NICARAGUA
 PANAMA
 ANTIGUA & BARBUDA
 THE BAHAMAS
 BARBADOS
 CUBA
 DOMINICA
DOMINICAN REPUBLIC

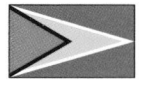 GUYANA
 SURINAME
 VENEZUELA
 BOLIVIA
 ECUADOR
 PERU
 BRAZIL
 ARGENTINA

 BENIN
 BURKINA FASO
 CAPE VERDE
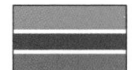 GAMBIA
 GHANA
 GUINEA
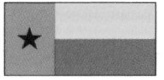 GUINEA-BISSAU
 IVORY COAST

 CAMEROON
 CENTRAL AFRICAN REPUBLIC
 CHAD
 CONGO
 DEM. REP. CONGO
 EQUATORIAL GUINEA
 GABON
 SAO TOME & PRINCIPE

 SOUTH SUDAN
 TANZANIA
 UGANDA
 ANGOLA
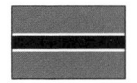 BOTSWANA
 LESOTHO
 MALAWI
 MOZAMBIQUE

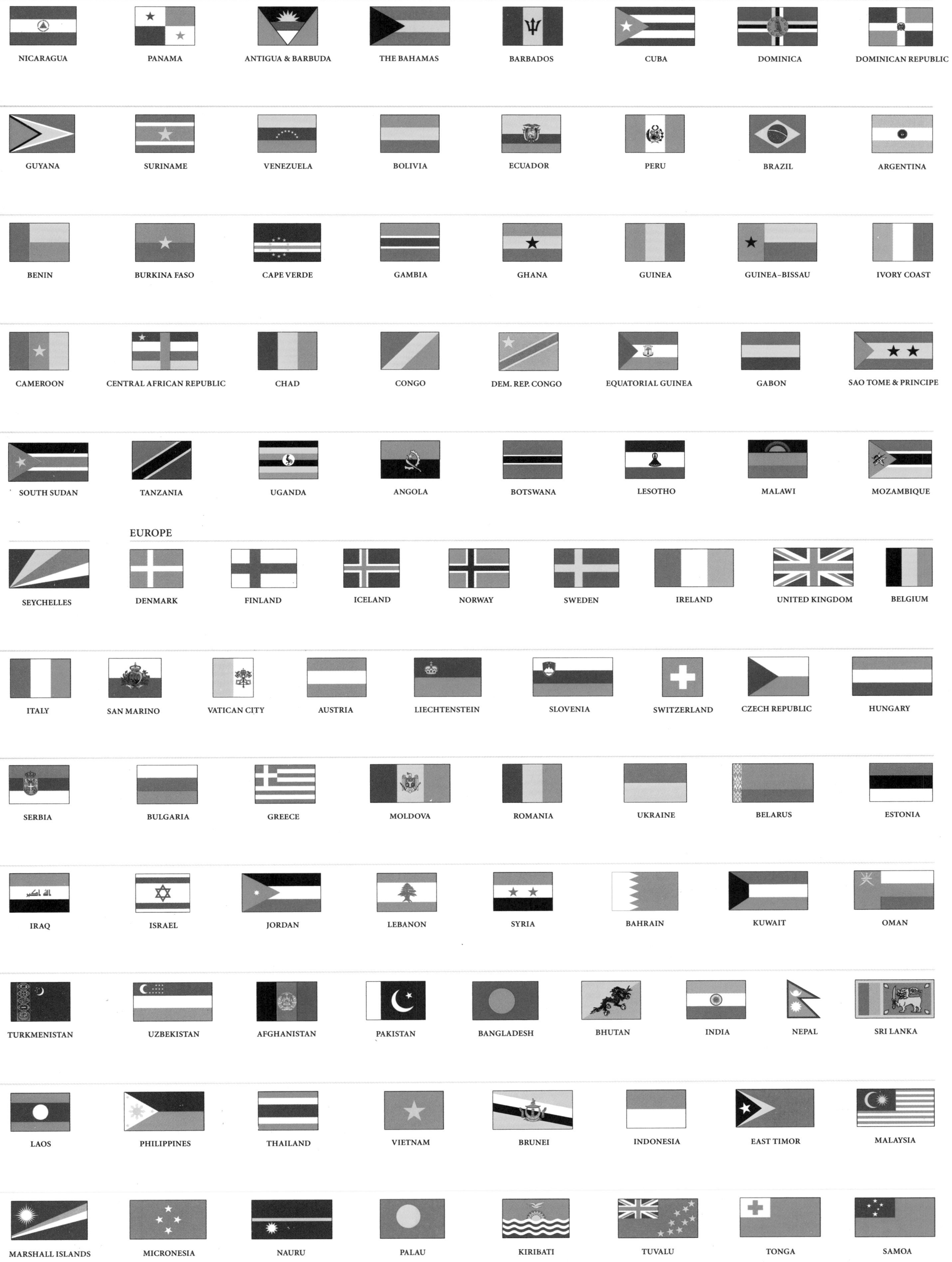

EUROPE

 SEYCHELLES
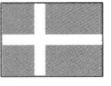 DENMARK
 FINLAND
 ICELAND
 NORWAY
 SWEDEN
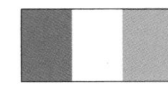 IRELAND
 UNITED KINGDOM
 BELGIUM

 ITALY
 SAN MARINO
 VATICAN CITY
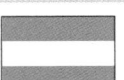 AUSTRIA
 LIECHTENSTEIN
 SLOVENIA
 SWITZERLAND
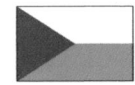 CZECH REPUBLIC
 HUNGARY

 SERBIA
 BULGARIA
 GREECE
 MOLDOVA
 ROMANIA
 UKRAINE
 BELARUS
 ESTONIA

 IRAQ
 ISRAEL
 JORDAN
 LEBANON
 SYRIA
 BAHRAIN
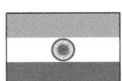 KUWAIT
 OMAN

 TURKMENISTAN
 UZBEKISTAN
 AFGHANISTAN
PAKISTAN
 BANGLADESH
 BHUTAN
 INDIA
 NEPAL
SRI LANKA

 LAOS
PHILIPPINES
THAILAND
VIETNAM
BRUNEI
INDONESIA
EAST TIMOR
MALAYSIA

MARSHALL ISLANDS
MICRONESIA
NAURU
PALAU
KIRIBATI
TUVALU
TONGA
SAMOA

COMPLETE
ATLAS
OF THE WORLD

COMPLETE
ATLAS
OF THE WORLD

FOR THE THIRD EDITION

Senior Cartographic Editor Simon Mumford

Cartographers Encompass Graphics Ltd, UK

Producer, Pre-Production Luca Frassinetti **Producer** Vivienne Yong

Jacket Design Development Manager Sophia MTT

Publishing Director Jonathan Metcalf **Associate Publishing Director** Liz Wheeler **Art Director** Karen Self

FOR PREVIOUS EDITIONS

Cartographic Editors
Tony Chambers, John Dear, Ruth Hall, Andrew Johnson, Belinda Kane, Lynn Neal, Ann Stephenson

Cartographers
Paul Eames, Edward Merritt, John Plumer, Rob Stokes, Iorwerth Watkins

Digital Map Suppliers
Advanced Illustration, Congleton, UK • Cosmographics, Watford, UK
Encompass Graphics, Brighton, UK • Lovell Johns Ltd., Long Hanborough, UK
Netmaps, Barcelona, Spain

Digital Terrain Data
Digital terrain data and continental panoramic images created by Planetary Visions Ltd, Farnham, UK

Editor
Robert Dinwiddie

Designers
Nicola Liddiard, Yak El-Droubie

Picture Research
Louise Thomas, Jenny Baskaya

Indexing and Database
T-Kartor, Sweden
Francesca Albini, Eleanor Arkwright, Renata Dyntarova, Edward Heelas, Britta Hansesgaard

Systems Coordinator
Philip Rowles

Flags courtesy of The Flag Institue, Cheshire, UK

First American Edition, 2007
This revised edition 2016
Published in the United States by DK Publishing, 345 Hudson Street, New York, New York 10014

16 17 18 19 20 10 9 8 7 6 5 4 3 2 1
265177—May 2016

Reprinted with revisions 2009, Second edition 2012, Third edition 2016
Copyright © 2007, 2009, 2012, 2016 Dorling Kindersley Limited. All rights reserved.

Published in Great Britain by Dorling Kindersley Limited. A Penguin Random House company.

A catalog record for this book is available from the Library of Congress.

ISBN: 978-1-4654-4401-1

DK books are available at special discounts when purchased in bulk for sales promotions, premiums, fund-raising, or educational use.
For details, contact: DK Publishing Special Markets, 345 Hudson Street, New York, New York 10014 or SpecialSales@dk.com

Printed and bound in Italy by L.E.G.O. S.p.A.

A WORLD OF IDEAS:
SEE ALL THERE IS TO KNOW
www.dk.com

Introduction

The World at the beginning of the 21st Century would be a place of unimaginable change to our forefathers. Since 1900 the human population has undergone a fourfold growth coupled with an unparalleled development in the technology at our disposal. The last vestiges of the unknown World are gone, and previously hostile realms claimed for habitation. The advent of aviation technology and the growth of mass tourism have allowed people to travel further and more frequently than ever before.

Allied to this, the rapid growth of global communication systems mean that World events have become more accessible than ever before and their knock on effects quickly ripple across the whole planet. News broadcasts bring the far-flung corners of the world into everyone's lives, and with it, a view of the people and places that make up that region. The mysteries of the World that once fueled global exploration and the quest to discover the unknown are behind us; we inhabit a world of mass transportation, a world where even the most extreme regions have been mapped, a world with multi faceted view points on every event, a World of communication overload.

However, does this help us make sense of the World? It is increasingly important for us to have a clear vision of the World in which we live and such a deluge of information can leave us struggling to find some context and meaning. It has never been more important to own an atlas; the *DK Complete Atlas of the World* has been conceived to meet this need. At its core, like all atlases, it seeks to define where places are, to describe their main characteristics, and to locate them in relation to other places. By gathering a spectacular collection of satellite imagery and draping it with carefully selected and up-to-date geographic information, this atlas filters the World's data into clear, meaningful and user-friendly maps.

The World works on different levels and so does the *DK Complete Atlas of the World*. Readers can learn about global issues of many kinds or they can probe in a little further for the continental context. Delving even further they can explore at regional, national or even sub-national level. The very best available satellite data has been used to create topography and bathymetry that reveal the breathtaking texture of landscapes and sea-floors. These bring out the context of the places and features selected to appear on top of them.

This third edition of the *DK Complete Atlas of the World* incorporates hundreds of revisions and updates affecting every map and every page, distilling the burgeoning mass of information available through modern technology into an extraordinarily detailed and reliable view of our World.

Contents

The atlas is organized by continent, moving eastward from the International Date Line. The opening section describes the world's structure, systems and its main features. The Atlas of the World which follows, is a continent-by-continent guide to today's world, starting with a comprehensive insight into the physical, political, and economic structure of each continent, followed by detailed maps of carefully selected geopolitical regions.

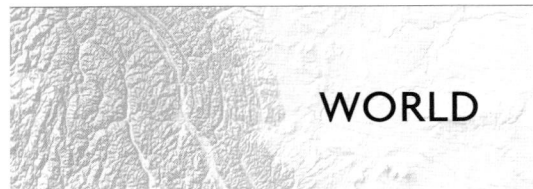

WORLD

NORTH AMERICA

SOUTH AMERICA

AFRICA

EUROPE

ASIA

AUSTRALASIA & OCEANIA

INDEX & GAZETTEER

Key to regional maps

Physical features

elevation

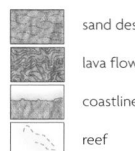

6000m / 19,686ft

4000m / 13,124ft

3000m / 9843ft

2000m / 6562ft

1000m / 3281ft

500m / 1640ft

250m / 820ft

100m / 328ft

sea level

below sea level

▲ elevation above sea level (mountain height)

▲ volcano

✕ pass

▼ elevation below sea level (depression depth)

sand desert

lava flow

coastline

reef

atoll

sea depth

sea level

-250m / -820ft

-2000m / -6562ft

-4000m / -13,124ft

▲ seamount / guyot symbol

▼ undersea spot depth

Drainage features

main river

secondary river

tertiary river

minor river

main seasonal river

secondary seasonal river

canal

waterfall

rapids

dam

perennial lake

seasonal lake

perennial salt lake

seasonal salt lake

reservoir

salt flat / salt pan

marsh / salt marsh

mangrove

wadi

○ spring / well / waterhole / oasis

Ice features

ice cap / sheet

ice shelf

glacier / snowfield

• • • • summer pack ice limit

• • • • winter pack ice limit

Graticule features

lines of latitude and longitude / Equator

Tropics / Polar circles

45° degrees of longitude / latitude

Communications

motorway / highway

motorway / highway (under construction)

major road

minor road

tunnel (road)

main railroad

minor railroad

tunnel (railroad)

✈ international airport

Borders

full international border

undefined international border

disputed de facto border

disputed territorial claim border

indication of country extent (Pacific only)

indication of dependent territory extent (Pacific only)

demarcation/ cease fire line

autonomous / federal region border

other 1st order internal administrative border

2nd order internal administrative border

Miscellaneous features

ancient wall

◇ site of interest

○ scientific station

Settlements

built up area

settlement population symbols

■ more than 5 million

◉ 1 million to 5 million

◉ 500,000 to 1 million

◎ 100,000 to 500,000

⊕ 50,000 to 100,000

○ 10,000 to 50,000

○ fewer than 10,000

■ ● ● country/dependent territory capital city

■ ● ● autonomous / federal region / other 1st order internal administrative center

■ ● ● 2nd order internal administrative center

Typographic key

Physical features

landscape features ... *Namib Desert*

Massif Central

ANDES

headland .... *Nordkapp*

elevation / volcano / pass .... Mount Meru 4556 m

drainage features .... *Lake Geneva*

rivers / canals spring / well / waterhole / oasis / waterfall / rapids / dam .... *Mekong*

ice features .... *Vatnajökull*

Physical features (continued)

sea features........... *Golfe de Lion*

Andaman Sea

INDIAN OCEAN

undersea features ... *Barracuda Fracture Zone*

Regions

country .... ARMENIA

dependent territory with parent state ...... NIUE (to NZ)

autonomous / federal region .... MINAS GERAIS

other 1st order internal administrative region .... MINSKAYA VOBLASTS'

2nd order internal administrative region .... Vaucluse

cultural region .... New England

Settlements

capital city .... BEIJING

dependent territory capital city .... FORT-DE-FRANCE

other settlements .... Chicago

Adana

Tizi Ozou

Yonezawa

Farnham

Miscellaneous

sites of interest / miscellaneous .... Valley of the Kings

Tropics / Polar circles .... *Antarctic Circle*

The Solar System

The Solar System consists of our local star, the Sun, and numerous objects that orbit the Sun – eight planets, five currently recognized dwarf planets, over 165 moons orbiting these planets and dwarf planets, and countless smaller bodies such as comets and asteroids. Including a vast outer region that is populated only by comets, the Solar System is about 9,300 billion miles (15,000 billion km) across. The much smaller region containing just the Sun and planets is about 7.5 billion miles (12 billion km) across. The Sun, which contributes over 99 percent of the mass of the entire Solar System, creates energy from nuclear reactions deep within its interior, providing the heat and light that make life on Earth possible.

THE MOON'S PHASES

As the Moon orbits Earth, the relative positions of Moon, Sun and Earth continuously change. Thus, the angle at which the Moon's sunlit face is seen by an observer on Earth varies in a cyclical fashion, producing the Moon's phases, as shown at right. Each cycle takes 29.5 days.

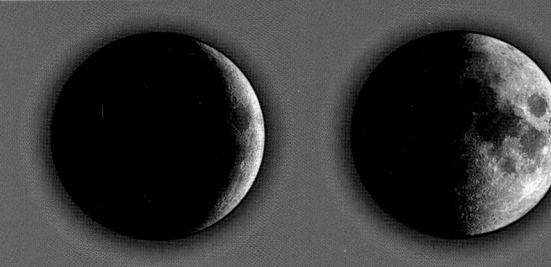

1. WAXING CRESCENT   2. FIRST QUARTER

The Moon

Earth's only satellite, the Moon, is thought to have formed 4.5 billion years ago from a cloud of debris produced when a large asteroid hit the young Earth. The Moon is too small to have retained an atmosphere, and is therefore a lifeless, dusty and dead world. However, although the Moon has only about 1 percent of the mass of the Earth, its gravity exerts an important influence on Earth's oceans, manifest in the ebb and flow of the tides.

The Earth and Moon's relative sizes are clear in this long-range image from space.

What is a Planet?

The International Astronomical Union defines a Solar System planet as a near-spherical object that orbits the Sun (and no other body) and has cleared the neighborhood around its orbit of other bodies. A dwarf planet is a planet that is not big enough to have cleared its orbital neighborhood. Extra-solar planets are objects orbiting stars other than the Sun.

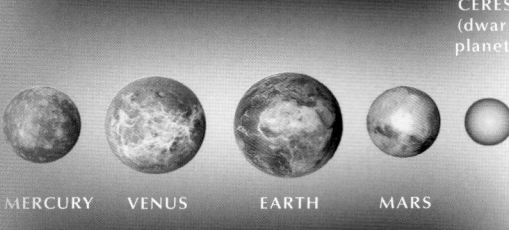

MERCURY   VENUS   EARTH   MARS

CERES (dwarf planet)

JUPITER

The Sun

The Sun is a huge sphere of exceedingly hot plasma (ionized gas), consisting mainly of the elements hydrogen and helium. It formed about 4.6 billion years ago, when a swirling cloud of gas and dust began to contract under the influence of gravity. When the center of this cloud reached a critically high temperature, hydrogen nuclei started combining to form helium nuclei – a process called nuclear fusion – with the release of massive amounts of energy. This process continues to this day.

SOLAR ECLIPSE

A solar eclipse occurs when the Moon passes between Earth and the Sun, casting its shadow on Earth's surface. During a total eclipse (below), viewers along a strip of Earth's surface, called the area of totality, see the Sun totally blotted out for a short time, as the umbra (Moon's full shadow) sweeps over them. Outside this area is a larger one, where the Sun appears only partly obscured, as the penumbra (partial shadow) passes over.

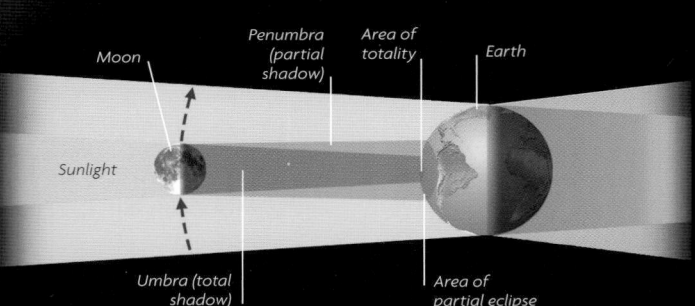

Moon

Penumbra (partial shadow)

Area of totality

Earth

Sunlight

Umbra (total shadow)

Area of partial eclipse

INSIDE THE SUN

The Sun has three internal layers. At its center is the core, where temperatures reach 27 million°F (15 million°C) and nuclear fusion occurs. The radiative zone is a slightly cooler region through which energy radiates away from the core. Further out, in the convective zone, plumes of hot plasma carry the energy towards the Sun's visible surface layer, called the photosphere. Once there, the energy escapes as light, heat and other forms of radiation.

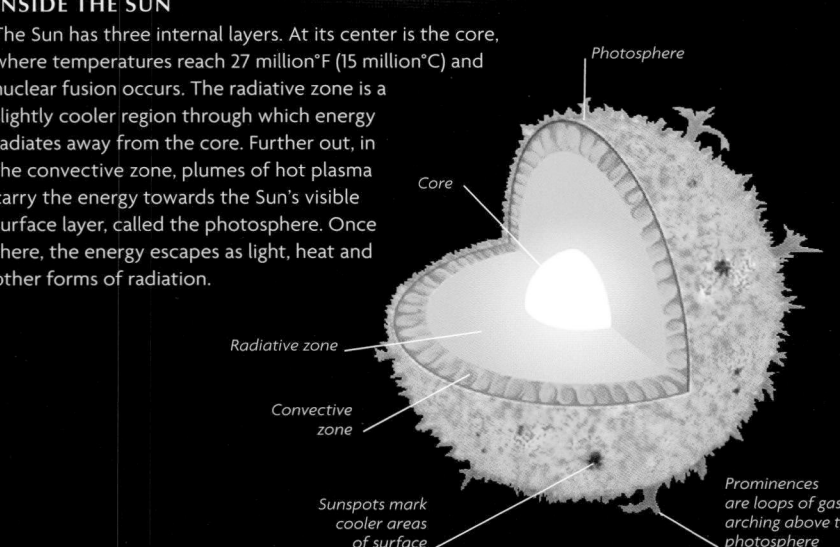

Photosphere

Core

Radiative zone

Convective zone

Sunspots mark cooler areas of surface

Prominences are loops of gas arching above the photosphere

| 3. WAXING GIBBOUS | 4. FULL MOON | 5. WANING GIBBOUS | 6. LAST QUARTER | 7. WANING CRESCENT | 8. NEW MOON |

PLANETS

MAIN DWARF PLANETS

	MERCURY	VENUS	EARTH	MARS	JUPITER	SATURN	URANUS	NEPTUNE	CERES	PLUTO	ERIS
DIAMETER	3029 miles (4875 km)	7521 miles (12,104 km)	7928 miles (12,756 km)	4213 miles (6780 km)	88,846 miles (142,984 km)	74,898 miles (120,536 km)	31,763 miles (51,118 km)	30,775 miles (49,528 km)	590 miles (950 km)	1432 miles (2304 km)	1429-1553 miles (2300-2500 km)
AVERAGE DISTANCE FROM THE SUN	36 mill. miles (57.9 mill. km)	67.2 mill. miles (108.2 mill. km)	93 mill. miles (149.6 mill. km)	141.6 mill. miles (227.9 mill. km)	483.6 mill. miles (778.3 mill. km)	889.8 mill. miles (1431 mill. km)	1788 mill. miles (2877 mill. km)	2795 mill. miles (4498 mill. km)	257 mill. miles (414 mill. km)	3675 mill. miles (5,915 mill. km)	6344 mill. miles (10,210 mill. km)
ROTATION PERIOD	58.6 days	243 days	23.93 hours	24.62 hours	9.93 hours	10.65 hours	17.24 hours	16.11 hours	9.1 hours	6.38 days	not known
ORBITAL PERIOD	88 days	224.7 days	365.26 days	687 days	11.86 years	29.37 years	84.1 years	164.9 years	4.6 years	248.6 years	557 years
SURFACE TEMPERATURE	-292°F to 806°F (-180°C to 430°C)	896°F (480°C)	-94°F to 131°F (-70°C to 55°C)	-184°F to 77 °F (-120°C to 25°C)	-160°F (-110°C)	-220°F (-140°C)	-320°F (-200°C)	-320°F (-200°C)	-161°F (-107°C)	-380°F (-230°C)	-405°F (-243°C)

DWARF PLANETS

In 2006 a new type of dwarf planet was defined in an attempt to classify the numerous smaller bodies within the solar system that behave like planets physically but which are only a fraction of the size of the major planets. Currently, there are five dwarf planets recognized under this system, Ceres, Pluto, Haumea, Makemake, and Eris.

ERIS (dwarf planet)

PLUTO (dwarf planet)

NEPTUNE

URANUS

THE OUTER PLANETS

SATURN

Orbits

All the Solar System's planets and dwarf planets orbit the Sun in the same direction and (apart from Pluto) roughly in the same plane. All the orbits have the shapes of ellipses (stretched circles). However in most cases, these ellipses are close to being circular: only Pluto and Eris have very elliptical orbits. Orbital period (the time it takes an object to orbit the Sun) increases with distance from the Sun. The more remote objects not only have further to travel with each orbit, they also move more slowly.

THE OUTER PLANETS

The four gigantic outer planets – Jupiter, Saturn, Uranus and Neptune – consist mainly of gas, liquid and ice. All have rings and many moons. The dwarf planet Pluto is made of rock and ice.

THE INNER PLANETS

The four planets closest to the Sun – Mercury, Venus, Earth and Mars – are composed mainly of rock and metal. They are much smaller than the outer planets, have few or no moons, and no rings.

THE INNER PLANETS

AVERAGE DISTANCE FROM THE SUN

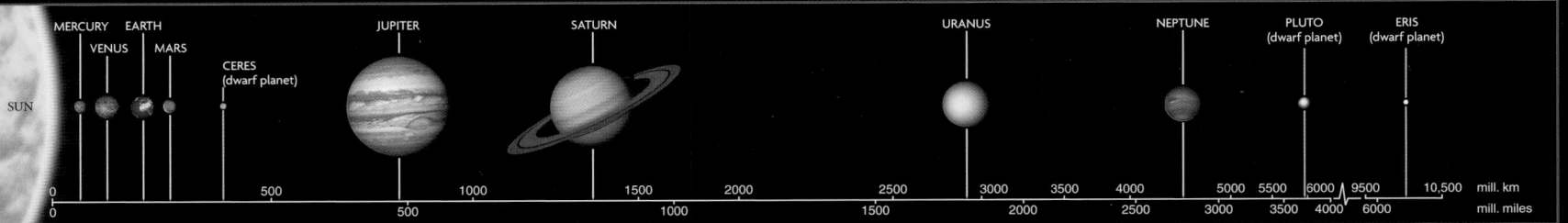

| SUN | MERCURY | EARTH | | CERES (dwarf planet) | JUPITER | SATURN | URANUS | NEPTUNE | PLUTO (dwarf planet) | ERIS (dwarf planet) |

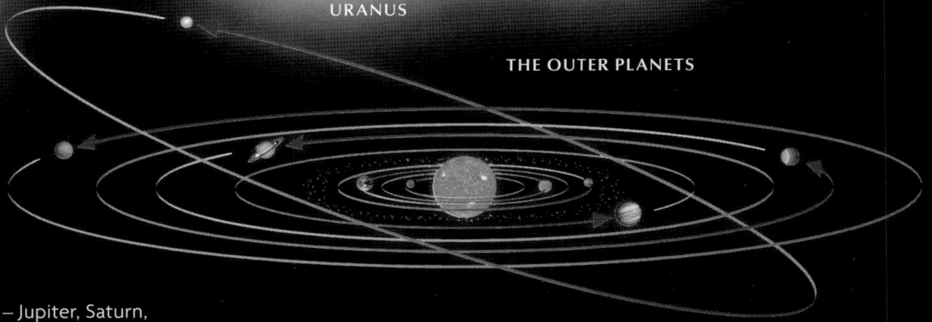

VENUS MARS

| 0 | 500 | 1000 | 1500 | 2000 | 2500 | 3000 | 3500 | 4000 | 5000 5500 | 6000 9500 | 10,500 mill. km |
| 0 | 500 | 1000 | 1500 | 2000 | 2500 | 3000 | 3500 | 4000 | 5000 | 6000 | mill. miles |

The Physical World

Earth's surface is constantly being transformed. Movements of the rigid tectonic plates that make up this surface are continuously, if slowly, shifting its landmasses around, while the land itself is constantly weathered and eroded by wind, water, and ice. Sometimes change is dramatic, the spectacular results of earthquakes or floods. More often it is a slow process lasting for millions of years. A physical map of the world represents a snapshot of Earth's ever-evolving architecture. The maps below and at right show the planet's whole surface, including variations in ocean depth as well as the mountain-rippled texture of Earth's continents.

THE WORLD'S OCEANS
Earth's surface is dominated by water. The hemisphere shown here, centered around the southwest Pacific, is nearly all ocean, with the waters interrupted only by Antarctica, a part of South America, Australia, and the numerous islands of Australasia & Oceania, and southeast Asia.

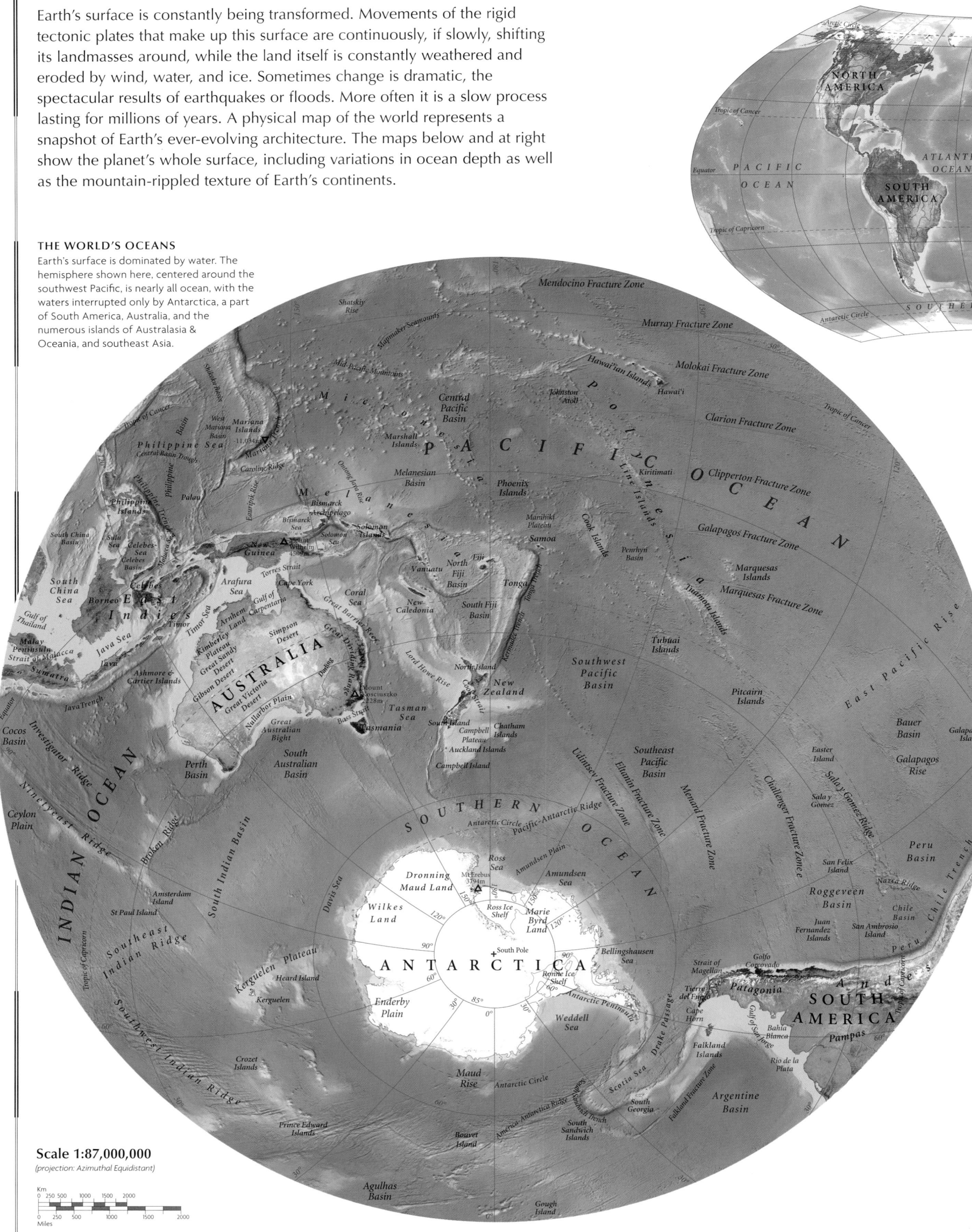

Scale 1:87,000,000
(projection: Azimuthal Equidistant)

Km
0 250 500   1000   1500   2000

0   250   500   1000   1500   2000
Miles

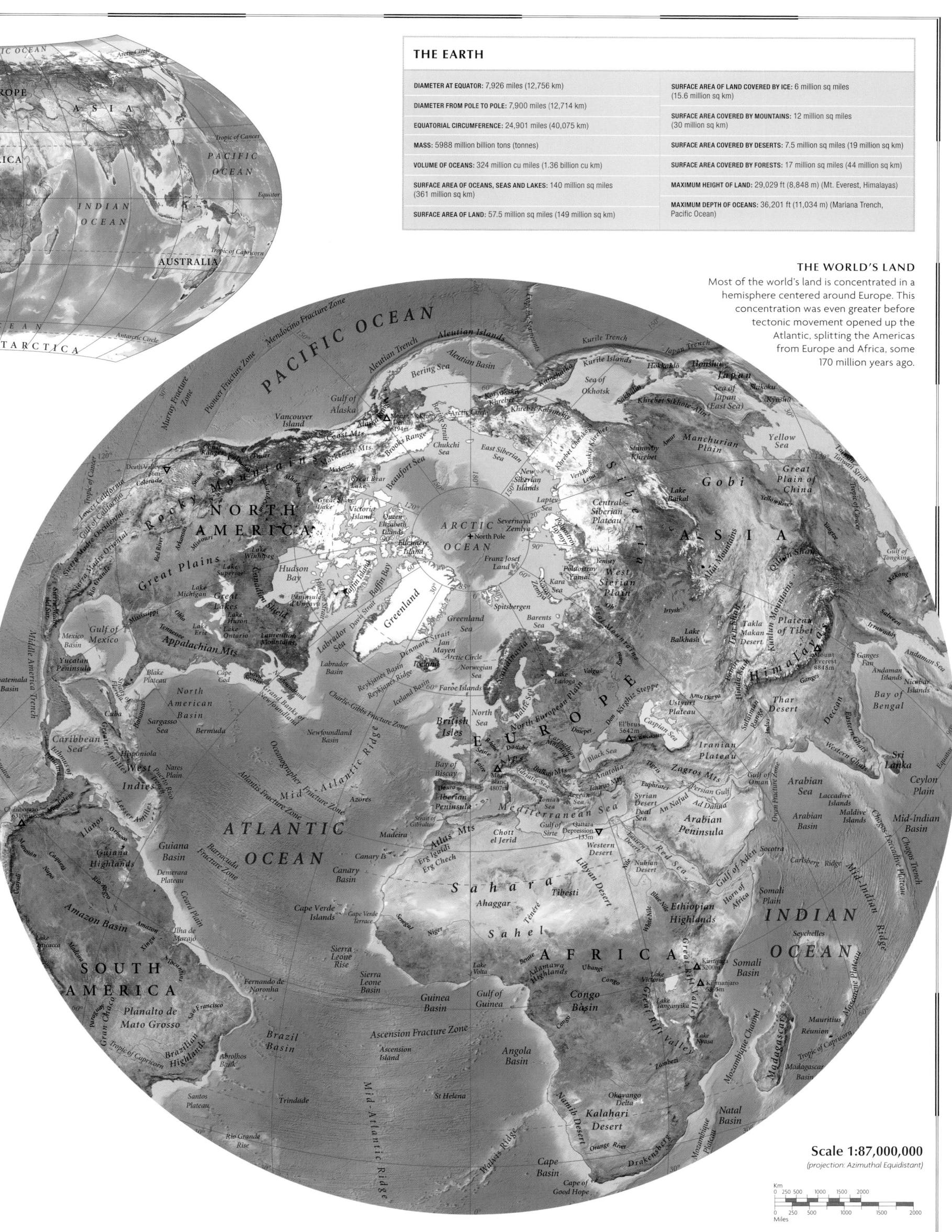

THE EARTH

DIAMETER AT EQUATOR: 7,926 miles (12,756 km)

DIAMETER FROM POLE TO POLE: 7,900 miles (12,714 km)

EQUATORIAL CIRCUMFERENCE: 24,901 miles (40,075 km)

MASS: 5988 million billion tons (tonnes)

VOLUME OF OCEANS: 324 million cu miles (1.36 billion cu km)

SURFACE AREA OF OCEANS, SEAS AND LAKES: 140 million sq miles (361 million sq km)

SURFACE AREA OF LAND: 57.5 million sq miles (149 million sq km)

SURFACE AREA OF LAND COVERED BY ICE: 6 million sq miles (15.6 million sq km)

SURFACE AREA COVERED BY MOUNTAINS: 12 million sq miles (30 million sq km)

SURFACE AREA COVERED BY DESERTS: 7.5 million sq miles (19 million sq km)

SURFACE AREA COVERED BY FORESTS: 17 million sq miles (44 million sq km)

MAXIMUM HEIGHT OF LAND: 29,029 ft (8,848 m) (Mt. Everest, Himalayas)

MAXIMUM DEPTH OF OCEANS: 36,201 ft (11,034 m) (Mariana Trench, Pacific Ocean)

THE WORLD'S LAND
Most of the world's land is concentrated in a hemisphere centered around Europe. This concentration was even greater before tectonic movement opened up the Atlantic, splitting the Americas from Europe and Africa, some 170 million years ago.

Scale 1:87,000,000
(projection: Azimuthal Equidistant)

Km
0 250 500 1000 1500 2000

0 250 500 1000 1500 2000
Miles

The Structure of the Earth

Earth is an almost perfect sphere consisting of a partly liquid core overlain by a deep, semisolid layer, called the mantle, and two types of surface crust, known as continental and oceanic crust. Our planet has constantly evolved since it formed some 4.5 billion years ago. Its continents are neither fixed nor stable. Over the course of history, gradual movements of rocky material within Earth's mantle, resulting from massive internal flows of heat, have caused the great slabs of material that make up the planet's surface, known as tectonic plates, to shift around. The plates have moved, collided, joined together, and sometimes split apart. These processes continue to mold Earth's surface, causing earthquakes and volcanic eruptions, and creating oceans, mountain ranges, rift valleys, deep ocean trenches, and island chains.

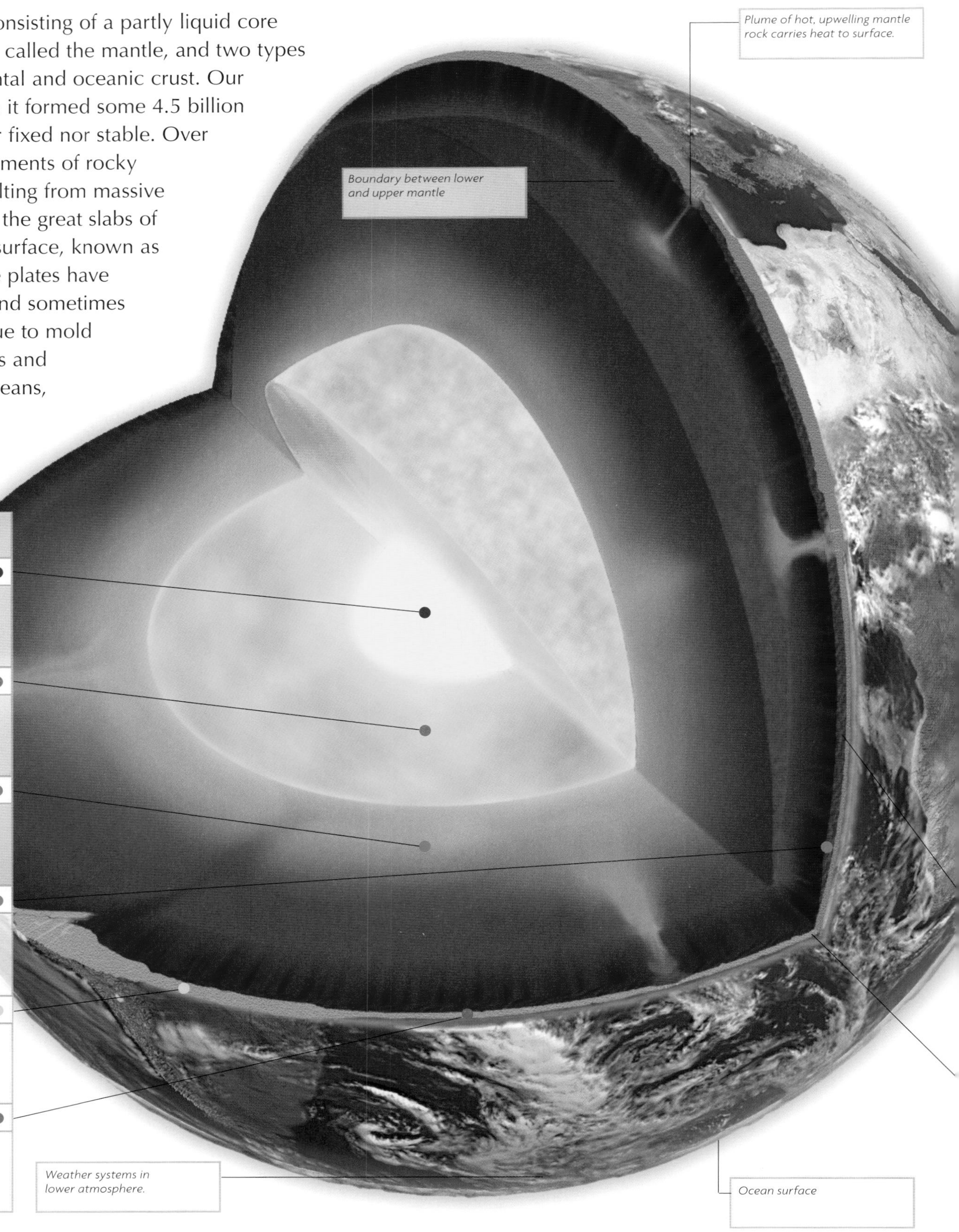

Plume of hot, upwelling mantle rock carries heat to surface.

Boundary between lower and upper mantle

EARTH FACTS & FIGURES

INNER CORE
COMPOSITION: Solid iron, with some nickel
DENSITY: 7.0 oz/in³ (12 g/cm³)
DEPTH: 3200-3963 miles (5150-6378 km) below surface
TEMPERATURE: 7200-8500°F (4000-4700°C)

OUTER CORE
COMPOSITION: Liquid iron and nickel
DENSITY: 5.7 oz/in³ (10 g/cm³)
DEPTH: 1907-3200 miles (2990-5150 km) below surface
TEMPERATURE: 6300-7200°F (3500-4000°C)

LOWER MANTLE
COMPOSITION: Semisolid high-density silicates
DENSITY: 3.2 oz/in³ (5.5 g/cm³)
DEPTH: 48-1907 miles (75-2990 km) below surface
TEMPERATURE: 1800-6300°F (1000-3500°C)

UPPER MANTLE
COMPOSITION: Semisolid rock, primarily peridotite
DENSITY: 2.0 oz/in³ (3.5 g/cm³)
DEPTH: 3-48 miles (5-75 km) below surface
TEMPERATURE: 1800°F (Less than 1000°C)

CONTINENTAL CRUST
COMPOSITION: Solid, relatively light rock such as granite
DENSITY: 1.6 oz/in³ (2.7 g/cm³)
DEPTH: 0-48 miles (0-75 km) below surface
TEMPERATURE: 1800°F (Less than 1000°C)

OCEANIC CRUST
COMPOSITION: Solid, relatively dense basaltic lava
DENSITY: 1.7 oz/in³ (3 g/cm³)
DEPTH: 2-7 miles (3-11 km) below surface
TEMPERATURE: 1800°F (Less than 1000°C)

Weather systems in lower atmosphere.

Ocean surface

FROM THE BIG BANG TO THE PRESENT DAY

The Big Bang | first galaxies form | Milky Way galaxy forms

13,700 million years ago (mya) | 12,000 mya | 11,000 mya | 10,000

1000 mya | 2000 mya first multi-celled organisms | 3000 mya first landmasses

Phanerozoic Eon
543 mya | present day

Phanerozoic Eon *(right)* has been enlarged to show geological eras, periods and epochs

ERA	Paleozoic - age of ancient life						
PERIOD	Cambrian	Ordovician	Silurian	Devonian	Carboniferous		Permian
EPOCH						Mississippian	Pennsylvanian
	543	490	443	418	354	323	290

Continental drift

Although Earth's tectonic plates move only a few inches (centimeters) each year, over hundreds of millions of years, its landmasses have moved many thousands of miles (kilometers), to create new continents, oceans, and mountain chains.

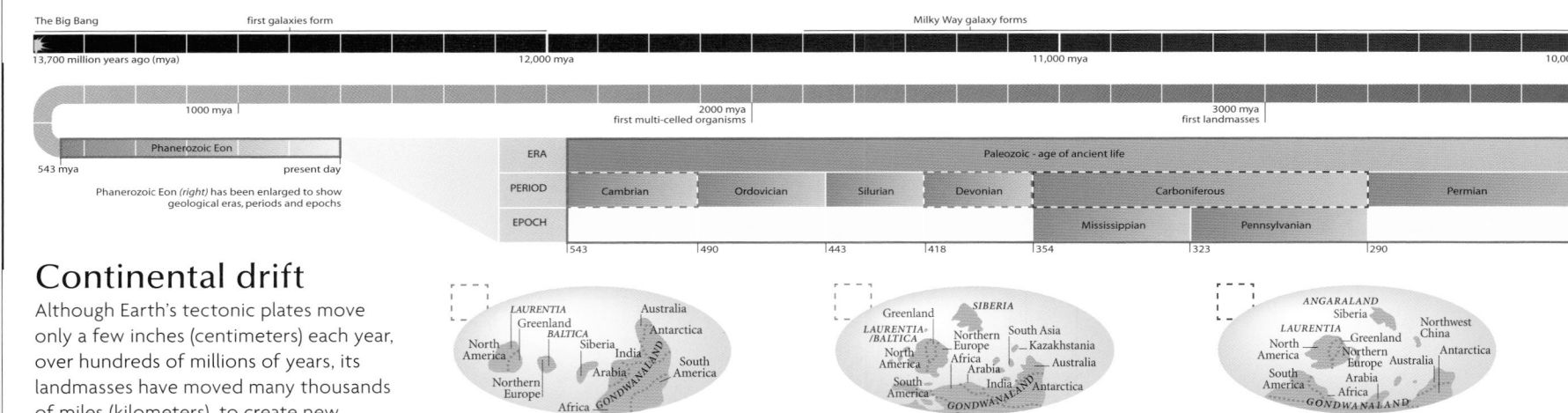

Cambrian 543–490 million years ago

Devonian 418–354 million years ago

Carboniferous 354–290 million years ago

Dynamic Earth

Earth's surface is split up into several rigid, closely-fitting sections, called tectonic plates. Each of the plates contains some oceanic crust, and most also contain some continental crust. The plates constantly move relative to one another. Movements at different types of plate boundary produce various types of geological structure and activity.

Tectonic activity and geological regions

Plate boundaries
- Convergent
- Divergent
- Transform
- Uncertain

Tectonic activity
- volcanic zone
- earthquake zone
- hot spot
- rift valley
- Sedimentary cover
- Mesozoic & Cenozoic volcanic rock
- Cenozoic (65 mya – present)
- Mesozoic (252 mya – 65 mya)
- Paleozoic (543 mya – 252 mya)
- pre-Cambrian Shields

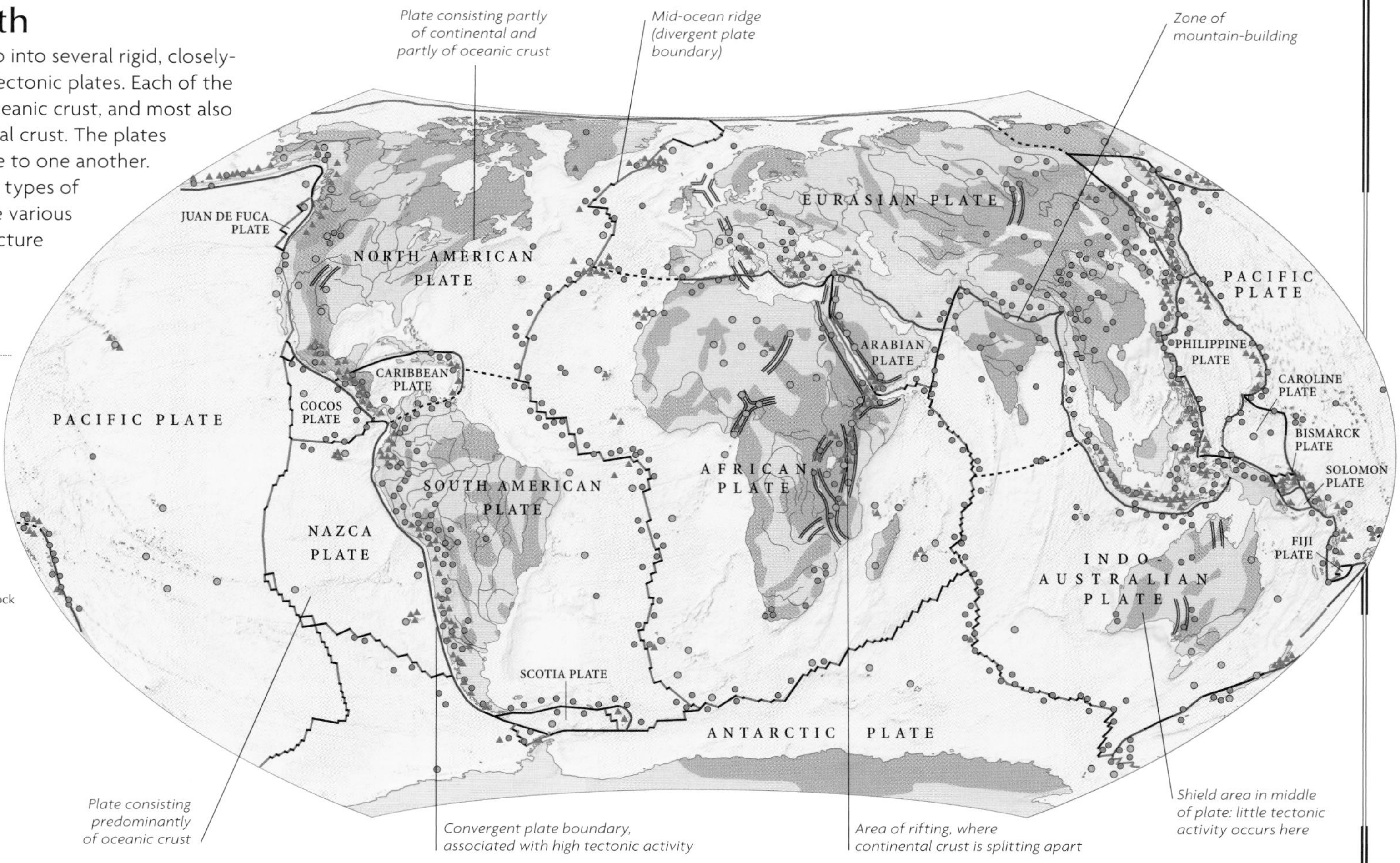

Plate consisting partly of continental and partly of oceanic crust
Mid-ocean ridge (divergent plate boundary)
Zone of mountain-building
JUAN DE FUCA PLATE
NORTH AMERICAN PLATE
EURASIAN PLATE
PACIFIC PLATE
CARIBBEAN PLATE
COCOS PLATE
PACIFIC PLATE
ARABIAN PLATE
PHILIPPINE PLATE
CAROLINE PLATE
BISMARCK PLATE
SOLOMON PLATE
SOUTH AMERICAN PLATE
AFRICAN PLATE
NAZCA PLATE
INDO-AUSTRALIAN PLATE
FIJI PLATE
SCOTIA PLATE
ANTARCTIC PLATE
Plate consisting predominantly of oceanic crust
Convergent plate boundary, associated with high tectonic activity
Area of rifting, where continental crust is splitting apart
Shield area in middle of plate: little tectonic activity occurs here

EFFECTS AT PLATE BOUNDARIES

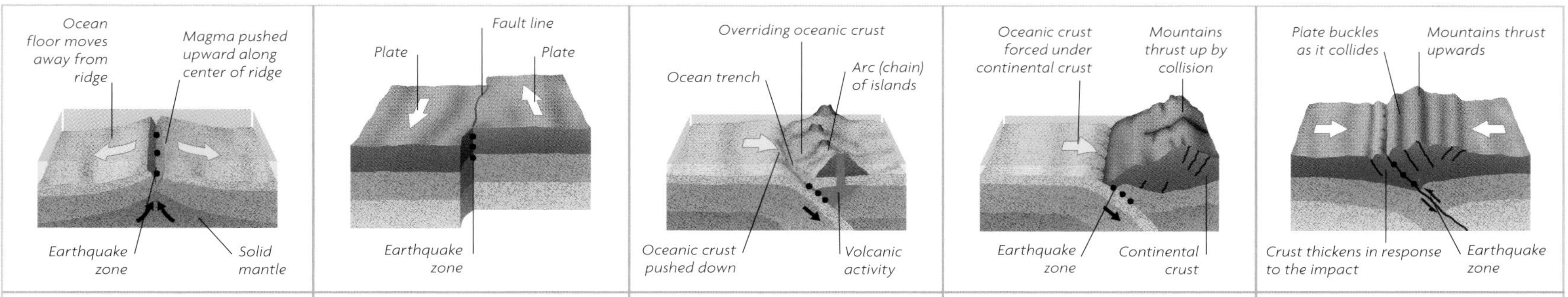

Ocean floor moves away from ridge
Magma pushed upward along center of ridge
Earthquake zone
Solid mantle
Fault line
Plate
Plate
Earthquake zone
Overriding oceanic crust
Ocean trench
Arc (chain) of islands
Oceanic crust pushed down
Volcanic activity
Oceanic crust forced under continental crust
Mountains thrust up by collision
Earthquake zone
Continental crust
Plate buckles as it collides
Mountains thrust upwards
Crust thickens in response to the impact
Earthquake zone

FORMATION OF A MID-OCEAN RIDGE	SLIDING PLATES (TRANSFORM BOUNDARY)	FORMATION OF ISLAND ARC AND OCEAN TRENCH	SUBDUCTION OF OCEANIC CRUST UNDER CONTINENTAL CRUST	BLOCKS OF CONTINENTAL CRUST COLLIDE TO FORM MOUNTAINS

Boundary between upper mantle and crust

Sea floor made of oceanic crust

CONVECTION CURRENTS

Deep within Earth's core, temperatures may exceed 8100°F (4500°C). The heat from the core warms rocks in the mantle, which become semimolten and rise upwards, displacing cooler rock below the solid oceanic and continental crust. This rock sinks and is warmed again by heat given off from the core. The process continues in a cyclical fashion, producing convection currents below the crust. These currents lead, in turn, to gradual movements of the tectonic plates over the planet's surface.

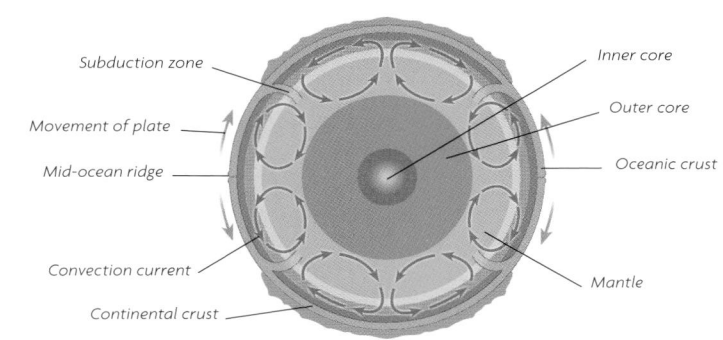

Subduction zone
Movement of plate
Mid-ocean ridge
Convection current
Continental crust
Inner core
Outer core
Oceanic crust
Mantle

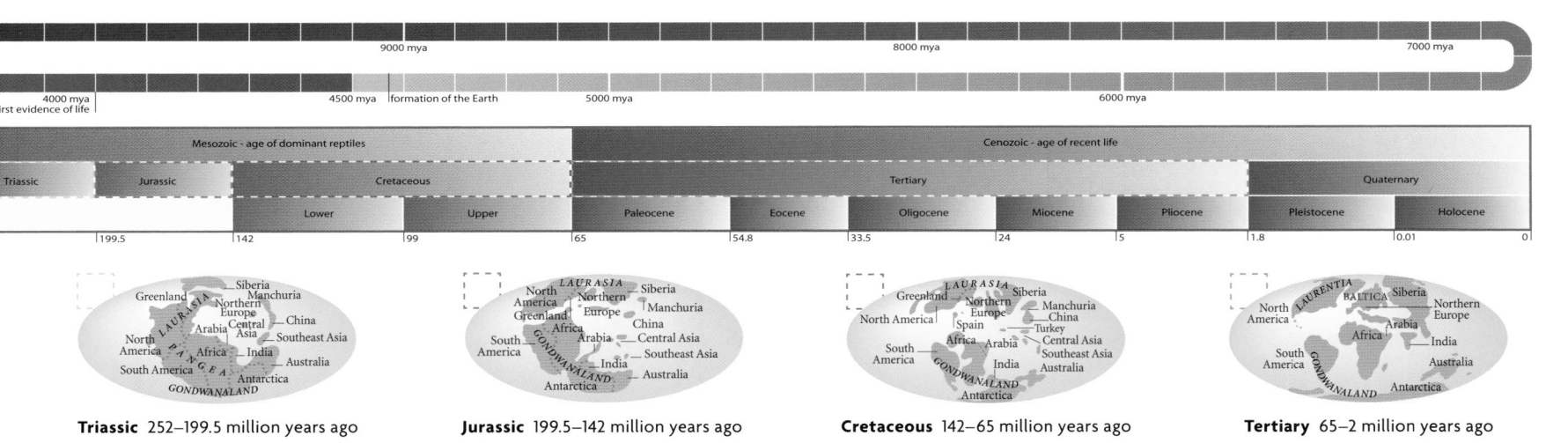

9000 mya
8000 mya
7000 mya
4000 mya
first evidence of life
4500 mya
formation of the Earth
5000 mya
6000 mya
Mesozoic - age of dominant reptiles
Cenozoic - age of recent life
Triassic
Jurassic
Cretaceous
Tertiary
Quaternary
Lower
Upper
Paleocene
Eocene
Oligocene
Miocene
Pliocene
Pleistocene
Holocene
199.5
142
99
65
54.8
33.5
24
5
1.8
0.01
0

Siberia
Greenland
Manchuria
Northern Europe
LAURASIA
Arabia
Central Asia
China
North America
PANGEA
Africa
Southeast Asia
India
South America
Australia
Antarctica
GONDWANALAND

Triassic 252–199.5 million years ago

LAURASIA
North America
Northern Europe
Siberia
Greenland
Manchuria
Africa
China
South America
Arabia
Central Asia
Southeast Asia
GONDWANALAND
India
Australia
Antarctica

Jurassic 199.5–142 million years ago

LAURASIA
Greenland
Northern Europe
Siberia
Manchuria
China
North America
Spain
Turkey
Africa
Arabia
Central Asia
Southeast Asia
South America
India
Australia
GONDWANALAND
Antarctica

Cretaceous 142–65 million years ago

LAURENTIA
BALTICA
Siberia
North America
Northern Europe
Arabia
Africa
India
South America
Australia
GONDWANALAND
Antarctica

Tertiary 65–2 million years ago

Shaping the Landscape

The basic material of Earth's surface is solid rock: valleys, deserts, soil, and sand are all evidence of the powerful agents of weathering, erosion and deposition that constantly transform Earth's landscapes. Water, whether flowing in rivers or grinding the ground in the form of glaciers, has the most clearly visible impact on Earth's surface. Also, wind can transport fragments of rock over huge distances and strip away protective layers of vegetation, exposing rock surfaces to the impact of extreme heat and cold. Many of the land-shaping effects of ice and water can be seen in northern regions such as Alaska *(below)*, while the effects of heat and wind are clearly visible in the Sahara *(far right)*.

● FJORD
A valley carved by an ancient glacier and later flooded by the sea is called a fiord.

Ice and water

Some of the most obvious and striking features of Earth's surface are large flows and bodies of liquid water, such as rivers, lakes, and seas. In addition to these are landforms caused by the erosional or depositional power of flowing water, which include gullies, river valleys, and coastal features such as headlands and deltas. Ice also has had a major impact on Earth's appearance. Glaciers—rivers of ice formed by the compaction of snow—pick up and carry huge amounts of rocks and boulders as they pass over the landscape, eroding it as they do so. Glacially-sculpted landforms range from mountain *cirques* and U-shaped valleys to fiords and glacial lakes.

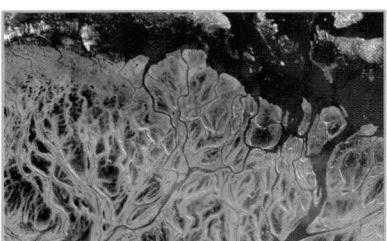

● DELTA
A delta, such as that of the Yukon River (above), is a roughly triangular or fan-shaped area of sediment deposited by a river at its mouth.

● PINGO
These blister-like mounds, seen in regions of Arctic tundra, are formed by the upward expansion of water as it freezes in the soil.

● TIDEWATER GLACIER
Glaciers of this type flow to the sea, where they calve (disgorge) icebergs. Like all glaciers, they erode huge amounts of rock from the landscape.

● LANDSLIDE
The freezing and later thawing of water, which occurs in a continuous cycle, can shatter and crumble rocks, eventually causing landslides.

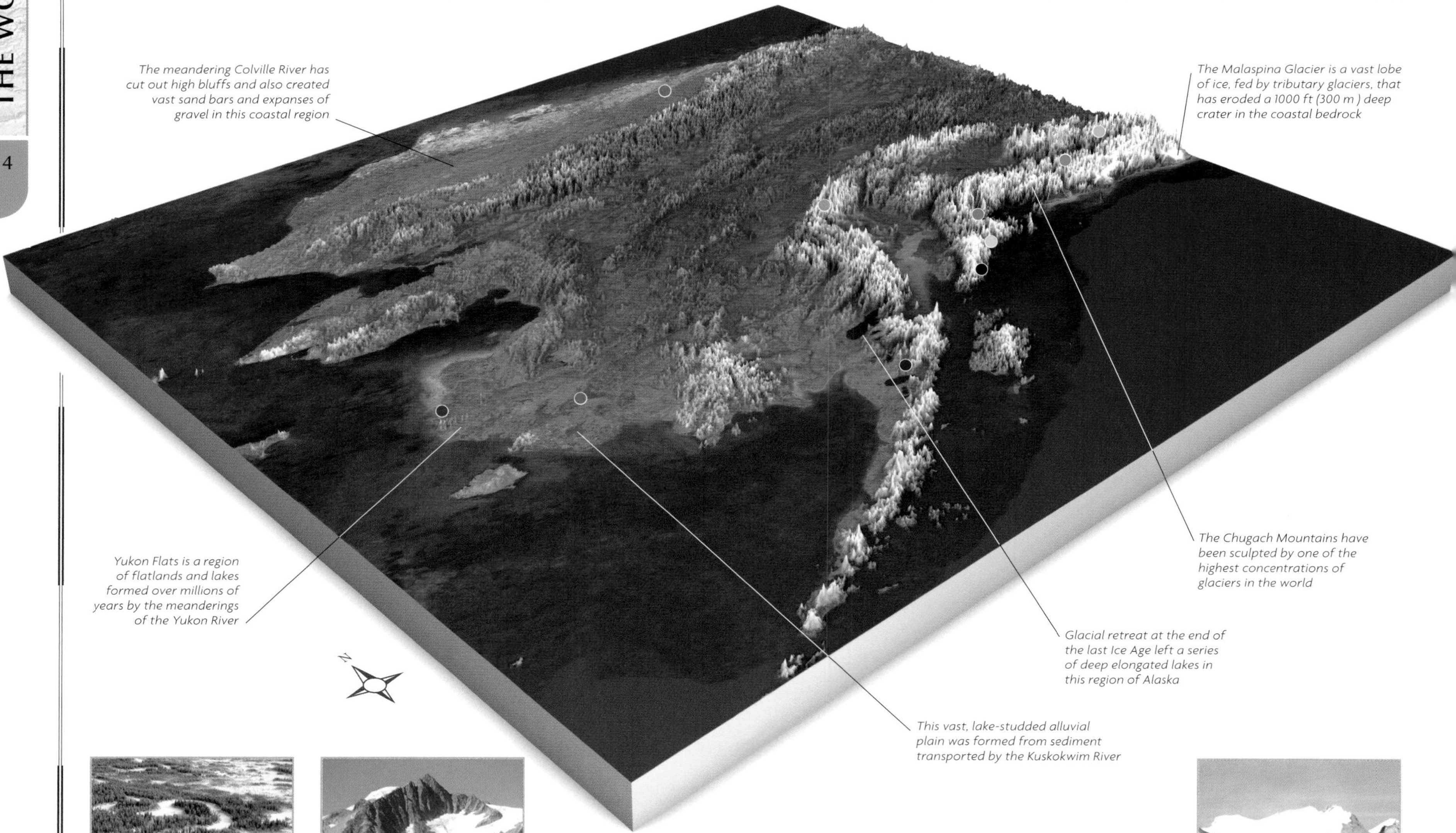

The meandering Colville River has cut out high bluffs and also created vast sand bars and expanses of gravel in this coastal region

The Malaspina Glacier is a vast lobe of ice, fed by tributary glaciers, that has eroded a 1000 ft (300 m) deep crater in the coastal bedrock

Yukon Flats is a region of flatlands and lakes formed over millions of years by the meanderings of the Yukon River

The Chugach Mountains have been sculpted by one of the highest concentrations of glaciers in the world

Glacial retreat at the end of the last Ice Age left a series of deep elongated lakes in this region of Alaska

This vast, lake-studded alluvial plain was formed from sediment transported by the Kuskokwim River

● MEANDERING RIVER
In their lower courses, some rivers carve out a series of looping bends called meanders.

● CIRQUE
A cirque is a hollow formed high on a mountain by glacial action. It may be ice-filled.

● POSTGLACIAL FEATURES
Glacially-polished cliffs like these are a tell-tale sign of ancient glacial action. Other signs include various forms of sculpted ridge and hummock.

● RIVER VALLEY
Over thousands of years, rivers erode uplands to form characteristic V-shaped valleys, with flat narrow floors and steeply-rising sides.

● GULLIES
Gullies are deep channels cut by rapidly flowing water, as here below Alaska's Mount Denali.

Heat and wind

Marked changes in temperature—rapid heating caused by fierce solar radiation during the day, followed by a sharp drop in temperature at night—cause rocks at the surface of hot deserts to continually expand and contract. This can eventually result in cracking and fissuring of the rocks, creating thermally-fractured desert landscapes. The world's deserts are also swept and scoured by strong winds. The finer particles of sand are shaped into surface ripples, dunes, or sand mountains, which can rise to a height of 650 ft (200 m). In other areas, the winds sweep away all the sand, leaving flat, gravelly areas called desert pavements.

DESERT LANDSCAPES

In desert areas, wind picks up loose sand and blasts it at the surface, creating a range of sculpted landforms from faceted rocks to large-scale features such as *yardangs*. Individually sculpted-rocks are called ventifacts. Where the sand abrasion is concentrated near the ground, it can turn these rocks into eccentrically-shaped "stone mushrooms." Other desert features are produced by thermal cracking and by winds continually redistributing the vast sand deposits.

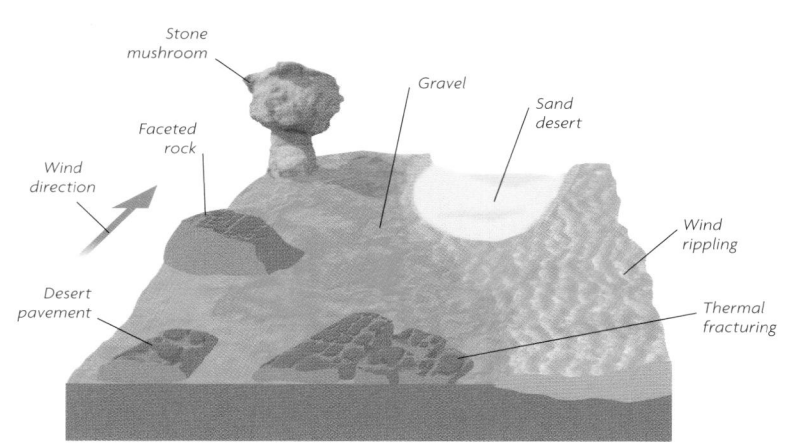

Stone mushroom

Gravel

Faceted rock

Sand desert

Wind direction

Wind rippling

Desert pavement

Thermal fracturing

FEATURES OF A DESERT SURFACE

● **DUST STORM**
A common phenomenon in some deserts, dust storms result from intense heating of the ground creating strong convection currents.

● **LOESS DEPOSIT**
A deposit of silt that has been transported over long distances by wind, then compacted. Loess is found in a few marginal areas of the Sahara.

● **YARDANG**
A yardang is a ridge of rock produced by wind erosion, usually in a desert. Large yardangs can be many miles long.

● **DESERT PAVEMENT**
Dark, gravelly surfaces like this result from wind removing all the sand from an area of desert.

Part of the Grand Erg Oriental, this region is a vast wind-sculpted sea of sand, much affected by sand storms

This area of complex dune morphology has resulted from two different types of dunes overlapping and coalescing

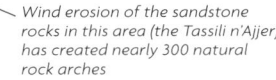

Wind erosion of the sandstone rocks in this area (the Tassili n'Ajjer) has created nearly 300 natural rock arches

The Tefedest is an impressive, sun-baked, wind-eroded, granite massif located in southern Algeria

This highland region, called the Ahaggar Mountains, has largely been blasted free of sand and is heavily eroded throughout

● **TRANSVERSE DUNES**
This series of parallel sand ridges lies at right angles to the prevailing wind direction.

● **VENTIFACT**
A ventifact is a rock that has been heavily sculpted and abraded by wind-driven sand.

● **CRACKED DESERT**
Intensely heated and dried-out desert areas often developed geometrically-patterned surface cracking.

● **WADI**
Wadis are dried out stream beds, found in some desert regions, that carry water only during occasional periods of heavy rain.

● **BARCHAN DUNE**
This arc-shaped type of dune migrates across the desert surface, blown by the wind.

The World's Oceans

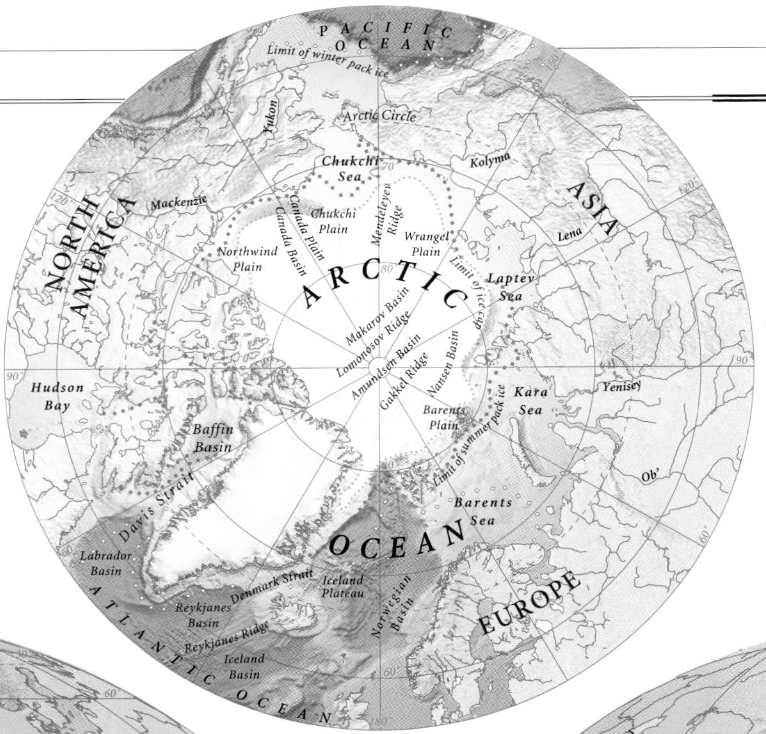

Two-thirds of Earth's surface is covered by the five oceans: the Pacific, Atlantic, Indian, Southern (or Antarctic), and Arctic. The basins that form these oceans, and the ocean floor landscape, have formed over the past 200 million years through volcanic activity and gradual movements of the Earth's crust. Surrounding the continents are shallow flat regions called continental shelves. These shelves extend to the continental slope, which drops steeply to the ocean floor. There, vast submarine plateaus, known as abyssal plains, are interrupted by massive ridges, chains of seamounts, and deep ocean trenches.

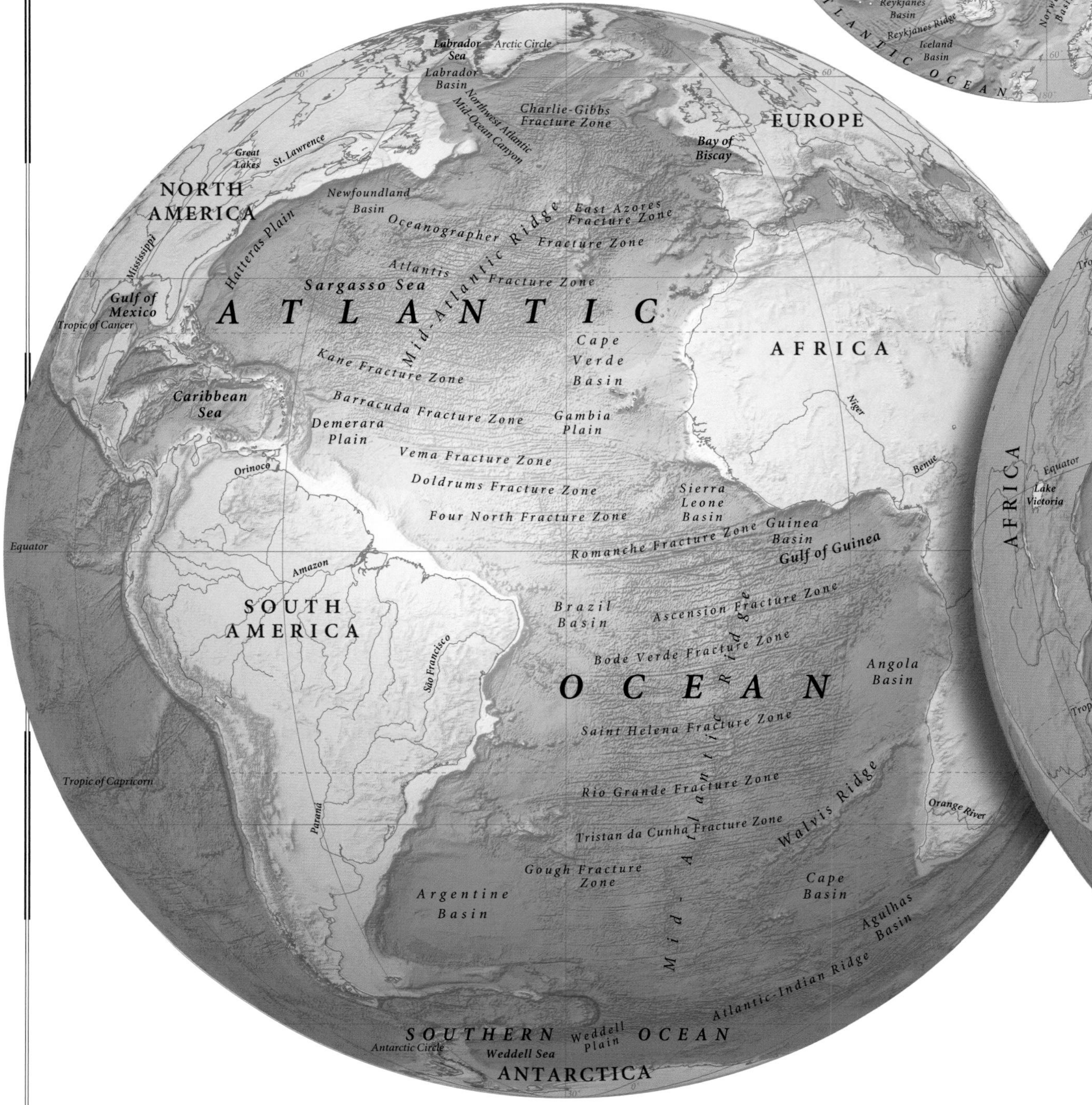

Arctic globe labels
PACIFIC OCEAN
Limit of winter pack ice
Arctic Circle
NORTH AMERICA
Mackenzie
Chukchi Sea
Kolyma
ASIA
Lena
Northwind Plain
Canada Plain
Canada Basin
Mendeleyev Ridge
Wrangel Plain
Laptev Sea
A R C T I C
Makarov Basin
Lomonosov Ridge
Amundsen Basin
Gakkel Ridge
Nansen Plain
Kara Sea
Yenisey
Ob'
Hudson Bay
Baffin Basin
Barents Plain
Barents Sea
O C E A N
EUROPE
Davis Strait
Labrador Basin
Reykjanes Basin
Denmark Strait
Iceland Plateau
Norwegian Basin
Reykjanes Ridge
Iceland Basin
ATLANTIC OCEAN

Atlantic globe labels
Labrador Sea
Arctic Circle
Labrador Basin
Northwest Atlantic Mid-Ocean Canyon
Charlie-Gibbs Fracture Zone
EUROPE
Great Lakes
St. Lawrence
NORTH AMERICA
Newfoundland Basin
Oceanographer Fracture Zone
East Azores Fracture Zone
Bay of Biscay
Mississippi
Hatteras Plain
Atlantis Fracture Zone
Fracture Zone
Sargasso Sea
Gulf of Mexico
Tropic of Cancer
A T L A N T I C
Cape Verde Basin
AFRICA
Caribbean Sea
Kane Fracture Zone
Barracuda Fracture Zone
Gambia Plain
Demerara Plain
Vema Fracture Zone
Niger
Orinoco
Doldrums Fracture Zone
Sierra Leone Basin
Four North Fracture Zone
Romanche Fracture Zone
Guinea Basin
Gulf of Guinea
Benue
Equator
Amazon
SOUTH AMERICA
Brazil Basin
Ascension Fracture Zone
O C E A N
Angola Basin
São Francisco
Bode Verde Fracture Zone
Saint Helena Fracture Zone
Tropic of Capricorn
Paraná
Rio Grande Fracture Zone
Walvis Ridge
Orange River
Tristan da Cunha Fracture Zone
Cape Basin
Gough Fracture Zone
Argentine Basin
Agulhas Basin
Mid-Atlantic Ridge
Atlantic-Indian Ridge
SOUTHERN OCEAN
Weddell Plain
Antarctic Circle
Weddell Sea
ANTARCTICA

Indian globe labels
Caspian Sea
Brahmaputra
Ganges
Tropic of Cancer
Arabian Sea
Arabian Basin
Bay of Bengal
Indus
Ganges
Alula-Fartak Trench
Owen Fracture Zone
Ceylon Plain
Mid-Indian Ridge
Chagos-Laccadive Plateau
Chagos Trench
AFRICA
Equator
Lake Victoria
Somali Basin
I N D I A N
Ninetyeast Ridge
Mascarene Plateau
Carlsberg Ridge
Vema Fracture Zone
Mid-Indian Basin
Argo Fracture Zone
Egeria Fracture Zone
Tropic of Capricorn
Madagascar Basin
O C E A N
Madagascar Plateau
Prince Edward Fracture Zone
Crozet Basin
Southeast
Natal Basin
Southwest Indian Ridge
Agulhas Plateau
Crozet Plateau
Kerguelen Plateau
Agulhas Basin
Atlantic-Indian Basin
Enderby Plain
SOUTHERN
Antarctic Circle
ANT

Ocean currents

Surface currents are driven by winds and by the Earth's rotation. Together these cause large circular flows of water over the surface of the oceans, called gyres. Deep sea currents are driven by changes in the salinity or temperature of surface water. These changes cause the water to become denser and sink, forcing horizontal movements of deeper water.

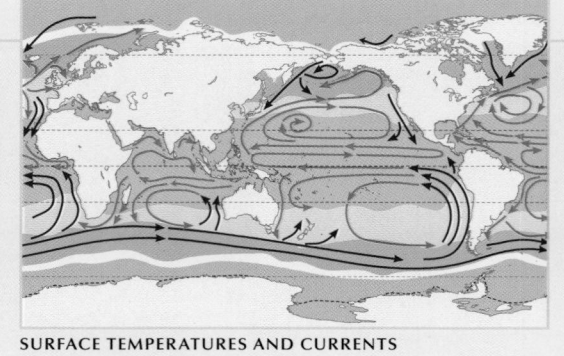

Surface temperature and currents

---- ice-shelf (below 32°F / 0°C)

▓ sea-ice* (average) below 28°F / -2°C

□ sea-water 28–32°F / -2 to 0°C
* sea-water freezes at 28.4°F / -1.9°C

32–50°F / 0–10°C
50–68°F / 10–20°C
68–86°F / 20–30°C

→ warm current
→ cold current

SURFACE TEMPERATURES AND CURRENTS

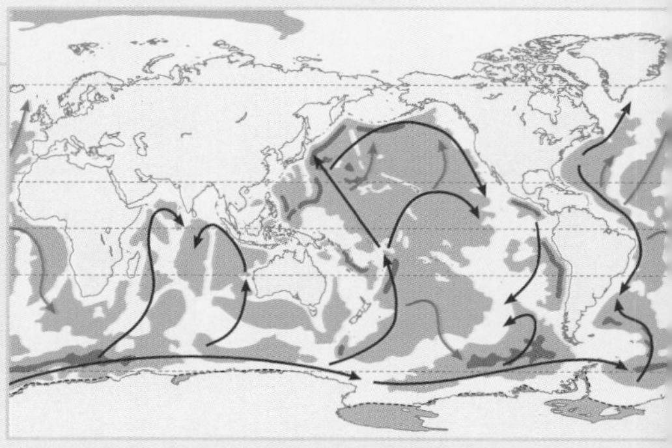

DEEP SEA TEMPERATURES AND CURRENTS

The ocean floor

The ages of seafloor rocks increase in parallel bands outward from central ocean ridges. At these ridges, new oceanic crust is continuously created from lava that erupts from below the seafloor and then cools to form solid rock. As this new crust forms, it gradually pushes older crust away from the ridge.

Ages of the ocean crust

- 0–5 million years
- 5–21 million years
- 21–38 million years
- 38–65 million years
- 65–140 million years
- 140–190 million years
- continental shelf
- no data

Tides

Tides are caused by gravitational interactions between the Earth, Moon, and Sun. The strongest tides occur when the three bodies are aligned and the weakest when the Sun and Moon align at right angles

Strongest tides

Weakest tides

Gravitational pull from the Sun

Tidal bulges created by gravitational interactions

Earth

Moon

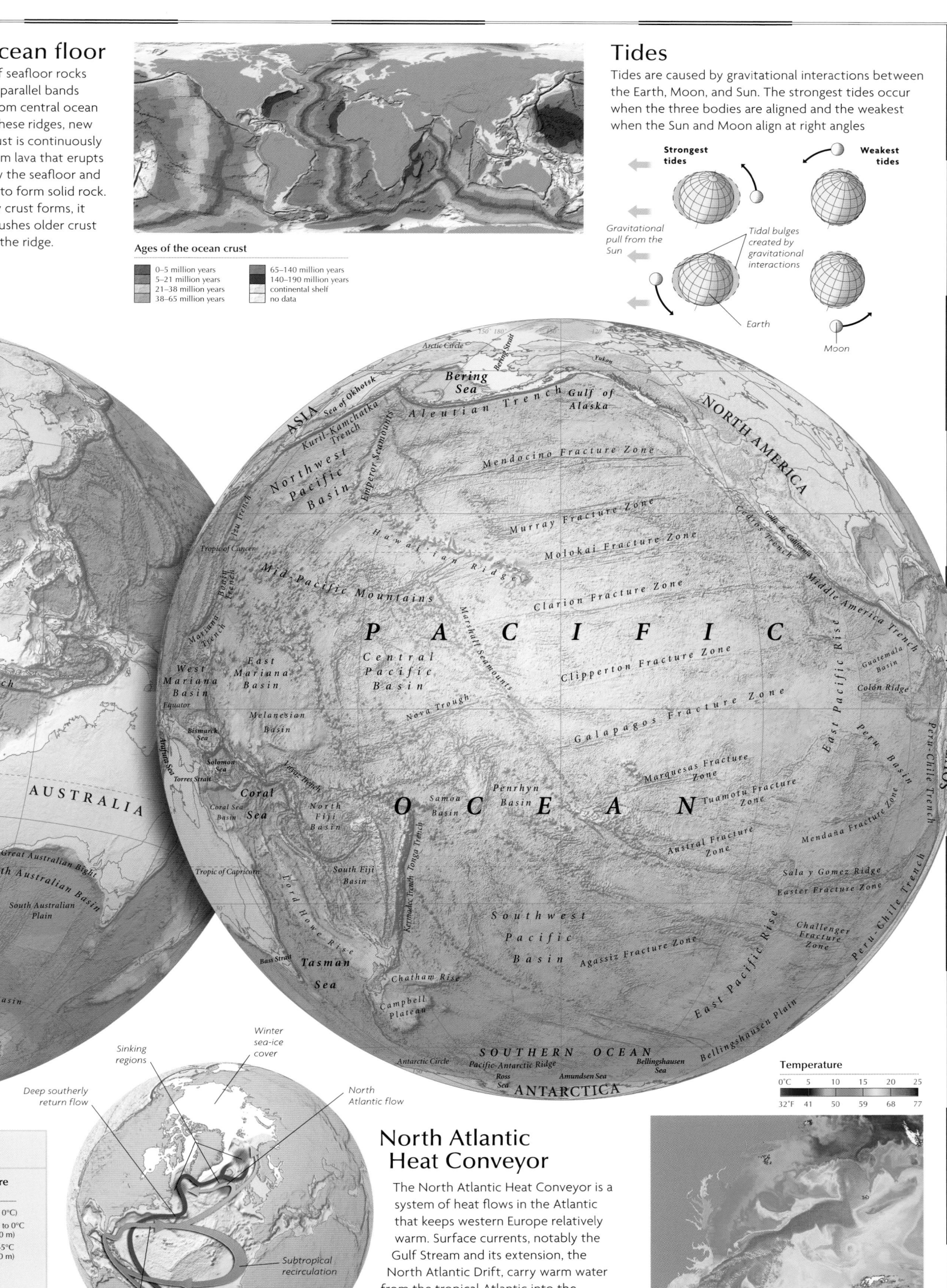

ASIA
Yellow River
Yangtze
Mekong
Gulf of Thailand
Strait of Malacca
Investigator Ridge
Java Trench
Wharton Basin
Indiaman Ridge
Perth Basin
Hatton Ridge
antena Fracture Zone
Great Australian Bight
South Australian Basin
AUSTRALIA
South Australian Plain
dian Ridge
South Indian Basin
CEAN
ICA

Arctic Circle
Bering Strait
Yukon
ASIA
Sea of Okhotsk
Kuril–Kamchatka Trench
Northwest Pacific Basin
Emperor Seamounts
Bering Sea
Aleutian Trench
Gulf of Alaska
NORTH AMERICA
Mendocino Fracture Zone
Murray Fracture Zone
Izu Trench
Tropic of Cancer
Hawaiian Ridge
Molokai Fracture Zone
Golfo de California
Cedros Trench
Mid-Pacific Mountains
Marshall Seamounts
Clarion Fracture Zone
Middle America Trench
Mariana Trench
West Mariana Basin
East Mariana Basin
Central Pacific Basin
PACIFIC
Clipperton Fracture Zone
Guatemala Basin
Colón Ridge
Equator
Nova Trough
Galapagos Fracture Zone
East Pacific Rise
Peru Basin
SOUTH AMERICA
Bismarck Sea
Melanesian Basin
Solomon Sea
Torres Strait
Coral Sea Basin
Coral Sea
Vanua Trench
North Fiji Basin
OCEAN
Samoa Basin
Penrhyn Basin
Marquesas Fracture Zone
Tuamotu Fracture Zone
Mendaña Fracture Zone
Peru–Chile Trench
Tropic of Capricorn
South Fiji Basin
Kermadec Trench Tonga Trench
Austral Fracture Zone
Sala y Gomez Ridge
Easter Fracture Zone
Lord Howe Rise
Southwest Pacific Basin
Agassiz Fracture Zone
East Pacific Rise
Challenger Fracture Zone
Bass Strait
Tasman Sea
Chatham Rise
Campbell Plateau
East Pacific Rise
Bellingshausen Plain
Antarctic Circle
SOUTHERN OCEAN
Pacific-Antarctic Ridge
Ross Sea
Amundsen Sea
Bellingshausen Sea
ANTARCTICA

Sinking regions

Winter sea-ice cover

Deep southerly return flow

North Atlantic flow

Subtropical recirculation

Gulf Stream

Deep sea temperature and currents

- ice-shelf (below 32°F / 0°C)
- sea-water 28–32°F / -2 to 0°C (below 16,400 ft / 5000 m)
- sea-water 32–41°F / 0–5°C (below 13,120 ft / 4000 m)
- → primary currents
- → secondary currents

North Atlantic Heat Conveyor

The North Atlantic Heat Conveyor is a system of heat flows in the Atlantic that keeps western Europe relatively warm. Surface currents, notably the Gulf Stream and its extension, the North Atlantic Drift, carry warm water from the tropical Atlantic into the northeastern Atlantic. There, the heat they supply is released, warming Europe, while the water itself cools and sinks. This cold water then returns at depth towards the equator.

Temperature

0°C	5	10	15	20	25
32°F	41	50	59	68	77

A key part of the North Atlantic Heat Conveyor is the warm Gulf Stream, visible as the dark red ribbon in this Atlantic sea-surface temperature map.

Global Climate

The climates of different regions on Earth are the typical long-term patterns of temperature and humidity in those regions. By contrast, weather consists of short-term variations in factors such as wind, rainfall, and sunshine. Climates are determined primarily by the Sun's variable heating of different parts of Earth's atmosphere and oceans, and by Earth's rotation. These factors drive the ocean currents and prevailing winds, which in turn redistribute heat energy and moisture between the equator and poles, and between sea and land. Most scientists think that major changes are currently occurring in global climate due to the effects of rising carbon dioxide levels in the atmosphere.

The atmosphere

Earth's atmosphere is a giant ocean of air that surrounds the planet. It extends to a height of about 625 miles (1000 km) but has no distinct upper boundary. The Sun's rays pass through the atmosphere and warm Earth's surface, causing the air to move and water to evaporate from the oceans.

Global air circulation

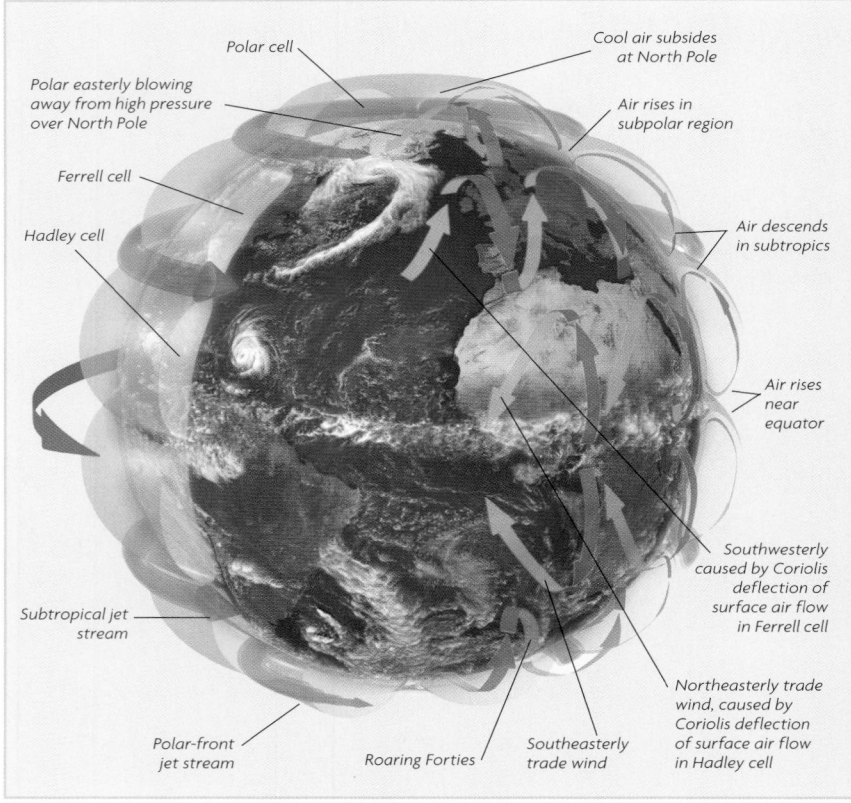

Polar cell

Polar easterly blowing away from high pressure over North Pole

Ferrell cell

Hadley cell

Cool air subsides at North Pole

Air rises in subpolar region

Air descends in subtropics

Air rises near equator

Southwesterly caused by Coriolis deflection of surface air flow in Ferrell cell

Northeasterly trade wind, caused by Coriolis deflection of surface air flow in Hadley cell

Subtropical jet stream

Polar-front jet stream

Roaring Forties

Southeasterly trade wind

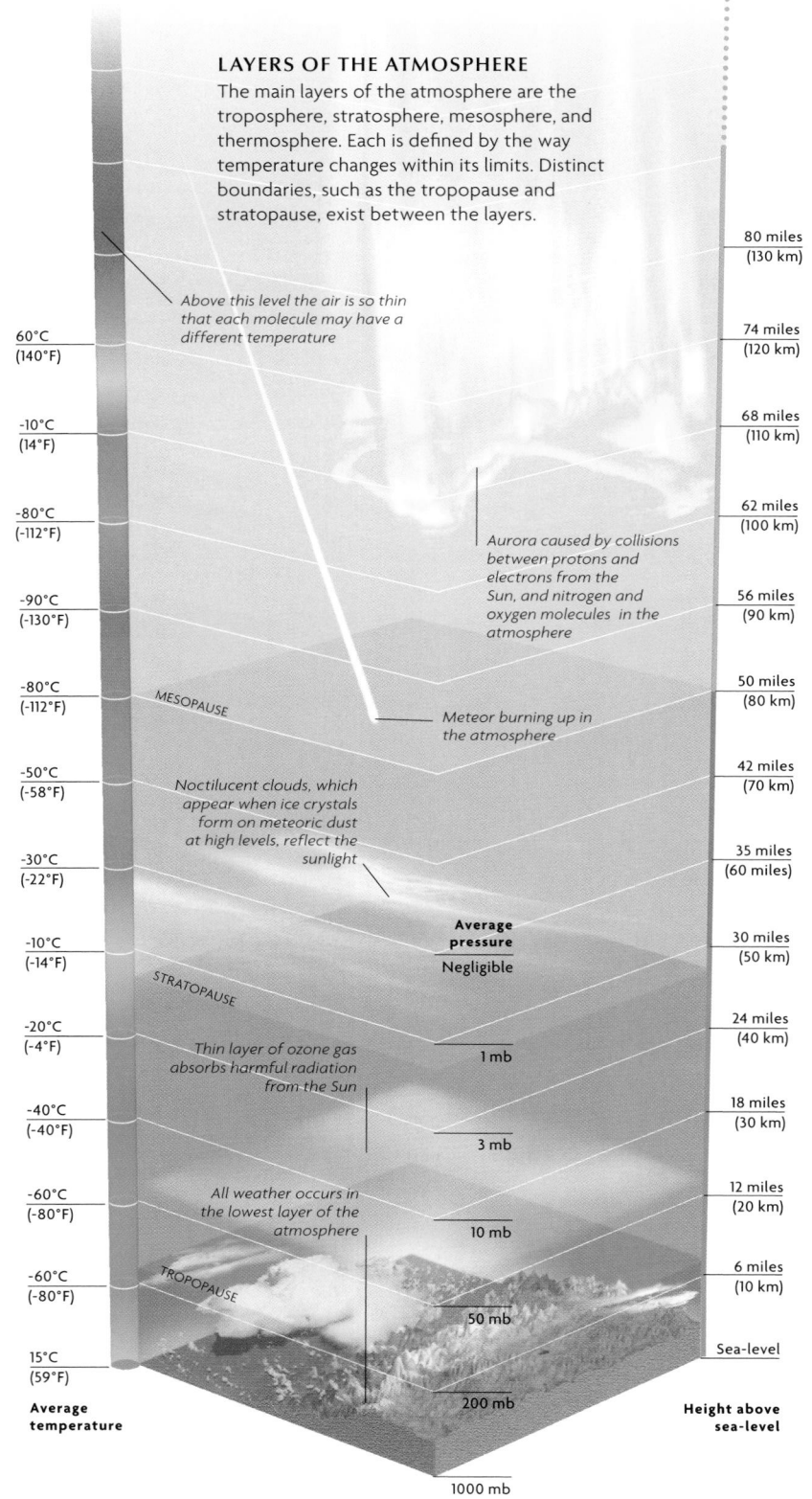

LAYERS OF THE ATMOSPHERE

The main layers of the atmosphere are the troposphere, stratosphere, mesosphere, and thermosphere. Each is defined by the way temperature changes within its limits. Distinct boundaries, such as the tropopause and stratopause, exist between the layers.

Above this level the air is so thin that each molecule may have a different temperature

Aurora caused by collisions between protons and electrons from the Sun, and nitrogen and oxygen molecules in the atmosphere

MESOPAUSE

Meteor burning up in the atmosphere

Noctilucent clouds, which appear when ice crystals form on meteoric dust at high levels, reflect the sunlight

STRATOPAUSE

Thin layer of ozone gas absorbs harmful radiation from the Sun

All weather occurs in the lowest layer of the atmosphere

TROPOPAUSE

Average temperature		Height above sea-level
60°C (140°F)		80 miles (130 km)
-10°C (14°F)		74 miles (120 km)
-80°C (-112°F)		68 miles (110 km)
-90°C (-130°F)		62 miles (100 km)
-80°C (-112°F)		56 miles (90 km)
-50°C (-58°F)		50 miles (80 km)
-30°C (-22°F)		42 miles (70 km)
-10°C (-14°F)	Average pressure Negligible	35 miles (60 miles)
-20°C (-4°F)	1 mb	30 miles (50 km)
-40°C (-40°F)	3 mb	24 miles (40 km)
-60°C (-80°F)	10 mb	18 miles (30 km)
-60°C (-80°F)	50 mb	12 miles (20 km)
15°C (59°F)	200 mb	6 miles (10 km)
	1000 mb	Sea-level

Winds, currents, and climate

Earth has 12 climatic zones, ranging from ice-cap and tundra to temperate, arid (desert), and tropical zones. Each of these zones features a particular combination of temperature and humidity. The effects of prevailing winds, ocean currents of both the warm and cold variety, as well as latitude and altitude, all have an important influence on a region's climate. For example, the climate of western Europe is influenced by the effects of the warm North Atlantic Drift current.

● **THERMOSPHERE**
This layer extends from a height of 50 miles (80 km) upward. Its temperature increases rapidly above a height of 60 miles (90 km), due to absorption of highly energetic solar radiation.

● **MESOSPHERE**
The temperature of the lower part of this layer stays constant with height; but above 35 miles (55 km), it drops, reaching -112°F (80°C) at the mesopause.

● **STRATOSPHERE**
The temperature of the stratosphere is a fairly constant -76° F (-60°C) up to an altitude of about 12 miles (20 km), then increases, due to absorption of ultraviolet radiation.

● **TROPOSPHERE**
This layer extends from Earth's surface to a height of about 10 miles (16 km) at the equator and 5 miles (8 km) at the poles. Air temperature in this layer decreases with height.

Arctic Circle January

July

Alaska Current

North Pacific Current

WESTERLIES

NORTH

EAST

North Equatorial Current

TRADES

Equatorial Counter Current

Doldrums · El Niño

Equator

South Equatorial Current

SOUTH

EAST

TRADES

Tropic of Capricorn

California Current

Northern Equatorial Current

Tropic of Cancer

January

July

WESTERLIES

West Wind Drift

Antarctic Circle

Air moves within giant atmospheric cells called Hadley, Ferrell, and polar cells. These cells are caused by air being warmed and rising in some latitudes, such as near the equator, and sinking in other latitudes. This north-south circulation combined with the Coriolis effect *(below)* produces the prevailing surface winds.

THE CORIOLIS EFFECT

Air moving over Earth's surface is deflected in a clockwise direction in the northern hemisphere and counterclockwise in the south. Known as the Coriolis effect, and caused by Earth's spin, these deflections to the air movements produce winds such as the trade winds and westerlies.

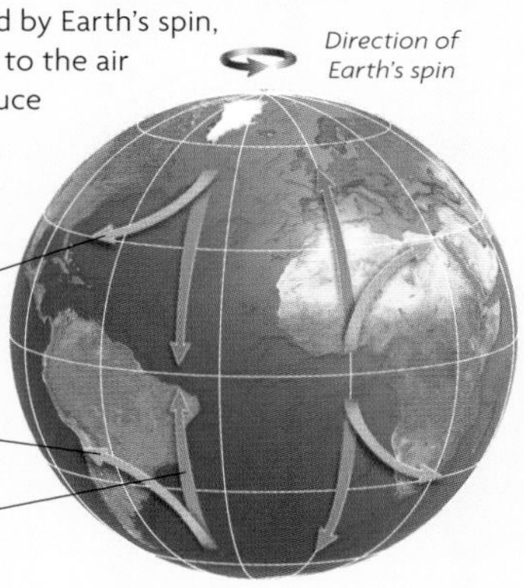

Direction of Earth's spin

Deflected clockwise

Deflected counterclockwise

Initial direction

Temperature and precipitation

The world divides by latitude into three major temperature zones: the warm tropics, the cold polar regions; and an intermediate temperate zone. In addition, temperature is strongly influenced by height above sea level. Precipitation patterns are related to factors such as solar heating, atmospheric pressure, winds, and topography. Most equatorial areas have high rainfall, caused by moist air being warmed and rising, then cooling to form rain clouds. In areas of the subtropics and near the poles, sinking air causes high pressure and low precipitation. In temperate regions rainfall is quite variable.

AVERAGE JANUARY TEMPERATURE

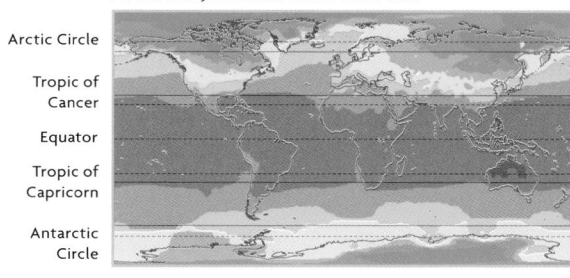

AVERAGE JANUARY RAINFALL

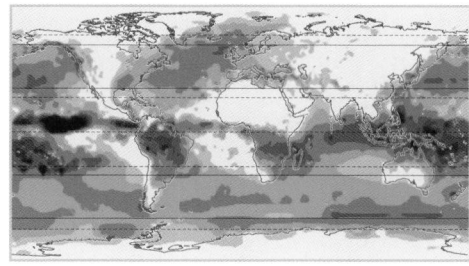

AVERAGE JULY TEMPERATURE

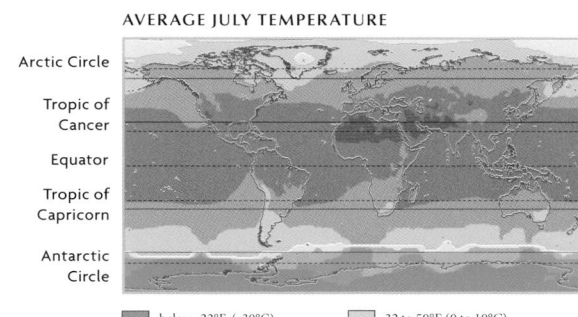

AVERAGE JULY RAINFALL

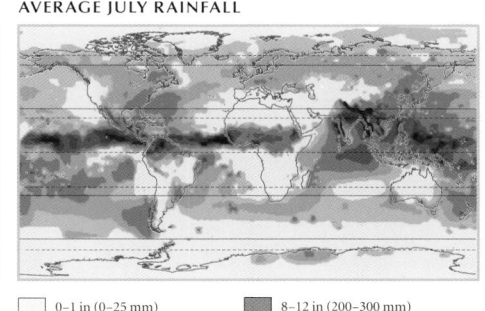

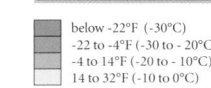
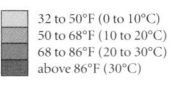

below -22°F (-30°C)	32 to 50°F (0 to 10°C)
-22 to -4°F (-30 to - 20°C)	50 to 68°F (10 to 20°C)
-4 to 14°F (-20 to - 10°C)	68 to 86°F (20 to 30°C)
14 to 32°F (-10 to 0°C)	above 86°F (30°C)

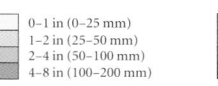
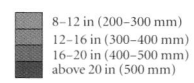

0–1 in (0–25 mm)	8–12 in (200–300 mm)
1–2 in (25–50 mm)	12–16 in (300–400 mm)
2–4 in (50–100 mm)	16–20 in (400–500 mm)
4–8 in (100–200 mm)	above 20 in (500 mm)

Ocean currents, winds and climatic regions

Climate zones

ice-cap	temperate
subarctic	warm temperate
tundra	mediterranean
continental	semi-arid
	arid
	hot humid
	humid-equatorial
	tropical

Ocean currents

- warm
- cold

Prevailing winds

→ warm
→ cold

Local winds

→ warm
→ cold
⇢ *July* seasonal winds (cold or warm)

POLAR EASTERLIES

WESTERLIES

NORTH EAST TRADES

SOUTH EAST TRADES

SOUTH EAST TRADES

NORTH EAST TRADES

WESTERLIES

POLAR EASTERLIES

Doldrums

Doldrums

Doldrums

Arctic Circle

Tropic of Cancer

Equator

Tropic of Capricorn

Antarctic Circle

Gulf Stream
North Atlantic Drift
Labrador Current
North Equatorial Current
Canary Current
Brazil Current
Falkland Current
Benguela Current
Peru (Humboldt)
Pamperos
South Equatorial Current
West Wind Drift
Equatorial Counter Current
South Equatorial Current
Northeast Monsoon October
Southwest Monsoon
Monsoon Drift
Kuro Siwo Current
Typhoon July-October
North Equatorial Current
Equatorial Counter Current
South Equatorial Current
Southeast Monsoon October-March
West Australian Current
Queensland
Hurricanes January
West Wind Drift

Mistral
Föhn
Bora
Sirocco
Khamsin
Harmattan
Haboob
Buran
Willy Willies January

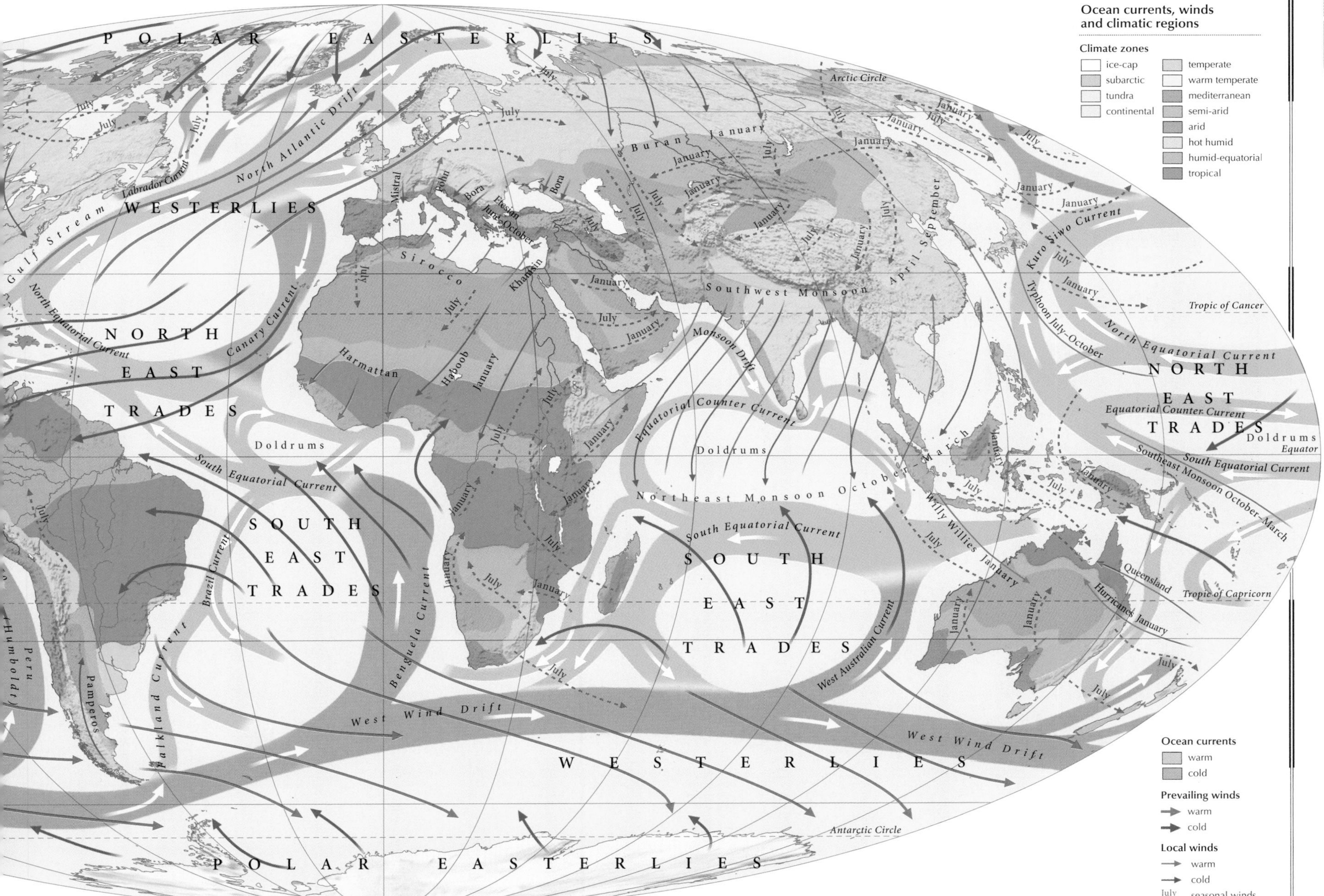

Life on Earth

A unique combination of an oxygen-rich atmosphere and plentiful surface water is the key to life on Earth, where few areas have not been colonized by animals, plants, or smaller life-forms. An important determinant of the quantity of life in a region is its level of primary production—the amount of energy-rich substances made by organisms living there, mainly through the process of photosynthesis. On land, plants are the main organisms responsible for primary production; in water, algae fulfil this role. These primary producers supply food for animals. Primary production is affected by climatic, seasonal, and other local factors. On land, cold and aridity restrict the quantity of life in a region, whereas warmth and regular rainfall allow a greater diversity of species. In the oceans, production is mainly affected by sunlight levels, which reduce rapidly with depth, and by nutrient availability.

POLAR REGIONS
Ice restricts life in these regions to just a few species, such as polar bears in the Arctic.

Biogeographical regions

Earth's biogeographical regions, or biomes, are communities where certain species of plants and animals coexist within the constraints of particular climatic conditions. They range from tundra to various types of grassland, forest, desert, and marine biomes such as coral reefs. Factors like soil richness, altitude, and human activities such as deforestation can affect the local distribution of living species in each biome.

TEMPERATE GRASSLAND
Also known as steppe or prairie, grassland of this type occurs mainly in the northern hemisphere and in South America (the Pampas).

NEEDLELEAF FOREST
These vast forests of coniferous trees cover huge areas of Canada, Siberia, and Scandinavia.

TROPICAL GRASSLAND
This type of grassland is widespread in Africa and South America, supporting large numbers of grazing animals and their predators.

World biomes

- ice
- tundra
- temperate coniferous forest
- temperate broadleaf mixed forest
- temperate grassland
- mediterranean
- desert and shrubland
- boreal forest/taiga

Animal diversity

The number of animal species, and the range of genetic diversity within the populations of those species, determines the level of animal diversity within each country or other region of the world. The animals that are endemic to a region—that is, those found nowhere else on the planet—are also important in determining its level of animal diversity.

Number of animal species per country

- more than 2,000
- 1000–1999
- 700–999
- 400–699
- 200–399
- 100–199
- 0–99
- data not available

TUNDRA
With little soil and large areas of frozen ground, the tundra is largely treeless, though briefly clothed by small flowering plants in summer.

TEMPERATE RAIN FOREST
Occurring in mid-latitudes in areas of high rainfall, these forests may be predominantly coniferous or mixed with deciduous species.

CORAL REEFS
Occurring in clear tropical waters, coral reefs support an extraordinary diversity of species, especially fish and many types of invertebrate.

MOUNTAINS
In high mountain areas only a few hardy species of plant will grow above the tree-line.

TROPICAL RAINFOREST
Characterized by year-round warmth and high rainfall, tropical rainforests contain the highest diversity of plant and animal species on Earth.

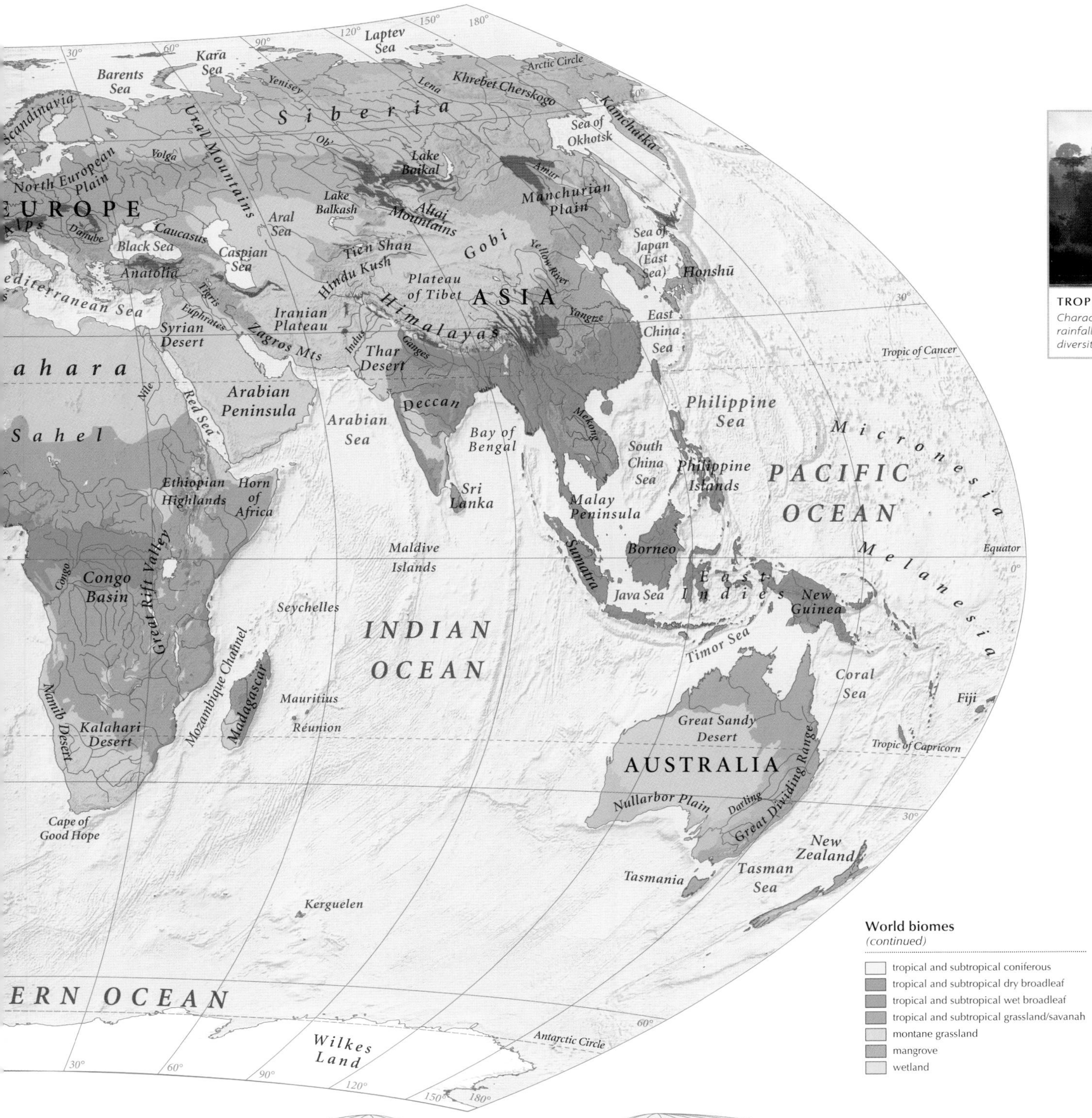

HOT DESERT
Only a few highly adapted species can survive in hot deserts, which occur mainly in the tropics.

World biomes
(continued)

- tropical and subtropical coniferous
- tropical and subtropical dry broadleaf
- tropical and subtropical wet broadleaf
- tropical and subtropical grassland/savanah
- montane grassland
- mangrove
- wetland

OPEN OCEAN
Earth's largest biome, the oceans are home to a vast diversity of fish, mammals, invertebrates, and algae.

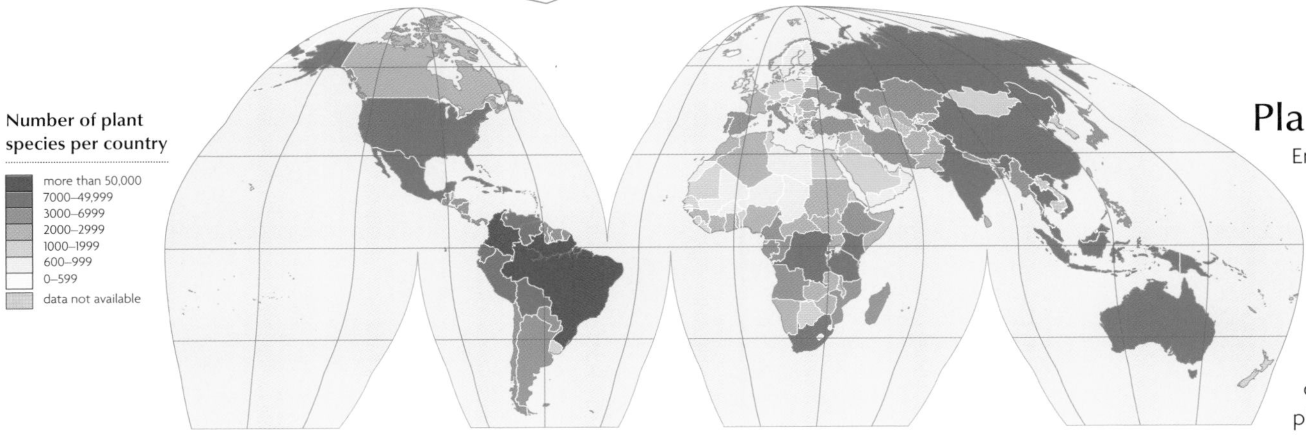

Number of plant species per country

- more than 50,000
- 7000–49,999
- 3000–6999
- 2000–2999
- 1000–1999
- 600–999
- 0–599
- data not available

Plant diversity

Environmental conditions, particularly climate, soil type, and the extent of competition with other living organisms, influence the development of plants into distinctive forms and thus also the extent of plant diversity. Human settlement and intervention has considerably reduced the diversity of plant species in many areas.

Man and the Environment

The impact of human activity on the environment has widened from being a matter of local concern (typically over the build-up of urban waste, industrial pollution, and smog) to affect whole ecosystems and, in recent decades, the global climate. Problems crossing national boundaries first became a major issue over acid rain, toxic waste dumping at sea, and chemical spillages polluting major rivers. Current concerns center on loss of biodiversity and vital habitat including wetlands and coral reefs, the felling and clearance of great tropical and temperate forests, overexploitation of scarce resources, the uncontrolled growth of cities and, above all, climate change.

OZONE HOLE
Man-made chlorofluorocarbons (CFCs), used in refrigeration and aerosols, damaged the ozone layer in the stratosphere which helps filter out the sun's harmful ultraviolet rays. When a seasonal ozone hole first appeared in 1985 over Antarctica, a shocked world agreed to phase out CFC use.

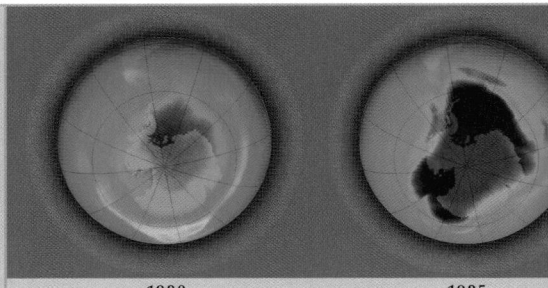
1980   1985

CO_2 emissions in 2008
(million tons)

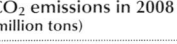
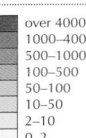

- over 4000
- 1000–4000
- 500–1000
- 100–500
- 50–100
- 10–50
- 2–10
- 0–2
- no data

Climate change

Global warming is happening much faster than Earth's normal long-term cycles of climate change. The consequences include unpredictable extreme weather and potential disruption of ocean currents. Melting ice-caps and glaciers, and warmer oceans, will raise average sea levels and threaten coastlines and cities. Food crops like wheat are highly vulnerable to changes in temperature and rainfall. Such changes can also have a dramatic affect on wildlife habitats.

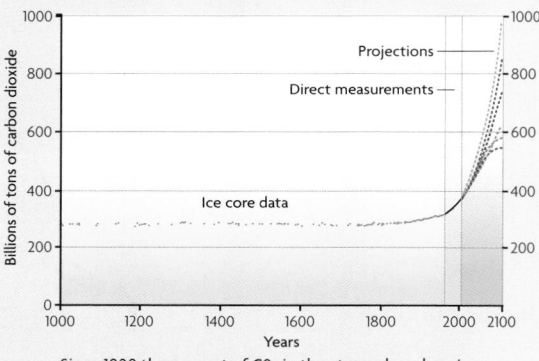

Since 1800 the amount of CO_2 in the atmosphere has risen sharply. Urgent worldwide action to control emissions is vital to stabilize the level by the mid 21st century.

THE GREENHOUSE EFFECT
Some solar energy, reflected from the Earth's surface as infra red radiation, is reflected back as heat by "greenhouse gases" (mainly carbon dioxide and methane) in the atmosphere. Nearly all scientists now agree that an upsurge in emissions caused by humans burning fossil fuel has contributed to making the resultant warming effect a major problem.

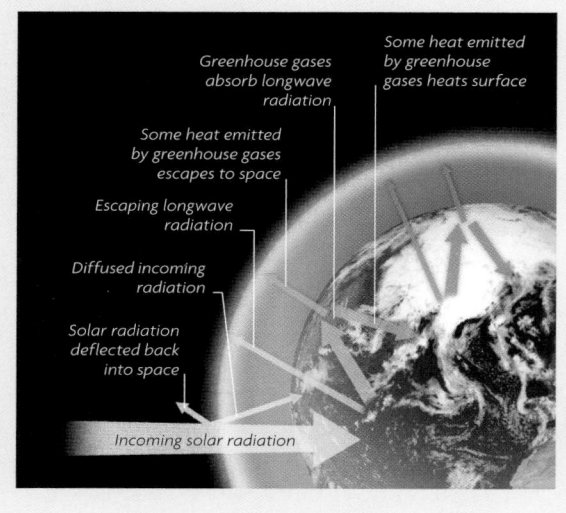

Greenhouse gases absorb longwave radiation

Some heat emitted by greenhouse gases heats surface

Some heat emitted by greenhouse gases escapes to space

Escaping longwave radiation

Diffused incoming radiation

Solar radiation deflected back into space

Incoming solar radiation

CO_2 emissions by country in 2014
(million tons)

TOTAL: 35,669

China 10,540
Other 11,349
United States 5334
Brazil 501
Canada 565
South Korea 610
Iran 618
Germany 767
Japan 1278
India 2341
Russian Federation 1766

FOOD AND LAND USE
The world has about five billion hectares of agriculturally useful land, well under one hectare per person. The majority of this is pasture for grazing. Crops are grown on about 30 percent (and nearly a fifth of cropland is artificially irrigated). Mechanized farming encouraged vast single crop "monocultures," dependent on fertilizers and pesticides. North America's endless prairies of wheat and corn, huge soybean plantations, and southern cotton fields are mirrored in Ukraine (wheat), Brazil and Argentina (soya) and Uzbekistan (cotton). Elsewhere, scarce farmland can be squeezed by the housing needs of growing urban populations. Current interest in crop-derived "biofuels" means further pressure to grow food more productively on less land.

Intensive farming. Satellite photography picks up the greenhouses that now cover almost all the land in this Spanish coastal area southwest of Almeria.

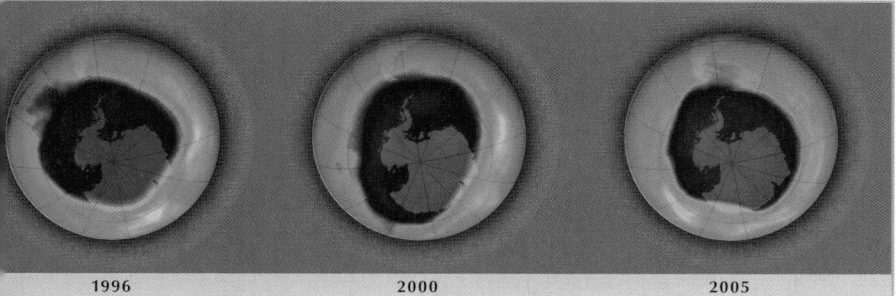

DEFORESTATION

At current rates of destruction, all tropical forests, and most old-growth temperate forest, will be gone by 2090. The Amazon rain forest is a valuable genetic resource, containing innumerable unique plants and animals, as well as acting as a crucial natural "sink" for absorbing climate-damaging carbon dioxide. Stemming the loss of these precious assets to logging and farming is one of the major environmental challenges of modern times.

Over 25,000 sq miles (60,000 sq km) of virgin rain forest are cleared annually by logging and agricultural activities, destroying an irreplaceable natural resource.

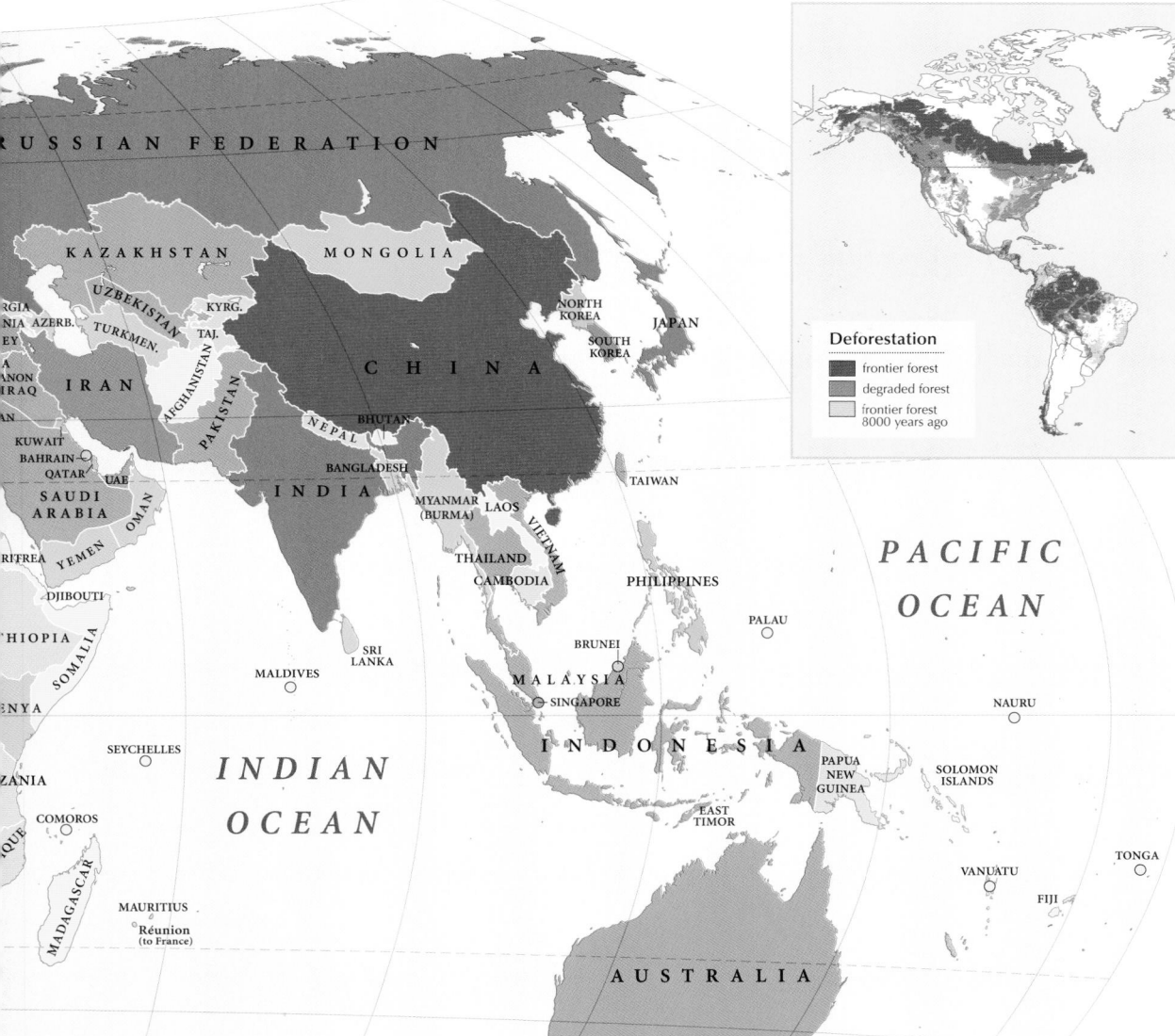

Deforestation
- frontier forest
- degraded forest
- frontier forest 8000 years ago

GLACIATION

The world's glaciers and ice sheets have been in retreat for decades, forming less new ice at high altitudes than they lose by melting lower down. The loss of ice from Greenland doubled between 1996 and 2005, with alarming implications for rising sea levels. Other dramatic evidence of global warming includes the rapid thinning of ice in the Himalayas, and the highly symbolic loss of the snowcap on Africa's Mount Kilimanjaro.

Helheim Glacier 2001
The Helheim glacier *(above)* almost completely fills this image, with the leading edge visible on the righthand side, and was in a relatively stable condition.

Helheim Glacier 2005
By 2005 *(right)* it had retreated by 2.5 miles (4 km).

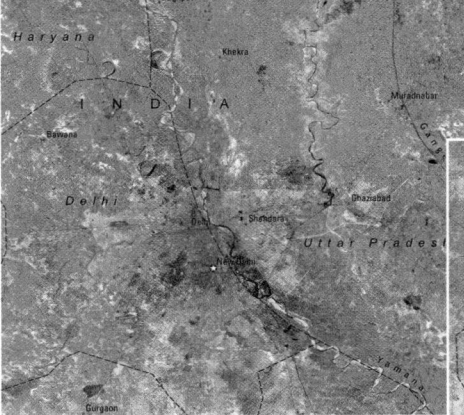

Delhi 1971
In 1971 Delhi *(above)* occupied an area of about 190 sq miles (500 sq km).

Delhi 1999
By 1999 *(right)* it had sprawled to cover 500 sq miles (1300 sq km). It vies with Mumbai in the southwest to be the sub-continent's most populous city, fast approaching 20 million people.

CITY GROWTH

The world in 2015 had 15 cities with populations over 20 million. The number of cities with populations between 10 and 20 million has surpassed 20 and continues to rise. The search for work, and the hope of escape from rural poverty, drives migration from rural to urban areas across the developing world. Urban dwellers now amount to more than half the world's population, and consume more resources than their rural counterparts.

Population and Settlement

Earth's human population is projected to rise from its current level of 7.3 billion to between 8.1 and 11 billion by the year 2050. The distribution of this population is very uneven and is dictated by climate, terrain, and by natural and economic resources. Most people live in coastal zones and along the valleys of great rivers such as the Ganges, Indus, Nile, and Yangtze. Deserts cover over 20 percent of Earth's surface but support less than 5 percent of its human population. Over half the world's population live in cities—most of them in Asia, Europe, and North America—as a result of mass migrations that have occurred from rural areas as people search for jobs. Many of these people live in so-called "megacities"—sprawling urban areas that have populations higher than 10 million.

Population density by country (population per sq mile)

over 2600	260-389	65–129
775-2599	195–259	26–64
390-774	130–194	0–25

Population density

A few regions, including Europe, India, and much of eastern Asia, have extremely high population densities. Within these areas, a few spots, such as Monaco and Hong Kong, have densities of over 12,900 per sq mile (5000 people per sq km). Other regions (mostly desert, mountain, ice cap, tundra, or thickly forested areas) have densities close to zero –examples include large areas of Australia, western China, Siberia, North Africa, Canada, Greenland, and much of the Amazon rain forest region.

NORTH AMERICA

World population World land area
8% 17.0%

EUROPE

World population World land area
10% 7.1%

Million-person cities

In the year 1900 there were fewer than 20 cities in the world with a population that exceeded one million. By 1950 there were 83 such cities, and by the year 2015 there were more than 500 such cities, 100 of them in China alone, with another 54 in India, 20 in Brazil, and 14 in Japan.

Million-cities in 1900

• Cities over 1 million in population

Population density (persons per sq mile)

520–2600
260–520
130–260
52–130
26–52
13-26
3–13
0–3

SOUTH AMERICA

World population World land area
6% 11.8%

Million-cities in 1950

ANTARCTICA

World population World land area
0.0% 8.9%

Million-cities in 2006

Tokyo urban sprawl

—— City boundary, 1860   —— City boundary, 1964

GREATER TOKYO

The Greater Tokyo Area is the most populous urban area in the world, with an estimated head count in 2015 of 37.8 million. It includes Tokyo City, which has a population of about 12 million, and adjoining cities such as Yokohama. This satellite photograph shows the Greater Tokyo Area today, and also the boundaries of Tokyo City in 1860 (red) and 1964 (yellow).

Migration

Every year about 200 million people – 3 percent of the global population – change their country of residence. Emigration rates are generally highest in countries affected by wars or that have suffered economic woes, natural disasters, or where groups of people have been oppressed or persecuted. Immigration rates are generally highest in stable, developed countries, such as the USA, Spain and Canada, which usually have low birth rates and need immigrants to provide economic support to their ageing populations.

NET MIGRATION

The map at right shows the net migration rate for each country. A positive value means more people are entering the country than leaving it (net immigration) and a negative value the reverse (net emigration).

Net migration
average annual migration
(thousands)

- over 300
- 200 to 300
- 100 to 200
- 0 to 100
- -100 to 0
- -200 to -100
- -300 to -200
- -400 to -300
- data unavailable

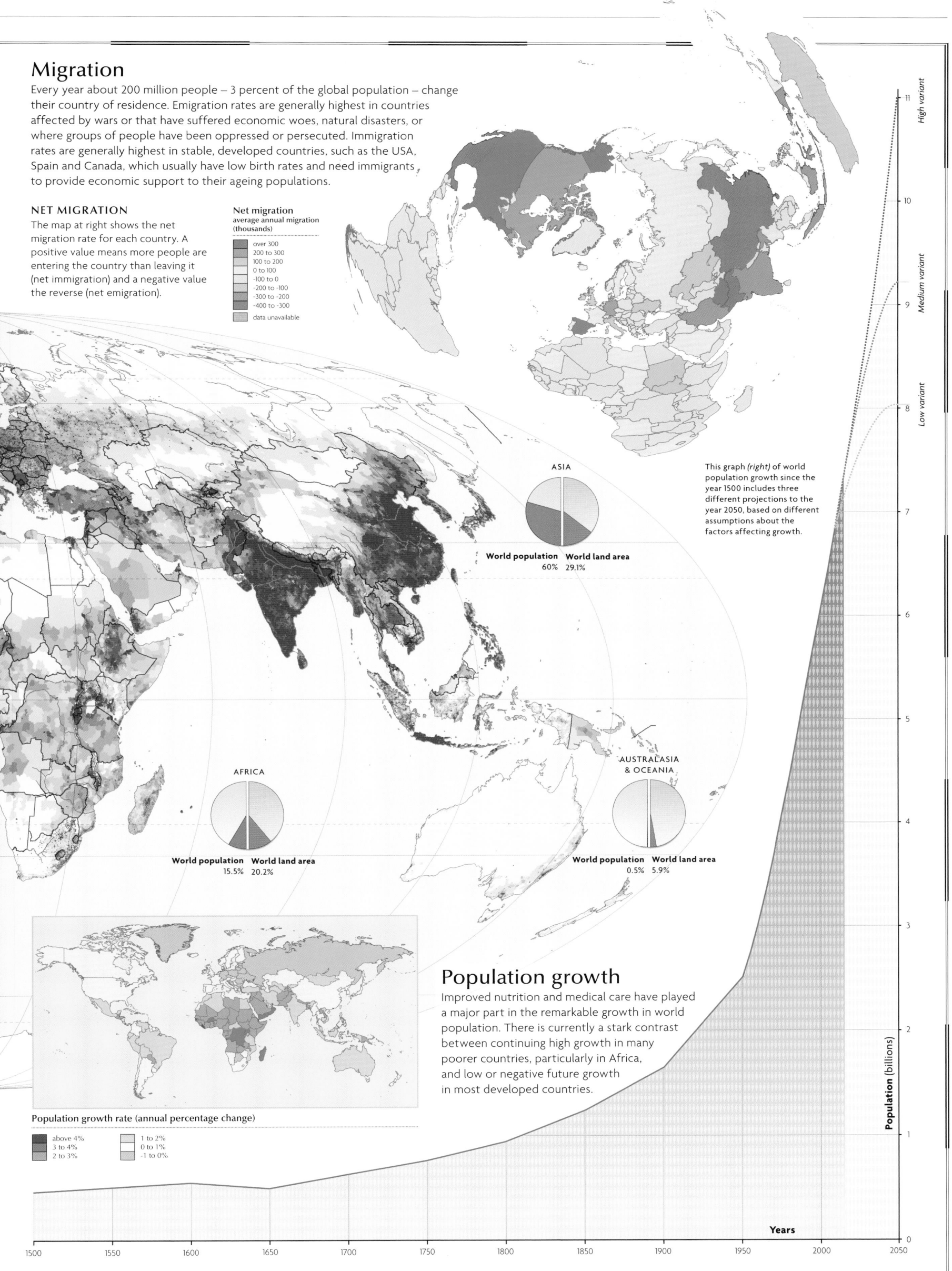

ASIA

World population 60%   **World land area** 29.1%

AFRICA

World population 15.5%   **World land area** 20.2%

AUSTRALASIA & OCEANIA

World population 0.5%   **World land area** 5.9%

This graph *(right)* of world population growth since the year 1500 includes three different projections to the year 2050, based on different assumptions about the factors affecting growth.

High variant
Medium variant
Low variant

Population growth

Improved nutrition and medical care have played a major part in the remarkable growth in world population. There is currently a stark contrast between continuing high growth in many poorer countries, particularly in Africa, and low or negative future growth in most developed countries.

Population growth rate (annual percentage change)

- above 4%
- 3 to 4%
- 2 to 3%
- 1 to 2%
- 0 to 1%
- -1 to 0%

Population (billions)

11
10
9
8
7
6
5
4
3
2
1
0

Years

1500   1550   1600   1650   1700   1750   1800   1850   1900   1950   2000   2050

Language

Over 6800 different languages exist throughout the world, each one with its own unique evolutionary history and cultural connotations. Most of these languages are spoken only by small groups of people in remote regions. Sadly these minority tongues are dying out—it is estimated that about a third will have disappeared by the year 2100. The relatively small number of widely-spoken languages have gained their current predominance and pattern of distribution through a variety of historical factors. Among these have been the economic, military, or technological success of certain peoples and cultures, differing population growth rates, and the effects of migrations and colonization.

The European Union (EU) embraces the diversity of its 28 countries and 24 official languages by providing a translation and interpretation service for the majority of its meetings and documentation. This costs around US$ 650 million per year, which equates to 1 percent of the EU budget.

Main international languages

- Chinese
- Spanish
- Arabic
- Hindi
- English
- French
- Russian
- Portuguese
- English/Spanish
- Spanish/other
- Arabic/French
- French/other
- English/other
- Arabic/other
- Hindi/English/other
- Chinese/other
- Russian/other
- English/French
- Portuguese/other
- other language

Bantu language group
Mari other language

uninhabited land

The colonial powers

Colonialism between the 15th and 20th centuries had a major influence in establishing the world prevalence of various, mainly European, languages. Britain, for example, was the colonial power in Canada, the USA (until 1776), the Indian subcontinent, Australia, and parts of Africa and the Caribbean. Hence, English is still the main (or a major) language in these areas. The same applies to France and the French language in parts of Africa and southeast Asia, and to Spain and the Spanish language in much of Latin America. For similar reasons, Portuguese is the main language in Brazil and parts of Africa, and there are many Dutch speakers in Indonesia.

This dual language sign, written in both in Hindi and English, stands outside Shimla railway station in northern India. The sign reflects India's past—the British used Shimla as their summer capital during the colonial period.

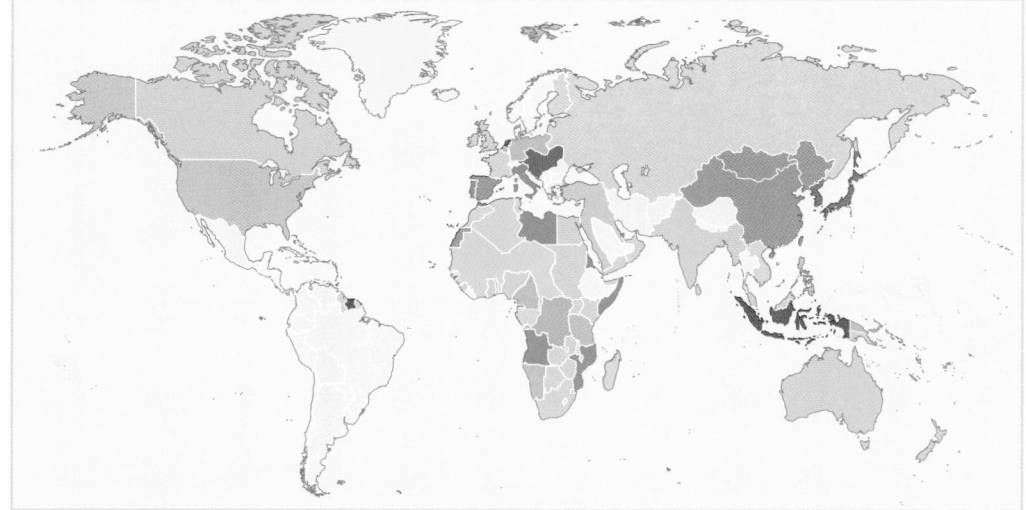

TOP TEN LANGUAGES

About 45 percent of people speak one of just ten languages as their native tongue. Mandarin Chinese is spoken by far the largest number—a situation likely to persist, as minority language speakers in China are encouraged to switch to Mandarin. English usage is also increasing, as it is the most favored language on the internet and in business circles. Wherever English is not the mother tongue, it is often the second language.

THE TEN MOST SPOKEN LANGUAGES
(number of native speakers)

- Mandarin Chinese (1.2 billion)
- Spanish (470 million)
- English (340 million)
- Hindi/Urdu (260 million)
- Arabic (240 million)
- Portuguese (200 million)
- Bengali (190 million)
- Russian (170 million)
- Japanese (130 million)
- Javanese (84 million)

Colonial Empires in 1914

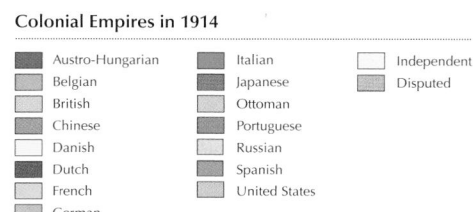

- Austro-Hungarian
- Belgian
- British
- Chinese
- Danish
- Dutch
- French
- German
- Italian
- Japanese
- Ottoman
- Portuguese
- Russian
- Spanish
- United States
- Independent
- Disputed

Religion

The spread of religion

By their nature, religions usually start off in small geographical areas and then spread. For Christianity and Islam, this spread was rapid and extensive. Buddhism diffused more slowly from around 500 BCE into a large part of Asia. The oldest religion, Hinduism, has always been concentrated in the Indian subcontinent, although its adherents in other parts of the world now number millions following migrations from India.

1ST– 7TH CENTURY
During this period, Christianity spread from its origins in the eastern Mediterranean, while Hinduism and forms of Buddhism spread in Asia. Islam became established in Arabia.

Rise and spread of the classical religions to 650 CE

- Buddhist heartland
- Chinese Confucianism/ Daoism and indigenous primal traditions
- Converted to Christianity by 600 CE
- Hinduism
- Islam under Muhammad
- Mahayana Buddhism
- Shintoism
- Zoroastrianism
- → spread of Buddhism
- → spread of Christianity
- → spread of Hinduism
- → dispersion of Jews, to 500 CE

7TH–16TH CENTURY
Islam later spread further through Asia and into parts of Africa and Europe. Christianity diffused through Europe and was then carried to many other parts of the world by colonialists and missionaries. Buddhism spread further in Asia.

About 83 percent of the world's population adheres to a religion. The remainder adopt irreligious stances such as atheism. In terms of broad similarities of belief, there are about 20 different religions in the world with more than 1 million adherents. However, the larger of these are split into several denominations, which differ in their exact beliefs and practices. Christianity, for example, includes three major groupings that have historically been in conflict—Roman Catholicism, Protestantism, and Orthodox Christianity—as well as hundreds of separate smaller groups. Many of the world's other main religious, such as Islam and Buddhism, are also subdivided.

Each year millions of Muslims visit Mecca during the the Islamic pilgrimage known as the *Hajj*

World religions c.1500 CE

- Catholic Christianity
- area converted to Catholic Christianity
- Hinduism
- Islam
- Mahayana Buddhism and Confucianism, Daoism and Shinto
- Mahayana Buddhism and Confucianism, Daoism
- Russian Orthodoxy
- Theravada Buddhism
- Tibetan Buddhism
- Aztec Empire
- Inca Empire
- → spread of Catholicism
- → spread of Islam
- → spread of Protestantism
- → spread of Russian Orthodoxy

RELIGION AROUND THE WORLD

About 72 percent of humanity adheres to one of five religions: Christianity, Islam, Hinduism, Buddhism, and Chinese traditional religion (which includes Daoism and Confucianism). Of the remainder, many are adherents of primal indigenous religions (a wide range of tribal or folk religions such as shamanism).

Buddhist (0.36 billion)
Judaism (15 million)
Sikhism (23 million)
Chinese traditional (0.36 billion)
Christianity (1.9 billion)
Primal indigenous (0.36 billion)
Hindu (0.84 billion)
Not religious (0.96 billion)
Islam (1.06 billion)

Majority religions

- Protestant Christianity
- Catholic Christianity
- Orthodox Christianity
- Shi'a Islam
- Sunni Islam
- Hinduism
- Judaism
- Theravada Buddhism
- Mahayana Buddhism
- Tibetan Buddhism
- other
- Marxism / Maoism

State policy

- ▲ secular ideologies governing
- ● communist states during 20th century
- ■ non-pluralist states

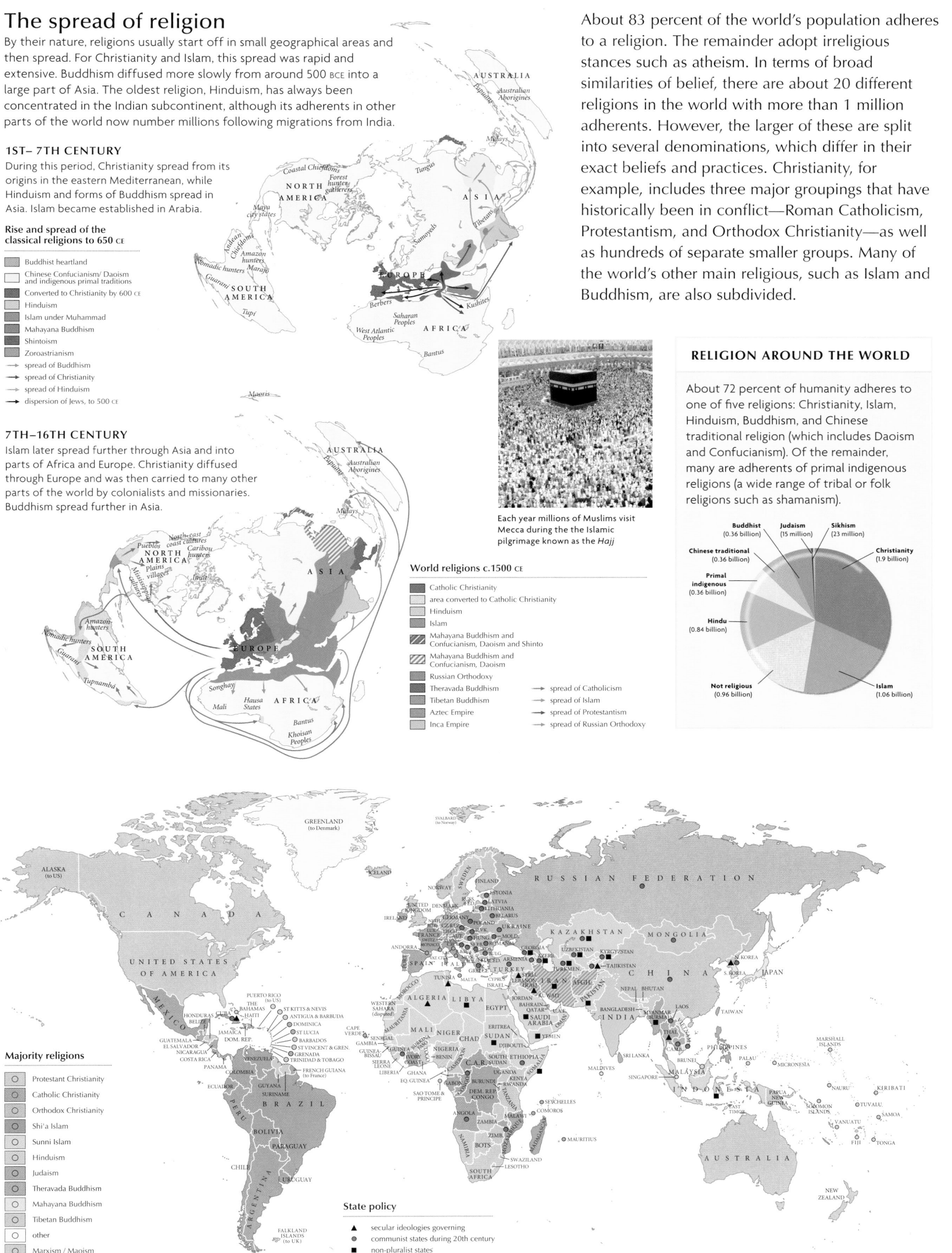

Health

On most health parameters, the countries of the world split into two distinct groups. The first of these encompass the richer, developed, countries, where medical care is good to excellent, infant mortality and the incidence of deadly infectious diseases is low, and life expectancy is high and rising. Some of the biggest health problems in these countries arise from overeating, while the two main causes of death are heart disease and cancer. The second region consists of the poorer developing countries, where medical care is much less adequate, infant mortality is high, many people are undernourished, and infectious diseases such as malaria are major killers. Life expectancy in these countries is much lower and in some cases is falling.

Life expectancy

Life expectancy has risen remarkably in developed countries over the past 50 years and has now topped 80 years in many of them. In contrast, life expectancy in many of the countries of sub-Saharan Africa has fallen well below 50, in large part due to the high prevalence of HIV/AIDS.

Many people in developed countries are now living for 15–20 years after retirement, putting greater pressure on welfare and health services.

Infant deaths and births

Infant mortality is still high in many developing nations, especially some African countries, due in part to stretched medical services. As well as lower infant mortality, the world's developed countries have much lower birth rates—greater female emancipation and easier access to contraceptives are two causative factors.

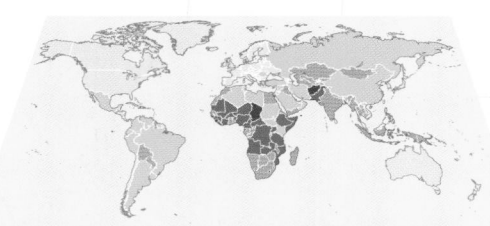

World infant mortality rates (deaths per 1000 live births)

■ above 125 ■ 75–124 ■ 35–74 ■ 15–34 □ below 15

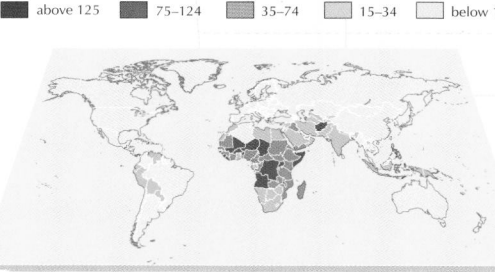

Number of births (per 1000 people)

■ above 40 ■ 30–39 ■ 20–29 □ below 20

Nutrition

Two-thirds of the world's food is consumed in developed nations, many of which have a daily calorific intake far higher than is needed by their populations. By contrast, about 800 million people in the developing world do not have enough food to meet basic nutritional needs.

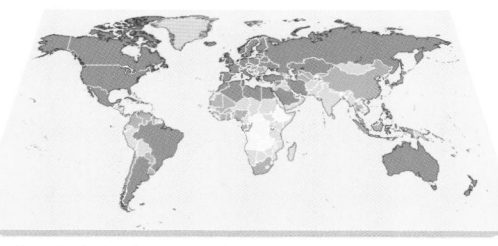

Daily calorie intake per capita

■ above 3000 ■ 2500–2999 ■ 2000–2499 □ below 2000

Life expectancy

◉ above 80 years
◉ 75–80 years
◉ 70–75 years
◉ 60–70 years
◉ 50–60 years
◉ below 50 years

United States of America: has an average life expectancy of about 78 years, with women living about 5 years longer than men.

Liberia: currently has one of the lowest life expectancies in West Africa, at about 45 years, owing to factors such as high rates of infectious disease, recent conflict, and poverty.

The extensive public healthcare system in Cuba provides for around 6 doctors per 1000 people, one of the highest ratios in the world.

Healthcare

An indicator of the strength of healthcare provision in a country is the number of doctors per 1000 population. Some communist and former communist countries such as Cuba and Russia score well in this regard. In general, healthcare provision is good or adequate in most of the world's richer countries but scanty throughout much of Africa and in parts of Asia and Latin America.

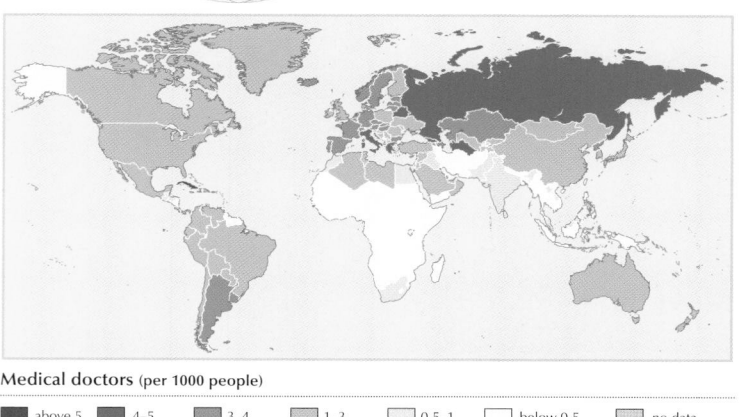

Medical doctors (per 1000 people)

■ above 5 ■ 4–5 ■ 3–4 ■ 1–3 □ 0.5–1 □ below 0.5 ■ no data

Smoking

Cigarette smoking—one of the most harmful activities to health—is common throughout much of the world. Smoking prevalence is generally highest in the richer, developed countries. However, awareness of the health risks has seen cigarette consumption in most of these countries stabilize or begin to fall. By contrast, more and more people, especially males, are taking up the habit in poorer developing countries.

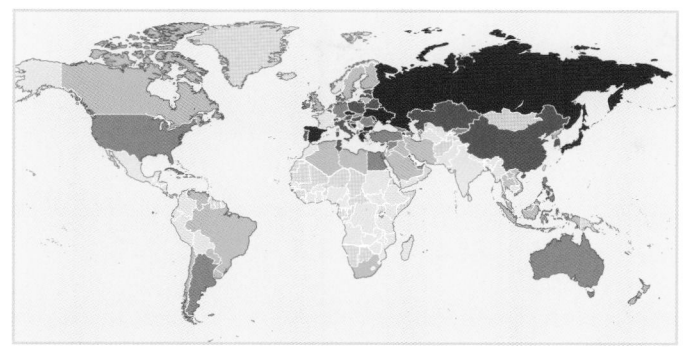

Annual cigarette consumption (per person)

- above 2000
- 1500–2000
- 1000–1499
- 500–999
- 0–499
- no data

Japan: has one of the world's highest life expectancies, at over 81 years—a fact commonly put down to the typical Japanese low-fat diet of rice, fish, and soy products.

Swaziland: currently has the lowest life expectancy in the world, at about 40 years, due to widespread HIV/AIDS.

Communicable diseases

Despite advances in their treatment and prevention, infectious diseases remain a huge problem, especially in developing countries. Three of the most common and deadly are tuberculosis (TB), HIV/AIDS, and malaria. Of these, active TB affects about 25 million people (often as a complication of AIDS), with a particularly high prevalence in parts of Africa. HIV/AIDS has spread since 1981 to become a global pandemic. Malaria affects about 225 million people every year.

Estimated tuberculosis cases (per 100,000 per year)

- above 300
- 100–300
- 50–100
- 10–50
- below 10

Adult (15-49) HIV prevalence rate (percent of population)

- 15–34
- 5–15
- 1–5
- 0.5–1
- 0.1–0.5
- below 0.1
- no data

Malaria cases (per 100,000 per year)

- above 25,000
- 10,000–25,000
- 1000–10,000
- 100–1000
- 10–100
- below 10
- low risk

Preventive medicine

Throughout the world, doctors recognize that the prevention of disease and disease transmission is just as important as the treatment of illness. Preventive medicine has many aspects and includes advice about diet and nutrition; education about the avoidance of health-threatening behaviors such as smoking, excess alcohol consumption, and unprotected sex; and the use of vaccines against diseases such as typhoid, polio and cholera. In developing countries, some of the main priorities in preventive medicine are the provision of pure water supplies and proper sanitation, as well as measures against malaria, including the use of antimalarial drugs and mosquito nets.

The use of mosquito nets greatly reduces the transmission of malaria and the risk of infection.

TOP TEN KILLER DISEASES

The world's biggest killer diseases fall into two main groups. One group, which includes HIV/AIDS, malaria, tuberculosis, and childhood diseases such as measles, mainly kills people in poor countries. The other group includes cardiovascular diseases and cancer, the big killers in rich countries.

- Cardiovascular diseases (heart disease and stroke) (16.7 million)
- Cancer (7.1 million)
- Respiratory infection (4 million)
- Other respiratory diseases (3.7 million)
- HIV/AIDS (2.9 million)
- Digestive diseases (1.9 million)
- Tuberculosis (1.5 million)
- Childhood cluster diseases (1.1 million)
- Diabetes (0.9 million)
- Malaria (0.9 million)

Water Resources

Water covers 71 percent of Earth's surface, but only 2.5 percent of this is fresh water, and two thirds of that is locked up in glaciers and polar ice sheets. Patterns of human settlement have developed around fresh water availability, but increasing numbers of people are now vulnerable to chronic shortage or interruptions in supply. Worldwide, fresh water consumption multiplied more than sixfold during the 20th century as populations increased and agriculture became more dependent on irrigation, much of it hugely wasteful because of evaporation and run-off. Industrial water demand also rose, as did use in the home, for washing, flushing, cooking, and gardening.

Amid the desert of Wadi Rum, Jordan, crops grow on circular patches of land irrigated with water from an underground aquifer.

Water withdrawal

Agriculture accounts for 70 percent of water consumption worldwide. Industry and domestic use each account for 15 percent. Excessive withdrawal of water affects the health of rivers and the needs of people. China's Yellow River now fails to reach the sea for most of the year.

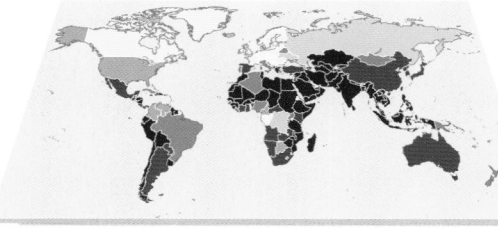

Percentage of freshwater withdrawal by agriculture

| 79–100 | 66–79 | 47–66 | 31–47 | 16–31 | 0–16 |

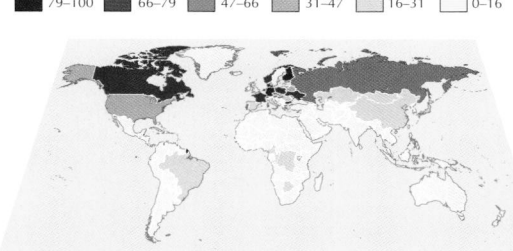

Percentage of freshwater withdrawal by industry

| 79–100 | 66–79 | 47–66 | 31–47 | 16–31 | 0–16 |

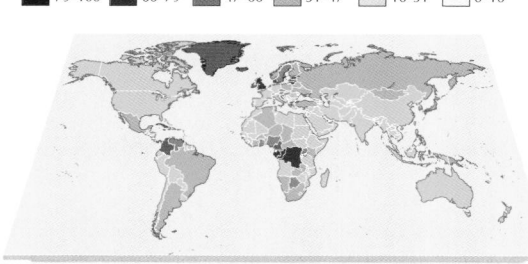

Percentage of freshwater withdrawal by domestic use

| 60–81 | 45–60 | 30–45 | 15–30 | 0–15 | no data |

Availability of fresh water
total renewable
(cubic yards/capita/per year)

- less than 1300 (water scarcity)
- 1300–2221 (water stress)
- 2222–3921 (insufficient water)
- 3922–12,999 (relatively sufficient)
- 13,000 or more (plentiful supplies)

▬ major drainage basin

▼ over 50% of water resource originating from outside country

Drought

The disruption of normal rainfall patterns can cause drought problems even in temperate zones, with consequences ranging from domestic water usage restrictions to low crop yields to forest fires. In regions of the developing world where monsoon rains fail, or water is perennially scarce, drought is a life or death issue. Parts of central and east Africa, for instance, have suffered severe and recurring droughts in recent decades, with disastrous results including destruction of livestock, desertification, famine, and mass migration.

In a severe drought, river beds may dry up (above left), leaving stranded fish to die, as here in Florida.

A Chinese farmer waters dry fields (above) in China's southern province of Guangdong. This picture was taken in May 2002, but the image is timeless; it could be August 2006 in Sichuan province, to the northwest of here—or almost any year in water-stressed northern China.

Water stress

A region is under "water stress" when the rate of water withdrawal from its rivers and aquifers exceeds their natural replenishment, so that people living there are subject to frequent shortages. Currently 1.7 billion people live in "highly stressed" river basins worldwide. This is a major potential cause of conflict, particularly when several countries share one river; the Euphrates, running through Turkey, Syria, and Iraq, or the rivers of southern China running south into Korea, are just two examples.

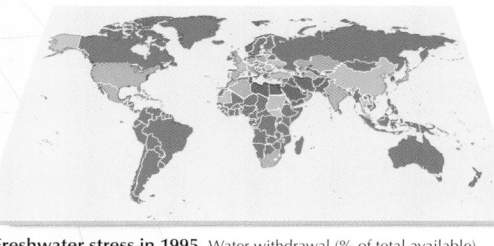

Freshwater stress in 1995 Water withdrawal (% of total available)

| above 40 | 20–40 | 10–20 | below 10 |

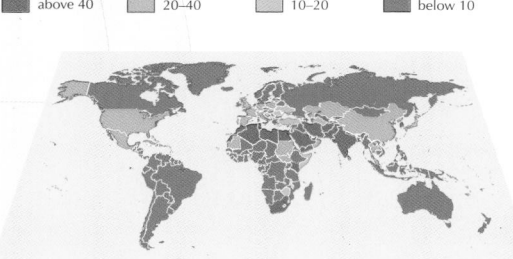

Freshwater stress in 2025 Water withdrawal (% of total available)

| above 40 | 20–40 | 10–20 | below 10 |

WATER AVAILABILITY

(by percentage of world's population)

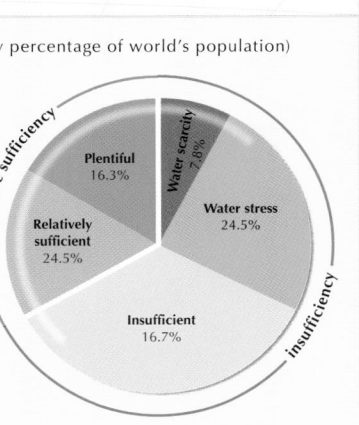

relative sufficiency

Plentiful 16.3%

Water scarcity 7.8%

Water stress 24.5%

Relatively sufficient 24.5%

Insufficient 16.7%

insufficiency

Mozambican children *(above)* fetch precious water in metal pans.

Gujarati villagers gather to draw water from a huge well *(above left)* in Natwarghad, western India. Many wells and village ponds ran dry in the severe drought of 2003, leaving local people to wait for irregular supplies brought in by state-run tankers.

Clean drinking water

Sub-Saharan Africa is among the most deprived regions for lack of access to safe drinking water. Worldwide, this terrible health hazard affects over a billion people—at least 15 percent of the population. One of the agreed United Nations "millennium goals" for international development is to halve this proportion by 2015, by tackling chemical pollution from agriculture and industry, and by introducing essential purification facilities and local supply systems. In the industrialized world, people have come to expect clean drinking water on tap, even if they face rising prices for its treatment and supply.

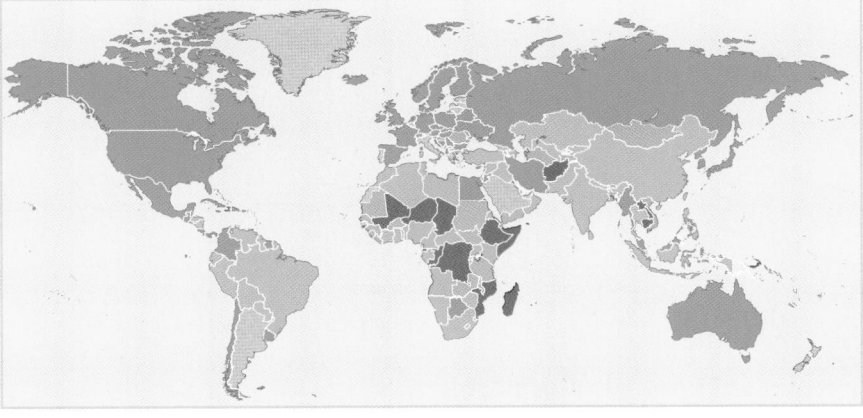

Access to safe drinking water source (percentage of population)

| 91%–100% | below 50% |
| 76%–90% | no data |
| 50%–75% |

Economic Systems

The world economy is now effectively a single global system based on "free market" capitalist principles. Few countries still cling, like North Korea, to the "command economy" formula developed in the former communist bloc, where centralized state plans set targets for investment and production. In the West, state ownership of companies has greatly diminished thanks to the wave of privatization in the last 25 years. Major companies move capital and raw materials around the globe to take advantage of different labor costs and skills. The World Trade Organization (WTO) promotes free trade, but many countries still use subsidies, and protect their markets with import tariffs or quotas, to favor their own producers.

Enormous volumes of trade pass through the world's stock markets making them key indicators of the strength of the global economy.

Balance of trade

Few countries earn from their exports exactly as much as they spend on imports. If the imbalance is persistently negative, it creates a potentially serious problem of indebtedness. The European Union's (EU) external trade is broadly in balance, but the US balance of trade has been in deficit since the 1970s, partly because it imports so many consumer goods. This deficit now stands at around US$ 500 billion a year.

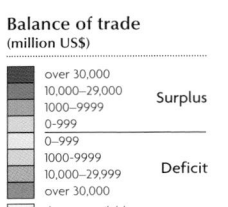

Balance of trade
(million US$)

- over 30,000
- 10,000–29,000
- 1000–9999
- 0–999   **Surplus**
- 0–999
- 1000–9999
- 10,000–29,999
- over 30,000   **Deficit**
- data unavailable

TOP TEN GLOBAL COMPANIES (2015)

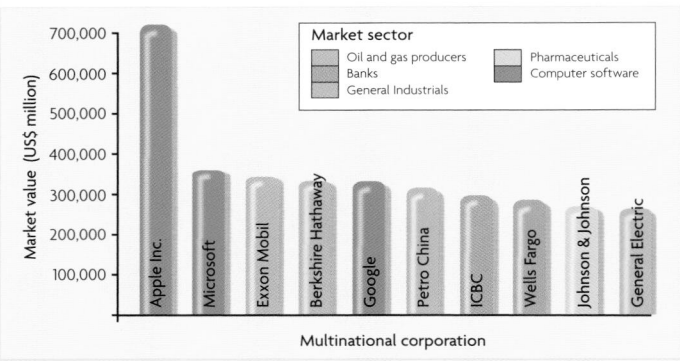

Market sector
- Oil and gas producers
- Banks
- General Industrials
- Pharmaceuticals
- Computer software

Market value (US$ million): Apple Inc., Microsoft, Exxon Mobil, Berkshire Hathaway, Google, Petro China, ICBC, Wells Fargo, Johnson & Johnson, General Electric

Multinational corporation

Energy

Countries with oil and gas to sell (notably in the Middle East and Russia) can charge high prices; trade in fuel was worth US$ 1.4 trillion in 2005. The US and others are turning back to nuclear power (despite safety fears) for generating electricity. China relies heavily on (polluting) coal. Renewable technologies promise much, but so far make relatively minor contributions.

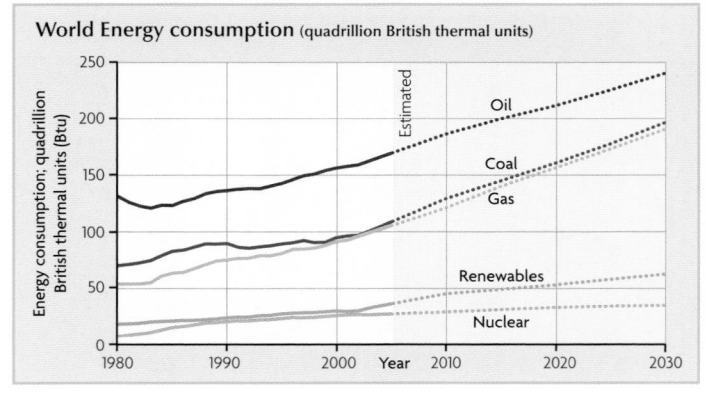

World Energy consumption (quadrillion British thermal units)

Oil, Coal, Gas, Renewables, Nuclear — Estimated

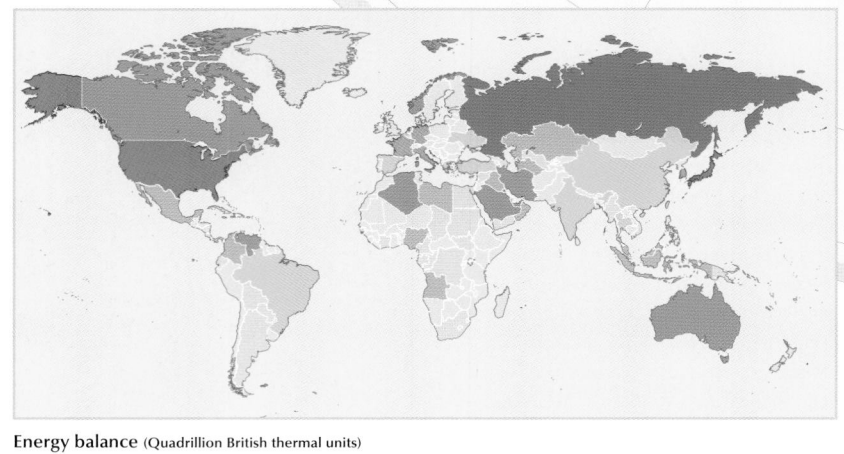

Energy balance (Quadrillion British thermal units)

net producer: 10 and above | 5 to 10 | 1 to 5 | 0 to 1 | data unavailable

net consumer: 0 to -1 | -1 to -5 | -5 to -10 | -10 and below

SOUTH AMERICA

New York

London

EUROPE

AFRICA

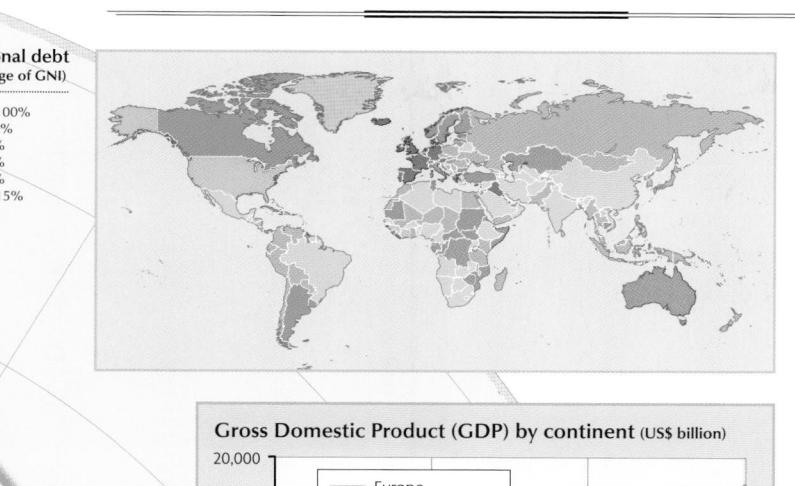

International debt

Saddled with crippling debts from past borrowing, the world's poorest countries are still paying off US $100 million a day. This is despite recent successful campaigns to get some of their debts cancelled to allow them to use their limited resources for development. Most international debt, however, is owed by developed countries to one another. The US owes just over a trillion dollars, around 7% of its total debt, to China.

Trade sector

World trade in merchandise tops US$ 10 trillion a year. The global pattern is uneven. Latin America, Africa, the Middle East, and Russia principally export "primary" goods (agricultural produce, mining and fuel). The "secondary" manufacturing sector includes iron and steel, machine tools, chemicals, clothing and textiles, cars and other consumer goods. The West still dominates the "tertiary" or non-merchandise sector, worth US$ 2.4 trillion, in services such as insurance and banking.

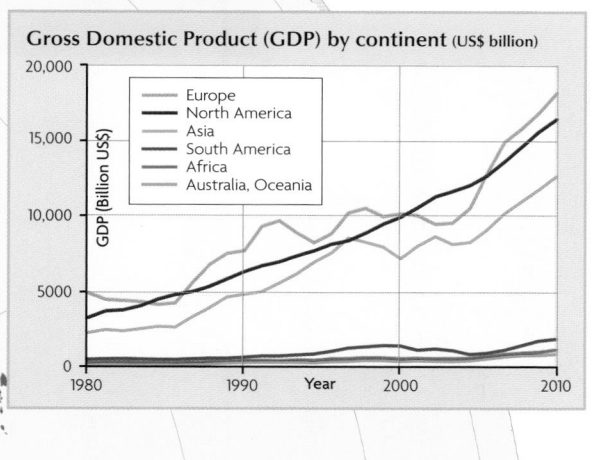

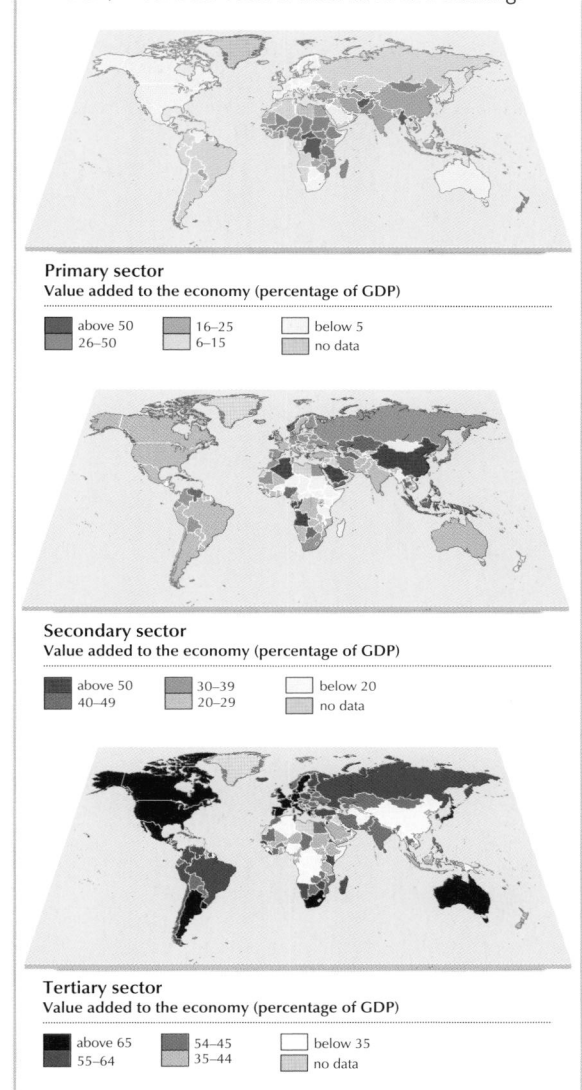

Primary sector
Value added to the economy (percentage of GDP)

above 50   16–25   below 5
26–50   6–15   no data

Secondary sector
Value added to the economy (percentage of GDP)

above 50   30–39   below 20
40–49   20–29   no data

Tertiary sector
Value added to the economy (percentage of GDP)

above 65   54–45   below 35
55–64   35–44   no data

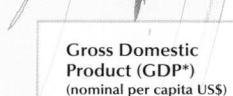

Gross Domestic Product (GDP) by continent (US$ billion)

Europe
North America
Asia
South America
Africa
Australia, Oceania

GDP (Billion US$)

20,000

15,000

10,000

5000

1980   1990   **Year**   2000   2010

NORTH
AMERICA

ASIA

AUSTRALIA

Tokyo

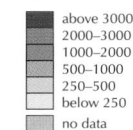

Labor

China's huge low-cost labor force promotes its conquest of world markets for manufactured goods. India's educated workforce attracts call centers and other service sector jobs, while the more economically developed countries's (MEDC) caring professions, and low-wage agriculture, draw in immigrant labor.

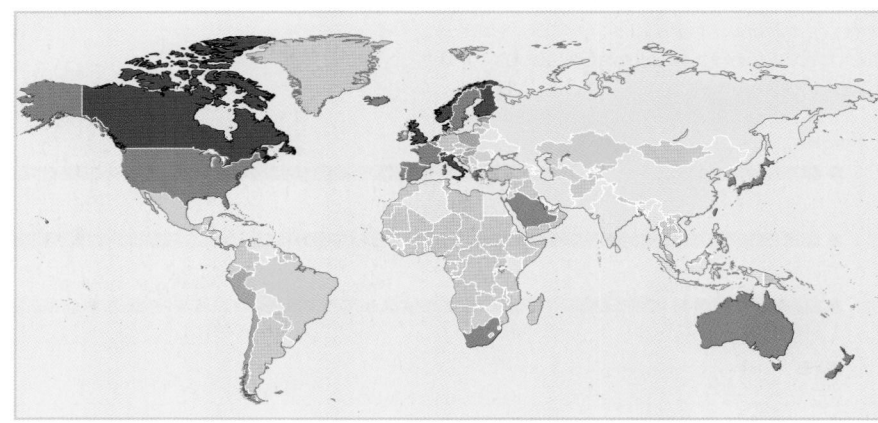

Travel

Mass travel is now a ubiquitous feature of all developed countries, and the provision of transport and tourism facilities one of the world's biggest industries, employing well over 100 million people. The travel explosion has come about, first, through major improvements in transportation technology; and second, as a result of increasing amounts of disposable income and leisure time in the world's wealthier countries. The main reasons for travel today include leisure pursuits and tourism (accounting for well over half of the total financial outlay), work and business, pilgrimage, migration, and visits to family and friends.

There are currently around 4.2 billion air travelers a year passing through over 1600 international and domestic airports. This figure is forecast to grow by 4 percent each year, leading to increased pressure on air traffic control and ground handling systems that, in many areas, are already close to maximum capacity.

Major modes of transportation

The major transport modes for people in the 21st century are road, rail, and air travel. The most popular air routes are highly concentrated within and between the USA, western Europe, and Asia. Major roads and railroads are more evenly spread, following the general distribution of the world's population.

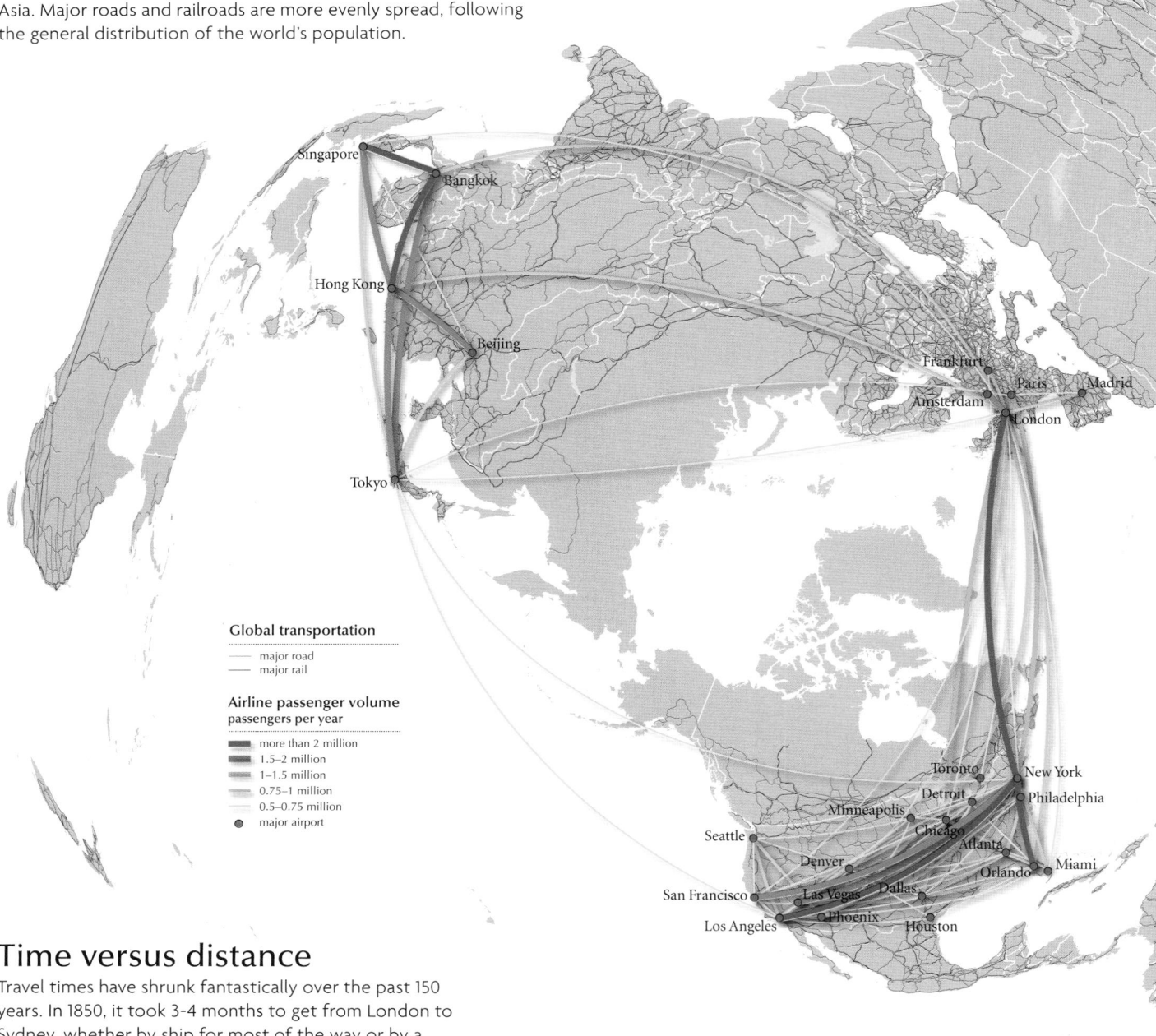

Global transportation

— major road
— major rail

Airline passenger volume
passengers per year

— more than 2 million
— 1.5–2 million
— 1–1.5 million
— 0.75–1 million
— 0.5–0.75 million
● major airport

Time versus distance

Travel times have shrunk fantastically over the past 150 years. In 1850, it took 3-4 months to get from London to Sydney, whether by ship for most of the way or by a series of different transports. By 1930, trains and faster ships had reduced the journey to about 40 days. In 2005, the trip took just 21 hours by air.

London

| 1850 | by coach to Portsmouth and thence ship around the Cape of Good Hope | | | | | | Basra |

Istanbul

| 1850 | *coach . ferry . coach . horseback* | | | *horseback . river boat* | | *river b* |

Istanbul | **Basra** | **Bombay** | **Calcutta** | **Singapore** | **Sydney**

| 1930 | *train . ferry . train* | | *train . river boat* | *river boat . steamship* | *train* | *steamship* | *steamship* |

| 2005 | ●I● London–Sydney by air including one refueling stop |

DAYS 1 2 3 4 5 6 7 8 9 10 11 12 13 14 15 16 17 18 19 20 21 22 23 24 25 26 27 28 29 30 31 32 33 34 35 36 37 38 39 40 41 42 43 44 45 46 47 48 49 50 51 52 53 54 5

Media and Communications

Over the past 50 years, the term "media" has come to denote various means of communicating information between people at a distance. These include mass media—methods such as newspapers, radio, and television that can be used to rapidly disseminate information to large numbers of people—and two-way systems, such as telephones and e-mail. Currently, the communication systems undergoing the most rapid growth worldwide include mobile telephony and various Internet-based applications, such as web sites, blogs, and podcasting, which can be considered forms of mass media.

Internet usage

Internet usage has grown extremely rapidly since the early 1990s, largely as a result of the invention of the World Wide Web. Usage rates are highest in the USA (where about 80 percent of people were using the Internet in 2006), Australia, Japan, South Korea, and Finland. They are lowest in Africa, where on average less than 5 percent of the population were Internet users in 2006.

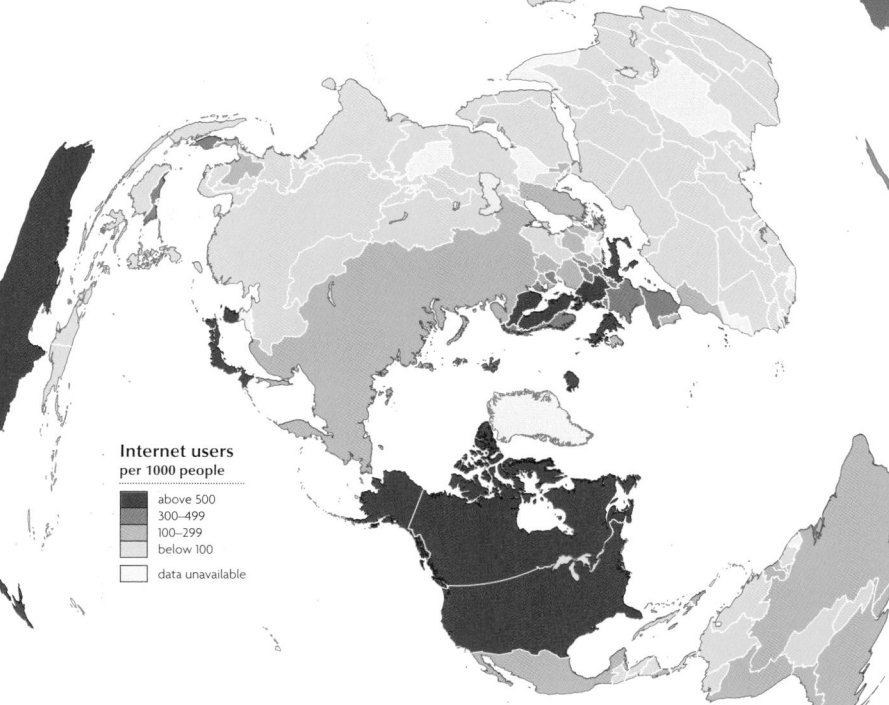

Internet users
per 1000 people
- above 500
- 300–499
- 100–299
- below 100
- data unavailable

The internet emerged in the early 1990s as a computer-based global communication system. Since then massive growth has seen user numbers increase to around 1.1 billion people, or roughly 17 percent of the world's population.

Mobile phone usage

By 2006, there were more than 2.5 billion mobile phone users worldwide. In some parts of Europe, such as Italy, almost everyone owns and uses a mobile—many possess more than one phone. In contrast, throughout much of Southern Asia and Africa, less than 10 percent of the population are users. As well as utilizing them as telephones, most users now employ the devices for the additional functions they offer, such as text messaging and e-mail.

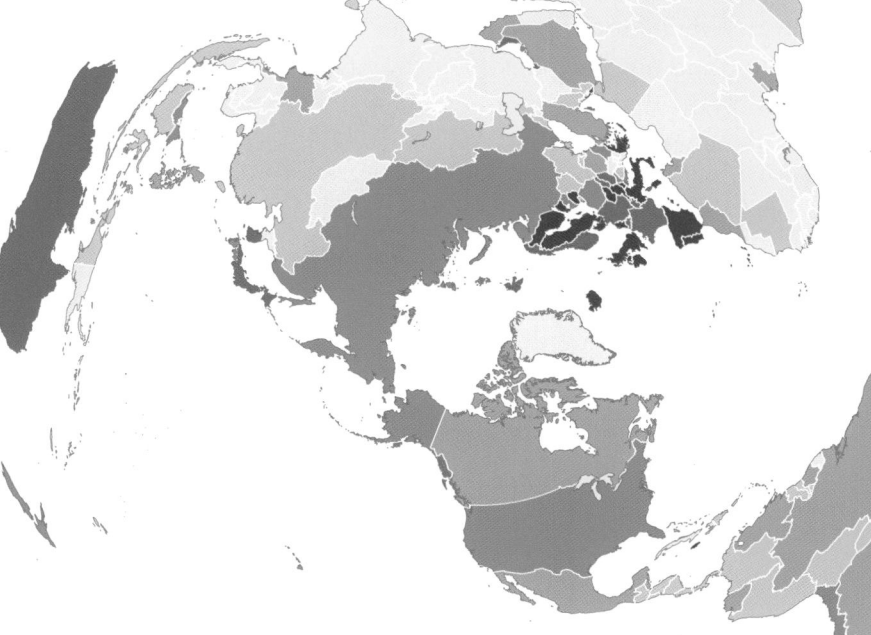

Mobile phone users
per 1000 people
- above 900
- 700–899
- 500–699
- 300–499
- 100–299
- below 100
- data unavailable

Satellite Communications

Modern communications satellites are used extensively for international telephony, for television and radio broadcasting, and to some extent for transmitting Internet data. Many of these satellites are deployed in clusters or arrays, often in geostationary orbits—that is, in positions that appear fixed to Earth-based observers.

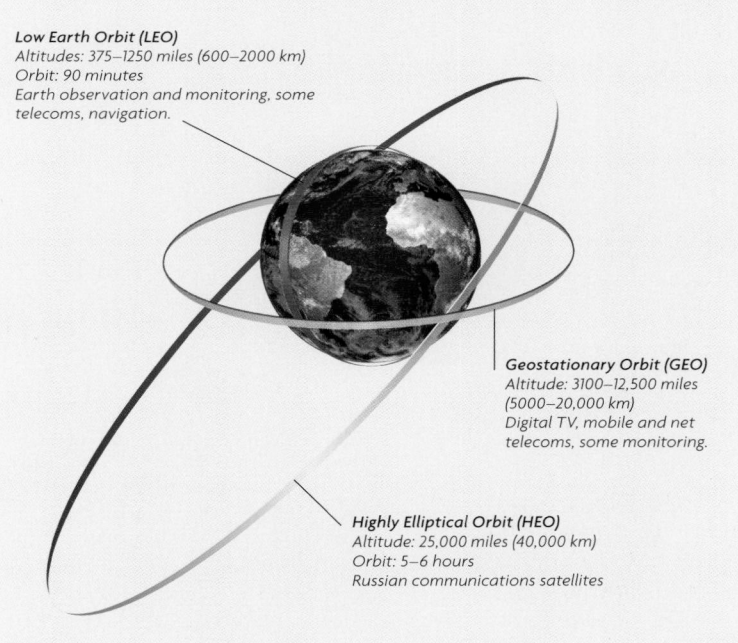

Low Earth Orbit (LEO)
Altitudes: 375–1250 miles (600–2000 km)
Orbit: 90 minutes
Earth observation and monitoring, some telecoms, navigation.

Geostationary Orbit (GEO)
Altitude: 3100–12,500 miles (5000–20,000 km)
Digital TV, mobile and net telecoms, some monitoring.

Highly Elliptical Orbit (HEO)
Altitude: 25,000 miles (40,000 km)
Orbit: 5–6 hours
Russian communications satellites

Sydney

Bombay · · · Calcutta · Singapore · Sydney

steamship · horseback · steamship · steamship

59 60 61 62 63 64 65 66 67 68 69 70 71 72 73 74 75 76 77 78 79 80 81 82 83 84 85 86 87 88 89 90 91 92 93 94 95 96 97 98 99 100 101 102 103 104 105 106 107 108 109 110 111 112 113 114 115

The Political World

Today's world map shows nearly 200 independent states, compared with about 80 after World War II. The transformation is mainly due to the withdrawal of European powers from huge colonial empires; their remaining overseas dependencies are tiny by comparison. The late 20th century also saw the collapse of communism, realignment in Europe, and fragmentation in former Yugoslavia. Globally, the Soviet Union's demise left the USA as the sole superpower, though with fast-growing China and India emerging as economic giants of the future. US security preoccupations switched to combating terrorism, while looming oil and other resource shortages, and environmental constraints, underlined the need for more effective international cooperation.

CONTINENTAL FACTFILE

	Total area: sq miles	Total area: sq km	Total population
North & Central America	9,358,340	24,238,000	565.3 million
South America	6,886,000	17,835,000	406.7 million
Africa	11,712,434	30,335,000	1110.6 million
Europe	4,053,309	10,498,000	742.4 million
Asia	16,838,365	43,608,000	4298.7 million
Australia & Oceania	3,285,048	8,508,238	38.3 million

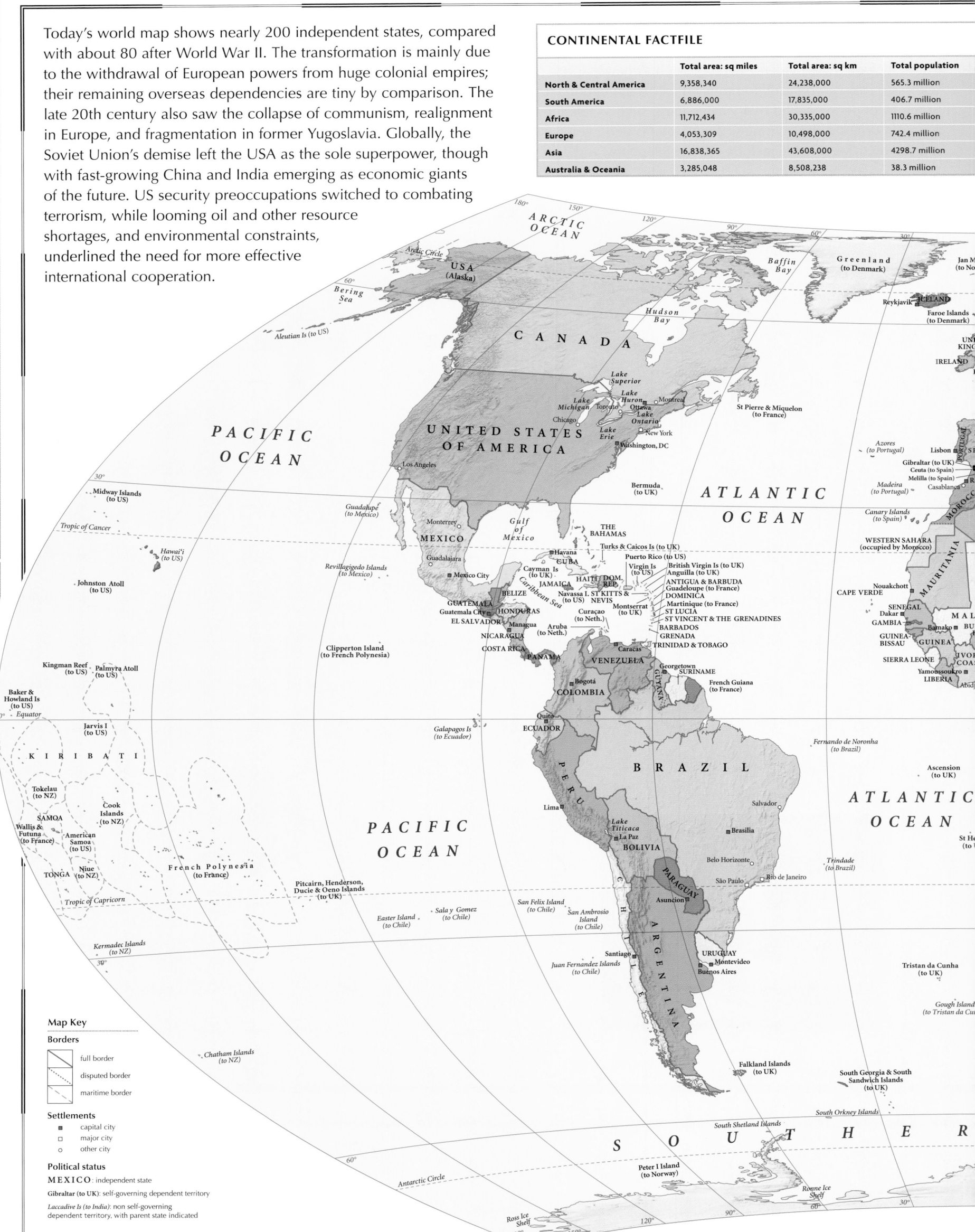

Map Key

Borders

full border

disputed border

maritime border

Settlements

■ capital city

□ major city

○ other city

Political status

MEXICO: independent state

Gibraltar (to UK): self-governing dependent territory

Laccadive Is (to India): non self-governing dependent territory, with parent state indicated

International borders

The world political map of today displays a complex pattern of boundaries that has evolved through history, and is still constantly changing as new countries emerge and disputes and territorial claims are slowly resolved. The map shows two main types of border. Full borders represent internationally agreed and recognized territorial boundaries. A disputed border is indicated where a *de facto* territorial boundary exists, which is not agreed or is still subject to arbitration.

Countries	Largest country	Country with largest population
23	Canada 3,855,171 sq miles (9,984,670 sq km)	United States 323 million
12	Brazil 3,286,470 sq miles (8,511,965 sq km)	Brazil 202 million
54	Algeria 919,590 sq miles (2,381,740 sq km)	Nigeria 178 million
46	European Russia 1,527,341 sq miles (3,955,818 sq km)	European Russia 110 million
49	Asiatic Russia 5,065,394 sq miles (13,119,382 sq km)	China 1394 million
14	Australia 2,967,893 sq miles (7,686,850 sq km)	Australia 24 million

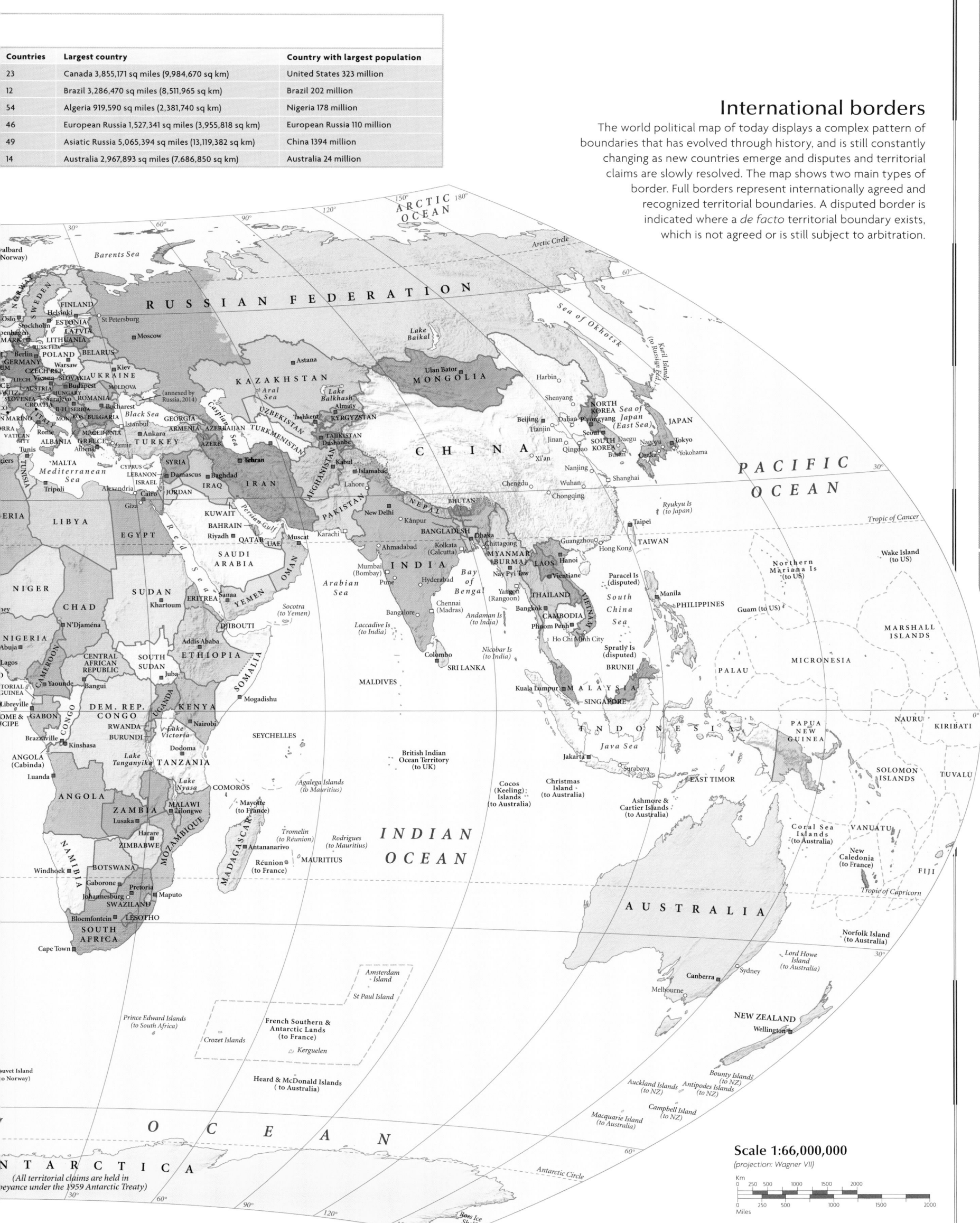

Scale 1:66,000,000

(projection: Wagner VII)

Borders, conflicts and disputes

Conflict evolved in the 20th century from conventional land- or sea-based warfare to increasingly long-range airborne attacks. Nuclear arms from 1945 took this to the intercontinental scale. The Cold War presented a new type of conflict, underlined by the race for weapons capabilities between the US and the Soviet Union. In Korea, Vietnam, the Middle East and elsewhere, soldiers and civilians were exposed to deadly chemicals. International treaties aimed to prevent the spread of nuclear, biological and chemical "weapons of mass destruction". Intercommunal conflict and "ethnic cleansing" reminded the world that horror needed no sophisticated weaponry. After 9/11, the US-led "war on terror" perceived conflict in a new light, where international terrorism knew no borders.

THE PEACEKEEPERS

Over 130 countries have contributed around a million troops to UN missions to monitor peace processes and help implement peace accords since 1948. Regional alliances such as NATO and the African Union (AU) are increasingly deploying their own multinational forces in trouble-spots, while Australia has intervened in a similar manner in nearby Pacific island states. Peacekeepers oversaw East Timor's elections in 2001 and subsequent celebration of independence *(above)*. The US defines many of its activities as peacekeeping, despite the confrontational nature of some of its interventions.

DARFUR

African ethnic minorities in Darfur in western Sudan have suffered appalling violence since 2003 at the hands of genocidal Arab Janjaweed militias, for which the government in Khartoum denies responsibility. Displaced in their hundreds of thousands, refugees receive inadequate protection and aid from an international community unwilling to commit to full-scale intervention.

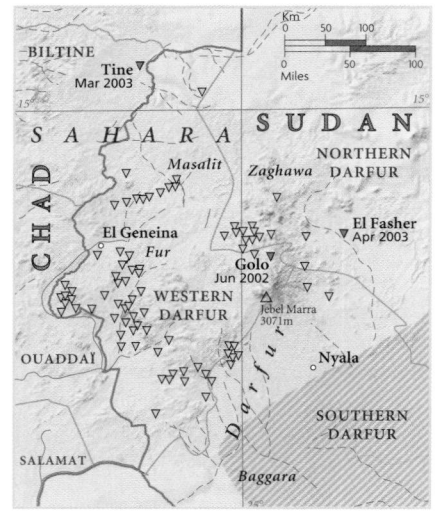

Darfur conflict

Fur ethnic group

░ arabic speaking area

▽ villages destroyed by Janjaweed

▼ towns that have been attacked by rebels opposing the Sudanese government

ISRAEL

Since its creation in 1948, Israel has been at war with its Arab neighbors. The Palestinians are fighting for a separate, viable state, comprising of at least East Jerusalem, and the West Bank and Gaza Strip, territories occupied by Israel in 1967. Their struggle *(intifada)* has attracted international support, but has been met by a hard-line response from Israel, which is backed by the US.

Arab-Israeli Wars 1947-2006

MAIN MAP: Arab-Israeli Wars

▨ Israel in 1949

▨ occupied by Israel after 1967 war

▨ occupied by Israel after 1973 war

▨ occupied by Israel after 1967 war reoccupied by Egypt after 1973 war

▭ demilitarized zone held by UN after Israel-Syria agreement, 1974, and 2nd Sinai agreement, 1975

▽ Hezbollah rocket attacks 2006

▼ Israeli rocket attacks 2006

— · disputed border

INSET MAP 1: UN Partition plan in 1947

— border of British mandate 1923

▨ proposed Arab State

▨ proposed Jewish State

▨ proposed international zone

INSET MAP 2: West Bank security

▨ Palestinian responsibility for civil affairs and internal security

▨ Palestinian responsibility for civil affairs; Israel responsible for security

— Security Wall (existing and planned)

Conflicts and international disputes

- Major active territorial or border disputes
- Countries involved in internal conflict
- Active territorial or border disputes and internal conflict

Types of government

- Multiparty democracy for more than 10 yrs
- Multiparty democracy within last 10 yrs
- Single-party government
- Military regime
- Theocracy
- Monarchy
- Non-party system
- Transitional regime

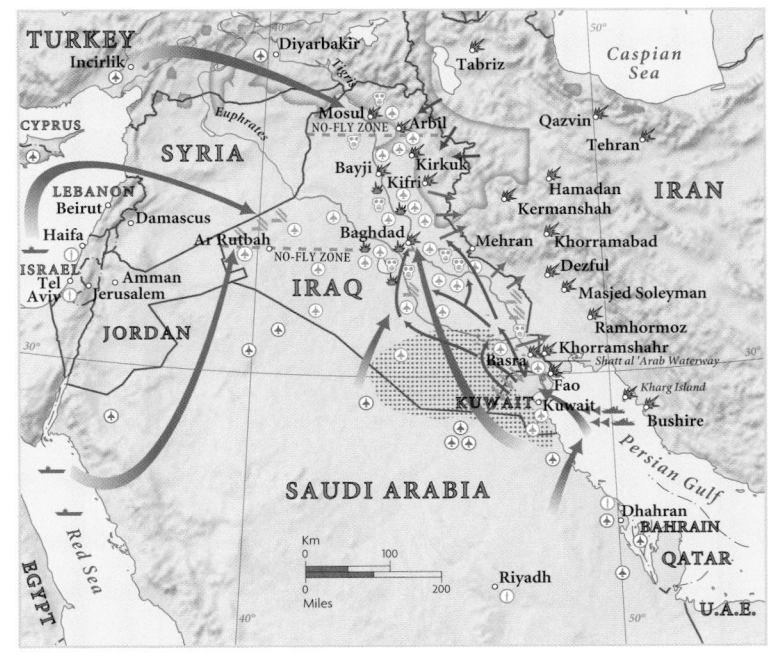

RUSSIAN FEDERATION
KAZAKHSTAN
MONGOLIA
RUS
INE
UVA
RIA
GEORGIA
ARMENIA AZERB.
RKEY
TURKMEN.
UZBEKISTAN
KYRG.
TAJ.
SYRIA
IRAN
AFGHANISTAN
PAKISTAN
NEPAL
BHUTAN
CHINA
NORTH KOREA
SOUTH KOREA
JAPAN
ION
JORDAN
WEST BANK
IRAQ
KUWAIT
BAHRAIN
QATAR
UAE
SAUDI ARABIA
OMAN
INDIA
BANGLADESH
MYANMAR (BURMA)
LAOS
THAILAND
VIETNAM
CAMBODIA
TAIWAN
PHILIPPINES
MICRONESIA
MARSHALL ISLANDS
ERITREA
YEMEN
DJIBOUTI
ETHIOPIA
TH
AN
SRI LANKA
MALDIVES
BRUNEI
MALAYSIA
SINGAPORE
PALAU
NAURU
KIRIBATI
UGANDA
KENYA
RWANDA
BURUNDI
SEYCHELLES
INDONESIA
EAST TIMOR
PAPUA NEW GUINEA
SOLOMON ISLANDS
TUVALU
SAMOA
TANZANIA
COMOROS
MOZAMBIQUE
MADAGASCAR
MAURITIUS
INDIAN OCEAN
VANUATU
FIJI
TONGA
ABWE
SWAZILAND
OTHO
AUSTRALIA
NEW ZEALAND
PACIFIC OCEAN

Lines on the map

The determination of international boundaries can use a variety of criteria. Many borders between older states follow physical boundaries, often utilizing natural defensive features. Others have been determined by international agreement or arbitration, or simply ended up where the opposing forces stood at the end of a conflict.

WORLD BOUNDARIES

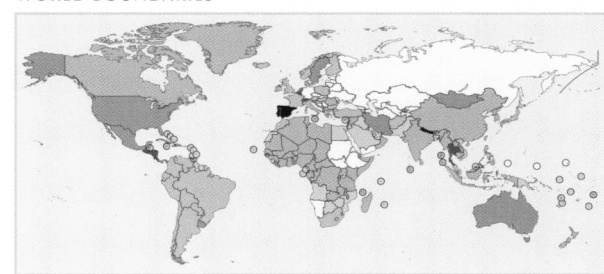

Dates from which current boundaries have existed

- 1990–present
- 1966–1989
- 1946–1965
- 1915–1945
- 1850–1914
- 1800–1849
- Pre-1800

POST-COLONIAL BORDERS

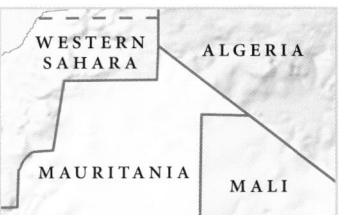

WESTERN SAHARA
ALGERIA
MAURITANIA
MALI

Independent African countries have largely inherited the earlier carve up of the continent by European colonial powers. These often arbitrarily divided or grouped differing ethnic and religious groups which has, in turn, contributed to the tensions that underlie the many civil conflicts that have plagued post-colonial Africa.

ENCLAVES

Changes to international boundaries occasionally create pockets of land cut off from the main territory of the country they belong to. In Europe, Kaliningrad has been separated from the rest of the Russian Federation since the independence of the Baltic States. Likewise, when Morocco was granted independence, Spain retained the coastal enclaves of Ceuta and Melilla.

Baltic Sea
ESTONIA
RUSSIAN FED.
LATVIA
LITHUANIA
RUSS. FED. (Kaliningrad)
BELARUS
POLAND

GEOMETRIC BORDERS

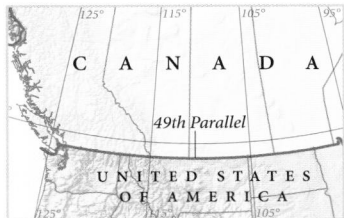

CANADA
49th Parallel
UNITED STATES OF AMERICA

Straight lines and lines of longitude and latitude have occasionally been used to determine international boundaries: the 49th Parallel forms a large section of the Canada–US border, while the 38th Parallel roughly divides the Korean Peninsula. Internal administrative divisions within Canada, the US, and Australia also use geometric boundaries.

PHYSICAL BORDERS

Rivers account for one-sixth of the world's borders: the Danube forms part of the boundaries for nine European nations. Changes in a river's course or disruption of its flow can lead to territorial disputes. Lakes and mountains also form natural borders.

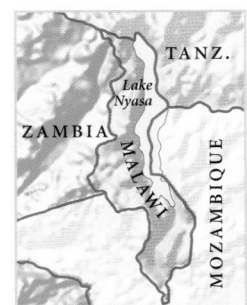

TANZ.
Lake Nyasa
ZAMBIA
MALAWI
MOZAMBIQUE

Lake border (right)
Mountain border (below left)
River border (below right)

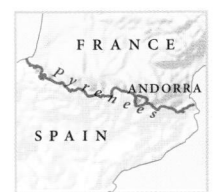

FRANCE
ANDORRA
SPAIN

GERMANY
Danube
UKRAINE
SLOVAKIA
AUSTRIA
HUNGARY
CROATIA
ROMANIA
SERBIA
Danube
BULGARIA

GULF CONFLICTS

Although the West armed Saddam Hussein in the brutal 1980s Iran-Iraq War, his unprovoked invasion of Kuwait in 1990 was decried the world over. A US-led coalition, including Arab states, repelled his troops but left him in power. A decade of sanctions followed until, in 2003, Saddam was finally toppled by US-led forces. Following elections in 2005, Iraq has struggled to contain a violent insurgency.

TURKEY
Incirlik
Diyarbakir
Tabriz
Caspian Sea
CYPRUS
Mosul
NO-FLY ZONE
Arbil
Qazvin
SYRIA
Bayji
Kirkuk
Kifri
Tehran
LEBANON
Beirut
Damascus
Haifa
Ar Rutbah
Baghdad
NO-FLY ZONE
Hamadan
Kermanshah
Mehran
IRAN
Khorramabad
Dezful
Masjed Soleyman
ISRAEL
Tel Aviv
Amman
Jerusalem
IRAQ
JORDAN
Ramhormoz
Khorramshahr
Basra
Shatt al 'Arab Waterway
Fao
Kharg Island
KUWAIT
Kuwait
Bushire
SAUDI ARABIA
Persian Gulf
EGYPT
Red Sea
Dhahran
BAHRAIN
QATAR
Riyadh
U.A.E.

Conflict in the Persian Gulf

Iran–Iraq War (1980–88)
- Iraqi invasion force Sep–Nov 1980
- Iranian invasion force Oct 1984
- air strike 1980–88

Persian Gulf War (1990–91)
- Iraqi invasion of Kuwait 1–2 Aug 1990
- Iraqi air base
- Allied air base
- SCUD installation
- Iraqi SCUD missile attacks
- Iraqi weapons plant
- US battleship
- US aircraft carrier
- Allied amphibious attack
- Allied airborne attack
- area of Allied ground combat
- Kurdish region

Iraq War from 2003
- Allied air exclusion zone (1991–2003)
- Allied land campaign 2003
- main centres of insurgency 2003–04

The World's Standard Time Zones

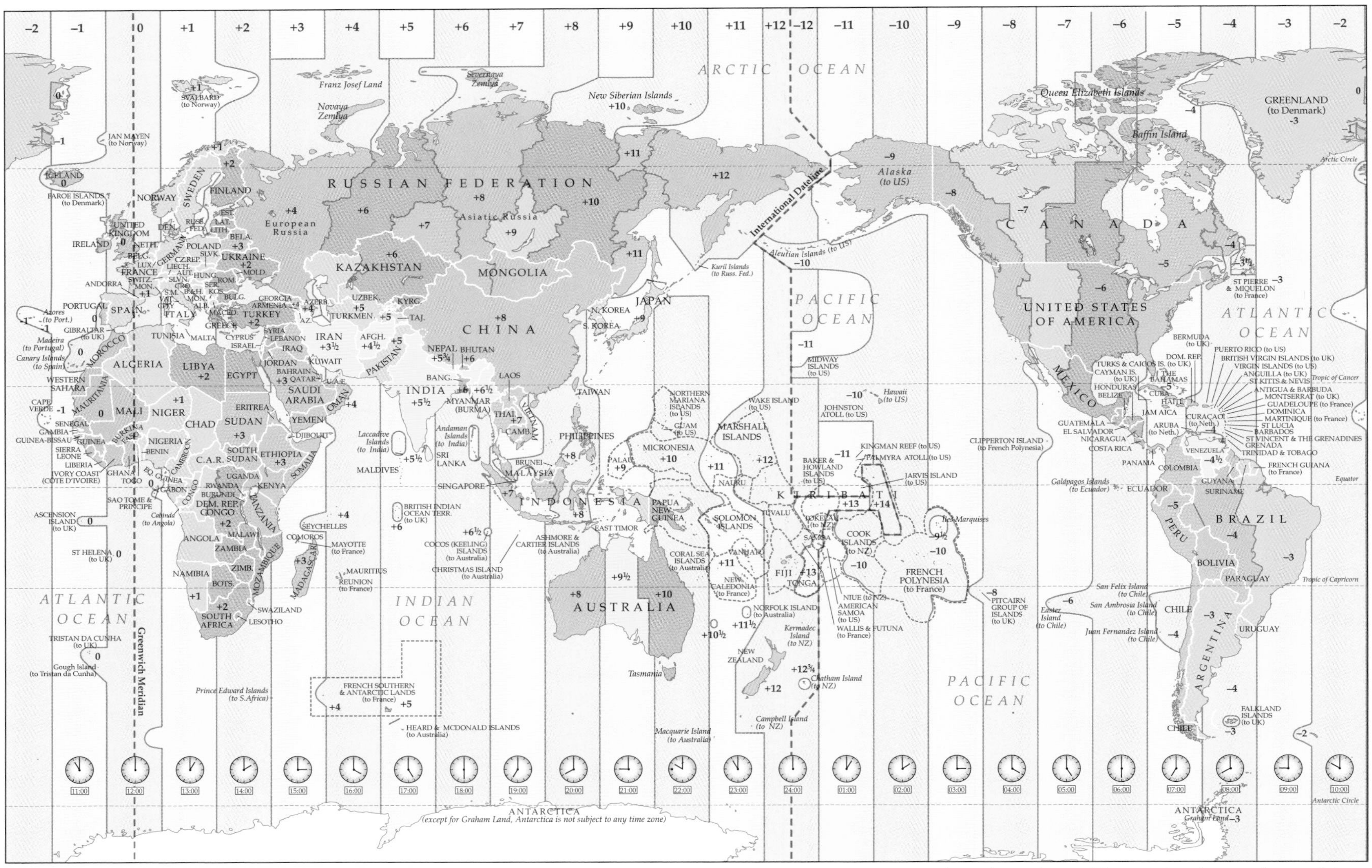

The numbers at the top of the map indicate how many hours each time zone is ahead or behind Coordinated Universal Time (UTC). The row of clocks indicate the time in each zone when it is 12:00 noon UTC.

TIME ZONES

Because Earth is a rotating sphere, the Sun shines on only half of its surface at any one time. Thus, it is simultaneously morning, evening and night time in different parts of the world (see diagram below). Because of these disparities, each country or part of a country adheres to a local time. A region of Earth's surface within which a single local time is used is called a time zone. There are 24 one hour time zones around the world, arranged roughly in longitudinal bands.

STANDARD TIME

Standard time is the official local time in a particular country or part of a country. It is defined by the time zone or zones associated with that country or region. Although time zones are arranged roughly in longitudinal bands, in many places the borders of a zone do not fall exactly on longitudinal meridians, as can be seen on the map (above), but are determined by geographical factors or by borders between countries or parts of countries. Most countries have just one time zone and one standard time, but some large countries (such as the USA, Canada and Russia) are split between several time zones, so standard time varies across those countries. For example, the coterminous United States straddles four time zones and so has four standard times, called the Eastern, Central, Mountain and Pacific standard times. China is unusual in that just one standard time is used for the whole country, even though it extends across 60° of longitude from west to east.

COORDINATED UNIVERSAL TIME (UTC)

Coordinated Universal Time (UTC) is a reference by which the local time in each time zone is set. For example, Australian Western Standard Time (the local time in Western Australia) is set 8 hours ahead of UTC (it is UTC+8) whereas Eastern Standard Time in the United States is set 5 hours behind UTC (it is UTC-5). UTC is a successor to, and closely approximates, Greenwich Mean Time (GMT). However, UTC is based on an atomic clock, whereas GMT is determined by the Sun's position in the sky relative to the 0° longitudinal meridian, which runs through Greenwich, UK.

In 1884 the Prime Meridian (0° longitude) was defined by the position of the cross-hairs in the eyepiece of the "Transit Circle" telescope in the Meridian Building at the Royal Observatory, Greenwich, UK.

DAY AND NIGHT AROUND THE WORLD

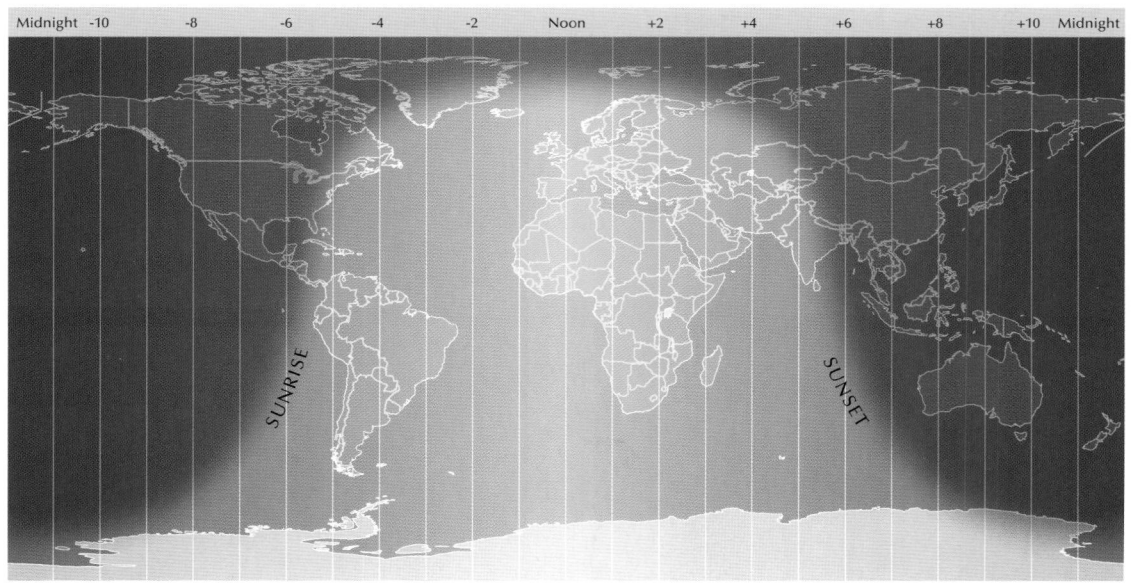

THE INTERNATIONAL DATELINE

The International Dateline is an imaginary line from pole to pole that roughly corresponds to the 180° longitudinal meridian. It is an arbitrary marker between calendar days. The dateline is needed because of the use of local times around the world rather than a single universal time. When moving from west to east across the dateline, travelers have to set their watches back one day. Those traveling in the opposite direction, from east to west, must add a day.

DAYLIGHT SAVING TIME

Daylight saving is a summertime adjustment to the local time in a country or region, designed to cause a higher proportion of its citizens' waking hours to pass during daylight. To follow the system, timepieces are advanced by an hour on a pre-decided date in spring and reverted back in the fall. About half of the world's nations use daylight saving.

COMPLETE ATLAS OF THE WORLD

THE MAPS IN THIS ATLAS ARE ARRANGED CONTINENT BY CONTINENT, STARTING

FROM THE INTERNATIONAL DATE LINE, AND MOVING EASTWARD. THE MAPS PROVIDE

A UNIQUE VIEW OF TODAY'S WORLD, COMBINING TRADITIONAL CARTOGRAPHIC

TECHNIQUES WITH THE LATEST REMOTE-SENSED AND DIGITAL TECHNOLOGY.

North America is the world's third largest continent with
a total area of 9,358,340 sq miles (24,238,000 sq km)
including Greenland and the Caribbean islands.
It lies wholly within the Northern Hemisphere.

FACTFILE

N **Most Northerly Point:** Kap Morris Jesup, Greenland 83° 38' N
S **Most Southerly Point:** Peninsula de Azuero, Panama 7° 15' N
E **Most Easterly Point:** Nordostrundingen, Greenland 12° 08' W
W **Most Westerly Point:** Attu, Aleutian Islands, USA 172° 30' E

Largest Lakes:
1 Lake Superior, Canada/USA 31,151 sq miles (83,270 sq km)
2 Lake Huron, Canada/USA 23,436 sq miles (60,700 sq km)
3 Lake Michigan, USA 22,402 sq miles (58,020 sq km)
4 Great Bear Lake, Canada 12,274 sq miles (31,790 sq km)
5 Great Slave Lake, Canada 10,981 sq miles (28,440 sq km)

Longest Rivers:
1 Mississippi-Missouri, USA 3710 miles (5969 km)
2 Mackenzie, Canada 2640 miles (4250 km)
3 Yukon, Canada/USA 1978 miles (3184 km)
4 St Lawrence/Great Lakes, Canada/USA 1900 miles (3058 km)
5 Rio Grande, Mexico/USA 1900 miles (3057 km)

Largest Islands:
1 Greenland 849,400 sq miles (2,200,000 sq km)
2 Baffin Island, Canada 183,800 sq miles (476,000 sq km)
3 Victoria Island, Canada 81,900 sq miles (212,000 sq km)
4 Ellesmere Island, Canada 75,700 sq miles (196,000 sq km)
5 Newfoundland, Canada 42,031 sq miles (108,860 sq km)

Highest Points:
1 Mount McKinley (Denali), USA 20,332 ft (6194 m)
2 Mount Logan, Canada 19,550 ft (5959 m)
3 Volcán Pico de Orizaba, Mexico 18,700 ft (5700 m)
4 Mount St Elias, USA 18,008 ft (5489 m)
5 Popocatépetl, Mexico 17,887 ft (5452 m)

Lowest Point:
▼ Death Valley, USA -282 ft (-86 m) below sea level

Highest recorded temperature:
＋ Death Valley, USA 135°F (57°C)

Lowest recorded temperature:
－ Northice, Greenland -87°F (-66°C)

Wettest Place:
≋ Vancouver, Canada 262 in (6650 mm)

Driest Place:
⊖ Death Valley, USA 2 in (50 mm)

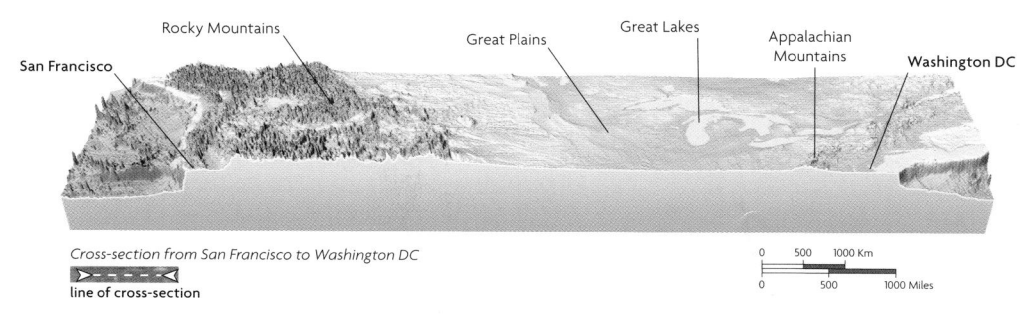

San Francisco · Rocky Mountains · Great Plains · Great Lakes · Appalachian Mountains · Washington DC

Cross-section from San Francisco to Washington DC
⟩⟩⟩—⟩—
line of cross-section

0   500   1000 Km
0   500   1000 Miles

Political

Democracy is well established in some parts of the continent but is a recent phenomenon in others. The economically dominant nations of Canada and the USA have a long democratic tradition but elsewhere, notably in the countries of Central America, political turmoil has been more common. In Nicaragua and Haiti, harsh dictatorships have only recently been superseded by democratically-elected governments. North America's largest countries—Canada, Mexico, and the USA—have federal state systems, sharing political power between national and state or provincial governments. The USA has intervened militarily on several occasions in Central America and the Caribbean to protect its strategic interests.

Transportation

In the 19th century, railroads were used to open up the North American continent. Air transport is now more common for long distance passenger travel, although railroads are still extensively used for bulk freight transport. Waterways, like the Mississippi River, are important for the transport of bulk materials, and the Panama Canal is a vital link between the Pacific Ocean and the Caribbean. In the 20th century, road transportation increased massively in North America, with the introduction of cheap, mass-produced cars and extensive highway construction.

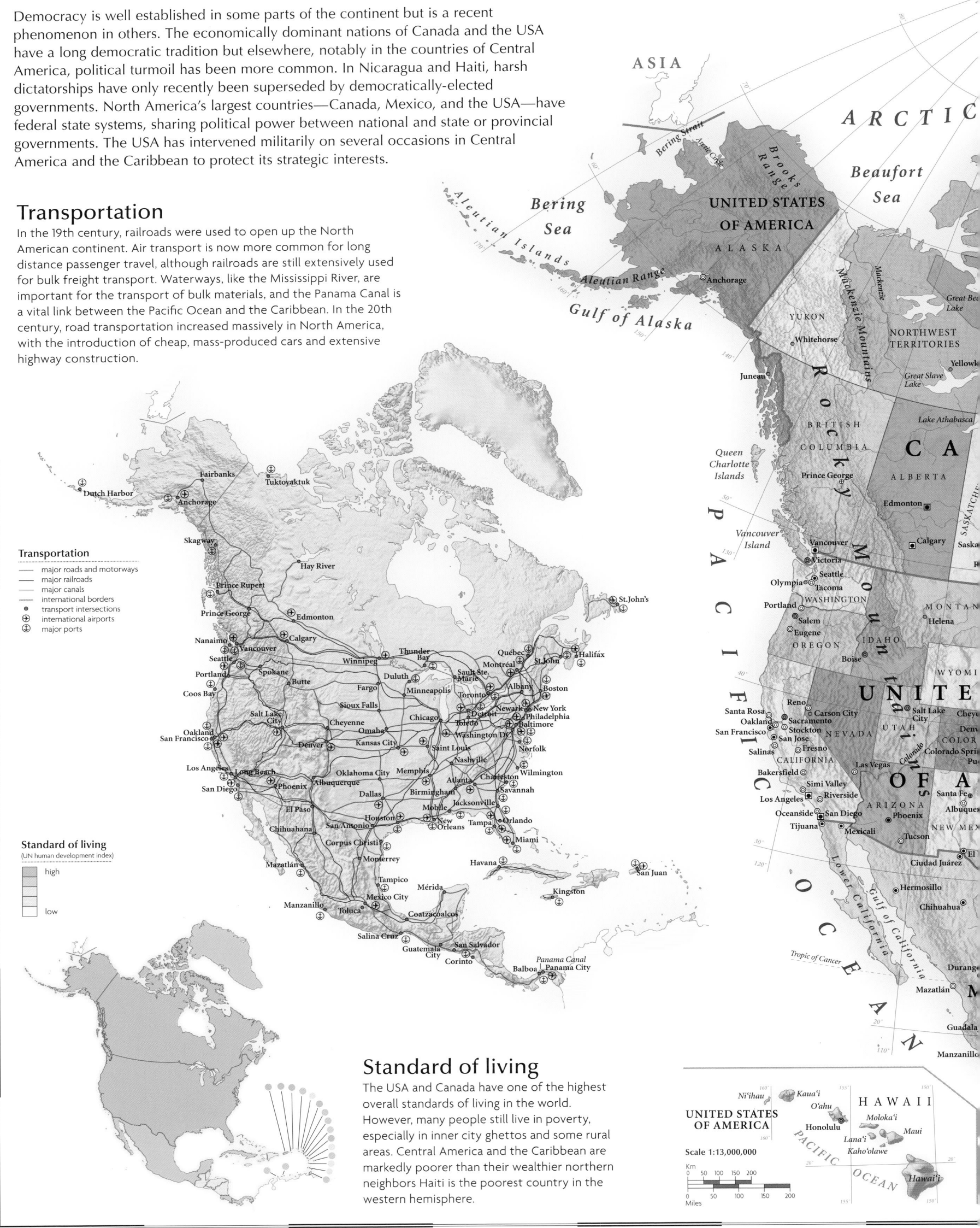

Transportation
- major roads and motorways
- major railroads
- major canals
- international borders
- transport intersections
- international airports
- major ports

Standard of living
(UN human development index)
high → low

Standard of living

The USA and Canada have one of the highest overall standards of living in the world. However, many people still live in poverty, especially in inner city ghettos and some rural areas. Central America and the Caribbean are markedly poorer than their wealthier northern neighbors Haiti is the poorest country in the western hemisphere.

UNITED STATES OF AMERICA

Scale 1:13,000,000

Languages

The three major official languages of North America are of European origin, brought by settlers in the 16th century. In Canada, French and English are spoken; in the USA, English is the main language, with large Spanish-speaking areas in the southwest; Mexicans are Spanish-speaking; while the Caribbean islands use French, English, and Spanish as well as the hybrid Creole tongues. In isolated areas, languages of the indigenous peoples still exist, such as Inuit in the far north of the continent.

Language groups

- American Indian
- Germanic
- Romance
- Eskimo-Aleut
- Uninhabited

ESKIMO-ALEUT
ATHABASCAN
ALGONQUIN
FRENCH
ENGLISH
ENGLISH/SPANISH
UTO-AZTECAN
FRENCH/ENGLISH
ENGLISH/SPANISH
ENGLISH
SPANISH-FRENCH
CREOLE
CREOLE
CREOLE
MAYAN
SPANISH

Population

Much of North America is almost empty, especially the frozen far north. Population densities are highest in the highlands of Mexico and Central America; the coastal plain stretching from the Gulf of Mexico along the Atlantic coast; the Great Lakes area; and the Pacific coast. Large conurbations have developed, notably the San-San (San Francisco–San Diego), Boswash (Boston–Washington) and Main Street (Toronto–Montréal). The populations of the Caribbean islands are small, but settlement is dense, due to the limited amount of land available.

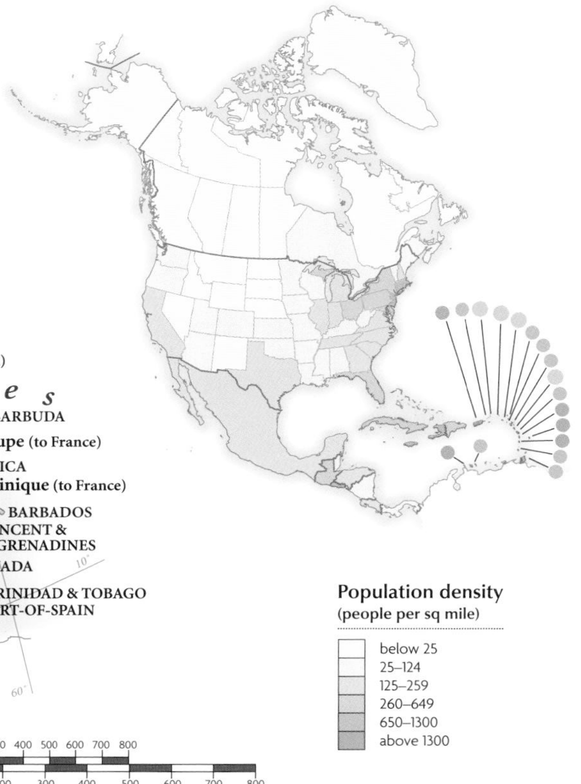

Population

- ■ above 5 million
- ◉ 1 million to 5 million
- ◎ 500,000 to 1 million
- ◍ 100,000 to 500,000
- ⊕ 50,000 to 100,000
- ○ 10,000 to 50,000
- ∘ below 10,000
- State / Province capital
- Country capital

Borders

- full international border
- state border

Population density
(people per sq mile)

- below 25
- 25–124
- 125–259
- 260–649
- 650–1300
- above 1300

Km
0 100 200 300 400 500 600 700 800
Miles
0 100 200 300 400 500 600 700 800

Scale 1:30,750,000
(projection: Lambert Azimuthal Equal Area)

OCEAN
Ellesmere Island
Greenland (to Denmark)
Baffin Bay
NUUK
Baffin Island
Davis Strait
toria and
Foxe Basin
NUNAVUT
Iqaluit
Labrador Sea
Hudson Strait
Hudson Bay
DA
Reindeer Lake
MANITOBA
Lake Winnipeg
ONTARIO
QUÉBEC
NEWFOUNDLAND AND LABRADOR
Newfoundland
St.John's
Winnipeg
Thunder Bay
St Pierre & Miquelon (to France)
PRINCE EDWARD ISLAND
Charlottetown
NORTH DAKOTA
Bismarck
MINNESOTA
Lake Superior
Lake Huron
OTTAWA
Québec
Montréal
St.Lawrence
NEW BRUNSWICK
Fredericton
NOVA SCOTIA
Halifax
MAINE
Augusta
TH DAKOTA
erre
Saint Paul
MICHIGAN
Oshawa
Toronto
Lake Ontario
Rochester
VERMONT
NEW HAMPSHIRE
Montpelier
Concord
Minneapolis
WISCONSIN
Lansing
Hamilton
Buffalo
NEW YORK
Albany
Boston
MASSACHUSETTS
Sioux Falls
Madison
Milwaukee
Detroit
Lake Erie
Hartford
Providence
RHODE ISLAND
Missouri
Lincoln
Chicago
Lake Michigan
Cleveland
PENNSYLVANIA
CONNECTICUT
New York
BRASKA
Des Moines
IOWA
ILLINOIS
INDIANA
OHIO
Toledo
Pittsburgh
Harrisburg
Trenton
NEW JERSEY
Philadelphia
Dover
DELAWARE
STATES
Omaha
Springfield
Indianapolis
Columbus
Cincinnati
WEST VIRGINIA
MARYLAND
Baltimore
Annapolis
WASHINGTON DC
Topeka
Kansas City
saint Louis
Frankfort
Charleston
Richmond
VIRGINIA
KANSAS
Jefferson City
MISSOURI
Louisville
KENTUCKY
Norfolk
Wichita
Springfield
Evansville
Nashville
Appalachian Mountains
Raleigh
NORTH CAROLINA
RICA
Arkansas
TENNESSEE
Charlotte
Columbia
SOUTH CAROLINA
rillo
Oklahoma City
OKLAHOMA
Little Rock
ARKANSAS
Memphis
Birmingham
Atlanta
GEORGIA
ubbock
Mississippi
ALABAMA
Savannah
Fort Worth
Dallas
Shreveport
Jackson
MISSISSIPPI
Montgomery
Columbus
TEXAS
Jacksonville
Austin
LOUISIANA
Baton Rouge
Mobile
Tallahassee
San Antonio
Houston
New Orleans
Mississippi Delta
Orlando
FLORIDA
Saint Petersburg
Tampa
Corpus Christi
Rio Grande
Fort Lauderdale
Miami
Gulf of Mexico
Monterrey
NASSAU
THE BAHAMAS
Tropic of Cancer
Virgin Islands (to US)
British Virgin Islands (to UK)
Anguilla (to UK)
HAVANA
Guantanamo Bay (to US)
Turks & Caicos Islands (to UK)
Puerto Rico (to US)
ANTIGUA & BARBUDA
Santa Clara
CUBA
SAN JUAN
Guadeloupe (to France)
ICO
Tampico
San Luis Potosí
eón
Santiago de Cuba
HAITI
Greater
DOMINICAN REPUBLIC
SANTO DOMINGO
ST KITTS & NEVIS
Montserrat (to UK)
DOMINICA
Martinique (to France)
Mérida
Cayman Islands (to UK)
PORT-AU-PRINCE
Navassa Island (to US)
Antilles
ST LUCIA
Querétaro
Yucatan Peninsula
JAMAICA
KINGSTON
BARBADOS
ST VINCENT & THE GRENADINES
Morelia
oluca
MEXICO CITY
Puebla
West
Indies
GRENADA
Acapulco
Villahermosa
BELIZE
BELMOPAN
Aruba (Neth.)
Lesser Antilles
TRINIDAD & TOBAGO
PORT-OF-SPAIN
Caribbean Sea
Bonaire (to Neth.)
GUATEMALA
HONDURAS
San Pedro Sula
TEGUCIGALPA
Curaçao (Neth.)
GUATEMALA CITY
SAN SALVADOR
NICARAGUA
EL SALVADOR
Lake Nicaragua
MANAGUA
SOUTH AMERICA
SAN JOSÉ
PANAMA CITY
COSTA RICA
PANAMA

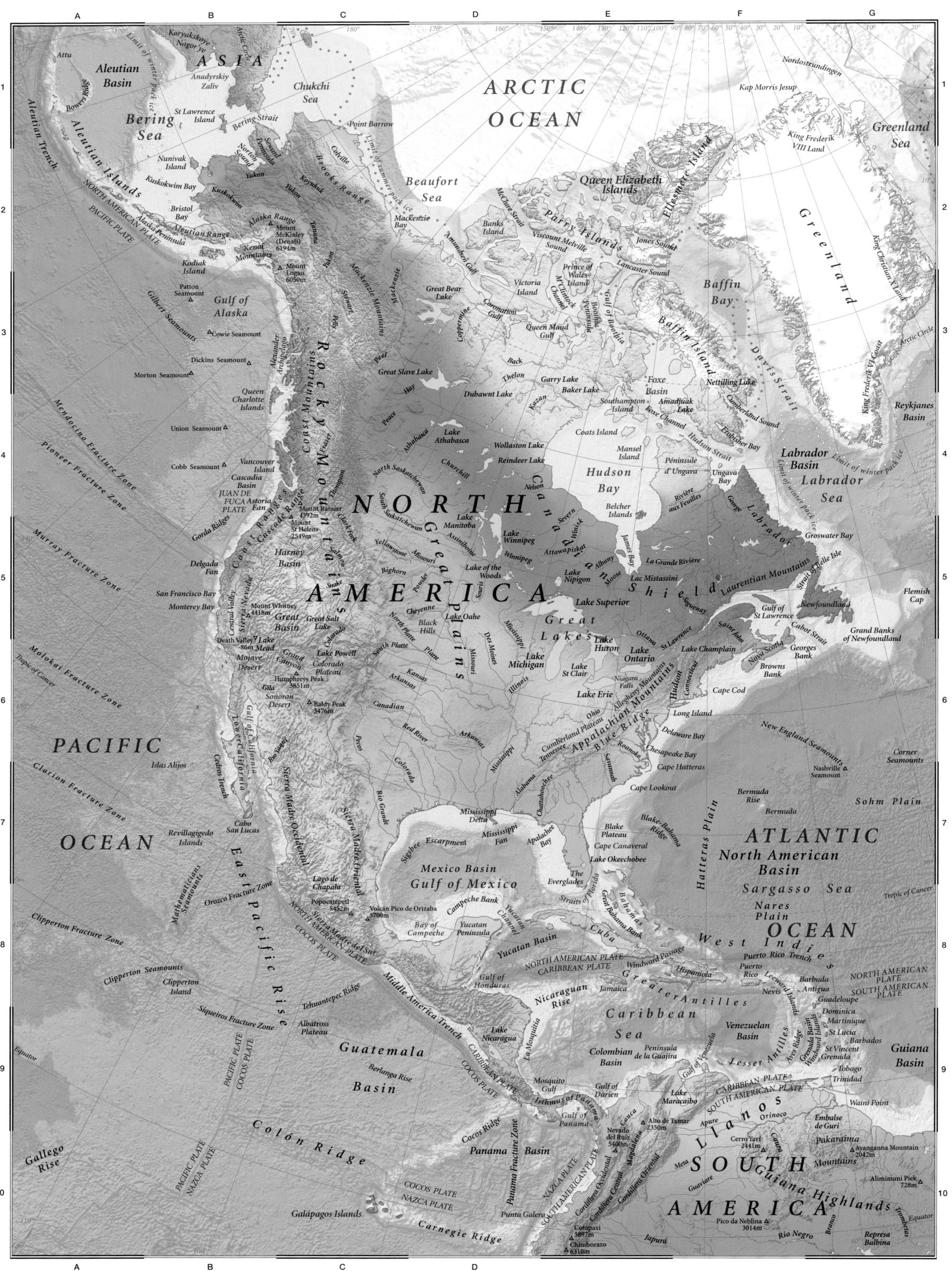

NORTH AMERICA

46

Scale 1:33,500,000
(projection: Lambert Conformal Conic)

Environmental Issues

Many fragile environments are under threat throughout the region. In Haiti, all the primary rain forest has been destroyed, while air pollution from factories and cars in Mexico City is among the worst in the world. Elsewhere, industry and mining pose threats, particularly in the delicate arctic environment of Alaska where oil spills have polluted coastlines and decimated fish stocks.

Climate

North America's climate includes extremes ranging from freezing Arctic conditions in Alaska and Greenland, to desert in the southwest, and tropical conditions in southeastern Florida, the Caribbean, and Central America. Central and southern regions are prone to severe storms including tornadoes and hurricanes.

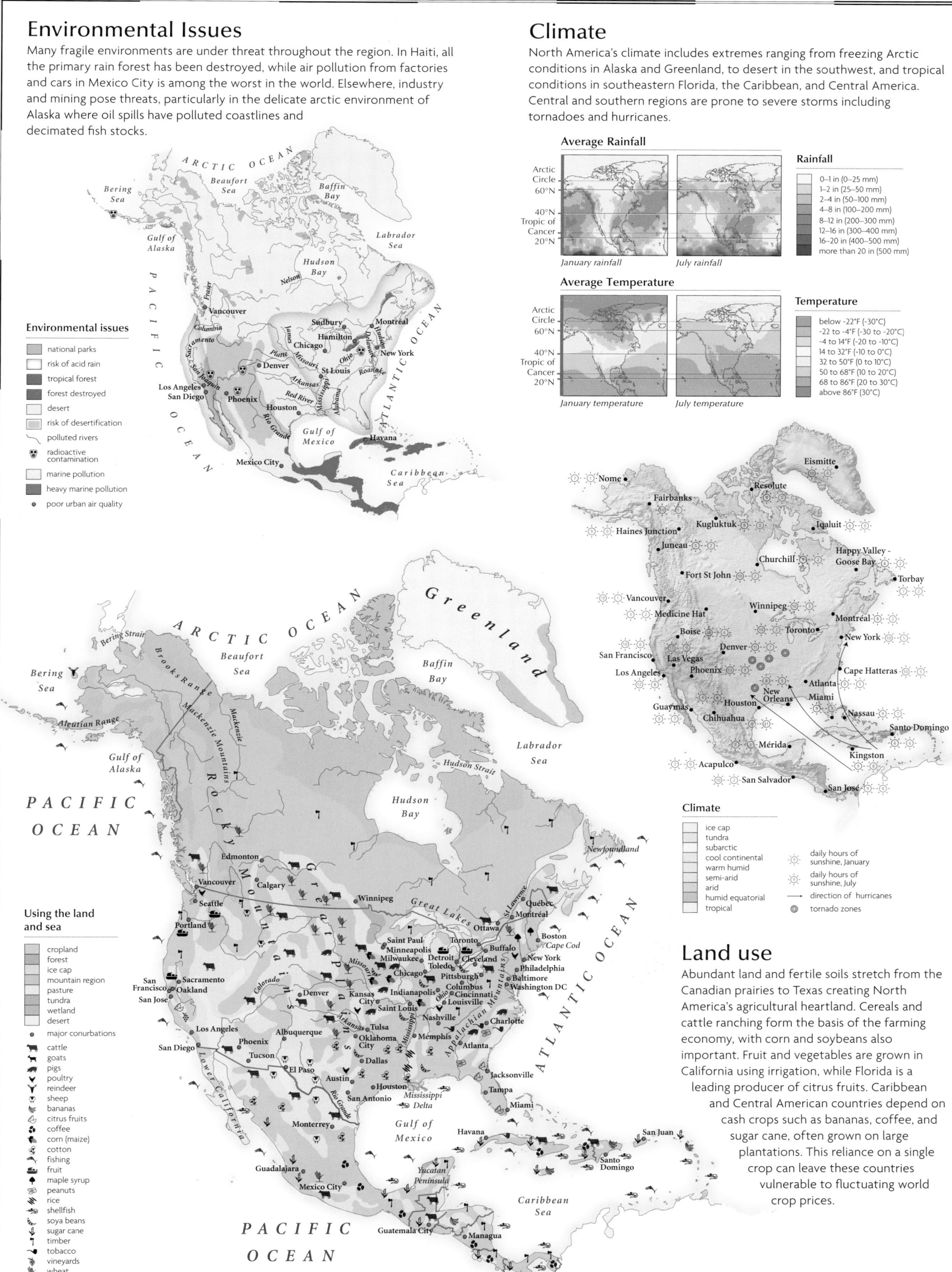

Average Rainfall

January rainfall   *July rainfall*

Rainfall
- 0–1 in (0–25 mm)
- 1–2 in (25–50 mm)
- 2–4 in (50–100 mm)
- 4–8 in (100–200 mm)
- 8–12 in (200–300 mm)
- 12–16 in (300–400 mm)
- 16–20 in (400–500 mm)
- more than 20 in (500 mm)

Average Temperature

January temperature   *July temperature*

Temperature
- below -22°F (-30°C)
- -22 to -4°F (-30 to -20°C)
- -4 to 14°F (-20 to -10°C)
- 14 to 32°F (-10 to 0°C)
- 32 to 50°F (0 to 10°C)
- 50 to 68°F (10 to 20°C)
- 68 to 86°F (20 to 30°C)
- above 86°F (30°C)

Environmental issues
- national parks
- risk of acid rain
- tropical forest
- forest destroyed
- desert
- risk of desertification
- polluted rivers
- radioactive contamination
- marine pollution
- heavy marine pollution
- poor urban air quality

Climate
- ice cap
- tundra
- subarctic
- cool continental
- warm humid
- semi-arid
- arid
- humid equatorial
- tropical
- daily hours of sunshine, January
- daily hours of sunshine, July
- direction of hurricanes
- tornado zones

Land use

Abundant land and fertile soils stretch from the Canadian prairies to Texas creating North America's agricultural heartland. Cereals and cattle ranching form the basis of the farming economy, with corn and soybeans also important. Fruit and vegetables are grown in California using irrigation, while Florida is a leading producer of citrus fruits. Caribbean and Central American countries depend on cash crops such as bananas, coffee, and sugar cane, often grown on large plantations. This reliance on a single crop can leave these countries vulnerable to fluctuating world crop prices.

Using the land and sea
- cropland
- forest
- ice cap
- mountain region
- pasture
- tundra
- wetland
- desert
- major conurbations
- cattle
- goats
- pigs
- poultry
- reindeer
- sheep
- bananas
- citrus fruits
- coffee
- corn (maize)
- cotton
- fishing
- fruit
- maple syrup
- peanuts
- rice
- shellfish
- soya beans
- sugar cane
- timber
- tobacco
- vineyards
- wheat

NORTH AMERICA

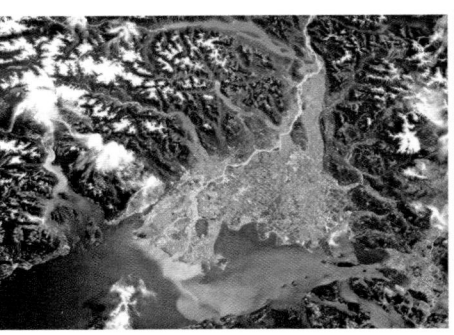

1 VANCOUVER, BRITISH COLUMBIA, CANADA
Canada's premier west coast city occupies the delta of the Fraser river, formed among the Coast Mountains.

2 MOUNT SAINT HELENS, WASHINGTON, USA
In 1980, this volcano's catastrophic eruption devastated 270 sq miles (700 sq km) of forest almost instantly.

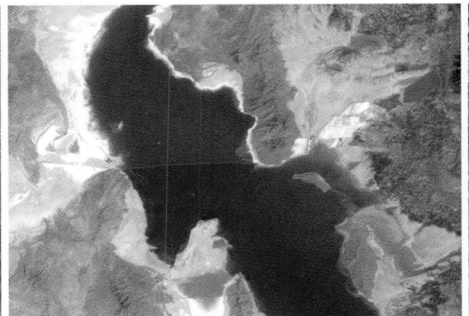

3 GREAT SALT LAKE, UTAH, USA
A causeway carries a railroad, blocking circulation between the northern and southern parts, the water reddened by bacteria in the more saline north.

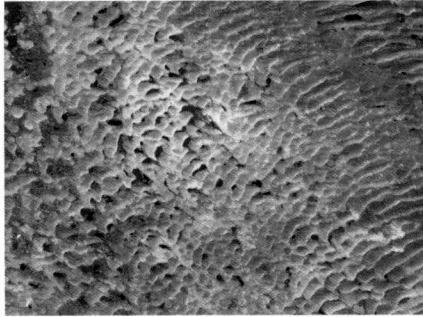

4 SAND HILLS, NEBRASKA, USA
Forming the largest sand sea in the Western Hemisphere, these hills are not classified as desert because today's relatively wet climate has allowed grasses to take hold.

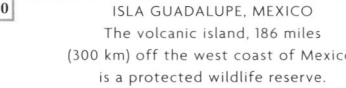

9 LOS ANGELES AND LONG BEACH, CALIFORNIA, USA
Taken together, these west coast cities constitute the busiest seaport in the United States.

10 ISLA GUADALUPE, MEXICO
The volcanic island, 186 miles (300 km) off the west coast of Mexico, is a protected wildlife reserve.

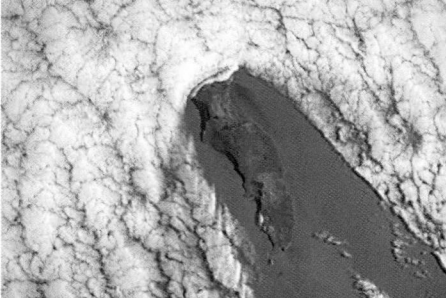

11 GRAND CANYON, ARIZONA, USA
The 5250 ft (1600 m) deep canyon cuts through the Kaibab Plateau in this southwest-looking view.

12 DENVER, COLORADO, USA
Colorado's state capital nestles under the Rocky Mountains with the South Platte River running through its center.

BELCHER ISLANDS, NUNAVUT, CANADA 5
These low-lying, treeless, and sparsely-
populated islands lie icebound in Hudson Bay
for much of the year.

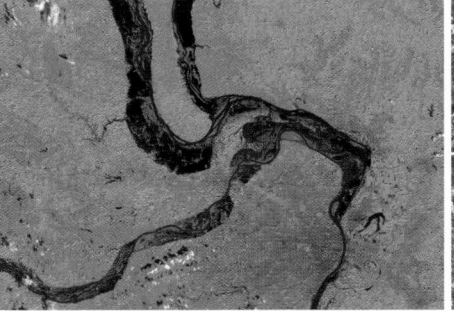

MISSISSIPPI, MISSOURI, AND ILLINOIS RIVERS, USA 6
This Infrared image shows how these rivers burst their
banks in many places after heavy rains in the summer
of 1993, leading to the area's worst floods on record.

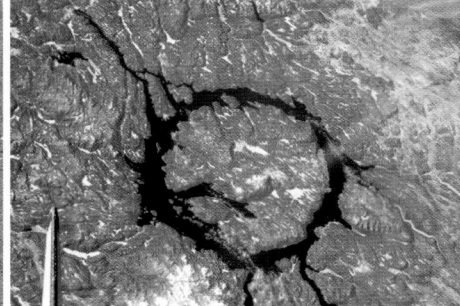

RÉSERVOIR MANICOUAGAN, QUÉBEC, CANADA 7
This unusual 62 mile (100 km) diameter annular
lake occupies the low ground between the rim
and central uplift of an ancient meteorite crater.

NEW YORK CITY, USA 8
The largest city in the United States, with
a population of over 8 million, it is also the
country's main financial center.

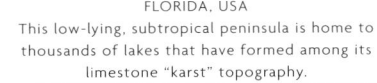

MISSISSIPPI RIVER DELTA, LOUISIANA, USA 13
This delta has developed a "bird's foot" shape
due to the shifting course of the river over
the last 6000 years.

FLORIDA, USA 14
This low-lying, subtropical peninsula is home to
thousands of lakes that have formed among its
limestone "karst" topography.

HAVANA, CUBA 15
Cuba's capital city is home to 2 million
people and was founded by the Spanish in
1519 around a natural harbor.

BARRIER REEF, BELIZE 16
The world's second-longest barrier reef lies
about 12 miles (20 km) off the coast of Belize.

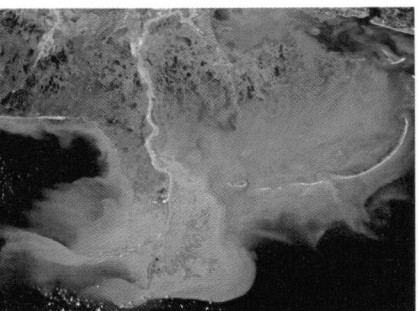

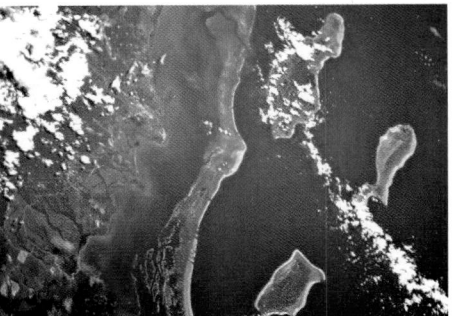

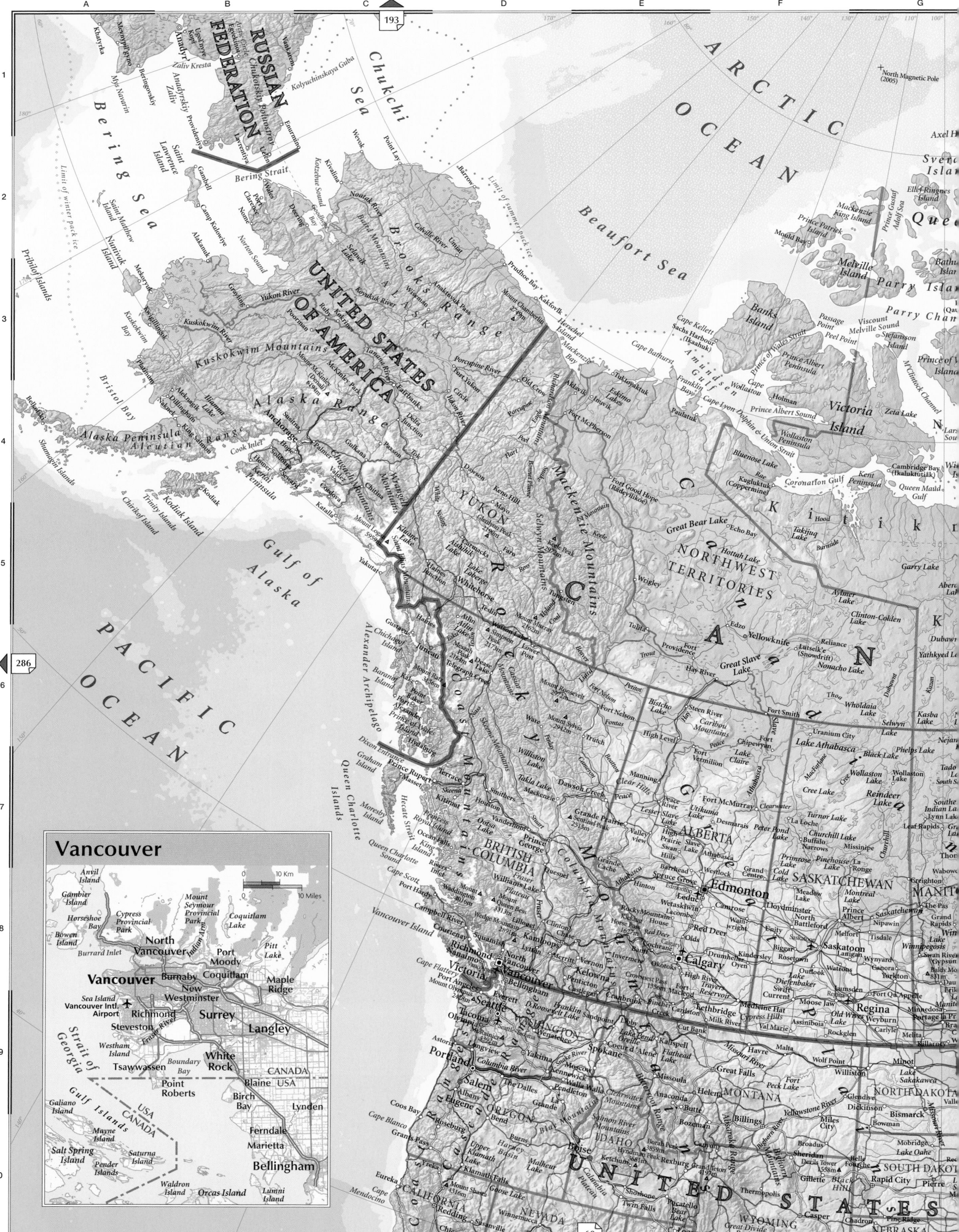

Northern Canada

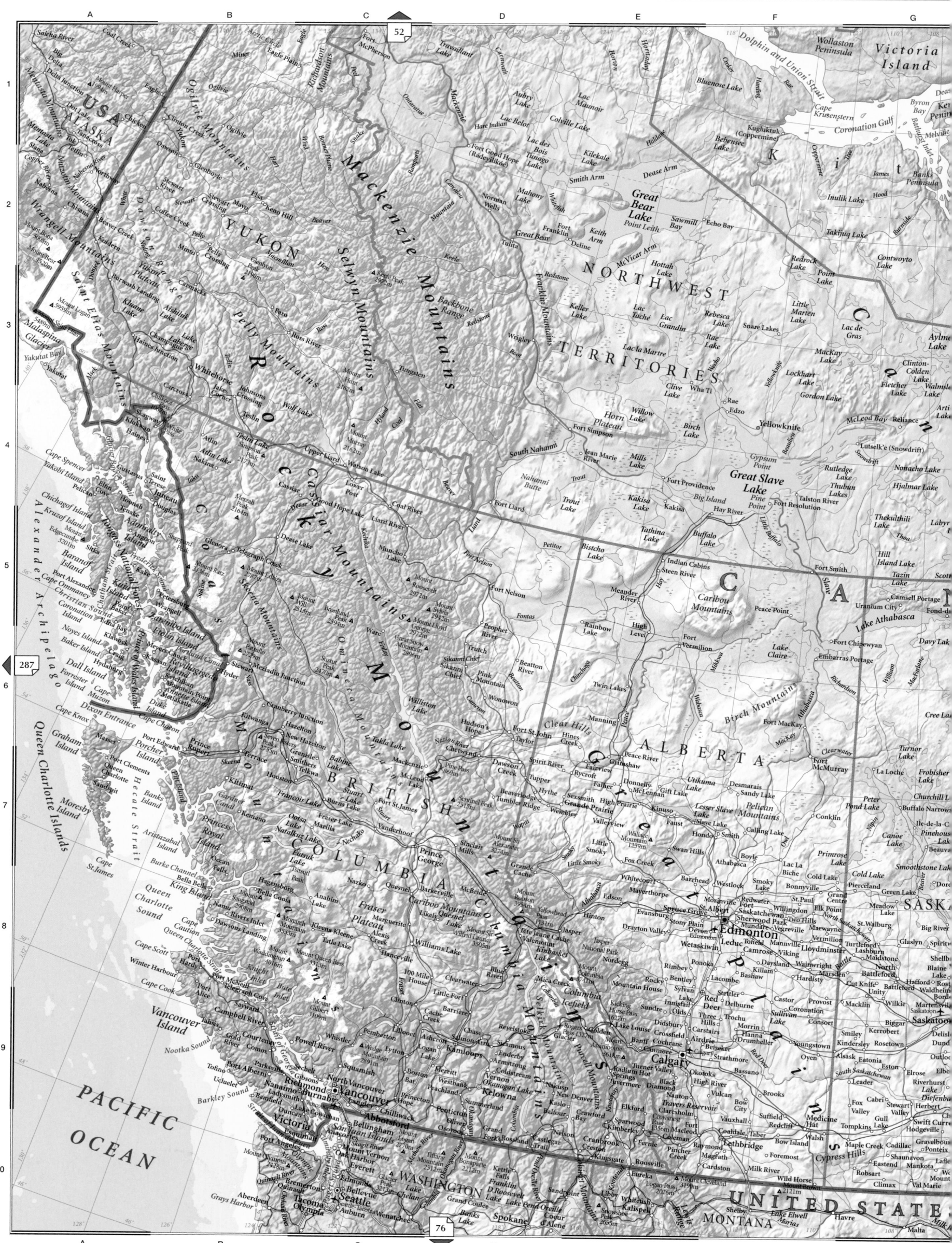

Scale 1:7,500,000
(projection: Lambert Conformal Conic)

0 25 50 75 100 125 150 175 200 Km
0 25 50 75 100 125 150 175 200 Miles

Population
▪ above 5 million
◉ 100,000 to 500,000
▪ 1 million to 5 million
⊕ 50,000 to 100,000
◉ 500,000 to 1 million
○ 10,000 to 50,000
○ below 10,000

NUNAVUT
Kivalliq
QUÉBEC
Péninsule d'Ungava
MANITOBA
ONTARIO
SASKATCHEWAN
MINNESOTA
NORTH DAKOTA
MICHIGAN
OF AMERICA

Hudson Bay
James Bay
Hudson Strait
Foxe Basin
Foxe Channel
Foxe Peninsula
Roes Welcome Sound
Southampton Island
Melville Peninsula
Prince Charles Island
Coats Island
Mansel Island
Belcher Islands
Sleeper Islands
King George Islands
North Belcher Islands
Ottawa Islands
Akimiski Island
Lake Winnipeg
Lake Winnipegosis
Lake Manitoba
Lake of the Woods
Lake Superior
Lake Nipigon
Lake Seul
Reindeer Lake
Wollaston Lake
Baker Lake (Qamanittuaq)
Dubawnt Lake
Aberdeen Lake
Garry Lake
Chesterfield Inlet
Rankin Inlet
Churchill
Winnipeg
Regina
Thunder Bay
Sault Ste Marie
York Factory
Fort Severn
Fort Albany
Attawapiskat
Moosonee
Kuujjuaq

Elevation scale:
19,686ft
13,124ft
9843ft
6562ft
3281ft
1640ft
820ft
328ft
Sea Level
-820ft
-6562ft
-13,124ft

53
58
72

NORTH AMERICA

58

54
76
287

PACIFIC OCEAN

BRITISH COLUMBIA

ROCKY MOUNTAINS

Coast Mountains

Columbia Mountains

Selkirk Mountains

Cariboo Mountains

Fraser Plateau

Skeena Mountains

Omineca Mountains

Clear Hills

Vancouver Island

Olympic Mountains

UNITED STATES

WASHINGTON

Queen Charlotte Sound

Queen Charlotte Strait

Strait of Georgia

Strait of Juan de Fuca

USA ALASKA

Elevation scale:
- 6000m
- 4000m
- 3000m
- 2000m
- 1000m
- 500m
- 250m
- 100m
- Sea Level
- -250m
- -2000m
- -4000m

Wrangell, Ketchikan, Metlakatla, Annette Island, Revillagigedo Island, Duke Island, Hyder, Stewart, Meziadin Junction, Mount Pattullo 2739m, Cranberry Junction, Kitwanga, Hazelton, New Hazelton, Terrace, Kitimat, Prince Rupert, Port Edward, Porcher Island, Pitt Island, Banks Island, Princess Royal Island, Aristazabal Island, Ocean Falls, Bella Bella, King Island, Namu, Rivers Inlet, Dawsons Landing, Hagensborg, Bella Coola, Mount Saugstad 2908m, Burke Channel, Monarch Mountain 3533m, Klecna Kleene, Mount Waddington 4016m, Mount Queen Bess 3313m, Knight Inlet, Bute Inlet, Mount Gilbert 3109m, Cape Caution, Cape Scott, Cape Cook, Winter Harbour, Port Hardy, Port McNeill, Port Alice, Telegraph Cove, Sayward, Campbell River, Gold River, Tahsis, Nootka Sound, Tofino, Ucluelet, Bamfield, Barkley Sound, Port Alberni, Parksville, Ladysmith, Lake Cowichan, Courtenay, Comox, Powell River, Sechelt, Gibsons, Nanaimo, Duncan, Swarts Bay, Victoria, Esquimalt, Port Renfrew, Cape Flattery, Neah Bay, Clallam Bay

Ware, Great Snow Mountain 2896m, Trutch, Beatton River, Sikanni Chief, Pink Mountain, Wonowon, Cameron, Williston Lake, Fort St.John, Hudson's Hope, Taylor, Peace, Hines Creek, Fairview, Grimshaw, Chetwynd, Dawson Creek, Spirit River, Rycroft, Mackenzie, Pine Pass 869m, McLeod Lake, Tupper, Hythe, Beaverlodge, Wembley, Grande Prairie, Sexsmith, Tumbler Ridge, Sentinel Peak 2515m, Wapiti, Sinclair Mills, Mount Sir Alexander 3274m, Grande Cache, Prince George, Vanderhoof, Fraser Lake, Fort St.James, Stuart Lake, Fort, Nechako, Takla Lake, Babine Lake, Granisle, Smithers, Telkwa, Houston, Burns Lake, Ootsa Lake, Eutsuk Lake, Nazko, Quesnel, Barkerville, Mount Robson 3954m, Tete Jaune Cache, Yellowhead Pass 1131m, Jasper, Jasper National, Likely, Mount Sir Wilfrid Laurier 3505m, Valemount, McBride, Marguerite, Williams Lake, Blue River, Mica Creek, Mount Columbia 3741m, Kinbasket Lake, 100 Mile House, Hanceville, Alexis Creek, Tatla Lake, Clearwater, Little Fort, Clinton, Barriere, Cache Creek, Thompson, Ashcroft, Lillooet, Kamloops, Chase, Salmon Arm, Enderby, Sicamous, Revelstoke, Rogers Pass 1327m, Logan Lake, Armstrong, Vernon, Coldstream, Pemberton, Whistler, Wedge Mountain 2891m, Lytton, Merritt, Okanagan Lake, Westbank, Kelowna, Peachland, Summerland, Nakusp, New Denver, Boston Bar, Squamish, North Vancouver, Vancouver, Burnaby, Richmond, Langley, Chilliwack, Hope, Princeton, Penticton, Abbotsford, Blaine, Ferndale, Lynden, Sumas, Deming, Bellingham, Newhalem, Oliver, Osoyoos, Oroville, Grand Forks, Rossland, Castlegar, Mount Bonaparte 2213m, Orient, Northport, Tiffany Mountain 2512m, Tonasket, Republic, Kettle Falls, Colville, San Juan Islands, Oak Harbor, Port Townsend, Coupeville, Anacortes, Mount Vernon, Skagit River, Rockport, Mount Logan 2770m, Mazama, Winthrop, Lake Chelan, Chewelah, Sequim, Port Angeles, Forks, Queets, Quinault, Moclips, Taholah, Pacific Beach, Copalis Beach, Humptulips, Quinault River, Mount Olympus 2428m, Bremerton, Port Orchard, Seattle, Bellevue, Redmond, Edmonds, Monroe, Everett, Marysville, Arlington, Darrington, Glacier Peak 3213m, Leavenworth, Skykomish, Entiat, Wenatchee, Kent, Auburn, Tacoma, Puyallup, Shelton, Roslyn, Ephrata, Odessa, Sprague, Davenport, Coulee City, Grand Coulee, Franklin D. Roosevelt Lake, Banks Lake, Wilbur, Crab Creek, Spokane River, Spokane, Deer Park, Green River

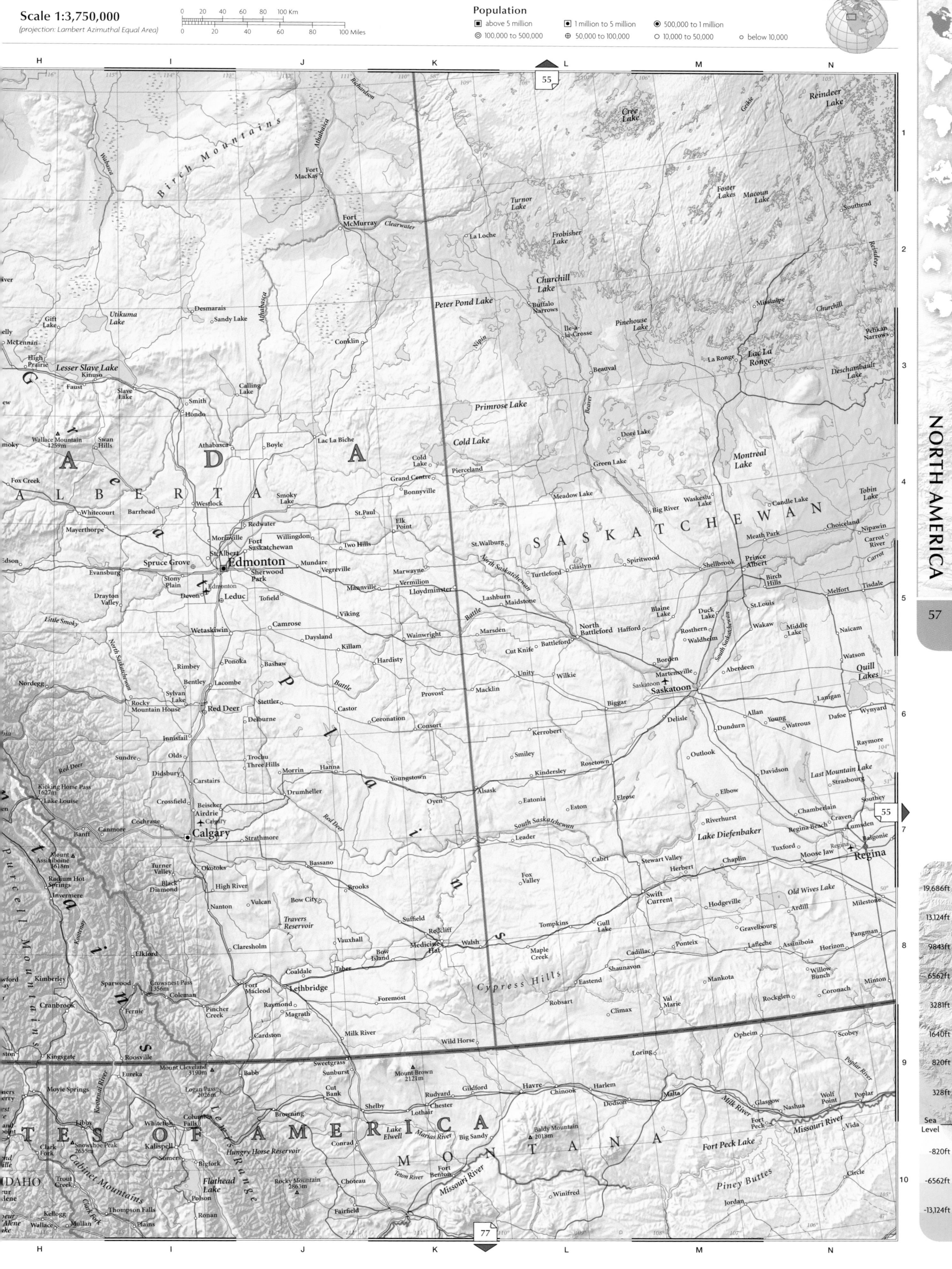

Map Content

Scale and Projection

Scale 1:3,750,000
(projection: Lambert Azimuthal Equal Area)

0 20 40 60 80 100 Km
0 20 40 60 80 100 Miles

Population

■ above 5 million	◾ 1 million to 5 million	◉ 500,000 to 1 million
◎ 100,000 to 500,000	⊕ 50,000 to 100,000	○ 10,000 to 50,000	∘ below 10,000

Elevation Scale

19,686ft
13,124ft
9843ft
6562ft
3281ft
1640ft
820ft
328ft
Sea Level
-820ft
-6562ft
-13,124ft

Grid references (top)

H I J K L M N
55
77

Place Names

CANADA

ALBERTA

SASKATCHEWAN

Great Plains

Rocky Mountains

Purcell Mountains

UNITED STATES OF AMERICA

MONTANA

IDAHO

Birch Mountains

Cypress Hills

Reindeer Lake
Southend
Foster Lakes
Macoun Lake
Cree Lake
Churchill
Missinipe
Pelikan Narrows
Lac La Ronge
La Ronge
Deschambault Lake
Turnor Lake
Frobisher Lake
La Loche
Churchill Lake
Buffalo Narrows
Ile-a-la-Crosse
Pinehouse Lake
Beauval
Green Lake
Dore Lake
Montreal Lake
Peter Pond Lake
Nipin
Primrose Lake
Beaver

Fort MacKay
Fort McMurray
Clearwater

Desmarais
Sandy Lake
Athabasca
Conklin

Gift Lake
Utikuma Lake
McLennan
High Prairie
Lesser Slave Lake
Kinuso
Faust
Slave Lake
Smith
Hondo
Calling Lake

Athabasca
Boyle
Lac La Biche
Cold Lake
Pierceland
Grand Centre
Cold Lake
Bonnyville
Meadow Lake
Big River
Waskesiu Lake
Candle Lake
Choiceland
Carrot River
Carrot
Nipawin
Tisdale
Melfort
Watson
Quill Lakes
Lanigan
Dafoe
Wynyard
Raymore
Strasbourg
Last Mountain Lake
Southey
Chamberlain
Craven
Lumsden
Regina Beach
Balgonie
Regina
Milestone
Pangman
Horizon
Assiniboia
Willow Bunch
Coronach
Minton
Rockglen
Val Marie
Climax
Robsart
Eastend
Shaunavon
Mankota
Cadillac
Ponteix
Maple Creek
Gull Lake
Tompkins
Swift Current
Hodgeville
Gravelbourg
Lafleche
Ardill
Old Wives Lake
Cabri
Stewart Valley
Herbert
Chaplin
Moose Jaw
Tuxford
Leader
Lake Diefenbaker
Riverhurst
Elbow
Outlook
Davidson
Elrose
Eston
Kerrobert
Kindersley
Rosetown
Smiley
Youngstown
Alsask
Oyen
Eatonia
Delisle
Dundurn
Young
Watrous
Allan
Aberdeen
Saskatoon
Biggar
Martensville
Borden
Wilkie
Unity
Macklin
Provost
Cut Knife
Battleford
North Battleford
Hafford
Blaine Lake
Duck Lake
Rosthern
Waldheim
St.Louis
Birch Hills
Prince Albert
Shellbrook
Meath Park
Turtleford
Glaslyn
Spiritwood
St.Walburg
Marsden
Lashburn
Maidstone
Lloydminster
Vermilion
Marwayne
St.Paul
Elk Point
Two Hills
Willingdon
Vegreville
Mundare
Mannville
Wainwright
Hardisty
Killam
Daysland
Camrose
Wetaskiwin
Viking
Ponoka
Bashaw
Stettler
Castor
Coronation
Consort
Provost
Rimbey
Bentley
Lacombe
Sylvan Lake
Red Deer
Innisfail
Delburne
Trochu
Three Hills
Hanna
Morrin
Didsbury
Carstairs
Drumheller
Sundre
Olds
Crossfield
Beiseker
Airdrie
Cochrane
Calgary
Strathmore
Bassano
Brooks
Okotoks
High River
Turner Valley
Black Diamond
Nanton
Vulcan
Bow City
Travers Reservoir
Claresholm
Vauxhall
Taber
Bow Island
Redcliff
Medicine Hat
Walsh
Suffield
Fox Valley
Leader

Edmonton
Spruce Grove
Stony Plain
Devon
Leduc
St.Albert
Morinville
Fort Saskatchewan
Sherwood Park
Redwater
Westlock
Barrhead
Whitecourt
Mayerthorpe
Smoky Lake
Evansburg
Drayton Valley
Little Smoky
North Saskatchewan
Nordegg
Rocky Mountain House
Fox Creek
Edson
Swan Hills
Wallace Mountain 1259m

Banff
Canmore
Kicking Horse Pass 1627m
Lake Louise
Mount Assiniboine 3618m
Radium Hot Springs
Invermere
Kimberley
Cranbrook
Fernie
Sparwood
Crowsnest Pass 1356m
Coleman
Fort Macleod
Lethbridge
Coaldale
Pincher Creek
Raymond
Magrath
Cardston
Foremost
Milk River
Wild Horse
Kingsgate
Roosville
Eureka
Babb
Sweetgrass
Sunburst
Mount Cleveland 3190m
Logan Pass 2026m
Cut Bank
Shelby
Lothair
Chester
Conrad
Browning
Columbia Falls
Whitefish
Kalispell
Somers
Bigfork
Hungry Horse Reservoir
Lake Elwell
Marias River
Choteau
Fairfield
Teton River
Fort Benton
Missouri River
Big Sandy
Baldy Mountain 2018m
Winifred
Rudyard
Gildford
Havre
Chinook
Harlem
Malta
Dodson
Glasgow
Nashua
Wolf Point
Poplar
Scobey
Opheim
Loring
Milk River
Poplar River
Missouri River
Vida
Circle
Jordan
Piney Buttes
Fort Peck Lake
Fort Peck
Mount Brown 2121m
Flathead Lake
Polson
Ronan
Plains
Thompson Falls
Snowshoe Peak 2665m
Cabinet Mountains
Clark Fork
Trout Creek
Libby
Moyie Springs
Kellogg
Wallace
Mullan
Rocky Mountain 2663m
Roosevelt

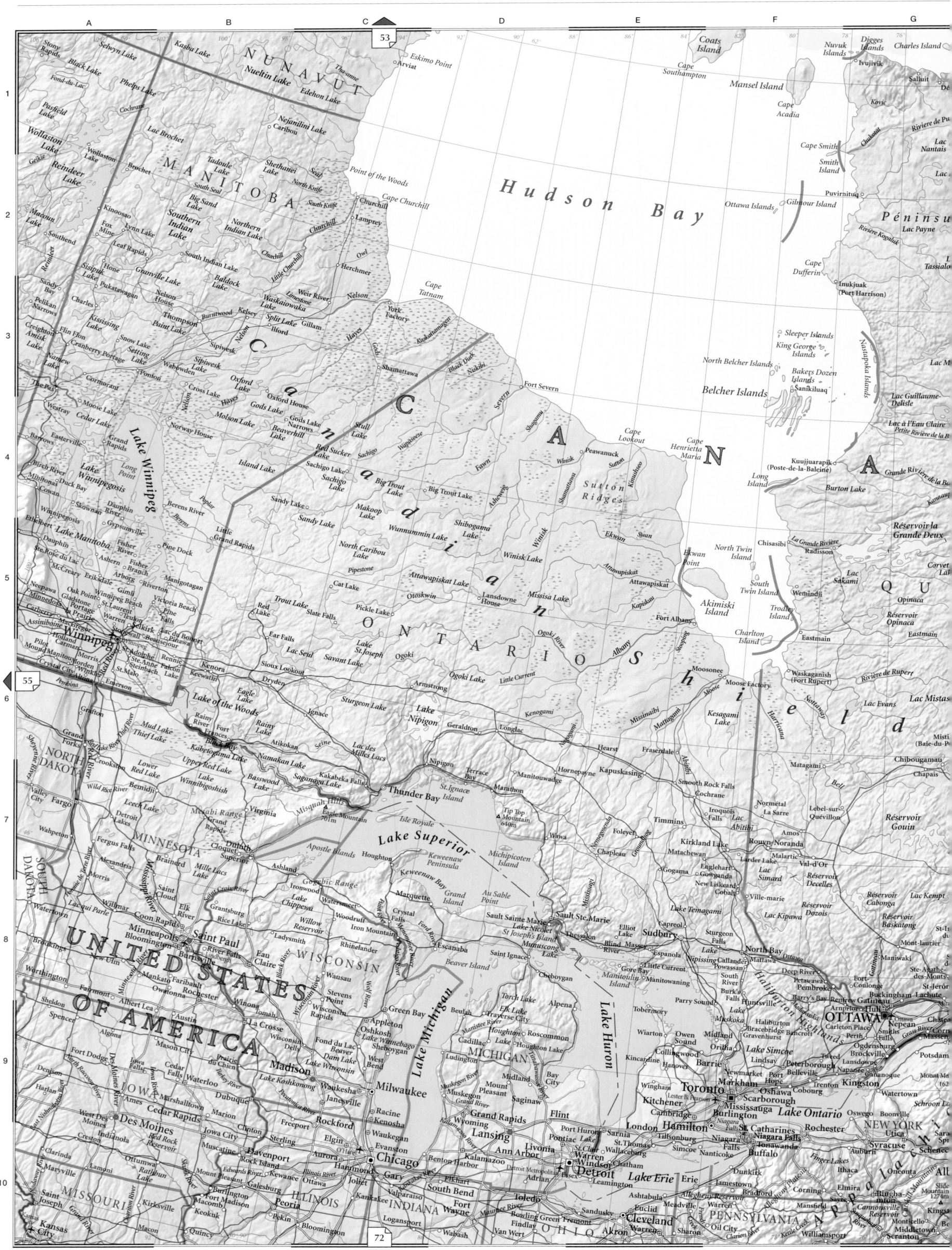

Scale 1:10,250,000
(projection: Lambert Azimuthal Equal Area)

| 0 | 50 | 100 | 150 | 200 | 250 | 300 Km |
| 0 | 50 | 100 | 150 | 200 | 250 | 300 Miles |

Population

- ■ above 5 million
- ■ 1 million to 5 million
- ◉ 500,000 to 1 million
- ◎ 100,000 to 500,000
- ⊕ 50,000 to 100,000
- ○ 10,000 to 50,000
- ∘ below 10,000

51

290

90

19,686ft
13,124ft
9843ft
6562ft
3281ft
1640ft
820ft
328ft
Sea Level
-820ft
-6562ft
-13,124ft

ATLANTIC OCEAN

Gulf of Mexico

THE BAHAMAS

CUBA

LA HABANA (HAVANA)

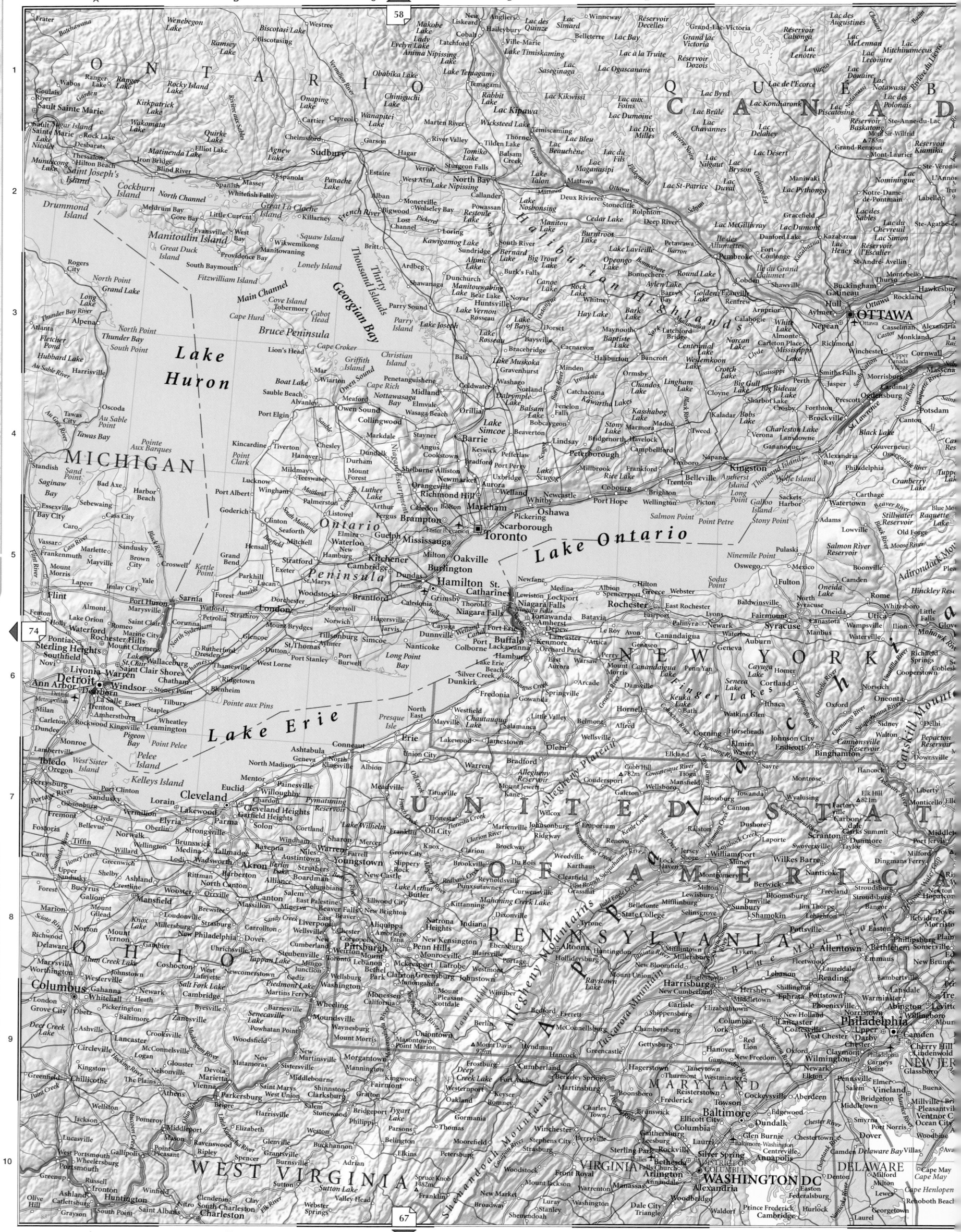

Scale 1:3,000,000
(projection: Lambert Conformal Conic)

0 20 40 60 80 100 Km
0 20 40 60 80 100 Miles

Population
■ above 5 million
◉ 1 million to 5 million
◉ 500,000 to 1 million
◎ 100,000 to 500,000
⊙ 50,000 to 100,000
○ 10,000 to 50,000
○ below 10,000

59

290

91

Boston

Pinehurst
Reading
Beverly
Salem
Salem Maritime N.H.S.
Peabody
Wakefield
Lynn Woods Reserve
Woburn
Marblehead
Melrose
Saugus
Tinkers Island
Lynn
Swampscott
Lexington
Winchester
Middlesex Park Reserve
Nahant Bay
Nahant
Arlington
Medford
Malden
Revere
Broad Sound
East Point
Lincoln
Somerville
Everett
Chelsea
Winthrop
Massachusetts Bay
Waltham
Cambridge
Bunker Hill Monument
Watertown
Logan International Airport
John F. Kennedy N.H.S.
Quincy Market
Newton
Boston
Boston Harbor
Brewster Islands
Point Allerton
Wellesley
Brookline
Franklin Park
Needham
Quincy Bay
Hingham Bay
Norwood
Charles River
Dedham
Neponset River
Milton
Adams N.H.S.
Quincy
Weymouth
Braintree

4 Km
4 Miles

19,686ft
13,124ft
9843ft
6562ft
3281ft
1640ft
820ft
328ft
Sea Level
-820ft
-6562ft
-13,124ft

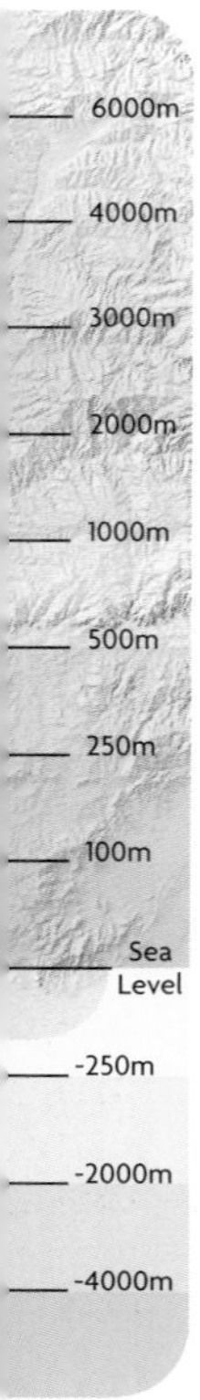

Scale 1:1,500,000
(projection: Lambert Conformal Conic)

0 10 20 30 40 50 Km
0 10 20 30 40 50 Miles

Population
- ■ above 5 million
- ◉ 100,000 to 500,000
- ▣ 1 million to 5 million
- ⊕ 50,000 to 100,000
- ◉ 500,000 to 1 million
- ○ 10,000 to 50,000
- ○ below 10,000

VERMONT

NEW HAMPSHIRE

MASSACHUSETTS

Gulf of Maine

Boston

Massachusetts Bay

CONNECTICUT

RHODE ISLAND

Hartford

Providence

Cape Cod Bay

Cape Cod

Nantucket Sound

Nantucket Island

Martha's Vineyard

Albany

Long Island Sound

New York

Long Island

Block Island Sound

Montauk Point

ATLANTIC OCEAN

New York

Passaic
Rutherford
North Arlington
North Bergen
Fort Lee
Bronx
Kings Point
Whitestone
Flushing
East Orange
Kearny
Union City
Manhattan
Central Park
Rikers Island
La Guardia Airport
Jackson Heights
Queens
Newark
Hoboken
Met. Museum of Art
Empire State Building
Maspeth
Jamaica
Jersey City
New York
Woodhaven
Newark Intl. Airport
Statue of Liberty
Upper New York Bay
Prospect Heights
Brooklyn
Elizabeth
Bayonne
Sunset Park
Flatlands
John F. Kennedy Intl. Airport
Staten Island
Dongan Hills
Fort Hamilton
Jamaica Bay
Oakwood
Lower New York Bay
Coney Island
Brighton Beach
Rockaway Beach
Rockaway Point
Atlantic Ocean
Hudson River
East River
Long Island Sound

0 5 10 Km
0 5 10 Miles

19,686ft
13,124ft
9843ft
6562ft
3281ft
1640ft
820ft
328ft
Sea Level
-820ft
-6562ft
-13,124ft

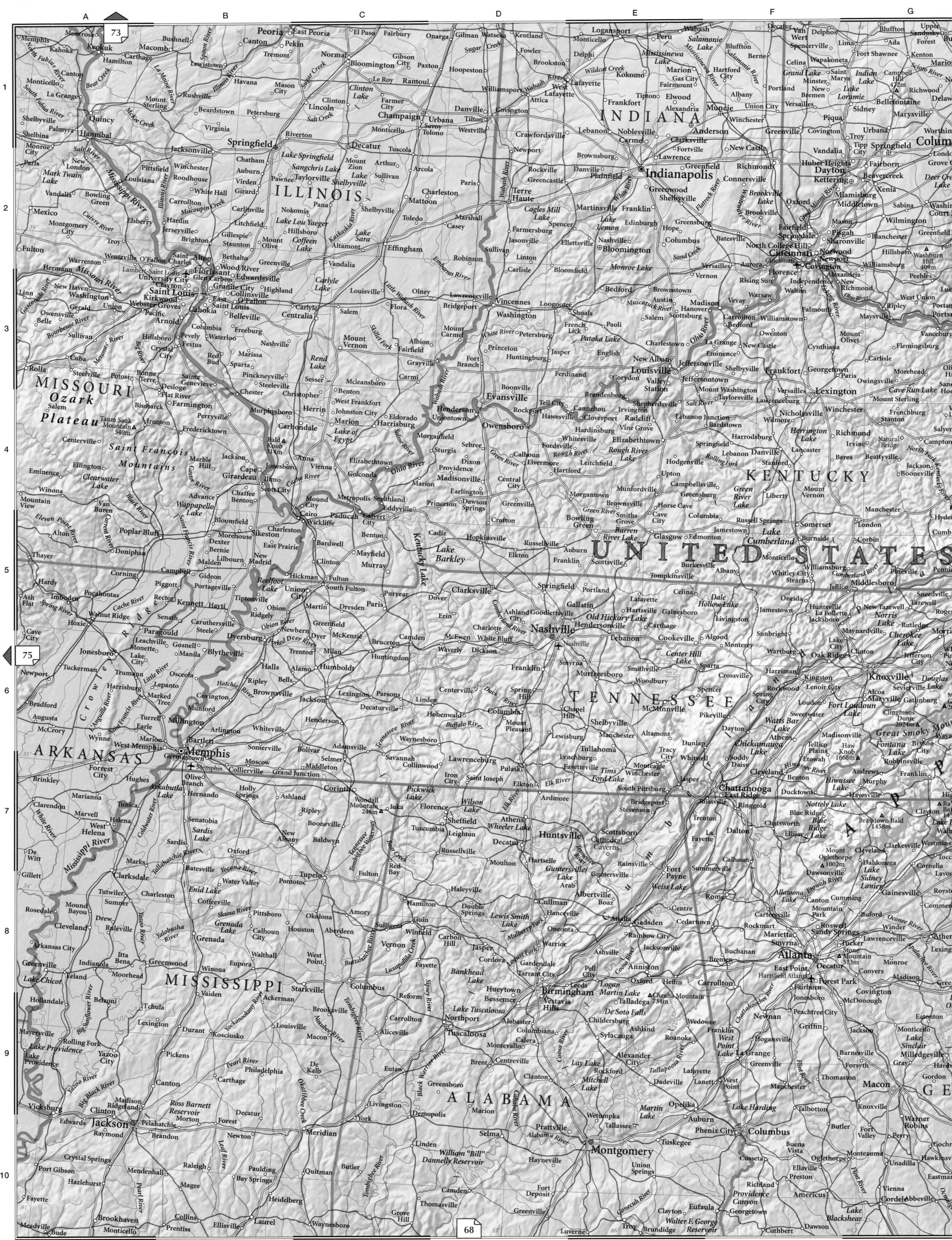

Scale 1:3,000,000
(projection: Lambert Conformal Conic)

Population
■ above 5 million
◉ 1 million to 5 million
◉ 500,000 to 1 million
◉ 100,000 to 500,000
⊕ 50,000 to 100,000
○ 10,000 to 50,000
○ below 10,000

0 20 40 60 80 100 Km
0 20 40 60 80 100 Miles

62

290

PENNSYLVANIA

WEST VIRGINIA

VIRGINIA

NORTH CAROLINA

SOUTH CAROLINA

MARYLAND

DELAWARE

NEW JERSEY

WASHINGTON DC
DISTRICT OF COLUMBIA

ATLANTIC OCEAN

Pittsburgh
Philadelphia
Baltimore
Richmond
Charlotte
Columbia
Savannah
Charleston
Raleigh
Durham
Greensboro
Winston Salem
Roanoke
Lynchburg
Norfolk
Virginia Beach
Newport News
Hampton
Portsmouth
Chesapeake Bay
Albemarle Sound
Pamlico Sound
Cape Hatteras
Hatteras Island
Ocracoke Island
Roanoke Island
Cape Lookout
Cape Fear
Onslow Bay
Raleigh Bay
Long Bay
Myrtle Beach
Wilmington
Fayetteville
Pinehurst
Asheville
Mount Mitchell 2037m
Mount Rogers 1746m

UNITED STATES OF AMERICA

Chesapeake Bay
Delaware Bay
Cape May
Cape Henlopen
Assateague Island
Chincoteague
Tangier Island
Cape Charles
Cape Henry

Lake Gaston
John H. Kerr Reservoir
B. Everett Jordan Reservoir
Smith Mountain Lake
Lake Norman
Lake Moultrie
Lake Marion
Lake Murray
Clark Hill Lake

BERMUDA (to UK)
Scale 1:500,000
Hamilton
St George's Island
St Catherine Point
St George
St David's Island
Ireland Island North
Ireland Island South
Somerset Island
Spanish Point
Great Sound
Little Sound
Gibbs Hill 73m
Kindley Field
Commissioner's Point
Castle Harbour
Harrington Sound
Tucker's Town
Flatts Village
Hamilton Harbour
ATLANTIC OCEAN

69

Elevation:
19,686ft
13,124ft
9843ft
6562ft
3281ft
1640ft
820ft
328ft
Sea Level
-820ft
-6562ft
-13,124ft

Scale 1:3,000,000
(projection: Lambert Conformal Conic)

0 20 40 60 80 100 Km
0 20 40 60 80 100 Miles

Population
■ above 5 million
◉ 100,000 to 500,000
■ 1 million to 5 million
⊕ 50,000 to 100,000
◉ 500,000 to 1 million
○ 10,000 to 50,000
○ below 10,000

67

88

290

ATLANTIC OCEAN

GEORGIA

ALABAMA

STATES OF AMERICA

FLORIDA

SOUTH CAROLINA

Piedmont

Miami

Birmingham
Vestavia Hills
Bessemer
Columbiana
Calera
Sylacauga
Talladega
Childersburg
Ashland
Wedowee
Cheaha Mountain
De Soto Falls
Roanoke
Newnan
Franklin
McDonough
Peachtree City
Griffin
Jackson
Monticello
Eatonton
Sparta
Warrenton
Lake Oconee
Milledgeville
Hardwick
Louisville
Wadley
Millen
Swainsboro
Denmark
Barnwell
Branchville
Saint George
Moncks Corner
Lake Moultrie
McClellanville
Cape Island
Cape Romain
Bull Island
North Charleston
Charleston
Mount Pleasant
Folly Beach

Montgomery
Tallassee
Auburn
Opelika
Phenix City
Columbus
Macon
Warner Robins
Perry
Cochran
Dublin
Soperton
Vidalia
Lyons
Claxton
Pembroke
Metter
Statesboro
Reidsville
Glennville
Hinesville
Savannah
Hilton Head Island
Hilton Head Island
Tybee Island

Tallahassee
Panama City
Apalachicola
Jacksonville
Gainesville
Ocala
Orlando
Tampa
Saint Petersburg
Lakeland
Sarasota
Fort Myers
Naples
West Palm Beach
Fort Lauderdale
Hollywood
Miami
Key West
Homestead

Lake Okeechobee
The Everglades
Big Cypress Swamp
Florida Keys
Straits of Florida
Dry Tortugas
Key Largo

Miramar
Hallandale
Carol City
Norland
North Miami Beach
Golden Glades
Westview
Miami Shores
Hialeah
Hialeah Park
Gladeview
North Bay Village
Virginia Gardens
Miami International Airport
Brownsville
West Miami
Miami Beach
Westchester
Florida Intl. University
Miami
Olympia Heights
Coral Gables
Kendall
Pinecrest
Richmond Heights
Westwood Lake
Goulds
Cutler Ridge
Redland
Naranja
South Allapattah
Leisure City
Homestead
Florida City
Virginia Key
Miami Seaquarium
Key Biscayne
Cape Florida
Biscayne Bay
Sands Key
Elliot Key
Atlantic Ocean

0   5 Km
0   5 Miles

19,686ft
13,124ft
9843ft
6562ft
3281ft
1640ft
820ft
328ft
Sea Level
-820ft
-6562ft
-13,124ft

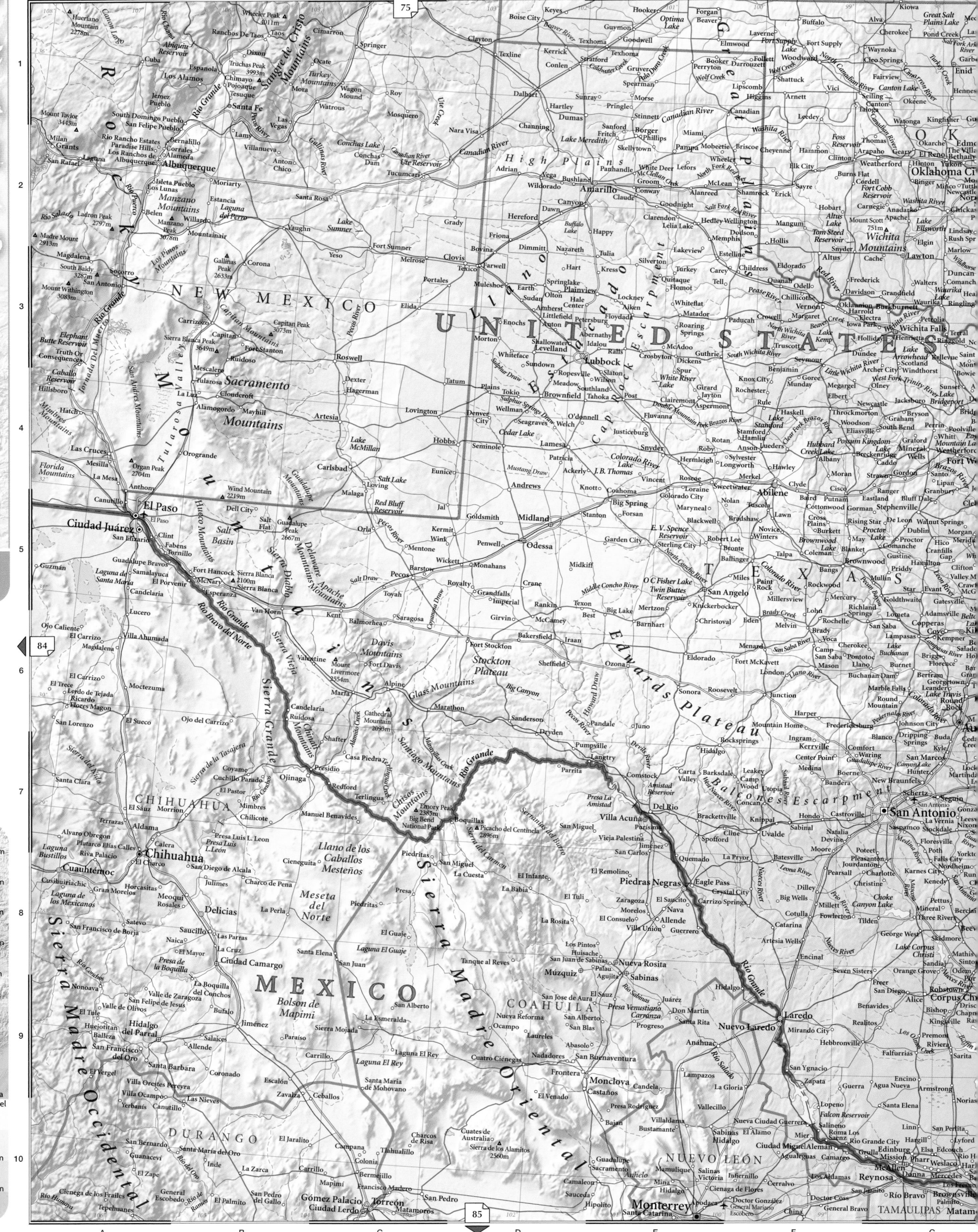

Scale 1:3,750,000
(projection: Lambert Conformal Conic)

0 20 40 60 80 100 Km
0 20 40 60 80 100 Miles

Population

■ above 5 million ■ 1 million to 5 million ◉ 500,000 to 1 million
◎ 100,000 to 500,000 ⊕ 50,000 to 100,000 ○ 10,000 to 50,000 ○ below 10,000

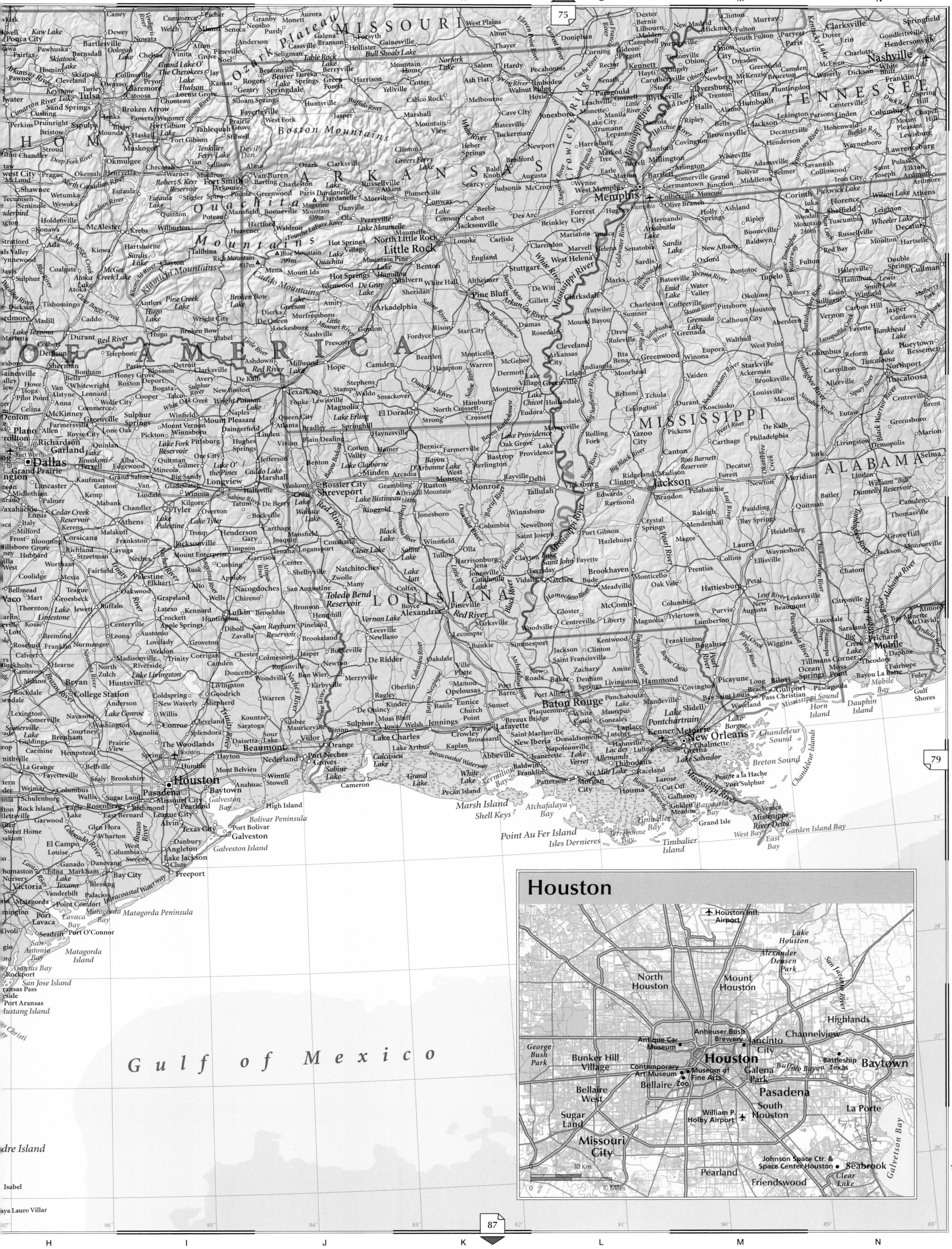

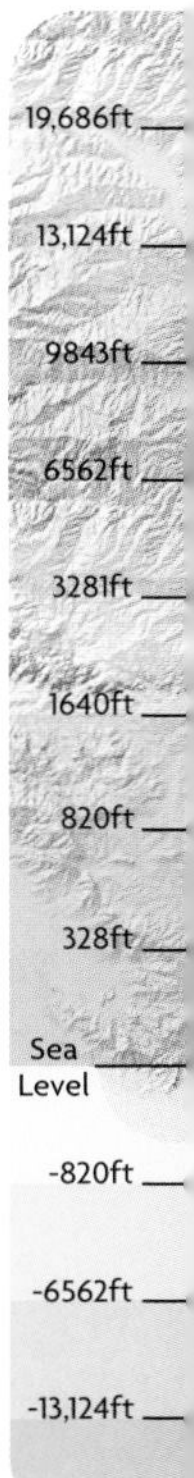

MISSOURI

TENNESSEE

ARKANSAS

Ozark Plateau

Boston Mountains

Ouachita Mountains

Kiamichi Mountains

OF AMERICA

MISSISSIPPI

ALABAMA

LOUISIANA

Little Rock
North Little Rock
Memphis
Tulsa
Dallas
Jackson
Shreveport
Bossier City
Monroe
Alexandria
Baton Rouge
Lafayette
Lake Charles
Beaumont
Houston
Pasadena
Baytown
Galveston
New Orleans
Metairie
Kenner

Gulf of Mexico

Houston

Houston Intl. Airport
North Houston
Mount Houston
Lake Houston
Alexander Deussen Park
Channelview
Highlands
Baytown
George Bush Park
Bunker Hill Village
Antique Car Museum
Anheuser Busch Brewery
Jacinto City
Galena Park
Houston
Contemporary Art Museum
Museum of Fine Arts
Houston Zoo
Pasadena
Bellaire West
Bellaire
South Houston
La Porte
Sugar Land
William P. Holby Airport
Missouri City
Pearland
Friendswood
Johnson Space Ctr. & Space Center Houston
Seabrook
Clear Lake
Galveston Bay

19,686ft
13,124ft
9843ft
6562ft
3281ft
1640ft
820ft
328ft
Sea Level
-820ft
-6562ft
-13,124ft

75

79

87

USA – Great Lake States

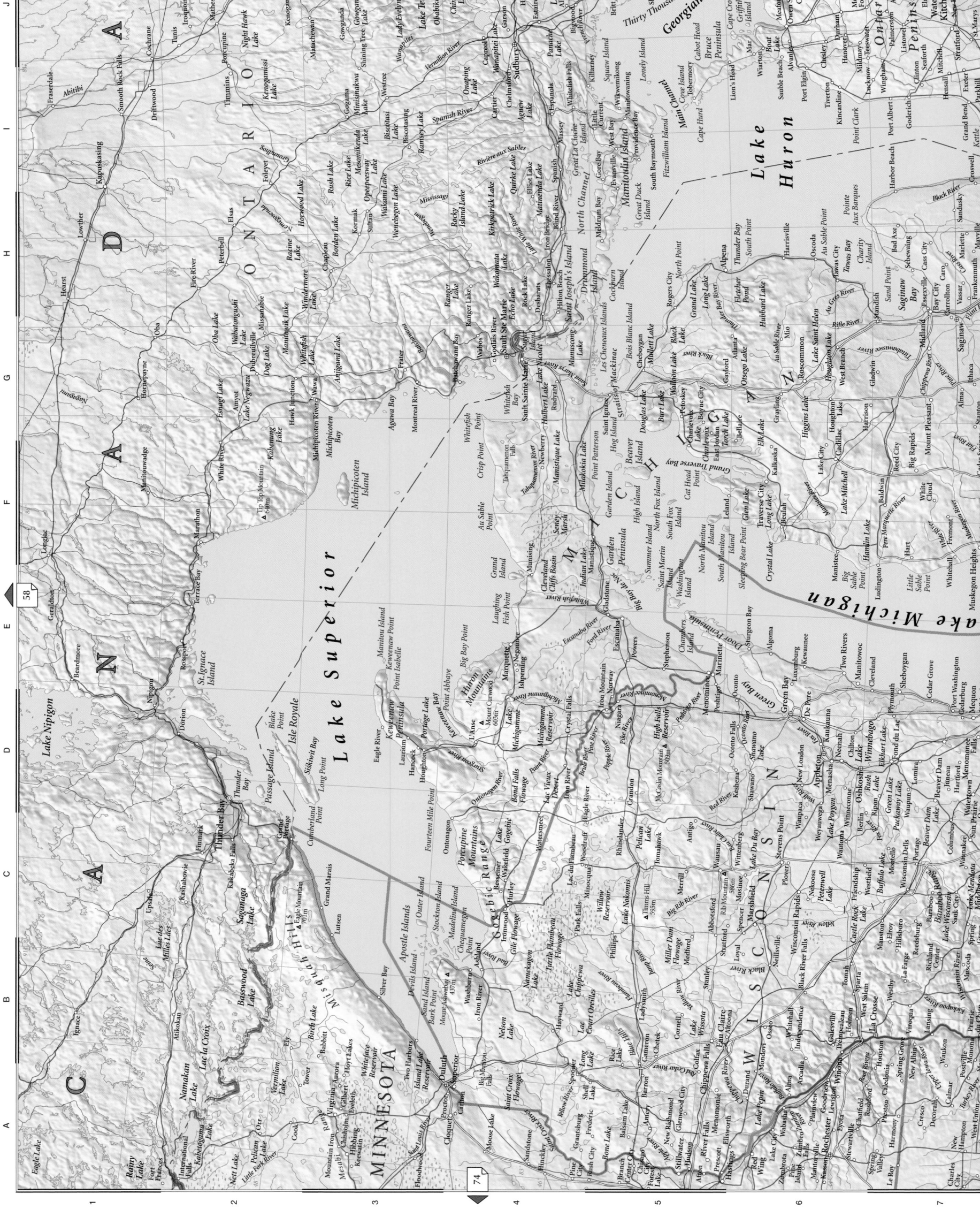

Scale 1:3,000,000
(projection: Lambert Conformal Conic)

0   20   40   60   80   100 Km
0   20   40   60   80   100 Miles

Population

■ above 5 million
◉ 1 million to 5 million
◉ 500,000 to 1 million
◎ 100,000 to 500,000
⊕ 50,000 to 100,000
○ 10,000 to 50,000
∘ below 10,000

Lake Erie

OHIO

INDIANA

ILLINOIS

MISSOURI

KENTUCKY

TENNESSEE

WEST VIRGINIA

VIRGINIA

NORTH CAROLINA

ARKANSAS

IOWA

UNITED STATES OF AMERICA

Detroit

Cleveland

Columbus

Cincinnati

Indianapolis

Chicago

Springfield

Nashville

Louisville

Lexington

19,686ft
13,124ft
9843ft
6562ft
3281ft
1640ft
820ft
328ft
Sea Level
-820ft
-6562ft
-13,124ft

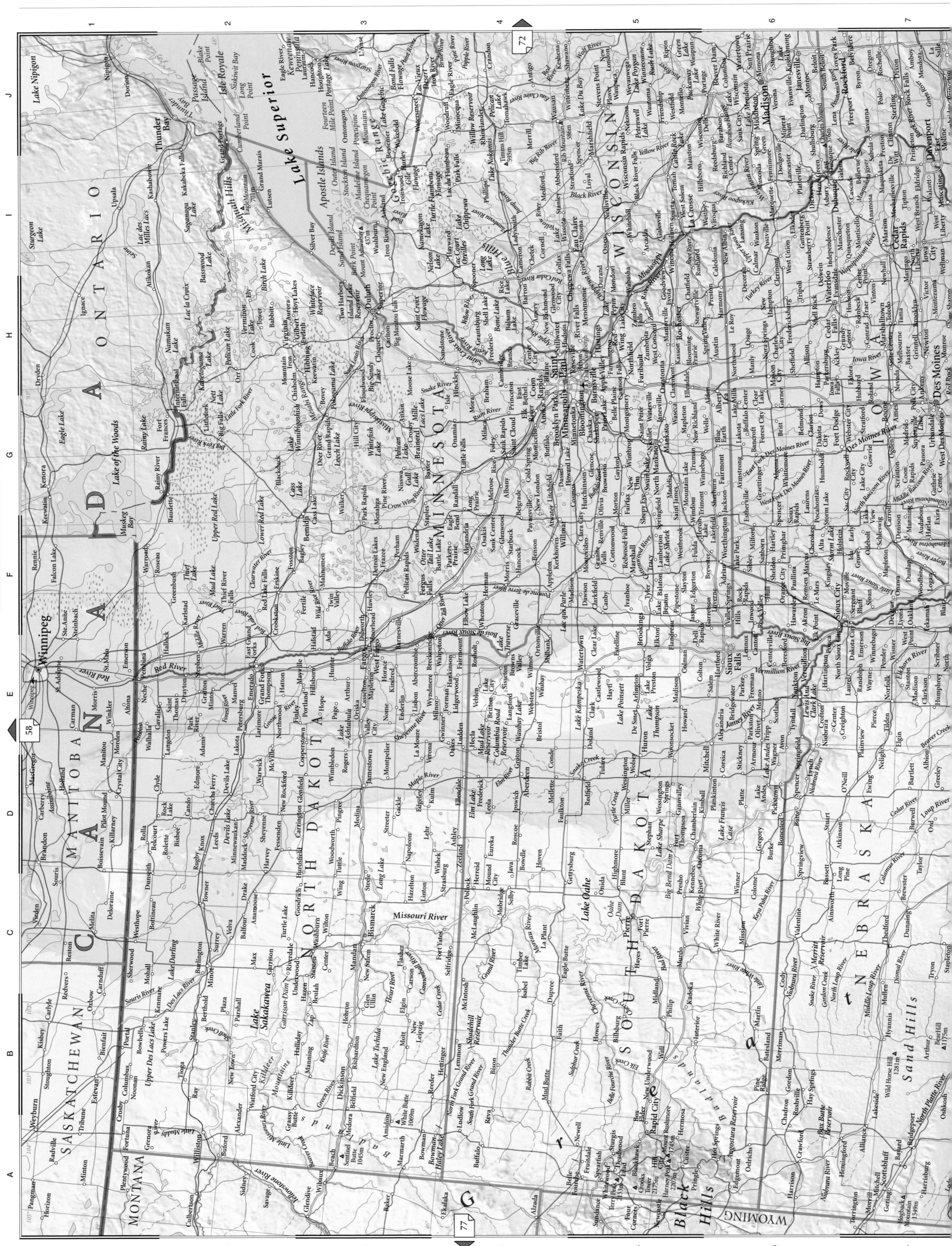

Scale 1:3,750,000
(projection: Lambert Conformal Conic)

```
0   20   40   60   80   100 Km
0   20    40    60    80   100 Miles
```

Population
- ■ above 5 million
- ◉ 500,000 to 1 million
- ⊕ 50,000 to 100,000
- ○ 10,000 to 50,000
- □ 1 million to 5 million
- ◎ 100,000 to 500,000
- ○ below 10,000

Elevation scale:
- 19,686ft
- 13,124ft
- 9843ft
- 6562ft
- 3281ft
- 1640ft
- 820ft
- 328ft
- Sea Level
- -820ft
- -6562ft
- -13,124ft

ILLINOIS
MISSOURI
KANSAS
ARKANSAS
OKLAHOMA
TEXAS
COLORADO
NEW MEXICO
MISSISSIPPI
LOUISIANA
UNITED STATES OF AMERICA

Kansas City
Topeka
Wichita
Lincoln
Springfield
Little Rock
Oklahoma City
Dallas
Fort Worth
Arlington
Memphis
Shreveport
Amarillo
Lubbock

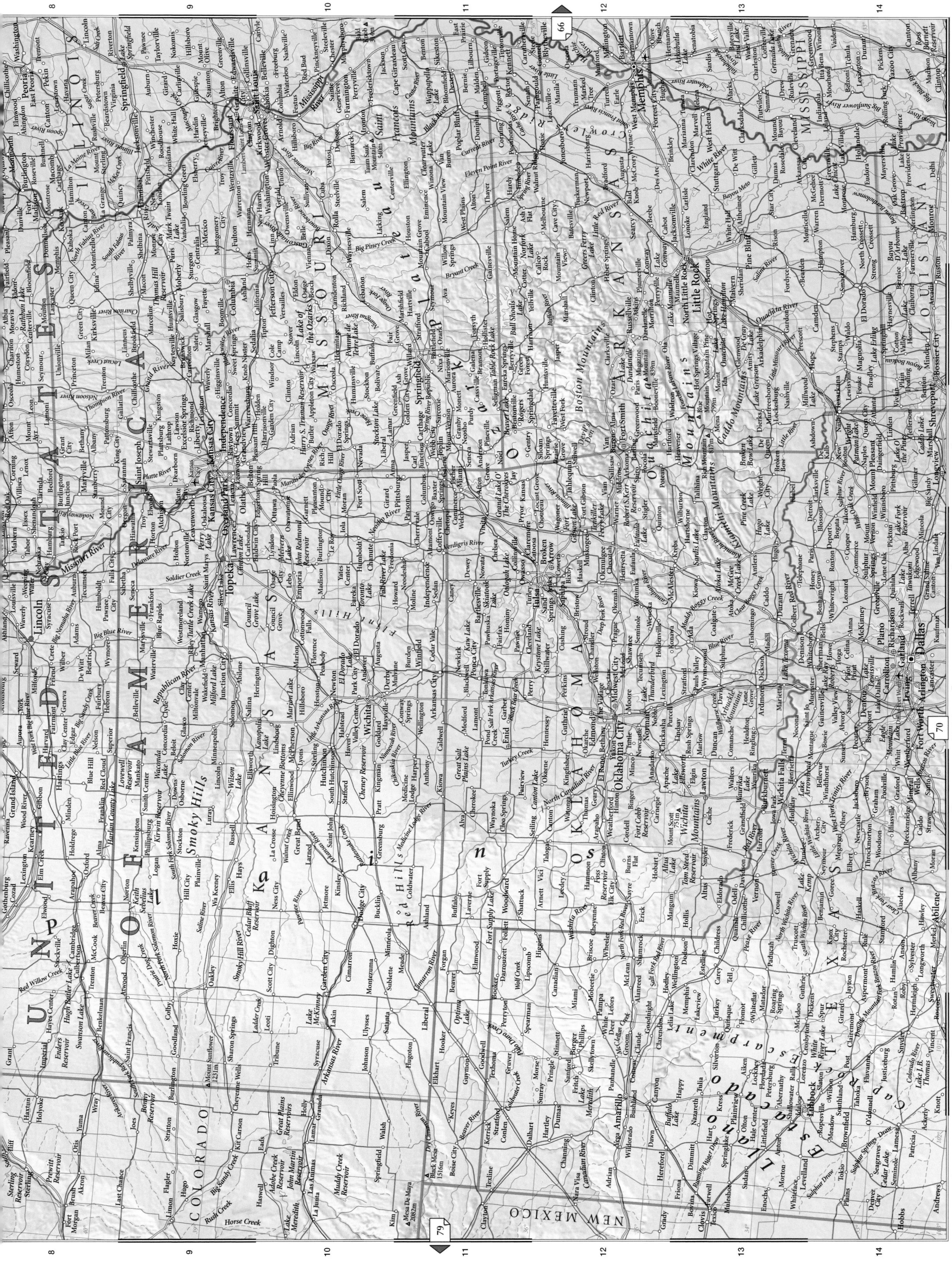

Scale 1:3,750,000
(projection: Lambert Conformal Conic)

0 20 40 60 80 100 Km
0 20 40 60 80 100 Miles

Population
■ above 5 million
◉ 1 million to 5 million
◉ 500,000 to 1 million
◎ 100,000 to 500,000
⊕ 50,000 to 100,000
○ 10,000 to 50,000
○ below 10,000

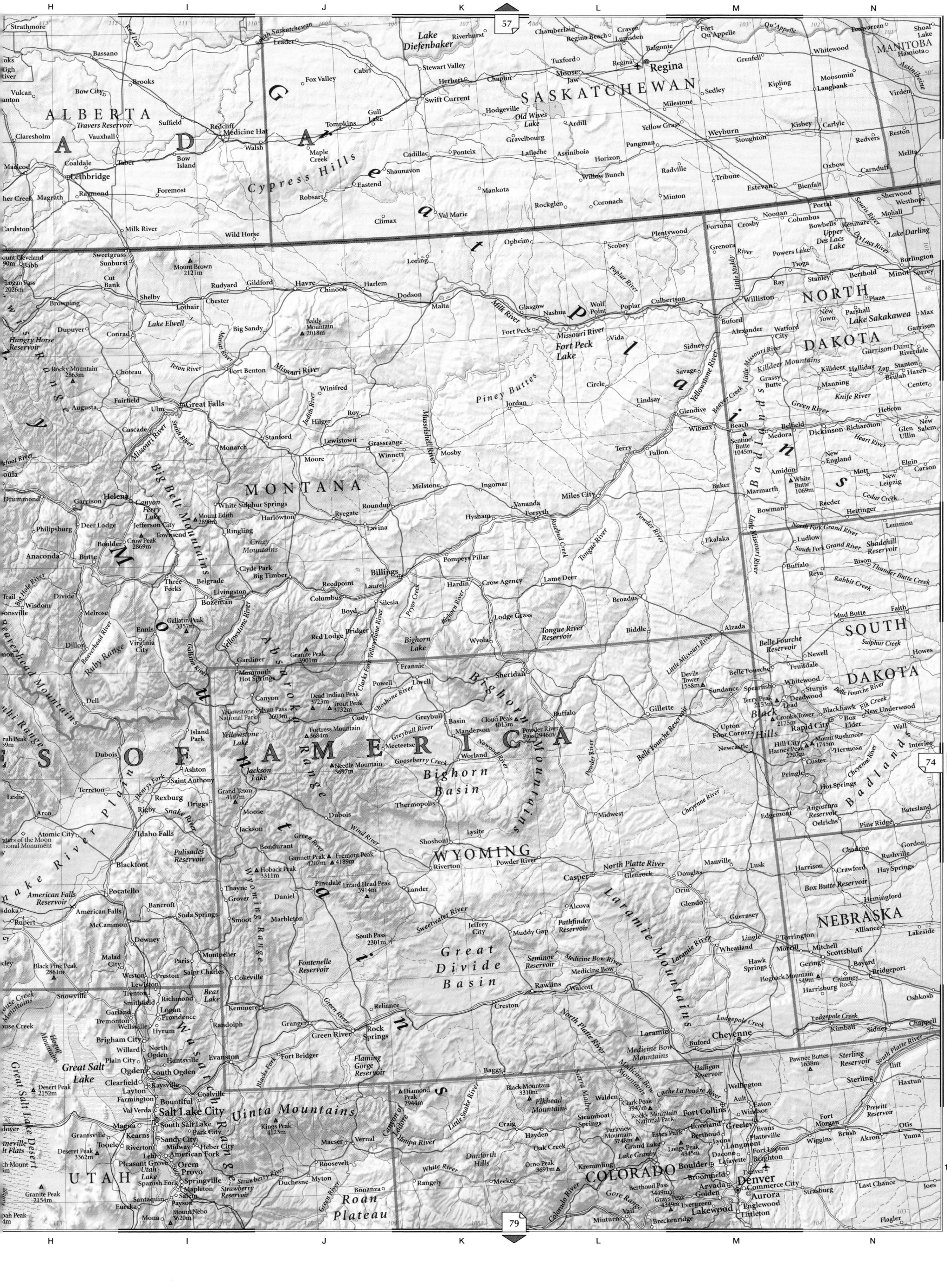

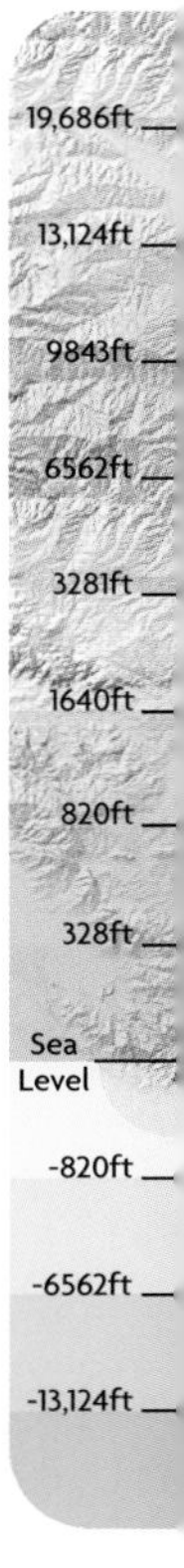

H I J K L M N

57

79

ALBERTA

SASKATCHEWAN

MANITOBA

NORTH DAKOTA

SOUTH DAKOTA

MONTANA

WYOMING

NEBRASKA

UTAH

COLORADO

C A N A D A

U N I T E D   S T A T E S   O F   A M E R I C A

Great Plains

Rocky Mountains

Badlands

Great Salt Lake

Great Divide Basin

Bighorn Basin

Yellowstone National Park

Missouri River
Yellowstone River
Milk River
North Platte River
Green River
Snake River
Bear Lake
Fort Peck Lake
Lake Sakakawea

Regina
Medicine Hat
Lethbridge
Swift Current
Great Falls
Helena
Butte
Bozeman
Billings
Miles City
Williston
Bismarck area — Dickinson
Rapid City
Denver
Aurora
Lakewood
Boulder
Fort Collins
Greeley
Cheyenne
Laramie
Casper
Salt Lake City
Ogden
Provo
Idaho Falls
Pocatello

Gannett Peak 4207m
Grand Teton 4197m
Harney Peak 2207m
Kings Peak 4123m
Gannett Peak
Cloud Peak 4013m
Granite Peak 3901m

19,686ft
13,124ft
9843ft
6562ft
3281ft
1640ft
820ft
328ft
Sea Level
-820ft
-6562ft
-13,124ft

74

79

Scale 1:3,750,000
(projection: Lambert Conformal Conic)

0 20 40 60 80 100 Km
0 20 40 60 80 100 Miles

Population
■ above 5 million   ◉ 1 million to 5 million   ◉ 500,000 to 1 million
◎ 100,000 to 500,000   ⊕ 50,000 to 100,000   ○ 10,000 to 50,000   ∘ below 10,000

NORTH AMERICA

79

75
70

19,686ft
13,124ft
9843ft
6562ft
3281ft
1640ft
820ft
328ft
Sea Level
-820ft
-6562ft
-13,124ft

NEVADA

UNITED STATES OF AMERICA

Sierra Nevada

Central Valley

San Joaquin Valley

Sacramento Valley

Coast Range

Diablo Range

Santa Lucia Range

Toiyabe Range

Shoshone Mountains

Stillwater Range

Monitor Range

Grant Range

Monte Cristo Range

White Mountains

Gabbs Valley Range

Excelsior Mountains

Wassuk Range

San Francisco

Sacramento

Oakland

San Jose

Fresno

Reno

Carson City

Lake Tahoe

Mono Lake

Walker Lake

Pyramid Lake

Humboldt Lake

Carson Sink

Monterey Bay

Point Reyes

Sea Level

6000m
4000m
3000m
2000m
1000m
500m
250m
100m
-250m
-2000m
-4000m

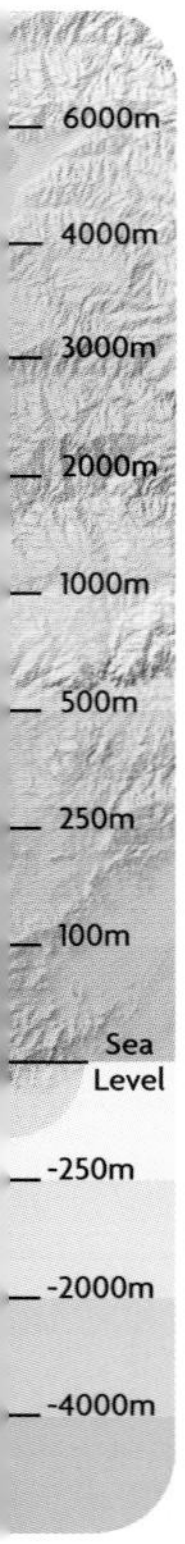

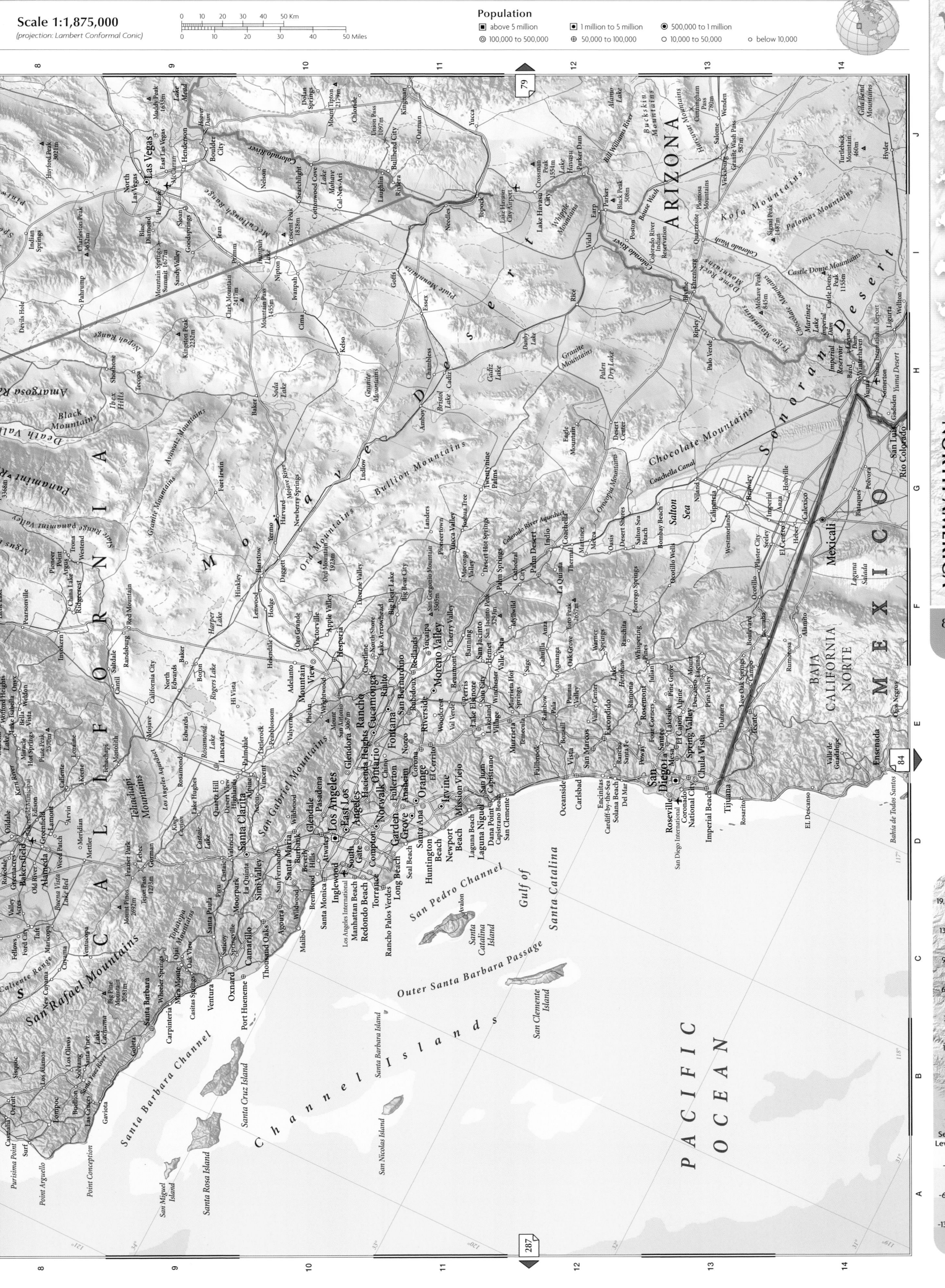

Scale 1:1,875,000
(projection: Lambert Conformal Conic)

Population
■ above 5 million   ■ 1 million to 5 million   ◉ 500,000 to 1 million
◎ 100,000 to 500,000   ⊕ 50,000 to 100,000   ○ 10,000 to 50,000   ○ below 10,000

19,686ft
13,124ft
9843ft
6562ft
3281ft
1640ft
820ft
328ft
Sea
Level
-820ft
-6562ft
-13,124ft

A R C T I C

Chukchi

Sea

Icy Cape

Wain

294

Point Lay

Cape Lisburne

Point Hope
Point Hope

Wevok

Misheguk Mountain
1350m

Nuvik

Noatak

Tutuilakat Mou
1364m

Kivalina

Kotzebue Sound

Kotzebue

Noorvik

Sel

Enurmino

Kougarok Mountain
875m

Seward Peninsula

Koyuk

Cape
Espenberg

Goodhope
Bay

Deering
Candle

Buck

Kiwalik

Arctic Circle

Shishmaref

Poluostrov
Chukotskiy

Uelen

Little Diomede
Island

Brooks
Mountain
863m

Brevig
Mission

Wales

Teller

Council

Lavrentiya

Cape Prince
of Wales

Port Clarence

Kauitrin River

Solomon
Golovin

Elim

Koyuk

Providentya

Port Clarence

Teller

Nome
Cape
Nome

Solomon

Norton

Mys Chukotskiy

Cape Rodney

Cape
Darby

Shaktoolik

Anadyrskiy Zaliv

Northwest Cape

Gambell

Savoonga

Unalakleet

Southwest Cape

Saint Lawrence
Island

Northeast Cape

Camp Kulowiye

Stuart
Island

Norton Sound

Stebbins

Saint
Michael

Southeast Cape

Pastol
Bay

Hamilton

Kotlik

Grayl
Sha

CHUKOTSKIY
AVTONOMNYY OKRUG

Velikaya

Emmonak
Alakanuk

Yukon
River

Anv
Parad
551m

Kamenskoye

Koryakskoye

Nagor'ye

Sheldons Point

Mountain Village

Saint Marys

Russian
Mission

K o r y a k s k o y e   N a g o r' y e

Meynypil'gyno

Scammon Bay

Pitkas Point

Pilot Station

KORYAKSKIY
AVTONOMNYY OKRUG

Gora Ledyanaya
2562m

Hooper Bay

Chevak

Marshall

Lower Kalskag

Khatyrka

Newtok

Aropuk Lake

Kasigluk
Akiachak,
Akiak

Bethel
Napakiak
Napaskiak

Kweth
Kwethi

Ol'yutorskiy
Zaliv

Mys Navarin

Hazen Bay

Tanunak

Toksook Bay
Nightmute

Tuntutuliak

Eek River

Korf

Ulichiki

Pakhachi

Mekoryuk

Nunivak
Island

Dall
Lake
Chefornak

Eek

Beringovskiy

RUSSIAN
FEDERATION

Roberts Mountain
510m

Kipnuk

Kwigillingok

Mys Ol'yutorskiy

Hall Island

Glory of Russia Cape

Cape Mendenhall

Quinhagak

Saint Matthew Island

Upright Cape

Kuskokwim Bay

Pinnacle Island

Cape Mohican

Goodnews

Platinum

Twi
Togiak

Limit of winter pack ice

Cape Newenham

Hagemeiste
Island

B e r i n g

S e a

Bristol Ba

Saint Paul Island

Saint Paul

Pribilof Islands

Port Moller

Saint George Island

Saint George

193

Cape Wrangell

Near Islands

Amak Island

Alask

Attu

Ivan

Attu Island

Agattu Strait

Shemya Island

Cold Bay

Korovin Island
Sand Point

Agattu Island

Krugloi Point

Buldir
Island

King Cove

Belkofski

Squaw
Harbor

Cape Sabak

Unimak Island

Shishaldin Volcano

False Pass

Deer
Island

Nagai
Island

Shum

Pogromni Volcano
2002m

Paulof Harbor

Sanak Islands

Kiska Island

Segula
Island

Little
Sitkin
Island

Semisopochnoi
Island

Akun Island

Tigalda Island

Vega Point

Anvil Peak
1221m

Makushin Volcano
2306m

Akutan Island

Avatanak Island

Rat Island

A l e u t i a n   I s l a n d s

Akutan

Krenitzin Islands

R a t   I s l a n d s

Amchitka
Island

Cape Kanaga

Tanaga
Volcano
1806m

Kanaga
Volcano

Great Sitkin
Volcano
1307m

Atka
Island

Unalaska Island

Dutch
Harbor

Delarof
Islands

Gareloi
Island

Seguam
Island

Fox Islands

Amchitka Pass

Tanaga Island

Kagalaska

Adak Island

Atka

Fenimore Pass

Seguam Pass

Yunaska Island

Umnak Island

Nikolski

Sasmik Island

Amlia Island

Islands of Four Mountains

Amukta Pass

Amukta
Island

Herbert
Island

A n d r e a n o f   I s l a n d s

UNITED STATES OF AMERICA
HAWAII

Kaua'i

Hanalei Kilauea

Lehua Island

Kahala Point

Kaumakali Channel

Pu'uwai

'Ekeha

Kapa'a

Ni'ihau

Lihu'e
Koloa

Kawaihoa
Point

'Ele'ele
Makahü'ena
Point

O'ahu

Kahuku Point

Kahuku

Kaua'i Channel

Ka'ena Point

Hau'ula

Wahiawa

Wai'anae

Pearl City

Makakilo City

Honolulu

Diamond
Head

'Ewa Beach

Pearl Harbor

Kaunakakai

Kalohi Channel

Kailua
Pa'ia

Moloka'i

Kalaupapa

Päilolo Channel

Näkälele Point

Lahaina

Maui

Läna'i City

Kihei

Haleakalä

Läna'i

Kaho'olawe

'Alaläkeiki Channel

Hāna

Pu'u 'Ula'ula (Red Hill)
3055m

'Alenuihähä Channel

'Upolu Point

Hāwī

Honoka'a

Laupähoehoe

PACIFIC

Waimea

Wailea

OCEAN

Keāhole Point

Hawai'i

Mauna Kea
4205m

Päpa'ikou
Hilo

Kailua-Kona

Kalaoa

Ke'a'au

Kahalu'u

Kealakekua

Mountain View

Captain Cook

Mauna Loa
4169m

Kilauea Pähoa

Caldera

Cape Kumukahi

Pähala

'Āpua Point

Kaunā Point

Na'älehu

Ka Lae
(South Point)

Scale 1:5,000,000

0 20 40 60 80 100 120 Km

0 20 40 60 80 100 120 Miles

H a w a i' i a n   I s l a n d s

286

Northern Mexico

0 20 40 60 80 100 Km
0   20   40   60   80 100 Miles

Population

- ■ above 5 million
- ▣ 1 million to 5 million
- ◉ 500,000 to 1 million
- ◎ 100,000 to 500,000
- ⊕ 50,000 to 100,000
- ○ 10,000 to 50,000
- ∘ below 10,000

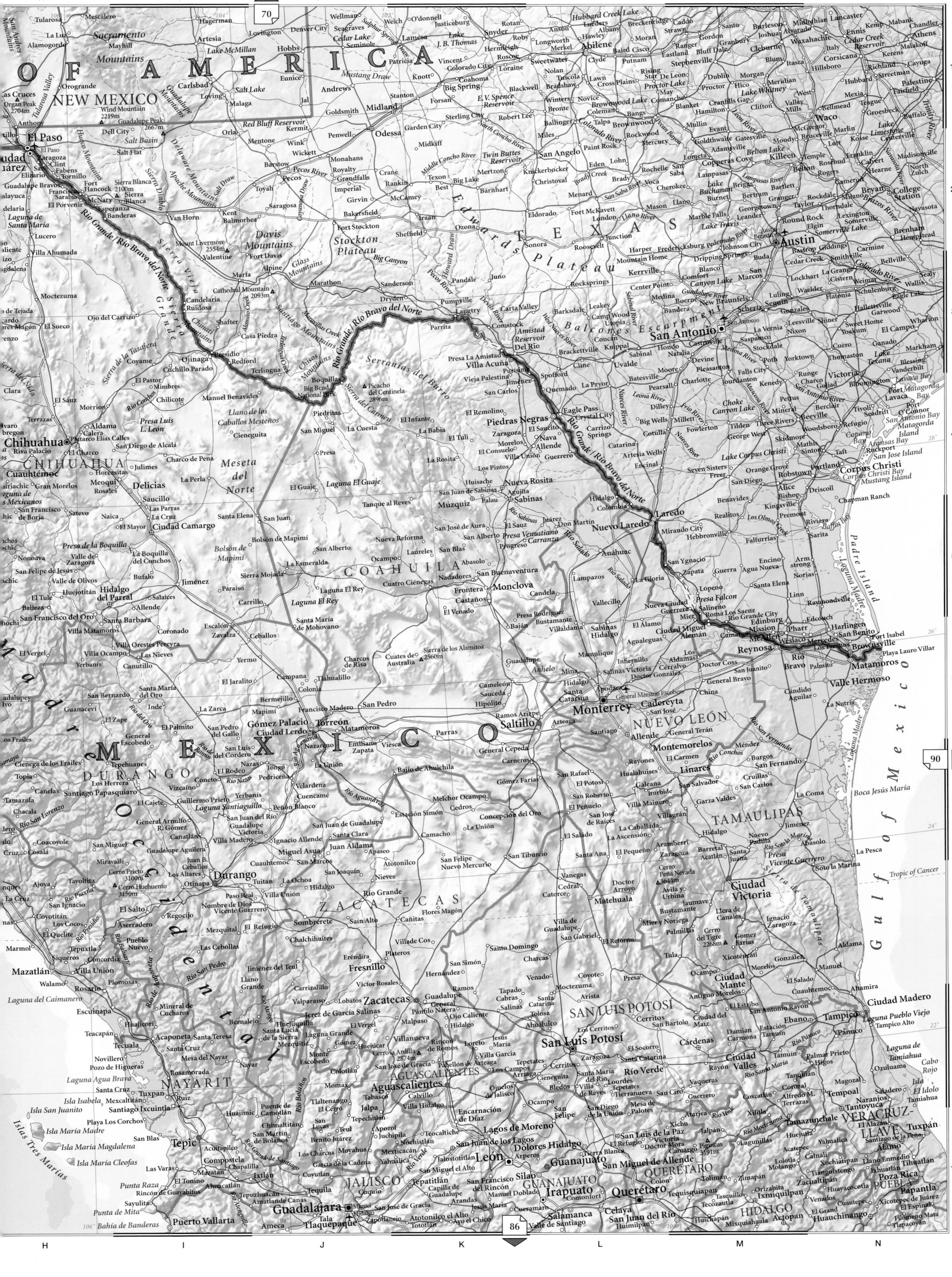

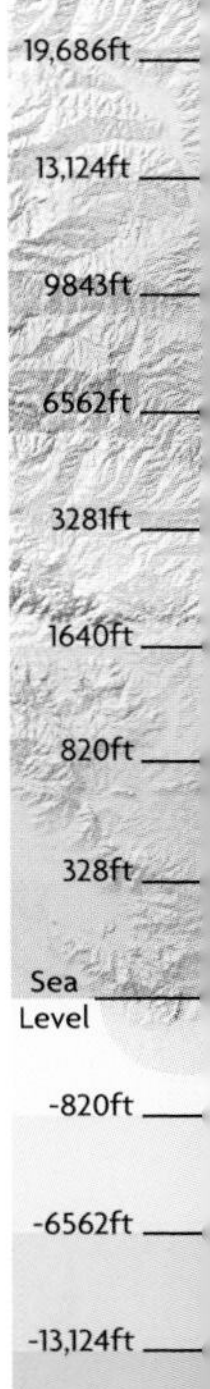

70

90

86

19,686ft
13,124ft
9843ft
6562ft
3281ft
1640ft
820ft
328ft
Sea
Level
-820ft
-6562ft
-13,124ft

OF AMERICA

NEW MEXICO

TEXAS

Edwards Plateau

Stockton Plateau

Balcones Escarpment

San Antonio

Austin

Waco

Chihuahua

CHIHUAHUA

COAHUILA

Meseta del Norte

Bolsón de Mapimí

NUEVO LEÓN

Monterrey

Saltillo

Nuevo Laredo

Laredo

Corpus Christi

Matamoros

Reynosa

TAMAULIPAS

Gulf of Mexico

Padre Island

Laguna Madre

MEXICO

DURANGO

Durango

Sierra Madre Occidental

ZACATECAS

Zacatecas

Fresnillo

Torreón

Gómez Palacio

SAN LUIS POTOSÍ

San Luis Potosí

Ciudad Victoria

Tampico

Ciudad Madero

NAYARIT

Tepic

Mazatlán

AGUASCALIENTES

Aguascalientes

JALISCO

Guadalajara

GUANAJUATO

Irapuato

Guanajuato

León

QUERÉTARO

Querétaro

HIDALGO

VERACRUZ-LLAVE

Tropic of Cancer

El Paso

Ciudad Juárez

Río Grande / Río Bravo del Norte

Presa Falcón

PACIFIC

OCEAN

COAHUILA

DURANGO

SINALOA

NUEVO LEON

TAMAULIPAS

ZACATECAS

SAN LUIS POTOSÍ

NAYARIT

AGUASCALIENTES

VERACRUZ-LLAVE

JALISCO

GUANAJUATO

QUERÉTARO

HIDALGO

MICHOACÁN

MEXICO

TLAXCALA

PUEBLA

COLIMA

MORELOS

GUERRERO

Monterrey

Saltillo

Durango

Mazatlán

Ciudad Victoria

Ciudad Madero

Tampico

Tepic

Zacatecas

San Luis Potosí

Aguascalientes

Ciudad Valles

Guadalajara

León

Guanajuato

Querétaro

Morelia

Pachuca

MÉXICO (MEXICO CITY)

Toluca

Puebla

Cuernavaca

Colima

Manzanillo

Lázaro Cárdenas

Chilpancingo

Acapulco

Puerto Escondido

Brownsville

Matamoros

Scale 1:4,250,000
(projection: Lambert Conformal Conic)

0 20 40 60 80 100 Km
0 20 40 60 80 100 Miles

Population
- ◼ above 5 million
- ◼ 1 million to 5 million
- ◉ 500,000 to 1 million
- ◎ 100,000 to 500,000
- ⊞ 50,000 to 100,000
- ◌ 10,000 to 50,000
- ○ below 10,000

Mexico City

Atizapán de Zaragoza
Tlalnepantla
Naucalpan de Juárez
Azcapotzalco
Gustavo A. Madero
Basílica de Guadalupe
Parque San Juan de Aragón
Zoo
Catedral
Lago Nabor Carrillo
Benito Juárez Airport
Mexico City (México)
Castillo de Chapultepec
Nezahualcóyotl
Bosque de Chapultepec
Miguel Hidalgo
Moderna
Iztacalco
La Perla
Benito Juárez
Iztapalapa
Coyoacán
Parque Nacional Cerro de Estrella
Álvaro Obregón
Pirámide de Cuicuilco
Estadio Azteca
Tlahuac
Reserva Ecológica del Pedregal
Tlalpan
Parque de Xochimilco
Xochimilco

0 5 Km
0 5 Miles

Gulf of Mexico

Tropic of Cancer

Cabo Catoche
Isla Contoy
Punta Yalkabul   Isla Holbox
San Felipe   Río Lagartos   Laguna de Yalahau   Isla Mujeres
Telchac Puerto   El Cuyo   Chiquilá
Chicxulub   Dzilam de Bravo   Panabá   Cancún
Progreso   Dzidzantún   Tizimín   Yucatán   Puerto Juárez   Isla Cancún
Sisal   Motul   Temax   Buctzotz   Kantunilkin   Punta Molas del Norte
Conkal   Izamal   Cenotillo   Espita   San Francisco   Puerto Morelos   Molas
Hunucmá   **Mérida**   Tunkas   Piste   Valladolid   Chemax   Cozumel
Kinchil   Xocchel   Kantunil   Chichén-Itzá   Chichimilá   Playa del Carmen
Umán   Hoctún   Sotuta   Yaxcabá   Coba   Xel-Há   Akumal
Celestún   San Rafael   Maxcanú   Soruta   Ruinas   Isla Cozumel
La Costa   Halachó   Muna   Teabo   Oxkutzcab   Ruinas de Tulum   Tulum
Becal   Ruinas Calkiní   Ticul   Maní   Tihosuco   San Ramón   Muyil
Jaina   Tenabo   Dzibalché   Xul   Tekax   Tzucacab   Santa Rosa   Filomeno Mata   Vigía Chico   Punta Allen
Hecelchakán   Bolónchén de Rejón   Xmaben   Xiatil   Bahía de la Ascensión
Hampolol   Becanchén
Campeche   Chencoyi   Hopelchén   Yucatan   Felipe Carrillo Puerto
Lerma   Seybaplaya   Pich   Dzibalchen   Iturbide   Chunhuhub   QUINTANA ROO   Bahía del Espíritu Santo
Sihochac   Chenoyh   Naranjal Poniente   Polyuc   Punta Herrero
Champotón   Arellano   Valle Hermoso   Pocboc   Chacchoben

Bahía de Campeche

Ulumal   Pustunich   Peninsula   Majahual
Sabancuy   Chekubul   Reforma   Laguna Bacalar   Bancó Chinchorro
Isla del Carmen   Puerto Real   Francisco Escárcega   Xpujil   Bacalar   Chetumal   Banco Chinchorro   Cayo Centro
Álvaro Obregón   Isla Aguada   Tres Garantías   Kohunlich   Chetumal   Bay   Cayo Lobos
Frontera   Zacatal   Laguna de Términos   Mamantel   El Corozal   Pucte   Corozal   Rocky Point   Santa Cecilia
Paraíso   Candelaria   Nuevo Coahuila   Tomás Garrido   Hondo   Corozal   Boca Bacalar Chico
Comalcalco   TABASCO   Palizada   Nuevo Coahuila   Orange Walk   Ambergris Cay
Cunduacán   San Joaquín   Carmelita   Indian Church   New River   Belize   San Pedro
Cárdenas   Reforma   Villahermosa   Morelos   Hill Bank   Turneffe Islands
Las Choapas   Huimanguillo   Jalapa   Chablé   Balancán   Orange Walk   Belize City
Filisola   Macuspana   Emiliano Zapata   Benito Juárez   Lighthouse Reef
Medias Aguas   Teapa   Salto de Agua   Estapilla   BELIZE   BELMOPAN
Pichucalco   Ixtapangajoya   Tila   Palenque   Tenosique   Río San Pedro   San Ignacio   Middlesex   Glovers Reef
Volcán El Chichonal 1064m   Tapijulapa   Carmelita   Tikal   Benque Viejo   Dangriga
Ostuacán   Cuauhtémoc   Simojovel   Yaxchilán   PETÉN   del Carmen   CAYO   Maya Mountains
Raudales   Tapilula   Copainalá   Lago Petén Itzá   Ciudad Melchor de Mencos   Victoria Peak 1120m   Richardson Peak   CREEK
Tecpatán   Pantelhó   Ocosingo   Flores   Maya Mountains 1120m   Monkey River Town
OAXACA   Cerro Zempoaltepec 3395m   Apic-Pac   Las Rosas   La Libertad   San Benito   Dolores   TOLEDO
Presa Netzahualcóyotl   Benito Juárez   San Cristóbal de Las Casas   Bonampak   Sayaxché   San Luis   Punta Gorda
Ocozocuautla   Altamirano   Río de la Pasión   Nueva Santa Rosa   San Antonio   Gulf of Honduras
Istmo de Tehuantepec   **Tuxtla**   Chiapa de Corzo   CHIAPAS   Venustiano Carranza   Las Margaritas   Sayaxché   Sarstoon   Bahía de Amatique   Punta Manabique   Punta Sal
Cintalapa   Suchiapa   Teopisca   Comitán   Francisco Saravia   Modesto Méndez   Santo Tomás de Castilla   Puerto Cortés
Matías Romero   San Miguel Chimalapa   Venustiano Carranza   La Trinitaria   Livingston   CORTÉS   ATLÁNTIDA
Ixtepec   Unión Hidalgo   Santo Domingo   Villa Flores   Chisec   Morales   Puerto Barrios   San Pedro Sula   El Progreso
Juchitán   Zanatepec   Villa Corzo   La Concordia   Barillas   ALTA VERAPAZ   El Estor   Lago de   Izabal   Macuelizo   Villanueva
Tehuantepec   Laguna Superior   Arriaga   Presa de la Angostura   San Mateo   IZABAL   Izabal   YORO
Salina Cruz   Laguna Inferior   Rincón Juárez   Ixtatán   Cobán   Carchá   Los Amates   SANTA BÁRBARA   Yojoa   Victoria
Santiago Astata   Mar Muerto   Paredón   Tonalá   Cerro Tres Picos 2952m   HUEHUETENANGO   Jacaltenango   QUICHÉ   Cahabón   San Cristóbal Verapaz   COPÁN   Represa El Cajón
Puerto Arista   La Angostura   Huehuetenango   Rabinal   Salamá   Copán   Santa Rosa de Copán   COMAYAGUA
Golfo de Tehuantepec   El Manguito   Cerro Tres Cruces 2952m   Ciudad Cuauhtémoc   Comalapa   Río Cuilco   BAJA VERAPAZ   Río Cahabón   Zacapa   Camotán   Santa Rita   HONDURAS   Comayagua
Tres Picos   Motozintla de Mendoza   Colomba   Río Chixoy   EL PROGRESO   Sierra de las Minas   Chiquimula   Santa Rosa   Sensuntepeque   La Paz
Mapastepec   Amatenango   Volcán Tajumulco 4093m   San Pedro Carchá   Guastatoya   QUICHÉ   Río Motagua   Chiquimula   SANTA BÁRBARA   Villa de
Escuintla   Tuxtla   SAN MARCOS   Huehuetenango   CHIMALTENANGO   GUATEMALA   Jalapa   CHIQUIMULA   INTIBUCA
Tapachula   Quezaltenango   Sololá   Chichicastenango   **CIUDAD DE GUATEMALA**   JALAPA   Esquipulas   LEMPIRA   Marcala
Puerto Madero   Tilapa   Lago de Atitlán   Santiago Atitlán   Ciudad Vieja   Cuilapa   SANTA ROSA   JUTIAPA   EL SALVADOR
Ciudad Hidalgo   Retalhuleu   Escuintla   GUATEMALA CITY   Santa Ana   Nueva Ocotepeque   La Paz
Champerico   Pueblo Nuevo Tiquisate   Chiquimulilla   Escuintla   Chalatenango

GUATEMALA: ADMINISTRATIVE REGIONS:
1. RETALHULEU
2. QUEZALTENANGO
3. TOTONICAPÁN
4. SOLOLÁ
5. SUCHITEPÉQUEZ
6. ESCUINTLA
7. CHIMALTENANGO
8. SACATEPÉQUEZ
9. GUATEMALA
10. BAJA VERAPAZ
11. EL PROGRESO
12. ZACAPA
13. CHIQUIMULA
14. JALAPA
15. SANTA ROSA
16. JUTIAPA

Veracruz   Boca del Río   Punta de Antón Lizardo   Antón Lizardo
Punta Villa Rica   Punta Zempoala   La Antigua   Jamapa
Alvarado   Laguna Alvarado   Punta Roca Partida
San Andrés Tuxtla   Cosamaloapan   Coatzacoalcos   Minatitlán
Acayucan   Sayula de Alemán   Azueta   Villa

Elevation scale:
19,686ft
13,124ft
9843ft
6562ft
3281ft
1640ft
820ft
328ft
Sea Level
-820ft
-6562ft
-13,124ft

71
88
90
87

Central America

GUATEMALA:
ADMINISTRATIVE REGIONS:

① RETALHULEU
② QUEZALTENANGO
③ TOTONICAPÁN
④ SOLOLÁ
⑤ SUCHITEPÉQUEZ
⑥ ESCUINTLA
⑦ CHIMALTENANGO
⑧ SACATEPÉQUEZ
⑨ GUATEMALA
⑩ BAJA VERAPAZ
⑪ EL PROGRESO
⑫ ZACAPA
⑬ CHIQUIMULA
⑭ JALAPA
⑮ SANTA ROSA
⑯ JUTIAPA

EL SALVADOR:
ADMINISTRATIVE REGIONS:

① AHUACHAPÁN
② SANTA ANA
③ SONSONATE
④ CHALATENANGO
⑤ LA LIBERTAD
⑥ SAN SALVADOR
⑦ CUSCATLÁN
⑧ CABAÑAS
⑨ LA PAZ
⑩ SAN VICENTE
⑪ USULUTÁN
⑫ SAN MIGUEL
⑬ MORAZÁN
⑭ LA UNIÓN

Guatemala City

6000m
4000m
3000m
2000m
1000m
500m
250m
100m
Sea Level
-250m
-2000m
-4000m

0 20 40 60 80 100 Km
0 20 40 60 80 100 Miles

Population
■ above 5 million   ■ 1 million to 5 million   ● 500,000 to 1 million
◎ 100,000 to 500,000   ⊕ 50,000 to 100,000   ○ 10,000 to 50,000   ○ below 10,000

JAMAICA

Montego Bay · Falmouth · Port Maria · Port Antonio · North East Point
Lucea · Sangster · The Cockpit Country · Ocho Rios · Annotto Bay · Blue · Jamaica Channel
Grange Hill · Denham · Christiana · Spanish Town · Mountain Peak 2256m
South Negril Point · Savanna-La-Mar · Mandeville · Old Harbour · Norman Manley
Crab Pond Point · Black River · Malvern · May Pen · KINGSTON
Great Pedro Bluff · Long Bay · Lionel Town · Morant Bay
Portland Point · Wreck Point · Morant Point

C a r i b b e a n

S e a

Laguna de Caratasca
Puerto Lempira
ACIAS A DIOS
Cabo de Gracias a Dios · Arrecife Edinburgh
Río Coco · Boom
Waspam · Laguna Bismuna
Arrecifes de la Media Luna
Río Wawa
REGIÓN AUTÓNOMA
Cayo Muerto
Dákura · Cayos Miskitos
Yablis · Cayos Londres
San Luis esita
ATLÁNTICO · Tuapi
Puerto Cabezas
Wounta
Bambana · Prinzapolka
NORTE · Makantaka
Isla Santa Catalina
Isla de Providencia
La Cruz de Río Grande
Cayos Guerrero
REGIÓN AUTÓNOMA · Barra de Río Grande
SAN ANDRÉS Y PROVIDENCIA
(to Colombia)
Cayos King
ATLÁNTICO · Laguna de Perlas
Cayos de Perlas
SUR · Punta de Perlas · Punta Mosquito
Isla de San Andrés
Bahía de Bluefields · Islas del Maíz
Cayos del Este Sudeste'
ondido · Kama · El Bluff · Cayos de Albuquerque
Bluefields

Monkey Point
Punta Gorda

info
nda · San Juan del Norte
Guinea · Barra del Colorado
a Gorda
lo San Juan
HEREDIA · Río Colorado

Panama City

Villa · Sara Sotillo · San · Santa
Caceres · Miguelito · Clara
Río Abajo · Miraflores
Clayton · Río Abajo · Juan Díaz
Villa de · Pueblo · Nuevo · Reparto
Fuentes · Nuevo · Paitilla · Nuevo
Parque · Curundú · Los · El Cangrejo · Panamá
Nacional · Angeles · Parque
Metropolitano · Lo cería · San · Omar Torrijos
· Francisco · Herrera
Museo Reina · Coco de · Bahía de
Torres de Araúz · Iglesia · Mar · Panamá
del Carmen · **Panama City**
Corozal · Marcos A. · **(Panamá)**
· Gelabert · Marbella
· International · Museo de las
Altos de · Airport · Ciencias Naturales · Pacific
Diablo · Calidonia · Ocean
· Anacón · Bahía de
· Palacio Presidencial · Panamá
· Balboa Santa · Iglesia de San José
· Ana

0   1 Km
0   1 Miles

COSTA RICA
Puerto Viejo
LIMÓN
Volcán Barva 2906m
Guápiles
Volcán Irazú 3339m
Heredia · Siquirres · Matina
HEREDIA · Turrialba · Limón
JOSÉ · Paraíso
SAN JOSÉ · Cartago · Punta Mona
CARTAGO
SAN JOSÉ · Cerro La Muerte 3491m
Cerro Chirripó Grande 3819m · Bribrí
Dominical · San Isidro · Guabito
ARENAS · Cerro Kamuk 3554m · Changuinola
Cortés · Almirante · Bocas del Toro
Bahía de Coronado · Buenos Aires · Península Valiente
Palmar Sur · Río Grande de Térraba
Península de Osa · San Vito · Volcán Barú 3475m · Boquete
Golfo Dulce · Volcán · BOCAS DEL TORO · Chiriquí Grande
Puerto Armuelles · Golfito · La Concepción · David
Punta Burica · Alanje · Cerro Chorcha 2298m
Isla Sevilla · Pedregal · NGÖBE BUGLÉ
Isla Parida · Remedios · Cerro Santiago 2121m
Golfo de Chiriquí · Horconcitos · Gualaca
Isla Coiba · Las Palmas · Quebrada Guabo
VERAGUAS · Canazas · Calobre
Soná · Santiago · San Francisco · Santa Fé
Río de Jesús · Río Santa María · Ponuga
Guarumal · Parita · Ocú · Chitré · Los Santos
HERRERA · Montijo · Macaracas · Monagrillo
Península de Azuero · Las Tablas
LOS SANTOS · Cerro Hoya 1560m · Tonosí · Pedasí
Isla Cébaco · Punta Mala

Portobelo · Santa Isabel · El Porvenir
Colón · KUNA YALA
Cristóbal · Cordillera de San Blas · Ailigandí
Nuevo Chagres · Palma de · Río Chagres · PANAMÁ · KUNA DE MADUNGANDI
Miguel de la Borda · Lago · Panama City · Lago · Punta Mosquito
Coclé del Norte · Gatún · San Miguelito · Cerro Chucanti 1439m · KUNA DE WARGANDI · Punta Escocés
COLÓN · Balboa · PANAMÁ · Serranía de Majé
La Chorrera · WEST PANAMÁ · Chepo · Río Bayano · Gulf of Darien
Capira · Bayano · Chimán · Puerto Obaldía · Lorica
Cerro Peña Blanca 1314m · El Valle · Cerro Gaital 1173m · Bahía de Panamá · Acandí · Golfo de Urabá · Montería
COCLÉ · Penonomé · San Carlos · Punta Chame · Archipiélago de las Perlas · San Miguel · La Palma · EMBERÁ WOUNAAN
Aguadulce · Río Hato · Isla del Rey · Punta Brava · DARIÉN · Tierralta
Isla San José · Garachiné · Yaviza · El Real · Unión Chocó · Turbo
PANAMÁ · Golfo de San Miguel · Cerro Tacarcuna 1875m · CÓRDOBA
Golfo de Panamá · Punta Garachiné · Cerro Pirre 1200m · ANTIOQUIA
Jaqué · Cerro Setetule 1220m · CHOCÓ · COLOMBIA
Ríosucio · Mutatá · Dabeiba
Jurado · Alto Musinga 3850m · Frontino

19,686ft
13,124ft
9843ft
6562ft
3281ft
1640ft
820ft
328ft
Sea Level
-820ft
-6562ft
-13,124ft

UNITED STATES
OF AMERICA

FLORIDA

Saint Petersburg
Bradenton
Sarasota
Venice
Port Charlotte
Fort Myers
Cape Coral
Bonita Springs
Naples

Lake Wales
Avon Park
Wauchula
Arcadia
Okeechobee
Punta Gorda
La Belle
Immokalee
Big Cypress Swamp
The Everglades
Cape Romano
Everglades City
Cape Sable

Sebastian
Vero Beach
Fort Pierce
Jensen Beach
Stuart
Hobe Sound
Indiantown
West Palm Beach
Lake Worth
Belle Glade
Boynton Beach
Boca Raton
Pompano Beach
Fort Lauderdale
Hollywood
North Miami
Miami Beach
Hialeah
Kendall
Miami
Homestead

Gulf of Mexico

Dry Tortugas
Marquesas Keys
Key West
Marathon
Florida Bay
Key Largo
Key Largo
Islamorada
Florida Keys

Straits of Florida

Tropic of Cancer

THE BAHAMAS

Grand Bahama Island
Great Sale Cay
Little Abaco
Coopers Town
West End
Pelican Point
Marsh Harbour
Great Abaco
Freeport
Eight Mile Rock
Moores Island
Cherokee Sound
Freeport

Northwest Providence Channel

Bimini Islands
Berry Islands
Northeast Providence Channel
Eleuthera Island
Nicholls Town
Linden Pindling
Adelaide
NASSAU
New Providence
Current
Governor's Harbour
Rock Sound
Behring Point
Andros Town
Andros Island
Kemp's Bay
Exuma Sound
Exuma Cays
Great Guana Cay
Bannerman Town
Cat Island
Arthur's Town
Cockburn Town
San Salvador
Columbus Point
Cape Santa Maria
Conception Island
George Town
Rum Cay
Long Island
Samana Cay
Clarence Town
Great Exuma Island
Little Exuma
Deadman's Cay
Cape Verde
Crooked Island
Colonel Hill
Long Cay
Plana Cays
Snug Corner
Maya
The Ca
Acklins Island
Salina Point
Ragged Island Range
Mayaguana Passage
Caicos
Southeast I

Cay Sal
Anguilla Cays
Nicholas Channel
Santaren Channel
Archipiélago de Sabana
Old Bahama Channel
Archipiélago de Camagüey

LA HABANA (HAVANA)
Mariel
Guanabacoa
Matanzas
Cárdenas
Artemisa
San Cristóbal
Güines
Jovellanos
Minas de Matahambre
Los Palacios
Consolación del Sur
Güira de Melena
Colón
Santo Domingo
Santa Clara
Caibarién
Cayo Fragoso
Jagüey Grande
Cruces
Sagua la Grande
Cayo Coco
Great Inagua
Pinar del Río
Viñales
Cienfuegos
Placetas
Cayo Romano
Cayo Guajaba
Peninsula Aguada de Pasajeros
Pico San Juan 1156m
Sancti Spíritus
Morón
Esmeralda
Sierra de los Órganos
Archipiélago de los Colorados
Cabo de San Antonio
Nueva Gerona
Santa Fé
Cayo Largo
Trinidad
Ciego de Ávila
Florida
Camagüey
Nuevitas
Golfo de Guanahacabibes
Golfo de Batabanó
Cabo Corrientes
Cayos de San Felipe
Isla de la Juventud
Archipiélago de los Canarreos
CUBA
Golfo de Ana María
San Pedro
Vertientes
Puerto Padre
Gibara
Cabo Lucrecia
Archipiélago de los Jardines de la Reina
Santa Cruz del Sur
Las Tunas
Holguín
Banes
Mayarí
Moa
Matthew Town
Golfo de Guacanayabo
Campechuela
Cauto
Bayamo
Jiguaní
Manzanillo
Sierra de Nipe
Sagua de Tánamo
Punta Guarico
La Maya
Baracoa
Maisí
Punta de Quemado
Guantánamo
Sierra Maestra
Palma Soriano
Pilón
Cabo Cruz
Pico Turquino 1944m
Santiago de Cuba
Bahía de Guantánamo (to US)

Greater

CAYMAN ISLANDS (to UK)
Little Cayman
Cayman Brac
GEORGE TOWN
Owen Roberts
Bodden Town
Grand Cayman

HAITI
Cap Dame Marie
Dame-Marie
Jérémie
Corail
Môle- St-Nicolas
Gros-N
Gona
Golfe de la Gona
St-
Port-de-Pa
Île de la Gonâve
Windward Passage
Canal de la
Miragoâne
Chardonnières
Port Salut
Pointe à Gravois
Cayes
Île à Vache
Aquin
Ja

JAMAICA
Montego Bay
Sangster
Port Maria
South Negril Point
Christiana
Port Antonio
Savanna-La-Mar
Mandeville
Spanish Town
Black River
May Pen
Blue Mountain Peak 2256m
Morant Bay
Portland Point
KINGSTON
Norman Manley
Port Royal
NAVASSA ISLAND (to US)
Jamaica Channel
Massif de la Hatte

ISLAS DE LA BAHÍA
Isla de Guanaja
Roatán
Isla de Roatán
Punta Caxinas
Trujillo
Limón
Balfate
Iriona
Río Aguán
COLÓN
Río Sico Tinto
Brus Laguna
San Esteban
Nava
Gualaco
Laguna de Caratasca
Dulce Nombre de Culmí
Puerto Lempira
HONDURAS
GRACIAS A DIOS
Juticalpa
Catacamas
Arrecifes de la Media Luna
OLANCHO
Cabo de Gracias a Dios
Río Coco
Arrecife Edinburgh
Bonanza
La Rosita
Waspam
Boom
Bocay
Ulmukhuás
Yablis
Cayos Miskitos
Dakura
Wiwilí
San Luis
REGIÓN AUTÓNOMA ATLÁNTICO NORTE
Siuna
Tuapi
Cayos Londres
Cerro Chachagua 1504m
Puerto Cabezas
Jinotega
Río Tuma
Wounta
Matagalpa
Río Grande de Matagalpa
Prinzapolka
Isla Santa Catalina
MATAGALPA
Río Prinzapolka
Isla de Providencia
La Sirena
Cayos Guerrero
BOACO
Río Grande
SAN ANDRÉS Y PROVIDENCIA (to Colombia)
CHONTALES
Muelle de los Bueyes
Kara
Boaco
El Rama
Barra de Río Grande
Juigalpa
Santo Tomás
NICARAGUA
Bluefields
Isla de San Andrés
Lago de Nicaragua
REGIÓN AUTÓNOMA ATLÁNTICA SUR
Cayos King
Volcán Concepción 1610m
Rivas
Río San Juan
Monkey Point
Isla de Ometepe
Morrito
RÍO SAN JUAN
Río Punta Gorda
Cayos de Perlas
Cayos del Este Sudeste
Punta Gorda
Islas del Maíz
San Miguelito
Cayos del Albuquerque
San Carlos
La Cruz
Upala
El Castillo de La Concepción
San Juan del Norte
Liberia
Volcán Miravalles 2028m
San Juan del Sur
San Juan
Barra del Colorado
GUANACASTE
Cañas
Volcán Arenal
ALAJUELA
Río San Juan
Santa Cruz
Quesada
HEREDIA
Puerto Viejo
Puntarenas
Volcán Poás 2704m
Volcán Irazú
LIMÓN
PUNTARENAS
Atenas
Heredia
Alajuela
SAN JOSÉ
COSTA RICA
Siquirres
Guápiles
Limón

Caribbean Sea

Caribbean Sea

JAMAICA inset

JAMAICA
Caribbean Sea
Montego Bay
Sangster
Falmouth
Discovery Bay
St Ann's Bay
Port Maria
Lucea
Clark's Town
Browns Town
Ocho Rios
Don Christophers Point
Negril
Dolphin Head 545m
The Cockpit Country
Cambridge
Alexandria
Claremont
Annotto Bay
Buff Bay
North East Point
Little London
Grange Hill
Savanna-La-Mar
Mount Denham 886m
Frankfield
Linstead
Ewarton
Highgate
Port Antonio
Crab Pond Point
Maggotty
Christiana
Mandeville
Chapelton
Spanish Town
Bog Walk
Blue Mountain Peak 2258m
Black River
Santa Cruz
May Pen
Port Morant
Golden Grove
Malvern 725m
Old Harbour
Portmore
Norman Manley
KINGSTON
Bath
Yallahs Hill 730m
Morant Bay
Great Pedro Bluff
Alligator Pond
Lionel Town
Long Bay
Port Royal
Wreck Point
Portland Bight
Portland Point
Caribbean Sea

Scale 1:2,500,000
0 5 10 20 Km
0 5 10 20 Miles

SAN ANDRÉS Y PROVIDENCIA (to Colombia)

Caribbean

Sea

COLOMBIA
Ríohac
Santa Marta
Dibulla
Barranquilla
MAGDALENA
Puerto Colombia
Ciénaga
Pico Cristóbal Colón 5775m
Ernesto Cortissoz
Soledad
Sitionuevo
Sierra Nevada de Santa Marta
ATLÁNTICO
Sabanalarga
Piojay
Aracataca
Santa Catalina
Salamina
Fundación
Fo
Cartagena
Campo de la Cruz
Turbaco
Vallledupar

6000m
4000m
3000m
2000m
1000m
500m
250m
100m
Sea Level
-250m
-2000m
-4000m

Scale 1:6,250,000
(projection: Lambert Conformal Conic)

0 25 50 75 100 125 150 175 200 Km
0 25 50 75 100 125 150 175 200 Miles

Population

■ above 5 million
◉ 1 million to 5 million
◉ 500,000 to 1 million
◎ 100,000 to 500,000
⊕ 50,000 to 100,000
○ 10,000 to 50,000
○ below 10,000

H I J K L M N

290

GUADELOUPE (to France)

Caribbean Sea
Guadeloupe Passage
Pointe de la Grande Vigie
Anse-Bertrand
Ste-Rose
Baie-Mahault
Lamentin
Pointe Noire
Morne-à-l'Eau
le Moule
Grande Terre
Port-Louis
les Abymes
St-François
Basse Terre
Petit-Bourg
Ste-Anne
Vieux-Habitants
Soufrière 1467m
Petit Cul-de-Sac Marin
Canal de Marie-Galante
St-Claude
Capesterre-Belle-Eau
Canal des Saintes

BASSE-TERRE
Scale 1:2,500,000
0 5 10 20 Km
0 5 10 20 Miles

DOMINICA

Dominica Passage
Pointe Jaco
Vieille Case
Portsmouth
Melville Hall
Marigot
Morne Diablotins 1447m
Salisbury
Castle Bruce
St. Joseph
Canefield
Rosalie
La Plaine
ROSEAU
Berekua
Scotts Head Village
Martinique Passage

Scale 1:2,000,000
0 5 10 Km
0 5 10 Miles

MARTINIQUE (to France)

Martinique Passage
Grand' Rivière
Basse-Pointe
le Prêcheur
Montagne Pelée 1397m
Ste-Marie
St-Pierre
la Trinité
le Robert
Schœlcher
le Lamentin
FORT-DE-FRANCE
le François
Baie de Fort-de-France
Rivière-Pilote
les Anses-d'Arlets
le Diamant
Ste-Anne
Saint Lucia Channel

Scale 1:2,500,000
0 5 10 20 Km
0 5 10 20 Miles

ST LUCIA

Caribbean Sea
Pointe Du Cap
Gros Islet
George F.L Charles
CASTRIES
Anse La Raye
Dennery
Soufrière
Petit Piton 743m
Micoud
Mount Gimie 950m
Gros Piton 798m
Laborie
Hewanorra
Vieux Fort
Ministre Point
Saint Vincent Passage

Scale 1:2,000,000
0 5 10 Km
0 5 10 Miles

BARBADOS

North Point
Crab Hill
ATLANTIC OCEAN
Speightstown
Bathsheba
Mount Hillaby 340m
Welchman Hall
Holetown
BRIDGETOWN
The Crane
Oistins
Grantley Adams

Scale 1:2,000,000
0 5 10 Km
0 5 10 Miles

ST VINCENT & THE GRENADINES

Caribbean Sea
Saint Vincent Passage
Porter Point
Fancy
La Soufrière 1234m
Chateaubelair
Georgetown
St Vincent
Layou
North Union
Barrouallie
KINGSTOWN
Stubbs
Arnos Vale
ATLANTIC OCEAN

Scale 1:2,000,000
0 5 10 Km
0 5 10 Miles

GRENADA

Caribbean Sea
Sauteurs
Victoria
Mount St Catherine 840m
Gouyave
Grenville
ST.GEORGE'S
St. David's
Point Salines
Grand Anse
ATLANTIC OCEAN

Scale 1:2,000,000
0 5 10 Km
0 5 10 Miles

Tropic of Cancer

ATLANTIC OCEAN

WKS & CAICOS ISLANDS (to UK)
s Caicos Islands
Grand Caicos
ncials East Caicos
Caicos
Grand Turk Island
Cockburn **COCKBURN TOWN**
Harbour Turks Islands

Mouchoir Passage

ispaniola
DOMINICAN REPUBLIC
Monte Cristi Cabo Isabela
-Haitien
Fort-Liberté Mao
Dajabón Sabaneta Nagua
Santiago Moca Cabrera
La Vega Salcedo
San Francisco de Macorís
Bahía Escocesa
Hinche
Río Yuna Miches
Bonao Monte Bahía de Samaná
Pico Duarte 3175m Hato Mayor El Seibo
Higüey Engano
Comendador Azua Cristóbal
SANTO DOMINGO La Romana
San Pedro de Macorís
Barahona
Enriquillo
Oviedo
Cabo Beata

PUERTO RICO (to US)
BRITISH VIRGIN ISLANDS (to UK)
Anegada
Aguadilla
Arecibo SAN JUAN Carolina
Sombrero (to Anguilla)
Luis Muñoz Marín
St Thomas ROAD TOWN
Bayamón Caguas Fajardo Tortola
Beef Island
Mayagüez Cordillera Central 1065m CHARLOTTE AMALIE
Yauco Ponce Guayama VIRGIN ISLANDS (to US)
Isla de Vieques
Frederiksted St Croix Christiansted

ANGUILLA (to UK)
THE VALLEY
Marigot St Martin (to France)
St Barthélemy (to France)
PHILIPSBURG Sint Maarten (Netherlands)
Saba (to Neth.) St Eustatius (to Neth.)
St Kitts
Robert L. Bradshaw Vance W. V.C.Bird
Nevis Charlestown ST JOHN'S
ANTIGUA & BARBUDA
Codrington Barbuda
Antigua
ST KITTS & NEVIS
Redonda Falmouth
BRADES John A. Osborne
MONTSERRAT (to UK)
Ste-Rose le Raizet Port-Louis
la Désirade
Basse Terre Pointe-à-Pitre GUADELOUPE (to France)
Soufrière 1467m Marie-Galante
BASSE-TERRE les Saintes Grand-Bourg
Dominica Passage
Portsmouth Marigot
ROSEAU Canefield La Plaine DOMINICA
Martinique Passage
Montagne Pelée 1397m Ste-Marie MARTINIQUE
St-Pierre
FORT-DE-FRANCE Lamentin
Aimé Césaire
Rivière-Pilote
St Lucia Channel
CASTRIES Mount Gimie 950m
ST LUCIA Soufrière Vieux Fort
St Vincent Passage
St Vincent La Soufrière 1234m **BARBADOS**
Chateaubelair Georgetown Speightstown
ST VINCENT & THE GRENADINES KINGSTOWN BRIDGETOWN
Arnos Vale Grantley Adams
Bequia
Port Elizabeth Mustique
Canouan The Grenadines
Union Island
Hillsborough Carriacou
GRENADA Victoria
ST GEORGE'S Grenville
Point Salines

Lesser Antilles
Leeward Islands
Windward Islands

PUERTO RICO (to US)

ATLANTIC OCEAN
Isabela
Aguadilla Arecibo Laguna Tortuguero
Vega Baja Manati SAN JUAN
Punta Higüero San Sebastián Lago Dos Bocas Cataño Rio Grande
Bahía de Mayagüez Río Grande de Añasco Utuado Orocovis Bayamón Carolina Cabezas de San Juan
Mayagüez Adjuntas Guaynabo Caguas Sonda de Vieques
Cerro de Punta 1338m El Yunque 1065m Isla de Culebra
San Germán Yauco Juana Díaz Cayey Humacao Yabucoa Monte Pirata 301m Isla de Vieques
Embalse Salinas Punta Puerca
Cabo Rojo Ponce Guayama Punta Guayanés
Punta Brea Punta Petrona

Caribbean Sea
Scale 1:2,500,000
0 5 10 20 Km
0 5 10 20 Miles

TRINIDAD & TOBAGO

Caribbean Sea
Galera Point
Tobago
Blanchisseuse Matelot
Redhead
The Dragon's Mouth **PORT-OF-SPAIN** Arima
Tunapuna Piarco Sangre Grande
Chaguanas Caroni River Caroni Arena Dam
Couva Guatuaro Point
Gulf of Paria San Fernando Trinidad Rio Claro
La Brea Princes Town Killdeer River Rushville
Point Fortin Siparia Galeota Point
Bonasse Moruga
The Serpent's Mouth
VENEZUELA

Scale 1:2,500,000
0 5 10 20 Km
0 5 10 20 Miles

290

Lesser Antilles
ARUBA (Netherlands)
ORANJESTAD Reina Beatrix Aruba
CURAÇAO (Netherlands)
Noordpunt Sint Nicolaas
Hato Airport BONAIRE (to Netherlands)
Santa Catherina Malmok **KRALENDIJK**
Curaçao Islas
WILLEMSTAD Las Aves
Islas Los Roques
Isla La Orchila
Isla Blanquilla
Charlotteville Tobago
Scarborough
TRINIDAD & TOBAGO
Galera Point
PORT-OF-SPAIN
Trinidad San Fernando

VENEZUELA
Peninsula de la Guajira
s Gallinas
Uribia
ZULIA Puerto López
Punta López Cabo San Román
Pueblo Nuevo
Golfo de Venezuela Los Taques Punta Fijo
Paraguaipoa
Puerto Cumarebo
Coro San Felipe La Vela de Coro Mirimire
Sabaneta Tucuyo de La Costa Chichiriviche
Capatárida FALCÓN
Pedregal Churuguara Maparari
Altagracia Mene de Mauroa
San Francisco Cabimas Siquisique Aroa
Concepción Tía Juana LARA San Luis YARACUY
Maracaibo
Churuguara
San Juan de los Cayos
La Tortuga
NUEVA ESPARTA La Asunción
Boca de Pozo Juangriego Pampatar
Isla de Margarita Porlamar
Punta de Piedras
Río Caribe Puerto de Hierro
Carúpano Güiria
Araya SUCRE Cariaco Casanay El Pilar Irapa
Cumaná Cumanacoa San Antonio Caripe
VARGAS Barcelona Santa Inés Aragua de Maturín Caicara
La Guaira **CARACAS** ANZOÁTEGUI Quiriquire
Petare Puerto Píritu Clarines Punta de Mata MONAGAS
Los Teques MIRANDA Maturín
Río Chico
Morón Puerto Cabello
Maracay Cagua Caño Manamo DELTA AMACURO
Valencia Villa de Cura del Tuy Valle de Guanape Punta Baja
YARACUY Ocumare Pedernales
San Felipe Guigue Turmero The Serpent's
Chivacoa San Luis San Francisco de Yare

102

Elevation

19,686ft
13,124ft
9843ft
6562ft
3281ft
1640ft
820ft
328ft
Sea Level
-820ft
-6562ft
-13,124ft

NORTH AMERICA

91

Atlanta

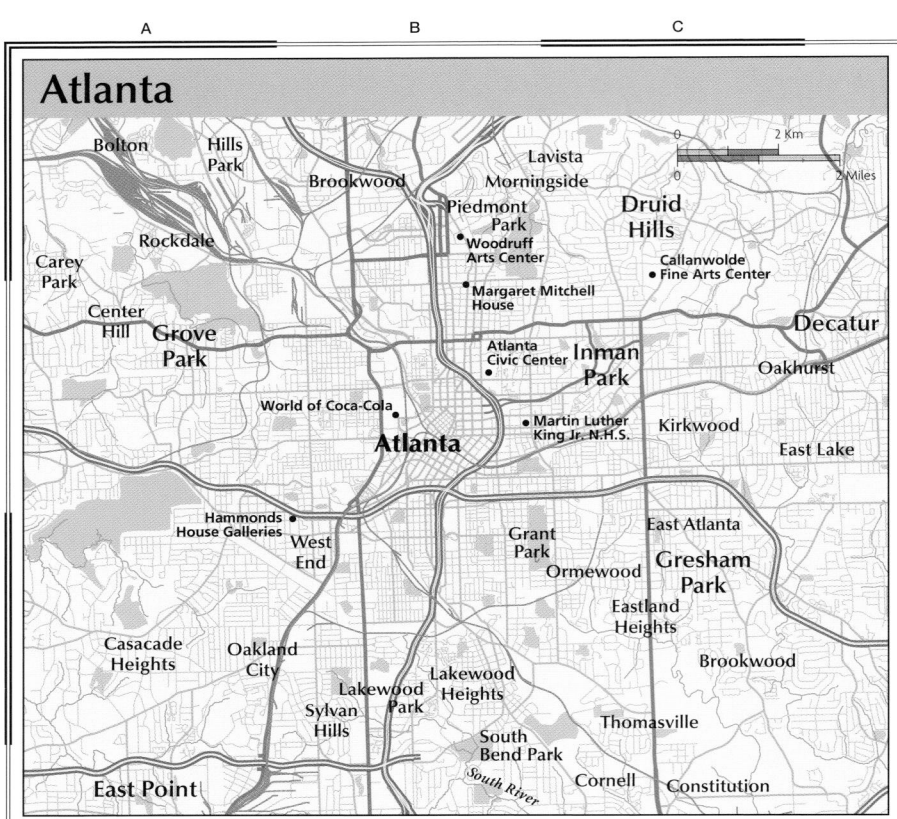

Bolton
Hills Park
Brookwood
Lavista
Morningside
Piedmont Park
Woodruff Arts Center
Druid Hills
Rockdale
Callanwolde Fine Arts Center
Carey Park
Margaret Mitchell House
Center Hill
Grove Park
Decatur
Atlanta Civic Center
Inman Park
Oakhurst
World of Coca-Cola
Atlanta
Martin Luther King Jr. N.H.S.
Kirkwood
East Lake
Hammonds House Galleries
West End
Grant Park
East Atlanta
Ormewood
Gresham Park
Casacade Heights
Oakland City
Eastland Heights
Lakewood Park
Lakewood Heights
Brookwood
Sylvan Hills
South Bend Park
Thomasville
East Point
South River
Cornell
Constitution

Chicago

Chicago O'Hare Intl. Airport
Harwood Heights
Lake Michigan
Uptown
Lincoln Park Zoo
Addison
Elmwood Park
Melrose Park
Avondale
Lincoln Park
Elmhurst
Maywood
Oak Park
Bucktown
Lombard
Sears Tower
Chicago
Westchester
Berwyn
Cicero
Chinatown
Oak Brook
Pilsen
Bronzeville
Bridgeport
La Grange
Downers Grove
Summit
Chicago Midway Airport
Elsdon
Englewood
Kenwood
Bedford Park
Darien
Forest Hill
Burbank
Ashburn
Evergreen Park
Waterfall Glen Forest Preserve
Des Plaines
Hickory Hills
Oak Lawn
Chicago State University

Dallas

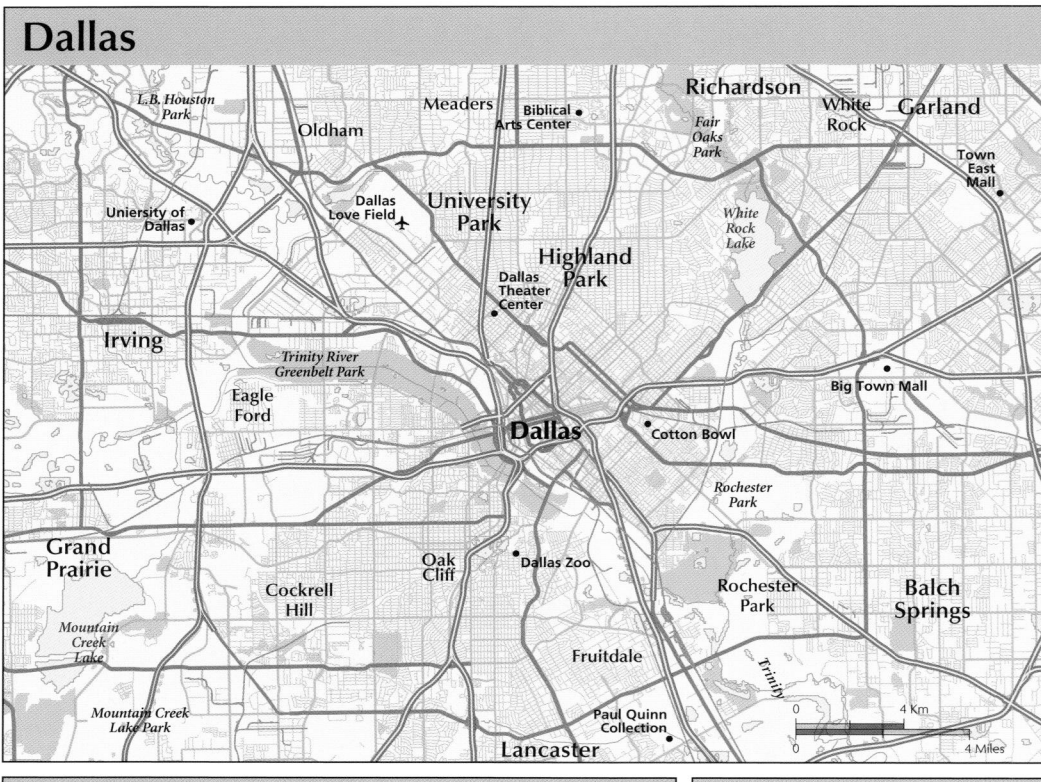

L.B. Houston Park
Meaders
Biblical Arts Center
Richardson
White Rock
Garland
Oldham
Fair Oaks Park
Town East Mall
University of Dallas
Dallas Love Field
University Park
White Rock Lake
Irving
Highland Park
Dallas Theater Center
Trinity River Greenbelt Park
Big Town Mall
Eagle Ford
Dallas
Cotton Bowl
Grand Prairie
Rochester Park
Cockrell Hill
Oak Cliff
Dallas Zoo
Rochester Park
Balch Springs
Mountain Creek Lake
Fruitdale
Trinity
Mountain Creek Lake Park
Paul Quinn Collection
Lancaster

Denver

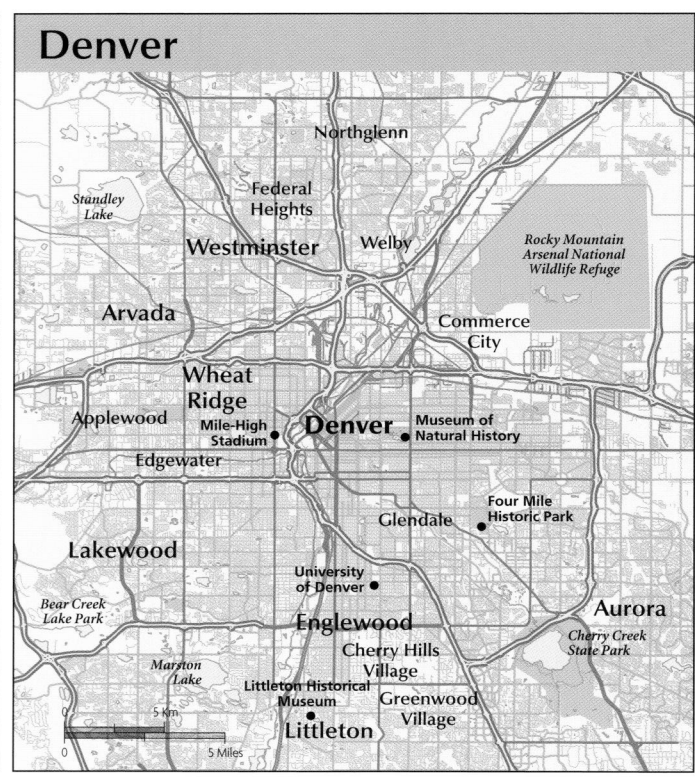

Northglenn
Standley Lake
Federal Heights
Westminster
Welby
Rocky Mountain Arsenal National Wildlife Refuge
Arvada
Commerce City
Wheat Ridge
Applewood
Mile-High Stadium
Denver
Museum of Natural History
Edgewater
Glendale
Four Mile Historic Park
Lakewood
University of Denver
Bear Creek Lake Park
Marston Lake
Englewood
Cherry Hills Village
Cherry Creek State Park
Aurora
Littleton Historical Museum
Greenwood Village
Littleton

Detroit

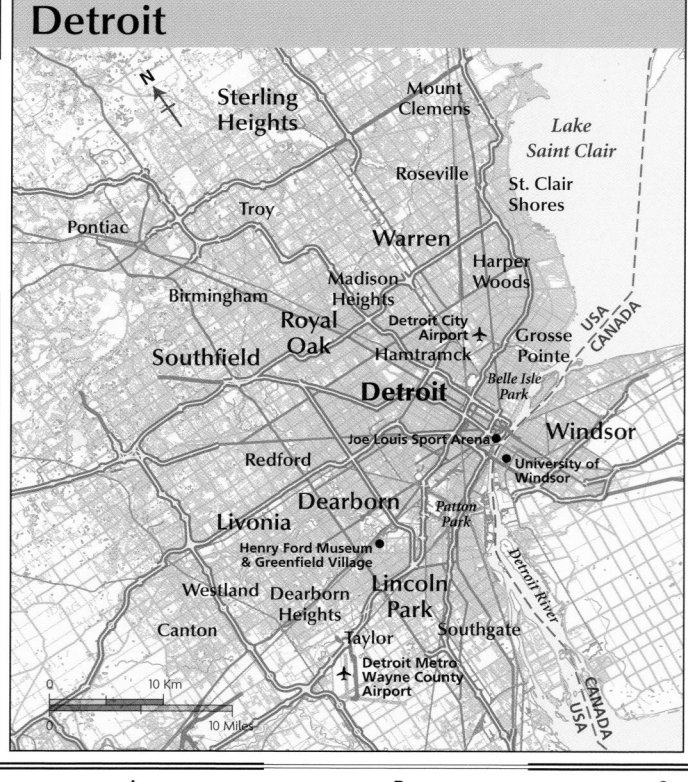

N
Sterling Heights
Mount Clemens
Lake Saint Clair
Roseville
St. Clair Shores
Pontiac
Troy
Warren
Birmingham
Madison Heights
Harper Woods
Southfield
Royal Oak
Detroit City Airport
Grosse Pointe
USA CANADA
Hamtramck
Belle Isle Park
Detroit
Joe Louis Sport Arena
Windsor
Redford
University of Windsor
Livonia
Dearborn
Patton Park
Henry Ford Museum & Greenfield Village
Lincoln Park
Detroit River
Westland
Dearborn Heights
Southgate
CANADA USA
Canton
Taylor
Detroit Metro Wayne County Airport

Las Vegas

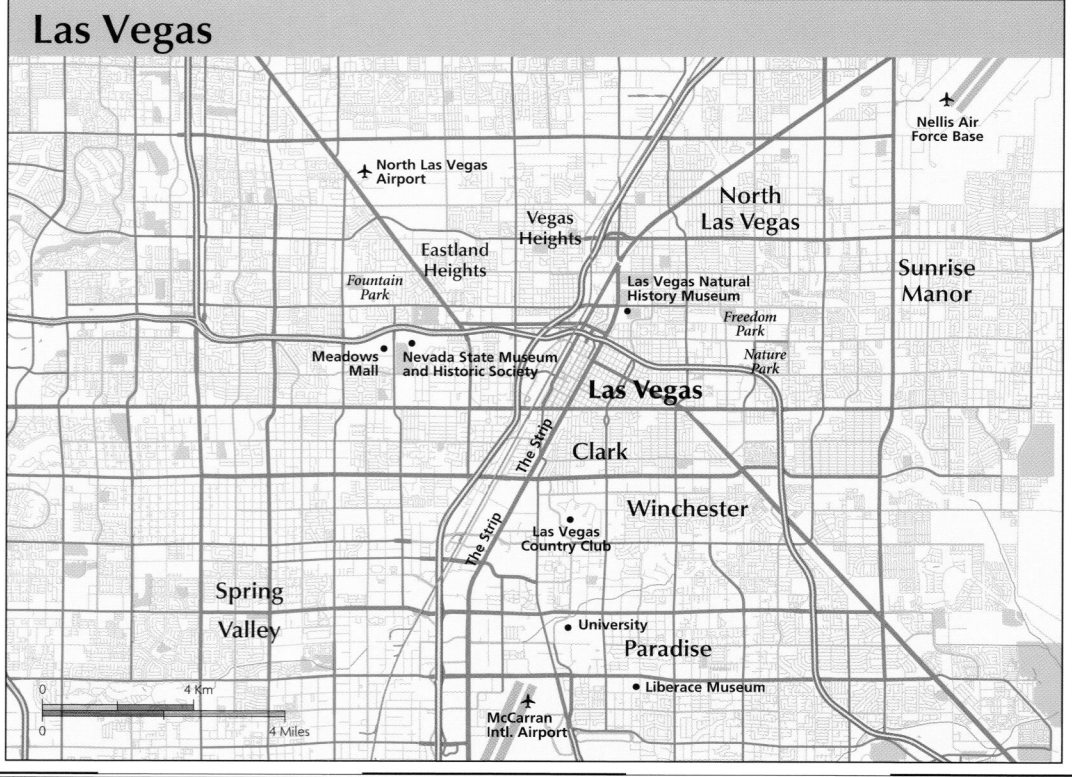

Nellis Air Force Base
North Las Vegas Airport
Vegas Heights
North Las Vegas
Eastland Heights
Sunrise Manor
Fountain Park
Las Vegas Natural History Museum
Freedom Park
Meadows Mall
Nevada State Museum and Historic Society
Nature Park
Las Vegas
Clark
The Strip
Winchester
The Strip
Las Vegas Country Club
Spring Valley
University
Paradise
Liberace Museum
McCarran Intl. Airport

Los Angeles

Montréal

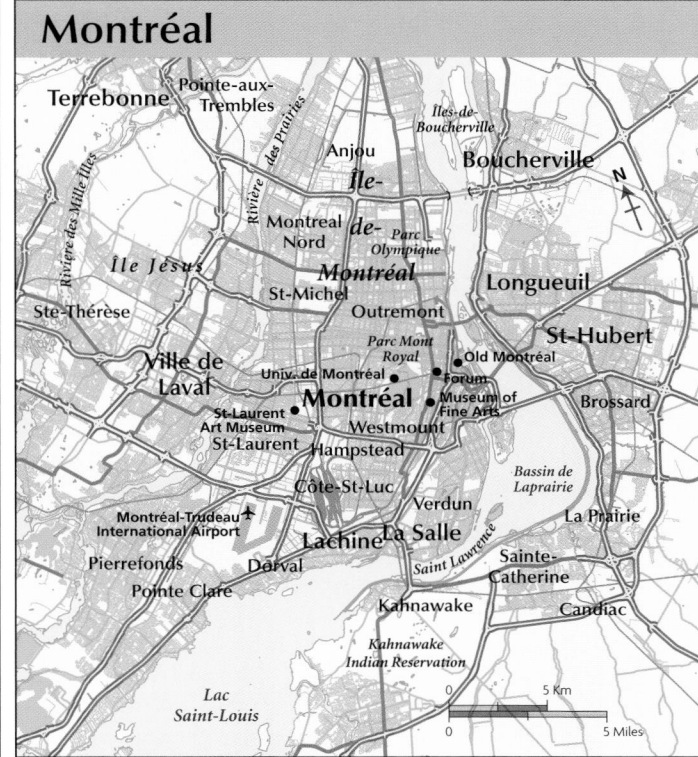

Philadelphia

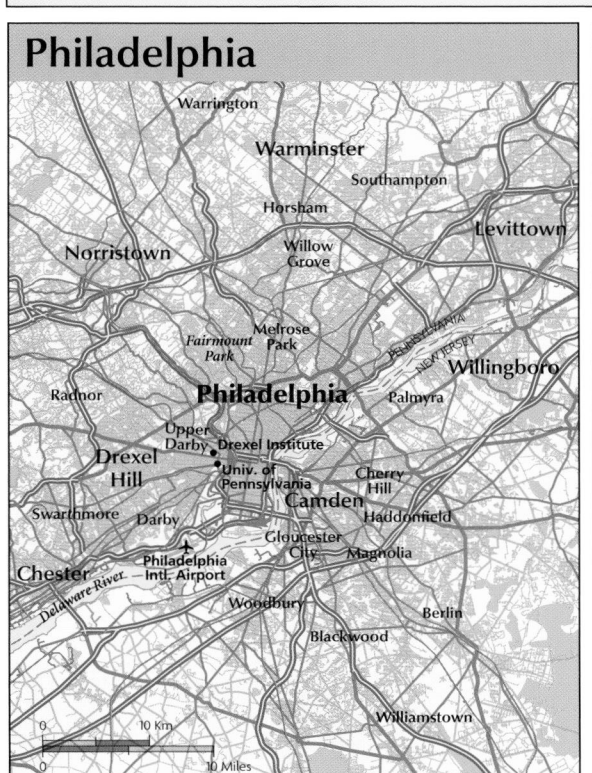

Seattle

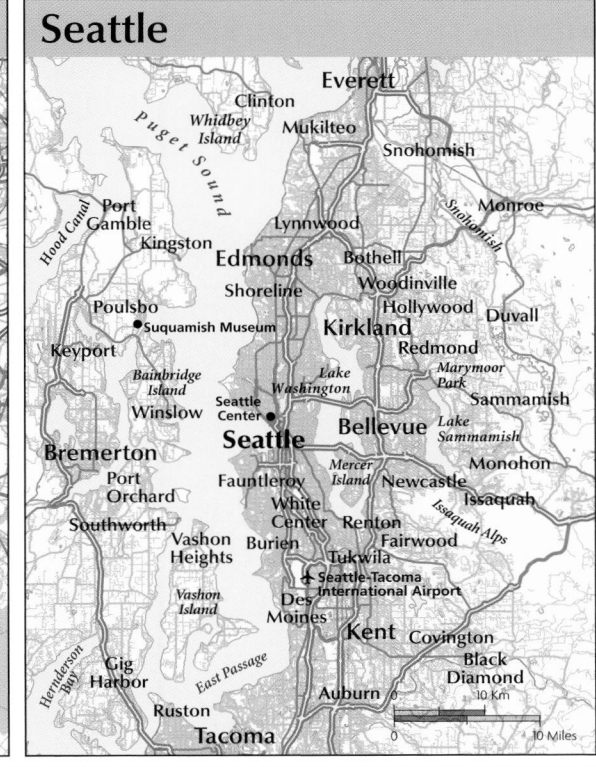

Toronto

San Francisco

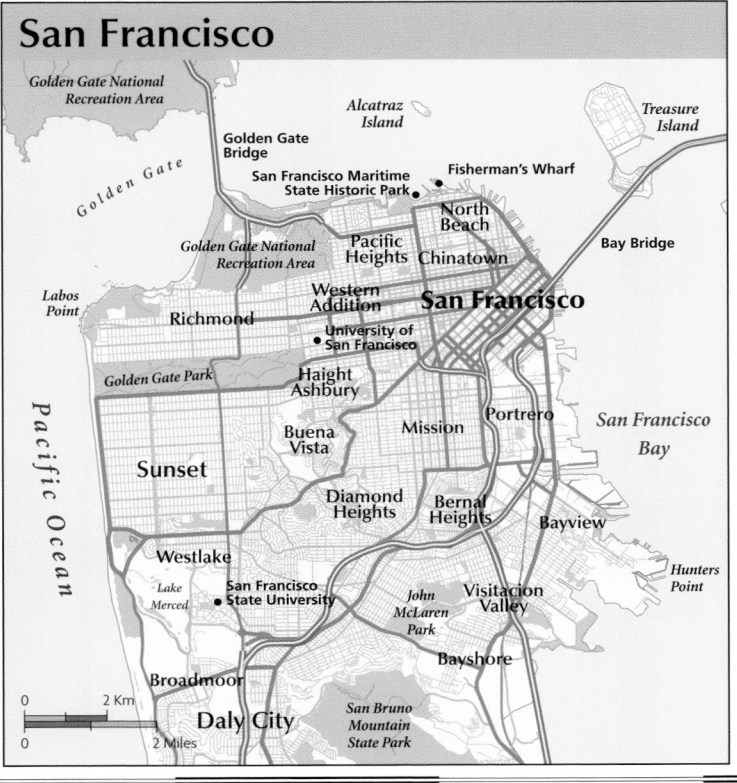

Washington D.C.

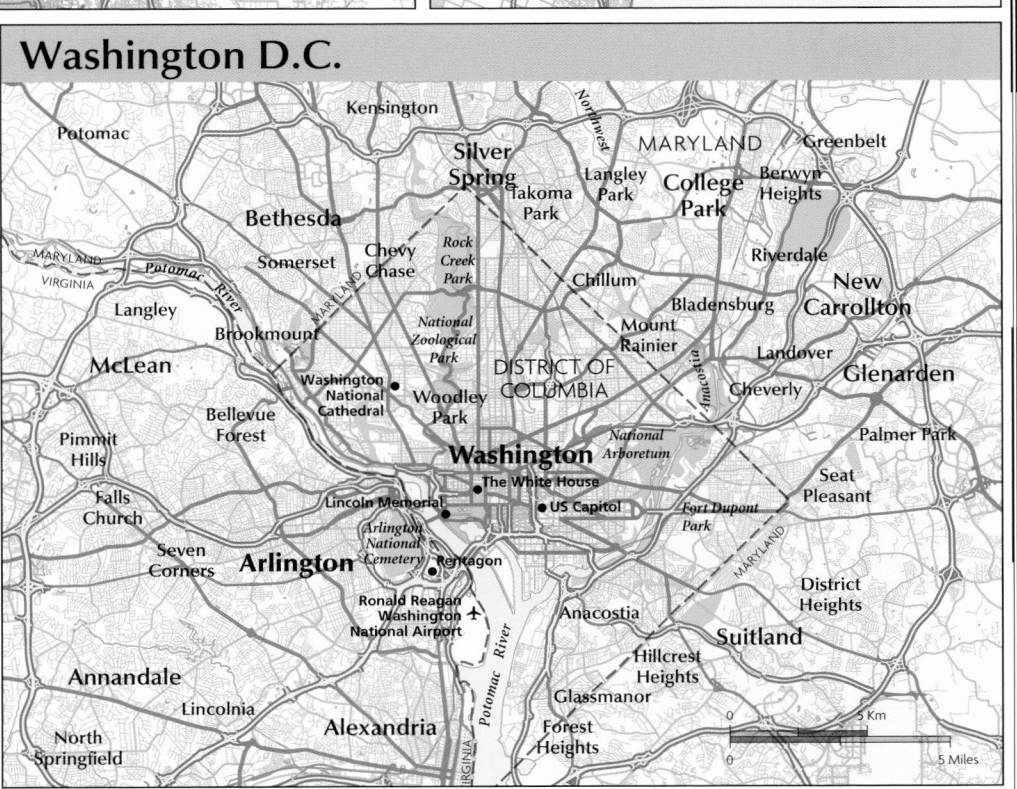

South America reaches from the humid tropics down into the cold South Atlantic, with a total area of 6,886,000 sq miles (17,835,000 sq km). It comprises 12 separate countries, with the largest, Brazil, covering almost half the continent.

FACTFILE

N Most Northerly Point: Punta Gallinas, Colombia 12° 28' N
S Most Southerly Point: Cape Horn, Chile 55° 59' S
E Most Easterly Point: Ilhas Martin Vaz, Brazil 28° 51' W
W Most Westerly Point: Galapagos Islands, Ecuador 92° 00' W

Largest Lakes:
1. Lake Titicaca, Bolivia/Peru 3141 sq miles (8135 sq km)
2. Mirim Lagoon, Brazil/Uruguay 1158 sq miles (3000 sq km)
3. Lago Poopó, Bolivia 976 sq miles (2530 sq km)
4. Lago Buenos Aires, Argentina/Chile 864 sq miles (2240 sq km)
5. Laguna Mar Chiquita, Argentina 695 sq miles (1800 sq km)

Longest Rivers:
1. Amazon, Brazil/Colombia/Peru 4049 miles (6516 km)
2. Paraná, Argentina/Brazil/Paraguay 2920 miles (4700 km)
3. Madeira, Bolivia/Brazil 2100 miles (3379 km)
4. Purus, Brazil/Peru 2013 miles (3239 km)
5. São Francisco, Brazil 1802 miles (2900 km)

Largest Islands:
1. Tierra del Fuego, Argentina/Chile 18,302 sq miles (47,401 sq km)
2. Ilha de Marajo, Brazil 15,483 sq miles (40,100 sq km)
3. Isla de Chiloé, Chile 3241 sq miles (8394 sq km)
4. East Falkland, Falkland Islands 2550 sq miles (6605 sq km)
5. Isla Wellington, Chile 2145 sq miles (5556 sq km)

Highest Points:
1. Cerro Aconcagua, Argentina 22,831 ft (6959 m)
2. Cerro Ojos del Salado, Argentina/Chile 22,572 ft (6880 m)
3. Cerro Bonete, Argentina 22,546 ft (6872 m)
4. Monte Pissis, Argentina 22,224 ft (6774 m)
5. Cerro Mercedario, Argentina 22,211 ft (6768 m)

Lowest Point:
▼ Laguna del Carbón, Argentina -344 ft (-105 m) below sea level

Highest recorded temperature:
⊕ Rivadavia, Argentina 120°F (49°C)

Lowest recorded temperature:
⊖ Sarmiento, Argentina -27°F (-33°C)

Wettest Place:
≋ Quibdó, Colombia 354 in (8990 mm)

Driest Place:
⊖ Arica, Chile 0.03 in (0.8 mm)

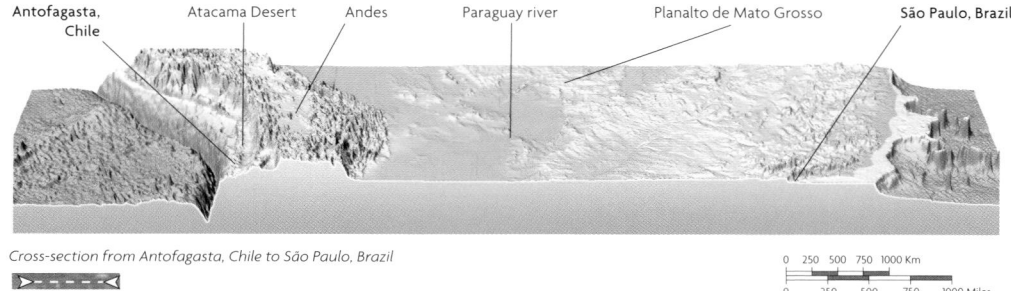

Antofagasta, Chile | Atacama Desert | Andes | Paraguay river | Planalto de Mato Grosso | São Paulo, Brazil

Cross-section from Antofagasta, Chile to São Paulo, Brazil

line of cross-section

0 250 500 750 1000 Km
0 250 500 750 1000 Miles

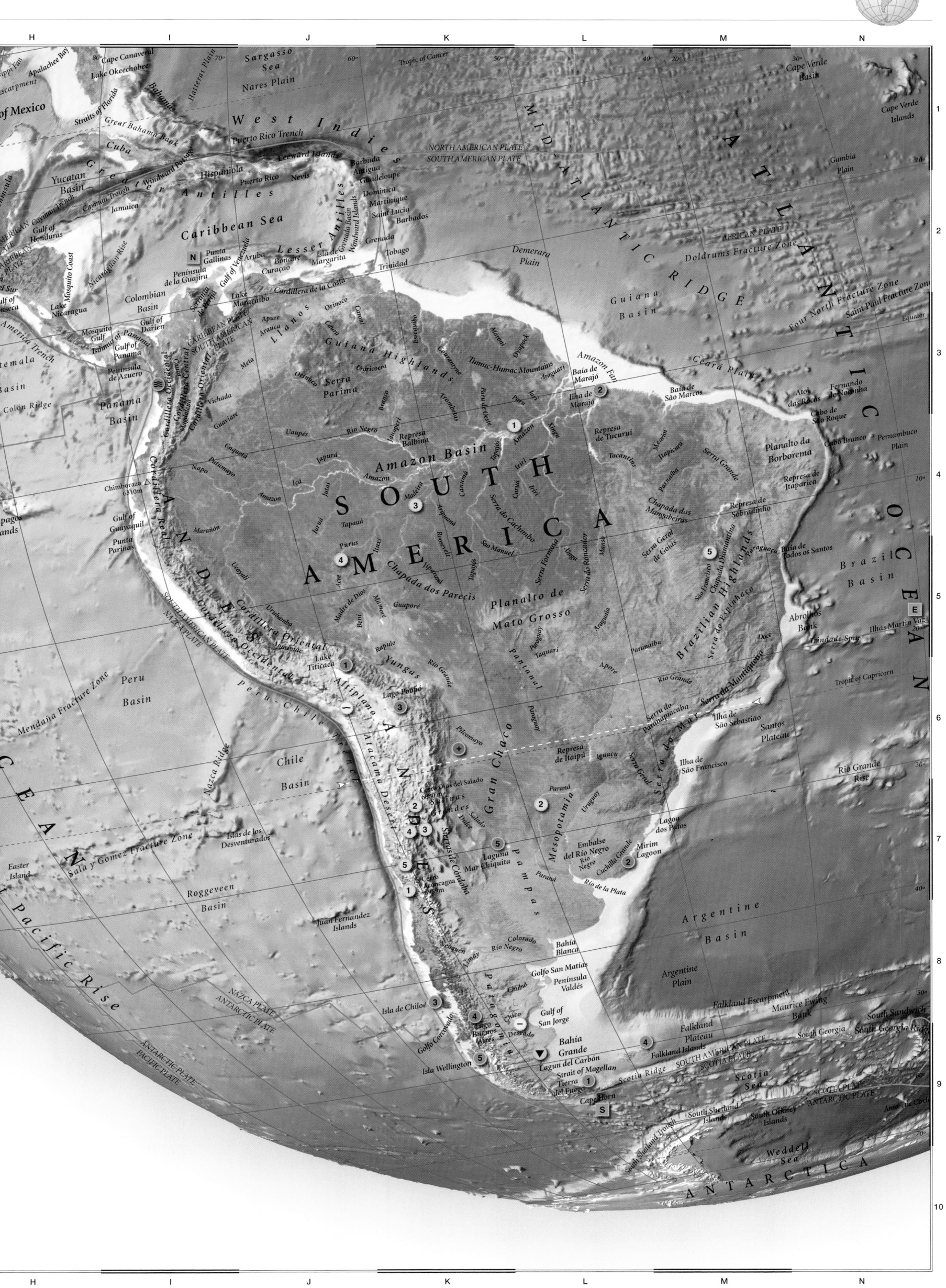

Km
0 100 200 300 400 500 600 700 800

Miles
0   100   200   300   400   500   600   700   800

Scale 1:24,000,000
(projection: Lambert Azimuthal Equal Area)

Population

- ■ above 5 million
- ▣ 1 million to 5 million
- ◉ 500,000 to 1 million
- ◎ 100,000 to 500,000
- ⊕ 50,000 to 100,000
- ○ 10,000 to 50,000
- ∘ below 10,000
- ● Country capital
- ● State capital

Borders

- ⬩ full international border
- ⬩ disputed de facto border
- ⬩ disputed territorial claim border
- ⬩ state border

Political

Modern South America's political boundaries have their origins in the territorial endeavors of explorers during the 16th century, who claimed almost the entire continent for Portugal and Spain. The Portuguese land in the east later evolved into the federal states of Brazil, while the Spanish vice-royalties eventually emerged as separate independent nation-states in the early 19th century. South America's growing population has become increasingly urbanized, with the expansion of coastal cities into large conurbations like Rio de Janeiro and Buenos Aires. In Brazil, Argentina, Chile, and Uruguay, a succession of military dictatorships has given way to fragile, but strengthening, democracies.

Languages

Prior to European exploration in the 16th century, a diverse range of indigenous languages were spoken across the continent. With the arrival of Iberian settlers, Spanish became the dominant language, with Portuguese spoken in Brazil, and Native American languages, such as Quechua and Guaraní, becoming concentrated in the continental interior. Today this pattern persists, although successive European colonization has led to Dutch being spoken in Suriname, English in Guyana, and French in French Guiana, while in large urban areas, Japanese and Chinese are increasingly common.

Language groups

- American Indian
- Germanic
- Romance

Standard of living

Wealth disparities throughout the continent create a wide gulf between affluent landowners and those afflicted by chronic poverty in inner-city slums. The illicit production of cocaine, and the hugely influential drug barons who control its distribution, contribute to the violent disorder and corruption which affect northwestern South America, destabilizing local governments and economies.

Standard of living
(UN human development index)

- low
- high

Population

Almost half of South America's population lives in Brazil but, due to the large uninhabited expanses of the Amazon Basin, its overall population density is much lower than in other countries. During the 20th century the most important population trend was the movement from rural to urban areas, giving rise to great population concentrations in cities like São Paulo, Rio de Janeiro, Caracas, Lima, Bogotá, and Buenos Aires.

Population density
(people per sq mile)

- below 10
- 11–23
- 24–36
- 37–49
- 50–75
- above 75

Transportation

- major roads and motorways
- major railroads
- international borders
- transport intersections
- international airports
- major ports

Transportation

Most major road and rail routes are confined to the coastal regions by the forbidding natural barriers of the Andes mountains and the Amazon Basin. Few major cross-continental routes exist, although Buenos Aires serves as a transport center for the main rail links to La Paz and Valparaíso, while the construction of the Trans-Amazon and Pan-American Highways have made direct road travel possible from Recife to Lima and from Puerto Montt up the coast into central America. A new waterway project is proposed to transform the Paraguay river into a major shipping route, although it involves considerable wetland destruction.

SOUTH AMERICA

97

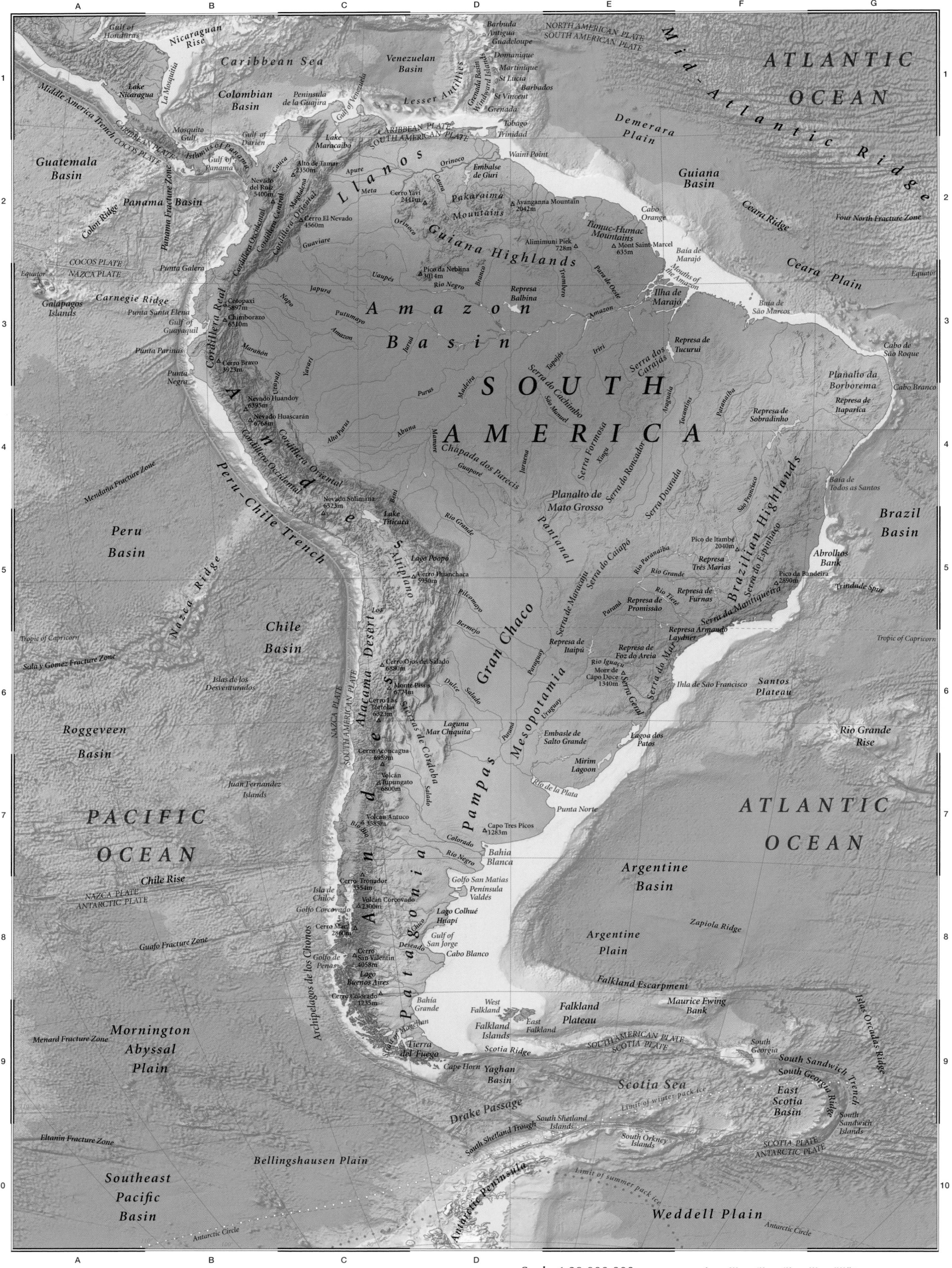

Scale 1:29,000,000
(projection: Lambert Azimuthal Equal Area)

0	200	400	600	800	1000 Km

0	200	400	600	800	1000 Miles

Climate

The climate of South America is influenced by three principal factors: the seasonal shift of high pressure air masses over the tropics, cold ocean currents along the western coast, affecting temperature and precipitation, and the mountain barrier produced by by the Andes, which creates a rain shadow over much of the south.

Climate

- tundra
- cool continental
- warm humid
- semi-arid
- arid
- humid equatorial
- tropical

- ☼ daily hours of sunshine, January
- ☼ daily hours of sunshine, July
- → cold wind

Average Rainfall

January rainfall

July rainfall

Rainfall

- 0–1 in (0–25 mm)
- 1–2 in (25–50 mm)
- 2–4 in (50–100 mm)
- 4–8 in (100–200 mm)
- 8–12 in (200–300 mm)
- 12–16 in (300–400 mm)
- 16–20 in (400–500 mm)
- more than 20 in (500 mm)

Average Temperature

January temperature

July temperature

Temperature

- below -22°F (-30°C)
- -22 to -4°F (-30 to -20°C)
- -4 to 14°F (-20 to -10°C)
- 14 to 32°F (-10 to 0°C)
- 32 to 50°F (0 to 10°C)
- 50 to 68°F (10 to 20°C)
- 68 to 86°F (20 to 30°C)
- above 86°F (30°C)

Land use

Many foods now common worldwide originated in South America. These include the potato, tomato, squash, and cassava. Today, large herds of beef cattle roam the temperate grasslands of the Pampas, supporting an extensive meat-packing trade in Argentina, Uruguay and Paraguay. Corn (maize) is grown as a staple crop across the continent and coffee is grown as a cash crop in Brazil and Colombia. Coca plants grown in Bolivia, Peru and Colombia provide most of the western world's cocaine. Fish and shellfish are caught off the western coast, especially anchovies off Peru, shrimps off Ecuador and sardines off Chile.

Environmental Issues

The Amazon Basin is one of the last great wilderness areas left on Earth. The tropical rainforests which grow there are a valuable genetic resource, containing innumerable unique plants and animals. The forests are increasingly under threat from new and expanding settlements and "slash and burn" farming techniques, which clear land for the raising of beef cattle, causing land degradation and soil erosion.

Environmental Issues

- national parks
- tropical forest
- forest destroyed
- desert
- desertification
- polluted rivers
- marine pollution
- heavy marine pollution
- poor urban air quality

Using the Land and Sea

- barren land
- cropland
- desert
- forest
- mountain region
- pasture
- major conurbations
- cattle
- pigs
- sheep
- bananas
- corn (maize)
- citrus fruits
- cocoa
- cotton
- coffee
- fishing
- oil palms
- peanuts
- rubber
- shellfish
- soya beans
- sugar cane
- vineyards
- wheat

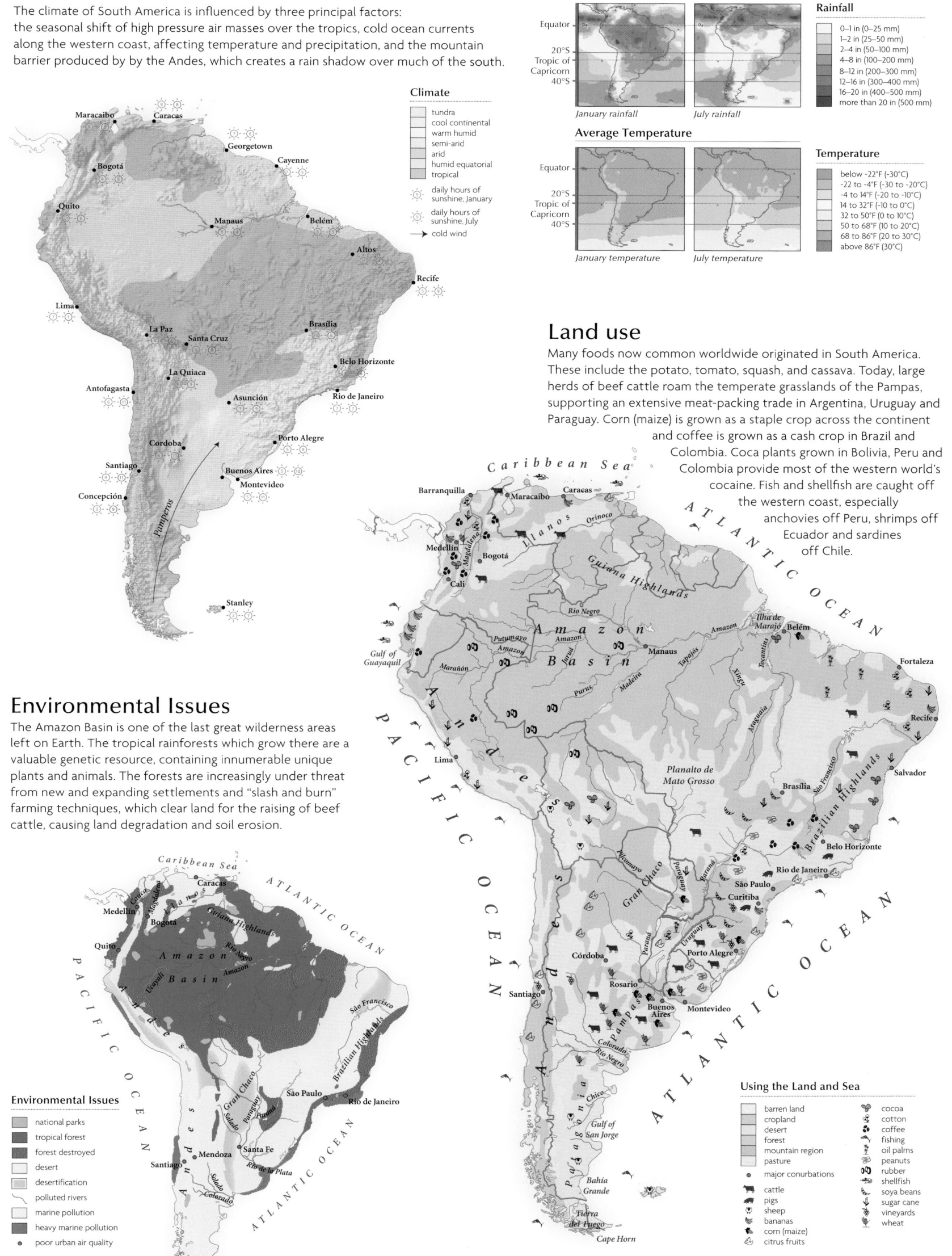

SOUTH AMERICA

1 SANTIAGO, CHILE
Chile's capital city was founded in 1541 by Pedro de Valdivia who chose the location because it had a Mediterranean climate and was easy to defend.

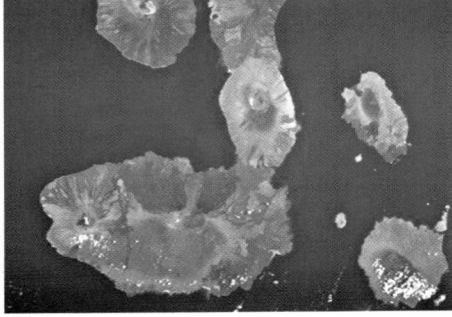

2 GALAPAGOS ISLANDS, ECUADOR
These islands are a collection of volcanoes rising from the ocean floor 621 miles (1000 km) west of the South American mainland.

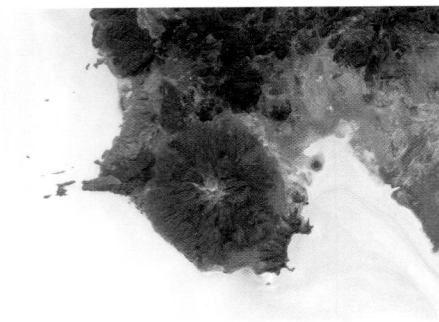

3 SALAR DE UYUNI, BOLIVIA
Occupying a depression high up on the Altiplano between the volcanoes of the western Andes and the fold belts of the eastern Andes, this is the world's largest salt flat.

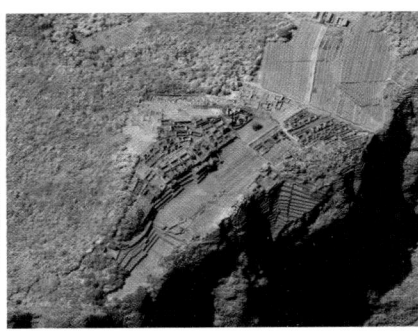

4 MACHU PICCHU, PERU
Perched precariously above the Urubamba valley, the lost Inca retreat was rediscovered in 1911 by Hiram Bingham, an American archaeologist.

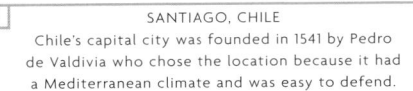

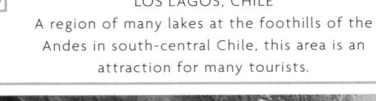

9 LAGO VIEDMA, ARGENTINA
Lago Viedma enjoys a milky-blue appearance due to the glacial sediment suspended in its waters.

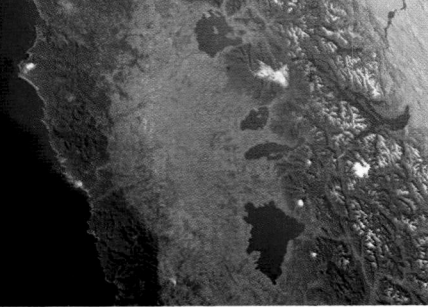

10 LOS LAGOS, CHILE
A region of many lakes at the foothills of the Andes in south-central Chile, this area is an attraction for many tourists.

11 ROSARIO, ARGENTINA
Located on the west bank of the Paraná river, Rosario lies at that heart of Argentina's industrial corridor, centered on the river.

12 RIVER PLATE, ARGENTINA/URUGUAY
Fed by the Paraná and Uruguay rivers, this Atlantic Ocean inlet separates Argentina and Uruguay.

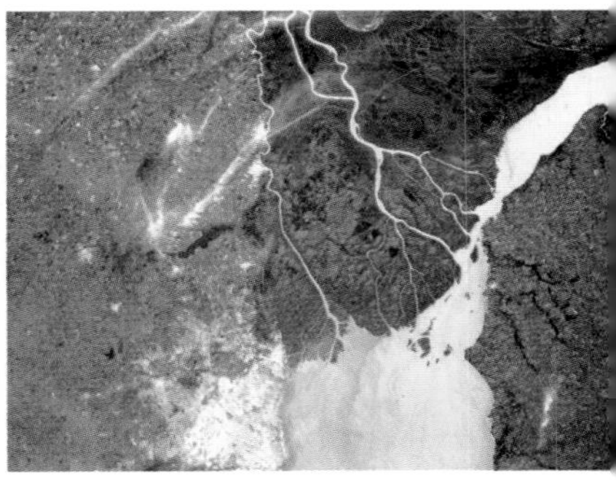

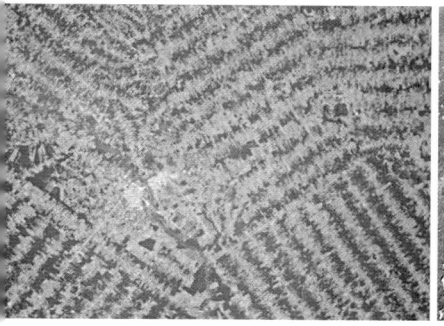

RONDÔNIA, BRAZIL 5
Pale strips of forest clearance can be seen along perpendicular tracks in this region of the Amazon Basin.

MARACAIBO, VENEZUELA 6
Maracaibo is the center of Venezuela's oil industry and its second largest city with a population of 1.6 million.

AMAZON RIVER/RIO NEGRO, BRAZIL 7
The dark, plant debris-stained waters of the Rio Negro join the beige Amazon near the city of Manaus.

EMBALSE DE GURI, VENEZUELA 8
This enormous reservoir, on the Caroni river, was completed in 1986 and its hydroelectric plant was the first to produce more than 10 gigawatts of electricity.

FOREST CLEARANCE IN SANTA CRUZ STATE, BOLIVIA 13
This infrared image shows the distinctive radial clearance patterns of original tropical dry forest with a small settlement at each center.

LAGOA DOS PATOS AND MIRIM LAGOON, BRAZIL/URUGUAY 14
These two lagoons are separated from the Atlantic Ocean by 248 miles (400 km) of sandbar.

ITAIPU DAM, BRAZIL/PARAGUAY 15
With an installed capacity of 14 gigawatts this is the world's largest hydroelectric power scheme, delivering 95% of Paraguay's energy needs and 24% of Brazil's.

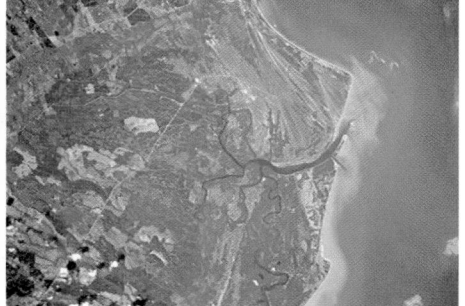

POINT BALEIA, BRAZIL 16
This headland has built up through steady accumulation of silt and sediment, shaped by tides and ocean currents.

Scale 1:6,500,000
(projection: Lambert Azimuthal Equal Area)

0 25 50 75 100 125 150 175 200 Km

0 25 50 75 100 125 150 175 200 Miles

Population

■ above 5 million   ▪ 1 million to 5 million   ◉ 500,000 to 1 million

◎ 100,000 to 500,000   ⊕ 50,000 to 100,000   ○ 10,000 to 50,000   ○ below 10,000

Bogotá

Suba

Usaquén

Engativá

Molinos

Aeropuerto Internacional El Dorado

Barrios Unidos

Chapinero

Fontibón

Teusaquillo

Bogotá

Kennedy

Puente Aranda

Los Mártires

• Museo Nacional de Colombia

Catedral •

La Candelaria

Bosa

Antonio Mariño

Santa Fe

Rafael Uribe

San Cristóbal

Tunjuelito

ATLANTIC OCEAN

CASTRIES
George F L Charles
Mount Gimie
950m
ST LUCIA
Hewanorra
Vieux Fort
Saint Vincent Passage
St Vincent
KINGSTOWN
Arnos Vale
BARBADOS
Bequia
BRIDGETOWN
Grantley Adams
ST VINCENT &
THE GRENADINES
Mustique
Canouan
Union Island
Carriacou
ST. GEORGE'S
Point Salines
GRENADA

la Blanquilla

Islas los Testigos

Tobago

Charlotteville

Isla de Margarita

Scarborough

Galera Point

VA ESPARTA

Juangriego La Asunción
Pampatar
Punta de
Porlamar
Piedras
Carúpano
Carúpano
Río Caribe
to
Cumaná
El Pilar
Trapa
Puerto de
Sangre Grande
Casanay
Güiria Hierro
Piaco
195m
Cumanacoa
San Fernando
San Antonio
Río Claro
SUCRE
Caripe
Point Fortin
Siparia
Rushville
celona
San Antonio
Bonasse
Galeota Point
Aragua de Maturín
Quiriquire

ATLANTIC
OCEAN

Ariima
PORT-OF-SPAIN
Trinidad
TRINIDAD
& TOBAGO
Gulf of Paria

Maturín
MONAGAS
Santa Rosa
Punta
de Mata
Anaco
Aguasay
aquin
Cantaura

San Tomé
San José de Guanipa
ANZOÁTEGUI
Ciudad Guayana
Soledad
dad Bolívar
Borbón
pire
Ciudad Piar
Embalse
de Guri
El Manteco

Pedernales
Punta Baja

Tucupita
DELTA AMACURO

Guayabones
Curiapo

Waini Point

Walni

Port Kaituma

Arakaka
Matthews
Ridge
Barama River

Charity

Spring Garden

Essequibo Islands

El Pao
El Palmar
Opata

Río Orinoco
Barrancas
Temblador
La Horqueta
Cano Mánaca

Río Cuchivero

Guasipati
Tumeremo
El Dorado

Kuracki
Aurora
Parika
Bartica
Chedd-Jagan
GEORGETOWN

Cerro Turagua
4838m
ancheras

Caruana de Montana

Cano Negro
Canaima

Enachu Landing

Peters Mine
Rockstone
Linden

Nieuw Amsterdam
Rose Hall

BOLÍVAR

La Paragua

Salto Angel
Auyán Tepuy
2950m
Uruyén
Cerro Venamo
1563m

Kamarang
Imbaimadai

Issano
Mahdia
Ituni

Nieuw Nickerie
Wageningen
Erlendship
Corriverton
Totness
WANICA
Paramaribo
PARAMARIBO
COMMEWIJNE

Cerro Guaiquinima
2100m

Mount Roraima
2810m

Ayanganna Mountain
2042m

Kaieteur Falls

GUYANA

Matarara River

Demerara River

Berbice River

Corantijn River

CORONIE
NICKERIE
Corneliskondre
Groningen
SARA-MACCA
PARA
Onverwacht
Nieuw Amsterdam
Albina
St-Laurent-du-Maroni
Iracoubo

Iles du Salut
Ile du Diable
Centre Spatial Guyanais
Kourou

Carapo

Santa María
Erebato

Uonán

Santa Elena de Uairén

Glendor
Mountains

Orealla
Wasjabo
Kwakoegron
Apoera
Donderkamp
Kaaimanston

Bergen Dal
Brokopondo
Brownsweg

MAROWIJNE

St-Jean
Apatou
Citron
Délices

Tonate
Cayenne
CAYENNE

adisocana

Catisimina

Pakaraima Mountains

Normandia

Kurupukari

Hendrik Top
957m

Tafelberg
1026m

W.J. van
Blommesteinmeer
Bergi

Pokigron

Grand-Santi
Herminadorp

St-Elie
Cacao

Rémire
Matoury

Sararina

Conceição do Maú

Lethem

Rupununi River

Lucie River

Boti-Pasi
Djoemoe

Poeketi

FRENCH
GUIANA
(to France)

Régina

Ouanary

Cabo Orange

Uaiacás
Urairicoera

Kanuku Mountains
Saudwaunawa

Juliana Top
1230m

Wilhelmina Gebergte

Apetina

Maripasoula
Saul

Buie de l'Oyapok

Boa Vista

Jacobs
Ladder
Falls

SURINAME

SIPALIWINI

Pédima

Montagnes
Bellevue de l'Inini

L'Oyapok River

Pointe Béhague

RORAIMA

Santa Rosa

Conceição do Maú

Essequibo River

New River
(Claimed by Suriname)

Appikalo

Alimimuni Pick
728m

Tumuc-Humac Mountains

Mont Saint-Marcel
635m

Massif du Mitaraka
690m

Trois
Sauts

Rio Oiapoque

Calçoene

290

Orinoco
Horqueta Minas
de Unturán

Rio Catrimoni

Kuyuwini
Landing

Johi Village

(Venezuela claims all of Guyana
west of the Essequibo river)

Acarai Mountains

(Claimed by Suriname)

AMAPÁ

Amapá

rapeco

Missão Catrimani

Caracarai

São Luís

Serra do Jatapu

Sete Ilhas

Rio Araguari

Catrimani

Rio Demini

Rio Branco

Rio Jauaperi

Boiaçu

Represa
Balbina

Rio Trombetas

Rio Nhamundá

Rio Paru do Oeste

Rio Paru

Planalto
Maracanaquará

Monte Dourado

Macapá
Equator

Ilha Grande
de Gurupá

o Negro

Tapurucuará

Barcelos

Moura

PARÁ

Oriximiná
Óbidos

Alenquer

Amazon

Porto de Moz

Portel

B R A Z I L

AMAZONAS

Novo Airão

Eduardo
Gomes

Manacapuru
Caldeirão

Iranduba

Itacoatiara

Parintins

Santarém

Rurópolis Presidente

Altamira

Codajás
Beruri

Manaus

Manaquiri

Autazes

Careiro

Amazon Basin

Coari

Rio Purus

Itaituba
Pimenta

Rio Iriri

107

19,686ft
13,124ft
9843ft
6562ft
3281ft
1640ft
820ft
328ft
Sea Level
-820ft
-6562ft
-13,124ft

290

SOUTH AMERICA

106

102

287

VENEZUELA

COLOMBIA

GUAVIARE

GUAINÍA

VAUPÉS

CAQUETÁ

PUTUMAYO

AMAZONAS

B r a s i l B a s i n

B R A Z I L

AMAZONAS

ACRE

PANDO

MADRE DE DIOS

UCAYALI

A m a z o n

Río Napo

Río Napo

Río Marañón

Río Juruá

Río Purus

ECUADOR

ESMERALDAS

MANABÍ

CARCHI

IMBABURA

PICHINCHA

COTOPAXI

LOS RÍOS

GUAYAS

BOLÍVAR

TUNGURAHUA

CHIMBORAZO

CAÑAR

AZUAY

EL ORO

LOJA

ZAMORA CHINCHIPE

MORONA SANTIAGO

NAPO

SUCUMBÍOS

ORELLANA

PASTAZA

LORETO

P E R U

TUMBES

PIURA

LAMBAYEQUE

CAJAMARCA

AMAZONAS

SAN MARTÍN

LA LIBERTAD

ANCASH

HUÁNUCO

PASCO

JUNÍN

LIMA

Cordillera Blanca

Cordillera Azul

C o r d i l l e r a   A n d e s

Quito

Guayaquil

Cuenca

Loja

Machala

Esmeraldas

Manta

Portoviejo

Santo Domingo de los Colorados

Ambato

Riobamba

Latacunga

Bahahoyo

Iquitos

Pucallpa

Leticia

Benjamín Constant

Tabatinga

Río Branco

Trujillo

Chiclayo

Chimbote

Piura

Sullana

Tumbes

Paita

Talara

Lima

Callao

Equator

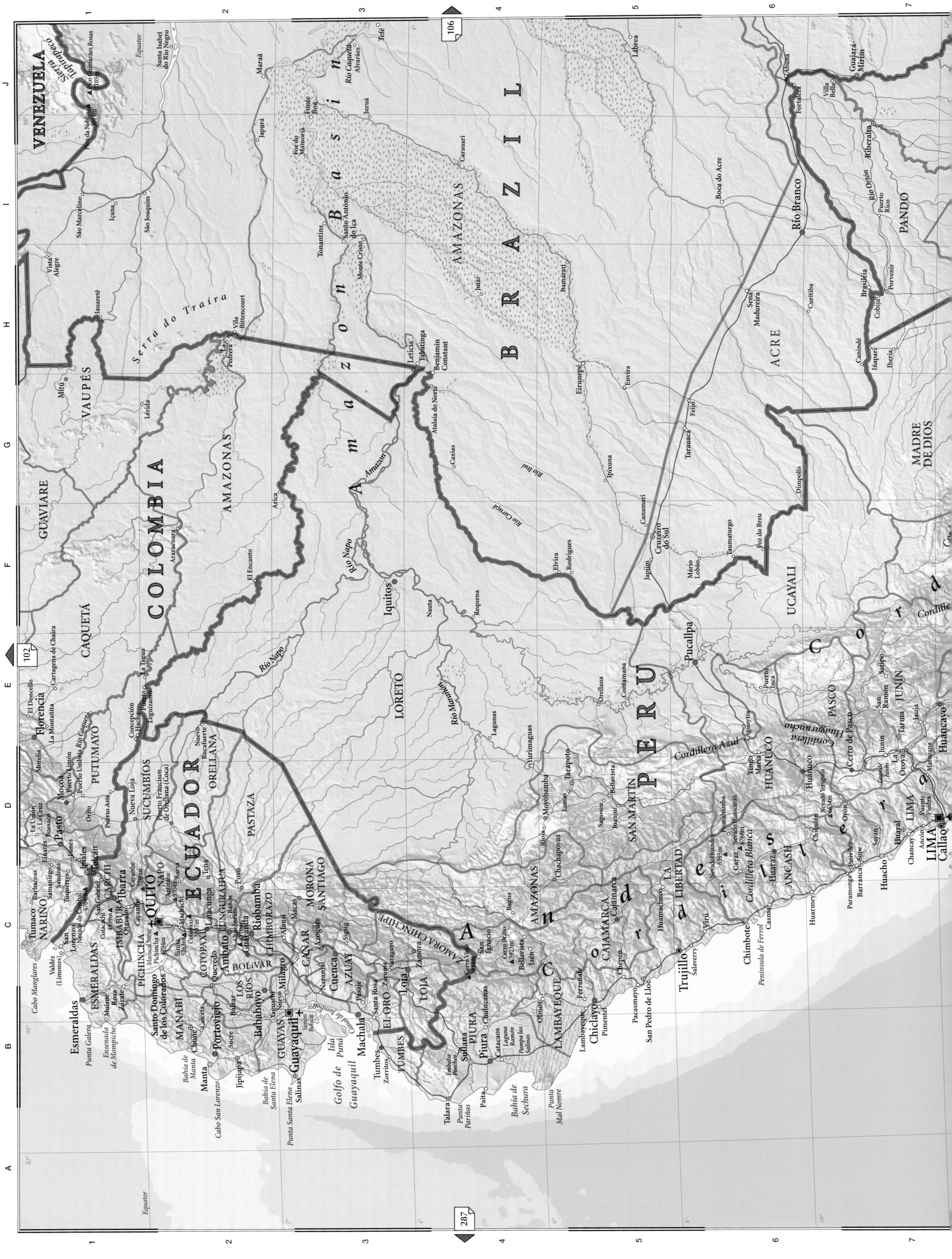

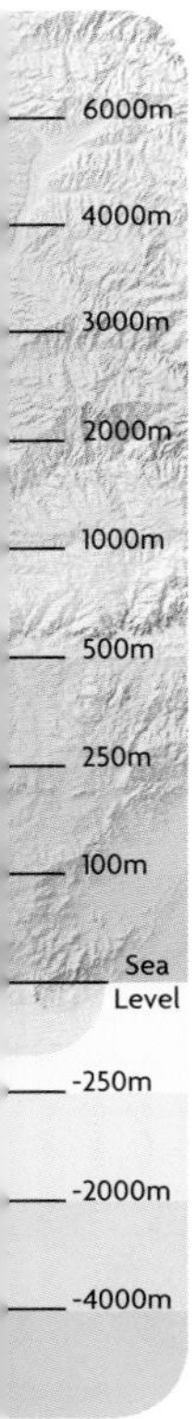

Scale 1:6,500,000
(projection: Lambert Azimuthal Equal Area)

| 0 | 25 | 50 | 75 | 100 | 125 | 150 | 175 | 200 Km |
| 0 | 25 | 50 | 75 | 100 | 125 | | 175 | 200 Miles |

Population

- ■ above 5 million
- ◉ 1 million to 5 million
- ◉ 500,000 to 1 million
- ◎ 100,000 to 500,000
- ⊕ 50,000 to 100,000
- ○ 10,000 to 50,000
- ○ below 10,000

Map labels

BOLIVIA
PERU
CHILE
ARGENTINA

EL BENI
LA PAZ
ORURO
POTOSÍ
CHUQUISACA
TARIJA
COCHABAMBA
CUSCO
APURÍMAC
AYACUCHO
HUANCAVELICA
ICA
PUNO
MOQUEGUA
AREQUIPA
TACNA
ARICA Y PARINACOTA
TARAPACÁ
ANTOFAGASTA
ATACAMA
JUJUY
SALTA
TUCUMÁN
CATAMARCA

La Paz
Sucre
Cochabamba
Oruro
Potosí
Tarija
Cusco
Arequipa
Puno
Juliaca
Tacna
Arica
Iquique
Antofagasta
Salta
San Salvador de Jujuy
San Miguel de Tucumán

Lake Titicaca

PACIFIC OCEAN

Tropic of Capricorn

Galápagos Islands inset

Galápagos Islands
(Archipiélago de Colón)
(to Ecuador)

Isla Pinta
Isla Marchena
Isla Genovesa
Isla San Salvador
Isla Santa Cruz
Isla San Cristóbal
Isla Española
Isla Santa María
Isla Isabela
Isla Fernandina
Isla Santa Fe
Isla Pinzón
Puerto Ayora
Puerto Baquerizo Moreno
Puerto Villamil

Volcán Wolf 1646m
Volcán Darwin 1280m
Volcán Alcedo 1097m
Volcán La Cumbre 1463m
Volcán Santo Tomás 1490m

Equator

PACIFIC OCEAN

Scale 1:7,750,000

| 0 | 25 | 50 | 75 | 100 |

Bolivia capital cities inset

BOLIVIA: CAPITAL CITIES

LA PAZ – seat of government
SUCRE – legal capital

La Paz inset

La Paz

Palacio de Comunicaciones
Museo Nacional de Arte
Museo Nacional de Arqueología de Bolivia
Museo de Marina Núñez del Prado
El Alto International Airport
El Alto
Tupac Katari
Villa Adela
Caluyo
Santa Rosa
Magisterio
Irpaví
Los Pinos

Río Choqueyapu
Río Seco

| 0 | 3 Km |
| 0 | 3 Miles |

Lima inset

Lima

Los Olivos
San Juan de Lurigancho
Independencia
Santa Anita
Pederos
Vitarte
La Molina
Universidad de Lima
Hipódromo de Monterrico
Villa María Del Triunfo
Villa El Salvador
Santiago de Surco
Chorrillos
Miraflores
San Isidro
Magdalena del Mar
La Victoria
San Borja
San Luis
Estadio Nacional
Congreso de la República del Perú
Univ. Nacional de Ingeniería
Rímac
Cerro Altillo
Callao
Bocanegra
Aeropuerto Internacional Jorge Chávez
Museo Histórico del Real Felipe
Palacio del Gobierno
San Martín de Porras
Matasango
Playa Agua Dulce
Playa Villa
Morro Solar

Río Rímac

Pacific Ocean

| 0 | 5 Km |
| 0 | 5 Miles |

Elevation scale

19,686ft
13,124ft
9843ft
6562ft
3281ft
1640ft
820ft
328ft
Sea Level
-820ft
-6562ft
-13,124ft

Amazon Basin

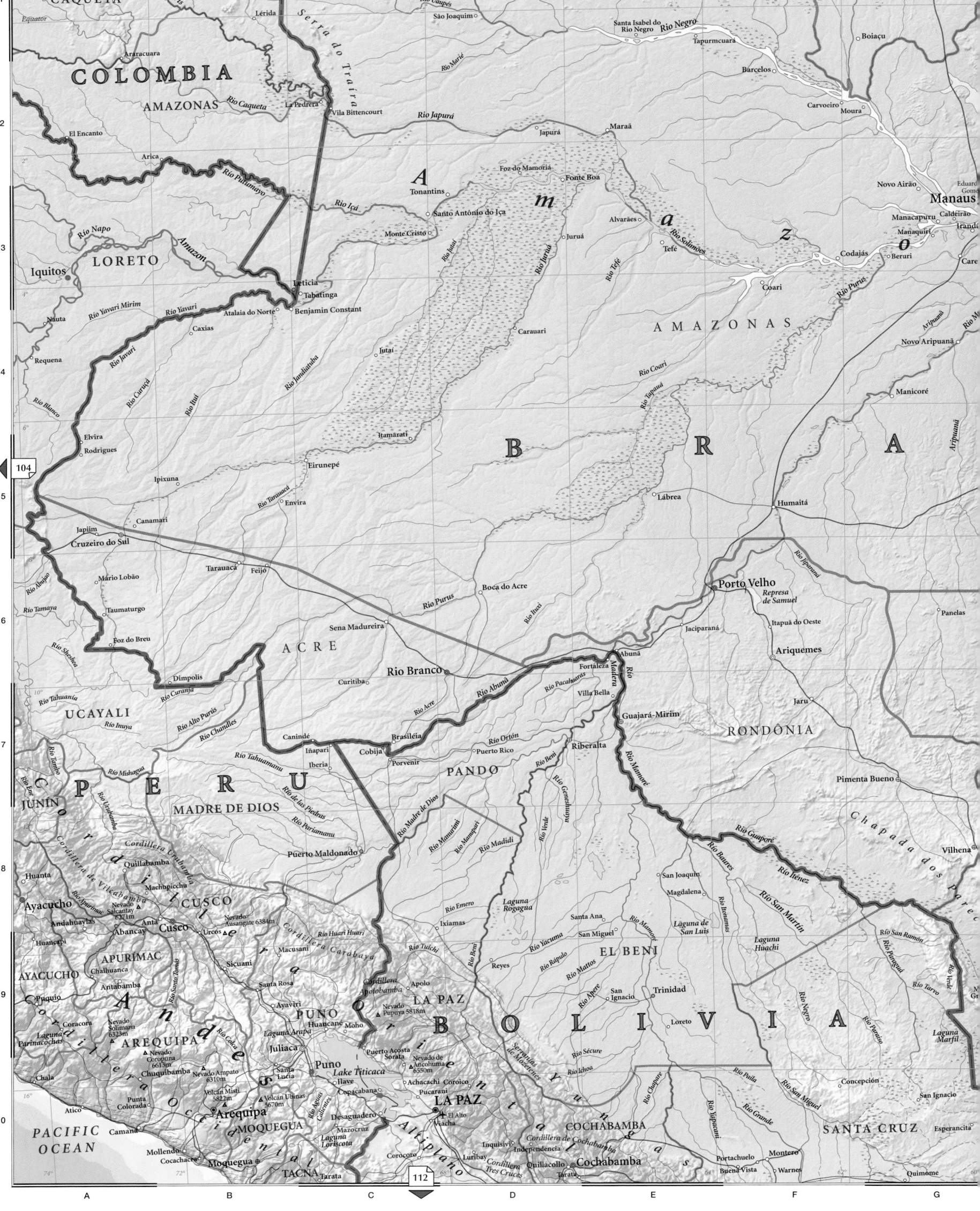

102

104

112

RORAIMA

COLOMBIA

CAQUETÁ

VAUPÉS

AMAZONAS

LORETO

Iquitos

Equator

PERU

JUNIN

UCAYALI

MADRE DE DIOS

CUSCO

APURIMAC

AYACUCHO

AREQUIPA

MOQUEGUA

TACNA

PACIFIC
OCEAN

PUNO

LA PAZ

EL BENI

BOLIVIA

COCHABAMBA

SANTA CRUZ

RONDÔNIA

PANDO

ACRE

B R A
A M A Z O N A S

Manaus

Porto Velho

Rio Branco

Amazonas

Cruzeiro do Sul

6000m
4000m
3000m
2000m
1000m
500m
250m
100m
Sea
Level
-250m
-2000m
-4000m

Scale 1:6,500,000
(projection: Lambert Azimuthal Equal Area)

0 25 50 75 100 125 150 175 200 Km
0 25 50 75 100 125 150 175 200 Miles

Population
- ■ above 5 million
- ◉ 100,000 to 500,000
- ■ 1 million to 5 million
- ⊕ 50,000 to 100,000
- ◉ 500,000 to 1 million
- ○ 10,000 to 50,000
- ○ below 10,000

103

113

108

ATLANTIC OCEAN

Equator

AMAPÁ

Sete Ilhas
Ilha Bailique
Ilha do Curuá
Ilha Janaucu
Ilha Caviana de Fora
Ilha Mexiana

Mouths of the Amazon

Macapá

Planalto
Maracanaquará

Monte Dourado

Ilha Grande de Gurupá

Ilha de Marajó

Baía de Marajó

Marudá
Vigia

Belém
Castanhal
Capanema
Viseu
Carutapera
Turiaçu
Alto Bonito

Rio Araguari
Rio Jari

Amazon

Oriximiná
Óbidos
Alenquer
Porto de Moz
Portel

Ilha Siriuba

Tomé-Açu

Rio Trombetas
Rio Paru de Oeste
Rio Nhamundá

Urucará
Parintins
Itacoatiara

Santarém

Altamira

B **a** **s** **i** **n**

Rio Tocantins

Rio Gurupi

Serra do Tiracambu

MARANHÃO

Dom Eliseu

Açailândia

Imperatriz

Grajaú

Rio Gnapu

Itaituba
Pimenta

Rurópolis Presidente Medici

PARÁ

Represa de Tucuruí

São Félix
Marabá

Tucunaré

Rio Tapajós
Rio Jamanxim
Rio Iriri

José Rodrigues

Serra dos Carajás

Parauapebas

Estreito

São Raimundo das Mangabeiras

Jacaré-a-Canga

Araras
Bom Futuro

Z

B **R** **I** **A** **Z** **I** **L**

São Félix do Xingu

Rio Xingu

Araguaína

Carolina

Balsas

Manuel Zinho

Barra do São Manuel

Serra do Cachimbo

Recreio
Pereirinha

Serra dos Gradaús

Conceição do Araguaia

Craolândia

Tasso Fragoso

Chapada das Mangabeiras

Rio São Manuel

Bandeirantes

Cachimbo

Rio Araguaia

Alto Parnaíba
Santa Filomena

PIAUÍ

Paranaíta

Vila Rica

Rio Tocantins

TOCANTINS

Palmas do Tocantins

Corrente

Juruena

Peixoto de Azevedo

Porto Nacional

Serra Geral de Goiás

Espigão Mestre

Rio Juruena

Juará
Novo Horizonte

Marcelândia
Campo de Diauarum

São Félix do Araguaia

Ilha do Bananal

Gurupi

Porto dos Gaúchos

Serra Formosa

Sinop

Pôsto Jacaré

Rio Arinos
Rio do Sangue

Rio das Mortes

Serra do Roncador

Taguatinga

BAHIA

Campos Belos

MATO GROSSO

Porangatu

Planalto de
Mato Grosso

Cocalinho

Alto Paraíso de Goiás

Arenápolis
Nobres
Rosário Oeste

Rio Manso

Itacaiu

Uruaçu

Tupiraçaba

Serra Dourada

Pontes e Lacerda

Várzea Grande
Cuiabá

Ceres
Rialma
Barro Alto
Goianésia

GOIÁS

Cáceres

Jaciara

Pirenópolis

BRASÍLIA

Planaltina

DISTRITO FEDERAL

Rio São Francisco

San Matías

Rondonópolis

Aragarças
Goiás

Anápolis

Unaí

MINAS GERAIS

Pantanal

Laguna Uberaba

Rio Cuiabá
Rio Piquiri

Piranhas

Alto Araguaia
Santa Rita de Araguaia
Mineiros

Indiara

Goiânia

Cristalina

Paracatu

Planalto Central

Rio São Lourenço

19,686ft
13,124ft
9843ft
6562ft
3281ft
1640ft
820ft
328ft
Sea Level
-820ft
-6562ft
-13,124ft

SOUTH AMERICA

108

ATLANTIC OCEAN

Equator

Mouths of the Amazon

Amazon Basin

AMAPÁ

PARÁ

MARANHÃO

PIAUÍ

CEARÁ

RIO GRANDE DO NORTE

PARAÍBA

PERNAMBUCO

ALAGOAS

SERGIPE

TOCANTINS

Planalto da Borborema

Serra Grande

Chapada das Mangabeiras

Serra Geral de Goiás

Serra dos Grades

Serra do Tiracambu

Serra dos Carajás

Espigão

BRAZIL

6000m
4000m
3000m
2000m
1000m
500m
250m
100m
Sea Level
-250m
-2000m
-4000m

Macapá
Belém
Castanhal
Capanema
Vigia
Viseu
Turiaçu
São Luís
São João de Cortes
Parnaíba
Fortaleza
Caucaia
Camocim
Acaraú
Sobral
Itapipoca
Araras
Crateús
Quixadá
Natal
Ceará Mirim
Touros
Macau
Mossoró
Assú
Areia Branca
Currais Novos
Caicó
Aracati
Cascavel
João Pessoa
Campina Grande
Olinda
Recife
Jaboatão
Caruaru
Arcoverde
Garanhuns
Arapiraca
Maceió
Aracaju
São Cristóvão
Estância
Propriá
Petrolina
Juazeiro
Paulo Afonso
Salgueiro
Ouricuri
Juazeiro do Norte
Campos Sales
Marcolândia
Picos
Oeiras
Floriano
Teresina
Timon
Caxias
Codó
Bacabal
Presidente Dutra
Colinas
São João dos Patos
São Raimundo das Mangabeiras
Balsas
Carolina
Estreito
Imperatriz
Açailândia
São Félix
Marabá
Araguaína
Palmas
Conceição do Araguaia
Xique-Xique
Barra
Piripiri
Campo Maior
Barro Duro
Valença do Piauí
Gaturiano
Canto do Burití
Alto Parnaíba
Corrente
Santa Filomena
Tasso Fragoso
Grajaú
Roncador
Chapadinha
Itapecuru-Mirim
Senador Pompeu
Taua
Afrânio
Sobradinho
Santa Rita de Cássia
Campo Alegre de Lourdes
Mazagão
Sento Sé
Crixás
Carolândia

Rio Parnaíba
Rio Itapicuru
Rio Gurupi
Rio Grajaú
Rio Mearim
Rio Tocantins
Rio Araguaia
Rio Xingu
Rio São Francisco
Rio Anguaia

Ilha de Marajó
Baía de Marajó
Ilha do Caju
Baía de São Marcos
Recife Manuel Luís
Recife do Silva
Atol das Rocas
Cabo de São Roque

Açude Banabuiú
Açude Orós
Represa de Itaparica
Represa de Sobradinho
Represa de Tucuruí
Açude Poço da Cruz

290

291

107

Scale 1:6,500,000
(projection: Lambert Azimuthal Equal Area)

0 25 50 75 100 125 150 175 200 Km
0 25 50 75 100 125 150 175 200 Miles

Population

| ■ above 5 million | ▣ 1 million to 5 million | ◉ 500,000 to 1 million |
| ◎ 100,000 to 500,000 | ⊕ 50,000 to 100,000 | ○ 10,000 to 50,000 | ○ below 10,000 |

291

113

19,686ft
13,124ft
9843ft
6562ft
3281ft
1640ft
820ft
328ft
Sea Level
-820ft
-6562ft
-13,124ft

ATLANTIC OCEAN

Salvador
Baía de Todos os Santos
Ilha de Boipeba
Ponta do Mutá
Maraú
Valença
Itaberaba
Jequié
Ilhéus
Itabuna
Comandatuba
Canavieiras
Belmonte
Santa Cruz Cabrália
Porto Seguro
Prado
Caravelas
Ilha Caçumba
Lençóis
Itapetinga
Itapetinga
Rio Jequitinhonha
Eunápolis
Itamaraju
São Mateus
Chapada Diamantina
Bom Jesus da Lapa
Gaeté
Brumado
Vitória da Conquista
Pedra Azul
Nanuque
Espírito Santo
Linhares
Santa Maria da Vitória
Espinosa
Monte Azul
Itaobim
Aracuaí
Teófilo Otoni
Governador Valadares
Rio Doce
Colatina
Santa Teresa
Vitória
Guarapari
Cachoeiro de Itapemirim
ESPÍRITO SANTO
São Mateus
Januária
Jamaúba
Diamantina
Serro
Pico da Bandeira 2890m
São João da Barra
Pirapora
MINAS GERAIS
Pico do Itambé 2060m
Ipatinga
Ponte Nova
Manhuaçu
Bom Jesus do Itabapoana
Campos
Montes Claros
Curvelo
Guanhães
Conselheiro Lafaiete
Barbacena
Juiz de Fora
Três Rios
Miracema
Itaperuna
Macaé
Cordeiro
Nova Friburgo
RIO DE JANEIRO
Arraial do Cabo
Cabo Frio
Brazilian Highlands (Planalto Central)
Sete Lagoas
Belo Horizonte
Betim
Ouro Preto
São João del Rei
Teresópolis
Petrópolis
São Gonçalo
Niterói
Rio de Janeiro
Rio São Francisco
Abaeté
Três Pontas
Pouso Alegre
Barra Mansa
Volta Redonda
Angra dos Reis
Ilha Grande
Ilha de São Sebastião
Boa Vinopolis
Represa Três Marias
Passos
Poços de Caldas
São José do Rio Preto
Represa de Furnas
Guaratinguetá
São José dos Campos
Caraguatatuba
Ubatuba
Ilha de São Sebastião
Brasília
DISTRITO FEDERAL
Unaí
Ibiá
Araxá
Franca
Represa de Furnas
Mococa
Casa Branca
Piracaia
Atibaia
Campinas
Jundiaí
Osasco
São Paulo
Santos
São Vicente
Ilha de Santo Amaro
Alto Paraíso de Goiás
Cristalina
Patos de Minas
Uberlândia
Uberaba
Ribeirão Preto
Batatais
Sertãozinho
São Joaquim da Barra
Guará
Igarapava
Araraquara
Matão
Limeira
Americana
Indaiatuba
Sorocaba
Itapira
Mogi-Mirim
Amparo
Bragança Paulista
São Caetano do Sul
Pedro Barros
Registro
Iguape
Jacupiranga
Ilha Comprida
BRASÍLIA
GOIÁS
Goiás
Anápolis
Pirenópolis
Anhanguera
Araguari
Rio Grande
Rio Paranaíba
São Carlos
Jaú
Bauru
Rio Tietê
Piracicaba
Botucatu
Tatuí
Itapetininga
Capão Bonito
Apiaí
Ilha do Mel
Antonina
Paranaguá
SÃO PAULO
Goiânia
Ceres
Porangatu
Uruaçu
Barro Alto
Goianésia
Rialma
Itumbiara
Anápolis
Araguari
Itumbiara
Serra dos Pireneus
Serra Dourada
MATO GROSSO DO SUL
Indiara
Rio Verde
Represa de Água Vermelha
São José do Rio Preto
Catanduva
Novo Horizonte
Araraquara
Jaboticabal
Taquaritinga
Ibitinga
Marília
Garça
Santa Cruz do Rio Pardo
Avaré
Ourinhos
Itapeva
Itararé
Jaguariaíva
Castro
Ponta Grossa
Lapa
Campo Largo
Curitiba
São Mateus do Sul
Rio Negro
Mafra
Itaici
Caçu
Jataí
Piranhas
Aragarças
Barra do Garças
Mineiros
Rio Verde
Serra do Caiapó
Rio Aporé
Represa de Ilha Solteira
Santa Fé do Sul
Jales
Fernandópolis
Votuporanga
Nhandeara
Araçatuba
Penápolis
Birigüi
Lins
Promissão
Novo Horizonte
Tupã
Garça
Pompéia
Cândido Mota
Londrina
Maringá
Apucarana
Serra Geral
Ivaiporã
PARANÁ
Ibaiti
Arapongas
Cornélio Procópio
Jacarezinho
Rio Itararé
Tibagi
Telêmaco Borba
Ponta Grossa
Pinhão
Prudentópolis
Guarapuava
Represa de Foz do Areia
Três Lagoas
Andradina
Pereira Barreto
Mirandópolis
Adamantina
Dracena
Presidente Prudente
Presidente Venceslau
Presidente Epitácio
Teodoro Sampaio
Rio Paraná
Rio Paranapanema
Paranavaí
Nova Esperança
Cruzeiro do Oeste
Campo Mourão
Goioerê
Cianorte
Umuarama
Ivaí
Rio Ivaí
Rio Piquiri
Guarapuava
Lapa
Irati
Três Pinheiros
Pato Branco
Dois Vizinhos
Tropic of Capricorn
MATO GROSSO
Serra do Roncador
Rio das Mortes
Cocalinho
Itacaiú
Cocalinho
Aragarças
Água Clara

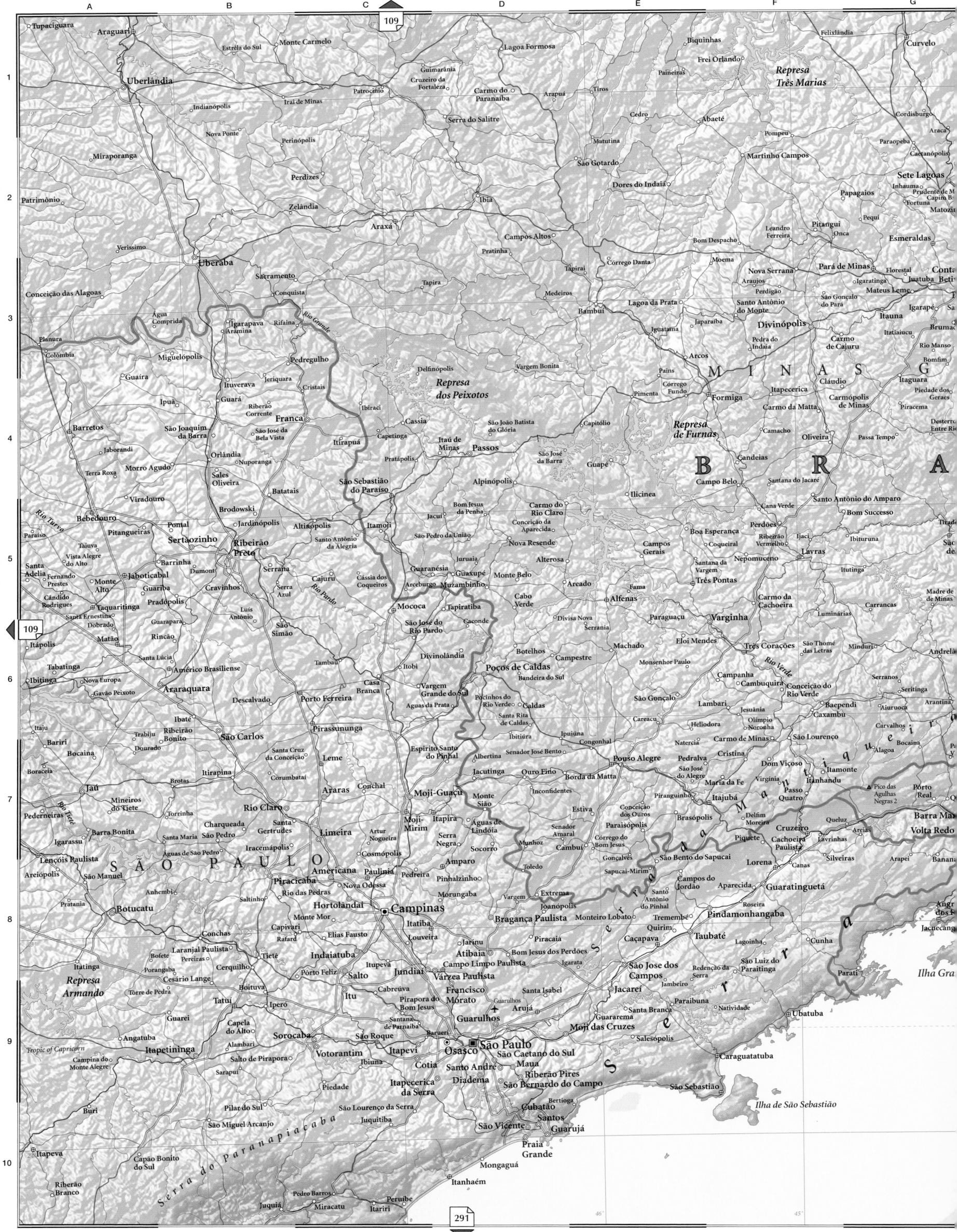

Central South America

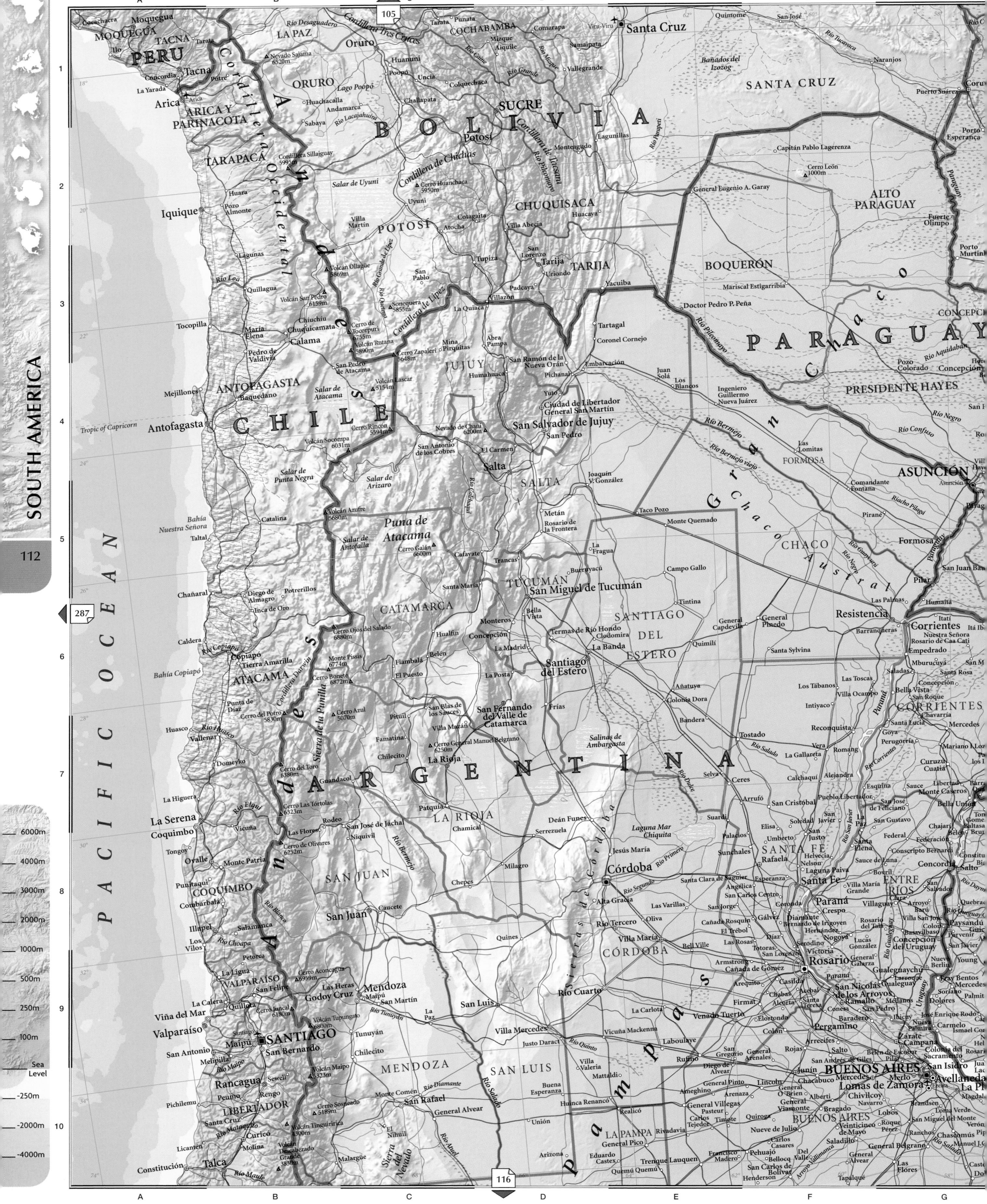

SOUTH AMERICA

Scale 1:6,500,000
(projection: Lambert Azimuthal Equal Area)

0 25 50 75 100 125 150 175 200 Km
0 25 50 75 100 125 150 175 200 Miles

Population

- ■ above 5 million
- ▣ 1 million to 5 million
- ◉ 500,000 to 1 million
- ◎ 100,000 to 500,000
- ⊕ 50,000 to 100,000
- ○ 10,000 to 50,000
- ∘ below 10,000

109

291

291

ATLANTIC

OCEAN

Tropic of Capricorn

Brazil / region labels

GOIÁS

MINAS GERAIS

MATO GROSSO DO SUL

SÃO PAULO

B R A Z I L

PARANÁ

SANTA CATARINA

RIO GRANDE DO SUL

RIO DE JANEIRO

URUGUAY

Pantanal

Serra do Mar

Selected cities

Belo Horizonte, São Paulo, Rio de Janeiro, Campinas, Curitiba, Porto Alegre, Florianópolis, Uberlândia, Ribeirão Preto, Londrina, Maringá, Campo Grande, Santos, Niterói, Nova Iguaçu, São Gonçalo, Montevideo, Ciudad del Este, Posadas, Encarnación, Pelotas, Rio Grande, Bagé, Criciúma, Blumenau, Joinville, Caxias do Sul, Novo Hamburgo, Canoas, Gravataí, Santa Maria, Passo Fundo, Ponta Grossa, Juiz de Fora, Petrópolis, Teresópolis, Volta Redonda, Barra Mansa, Taubaté, São José dos Campos, Jundiaí, Sorocaba, Bauru, Araraquara, São Carlos, Limeira, Piracicaba, Marília, Presidente Prudente, Araçatuba, São José do Rio Preto, Franca, Barretos, Araguari, Uberaba, Divinópolis, Ouro Preto, Barbacena, Governador Valadares, Ipatinga, Itabira, Conselheiro Lafaiete, Diamantina, Piripora, Luislândia do Oeste

Buenos Aires (inset)

Buenos Aires

Las Conchas, Tigre, Pilar, Garín, San Isidro, Vicente López, Belgrano, Palermo, Hippodrome, Zoo, Teatro Colón, Cathedral, Plaza de Mayo, San Miguel, General San Martín, Moreno, Sáenz Peña, Floresta, Barracas, Merlo, San Justo, Villa Madero, Avellaneda, Villa Alsina, Quilmes, Lanús, Lomas de Zamora, Berazategui, Florencio Varela, Longchamps, Mariano Acosta, Pontevedra, González Catán, Morón, Aeropuerto Internacional de Ezeiza

Río de la Plata

Rectificación del Riachuelo

0 10 Km
10 Miles

Elevation scale

19,686ft
13,124ft
9843ft
6562ft
3281ft
1640ft
820ft
328ft
Sea Level
-820ft
-6562ft
-13,124ft

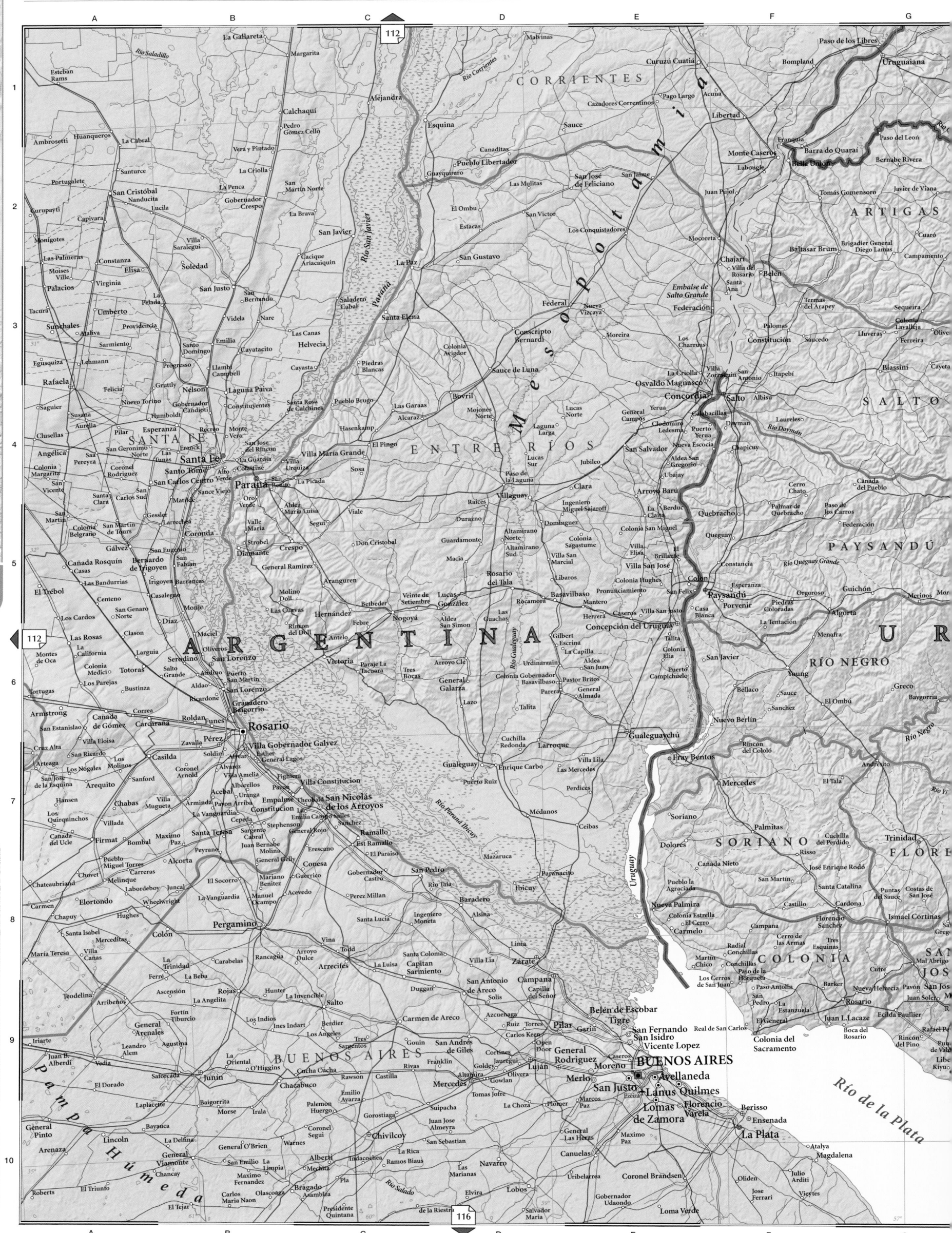

112
116

Scale 1:2,000,000
(projection: Lambert Conformal Conic)

0 10 20 30 40 50 60 70 80 Km

0 10 20 30 40 50 60 70 80 Miles

Population
- ■ above 5 million
- ▪ 1 million to 5 million
- ◉ 500,000 to 1 million
- ◎ 100,000 to 500,000
- ⊕ 50,000 to 100,000
- ○ 10,000 to 50,000
- ○ below 10,000

113

116

291

Elevation scale
- 19,686ft
- 13,124ft
- 9843ft
- 6562ft
- 3281ft
- 1640ft
- 820ft
- 328ft
- Sea Level
- -820ft
- -6562ft
- -13,124ft

BRAZIL

RIO GRANDE DO SUL

Serra das Encantadas

Santa Cruz do Sul
Santa Maria
Silveira Martins
Águdo
Candelária
General Câmara
Triunfo
Charqueadas
São Jerônimo
Passo do Sobrado
Mina do Loao
Butiá
Rio Pardo
Cachoeira do Sul
Capane
Cordilheira
Pántano Grande
Capivarita
Mariana Pimentel
Barão do Triunfo
Quitéria
Cerro Grande
Sertão de Santana
Camaquã
Dom Feliciano
Rio Camaquã
Encruzilhada do Sul
Boqueirão
São Lourenço do Sul
Canguçu
Cerrito Alegre
Quilombo
Pacheca

Passo Novo
Jacaquá
Rio Ibicuí
Loreto
São Vicente do Sul
Restinga Seca
Três Vendas
Ferreira
Barro Vermelho
Formigueiro
São Sepé
Cacequi
Azevedo Sodré
Vacacaí
Tiaraju
São Gabriel
Cacapava do Sul
Lavras do Sul
Dom Pedrito
Santana da Boa Vista
Pedras Altas
Candiota
Piratini
Pinheiro Machado
Capao do Leao
Alegrete
Rosário do Sul
Pampeiro
Santana do Livramento
Palomas
Rivera
Masoller
Tranqueiras
Zanja Honda
Quintana
Paso del Cerro
Banado de Rocha
Cerro Pelado del Este
Lapuente
Cerrillada
Tacuarembó
Los Rosanos
Pueblo de Arriba
Ansina
Minas de Corrales
Caraguatá
Vichadero
Arroyo Blanco
Abrojal
Zanja Honda
Coronilla
Acegua
Maria Isabel
Isidoro Noblia
Cruz de Piedra
Buena Vista
Pedreiras
Lagoa dos Patos
Pelotas
Cerrito
Quinta
São José de Norte
Rio Grande
Estreito
Cassimo

Tambores
Quiebra Yugos
Zapata
Los Toros
Rincón del Bonete
Peralta
Achar
Curtina
Clara
Blanquillo
Cerro Convento
Blanquillo
La Paloma
Cerrezuelo
Capilla Farruco
Carpintería
Santa Clara de Olimar
Tupambaé
Paso Pereira
Rio Negro
Arévalo
Fraile Muerto
Cerro de las Cuentas
Toledo
Tres Islas
Melo
La Pedrera
Banado de Medina
CERRO LARGO
Uruguay
Rio Yaguarón
Jaguarão
Rio Branco
Arbolito
Placido Rosas
Rincón
Mirim Lagoon
Taim

URUGUAY
Embalse del Río Negro
San Gregorio de Polanco
Verdún
Cardoso
Ombúes de Oribe
Carlos Reyles
DURAZNO
Cerro Chato
Sarandí del Yí
Capilla del Sauce
José Batlle y Ordoñez
Nico Pérez
Zapicán
José Pedro Varela
Maria Albina
Villa Sara
Julio Maria Sanz
Mendizabal
TREINTA Y TRES
General Enrique Martínez
Treinta y Tres
Cebollatí
Arrozal Victoria
La Coronilla
Santa Vitória do Palmar
Lagoa Mangueira

Cuchilla Grande
Valentines
Illescas
Sarandí Grande
FLORIDA
Pintado
La Cruz
Reboledo
Florida
Mendoza Chico
Chamizo
San Ramón
Tala
Solís
Minas
Pirarajá
Rio Cebollati
Lascano
Diez y Ocho de Julio
Chuí
Chuy
Maria Isabel
Velázquez
LAVALLEJA
Polanco
Polanco Sur
Mariscala
Los Talas
ROCHA
La Coronilla
Aigua
Villa Serrana
Parallé
Castillos
La Barra
Cabo Polonio
Rocha
La Paloma

Mendoza
San Antonio
San Bautista
Casupá
Fray Marcos
Bolívar
Veintcinco de Agosto
Santa Lucía
Santa Rosa
San Jacinto
CANELONES
Migues
Montes
Tapia
Solís de Matajo
Gregorio Aznarez
Nueva Carrara
Pan de Azúcar
San Carlos
MALDONADO
Independencia
Corrientes
Canelones
Los Cerrillos
Cruz de los Caminos
Totoral
Piedra del Toro
Soca
La Querencia
Las Flores
Piedras de Afilar
Pando
Joaquín Suárez
La Paz
Toledo
Las Piedras
Carrasco
Barra de Carrasco
Piriápolis
Maldonado

MONTEVIDEO

Punta del Este

ATLANTIC OCEAN

TACUAREMBÓ
RIVERA
Santana
Cordilheira
Bagé
Pintado Grande
Artigas
Rio Cuareim
Urupey Grande
Masoller
Quinta

Southern South America

SOUTH AMERICA

116

BRAZIL

URUGUAY

CORRIENTES

ENTRE RIOS

SANTIAGO DEL ESTERO

CATAMARCA

LA RIOJA

SAN JUAN

SAN LUIS

CÓRDOBA

SANTA FE

MENDOZA

LA PAMPA

BUENOS AIRES

RIO NEGRO

NEUQUÉN

ARGENTINA

CHILE

ATACAMA

COQUIMBO

VALPARAISO

LIBERTADOR

MAULE

BIO BIO

ARAUCANIA

LOS RIOS

LOS LAGOS

Andes

RIO GRANDE DO SUL

MONTEVIDEO

BUENOS AIRES

Avellaneda

La Plata

Mar del Plata

Córdoba

Rosario

Santa Fe

Paraná

Mendoza

San Juan

San Luis

La Rioja

Santiago del Estero

Corrientes

Posadas

SANTIAGO

Valparaíso

Viña del Mar

Concepción

Talcahuano

Temuco

Valdivia

Neuquén

General Roca

Viedma

Río de la Plata

PACIFIC OCEAN

ATLANTIC OCEAN

Sea Level
6000m
4000m
3000m
2000m
1000m
500m
250m
100m
-250m
-2000m
-4000m

291
112
287

Scale 1:6,500,000
(projection: Lambert Azimuthal Equal Area)

0 25 50 75 100 125 150 175 200 Km
0 25 50 75 100 125 150 175 200 Miles

Population
- ■ above 5 million
- ■ 1 million to 5 million
- ● 500,000 to 1 million
- ◉ 100,000 to 500,000
- ⊕ 50,000 to 100,000
- ○ 10,000 to 50,000
- ○ below 10,000

291
292
287

Montevideo

Joaquín Suárez
Toledo
Villa García
Aeropuerto de Carrasco
Ciudad de la Costa
Flor de Maroñas
Villa Española
Las Piedras
La Paz
Aeródromo Melilla
Villa García
Museo de Arte Contemporáneo
Estadio Centenario
Santiago Vázquez
Cerro
Capurro
Terminal Marítima
Iglesia Matriz
Ciudad Vieja
Museo
Montevideo
Pajas Blancas
Punta de Lobos
Punta Carretas
Delta del Tigre
Delta del Tigre
Punta Pajas Blancas
Punta Yeguas

Río de la Plata

5 Km
5 Miles

FALKLAND ISLANDS (to UK)

ATLANTIC OCEAN

Cape Bougainville
Cape Dolphin
North Fall
Salvador
Port Salvador
Berkeley Sound
Macbride Head
Menguera Point
STANLEY
Falkland Sound
Port Howard
San Carlos
San Carlos Settlement
Mount Adam
700m
Port Stephens
Mount Usborne
705m
Bluff Cove
Darwin
Goose Green
Mount Pleasant
Fox Point
Glorious Hill
Choiseul Sound
Lively Island
North East Island
Speedwell Island
North Arm
Bleaker Island
Eagle Passage
Bay of Harbours
Sea Lion Islands
George Island

West Falkland
East Falkland

Keppel Sound
Pebble Island
Saunders Island
Keppel Island
Byron Sound
King George Bay
Queen Charlotte Bay
Mount Alice
Mount Meredith
Port Stephens Settlement
Cape Meredith

Jason Islands
Grand Jason
Sedge Island
Carcass Island
New Island
Weddell Island
Beaver Settlement
Beaver Island
Port Orford
Passage Island
Roy Cove Settlement

Scale 1:3,000,000

0 10 20 30 40 50 60 Km
0 10 20 30 40 50 60 Miles

FALKLAND ISLANDS (to UK)

Cape Dolphin
STANLEY
Bluff Cove
Mount Pleasant
Mt Usborne 705m
Darwin
Goose Green
Glorious Hill 204m
East Falkland
West Falkland
Jason Islands
Keppel Island
King George Bay
Weddell Island
Mt Adam 700m
Port Stephens 361m
Cape Meredith

Estrecho de Le Maire

South America mainland

Salinas Grandes
Punta Delgada
Golfo Nuevo
Puerto Madryn
Trelew
Rawson
Gaimán
Dolavon
Golfo de San Matías
Bahía Vera
Bahía Camarones
Camarones
Bahía Bustamante
Golfo San Jorge
Comodoro Rivadavia
Rada Tilly
Caleta Olivia
Cabo Blanco
Puerto Deseado
Punta Peñas
Bahía de los Nodales
Telsen
Península de Valdés
Bahía Engaño

CHUBUT
Río Chubut
Las Plumas
Paso de Indios
Lago Colhué Huapi
Lago Musters
Sarmiento
Río Chico
Río Senguerr
Gran Bajo
Pico Truncado
Río Deseado

Esquel
Trevelin
Gastre
José de San Martín
Alto Río Senguer
Río Mayo
Perito Moreno
Gobernador Gregores
Puerto San Julián
Laguna del Carbón -105m
Comandante Luis Piedra Buena
Puerto Santa Cruz
Punta Entrada

SANTA CRUZ
Gran Bajo de San Julián
Gran Altiplanicie Central
Río Chico
Río Santa Cruz
Río Coig
Río Gallegos
Río Coyle

Bahía Grande
Punta Sur
Bahía Bustamante

Esquel
Cholila
Cerro Tres Picos 2492m
El Bolsón
El Maitén
Río Pico
Cerro Cónico 2771m

Chile Chico
Lago Buenos Aires
Los Antiguos
Lago Pueyrredón
Monte Zeballos 2743m
Lago Posadas
Lago San Martín
Lago Cardiel
Lago Strobel
Lago Viedma
Lago Argentino
El Calafate
El Chaltén
Cerro Fitzroy 3405m
Cerro Torre 3128m

Cerro Murallón 3600m
Cerro Bertrand 2900m
Cerro Payne 2600m

Balmaceda
Coihaique
Puerto Aisén
Puerto Chacabuco

AISÉN

Cerro Maca 2960m
Cerro Melimoyu 2400m
Puerto Cisnes
Cerro Hudson 1905m

Puerto Natales
Puerto Bories
Cerro Balmaceda 2035m
Río Turbio
Cerro Castillo

MAGALLANES

Punta Arenas
Península Brunswick
Isla Riesco
Isla Santa Inés
Península Muñoz Gamero
Seno Skyring
Seno Otway
Punta Dungeness
Río Gallegos
Bahía Lomas
Porvenir
Bahía Inútil
Bahía San Sebastián
Punta de Arenas

TIERRA DEL FUEGO
Río Grande
Monte Sarmiento 2300m
Lago Fagnano
Ushuaia
Canal Whiteside
Canal Beagle
Beagle Channel

Cabo de Hornos
Isla de los Estados
Isla Navarino
Isla Hoste
Isla Lennox
Isla Wollaston
Isla Nueva
Isla Gordon
Isla Londonderry
Isla Stewart
Isla Desolación
Isla Clarence
Isla Dawson
Estrecho de Magallanes
Strait of Magellan

ANDES

Isla de Chiloé
Ancud
Castro
Quellón
Chonchi
Golfo Corcovado
Golfo de Ancud
Volcán Corcovado 2300m
Canal Moraleda
Archipiélago de los Chonos
Península de Taitao
Golfo de Penas
Península Tres Montes
Isla Byron
Isla Javier
Isla Campana
Isla Patricio Lynch
Isla Esmeralda
Isla Wellington
Isla Madre de Dios
Isla Duque de York
Golfo Trinidad
Bahía Salvación
Isla Diego de Almagro
Isla Hanover
Archipiélago Reina Adelaida
Península Nelson

RÍO NEGRO
Golfo de Arcud
Isla Guafo
Isla Guaiteca
Península Taitao

Elevation legend

19,686ft
13,124ft
9843ft
6562ft
3281ft
1640ft
820ft
328ft
Sea Level
-820ft
-6562ft
-13,124ft

Scale 1:2,600,000
(projection: Lambert Conformal Conic)

0 10 20 30 40 50 Km
0 10 20 30 40 50 Miles

SOUTH AMERICA

118

Easter Island
(Isla de Pascua)
(to Chile)

Punta San Juan
Cabo Norte
Playa de Anakena
Punta Rosalia
Bahía de La Pérouse
Cabo O'Higgins
Maunga Terevaka 506m
Naunau
Ahu Rarakui
Cabo Roggewein
Ahu Tepeu
Motu Tautara
Maunga Pukatikei 370m
Hanga Roa
Maunga Tangaroa 270m
Ahu Akivi
Rano Raraku
Mataveri
Vaihu
Punta Akahanga
Punta Cuidado
Ahu Vinapu
Orongo
Punta Baja
Rano Kau
Motu Nui

Scale 1:500,000
0 2.5 5 Km
0 2.5 5 Miles

PACIFIC OCEAN

CHILE

ARGENTINA

SANTIAGO

VALPARAÍSO

COQUIMBO

SAN JUAN

MENDOZA

LIBERTADOR

MAULE

BIO BIO

NEUQUÉN

LA PAMPA

ARAUCANÍA

Cerro Aconcagua 6959m
Volcán Tupungato 6800m
Volcán Maipo 5323m
Cerro Juncal 6180m
Volcán Descabezado Grande 3830m
Cerro Campanario 4049m
Volcán Domuyo 4709m
Cerro Payún 3680m
Cerro Nevado 3810m
Volcán Antuco 3585m
Volcán Copahue 2980m

6000m
4000m
3000m
2000m
1000m
500m
250m
100m
Sea Level
-250m
-2000m
-4000m

112
287
116

SOUTH AMERICA – Cities

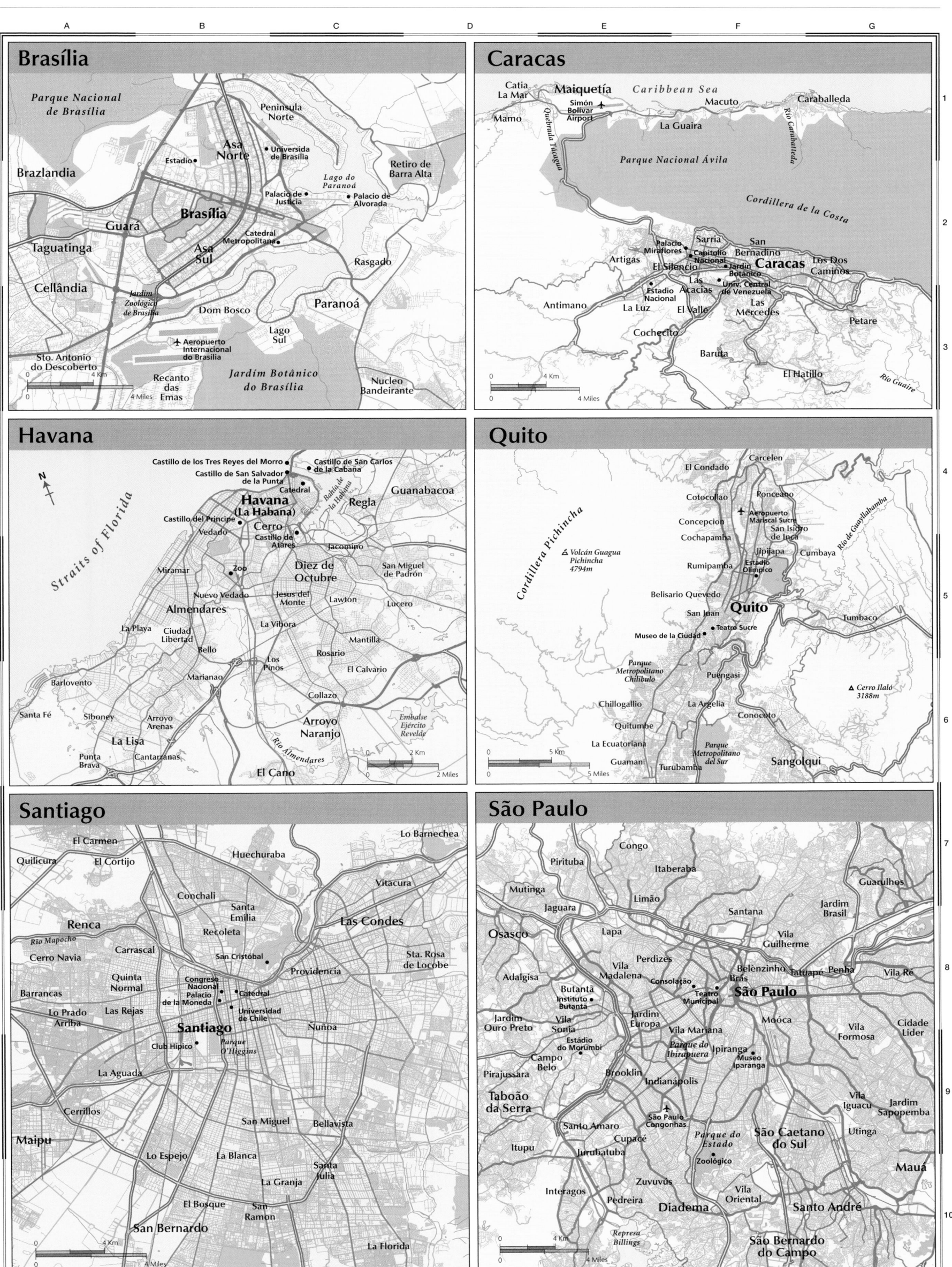

Brasília

Parque Nacional de Brasília
Peninsula Norte
Asa Norte
Brazlandia
Estadio
Universida de Brasília
Lago do Paranoá
Retiro de Barra Alta
Palacio de Justicia
Palacio de Alvorada
Brasília
Guará
Catedral Metropolitana
Taguatinga
Asa Sul
Rasgado
Cellândia
Jardim Zoológico de Brasília
Dom Bosco
Paranoá
Lago Sul
Aeropuerto Internacional do Brasília
Sto. Antonio do Descoberto
Recanto das Emas
Jardím Botânico do Brasília
Nucleo Bandeirante
0   4 Km
0   4 Miles

Caracas

Catia La Mar
Maiquetía
Simón Bolívar Airport
Caribbean Sea
Macuto
Caraballeda
Mamo
Quebrada Tacagua
La Guaira
Rio Caraballeda
Parque Nacional Ávila
Cordillera de la Costa
Palacio Miraflores
Sarría
San Bernadino
Artigas
Capitolio Nacional
Los Dos Caminos
El Silencio
Jardin Botánico
Caracas
Las Acacias
Univ. Central de Venezuela
Antimano
Estadio Nacional
Las Mercedes
La Luz
El Valle
Cochecito
Petare
Baruta
El Hatillo
Rio Guaire
0   4 Km
0   4 Miles

Havana

N
Castillo de los Tres Reyes del Morro
Castillo de San Carlos de la Cabaña
Castillo de San Salvador de la Punta
Catedral
Havana (La Habana)
Bahía de la Habana
Regla
Guanabacoa
Castillo del Principe
Cerro
Castillo de Atares
Straits of Florida
Vedado
Jacomino
Miramar
Zoo
Diez de Octubre
San Miguel de Padrón
Nuevo Vedado
Jesus del Monte
Lawton
Lucero
Almendares
La Playa
Ciudad Libertad
Bello
La Vibora
Mantilla
La Lisa
Marianao
Rosario
El Calvario
Barlovento
Los Pinos
Collazo
Santa Fé
Siboney
Arroyo Arenas
Arroyo Naranjo
Embalse Ejército Revelde
Punta Brava
Cantarranas
Rio Almendares
El Cano
0   2 Km
0   2 Miles

Quito

Carcelen
El Condado
Cotocollao
Ponceano
Aeropuerto Mariscal Sucre
Concepcion
San Isidro de Inca
Cordillera Pichincha
Volcán Guagua Pichincha 4794m
Cochapamha
Jipijapa
Cumbaya
Rio de Guayllabamba
Rumipamba
Estadio Olímpico
Belisario Quevedo
Quito
Tumbaco
San Juan
Teatro Sucre
Museo de la Ciudad
Parque Metropolitano Chilibulo
Puengasi
Cerro Ilaló 3188m
Chillogallio
La Argelia
Conocoto
Quitumbe
La Ecuatoriana
Parque Metropolitano del Sur
Sangolqui
Guamani
Turubamba
0   5 Km
0   5 Miles

Santiago

El Carmen
Lo Barnechea
Quilicura
El Cortijo
Huechuraba
Conchali
Vitacura
Renca
Santa Emilia
Las Condes
Rio Mapocho
Recoleta
Cerro Navia
Carrascal
San Cristóbal
Sta. Rosa de Locobe
Quinta Normal
Providencia
Barrancas
Congreso Nacional Palacio de la Moneda
Catedral
Lo Prado Arriba
Las Rejas
Universidad de Chile
Santiago
Nuñoa
Club Hípico
Parque O'Higgins
La Aguada
Cerrillos
San Miguel
Bellavista
Maipu
Lo Espejo
La Blanca
Santa Julia
El Bosque
San Ramon
La Granja
San Bernardo
La Florida
0   4 Km
0   4 Miles

São Paulo

Congo
Pirituba
Itaberaba
Mutinga
Limão
Guarulhos
Jaguara
Santana
Jardim Brasil
Osasco
Lapa
Vila Guilherme
Adalgisa
Vila Madalena
Perdizes
Belénzinho
Penha
Vila Ré
Butantã
Consolação
Brás
Tatuapé
Instituto Butantã
Teatro Municipal
São Paulo
Jardim Ouro Preto
Vila Sonia
Jardim Europa
Moóca
Vila Formosa
Cidade Lider
Campo Belo
Estádio do Morúmbi
Vila Mariana
Ipiranga
Pirajussara
Parque do Ibirapuera
Museo Iparanga
Brooklin
Indianápolis
Vila Iguacu
Jardim Sapopemba
Taboão da Serra
São Paulo Congonhas
Santo Amaro
Cupacé
Parque do Estado
São Caetano do Sul
Utinga
Itupu
Jurubatuba
Zoológico
Maua
Zuvuvus
Vila Oriental
Interagos
Pedreira
Diadema
Santo André
Represa Billings
São Bernardo do Campo
0   4 Km
0   4 Miles

19,686ft
13,124ft
9843ft
6562ft
3281ft
1640ft
820ft
328ft
Sea Level
-820ft
-6562ft
-13,124ft

Africa is the world's second largest continent with a total area of 11,712,434 sq miles (30,335,000 sq km). It has 54 separate countries, including Madagascar in the Indian Ocean. It straddles the equator and is the only continent to stretch from the northern to southern temperate zones.

FACTFILE

N **Most Northerly Point:** Jalta, Tunisia 37° 31' N
S **Most Southerly Point:** Cape Agulhas, South Africa 34° 52' S
E **Most Easterly Point:** Raas Xaafuun, Somalia 51° 24' E
W **Most Westerly Point:** Santo Antão, Cape Verde, 25° 11' W

Largest Lakes:
1. Lake Victoria, Kenya/Tanzania/Uganda 26,828 sq miles (69,484 sq km)
2. Lake Tanganyika, Dem. Rep. Congo/Tanzania 12,703 sq miles (32,900 sq km)
3. Lake Nyasa, Malawi/Mozambique/Tanzania 11,600 sq miles (30,044 sq km)
4. Lake Turkana, Ethiopia/Kenya 2473 sq miles (6405 sq km)
5. Lake Albert, Dem. Rep. Congo/Uganda 2046 sq miles (5299 sq km)

Longest Rivers:
1. Nile, NE Africa 4160 miles (6695 km)
2. Congo, Angola/Congo/Dem. Rep. Congo 2900 miles (4667 km)
3. Niger, W Africa 2589 miles (4167 km)
4. Zambezi, Southern Africa 1673 miles (2693 km)
5. Ubangi-Uele, C Africa 1429 miles (2300 km)

Largest Islands:
1. Madagascar, 229,300 sq miles (594,000 sq km)
2. Réunion, 970 sq miles (2535 sq km)
3. Tenerife, Canary Islands 785 sq miles (2034 sq km)
4. Isla de Bioco, Equatorial Guinea 779 sq miles (2017 sq km)
5. Mauritius, 709 sq miles (1836 sq km)

Highest Points:
1. Kilimanjaro, Tanzania 19,340 ft (5895 m)
2. Kirinyaga, Kenya 17,058 ft (5199 m)
3. Mount Stanley, Dem. Rep. Congo/Uganda 16,762 ft (5109 m)
4. Mount Speke, Uganda 16,043 ft (4890 m)
5. Mount Baker, Uganda 15,892 ft (4844 m)

Lowest Point:
▼ Lac 'Assal, Djibouti -512 ft (-156 m) below sea level

Highest recorded temperature:
✛ Al'Aziziyah, Libya 136°F (58°C)

Lowest recorded temperature:
⊖ Ifrane, Morocco -11°F (-24°C)

Wettest Place:
≋ Cape Debundsha, Cameroon 405 in (10,290 mm)

Driest Place:
⌣ Wadi Halfa, Sudan <0.1 in (<2.5 mm)

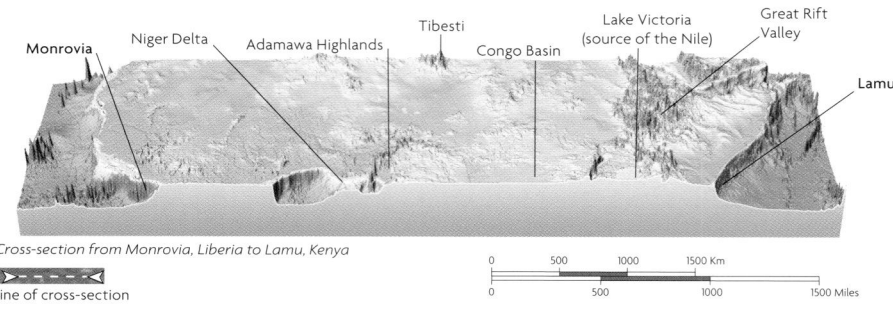

Cross-section from Monrovia, Liberia to Lamu, Kenya

line of cross-section

0   500   1000   1500 Km
0   500   1000   1500 Miles

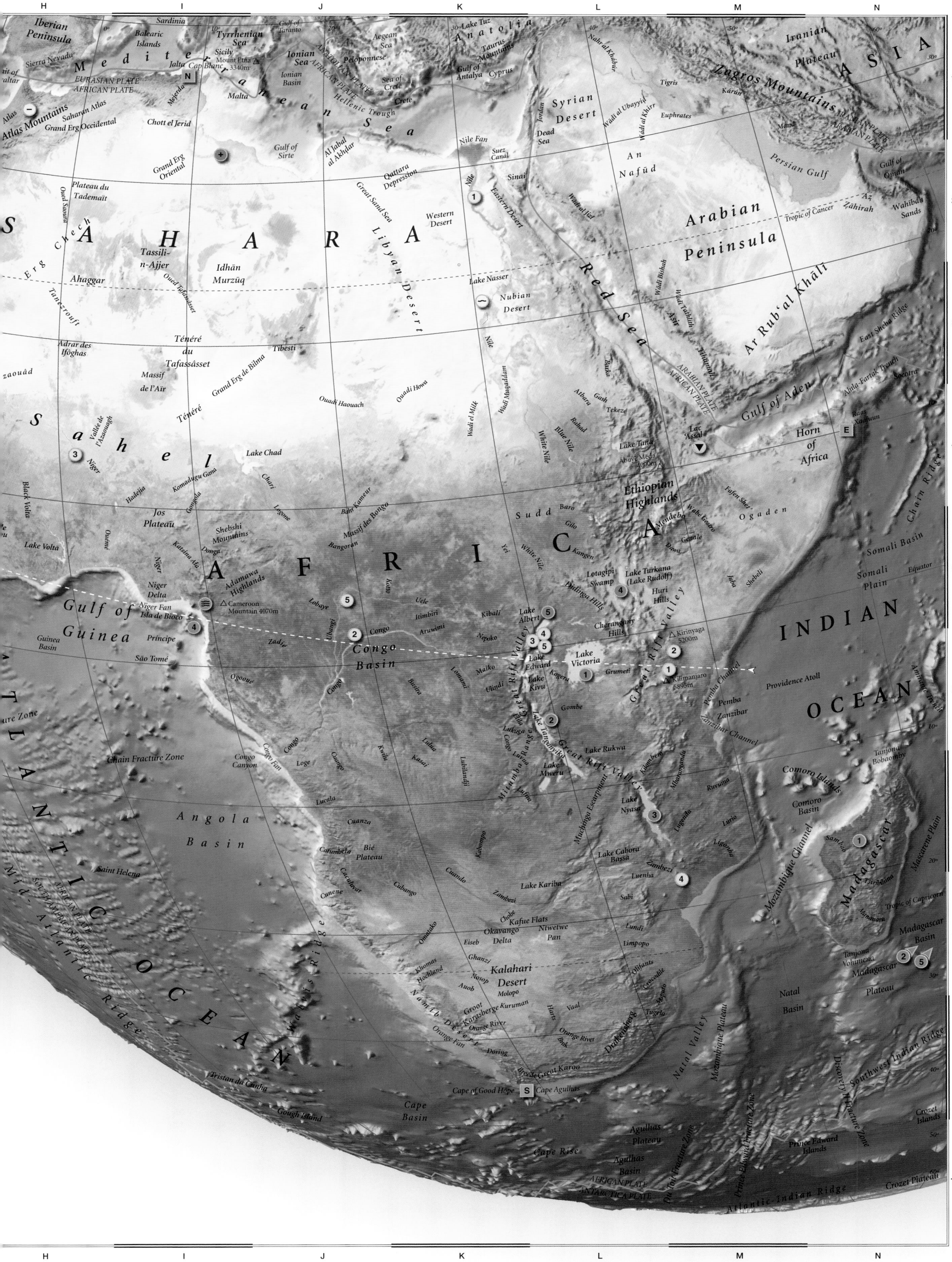

H | I | J | K | L | M | N

Seas, Gulfs, Oceans and Water Bodies

Mediterranean Sea, Tyrrhenian Sea, Ionian Sea, Ionian Basin, Aegean Sea, Sea of Crete, Hellenic Trough, Gulf of Taranto, Gulf of Sirte, Gulf of Antalya, Lake Tuz, Nile Fan, Suez Canal, Dead Sea, Red Sea, Persian Gulf, Gulf of Oman, Gulf of Aden, Lac Assal, Lake Tana, Lake Nasser, Lake Chad, Lake Volta, Lake Turkana (Lake Rudolf), Lotagipi Swamp, Lake Albert, Lake Edward, Lake Kivu, Lake Victoria, Lake Tanganyika, Lake Rukwa, Lake Mweru, Lake Nyasa, Lake Cabora Bassa, Lake Kariba, Kafue Flats, Okavango Delta, Ntwetwe Pan, Gulf of Guinea, Guinea Basin, Angola Basin, Congo Basin, Cape Basin, Cape Rise, Agulhas Basin, Agulhas Plateau, Natal Basin, Natal Valley, Mozambique Basin, Comoro Basin, Mascarene Plain, Somali Basin, Somali Plain, Indian Ocean, Atlantic Ocean, Zanzibar Channel, Mozambique Channel, Pemba Channel

Deserts and Plateaux

Sahara, Sahel, Libyan Desert, Nubian Desert, Western Desert, Eastern Desert, Syrian Desert, An Nafūd, Ar Rub' al Khālī, Wahibah Sands, Ténéré, Ténéré du Tafassâsset, Grand Erg Oriental, Grand Erg Occidental, Grand Erg de Bilma, Great Sand Sea, Qattara Depression, Plateau du Tademaït, Idhân Murzûq, Ahaggar, Tassili-n-Ajjer, Massif de l'Aïr, Adrar des Ifôghas, Tanezrouft, Erg Chech, Ouad Toudoukset, Ouadi Haouach, Ouadi Howa, Wadi al Milk, Kalahari Desert, Namib Desert, Bié Plateau, Angola Basin, Jos Plateau, Ethiopian Highlands, Adamawa Highlands, Ogaden, Sudd, Great Karoo, Khomas Hochland, Madagascar Plateau

Mountains

Atlas Mountains, Saharan Atlas, Sierra Nevada, Taurus Mountains, Zagros Mountains, Iranian Plateau, Shebshi Mountains, Cameroon Mountain 4070m, Kilimanjaro 5895m, Kirinyaga 5200m, Mount Etna 3340m, Cape Blanc, Drakensberg, Mulanje, Kinyeti, Dialigu Hills, Cherangany Hills, Huri Hills

Rivers

Nile, White Nile, Blue Nile, Niger, Congo, Zambezi, Limpopo, Orange River, Vaal, Okavango, Cunene, Cuanza, Kasai, Lualaba, Lomami, Ubangi, Uele, Aruwimi, Lulonga, Lomani, Kwango, Kwilu, Logone, Chari, Komadugu Gana, Benue, Baro, Gilo, Sobat, Atbara, Tekeze, Rahad, Dinder, Wabe Shebele, Gebele, Jubba, Shebeli, Tigris, Euphrates, Karun, Wadi al Hasa, Wadi ar Rumah, Orange Fan, Congo Canyon, Congo Fan, Niger Fan, Nile Fan

Capes and Islands

Iberian Peninsula, Balearic Islands, Sardinia, Sicily, Malta, Cyprus, Crete, Peloponnese, Anatolia, Arabian Peninsula, Sinai, Horn of Africa, Socotra, Zanzibar, Pemba, Providence Atoll, Comoro Islands, Madagascar, Isla de Bioco, Príncipe, São Tomé, Saint Helena, Tristan da Cunha, Gough Island, Prince Edward Islands, Crozet Islands, Crozet Plateau, Cape of Good Hope, Cape Agulhas

Ridges and Fracture Zones

Mid-Atlantic Ridge, South Atlantic Ridge, Walvis Ridge, Southwest Indian Ridge, Atlantic-Indian Ridge, Chain Fracture Zone, Fracture Zone, Andrew Bain Fracture Zone, Prince Edward Fracture Zone, Discovery Fracture Zone, East Sheba Ridge, Alula-Fartak Trench, Carlsberg Ridge

Tectonic Plates

EURASIAN PLATE, AFRICAN PLATE, ANATOLIAN PLATE, ARABIAN PLATE, ANTARCTICA PLATE

Tropic of Cancer, Tropic of Capricorn, Equator

Great Rift Valley, Mitumba Escarpment, Muchinga Escarpment

Political

The political map of modern Africa only emerged following the end of World War II. Over the next half-century, all of the countries formerly controlled by European powers gained independence from their colonial rulers—only Liberia and Ethiopia were never colonized. The post-colonial era has not been an easy period for many countries, but there have been moves toward multi-party democracy across much of the continent. In South Africa, democratic elections replaced the internationally-condemned apartheid system only in 1994. Other countries have still to find political stability; corruption in government and ethnic tensions are serious problems. National infrastructures, based on the colonial transportation systems built to exploit Africa's resources, are often inappropriate for independent economic development.

Scale 1:30,500,000
(projection: Lambert Azimuthal Equal Area)

Km
0 200 400 600 800 1000
Miles
0 200 400 600 800 1000

Population
- ■ above 5 million
- ■ 1 million to 5 million
- ● 500,000 to 1 million
- ◎ 100,000 to 500,000
- ⊕ 50,000 to 100,000
- ○ 10,000 to 50,000
- ● Country capital

Borders
- full international border
- disputed de facto border
- ceasefire line

Standard of living

Since the 1960s most countries in Africa have seen significant improvements in life expectancy, healthcare, and education. However, 28 of the 30 most deprived countries in the world are African, and the continent as a whole lies well behind the rest of the world in terms of meeting many basic human needs.

Standard of living
(UN human development index)
- high
- low

AFRICA

122

Transportation

African railroads were built to aid the exploitation of natural resources, and most offer passage only from the interior to the coastal cities, leaving large parts of the continent untouched—five land-locked countries have no railroads at all. The Congo, Nile, and Niger river networks offer limited access to land within the continental interior, but have a number of waterfalls and cataracts which prevent navigation from the sea. Many roads were developed in the 1960s and 1970s, but economic difficulties are making the maintenance and expansion of the networks difficult.

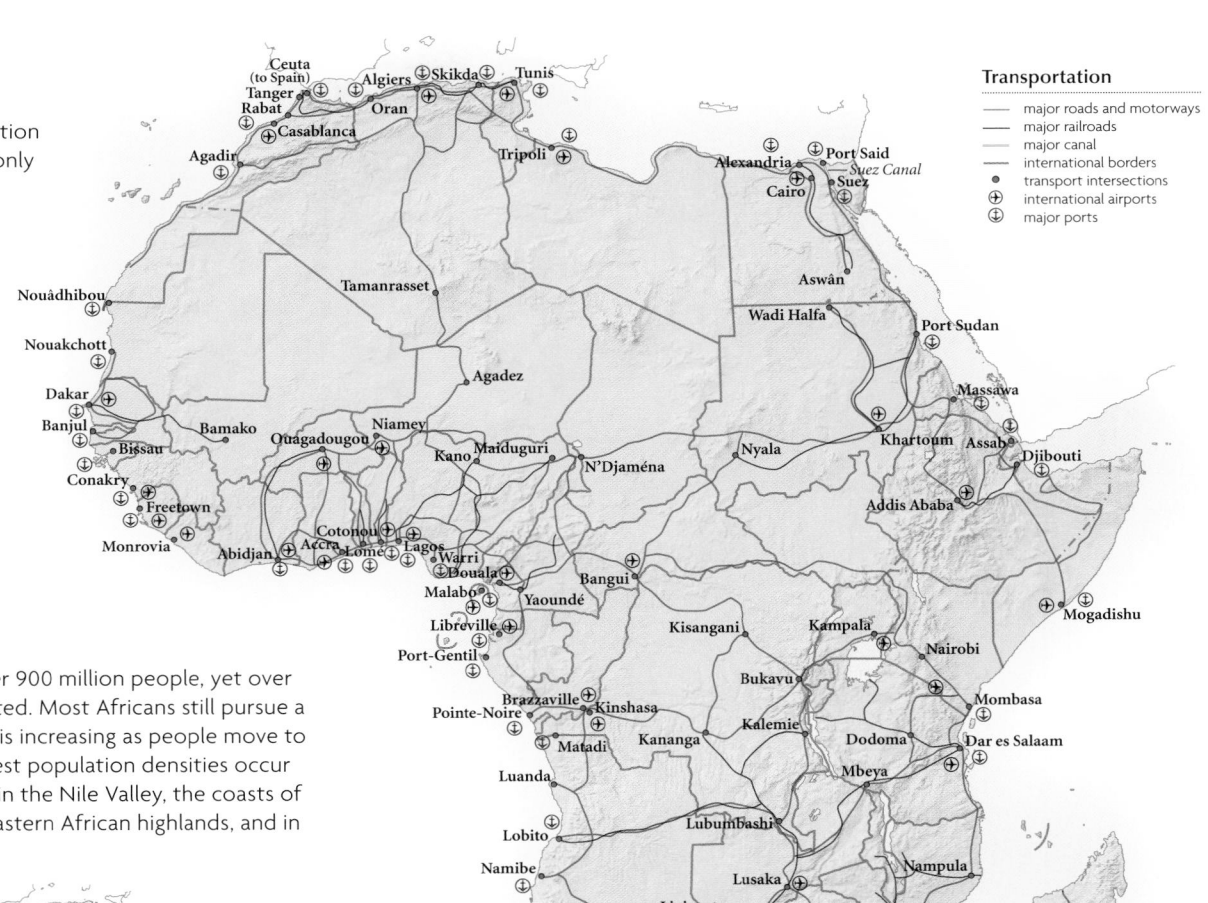

Transportation

- major roads and motorways
- major railroads
- major canal
- international borders
- ● transport intersections
- ⊕ international airports
- ⊕ major ports

Population

Africa has a rapidly-growing population of over 900 million people, yet over 75% of the continent remains sparsely populated. Most Africans still pursue a traditional rural lifestyle, though urbanization is increasing as people move to the cities in search of employment. The greatest population densities occur where water is more readily available, such as in the Nile Valley, the coasts of North and West Africa, along the Niger, the eastern African highlands, and in South Africa.

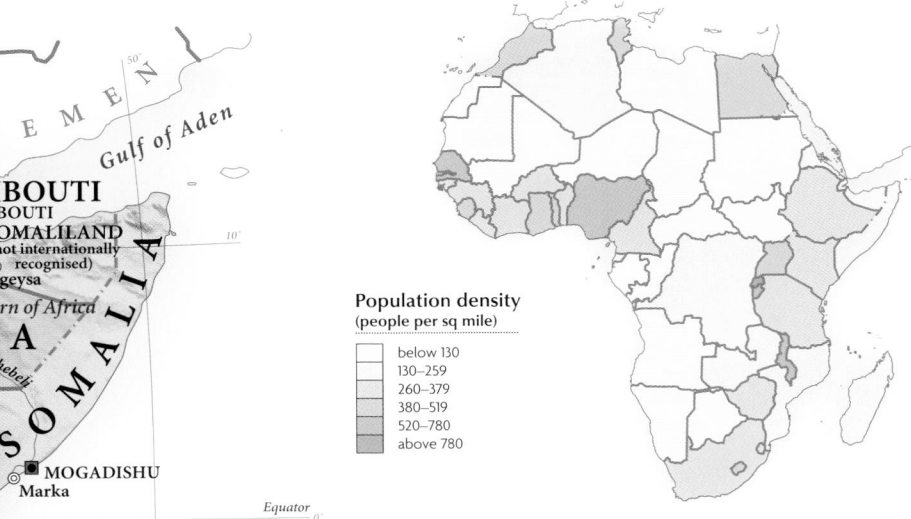

Population density
(people per sq mile)

- below 130
- 130–259
- 260–379
- 380–519
- 520–780
- above 780

Languages

Three major world languages act as *lingua francas* across the African continent: Arabic in North Africa; English in southern and eastern Africa and Nigeria; and French in Central and West Africa, and in Madagascar. A huge number of African languages are spoken as well—over 2000 have been recorded, with more than 400 in Nigeria alone—reflecting the continuing importance of traditional cultures and values. In the north of the continent, the extensive use of Arabic reflects Middle Eastern influences while Bantu languages are widely-spoken across much of southern Africa.

Language groups

- Afro-Asiatic (Hamito-Semitic)
- Niger-Congo
- Nilo-Saharan
- Khoisan
- Indo-European
- Austronesian

Official African Languages

- French
- English
- Arabic
- Portuguese
- Swahili
- Amharic
- Spanish
- French/English
- French/Arabic
- French/Malagasy
- English/Swahili
- Arabic/Somali

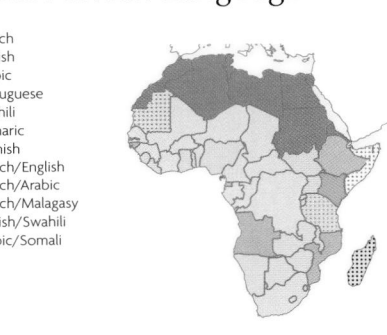

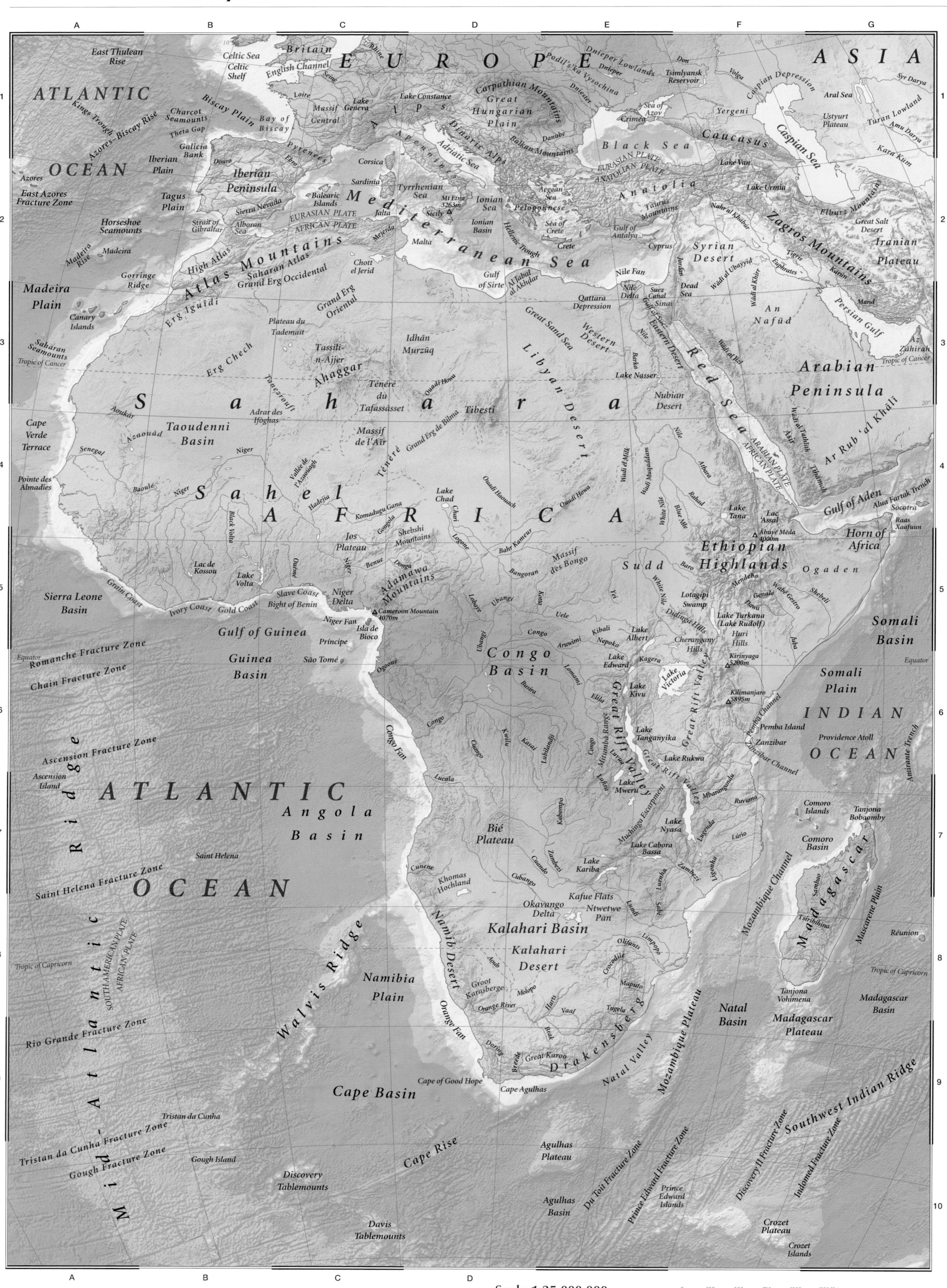

Scale 1:35,000,000
(projection: Lambert Azimuthal Equal Area)

Climate

The climates of Africa range from mediterranean to arid, dry savannah and humid equatorial. In East Africa, where snow settles at the summit of volcanoes such as Kilimanjaro, climate is also modified by altitude. The winds of the Sahara export millions of tons of dust a year both northward and eastward.

Climate

- arid
- humid equatorial
- mediterranean
- semi-arid
- tropical
- warm humid
- ☼ daily hours of sunshine, January
- ☼ daily hours of sunshine, July
- → cold wind
- → hot wind

Average Rainfall

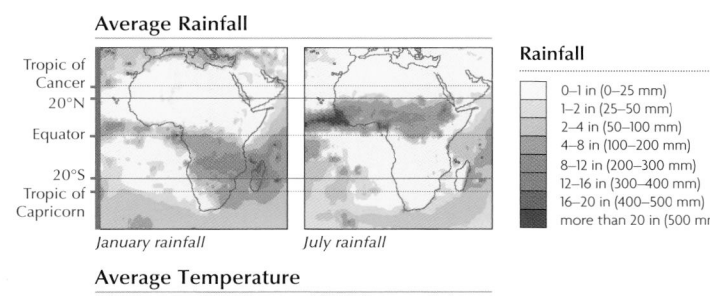

January rainfall *July rainfall*

Rainfall

- 0–1 in (0–25 mm)
- 1–2 in (25–50 mm)
- 2–4 in (50–100 mm)
- 4–8 in (100–200 mm)
- 8–12 in (200–300 mm)
- 12–16 in (300–400 mm)
- 16–20 in (400–500 mm)
- more than 20 in (500 mm)

Average Temperature

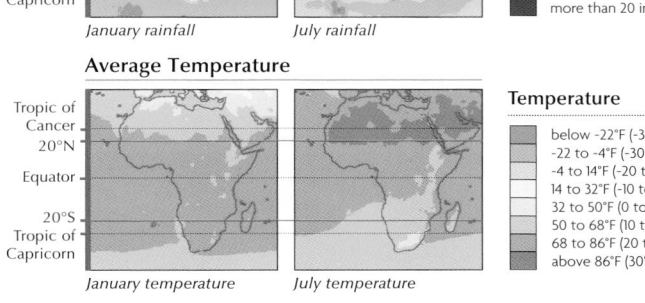

January temperature *July temperature*

Temperature

- below -22°F (-30°C)
- -22 to -4°F (-30 to -20°C)
- -4 to 14°F (-20 to -10°C)
- 14 to 32°F (-10 to 0°C)
- 32 to 50°F (0 to 10°C)
- 50 to 68°F (10 to 20°C)
- 68 to 86°F (20 to 30°C)
- above 86°F (30°C)

Land use

Some of Africa's most productive agricultural land is found in the eastern volcanic uplands, where fertile soils support a wide range of valuable export crops including vegetables, tea, and coffee. The most widely-grown grain is corn and peanuts (groundnuts) are particularly important in West Africa. Without intensive irrigation, cultivation is not possible in desert regions and unreliable rainfall in other areas limits crop production. Pastoral herding is most commonly found in these marginal lands. Substantial local fishing industries are found along coasts and in vast lakes such as Lake Nyasa and Lake Victoria.

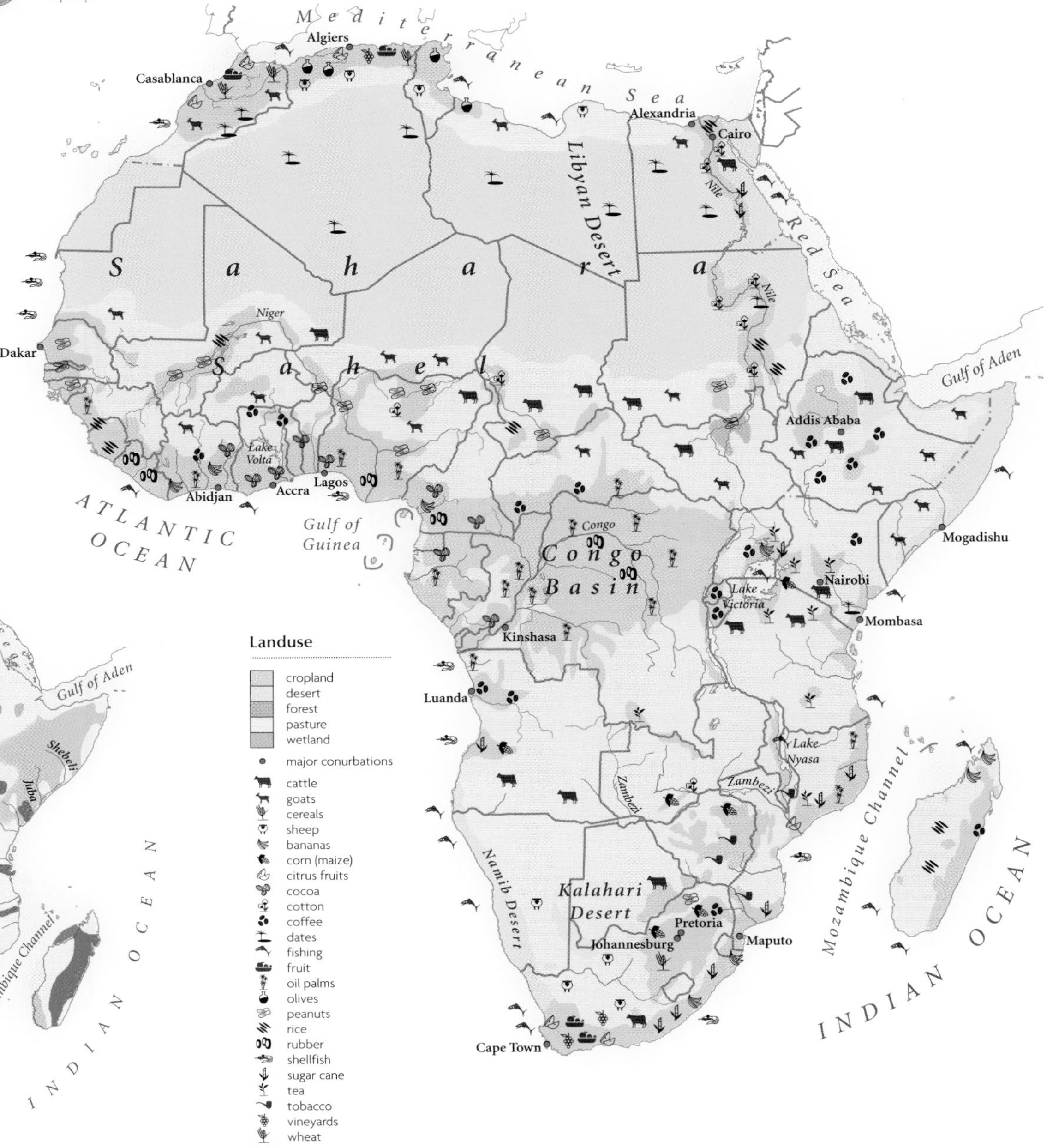

Environmental issues

One of Africa's most serious environmental problems occurs in marginal areas such as the Sahel where scrub and forest clearance, often for cooking fuel, combined with overgrazing, are causing desertification. Game reserves in southern and eastern Africa have helped to preserve many endangered animals, although the needs of growing populations have led to conflict over land use, and poaching is a serious problem.

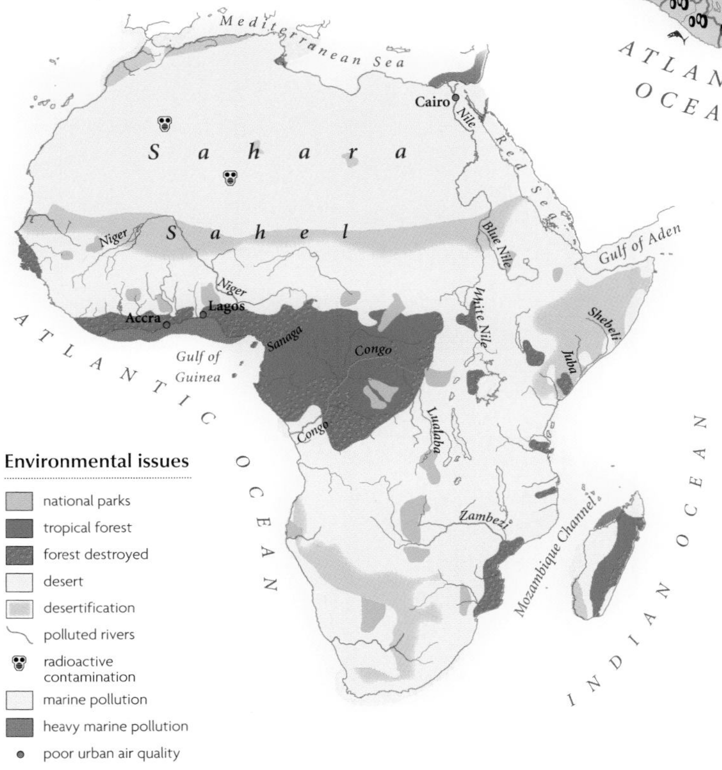

Environmental issues

- national parks
- tropical forest
- forest destroyed
- desert
- desertification
- polluted rivers
- ☢ radioactive contamination
- marine pollution
- heavy marine pollution
- ● poor urban air quality

Landuse

- cropland
- desert
- forest
- pasture
- wetland
- ● major conurbations
- cattle
- goats
- cereals
- sheep
- bananas
- corn (maize)
- citrus fruits
- cocoa
- cotton
- coffee
- dates
- fishing
- fruit
- oil palms
- olives
- peanuts
- rice
- rubber
- shellfish
- sugar cane
- tea
- tobacco
- vineyards
- wheat

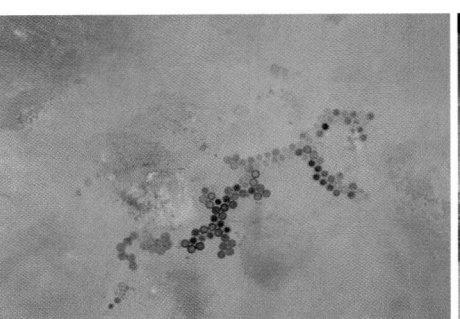

1 AL KHUFRAH, LIBYA
The circular irrigation patterns at this oasis have developed through the use of sprinkler units sweeping around a central point.

2 ERG DU DJOURAB, CHAD
Looking southwest, the pale area, just south of the darker Tibesti mountains on the right and the Ennedi plateau on the left, shows a desert sandstorm in motion.

3 ASWAN HIGH DAM, EGYPT
Completed in 1970 the dam controls flooding along the lower stretches of the Nile river.

4 KHARTOUM, SUDAN
The capital of Sudan lies at the junction of the Blue Nile, flowing from the east, and the broad White Nile, flowing from the south.

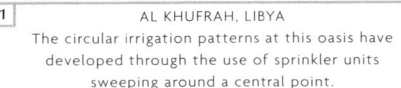

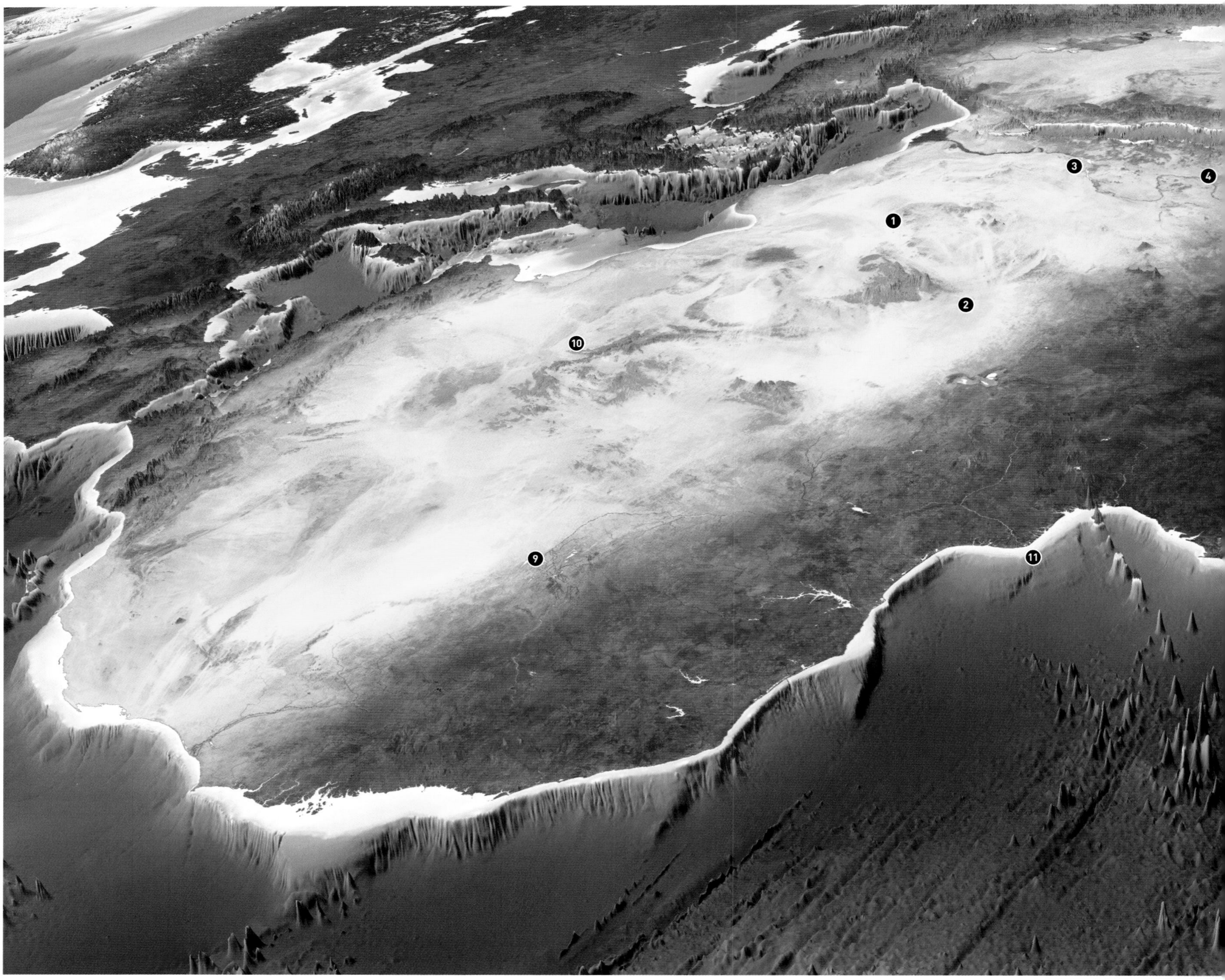

9 LAKE FAGUIBINE, MALI
Part of the Niger river's "inland delta," a region of lakes, creeks and backwaters near Tombouctou.

10 TASSILI-N-AJJER, ALGERIA
These sand dunes, one of a variety found in the Sahara, overlie the darker sandstone bedrock of the Tassili-n-Ajjer plateau.

11 NIGER DELTA, NIGERIA
At this point lies the vast, low-lying region through which the waters of the Niger river drain into the Gulf of Guinea.

12 CONGO/UBANGI RIVERS, DR CONGO
The confluence of these two rivers lies at the heart of the Congo Basin.

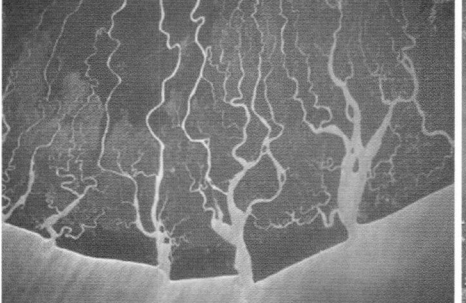

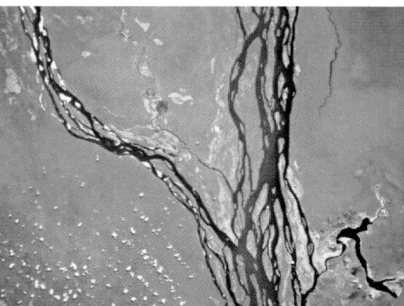

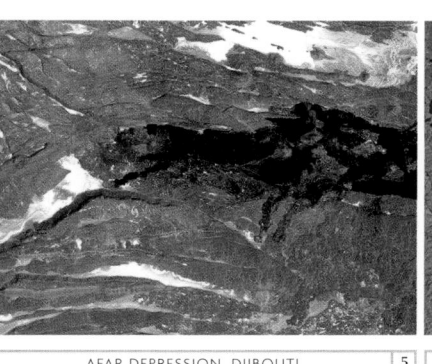

AFAR DEPRESSION, DJIBOUTI | 5
This low point is located at the junction of three
ectonic plates—the Gulf of Aden to the east, the Red
ea to the north and the Great Rift Valley to the south.

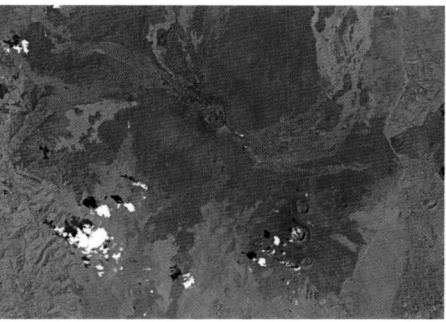

NYIRAGONGO AND NYAMURAGIRA VOLCANOES, | 6
DR CONGO
These two volcanoes, lying to the west of the Great Rift
Valley, last erupted in 2002 and 2001 respectively.

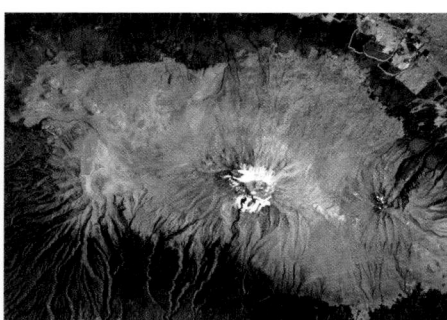

KILIMANJARO, TANZANIA | 7
An extinct volcano, its great height modifies the local
climate, forcing moist air streams from the Indian Ocean
to rise, inducing rain and, higher up, snow.

BETSIBOKA RIVER, MADAGASCAR | 8
The waters of Madagascar's second longest river are
red with sediment as it carries eroded topsoil from the
interior and deposits it in the Indian Ocean.

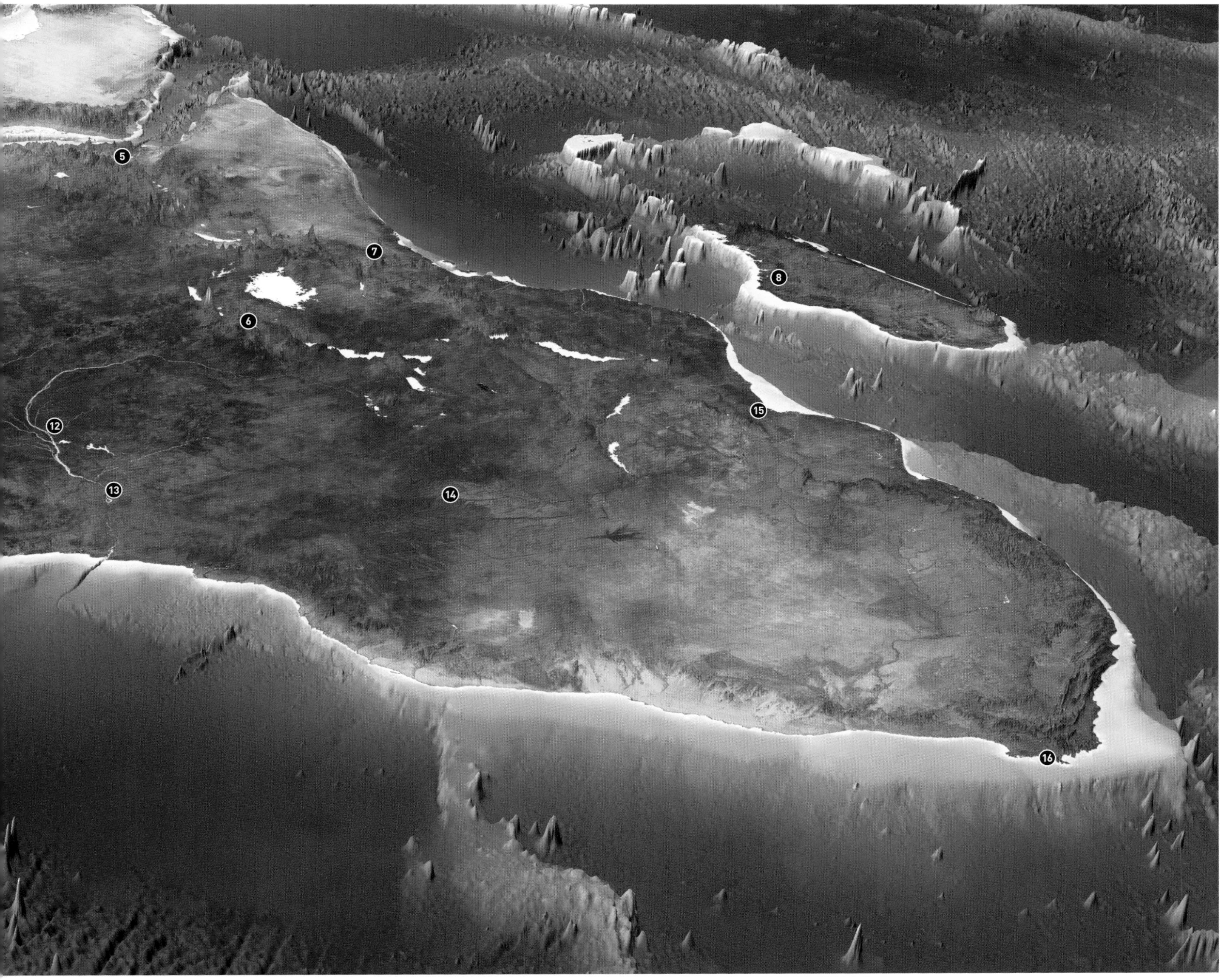

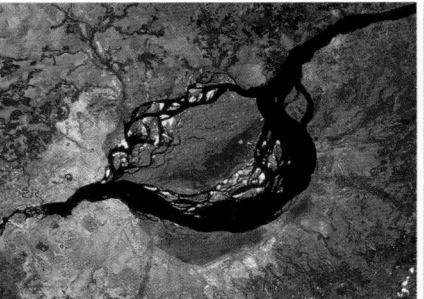

MALEBO POOL, CONGO/DR CONGO | 13
ake in the lower reaches of the Congo river, it hosts
wo capital cities on its banks, Brazzaville, Congo to
the north and Kinshasa, DR Congo to the south.

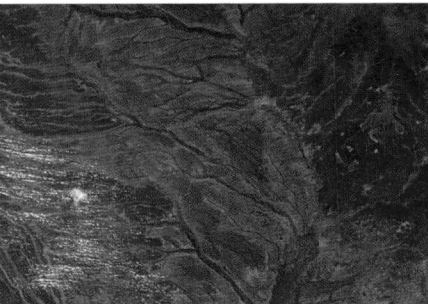

ZAMBEZI RIVER, ZAMBIA | 14
Seasonal flooding of the river and its tributaries turned
the Mulonga and Liuwa plains on the Zambia-Angola
border into a vast wetland in April 2004

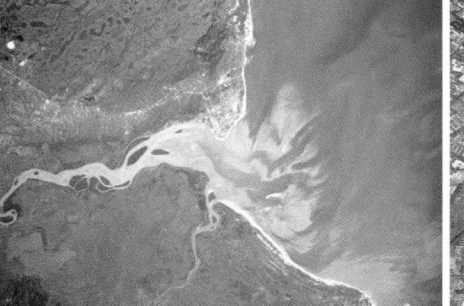

BEIRA, MOZAMBIQUE | 15
This port and beach resort lies on the north
side of the mouth of the Pungoé river.

CAPE TOWN, SOUTH AFRICA | 16
South Africa's third largest city with a
population of 2.9 million, it is also the seat of
the country's parliament.

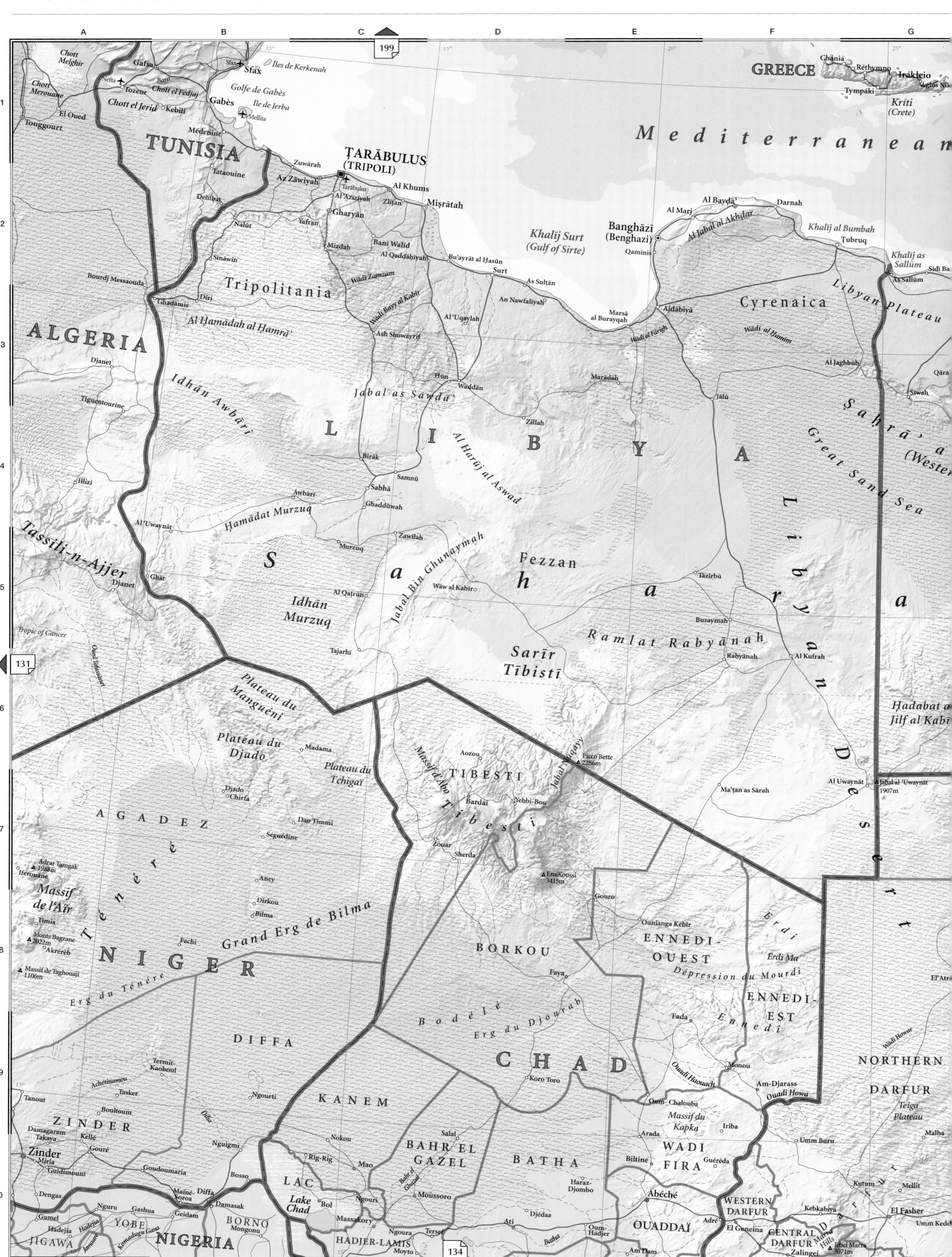

Northeast Africa

GREECE

Mediterranean

TUNISIA

TARĀBULUS (TRIPOLI)

ALGERIA

Tripolitania

Al Ḥamādah al Ḥamrā'

Khalīj Surt (Gulf of Sirte)

Banghāzī (Benghazi)

Cyrenaica

Libyan Plateau

Ṣaḥrā' a (Wester

L   I   B   Y   A

Jabal as Sawdā'

Al Harūj al Aswad

Great Sand Sea

Tassili-n-Ajjer

Idhān Awbāri

Ḥamādat Murzuq

S   a   h

Fezzan

a   r   y   a   n   De

Ramlat Rabyānah

Sarīr Tībistī

Ḥadabat a Jilf al Kabī

Idhān Murzuq

Plateau du Manguéni

Plateau du Djado

Plateau du Tchigaï

Massif du Tibesti

TIBESTI

Aozou

Bardaï

Ennedi Tibesti

Picco Bette 2266m

Ma'tan as Sārah

Al Uwaynāt

Jabal al 'Uwaynāt 1907m

Ténéré

AGADEZ

Massif de l'Aïr

NIGER

Erg du Ténéré

Grand Erg de Bilma

BORKOU

Faya

ENNEDI-OUEST

ENNEDI-EST

Ennedi

Erdi

Erdi Ma

Dépression du Mourdi

Bodélé

Erg du Djourab

CHAD

DIFFA

KANEM

BAHR EL GAZEL

BATHA

WADI FIRA

Massif du Kapka

NORTHERN DARFUR

Teiga Plateau

LAC

Lake Chad

BORNO

NIGERIA

HADJER-LAMIS

OUADDAÏ

WESTERN DARFUR

CENTRAL DARFUR

ZINDER

YOBE

JIGAWA

0 25 50 75 100 125 150 175 200 Km
0 25 50 75 100 125 150 175 200 Miles

Population
- ■ above 5 million
- ▣ 1 million to 5 million
- ◉ 500,000 to 1 million
- ◎ 100,000 to 500,000
- ⊙ 50,000 to 100,000
- ○ 10,000 to 50,000
- ∘ below 10,000

AFRICA

129

216

136

220

Elevation scale:
19,686ft
13,124ft
9843ft
6562ft
3281ft
1640ft
820ft
328ft
Sea Level
-820ft
-6562ft
-13,124ft

H   I   J   K   L   M   N

1 2 3 4 5 6 7 8 9 10

TURKISH REPUBLIC OF NORTHERN CYPRUS
(recognized only by Turkey)
Nicosia (Kerýneia)
Girne
Gazimağusa (Famagusta)
Olympos 1951m
Páfos
Lemesós (Limassol)
Lárnaka
CYPRUS

Sea

SYRIA
Al Lādhiqīyah (Latakia)
Ḥamāh
Ṭarṭūs
Tripoli
Ḥimṣ (Homs)
Jabal Abū Rahbah
Al Jazīrah
Tikrīt
IRAQ
BAGHDĀD
Ba'qūbah
Al Fallujah
Euphrates
Tigris
Kermānshāh (Bākhtarān)
KERMĀNSHĀH
IRAN
LORESTĀN
ĪLĀM

LEBANON
BEYROUTH (BEIRUT)
DIMASHQ (DAMASCUS)
Douma
Jabal Lubnān
Syrian Desert
Buḥayrat ath Tharthār
Buḥayrat ar Razāzah
Karbalā'
Al Ḥillah
An Najaf
Al Kūt
Al 'Amārah
An Nāṣirīyah
Ar Rihāb
Ḥawr al Ḥammār

Al Quṇayṭirah
Nahariyya
Hefa (Haifa)
Lake Tiberias
Irbid
Al Mafraq
Az Zarqā'
'As Suwaydā'
Umm al Jimāl
Jabal 'Unayzah 940m
Ar Rutbah
Ar Ruthīyah
'Ar'ar
Zahrat al Baṭn

ISRAEL
Netanya
Tel Aviv-Yafo
Holon
Rechovot
Ashdod
West Bank
'AMMĀN
Mādabā
Jabal al 'Amūd 1077m
'Adilfa'
Sakākah
Rafhah
Bi'r Juraybiyāt
Wadi al Baṭin

JERUSALEM
Gaza Strip Gaza
Be'ér Sheva'
Hebron
Arad
Dead Sea
Al Karak
Ma'ān
Al Qurayyāt
Ṭurayf
Al Jarāwī
Al Murayr
Al Jawf
Al Labbah
Al Ḥudūd Ash Shamālīyah
Niṣāb
Ḥafar al Bāṭin
Ath Thamāmī

Nile Delta
Alexandria (Al Iskandarīyah)
Rashīd (Rosetta)
Kafr ash Shaykh
Dumyāṭ (Damietta)
Būr Sa'īd (Port Said)
Al 'Arīsh
Damanhūr
Az Zaqāzīq
Ṭanṭā
Suez Canal (Qanāt as Suways)
Al Ismā'īliya
Al Manṣūrah
Shibīn al Kawm
Banha
CAIRO (AL QĀHIRAH)
Giza (Al Jīzah)
Pyramids of Giza
Hilwān
Suez (As Suways)
As Saff
Sinai (Sīnā')
Za'farānah
Abu Zenima
Haql
Al 'Aqabah
Al Bi'r
Bi'r Fajr
El 'Arīsh
'Isāwiyah
Al Qurayyāt

JORDAN
Al Fayyūm
Banī Suwayf
Al Fashn
Khalīj as Suways (Gulf of Suez)
Jabal Mūsā 2285m
Al Ṭūr
Sharm ash Shaykh
Ra's Muḥammad
Jabal al Lawz 2580m
Ṭabūk
Al Qalībah
Bi'r al Murrā
An Nafūd
Al Mayyāh
Al Ḥamūdīyah

GYPT
Baḥarīyah
Qaṣr al Farāfirah
Asyūṭ
Mallawi
Dayrūt
Abnūb
Bani Mazār
Al Minyā
Al Ghardaqah (Hurghada)
Būr Safāga
Jazīrat Tīrān
Al Muwayliḥ
Dubā
Ra's Ghārib
Tabūk
Ad Dār al Ḥamrā'
Taymā'
Bi'r al Murrā
HĀ'IL
Al Ghazālah
Buraydah
Az Zilfī
Al Arṭāwīyah

Sawhāj
Akhmīm
Jirjā
Tahtā
Qinā
Luxor (Al 'Uqṣur)
Isnā
Idfū
Al Quṣayr
Marsā al 'Alam
Umm Laj
Al 'Ulā
Khaybar
Ḥaḍlīyah
Al Ḥanākīyah
'Uqlat aṣ Ṣuqūr
Wadi ar Rimāh
AL QAṢĪM
Najd
'Unayzah
Shaqrā'
Marāh
Durmā
Ad Dawādimī
Al Quwayyah

Al Qaṣr
Mūt
Al Khārijah
Bāris
Kom Ombo (Kawm Umbū)
Aswān
Aswan Dam (Khazzān Aswān)
Baranis
Ra's Banās
Wādī al Ḥamḍ
Jabal Raḍwā 1814m
Yanbu' al Baḥr
AL MADĪNAH
Al Madīnah (Medina)
Mahd adh Dhahab
Afif
Ar Ruwaydah
Ḥarrat Rahat
SAUDI ARABIA
AR RIYĀḌ
Durmā
Al Miṣlah
Tropic of Cancer

Lake Nasser (Buḥayrat Nāṣir)
Abu Simbel (Abū Sunbul)
Wadi Halfa
(Administrative border)
(Political border)
Hala'ib Triangle
Halaib
Badr Ḥunayn
Rābigh
Buḍayy'ah
Ḥalabān

Red Sea

MAKKAH
Makkah (Mecca)
King Abdul Aziz
Jiddah (Jedda)
At Ṭā'if
Turabah
Az Rawdah
Al Khurmah
220

Selima Oasis
Akasha
Wādī Ḥalfāyā
Delgo
Laqiya Arba'in
NORTHERN
Argo
Dongola
El Khandaq
Korti
Merowe
Shereik
Abū Ḥamed
Wādī 'Amur
Nubian Desert
RED SEA
Port Sudan
Sallom
Suakin
Sinkat
Ekowit
Tokar
Dungúnab
Ras Abu Shagara
Muhammad Qol
Salala
Al Lith
Daws
AL BĀHAH
Al Bāhah
Qal'at Bishah
Tathlith
Wādī Tathlīth
'ASĪR
Muhāyil
Khamīs Mushayṭ
Al Qunfudhah
Al Birk
Jabal Sawda' 3133m
Abhā
Zahrān
Ad Darb
JĪZĀN
NAJRĀN
Najrān

SUDAN
RIVER NILE
Ed Damer
Atbara
Berber
Kabushiya
Shendi
Efile
Abu 'Uruq
Umm Inderab
KHARTOUM NORTH
Omdurman
KHARTOUM
Wadi al Milk
NORTHERN KORDOFAN
Hamrat esh Sheikh
Sodiri
Bara
WESTERN KORDOFAN
Wad Banda
Khuwei
El Obeid
Umm Ruwaba
Tendelti
Musmar
Haiya
Derudeb
Wādī Langeb
Ras Shakal
Göz Regeb
KASSALA
Kassala
Khashm el Girba
Teseney
Barentu
Aroma
ERITREA
Keren
Massawa
Akurdet
Engbershatu 2576m
Nakfa
Banda
Massawā Channel
Dahlak Archipelago
Jazā'ir Farasān
Midi
'Abs
Khamir
Al Luḥayyah
Wādī Mawr
Ḥajjah
Kamarān
Az Zaydīyah
Al Ḥudaydah (Hodeida)
SAN'Ā (SANA)
YEMEN
'Amrān
Jabal an Nabī Shu'ayb 3760m
Jīzān
Ṣabyā
Sa'dah
Midi

WHITE NILE
GEZIRA
Wad Medani
Ed Dueim
Rufa'a
El Kamlin
Hag 'Abdullah
El Manaqil
Al Barakat
Sennar
Singa
Rabak
Kosti
SINNAR
GEDAREF
Gedaref
Doka
El Hawata
Gallabat
Metema
Aykel
Gonder
Sek'ot'a
AMHARA
TIGRAY
Adīgrat
'Ādwa
Āksum
Ādī Ārk'ay
Terara 4620m
Ras Dashen 4620m
Mek'elē
Maych'ew
Korem
Semēn
ETHIOPIA
AFAR
Danakil Desert
Om Hajer
Barka
Ṣawrā 3018m
Mendefera
ASMARA
Adi Ark'ay
Zula
Mersa Fatma
Jazīrat al Hanish al Kabīr
Jazīrat Zuqar
Ed
Beylul
Aseb
Mouhlouri
DJIBOUTI
Bāb al Mandab
Moka (Al Mukhā)
Madīnat ash Sha'b
Turbah
Ta'izz
Ibb
Dhamār
Radā'
Yarīm
Qa'ṭabah
Lahij
Ras Shakal

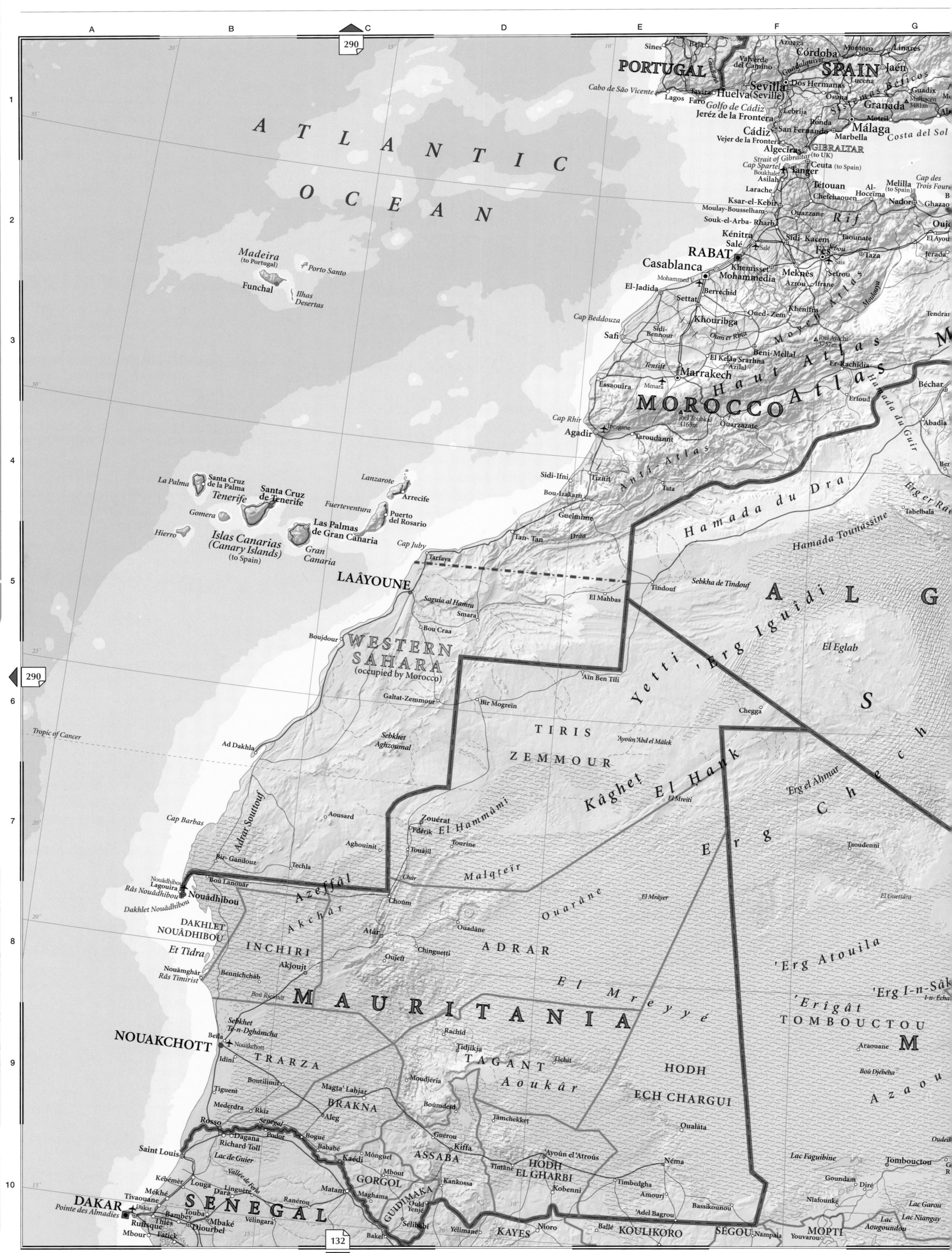

ATLANTIC

OCEAN

PORTUGAL

SPAIN

Cabo de São Vicente

Sines
Beja
Azuaga
Córdoba
Montoro
Linares

Évora
Valverde del Camino
Guadalquivir
Jaén

Lagos
Faro
Huelva
Sevilla
Dos Hermanas
Lucena
Guadix

Golfo de Cádiz
Seville
Osuna
Antequera
Granada
3481m

Jerez de la Frontera
Lebrija
Ronda
Motril
Costa del Sol

Cádiz
San Fernando
Marbella

Vejer de la Frontera
Algeciras
GIBRALTAR

Strait of Gibraltar
Cap Spartel
Tanger
Ceuta (to Spain)

Cap des Trois Fourches

Larache
Tétouan
Al-Hoceïma
Melilla (to Spain)

Asilah
Chefchaouen
Nador
Ghazao

Ksar-el-Kébir
Rif

Moulay-Bousselham
Souk-el-Arba-Rharb

Kénitra
Sidi-Kacem
Taounate
El Ayou

Salé
Taza
Jerada

RABAT
Meknès
Sefrou
Fès

Casablanca
Khemisset
Azrou
Moyen Atlas

Mohammedia
Ifrane

El-Jadida
Berrechid
Khénifra

Settat
Oued-Zem

Khouribga

Cap Beddouza
Sidi-Bennour
Beni-Mellal
Er-Rachidia

Safi
El Kelaa Srarhna
Azilal
Jbel Ayachi

Tensift
Beni-Mellal
Errachidia

Marrakech
Erfoud
Abadla

Essaouira
Menara
MOROCCO
Atlas
Béchar

Cap Rhir
Jbel Toubkal 4165m
Ouarzazate

Imezgane
Agadir
Taroudannt

Anti-Atlas
Hamada du Guir

Sidi-Ifni
Tiznit
Tata

Bou-Izakarn
Guelmime
Erg er Ra

Tan-Tan
Draa
Hamada du Dra
Tabelbala

Tarfaya
Hamada Tounassine

Cap Juby

LAÂYOUNE
Sebkha de Tindouf
El Mahbas
Tindouf
A L G

Saguia al Hamra
Smara

Bou Craa

WESTERN SAHARA
(occupied by Morocco)
'Aïn Ben Tili
Yetti
'Erg Iguîdi
El Eglab

Boujdour

Galtat-Zemmour
Bir Mogreïn
Chegga

TIRIS
'Ayoûn 'Abd el Málek

Tropic of Cancer
Ad Dakhla
ZEMMOUR
El Mreïti
'Erg el Ahmar

Sebkhet Aghzoumal
Erg Ch

Cap Barbas
Aousard
El Hammâmi
Kâghet
El Hank
Erg

Zouérat
Taoudenni

Fdérik
Touîne
El Guettâra

Adrar Souttouf
Aghouinit
Iouâjil
'Erg Atouila

Bir-Gandouz
Char
Malqteïr

Techla
Choûm
El Mrâyer
'Erg I-n-Sâk

Nouâdhibou
Boû Lanouar
Ouarâne
'Erîgât
TOMBOUCTOU

Lagouira
Azeffâl
Ouadâne
'Erg I-n-Echa

Râs Nouâdhibou
Nouâdhibou
Atâr
Chinguetti

Dakhlet Nouâdhibou
Akchâr
Oujeft
ADRAR
Boû Djébéha
M

DAKHLET NOUÂDHIBOU
INCHIRI
Akjoujt
El Mreyyé
Araouane

Et Tidra
Bennichchâb
TOMBOUCTOU

Nouâmghâr
Râs Timirist
Boû Richât
Azaou

MAURITANIA

Sebkhet Ta-n-Dghâmcha
Rachid
HODH

NOUAKCHOTT
Idini
Nouakchott
Bella
Tidjikja
Tichit
Boû Djébéha

TRARZA
Moudjéria
TAGANT
Aoukâr
HODH ECH CHARGUI

Tiguent
Boutilimit
Magta' Lahjar
Boûmdeïd
Jâmchekker
Oualâta

Mederdra
Rkiz
BRAKNA
Guérou
Kiffa
'Ayoûn el 'Atroûs
Néma

Rosso
Aleg
Kaédi
Tîntâne
Bou Djébéha
Tombouctou

Dagana
Padot
Bogué
Jâmchekker

Richard Toll
Bababé
Mônguel
Kankossa
Bassikounou
Lac Faguibine

Saint Louis
Lac de Guier
Senegal
ASSABA
HODH EL GHARBI
Goundam
Diré

Kébémer
Louga
Maghama
Timbedgha
Lac Garou

Mékhé
Matam
Kobenni
Amourj
Lac Tanma
Niafunké

DAKAR
Tivaouane
GORGOL
Maghama
Bakel
Amourj
Diré

Pointe des Almadies
Thiès
Dara
Linguère
Ranérou
Yélimané
KAYES
Youvarou
MOPTI

Rufisque
Bambey
Vélingara
GUIDIMAKA
'Adel Bagrou

SENEGAL
Diourbel
Ouîd Yenjé
Sélibabi
Mbour
Fatick
Sélibabi
Ballé
KOULIKORO
SÉGOU
Nampala

Mbour
Nioro

Touba
Maghama

Dakar
Bababé

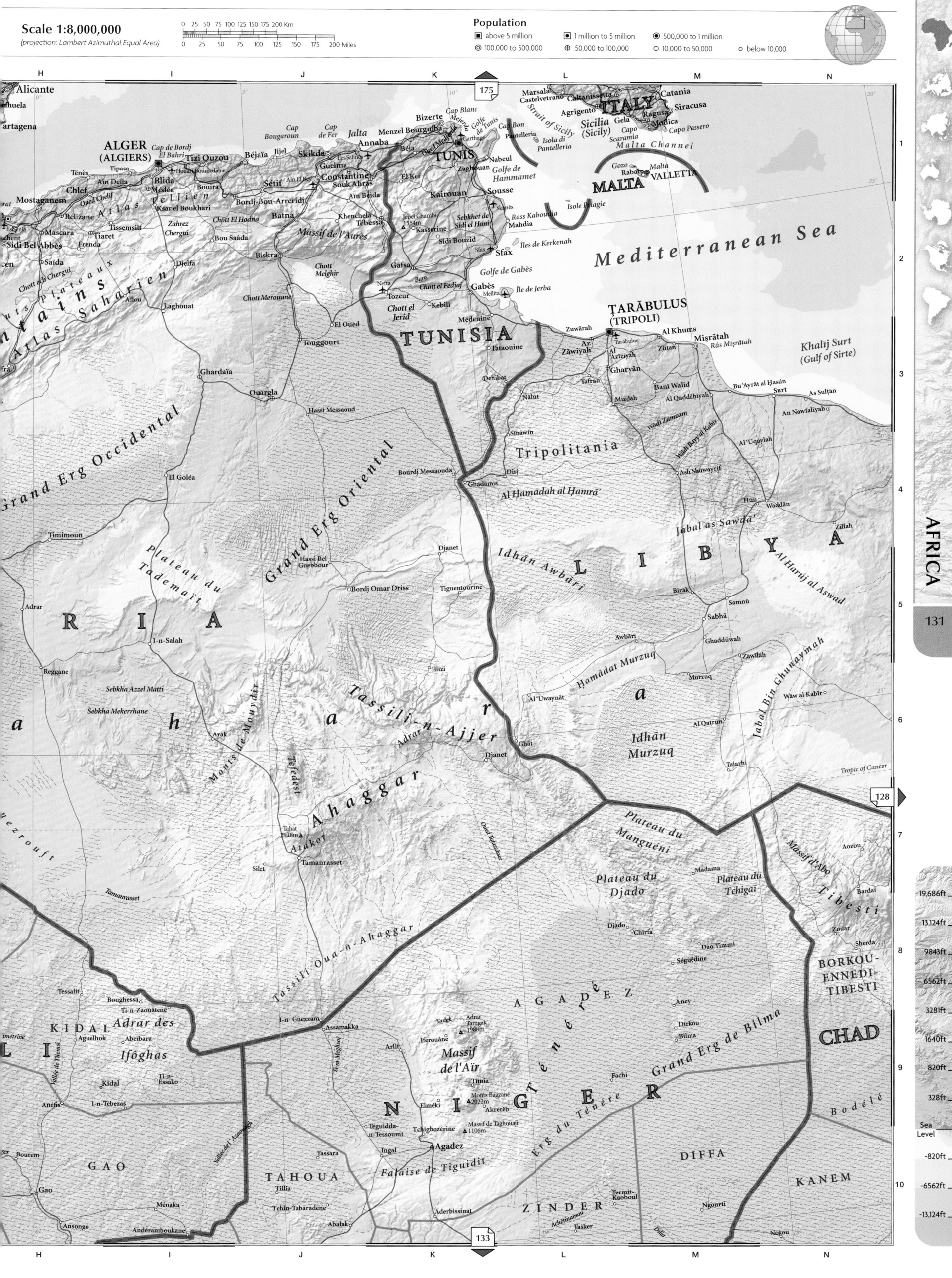

Population

- ■ above 5 million
- ▣ 1 million to 5 million
- ◉ 500,000 to 1 million
- ◎ 100,000 to 500,000
- ⊕ 50,000 to 100,000
- ○ 10,000 to 50,000
- ∘ below 10,000

0 25 50 75 100 125 150 175 200 Km
0 25 50 75 100 125 150 175 200 Miles

AFRICA

131

128

Scale bar / elevation:
19,686ft
13,124ft
9843ft
6562ft
3281ft
1640ft
820ft
328ft
Sea Level
-820ft
-6562ft
-13,124ft

Map labels:

Alicante · huela · artagena

ALGER (ALGIERS) · Cap de Bordj El Bahri · Tizi Ouzou · Béjaïa · Jijel · Skikda · Annaba · Béja · TUNIS · Nabeul · Zaghouan · Golfe de Hammamet · Sousse · Skanès · Monastir · Mahdia · Îles de Kerkenah · Sfax

Ténès · Tipasa · Blida · Médéa · Bouira · Hoàra Boumedienne · Aïn Defla · Oued Chelif · Sétif · Constantine · Souk Ahras · Guelma · Les Salines · Carthage · Cap Bon

Chlef · Mostaganem · Relizane · Ksar el Boukhari · Bordj-Bou-Arréridj · Aïn Beïda · El Kef · Kairouan · Sidi Bouzid

Mascara · Saïda · Tiaret · Frenda · Zahrez Chergui · Batna · Khenchela · Tébessa · Kasserine · Sebkhet el Hani · Sfax · Sidi El Hani

Sidi Bel Abbès · ent · cen · Chott ech Chergui · Djelfa · Chott El Hodna · Massif de l'Aurès · Jebel Chambi 1544m · Sbeïtla

Atlas Saharien · Plateaux · Laghouat · Bou Saâda · Biskra · Gafsa · Golfe de Gabès · Gabès · Île de Jerba

Aflou · Chott Melrhir · Nefta · Chott el Fedjaj · Mellita

El Oued · Chott Merouane · Tozeur · Chott el Jerid · Kebili · Médenine

Touggourt · Ghardaïa · El Oued · Tataouine

Ouargla · Hassi Messaoud · Dehibat · Nalūt

MALTA · Marsala · Castelvetrano · Agrigento · ITALY · Caltanissetta · Catania · Gela · Ragusa · Siracusa · Modica · Capo Scaramia · Capo Passero

Bizerte · Cap Blanc · Golfe de Tunis · Strait of Sicily · Sicilia (Sicily) · Pantelleria · Isola di Pantelleria · Malta Channel · Gozo · Rabat · VALLETTA · Malta · Isole Pelagie · Île de Pantelleria

Mediterranean Sea

TARĀBULUS (TRIPOLI) · Zuwārah · Az Zāwiyah · Al 'Azīzīyah · Gharyān · Tarābulus · Al Khums · Zlitān · Mişrātah · Rās Mişrātah · Khalīj Surt (Gulf of Sirte)

Yafran · Banī Walīd · Bu 'Ayrāt al Hasūn · Surt · As Sulţān · An Nawfalīyah · Al 'Uqaylah

Mizdah · Al Qaddāhiyah · Ash Shuwayrif · Hūn · Waddān · Zillah

TUNISIA · Tripolitania · Wādī Zamzam · Wādī Bayy al Kabīr · Jabal as Sawdā' · Al Harūj al Aswad

Sināwin · Dirj · Ghadāmis · Al Ḩamādah al Ḩamrā' · Birāk · Samnū · Sabhā · Ghaddūwah

LIBYA · Idhān Awbārī · Awbārī · Murzuq · Zawīlah · Jabal Bin Ghunaymah · Wāw al Kabīr

Bourj Messaouda · Djanet · Tiguentourine · Hamādat Murzuq · Al Qaţrūn · Tajarhī

Grand Erg Occidental · Timimoun · El Goléa · Plateau du Tademaït · Grand Erg Oriental · Hassi Bel Guebbour · Bordj Omar Driss · Illizi · Idhān Murzuq · Al 'Uwaynāt · Ghāt

RIA · Adrar · I-n-Salah · Reggane · Sebkha Azzel Matti · Sebkha Mekerrhane · Tassili-n- · Adrar-n-Ajjer · Tassili Oua-n-Ahaggar · Monts de Mouydir · Arāk · Tefedest · Ahaggar · Tahat 2918m · Atakor · Silet · Tamanrasset

Tropic of Cancer

ezrouft · Plateau du Manguéni · Aozou · Massif d'Abo · Tibesti · Bardaï

Plateau du Djado · Madama · Plateau du Tchigaï · Zouar · Sherda

Djado · Chīrfa · Dao Timmi · Séguédine · BORKOU-ENNEDI-TIBESTI

KIDAL · Adrar des Ifôghas · Tessalit · Boughessa · Ti-n-Zaouâtene · I-n-Guezzam · Assamakka · Tadek · Adrar Tamgak 1988m · Iferouâne · AGADEZ · Aney · Dirkou · Bilma · Grand Erg de Bilma · CHAD

métrine · Aguelhok · Abeïbara · Arlit · Tin-Méghsoi · Fachi

Kidal · Ti-n-Essako · Tunia · Massif de l'Aïr · Erg du Ténéré · Diffa

Anéfis · I-n-Tebezas · Elméki · Monts Bagzane 2022m · Akréréb · Teguidda-n-Tessoumt · Tchighozérine · Massif de Taghouaji 1106m · Ingal · Agadez · Falaise de Tiguidit · DIFFA · Bodélé · KANEM

Bourem · GAO · Tassara · TAHOUA · Tillia · ZINDER · Termit-Kaoboul · Ngourti

Gao · Ménaka · Tchin-Tabaradene · Abalak · Aderbissinat · Achégour · Tasker · Ngouri · Nokou

Ansongo · Andéramboukane · NIGER

AFRICA

132

MAURITANIA

SENEGAL

GAMBIA

GUINEA-BISSAU

GUINEA

SIERRA LEONE

LIBERIA

IVORY COAST

MALI

BURKINA FASO

ATLANTIC OCEAN

Cap Barbas
Bir-Gandouz
Aghouinit
Touâjil
Tourine
TIRIS ZEMMOUR
Taoudenni
El Guettâra
Nouâdhibou
Lagouira
Râs Nouâdhibou
Dakhlet Nouâdhibou
Bou Lanouâr
Techla
Châr
Choûm
Ouadâne
Malqteïr
El Mrâyer
Azeffâl
Akchâr
Ouarâne
ADRAR
'Erg Atouila
'Erg I-n-Sâkân
I-n-Echaï
DAKHLET NOUÂDHIBOU
INCHIRI
Atâr
Chinguetti
Oujeft
'Erîgât
Azaouâd
Nouâmghâr
Bennichchâb
Râs Timirist
Boû Rjeimat
Akjoujt
EL MRÂYYÉ
TOMBOUCTOU
Araouane
Boû Djébéha
NOUAKCHOTT
Sebkhet Ten-Dghâmcha
Bella
Nouâkchott
Idini
TAGANT
Tidjikja
Tichit
Aoukâr
HODH
ECH CHARGUI
h
TRARZA
Boutilimit
Magta'Lahjar
Moudjéria
Boûmdeïd
Tâmchekket
Oualâta
Oudeïka
Rkîz
BRAKNA
Aleg
Guérou
Kiffa
'Ayoûn el 'Atroûs
Néma
Diré
M
Rosso
Sénégal
Padar
Bogué
Bababé
Mônguel
Kankossa
Timbedgha
Amourj
'Adel Bagrou
Lac Faguibine
Tombouctou
Gourma-Rharous
Goundam
Dagana
Richard Toll
Kaédi
Mbout
Maghama
Ould Yenjé
Selibabi
Kobenni
Bassikounou
Nampala
Niafounké
Lac Garou
Lac Niangay
Saint Louis
Lac de Guier
Louga
Kébémèr
Dara
Matam
ASSABA
Tintâne
HODH EL GHARBI
Nara
Youvarou
Lac Aougoundou
Pointe des Almadies
Dakar
Mékhé
Tivaouane
Louga
Ranérou
Linguère
GORGOL
GUIDIMAKA
Yélimané
Nioro
Balé
Sokolo
Dioura
Ténenkou
Konna
Sévaré
Mopti
Douenza
DAKAR
Thiès
Rufisque
Touba
Mbaké
Diourbel
Bambey
SENEGAL
Bakel
Kidira
Ambidédi
Kayes
Maréna
Sandaré
Diéma
Mourdiah
Niono
Massina
Diafarabé
MOPTI
Bandiagara
Koro
Mbour
Fatick
Kaolack
Joal-Fadiout
Sokone
Kaffrine
Koungheul
Nioro du Rip
Goudiri
Diamou
Sadiola
Baoulabé
KAYES
Didiéni
Banamba
Djénné
Bankass
Tominian
GAMBIA
BANJUL
Brikama
Banjul
Mansa Konko
Georgetown
Maka
Tambacounda
Dialakoto
KAYES
Kita
Sébékoro
Kolokani
KOULIKORO
Ségou
Markala
SÉGOU
San
Bénéna
Tougan
S
Ouahigouya
Yako
Dioulouiou
Bignona
Ziguinchor
Sédhiou
Farim
Kolda
Vélingara
Medina Gounas
Saraya
Kédougou
Satadougou
Kéniéba
Bafing
Kokofata
Fana
Diolla
Ouéléssébougou
Koutiala
Yorosso
Nouna
Dédougou
Réo
Koudougou
OUAGADOUGOU
BURKINA FASO
Cacheu
Bissorã
Mansôa
Bafatá
Gabú
BISSAU
Bolama
Fulacunda
Buba
Koundara
Mali
Támgue 1538m
Labé
Tougué
Siguiri
Kangaba
Lac de Sélingué
Bougouni
SIKASSO
Kadiolo
Koloko
Sindou
Orodara
Bobo-Dioulasso
Diébougou
Houndé
Boromo
Boundoukui
GUINEA-BISSAU
Quinhámel
Bolama
Catió
Gaoual
Fouta Djallon
Pita
Kavendou 1421m
Dalaba
Dinguiraye
Dabola
Niger
Kouroussa
Mandiana
Manankoro
SIKASSO
Kolondiéba
Garalo
Yanfolila
Ouangolodougou
Niangoloko
Banfora
Gaoua
Kampti
Wa
290
Kamsar
Boké
Télimélé
Konkouré
Fria
Boffa
Dubréka
Kindia
Cap Verga
Mamou
Fria
GUINEA
Faranah
Kissidougou
Kankan
Samatiguila
Madinani
Odienné
Kouto
Boundiali
Korhogo
Ferkessédougou
Tehini
Bouna
Sawla
Damongo-Bole
CONAKRY
Conakry
Forécariah
Coyah
Falaba
Mongo
Kabala
Bintimani 1948m
Tokounou
Kérouané
Bako
Niellé
Orodara
Tafiré
Katiola
Dabakala
FREETOWN
Kambia
Pendembu
Port Loko
Pepel
Lungi
Lunsar
Makeni
Magburaka
Koidu
Guéckédou
Macenta
Voinjama
Beyla
Pic de Tibé 1504m
Madinani
Séguéla
Mankono
IVORY COAST
Béoumi
Bouaké
Mbahiakro
Agnibilékrou
Wenchi
Berekum
Sunyani
Mampong
SIERRA LEONE
Bo
Kenema
Moyamba
Matru
Shenge
Bonthe
Pujehun
Sulima
Zimmi
Kolahun
Zorzor
Nzérékoré
Yomou
Lola
Sinfra
Zuénoula
Daloa
Lac de Kossou
Tiébissou
YAMOUSSOUKRO
Dimbokro
Abengourou
Goaso
Bibiani
Awaso
Obuasi
GH
Kumasi
Robertsport
Tubmanburg
Gbanga
Sanniquellie
Ganta
Danané
Man
Biankouma
Duékoué
Guiglo
Bangolo
Bouaflé
Oumé
Issia
Gagnoa
Lakota
Divo
Agboville
Adzopé
Enchi
MONROVIA
Marshall
LIBERIA
Kakata
Harbel
Buchanan
River Cess
Zwedru
Taï
Toulepleu
Guiglo
Soubré
Guéyo
Sassandra
Fresco
Grand-Lahou
Grand-Bassam
Aboisso
Half Assini
Axim
Takoradi
Greenville
Grabo
San Pédro
Sassandra
ABIDJAN
Port-Bouet
Grand Cess
Plibo
Tabou
Grand-Béréby
Cape Three Points
Harper
Cape Palmas

Elevation scale:
6000m
4000m
3000m
2000m
1000m
500m
250m
100m
Sea Level
-250m
-2000m
-4000m

CAPE VERDE

Santo Antão
Pombas
Mindelo
São Vicente
Ilhas de Barlavento
Ribeira Brava
Amílcar Cabral
Pedra Lume
Sal
São Nicolau
Boa Vista
João Barrosa
ATLANTIC OCEAN
Tarrafal
Fogo
São Filipe
Maio
Maio
Santiago
PRAIA
Ilhas de Sotavento

Scale 1:8,000,000
0   50   100 Km
0   50   100 Miles

ASCENSION ISLAND (to UK)

North Point
Porpoise Point
North East Bay
Clarence Bay
Sisters Peak 446m
South East Point
GEORGETOWN
The Peak 859m
South East Bay
South West Bay
Portland Point
Mars Bay
South Point
Widewake Airfield
Pillar Bay
ATLANTIC OCEAN

Scale 1:750,000
0   5   10 Km
0   5   10 Miles

TRISTAN DA CUNHA (to UK)

ATLANTIC OCEAN
Big Point
Rookery Point
EDINBURGH
Anchorstock Point
Sandy Point
Queen Mary's Peak 2060m
Lyon Point
Longbluff
Cave Point
Stonybeach Bay
Stonyhill Point

Scale 1:750,000
0   5   10 Km
0   5   10 Miles

SAINT HELENA (to UK)

Sugar Loaf Point
Flagstaff Bay
JAMESTOWN
The Haystack 616m
Horse Pasture Point
Longwood
Egg Island
Diana's Peak 823m
Gill Point
South West Point
Long Range Point
Speery Island
Castle Rock Point
ATLANTIC OCEAN

Scale 1:750,000
0   5   10 Km
0   5   10 Miles

130
290
291

Elevation scale

6000m
4000m
3000m
2000m
1000m
500m
250m
100m
Sea Level
-250m
-2000m
-4000m

Countries and major regions:

NIGER — DIFFA, ZINDER, AGADEZ

SUDAN — NORTHERN DARFUR, WESTERN DARFUR, CENTRAL DARFUR, SOUTHERN DARFUR, EASTERN DARFUR, NORTHERN KORDOFAN, WESTERN KORDOFAN, SOUTHERN KORDOFAN

SOUTH SUDAN — NORTHERN BAHR EL GHAZAL, WESTERN BAHR EL GHAZAL, EL GHAZAL, UNITY, WARRAP, LAKES, WESTERN EQUATORIA, CENTRAL EQUATORIA

UGANDA

CHAD — BORKOU, KANEM, BAHR EL GAZEL, BATHA, WADI-FIRA, OUADDAÏ, SILA, GUÉRA, CHARI-BAGUIRMI, HADJER-LAMIS, SALAMAT, MOYEN-CHARI, MANDOUL, TANDJILÉ, LOGONE OCCIDENTAL, LOGONE ORIENTAL, MAYO-KEBBI EST, MAYO-KEBBI OUEST

CENTRAL AFRICAN REPUBLIC — VAKAGA, BAMINGUI-BANGORAN, HAUT-MBOMOU, MBOMOU, HAUTE-KOTTO, BASSE-KOTTO, NANA-GRÉBIZI, KÉMO, OUAKA, OMBELLA-MPOKO, NANA-MAMBÉRÉ, MAMBÉRÉ-KADÉÏ, SANGHA-MBAÉRÉ, LOBAYE, OUHAM, OUHAM-PENDE, BANGUI, LA LIKOUALA, LA SANGHA, CUVETTE-OUEST, OGOOUÉ-IVINDO

DEMOCRATIC REPUBLIC OF THE CONGO — ORIENTALE, ÉQUATEUR, BAS-UELE

CAMEROON — EXTRÊME-NORD, NORD, ADAMAWA, NORD-OUEST, SUD-OUEST, CENTRE, EST, LITTORAL, OUEST, SUD

NIGERIA — YOBE, BORNO, JIGAWA, KANO, KADUNA, BAUCHI, GOMBE, PLATEAU, TARABA, ADAMAWA, BENUE, CROSS RIVER

EQUATORIAL GUINEA

GABON — ESTUAIRE, WOLEU-NTEM

Major cities: N'DJAMENA, YAOUNDÉ, BANGUI, LIBREVILLE, MALABO, Kano, Maiduguri, El Obeid, El Fasher, Nyala, Wau, Bertoua, Douala

Physical features: Lake Chad, Congo Basin, Massif de l'Adamaoua, Chari, Logone, Congo, Sudd, Lake Albert

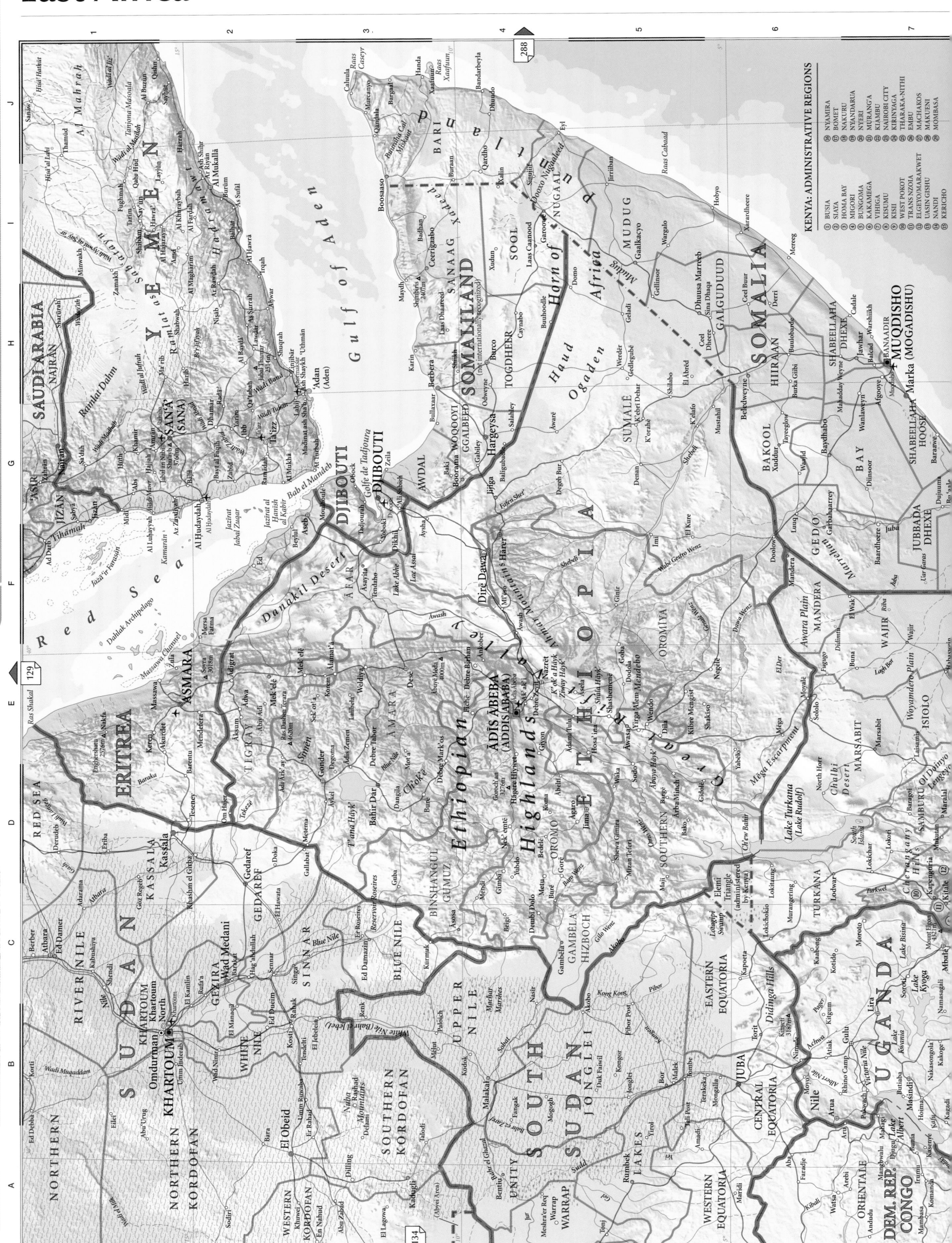

SOUTH AFRICA: CAPITAL CITIES

PRETORIA – administrative capital
CAPE TOWN – legislative capital
BLOEMFONTEIN – judicial capital

Scale 1:8,000,000
(projection: Lambert Azimuthal Equal Area)

0 25 50 75 100 125 150 175 200 Km
0 25 50 75 100 125 150 175 200 Miles

Population
■ above 5 million
◉ 100,000 to 500,000
▣ 1 million to 5 million
⊕ 50,000 to 100,000
◉ 500,000 to 1 million
○ 10,000 to 50,000
○ below 10,000

137

NORTHERN
EASTERN
CENTRAL
Mwanya
Kapandashila
Mfuwe
Kamoto
Kasungu
Lichinga
Metangula
Macaloge
Mataca
Unango
Quissanga
Mutsamudu
Nzwani
Comoro Islands
COMOROS
Mwali

CABO DELGADO
Nantulo
Natiroto
Macomia
Ancuabi
Baia de Pemba
Pemba
Mecúfi
Pamandzi
MAMOUDZOU
MAYOTTE
(to France)

LILONGWE
MALAWI
SOUTHERN
Monkey Bay
Dedza Mountain
2198m
NIASSA
Marrupa
Montepuez
Montepuez
Namuno
Ocua
Namapa
Lúrio
Memba
Baia de Memba

Lake Chilwa
Zomba
Blantyre
Mulanje
NAMPULA
Malema
Ribáuè
Nacala
Minguri
Baia de Fernão Veloso

MOZAMBIQUE
ZAMBÉZIA
Nampula
Murrupula
Liúpo
Quixaxe
Lumbo
Mogincual

HARARE
MASHONALAND
CENTRAL
Bindura
Mutoko
SOFALA
Gorongosa
Beira
Baia de Sofala

COMOROS
Comoro Islands
Mahajanga
Boriziny
Mampikony
MAHAJANGA
Tanjona
Vilanandro
Soalala
Mitsinjo
Marovoay
Besalampy
Morafenobe
Maintirano

MADAGASCAR
Nosy Glorieuses
Tanjona Bobaomby
Antsiraňana
ANTSIRAŇANA
Nosy Be
Ambilobe
Iharaùa
Ambanja
Maromokotro
2876m
Sambava
Analalava
Bealanana
Andapa
Antalaha
Antsohihy
Befandriana
Avaratra
Ambohitralanana
Mahajanga
Mahajanga
Boriziny
Mandritsara
MAHAJANGA
Maroantsetra
Tanjona
Masoala
Tanjona
Vilanandro
Soalala
Mitsinjo
Tsaratanana
Andilamena
Mananara Avaratra
Besalampy
Kandreho
Maevatanana
Soanierana-
Ivongo
Nosy Sainte Marie
Morafenobe
Ambatomainty
Farihy Alaotra
Amparafaravola
Fenoarivo
Atsinanana
Ambodifotatra
Maintirano
Ankazobe
Anjozorobe
Vavatenina
Ambatondrazaka
ANTANANARIVO
Antsalova
Tsiroanomandidy
Ambohidratrimo
Moramanga
Ampasimanolotra
Toamasina
Soavinandriana
Arivonimamo
Antananarivo
TOAMASINA
Belo Tsiribihina
Miandrivazo
Ambatolampy
Anosibe
An'ala
Vatomandry
Faratsiho
Antanambao
Manampotsy
Betafo
Fandriana
Mahanoro
Morondava
Mahabo
Ambato Finandrahana
Marolambo
Ambositra
Ambohimahasoa
Nosy Varika
Mandabe
Ikalamavony
Ifanadiana
Mananjary
Beroroha
Fianarantsoa
Ambalavao
Ikongo
FIANARANTSOA
Manakara
Ankazoabo
Ihosy
Ivohibe
Vohipeno
Morombe
Tanjona
Ankaboa
Sakaraha
Betroka
Iakora
Farafangana
Toliara
Onilahy
Benenitra
Midongy
Atsimo
Vangaindrano
Betioky
TOLIARA
Befotaka
Bekily
Ampanihy
Amboasary
Tôlaňaro
Beloha
Ambovombe
Tsiombe
Tanjona
Vohimena

MOZAMBIQUE
CHANNEL

INDIAN
OCEAN

MAPUTO
SWAZILAND
MBABANE
Matola
KWAZULU/
NATAL
Pietermaritzburg
Durban

INDIAN
OCEAN

Tropic of Capricorn

Ngazidja (Grande Comore)
Scale 1:4,500,000
0 20 40 60 80 Km
0 20 40 60 80 Miles
Mitsamiouli
Saondzou
1087m
Hahaya
Mbéni
MORONI
Koimbani
Le Kartala
2361m
Mitsoudjé
Foumbouni
Dembéni
COMOROS
INDIAN
OCEAN
Mwali (Mohéli)
Nzwani
(Anjouan)
Miringoni
Fomboni
Sima
Quani
Domoni
Nioumachoua
Ouanani
Moya
Mramani
Mozambique
Channel
Comoro Islands
MAYOTTE
(to France)
MAMOUDZOU
Dzaoudzi
Pamandzi
Bandrélé

Scale 1:8,000,000
0 25 50 75 100 125 150 Km
0 25 50 75 100 125 150 Miles

288

19,686ft
13,124ft
9843ft
6562ft
3281ft
1640ft
820ft
328ft
Sea Level
-820ft
-6562ft
-13,124ft

South Africa

A B C D E F G

1

2

3

4

5

291

6

7

8

9

10

AFRICA

140

6000m
4000m
3000m
2000m
1000m
500m
250m
100m
Sea Level
-250m
-2000m
-4000m

Cape Cross
Hentiesbaai
Wlotzkasbaken
Swakopmund
Walvis Bay
Pelican Point
Rooibank
Ilhea Point
Tropic of Capricorn
Conception Bay
Black Reef
Hollandsbird Island
St Francis Bay
Dolphin Head

E R O N G O

Omaruru
Karibib
Ebony
Usakos
Okahandja
Otjinbingwe
K H O M A S
WINDHOEK
Nonidas
Windhoek
Rössing
Dordabis
Bergland
Rehoboth
Khan
Swakop

Steinhausen
Black Nossob
Omitara
Witvlei
Gobabis
O M A H E K E

Buitepos
Mammo
Karakubis
Tswaane
Okwa
G H A N Z I

Central Kalahari Game Reserve

B O T S W

Kule
Takatswaane
Palamakoloi

K a l a h a r i

Ukwi

Leonardville
Aminuis

D e s e r t

Gobabed
Nauchas
Klein Aub
Petrusdal
Hoachanas
Derm

Hukuntsi
Tshane
K G A L A G A D I
Kokong
Kang

Mabutsane
Sekoma

Dutlwe
Motokwe
Khakhea

Bullsport
Solitaire
Kalkrand
Narib
H A R D A P
Hardap Dam
Mariental
Stampriet
Aranos
Nossob

Lendepas

N A M I B I A

Maltahöhe
Nomtsas
Witbooisvlei
Gochas

Bossiesvlei
Gibeon
Asab
Tweerivier
Kalahari Gemsbok National Park
Gemsbok National Park
Mabuasehube Game Reserve

Werda
Makopong
Bray
Vergelee
Tosca

Terra Firma
Verstershoop

Namaqualand
Helmeringhausen
Berseba
Tses

Maralaleng
Tshabong
Senlac
Morokweng
Piet Plessis

N O R T

Bethanie
Tsawisis
Keetmanshoop
Koës

Ewbank
Severn
Tlhakgameng
Ganyesa

K A R A S
Fish
Twee Rivieren
Khuis
Aansluit
Van Zylsrust
Klein-Tswaing

Lüderitz
Garub
Goageb
Sandverhaar
Seeheim Noord
Lower
Narubis
Aroab
Rietfontein
Hakskeenpan
Addriesvale
Witdraai
Askham
Lower
Dikgathong
Tsineng
Lolwane

Grillenthal
Aus
Chamaites
Kuruman
Cramond
Ontemoeting
Sonstraal
Sutton
Hotazel

Possession Island
Klein Karas
Obobogorab
Abierkwasputs
Koopankuil
Noenieput
Wincanton
Dibeng
Kuruman
Kathu
Wesselsvlei

Witputz
Gorges
Grünau
Karasburg
Vrouenspan
Swartmodder
Moeswal
Langkloof Pass
Sishen
Dingleton
Blinkfontein

Chamais Bay
Rosh Pinah
Ai-Ais
Haib
Kums
Nakop
Langklip
Geluksprult
Upington
Karos
Grootdrink
Olifantshock
Danielskuil

Ai-Ais Hot Springs & Fish River Canyon Park
Ariamsvlei
Langkip
Postmasburg
Owendale
Silver Streams
Swartp

Affenrücken
Sendelingsdrif
Warmbad
Kokerboom
Lutzputs
Augrabies
Keimoes
Neilersdrif
Louisvale
Kanoneiland
Kalkwerf
Volop
Beeshock

Richtersveld National Park
Kuboes
Holgat
Onseepkans
Marchand
Kakamas
Wegdraai
Kleinbegin
Groblershoop
Griekwastad
Campbell
Barkly

Oranjemund
Vioolsdrif
Goodhouse
Dabenoris
Pella
Pofadder
Nabies
Koegrabie
Pussonderwater
Rooidam
Marydale
Westerberg
Niekerkshoop
Bucklands
Douglas

Alexander Bay
Eksteenfontein
Lekkersing
Steinkopf
Namies
Bossiekom
Kenhardt
Koegasbrug
Prieska
Salt lake

Wreck Point
Cliff Point
Aninauspos
Bulletrap
Concordia
Uitkyk
Koegabie
Marydale
N O R T H E R N   C A P E
Asbesberge
Hopetown

Port Nolloth
McDougall's Bay
Wedge Point
Nababeep
Okiep
Springbok
Gamoep
Jagt Drift
Groot Vloer
Kareeboskolk
Copperton
Omdraaisvlei
Poupan
Sodium

Grootmis
Kleinsee
Melkbospunt
Buffels
Messelpad Pass
Burke's Pass
Granaatboskolk
Halfweg
Swartkop
Ondersteorings
Van Wyksvlei
Vosburg
Britstown
De Aar

Skulpfonteinpunt
Soebatsfontein
Kamieskroon
Leliefontein
Stofvlei
Brandvlei
Vanwyksvleidam
Giesenskraal
Houwaterdam
Philipstown

Koingnaas
Hondeklipbaai
Wallekraal
Karkams
Platbakkies
Riet se Vloer
Kareebospoort
Smartt Syndicate Dam
Die Put

Strandfonteinpunt
Witwater
Garies
Kliprand
Sakrivier
Carnarvon
Groen
Deelfontein

Island Point
Groen
Nariep
Rietpoort
Loeriesfontein
Tontelbos
Kareeberge
Pampoenpoort

Kotzesrus
Bitterfontein
Nuwerus
Komkans

Hantams
Calvinia
Williston
Loxton
Meltonwold
Victoria West
Hutchinson

A T L A N T I C

Roodeduinen Point
Lambert's Bay
Strandfontein
Vredendal
Vanrhynsdorp
Klawer
Trawal
Heerenlogement
Botterkloof Pass
Middelpos
Bonekraal
Sutherland
Fraserburg
Beaufort West
Murraysburg
Restvale
Aberdeen

O C E A N

Baboon Point
Noordkuil
Redelinghuys
Clanwilliam
Cederberg
Wupperthal
Hondefontein
Merweville
Kamsberg
Koringplaas
Letjiesbos

St Helena Point
Elands Bay
Citrusdal
Grootwater
W E S T E R N   C A P E
Sutherland
Rooikloof Pass
Seekoegat
Willowmore

St Helena Bay
Paternoster
Piketberg
Porterville
Renosterveld
Laingsburg
Vleifontein
Prince Albert

Cape Columbine
Vredenburg
Saldanha
Hopefield
Moorreesburg
Tulbagh
Ceres
Witberg
Oudtshoorn

Saldanha Bay
Churchhaven
Yzerfontein
Yzerfonteinpunt
Dassen Island
Malmesbury
Riebeek-Wes
Wolseley
Tows River
Montagu
George
Knysna

Bokpunt
Atlantis
Darling
Abbotsdale
Wellington
Paarl
Worcester
Robertson
Swellendam
Mosselbaai

Robben Island
Table Bay
CAPE TOWN
Bellville
Stellenbosch
Somerset-West
Strand
Caledon
Protem
Vleesbaai

Simon's Town
False Bay
Kleinmond
Hermanus
Bredasdorp
St Sebastian Bay

Cape of Good Hope
Cape Point
Walker Bay
Danger Point
Elim
Skipskop
Struisbaai
L'Agulhas
Cape Agulhas

SOUTH AFRICA: CAPITAL CITIES

PRETORIA – administrative capital
CAPE TOWN – legislative capital
BLOEMFONTEIN – judicial capital

Scale 1:4,650,000
(projection: Lambert Azimuthal Equal Area)

| 0 | 20 | 40 | 60 | 80 | 100 Km |
| 0 | 20 | 40 | 60 | 80 | 100 Miles |

Population

- ■ above 5 million
- ■ 1 million to 5 million
- ◉ 500,000 to 1 million
- ◎ 100,000 to 500,000
- ⊕ 50,000 to 100,000
- ○ 10,000 to 50,000
- ○ below 10,000

ZIMBABWE

MOZAMBIQUE

INHAMBANE

GAZA

LIMPOPO
(NORTHERN)

CENTRAL

KGATLENG

GABORONE

SOUTHEAST

MPUMALANGA

PRETORIA
GAUTENG
Johannesburg
Krugersdorp
Soweto
Vereeniging

MATOLA
MAPUTO
MBABANE
SWAZILAND

FREE STATE

A F R I C A

BLOEMFONTEIN

MASERU
LESOTHO

KWAZULU/NATAL

Pietermaritzburg
Durban

EASTERN CAPE

INDIAN
OCEAN

Port Elizabeth
East London
Mdantsane
Bhisho

Tropic of Capricorn

| 19,686ft |
| 13,124ft |
| 9843ft |
| 6562ft |
| 3281ft |
| 1640ft |
| 820ft |
| 328ft |
| Sea Level |
| -820ft |
| -6562ft |
| -13,124ft |

139

288

293

AFRICA

142

Algiers

L'Ermitage

Mediterranean Sea

Cap de Bordj

Bab El Oued
Grande Mosquée
Kasbah
El Biar
Palais du Gouvernement
Chéraga
Cité Olympique
Ben Aknoûn
Agha
Algiers (Alger)
Bordj El Bahri
Bordj El Kiffan

Musée des Beaux Arts
Birmandreis
Hussein-Dey
Kouba

Draria
Birkhadem
El Harrach
Dar El Beïda

Douera
Baraki
Oued Smar
Algiers Airport
Oued Harrach

3 Km
3 Miles

Cairo

Abu Al Ghayt
Bahtim
El Matariya
Cairo International Airport

Nile
Shubra Al Amiriya
El Zeitûn
Masr el Gedida (Heliopolis)

Warrâq el Hadr
Shubra Al Khaymah

Warrâq el'Arab
Mâdinet Nasr

Imbâbah
Bûlâq
El Ezbekiya

Aguza
Âbdin
Cairo (Al Qâhirah)
Egyptian Antiquities Museum
Central Government Building

El Duqqi
The Citadel
Zoological Gardens
Garden City

Giza (Al Jîzah)
Masr el Qadima

El Basâtin

Cheops
Sphinx
Pyramids of Giza
El Ma'âdi
Nile

3 Km
3 Miles

Cape Town

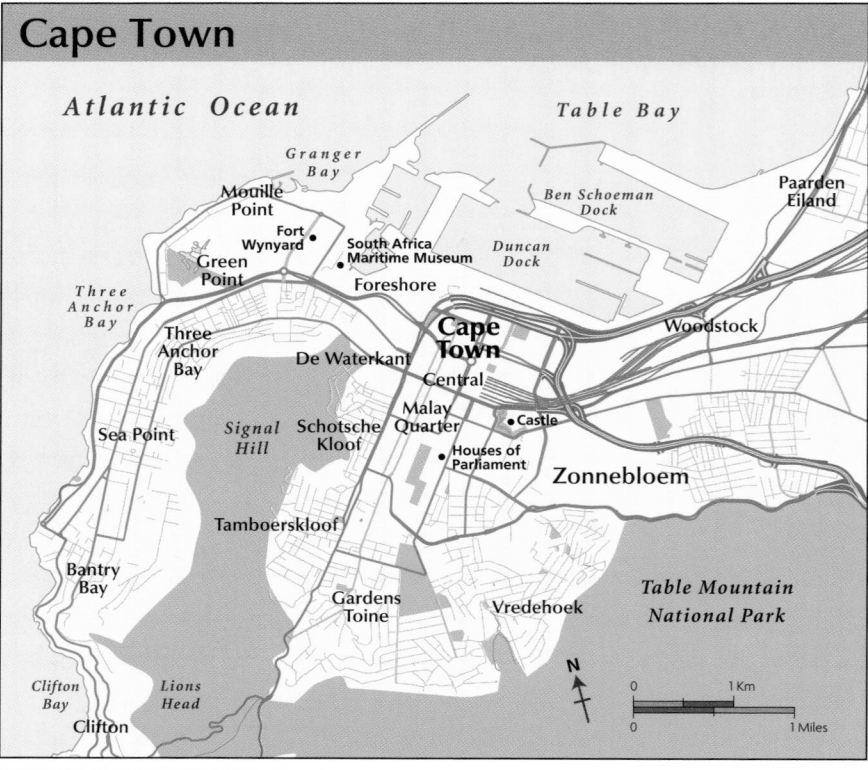

Atlantic Ocean

Table Bay

Granger Bay
Mouille Point
Fort Wynyard
Green Point
South Africa Maritime Museum
Foreshore

Ben Schoeman Dock
Duncan Dock
Paarden Eiland

Three Anchor Bay
Three Anchor Bay
De Waterkant
Cape Town
Central
Woodstock

Sea Point
Signal Hill
Schotsche Kloof
Malay Quarter
Castle
Houses of Parliament
Zonnebloem

Bantry Bay
Tamboerskloof

Clifton Bay
Lions Head
Gardens Toine
Vredehoek
Table Mountain National Park

Clifton

N

1 Km
1 Miles

Casablanca

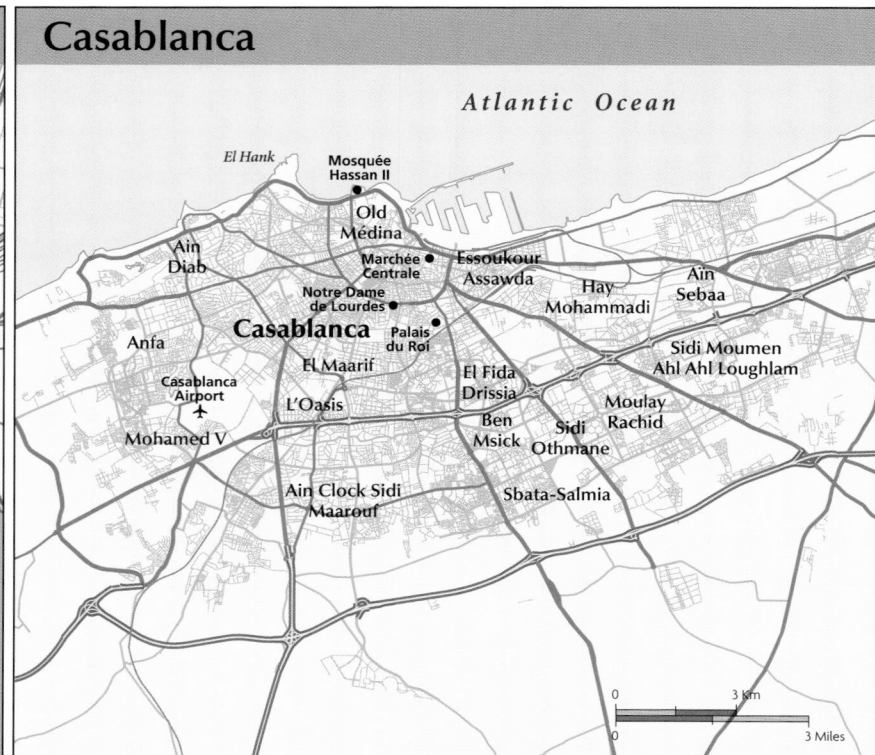

Atlantic Ocean

El Hank
Mosquée Hassan II
Old Médina
Ain Diab
Marchée Centrale
Essoukour Assawda
Hay Mohammadi
Aïn Sebaa

Anfa
Notre Dame de Lourdes
Casablanca
Palais du Roi
Sidi Moumen Ahl Ahl Loughlam

Casablanca Airport
El Maarif
El Fida Drissia
Moulay Rachid

Mohamed V
L'Oasis
Ben Msick
Sidi Othmane

Ain Clock Sidi Maarouf
Sbata-Salmia

3 km
3 Miles

Dakar

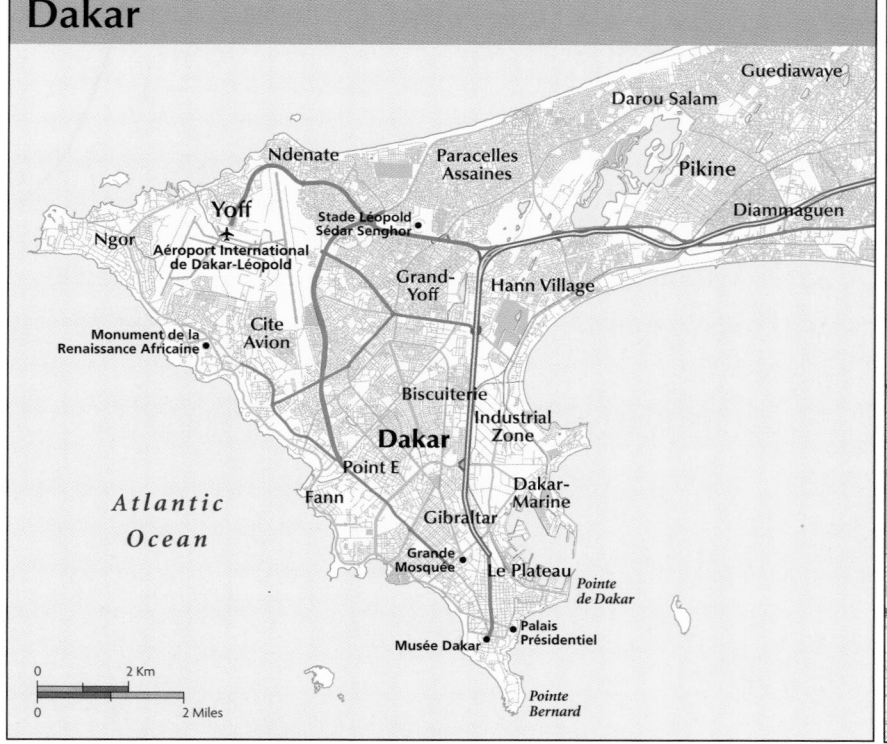

Guediawaye
Darou Salam

Ndenate
Paracelles Assaines
Pikine

Yoff
Stade Léopold Sédar Senghor
Diammaguen

Ngor
Aéroport International de Dakar-Léopold
Grand-Yoff
Hann Village

Monument de la Renaissance Africaine
Cite Avion

Biscuiterie
Industrial Zone

Dakar
Point E
Dakar-Marine

Fann
Gibraltar

Atlantic Ocean

Grande Mosquée
Le Plateau
Pointe de Dakar

Musée Dakar
Palais Présidentiel

Pointe Bernard

2 Km
2 Miles

Harare

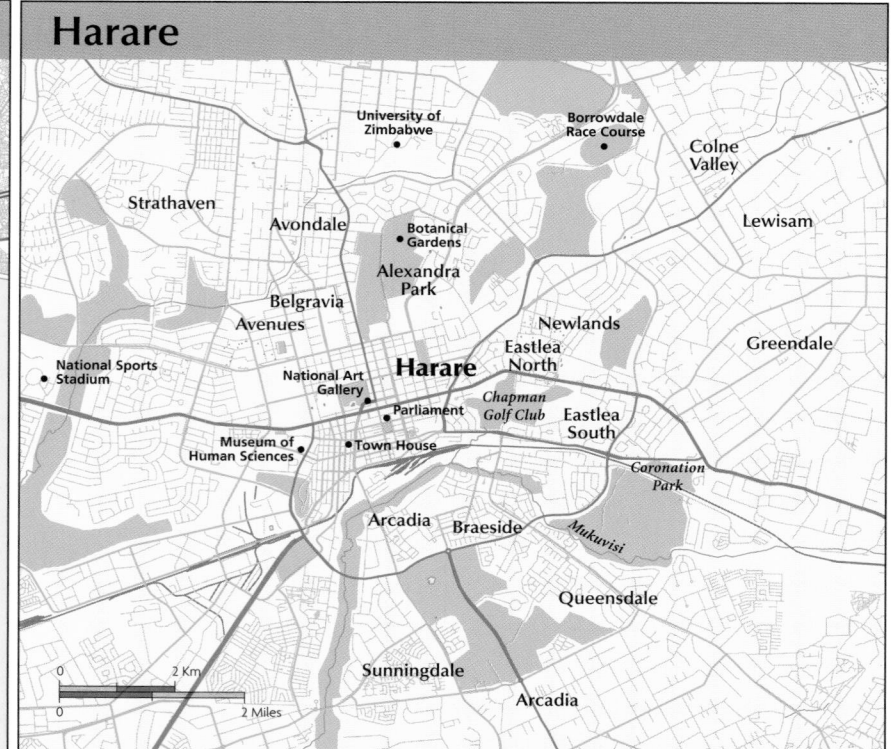

University of Zimbabwe
Borrowdale Race Course
Colne Valley

Strathaven
Avondale
Botanical Gardens
Lewisam

Belgravia Avenues
Alexandra Park
Newlands
Eastlea North
Greendale

National Sports Stadium
National Art Gallery
Harare
Parliament
Chapman Golf Club
Eastlea South

Museum of Human Sciences
Town House
Coronation Park

Arcadia
Braeside
Mukuvisi
Queensdale

Sunningdale
Arcadia

2 Km
2 Miles

Johannesburg

Diepsloot N.R.

Tembisa

Fourways

Olivedale

Ferndale

Modderfontein

Sandton

Kempton Park

Krugersdorp

Alexandra

Roodepoort

Edenvale

Johannesburg

O.R. Tambo International Airport

Florida

Bedfordview

Wits University

National School of Arts

National Exhibition Centre

Johannesburg Library

Museum of Africa

Boksburg

Germiston

Soweto

Klipriviersberg N.R.

Alberton

Elsburg

Lenasia

Klip

0 10 Km
0 10 Miles

Kinshasa

Congo

Palais de Nation

Palace de Justice

Gombe

Kinshasa

Mont Ngaliema

Musée de Kinshasa

Lingwala

Barumbu

Ngaliema

Kintambo

Binza Ozone

Kasa-Vubu

Bandalungwa

Kalamu

Binza Meteo

Ngiri-Ngiri

Limete

Bumbu

Binza Delvaux

Selembao

Makala

Ngaba

Matete

Masina

Kinsenso

Ndjili

Ngafula

Kimbanseke

0 3 Km
0 3 Miles

Lagos

Lagos Lagoon

Yaba

Ebute-Metta

Iganmu

National Theatre

Ijora

Oba's Palace

Central Mosque

Bamgboshe

Lagos Island

Lagos

Onikan National Museum

Moba

Apapa

Obalende

Ikoyi

Falomo

Lekki Peninsula

Lagos Harbour

Apapa Warf

Five Cowrie Creek

Maroko

Ogoyo

Porto Novo Creek

Victoria Island

Alaguntan

Atlantic Ocean

Tarqua Bay

0 2 Km
0 2 Miles

Nairobi

Ndenderu

Kasarini

Muthaiga

Kitisuru

Huruma

Westlands

Kileleshwa

Kenya National Theatre

Pumwani

Umoja

Maziwa

All Saints Cathedral

Parliament Buildings

Kimathi

Nairobi

Nairobi

Embakasi

Nairobi South

Kuwinda

Kawa Njenga

Nairobi Airport

Nairobi Hill

Nairobi National Park

0 2 Km
0 2 Miles

Tripoli

Mediterranean Sea

Harbour

Gurji Mosque

Assaraya Al Hamra

Mitiga Intl. Airport

Al Madinah

People's Palace

Sidi al Marsri

Suq al Juma'a

Tripoli (Tarābulus)

Annasr Forest

Al-Seyaheyya

University of Tripoli

Tajoura

Janzur

Abu Salim

Ain Zara

Hai Alsslam

Wadi al Migynin

0 5 Km
0 5 Miles

Tunis

Sebkhet Ariana

La Marsa

Bou Saïd

Tunis-Carthage Airport

Parc Archéologique

El Manar

El Aouine

Musée Océanographique

Université

Tunis

Cité Olympique

Carthage

Parc du Belvédère

Cité El Zhadra

Bardo

Lac du Tunis

La Goulette

Musée du Bardo

La Médina

El Bhira

Mediterranean Sea

Mégrine

Rades

Golfe de Tunis

El Bhira

Sebkhet Sejoumi

Ben Arous

Ez Zahra

Oued Méliane

Habeul

Hammam Lif

0 5 Km
0 5 Miles

Europe is the world's second smallest continent with a total area of 4,053,309 sq miles (10,498,000 sq km). It comprises 46 separate countries, including Turkey and the Russian Federation, although the greater parts of these nations lie in Asia.

FACTFILE

N **Most Northerly Point:** Ostrov Rudol'fa, Russian Federation 81° 47′ N
S **Most Southerly Point:** Gávdos, Greece 34° 51′ N
E **Most Easterly Point:** Mys Flissingskiy, Novaya Zemlya, Russian Federation 69° 03′ E
W **Most Westerly Point:** Bjargtangar, Iceland 24° 33′ W

Largest Lakes:
1. Lake Ladoga, Russian Federation 7100 sq miles (18,390 sq km)
2. Lake Onega, Russian Federation 3819 sq miles (9891 sq km)
3. Vänern, Sweden 2141 sq miles (5545 sq km)
4. Lake Peipus, Estonia/Russian Federation 1372 sq miles (3555 sq km)
5. Vättern, Sweden 737 sq miles (1910 sq km)

Longest Rivers:
1. Volga, Russian Federation 2265 miles (3645 km)
2. Danube, C Europe 1771 miles (2850 km)
3. Dnieper, Belarus/Russian Federation/Ukraine 1421 miles (2287 km)
4. Don, Russian Federation 1162 miles (1870 km)
5. Pechora, Russian Federation 1124 miles (1809 km)

Largest Islands:
1. Britain, 88,700 sq miles (229,800 sq km)
2. Iceland, 39,315 sq miles (101,826 sq km)
3. Ireland, 31,521 sq miles (81,638 sq km)
4. Ostrov Severny, Novaya Zemlya, Russian Federation 18,177 sq miles (47,079 sq km)
5. Spitsbergen, Svalbard 15,051 sq miles (38,981 sq km)

Highest Points:
1. El'brus, Russian Federation 18,510 ft (5642 m)
2. Dykhtau, Russian Federation 17,077 ft (5205 m)
3. Koshtantau, Russian Federation 16,877 ft (5144 m)
4. Gora Kazbek, Georgia/Russian Federation 16,647 ft (5074 m)
5. Gora Dzhangitau, Georgia/Russian Federation 16,571 ft (5051 m)

Lowest Point:
▼ Caspian Depression, Russian Federation -92 ft (-28 m) below sea level

Highest recorded temperature:
⊕ Seville, Spain 122°F (50°C)

Lowest recorded temperature:
⊖ Ust'-Shchuger, Russian Federation -72.6°F (-58.1°C)

Wettest Place:
≋ Crkvice, Bosnia and Herzegovina 183 in (4648 mm)

Driest Place:
⌒ Astrakhan', Russian Federation 6.4 in (162.5 mm)

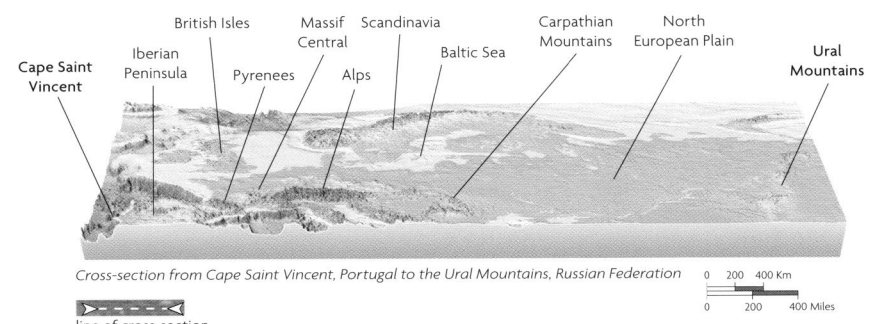

Cross-section from Cape Saint Vincent, Portugal to the Ural Mountains, Russian Federation

▷ ─ ▷ ─ ▷
line of cross-section

0   200   400 Km
0   200   400 Miles

Political

The political boundaries of Europe have changed many times, especially during the 20th century in the aftermath of two world wars, the break-up of the empires of Austria-Hungary, Nazi Germany, and, toward the end of the century, the collapse of communism in eastern Europe. The fragmentation of Yugoslavia has again altered the political map of Europe, highlighting a trend towards nationalism and devolution. In contrast, economic federalism is growing. In 1958, the formation of the European Economic Community (now the European Union or EU) started a move toward economic and political union and increasing internal migration. This process is still ongoing and the accession of Bulgaria and Romania in January 2007, and Croatia in 2013, brought the number of EU member states to twenty eight. Of these, nineteen have joined the Eurozone by adopting the Euro as their official currency.

Population
- ▪ above 5 million
- ▪ 1 million to 5 million
- ◉ 500,000 to 1 million
- ◎ 100,000 to 500,000
- ⊕ 50,000 to 100,000
- ○ 10,000 to 50,000
- ● Country capital

Borders
- full international border

Scale 1:17,250,000
(projection: Lambert Azimuthal Equal Area)

Km
0 100 200 300 400 500 600 700

Miles
0 100 200 300 400 500 600 700

AN
ION
Ural Mountains
Perm'
Ufa
Orenburg
Samara
Kazakhstan
Volga
Astrakhan'
Caspian Sea
Groznyy
Caucasus
Georgia
Azerbaijan

Languages

There are three main European language groups: Germanic languages predominate in central and northern Europe; Romance languages in western and Mediterranean Europe and Romania; while Slavic languages are spoken in eastern Europe and the Russian Federation. Isolated pockets of local languages, such as Basque and Gaelic, persist and frequently provide a focus for national identity.

Language groups
Turkic
Albanian
Finno-Ugric/Samoyed
Germanic
Slavic
Romance
Basque
Baltic
Celtic
Greek
Caucasian
Iranian
Mongol
ICELANDIC
FAROESE
NENETS
KOMI
LAPPISH (SAMI)
NORWEGIAN
SWEDISH
FINNISH
KARELIAN
SWEDISH
GAELIC
ENGLISH
IRISH
ENGLISH
ENGLISH
WELSH
VEPS
RUSSIAN
UDMURT
MARI
CHUVASH
TARTAR
BASHKIR
MORDVINIAN
SWEDISH
ESTONIAN
LATVIAN
KARELIAN
DANISH
LITHUANIAN
RUSSIAN
POLISH
FRISIAN
DUTCH
BRETON
FRENCH
GERMAN
POLISH
BELORUSSIAN
CZECH
SLOVAK
UKRAINIAN
KALMYK
KABARDIAN
FRENCH
GERMAN
HUNGARIAN
GALICIAN
BASQUE
SLOVENE
CROATIAN
ROMANIAN
ADYGHE
KARACHAY
KUMYK
CHECHEN
AVAR
LEZGHIAN
OSSETIAN
BALKAR
PORTUGUESE
SPANISH
CATALAN
FRENCH/CORSICAN
ITALIAN
BOSNIAN
SERBIAN
BULGARIAN
MACEDONIAN
TURKISH
ALBANIAN
CATALAN
SARDINIAN
ITALIAN
GREEK
MALTESE

Population

Europe is a densely populated, urbanized continent; in Belgium over 90% of people live in urban areas. The highest population densities are found in an area stretching east from southern Britain and northern France, into Germany. The northern fringes are only sparsely populated.

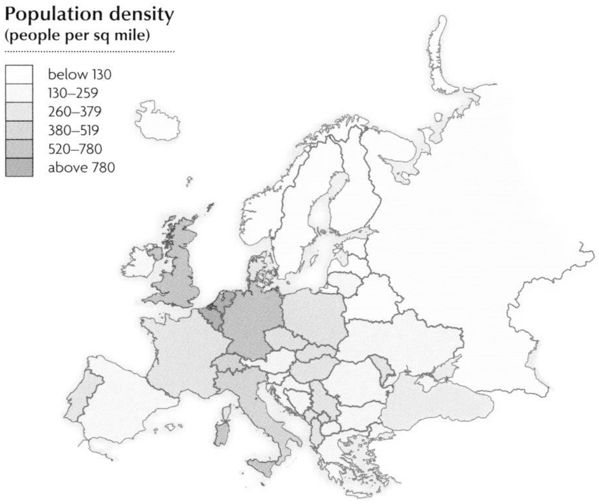
Population density
(people per sq mile)
below 130
130–259
260–379
380–519
520–780
above 780

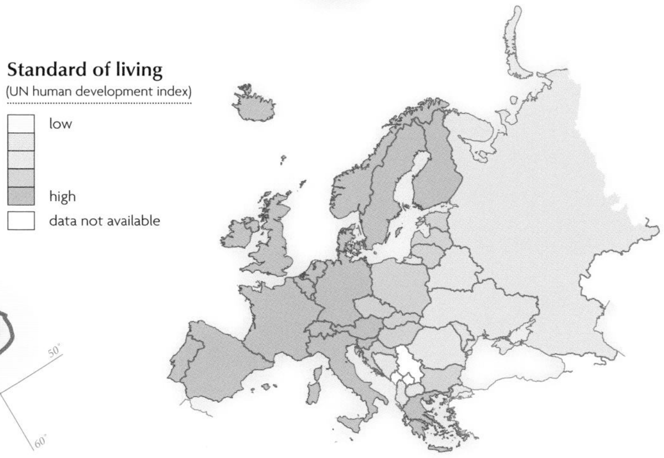
Standard of living
(UN human development index)
low
high
data not available

Standard of living

Living standards in western Europe are among the highest in the world, although there is a growing sector of homeless, jobless people. Eastern Europeans have lower overall standards of living—a legacy of stagnated economies.

Transportation

Despite its fragmented geography and many natural frontiers, communications in Europe are well developed. Extensive motorway links allow rapid road transportation, while high-speed rail connections like France's TGV *(Train à Grande Vitesse)*, and the Channel Tunnel have improved rail travel. Outdated communication infrastructures in parts of eastern Europe, and insufficient transport links across the Alps, however, remain weak parts of the network.

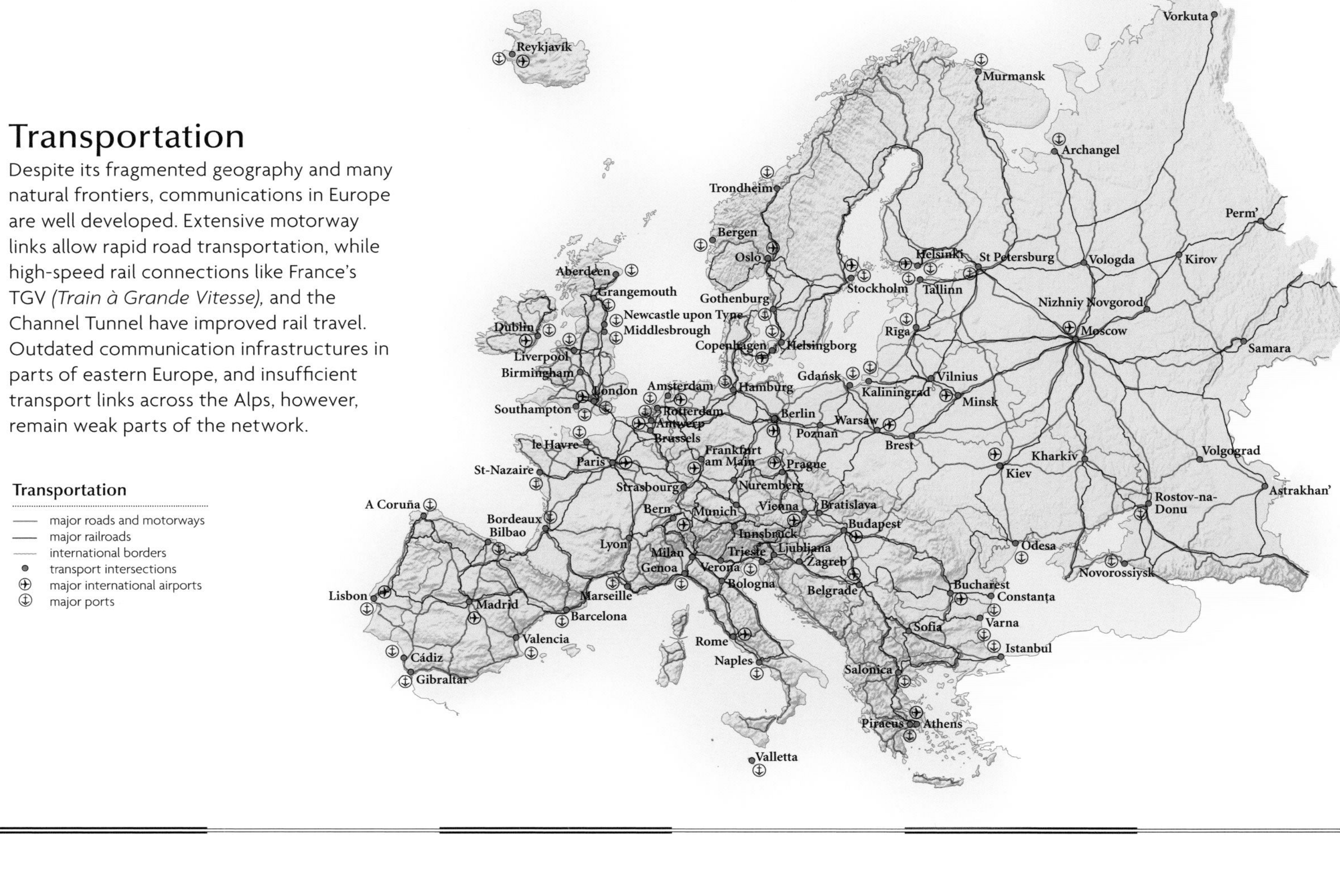
Transportation
major roads and motorways
major railroads
international borders
transport intersections
major international airports
major ports
Reykjavík
Vorkuta
Murmansk
Archangel
Trondheim
Bergen
Oslo
Aberdeen
Grangemouth
Newcastle upon Tyne
Middlesbrough
Gothenburg
Helsinki
St Petersburg
Vologda
Kirov
Perm'
Stockholm
Tallinn
Nizhniy Novgorod
Dublin
Liverpool
Birmingham
Copenhagen
Helsingborg
Riga
Moscow
Samara
London
Southampton
Amsterdam
Hamburg
Gdańsk
Kaliningrad
Vilnius
Minsk
Rotterdam
Antwerp
Brussels
Berlin
Poznań
Warsaw
Brest
Le Havre
St-Nazaire
Paris
Frankfurt am Main
Prague
Kharkiv
Kiev
Volgograd
Strasbourg
Nuremberg
Bratislava
A Coruña
Bern
Munich
Vienna
Rostov-na-Donu
Astrakhan'
Bordeaux
Bilbao
Lyon
Innsbruck
Budapest
Milan
Trieste
Ljubljana
Zagreb
Odesa
Genoa
Verona
Bologna
Novorossiysk
Lisbon
Madrid
Marseille
Belgrade
Bucharest
Constanța
Barcelona
Sofia
Varna
Valencia
Rome
Istanbul
Cádiz
Naples
Gibraltar
Salonica
Piraeus
Athens
Valletta

A B C D E F G

1

ARCTIC OCEAN

King Frederik
VIII Land
Ostrov
Rudol'fa

Franz Josef Land

Mys Flissingskiy

Greenland

Spitsbergen

Ostrov Severnyy

pyasina

*Putorama
Mountains*

North Siberian Lowland

2

*Greenland
Sea*

Limit of Summer Pack Ice

Bjørnøya

**Barents
Sea**

Ostrov
Yuzhnyy

Novaya
Zemlya

Ostrov Belyy

Gydanskiy Poluostrov

Yenisey

King Christian X Land

Kara Sea

Poluostrov Yamal

Gulf of Ob

Taz

Pur

*West
Siberian
Plain*

3

Denmark
Strait

Kolbeinsey Ridge

NORTH AMERICAN PLATE
EURASIAN PLATE

Jan Mayen Fracture Zone

Jan Mayen

Tromsøflaket

Limit of Winter Pack Ice

North Cape
Nordkinn

Nordkapp

Fugløya
Bank

Murmansk Rise

Ostrov
Kolguyev

Poluostrov
Kanin

Timanskiy Kryazh

Mezen'

Gora Narodnaya
1895m

Pechora

Kara Strait

Baydaratskaya Guba

Ob'

Ob'

Irtysh

ASIA

Arctic Circle

Iceland
Plateau

Jan Mayen Ridge

*Norwegian
Basin*

Inarijärvi

Kebnekaise
2117m

Kola
Peninsula

Ozero
Imandra

White
Sea

Northern Dvina

Yug

Vychegda

4

Bjargtangar

Iceland

Vatnajökull

Faroe–Iceland Ridge

Voring
Plateau

*Norwegian
Sea*

Traena
Bank

Vesterålen

Lofoten

S c a n d i n a v i a

Kemijoki
Oulujoki

Onega Bay

Ozero
Beloye
Ozero
Vygozero

Sukhona

Vaga

Tobol

Faroe
Islands

Galdhøpiggen
2469m

Glåma

Umeälven

Lake Onega

Yug

Uy

Insev'

Ishim

5

Hatton Ridge

Rockall
Rise

Feni Ridge

Rockall Trough

Bill Baileys
Bank

Outer
Hebrides

Shetland
Islands

Orkney
Islands

Viking
Bank

Norwegian Trench

Lule

Gøta

Dal

Vänern

Åland

Gulf of Bothnia

Gulf of Finland

Lake
Peipus

Lake Ladoga

Lake Pskov

Lake Ilmen

Msta

Ozero
Vygozero

Rybinsk
Reservoir

Gor'kiy
Reservoir

Mologa

Oka

Belaya

Chusovaya

Votkinsk
Reservoir

Kuybyshev
Reservoir

U r a l   M o u n t a i n s

Ural

6

Faroe–Shetland Trough

Grampian
Mountains

Ben Nevis
1343m

North
Channel

**British
Isles**

Ireland

Shannon

Irish
Sea

St. George's
Channel

Britain

Snowdon
1085m

Severn

The
Fens

Dogger
Bank

Great Fisher
Bank

Jutland
Bank

Skagerrak

Kattegat

Jutland

Sjælland

Vättern

Gotland

Gulf of
Riga

Baltic Sea

Neman

Western Dvina

North European Plain

Dnieper

**Central
Russian
Upland**

Don

Volga Uplands

Volga

Kirghiz Steppe

Aral Sea

*Caspian
Depression*

7

Porcupine
Plain

Celtic Sea

Celtic
Shelf

Bristol Channel

Land's
End

English Channel

Channel Islands

Strait of Dover

Frisian Islands

Elbe

Oder

Vistula

Warta

Rhine

E U R O P E

Bug

Western Bug

Pripet
Marshes

Byerezino

Desna

Dnieper Lowlands

Seym

Kiev
Reservoir

Dnieper

Kremenchuk
Reservoir

Podil's'ka Vysochina

Donets

Southern Bug

Tsimlyansk
Reservoir

Yergeni

Ustyurt
Plateau

Biscay
Plain

Theta Gap

Galicia
Bank

Bay of
Biscay

Massif
Central

Seine

Marne

Loire

Vienne

Cher

Lake
Geneva

Ardennes

Harz

Danube

Black Forest

Lake Constance

Danube

Matra

Bakony

Tisza

Carpathian Mountains

**Great
Hungarian
Plain**

Drava

Sava

Siret

Prut

Dniester

Danube

Black Sea Lowland

Sea of
Azov

Crimea

Kuban

Manych

Don

Terek

**Caspian
Sea**

EURASIAN PLATE

8

Mt Blanc
4807m

Anceto
3404m

Cévennes

Cordillera Cantabrica

Pyrenees

Sistema Ibérico

Ebro

A l p s

Po

Lot

Garonne

Rhône

Jura

Dolomites

Lake
Garda

A
p
e
n
n
i
n
e
s

D i n a r i c   A l p s

Transylvanian Alps

Balkan Mountains

Danube

Maritsa

Rhodope
Mountains

Vardar

B o s p o r u s

Sea of
Marmara

EURASIAN PLATE
ANATOLIAN PLATE

A n a t o l i a

Caucasus
El'brus
5642m

Kura

Lake
Van

Mount Ararat
5137m

Aras

Kizel Owzan

Elbruz
Mts

**Iberian
Peninsula**

Cubo
da Roca

Douro

Duero

Sistema Central

Guadiana

Sierra Morena

Guadalquivir

Sistemas Béticos

Sierra Nevada

Segura

Júcar

Gulf of
Valencia

Balearic
Islands

Algerian
Basin

Corsica

Strait of Bonifacio

Sardinia

Tyrrhenian
Sea

Tyrrhenian
Basin

Ligurian
Sea

Corno Grande
2912m

Vesuvius
1277m

Adriatic
Sea

Gulf of
Taranto

Strait of Messina

Strait of Otranto

Adriatic
Basin

Lake Scutari

Lake
Ohrid

Lake
Prespa

Pindus Mountains

Gulf of
Antalya

Taurus Mountains

Cyprus

Cyprus
Basin

Lake Tuz

Lake
Urmia

Nahr al Khabur

Zagros Mountains

EURASIAN PLATE
AFRICAN PLATE

Euphrates

Tigris

9

Cape
Saint
Vincent

Strait of
Gibraltar

Rif

Alboran
Sea

EURASIAN PLATE
AFRICAN PLATE

Oued Chelf

M e d i t e r r a n e a n

Strait of Sicily

Sicily

Etna
3263m

Malta

Ionian Sea

Ionian
Basin

Peloponnese

Aegean
Sea

Mirtoan
Sea

Sea of Crete

Gávdos

Crete

Karpathos Basin

Kasos Strait

Rhodes

ANATOLIAN PLATE
AFRICAN PLATE

Mediterranean Ridge

Levantine Basin

Nile Fan

Dead
Sea

**Syrian
Desert**

Buhayrat ath
Tharthar

Buhayrat ar
Razazah

Jordan

An Nafud

Omer Rbia

Sebou

Middle Atlas

High Atlas

**Atlas
Mountains**

Moulouya

Tell Atlas

Saharan Atlas

Chott el Jerid

Gulf of
Sirte

Qattara Depression
-133m

S e a

Nile

Suez
Canal

Sinai

Gulf of Suez

ARABIAN PLATE
AFRICAN PLATE

R e d   S e a

Eastern Desert

**Arabian
Peninsula**

Tropic of Cancer

10

Tropic of Cancer

'Erg Iguidi

A F R I C A

Erg Chech

S a h a r a

Grand Erg Occidental

Grand Erg Oriental

Libyan Desert

Western Desert

Scale 1:22,500,000
(projection: Lambert Conformal Conic)

0 200 400 600 800 1000 Km

0 200 400 600 800 1000 Miles

Climate

Europe experiences few extremes in either rainfall or temperature, with the exception of the far north and south. Along the west coast, the warm currents of the North Atlantic Drift moderate temperatures. Although east–west air movement is relatively unimpeded by relief, the Alpine Uplands halt the progress of north–south air masses, protecting most of the Mediterranean from cold, north winds.

Average Rainfall

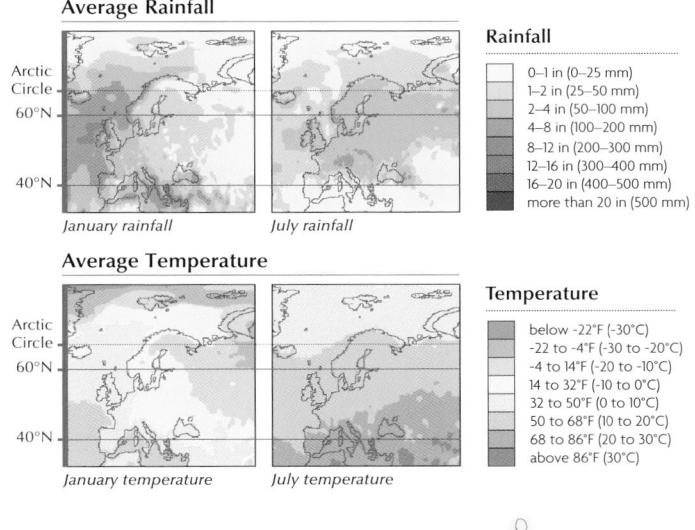

January rainfall   July rainfall

Rainfall

- 0–1 in (0–25 mm)
- 1–2 in (25–50 mm)
- 2–4 in (50–100 mm)
- 4–8 in (100–200 mm)
- 8–12 in (200–300 mm)
- 12–16 in (300–400 mm)
- 16–20 in (400–500 mm)
- more than 20 in (500 mm)

Average Temperature

January temperature   July temperature

Temperature

- below -22°F (-30°C)
- -22 to -4°F (-30 to -20°C)
- -4 to 14°F (-20 to -10°C)
- 14 to 32°F (-10 to 0°C)
- 32 to 50°F (0 to 10°C)
- 50 to 68°F (10 to 20°C)
- 68 to 86°F (20 to 30°C)
- above 86°F (30°C)

Climate

- tundra
- subarctic
- cool continental
- warm humid
- mediterranean
- semi-arid
- daily hours of sunshine, January
- daily hours of sunshine, July
- cold wind
- hot wind

Environmental issues

The partially enclosed waters of the Baltic and Mediterranean seas have become heavily polluted, while the Barents Sea is contaminated with spent nuclear fuel from Russia's navy. Acid rain, caused by emissions from factories and power stations, is actively destroying northern forests. As a result, pressure is growing to safeguard Europe's natural environment and prevent further deterioration.

Environmental issues

- national parks
- risk of acid rain
- polluted rivers
- radioactive contamination
- marine pollution
- heavy marine pollution
- poor urban air quality

Land use

Europe's swelling urban population and the outward expansion of many cities has created acute competition for land. Despite this, European resourcefulness has maximized land potential, and over half of Europe's land is still used for a wide variety of agricultural purposes. Land in northern Europe is used for cattle-rearing, pasture, and arable crops. Towards the Mediterranean, the mild climate allows the growing of grapes for wine; olives, sunflowers, tobacco and citrus fruits. EU subsidies, however, have resulted in massive overproduction and a land "set-aside" policy has been introduced.

Using the land and sea

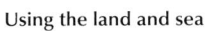
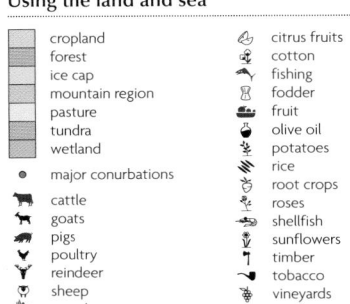

- cropland
- forest
- ice cap
- mountain region
- pasture
- tundra
- wetland
- major conurbations
- cattle
- goats
- pigs
- poultry
- reindeer
- sheep
- cereals
- citrus fruits
- cotton
- fishing
- fodder
- fruit
- olive oil
- potatoes
- rice
- root crops
- roses
- shellfish
- sunflowers
- timber
- tobacco
- vineyards

EUROPE

1 VATNAJÖKULL, ICELAND
Europe's largest ice cap is located in the southeast of this Atlantic island.

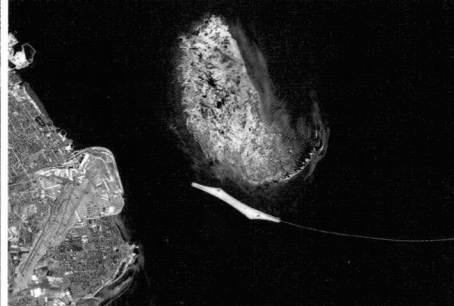

2 ORESUND LINK, DENMARK/SWEDEN
This link was opened to traffic in 2000, joining the Danish capital, Copenhagen, with the Swedish town of Malmö across the waters of the Oresund Strait.

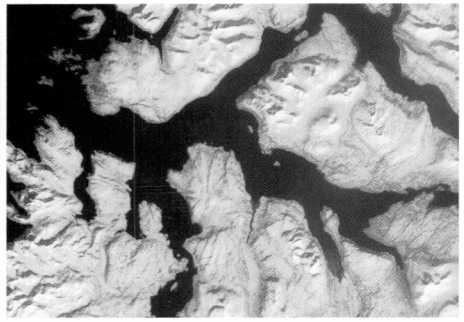

3 BALSFJORD, NORWAY
Fjords were cut into Norway's west coast by glaciers during the last ice age but as the ice retreated rising sea-levels flooded the valleys left behind.

4 PRAGUE, CZECH REPUBLIC
In August 2002 some parts of the capital were still under water after the worst floods in living memory.

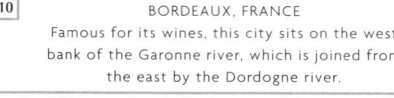

9 GIBRALTAR
A British colony since 1713, this rocky promontory commands a strategic position at the southern end of the Iberian Peninsula.

10 BORDEAUX, FRANCE
Famous for its wines, this city sits on the west bank of the Garonne river, which is joined from the east by the Dordogne river.

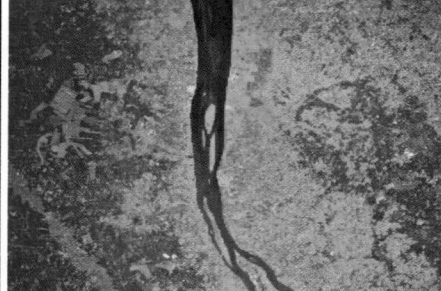

11 SOUTH FLEVOLAND, NETHERLANDS
This polder was reclaimed from the sea in the early 1970s and is now home to extensive farmland and small towns.

12 RHINE, GERMANY
The Rhine has been straightened in places, such as here, just south of Mannheim, to ease navigation.

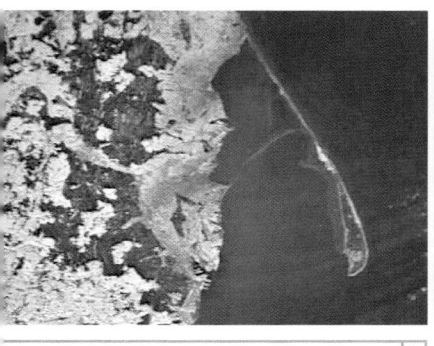

HEL PENINSULA, POLAND | 5
The long spit of this peninsula encloses Puck Bay and shelters the important port of Gdynia.

TALLINN, ESTONIA | 6
The capital and main port of Estonia has become a popular tourist destination in recent years.

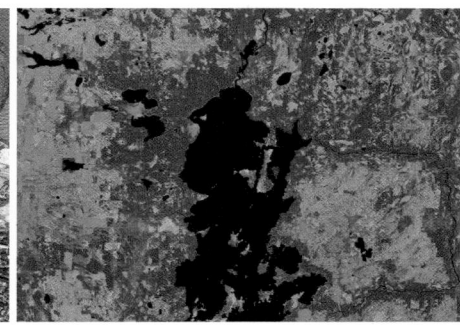

LAKE VODLOZERO, RUSSIAN FEDERATION | 7
The lake lies within a national park, which protects one of the most untouched wilderness areas in Europe and encompasses plains, taiga forests, and wetlands.

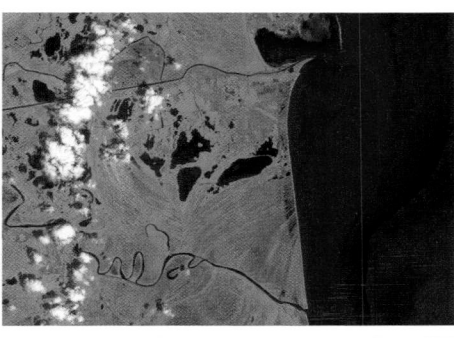

DANUBE DELTA, ROMANIA | 8
The Danube river splits into several channels as it flows into the Black Sea, forming one of Europe's most important wetland ecosystems.

VENICE, ITALY | 13
ccupying the largest island in a sheltered lagoon at the north end of the Adriatic, this city was founded in 52 CE and grew rich on an extensive trading network.

ISTRA PENINSULA, CROATIA | 14
This triangular peninsula marks the northern extent of Croatia's Dalmatian coastline.

MOUNT ETNA, SICILY, ITALY | 15
This combination of visible and thermal images shows the volcano erupting in July 2001 and clearly indicates the major lava flows.

KEFALLONIÁ, GREECE | 16
The largest of the Ionian Islands off Greece's west coast, Kefalloniá is mountainous with relatively high rainfall.

Scandinavia, Finland & Iceland

194

295

290

ARCTIC OCEAN

Barents Sea

RUSSIAN FEDERAT

MURMANSKAYA OBLAST'

RESPUBLIKA KARELIYA

KAINUU

POHJOIS-POHJANMAA

LAPPI

FINNMARK

TROMS

NORRBOTTEN

VÄSTERBOTTEN

NORDLAND

NORD-TRØNDELAG

ATLANTIC OCEAN

Greenland Strait

Denmark Strait

Greenland Sea

ICELAND

NORÐURLAND VESTRA

NORÐURLAND EYSTRA

VESTURLAND

VESTFIRÐIR

SUÐURLAND

AUSTURLAND

SUÐURNES

REYKJAVÍK

Scale 1:4,900,000

ARCTIC OCEAN

SVALBARD (to Norway)

Spitsbergen

LONGYEARBYEN

Barents Sea

Norwegian Sea

Greenland Sea

Scale 1:8,000,000

Lofoten

Vesterålen

Sea Level
6000m
4000m
3000m
2000m
1000m
500m
250m
100m
Sea Level
-250m
-2000m
-4000m

Southern Scandinavia

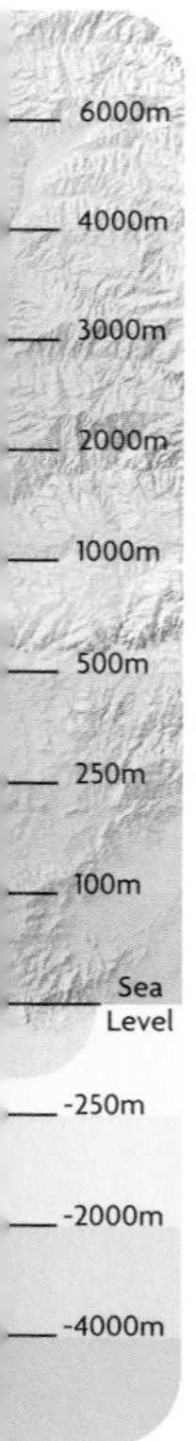

SWEDEN

NORWAY

NORDLAND

NORD-TRØNDELAG

SØR-TRØNDELAG

JÄMTLAND

VÄSTERBOTTEN

VÄSTERNORRLAND

GÄVLEBORG

DALARNA

HEDMARK

OPPLAND

BUSKERUD

HORDALAND

MØRE OG ROMSDAL

SOGN OG FJORDANE

AKERSHUS

UPPSALA

VÄSTMANLAND

ÅLAND

Norwegian Sea

Gulf of Bothnia

Ålands Hav

ATLANTIC OCEAN

FAROE ISLANDS (to Denmark)

Streymoy · Bordhoy · Fugloy · Vidhoy · Kunoy · Kalsoy · Svinoy · Eysturoy · Nolsoy · Sandoy · Skúvoy · Husavík · Sudhuroy · Vágur · Mykines · Vestmanna · TÓRSHAVN

Scale 1:2,500,000

Trondheim · Östersund · Umeå · Härnösand · Sundsvall · Gävle · Falun · Lillehammer · Hamar · Molde · Kristiansund · Steinkjer · Namsos · Bergen · Voss

Trondheimsfjorden · Storsjön · Siljan · Indalsälven · Ströms Vattudal · Femunden · Mjøsa · Dovrefjell · Jotunheimen · Hardangervidda · Hardangerfjorden · Sognefjorden

Scale 1:2,750,000
(projection: Lambert Conformal Conic)

0 10 20 30 40 50 60 70 80 Km
0 10 20 30 40 50 60 70 80 Miles

Population
■ above 5 million
◉ 100,000 to 500,000
■ 1 million to 5 million
⊕ 50,000 to 100,000
● 500,000 to 1 million
○ 10,000 to 50,000
○ below 10,000

19,686ft
13,124ft
9843ft
6562ft
3281ft
1640ft
820ft
328ft
Sea Level
-820ft
-6562ft
-13,124ft

United Kingdom & Ireland

UNITED KINGDOM

North Sea

ATLANTIC OCEAN

Shetland Islands

Herma Ness
Unst
Fetlar
Yell
Out Skerries
Whalsay
Sullom Voe
Hillswick
Bressay
Lerwick
St Magnus Bay
Mainland
Papa Stour
Scalloway
West Burra
Foula
Fitful Head
Sumburgh Head
Yell Sound

Fair Isle

Orkney Islands
Papa Westray
Westray
The North
North Ronaldsay
Sanday
Rousay
Eday
Stronsay
Sound
Mainland
Shapinsay
Kirkwall
Hoy
Scapa
Flow
Burray
St Margaret's Hope
South Ronaldsay
Duncansby Head
John o'Groats
Pentland Firth
Noss Head
Wick

Stromness

Dunnet Head
Dounreay
Thurso
Halkirk
Halladale
Kinbrace
Helmsdale
Brora
Golspie
Loch Fleet
Dornoch
Tarbat Ness
Tain
Dornoch Firth
Moray Firth
Lairg
Bonar Bridge
Cromarty
Nairn
Elgin
Forres
Lossiemouth
Buckie
Keith
Huntly
Macduff
Banff
Turriff
Kinnaird Head
Fraserburgh
Peterhead
Buchan Ness

Strathy Point
Tongue
Ben Klibreck
721m
Loch Erriboll
Loch Naver
Beinn Dearg
1081m
Fort Augustus
Loch Ness
Inverness
Beauly
Dingwall
Invergordon
Aviemore
Grantown-on-Spey
Cairngorm
Mountains
Cairn Gorm
1245m
Ben Macdui
1309m
Ellon
Inverurie
Dyce
Aberdeen
Girdle Ness
Stonehaven

Cape Wrath
Durness
Rhiconich
Scourie
Ben More Assynt
998m
Lochinver
Edrachillis Bay
Enard Bay
Ullapool
Loch Broom
Loch Maree
Loch Torridon
Kinlochewe

Ben Nevis
1344m
Fort William
Loch Linnhe
Loch Leven
Glen Coe
Ballachulish

GRAMPIAN MOUNTAINS
SCOTLAND
North West Highlands

Braemar
Ballater
Lochnagar
1154m
Kirriemuir
Blairgowrie
Forfar
Brechin
Montrose
Arbroath
Carnoustie
Firth of Tay
Dundee
Perth
St Andrews
Fife Ness
Cupar

Ben Lawers
1214m
Loch Tay
Crieff
Callander
Loch Katrine
Loch Lomond
Loch Rannoch
Loch Ericht
Loch Tummel
Kinross
Dunfermline
Kirkcaldy
Firth of Forth
North Berwick
Haddington
Dunbar

Stirling
Falkirk
Grangemouth
Livingston
Edinburgh
Musselburgh
Dalkeith
Penicuik

St Abb's Head
Eyemouth
Berwick-upon-Tweed
Holy Island

Alloa
Clydebank
Glasgow
Paisley
Coatbridge
Motherwell
Hamilton
East Kilbride
Greenock
Port Glasgow
Helensburgh
Dumbarton
Firth of Clyde

The Cheviot
815m
Wooler
Alnwick
Morpeth
Blyth
Tynemouth

Applecross
Inner Sound
Stromeferry
Kyle of Lochalsh
Kyleakin
Broadford
Portree
Isle of Skye
Raasay
Sea of the Hebrides

Canna
Rhum
Eigg
Muck
Coll
Tiree
Point of Ardnamurchan
Tobermory
Isle of Mull
Ben More
966m
Oban
Firth of Lorn
Colonsay
Jura
Islay
Port Askaig
Port Ellen
Mull of Oa

Sound of Jura
Loch Fyne
Gigha Island
Campbeltown
Mull of Kintyre
Isle of Arran
Brodick
Ailsa Craig

INNER HEBRIDES
OUTER HEBRIDES

Butt of Lewis
Port of Ness
Ness
Eye Peninsula
Broad Bay
Carloway
Stornoway
Tarbert
Isle of Lewis
Loch Roag
Scarp
Harris
Taransay
Sound of Harris
Pabbay
Flannan Isles
Berbecula
North Uist
Monach Islands
Lochmaddy
Benbecula
South Uist
Lochboisdale
Eriskay
Barra
Barra Head

St Kilda

Sula Sgeir
North Rona
Sula Sgeir
Sule Skerry
Stack Skerry

The Minch
The Little Minch

Shiant Islands

Kilmarnock
Troon
Ayr
Irvine
Prestwick
Cumnock
Girvan
Ballantrae
Merrick
842m
New Galloway
Thornhill
Moffat
Peebles
Galashiels
Selkirk
Hawick
Kelso
Coldstream
Jedburgh
Duns
Lauder
Newton St Boswells
Dumfries

Ireland
Tory Island
Bloody Foreland
Gweedore
Dunlewy
Errigal Mountain
752m
Dungloe
Donegal
Sheep Haven
Malin Head
Inishtrahull
Carndonagh
Buncrana
Lough Swilly
Letterkenny
Derry/Londonderry
Coleraine
Portstewart
Portrush
Ballymoney
Ballycastle
Rathlin Island
Giant's Causeway

Sea Level
6000m
4000m
3000m
2000m
1000m
500m
250m
100m
Sea Level
-250m
-2000m
-4000m

Scale 1:2,750,000
(projection: Lambert Conformal Conic)

| 0 10 20 30 40 50 60 70 80 Km |
| 0 10 20 30 40 50 60 70 80 Miles |

Population
- above 5 million
- 1 million to 5 million
- 500,000 to 1 million
- 100,000 to 500,000
- 50,000 to 100,000
- 10,000 to 50,000
- below 10,000

19,686ft
13,124ft
9843ft
6562ft
3281ft
1640ft
820ft
328ft
Sea
Level
-820ft
-6562ft
-13,124ft

FRANCE

PICARDIE

NORD-PAS-DE-CALAIS

HAUTE-NORMANDIE

BASSE-NORMANDIE

BRETAGNE

ENGLAND

WALES

IRELAND

NORTHERN IRELAND

LONDON

DUBLIN

East Anglia

Irish Sea

Celtic Sea

Bristol Channel

Cardigan Bay

Channel

English

Isle of Man
DOUGLAS
(British Crown Dependency)

ST PETER PORT
GUERNSEY
(British Crown Dependency)

ST HELIER
JERSEY
(British Crown Dependency)

Channel Islands

Dublin

Swords
Malahide
Portmarnock
Sutton
Howth
Finglas
Artane
Clontarf
Drumcondra
Dublin
Dún Laoghaire
Blackrock
Dalkey
Crumlin
Rathfarnham
Stillorgan
Dundrum
Stepaside
Bray
Clondalkin
Tallaght
Lucan
Palmerston
Blanchardstown
Clonee

Dublin Bay

Northern Britain & Ireland

ATLANTIC

OCEAN

Sea of the Hebrides

Inner Hebrides

IRELAND

NORTHERN IRELAND

UNITED

SCO...

Highland

Grampian Mts

Argyll

Stirlingshire

Ayrshire

North Channel

Irish Sea

Belfast

DUBLIN

Dublin Bay

ISLE OF MAN
(British Crown Dependency)

Isle of Man

DOUGLAS

Anglesey

DONEGAL

SLIGO

LEITRIM

MAYO

ROSCOMMON

LONGFORD

WEST MEATH

MEATH

MONAGHAN

CAVAN

LOUTH

KILDARE

OFFALY

GALWAY

CONNAUGHT

Sea Level
6000m
4000m
3000m
2000m
1000m
500m
250m
100m
Sea Level
-250m
-2000m
-4000m

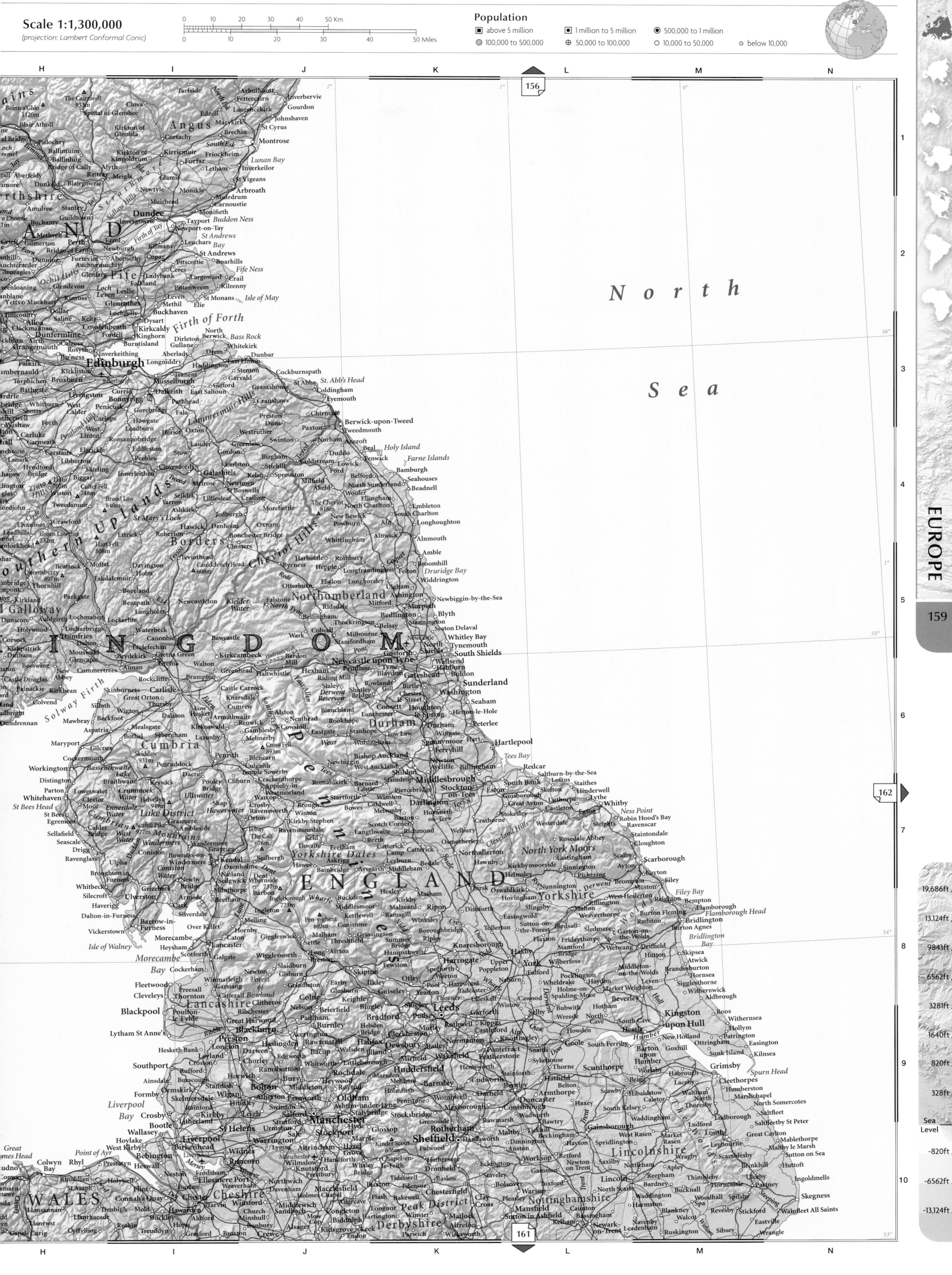

Scale 1:1,300,000
(projection: Lambert Conformal Conic)

Population
■ above 5 million
◉ 100,000 to 500,000
◪ 1 million to 5 million
⊕ 50,000 to 100,000
◉ 500,000 to 1 million
○ 10,000 to 50,000
○ below 10,000

0 10 20 30 40 50 Km
0 10 20 30 40 50 Miles

156
162
161
159

EUROPE

North
Sea

KINGDOM

ENGLAND

WALES

Edinburgh

Fife

Firth of Forth

Southern Uplands

Borders

Cheviot Hills

Northumberland

Newcastle upon Tyne

Sunderland

Durham

Cumbria

Pennines

Lake District

Cumbrian Mountains

Carlisle

Middlesbrough

Darlington

North York Moors

Yorkshire

Yorkshire Dales

Scarborough

York

Leeds

Bradford

Kingston upon Hull

Grimsby

Blackpool

Preston

Lancashire

Bolton

Manchester

Liverpool

Birkenhead

Stockport

Sheffield

Rotherham

Doncaster

Lincolnshire

Lincoln

Nottinghamshire

Peak District

Derbyshire

Cheshire

Solway Firth

Morecambe Bay

19,686ft
13,124ft
9843ft
6562ft
3281ft
1640ft
820ft
328ft
Sea
Level
-820ft
-6562ft
-13,124ft

Southern Britain

157

Irish Sea

Celtic Sea

St George's Channel

Cardigan Bay

Bristol Channel

Lyme Bay

IRELAND

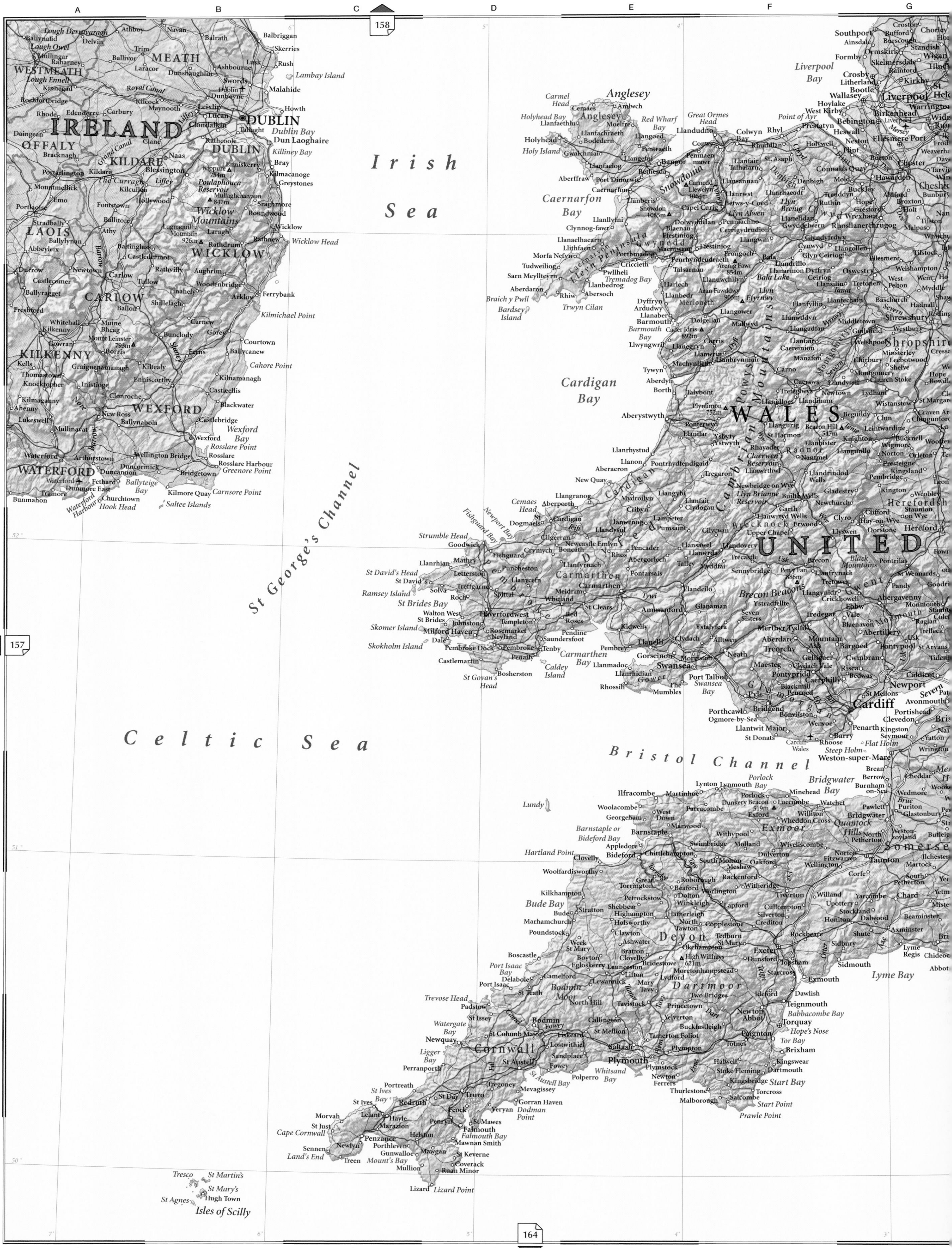

WALES

UNITED

WESTMEATH

MEATH

OFFALY

KILDARE

LAOIS

WICKLOW

CARLOW

KILKENNY

WEXFORD

WATERFORD

DUBLIN

Anglesey

Snowdonia

Cambrian Mountains

Cornwall

Devon

Dartmoor

Bodmin Moor

Exmoor

Cardiff

Swansea

Plymouth

Land's End

Lizard Point

Isles of Scilly

Liverpool

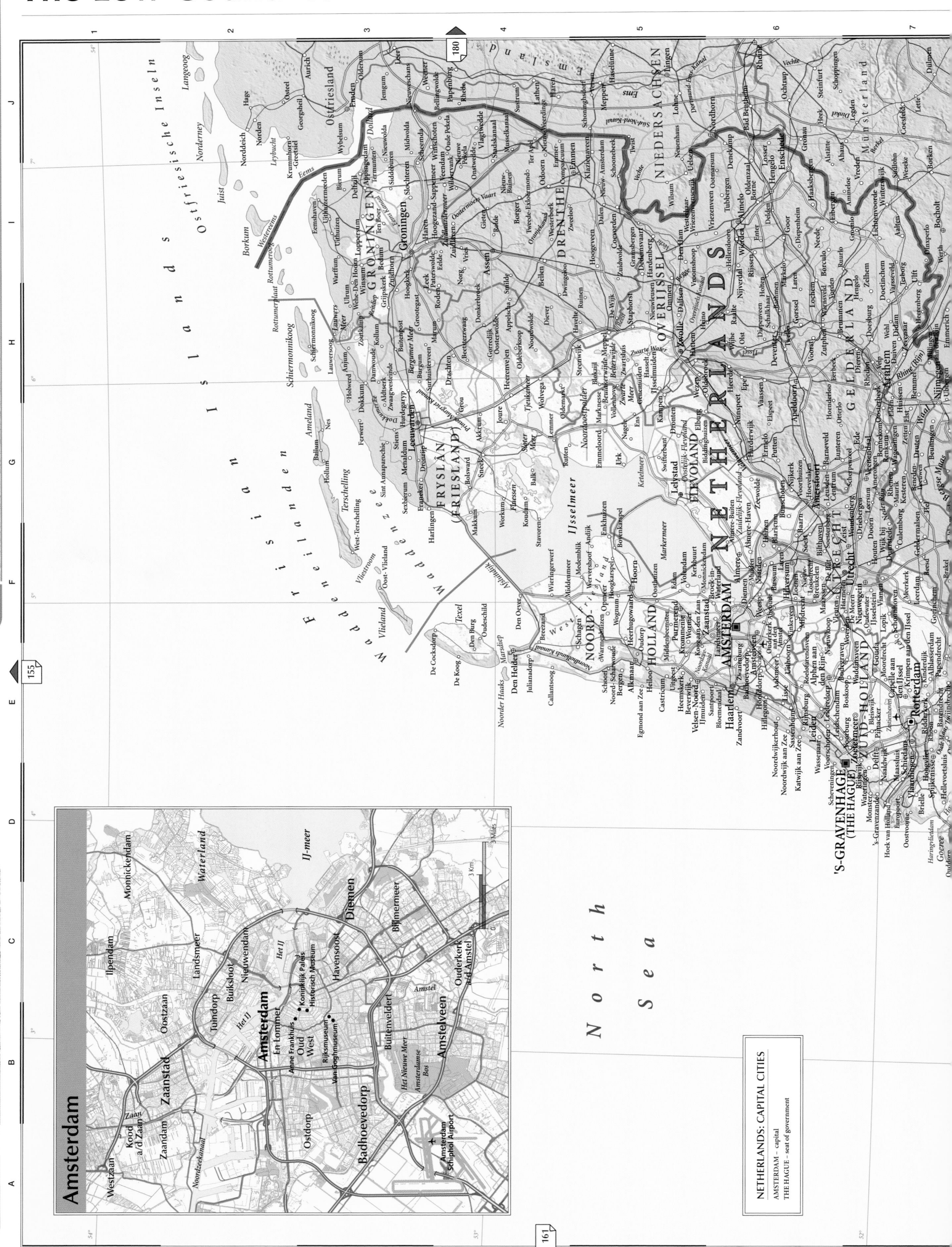

Scale 1:1,125,000

(projection: Lambert Conformal Conic)

Population

■ above 5 million   ▫ 1 million to 5 million   ◉ 500,000 to 1 million

◎ 100,000 to 500,000   ⊕ 50,000 to 100,000   ○ 10,000 to 50,000   ○ below 10,000

19,686ft
13,124ft
9843ft
6562ft
3281ft
1640ft
820ft
328ft
Sea Level
-820ft
-6562ft
-13,124ft

France

160

UNITED
KINGDOM

Hartland Point Bideford Barnstaple Taunton Bridgwater Salisbury Andover Guildford Reigate
Bude Tiverton Sherborne Shaftesbury Winchester Eastleigh Crawley Royal Tunbridge Wells
Okehampton High Willhays Lyme Regis Dorchester Southampton Littlehampton Hove Brighton Bexhill
Newquay Bodmin Moor Torquay Paignton Poole Bournemouth Portsmouth Bognor Worthing Eastbourne
St Ives St Austell Dartmouth Weymouth Swanage Isle of Wight Beachy Head
Cape Cornwall Penzance Truro Falmouth Start Point Portland Bill St Catherine's Point
Land's End Helston Lizard Point Mount's Bay
Isles of Scilly

English Channel

ATLANTIC

OCEAN

Alderney Cap de la Hague Pointe de Barfleur Cap d'Antifer Fécamp Seine-Maritime
Cherbourg Barfleur Montivilliers Cap de la Hève le Havre
ST PETER PORT Herm Octeville Tourlaville Valognes Cotentin Baie de la Seine Trouville Deauville
GUERNSEY (British Crown Dependency) Sark Manche Isigny-sur-Mer Honfleur Pont-Audemer
Channel Islands Carentan Bayeux Calvados Pont-l'Évêque Lisieux
ST HELIER JERSEY (British Crown Dependency) St-Lô Caen Bernay Conches-en-Ouche
Passage de la Déroute Coutances BASSE-NORMANDIE Falaise
Îles Chausey Villedieu-les-Poêles Vire Collines de Normandie Flers Argentan Orne l'Orne
Granville Avranches Mortain Domfront Sées Alençon
Île de Batz Perros-Guirec Tréguier Pointe de l'Arcouest Le Mont-St-Michel Mortagne-au-Perche
Roscoff St-Pol-de-Léon Lannion Paimpol Golfe de St-Malo St-Malo Dinard Dol-de-Bretagne 417m Beaumont Mamers Nogent-le-Rotrou la Ferté-Bernard
Lesneven Île de Bréhat Morlaix Bégard Baie de St-Brieuc Dinan Combourg Avoirs Mont des Avaloirs
Landerneau Monts d'Arrée Plérin St-Brieuc Lamballe Ille-et-Vilaine Mayenne Mamers Sarthe
Brest Finistère Callac Côtes d'Armor Fougères Sillé-le-Guillaume
Iroise Châteaulin Guingamp Loudéac BRETAGNE Rennes Vitré Laval Mayenne Loué le Mans
Pointe St-Mathieu Crozon Carhaix-Plouguer Montagnes Noires Pontivy Josselin Ploërmel Ille-et-Vilaine Château-Gontier Segré Sablé-sur-Sarthe St-Calais
Douarnenez Châteauneuf Scaër Rosporden Mauron Guichen Janzé Craon La Flèche
Pointe du Raz Quimper Quimperlé Morbihan Redon Guémené-Penfao Châteaubriant Angers Noyant
Baie d'Audierne Pont-l'Abbé Concarneau Lorient Auray Vannes Questembert Loire-Atlantique Anjou Saumur Montreuil-Bellay
Pointe de Penmarch Îles Glénan Île de Groix Baie de Quiberon Pontchâteau Blain Nantes Cholet Montaigu Chemillé Vihiers Thouars Richelieu
Belle Île Quiberon Guérande St-Nazaire PAYS DE LA LOIRE Clisson Mortagne-sur-Sèvre Bressuire Châtellerault
Pointe du Croisic St-Gildas Pornic Lac de Grand-Lieu Beaupréau Mauléon Airvault Parthenay
Noirmoutier-en-l'Île Île de Noirmoutier Beauvoir-sur-Mer Challans la Roche-sur-Yon les Herbiers Deux-Sèvres Fontenay-le-Comte POITOU-CHARENTES Poitiers
St-Jean-de-Monts Île d'Yeu St-Gilles-Croix-de-Vie Vendée Lusignan Vienne
les Sables-d'Olonne la Mothe-Achard Luçon Niort Celle-l'Évescault Civray
Pertuis Breton Marans St-Maixent-l'École Melle
Île de Ré la Rochelle Surgères Vienne
Pertuis d'Antioche Pertuis Charente-Maritime Ruffec
Île d'Oléron Rochefort St-Jean-d'Angély Mansle Charente Rochechouart
le Château d'Oléron Marennes Matha Charente
Pointe de la Coubre Royan Saintes Cognac Jarnac Angoulême Rochefoucauld
Pointe de Grave Pons Barbezieux St-Hilaire Brantôme
Gironde Jonzac Montmoreau Nontron
Lesparre-Médoc Mirambeau Montendre
Médoc St-Laurent-Médoc Blaye Montlieu-la-Garde Coutras Périgueux
Lac d'Hourtin-Carcans Lacanau St-André-de-Cubzac Libourne St-Médard-en-Jalles Cenon Bordeaux AQUITAINE Sauveterre-de-Guyenne Dordogne
Bassin d'Arcachon Mérignac Pessac Gironde Miramont-de-Guyenne Bergerac
Arcachon la Teste Marmande Villeneuve-sur-Lot
Cap Ferret Gradignan
Étang de Biscarrosse et de Parentis Belin-Béliet Bazas Langon Casteljaloux
Mimizan Parentis-en-Born Captieux Agen Nérac
Labouheyre Sabres Roquefort Houeillès Mézin Condom
Léon Landes Morcenx Gers
Mont-de-Marsan Aire-sur-l'Adour Auch
St-Paul-lès-Dax Tartas Nogaro
Dax Grenade Gimont
St-Vincent-de-Tyrosse Capbreton Hagetmau Vic-en-Bigorre
Anglet Perchacate Orthez Lembeye Mirande Castelnau-Magnoac L'Isle-Jourdain
Biarritz Bayonne Hasparren Lacq Mourenx Pau Tarbes
Donostia/San Sebastián Hendaye la Rhune Orthez Pontacq Bagnères-de-Bigorre
Irún St-Jean-de-Luz Lourdes Pyrénées-Atlantiques Pic du Midi de Bigorre 2872m
Oloron-Ste-Marie Argelès-Gazost Arreau Bagnères-de-Luchon
PAÍS VASCO Tolosa Pic d'Anie Gavarnie 3144m

Bay of

Biscay

Paris

Seine Montmorency Aéroport Charles de Gaulle Aéroport Le Bourget Tremblay-en-France
Forêt de St-Germain Montmorency Aulnay-sous-Bois
Enghien St-Denis Drancy
Poissy Argenteuil Asnières Aubervilliers Le Raincy
St-Germain-en-Laye Nanterre Montmartre Sacré-Cœur Paris Montreuil Marne-la-Vallée
Rueil-Malmaison Arc de Triomphe Musée du Louvre Bastille Marne
Boulogne-Billancourt Tour Eiffel Notre Dame Vincennes
Château de Versailles Meudon Seine Champigny-sur-Marne
Versailles Vitry-sur-Seine Créteil
Trappes Sceaux Seine
Antony Orly
Chevreuse Palaiseau Aéroport d'Orly Brie-Comte-Robert
Orsay Mortgeron

0 5 Km 5 Miles

290

170

Cabo Ortegal Punta de Estaca de Bares Cabo Prior Ortigueira Cedeira Cabo de Peñas Luarca Avilés Lluanco Gijón (Xixón) Costa Verde Santander Cabo de Ajo
A Coruña (La Coruña) Ferrol Viveiro Cervo Tapia de Casariego Navia Candás ASTURIAS Villaviciosa Llanes Torrelavega Santoña
Carballo Pontedeume Ribadeo Castropol Grau Mieres del Camino Las Arriondas Cangas de Onís Castro-Urdiales CANTABRIA
Camariñas Betanzos Mondoñedo Vilalba Oviedo La Pola Llaviana Cabezón de la Sal Santurtzi
Vimianzo Ordes Lugo A Fonsagrada Picos de Europa GALICIA 2570m San Vicente de la Barquera Bilbao PAÍS VASCO
Muros Negreira Santiago de Compostela Sarria GALICIA Cordillera Cantábrica Reinosa Vitoria-Gasteiz NAVARRA
Noia Melide Palas de Rei O Corgo Becerreá Villafranca del Bierzo La Robla León Sahagún Carrión de los Condes Pamplona (Iruña)
A Estrada Lalín Monforte de Lemos Ponferrada Astorga Palencia Burgos Estella Tafalla
Pontevedra Ourense (Orense) O Barco León La Bañeza Benavente Valderas Villada Villalón de Campos Valladolid Castilla y León Soria ARAGÓN
Vigo Ribadavia Xinzo de Limia Verín Puebla de Sanabria Toro Medina de Rioseco Zamora Tordesillas SPAIN Sistema Ibérico Zaragoza
PORTUGAL Braga Chaves Bragança Mirandela Rueda Medina del Campo Segovia Ávila Aranda de Duero Calatayud

6000m
4000m
3000m
2000m
1000m
500m
250m
100m
Sea Level
-250m
-2000m
-4000m

Population
■ above 5 million   ▣ 1 million to 5 million   ◉ 500,000 to 1 million
◎ 100,000 to 500,000   ⊕ 50,000 to 100,000   ○ 10,000 to 50,000   ○ below 10,000

EUROPE

165

Scale bar: 0 20 40 60 80 100 Km / 0 20 40 60 80 100 Miles

Elevation scale (right margin):
19,686ft
13,124ft
9843ft
6562ft
3281ft
1640ft
820ft
328ft
Sea Level
-820ft
-6562ft
-13,124ft

Grid references (top): H I J K L M N
Column references: 181 177 198

Mediterranean Sea

Ligurian Sea

Tyrrhenian Sea

Corse (Corsica)

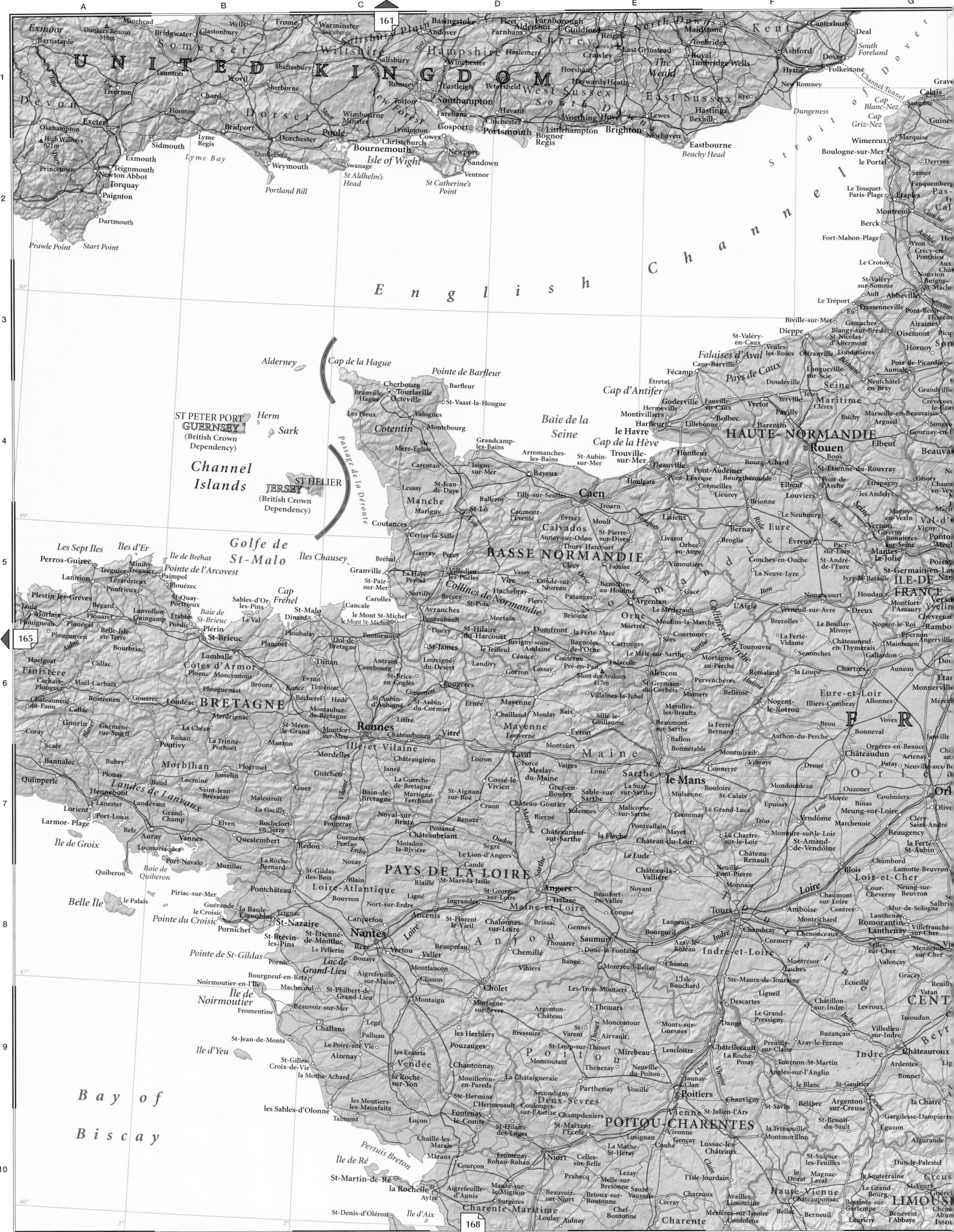

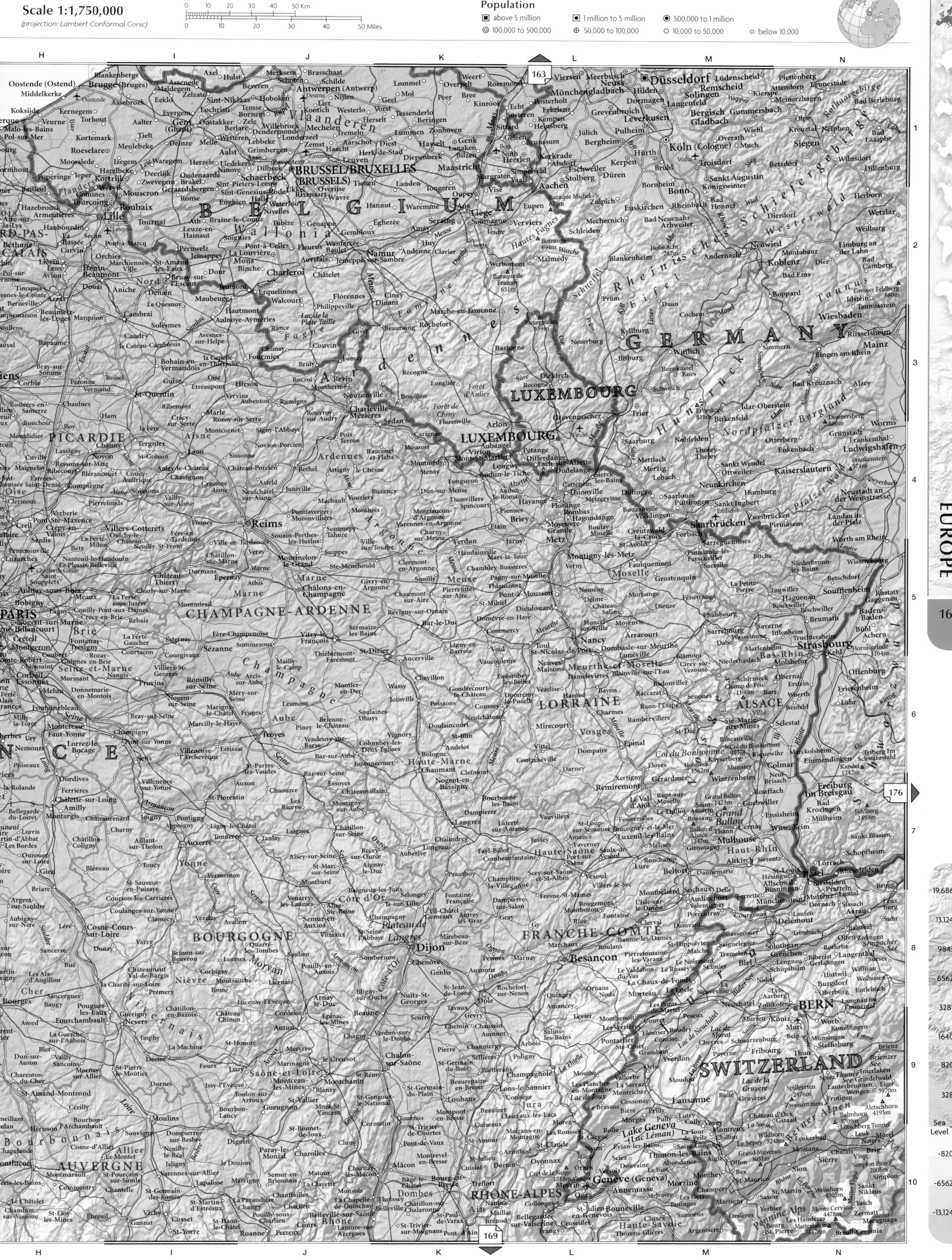

EUROPE

168

ATLANTIC OCEAN

Bay of Biscay

Costa Verde

POITOU-CHARENTES

LIMOUSIN
Plateau de Millevaches
Corrèze
Haute-Vienne
Limoges

AQUITAINE
Bordeaux
Dordogne
Gironde

MIDI-PYRÉNÉES
Tarn-et-Garonne
Montauban
Lot
Toulouse
Haute-Garonne
Tarn
Ariège

Landes
Armagnac
Gers
Auch

PAÍS VASCO
Bilbao/Bilbo
Bizkaia
Gipuzkoa
Donostia/San Sebastián
Biarritz
Bayonne
Pau
Béarn

Pyrénées-Atlantiques
Pyrénées
Hautes-Pyrénées
Lourdes
ANDORRA
ANDORRA LA VELLA

NAVARRA
Pamplona (Iruña)
LA RIOJA
Logroño
Sierra de la Demanda

Vitoria-Gasteiz
Álava

Zaragoza

ARAGÓN

SPAIN

Sierra del Moncayo
Moncayo 2316m

Sierra de Guara
Huesca

CATALUNA
Lleida (Lérida)
Barcelona
L'Hospitalet de Llobregat
Terrassa
Manresa
Tarragona

Golfo de Sant Jordi

Sea Level
6000m
4000m
3000m
2000m
1000m
500m
250m
100m
-250m
-2000m
-4000m

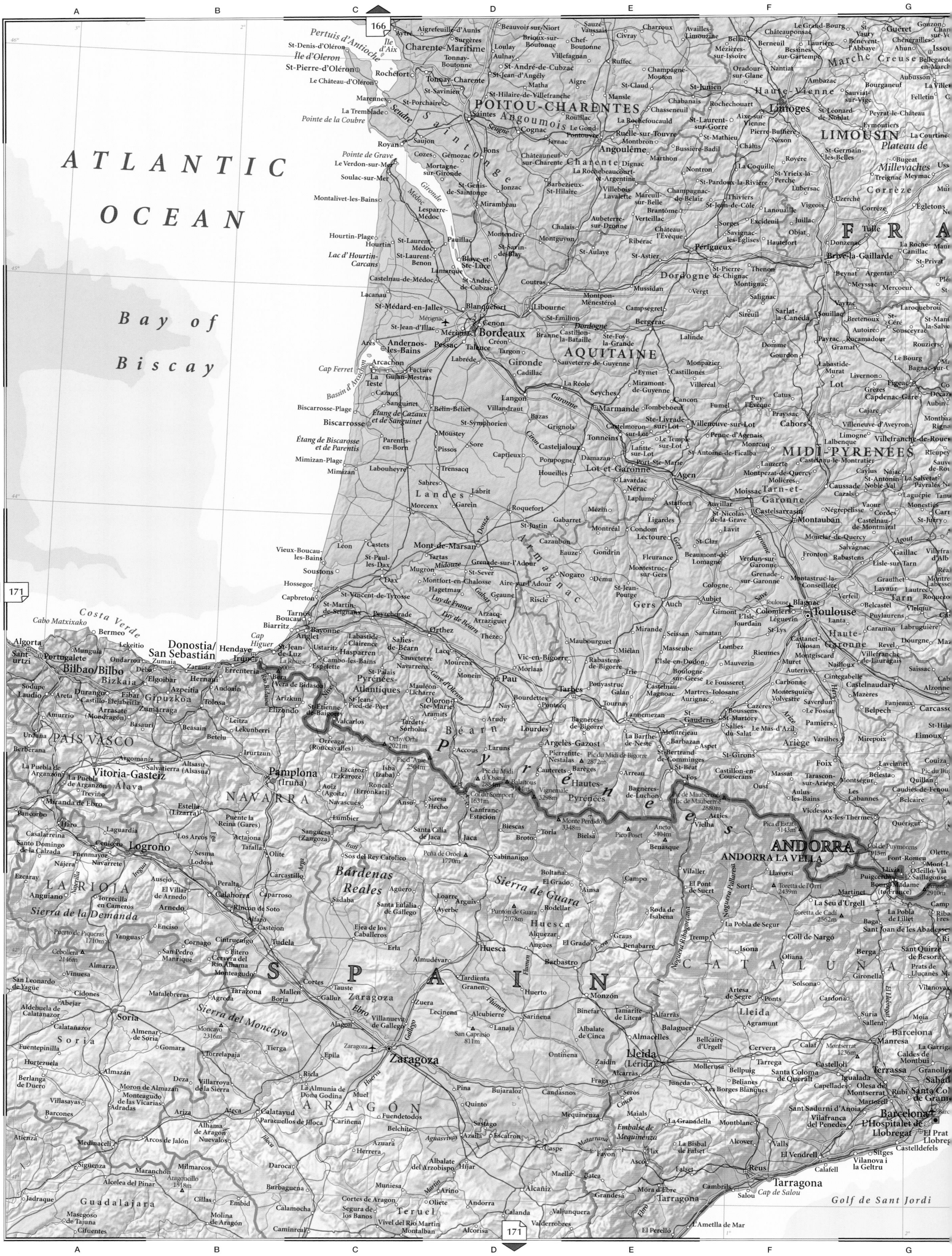

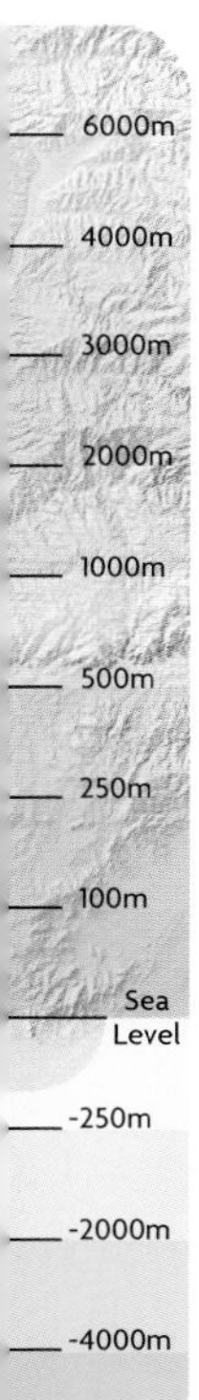

The Iberian Peninsula

The Italian Peninsula

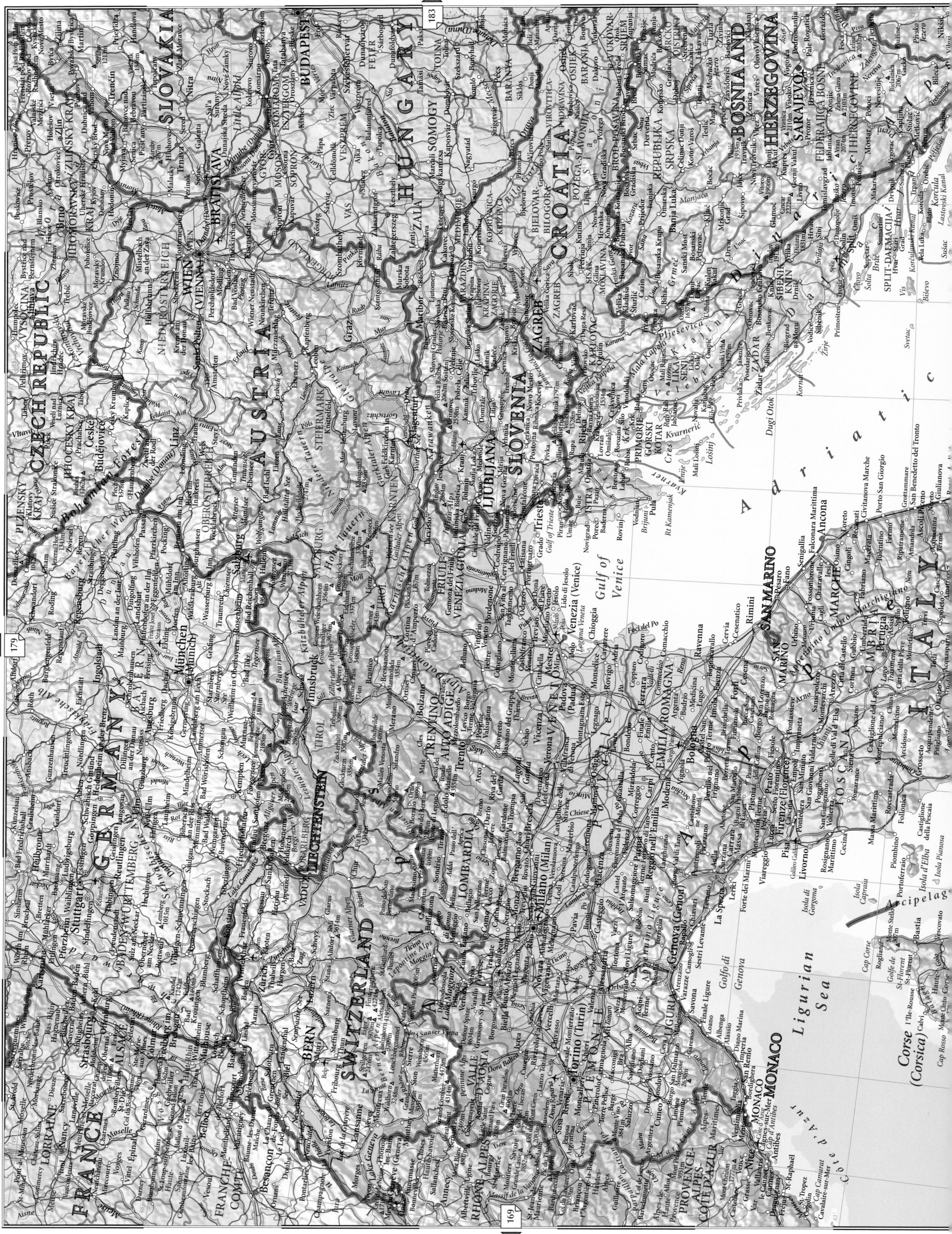

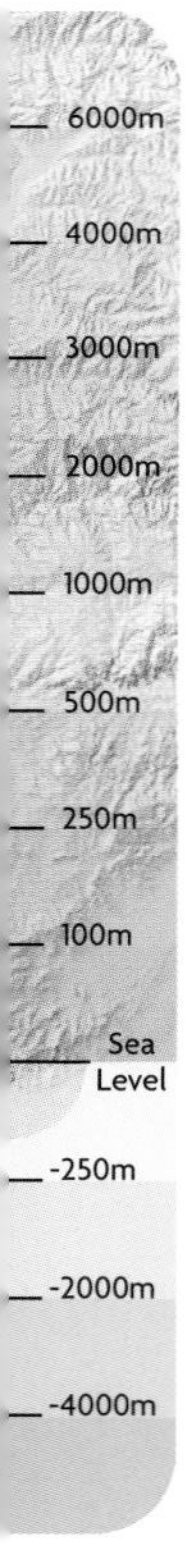

6000m
4000m
3000m
2000m
1000m
500m
250m
100m
Sea Level
-250m
-2000m
-4000m

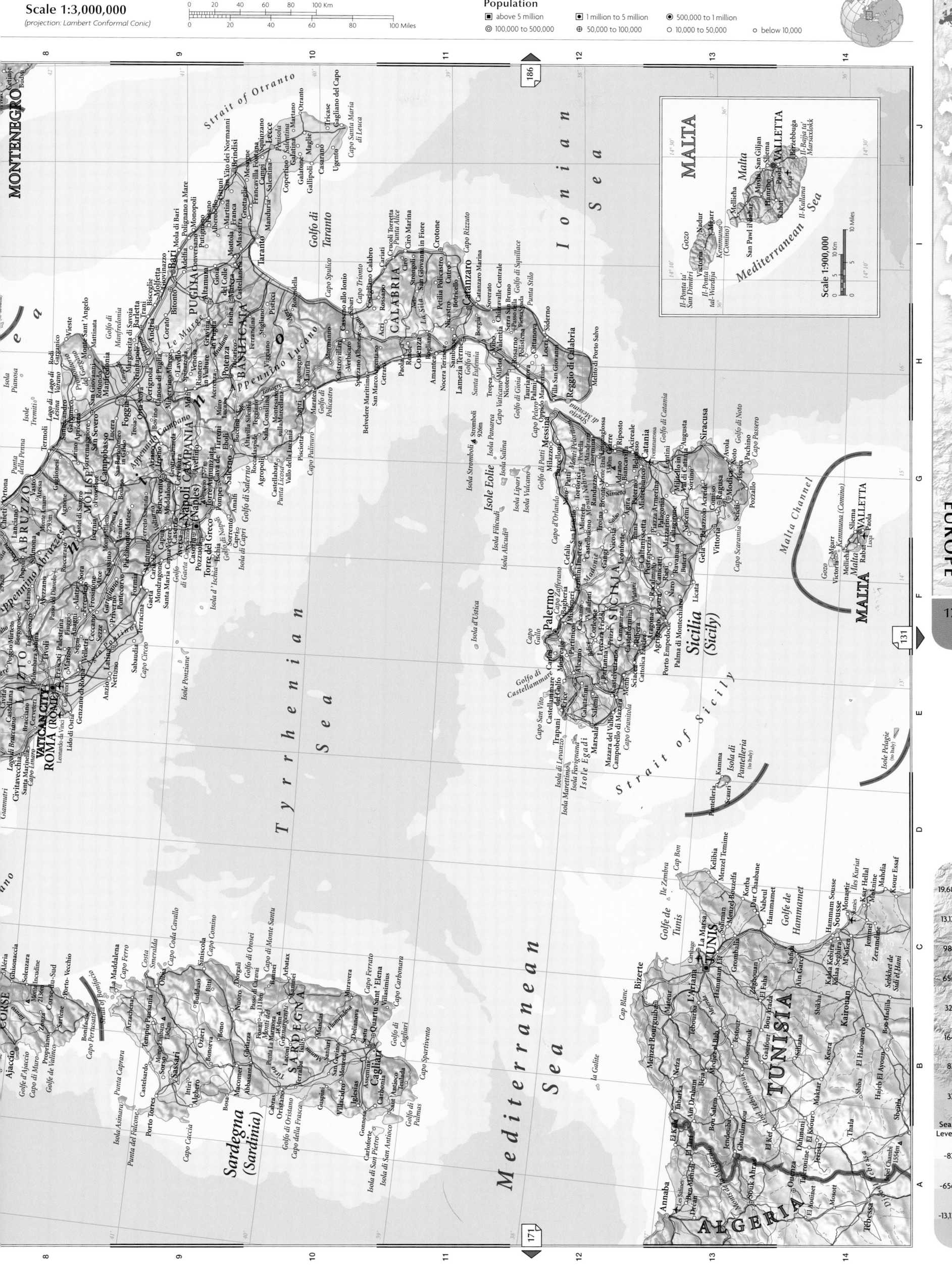

Scale 1:2,100,000
(projection: Lambert Conformal Conic)

0 10 20 30 40 50 60 70 80 Km
0 10 20 30 40 50 60 70 80 Miles

Population
- ■ above 5 million
- ▣ 1 million to 5 million
- ◉ 500,000 to 1 million
- ◎ 100,000 to 500,000
- ⊕ 50,000 to 100,000
- ○ 10,000 to 50,000
- ∘ below 10,000

183

175

184

19,686ft
13,124ft
9843ft
6562ft
3281ft
1640ft
820ft
328ft
Sea Level
-820ft
-6562ft
-13,124ft

CZECH REPUBLIC
JIHOČESKÝ KRAJ
JIHOMORAVSKÝ KRAJ

SLOVAKIA
BRATISLAVSKY KRAJ
TRNAVSKY KRAJ
NITRIANSKY KRAJ
BRATISLAVA

AUSTRIA
OBERÖSTERREICH
NIEDERÖSTERREICH
WIEN (VIENNA)
Linz
Salzburg
SALZBURG
TIROL
KÄRNTEN
STEIERMARK
Graz
Klagenfurt
BURGENLAND

HUNGARY
GYŐR-MOSON-SOPRON
VAS
VESZPRÉM
ZALA
SOMOGY
TOLNA
BARANYA
KOMÁROM-ESZTERGOM
FEJÉR
Pécs

SLOVENIA
LJUBLJANA
Maribor

CROATIA
ZAGREB
KARLOVAC
LIKA-SENJ
ISTRA
PRIMORJE-GORSKI KOTAR
SISAK-MOSLAVINA
BJELOVAR-BILOGORA
POŽEGA-SLAVONIJA
VIROVITICA-PODRAVINA
OSIJEK-BARANJA
BROD-POSAVINA
ZADAR
ŠIBENIK-KNIN
SPLIT-DALMACIJA
DUBROVNIK-NERETVA

BOSNIA & HERZEGOVINA
REPUBLIKA SRPSKA
FEDERACIJA BOSNE I HERCEGOVINE
Banja Luka

ITALY
VENETO
FRIULI-VENEZIA GIULIA
Venezia (Venice)
Trieste
MARCHE
UMBRIA
Ravenna
Ancona
Rimini

SAN MARINO
SAN MARINO

Gulf of Venice
Gulf of Trieste
Adriatic Sea

München (Munich)
Regensburg
Landshut

North Sea

Baltic Sea

Kattegat

SWEDEN

BLEKINGE

SKÅNE

DENMARK

JYLLAND

MIDTJYLLAND

SYDDANMARK

SJÆLLAND

FYN

HOVEDSTADEN

KØBENHAVN (COPENHAGEN)

BORNHOLM

Bornholm

LOLLAND

Falster

Langeland

Pomeranian Bay

Pomerania

POLAND

ZACHODNIO-POMORSKIE

LUBUSKIE

BERLIN

BRANDENBURG

MECKLENBURG-VORPOMMERN

SACHSEN-ANHALT

NIEDERSACHSEN

SCHLESWIG-HOLSTEIN

Hamburg

BREMEN

Hannover

Braunschweig

Magdeburg

Potsdam

NETHERLANDS

GRONINGEN

FRYSLÂN (FRIESLAND)

DRENTHE

OVERIJSSEL

GELDERLAND

FLEVOLAND

North Frisian Islands (Nordfriesische Inseln)

Ostfriesische Inseln

Helgoländer Bucht

Kieler Bucht

Mecklenburger Bucht

Lübecker Bucht

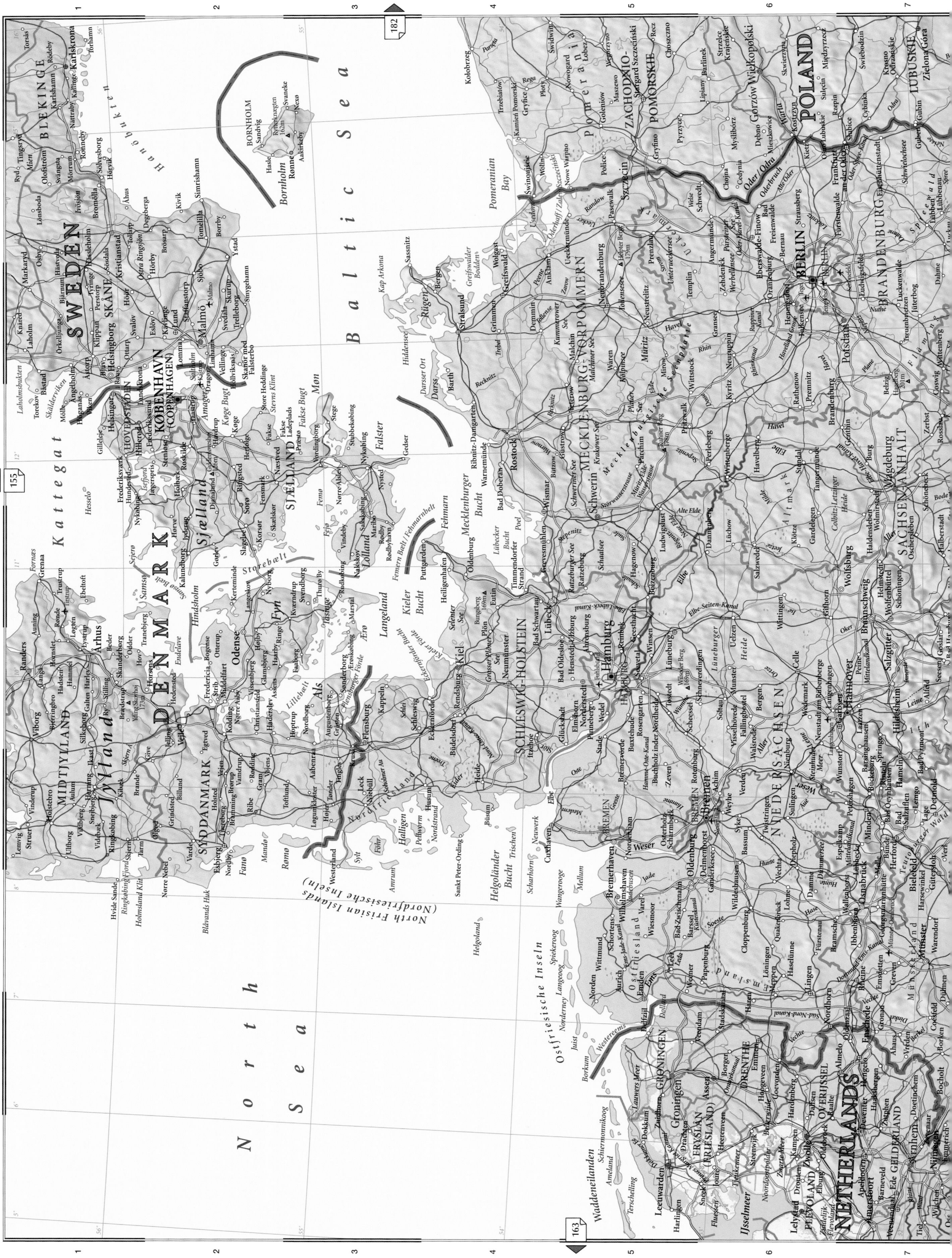

6000m
4000m
3000m
2000m
1000m
500m
250m
100m
Sea Level
-250m
-2000m
-4000m

Scale 1:2,250,000
(projection: Lambert Conformal Conic)

0 10 20 30 40 50 60 70 80 Km
0 10 20 30 40 50 60 70 80 Miles

Population
- ■ above 5 million
- ◉ 100,000 to 500,000
- ▣ 1 million to 5 million
- ⊕ 50,000 to 100,000
- ◉ 500,000 to 1 million
- ○ 10,000 to 50,000
- ○ below 10,000

GERMANY

CZECH REPUBLIC

AUSTRIA

SWITZERLAND

LIECHTENSTEIN

ITALY

FRANCE

BELGIUM

LUXEMBOURG

PRAHA (PRAGUE)

München (Munich)

Nürnberg (Nuremberg)

SACHSEN

THÜRINGEN

HESSEN

RHEINLAND-PFALZ

SAARLAND

BADEN-WÜRTTEMBERG

BAYERN

OBERÖSTERREICH

NIEDERÖSTERREICH

SALZBURG

STEIERMARK

KÄRNTEN

TIROL

VORARLBERG

LIBERECKÝ KRAJ

ÚSTECKÝ KRAJ

KARLOVARSKÝ KRAJ

PLZEŇSKÝ KRAJ

STŘEDOČESKÝ KRAJ

JIHOČESKÝ KRAJ

Bohemian Forest

Bohemian Forest

Lake Constance

Rhine (Rhein)

Rhône (Rhône)

ALSACE

LORRAINE

Meurthe-et-Moselle

Moselle

FRANCHE-COMTÉ

HAUTE-SAÔNE

Vosges

Doubs

LIMBURG

Sea Level

19,686ft
13,124ft
9843ft
6562ft
3281ft
1640ft
820ft
328ft
-820ft
-6562ft
-13,124ft

Rhineland & Hamburg

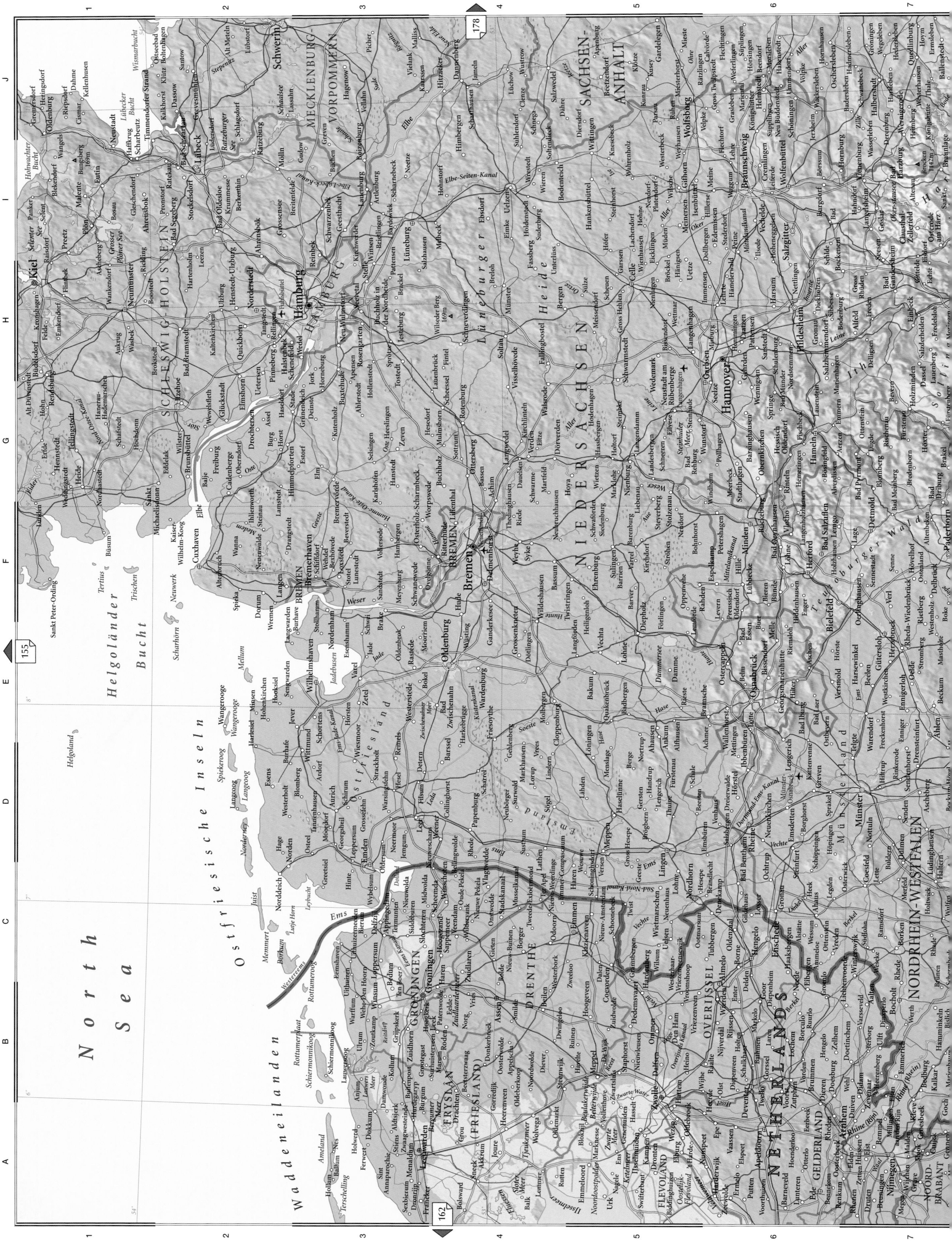
North Sea
Helgoländer Bucht
Ostfriesische Inseln
Waddeneilanden
NETHERLANDS
NIEDERSACHSEN
SCHLESWIG-HOLSTEIN
MECKLENBURG-VORPOMMERN
SACHSEN-ANHALT
NORDRHEIN-WESTFALEN
Lüneburger Heide
Hamburg
Bremen
Hannover
Kiel
Lübeck
Schwerin
Bremerhaven
Cuxhaven
Wilhelmshaven
Emden
Oldenburg
Osnabrück
Münster
Bielefeld
Braunschweig
Wolfsburg
Groningen
178
155
162

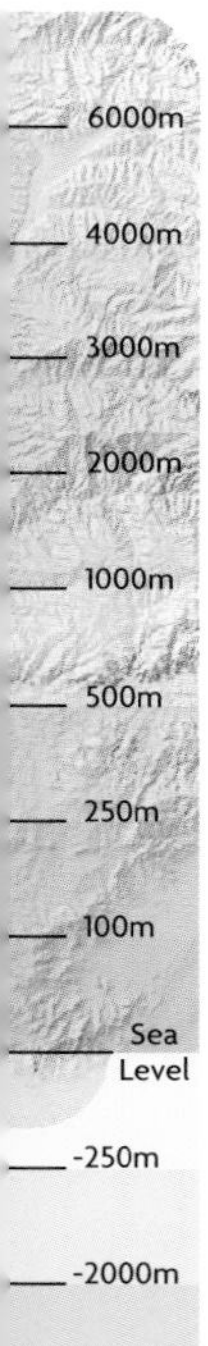
6000m
4000m
3000m
2000m
1000m
500m
250m
100m
Sea Level
-250m
-2000m
-4000m

Scale 1:1,200,000

(projection: Lambert Conformal Conic)

Population		
■ above 5 million	■ 1 million to 5 million	● 500,000 to 1 million
◉ 100,000 to 500,000	⊕ 50,000 to 100,000	○ 10,000 to 50,000	∘ below 10,000

19,686ft
13,124ft
9843ft
6562ft
3281ft
1640ft
820ft
328ft
Sea Level
-820ft
-6562ft
-13,124ft

THÜRINGEN

BAYERN

HESSEN

GERMANY

RHEINLAND-PFALZ

BADEN-WÜRTTEMBERG

SAARLAND

BELGIUM

LUXEMBOURG

FRANCE

ALSACE

LORRAINE

LIMBURG

Erfurt

Göttingen

Kassel

Würzburg

Nürnberg

Frankfurt am Main

Wiesbaden

Mainz

Mannheim

Heidelberg

Karlsruhe

Koblenz

Köln Cologne

Düsseldorf

Dortmund

Essen

Duisburg

Saarbrücken

Luxembourg

Rhine (Rhein)

Main

Mosel-Moselle

179

176

163

Elevation scale:
6000m
4000m
3000m
2000m
1000m
500m
250m
100m
Sea Level
-250m
-2000m
-4000m

Countries and regions:

LATVIA
LITHUANIA
ŽEMAITIJA
AUKŠTAITIJA
TELŠIAI
KLAIPĖDA
TAURAGĖ
KAUNAS
MARIJAMPOLĖ
ALYTUS
SWEDEN
KRONOBERG
KALMAR
BLEKINGE
SKÅNE
DENMARK
SJÆLLAND
BORNHOLM
RUSSIAN FEDERATION
KALININGRADSKAYA OBLAST
BELARUS
PODLASKIE
WARMIŃSKO-MAZURSKIE
MAZOWIECKIE
LUBELSKIE
POLAND
POMORSKIE
KUJAWSKO-POMORSKIE
ŁÓDZKIE
ZACHODNIO-POMORSKIE
WIELKOPOLSKIE
LUBUSKIE
DOLNOŚLĄSKIE
GERMANY
MECKLENBURG-VORPOMMERN
BRANDENBURG
SACHSEN
SACHSEN-ANHALT
THÜRINGEN

Water bodies:

BALTIC SEA
Gulf of Danzig
Kattegat
Pomeranian Bay
Courland Lagoon
Courland Spit

Selected cities:

Liepāja, Klaipėda, Kaliningrad, Gdańsk, Gdynia, Sopot, Elbląg, Olsztyn, Białystok, Grodno, Brest, WARSZAWA (WARSAW), Łódź, Płock, Włocławek, Toruń, Bydgoszcz, Poznań, Kalisz, Wrocław, Legnica, Zielona Góra, Gorzów Wielkopolski, Szczecin, Koszalin, Słupsk, Kołobrzeg, Świnoujście, BERLIN, Potsdam, Schwerin, Rostock, Dresden, Cottbus, Frankfurt an der Oder, KØBENHAVN COPENHAGEN, Malmö, Halmstad, Kalmar, Lublin, Kielce, Radom

Scale 1:2,750,000
(projection: Lambert Conformal Conic)

0 10 20 30 40 50 60 70 80 Km
0 10 20 30 40 50 60 70 80 Miles

Population

■ above 5 million
◉ 100,000 to 500,000

▣ 1 million to 5 million
⊕ 50,000 to 100,000

● 500,000 to 1 million
○ 10,000 to 50,000

○ below 10,000

UKRAINE

POLAND

L'VIVS'KA OBLAST'

PODKARPACKIE

MAŁOPOLSKIE

ZAKARPATS'KA OBLAST'

CZECH REPUBLIC

SLEZSKÝ KRAJ

OLOMOUCKÝ KRAJ

PARDUBICKÝ KRAJ

KRÁLOVÉ HRADECKÝ KRAJ

JIHOMORAVSKÝ KRAJ

VYSOČINA

JIHOČESKÝ KRAJ

STŘEDOČESKÝ KRAJ

PLZEŇSKÝ KRAJ

KARLOVARSKÝ KRAJ

PRAHA (PRAGUE)

SLOVAKIA

ŽILINSKÝ KRAJ

PREŠOVSKÝ KRAJ

KOŠICKÝ KRAJ

BANSKOBYSTRICKÝ KRAJ

TRENČIANSKY KRAJ

TRNAVSKÝ KRAJ

NITRIANSKY KRAJ

BRATISLAVSKÝ KRAJ

BRATISLAVA

Carpathians

Tatra Mountains

Malé Karpaty

Low Tatras

HUNGARY

BUDAPEST

PEST

NÓGRÁD

HEVES

BORSOD-ABAÚJ-ZEMPLÉN

SZABOLCS-SZATMÁR-BEREG

HAJDÚ-BIHAR

SZOLNOK

JÁSZ-NAGYKUN-SZOLNOK

BÉKÉS

CSONGRÁD

BÁCS-KISKUN

KOMÁROM-ESZTERGOM

FEJÉR

VESZPRÉM

GYŐR-MOSON-SOPRON

VAS

ZALA

SOMOGY

TOLNA

BARANYA

Great Hungarian Plain

Danube (Duna)

Little Alföld

AUSTRIA

WIEN (VIENNA)

NIEDERÖSTERREICH

OBERÖSTERREICH

STEIERMARK

KÄRNTEN

SALZBURG

BURGENLAND

Hohe Tauern

Niedere Tauern

SLOVENIA

LJUBLJANA

CROATIA

ZAGREB

MEĐIMURJE

VARAŽDIN

KOPRIVNICA-KRIŽEVCI

BJELOVAR-BILOGORA

VIROVITICA-PODRAVINA

KRAPINA-ZAGORJE

SISAK-MOSLAVINA

POŽEGA-SLAVONIA

BROD-POSAVINA

OSIJEK-BARANJA

KARLOVAC

LIKA-SENJ

PRIMORJE-GORSKI KOTAR

ISTRA

BOSNIA AND HERZEGOVINA

SERBIA

BEOGRAD (BELGRADE)

VOJVODINA

SRIJEM

ROMANIA

ALBA

CLUJ

BIHOR

SATU MARE

SĂLAJ

MARAMUREŞ

ARAD

TIMIŞ

CARAŞ-SEVERIN

HUNEDOARA

GORJ

MEHEDINŢI

ITALY

FRIULI-VENEZIA GIULIA

VENETO

Gulf of Venice

Danube (Donau)

San

Wisła

Odra

Tisza

Drava

Sava

Mur

Drina

EUROPE

183

188

184

177

19,686ft

13,124ft

9843ft

6562ft

3281ft

1640ft

820ft

328ft

Sea Level

-820ft

-6562ft

-13,124ft

Scale 1:2,500,000
(projection: Lambert Conformal Conic)

0 10 20 30 40 50 Km
0 10 20 30 40 50 Miles

Population
- ■ above 5 million
- ▣ 1 million to 5 million
- ◉ 500,000 to 1 million
- ◎ 100,000 to 500,000
- ⊕ 50,000 to 100,000
- ○ 10,000 to 50,000
- ∘ below 10,000

Elevation scale:
19,686ft / 13,124ft / 9843ft / 6562ft / 3281ft / 1640ft / 820ft / 328ft / Sea Level / -820ft / -6562ft / -13,124ft

MOLDOVA

UKRAINE

ROMANIA
BIHOR · CLUJ · MUREŞ · HARGHITA · NEAMŢ · BACĂU · VASLUI · GALAŢI · VRANCEA · COVASNA · BRAŞOV · SIBIU · ALBA · HUNEDOARA · CARAŞ-SEVERIN · GORJ · VÂLCEA · ARGEŞ · DÂMBOVIŢA · PRAHOVA · BUZĂU · BRĂILA · IALOMIŢA · ILFOV · CĂLĂRAŞI · GIURGIU · TELEORMAN · OLT · DOLJ · MEHEDINŢI · VIDIN · CONSTANŢA · TULCEA · DOBRICH

Cluj-Napoca · Târgu Mureş · Bacău · Braşov · Sibiu · Alba Iulia · Deva · Hunedoara · Petroşani · Târgu Jiu · Râmnicu Vâlcea · Piteşti · Târgovişte · Ploieşti · Buzău · Brăila · Galaţi · Focşani · Slobozia · Călăraşi · Craiova · Slatina · Caracal · Alexandria · Giurgiu · Drobeta-Turnu Severin · Reşiţa · BUCUREŞTI (BUCHAREST) · Constanţa · Tulcea · Mangalia

Carpaţii Meridionali · Carpathian Mountains · Delta Dunării · Vârful Moldoveanu 2544m · Vârful Negoiu 2535m

BULGARIA
Vidin · Montana · VRATSA · PLEVEN · LOVECH · VELIKO TARNOVO · RAZGRAD · SHUMEN · VARNA · DOBRICH · SILISTRA · RUSE · TARGOVISHTE · SOFIA · PERNIK · KYUSTENDIL · BLAGOEVGRAD · PAZARDZHIK · PLOVDIV · STARA ZAGORA · SLIVEN · YAMBOL · BURGAS · HASKOVO · KARDZHALI · SMOLYAN

SOFIA · Pernik · Kyustendil · Blagoevgrad · Pazardzhik · Plovdiv · Stara Zagora · Sliven · Yambol · Burgas · Haskovo · Kardzhali · Smolyan · Pleven · Lovech · Gabrovo · Veliko Tarnovo · Razgrad · Shumen · Varna · Dobrich · Silistra · Ruse · Vratsa · Montana · Vidin

Dunavska Ravnina · Balkan Mountains · Sredna Gora · Stara Planina · Rhodope Mountains · Pirin

GREECE
KENTRIKÍ MAKEDONÍA · ANATOLIKÍ MAKEDONÍA KAI THRÁKI · THESSALÍA · VÓREION AIGAÍON
Thessaloníki · Véroia · Kateríni · Sérres · Dráma · Kaválas · Kozáni · Lárisa · Vólos · Kými Olympus Mt Olympus 2917m · Chalkidikí · Lésvos (Lesbos) · Límnos · Samothráki · Thásos

TURKEY
KIRKLARELI · TEKIRDAĞ · EDİRNE · İSTANBUL · ÇANAKKALE · BALIKESİR · BURSA
İstanbul · Edirne · Tekirdağ · Lüleburgaz · Çanakkale · Bandırma · Balıkesir · Bursa
İstanbul Boğazı (Bosporus) · Çanakkale Boğazı (Dardanelles) · Marmara Denizi (Sea of Marmara)

MACEDONIA
Skopje region · Strumica · Kavadarci · Gevgelija

Black Sea

Thracian Sea

Aegean Sea

Marmara Denizi (Sea of Marmara)

Greece

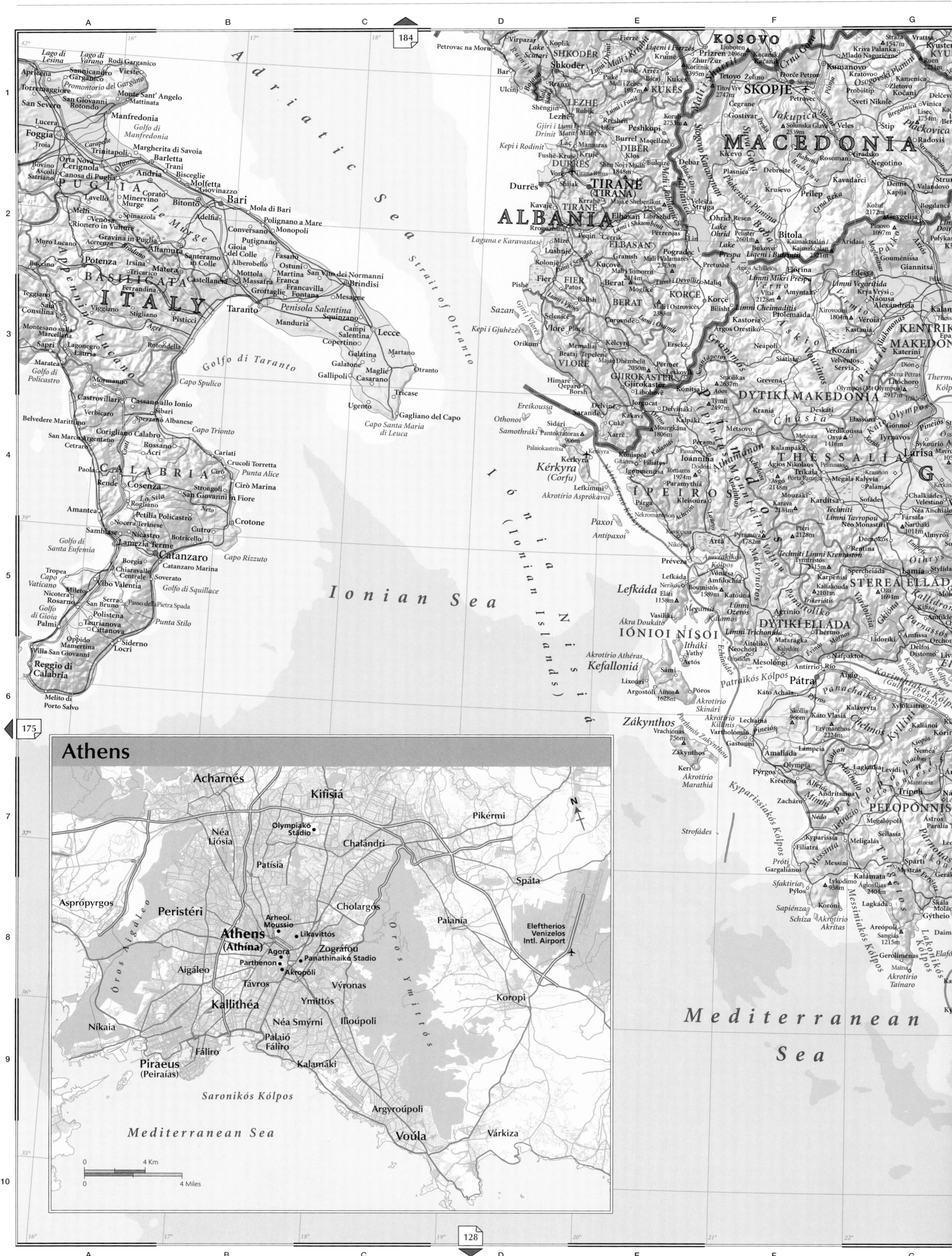

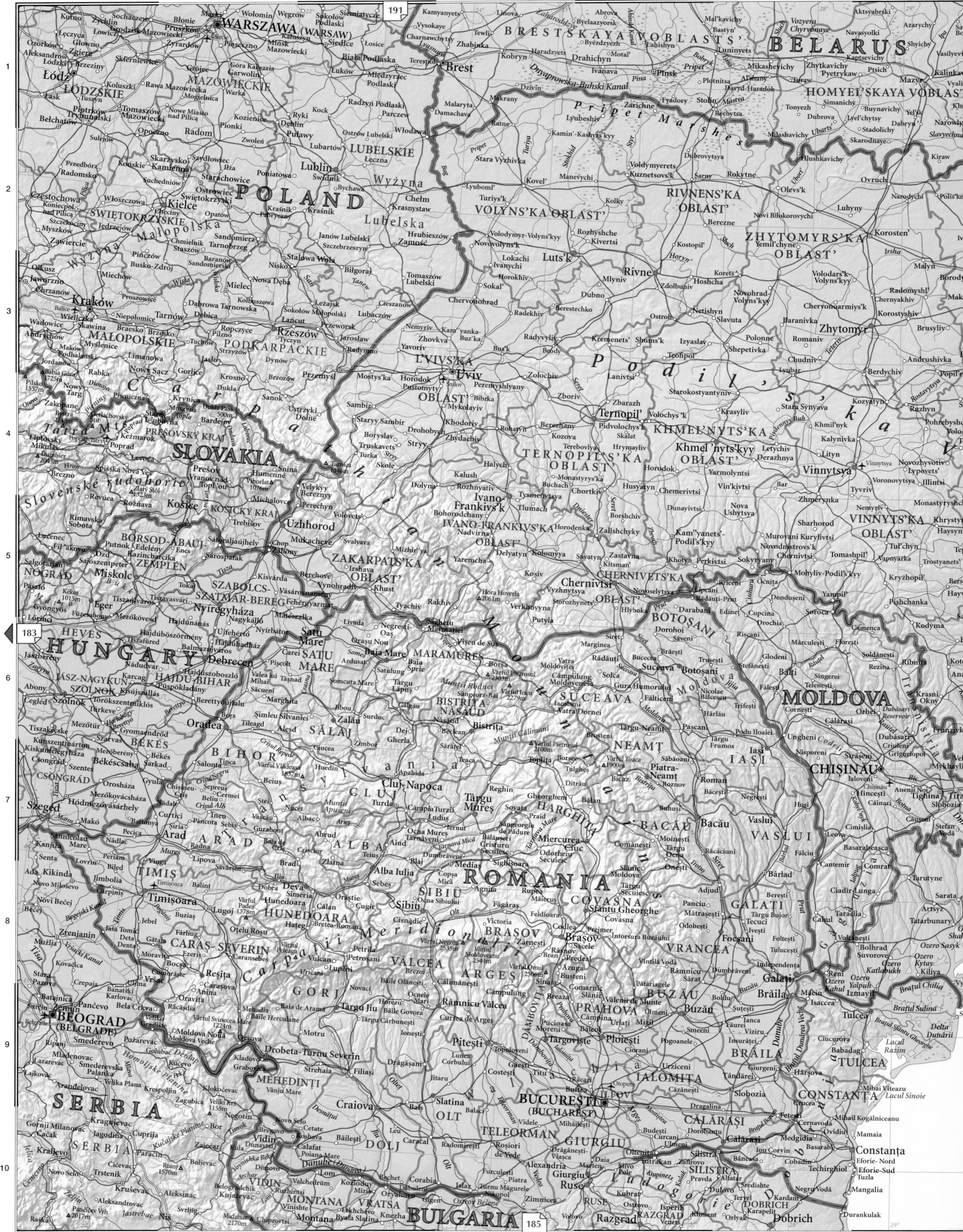

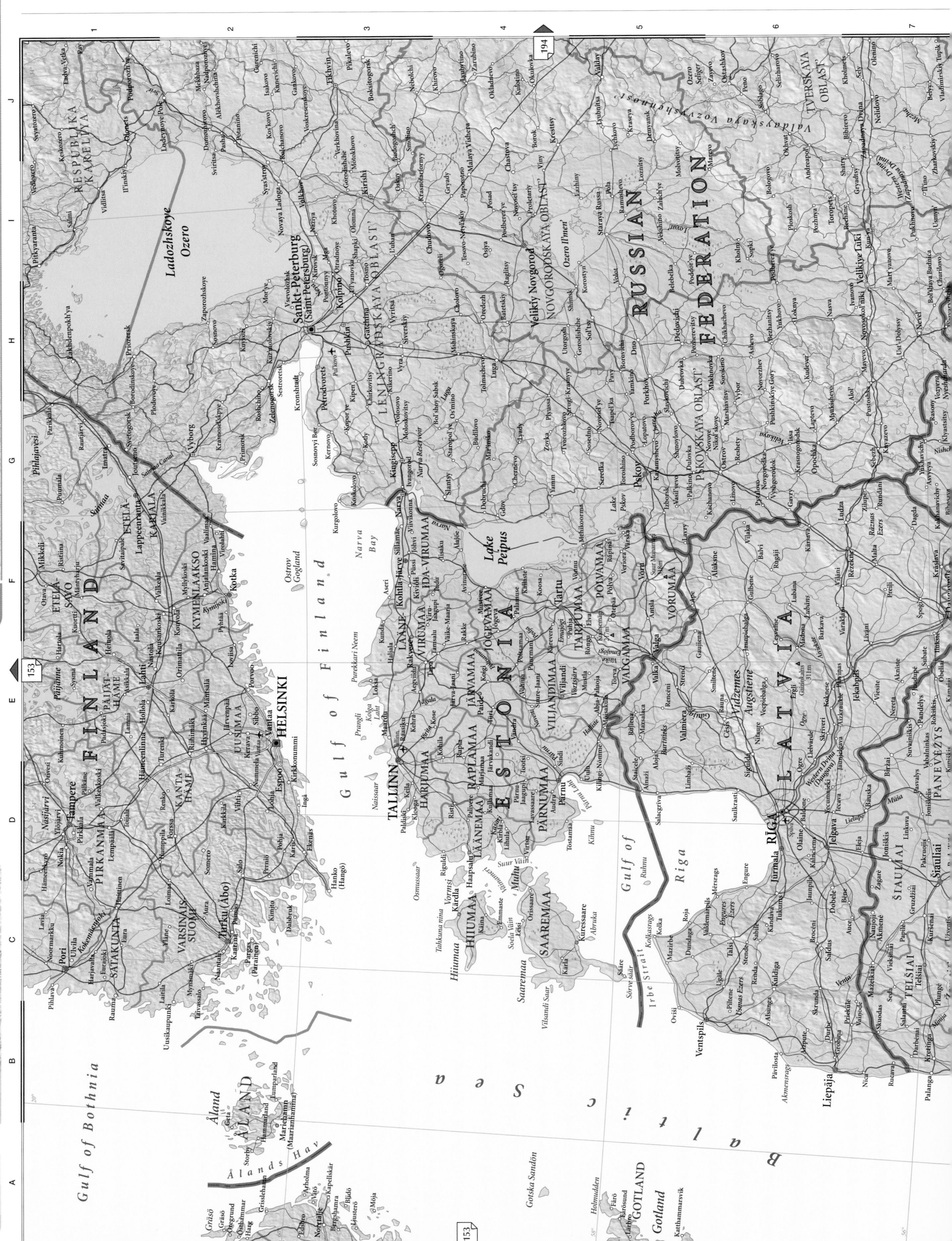

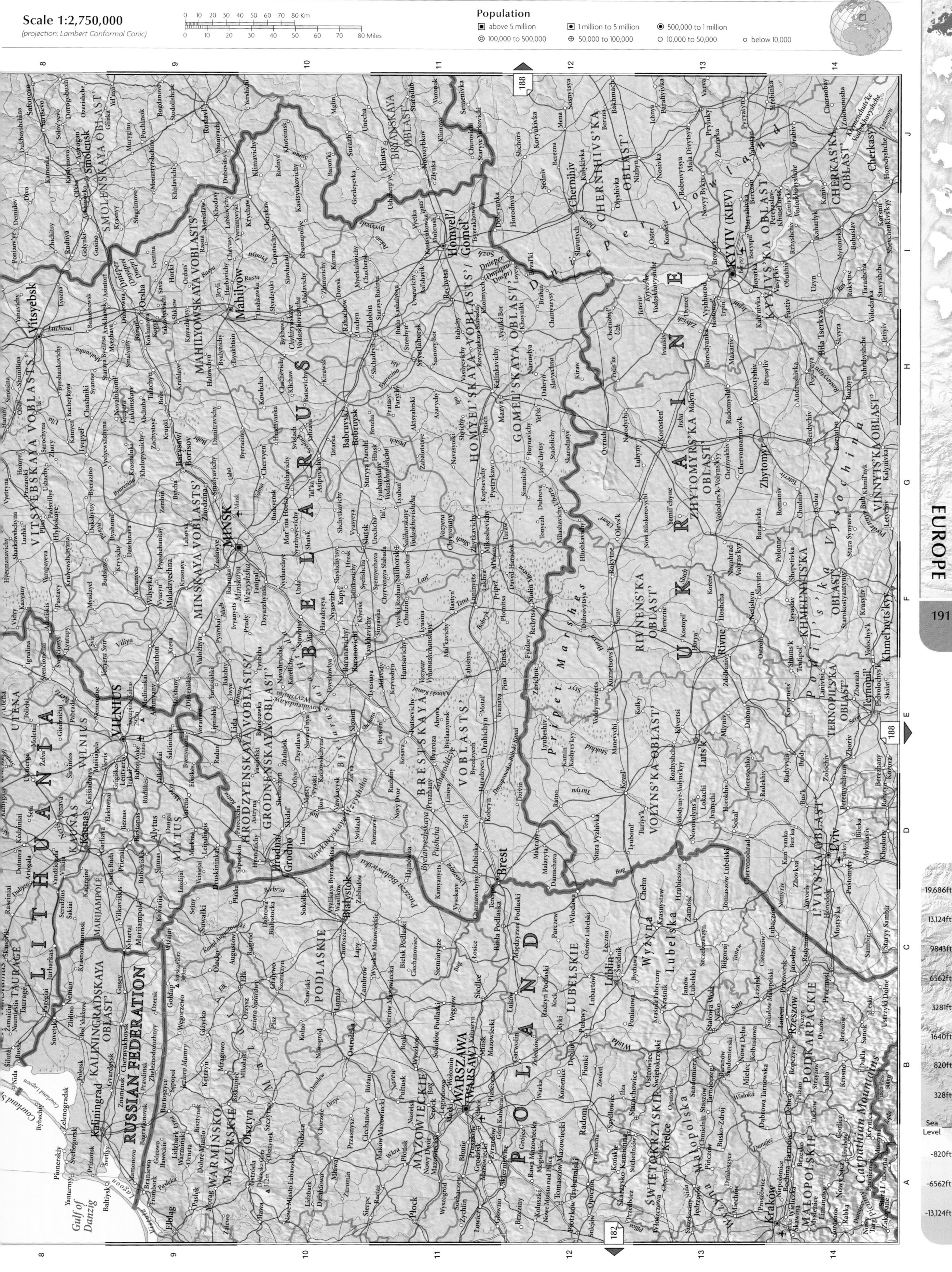

The Russian Federation

THE RUSSIAN FEDERATION: ADMINISTRATIVE REGIONS

The administrative area names in European Russia have been omitted west of the Ural Mountains. Please refer to pages 194-195 and 196-197 where these areas are shown at a larger scale.

Scale / elevation legend:

6000m
4000m
3000m
2000m
1000m
500m
250m
100m
Sea Level
-250m
-2000m
-4000m

Major labels:

Norwegian Sea

NORWAY

SWEDEN

FINLAND

RUSSIAN

ESTONIA

LATVIA

LITHUANIA

BELARUS

Baltic Sea

Gulf of Finland

Gulf of Bothnia

Barents Sea

Beloye More (White Sea)

Ladozhskoye Ozero

Onezhskoye Ozero

Lake Peipus

Lake Pskov

Kol'skiy Poluostrov

MURMANSKAYA OBLAST'

RESPUBLIKA KARELIYA

LENINGRADSKAYA OBLAST'

NOVGORODSKAYA OBLAST'

PSKOVSKAYA OBLAST'

VOLOGODSKAYA OBLAST'

YAROSLAVSKAYA OBLAST'

TVERSKAYA OBLAST'

VLADIMIRSKAYA OBLAST'

IVANOVSKAYA OBLAST'

ARKHANGEL'SKAYA

Lapland / LAPPI

NORDLAND

TROMS

FINNMARK

NORRBOTTEN

VÄSTERBOTTEN

VÄSTERNORRLAND

GÄVLEBORG

UPPSALA

POHJOIS-POHJANMAA

KESKI-POHJANMAA

ETELÄ-POHJANMAA

ÖSTERBOTTEN

KAINUU

POHJOIS-SAVO

POHJOIS-KARJALA

ETELÄ-SAVO

ETELÄ-KARJALA

KESKI-SUOMI

PIRKANMAA

SATAKUNTA

VARSINAIS-SUOMI

KANTA-HÄME

PÄIJÄT-HÄME

UUSIMAA

KYMENLAAKSO

ÅLAND

Cities:

Murmansk, Severomorsk, Apatity, Kandalaksha, Arkhangel'sk (Archangel), Severodvinsk, Petrozavodsk, Sankt-Peterburg (Saint Petersburg), HELSINKI, Espoo, Vantaa, Turku (Åbo), Tampere, TALLINN, Tartu, Velikiy Novgorod, Pskov, Vologda, Rybinsk, Yaroslavl', Ivanovo, Tver, Vladimir, RIGA, Kaliningrad, STOCKHOLM, Gävle, Luleå, Umeå, Oulu, Rovaniemi, Kuopio, Joensuu, Kiruna, Narvik, Bodø, Kemi, Vaasa, Pori, Kotka, Vyborg

Scale 1:5,750,000
(projection: Lambert Conformal Conic)

0 25 50 75 100 125 150 175 200 Km
0 25 50 75 100 125 150 175 200 Miles

Population
■ above 5 million
■ 1 million to 5 million
● 500,000 to 1 million
◉ 100,000 to 500,000
⊕ 50,000 to 100,000
○ 10,000 to 50,000
○ below 10,000

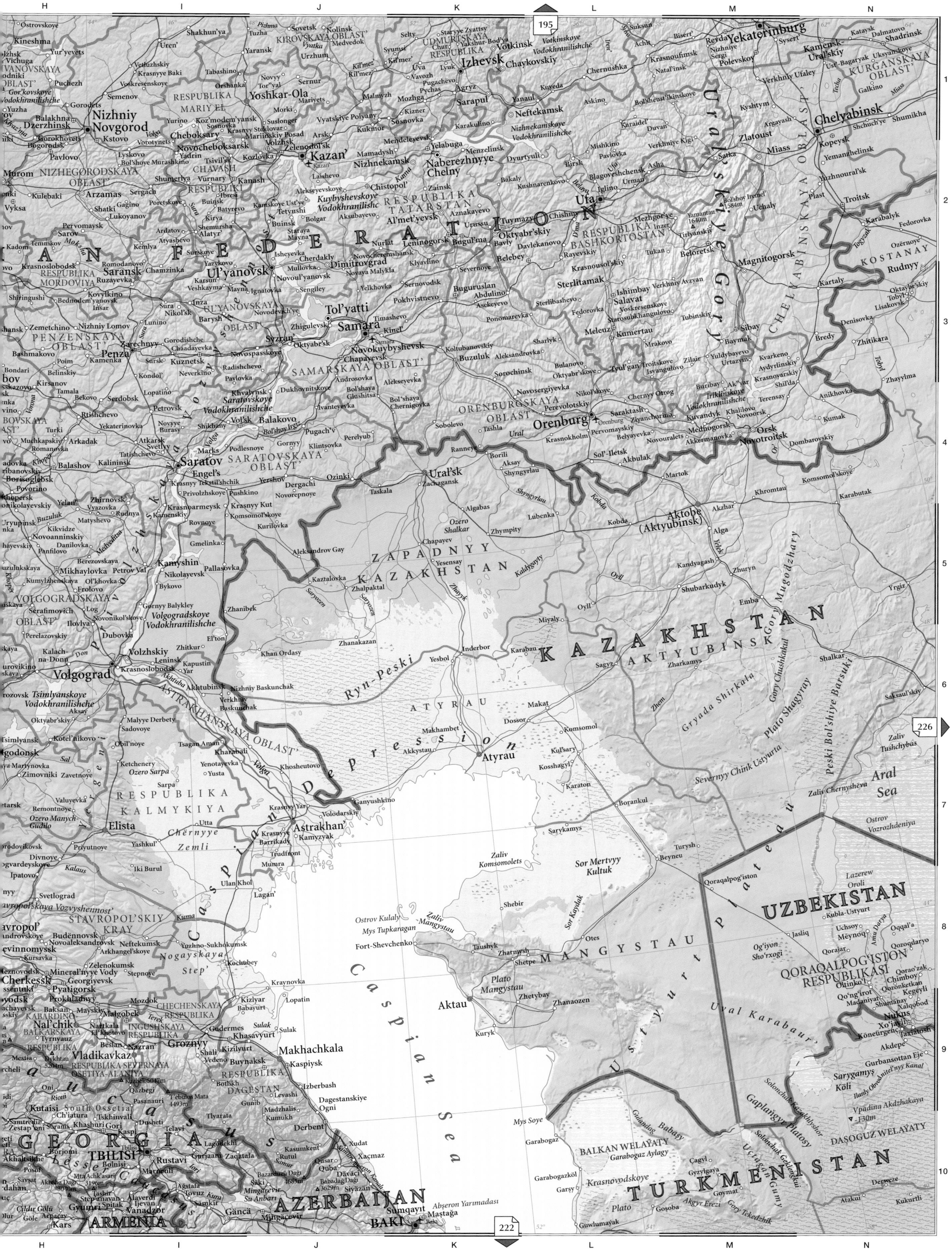

19,686ft
13,124ft
9843ft
6562ft
3281ft
1640ft
820ft
328ft
Sea Level
-820ft
-6562ft
-13,124ft

The Mediterranean

Scale 1:8,750,000
(projection: Lambert Conformal Conic)

0 25 50 75 100 125 150 175 200 Km
0 25 50 75 100 125 150 200 Miles

Population
■ above 5 million
◉ 100,000 to 500,000
■ 1 million to 5 million
⊕ 50,000 to 100,000
◉ 500,000 to 1 million
○ 10,000 to 50,000
○ below 10,000

196

129

216

19,686ft
13,124ft
9843ft
6562ft
3281ft
1640ft
820ft
328ft
Sea Level
-820ft
-6562ft
-13,124ft

POLAND
WARSZAWA (WARSAW)
BELARUS
UKRAINE
KYYIV (KIEV)
SLOVAKIA
BRATISLAVA
HUNGARY
BUDAPEST
ROMANIA
MOLDOVA
CHIŞINĂU
BUCUREŞTI (BUCHAREST)
SERBIA
BEOGRAD (BELGRADE)
BOSNIA AND HERZEGOVINA
SARAJEVO
MONTENEGRO
PODGORICA
KOSOVO
PRISHTINË (PRISTINA)
MACEDONIA
SKOPJE
ALBANIA
TIRANË (TIRANA)
BULGARIA
SOFIYA (SOFIA)
GREECE
ATHÍNA (ATHENS)
RUSSIAN FEDERATION
GEORGIA
TURKEY
ANKARA
Istanbul
İzmir
CYPRUS
NICOSIA
TURKISH REPUBLIC OF NORTHERN CYPRUS (recognised only by Turkey)
SYRIA
DIMASHQ (DAMASCUS)
IRAQ
LEBANON
BEYROUTH (BEIRUT)
ISRAEL
JERUSALEM
Tel Aviv-Yafo
JORDAN
'AMMĀN
SAUDI ARABIA
EGYPT
CAIRO (AL QĀHIRAH)
Alexandria (Al Iskandarīyah)
LIBYA
Banghāzī (Benghazi)
Cyrenaica
Libyan Plateau

Black Sea
Sea of Azov
Caucasus
Mediterranean Sea
Ionian Sea
Aegean Sea
Sea of Crete (Kritikó Pélagos)
Red Sea
Kríti (Crete)
Ródos (Rhodes)
Khalij Surt (Gulf of Sirte)
Nile Delta
Sinai (Sinaʾ)

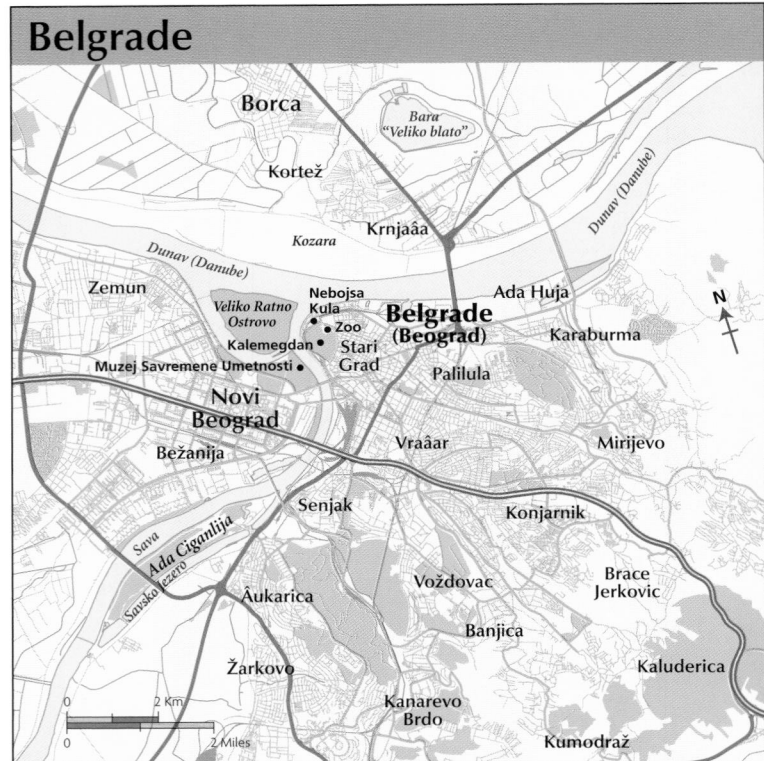

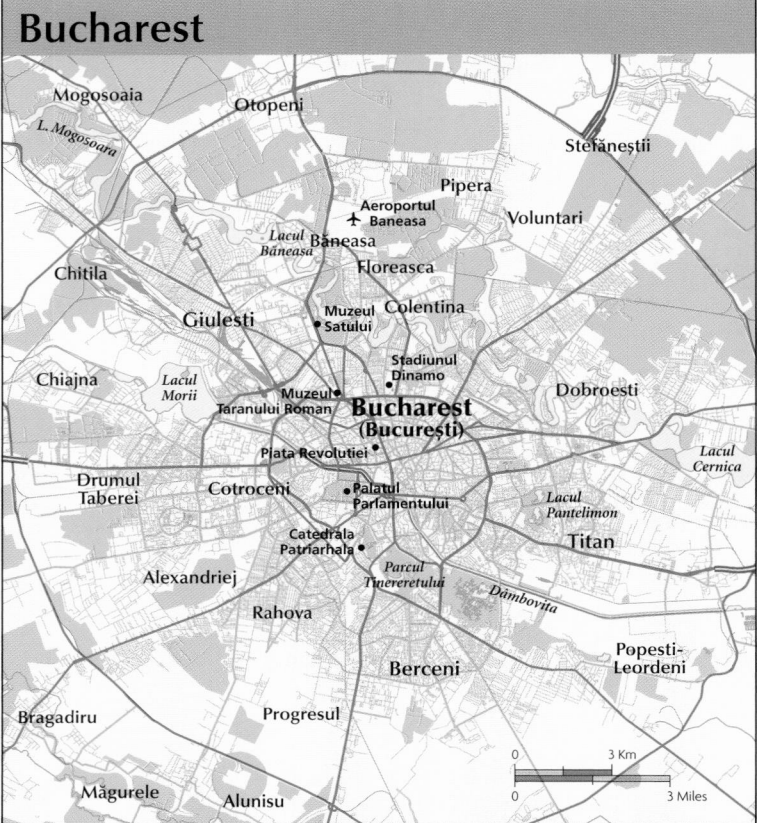

Copenhagen

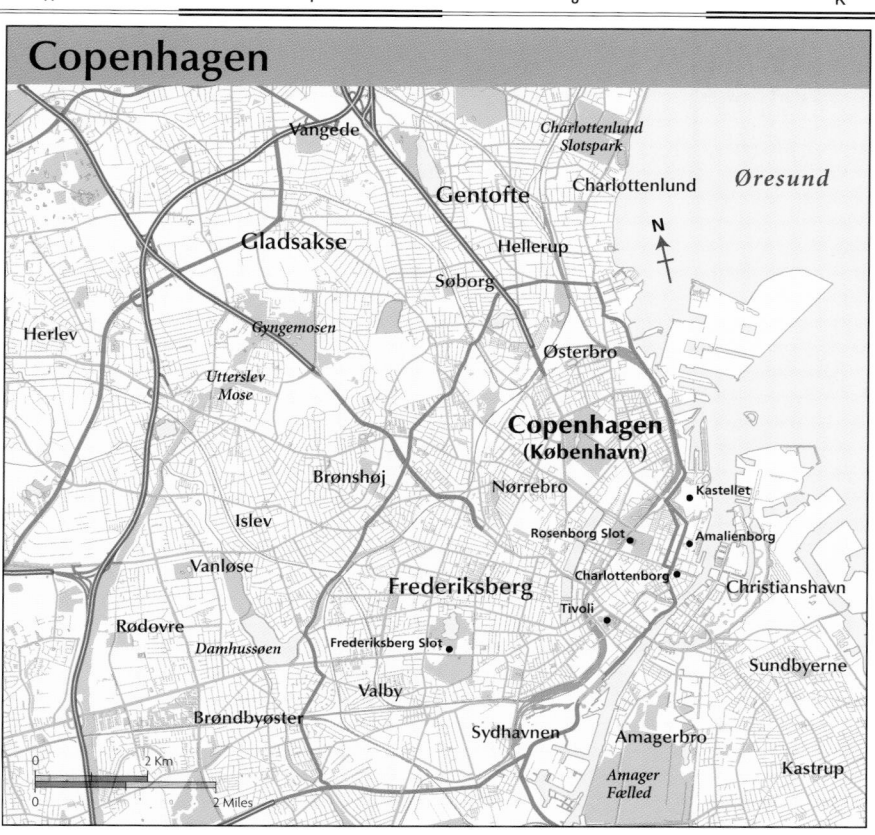

Kiev

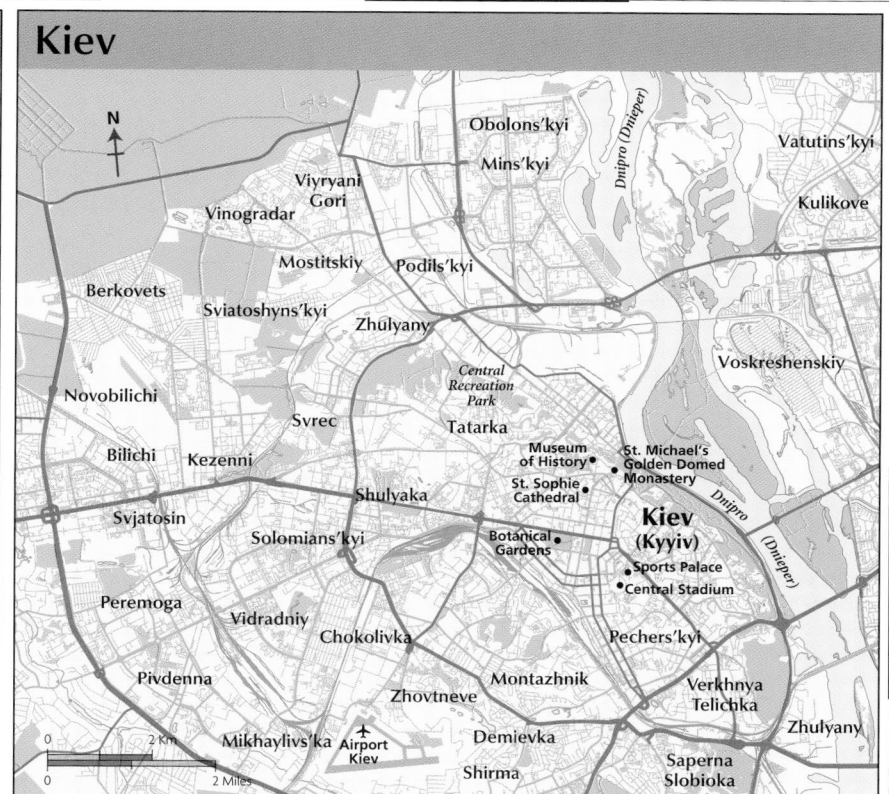

London

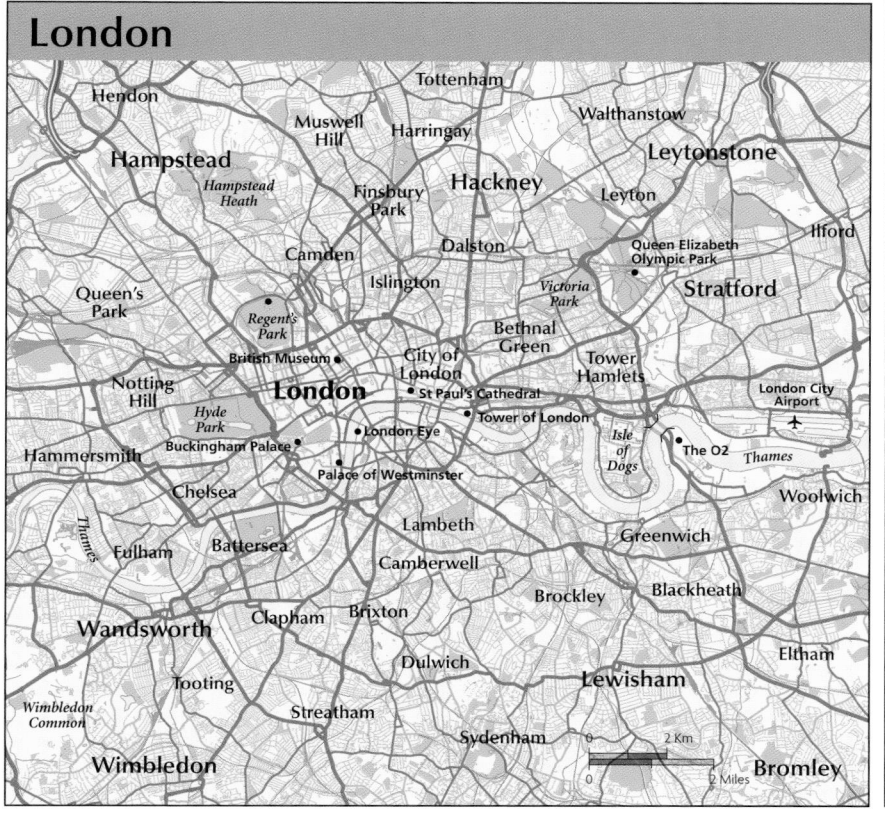

Lisbon

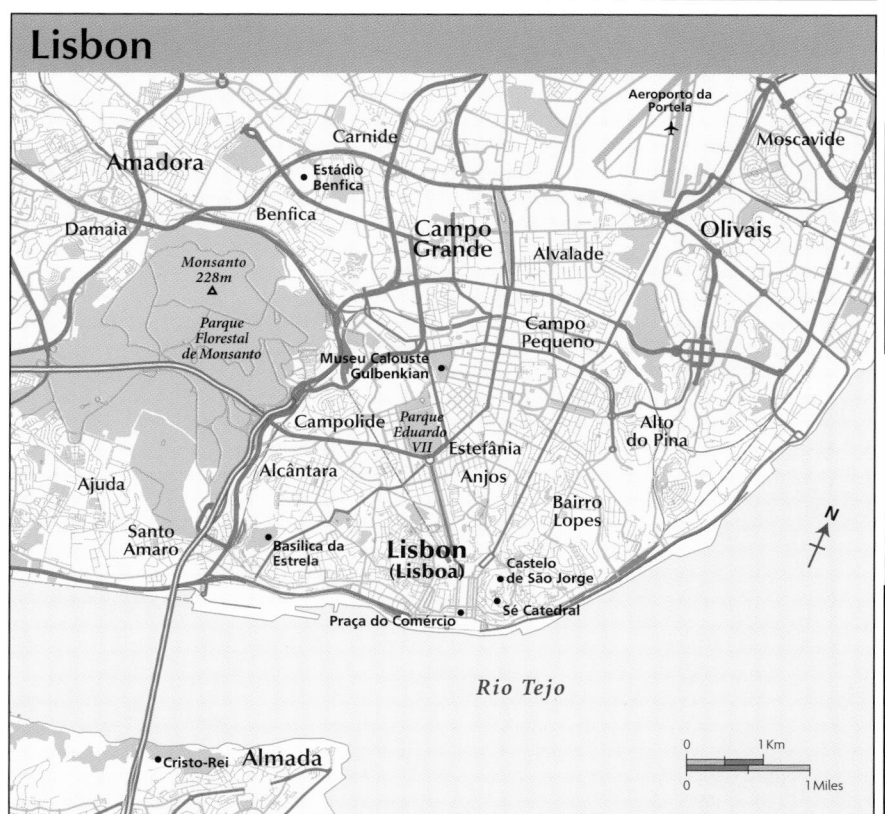

Madrid

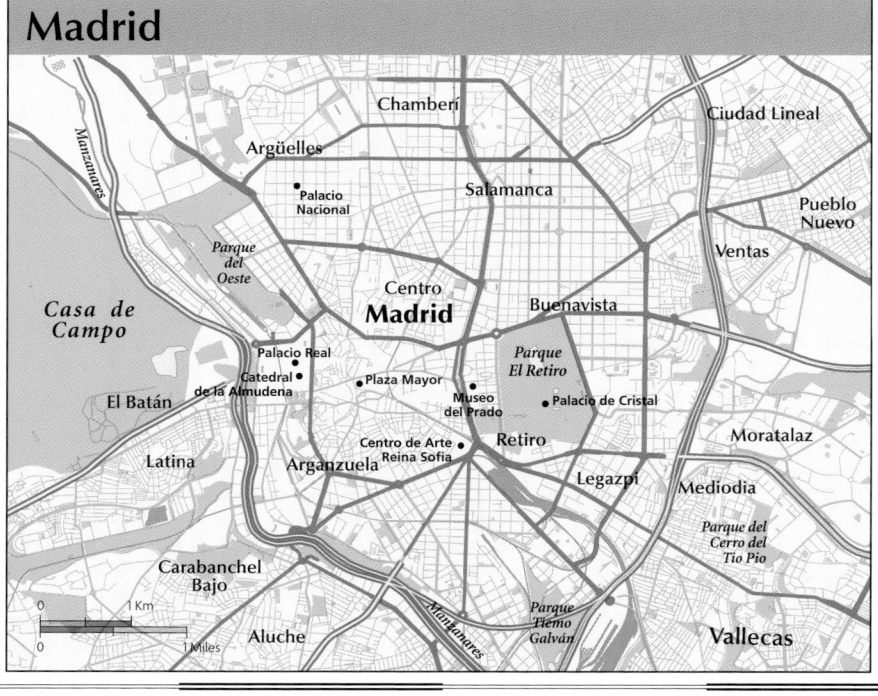

Minsk

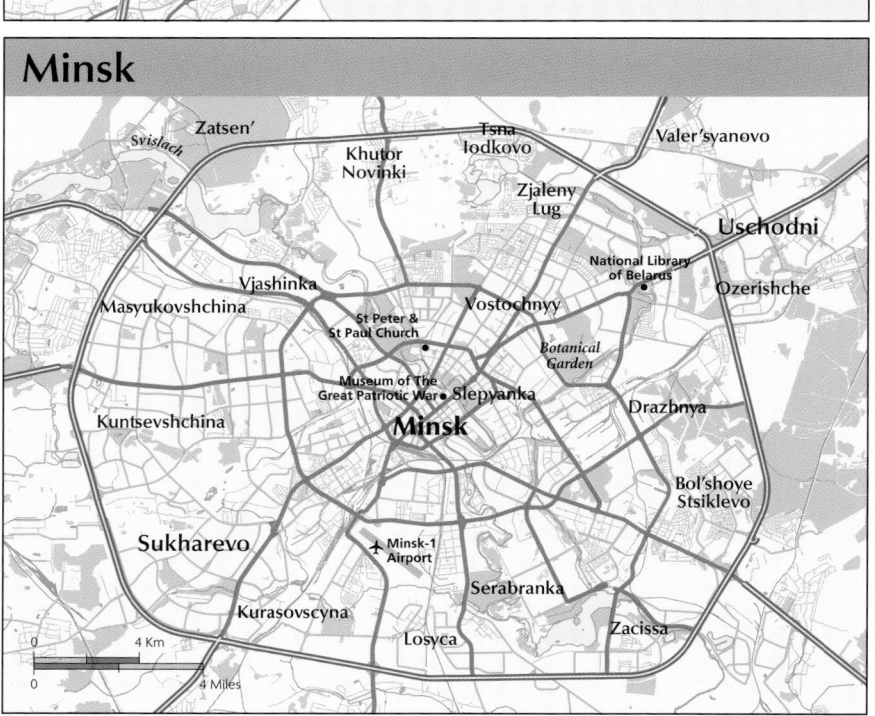

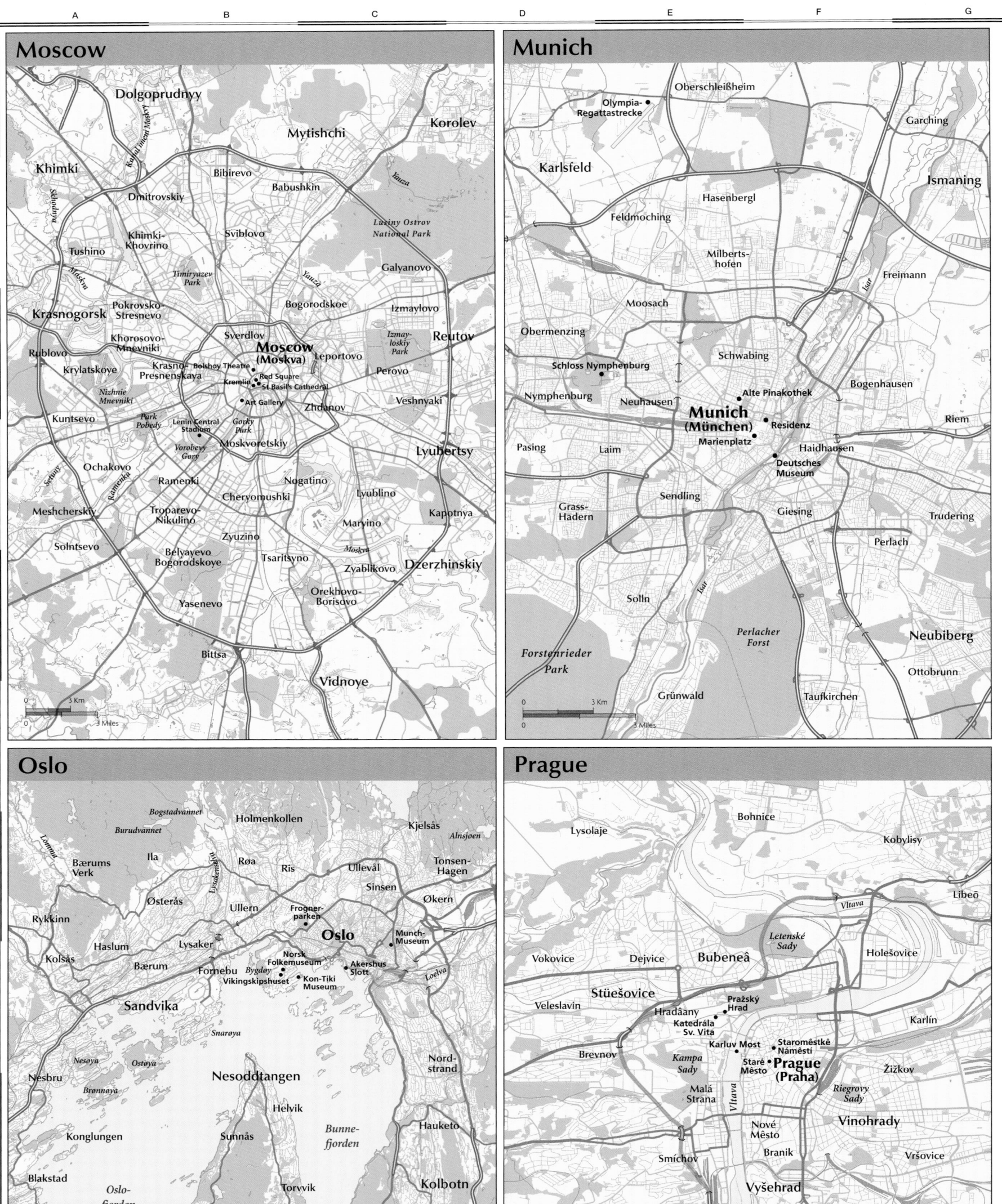

Moscow

Dolgoprudnyy
Mytishchi
Korolev
Khimki
Bibirevo
Babushkin
Dmitrovskiy
Kanal Imeni Moskvy
Sviblovo
Yauza
Lusiny Ostrov National Park
Khimki-Khovrino
Tushino
Galyanovo
Moskva
Timiryazev Park
Yauza
Izmaylovo
Krasnogorsk
Pokrovsko-Streснevo
Bogorodskoe
Reutov
Khorosovo-Mnevniki
Izmay-loskiy Park
Rublovo
Sverdlov
Leportovo
Perovo
Krylatskoye
Krasno-Presnenskaya
Bolshoy Theatre
Red Square
St Basil's Cathedral
Veshnyaki
Nizhnie Mnevniki
Kremlin
Art Gallery
Zhdanov
Kuntsevo
Park Pobedy
Lenin Central Stadium
Gorky Park
Moskvoretskiy
Lyubertsy
Ochakovo
Vorobevy Gory
Ramenki
Nogatino
Lyublino
Kapotnya
Meshcherskiy
Troparevo-Nikulino
Cheryomushki
Maryino
Dzerzhinskiy
Solntsevo
Zyuzino
Moskva
Belyayevo Bogorodskoye
Tsaritsyno
Zyablikovo
Yasenevo
Orekhovo-Borisovo
Bittsa
Vidnoye

0   3 Km
0   3 Miles

Munich

Olympia-Regattastrecke
Oberschleißheim
Garching
Karlsfeld
Hasenbergl
Ismaning
Feldmoching
Milberts-hofen
Freimann
Moosach
Isar
Obermenzing
Schwabing
Bogenhausen
Schloss Nymphenburg
Alte Pinakothek
Riem
Nymphenburg
Neuhausen
Munich (München)
Residenz
Pasing
Laim
Marienplatz
Haidhausen
Grass-Hadern
Sendling
Deutsches Museum
Giesing
Trudering
Perlach
Isar
Solln
Neubiberg
Perlacher Forst
Forstenrieder Park
Grünwald
Ottobrunn
Taufkirchen

0   3 Km
0   3 Miles

Oslo

Bogstadvannet
Holmenkollen
Kjelsås
Burudvannet
Alnsjøen
Bærums Verk
Ila
Røa
Ris
Ullevål
Tonsen-Hagen
Lysakerelva
Østerås
Sinsen
Økern
Rykkinn
Ullern
Frogner-parken
Oslo
Kolsås
Haslum
Lysaker
Munch-Museum
Bærum
Norsk Folkemuseum
Akershus Slott
Fornebu
Bygdøy
Kon-Tiki Museum
Loelva
Vikingskipshuset
Sandvika
Nordstrand
Snarøya
Nesøya
Ostøya
Nesoddtangen
Nesbru
Brønnøya
Helvik
Konglungen
Sunnås
Bunne-fjorden
Hauketo
Blakstad
Oslo-fjorden
Torvvik
Kolbotn
Bjetkås
Fjellstrand
Gjersjøen
Blylaget
Oppegård
Nebba

0   2 Km
0   2 Miles

Prague

Bohnice
Lysolaje
Kobylisy
Vokovice
Dejvice
Bubeneâ
Vltava
Libeô
Letenské Sady
Holešovice
Veleslavin
Stüešovice
Hradâany
Pražský Hrad
Karlín
Katedrála Sv. Víta
Staroměstké Náměstí
Brevnov
Karluv Most
Žižkov
Kampa Sady
Staré Město
Prague (Praha)
Malá Strana
Vltava
Riegrovy Sady
Nové Město
Vinohrady
Smíchov
Branik
Vršovice
Vyšehrad
Jinonice
Radlice
Michle
Podolí

0   1 Km
0   1 Miles

Rome

Tor di Quinto
Aeroporto di Roma - Urbe
Tufello
Torrevécchia
Stadio Olimpico
Flaminio
Villa Ada Savoia
Trieste
Monte Sacro
Primavalle
Trionfale
Parioli
Pietralata
Parco Regionale Urbano Pineto
Nomentano
Villa Borghese
Montespaccato
CITTÀ DEL VATICANO
Cappella Sistina
Rome (Roma)
Castel Sant'Angelo
Fontana di Trevi
Basilica di San Pietro
Piazza San Pietro
Pantheon
Palazzo di Quirinale
Aurelio
Foro Romano
Tiburtino
Trastevere
Colosseo
Tor Pignattara
Valcannuta
Villa Doria Pamphili
Fiume Tévere
Tuscolano
Montverde Nuovo
Cinecittà
Corviale
Parco della Caffaerella
Garbatella
Catacombe di Domitilla
Ostiense
Magliana

0   1 Km
0   1 Miles

St Petersburg

Ozero Lakhtinskiy Razliv
Olgino
Vdelnoe
Grazhdanka
Rzhevka

0   3 Km
0   3 Miles

Ostrova Krestovskiye
Stoyka
Polyustrovo
Petrogradskaya Storona
Cruiser Aurora
Nevá
Smolnyy Cathedral
Bolshaya-Okhta
Zhernovka
Vasilyevskiy Ostrov
Hermitage and Winter Palace
Kirov Palace of Culture
Admiralty
Alexander Nevsky Abbey
Malaya-Okhta
St Isaac's Cathedral
Finskiy Zaliv
St Petersburg (Sankt-Peterburg)
Ostrov Gutuyevskiy
Volynkina Derevnya
Volodanskoye
Vesolyy Posolok
Montespaccato
Obukhovo
Utkina Zavod
Avtovo
Aleksandrovskoye
Ulyanka
Kupchino
Ligovo
Srednaya Rogatka
Novoaleksandrovskoye

Sofia

Vrybnitsa
Benkouski
Nadezhda

0   2 Km
0   2 Miles

Serdika
Lyulin
Vasil Levski
Vrazdebna
Ilinden
Sofia
Iskǎr River
Krasna Polyana
National Art Gallery
Cathedral
Poduyane
Sofia Art Gallery
Ovcha Kupel
National Palace of Culture
National Stadium
Sofia Airport
Borisova Gradina
Slatina
Krasno Selo
Lozenets
Izgrev
Iskyr
Triaditsa
Drouzhba Li
Knyazhevo
Bakston
Studentski
Mladost
Vitosha National Park
Vitosha

Stockholm

Kista
Inverness
Sundbyberg
Friends Arena
Solna
Lidingö
Stockholm-Bromma Airport
Lidingö
Bromma
Stockholm
Lilla Värten
Älsten
Vasamuseet
Kärsön
Stadshuset
Skansen
Kungliga Slottet
Saltsjön
Ektorp
Södermalm
Nacka
Mälaren
Hägersten
Enskede
Bagarmossen
Skärholmen
Farsta

0   2 Km
0   2 Miles

Warsaw

Marcelin
Ząbkii
Tarchomin
Zacisze
Mlociny
Bródno
Rembertów
Wisla
Park Skaryszewski
Wawrzyszew
Praga
Grochów
Zoliborz
Zoo
Warsaw (Warszawa)
Warsaw Babice Airport
St John's Cathedral
Stadium
Saska Kępa
Górce
Royal Castle
Chopin Museum
Gocław
Palace of Culture and Science
Wola
Park Lazienkowski
Wisla
Mokotów
Ochota
Sadyba
Raków
Wierzbno
Augustówka
Opacz
Okęcie
Słuzew
Wilanów
Warsaw Frederic Chopin Airport

0   2 Km
2 Miles

Zagreb

0   2 Km
0   2 Miles

Granešina
Maksimir
Sesvete
Crnomerec
Croatian History Museum
Croatian National Theatre
Cathedral
Tresnjevka
Zagreb
Jarun
Sava
Lake Jarun
Novi Zagreb
Botinec
Zagreb Airport

Asia is the world's largest continent with a total area of 16,838,365 sq miles (43,608,000 sq km). It comprises 49 separate countries, including 97% of Turkey and 72% of the Russian Federation. Almost 60% of the world's population lives in Asia.

FACTFILE

N **Most Northerly Point:** Mys Articheskiy, Russia 81° 12′ N
S **Most Southerly Point:** Pulau Pamana, Indonesia 11° S
E **Most Easterly Point:** Mys Dezhneva, Russia 169° 40′ W
W **Most Westerly Point:** Bozzcaada, Turkey 26° 2′ E

Largest Lakes:
1. Caspian Sea, Asia/Europe 143,243 sq miles (371,000 sq km)
2. Lake Baikal, Russian Federation 11,776 sq miles (30,500 sq km)
3. Lake Balkhash, Kazakhstan/China 7115 sq miles (18,428 sq km)
4. Aral Sea, Kazakhstan/Uzbekistan 6625 sq miles (17,160 sq km)
5. Tonlé Sap, Cambodia 3861 sq miles (10,000 sq km)

Longest Rivers:
1. Yangtze, China 3915 miles (6299 km)
2. Yellow River, China 3395 miles (5464 km)
3. Mekong, SE Asia 2749 miles (4425 km)
4. Lena, Russian Federation 2734 miles (4400 km)
5. Yenisey, Russian Federation 2541 miles (4090 km)

Largest Islands:
1. Borneo, Brunie/Indonesia/Malaysia 292,222 sq miles (757,050 sq km)
2. Sumatra, Indonesia 202,300 sq miles (524,000 sq km)
3. Honshu, Japan 88,800 sq miles (230,000 sq km)
4. Sulawesi, Indonesia 73,057 sq miles (189,218 sq km)
5. Java, Indonesia 53,589 sq miles (138,794 sq km)

Highest Points:
1. Mount Everest, China/Nepal 29,029 ft (8848 m)
2. K2, China/Pakistan 28,253 ft (8611 m)
3. Kangchenjunga I, India/Nepal 28,210 ft (8598 m)
4. Lhotse, Nepal 27,939 ft (8516 m)
5. Makalu, China/Nepal 27,767 ft (8463 m)

Lowest Point:
▼ Dead Sea, Israel/Jordan -1401 ft (-427 m) below sea level

Highest recorded temperature:
+ Tirat Zevi, Israel 129°F (54°C)

Lowest recorded temperature:
− Verkhoyansk, Russian Federation -90°F (-68°C)

Wettest Place:
≈ Cherrapunji, India 450 in (11,430 mm)

Driest Place:
⌢ Aden, Yemen 1.8 in (46 mm)

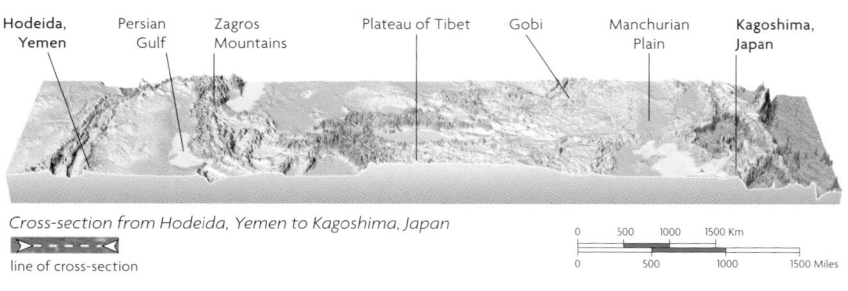

Cross-section from Hodeida, Yemen to Kagoshima, Japan

▷−▷−◁− line of cross-section

0   500   1000   1500 Km
0   500   1000   1500 Miles

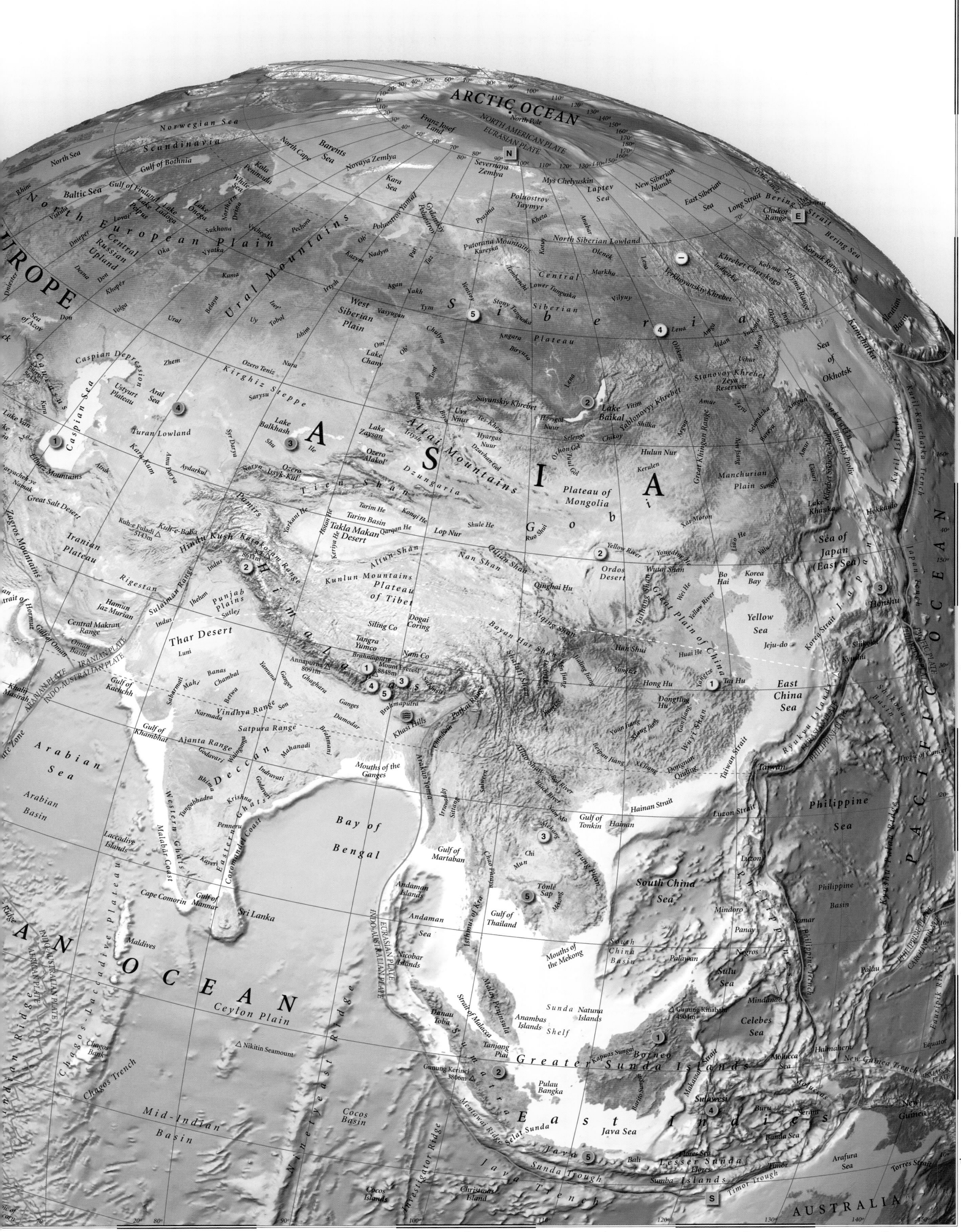

Political

Asia is the world's largest continent, encompassing many different and discrete realms, from the desert Arab lands of the southwest to the subtropical archipelago of Indonesia; from the vast barren wastes of Siberia to the fertile river valleys of China and South Asia, seats of some of the world's most ancient civilizations. The collapse of the Soviet Union has fragmented the north of the continent into the Siberian portion of the Russian Federation, and the new republics of Central Asia. Strong religious traditions heavily influence the politics of South and Southwest Asia. Hindu and Muslim rivalries threaten to upset the political equilibrium in South Asia where India—in terms of population—remains the world's largest democracy. China, another population giant, is reasserting its position as a world political and economic power, while on its doorstep, the dynamic Pacific Rim countries, led by Japan, continue to assert their worldwide economic force.

Population density
(people per sq mile)

- below 25
- 25–124
- 125–259
- 260–649
- 650–10,400
- above 10,400

Population

Some of the world's most populous and least populous regions are in Asia. The plains of eastern China, the Ganges River plains in India, Japan, and the Indonesian island of Java, all have very high population densities; by contrast parts of Siberia and the Plateau of Tibet are virtually uninhabited. China has the world's greatest population—20% of the globe's total—while India, with the second largest, has more than one billion.

Map labels

ARCTIC OCEAN
Beri... Sea
East Siberian Sea
Anadyr'
Laptev Sea
Kara Sea
Kolyma Range
Magadan
Indigirka
Arctic Circle
Noril'sk
Central Siberian Plateau
Siberia
Yakutsk
Sea of Okho...
Kureyka
Ob'
Lena
Vilyuy
Aldan
Stony Tunguska
RUSSIAN FEDERATION
West Siberian Plain
Yenisey
Yekaterinburg
Ural Mountains
Khabarovsk
Tobol
Irtysh
Chelyabinsk
Chulym
Amur
Tomsk
Krasnoyarsk
Lake Baikal
Omsk
Novosibirsk
Irkutsk
Novokuznetsk
Qiqihar
Harbin
Ural'sk
Zhayyk
KAZAKHSTAN
ASTANA
Karagandy
Semey
Sühbaatar
Choybalsan
Erdenet
ULAN BATOR
Changchun
Jilin
Vladivost...
Istanbul
Black Sea
Zhezkazgan
Balkhash
MONGOLIA
Shenyang
NOR...
ANKARA
Sokhumi
GEORGIA
Aktau
Aral Sea
Kyzylorda
Lake Balkhash
Gobi
Inner Mongolia
Anshan
Wonsan
TURKEY
Bat'umi
Kutaisi
TBILISI
Syr Darya
Taraz
Almaty
Ürümqi
Altai Mountains
PYONGYANG
Anatolia
Adana
Gaziantep
ARMENIA
Gäncä
AZERB.
BISHKEK
Karakol
Tien Shan
BEIJING
Dalian
SEOUL
CYPRUS
NICOSIA
Aleppo
YEREVAN
AZERB.
BAKU
UZBEKISTAN
Dasoguz
TASHKENT
KYRGYZSTAN
Osh
Baotou
Datong
Tangshan
Incheon
LEBANON
Tripoli
SYRIA
DAMASCUS
Mosul
Kirkuk
Amu Darya
DUSHANBE
Tarim He
Takla Makan Desert
Shijiazhuang
Baoding
Tianjin
SEJO...
BEIRUT
Haifa
Tel Aviv-Yafo
JERUSALEM
Gaza
ISRAEL
AMMAN
JORDAN
BAGHDAD
TEHRAN
Gorgan
AŞGABAT
TURKMENISTAN
Balkh
TAJIKISTAN
Qal'eh-ye Now
(line of control)
(claimed by India)
Kunlun Mountains
Taiyuan
Handan
Jinan
Qingdao
Yello...
An Najaf
Qom
Mashhad
Herat
Esfahan
IRAQ
IRAN
Iranian Plateau
AFGHANISTAN
KABUL
Jalalabad
(administered by China, claimed by India)
CHINA
Lanzhou
Zhengzhou
Xuzhou
Luoyang
Nanjing
Basra
Ahvaz
Kerman
Kandahar
Peshawar
Srinagar
ISLAMABAD
Jammu
Xi'an
Huainan
Hefei
Wuhan
Hangzhou
Nin...
KUWAIT
KUWAIT
Shiraz
Zahedan
Quetta
Faisalabad
Lahore
Ludhiana
Himalayas
Plateau of Tibet (Much of Arunachal Pradesh is claimed by China)
Salween
Mekong
Mianyang
Chengdu
Changsha
Nanchang
Fuzhou
SAUDI ARABIA
MANAMA
BAHRAIN
Bandar-e 'Abbas
PAKISTAN
Multan
Lhasa
Brahmaputra
Chongqing
Hengyang
TAIP...
Jedda
At Ta'if
RIYADH
QATAR
DOHA
ABU DHABI
UAE
Larkana
Shikarpur
Delhi
NEW DELHI
Bareilly
NEPAL
THIMPHU
Leshan
Guiyang
Liuzhou
Guangzhou
Gaoxi...
Ar Rub' al Khali (Empty Quarter)
Ar Rustaq
MUSCAT
Sur
Karachi
Hyderabad
Jaipur
Agra
Lucknow
KATHMANDU
BHUTAN
Rangpur
Guwahati
Kunming
Nanning
Hong Kong
SANA
YEMEN
OMAN
Arabian Sea
Ahmadabad
Vadodara
INDIA
Kanpur
Varanasi
Ganges
BANGLADESH
Brahmanbaria
MYANMAR
HANOI
Hai Phong
Ta'izz
Indore
Bhopal
Jamshedpur
Rajshahi
Khulna
DHAKA (BURMA)
Chittagong
Mandalay
Taunggyi
Pakokku
NAY PYI TAW
LAOS
VIENTIANE
Hainan Dao
Aden
Gulf of Aden
Socotra (to Yemen)
Narmada
Surat
Nagpur
Bhubaneswar
Bay of Bengal
Prome
Pegu
Bassein
Bogale
Chiang Mai
VINH
Da Nang
South China Sea
INDIAN OCEAN
Mumbai (Bombay)
Pune
Solapur
Godavari
Hyderabad
Yangon (Rangoon)
THAILAND
BANGKOK
Vijayawada
Krishna
Andaman Islands (to India)
CAMBODIA
Batdambang
Da Lat
Hubli
Bangalore
Chennai (Madras)
Coimbatore
Andaman Sea
PHNOM PENH
Ho Chi Minh City
Gulf of Thailand
Mysore
Kochi
Nicobar Islands (to India)
BANDAR SERI BEGAWAN
BRUNEI
Thiruvananthapuram
Jaffna
SRI LANKA
COLOMBO
SRI JAYAWARDENAPURA KOTTE
Kota Bharu
Taiping
MALAYSIA
Born...
Medan
KUALA LUMPUR
PUTRAJAYA
SINGAPORE
SINGAPORE
Pontianak
Balikpa...
Equator
Padang
Jambi
IN...
Sumatra
Palembang
Banjarmasin
Java Sea
Semarang
Sura...
JAKARTA
Bandung
Java
Malar...
AFRICA
Red Sea
Tropic of Cancer
Persian Gulf
Gulf of Oman
Caspian Sea
Euphrates
Tigris
Tabriz
EUROPE
Ural Mountains
Rudnyy
Yellow River
Yangtze
Mekong
Shantou
Arabian Peninsula
Indus
Thar Desert
Irrawaddy
Arabian Peninsula
Qingdao
Chongqing

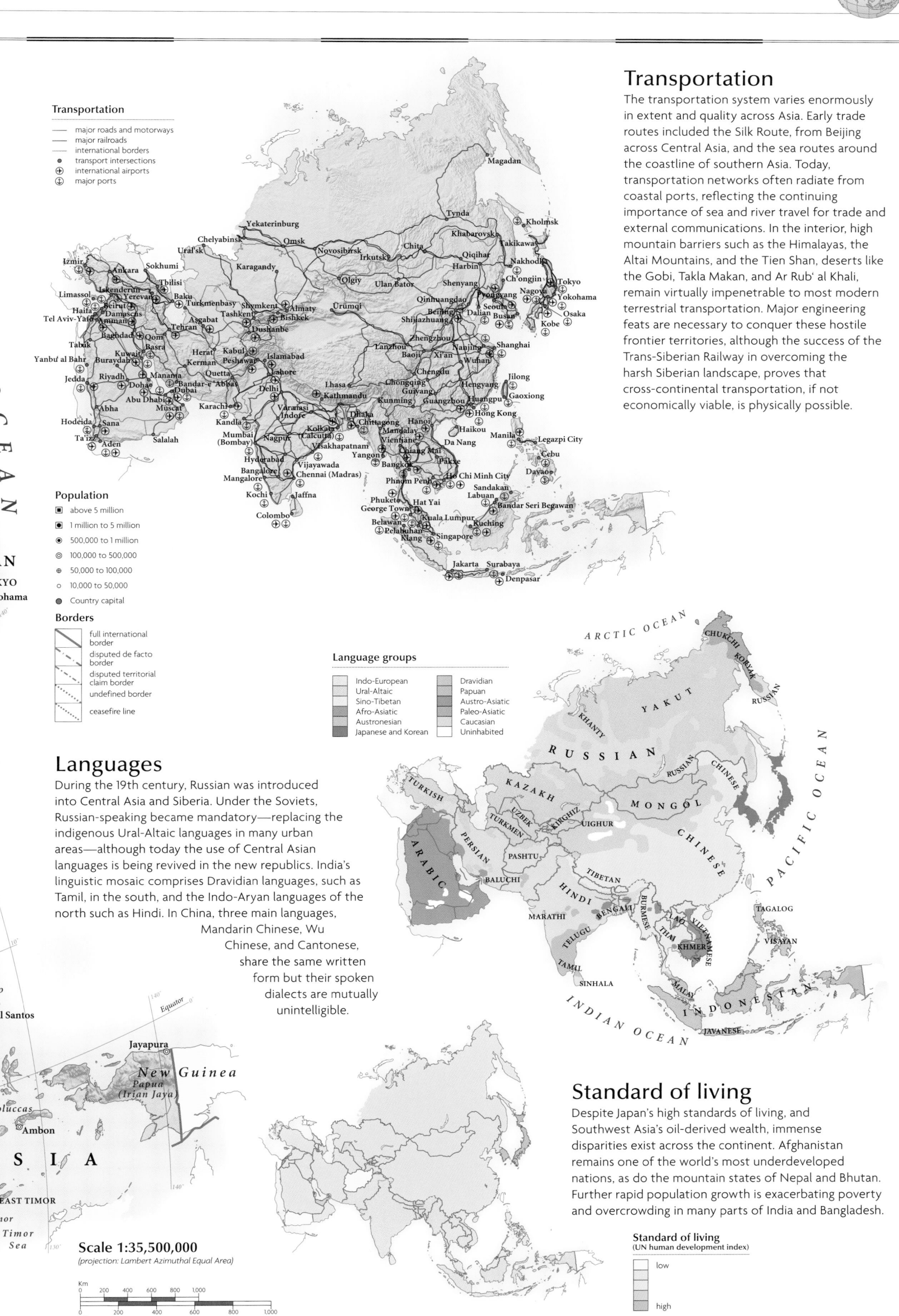

Transportation

The transportation system varies enormously in extent and quality across Asia. Early trade routes included the Silk Route, from Beijing across Central Asia, and the sea routes around the coastline of southern Asia. Today, transportation networks often radiate from coastal ports, reflecting the continuing importance of sea and river travel for trade and external communications. In the interior, high mountain barriers such as the Himalayas, the Altai Mountains, and the Tien Shan, deserts like the Gobi, Takla Makan, and Ar Rub' al Khali, remain virtually impenetrable to most modern terrestrial transportation. Major engineering feats are necessary to conquer these hostile frontier territories, although the success of the Trans-Siberian Railway in overcoming the harsh Siberian landscape, proves that cross-continental transportation, if not economically viable, is physically possible.

Transportation

— major roads and motorways
— major railroads
— international borders
• transport intersections
⊕ international airports
⊕ major ports

Population

■ above 5 million
■ 1 million to 5 million
◉ 500,000 to 1 million
◎ 100,000 to 500,000
⊕ 50,000 to 100,000
○ 10,000 to 50,000
● Country capital

Borders

full international border
disputed de facto border
disputed territorial claim border
undefined border
ceasefire line

Language groups

Indo-European
Ural-Altaic
Sino-Tibetan
Afro-Asiatic
Austronesian
Japanese and Korean
Dravidian
Papuan
Austro-Asiatic
Paleo-Asiatic
Caucasian
Uninhabited

Languages

During the 19th century, Russian was introduced into Central Asia and Siberia. Under the Soviets, Russian-speaking became mandatory—replacing the indigenous Ural-Altaic languages in many urban areas—although today the use of Central Asian languages is being revived in the new republics. India's linguistic mosaic comprises Dravidian languages, such as Tamil, in the south, and the Indo-Aryan languages of the north such as Hindi. In China, three main languages, Mandarin Chinese, Wu Chinese, and Cantonese, share the same written form but their spoken dialects are mutually unintelligible.

Standard of living

Despite Japan's high standards of living, and Southwest Asia's oil-derived wealth, immense disparities exist across the continent. Afghanistan remains one of the world's most underdeveloped nations, as do the mountain states of Nepal and Bhutan. Further rapid population growth is exacerbating poverty and overcrowding in many parts of India and Bangladesh.

Standard of living
(UN human development index)

low
high

Scale 1:35,500,000
(projection: Lambert Azimuthal Equal Area)

Km
0   200   400   600   800   1,000

Miles
0   200   400   600   800   1,000

Scale 1:47,500,000
(projection: Gall Stereographic)

0   500   1000   1500 Km
0   500   1000   1500 Miles

Climate

The climate of Asia exhibits marked differences from region to region, with freezing polar conditions in the north, hot and cold deserts in central regions and subtropical conditions throughout the south. Much of this variation can be attributed to enormous mountain barriers and internal depressions found across the continent. Monsoon winds, which reverse semi-annually, cause alternate wet and dry seasons across southern Asia. These air masses moving north from the ocean are stripped of their moisture over the Himalayas causing arid conditions across the Plateau of Tibet. Both the south and east are susceptible to tropical cyclones or typhoons.

Average Rainfall

January rainfall *July rainfall*

Rainfall

- 0–1 in (0–25 mm)
- 1–2 in (25–50 mm)
- 2–4 in (50–100 mm)
- 4–8 in (100–200 mm)
- 8–12 in (200–300 mm)
- 12–16 in (300–400 mm)
- 16–20 in (400–500 mm)
- more than 20 in (500 mm)

Average Temperature

January temperature *July temperature*

Temperature

- below -22°F (-30°C)
- -22 to -4°F (-30 to -20°C)
- -4 to 14°F (-20 to -10°C)
- 14 to 32°F (-10 to 0°C)
- 32 to 50°F (0 to 10°C)
- 50 to 68°F (10 to 20°C)
- 68 to 86°F (20 to 30°C)
- above 86°F (30°C)

Climate

- tundra
- subarctic
- cool continental
- warm humid
- mediterranean
- semi-arid
- arid
- humid equatorial
- tropical

- daily hours of sunshine, January
- daily hours of sunshine, July
- → cyclone
- ⇒ typhoon
- → cold/dry monsoon
- → warm/wet monsoon
- → cold wind

Environmental issues

The transformation of Uzbekistan by the former Soviet Union into the world's fifth largest producer of cotton led to the diversion of several major rivers for irrigation. Starved of this water, the Aral Sea diminished in volume by over 75% since 1960, irreversibly altering the ecology of the area. Heavy industries in eastern China have polluted coastal waters, rivers and urban air, while in Myanmar (Burma), Malaysia and Indonesia, ancient hardwood rainforests are felled faster than they can regenerate.

Environmental issues

- tropical forest
- forest destroyed
- desert
- desertification
- acid rain
- polluted rivers
- marine pollution
- heavy marine pollution
- radioactive contamination
- poor urban air quality

Using the land and sea

- cropland
- desert
- forest
- mountain region
- pasture
- tundra
- wetland

- fruit
- jute
- peanuts
- rice
- rubber
- shellfish
- soya beans
- sugar beet

- sugar cane
- tea
- timber
- wheat

- major conurbations
- cattle
- pigs
- goats
- sheep
- coconuts
- corn (maize)
- cotton
- dates
- fishing

Land use

Vast areas of Asia remain uncultivated as a result of unsuitable climatic and soil conditions. In favourable areas such as river deltas, farming is intensive. Rice is the staple crop of most Asian countries, grown in paddy fields on waterlogged alluvial plains and terraced hillsides, and often irrigated for higher yields. Across the black earth region of the Eurasian steppe in southern Siberia and Kazakhstan, wheat farming is the dominant activity. Cash crops, like tea in Sri Lanka and dates in the Arabian Peninsula, are grown for export, and provide valuable income. The sovereignty of the rich fishing grounds in the South China Sea is disputed by China, Malaysia, Taiwan, the Philippines and Vietnam, because of potential oil reserves.

ASIA

1 BOSPORUS, TURKEY
The Bosporus provides the only outlet for the Black Sea, linking it with the Sea of Marmara to the south and then with the Mediterranean Sea via the Dardanelles.

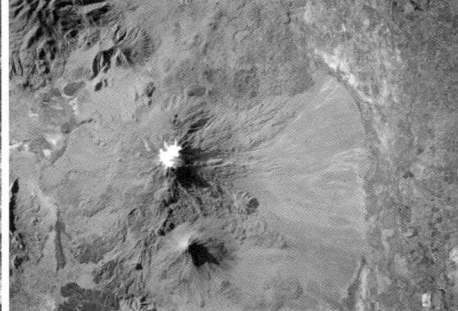

2 MOUNT ARARAT, TURKEY
Said to be the resting place for Noah's Ark, this extinct volcanic massif lies in the far east of Turkey.

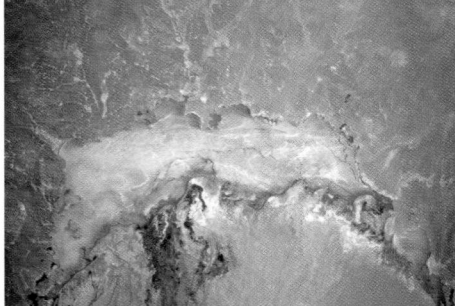

3 LAKE BALKASH, KAZAKHSTAN
Still covered in winter ice in this image, this lakes lies in a dry desert region and has no outlet.

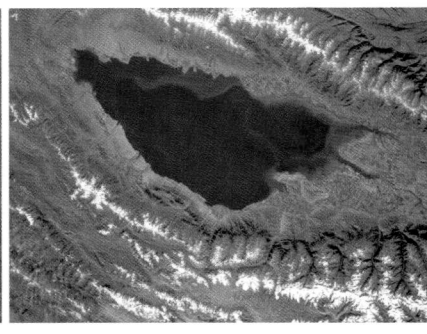

4 OZERO ISSYK-KUL', KYRGYZSTAN
Against the dry slopes of the Tien Shan mountains to the south this lake appears bright blue.

9 KUWAIT'S OILFIELDS, KUWAIT
The dark plumes are smoke rising from the 700 wells set alight by Iraqi forces during the Gulf War of 1991.

10 PALM ISLAND, UNITED ARAB EMIRATES
This luxury housing development and tourist resort, one mile (1.6 km) off the seafront of Dubai, is built from sediments dredged from the nearby port of Jebel Ali.

11 MALDIVES
The Maldives consist of 1300 coral formations in 19 atolls and stretch over 1491 miles (2400 km).

12 KARACHI, PAKISTAN
Pakistan's main seaport and former capital lies to the northwest of the delta of the Indus river.

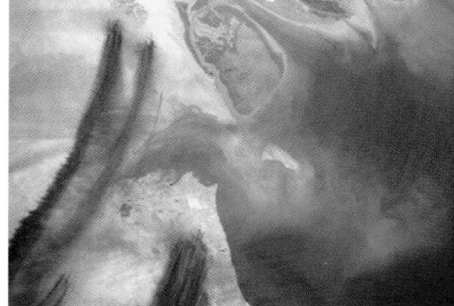

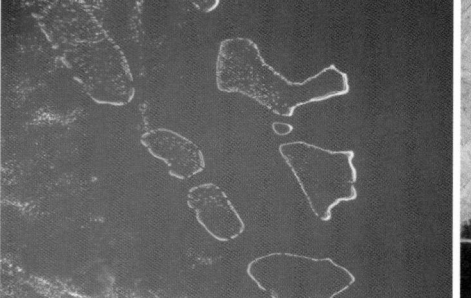

THREE GORGES DAM, CHINA 5
Seen here during its construction in 2000, the world's largest dam is designed to tame the Yangtze river which has regularly flooded.

BEIJING, CHINA 6
China's ancient capital was laid out on a grid pattern centred on the Forbidden City and its streets are picked out in this winter image by snowfall.

MOUNT FUJI, JAPAN 7
The steep, symmetrical, snow-capped volcano last erupted in 1707.

VULKAN KLYUCHEVSKAYA SOPKA, RUSSIAN FEDERATION 8
The Kamchatka Peninsula's highest and most active volcano last erupted in 1994.

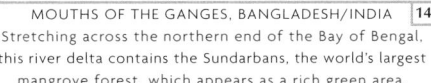

MOUNT EVEREST, CHINA/NEPAL 13
The world's highest mountain at 29,029 ft (8848 m) straddles the border between China and Nepal.

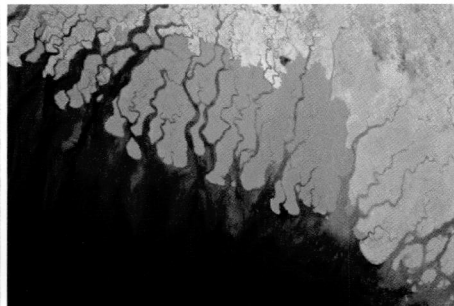

MOUTHS OF THE GANGES, BANGLADESH/INDIA 14
Stretching across the northern end of the Bay of Bengal, this river delta contains the Sundarbans, the world's largest mangrove forest, which appears as a rich green area.

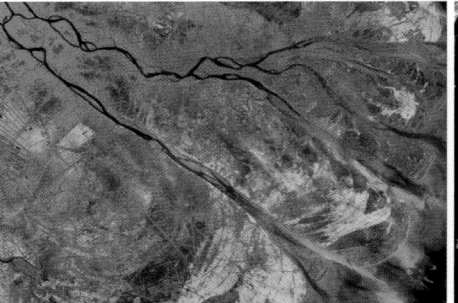

MEKONG DELTA, VIETNAM 15
The Mekong river flows over 2494 miles (4000 km) from the Plateau of Tibet before crossing Vietnam to reach the South China Sea.

HONG KONG, CHINA 16
This image was taken around the time that Hong Kong was handed back to China by the British in 1997, this city remains east Asia's trade and finance center.

ASIA

212

185

131

136

Elevation scale (m):
6000m
4000m
3000m
2000m
1000m
500m
250m
100m
Sea Level
-250m
-2000m
-4000m

Seas and water bodies:
Adriatic Sea
Tyrrhenian Sea
Ionian Sea (Iónia Nisiá / Ionian Islands)
Aegean Sea
Thracian Sea
Sea of Marmara (Marmara Denizi)
Black Sea
Mediterranean Sea
Kritikó Pélagos (Sea of Crete)
Mirtóo Pélagos
Red Sea
Dead Sea
Lake Nasser (Buhayrat Nāşir)
Khalīj Surt (Gulf of Sirte)
Gulf of Suez (Khalīj as Suways)
Gulf of Aqaba (Khalīj al 'Aqaba)

Countries / Regions:
ITALY, CROATIA, BOSNIA-HERZEGOVINA, MONTENEGRO, SERBIA, KOSOVO, MACEDONIA, ALBANIA, GREECE, BULGARIA, ROMANIA, TURKEY, CYPRUS, TURKISH REPUBLIC OF NORTHERN CYPRUS (recognised only by Turkey), SYRIA, LEBANON, ISRAEL, JORDAN, IRAQ, SAUDI ARABIA, EGYPT, LIBYA, CHAD, SUDAN, SOUTH SUDAN, ERITREA, ETHIOPIA, DJIBOUTI, CENTRAL AFRICAN REPUBLIC, RUSSIAN FEDERATION, GEORGIA, ARMENIA

Cities and towns:
Napoli (Naples), Salerno, Potenza, Cosenza, Catanzaro, Reggio di Calabria, Messina, Catania, Siracusa, Ragusa, Taranto, Lecce, Bari, Foggia, Campobasso, Pescara, L'Aquila, Terni, Perugia
Split, Podgorica, Shkodër, Tirane, Prizren, Priština, Pristina, Skopje, Sofia, Bitola, Thessaloniki (Saloníca), Kavala, Plovdiv, Stara Zagora, Sliven, Burgas, Varna, Shumen, Razgrad, Pleven, Vratsa, Niš, Kraljevo, Vidin, Giurgiu, Constanta, Dobrich
Athína (Athens), Pátra, Lárisa, Vólos, Akhísar, Manisa, Izmir, Aydin, Denizli, Marmaris, Ródos (Rhodes), Irákleio, Ágios Nikólaos, Kríti (Crete)
Istanbul, Izmit, Tekirdağ, Edirne, Çanakkale, Balıkesir, Bursa, Eskişehir, Kütahya, Afyon, Uşak, Isparta, Antalya, Alanya, Fínike, Konya, Karaman, Ereğli, Aksaray, Niğde, Mersin, Tarsus, Adana, Osmaniye, İskenderun, Antakya, Silifke, ANKARA, Polatlı, Kırıkkale, Çankırı, Kastamonu, Karabük, Zonguldak, İnebolu, Sinop, Samsun, Ordu, Giresun, Trabzon, Rize, Hopa, Bafra, Çorum, Tokat, Sivas, Kayseri, Erzincan, Erzurum, Ağrı, Muş, Bingöl, Elazığ, Malatya, Kahramanmaraş, Gaziantep, Adıyaman, Şanlıurfa, Diyarbakır, Batman, Siirt, Mardin, Van, Tatvan
Kırşehir, Tuz Gölü
Halab (Aleppo), Hamah, Hims (Homs), Tartūs, Latakia (Al Ladhiqiyah), Tripoli, Beyrouth (Beirut), Mount Hermon 2814m, Dimashq (Damascus), Dūmā, Irbid, Tudmur (Palmyra), Ar Raqqah, Dayr az Zawr, Manbij, Al Hasakah, Al Mawşil (Mosul), Al Fuhaymi, As Sulaymānīyah, Kirkūk, Bayji, Ba'qūbah, BAGHDAD, Karbalā', Al Hillah, An Najaf, As Samāwah
NICOSIA, Lemesós, Lárnaka
Hefa (Haifa), Netanya, Tel Aviv-Yafo, Ashdod, Gaza, Mizpe Ramon, Be'er Sheva, JERUSALEM, AMMAN, Az Zarqā', Al Karak, Ma'ān, Al Harrah, Ar'ar, Turayf, Al Jawf, An Nafud, Tabūk, Tabuk, Taymā', Hā'il, Buraydah, Al Majma'ah, Khaybar, Al Madīnah (Medina), Yanbu' al Bahr, Makkah (Mecca), Jiddah (Jedda), At Tā'if, Al Bāhah, Zalim, Abhā, Najrān, Sa'dah, Al Hudaydah (Hodeida), Zabid, SAN'Ā'
Banghāzī (Benghazi), Al Bayda', Al Jabal al Akhḍar, Tubruq, Ajdābiyā, Surt, Marādah, Jālū, Al Jaghbūb, Siwah, Al Kufrah, Picco Bette 2286m, Emi Koussi 3415m, Ounianga Kébir, Faya, Fada, Ennedi, Erg du Djourab, Abou-Déïa, Am Timan, Biltine, Abéché, Birao, Goz-Beïda, Kebkabiya, El Geneina, El Fasher, Nyala, En Nahud, Sodiri, El Obeid, Umm Ruwaba, Er Rahad, Kadugli, El Muglad, Ed Da'ein, Kadugli, Sumeih, Paloich
Alexandria (Al Iskandarīyah), Sīdī Barrāni, Al 'Alamayn, Tanta, Damietta (Dumyāt), Būr Sa'īd (Port Said), Az Zaqāzīq, Al Ismā'īlīya, Suez (As Suways), CAIRO (AL QĀHIRAH), Giza (Al Jīzah), Banī Suwayf, Bawītī, Al Minyā, Mallawi, Asyūt, Qasr al Farāfirah, Al Khārijah, Mūt, Sawhāj, Qinā, Luxor (Al 'Uqşur), Aswān, Al Ghurdaqah (Hurghada), Būr Safājah, Al Wajh, Wadi Halfa, Akasha, Delgo, Dongola, Ed Debba, El 'Atrun, Abu Hamed, Shereik, Atbara, Ed Damer, Port Sudan, Suakin, Haiya, Kassala, Keren, ASMARA, Teseney, Gedaref, Wad Medani, Sennar, Rabak, KHARTOUM, Khartoum North, Omdurman, Bahir Dar, Gonder, Mek'ele, Āksum, Ras Dashen Terara 4620m, DJIBOUTI, Obock, Aseb, Desē, Weldiya, Ta'izz, Mekelle, Roseires Reservoir, Ed Damazin

Physical features and deserts:
Cyrenaica, Libyan Plateau, Munkhafad al-Qattārah (Qattara Depression), Şahrā' al Gharbīyah (Western Desert), Great Sand Sea, Libyan Desert, Sahara, Ramlat Rabyānah, Hadabat al Jilf al Kabīr, Jabal al 'Uwaynāt 1907m, Nubian Desert, Sahel, Darfur, Kordofan, Ethiopian Highlands, Danakil Desert, Great Rift Valley, Sinai (Sīnā), Jabal Mūsā 2285m, Jabal ash Shifā, Şahrā' ash Sharqīyah (Eastern Desert), An Nafud, Al Hijaz, Nile, Blue Nile (Bahr el Azraq), White Nile (Bahr el Abyad), Atbara, Wadi Howar, Wadi el Milk, Nile Delta, Bir Ghaleb, Hala'ib Triangle, Syrian Desert, Euphrates, Tigris, Zāhir al Jafr, Tropic of Cancer, Kür Daği, Köroğlu Dağları, Toros Dağları (Taurus Mts), Anatolia, Pontic Mts, Lesser Caucasus, Mount Ararat 5165m, Van Gölü, Bosporus (İstanbul Boğazı), Dardanelles (Çanakkale Boğazı)

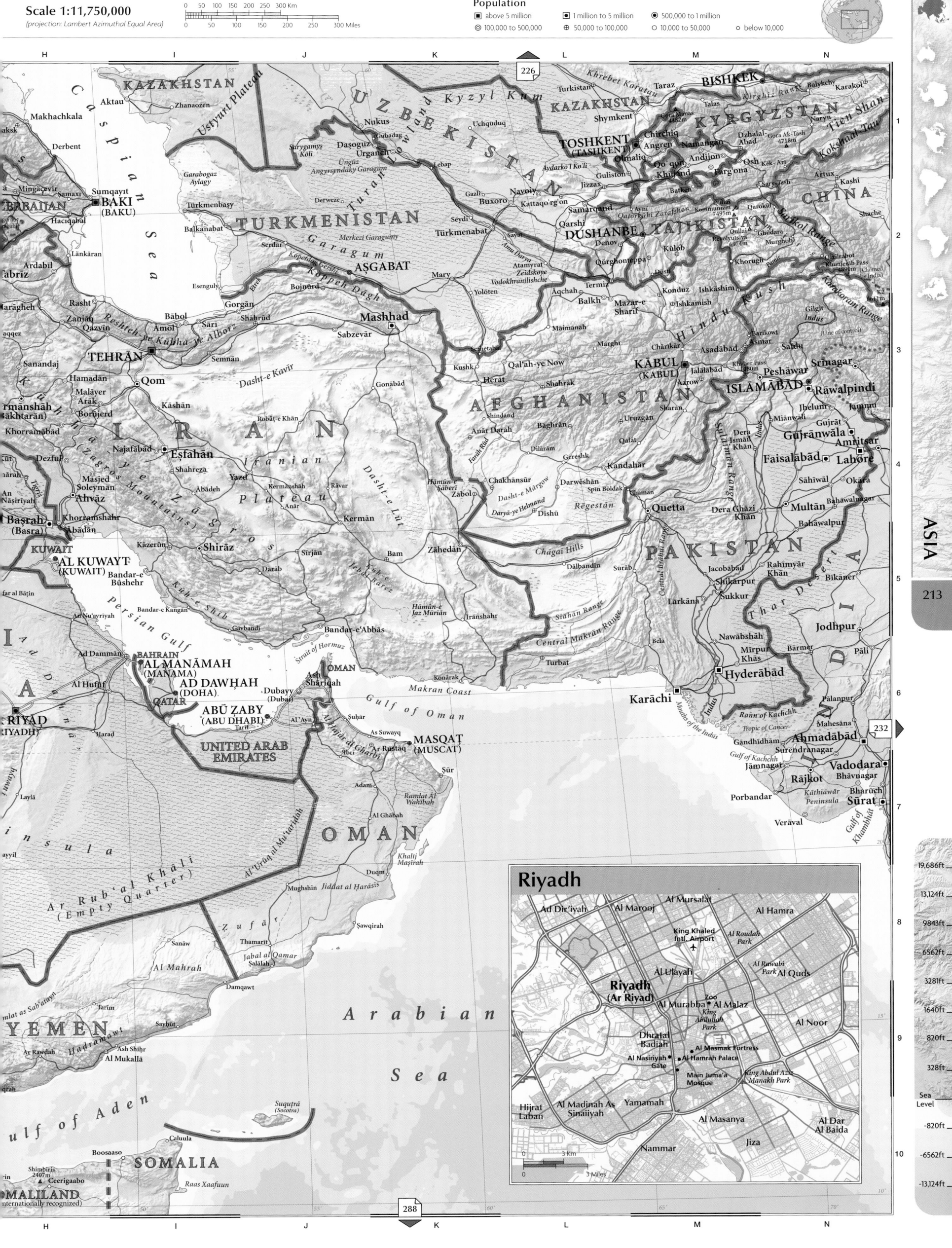

Scale 1:11,750,000
(projection: Lambert Azimuthal Equal Area)

0 50 100 150 200 250 300 Km

0   50   100   150   200   250   300 Miles

Population

■ above 5 million
◉ 100,000 to 500,000

▪ 1 million to 5 million
⊕ 50,000 to 100,000

● 500,000 to 1 million
○ 10,000 to 50,000

∘ below 10,000

19,686ft
13,124ft
9843ft
6562ft
3281ft
1640ft
820ft
328ft
Sea Level
-820ft
-6562ft
-13,124ft

KAZAKHSTAN

Aktau
Zhanaozen

Makhachkala
aks

Derbent

C a s p i a n   S e a

Ustyurt Plateau

Nukus
Uchquduq

UZBEKISTAN

Daşoguz
Urganch
Gubadag

Kyzyl Kum

Turkistan
Taraz

BISHKEK
Balykchy
Karakol

Talas
Kirghiz Range
Tien Shan

KAZAKHSTAN

Shymkent

KYRGYZSTAN

Narynn

Artux
Kashi

Derweze

Lebap

Gazli
Navoiy

TOSHKENT
(TASHKENT)
Chirchiq
Angren
Namangan
Olmaliq
Qo'qon
Andijon
Farg'ona

Qarshi

DUSHANBE
TAJIKISTAN

Denov
Ghudara
Murghob

Qürghonteppa

Kǔlob

Khorugh

Pamir

Karakoram Range

CHINA

K2

Köpetdag Gershi

Köpeh Dag

Bojnürd

Gorgān

Kuh-e Shib

Garagum

Mary
Yolöten

Aqchah
Balkh
Mazār-e Sharīf

Mäimanah

Marghi
Chārikār
Asadābād
Asmar
Saidu

Hindu Kush

Gilgit
Indus
Srinagar

AFGHANISTAN

Herāt

Shindand

Qal'ah-ye Now
Shahrak

Bāghrān
Dīlārām

Baghrān

KĀBUL
(KABUL)
Jalālābād
Khyber Pass
Peshāwar

ISLĀMĀBĀD
Rāwalpindi

Uruzgān
Sharan
Qalāt

Jhelum
Miānwāli
Jammu

Gujrānwāla
Amritsar

Chakhānsūr
Zābol

Darwēshān
Spin Bōldak

Régestān

Chaman

Kandahār
Dera Ghāzī Khān

Lahore
Faisalābād
Sāhiwāl
Okāra

Quetta

Dīshū
Daryā-ye Helmand

Dālbandin

Multān
Bahāwalnagar

Bahāwalpur

Jacobābād

Sūrab
Lārkāna

Rahīmyār Khān

Shikārpur
Sukkur

Bīkāner

P A K I S T A N

Central Makran Range

Bela
Nawābshāh

Mīrpur Khās

Bārmer

Jodhpur

I N D I A

Pāli

Karāchi
Mouths of the Indus

Rann of Kachchh
Tropic of Cancer

Palanpur

Mahesāna

Ahmadābād

Gāndhīdhām
Surendranagar

Gulf of Kachchh

Jāmnagar

Vadodara

Rājkot
Bhāvnagar

Porbandar

Kāthiāwār Peninsula

Bhāruch
Sūrat

Ve'rāval

Gulf of Khambhāt

Riyadh

Ad Dir'iyah
Al Marooj
Al Mursalat

King Khaled Intl. Airport

Al Hamra

Al Roudah Park

Al Ruwabi Park

Riyadh
(Ar Riyad)

Al Ulayah
Al Quds

Al Murabba
Al Malaz

King Abdullah Park

Zoo

Al Noor

Dhirat al Badiah

Al Masmak Fortress

Al Nasiriyah Gate
Al Hamrah Palace

Main Juma'a Mosque

King Abdul Aziz Manakh Park

Hijrat Laban

Al Madinah As Sinaiiyah

Yamamah

Al Masanya

Al Dar Al Baida

Nammar
Jiza

0   3 Km
0   3 Miles

ASIA

214

189
187
129

UKRAINE

Sea of Azov

Karkinits'ka Zatoka

AVTONOMNA RESPUBLIKA
KRYM (Crimea)
(annexed by Russian Federation, 2014)

Simferopol'
Sevastopol'
Yalta

ROMANIA

Wallachia

BUCUREŞTI
(BUCHAREST)

Constanţa

BULGARIA

Plovdiv
Burgas
Varna
Dobrich
Shumen

Edirne
KIRKLARELİ

Black   Sea

İSTANBUL
İstanbul
Marmara Denizi
(Sea of Marmara)
TEKİRDAĞ
ÇANAKKALE

Zonguldak
BARTIN
KASTAMONU
SİNOP
Sinop
SAMSUN
Samsun

KOCAELİ
İzmit
Adapazarı
SAKARYA
YALOVA
Bursa
BURSA
BİLECİK
BOLU
ÇANKIRI
ÇORUM
AMASYA
TOKAT

Köroğlu Dağları

ANKARA
ANKARA
KIRIKKALE
KIRŞEHİR
YOZGAT
SİVAS

BALIKESİR
Manisa
MANİSA
İzmir
İZMİR
KÜTAHYA
ESKİŞEHİR
NEVŞEHİR
KAYSERİ

Aegean Islands

Lésvos (Lesbos)
Chíos

UŞAK
AFYON
A   n   a   t   o   l   i   a
KONYA
Konya
AKSARAY
KARAMAN
ADANA
Adana

AYDIN
DENİZLİ
Denizli
ISPARTA
İsparta
BURDUR
ANTALYA
Antalya
MERSİN
Mersin
KAHRAMANMARAŞ
GAZİANTEP
Gaziantep

MUĞLA

Ródos (Rhodes)

Dodekánisa

GREECE

Kríti

TURKISH REPUBLIC OF NORTHERN CYPRUS
(recognised only by Turkey)

CYPRUS
NICOSIA
Lárnaka
Lemesós (Limassol)

Sovereign Base Area (to UK)

Al Lādhiqīyah
(Latakia)
HATAY
Antakya
İskenderun
HALAB
Halab
(Aleppo)
IDLIB

HAMAH
Ḥamāh
ṬARṬŪS
Ṭarṭūs
Ḥimş
(Homs)

Tripoli

LEBANON
BEYROUTH
(BEIRUT)
DIMASHQ
(DAMASCUS)

M e d i t e r r a n e a n   S e a

6000m
4000m
3000m
2000m
1000m
500m
250m
100m
Sea Level
-250m
-2000m
-4000m

The Near East

TURKEY

CYPRUS

TURKISH REPUBLIC OF NORTHERN CYPRUS
(recognized only by Turkey)

NICOSIA

SYRIA

LEBANON

BEYROUTH (BEIRUT)

DIMASHQ (DAMASCUS)

ISRAEL

JERUSALEM

TEL AVIV

JORDAN

AMMĀN

EGYPT

SAUDI

Mediterranean Sea

Gulf of Aqaba

Dead Sea

Sinai (Sīnā)

Syrian Desert

Al Hamād

6000m
4000m
3000m
2000m
1000m
500m
250m
100m
Sea Level
-250m
-2000m
-4000m

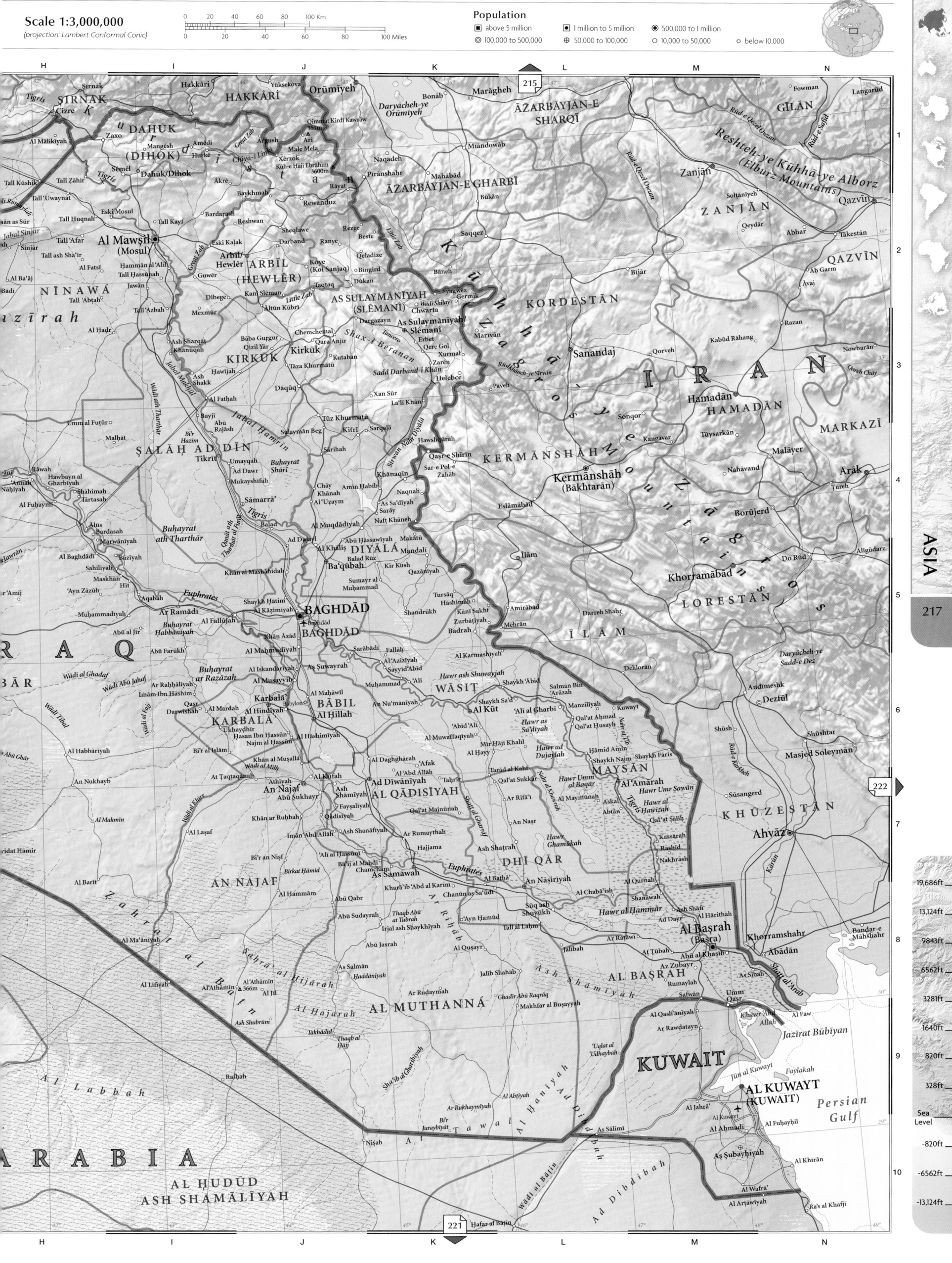

Israel & Lebanon

ASIA

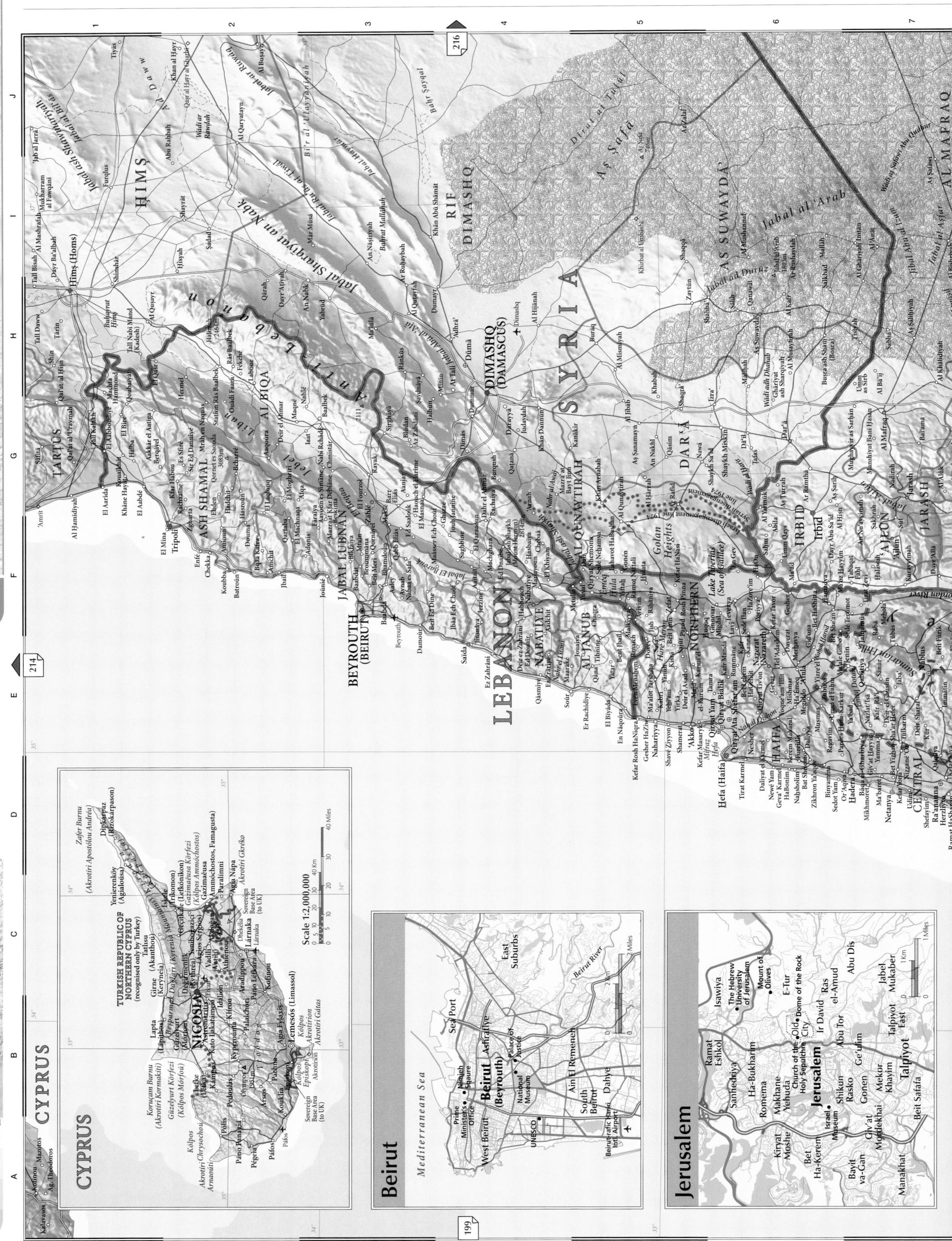

Scale 1:1,500,000
(projection: Lambert Conformal Conic)

0 10 20 30 40 50 Km
0 10 20 30 40 50 Miles

Population
- □ above 5 million
- ◉ 100,000 to 500,000
- ▣ 1 million to 5 million
- ⊕ 50,000 to 100,000
- ◉ 500,000 to 1 million
- ○ 10,000 to 50,000
- ○ below 10,000

Mediterranean Sea

SAUDI ARABIA

AL JAWF

AZ ZARQA'

'AMMAN

AMMAN

JORDAN

MA'AN

TABŪK

SAUDI ARABIA

WEST BANK

JERUSALEM

Dead Sea

ISRAEL

Gaza Strip

MADABA

AL KARAK

AT TAFILAH

Ash Sharah

AL 'AQABAH

Jabal al Khashsh

Gulf of Aqaba

Jibal al 'Adhriyat

Tuwayyil ash Shihaq

HaNegev

SOUTHERN

EGYPT

Sinai (Sinā)

Gebel el Tih

19,686ft
13,124ft
9843ft
6562ft
3281ft
1640ft
820ft
328ft
Sea Level
-820ft
-6562ft
-13,124ft

The Arabian Peninsula

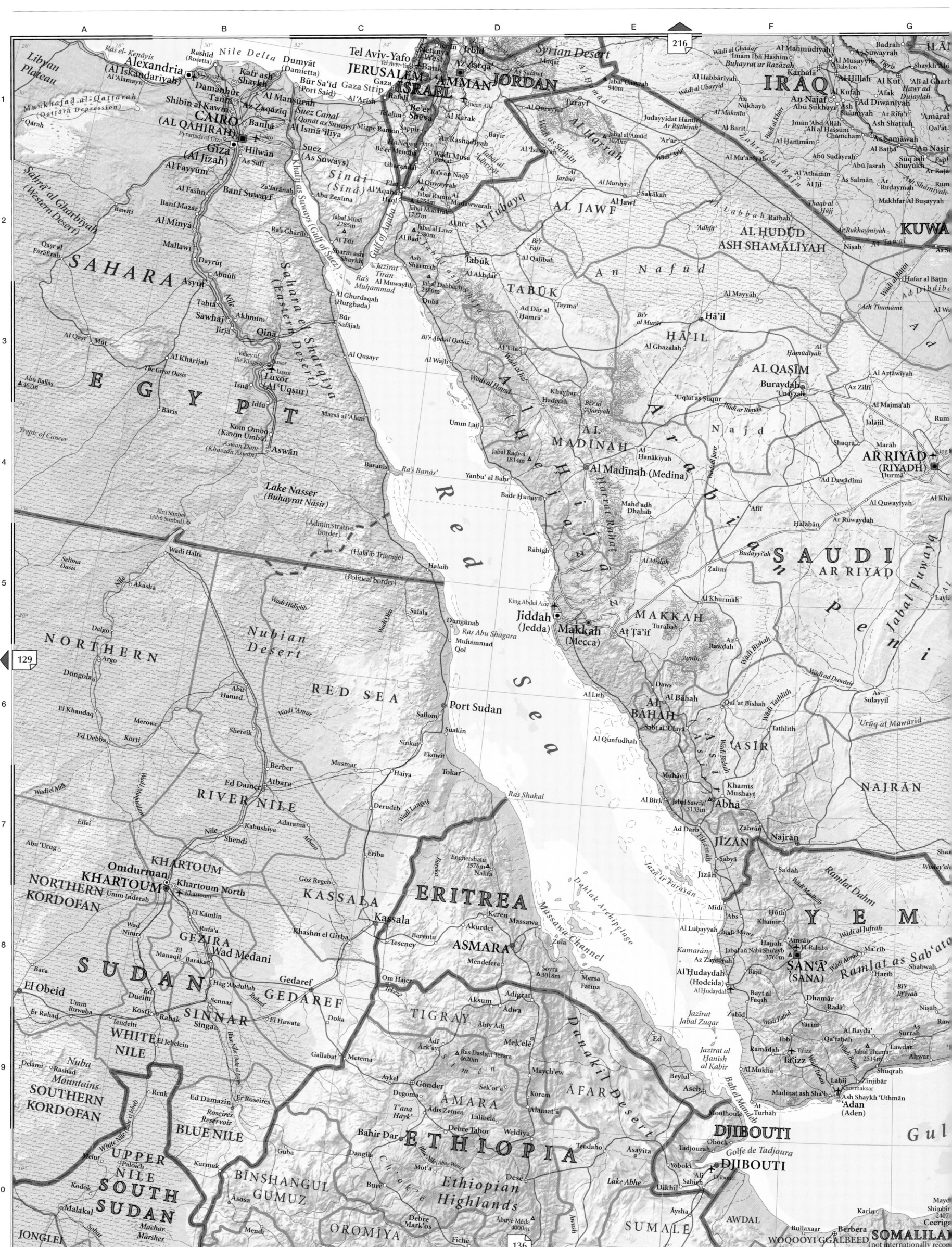

ASIA

222

197

220

220

Countries / major regions:

TURKEY

ARMENIA

AZERBAIJAN

AZER.

SYRIA

IRAQ

IRAN

SAUDI ARABIA

KUWAIT

BAHRAIN

QATAR

UNITED ARAB EMIRATES

Seas:

Caspian Sea

Persian Gulf

Elevation scale:

6000m
4000m
3000m
2000m
1000m
500m
250m
100m
Sea Level
-250m
-2000m
-4000m

Selected place names:

YEREVAN, BAKI (BAKU), TABRIZ, TEHRĀN, BAGHDAD, AL KUWAYT (KUWAIT), AL MANĀMAH (MANAMA), AD DAWHAH (DOHA), AR RIYĀD (RIYADH), ABŪ ZABY (ABU DHABI)

Erzincan, Erzurum, AGRI, MUŞ, BINGÖL, TUNCELI, Elazığ, Malatya, DİYARBAKIR, ŞANLIURFA, MARDIN, SİİRT, ŞIRNAK, HAKKARI, VAN, BİTLİS

Al Mawşil (Mosul), AR RAQQAH, AL HASAKAH, DAYR AZ ZAWR, ḤIMŞ

Kirkūk, As Sulaymānīyah/Slémāni, Sāmarrā, Tikrīt, Al Fallūjah, Karbalā, An Najaf, Al Baṣrah (Basra), An Nāṣirīyah, Ad Dīwānīyah

ORŪMĪYEH, Mahābād, Saqqez, Sanandaj, Bījār, KORDESTĀN, KERMĀNSHĀH, Kermānshāh, ĪLĀM, Khorramābād, LORESTĀN

Qazvīn, QAZVĪN, Rasht, GĪLĀN, MĀZANDARĀN, ĀZARBĀYJĀN-E SHARQĪ, ĀZARBĀYJĀN-E GHARBĪ, Zanjān, ZANJĀN, Ardabīl

Qom, QOM, Kāshān, MARKAZI, Arāk, ESFAHĀN, Esfahān, Najafābād, CHAHĀR MAHĀL VA BAKHTĪĀRĪ

Ahvāz, KHŪZESTĀN, Dezfūl, Shūshtar, Ābādān, Khorramshahr, Bandar-e Māhshahr, KOHGĪLŪYEH VA BOWYER AHMAD

Bandar-e Būshehr, BŪSHEHR, Shīrāz, FĀRS, Kāzerūn, Jahrom, Fīrūzābād

AL JAWF, AL ḤUDŪD ASH SHAMĀLĪYAH, AN NAFŪD, HĀ'IL, AL QAŞĪM, Buraydah, Najd, AR RIYĀD, ASH SHARQĪYAH, AD DAMMĀN, Al Hufūf

Tropic of Cancer

Syrian Desert

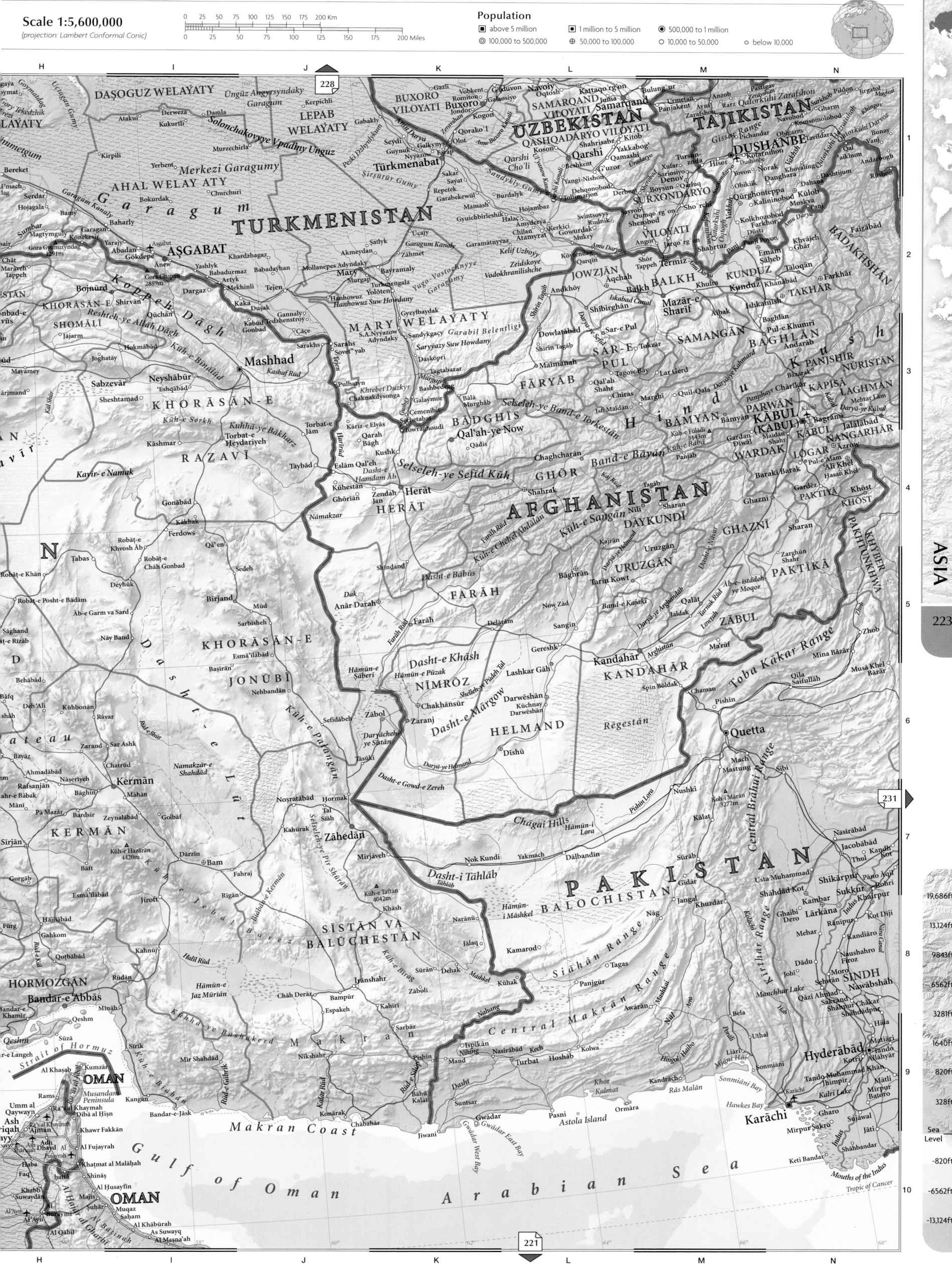

Scale 1:5,600,000
(projection: Lambert Conformal Conic)

| 0 | 25 | 50 | 75 | 100 | 125 | 150 | 175 | 200 Km |
| 0 | 25 | 50 | 75 | | 100 | 125 | 150 | 175 | 200 Miles |

Population

- ■ above 5 million
- ▣ 1 million to 5 million
- ◉ 500,000 to 1 million
- ◎ 100,000 to 500,000
- ⊕ 50,000 to 100,000
- ○ 10,000 to 50,000
- ∘ below 10,000

19,686ft
13,124ft
9843ft
6562ft
3281ft
1640ft
820ft
328ft
Sea Level
-820ft
-6562ft
-13,124ft

228
221
231

LAYATY
DAŞOGUZ WELAÝATY
Üñüz Angyrsyndaky Garagum
LEPAB WELAÝATY
BUXORO VILOYATI
Buxoro
UZBEKISTAN
SAMARQAND VILOYATI Samarqand
QASHQADARYO VILOYATI
TAJIKISTAN
DUSHANBE
BADAKHSHAN
AHAL WELAÝATY
Merkezi Garagumy
TURKMENISTAN
AŞGABAT
SURXONDARYO VILOYATI
Termiz
KUNDUZ
TAKHAR
KHORĀSĀN-E SHOMĀLĪ
JOWZJĀN
BALKH
Mazār-e Sharīf
SAMANGĀN
BAGHLĀN
PANJSHIR
NŪRISTĀN
Mashhad
MARY WELAÝATY
SAR-E PUL
FĀRYĀB
Hindu Kush
BĀMYĀN
KĀPISĀ LAGHMĀN
KHORĀSĀN-E RAZAVĪ
BĀDGHĪS
Qal'ah-ye Now
GHŌR
AFGHANISTAN
PARWĀN
KĀBUL (KABUL)
WARDAK
NANGARHĀR
HERĀT
Herāt
DĀYKUNDĪ
GHAZNĪ
LŌGAR
PAKTIYĀ
KHŌST
KHORĀSĀN-E JONŪBĪ
URUZGĀN
Kermān
FARĀH
ZĀBUL
PAKTĪKĀ
KHYBER PAKHTUNKHWA
Zābol
NĪMRŌZ
HELMAND
Kandahār
KANDAHĀR
Toba Kakar Range
Zāhedān
SĪSTĀN VA BALŪCHESTĀN
Quetta
HORMOZGĀN
Bandar-e 'Abbās
KERMĀN
OMAN
PAKISTAN
BALOCHISTAN
SINDH
Hyderābād
Karāchi
OMAN
Makran Coast
Gulf of Oman
Arabian Sea
Strait of Hormuz
Tropic of Cancer

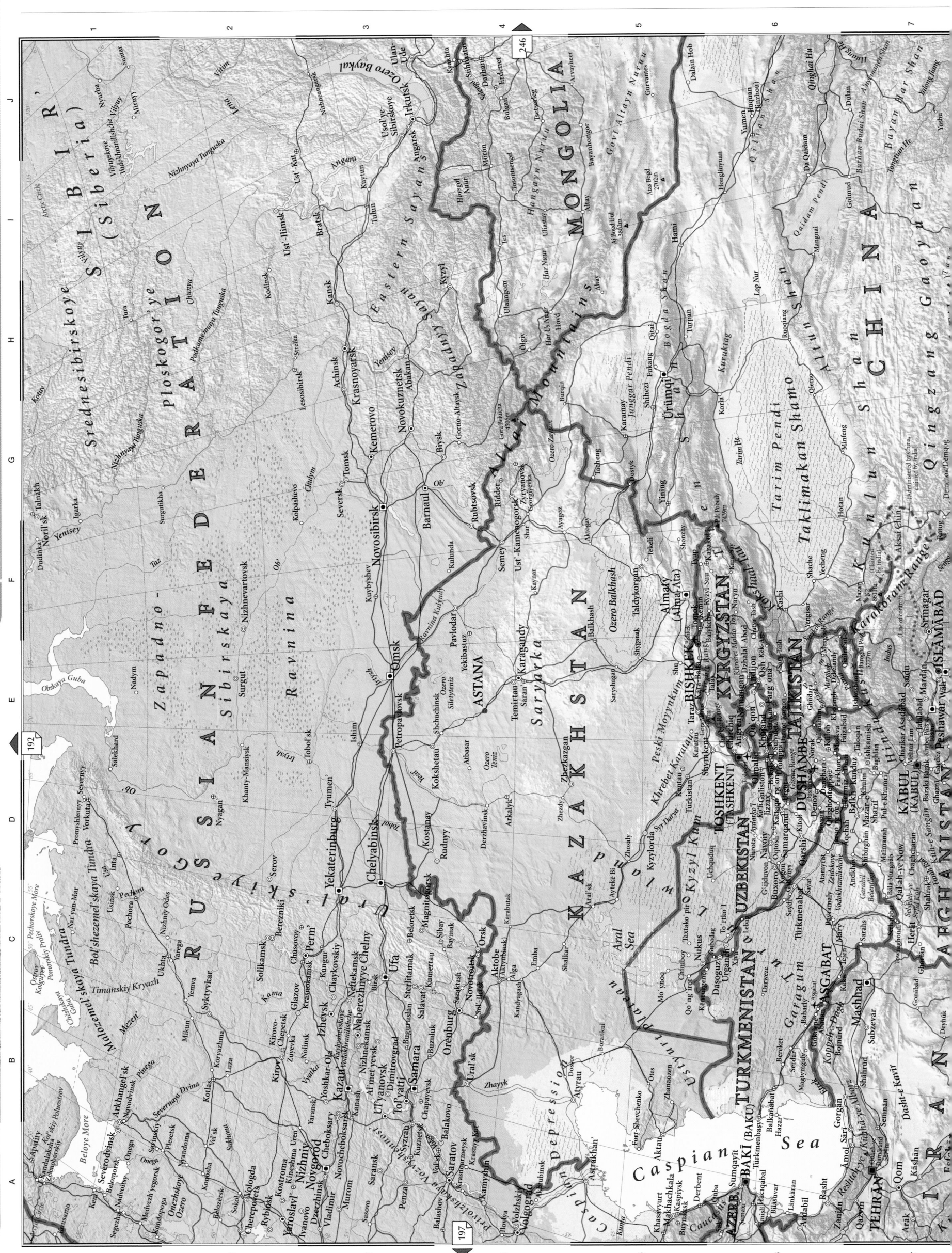

S I B E R I A (Siberia)

R U S S I A N   F E D E R A T I O N

Srednesibirskoye

Ploskogor'ye

Zapadno-Sibirskaya   Ravnina

Ozero Baykal

Nizhnyaya Tunguska

Ust'-Ilimsk

Bratsk

Kodinsk

Lesosibirsk

Achinsk

Krasnoyarsk

Kansk

Tomsk

Kemerovo

Novosibirsk

Novokuznetsk

Abakan

Biysk

Barnaul

Rubtsovsk

Ob'

Seversk

Yekaterinburg

Chelyabinsk

Magnitogorsk

Orsk

Orenburg

Ufa

Sterlitamak

Perm'

Izhevsk

Kazan'

Ul'yanovsk

Samara

Saratov

Volgograd

Astrakhan'

Omsk

Petropavlovsk

ASTANA

Pavlodar

Semey

Ust'-Kamenogorsk

MONGOLIA

Altai Mountains

Gobi

K A Z A K H S T A N

Ozero Balkhash

Almaty (Alma-Ata)

KYRGYZSTAN

BISHKEK

TAJIKISTAN

DUSHANBE

UZBEKISTAN

TOSHKENT (TASHKENT)

Samarqand

TURKMENISTAN

AŞGABAT

Aral Sea

Caspian Sea

Kyzyl Kum

Garagum

Mashhad

AFGHANISTAN

KABUL (KABUL)

C H I N A

Taklimakan Shamo

Tarim Pendi

Kunlun Shan

Karakoram Range

ISLAMĀBĀD

IRAN

TEHRAN

BAKÍ (BAKU)

AZERB.

Caucasus

Sea Level scale
6000m
4000m
3000m
2000m
1000m
500m
250m
100m
Sea Level
-250m
-2000m
-4000m

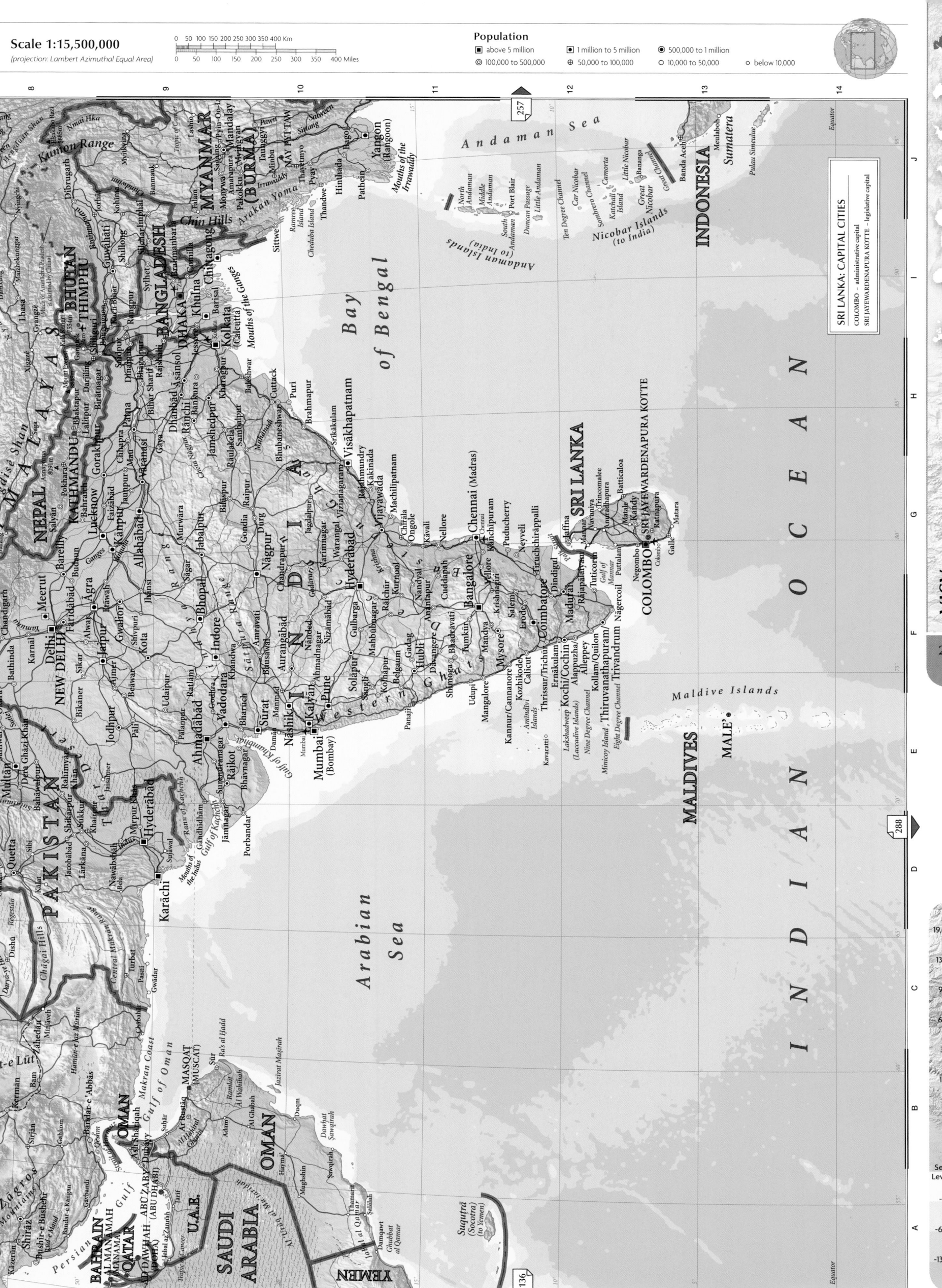

Kazakhstan

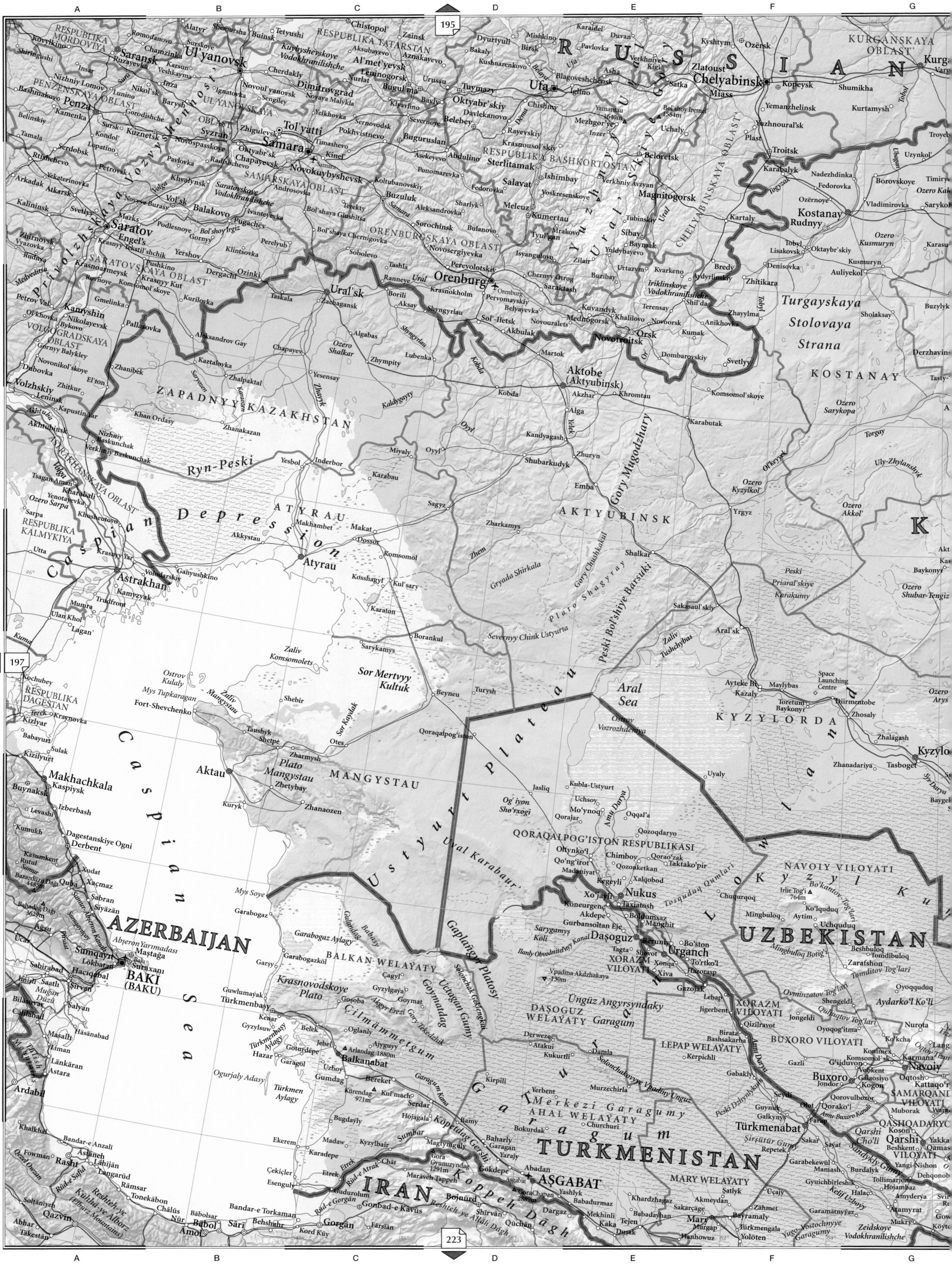

195

197

223

RUSSIAN

KAZAKHSTAN

ZAPADNYY KAZAKHSTAN

AZERBAIJAN

UZBEKISTAN

TURKMENISTAN

IRAN

KYZYLORDA

KOSTANAY

AKTYUBINSK

MANGYSTAU

Caspian Sea

Aral Sea

Caspian Depression

Ustyurt Plateau

Turgayskaya Stolovaya Strana

RESPUBLIKA MORDOVIYA

RESPUBLIKA TATARSTAN

RESPUBLIKA BASHKORTOSTAN

ORENBURGSKAYA OBLAST'

SAMARSKAYA OBLAST'

ULYANOVSKAYA OBLAST'

SARATOVSKAYA OBLAST'

VOLGOGRADSKAYA OBLAST'

ASTRAKHANSKAYA OBLAST'

RESPUBLIKA KALMYKIYA

RESPUBLIKA DAGESTAN

PENZENSKAYA OBLAST'

CHELYABINSKAYA OBLAST'

KURGANSKAYA OBLAST'

QORAQALPOG'ISTON RESPUBLIKASI

NAVOIY VILOYATI

XORAZM VILOYATI

BUXORO VILOYATI

SAMARQAND VILOYATI

QASHQADARYO VILOYATI

BALKAN WELAÝATY

DAŞOGUZ WELAÝATY

AHAL WELAÝATY

LEPAP WELAÝATY

MARY WELAÝATY

Samara
Saratov
Ul'yanovsk
Penza
Tol'yatti
Orenburg
Ural'sk
Atyrau
Aktobe (Aktyubinsk)
Astrakhan
Makhachkala
Aktau
Nukus
Urganch
Buxoro
Navoiy
Türkmenabat
Aşgabat
BAKI (BAKU)
Chelyabinsk
Magnitogorsk
Kostanay
Orsk

Sea Level
6000m
4000m
3000m
2000m
1000m
500m
250m
100m
-250m
-2000m
-4000m

Scale 1:6,250,000
(projection: Lambert Conformal Conic)

0 25 50 75 100 125 150 175 200 Km

0 25 50 75 100 125 150 175 200 Miles

Population
- ■ above 5 million
- ◼ 1 million to 5 million
- ◉ 500,000 to 1 million
- ◎ 100,000 to 500,000
- ⊕ 50,000 to 100,000
- ⊙ 10,000 to 50,000
- ○ below 10,000

192

238

231

19,686ft
13,124ft
9843ft
6562ft
3281ft
1640ft
820ft
328ft
Sea Level
-820ft
-6562ft
-13,124ft

ASIA

228

215

KAZA (KAZAKHSTAN)

Atyrau
ATYRAU
Komsomol
Kul'sary
Kosshagyl
Karaton
Zaliv Komsomolets
Shebir
Sarykamys
Borankul
Beyneu
Turysh
Gryada Shirkala
Gory Chushkakul
Plato Shagyray
Peski Bol'shiye Barsuki
Severnyy Chink Ustyurta
AKTYUBINSK
Shalkar
Sakasaul'skiy
Aral'sk
Ayteke Bi
Maylybas
Space Launching Centre
Diirmentobe
Toretum
Baykonyr
Syr Darya
Zhosaly
Zhalagas
Zhanadariya
Priaral'skiy Karakum
Ozero Shubar-Tengiz
Zaliv Tushchybas

Ostrov Kulaly
Mys Tupkaragan
Zaliv Mangystau
Fort-Shevchenko
Taushyk
Shetpe
Otes
Zharmysh
Sor Mertvyy Kultuk
Aktau
Plato Mangystau
Zhetybay
Kuryk
Zhanaozen

MANGYSTAU

Ustyurt Plateau

Qoraqalpog'iston

Aral Sea

KYZYLORDA
Kazaly
Qozoq'zak

Kyzy...

Mys Soye
Garabogaz

Jaslyq
Og'iyon Sho'rxogi
Kubla-Ustyurt
Uchsoy
Mo'ynoq
Qorajar
Amu Darya
Oqqal'a
Qozoqdaryo
Qozonketkan
Taxtako'pir
Uyaly

QORAQALPOG'ISTON RESPUBLIKASI

Ural Karabauri

Garabogaz
Garabogazköl
Garsy

Gaplaňgyr Tekiztigi
Solonchak Kaydaklakyr

Oltynko'l
Chimboy
Xo'jayli
Taxiatosh
Köneürgench
Boldumsaz
Akdepe
Madaniyat
Shumanay
Xalqobod
Nukus
Mang'it
Gurbansoltan Eje
Dasoguz
Gürlän
Berüni
Urganch
Tagta
Hanly Obvodnitel'nyy Kanal
XORAZM VILOYATI
Xonqa
Türtkül
Hazorasp
Xiva
Hazorasp, qishlaq
Gazojak
Lebap

NAVOIY VILOYATI
Bo'kantov Tog'lari
Idir Tog'i 764m
Ko'lquduq
Mingbuloq
Chuqurqoq
Beshbuloq
Tomdibuloq
Zarafshon
Tomditov-Tog'lari
Oymenzalov Tog'lari
Oyoq...
Shengeldi

UZBEKISTAN

Lowland

Kyzyl

Krasnovodskoye Plato
Guwlumaýak
Türkmenbaşy
Gyzylsuw
Kenar
Türkmenbaşy Aýlagy
Çagyl
Gyzylgaya
Akgyr Erezi
Goýmat
Gory Tekedzhik
Goýmatdag
Goşoba
Uçtagan Gumy
Solonchak Gektengkir
Goýmatdag

DAŞOGUZ WELAÝATY
Vpadina Akdzhakaya -130m▽
Ünguz Angyrsyndaky
Garagum

Derweze
Atakui
Kukurtli
Damla
Kerpichli

Solonchakovyye Vpadiny Unguz

Tuproqqal'a
Jigerbent
Qizilravot
Biratk
Bashsakarba

BUXORO VILOYATI
Gazli
G'ijduvon
Vobkent
Romiton
Buxoro
Galaosiyo
Jondor
Kogon

Nav...

Ogurjaly Adasy
Türkmen Aýlagy

Uzboý
Balkanabat
BALKAN WELAÝATY
Gumdag
Bereket
Künedag 971m
Kul'mach

Arlandag 1880m
Ajyguyy
Jebel
Oglaňly
Belek
Hazar
Goturdepe
Garagöl

Çilmämmetgum

Kirpili

Derweze

T u r a n

Gabakly
Murzechirla
Kerpichli

Guynük
Galkynys
Nyýazow
Farap
Seýdi
Zeravshan
Qorako'l
Olot

LEPAP WELAÝATY

Türkmenabat
Qarshi Çüli
Mubora...

Garaboýaz
Madaw
Kyzylbair
Gora Gyunuzyndag 1291m
Könekesir
Garagum Kanaly
Yerbent
Merkezi Garagumy

TURKMENISTAN

G a r a g u m

Bokurdak
Churchuri

Peski Dzhynlykum

Sakar
Sayat
Repetek

Şirşütür Gumy

Mamash
Burdalyk
Gyuichbirleshik
Halaç
Çilan
Hoja...
Atamyrat

Kelif Uzboýy

Çekiçler
Karadepe
Bugdaýly

Baharly
Serdar
Hojagala
Bamy
Garagan
Yarajy
Garaýalykum

Magtymguly
Sumbar
Etrek
Chät
Maraveh Tappeh

AHAL WELAÝATY

Gökdepe
Abadan
Anew
Gora Çapan 2889m
Aşgabat
Yashlyk
Babadurmaz
Artyk

MARY WELAÝATY
Şatlyk
Akmeýdan
Adyndaky
Garagum Kanaly
Garamätnýaz

Yugo-Vostochnyye Garagumy

Zeidskoye Vodokhranilishce

Koppeh Dagh

Esenguly
Gudurolum
Etrek
Rüd-e Atrak
Rüd-e Gorgan

Bojnürd
Shirvän
Reshteh-ye Allah Dägh

Dargaz
Mekhinli
Kaka
Dusak
Kabud Gonbad
Tedzhenstroy
Çäçe

Khardzhagaz
Babadayhan
Tejen

Sarakçäge
Mollanepes
Zähmet

Murgap

Mary
Turkmengala
Yolöten
Bayramaly

Hanhowuz
Hanhowuz Suw Howdany
Gannaly

Garabekewül

Bandar-e Torkaman
Gonbad-e Kavüs
GOLESTĀN
Kord Küy
Gorgan
Farsiän

KHORĀSĀN-E SHEMĀLĪ
Jäjarm
Shirvän
Qüchän

Hokmabad
Joghatay
Biärjümand

MĀZANDARĀN
Nür
Bābolsar
Behshahr
Sāri
Āmol
Qā'emshahr
Pol-e Safīd
Bābol

Reshteh-ye Kühhā-ye Alborz
(Elburz Mountains)
Damāvand 5671m

Dāmghān
Mayamey
Shāhrüd

Sabzevär
Neyshäbur
Sheshtamad
'Eshqäbad

Küh-e Bīnālūd
Mashhad
Kashaf Rüd

Sarahs
'Sovet'yab
S.A. Nyýazow Adyndaky
Gyzylbaydak

Garabil Belentligi

Sarayyazy Suw Howdany
Daşköpri

Murgap
Galaymor
Serhetabat

Pulhatyn
Tagtabazar
Çaknakdysonga

Khrebet Duzkyr

Bala Murghab

Maimanah

FĀRYĀB

Qal'...

SEMNĀN
Semnän
Gärmsär

I R A N

IRAN

Dasht-e Kavir

Kavïr-e Namak

Kavïr-e Sorkh
Küh-e Bakharz

Käshmar
Gonäbäd
Torbat-e Heydariyeh
Torbat-e Jām
Käriz-e Elyäs
Cemenibit
Towraghoudi

Kühestän
Eslām Qal'eh
Dasht-e Hamdam Āb
Herāt

BĀDGHĪS
Qarah Bägh
Qizil Murghab
Kushk

Qal'ah-ye Now
Qädis

AFG...

Selseleh-ye Band-e Torkestän
Selseleh-ye Sefïd Küh
Ghürian
Zindah Jän

ESFAHĀN
Anärak
Na'in
Jandaq

KHORĀSĀN-E JONŪBĪ
Tabas
Robaṭ-e Khosh Āb
Ferdows
Qä'en
Tāybäd

KHORĀSĀN-E RAZAVĪ

Robaṭ-e Posht-e Bādām
Robaṭ-e Khan
Deyhük
Robaṭ-e Chāh Gonbad

Kakhak
Sedeh

Namakzar
Ghōrian

HERĀT
Herāt

Zindah Jän

GHŌR
Chaghcharān
Shahrak

Caspian Sea

Elevation scale:
6000m
4000m
3000m
2000m
1000m
500m
250m
100m
Sea Level
-250m
-2000m
-4000m

226

223

215

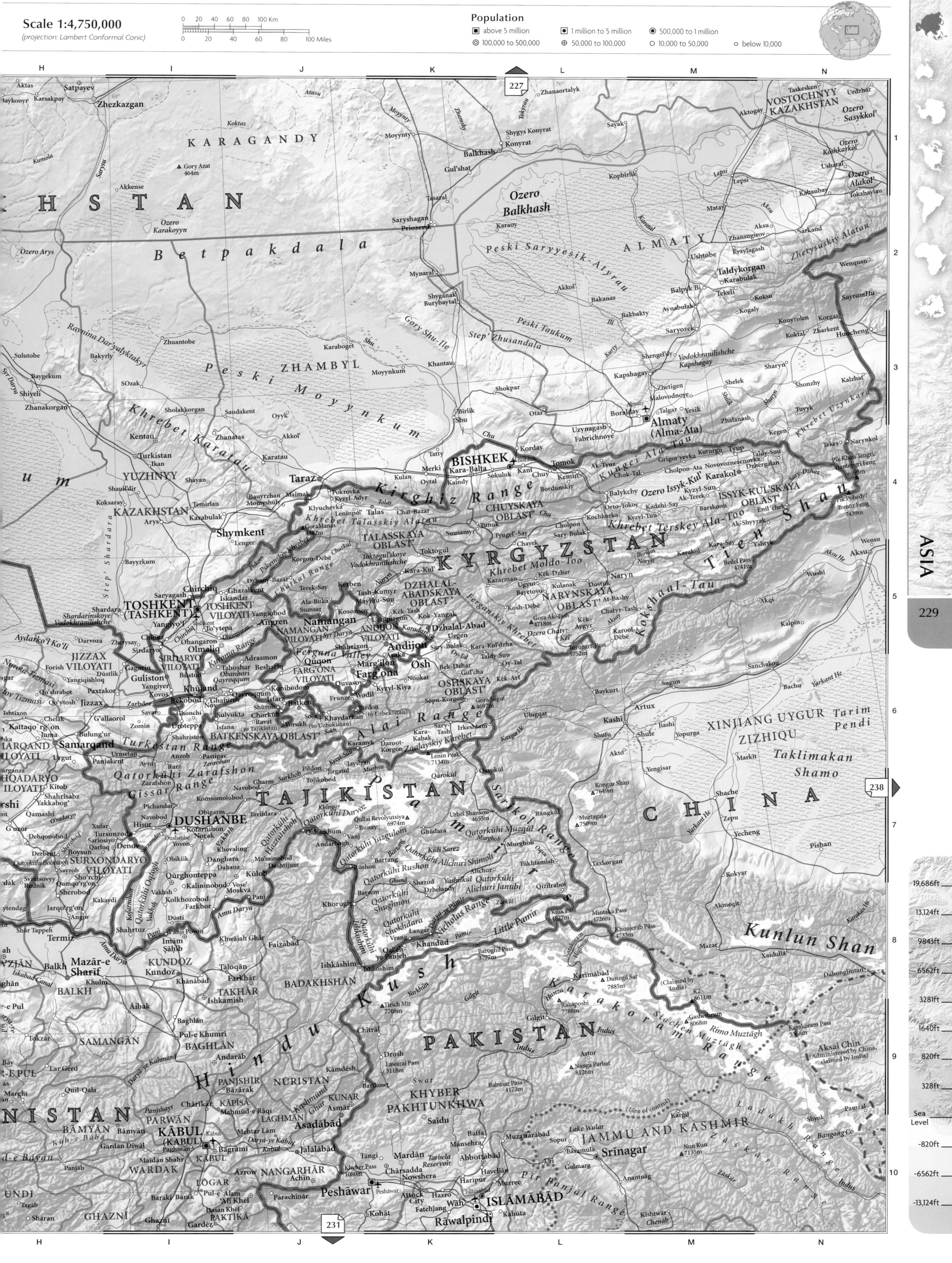

Afghanistan & Pakistan

228
223
288

ASIA

TURKMENISTAN

UZBEKISTAN

DUSHANBE

AFGHANISTAN

IRAN

PAKISTAN

OMAN

U.A.E.

Dasht-e Kavir

Dasht-e Lut

SEMNĀN

GOLESTĀN

KHORĀSĀN-E SHEMĀLĪ

KHORĀSĀN-E RAZAVĪ

KHORĀSĀN-E JANŪBĪ

YAZD

KERMĀN

HORMOZGĀN

SĪSTĀN VA BALŪCHESTĀN

Iranian Plateau

Makran

Makran Coast

Gulf of Oman

Strait of Hormuz

Arabian Sea

AHAL WELAŶATY

MARY WELAŶATY

LEBAP WELAŶATY

Garagum Kanaly

Yugo-Vostochnyye Garagumy

Garabil Belentligi

JOWZJĀN

BALKH

SAMANGĀN

FĀRYĀB

SAR-E PUL

BĀDGHĪS

GHOR

BĀMYĀN

PARWĀN

WARDAK

GHAZNĪ

DAYKUNDĪ

URŪZGĀN

ZĀBUL

PAKTĪ

HERĀT

FARAH

NIMRŌZ

HELMAND

KANDAHĀR

BALOCHISTAN

SINDH

Dasht-e Mārgow

Dasht-e Khash

Rēgestān

Chāgai Hills

Central Makrān Range

Siāhān Range

Siāhān Range

Kirthar Range

Central Brāhui Range

Toba Kākar Range

Selseleh-ye Sefid Kūh

Kūh-e Bābā

Band-e Bāyān

Kūh-e Chehel Abdālān

Kūh-e Sangān

Asgabat (AŞGABAT)

Mashhad

Mazār-e Sharīf

Herāt

Quetta

Kandahār

Lashkar Gāh

Zāhedān

Kermān

Bandar-e 'Abbās

Karāchi

Hyderābād

MASQAT (MUSCAT)

OMAN

Gwādar

Gulf of Oman

Elevation Scale

- 6000m
- 4000m
- 3000m
- 2000m
- 1000m
- 500m
- 250m
- 100m
- Sea Level
- -250m
- -2000m
- -4000m

Tropic of Cancer

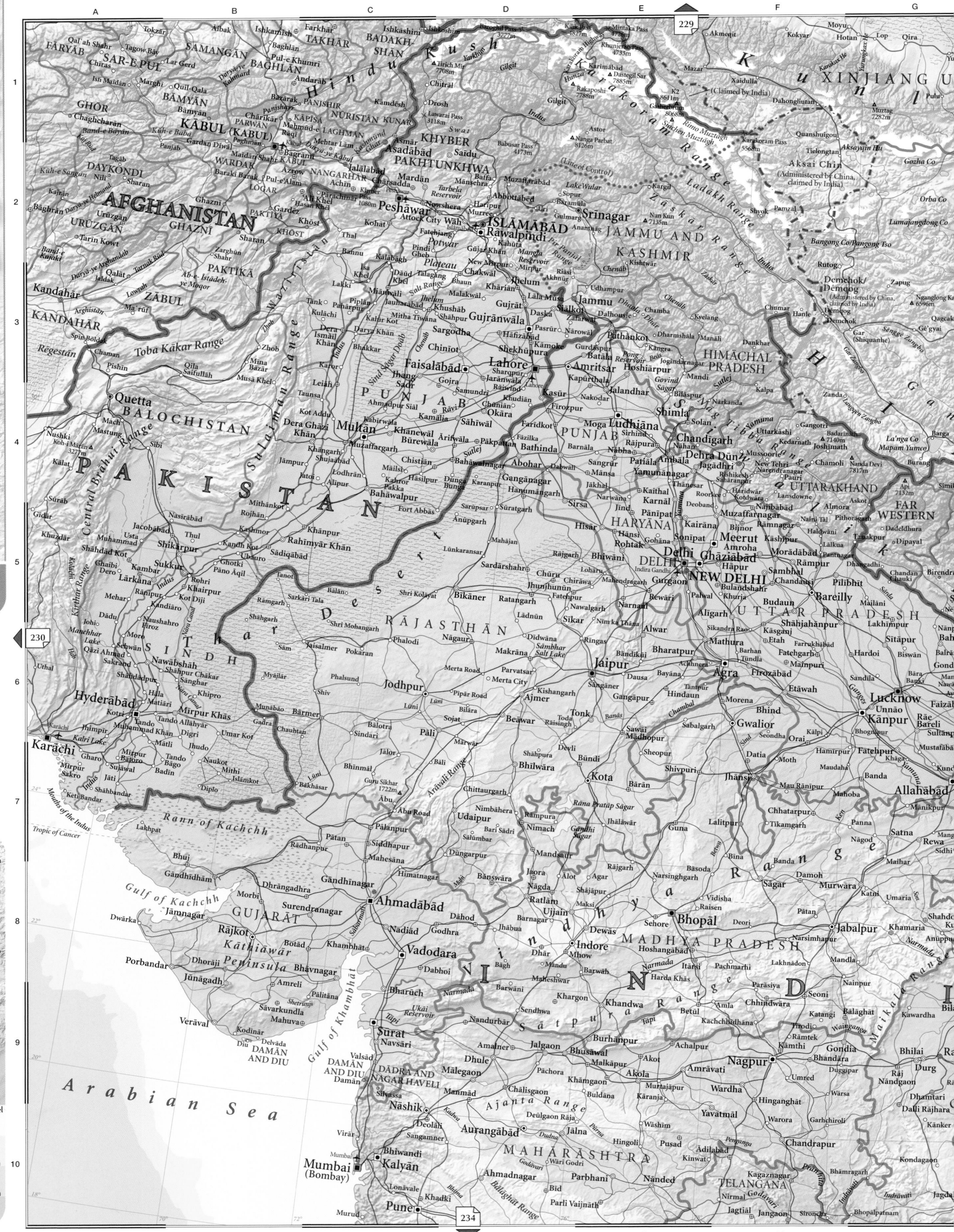

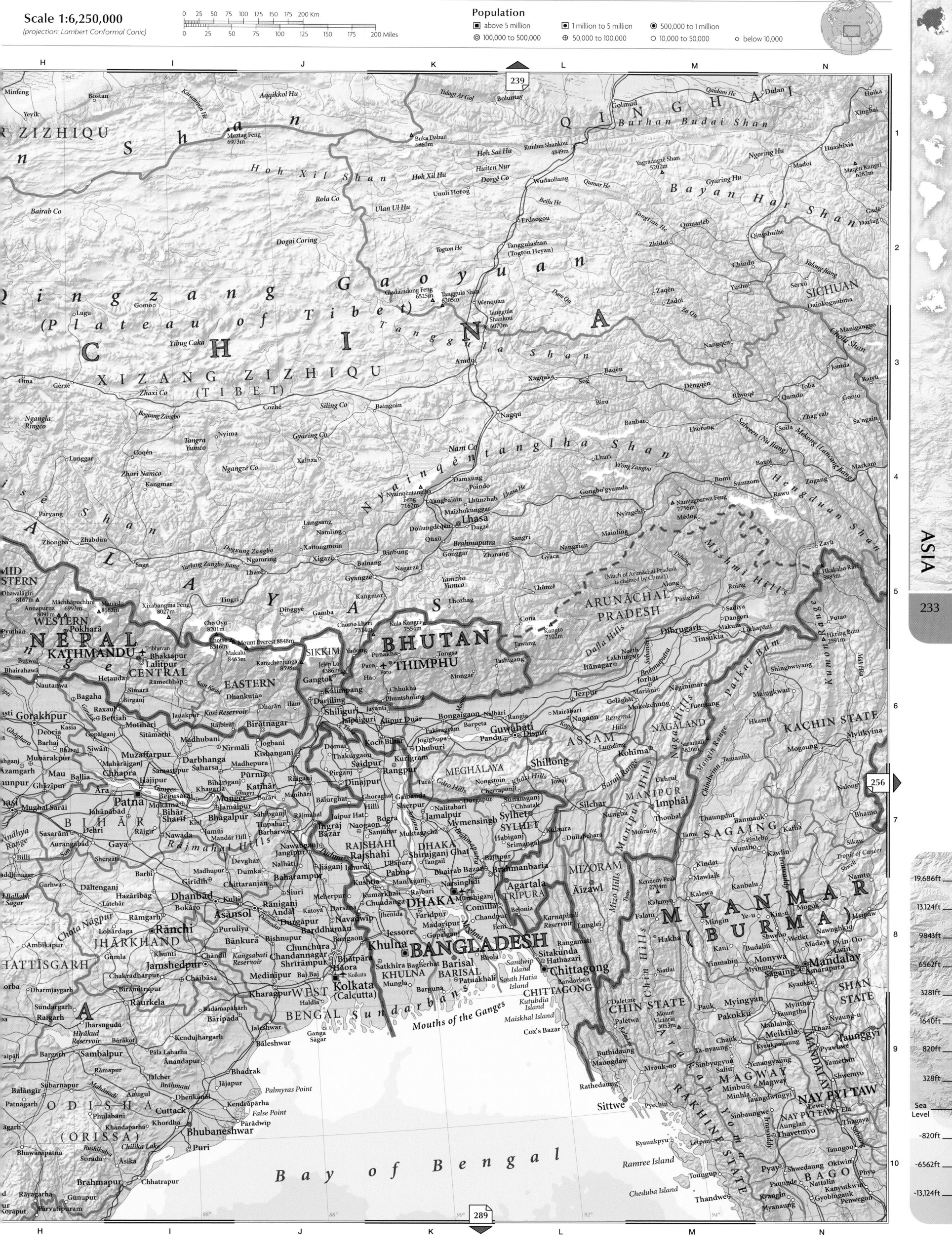

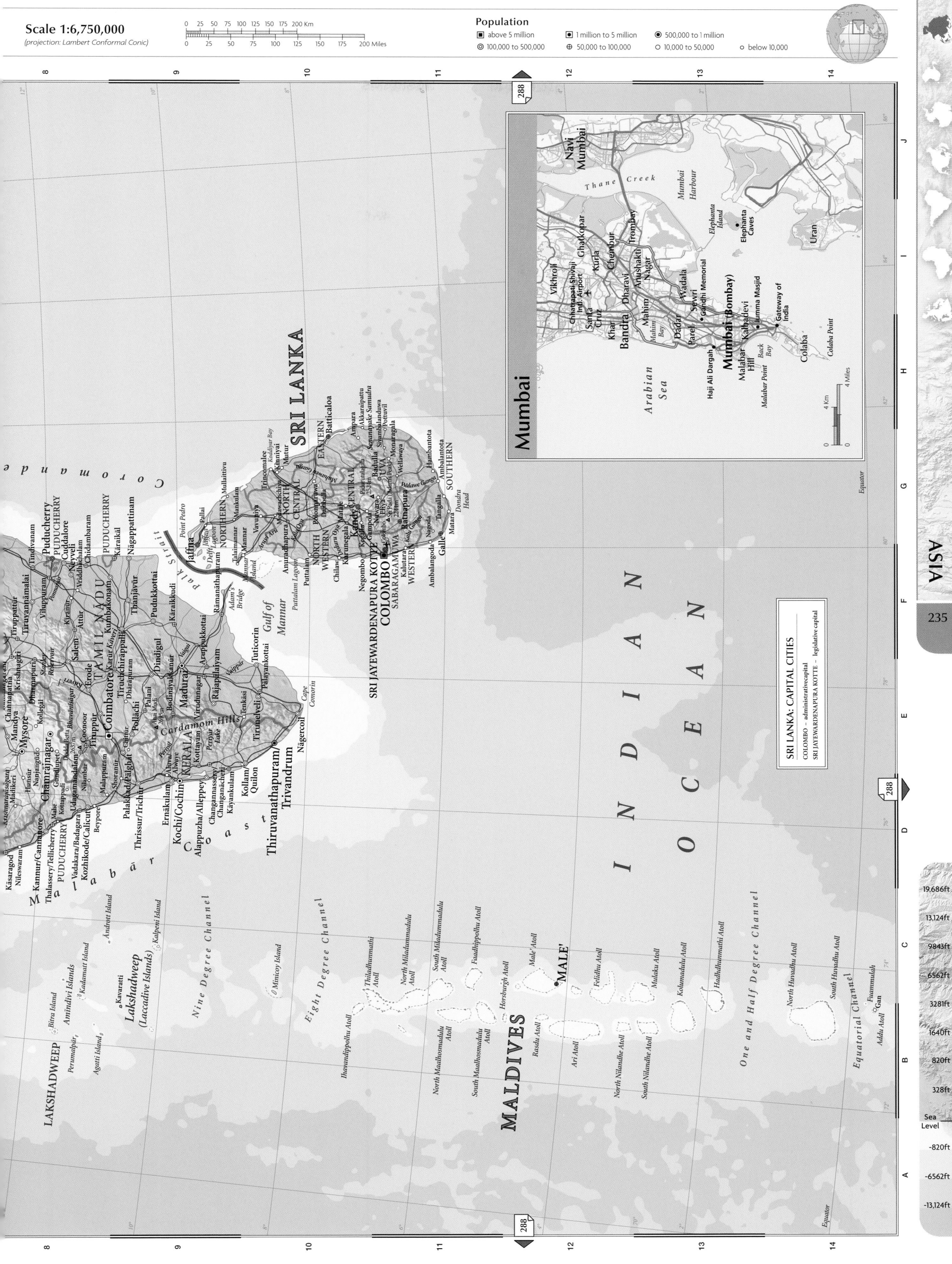

ASIA

236

192
231
256

Kostanay
Rudnyy
Omsk
Irtysh
Tatarsk
Novosibirsk
Tomsk
Yurga
Achinsk
Krasnoyarsk
Kansk
Ust'-Ilimsk

Derzhavinsk
Atbasar
Petropavlovsk
Yesil
Ravnina Kulundy
Berdsk
Leninsk-Kuznetskiy
Nizhneudinsk
Tayshet
Bratsk
Ust'-Kut

Arkalyk
Kokshetau
Shchuchinsk
Pavlodar
Karasuk
Kemerovo
Novokuznetsk
Abakan
Zima
Tulun
Angara
Lena

Zhezkazgan
ASTANA
Temirtau
Karagandy
Sharbakty
Barnaul
Novoaltaysk
Mezhdurechensk
Minusinsk
Usol'ye-Sibirskoye
Irkutsk

Saryarka
Abay
Karaganda
Semey
Ridder
Biysk
Gorno-Altaysk
Eastern Sayans
Cheremkhovo
Angarsk
Selenginsk
Ulan-Ude

KAZAKHSTAN
Kyzylzhar
Ust'-Kamenogorsk
Shar
Zyryanovsk
Gora Belukha 4506m
Zapadnyy Sayan
Kyzyl
Ak-Dovurak
Abaza
Tsetserleg
Moron
Erdenet
Darhan

RUSSIAN

MONGOLIA
ULAANBAATAR (ULAN BATOR)

Betpakdala
Balkhash
Konyrat
Ozero Balkhash
Ayagoz
Tacheng
Burqin
Uvs Nuur
Har Us Nuur
Hyargas Nuur
Hangayn Nuruu
Tosontsengel
Bulgan
Bayanhongor

Saryshagan
Ozero Alakol'
Zaysan
Ozero Zaysan
Altai Mountains
Olgiy
Tolbo
Dzereg
Uliastay
Altay
Bayanbulag
Arvayheer
Govi Altayn Nuruu
Mandalgovi

TIEN SHAN
Taraz
BISHKEK
KYRGYZSTAN
Almaty
Yining
Shonzhy
Urumqi
Qitai
Aj Bogd Uul 3802m
Atas Bogd 2702m
Dalandzadgad
Saynshand

Kashi
XINJIANG UYGUR ZIZHIQU
Tarim Pendi
Taklimakan Shamo
Korla
Kuruktag
Lop Nur
Hami
Anxi
Yumen
Honglyuyuan
Dalain Hob
Linhe
Baotou

Srinagar
Aksai Chin
Kunlun Shan
Altun Shan
Qilian Shan
Golmud
Qinghai Hu
Xining
GANSU
Lanzhou
NINGXIA
Wuzhong
Yinchuan

CHINA

Qingzang Gaoyuan (Plateau of Tibet)
QINGHAI
Tianshui
Xianyang
Baoji
Xi'an
SHAANXI

HIMACHAL PRADESH
Lahore
Amritsar
NEPAL
KATHMANDU
Lhasa
Xigazê
HIMALAYAS
BHUTAN
THIMPHU
SICHUAN
Chengdu
Mianyang
Nanchong
CHONGQING

NEW DELHI
Delhi
Meerut
Lucknow
Kanpur
Patna
BANGLADESH
DHAKA
Kolkata
MYANMAR (BURMA)
NAY PYI TAW
Mandalay
YUNNAN
Kunming
GUIZHOU
Guiyang
GUANGXI ZHUANG ZIZHIQU
Nanning

INDIA
Bhopal
Jabalpur
Nagpur
Raipur
Bay of Bengal
THAILAND
LAOS
VIANGCHAN (VIENTIANE)
HANOI
Hai Phong
Gulf of Tonkin
Hainan Dao
VIETNAM

Hyderabad
Vijayawada
ANDHRA PRADESH
Visakhapatnam
Yangon (Rangoon)
Mouths of the Irrawaddy
Chiang Mai
Bago
Da Nang

6000m
4000m
3000m
2000m
1000m
500m
250m
100m
Sea Level
-250m
-2000m
-4000m

0 50 100 150 200 250 300 350 400 Km
0 50 100 150 200 250 300 350 400 Miles

Population
■ above 5 million   ▣ 1 million to 5 million   ⦿ 500,000 to 1 million
◉ 100,000 to 500,000   ⊕ 50,000 to 100,000   ○ 10,000 to 50,000   ∘ below 10,000

193
263
286

Russian Federation / Sea of Okhotsk region

Bodaybo, Neryungri, Stanovoy Khrebet, Olekma, Tynda, Khrebet Dzhugdzhur, Peluostrov Kamchatka, Petropavlovsk-Kamchatskiy

FEDERATION

Vitim, Yablonovyy Khrebet, Shilka, Skovorodino, Never, Zeyskoye Vodokhranilishche, Pervyy Kuril'skiy Proliv, Ostrov Paramushir

Chita, Karymskoye, Mogocha, Shimanovsk, Svobodnyy, Berezovyy, Nogliki, Ostrov Sakhalin

Onon Gol, Olovyannaya, Krasnokamensk, Argun, Blagoveshchensk, Nikolayevsk-na-Amure, Komsomol'sk-na-Amure, Tatarskiy Proliv, Sea of Okhotsk, Kuril'skiye Ostrova (Kuril Islands), Ostrov Urup

Manzhouli, Zabaykalsk, Hulun Buir, Jagdaqi, Yichun, Hegang, Amur, Khor, Khabarovsk, Obluch'ye, Birobidzhan, Heilong Jiang, Bikin, Yuzhno-Sakhalinsk, Ostrov Iturup (Administered by Russian Federation, claimed by Japan)

Choybalsan, Baruun-Urt, Kerulen, Menengiyn Tal, Hulun Nur, Zalantun, HEILONGJIANG, Qiqihar, Daqing, Suihua, Harbin, Jiamusi, Jixi, Dal'nerechensk, Spassk-Dal'niy, Lake Khanka, Dal'negorsk, Khrebet Sikhote-Alin', Wakkanai, Rebun-tō, Rishiri-tō

Xiao Hinggan Ling, Da Hinggan Ling

Zamin-Üüd, Erenhot, Ulanhot, Bayan Ul, Baicheng, Shangzhi, Ussuriysk, Nakhodka, Vladivostok, La Pérouse Strait, Asahikawa, Takikawa, Asahi-dake, Kitami, Abashiri, Nemuro, Ostrov Kunashir

ONGOL ZIZHIQU, Hulingol, JILIN, Changchun, Jilin, Tongliao, Dunhua, Yanji, Mudanjiang, Tumen, Hoeryŏng, Najin, Ch'ŏngjin, Otaru, Ebetsu, Sapporo, Toma-komai, Muroran, Hakodate, Okushiri-tō, Hokkaidō, Kushiro, Horoshiri-dake

Xilinhot, Linxi, Liaoyuan, Siping, Baishan, Huch'ang, Hyesan, Kanggye, Tsugaru-kaikyō, Aomori, Hachinohe

Chifeng, Fuxin, Tieling, Fushun, Namsan-ni, Hŭich'ŏn, Kimch'aek, Hirosaki, Noshiro, Kuji, Miyako

Beipiao, Chaoyang, Shenyang, LIAONING, Anshan, Haicheng, Dandong, Sinŭiju, Chŏngju, Hamhŭng, NORTH KOREA, Akita, Yokote, Iwate, Morioka, Kesennuma

Baochang, Jinzhou, Qinhuangdao, P'YŎNGYANG, Namp'o, Sariwŏn, Wŏnsan, Sakata, Shinjō, Sendai, Ishinomaki

Hohhot, Datong, Zhangjiakou, Chengde, BEIJING, Tianjin, TIANJIN SHI, Dalian, Korea Bay, Haeju, Kaesŏng, Sŏkch'o, Sea of Japan (East Sea), Tsuruoka, Yamagata, Fukushima, Hitachi, JAPAN

Taiyuan, Shijiazhuang, HEBEI, Cangzhou, Bo Hai, Ongjin, Sunan, Ch'unch'ŏn, Gangneung, Niigata, Nagaoka, Kōriyama, Iwaki

Pingyang, Binzhou, Dezhou, Weifang, Yantai, Incheon, Incheon, SEOUL, Suwon, Wŏnju, Donghae, Jōetsu, Toyama, Nagano, Utsunomiya, Mito, Honshū

Xingtai, Anyang, Jinan, SHANDONG, Zibo, Qingdao, Cheonan, SEJONG CITY, Andong, Takaoka, Kanazawa, Komatsu, Matsumoto, Gifu, TŌKYŌ, Chiba, Haneda

Linyi, Kaifeng, Zhengzhou, Jining, Zaozhuang, SOUTH KOREA, Gunsan, Jeonju, Daejeon, Daegu, Pohang, Ulsan, Fukui, Maebashi, Fuji-san, Nagoya, Yokohama

Luoyang, HENAN, Xuzhou, JIANGSU, Namwon, Gwangju, Busan, Gyeongju, Tottori, Matsue, Kyōto, Kōbe, Ōsaka, Ise, Hamamatsu, Sagami-nada, O-shima

Nanyang, Xinyang, Bengbu, Yellow Sea, Mokpo, Suncheon, Gwangju, Geumje-do, Gōtsu, Okayama, Hiroshima, Yamaguchi, Hōfu, Tsu, Owase, Nii-jima, Mikura-jima, Izu-shotō

Suizhou, Hefei, ANHUI, Nanjing, Wuxi, Jeju-haehyeop, Korea Strait, Kitakyūshū, Kurume, Shikoku, Kōchi, Tanabe, Kii-suidō, Hachijō-jima

Xiangfan, Maohekou, HUBEI, Wuhan, Huangshi, Hangzhou, Jiaxing, SHANGHAI SHI, Shanghai, Hongqiao, Jeju-do, Fukuoka, Saga, Sasebo, Ōita, Nakamura, Hachijō-jima

Jingdezhen, Jiujiang, Anqing, Wuhu, Nagasaki, Yatsushiro, Kumamoto, Nobeoka, PACIFIC OCEAN

Changsha, Xiangtan, Nanchang, Shangrao, ZHEJIANG, Jinhua, Ningbo, Satsuma-Sendai, Miyazaki, Miyakonojō, Tanega-Shima, Yaku-Shima

East China Sea, Taizhou, Kagoshima, Kyūshū, Ogasawara-shotō

Shaoyang, Hengyang, JIANGXI, Ji'an, HUNAN, Wenzhou, Nansei-shotō (Ryukyu Islands), Naze, Amami-Ō-shima, Amami-guntō

Chenzhou, Ganzhou, Nanping, FUJIAN, Fuzhou, Yong'an, Mazu Dao (to Taiwan), (China and Taiwan claim all of each other's territory), Senkaku-shotō (Claimed by China, Japan and Taiwan), Okinawa, Ishigaki-jima, Miyake-jima

Shaoguan, Longyan, Zhangzhou, Quanzhou, Xiamen, Taiwan Strait, Taoyuan, Jilong, TAIBEI (TAIPEI), Jinmen Dao (to Taiwan), Iriomote-jima, Sakishima-shotō, Tropic of Cancer

GUANGDONG, Chaozhou, Shantou, Taizhong, Jiayi, Tainan, Taiwan, Okinawa, Naha

Zhaoqing, Guangzhou, Dongguan, Foshan, Gaoxiong, Gaoxiong, TAIWAN

Jiangmen, Hong Kong (Special Administrative Region), Chek Lap Kok, Macau (Special Administrative Region)

Luzon Strait, Babuyan Islands, Babuyan Channel, Philippine Sea

South China Sea, Luzon, Laoag, Tuguegarao, Ilagan, PHILIPPINES

Paracel Islands (disputed), Baguio, Dagupan, Angeles, Cabanatuan, Cordillera Central, Quezon City, MANILA, Ninoy Aquino

SOUTH KOREA: CAPITAL CITIES
SEOUL – capital
SEJONG CITY – administrative capital

Shanghai (inset map)

Gucun, Baoshan, Kailu Xincun, Huangpu Jiang, Shanghai University, Miaphang, Gaojing, Wujiao Chang, Dachang, Pengpu, Jiangwan, Hongkou Stadium, Lu Xun Tomb, Yangpu, Gaohang, Yichuan, Zhabei, Hongkou Tilan Qiao, Jinqiao, Zhengnu, Shanghai, Huangshan Xincun, Putuo, Temple of the Jade Buddha, Huangpu, Pudong, Yanguiadu, Zhangjiang, Jiaodong University, People's Square, Jing'an, Shanghai Museum, Humao Zhen, Beixinjing, Sun Yat Sen's Former Residence, Dapu, Muamu, Changning, Nanshi, Xujiahui, Luwan, Zhoujiadu, Yugiao, Beicai, Hongqiao, Longua, Qibao Zhen, Caohe, Sanlin, Liuliqiao

Elevation legend: 19,686ft, 13,124ft, 9843ft, 6562ft, 3281ft, 1640ft, 820ft, 328ft, Sea Level, -820ft, -6562ft, -13,124ft

Western China

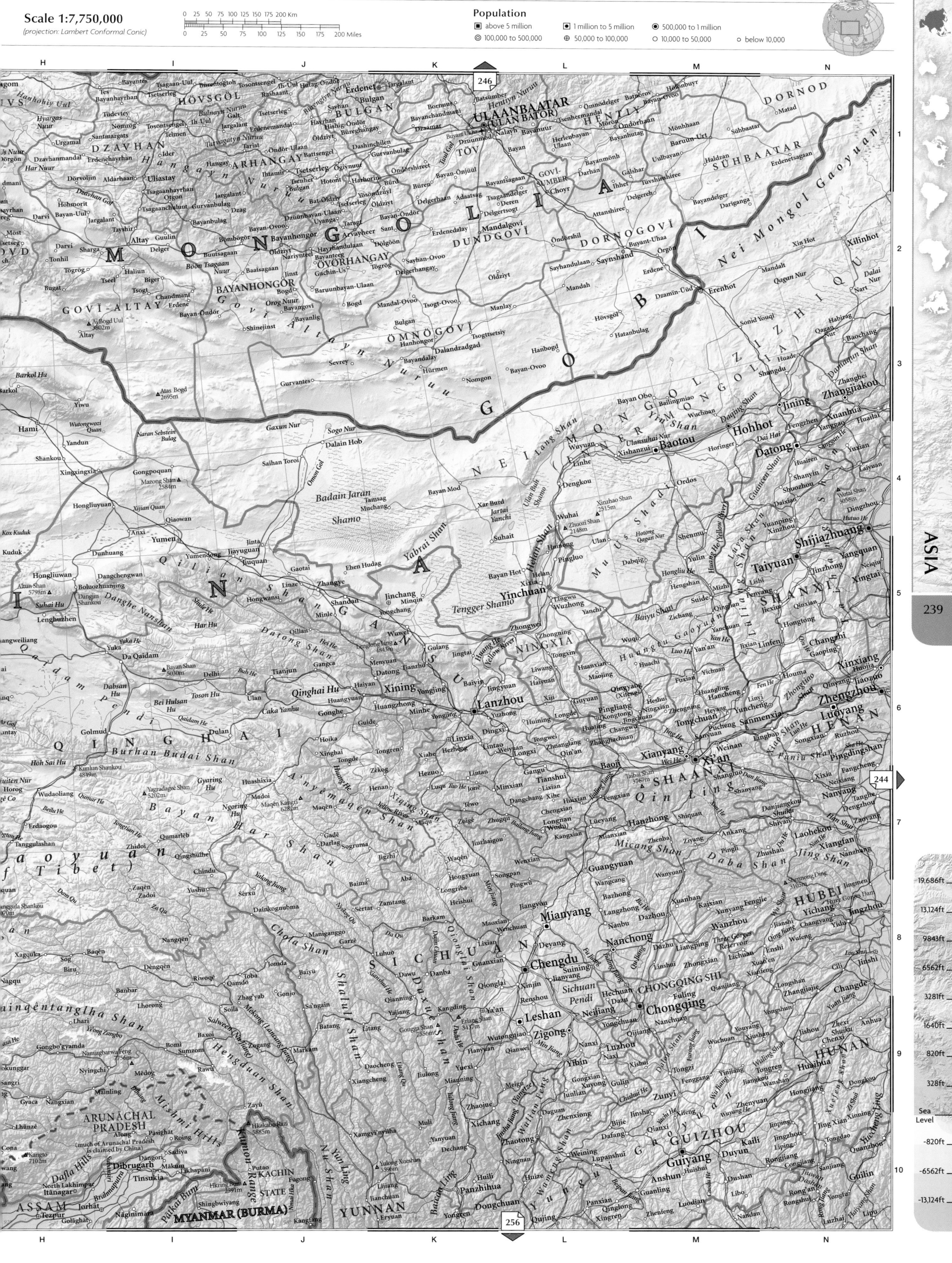

Scale 1:7,750,000
(projection: Lambert Conformal Conic)

0 25 50 75 100 125 150 175 200 Km
0 25 50 75 100 125 150 175 200 Miles

Population
- ■ above 5 million
- ■ 1 million to 5 million
- ⬤ 500,000 to 1 million
- ⊙ 100,000 to 500,000
- ⊕ 50,000 to 100,000
- ○ 10,000 to 50,000
- ○ below 10,000

19,686ft
13,124ft
9843ft
6562ft
3281ft
1640ft
820ft
328ft
Sea
Level
-820ft
-6562ft
-13,124ft

MONGOLIA

ULAANBAATAR
(ULAN BATOR)

GOBI

Nei Mongol Gaoyuan

DORNOD

SÜHBAATAR

HÖVSGÖL

BULGAN

ARHANGAY

ÖVÖRHANGAY

DZAVHAN

GOVI-ALTAY

BAYANHONGOR

TÖV

HENTIY

DUNDGOVI

DORNOGOVI

ÖMNÖGOVI

GOVI-SÜMBER

NEI MONGOL ZIZHIQU

Badain Jaran Shamo

Tengger Shamo

Muus Shadi

Ordos

Helan Shan

Hohhot

Baotou

Datong

Yinchuan

NINGXIA

SHANXI

Taiyuan

Shijiazhuang

Zhangjiakou

Jining

Xuanhua

GANSU

Lanzhou

Xining

QINGHAI

Qinghai Hu

Burhan Budai Shan

Bayan Har Shan

GAOYUAN
(Tibet)

Anyêmaqên Shan

SICHUAN

Chengdu

Chongqing

CHONGQING SHI

Sichuan Pendi

Mianyang

Nanchong

SHAANXI

Xi'an

Xianyang

Baoji

Qin Ling

Daba Shan

Micang Shan

HUBEI

Yichang

Jingzhou

HENAN

Zhengzhou

Luoyang

Nanyang

Hengduan Shan

Shaluli Shan

Daxue Shan

YUNNAN

Guiyang

GUIZHOU

HUNAN

Changde

Huaihua

Zunyi

Leshan

Zigong

Luzhou

Yibin

Panzhihua

Xichang

Lijiang

MYANMAR (BURMA)

KACHIN STATE

ARUNACHAL PRADESH
(much of Arunáchal Pradesh
is claimed by China)

ASSAM

Dibrugarh

Nu Shan

Yun Ling

Southeast China

ASIA

0 25 50 75 100 125 150 175 200 Km

0 25 50 75 100 125 150 175 200 Miles

Population

■ above 5 million   ▣ 1 million to 5 million   ◉ 500,000 to 1 million

◉ 100,000 to 500,000   ⊕ 50,000 to 100,000   ○ 10,000 to 50,000   ∘ below 10,000

ASIA

241

286

| H | I | J | K | L | M | N |

HEBEI

Xinji
Hengshui
Wuqiao
Huimin
Dongying
Laizhou Wan
Laizhou
Wendeng
SEJONG CITY
Gimcheon
Pohang

Nangong
Dezhou
Linyi
Binzhou Boxing
Rushan
Daejeon
Yeongcheon
Gyeongju

Handan
Yucheng
Jinan
Zibo
Qingzhou
Weifang
Pingdu
Laoshan Wan
Gunsan
Jeonju
Miryang
Daegu
Ulsan
Honshu

Anyang
Liaocheng
Laiwu
Boshan
SHANDONG
Yishui
Zhucheng
Weihai
Qingdao
Jin-do
SOUTH KOREA
Gwangju
Suncheon
Namwon
Jinju
Jinhae
Busan
Hiroshima
Kure

SEA OF JAPAN (EAST SEA)
Matsue
Gōtsu

PHILIPPINES   Luzon

South China Sea

Philippine Sea

| 19,686ft |
| 13,124ft |
| 9843ft |
| 6562ft |
| 3281ft |
| 1640ft |
| 820ft |
| 328ft |
| Sea Level |
| -820ft |
| -6562ft |
| -13,124ft |

ASIA

242

244

239

256

Provinces/Regions: SHAANXI, HUBEI, SICHUAN, CHONGQING, CHINA, HUNAN, GUIZHOU, YUNNAN, GUANGXI ZHUANGZU ZIZHIQU, VIETNAM, HA GIANG, CAO BANG, TUYEN QUANG, BAC CAN, YEN BAI, PHU THO, SON LA, LANG SON

Elevation legend:
6000m
4000m
3000m
2000m
1000m
500m
250m
100m
Sea Level
-250m
-2000m
-4000m

Tropic of Cancer

Selected place names:

Guangyuan, Mianyang, Chengdu, Nanchong, Chongqing, Leshan, Zigong, Yibin, Luzhou, Neijiang, Zunyi, Guiyang, Anshun, Duyun, Kaili, Tongren, Huaihua, Changde, Yiyang, Loudi, Shaoyang, Hengyang, Yongzhou, Guilin, Liuzhou, Liujiang, Nanning, Qinzhou, Fangchenggang, Mong Cai, Lang Son, Lao Cai, Yichang, Jingzhou, Xiangfan, Laohekou, Dengzhou, Three Gorges Reservoir, Three Gorges Dam, Chang Jiang (Yangtze)

Mountain ranges: Longmen Shan, Daba Shan, Wushan, Dalou Shan, Wuling Shan, Xuefeng Shan, Miao Ling, Duyang Shan, Jiuwan Dashan, Daming Shan, Shiwan Dashan, Wumeng Shan, Wulian Feng

Population
■ above 5 million   ■ 1 million to 5 million   ◉ 500,000 to 1 million
◎ 100,000 to 500,000   ⊕ 50,000 to 100,000   ○ 10,000 to 50,000   ∘ below 10,000

ASIA

243

286

HENAN
Xincai
Funan
Yingshang
Shouxian
Dingyuan
Chihe
Lai'an
Jiangdu
Taizhou
Jiangyan
Rugao
Rudong
Zhengyang
Huainan
Changfeng
Chuzhou
Luhe
Yangzhou
Yangzhong
Tongzhou
Xinyang
Xixian
Huoqiu
Feidong
Zhuxiang
Yizheng
Zhenjiang
Danyang
JIANGSU
Nantong
Haimen
Qidong
Liuchia-ho
Huangchuan
Gushi
Chengdong Hu
Wushan
Liangyuan
Quanjiao
Zhujiang
Nanjing
Jurong
Jintan
Changshu
Shaxi
Chongming Dao

ANHUI
Hefei
Feixi
Guozhen
Zhegao
Xibu
Dangtu
Liyang
Xueyangzhen
Wuxi
Suzhou
Kunshan
SHANGHAI SHI
Shanghai

HENAN
Wuhan
Dong Hu

ZHEJIANG
Hangzhou
Ningbo
Shaoxing
Zhoushan

JIANGXI
Nanchang
Jingdezhen
Shangrao
Jinhua

FUJIAN
Fuzhou
Nanping
Xiamen
Quanzhou
Zhangzhou
Shantou

GUANGDONG
Guangzhou
Dongguan
Shenzhen
Kowloon
Hong Kong (Special Administrative Region)
Macau (Special Administrative Region)
Zhuhai

TAIWAN
TAIBEI (TAIPEI)
Taoyuan
Xinzhu
Taizhong
Zhanghua
Jiayi
Tainan
Gaoxiong
Pingdong
Taidong
Hualian

East China Sea

Taiwan Strait

South China Sea

Bashi Channel

(China and Taiwan claim all of each other's territory)

Tropic of Cancer

19,686ft
13,124ft
9843ft
6562ft
3281ft
1640ft
820ft
328ft
Sea Level
-820ft
-6562ft
-13,124ft

ASIA

244

NEI MONGOL ZIZHIQU (INNER MONGOLIA)

QINGHAI

GANSU

NINGXIA

SHAANXI

SHANXI

SICHUAN

CHONGQING

HUBEI

Baotou
Hohhot
Yinchuan
Lanzhou
Xining
Xi'an
Xianyang
Baoji
Tianshui
Hanzhong
Mianyang
Guangyuan
Luoyang
Sanmenxia
Linfen

Huang He (Yellow River)
Wei He
Qin Ling
Taibai Shan 3767m
Helan Shan
Tengger Shama
Badain Jaran Shamo
Mu Us Shadi
Huangtu Gaoyuan

246

239

242

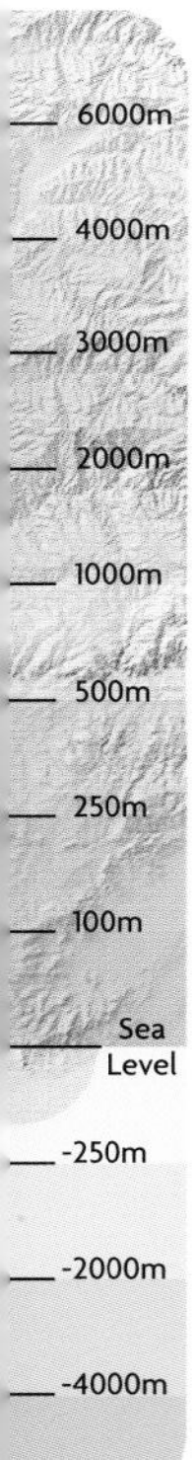

6000m
4000m
3000m
2000m
1000m
500m
250m
100m
Sea Level
−250m
−2000m
−4000m

Northeast China

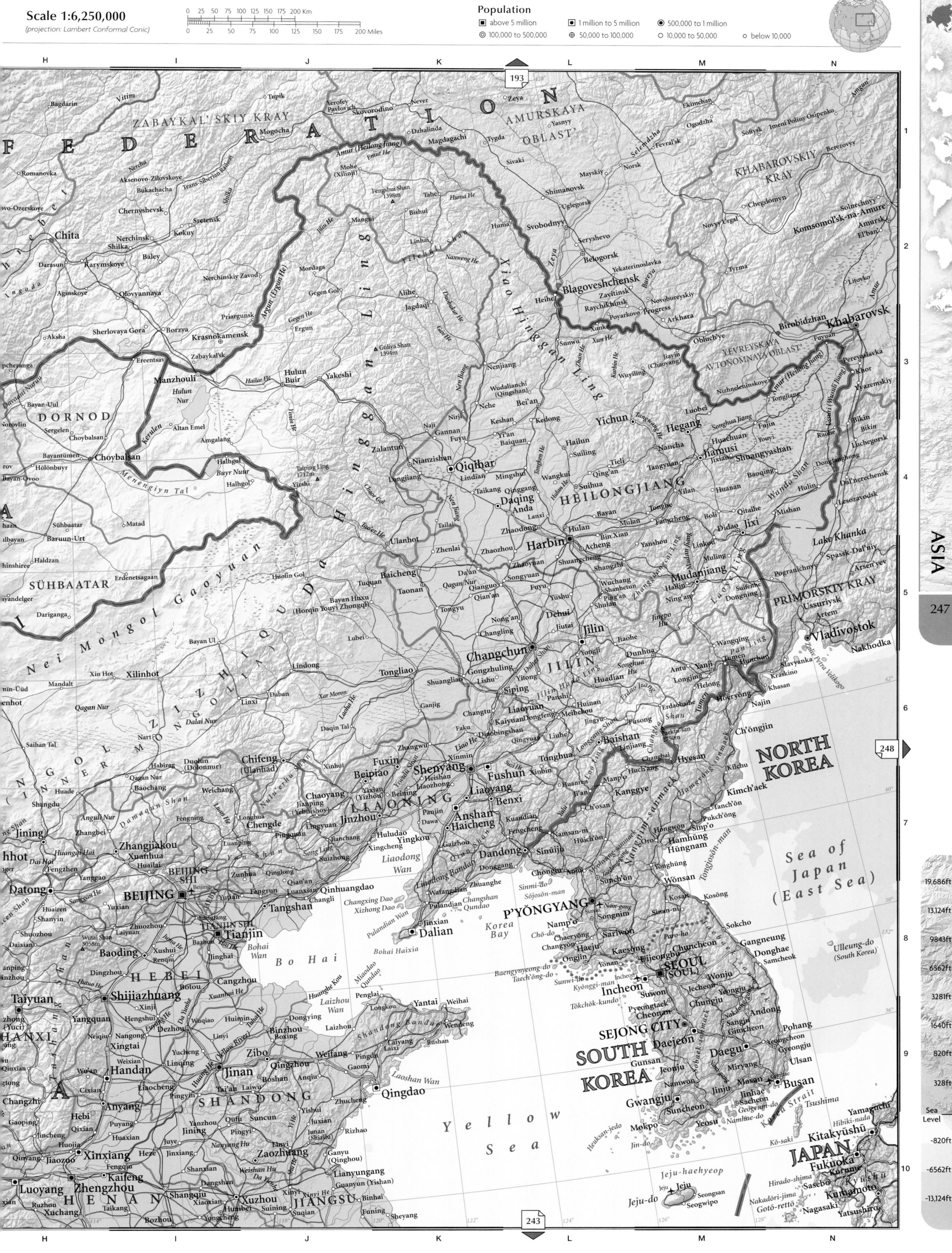

Korea & Japan

Sea of Japan (East Sea)

Liancourt Rocks
(Under South Korean control,
claimed by Japan)

Oki-shotō
Dōgo
Dōzen   Saigō
Nakano-shima
Chiburi-jima

Oki-kaikyō

Shimane-hantō   Jizō-zaki   Aoya
Hino-misaki   Hirata   Sakaiminato   Kurayoshi   Hamasak
Taisha   Izumo   Nakano-yama   Yonago   TOTTORI   Tott
Matsue   Yasugi   Dai-sen   Tōyō-   Tōt
Izumo   Kisuki   1729m   Wakasa   Chizu
Gōtsu   Oda   Sanbe-san   Tombara   Yokota   Katsuyama   Tsuyama   H
Hamada   1126m   Dōgo-yama   Niimi
SHIMANE   Miyoshi   1269m   Tōjō   Tsuyama
Susa   Masuda   Garyu-san   Shōbara   OKAYAMA   Bizen
Abu   1221m   Kake   Ashina-shi-gawa   Takahashi
Ōmi-shima   Ato   Kanmuri-yama   HIROSHIMA   Kōzan   Fuchū   Ibara   Okayama
Tsuno-shima   Tsuwano   1339m   Higashi-Hiroshima   Kurashiki
Hagi   YAMAGUCHI   Shinmanyō   Mihara   Tamano   N
Hōhoku   Mine   Ogōri   Hiroshima   Takehara   Onomichi   Bitchu-shima   Sakaide   T
Hibiki-nada   Yamaguchi   Kure   Fukuyama   Marugame   Takamatsu
Toyoura   Nōmi   Kan'onji   KAGAWA
Shimonoseki   Hōfu   Iwakuni   Imabari   Kawanoe   N
Onoda   Kudamatsu   Naga-shima   Yashiro-jima   Iyomishima   Niihama   Ikeda   TOKUSHIMA
Kitakyūshū   Ube   Hikari   Saijō   Kamega-mori   Komatsus
O-shima   Nakama   Suō-nada   Hime-jima   Heigun-tō   Matsuyama   1896m   TOKUSHIMA
Genkai-nada   Nōgata   Yukuhashi   Ya-shima   EHIME   Ishizuchi-san   Tsurugi-san
Iki   Iki-suidō   Tagawa   Nakatsu   Kuma   1982m   1955m
Katsumoto   Yobuko   Iizuka   Buzen   Bungo-Takada   Iyo   KŌCHI   Otoyo
Fukuoka   Karatsu   FUKUOKA   Usa   Nagahama   Yawatahama   Sukumo-shima   Kōchi
Azuchi-Ō-shima   Matsuura   Kasuga   Amagi   Hiko-san   Kitsuki   Ozu   SHIKOKU   Aki
SAGA   Onojo   1200m   Hita   Beppu-   Uwa   Sasa   Toyo
Ikitsuki-shima   Hirado   Taku   Ogōri   Kurogi   wan   Sadamisaki   Susaki   Muroto
Uku-jima   Hirado-   Imari   Saga   Yame   Hita   Beppu   Iyo-nada   Misaki   Uwajima   Onigaha-yama   Kubokawa   Tosa-wan
shima   Takeo   Kashima   Yanagawa   Ōita   1151m   Nakamura   Muroto-zaki
Ojika-jima   Kashima   Yamaga   Kikuchi   Kujūsan   Bungo-Ōno   Saiki   Tsurumi-   Sukumo   865m   Tosa-Shimizu
NAGASAKI   Omura   1791m   Taketa   zaki   Ashizuri-misaki
Sasebo   Matsubara   Ōmuta   Arao   Takamori   Sobo-san   Okino-shima
Gotō-rettō   Nakadōri-jima   Isahaya   Tamana   1757m   Bungo-suidō
Narao   Omura   Shimabara   Uto   Hinokage
Ōse-zaki   Fukue   Nagasaki   Kumamoto   Nobeoka
Fukue-jima   Minamishimabara   Kikuchi   Kumamoto-   Hyūga
Kuchinotsu   Yatsushiro   Kunimi-dake   Kyūshū-sanchi
Nōmo-zaki   Shimabara-wan   KUMAMOTO   1739m   Tsuno
Amakusa   Misumi   Kyūshū   Ichifusa-yama   Hyūga
Shimo-jima   Yatsushiro-kai   1727m
Amakusa-   Minamata   MIYAZAKI   Saito
nada   Izumi   Kirishima-yama   Takanabe
Ushibuka   Isa   Miyanojo   1700m
Naga-shima   Akune   Hitoyoshi   Kobayashi   Miyazaki
Kami-Koshiki-jima   Satsuma-Sendai   KAGOSHIMA   Miyakonojō
Shimo-Koshiki-jima   Kokubu   Nichinan
Koshikijima-rettō   Kushikino   Miyakonojō
Kagoshima   Tarumizu   Shibushi   Kushima
On-take   Kanoya   Shibushi-
1117m   wan   Toi-misaki
Noma-zaki   Satsuma-   Kagoshima-   Uchinoura
Minamisatsuma   hantō   wan
Makurazaki   Ibusuki   Yamagawa
Kusagaki-guntō   Sata-misaki   Ōsumi-kaikyō

East China Sea

Philippine Sea

Uji-guntō
Kuro-shima
Iō-jima
Take-shima
Mage-shima   Nishinoomote
Kuchinoerabu-jima   Tanega-shima
Kamiyaku
Yaku-shima   Minamitane

Inset Map 1

East China Sea

Sakishima-shotō   Miyako-shotō
Irabu-jima   Miyako-jima
Minna-jima   Hirara
Yaeyama-shotō   Tarama-jima
Yonaguni   OKINAWA
Yonaguni-jima   Hirakubo-saki
Iriomote-jima   Ishigaki-jima
Paimi-saki   Ishigaki
Kuro-shima

Philippine Sea

Hateruma-jima

Scale 1:3,250,000
0   10   20   40 Km
0   10   20   40 Miles

Elevation scale
6000m
4000m
3000m
2000m
1000m
500m
250m
100m
Sea Level
-250m
-2000m
-4000m

ASIA
250
245

Scale 1:2,500,000
(projection: Lambert Conformal Conic)

0 10 20 30 40 50 Km
0 10 20 30 40 50 Miles

Population
■ above 5 million   ▪ 1 million to 5 million   ● 500,000 to 1 million
◎ 100,000 to 500,000   ⊕ 50,000 to 100,000   ○ 10,000 to 50,000   ∘ below 10,000

ASIA

Inset maps locator

Scale 1:3,250,000
Scale 1:12,250,000

19,686ft
13,124ft
9843ft
6562ft
3281ft
1640ft
820ft
328ft
Sea Level
-820ft
-6562ft
-13,124ft

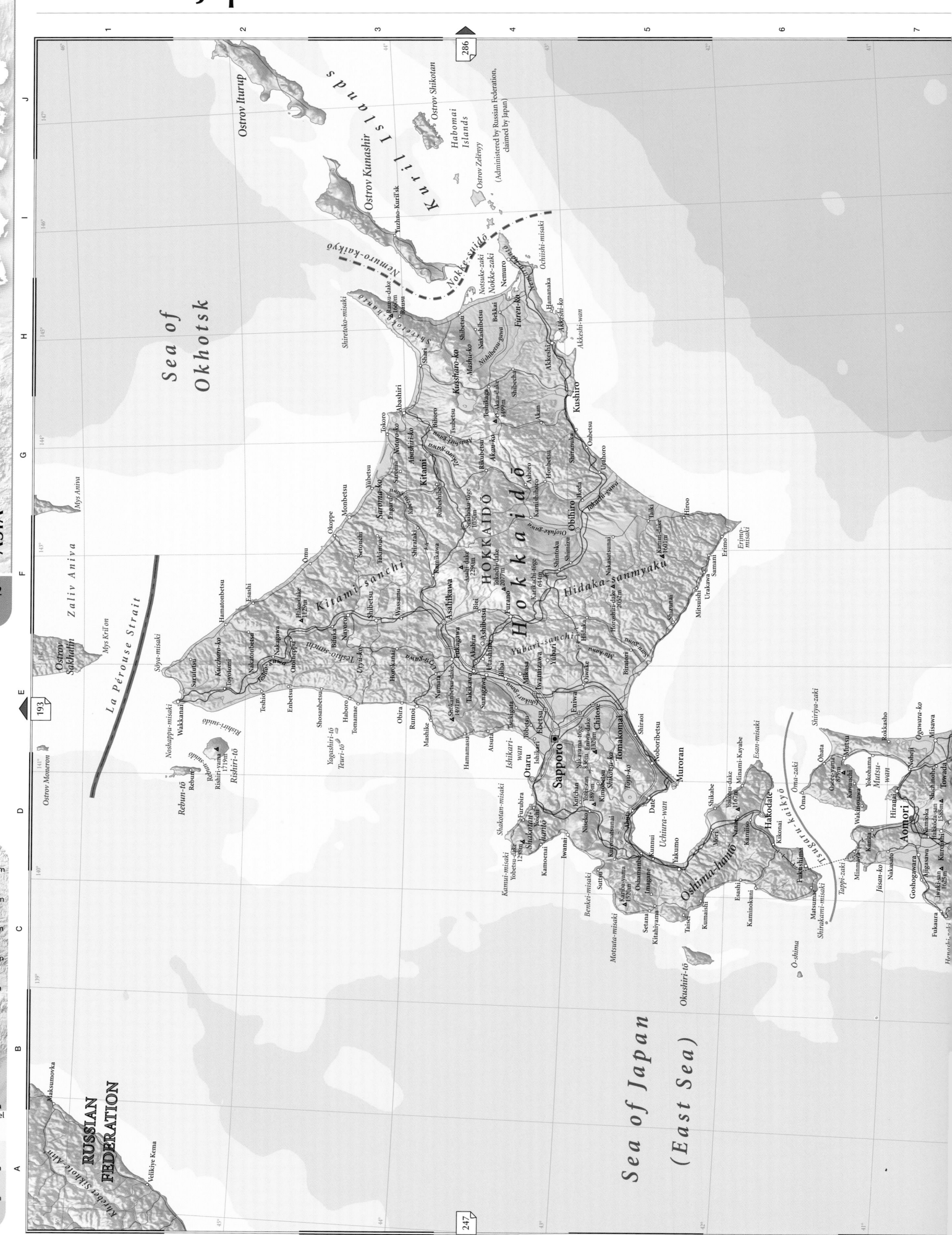

286
193
247

ASIA

6000m
4000m
3000m
2000m
1000m
500m
250m
100m
Sea Level
-250m
-2000m
-4000m

RUSSIAN FEDERATION

Khrebet Sikhote-Alin'

Velikiye Kema

Maksumovka

Sea of Japan (East Sea)

Sea of Okhotsk

Kuril Islands

Ostrov Iturup

Ostrov Kunashir

Ostrov Shikotan

Habomai Islands

Ostrov Zelënyy

(Administered by Russian Federation, claimed by Japan)

Yuzhno-Kurilsk

Nemuro-kaikyō

Nokke-suidō

Ostrov Sakhalin

Mys Aniva

Zaliv Aniva

Mys Krilon

La Pérouse Strait

Ostrov Moneron

Rebun-tō
Rebun
Rishiri-tō
Rishiri-yama 1719m

Sōya-misaki
Noshappu-misaki
Wakkanai
Sōya-suidō
Rishiri-suidō

HOKKAIDŌ

Hokkaido

Sapporo

Otaru

Hakodate

Tsugaru-kaikyō

Aomori

Oshima-hantō

Shakotan-hantō

Kitami-sanchi

Hidaka-sanmyaku

Teshio-sanchi

Yubari-sanchi

Shiretoko-hantō

Kushiro

Nemuro

Abashiri

Obihiro

Tomakomai

Muroran

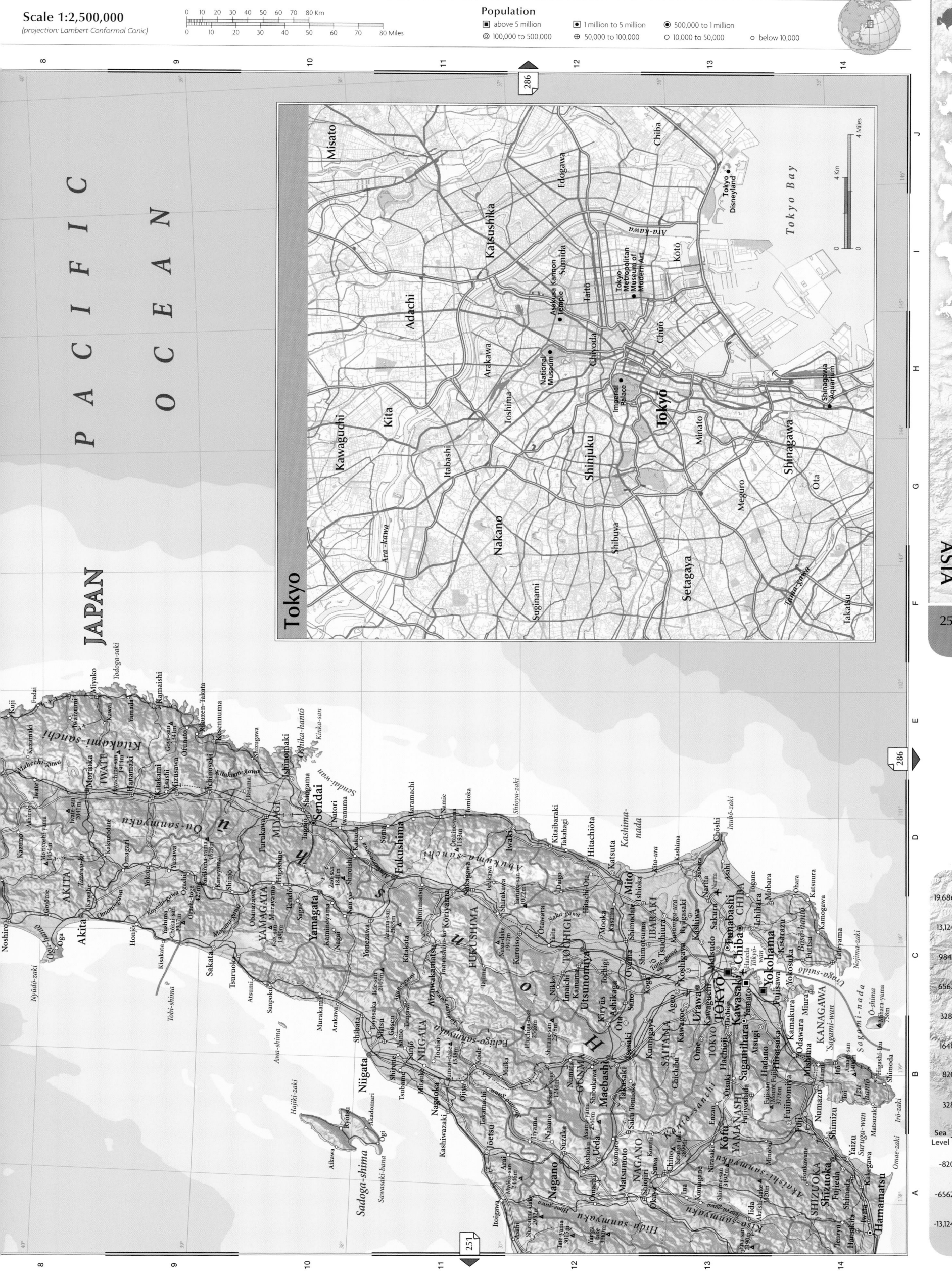

Southeast Asia

CHINA

BHUTAN
THIMPHU

BANGLADESH
DHAKA

INDIA

Himalayas
Mount Everest 8848m

Lhasa, Xigazê, Gyangzê, Nangxian, Mainling, Rawu
Kula Kangri, Lhünzê, Cona
Darjiling, Gangtok, Itanagar, Dibrugarh
Biratnagar, Shiliguri, Bongaigaon, Tezpur, Jorhat
Koch Bihar, Guwāhāti, Shillong, Kohima
Bhāgalpur, Dinājpur, Silchar, Imphāl
Kharagpur, Rajshahi, Pabna, Brahmanbaria, Comilla
Haora, Khulna, Jessore, Barisal
Kolkata, Chittagong

Tropic of Cancer
Bānkura, Baleshwar
Mouths of the Ganges

MYANMAR (BURMA)
NAY PYI TAW

Chin Hills, Falam, Monywa, Sagaing, Amarapura, Mandalay
Pakokku, Myingyan, Taunggyi
Sittwe, Minbu, Pyay, Thayetmyo
Ramree Island, Thandwe, Hinthada
Cheduba Island, Bago, Thaton
Pathein, Yangon (Rangoon), Mawlamyine, Kyaikkami
Mouths of the Irrawaddy
Gulf of Martaban

Kunming, Dali, Weishan, Baoshan, Luxi, Fengqing, Lincang
Lijiang, Panzhihua, Dongchuan, Qujing, Guanlei, Anshun
Xichang, Zhaotong, Bijie, Weining
Yibin, Zigong, Neijiang, Chongqing, Zhangjiajie, Yueyang
Tongzi, Zunyi, Guiyang, Kaili, Duyun, Dushan
Hechi, Tiandong, Nanning, Qinzhou, Beihai
Kaiyuan, Wenshan, Bose, Funing, Jingxi, Binyang, Litang
Gejiu, Mengzi, Thai Nguyen

LAOS
VIANGCHAN (VIENTIANE)

THAILAND
KRUNG THEP (BANGKOK)

Chiang Mai, Louangphabang, Xiangkhoang, Vinh
Udon Thani, Thakhek, Savannakhét, Huê
Phitsanulok, Khon Kaen, Roi Et, Ban Nadou
Nakhon Sawan, Nakhon Ratchasima, Ubon Ratchathani, Champasak, Pakxé
Khao Laem Reservoir, Dawei, Myeik
Phetchaburi, Rayong
Sirikit Reservoir, Korat Plateau, Plateau de Xiangkhoang
Mae Nam Ping, Ang Nam Ngum

VIETNAM
HÀ NỘI, Hải Phòng, Nam Định, Hạ Long
Cẩm Phả, Đà Nẵng, Quảng Ngãi
Plei Ku, Quy Nhơn, Tuy Hoa
Đà Lạt, Nha Trang, Cam Ranh
Biên Hòa, Hồ Chí Minh, Vung Tau
Châu Đốc, Mỹ Tho, Rạch Giá, Cần Thơ, Cà Mau
Mouths of the Mekong

CAMBODIA
PHNUM PENH (PHNOM PENH)
Chuŏr Phnum Dângrêk, Bătdâmbâng, Tônlé Sap
Kâmpóng Cham, Sihanoukville

GULF OF TONGKING
Hainan Dao, Dongfang, Qionghai, Sanya, Haikou, Zhanjiang, Maoming

SOUTH CHINA SEA

PARACEL ISLANDS (disputed)

SPRATLY ISLANDS (disputed)

Yiyang, Changde, Nanchang, Jingdezhen, Jiujiang, Jinhua
Xiangtan, Zhuzhou, Pingxiang, Yingtan, Shangr...
Changsha, Liling, Ji'an, Fuzhou, Nanping
Hengyang, Yong'an, Fuzhou, Putian, Quanzhou
Shaoyang, Loudi, Jingzhou, Guilin, Liuzhou
Hezhou, Longchuan, Meizhou, Xiamen, Gaoxio...
Quanzhou, Lianzhou, Zhaoqing, Foshan, Zhangzhou
Yulin, Yangjiang, Jiangmen, Dongguan, Chenghai, Shantou
Nanning, Lingshan, Guangzhou, Macau (SAR), Chaozhou
Kowloon, Hong Kong (Special Administrative Region)
Ai Jiang, Wuzhou, Jieyang
Taiwan

INDIAN OCEAN

Bay of Bengal

Andaman Islands (to India)
North Andaman, Middle Andaman, South Andaman, Port Blair, Little Andaman
Ten Degree Channel

Nicobar Islands (to India)
Car Nicobar, Camorta, Katchall Island, Little Nicobar, Great Nicobar, Bananga

Andaman Sea
Zadetkyi Kyun, Lanbi Kyun, Letsôk-aw Kyun, Mali Kyun
Isthmus of Kra

Ko Phangan, Ko Samui, Sichon, Nakhon Si Thammarat
Thung Song, Ko Phuket, Phuket
Songkhla, Hat Yai, Kota Bharu, Kuala Terengganu
George Town, Taiping, Ipoh, Kuantan
Langsa, Meulaboh, Banda Aceh
Medan, Pematangsiantar, Klang, Kuala Lumpur, Putrajaya

Strait of Malacca
Malay Peninsula

MALAYSIA

Kota Kinabalu, Gunung Kinabalu, Sandakan
BRUNEI
BANDAR SERI BEGAWAN
Miri, Bintulu, Sibu, Kuching, Singkawang

Borneo
Kalimantan
Pontianak, Samarinda, Balikpapan, Sampit, Amuntai, Kandangan, Banjarmasin

SINGAPORE
SINGAPORE
Melaka, Muar, Batu Pahat, Johor Bahru, Keluang

Pulau Simeulue, Pulau Nias, Pulau Siberut
Danau Toba, Sibolga
Kepulauan Banyak, Kepulauan Mentawai
Pekanbaru, Rengat, Jambi, Sungaipenuh
Padang, Bengkulu, Lahat, Palembang
Sumatera (Sumatra)
Pegunungan Barisan
Pangkalpinang, Pulau Bangka, Pulau Belitung

Selat Karimata
Kepulauan Lingga, Kepulauan Anambas, Kepulauan Natuna, Kepulauan Riau
Selat Sunda

Pulau Enggano, Bandar Lampung
JAKARTA
Serang, Bogor, Sukabumi, Bandung, Cirebon, Tegal, Pekalongan
Tasikmalaya, Cilacap, Magelang, Yogyakarta, Madiun, Kediri
Semarang, Surakarta, Surabaya, Probolinggo, Malang, Jember
Kudus, Pulau Madura
Jawa (Java)
Bali, Denpasar
Pulau Lombok, Pulau Sumba...
Nusa Tenggara

Java Sea (Laut Jawa)
Greater Sunda Islands

INDONESIA

Makassar, Parepare
Makassar Strait (Selat Makassar)
Pulau Laut
Flores Sea (Laut Flores)
Mataram

6000m
4000m
3000m
2000m
1000m
500m
250m
100m
Sea Level
-250m
-2000m
-4000m

239
234
289

Scale 1:15,500,000
(projection: Mercator)

0 50 100 150 200 250 300 350 400 Km
0   50   100   150   200   250   300   350 400 Miles

Population
- ■ above 5 million
- ◉ 1 million to 5 million
- ◉ 500,000 to 1 million
- ◎ 100,000 to 500,000
- ⊕ 50,000 to 100,000
- ○ 10,000 to 50,000
- ○ below 10,000

286

Taipei

Wuku
Shihlin
Luchou
Tanshui
Martyrs Shrine
Keelung
Nei Hu
Kuku
Sanchung
Confucius Temple
Datung
Hsingtien Temple
T'aipei Songshan Airport
Tiding
Sinzhuang
Sanchong
Sungshan
Taipei (Taibei)
Wanhua
Zhongcheng
Sinyi
Tai Shan
Lungshan Temple
National Theatre
Daan
Linguang
Shinjuang
National Museum of History
Banchiao
Banqiao
Hsin Chuang
Zhongher
Yungho
Wantang
T'aipei Zoo
Shu Lin
Tucheng
Fang Liao
Jhonghe
Wunshan
Sindian
0 Km
2 Miles

East China Sea
Nansei-shotō
Naze
Amami-ō-shima
Okinawa-shotō
Senkaku-shotō
Okinawa
Naha
Okinawa
Huangyan
nzhou
yang

PACIFIC OCEAN

Jilong
Sakishima-shotō
Miyake-jima
AIBEI
TAIPEI)
Iriomote-jima
Ishigaki-jima
Hualian
oyuan
TAIWAN
gdong

Tropic of Cancer

Philippine Sea

Babuyan Islands
on
ayan Channel
Tuguegarao
Ilagan
Central
Luzon
uio
pan
abanatuan
eles
MANILA
Lucena
Naga
Catanduanes Island
gas
Legazpi City
Sibuyan
oro
Calbayog
Samar
Sea
oxas City
PHILIPPINES
Tacloban
Panay Island
Cadiz
Leyte
Iloilo
Cebu
Negros
Bohol Sea
Butuan
Iligan
Bislig
Cagayan de Oro
Mindanao
Moro Gulf
Davao
Zamboanga
Lebak
Davao Gulf
General Santos
chipelago

Yap
COLONIA

HAGÅTÑA
(AGANA)
GUAM
(to USA)

Mariana Islands

MICRONESIA

Chuuk Islands

Babeldaob
MELEKEOK
PALAU

elebes Sea

Kepulauan Sangir
Kepulauan Talaud
Manado
Gorontalo
Ternate
Pulau Morotai
Pulau Halmahera

282

Equator

Molucca Sea
(Laut Maluku)
Pulau Waigeo
Selat Dampier
Sorong
Jazirah Doberai
Pulau Biak
Ninigo Group
Hermit Islands
Admiralty Islands
Manus Island
Lorengau
St. Matthias Group

awesi
elebes)
Laut Halmahera
Laut Seram
Pulau Misool
Pulau Yapen
Teluk Cenderawasih
Jayapura
Vanimo
Bismarck Archipelago
New Hanover
Kavieng
Lihir Group

Danau
Towuti
Waflia
Wahai
Teluk Berau
Fakfak
Sungai Mamberamo
Lumi
Green River
Wewak
Angoram
Bogia
Karkar Island
Kimbe
Bismarck Sea
Rabaul
New Ireland
Toriu
Taron

Kendari
Pulau
Buton
Ambon
Pulau Buru
Pulau Seram
Puncak Jaya
5040m
Pegunungan Maoke
Tembagapura
Amamapare
New Guinea
Central Range
Tabubil
Madang
Mount Wilhelm
4509m
Gloucester
Anepmete
Pomio
New Britain
Finschhafen
Gasmata
Solomon Sea

Kepulauan Sula
ESIA
A
Kepulauan Kai
Kiunga
Mendi
Mount Hagen
Goroka
Sialum
Lae
Kepulauan
Tukangbesi
Banda Sea
(Laut Banda)
Kepulauan Aru
Fly
Sungai Digul
Lake Murray
Huon Gulf
Kiriwina Islands

Kepulauan
Bonerate
Pulau Yamdena
Pulau Yos Sudarso
Emeti
Kerema
Manau
Popondetta
Tufi
Woodlark Island
Guasopa

Kepulauan
Alor
Pulau Wetar
Kepulauan Leti
Kepulauan Tanimbar
Weam
Kiwai Island
Oriomo
Daru
Hisiu
PAPUA NEW GUINEA
Owen Stanley Range
3676m
Mount Suckling
D'Entrecasteaux Islands
Magarida
Alotau
Louisiade Archipelago

es
ser Sunda Islands
DILI
EAST TIMOR
Timor
Arafura Sea
Mari
Torres Strait
PORT MORESBY
Kupiano

ulauan
awu
Pulau
Roti
Timor Sea
Savu Sea
Nikiniki
Kupang

Melville Island
Croker Island
South Goulburn Island
Wessel Islands
Prince of Wales Island
Cape York
Moa Island
Coral Sea
Tagula Island

Bathurst Island
Darwin
Van Diemen Gulf
Noonamah
Adelaide River
Arnhem Land
Nhulunbuy
Gulf of Carpentaria
AUSTRALIA
Cape York Peninsula

275

19,686ft
13,124ft
9843ft
6562ft
3281ft
1640ft
820ft
328ft
Sea Level
-820ft
-6562ft
-13,124ft

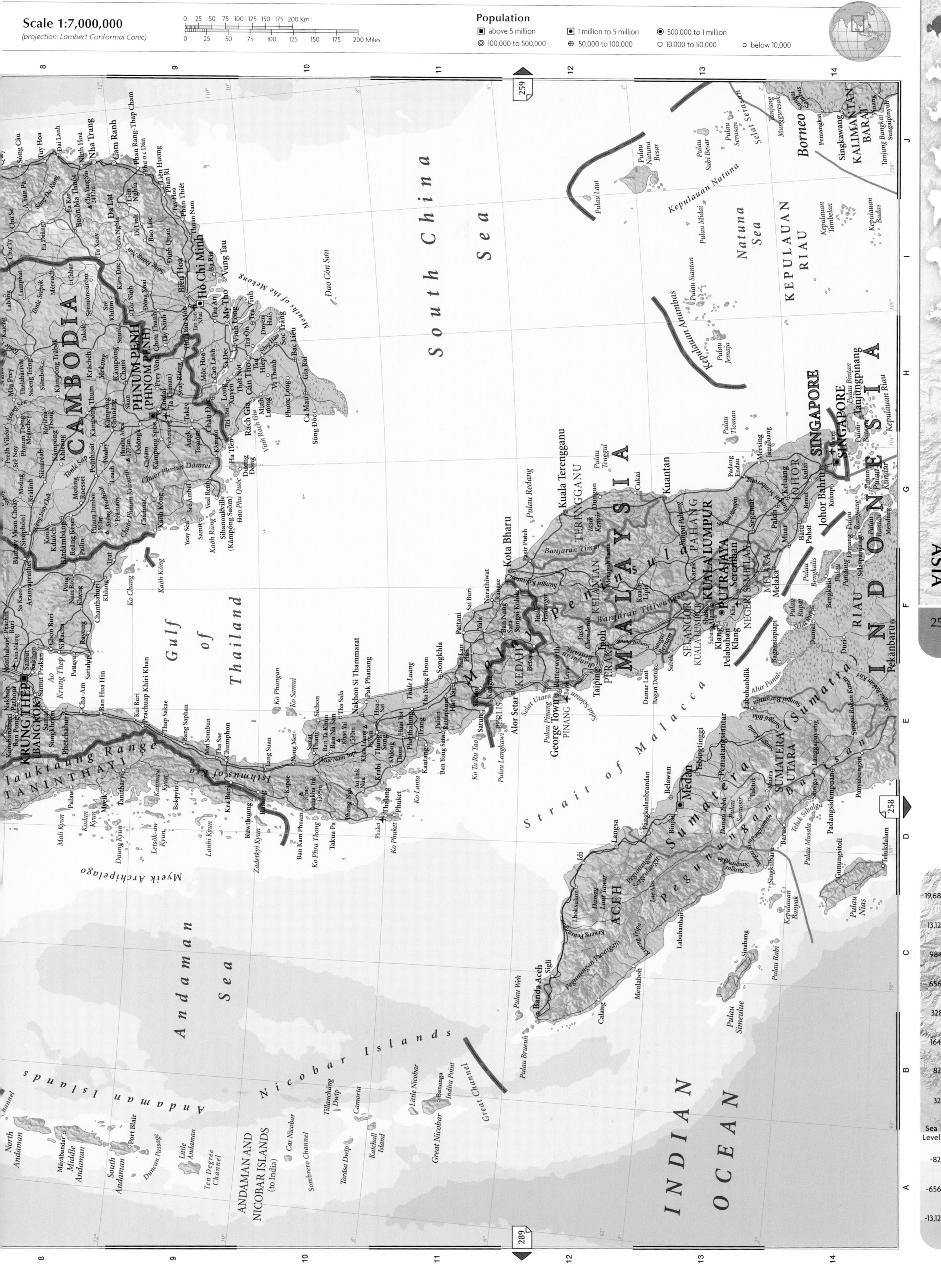

Scale 1:7,000,000
(projection: Lambert Conformal Conic)

0 25 50 75 100 125 150 175 200 Km

0 25 50 75 100 125 150 175 200 Miles

Population

■ above 5 million   ◉ 1 million to 5 million   ◉ 500,000 to 1 million
◎ 100,000 to 500,000   ⊕ 50,000 to 100,000   ○ 10,000 to 50,000   • below 10,000

259

CAMBODIA

PHNÙM PÉNH (PHNOM PENH)

Hồ Chí Minh

KRUNG THEP (BANGKOK)

THAILAND

Gulf of Thailand

Tanintharyi

TANINTHAYI

Tenasserim

Myeik Archipelago

Andaman Sea

ANDAMAN AND NICOBAR ISLANDS (to India)

Andaman Islands

North Andaman

Middle Andaman

South Andaman

Little Andaman

Nicobar Islands

Car Nicobar

Great Nicobar

Little Nicobar

Indira Point

Great Channel

South China Sea

Natuna Sea

Borneo

KALIMANTAN BARAT

Singkawang

KEPULAUAN RIAU

Kepulauan Natuna

Pulau Natuna Besar

MALAYSIA

Kuala Terengganu

TERENGGANU

PAHANG

Kota Bharu

KELANTAN

KEDAH

PERAK

Kuantan

KUALA LUMPUR
PUTRAJAYA

SELANGOR
Klang
Pelabuhan Klang

NEGERI SEMBILAN

MELAKA
Melaka

JOHOR

Johor Bahru

SINGAPORE
SINGAPORE

George Town
PINANG

PERLIS

Malay Peninsula

Banjaran Titiwangsa

Strait of Malacca

SUMATERA UTARA

Medan

ACEH

Banda Aceh

Sumatra

INDONESIA

Pekanbaru

RIAU

SUMATERA

Pegunungan Barisan

INDIAN OCEAN

258

289

19,686ft
13,124ft
9843ft
6562ft
3281ft
1640ft
820ft
328ft
Sea Level
-820ft
-6562ft
-13,124ft

Western Maritime Southeast Asia

ASIA

Andaman Sea

Nicobar Islands (to India)

Tarâsa Dwip
Camorta
Katchall Island
Little Nicobar
Great Nicobar
Bananga
Indira Point
Great Channel

THAILAND

Ko Luk Nua
Khao Luang 1835m
Nakhon Si Thammarat
Hua Sai
Pak Phanang
Phuket
Krabi
Thalang
Ko Phuket
Phuket
Trang
Khlong Thom
Ko Lanta
Kantang
Palian
Ban Yong Sata
Thung Song
Huai Yot
Phatthalung
Thale Luang
Tha Nong Phrom
Battaphum
Songkhla
Hat Yai
Pattani
Sai Buri
Ko Ta Ru Tao
Satun
Sadao
Ban Lam Phat
Yala
Narathiwat
Kangar
Ban Nang Sata
Rangae
Sungai Kolok
Kota Bharu
Pasir Puteh

PERLIS
Alor Setar
Pulau Langkawi
KEDAH
Betong
Pulau Redang

George Town
Pulau Pinang
Butterworth
PINANG
Bayan Lepas
Selat Selatan

KELANTAN
Tasik Temengor
Tasik Kenyir
Kuala Terengganu
TERENGGANU
Dungun
Pulau Tenggul

Taiping
Ipoh
PERAK
Tasik Chenderoh
Merapuh Lama
Kuala Lipis
Cukai

Damar Laut
Bagan Datuk
Sabak
Sungai Bernam
PAHANG
Sungai Pahang
Kuantan

Belawan
Pangkalanbrandan
MALAYSIA

Binjai
Medan
Tebingtinggi
Pematangsiantar
Labuhanbilik

SELANGOR
KUALA LUMPUR
Subang
Shah Alam
Klang
Pelabuhan Klang
Seremban
NEGERI SEMBILAN
KUALA LUMPUR
PUTRAJAYA
Karak
Padang Endau
Pulau Tioman

Labuhanhaji
Langgapayung

SUMATERA UTARA
Padangsidempuan
Panyabungan

Barus
Sungai Kualu
Muara
Sungai Bilah

Bagansiapiapi
MELAKA
Melaka
Muar
JOHOR
Paloh
Keluang
Mersing
Jamaluang

Pulau Rupat
Pulau Rupat
Dumai
Bengkalis
Batu Pahat
Benut
Kukup
Kulai
Johor Bahru
SINGAPORE
SINGAPORE
Pulau Bintan
Tanjungpinang

Duri
Pulau Padang
Selatpanjang
Pulau Rangsang
Pulau Batam
Pulau Kundur

KEPULAUAN RIAU

Kepulauan Anambas
Pulau Siantan
Pulau Jemaja
Pulau Midai

Natuna Sea

Kepulauan Tambelan
Kepulauan Badas
Pulau Pejantan

Pulau Simeulue
Sinabang
Singkilbaru
Pulau Rabi
Kepulauan Banyak
Pulau Musala
Teluk Sibolga
Sibolga
Danau Toba
Pulau Samosir
Tuktuk

Pekanbaru
Bangkinang
Sungai Rokan Kanan
Sungai Rokan Kiri
Sungai Kampar

RIAU
Rengat
Sapat
Pulau Lingga
Kelumu
Pulau Singkep
Selat Berhala
Labu

Gunungsitoli
Pulau Nias
Telukdalam

Kepulauan Batu
Pulau Pini
Lambak
Natal
Airbangis

Danau Maninjau
Bukittinggi
Tiku
Padangpanjang
Danau Singkarak
Taluk
Sungai Indragiri
Tungkal
Kualatungkal

Bawo Ofuloa
Pulau Tanahmasa
Pulau Tanahbela

Danau Singkarak
Solok
Sungaidareh
Kota Baru

JAMBI
Muarabungo
Muaratembesi
Jambi

Pulau Pagai
Kepulauan Mentawai

Muarasigep
Pulau Siberut
Selat Siberut
Padang
SUMATERA BARAT
Painan
Tarusan
Gunung Kerinci 3804m
Sungaipenuh
Danau Kerinci
Bangko
Sarolangun
Merangin

Taileleo
Selat Bungalaut
Pasirganting

Pulau Sipura
Pasapuat
Pulau Pagai Utara
Pulau Pagai Selatan
Tiop
Selat Sanding

Surulangun
Lubukhnggau
Muarabeliti
Air Musi
Perabumulih
Muaraenim
Palembang
SUMATERA SELATAN

Bengkulu
BENGKULU
Lahat
Tebingtinggi
Air Ogan

Toboali
Tanjung Kait
Pulau Lepar
Pangkalpinang
BANGKA BELITUNG
Pulau Bangka
Koba
Tanjungpandan
Belinyu
Mentok
Sungsang
Babat

Danau Ranau
Bukitkemuning
Kotabumi
Way Seputih
Way Sekampung
LAMPUNG
Krui
Bandarlampung

Pulau Enggano

South China Sea

Greater

Pulau Panaitan
Pulau Rakata
Serang
Rangkasbitung
Tangerang
JAKARTA
JAKARTA R.
Bekasi
Karawe
Kepulauan Seribu
Soekarno Hatta
Bogor
BANTEN
Cikawung
Citeureup
Cipanas
Purwakarta
Subang
Danau Jatiluhur
Sukabumi
Cianjur
JAWA BARAT
Ban
Pelabuhan Ratu
Teluk Pelabuhan Ratu
Genteng
Cik

INDIAN OCEAN

Christmas Island (to Australia)

Singapore

Zoological Gardens
Upper Peirce Reservoir
Central Catchment Nature Reserve
Ang Mo Kio
Buangkok
Selat Johor
Tampines
Bukit Timah Nature Reserve
MacRitchie Reservoir
Kallang
Chia Keng
Jurong East
Toa Payoh
Tan Tock Seng
K.G. Potong Pasir
Tai Seng
Bedok
Bukit Timah
Raffles Park
Kandang Kerbau
Geylang Serai
National Stadium
Clementi
University of Singapore
Queenstown
National Museum
Cathedral
City Hall
Raffles Hotel
Singapore River
Pasir Panjang
Singapore
Marina South
Buona Vista
Telok Blangah
Cable Car
Pulau Brani
Keppel Harbour
Sentosa Island
South China Sea
Straits of Singapore
Palau Bukum
Palau Tembakul
Palau Sakijang Bendera

0   3 Km
0   3 Miles

Elevation scale
6000m
4000m
3000m
2000m
1000m
500m
250m
100m
Sea Level
-250m
-2000m
-4000m

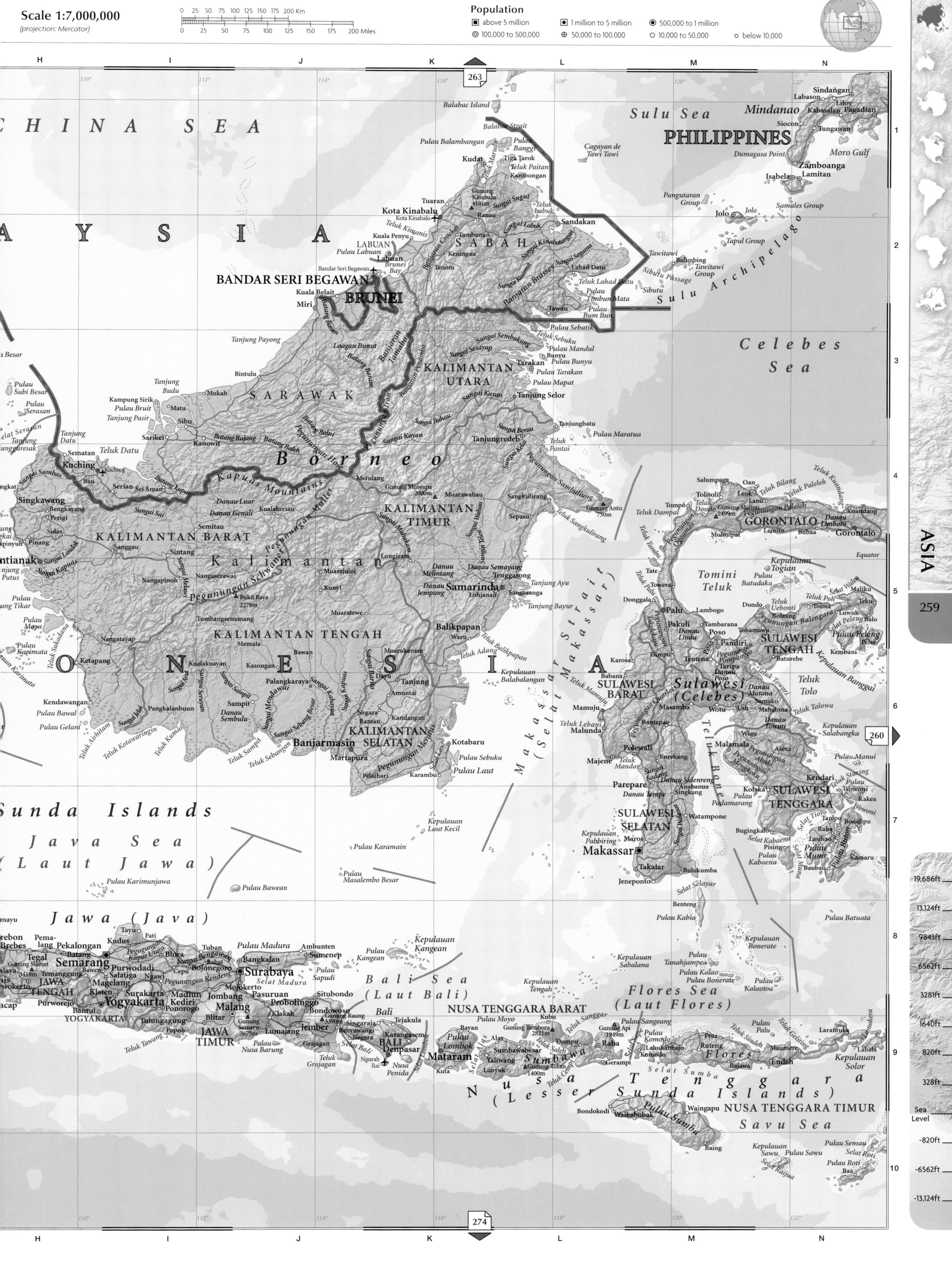

263
259
274

Scale 1:7,000,000
(projection: Mercator)

0 25 50 75 100 125 150 175 200 Km

0 25 50 75 100 125 150 175 200 Miles

Population
■ above 5 million
■ 1 million to 5 million
◉ 500,000 to 1 million
◎ 100,000 to 500,000
⊕ 50,000 to 100,000
○ 10,000 to 50,000
∘ below 10,000

282

PACIFIC OCEAN

Equator

Kepulauan
Asia

Kepulauan Mapia
Pulau Pegun   Pulau Bras

Kepulauan
Ayu

Pulau Waigeo
Selat Bougainville
Kabarei
Lamlam   Urbinasopon
Pulau   Koor   Warmandi
Besir Gam   Sausapor   Sau Korem
Selat Dampier   Napido   Pulau Supiori
Makbon   Todlo   Pegunungan Tamrau   Pulau Bepondi   Pulau Sansundi
Sorong   Asbakin   Sungai Kammundan   Manim Soyek   Sarwon
Segit   Megamo   Mubrani   Napam   Wardo   Sarwon
Pulau   Rawas   Mabui   Andoi   Mandori   Biak
Kofiau   Kwawi   Kouda   Teminabuan   Gunung Mebo   Oransbari   Pulau Biak
Atkri   SaiLeen   Baru   Mugoi   Ransiki   Saba
Tip   Pulau Misool   Bintuni   Mumi   Pulau Rumberpon   Rori
Kapocol   Koagas   Andamata   Sisember   Pulau Maswaar   Kuran
PAPUA BARAT   Teluk Bintuni   Sobiei   Pulau Roon   Serami
Kepulauan   Teluk Berau   Bombera   Teluk   Asori
Segaf   Kepulauan   Pakfak   Semenanjung   Cenderawasih   Pami
Kobi   Pisang   Jazirah   Bomberai   Napan-Yaur
Hoti   Rumbati   Weri   Tiwarra   Maki
Bolifar   Pegunungan Fakfak   Selassi   Wosimi   Nabire
Benu   Pulau Karas   Ibonawa   Warika   Kerai
Maswangi   Mas   Sopinusa   Lobo   Modowi
Waru   Teluk Sebaka   Obome   Pegunungan
Undur   Pulau   Wanggar   Danau
Kilwo   Gorong   Nusawulan   Kobowre   Paniai
Pulau Manawoka   Ilur   Manggawitu   Umari
Nama   Kepulauan   Pulau Adi   Yapa Kopra   Aiduna   Wanapiri
Kepulauan   Gorong   Uta
Banda   Pulau Kasiui   Timika
Gulir
Pulau Manuk   Kepulauan   Kokenau
Watubela   Amamapare

Jazirah
Doberai

New   Guinea

PAPUA

Pegunungan Van Rees
Rouffaer Reserves

Pegunungan Sudirman
Pegunungan Jayawijaya
Puncak Jaya
5040m
Tembagapura
Sahang

Pegunungan Maoke

Jayapura
Entrop
Nirahotong   Pue
Sentani
Demta

WEST SEPIK
Sissano   Aitape
Wutung   Vanimo
Imonda
Amanab
Green River
Ambunti
May River
Amisibil
Oksibil   3932m Mount Aiyang
Kawentinkim   3325m
Tabubil   Telefomin
Ok Tedi
Ningerum
Kiunga
Nomad

PAPUA
EAST SEPIK

NEW
GUINEA
WESTERN

Wewak
Schouten Islands
Kairiru Island
Muschu Island
Walis Island
Kaup
Angoram
Chambri
Lake   Yaminbot
Maprik   Sepik

Central Range
ENGA
Wabag   Kompiam
Laiagam   Wapenamanda
Koroba   Tari
Kandep   Margarima   Mount
Nipa   Mendi   Hagen
Poroma   Mount
SOUTHERN   Ialibu
HIGHLANDS   Kagua
Mount Bosavi   Erave
2397m

HELA

Simbai
Baiyer
River

Kikori
GULF   Kikori

Lake Copiago
Porgera
Kanggime
Wandip
Ketu
Ewe   Kaun
Tumu
Lake Murray   Strickland

Mayu   Odimmun
Yar
Tanjung De Jongs   Oreyabo
Heitske
Mapi   Keisak
Tusirah
Bado   Abemaree
Arak   Muting   Bupul
Yomuka   Kaba   Yodom   Kofarau
Pembre   Kurik
Pulau Yos   Wamal
Sudarso   Alotip
Solaka   Sungai Kumbe
Kladar   Monibum   Komoran
Tanjung Vals   Wan   Pulau Komoran
Merauke
Sakiramke
Kondomirat
Mari

Remoon   Wair
Pulau Kur   Har   Pulau Warilau   Warilau
Pulau   Piar   Pulau Lutu   Lutu
Tayandu   Watnil   Pulau   Gumzai
Kai Besar   Kai Ketil   Wamar   Komfane
Weduar   Dobo   Pulau Wokam   Pulau Jursian
Tanjung Weduar   Tanjung   Namalau
Kepulauan   Ngoni   Taberdate
Kai   Kepulauan
Aru
Pulau   Pulau Workai
Trangan   Baimun
Tanjung Ngabordamlu   Kepulauan Jin

Pulau Yamdena
Pulau Wuliaru
Koreare
Pulau Fordate
Larat
Pulau Larat
Amdassa
Saumlaki
Eliase
Tanjung   Pulau Selaru
Aro Usu
Yatoke
Pulau Babar
Amplawas
Kepulauan
Babar

Kepulauan Tanimbar

Arafura   Sea

Weam
Morehead
Sibidiri
Oriomo   Parama Island
Daru
Wabuda
Island   Purutu Island
Kiwai Island
Morigio
Island

Torres Strait
Badu Island   Moa Island

Prince of Wales
Island   Endeavour Strait   Cape York

Coral
Sea

Cape
York   Great Barrier Reef

AUSTRALIA

Bathurst
Island   Melville
Island
Van Diemen
Gulf
Beagle Gulf
Darwin
Noonaman
Adelaide River
Cooinda
Jabiru
Mount Evelyn
366m
Bulman

Croker Island
Cobourg Peninsula
South Goulburn Island
Marchinbar Island
Elcho
Island   Wessel Islands
Nhulunbuy

Arnhem
Land

Gulf of Carpentaria

Cape
York
Peninsula
Weipa

275
280

19,686ft
13,124ft
9843ft
6562ft
3281ft
1640ft
820ft
328ft
Sea
Level
-820ft
-6562ft
-13,124ft

ASIA

262

Countries and regions: CHINA, YUNNAN, GUANGXI ZHUANGZU ZIZHIQU, MYANMAR (BURMA), SHAN STATE, KAYAH STATE, KAYIN STATE, MON STATE, TANINTHAYI, THAILAND, LAOS, VIETNAM, CAMBODIA, MALAYSIA, INDONESIA, ACEH, KEDAH, PERLIS, PERAK, KELANTAN, TERENGGANU, PAHANG, PINANG, HAINAN

Seas and water bodies: Gulf of Tonkin, Gulf of Thailand, South China Sea, Hainan Dao, Qiongzhou Haixia, Weizhou Dao, Tropic of Cancer, Paracel Islands (disputed), Amphitrite Group, Crescent Group, Triton Island, Passu Keah, Spratly Island

Major cities: HA NOI, Hai Phong, Nanning, Haikou, KRUNG THEP (BANGKOK), VIANGCHAN (VIENTIANE), PHNUM PENH (PHNOM PENH), Hồ Chí Minh, Kuala Terengganu, George Town

Rivers and features: Mekong (Lancang Jiang), Red River, Black River, Salween, Annam Highlands, Korat Plateau, Tônlé Sap, Isthmus of Kra, Malay Peninsula, Myeik Archipelago, Mouths of the Mekong, Plateau de Xiangkhoang

Elevation scale: 6000m, 4000m, 3000m, 2000m, 1000m, 500m, 250m, 100m, Sea Level, -250m, -2000m, -4000m

Page references: 240, 258, 288

Grid references: A B C D E F G (columns), 1 2 3 4 5 6 7 8 9 10 (rows)

Population
■ above 5 million   ▣ 1 million to 5 million   ◉ 500,000 to 1 million
◎ 100,000 to 500,000   ⊕ 50,000 to 100,000   ⊙ 10,000 to 50,000   ∘ below 10,000

ASIA

263

243

286

260

Asia elevation scale (right margin):
19,686ft
13,124ft
9843ft
6562ft
3281ft
1640ft
820ft
328ft
Sea Level
-820ft
-6562ft
-13,124ft

Tropic of Cancer

GUANGDONG
Guangzhou
Dongguan
Huizhou
Jiangmen
Zhongshan
Zhuhai
Macau (SAR)
Shenzhen
Kowloon
Hong Kong (Special Administrative Region)
Chep Lap Kok
Shantou
Chaoyang
Haifeng
Lufeng
Puning
Nan'ao Dao
Honghai Wan
angchuan

TAIWAN
Tainan
Jiayi
Xinying
Yuli
Taodong
Pingdong
Gaoxing
Gaoxiong
Fangshan
Lu Dao
Hengchun
Eluan Pi
Lan Yu

Bashi Channel

Tungsha Tao

Batan Islands
Batan Island

Luzon Strait

Balintang Channel
Babuyan Island
Babuyan Islands

Babuyan Channel

Philippine Sea

Mayraira Point
Claveria
Escarpada Point
Laoag
Aparri
Mount Cagua 1133m
Cabugao
Dingras
Tuao
Tuguegarao
Vigan
Bangued
Jabuk
Candon
Bontoc
Ilagan
Cauayan
Bangar
San Fernando
Lagawe
Echague
La Trinidad
Bauang
Baguio
Bayombong
Baler
Bolinao
Dagupan
Lingayen Gulf
Lingayen
San Carlos
San Jose City
San Ildefonso Peninsula
Masinloc
Camiling
Palayan City
Iba
Tarlac
Cabanatuan
High Peak 2037m
Angeles
Mount Pinatubo 1485m
San Fernando
Olongapo
Malolos
Balanga
Quezon City
Caloocan
MANILA
Pasig
Imus
Laguna de Bay
Corregidor Island
Tagaytay
Nasugbu
Lipa
San Pablo
Batangas
Lucena
Lubang Island
Calapan
Boac
Catanauan
Cape Calavite
Mamburao
Marinduque
Pinamalayan
Mindoro
Sablayan
Mount Baco 2488m
Roxas
San Jose
Busuanga Island
Coron
Culion Island

Luzon

Cordillera Central
Sierra Madre
Cagayan

Polillo Islands
Lamon Bay
Tabo
Calauag
Daet
Caramoan
Naga
Pili
Iriga
Tabaco
Mayon Volcano 2425m
Legazpi City
Sorsogon
Donsol
Bulan
Catarman
Laoang
Samar
Dolores
Catbalogan
Borongan
Guiuan

Masbate
Calbayog
Naval
Biliran Island
Catanduanes Island
Virac

San Francisco
San Pascual
Ligao
Burias Island
Sibuyan Sea
Masbate
Tintotolo Channel
Placer
Cariagara
Galbiga
Leyte Gulf
Abuyog
Baybay
Leyte
Maasin
Dinagat
Dinagat Island
Siargao Island
Surigao

Tablas Strait
Odiongan
Tablas Island
Sibuyan Island
Cajidiocan
Balud
Masbate

Ibajay
Kalibo
Roxas City
Culasi
Passi
Cadiz
Bogo
Ormoc
Panay Island
Patnongon
San Jose de Buenavista
Iloilo
Miagao
Silay
Bacolod
San Carlos City
Cebu
Danao
Lapu-Lapu
Cebu
Toledo
Camotes Sea
Sogod
Ubay
Bohol Sea

La Carlota
Bago
Canlaon Volcano 2465m
Himamaylan
Negros
Sipalay
Argao
Bais
Bohol
Tagbilaran
Jagna
Bayawan
Dumaguete
Siquijor Island
Siaton
Siaton Point
Camiguin Island
Cabadbaran
Butuan
Surigao
Tandag

Cagayan Islands
Puerto Princesa
Quezon
Palawan
Brooke's Point
Palawan Passage

Sulu Sea

Balabac Island
Balabac Strait

Pulau Balambangan
Cagayan de Tawi Tawi
Pulau Banggi
Kudat
Tuaran
Gunung Kinabalu 4101m
Kota Kinabalu
Ranau
Teluk Labuk
Sandakan
Lahad Datu
Teluk Darvel
Tawau
Sebatik

MALAYSIA
SABAH
LABUAN
Pulau Labuan
Labuan
Teluk Marudu
Teluk Paitan
Kambongan
Tiga Tarok
Tambunan
Keningau
Tenom
Sungai Sugut
Sungai Labuk
Sungai Kinabatangan
Sungai Segama
Pulau Timbun Mata
Pulau Bum Bum

BRUNEI
BANDAR SERI BEGAWAN
Bandar Seri Begawan
Brunei Bay
Kuala Belait
Miri
SARAWAK
Belagan Crocker
Sungai Kuamut
Sungai Brassey
Banjaran Brassey

Pangutaran Group
Dumagasa Point
Isabela
Zamboanga
Lamitan
Basilan
Jolo
Jolo
Samales Group
Moro Gulf

Lebak
Palimbang
Tinaca Point
Sarangani Islands
Tapul Group
Tawitawi
Balimbing
Tawitawi Group
Sibutu
Sibutu Passage
Sulu Archipelago

Cagayan de Oro
Dipolog
Dapitan
Oroquieta
Tagoloan
Gingoog
Lianga
Prosperidad
Hinatuan
Bislig
Mangkayo
Nabunturan
Banganga
Tagum
Malaybalay
Maramag
Mindanao
Midsayap
Davao
Kidapawan
Digos
Mount Apo 2954m
Lupon
Manay
Mati
Tacurong
Isulan
Sultan Kudarat
Cotabato
Kabasalan
Tubod
Pagadian
Karomatan
Malabang
Sindangan
Labason
Liloy
Iligan
Iligan Bay
Ozamiz
Tangub
Mount Malindang 2425m
Lake Lanao
Marawi
Davao Gulf
Koronadal
Surallah
Mount Busa 2083m
Parker Volcano 1842m
General Santos
Glan
Jose Abad Santos
Kiamba
Cape San Agustin
Governor Generoso
Malita
Siocon
Tungawan

Agusan
Diuata Mountains

Celebes Sea

Spratly Islands
SPRATLY ISLANDS (disputed)
Northeast Cay
Southwest Cay
Sandy Cay
West York Island
Thitu Island
Loaita Island
Nanshan Island
Flat Island
Itu Aba Island
Sand Cay
Namyit Island
Sin Cowe Island
Sin Cowe East Island
Lansdowne Reef

El Nido
Taytay
Linapacan Island
Calamian Group
Cuyo West Pass
Cuyo East Pass
Linacan Island

PHILIPPINES

Visayan Sea

Kepulauan Nanusa
Pulau Karakelong
Kepulauan Talaud
Melanguane
Kepulauan Kawio

A B C D E F G

Baghdad

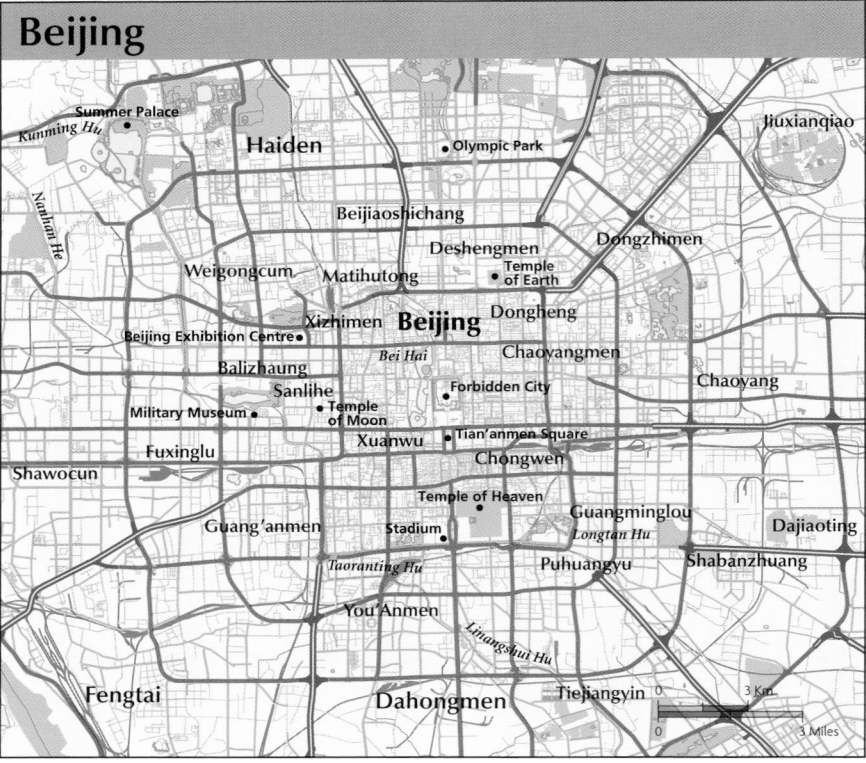

1

Tigris

Shaala

Quds

Zahrā   Tunis

Maghreb   Sadr City

Al'Azamiyah   Qanat Al Jaysh

Khansā'

Rusāfa

2

Adel   Arbataash   Shaikh   Aomar   Baghdād

Karkh   Baghdadi Museum

Iraqi National Museum   Al-Shaab Stadium

Khudrā   Aalām   Liberation Monument   Amin

Tishriyaa   Riyad   Muthana

Karrādah   Al-Rasheed Airport

Hamrā   Diyala

Firdows

3

Jihād   University   New Baghdad

Baghdad Intl. Airport

Amal Qādisiya   Jizīra

Dōra   Tigris

0   4 Km

0   4 Miles

Bangkok

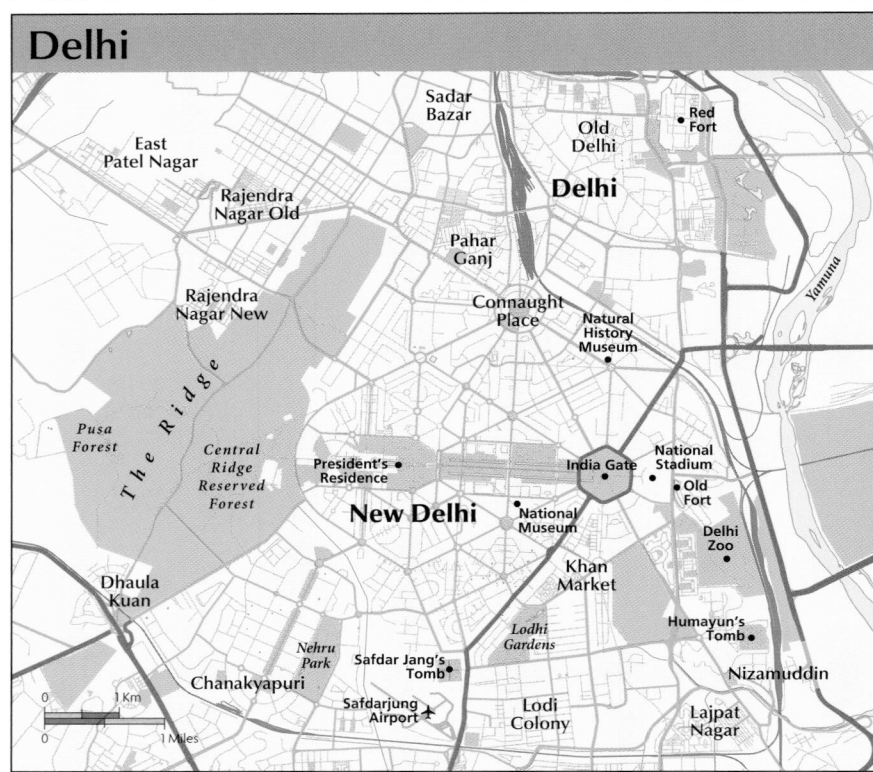

Bangkhen

Nonthaburi   Lad Phrao

Bangsu   Chatuchak

Chao Phraya   Huay Khwang   Bang Kapi

Dusit   Phaya Thai

Bangkok Noi   Chitralanda Palace

National Museum   Jim Thomson's House

Wat Phra Kaeo & Grand Palace   Bangkok (Krung Thep)   Khlong Toey

Wat Arun

Phasi Charoen   Khlong   Sathorn   Chao Phraya

Thonburi   Phra Khanong

Bang Kholaem

Chom Thong   Yannawa

Phra Pradaeng

0   5 Km

Ratburana   Samut Prakan

0   5 Miles

4

Beijing

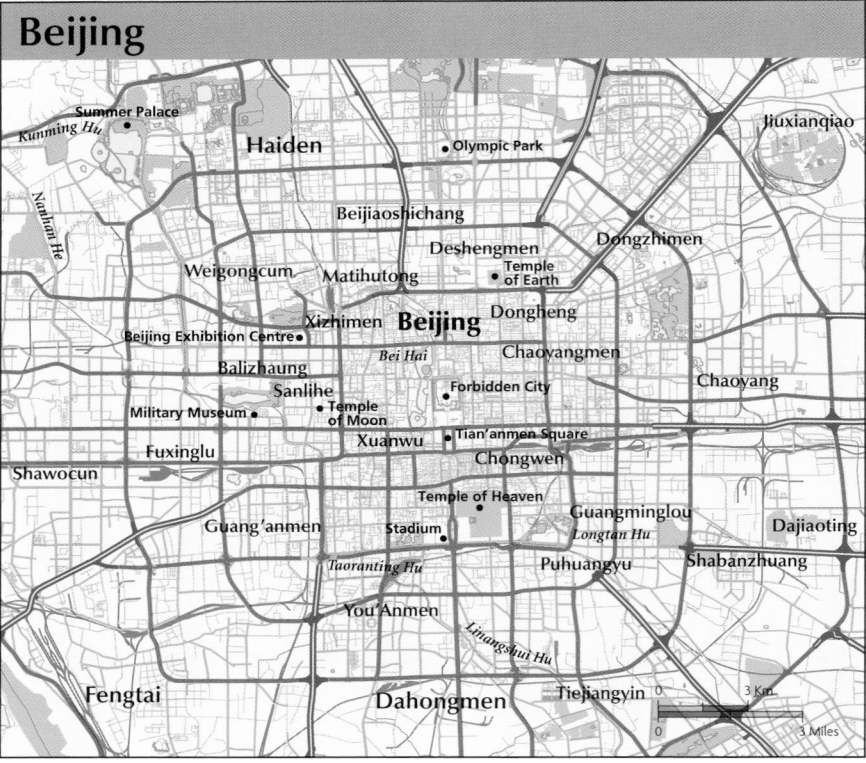

Summer Palace

Kunming Hu   Jiuxianqiao

Haiden   Olympic Park

Nanhai He

Beijiaoshichang

5

Deshengmen   Dongzhimen

Weigongcum   Matihutong   Temple of Earth

Xizhimen   Beijing   Dongheng

Beijing Exhibition Centre   Chaoyangmen

Balizhuang   Bei Hai   Chaoyang

Sanlihe   Forbidden City

Military Museum   Temple of Moon   Xuanwu   Tian'anmen Square

Fuxinglu   Chongwen

Shawocun

6

Guang'anmen   Temple of Heaven   Guangminglou   Dajiaoting

Stadium   Longtan Hu   Shabanzhuang

Taoranting Hu   Puhuangyu

You'Anmen

Lianhuashui Hu

Fengtai   Dahongmen   Tiejiangyin   0   3 Km

0   3 Miles

Delhi

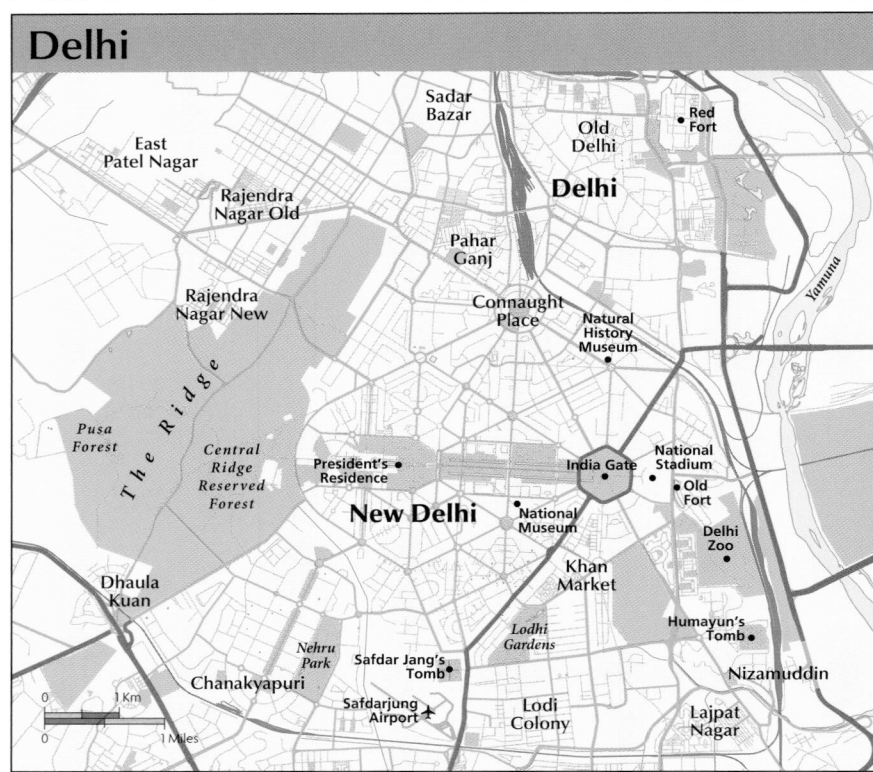

Sadar Bazar   Old Delhi   Red Fort

East Patel Nagar   Delhi

Rajendra Nagar Old

Pahar Ganj

Rajendra Nagar New   Connaught Place   Yamuna

Natural History Museum

The Ridge   India Gate   National Stadium

6

Pusa Forest   Central Ridge Reserved Forest   President's Residence   Old Fort

New Delhi   National Museum   Delhi Zoo

Dhaula Kuan   Khan Market

Humayun's Tomb

Nehru Park   Safdar Jang's Tomb   Lodhi Gardens   Nizamuddin

Chanakyapuri   Safdarjung Airport   Lodi Colony   Lajpat Nagar

0   1 Km

0   1 Miles

8

Dhaka

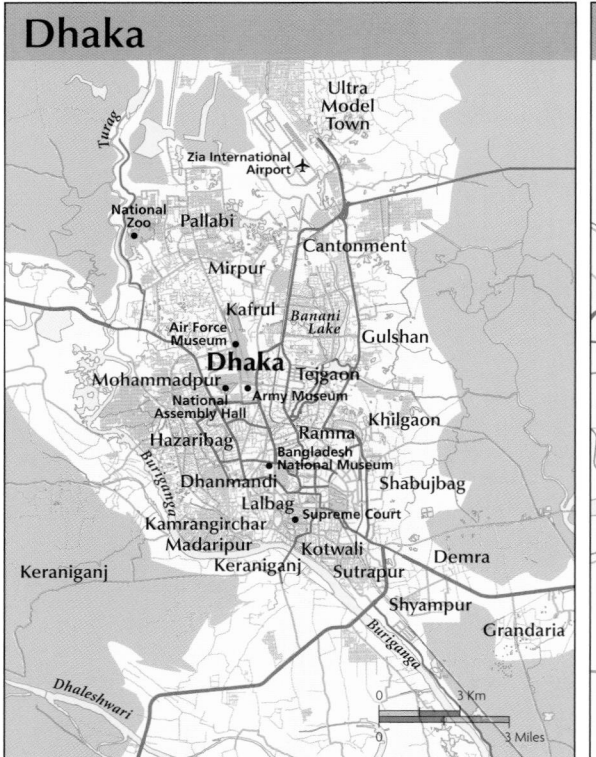

Turag   Ultra Model Town

Zia International Airport

National Zoo   Pallabi

Mirpur   Cantonment

Kafrul   Banani Lake   Gulshan

Air Force Museum   Dhaka   Tejgaon

Mohammadpur   Army Museum   Khilgaon

9

National Assembly Hall   Ramna

Hazaribag   Bangladesh National Museum   Shabujbag

Dhanmandi

Buriganga   Lalbag   Supreme Court

Kamrangirchar   Madaripur   Kotwali

Keraniganj   Keraniganj   Sutrapur   Demra

Shyampur   Grandaria

10

Dhaleswari   Buriganga   0   3 Km

0   3 Miles

Kolkata

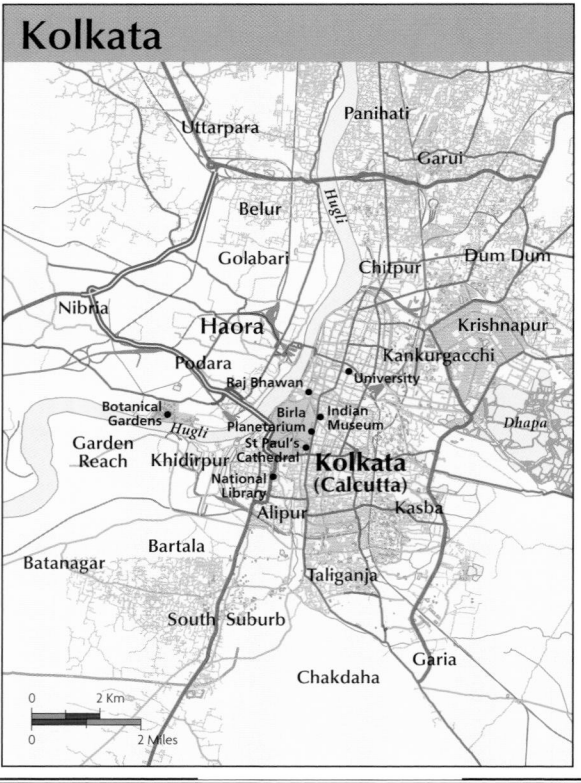

Uttarpara   Panihati

Garul

Hugli   Belur   Dum Dum

Golabari   Chitpur

Nibria   Krishnapur

Haora   Kankurgacchi

Podara   University

Raj Bhawan

Botanical Gardens   Birla Planetarium   Indian Museum   Dhapa

Hugli   St Paul's Cathedral

Garden Reach   Khidirpur   Kolkata (Calcutta)

National Library   Kasba

Alipur

Batanagar   Bartala   Taliganja

South Suburb   Garia

Chakdaha   0   2 Km

0   2 Miles

Kabul

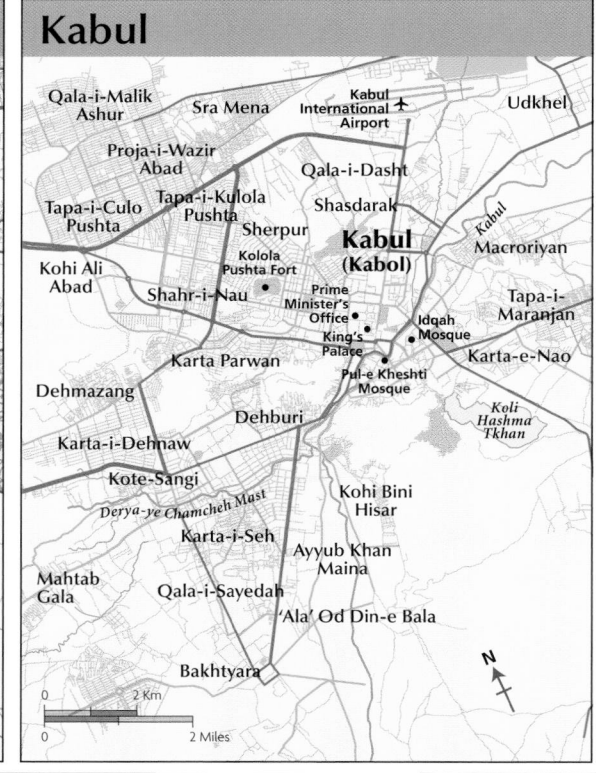

Qala-i-Malik Ashur   Sra Mena   Kabul International Airport   Udkhel

Proja-i-Wazir Abad   Qala-i-Dasht

Tapa-i-Culo Pushta   Tapa-i-Kulola Pushta   Shasdarak

Sherpur   Kabul   Kabul (Kabol)   Macroriyan

Kohi Ali Abad   Kolola Pushta Fort   Tapa-i-Maranjan

Shahr-i-Nau   Prime Minister's Office

Dehmazang   King's Palace   Idgah Mosque   Karta-e-Nao

Karta Parwan   Pul-e Kheshti Mosque

Dehburi   Koli Hashma Tkhan

Karta-i-Dehnaw

Kote-Sangi   Kohi Bini Hisar

Derya-ye Chamcheli Must   Karta-i-Seh   Ayyub Khan Maina

Mahtab Gala   Qala-i-Sayedah   'Ala' Od Din-e Bala

Bakhtyara   0   2 Km

N   0   2 Miles

A B C D E F G

Hong Kong

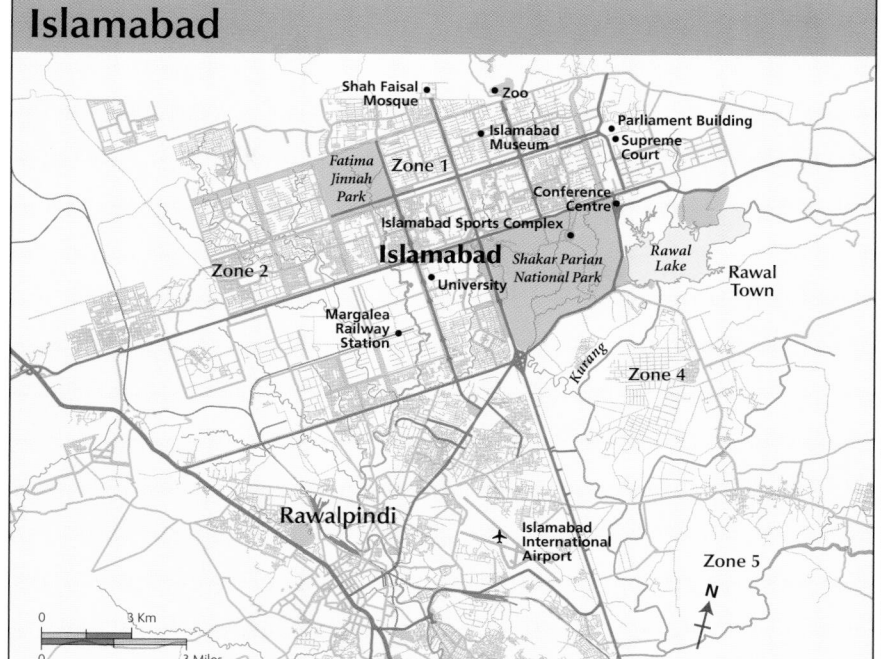

Mong Kok
Kowloon City
Kowloon
Ho Man Tin
Royal Observatory
Hung Hom
Kowloon Bay
Tsim Sha Tsui
Hong Kong Coliseum
Star Ferry
Space Museum & Planetarium
Sai Ying Pun
North Point
Kennedy Town
Victoria Harbour
Convention Centre
Jun Lo Wan
Tai Tam
University of Hong Kong
Sheung Wan
Government House
Hong Kong
Tai Hang
Mid Levels
Wan Chai
Victoria Peak 554m ▲
Happy Valley
Tiger Balm Garden
Hong Kong Island
0   1 Km
0   1 Miles

Istanbul

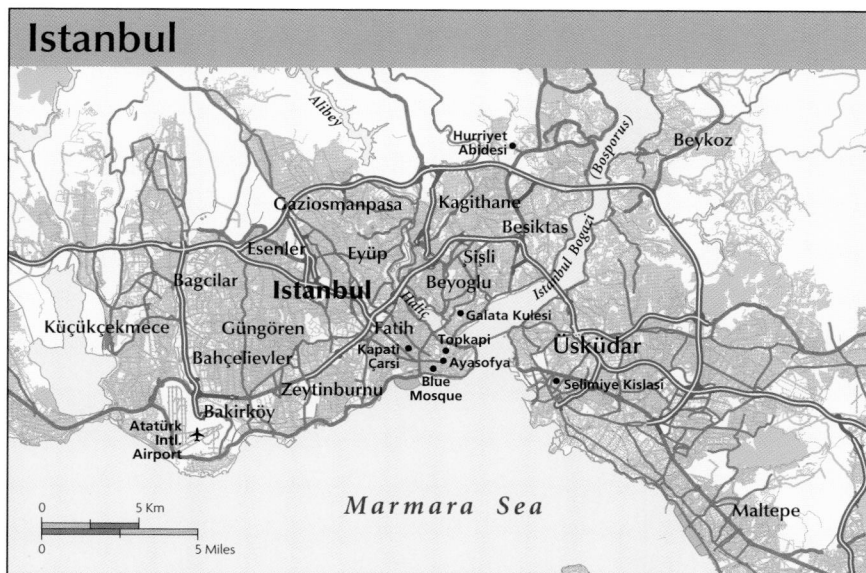

Alibey
Hürriyet Abidesi
Beykoz
Gaziosmanpasa
Kagithane
Besiktas
Esenler
Eyüp
Sisli
Beyoglu
Bagcilar
Istanbul
Galata Kulesi
Küçükcekmece
Güngören
Fatih
Topkapi
Üsküdar
Bahçelievler
Kapati Çarsi
Ayasofya
Zeytinburnu
Blue Mosque
Selimiye Kislasi
Bakirköy
Atatürk Intl. Airport
Maltepe
Marmara Sea
0   5 Km
0   5 Miles

Kuala Lumpur

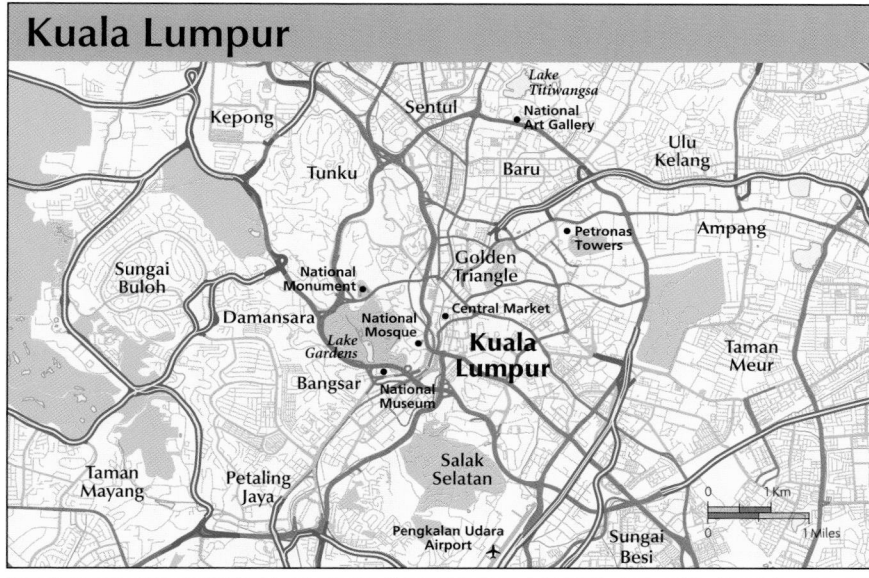

Lake Titiwangsa
Kepong
Sentul
National Art Gallery
Tunku
Baru
Ulu Kelang
Petronas Towers
Ampang
Sungai Buloh
National Monument
Golden Triangle
Central Market
Damansara
National Mosque
Kuala Lumpur
Lake Gardens
Bangsar
Taman Meur
National Museum
Taman Mayang
Petaling Jaya
Salak Selatan
Pengkalan Udara Airport
Sungai Besi
0   1 Km
0   1 Miles

Ulan Bator

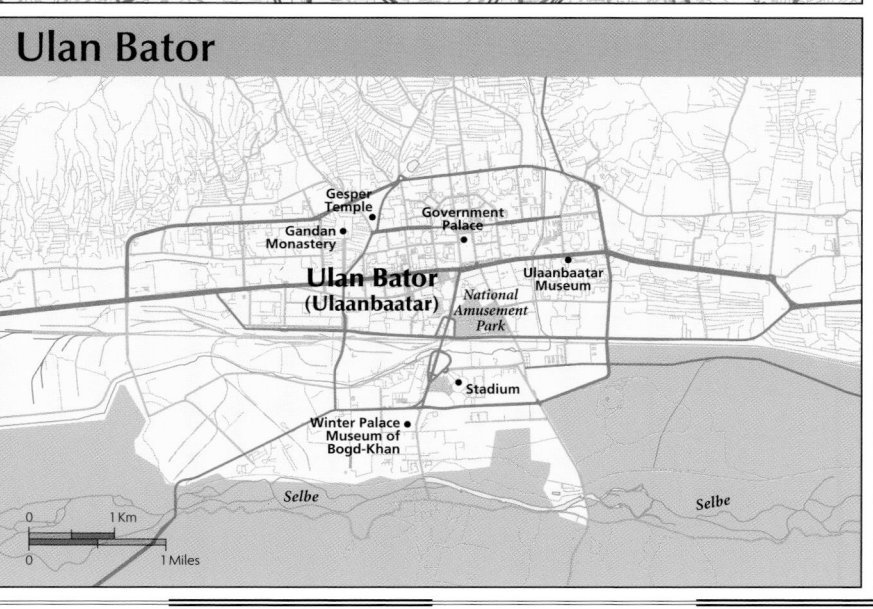

Gesper Temple
Government Palace
Gandan Monastery
Ulan Bator (Ulaanbaatar)
Ulaanbaatar Museum
National Amusement Park
Stadium
Winter Palace & Museum of Bogd-Khan
Selbe
Selbe
0   1 Km
0   1 Miles

Islamabad

Shah Faisal Mosque
Zoo
Parliament Building
Islamabad Museum
Supreme Court
Zone 1
Fatima Jinnah Park
Conference Centre
Islamabad Sports Complex
Zone 2
Islamabad
Shakar Parian National Park
Rawal Lake
Rawal Town
University
Margalea Railway Station
Zone 4
Kurang
Rawalpindi
Islamabad International Airport
Zone 5
N
0   3 Km
0   3 Miles

Jakarta

Java Sea
Teluk Jakarta
Soekarno-Hatta International Airport
Penjaringan
Pademangan
Ancol
Tanjung Priok
Kalideres
Taman Sari
Jakarta Museum
Sunter
Koja
Kembangan
Mardeka Palace
National Monument
Cempaka Putih
Kebon Jeruk
Welcome Monument
Jakarta
Kebayoran Lama
Parliament House
Menteng
University Rawamangun
Pulo Gadung
Matraman
Jatinegara
Kebayoran Baru
Pancoran
Manggarai
Kebayoran Baru
Ciliwung
Jakarta Halim Perdanakusuma Airport
Kramat Jati
Cilandak
Pasar Munggu
Ciracas
Pondok Cabe Airport
Jagakarsa
0   5 Km
0   5 Miles

Manila

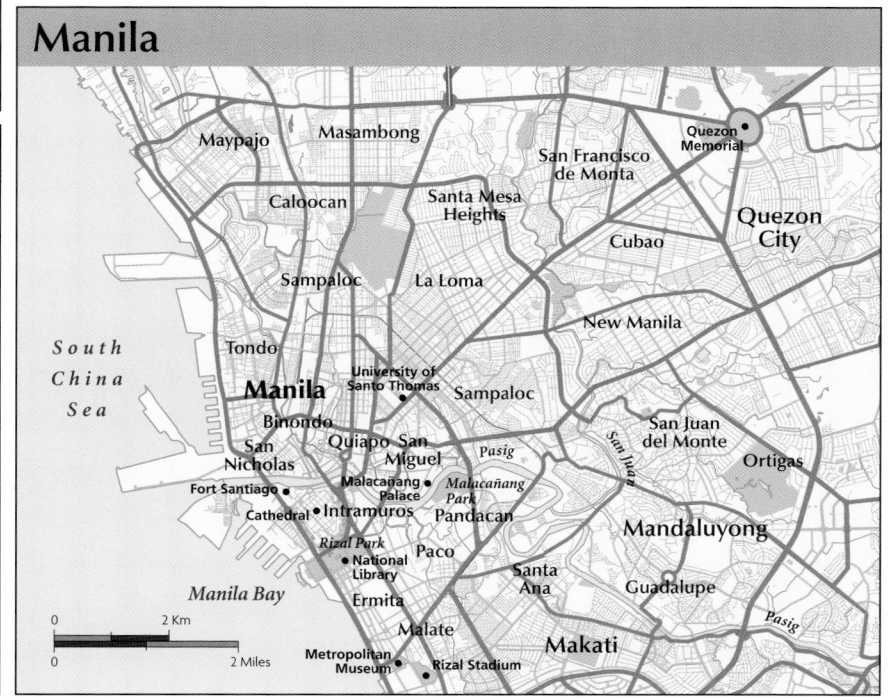

Maypajo
Masambong
San Francisco de Monta
Quezon Memorial
Caloocan
Santa Mesa Heights
Cubao
Quezon City
Sampaloc
La Loma
New Manila
Tondo
San China Sea
Manila
University of Santo Thomas
Sampaloc
Binondo
San Juan del Monte
Quiapo San Miguel
Pasig
Ortigas
San Nicholas
Malacañang Palace
Malacañang Park
Fort Santiago
Intramuros
Pandacan
Mandaluyong
Cathedral
Rizal Park
Paco
Santa Ana
National Library
Guadalupe
Manila Bay
Ermita
Pasig
Malate
Makati
Metropolitan Museum
Rizal Stadium
0   2 Km
0   2 Miles

AUSTRALASIA & OCEANIA

Australasia and Oceania with a total land area of 3,285,048 sq miles (8,508,238 sq km), takes in 14 countries including the continent of Australia, New Zealand, Papua New Guinea, and many island groups scattered across the Pacific Ocean.

FACTFILE

N **Most Northerly Point:** Eastern Island, Midway Islands 28° 15′ N

S **Most Southerly Point:** Macquarie Island, Australia 54° 30′ S

E **Most Easterly Point:** Clipperton Island, 109° 12′ W

W **Most Westerly Point:** Cape Inscription, Australia 112° 57′ E

Largest Lakes:
1 Lake Eyre, Australia 3430 sq miles (8884 sq km)
2 Lake Torrens, Australia 2200 sq miles (5698 sq km)
3 Lake Gairdner, Australia 1679 sq miles (4349 sq km)
4 Lake Mackay, Australia 1349 sq miles (3494 sq km)
5 Lake Argyle, Australia 800 sq miles (2072 sq km)

Longest Rivers:
1 Murray-Darling, Australia 2330 miles (3750 km)
2 Cooper Creek, Australia 880 miles (1420 km)
3 Warburton-Georgina, Australia 870 miles (1400 km)
4 Sepik, Indonesia/Papua New Guinea 700 miles (1126 km)
5 Fly, Indonesia/Papua New Guinea 652 miles (1050 km)

Largest Islands:
1 New Guinea, 312,000 sq miles (808,000 sq km)
2 South Island, New Zealand 56,308 sq miles (145,836 sq km)
3 North Island, New Zealand 43,082 sq miles (111,583 sq km)
4 Tasmania, Australia 24,911 sq miles (64,519 sq km)
5 New Britain, Papua New Guinea 13,570 sq miles (35,145 sq km)

Highest Points:
1 Mount Wilhelm, Papua New Guinea 14,793 ft (4509 m)
2 Mount Giluwe, Papua New Guinea 14,331 ft (4368 m)
3 Mount Herbert, Papua New Guinea 13,999 ft (4267 m)
4 Mount Bangeta, Papua New Guinea 13,520 ft (4121 m)
5 Mount Victoria, Papua New Guinea 13,360 ft (4072 m)

Lowest Point:
▼ Lake Eyre, Australia -53 ft (-16 m) below sea level

Highest recorded temperature:
➕ Bourke, Australia 128°F (53°C)

Lowest recorded temperature:
➖ Canberra, Australia -8°F (-22°C)

Wettest Place:
≋ Bellenden Ker, Australia 443 in (11,251 mm)

Driest Place:
➖ Mulka Bore, Australia 4.05 in (102.8 mm)

Cross-section from Dirk Hartog Island, Australia to Ducie Island, Pitcairn Islands

▷—▪—◁ line of cross-section

0 500 1000 1500 Km
0 500 1000 1500 Miles

H I J K L M N

150° 160° 170° 180° 170° 160° 150° 140° 130° 120°

Mapmaker Seamounts

Mid-Pacific Seamounts

Midway Islands

N

Wake Island

Hawaiian Islands

Murray Fracture Zone

40°

East Mariana Basin

Necker Ridge

Hawaiian Ridge

Molokai Fracture Zone

30°

Micronesia

Marshall Islands

Johnston Atoll

Schjetman Reef

Mauna Kea 4205m

Hawai'i

Tropic of Cancer

Magellan Seamounts

P A C I F I C

20°

Nauru

Banaba

Tungaru

Central Pacific Basin

Christmas Ridge

Clarion Fracture Zone

E

10°

Melanesian Basin

Phoenix Islands

Kiritimati

Clipperton Fracture Zone

Solomon Islands

Guadalcanal

Malaita

Santa Cruz Islands

Vityaz Trench

Tuvalu

O C E A N

Line Islands

Galapagos Fracture Zone

Equator 0°

South Solomon Trench

North New Hebrides Trench

Espíritu Santo

PACIFIC PLATE

FIJI PLATE

Robbie Ridge

Samoa Savaii Upolu

Northern Cook Islands

Manihiki Plateau

Marquesas Islands

Hiva Oa

Vanuatu

North Fiji Basin

Fiji

Vanua Levu

Samoa Basin

Penrhyn Basin

Polynesia

10°

Tanna

Viti Levu

Îles Loyauté

New Hebrides Ridge

Lau Basin

Tonga

Capricorn Tablemount

Southern Cook Islands

Rarotonga

Society Islands

Tahiti

Society Ridge

Tuamotu Islands

Tuamotu Ridge

Tiki Basin

Tuamotu Fracture Zone

New Caledonia

FIJI PLATE

Cook Fracture Zone

South Fiji Basin

Kermadec Ridge

Tonga Trench

Norfolk Ridge

Norfolk Island

Three Kings Rise

Kermadec Trench

Louisville Ridge

Les Australes

Îles Gambier

Austral Fracture Zone

20°

New Caledonia Basin

West Norfolk Ridge

Bay of Plenty

North Island

Pitcairn Island

Ducie Island

Henderson Island

Tropic of Capricorn

Lord Howe Rise

New Zealand

South Island

Taranaki (Mount Cook) 3714m

Southern Alps

Southwest Pacific Basin

NAZCA PLATE

East Pacific Rise

30°

South West Cape

Chatham Rise

Chatham Islands

Bounty Trough

Basin

Agassiz Fracture Zone

Macquarie Ridge

Campbell Plateau

S

Macquarie Island

Eltanin Fracture Zone

40°

S O U T H E R N   O C E A N

Udintsev Fracture Zone

PACIFIC PLATE

ANTARCTIC PLATE

ANTARCTICA

Pacific-Antarctic Ridge

130° 140° 150° 160° 170° 180° 170° 160° 150° 140° 130° 120°

Antarctic Circle

60°

H I J K L M N

1 2 3 4 5 6 7 8 9 10

Political

Vast expanses of ocean separate this geographically fragmented realm, characterized more by each country's isolation than by any political unity. Australia's and New Zealand's traditional ties with the United Kingdom, as members of the Commonwealth, are now being called into question as Australasian and Oceanian nations are increasingly looking to forge new relationships with neighboring Asian countries like Japan. External influences have featured strongly in the politics of the Pacific Islands; the various territories of Micronesia were largely under US control until the late 1980s, and France, New Zealand, the USA and the UK still have territories under colonial rule in Polynesia. Nuclear weapons-testing by Western superpowers was widespread during the Cold War period, but has now been discontinued.

Population
- above 5 million
- 1 million to 5 million
- 500,000 to 1 million
- 100,000 to 500,000
- 50,000 to 100,000
- 10,000 to 50,000
- below 10,000
- Country capital
- State capital

Borders
- full international border
- indication of maritime country extent
- indication of maritime dependent territory extent
- state border

Communications
- major roads
- major railroads

Scale 1:32,000,000
(projection: Lambert Azimuthal Equal Area)

Km
200 400 600 800
0
200 400 600 800
Miles

Map labels

Wake Island (to US)

Northern Mariana Islands (to US)

Saipan

Mariana Islands

Guam (to US)

HAGÅTÑA

Philippine Sea

Yap

Micronesia

MARSHALL ISLANDS

Bikini Atoll

Ratak Chain

Ralik Chain

Caroline Islands

Chuuk Pohnpei PALIKIR

Kosrae

MELEKEOK
Babeldaob

MICRONESIA

PALAU

Melanesia

Tarawa *Tungaru*

PACIFIC OCEAN

Baker & Howland Islands (to US)

NAURU

KIRIBATI

Phoenix Islands

Equator

PAPUA NEW GUINEA

TUVALU

Fongafale

Tokelau (to NZ)

Bismarck Sea

New Ireland

Wewak New Britain Rabaul

New Guinea Madang Ubai

Mount Hagen

Arawa
Bougainville Island

Solomon Islands

New Georgia Islands

HONIARA

Guadalcanal

SOLOMON ISLANDS

Lae

Solomon Sea

Tapini

PORT MORESBY

Amer Sam (to U

SAMOA APIA PA PA

Wallis and Futuna (to France)

Santa Cruz Islands

Banks Islands

Espiritu Santo

VANUATU

Malekula

Vanua Levu Labasa

Lautoka SUVA

Viti Levu *Lau Group*

TONGA N (to

Arafura Sea

Torres Strait

Coral Sea

Coral Sea Islands (to Australia)

Efate PORT-VILA

Erromango

Tanna *Iles Loyauté*

New Caledonia (to France)

NOUMÉA

FIJI

NUKU'ALOFA

PACIFIC

Timor Sea

Darwin

Arnhem Land

Joseph Bonaparte Gulf

Katherine

Gulf of Carpentaria

Cape York Peninsula

Great Barrier Reef

Cairns

Wyndham

Kimberley Plateau

Derby

Broome

NORTHERN

Normanton

Townsville

Hughenden Mackay

Rockhampton

Port Hedland

Tennant Creek

Tanami Desert

TERRITORY

Mount Isa

QUEENSLAND

Barcaldine

Norfolk Island (to Australia)

Kermadec Islands (to NZ)

INDIAN OCEAN

Great Sandy Desert

AUSTRALIA

Alice Springs

Charleville

Miles Brisbane

Toowoomba

Cunnamulla

Grafton

Lord Howe Island (to Australia)

Gibson Desert

WESTERN AUSTRALIA

Wilcannia

Grey Range

Bourke

NEW Dubbo

SOUTH WALES

Newcastle

North Island

Whangarei

Auckland

Hamilton Rotorua

New Plymouth

Hamersley Range

Carnarvon

Great Victoria Desert

SOUTH AUSTRALIA

Lake Eyre North

Flinders Ranges

Barwon

Darling

Wollongong

CANBERRA

AUSTRALIAN CAPITAL TERRITORY

Campbelltown

Port Augusta

Wagga Wagga

Hastings

Palmerston North

WELLINGTON

Tropic of Capricorn

Mount Magnet

Lake Torrens

Lake Everard Lake Gairdner

Whyalla

Murray

Ceduna

Adelaide

Kalgoorlie

Nullarbor Plain

Bendigo

VICTORIA

South Island

Southern Alps

Cook Strait

Geraldton

Great Australian Bight

Kangaroo Island

Horsham Ballarat

Melbourne

Geelong

Tasman Sea

Chatham Islands (to NZ)

Christchurch

NEW ZEALAND

Perth

Esperance

Mount Gambier

Albany

Bass Strait

Launceston TASMANIA

Tasmania Hobart

Dunedin

Invercargill

Stewart Island

Auckland Islands (to NZ)

SOUTHERN OCEAN

Sydney

Languages

English is spoken throughout Australia and New Zealand. In Australia, English has been superimposed on a mosaic of Aboriginal languages. In New Zealand, the indigenous language, Maori, is the official language besides English. In Papua New Guinea, Melanesian Pidgin has become a *lingua franca* alongside several hundred indigenous languages. Across the region, the indigenous languages can be grouped into (1) the Aboriginal languages of Australia, (2) the Papuan languages spoken mostly inland in Papua New Guinea, and (3) the widely dispersed Austronesian, which includes coastal languages of Papua New Guinea, New Zealand Maori, and languages of Oceania.

Language groups

- Australian
- Papuan
- Indo-European
- Austronesian

CHAMORRO

MARSHALLESE

GILBERTESE

EASTERN AUSTRONESIAN

TOK PISIN (PIDGIN)

PAPUAN

PIDGIN

ENGLISH

PIDGIN

ENGLISH

HINDI

FIJIAN

SAMOAN

TONGAN

TAHITIAN FRENCH

FRENCH

ENGLISH

MAORI

ENGLISH

Population

Density of settlement in the region is generally low. Australia is one of the least densely populated countries on Earth with over 80% of its population living within 25 miles (40 km) of the coast – mostly in the southeast of the country. New Zealand, and the island groups of Melanesia, Micronesia and Polynesia, are much more densely populated, although many of the smaller islands remain uninhabited.

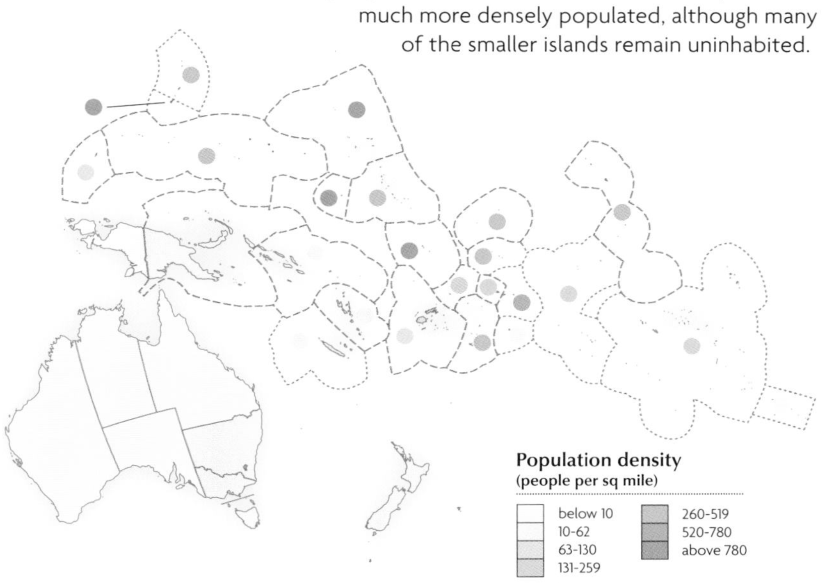

Population density
(people per sq mile)

- below 10
- 10–62
- 63–130
- 131–259
- 260–519
- 520–780
- above 780

Kingman Reef (to US)

Palmyra Atoll (to US)

· Teraina
· Tabuaran

KIRIBATI

Kiritimati

Jarvis Island (to US)

Line Islands

Equator

Malden Island

Starbuck Island

Northern Cook Islands

Penrhyn

Manihiki

Millennium Island

Flint Island

Cook Islands (to NZ)

Marquesas Islands

10°

Society Islands

PAPEETE

Tahiti

Tuamotu Islands

Southern Cook Islands

AVARUA

Rarotonga

Îles Australes

French Polynesia (to France)

Mururoa

Tropic of Capricorn

20°

Îles Gambier

Pitcairn, Henderson, Ducie & Oeno Islands (to UK)

· Pitcairn Island

O C E A N

30°

Standard of living

In marked contrast to its neighbor, Australia, with one of the world's highest life expectancies and standards of living, Papua New Guinea is one of the world's least developed countries. In addition, high population growth and urbanization rates throughout the Pacific islands contribute to overcrowding. The Aboriginal and Maori people of Australia and New Zealand have been isolated for many years. Recently, their traditional land ownership rights have begun to be legally recognized in an effort to ease their social and economic isolation, and to improve living standards.

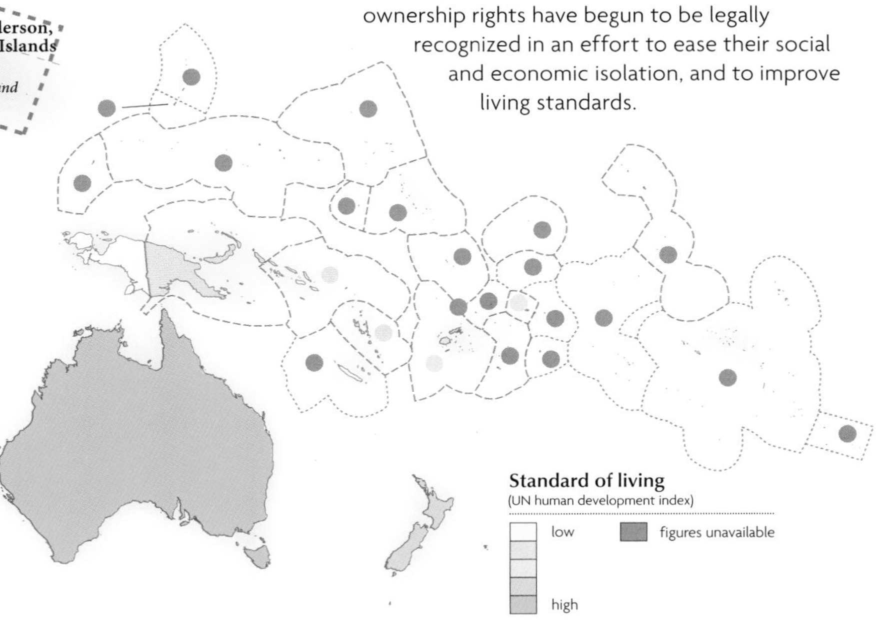

Standard of living
(UN human development index)

- low
- high
- figures unavailable

Transportation

While sea travel remains of paramount importance throughout the continent, well-developed regional and international air travel has reduced the region's global isolation. Internal air travel is particularly important in Australia, where distances are great and road systems are poorly developed or in some areas nonexistent. Australia's rail system, still operating on three different guages, a legacy of its piecemeal development, is being upgraded, particularly in the north-south links.

Scale 1:37,500,000
(projection: Lambert Azimuthal Equal Area)

0 200 400 600 800 1000 1200 Km

0 200 400 600 800 1000 1200 Miles

Climate

Surrounded by water, the climate of most areas is profoundly affected by the moderating effects of the oceans. Australia, however, is the exception. Its dry continental interior remains isolated from the ocean; temperatures soar during the day, and droughts are common. The coastal regions, where most people live, are cooler and wetter. The numerous islands scattered across the Pacific are generally hot and humid, subject to the different air circulation patterns and ocean currents that affect the area, including the El Niño ocean current anomaly, which produces extreme aridity.

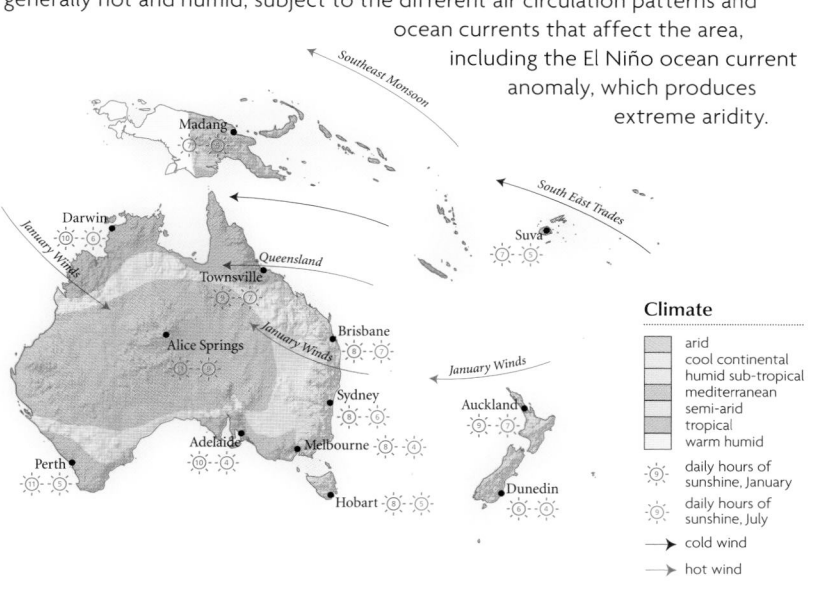

Southeast Monsoon
South East Trades
January Winds
Queensland
January Winds
January Winds
Madang
Darwin
Townsville
Suva
Alice Springs
Brisbane
Sydney
Auckland
Adelaide
Melbourne
Perth
Hobart
Dunedin
Climate
arid
cool continental
humid sub-tropical
mediterranean
semi-arid
tropical
warm humid
daily hours of sunshine, January
daily hours of sunshine, July
cold wind
hot wind

Average Rainfall

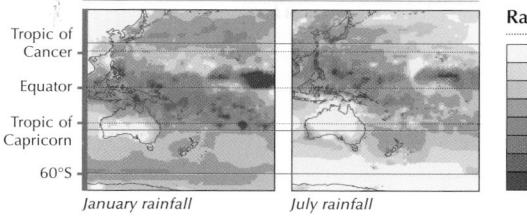

Tropic of Cancer
Equator
Tropic of Capricorn
60°S
January rainfall
July rainfall
Rainfall
0–1 in (0–25 mm)
1–2 in (25–50 mm)
2–4 in (50–100 mm)
4–8 in (100–200 mm)
8–12 in (200–300 mm)
12–16 in (300–400 mm)
16–20 in (400–500 mm)
more than 20 in (500 mm)

Average Temperature

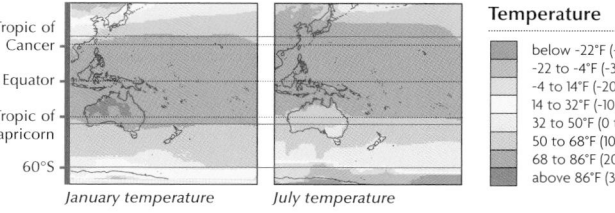

Tropic of Cancer
Equator
Tropic of Capricorn
60°S
January temperature
July temperature
Temperature
below -22°F (-30°C)
-22 to -4°F (-30 to -20°C)
-4 to 14°F (-20 to -10°C)
14 to 32°F (-10 to 0°C)
32 to 50°F (0 to 10°C)
50 to 68°F (10 to 20°C)
68 to 86°F (20 to 30°C)
above 86°F (30°C)

Environmental issues

The prospect of rising sea levels poses a threat to many low-lying islands in the Pacific. Nuclear weapons-testing, once common throughout the region, was finally discontinued in 1996. Australia's ecological balance has been irreversibly altered by the introduction of alien species. Although it has the world's largest underground water reserve, the Great Artesian Basin, the availability of fresh water in Australia remains critical. Periodic droughts combined with over-grazing lead to desertification and increase the risk of devastating bush fires, and occasional flash floods.

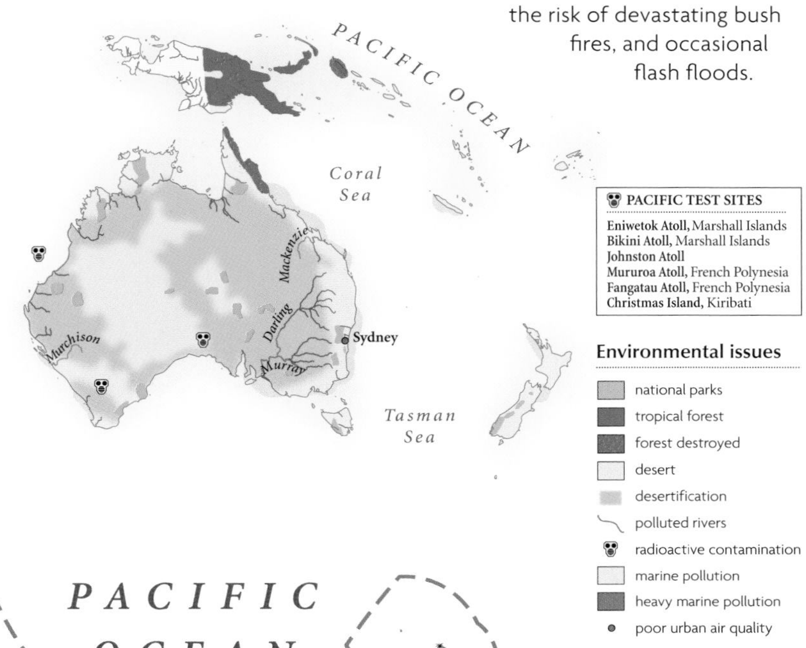

PACIFIC OCEAN
Coral Sea
Tasman Sea
Mackenzie
Darling
Murray
Murchison
Sydney
PACIFIC TEST SITES
Eniwetok Atoll, Marshall Islands
Bikini Atoll, Marshall Islands
Johnston Atoll
Mururoa Atoll, French Polynesia
Fangatau Atoll, French Polynesia
Christmas Island, Kiribati
Environmental issues
national parks
tropical forest
forest destroyed
desert
desertification
polluted rivers
radioactive contamination
marine pollution
heavy marine pollution
poor urban air quality

Land use

Much of the region's industry is resource-based: sheep farming for wool and meat in Australia and New Zealand; mining in Australia and Papua New Guinea and fishing throughout the Pacific islands. Manufacturing is mainly limited to the large coastal cities in Australia and New Zealand, like Sydney, Adelaide, Melbourne, Brisbane, Perth, and Auckland, although small-scale enterprises operate in the Pacific islands, concentrating on processing of fish and foods. Tourism continues to provide revenue to the area—in Fiji it accounts for 15 percent of GNP.

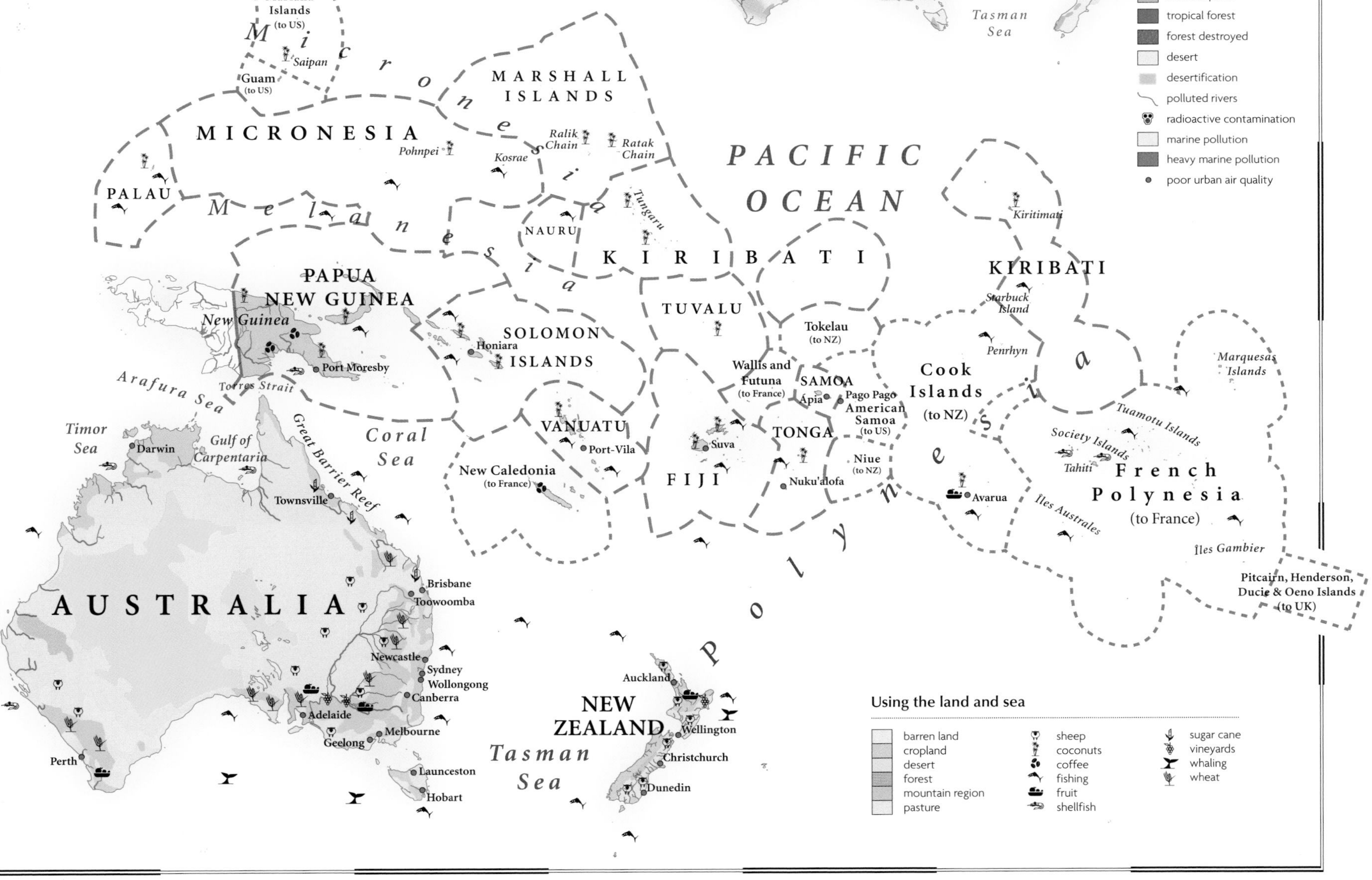

Northern Mariana Islands (to US)
Saipan
Guam (to US)
Micronesia
MARSHALL ISLANDS
MICRONESIA
Pohnpei
Kosrae
Ralik Chain
Ratak Chain
PALAU
Melanesia
NAURU
Tungaru
PACIFIC OCEAN
KIRIBATI
Kiritimati
KIRIBATI
Starbuck Island
PAPUA NEW GUINEA
New Guinea
Port Moresby
SOLOMON ISLANDS
Honiara
TUVALU
Tokelau (to NZ)
Wallis and Futuna (to France)
SAMOA
Apia
Pago Pago
American Samoa (to US)
Penrhyn
Cook Islands (to NZ)
Marquesas Islands
Tuamotu Islands
Society Islands
Tahiti
French Polynesia (to France)
Îles Australes
Îles Gambier
Pitcairn, Henderson, Ducie & Oeno Islands (to UK)
Arafura Sea
Torres Strait
Timor Sea
Darwin
Gulf of Carpentaria
Great Barrier Reef
Coral Sea
VANUATU
Port-Vila
New Caledonia (to France)
FIJI
Suva
TONGA
Nuku'alofa
Niue (to NZ)
Avarua
Polynesia
Townsville
AUSTRALIA
Brisbane
Toowoomba
Newcastle
Sydney
Wollongong
Canberra
Adelaide
Geelong
Melbourne
Perth
Launceston
Hobart
NEW ZEALAND
Auckland
Wellington
Christchurch
Dunedin
Tasman Sea
Using the land and sea
barren land
cropland
desert
forest
mountain region
pasture
sheep
coconuts
coffee
fishing
fruit
shellfish
sugar cane
vineyards
whaling
wheat

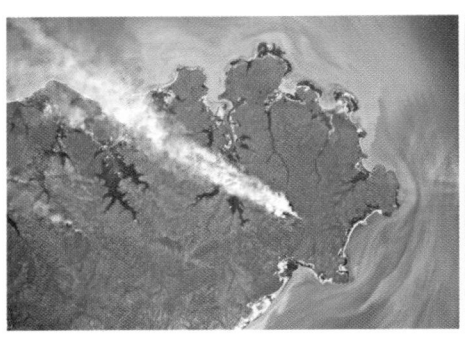

1 MELVILLE ISLAND, NORTHERN TERRITORY, AUSTRALIA
Lying off Australia's north coast, the island is sparsely populated consisting of sandy soils and mangrove swamps.

2 ANATAHAN, NORTHERN MARIANA ISLANDS
The volcano on Anatahan is one of 12 in the Mariana Islands and erupted on a large scale in April 2005.

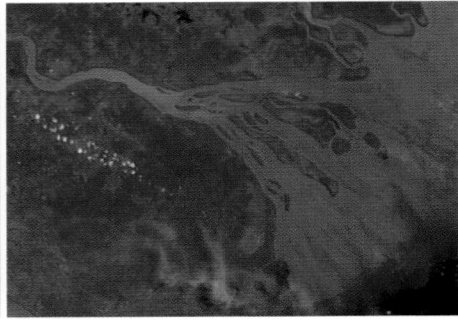

3 FLY RIVER, PAPUA NEW GUINEA
Flowing down from New Guinea's Central Range, the river carries a heavy load of sediment which it deposits in the Gulf of Papua, sometimes forming new islands.

4 RABAUL VOLCANO, NEW BRITAIN, PAPUA NEW GUIN
After erupting in 1994, this image shows how the high particles blew west causing condensation of water va over a wide area.

272

9 ULURU/AYERS ROCK, NORTHERN TERRITORY, AUSTRALIA
This enormous sandstone rock occupies Australia's heart, both physically and emotionally.

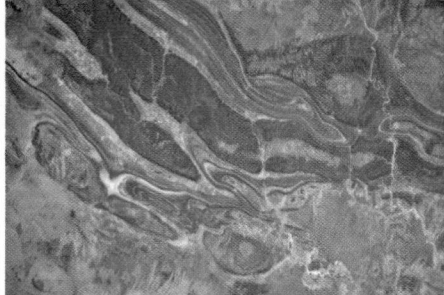

10 JAMES RANGES, NORTHERN TERRITORY, AUSTRALIA
A series of low ridges, these hills lie at the geographical center of Australia.

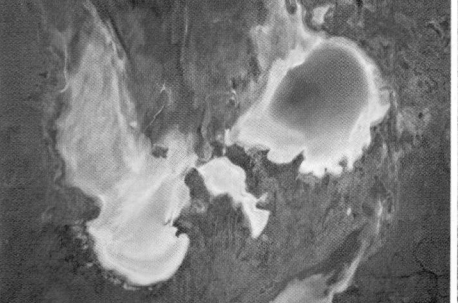

11 LAKE EYRE, SOUTH AUSTRALIA, AUSTRALIA
This great salt lake consists of north and south sections, joined by a narrow channel, Lake Eyre South being the smaller, elongated saltflat at the bottom of the image.

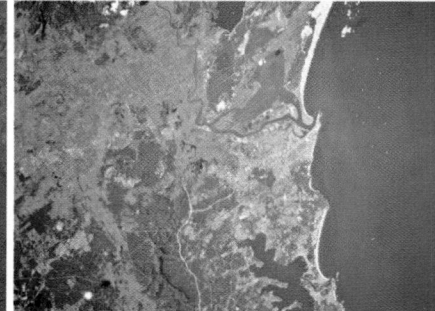

12 NEWCASTLE, NEW SOUTH WALES, AUSTRALIA
The industrial seaport of Newcastle lies on the south bank of Hunter river.

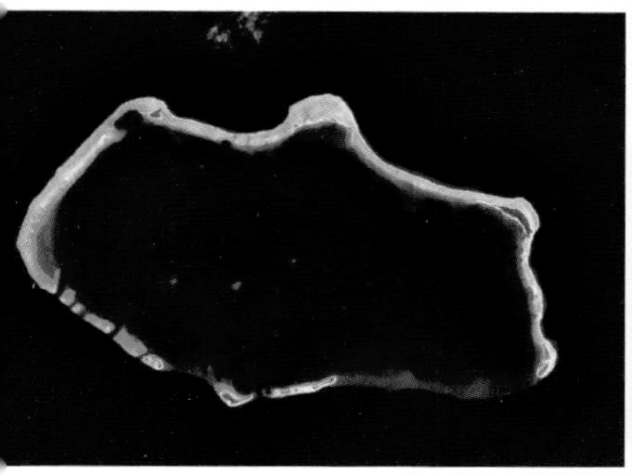

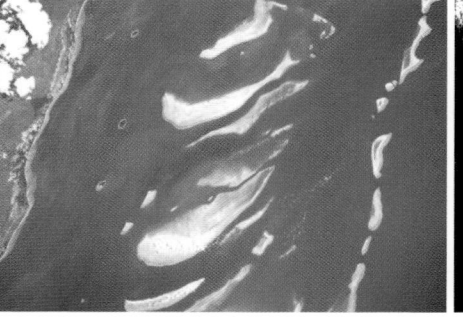

BIKINI ATOLL, MARSHALL ISLANDS 5
his atoll was the site of 23 atomic bomb tests in the
40s and 1950s, involving the intentional sinking of at
least 13 naval vessels in the shallow lagoon.

GREAT BARRIER REEF, QUEENSLAND, AUSTRALIA 6
The world's largest reef system is made up
of 3000 individual reefs and 900 islands and
stretches for 1600 miles (2600 km).

AMBRYM, VANUATU 7
Mount Marum, a 4166 ft (1270 m) volcano,
erupted in April 2004 producing an
extensive plume of ash.

KIRITIMATI, KIRIBATI 8
Kiritimati is the largest atoll in the
Pacific Ocean, its interior lagoon filled
in with coral growth.

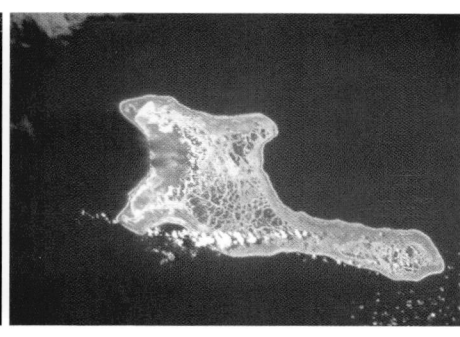

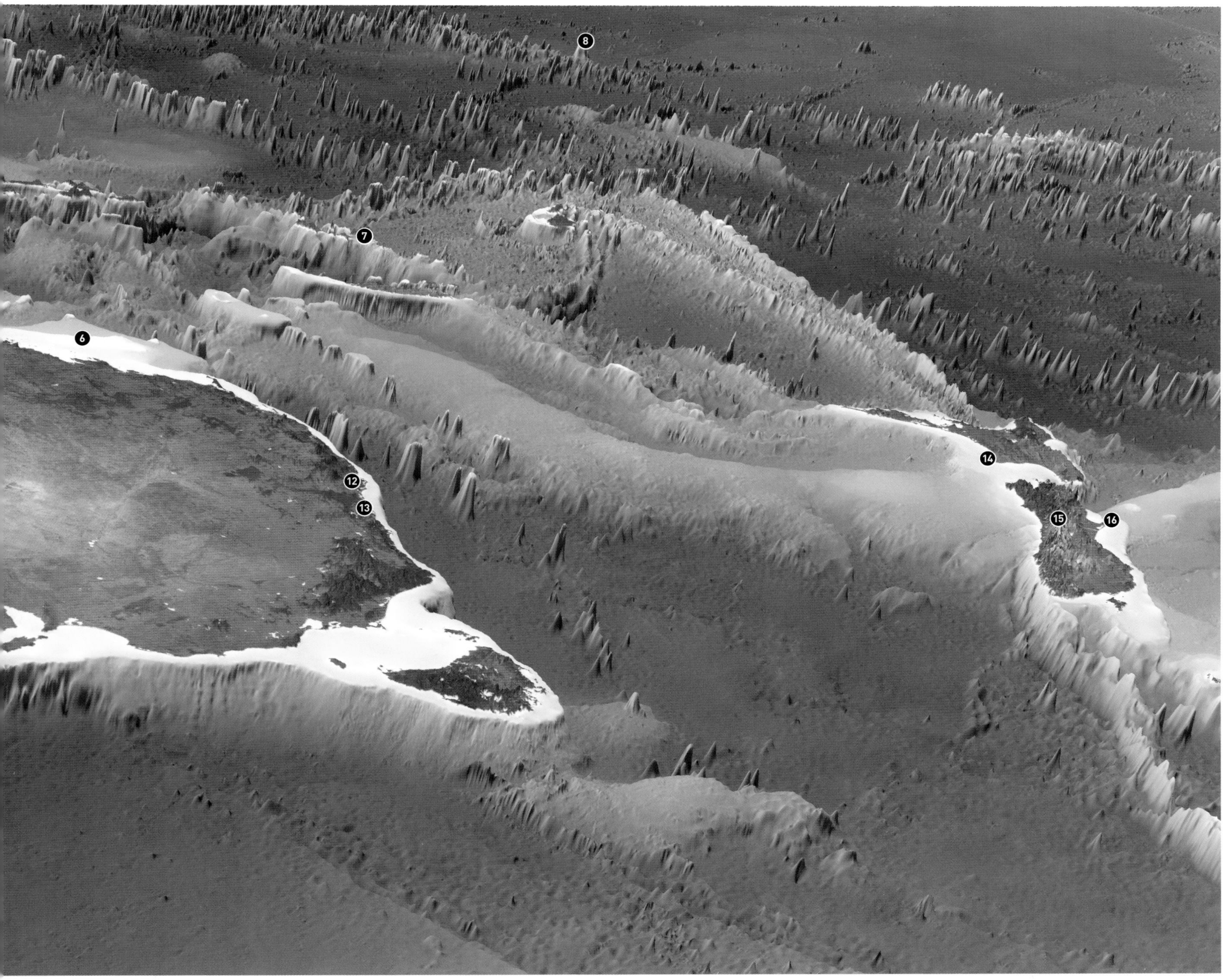

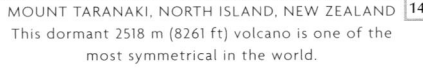

SYDNEY, NEW SOUTH WALES, AUSTRALIA 13
Expanding outward from the inlet of Port
Jackson, Australia's largest city was founded
in 1788.

MOUNT TARANAKI, NORTH ISLAND, NEW ZEALAND 14
This dormant 2518 m (8261 ft) volcano is one of the
most symmetrical in the world.

**AORAKI/MOUNT COOK,
SOUTH ISLAND, NEW ZEALAND** 15
New Zealand's highest peak rises 12,238 ft (3744 m)
and is surrounded by permanent ice fields.

BANKS PENINSULA, SOUTH ISLAND, NEW ZEALAND 16
With a circular drainage pattern typical of eroded
volcanoes, this is the only recognizably volcanic
feature on New Zealand's South Island.

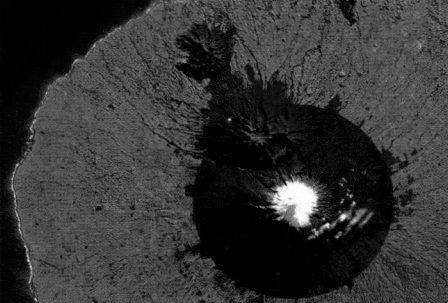

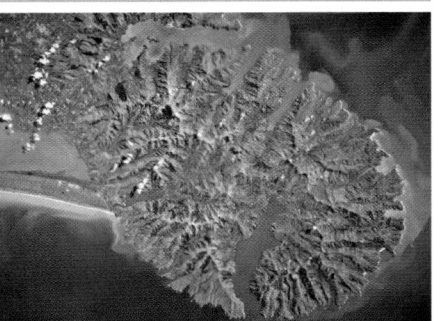

Australia

Perth

INDONESIA

INDIAN OCEAN

Java Sea

Bali Sea (Laut Bali)

Nusa Tenggara (Lesser Sunda Islands)

Savu Sea

EAST TIMOR

Timor Sea

CHRISTMAS ISLAND (to Australia)

JAKARTA
Serang Tangerang Bogor Cianjur Sukabumi Garut Ciamis Cilacap Bandung Cirebon Tegal Brebes Pekalongan Semarang Purwodadi Magelang Surakarta Yogyakarta Jombang Madiun Kediri Malang Jember Probolinggo Pasuruan Surabaya Denpasar Mataram

Kepulauan Seribu Sukarno-Hatta

Pulau Madura

Gunung Raung Bali Ngurah Rai Pulau Lombok

Sumbawa Raba Bima Endeh

Kepulauan Kangean

Kepulauan Tengah

Larantuka Labala Kabir Kalabahi

Kepulauan Alor Selat Wetar

Waikabubak Pulau Sumba Waingapu Kupang Baing

Pulau Roti Kepulauan Sawu Sawu Baa

DILI Manatuto Suai Kefamenanu Soe

Yatoke Amplawas Eliase

Kepulauan Tanimbe

Saumlaki

Kupang

Melville Island

Van Diemen Gulf

Bathurst Island

Beagle Gulf

Darwin

Noonamah

AdelaideRiver

PineCreek

Cape Bougainville Cape Londonderry

Joseph Bonaparte Gulf

Bonaparte Archipelago

Bigge Island

Heywood Islands

Adele Island

Collier Bay

Kalumburu

Mount Hann 779m

Kimberley Plateau

Kupingarri

King Leopold Ranges

Durack Range

Bungle Bungle

Mount Wells 970m

Lake Argyle

Turkey Creek

Lombadina

Derby

Broome

Fitzroy River

Fitzroy Crossing

Halls Creek

Kununurra

Wyndham

Victoria River Roadhouse

Timber Creek

Top Springs Roadhouse

Kalkarindji

NORTHERN

AUST

Tanami Desert

King Sound

Eighty Mile Beach

Great Sandy Desert

Port Hedland

De Grey River

Dampier Archipelago Dampier Wickham Karratha Roebourne Whim Creek

Barrow Island

MarbleBar

Percival Lakes

Tobin Lake

Lake Mackay

Mount Liebig 1274m Mount Ze 1531

Glen Hel

Yuend

Lake Dora Lake Auld

Lake Disappointment

Lake Macdonald

Fortescue River

Onslow

Witternoom

Hamersley Range

TomPrice Paraburdoo

Mount Meharry 1251m

Newman

Little Sandy Desert

Gibson Desert

Hopkins Lake

Lake Neale

Hermannsb James Rang

Lake Amadeu

North West Cape

Exmouth

Learmonth

Coral Bay

Ashburton River

Kenneth Range

Barlee Range

Mount Augustus 1105m Waldburg Range

Lake

Kumarina Roadhouse

Cafnarvon Range

Lake Gregory

WESTERN AUSTRALIA

Warburton

Kata Tjuta (Mount Olga) 1069m

Yulara Uluru (Ayers Ro 867m

Musgrave Ran Mount Morris 1288m

Erld Roadh

Tomkinson Ranges

Minilya

Lake Macleod

Gascoyne River

Robinson Range

Lake Carnegie

Everard Rang

Carnarvon

Gascoyne Junction

Wiluna

Lake Way

Lake Wells

Bernier Island

Dorre Island

Shark Bay

Denham

Murchison River

Lake Annean

Meekatharra

Lake Austin

Lake Throssell

LakeYeo

Great Victoria Desert

Dirk Hartog Island

SteepPoint

Yalgoo

Mongers Lake

Mount Magnet

Lake Carey

Lake Maurice

SOUTH

Tropic of Capricorn

Kalbarri

Geraldton

Lake Barlee

Leonora Lake Ballard

Menzies

Lake Rebecca

Nullarbor Plain

Watson

Moora

Wubin Pithara

Lake Moore

Kalgoorlie

Loongana Rawlinna

Kitchener

Reid

Eucla

Watson

The Pinnacles

Gingin Wanneroo

Merredin

Coolgardie Kambalda

Lake Lefroy

Madura Cocklebiddy

Boo

Southern Cross

Lake Cowan

Norseman Lake Dundas

Balladonia Caiguna

Perth

Fremantle Rockingham Mandurah

Bunbury Collie Busselton

Margaret River Cape Leeuwin Augusta

Darling Range

Northam York Brookton

Kondinin

Narrogin Wagin Katanning

Bridgetown Manjimup Pemberton

Mount Barker

Lake Hope

Lake King Ravensthorpe

Stirling Range

Tower Peak 594m

Esperance

Great Australian Bigh

Albany

259
289
293

Perth (inset map)

Indian Ocean

Wembley Downs Joondanna Bayswater

Herdsman Lake

City Beach

Lake Monger Jolimont North Perth

Alderbury Park Subiaco

Kings Park

Art Gallery of Western Australia **Perth** Belmont Park Redcliffe

Claremont

Zoo South Perth Kensington Carlisle

Cottesloe Dalkeith Swan River Cannington

Mosman Park Applecross Manning

Fremantle Myaree Partwood

Piney Lakes Reserve Canning River

Hilton Bull Creek

Murdoch

3 Miles

Km

Elevation scale:
6000m
4000m
3000m
2000m
1000m
500m
250m
100m
Sea Level
-250m
-2000m
-4000m

Scale 1:13,000,000
(projection: Lambert Conformal Conic)

0 50 100 150 200 250 300 350 400 Km
0 50 100 150 200 250 300 400 Miles

Population
■ above 5 million   ◘ 1 million to 5 million   ◉ 500,000 to 1 million
◎ 100,000 to 500,000   ⊕ 50,000 to 100,000   ○ 10,000 to 50,000   ∘ below 10,000

280

286

293

Arafura Sea

New Guinea

PAPUA NEW GUINEA

Torres Strait

PORT MORESBY

Gulf of Papua

Solomon Sea

SOLOMON ISLANDS

HONIARA
Guadalcanal

SOLOMON ISLANDS

Gulf of Carpentaria

Cape York Peninsula

Great Barrier Reef

Coral Sea

CORAL SEA ISLANDS
(to Australia)

Cairns

Townsville

Mount Isa

QUEENSLAND

Great Artesian Basin

Mackay

Rockhampton

Gladstone

Bundaberg

Hervey Bay
Fraser Island

Maryborough

Gympie
Sunshine Coast
Maroochydore-Mooloolaba

Toowoomba

Brisbane
Ipswich
Gold Coast
Surfers Paradise
Cape Byron

NEW CALEDONIA
(to France)

PACIFIC OCEAN

Lake Eyre Basin

Simpson Desert

Sturt Stony Desert

Lake Eyre North

Lake Eyre South

Coober Pedy

AUSTRALIA

Port Augusta

Whyalla

Port Pirie

Broken Hill

NEW SOUTH WALES

Dubbo

Bathurst

Orange

Newcastle

Sydney
Parramatta
Wollongong

CANBERRA
AUSTRALIAN CAPITAL TERRITORY
JERVIS BAY TERRITORY

Grafton

Coffs Harbour

Port Macquarie

Taree

Lord Howe Island
(to Australia)

Adelaide

Kangaroo Island

Mount Gambier

Warrnambool

Melbourne
Geelong

VICTORIA

Bass Strait

King Island

Flinders Island

Devonport
Launceston

TASMANIA

Hobart

Tropic of Capricorn

Brisbane

Chermside
Everton Park
The Gap
Red Hill
Lutwyche
Clayfield
Newstead
Hawthorne
Brisbane
Botanical Gardens
Queensland Art Gallery
Woolloongabba
Indooroopilly
Greenslopes
Carina Heights
Corinda
Mount Gravatt
Toombul
Brisbane Airport
Myrtletown
Wynnum
Tingalpa
Manly
Belmont
Tingalpa Reservoir
Burbank

19,686ft
13,124ft
9843ft
6562ft
3281ft
1640ft
820ft
328ft
Sea Level
-820ft
-6562ft
-13,124ft

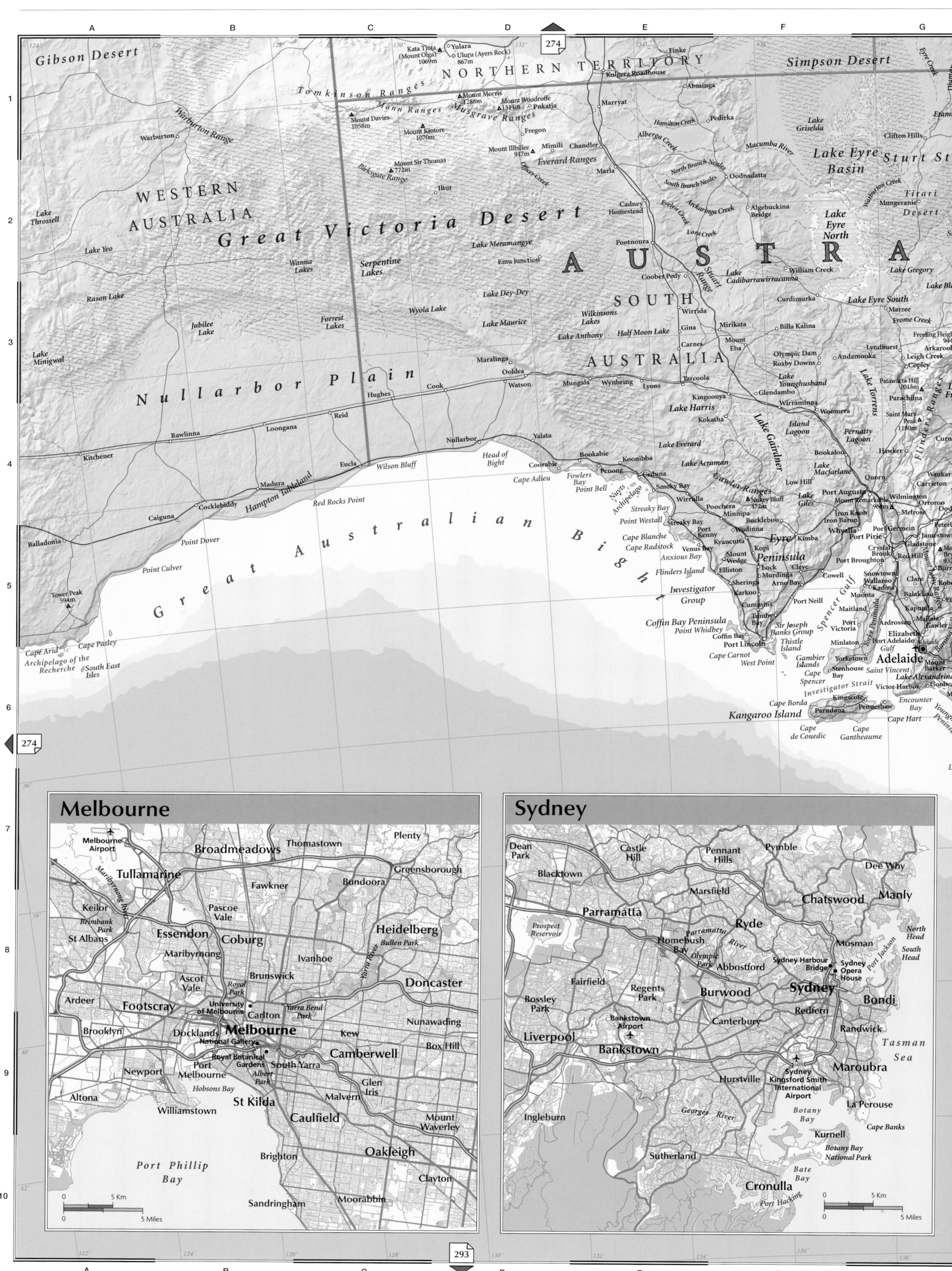

Southeast Australia

274

AUSTRALASIA & OCEANIA

276

274

Gibson Desert

NORTHERN TERRITORY

Simpson Desert

Kata Tjuta
(Mount Olga)
1069m

Yulara

Uluru (Ayers Rock)
867m

Finke

Eyre Creek

Tomkinson Ranges

Mount Morris
1288m

Mount Woodroffe
1514m

Pukatja

Kulgera Roadhouse

Mann Ranges

Musgrave Ranges

Marryat

Abminga

WESTERN
AUSTRALIA

Mount Davies
1058m

Mount Kintore
1070m

Fregon

Mimili

Chandler

Hamilton Creek

Pedirka

Lake
Griselda

Clifton Hills

Warburton Range

Mount Sir Thomas
772m

Mount Illbillee
917m

Everard Ranges

Marla

Alberga Creek

Macumba River

**Lake Eyre
Basin**

Sturt Sto

Warburton

Birksgate Range

Iltur

 Officer Creek

Cadney
Homestead

Evelyn Creek

North Branch Neales

Oodnadatta

Arckaringa Creek

Algebuckina
Bridge

Warburton Creek

Mungeranie

Eringa

Tirari
Desert

Lake
Throssell

Great Victoria Desert

Lake Meramangye

South Branch Neales

AUSTRA

**Lake Eyre
North**

Lake Yeo

Wanna
Lakes

Serpentine
Lakes

Emu Junction

Coober Pedy

Marla

SOUTH

William Creek

Lake Gregory

Strz

Rason Lake

Lake Dey-Dey

Wilkinsons
Lakes

Wirrida

Curdimurka

Lake Eyre South

Marree

Frome Creek

Jubilee
Lake

Forrest
Lakes

Wyola Lake

Lake Maurice

Lake Anthony

Half Moon Lake

Gina

Mirikata

Billa Kalina

AUSTRALIA

Freeling Heights
944m

Arkaroola

Lake Minigwal

Lake Dey-Dey

Carnes

Mount
Eba

Olympic Dam
Roxby Downs

Andamooka

Lyndhurst

Leigh Creek

Copley

Maralinga

Lake
Younghusband

Parachilna

Saint Mary
Peak
1180m

Patawarta Hill
1015m

Ooldea

Watson

Mungala

Wynbring

Lyons

Tarcoola

Kingoonya

Glendambo

Woomera

Marree

Hughes

Cook

Kokatha

Lake Harris

Island
Lagoon

Pernatty
Lagoon

Hawker

Curnam

Rawlinna

Loongana

Reid

Lake Everard

Low Hill

Bookaloo

Quorn

Nullarbor Plain

Yalata

Bookabie

Koonibba

Penong

Ceduna

Lake Acraman

Gawler Ranges

Nukey Bluff
472m

Lake
Gairdner

Lake
Macfarlane

Mount Remarkable
960m

Port Augusta

Wilmington

Orroroo

Melrose

Kitchener

Madura

Hampton Tableland

Eucla

Wilson Bluff

Head of
Bight

Coorabie

Fowlers
Bay
Point Bell

Smoky Bay

Wirrulla

Poochera

Minnipa

Buckleboo

Iron Knob

Iron Baron

Whyalla

Port Pirie

Red Hill

Jamestown

Peterbor

Caiguna

Cocklebiddy

Red Rocks Point

Cape Adieu

Nuyts
Archipelago

Point Westall

Streaky Bay

Streaky
Bay

Wudinna

Kopi

Kimba

Cowell

Crystal
Brook

Gladstone

Oodla

Balladonia

Point Dover

Cape Blanche
Cape Radstock

Port
Kenny

Venus Bay

Mount
Wedge

Eyre

Lock

Cleve

Arno Bay

Snowtown
Wallaroo

Clare

Burra

Great Australian Bight

Anxious Bay

Elliston

Murdinga

Karkoo

Moonta
Kadina

Balaklava

Pudu

Tower Peak
594m

Point Culver

Flinders Island

Sheringa

Cummins

Port Neill

Maitland

Kapunda

Cape Arid

Cape Pasley

Archipelago of the
Recherche

South East
Isles

**Investigator
Group**

Coffin Bay Peninsula
Point Whidbey

Coffin Bay

Port Lincoln

Tumby
Bay

Sir Joseph
Banks Group

Peninsula

Snowtown

Port
Victoria

Ardrossan

Mullaroo

Fawler St

Cape Carnot
West Point

Gambier
Islands

Minlaton

Port Adelaide

Spencer Gulf

Elizabeth

Cape
Spencer

Stenhouse
Bay

Yorketown

Saint Vincent

Adelaide

M

Cape Borda

Kingscote

Parndana

Penneshaw

Victor Harbor

Goolwa

Lake Alexandrina

Mount
Barker

Men

Kangaroo Island

Cape
de Couedic

Cape
Gantheaume

Cape Hart

Encounter
Bay

Younghusband
Peninsula

Lacu

Melbourne

Melbourne
Airport

Broadmeadows

Thomastown

Plenty

Tullamarine

Fawkner

Bondoora

Greensborough

Keilor

Maribyrnong Bay

Pascoe
Vale

Brimbank
Park

St Albans

Essendon

Coburg

Heidelberg

Bullen Park

Maribyrnong

Ivanhoe

Doncaster

Ascot
Vale

Royal
Park

Brunswick

Ardeer

Footscray

University
of Melbourne

Carlton

Yarra Bend
Park

Nunawading

Brooklyn

Docklands
National Gallery

Melbourne

Kew

Box Hill

Newport

Port
Melbourne

Royal Botanical
Gardens

South Yarra

Camberwell

Altona

Albert
Park

Glen
Iris

Williamstown

Hobsons Bay

Malvern

Mount
Waverley

St Kilda

Caulfield

Brighton

Oakleigh

Clayton

**Port Phillip
Bay**

Sandringham

Moorabbin

0   5 Km

0   5 Miles

Sydney

Dean
Park

Castle
Hill

Pennant
Hills

Pymble

Blacktown

Dee Why

Marsfield

Chatswood

Manly

Parramatta

Ryde

Parramatta

Homebush
Bay

Mosman

North
Head

Prospect
Reservoir

Olympic
Park

Abbotsford

Sydney Harbour
Bridge

Sydney
Opera
House

South
Head

Fairfield

Regents
Park

Burwood

Sydney

Port Jackson

Bossley
Park

Bankstown
Airport

Canterbury

Redfern

Bondi

Liverpool

Bankstown

Randwick

Tasman
Sea

Ingleburn

Sydney
Kingsford Smith
International
Airport

Hurstville

Maroubra

Georges River

La Perouse

Botany
Bay

Cape Banks

Kurnell

Sutherland

Botany Bay
National Park

Cronulla

Bate
Bay

Port Hacking

0   5 Km

0   5 Miles

6000m
4000m
3000m
2000m
1000m
500m
250m
100m
Sea
Level
-250m
-2000m
-4000m

293

Scale 1:6,500,000
(projection: Lambert Conformal Conic)

0 25 50 75 100 125 150 175 200 Km
0 25 50 75 100 125 150 175 200 Miles

Population
- ☐ above 5 million
- ☐ 1 million to 5 million
- ◉ 500,000 to 1 million
- ◎ 100,000 to 500,000
- ⊕ 50,000 to 100,000
- ○ 10,000 to 50,000
- ○ below 10,000

275

293

278

QUEENSLAND

NEW SOUTH WALES

VICTORIA

AUSTRALIAN CAPITAL TERRITORY

JERVIS BAY TERRITORY

TASMANIA

Tasmania

Great Dividing Range

Nandewar Range

Liverpool Range

Darling Downs

Riverina

Bass Strait

Tasman Sea

Brisbane
Ipswich
Toowoomba
Gold Coast
Surfers Paradise
Hervey Bay
Fraser Island
Maryborough
Gympie
Sunshine Coast
Caloundra
Bribie Island
Moreton Island
North Stradbroke Island
South Stradbroke Island
Tweed Heads
Byron Bay
Ballina
Casino
Lismore
Grafton
Coffs Harbour
Nambucca Heads
Port Macquarie
Taree
Forster-Tuncurry
Newcastle
Maitland
Cessnock
Gosford
Sydney
Parramatta
Penrith
Campbelltown
Wollongong
Port Kembla
Kiama
Nowra-Bomaderry
Goulburn
Canberra
Queanbeyan
Batemans Bay
Moruya
Narooma
Bega
Eden

Melbourne
Geelong
Ballarat
Bendigo
Wodonga
Albury
Wagga Wagga
Shepparton
Wangaratta
Warrnambool
Portland

Broken Hill
Mildura
Wentworth
Dubbo
Orange
Bathurst
Tamworth
Armidale
Moree
Bourke

Hobart
Launceston
Devonport
Burnie
Flinders Island
King Island
Cape Barren Island

Mount Kosciuszko 2228m

142° 144° 146° 148° 150° 152° 154° 156° 158°

Elevation scale:
19,686ft
13,124ft
9843ft
6562ft
3281ft
1640ft
820ft
328ft
Sea Level
-820ft
-6562ft
-13,124ft

AUSTRALASIA & OCEANIA

North Island

NEW ZEALAND

Auckland

Tasman Sea

NORTHLAND

AUCKLAND

WAIKATO

BAY OF PLENTY

GISBORNE

HAWKE'S BAY

TARANAKI

MANAWATU-WANGANUI

Coromandel Peninsula

Hawke Bay

Bay of Plenty

Elevation scale
- 6000m
- 4000m
- 3000m
- 2000m
- 1000m
- 500m
- 250m
- 100m
- Sea Level
- -250m
- -2000m
- -4000m

Auckland inset labels
Motutapu Island, Motuihe Island, Browns Island, Rangitoto Island, Waitemata Harbour, Kohimarama Bay, Karaka Bay, Mission Bay, Bucklands Beach, Glenn Innes, Howick, Highland Park, Pakuranga, Huntington Park, Otara, Clover Park, Papatoetoe, Manukau, Mangere, Ambury Park, Onehunga, Mount Wellington, Ellerslie, Remuera, Grafton, Auckland Museum, Maritime Museum, Skytower, Auckland, Auckland Zoo, Mount Eden, Mount Roskill, Royal Oak, Stardome & Auckland Observatory, One Tree Hill, St Johns Park, Eden Park, Devonport, Takapuna, Lake Pupuke, North Shore, Northcote, Kauri Park, Point Chevalier, Blockhouse Bay, Avondale, Green Bay, Te Atatu, Chester Park, Manukau Harbour, Auckland Intl. Airport, Tamaki River

3 Km / 3 Miles

Scale 1:3,200,000
(projection: Lambert Conformal Conic)

0 20 40 60 80 100 Km
0 20 40 60 80 100 Miles

Population
- ■ above 5 million
- ■ 1 million to 5 million
- ● 500,000 to 1 million
- ⊚ 100,000 to 500,000
- ⊕ 50,000 to 100,000
- ○ 10,000 to 50,000
- ○ below 10,000

PACIFIC OCEAN

South Island

MARLBOROUGH
TASMAN
WEST COAST
CANTERBURY
Canterbury Plains
Canterbury Bight
Christchurch
Banks Peninsula
Pegasus Bay
OTAGO
SOUTHLAND
Dunedin
Invercargill
Foveaux Strait
Stewart Island
Ruapuke Island

Wellington

Wellington Harbour
(Port Nicholson)

Karori
Chartwell
Wilton
Northland
Botanic Gardens
Lambton
Kelburn
Highbury
Thorndon
Parliament Buildings
Beehive
Museum of New Zealand
Te Aro
Wellington
Oriental Bay
Mount Victoria
Hataitai
Roseneath
Evans Bay
Point Halswell
Kau Bay
Mount Crawford
Scorching Bay
Karaka Bay
Worser Bay
Seatoun
Miramar
Strathmore Park
Point Dorset
Breaker Bay
Lyall Bay
Kilbirnie
Rongotai
Wellington Int. Airport
Melrose
Government House
Mitchelltown
Brooklyn
Kowhai Park
Wellington South
Mornington
Newtown
Wellington Zoo
Southgate
Houghton Bay
Lyall Bay
Kingston
Happy Valley
Island Bay
Owhiro Bay
Owhiro Bay

19,686ft
13,124ft
9843ft
6562ft
3281ft
1640ft
820ft
328ft
Sea Level
-820ft
-6562ft
-13,124ft

Papua New Guinea & Melanesia

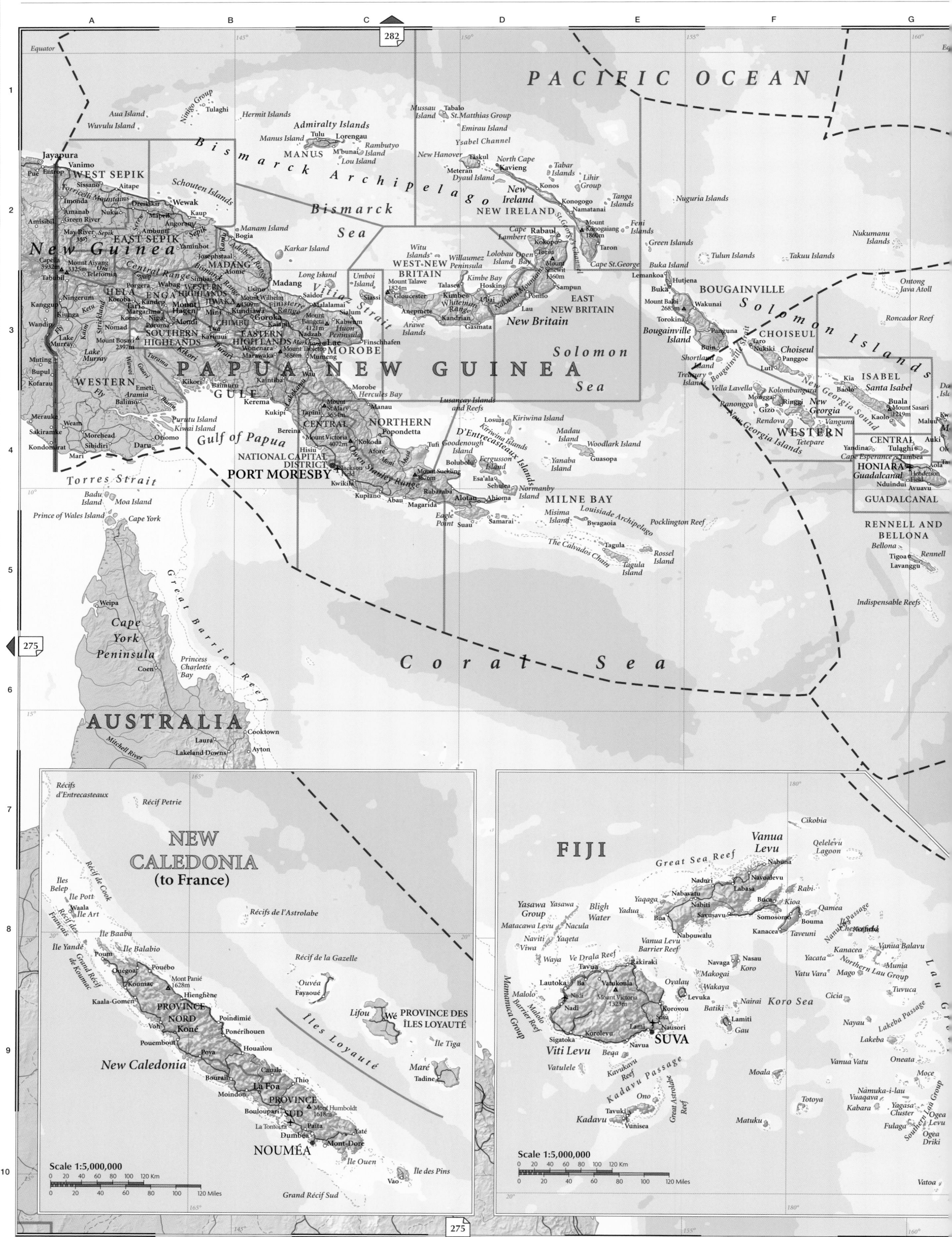

PACIFIC OCEAN

Equator

Bismarck Archipelago

MANUS

Admiralty Islands

Bismarck Sea

NEW IRELAND

BOUGAINVILLE

Solomon Islands

CHOISEUL

WEST SEPIK

EAST SEPIK

New Guinea

MADANG

WEST-NEW BRITAIN

EAST NEW BRITAIN

New Britain

Solomon Sea

WESTERN

ISABEL

CENTRAL

HONIARA
Guadalcanal

GUADALCANAL

RENNELL AND BELLONA

PAPUA NEW GUINEA

MOROBE

CENTRAL

NORTHERN

GULF

WESTERN

NATIONAL CAPITAL DISTRICT
PORT MORESBY

Gulf of Papua

Torres Strait

MILNE BAY

Louisiade Archipelago

AUSTRALIA

Cape York Peninsula

Great Barrier Reef

Coral Sea

NEW CALEDONIA
(to France)

New Caledonia

PROVINCE NORD

PROVINCE SUD

PROVINCE DES ÎLES LOYAUTÉ

NOUMÉA

Îles Loyauté

FIJI

Vanua Levu

Viti Levu

SUVA

Koro Sea

Northern Lau Group

Lau Group

Scale 1:5,000,000
0 20 40 60 80 100 120 Km
0 20 40 60 80 100 120 Miles

Scale 1:5,000,000
0 20 40 60 80 100 120 Km
0 20 40 60 80 100 120 Miles

6000m
4000m
3000m
2000m
1000m
500m
250m
100m
Sea Level
-250m
-2000m
-4000m

Scale 1:10,100,000
(projection: Mercator)

0 50 100 150 200 250 300 Km
0 50 100 150 200 250 300 Miles

Population

- ◼ above 5 million
- ◼ 1 million to 5 million
- ◉ 500,000 to 1 million
- ◎ 100,000 to 500,000
- ⊕ 50,000 to 100,000
- ○ 10,000 to 50,000
- ○ below 10,000

Main map (Solomon Islands inset)

CHOISEUL

Taro
Nukiki
Panggoe
Luti
Choiseul
Rob Roy
Vaghena
Kia
Baolo
Vella Lavella
Mongga
Kolombangara
Gizo
Ringgi
New Georgia
Munda
Rendova
Blanche Channel
Vangunu
Nggatokae
New Georgia Sound
Manning Strait

WESTERN
Tetepare
New Georgia Islands
Ranongga

ISABEL
Santa Isabel
Buala
Mount Sasari 1219m
Kaolo
San Jorge

SOLOMON ISLANDS

Roncador Reef

MALAITA

Dai Island
Mahu
Kwailibesi
Auki
Malaita
Olomburi
Baunani
Tarapaina
Maramasike
Apio

CENTRAL
Russell Islands
Florida Islands
Savo
Tulaghi
Cape Esperance
Tambea
Yandina
Iron Bottom Sound
Indispensable Strait

HONIARA
Henderson Field
Aola
Guadalcanal
Tangarare
Mount Popomanaseu 2330m
Nduindui
Avuavu

GUADALCANAL

Heuru
Kirakira
San Cristobal
Hauraha
Ulawa Island
Three Sisters Islands

RENNELL AND BELLONA
Bellona
Tigoa
Lavanggu
Rennell

MAKIRA-ULAWA

Scale 1:5,000,000
0 20 40 60 80 Km
0 20 40 60 80 Miles

Left lower area (repeated Solomon / Vanuatu)

ITA
kaiana
awa Island
Sisters Islands
akira
San Cristobal
Star Harbour
ha

SOLOMON ISLANDS

MAKIRA-ULAWA

Duff Islands
Reef Islands
Tinakula
Lata
Noka
Nendö
Santa Cruz Islands
Utupua
Vanikolo

TEMOTU

Tikopia

Torres Islands
Hiu
Toga
Ureparapara
Vanua Lava
Sola
Banks Islands
Gaua

VANUATU

Cape Cumberland
Nokuku
Espiritu Santo
Mount Tabwemasana 1879m
Malo
Bougainville Strait
Unmet
Malekula
Norsup
Lamap
Lamen Bay
Epi
Emae
Shepherd Islands
Nguna
Paonangisu
Bauer Field
Forari
PORT-VILA
Efate

Port-Olry
Luganville
Ambae
Navonda
Maéwo
Bwatnapne
Pentecost
Mount Marum 1270m
Ambrym
Toak
Tongoa

Erromango
Unpongkor
Ipota
Aniwa
Futuna
Tanna
Isangel

Huon
Récifs d'Entrecasteaux
Récif Petrie
Grand Passage
Récifs de Cook
Île'Art
Waala
Île Balabio
Poum
Pouébo
Mont Panié 1628m
Kaala-Gomen
Koumac
Hienghène
PROVINCE NORD
Voh
Koné
Ponérihouen
Houailou
Poya
Bourail
Canala
Thio
New Caledonia
La Foa
PROVINCE SUD
La Tontouta
Dumbéa
Yaté
NOUMÉA
Mont Dore
Vao
Île des Pins
Grand Récif Sud

NEW CALEDONIA (to France)

Récifs de l'Astrolabe
Fayaoué
Ouvéa
Lifou
We
Tadine
Maré
PROVINCE DES ÎLES LOYAUTÉ
Îles Loyauté
Récif Durand
Île Walpole

Tropic of Capricorn

Vanuatu inset (upper right)

Torres Islands
Hiu
Tegua
Loh
Toga
Ureparapara
Mota Lava
Vanua Lava
Mota
Sola
Banks Islands
Gaua
Mere Lava

VANUATU

Cape Cumberland
Nokuku
Big Bay
Port-Olry
Espiritu Santo
Naone
Maéwo
Mount Tabwemasana 1879m
Ambae
Navonda
Luganville
Malo
Bwatnapne
Bougainville Strait
Pentecost
Unmet
Norsup
Mount Marum 1270m
Ambrym
Malekula
Lamap
Paama
Toak
Lopevi
Lamen Bay
Epi
Emae
Tongoa
Shepherd Islands
Nguna
Emao
Bauer Field
Paonangisu
Forari
PORT-VILA
Efate

Unpongkor
Erromango
Ipota
Aniwa
Tanna
Isangel
Aneityum

Scale 1:5,000,000
0 20 40 60 80 100 120 Km
0 20 40 60 80 100 120 Miles

Fiji area (right)

Cikobia
Vanua Levu
Qelelevu Lagoon
Great Sea Reef
Nayolevu
Nabuna
Rabi
Naduri
Labasa
Bligh Water
Nabayatu
Bua
Somosomo
Bouma
Yasawa Group
Nabouwalu
Savusavu
Taveuni
Naitaba
Tavua
Rakiraki
Koro
Nasau
Vanua Balavu
Lautoka
Ba
Mount Victoria 1323m
Oyalau
Levuka
Mago
Northern Lau Group
Cicia
Mamanuca Group
Nadi
Korovou
Nausori
Lamiti
Nayau
Viti Levu
Navua
SUVA
Gau
Koro Sea
Lakeba Passage
Korolevu
Beqa
Lakeba
Oneata
FIJI
Vatulele
Kadavu Passage
Moala
Moce
Namuka-i-lau
Kabara
Vunisea
Ono
Totoya
Fulaga
Kadavu
Matuku
Lau Group
Southern Lau Group
Vatoa
Ono-i-lau

PACIFIC OCEAN

Tropic of Capricorn

Elevation scale

19,686ft
13,124ft
9843ft
6562ft
3281ft
1640ft
820ft
328ft
Sea Level
-820ft
-6562ft
-13,124ft

Micronesia

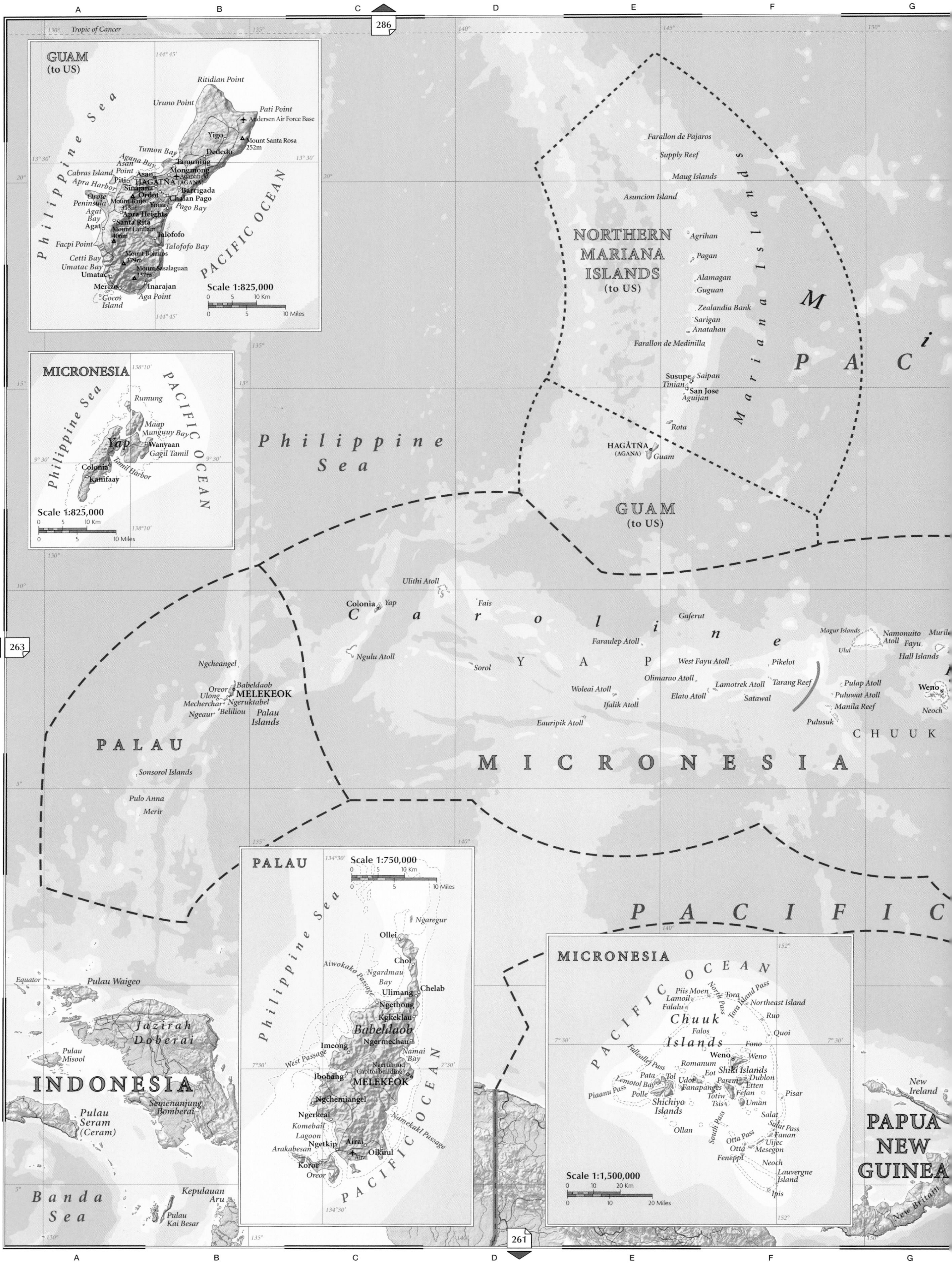

286

263

261

AUSTRALASIA & OCEANIA

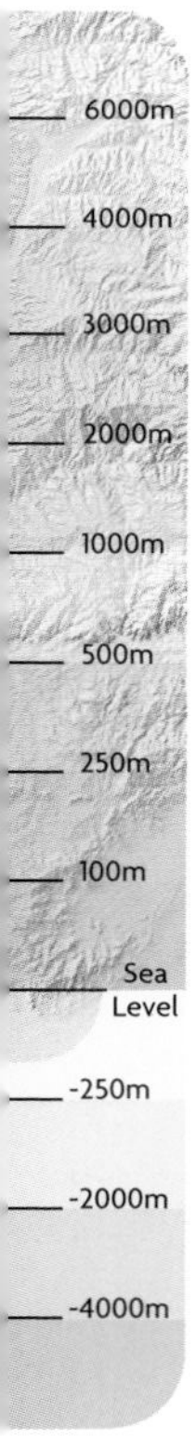

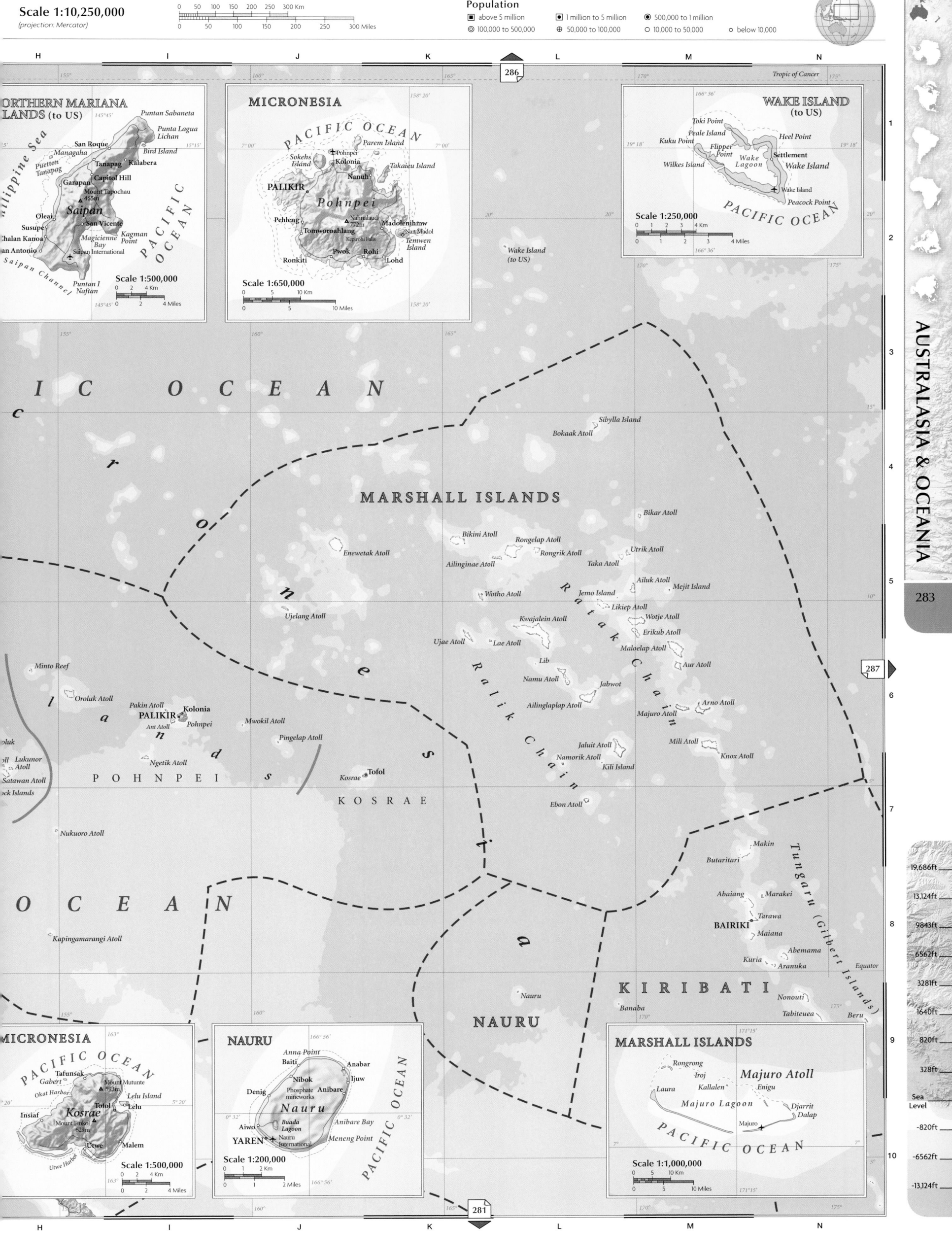

Scale 1:10,250,000
(projection: Mercator)

0 50 100 150 200 250 300 Km
0 50 100 150 200 250 300 Miles

Population
- ■ above 5 million
- ■ 1 million to 5 million
- ● 500,000 to 1 million
- ◉ 100,000 to 500,000
- ⊕ 50,000 to 100,000
- ○ 10,000 to 50,000
- · below 10,000

286

287

281

NORTHERN MARIANA ISLANDS (to US)

Philippine Sea

Puntan Sabaneta
Punta Lagua Lichan
Bird Island
San Roque
Managaha
Tanapag Kalabera
Puetton Tanapag
Capitol Hill
Garapan
Mount Tapochau 465m
Saipan
Oleai
San Vicente
Susupe
Magicienne Bay
Kagman Point
Chalan Kanoa
San Antonio
Saipan International
Puntan I Naftan

PACIFIC OCEAN

Saipan Channel

Scale 1:500,000
0 2 4 Km
0 2 4 Miles

MICRONESIA

PACIFIC OCEAN

Parem Island
Pohnpei
Sokehs Island
Kolonia
Nanuh
Takaieu Island
PALIKIR
Pohnpei
Pehleng
Nahnalaud 772m
Madolenihmw
Tomworoahlang
Kepirohi Falls
Nan Madol
Temwen Island
Ronkiti
Pwok Rohi
Lohd

Scale 1:650,000
0 5 10 Km
0 5 10 Miles

WAKE ISLAND (to US)

Toki Point
Peale Island
Kuku Point
Flipper Point
Heel Point
Wilkes Island
Wake Lagoon
Settlement
Wake Island
Wake Island
Peacock Point

PACIFIC OCEAN

Scale 1:250,000
0 1 2 3 4 Km
0 1 2 3 4 Miles

Tropic of Cancer

PACIFIC OCEAN

Sibylla Island
Bokaak Atoll

MARSHALL ISLANDS

Bikar Atoll
Bikini Atoll
Rongelap Atoll
Enewetak Atoll
Rongrik Atoll
Utrik Atoll
Ailinginae Atoll
Taka Atoll
Ailuk Atoll
Mejit Island
Wotho Atoll
Jemo Island
Likiep Atoll
Wotje Atoll
Ujelang Atoll
Erikub Atoll
Ujae Atoll
Lae Atoll
Kwajalein Atoll
Maloelap Atoll
Lib
Aur Atoll
Namu Atoll
Jabwot
Ailinglaplap Atoll
Majuro Atoll
Arno Atoll
Jaluit Atoll
Mili Atoll
Namorik Atoll
Knox Atoll
Kili Island
Ebon Atoll

Ratak Chain
Ralik Chain

Minto Reef
Oroluk Atoll
Pakin Atoll
Kolonia
PALIKIR
Pohnpei
Ant Atoll
Mwokil Atoll
Lukunor Atoll
Ngetik Atoll
Pingelap Atoll
Satawan Atoll
POHNPEI
Kosrae Tofol
Chuuk Islands
KOSRAE
Nukuoro Atoll
Ebon Atoll
Kapingamarangi Atoll

Makin
Butaritari
Abaiang Marakei
Tarawa
BAIRIKI
Maiana
Kuria
Abemama
Aranuka
Equator

Tungaru (Gilbert Islands)

Nauru
KIRIBATI
Banaba
Nonouti
NAURU
Tabiteuea
Beru

OCEAN

MICRONESIA

PACIFIC OCEAN

Tafunsak
Gabert
Mount Mutunte 593m
Okat Harbor
Lelu Island
Kosrae
Tofol
Lelu
Insiaf
Mount Finkol 629m
Utwe
Malem
Utwe Harbor

Scale 1:500,000
0 2 4 Km
0 2 4 Miles

NAURU

Anna Point
Baiti
Anabar
Nibok
Anabar
Denig
Ijuw
Phosphate mineworks
Anibare
Nauru
Buada Lagoon
Anibare Bay
YAREN
Nauru International
Meneng Point

PACIFIC OCEAN

Scale 1:200,000
0 1 2 Km
0 1 2 Miles

MARSHALL ISLANDS

Rongrong
Iroj
Majuro Atoll
Laura
Kallalen
Enigu
Majuro Lagoon
Djarrit
Majuro
Dalap

PACIFIC OCEAN

Scale 1:1,000,000
0 5 10 Km
0 5 10 Miles

19,686ft
13,124ft
9843ft
6562ft
3281ft
1640ft
820ft
328ft
Sea Level
-820ft
-6562ft
-13,124ft

MARSHALL ISLANDS

Erikub Atoll
Maloelap Atoll
Namu Atoll
Jabwot
Aur Atoll
Arno Atoll
Majuro Atoll
Mili Atoll
Jaluit Atoll
Kili Island
Knox Atoll
Ebon Atoll

Ralik Chain
Ratak Chain

Makin
Butaritari
Abaiang
Marakei
Tarawa
Maiana
Kuria
Abemama
Aranuka
Banaba
Nonouti
Tabiteuea
Beru Nikunau
Onotoa
Tamana
Arorae

Tungaru (Gilbert Islands)

Equator

KIRIBATI

Phoenix Islands
Kanton
Enderbury Island
McKean Island
Birnie Island
Rawaki
Orona
Manra
Nikumaroro

PACIFIC

Jarvis Island (to US)
Kingman Reef (to US)
Palmyra Atoll (to US)
Teraina
Tabuae
KIR

Howland Island (to US)
Baker Island (to US)

Nanumea Atoll
Niutao
Nanumaga
Nui Atoll
Vaitupu
Nukufetau Atoll
Funafuti Atoll
Nukulaelae Atoll
TUVALU
Niulakita

Atafu Atoll
Nukunonu Atoll
Fakaofo Atoll
TOKELAU (to NZ)

Rakahanga Manihiki
Northern Cook Is
Pukapuka Nassau
Swans Island
Suwarrow

WALLIS & FUTUNA (to France)
Îles Wallis
Île Futuna
Île Alofi

SAMOA
Savai'i
Sāmoa
'Upolu

AMERICAN SAMOA (to US)
Manua Islands
Tutuila

COOK ISLANDS (to NZ)

Mere Lava
Maéwo
Pentecost
Ambrym
Lopevi
Tongoa
Emao
PORT-VILA
Efate
Erromango
Ipota
Aniwa
Futuna
Tanna
Isangel
Aneityum
Maré
Île Walpole

VANUATU

Vanua Levu
Nabuna
Rabi
Yasawa Group
Bua
Koro
Taveuni
Lautoka
Mamanuca Group
Ovalau
Cicia
Lakeba
Viti Levu
SUVA
Gau
Kadavu
Moala
Nayau
Totoya
Matuku

Lau Group

FIJI
Vatoa

Niuatoputapu
Tafahi
Fonualei
Toku
Vava'u Group
Late
'Uta Vava'u
Kao
Ha'ano
Tofua
Kotu Group
Lifuka
Ha'apai Group
Nomuka Group
Otu Tolu Group
Tonumea
NUKU'ALOFA
Tongatapu
Tongatapu Group
'Eua

TONGA

ALOFI
Niue
NIUE (to NZ)

Palmerston
Southern Cook Isla
AVARUA
Aitu
Manu
Raro

Tropic of Capricorn

Inset maps

KIRIBATI
PACIFIC OCEAN
Iku Buariki
Taratai
Abaokoro
Marenanuka
Bikeman
Nabeina
Tabiteuea
Betio Bikenebu Bonriki
Eita Tarawa
Banraeaba
Bairiki
Tarawa
Scale 1:1,000,000
0 5 10 Km
0 5 10 Miles

TUVALU
Te Ava Te Lape
Amatuku
Fualifeke
Tepuka
Fualopa
Fongafale
Fuafatu
Funafuti
Atoll
Vaiaku Funafuti
Te Ava Fuagea
Vasafua
Fatato
Fuagea
Te Ava Pua Pua
Falefatu
Tefala
Funafara
Teafuafou
Telele
PACIFIC OCEAN
Scale 1:500,000
0 2 4 Km
0 2 4 Miles

WALLIS & FUTUNA (to France)
PACIFIC OCEAN
Pointe Fatua Pointe Matapu
Toloke Île Futuna
Leava Mont Puke 524m
Mala'e
Koliu Pointe Vele
Alofitai Pointe Matalesina
Mont Kolofau 417m
Île Alofi
Pointe Sauma
Scale 1:1,000,000
0 5 10 Km
0 5 10 Miles

WALLIS & FUTUNA (to France)
PACIFIC OCEAN
Nukuloa
Nukutapu
Hihifo
Île Luaniva
Alele Îles Wallis
Baie de l'Ouest
Ahoa Île
'Uvéa
MATÁ'UTU
Mala'atoli Baie de Mati'utu Nukuhifala
Tepa
Halalo
Nukuatea
Île Matala'a
Île Fenuafo'ou
Île Faioa
Scale 1:1,000,000
0 5 10 Km
0 5 10 Miles

TONGA
Niu 'Aunofa Atatá Maniloa Tau Ata
Kolovai Poloa Onevai Motu Nuku
Fafa Tapu Fukave
Piha Passage 'Eua Iki
NUKU'ALOFA Kolonga
Houma Fanga 'Uta Mui Hopohoponga
Pea Vaina Mu'a
Tongatapu Tongatapu
Houma Fua'amotu
Taloa
Houma 'Eua
'Ohonua
Ha'atua
Kalau
PACIFIC OCEAN
Scale 1:1,000,000
0 5 10 Km
0 5 10 Miles

COOK ISLANDS (to NZ)
Te Aiti Point
Nikao Te Aiti Ngatipa Avarua Harbour
Avarua Harbour
Rarotonga AVARUA Ikurangi 485m
Matavera
Arorangi Maungaroa 509m Te Manga 652m
Rarotonga Ngatangiia
Te Kou 564m Motutapu
Oneroa
Muri
Koromiri
Taakoka
Titikaveka
Scale 1:325,000
0 2 4 Km
0 2 4 Miles
COOK ISLANDS (to NZ)

PACIFIC OCEAN

Scale 1:15,500,000
(projection: Mercator)

0 50 100 150 200 250 300 350 400 Km
0 50 100 150 200 250 300 350 400 Miles

Population
- above 5 million
- 1 million to 5 million
- 500,000 to 1 million
- 100,000 to 500,000
- 50,000 to 100,000
- 10,000 to 50,000
- below 10,000

SAMOA (inset — Scale 1:3,000,000)

Savai'i
Faleālupo
Sātaua
Cape Puava
Fālelima
Mauga Silisili 1858m
Sala'ilua
Cape Asuisui
Taga
Fagamālo
Tuasivi
Pu'apu'a
Satupa'itea
Salelologa
Feleolo
Mauga Fito 1113m
ĀPIA 'Upolu
Apolima Strait
Matautu
Fagaloa Bay
Ti'avea
Palauli Bay
Lotofaga
Poutasi
Sāfata Bay
Salani

AMERICAN SAMOA (to US)
PAGO PAGO
Cape Matātula
'Aunu'u Island
Cape Taputapu
Steps Point
Tutuila
Manu'a Islands
Ofu
Olosega
Luma Ta'ū

Sàmoa

PACIFIC OCEAN

KIRIBATI (inset — Scale 1:1,175,000)

PACIFIC OCEAN
Northwest Point
Cape Manning
London
Banana
Northeast Point
Cook Island
Saint Stanislas
Paris
Kiritimati
Manulu Lagoon
Poland
Kiritimati (Christmas Island)
Bay of Wrecks
South West Point
Vaskess Bay
Isles Lagoon
Joe's Hill 12m
Aeon Point
Azur Lagoon
Pelican Lagoon
South East Point

FRENCH POLYNESIA (to France) (inset — Scale 1:1,000,000)

Baie d'Opunohu
Baie de Cook
Pointe Aroa
Papetoai
Îles du Vent
Baie de Matavai
Pointe Vénus
Mont Matorea 714m
Paopao
Mahina
Papenoo
Moorea
Afareaitu
PAPEETE
Pirea
Tiarei
Haapiti
Mont Tohiea 1207m
Faaa
Faaa
Hitiaa
Pointe Nuupere
Punaauia
Mont Aorai 2066m
Pointe Nuuroa
Mont Orohena 2241m
Tahiti
Passe Tamotoe
Paea
Mont Tetufera 1799m
Faaone
Baie de Taravao
Maraa
Taravao
Isthme de Taravao
Pointe Maraa
Mataiea
Afaahiti Toahotu
Tautira
Récif Tepaee
Vairao
Presqu'île de Taiarapu
Papara
Teahupoo
Mont Rooniu 1332m

PACIFIC OCEAN

Equator

Main map

OCEAN

ne Islands
Malden Island

Millennium Island
Vostok Island
Flint Island

Îles Marquises
Hatutu
Eiao
Nuku Hiva
Taiohae
Ua Huka
Ua Pu
Atuona
Hiva Oa
Tahuata
Motane
Fatu Hiva
Omoa

Îles du Roi Georges
Îles du Désappointement
Ahe
Manihi
Tepoto
Mataiva
Tikehau
Takaroa
Napuka
Pukapuka
Rangiroa
Takapoto
Tikei
Îles Palliser
Îles Sous le Vent
Aratika
Motu One
Toau
Kanehi
Takume
Fagatau
Makatea
Niau
Raraka
Raroia
Maupiti
Tupai
Fakarava
Katiu
Fakahina
Bora-Bora
Fare
Faaite
Makemo
Tahaa
Tetiaroa
Tahanea
Nihiru
Tehuata
Manuae
Raiatea
Huahine
Marūtea
Tauere
Tatakoto
Maupihaa
Moorea
PAPEETE
Anaa
Haraiki
Hikueru
Amanu
Maiao
Tahiti
Mehetia
Reitoru
Marokau
Hao
Îles du Vent
Ravahere
Akiaki
Pukarua
Archipel de la Société
Negonego
Vahitahi
Reao
Paraoa
Manuhagi
Vairaatea
Pinaki
Hereheretue
Ahunui
Îles du Duc de Gloucester
Vanavana
Tureia
FRENCH POLYNESIA (to France)
Tenararo
Groupe Actéon
Maturei
Maria
Moruroa
Maria
Rimatara
Tematagi
Fagataufa
Rurutu
Îles Gambier
Maria
Maria
Tubuai
Magareva
Temoe
Raevavae
Îles Australes

Tuamotu

PACIFIC OCEAN

PITCAIRN, HENDERSON, DUCIE & OENO ISLANDS (to UK)

Tropic of Capricorn
Oeno Island
Henderson Island
Ducie Island
Pitcairn Island

NIUE (to NZ) (inset — Scale 1:1,000,000)

Hikutavake
Toi
Mutalau
Makefu
Tuapa
Lakepa
Makapu Point
Alofi Bay
Niue
Liku
ALOFI
Halagigie Point
Tamakautoga
Avatele
Hakupu
Tepa Point
Mata Point

PACIFIC OCEAN

PITCAIRN ISLAND (to UK) (inset — Scale 1:125,000)

Young's Rock
Bounty Bay
ADAMSTOWN
Adam's Rock
Pitcairn Island
Point Christian
St Paul's Point

PACIFIC OCEAN

Rapa Iti
Marotiri

PACIFIC OCEAN

Elevation scale

19,686ft
13,124ft
9843ft
6562ft
3281ft
1640ft
820ft
328ft
Sea Level
-820ft
-6562ft
-13,124ft

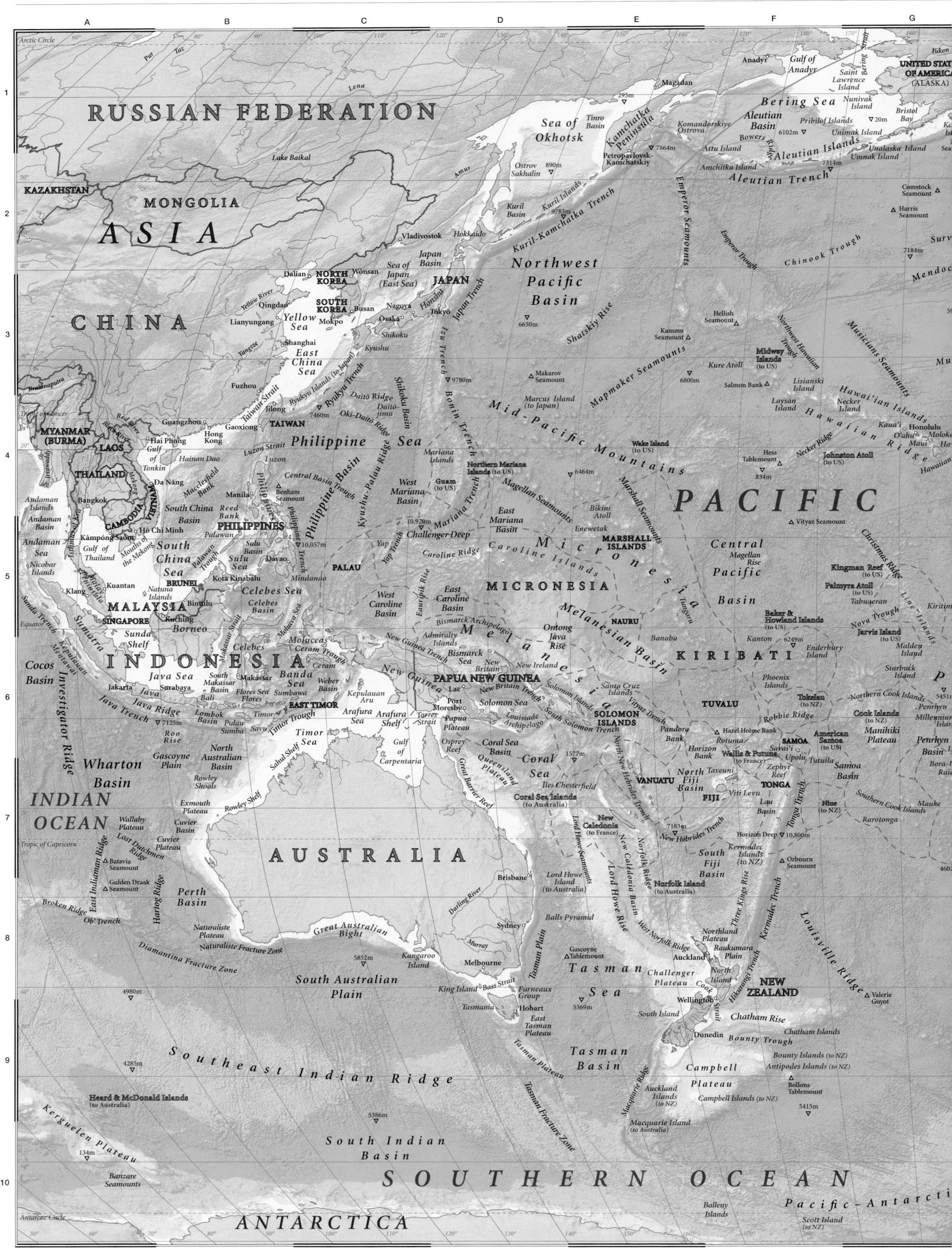

Pacific Ocean map labels:

Arctic Circle

RUSSIAN FEDERATION

KAZAKHSTAN

MONGOLIA

A S I A

CHINA

Lake Baikal

Lena

Amur

Ostrov Sakhalin   890m

Sea of Okhotsk   Tinro Basin   295m

Kamchatka Peninsula

Petropavlovsk-Kamchatskiy   7864m

Magadan

Kuril Basin   Kuril Islands   9783m

Kuril-Kamchatka Trench

Anadyr   Gulf of Anadyr

Saint Lawrence Island

Bering Strait   Yukon

UNITED STATES OF AMERICA (ALASKA)

Bering Sea   Nunivak Island   20m   Bristol Bay

Komandorskiye Ostrova

Aleutian Islands   6102m   Bowers Ridge   Attu Island   Amchitka Island   Unimak Island   Unalaska Island   Umnak Island   7314m   Pribilof Islands

Aleutian Trench

Comstock Seamount

Harris Seamount

7184m

Surv

Mendo

Chinook Trough

Emperor Seamounts   Emperor Trough

Vladivostok   Hokkaido

NORTH KOREA   Wonsan

Dalian   Sea of Japan (East Sea)   Japan Basin

SOUTH KOREA   Busan   Nagoya   Tokyo

Qingdao   Yellow Sea   Mokpo   JAPAN   Honshu   Osaka

Lianyungang   Shikoku   Japan Trench

Shanghai   Kyushu

East China Sea   6650m

Izu Trench

Northwest Pacific Basin

Shatskiy Rise

Hellish Seamount

Kammu Seamount   Midway Islands (to US)   Kure Atoll   6800m   Salmon Bank   Lisianski Island   Laysan Island   Necker Island   834m

Musicians Seamounts

Mu

Hawaiian Islands

Kaua'i   O'ahu   Honolulu   Moloke   Maui   Ha

Fuzhou

Yangtze

TAIWAN   Gaoxiong

Guangzhou   Hong Kong

Hai Phong   Gulf of Tonkin   Haidan Dao

MYANMAR (BURMA)

LAOS

THAILAND

Bangkok

VIETNAM   Da Nang

CAMBODIA   Kâmpóng Saôm   Hô Chi Minh

Andaman Islands   Andaman Basin   Andaman Sea   Nicobar Islands

Gulf of Thailand

Ryukyu Islands (to Japan)   Ryukyu Trench   7460m

Daitō Ridge   Daitō-jima   Oki-Daitō Ridge

Shikoku Basin   9780m

Bonin Trench

Mid-Pacific Mountains

6464m   Wake Island (to US)

Necker Ridge

Johnston Atoll (to US)

Hawaiian Ridge

Hawaii

Taiwan Strait   Jilong

Philippine Sea

Luzon Strait   Luzon

Macclesfield Bank

Central Basin Trough   Philippine Basin

Manila   Benham Bank

PHILIPPINES   Philippine Trench

South China Basin   Reed Bank

Palawan   Kyushu-Palau Ridge

West Mariana Basin

Mariana Islands   Guam (to US)

Northern Mariana Islands (to US)

Magellan Seamounts

East Mariana Basin

Bikini Atoll   Marshall Seamounts

Enewetak

MARSHALL ISLANDS

P A C I F I C

Vityaz Seamount

Central Magellan Rise

Central Pacific Basin

Kingman Reef (to US)

Palmyra Atoll (to US)

Tabuaeran

Christmas Ridge

Line Islands

Kiritim

South China Sea

Palawan Trough   Palawan

Kota Kinabalu   Sulu Basin   Sulu Sea

Davao   Mindanao

PALAU

Yap Trench   Yap Trench

10,057m   10,920m   Challenger Deep

Mariana Trench

Caroline Ridge   Caroline Islands

West Caroline Basin

East Caroline Basin

M i c r o n e s i a

MICRONESIA

NAURU

KIRIBATI

Kanton   6249m

Enderbury Island

5451m

Phoenix Islands   Tokelau (to NZ)

Baker & Howland Islands (to US)

Jarvis Island (to US)

Malden Island

Starbuck Island

MALAYSIA   SINGAPORE

BRUNEI   Bintulu

Kuching   Natuna Islands

Kuantan   Klang

Borneo   Celebes Sea   Celebes Basin

Sunda Shelf

Eauripik Rise

Bismarck Archipelago

Melanesian Basin

Santa Cruz Islands

Tuvalu Trench

TUVALU

Cocos Basin

Java Sea   Makassar Strait

Jakarta   Java   Surabaya   Bali

I N D O N E S I A

South Makassar Basin   Makassar   Flores Sea   Flores

Banda Sea

Moluccas   Ceram Trough   Ceram

Celebes

Weber Basin

Admiralty Islands   New Guinea Trench

New Guinea

Bismarck Sea   New Britain

New Ireland   New Britain Trench

Ontong Java Rise

Banaba

Robbie Ridge

Phoenix Islands

Northern Cook Islands (to NZ)

Penrhyn

Cook Islands (to NZ)

Manihiki Plateau

Penrhyn Basin

Bora-L   Rai

Equator

Sumatra   Kerimata Strait

Investigator Ridge

Java Ridge   7125m   Lombok Basin   Pulau Sumba

Timor   Savu   Timor Trough   East Timor

Roo Rise   Sumbawa

North Australian Basin

Kepulauan Aru

Arafura Sea   Arafura Shelf

Torres Strait

PAPUA NEW GUINEA   Lae

Port Moresby   Papua Plateau

Solomon Sea

Louisiade Archipelago

South Solomon Trench

SOLOMON ISLANDS

Solomon Islands

North New Hebrides Trench

Pandora Bank

Hazel Holme Bank

Rotuma   Wallis & Futuna (to France)

SAMOA   Savai'i   Upolu   Tutuila

American Samoa (to US)

Samoa Basin

Mauke

Rarotonga

Southern Cook Islands (to NZ)

Wharton Basin

Gascoyne Plain

Exmouth Plateau

Rowley Shoals   Rowley Shelf

Sahul Shelf

Timor Sea

Gulf of Carpentaria

Coral Sea Basin

Osprey Reef

Coral Sea

Queensland Plateau

Great Barrier Reef

1577m   Horizon Bank

North Fiji Basin

VANUATU

Zephyr Reef

North Fiji Basin

Taveuni   Viti Levu   FIJI

Lau Basin

TONGA

Niue (to NZ)

Tonga Trench

INDIAN OCEAN

Tropic of Capricorn

Wallaby Plateau

Lost Dutchmen

Cuvier Basin

Cuvier Plateau

A U S T R A L I A

Brisbane

Iles Chesterfield

Coral Sea Islands (to Australia)

New Caledonia (to France)

7183m

Lord Howe Seamounts

Lord Howe Rise

West Norfolk Ridge

Norfolk Basin

Norfolk Island (to Australia)

New Hebrides Trench

New Caledonia Basin

Lord Howe Rise

South Fiji Basin

Three Kings Rise

Kermadec Islands (to NZ)

Horizon Deep   10,800m

Kermadec Trench

Louisville Ridge

4602

Broken Ridge   Ob' Trench

Batavia Seamount

Gulden Draak Seamount

East Indiaman Ridge

Perth Basin

Hartog Ridge

Darling River

Sydney

Balls Pyramid

Lord Howe Island (to Australia)

Gascoyne Tablemount

Northland Plateau

Raukumara Plain

Auckland

North Island

Ozbourn Seamount

Valerie Guyot

Broken Ridge

Diamantina Fracture Zone

Naturaliste Plateau   Naturaliste Fracture Zone

Great Australian Bight

5852m

Kangaroo Island

Murray   Melbourne

King Island   Bass Strait

Tasmania   Hobart

Furneaux Group

East Tasman Plateau

5369m

Tasman Plain

Tasman Sea

Challenger Plateau   Cook Strait

North Island

Wellington   South Island

NEW ZEALAND

Dunedin

Chatham Rise   Chatham Islands

Bounty Trough

4980m

South Australian Plain

Southeast Indian Ridge

4285m

Tasman Basin

Tasman Plateau

Campbell Plateau

Bounty Islands (to NZ)   Antipodes Islands (to NZ)

Bollons Tablemount

Macquarie Ridge

Auckland Islands (to NZ)

Campbell Islands (to NZ)

5415m

Sea Level   -250m   -2000m   -4000m

Heard & McDonald Islands (to Australia)

Kerguelen Plateau

134m

Banzare Seamounts

South Indian Basin

5386m

Macquarie Island (to Australia)

S O U T H E R N   O C E A N

Balleny Islands

Scott Island (to NZ)

Pacific-Antarcti

Antarctic Circle

A N T A R C T I C A

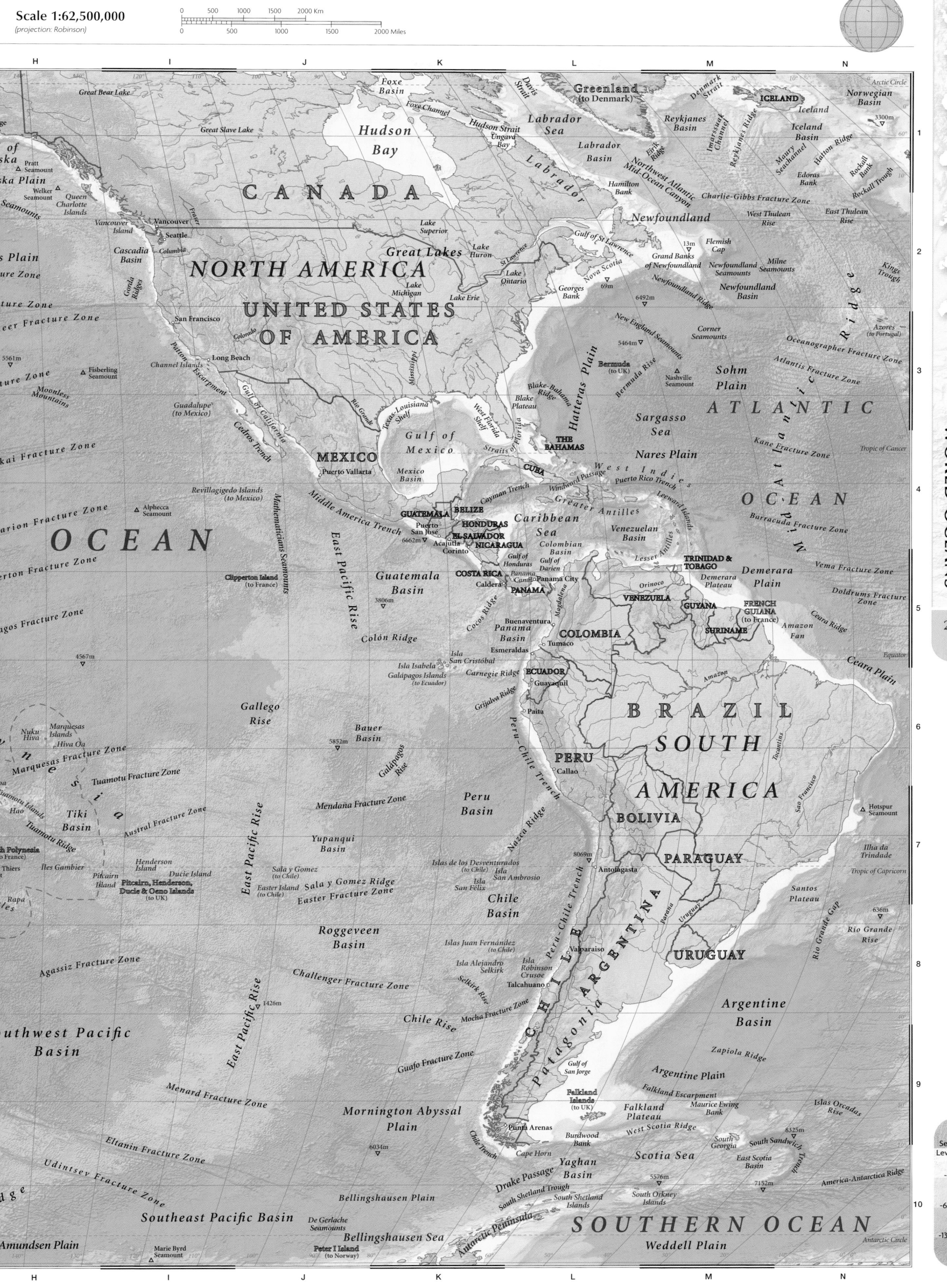

Indian Ocean

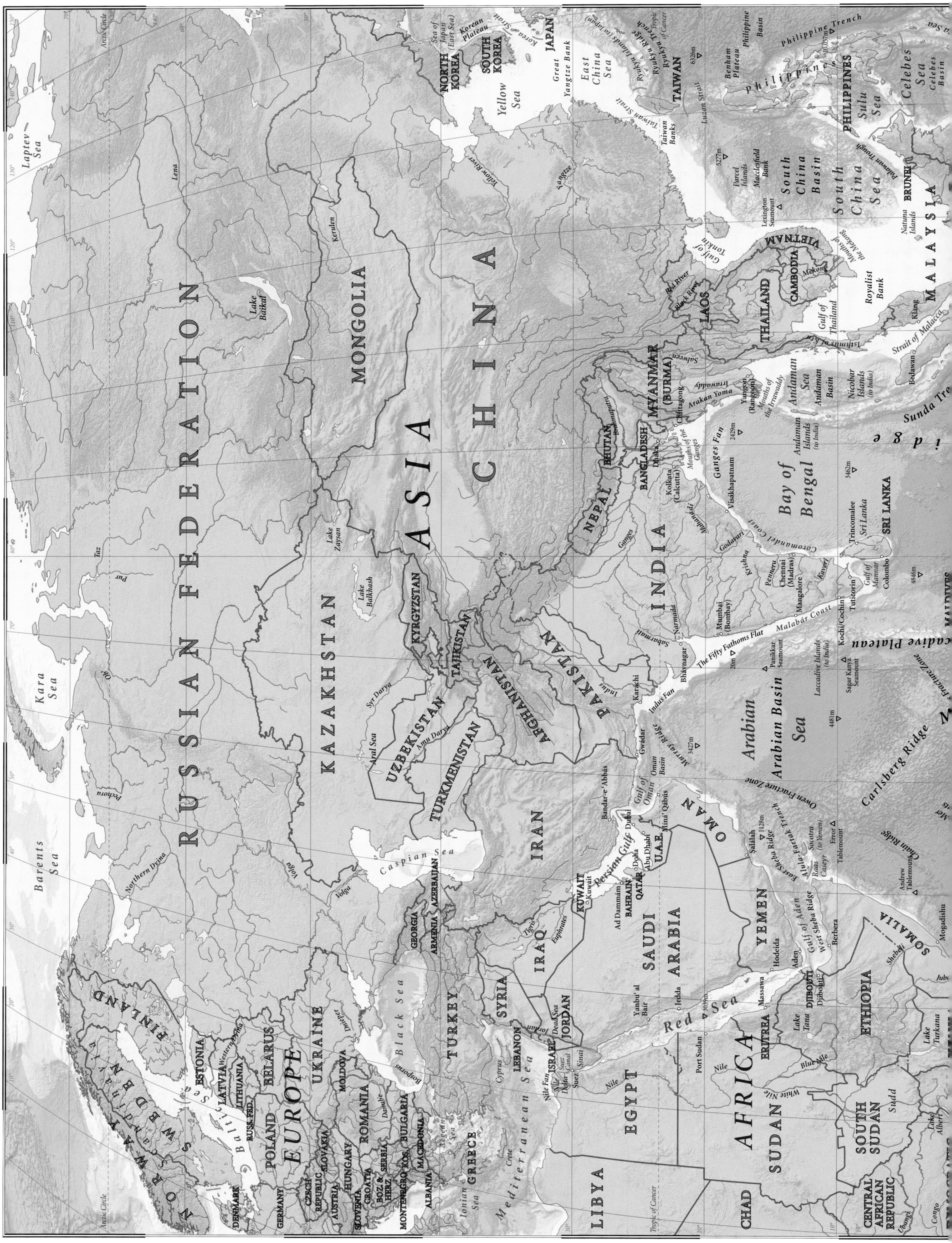

Sea Level

-250m

-2000m

-4000m

RUSSIAN FEDERATION

EUROPE

ASIA

CHINA

MONGOLIA

KAZAKHSTAN

KYRGYZSTAN

TAJIKISTAN

UZBEKISTAN

TURKMENISTAN

AFGHANISTAN

PAKISTAN

IRAN

IRAQ

SYRIA

TURKEY

LEBANON
ISRAEL
JORDAN

KUWAIT
BAHRAIN
QATAR
U.A.E.

SAUDI ARABIA

OMAN

YEMEN

EGYPT

LIBYA

SUDAN

SOUTH SUDAN

CHAD

CENTRAL AFRICAN REPUBLIC

AFRICA

ERITREA

DJIBOUTI

ETHIOPIA

SOMALIA

INDIA

NEPAL

BHUTAN

BANGLADESH

MYANMAR (BURMA)

SRI LANKA

THAILAND

LAOS

VIETNAM

CAMBODIA

MALAYSIA

BRUNEI

PHILIPPINES

TAIWAN

JAPAN

NORTH KOREA

SOUTH KOREA

FINLAND

SWEDEN

NORWAY

DENMARK

GERMANY

POLAND

BELARUS

UKRAINE

LITHUANIA

LATVIA

ESTONIA

RUSS. FED.

MOLDOVA

ROMANIA

BULGARIA

SERBIA

BOZ & HERZ

CROATIA

SLOVENIA

HUNGARY

SLOVAKIA

AUSTRIA

CZECH REPUBLIC

MONTENEGRO

KOS.

MACEDONIA

ALBANIA

GREECE

GEORGIA

ARMENIA

AZERBAIJAN

Laptev Sea

Barents Sea

Kara Sea

Caspian Sea

Black Sea

Mediterranean Sea

Red Sea

Arabian Sea

Bay of Bengal

Arabian Basin

Andaman Sea

Andaman Basin

South China Sea

South China Basin

East China Sea

Yellow Sea

Japan (East Sea)

Celebes Sea

Sulu Sea

Philippine Basin

Gulf of Thailand

Gulf of Tonkin

Persian Gulf

Gulf of Oman

Gulf of Aden

Lake Baikal

Lake Balkhash

Lake Zaysan

Aral Sea

Ionian Sea

Baltic Sea

Philippine Trench

Sunda Trench

Carlsberg Ridge

Nile

Volga

Ganges

Mekong

Yangtze

Indus

Lena

Scale 1:32,000,000
(projection: Robinson)

| 0 | 200 | 400 | 600 | 800 | 1000 | 1200 Km |
| 0 | 200 | 400 | 600 | 800 | 1000 | 1200 Miles |

Sea Level
-820ft
-6562ft
-13,124ft

INDIAN OCEAN

SOUTHERN OCEAN

AUSTRALIA

INDONESIA

Banda Sea
EAST TIMOR
Java Sea
Timor Sea
Timor Trough
Sumbawa
Bali
Lombok
Pulau Sumba
Savu
Ashmore & Cartier Islands (to Australia)
Java
Java Ridge
Java Trench
Christmas Island (to Australia)
Vening Meinesz Seamounts
Roo Rise
Sahul Shelf
Wyndham
Broome
Port Hedland
North Australian Basin
Rowley Shoals
Rowley Shelf
Exmouth Plateau
Gascoyne Plain
Cuvier Basin
Cuvier Plateau
Geraldton
Perth Basin
Fremantle
Bunbury
Albany
Naturaliste Plateau
Naturaliste Fracture Zone
Great Australian Bight
South Australian Basin
South Indian Basin
Wharton Basin
Wallaby Plateau
Lost Dutchmen Ridge
Batavia Seamount
Gulden Draak Seamount
Broken Ridge
Ob' Trench
East Indiaman Ridge
Hartog Ridge
Diamantina Fracture Zone
Southeast Indian Ridge
Amsterdam Fracture Zone
Amsterdam Island
St Paul Island
Investigator Ridge
Cocos Basin
Cocos Islands (to Australia)
Osborn Plateau
Ninetyeast Ridge
Mid-Indian Basin
Kerguelen Plateau
Heard & McDonald Islands (to Australia)
French Southern & Antarctic Lands (to France)
Kerguelen
Banzare Seamounts
Mid-Indian Ridge
Chagos Archipelago
British Indian Ocean Territory (to UK)
Diego Garcia
Chagos Trench
Chagos-Laccadive Fracture Zone
Vema Fracture Zone
Marie Celeste Fracture Zone
Egeria Fracture Zone
Rodrigues (to Mauritius)
Argo Fracture Zone
Central Indian Ridge
Crozet Basin
Crozet Plateau
Crozet Islands
Del Cano Rise
Southwest Indian Ridge
Indomed Fracture Zone
Lena Tablemount
Ob' Tablemount
Enderby Plain
Atlantic-Indian Basin
SEYCHELLES
Amirante Islands
Amirante Basin
Amirante Ridge
Amirante Trench
Seychelles Bank
Mahé
Mascarene Plateau
Saya de Malha Bank
Nazareth Bank
Cargados Carajos Bank
MAURITIUS
Mascarene Islands
Mascarene Basin
Réunion (to France)
Mascarene Plain
Mauritius Trench
Madagascar Basin
Madagascar Plateau
MADAGASCAR
Toamasina
Fred Seamount
Bardin Seamount
Agalega Islands
Farquhar Group
Aldabra Group
Giraud Seamount
Comoros
COMOROS
Comoro Basin
Mahajanga
Mozambique Channel
Davie Ridge
Bassas da India
Ile Europa
Jaguar Seamount
Mayotte (to France)
Natal Basin
Natal Valley
Mozambique Plateau
Mozambique Escarpment
Prince Edward Fracture Zone
Prince Edward Islands (to South Africa)
Walters Shoal
Transkei Basin
Africana Seamount
Agulhas Plateau
Agulhas Basin
Atlantic-Indian Ridge
SOUTH AFRICA
Cape Town
Cape of Good Hope
Cape Agulhas
Agulhas Bank
Protea Seamount
Mossel Baai
Port Elizabeth
East London
Durban
Drakensberg
Tugela
LESOTHO
SWAZILAND
MOZAMBIQUE
Maputo
Beira
Quelimane
Nacala
Pemba
Ruvuma
Cabo Delgado
Mafia
Zanzibar
Dar es Salaam
TANZANIA
Mombasa
BURUNDI
OF CONGO
Lake Tanganyika
Lake Rukwa
MALAWI
Lake Nyasa
ZAMBIA
Lake Mweru
Lake Cabora Bassa
ZIMBABWE
Zambezi
BOTSWANA
Limpopo
Orange River
Tropic of Capricorn
Antarctic Circle

5852m
5386m
498m
425m
184m
4684m
5386m
5819m
4035m
5641m
2078m
7023m
4936m
184m
497m
69m
3665m
7125m
+461m
5759m
5678m

10° 20° 30° 40° 50° 60° 70° 80° 90° 100° 110° 120° 130° 140° 150°

Atlantic Ocean

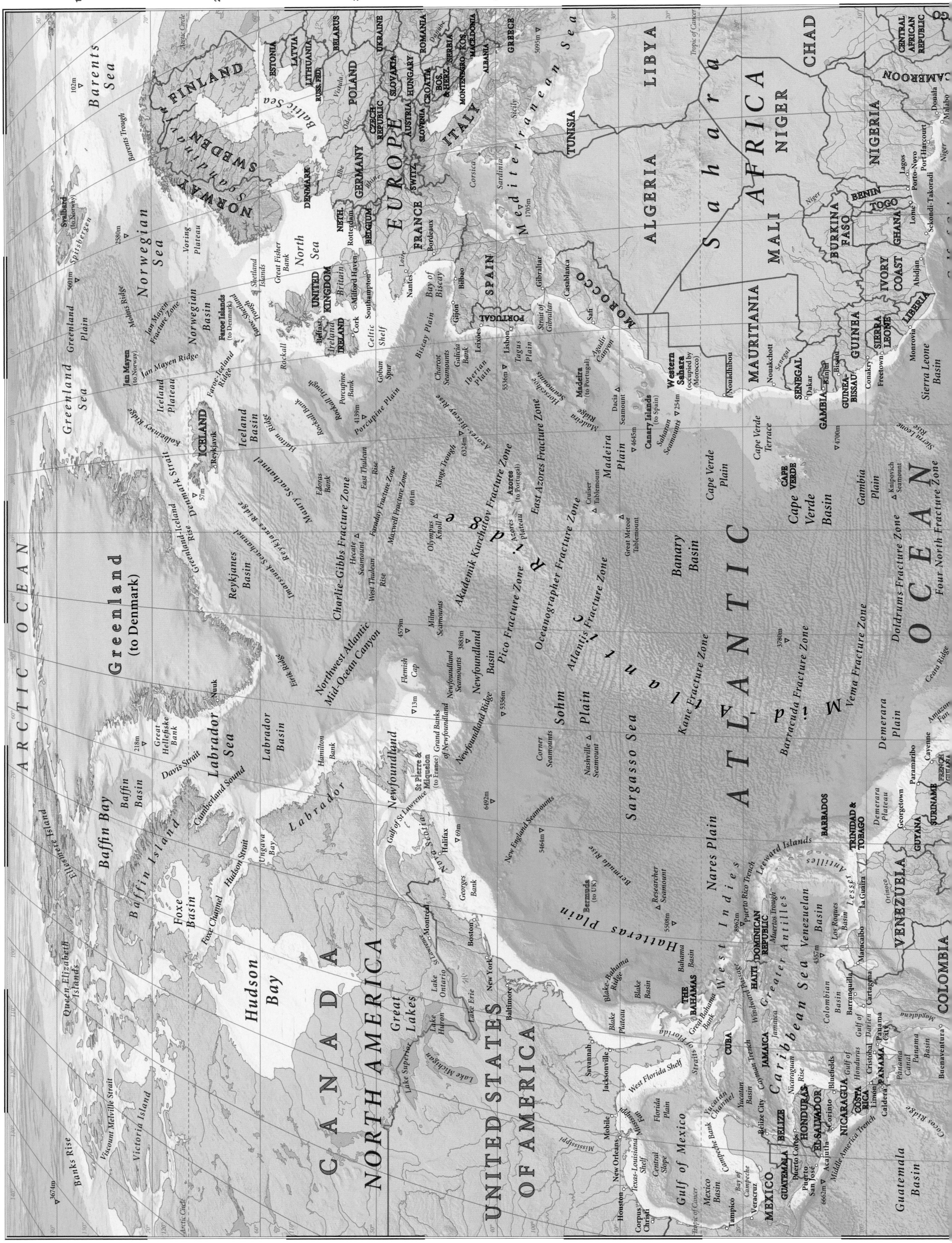

Sea Level
-250m
-2000m
-4000m

Scale 1:34,400,000
(projection: Robinson)

0 200 400 600 800 1000 1200 Km
0 200 400 600 800 1000 1200 Miles

Sea Level

-820ft

-6562ft

-13,124ft

CONGO

ANGOLA (Cabinda)
Matadi
Pointe-Noire
Congo Fan
Pierre Brazza Seamounts
Luanda
ANGOLA
Lobito
Namibe
NAMIBIA
Walvis Bay
Namib Desert
Lüderitz
Orange
Orange Fan
SOUTH AFRICA
Cape Town
Cape of Good Hope
Tropic of Capricorn

Angola Basin
Dampier Seamount
5042m
Zahov Seamount
6039m
Walvis Ridge
Namibia Plain
Agulhas Plateau
Protea Seamount
Atlantic-Indian Ridge
Bouvet Island (to Norway)

Chain Fracture Zone
6308m
Pernambuco Basin
Fernando do Noronha (to Brazil)
Parnaiba Ridge
Ascension Fracture Zone
Ascension Island (to UK)
Bode Verde Fracture Zone
5706m
Saint Helena Fracture Zone
Saint Helena (to UK)
Bonaparte Seamount
Rio Grande Fracture Zone
Tristan da Cunha (to UK)
1179m
Tristan de Cunha Fracture Zone
Gough Fracture Zone
Gough Island
Meteor Rise
Cape Rise
Schmidt-Ott Seamount
Vema Seamount
Cape Basin
5115m
Davis Seamount
Simmonds
America-Antarctica Ridge
Atlantic-Indian Basin
Maud Rise
Astrid Ridge
Lazarev Sea
Rüiser-Larsen Sea

SOUTHERN OCEAN

ANTARCTICA

Discovery Tablemount

M i d - A t l a n t i c   R i d g e

BRAZIL
SOUTH AMERICA
PERU
BOLIVIA
PARAGUAY
URUGUAY
ARGENTINA
CHILE
Patagonia

Amazon
Tocantins
São Francisco
Vitória
Rio de Janeiro
Santos
Paranaguá
Montevideo
Buenos Aires
Río de la Plata
Bahía Blanca
Paraná
Uruguay
Paraguay

Guayaquil
Paita
Callao
Antofagasta
Valparaíso
Talcahuano
Ponta Arenas
Tierra del Fuego
Cape Horn
Gulf of San Jorge
Gulf of San Matías

Recife
Stocks Seamount
Ferraz Ridge
Hotspur Seamount
Pernambuco Seamounts
Brazil Basin
Illa da Trindade
Ilhas Martin Vaz
Rio Grande Rise
636m
Rio Grande Gap
Santos Plateau
Argentine Basin
Zapiola Seamount
Zapiola Ridge
Argentine Plain
Argentine Escarpment
Falkland Escarpment
Falkland Plateau
Maurice Ewing Bank
Falkland Islands (to UK)
Patagonian Shelf
Burwood Bank
Yaghan Basin

Zahov Seamount

Islas Orcadas Rise
1738m
South Georgia Rise
3667m
South Georgia & the South Sandwich Islands (to UK)
South Georgia Ridge
West Scotia Ridge
Tehuelche Fracture Zone
Endurance Fracture Zone
Quest Fracture Zone
Protector Basin
South Scotia Ridge
South Orkney Islands
5575m
Pirie Deep
Orcadas Deep
Scotia Sea
South Sandwich Trench
8325m
East Scotia Ridge
East Scotia Basin
7152m
South Sandwich Fracture Zone
Weddell Plain
Weddell Sea
Ronne Ice Shelf
South Shetland Islands
South Scotia Ridge
South Sandwich Trough
Drake Passage
Antarctic Peninsula
SOUTHERN OCEAN
ANTARCTICA

PACIFIC OCEAN

Galápagos Ridge
Galápagos Rise
Salas y Gómez Ridge
Easter Fracture Zone
Mendaña Fracture Zone
5852m
Bauer Basin
Yupanqui Basin
Peru Basin
Nazca Ridge
Peru-Chile Trench
8069m
Chile Basin
Chile Rise
Roggeveen Basin
Challenger Fracture Zone
Islas Juan Fernández (to Chile)
Selkirk Rise
Isla Robinson Crusoe
Isla Alejandro Selkirk
Islas de los Desventurados (to Chile)
Isla San Ambrosio
Isla San Félix
Mocha Fracture Zone
Guafo Fracture Zone
Mornington Abyssal Plain
6034m
Menard Fracture Zone
Eltanin Fracture Zone
Southeast Pacific Basin
De Gerlache Seamounts
Peter I Island (to Norway)
Bellingshausen Plain
Bellingshausen Sea
Amundsen Sea
Antarctic Circle
Tropic of Capricorn

Orange

(to Ecuador)

DEM. 1

Antarctica

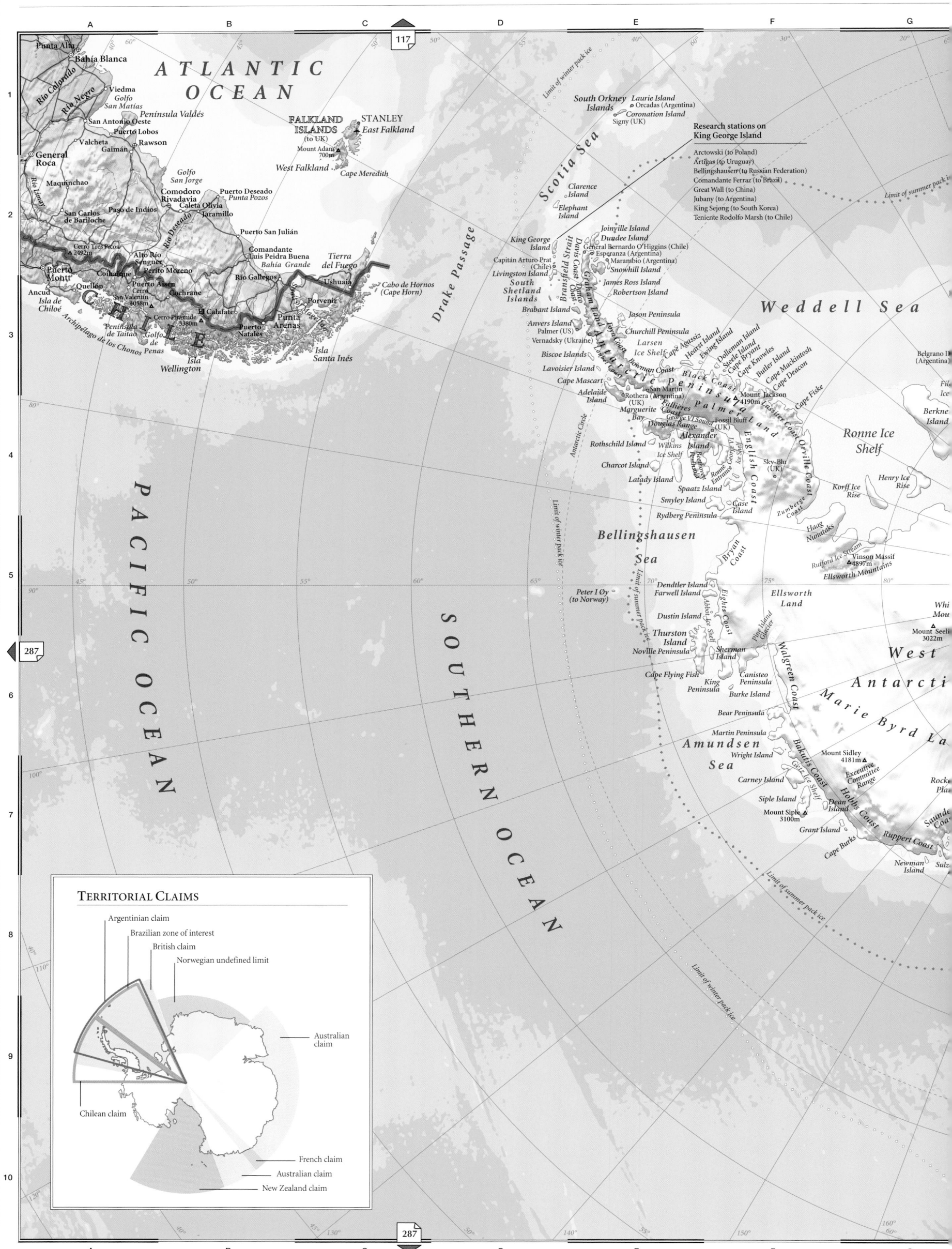

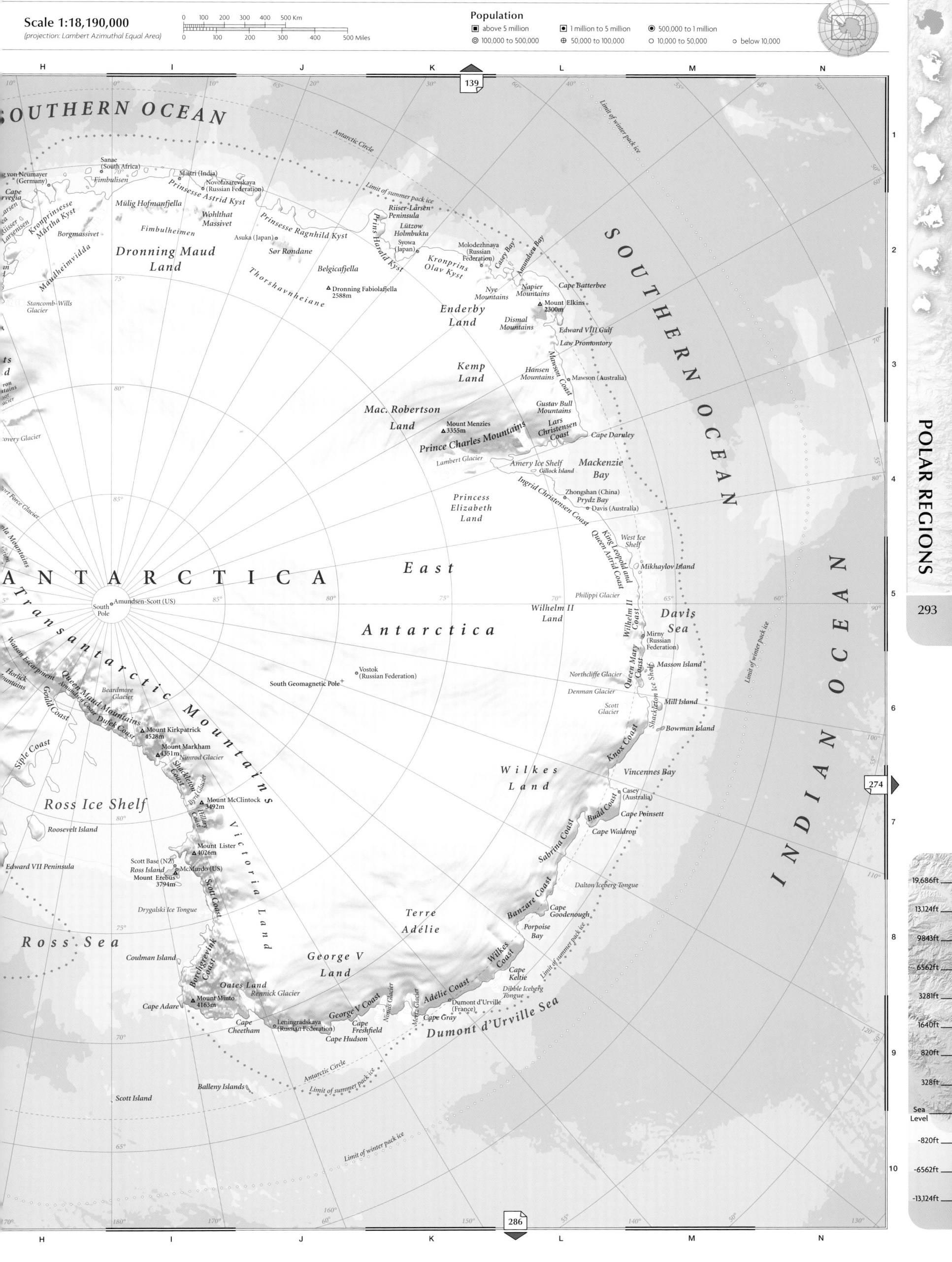

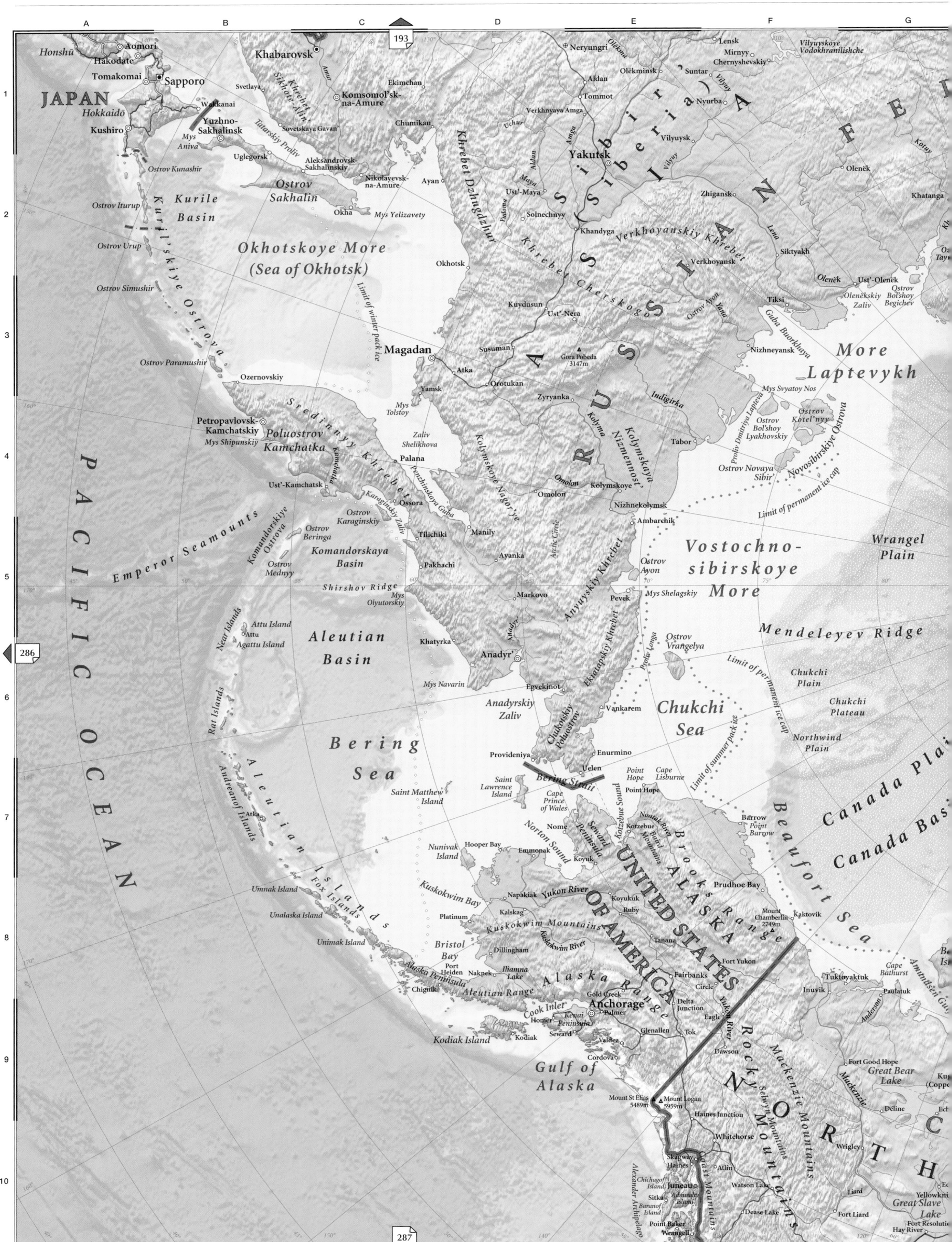

JAPAN

Honshū

Aomori
Hakodate
Tomakomai
Sapporo
Kushiro
Hokkaidō

Khabarovsk
Ekimchan
Svetlaya
Komsomol'sk-na-Amure
Chumikan
Wakkanai
Yuzhno-
Sakhalinsk
Mys Aniva
Uglegorsk
Aleksandrovsk-
Sakhalinskiy
Nikolayevsk-
na-Amure
Ayan
Okha
Mys Yelizavety
Okhotsk
Sovetskaya Gavan'
Ostrov Kunashir
Tatarskiy Proliv

*Ostrov
Sakhalin*

*Kurile
Basin*

Ostrov Iturup
Ostrov Urup
Ostrov Simushir
Ostrov Paramushir

Kuril'skiye Ostrova

*Okhotskoye More
(Sea of Okhotsk)*

Limit of winter pack ice

Ozernovskiy

Magadan
Atka
Yamsk
*Mys
Tolstoy*

Petropavlovsk-
Kamchatskiy
Mys Shipunskiy

*Poluostrov
Kamchatka*

Ust'-Kamchatsk

Sredinnyy Khrebet

*Komandorskiye
Ostrova*
*Ostrov
Beringa*
*Ostrov
Mednyy*
*Komandorskaya
Basin*
Shirshov Ridge
*Mys
Olyutorskiy*

*Ostrov
Karaginskiy*
Ossora

Karaginskiy Zaliv
Tilichiki
*Zaliv
Shelikhova*
Palana
Penzhinskaya Guba
Pakhachi

Kamchatka

Susuman
*Gora Pobeda
3147m*
Orotukan
Zyryanka
Omsukchan
Omolon
Omolon

Kolymskoye Nagor'ye

Kolyma

Indigirka

Nizhnekolymsk
Kolymskoye
Ambarchik

Anyuyskiy Khrebet

Nizhneyansk

*More
Laptevykh*

Tabor

Arctic Circle

Markovo
Pevek
Anadyr'
Mys Shelagskiy
*Ostrov
Ayon*

Khatyrka
Mys Navarin
Egvekinot
Vankarem

*Anadyrskiy
Zaliv*
*Chukotskiy
Poluostrov*
Enurmino

Providyeniya
Uelen
*Cape
Prince
of Wales*

*Bering
Sea*

Bering Strait

Saint
Lawrence
Island
*Saint Matthew
Island*

*Aleutian
Basin*

Attu Island
Attu
Agattu Island
Near Islands
Rat Islands
Atka
Andreanof Islands
Umnak Island
Unalaska Island
Unimak Island
Fox Islands
Aleutian Islands

PACIFIC OCEAN

Emperor Seamounts

*Vostochno-
sibirskoye
More*

Limit of permanent ice cap

*Ostrov
Vrangelya*

Proliv Longa

*Wrangel
Plain*

Mendeleyev Ridge

*Chukchi
Plain*

*Chukchi
Plateau*

*Northwind
Plain*

Limit of permanent ice cap

Limit of summer pack ice

*Chukchi
Sea*

Point
Hope
*Cape
Lisburne*
Point Hope
Kotzebue Sound
Kotzebue
Noatak River
Baird Mountains
Nome
*Seward
Peninsula*
Norton Sound
Koyuk
Emmonak
Hooper Bay
Nunivak
Island
Napakiak
Kalskag
Platinum
Kuskokwim Bay
Dillingham
*Bristol
Bay*
Port
Heiden
Naknek
Chignik
Alaska Peninsula
Aleutian Range
Kodiak
Kodiak Island

Kuskokwim River
Kuskokwim Mountains
Yukon River
Koyukuk
Ruby
Tanana
Fairbanks
Circle
Fort Yukon
Delta
Junction
Eagle
Tok
Glenallen
Yukon River
Dawson
Alaska Range
*Mount
Chamberlin
2749m*
Prudhoe Bay
Kaktovik
*Beaufort
Sea*
Cape
Bathurst
Tuktoyaktuk
Inuvik
Paulatuk
*Cape
Bathurst*
Amundsen Gulf

Brooks Range

UNITED STATES
OF AMERICA

ALASKA

Anchorage
Palmer
*Kenai
Peninsula*
Homer
Cook Inlet
Seward
Valdez
Cordova
Gold Creek
Iliamna
Lake

*Gulf of
Alaska*

*Mount St Elias
5489m*
*Mount Logan
5959m*
Haines Junction
Whitehorse
Skagway
Haines
Atlin
Juneau
*Chichagof
Island*
Sitka
*Baranof
Island*
*Alexander
Archipelago*
Point Baker
Wrangell
Watson Lake
Dease Lake

NORTH

Rocky Mountains
Mackenzie Mountains
Selwyn Mountains
Coast Mountains

Canada Plain
Canada Basin

Fort Good Hope
*Great Bear
Lake*
Déline
Wrigley
Fort Liard
Fort Resolution
Hay River
*Great Slave
Lake*
Yellowknife

Mackenzie

RUSSIA

Siberia

Khabarovsk
Amur
*Khrebet
Sikhote Alin'*
Amur

Yakutsk
Aldan
Maya
Maya
Yudoma
Uchur
Ust'-Maya
Solnechnyy
Khandyga
Lena
Aldan
Amga
Amga
Verkhnyaya Amga
Tommot
Aldan
Neryungri
Olëkminsk
Lensk
Olëkma
Chernyshevskiy
Suntar
Mirnyy
Nyurba
Vilyuy
Vilyuy
Vilyuysk
*Vilyuyskoye
Vodokhranilishche*
Zhigansk
Lena
Siktyakh
Verkhoyansk
Verkhoyanskiy Khrebet
Kuydusun
Ust'-Nera
Khrebet Cherskogo
Khrebet Dzhugdzhur

Olenëk
Olenëk
Ust'-Olenëk
Khatanga
Guba Buorkhaya
Tiksi
Nizhneyansk
Yana
Proliv Dmitriya Laptyeva
*Ostrov
Bol'shoy
Lyakhovskiy*
*Ostrov
Kotel'nyy*
Novosibirskiye Ostrova
*Ostrov Novaya
Sibir'*
Mys Svyatoy Nos
*Ostrov Bol'shoy
Begichev*
*Olenëkskiy
Zaliv*
*Kolymskaya
Nizmennost'*
Ekatapskiy Khrebet

193
287
286

6000m
4000m
3000m
2000m
1000m
500m
250m
100m
Sea
Level
-250m
-2000m
-4000m

Scale 1:18,190,000
(projection: Lambert Azimuthal Equal Area)

0 100 200 300 400 500 Km
0 100 200 300 400 500 Miles

Population
- ■ above 5 million
- ◙ 1 million to 5 million
- ◉ 500,000 to 1 million
- ◎ 100,000 to 500,000
- ⊕ 50,000 to 100,000
- ○ 10,000 to 50,000
- ○ below 10,000

19,686ft
13,124ft
9843ft
6562ft
3281ft
1640ft
820ft
328ft
Sea Level
-820ft
-6562ft
-13,124ft

Selected map labels:

RUSSIAN FEDERATION · EUROPE · BELARUS · UKRAINE · LATVIA · LITHUANIA · POLAND · ESTONIA · FINLAND · SWEDEN · NORWAY · DENMARK · GERMANY · UNITED KINGDOM · FAROE ISLANDS · ICELAND · CANADA · AMERICA · GREENLAND (to Denmark) · SVALBARD (to Norway) · JAN MAYEN (to Norway)

MINSK · L'viv · Brest · VILNIUS · RIGA · Kaliningrad · WARSZAWA (WARSAW) · Olsztyn · Poznań · Szczecin · BERLIN · Hamburg · Rostock · Malmö · KØBENHAVN (COPENHAGEN) · Göteborg · OSLO · Stavanger · Bergen · Trondheim · Ålesund · STOCKHOLM · Norrköping · Örebro · Gävle · Umeå · Vaasa · Oulu · HELSINKI · TALLINN · Tartu · Klaipeda · Gdańsk · Gotland · Öland · Tampere · Turku · Åland

Tver' · Torzhok · Vyshniy Volochek · Velikiye Luki · Velikiy Novgorod · Pskov · Sankt-Peterburg · Pul'kovo · Lake Peipus · Vologda · Cherepovets · Belozersk · Rybinskoye Vodokhranilishche · Vodokhranilishche · Petrozavodsk · Medvezh'yegorsk · Kem' · Belomorsk · Onezhskoye Ozero · Ladozhskoye Ozero · Sortavala · Nadvoitsy · Plesetsk · Arkhangel'sk (Archangel) · Severodvinsk · Mezen' · Severnaya Dvina · Vel'sk

Murmansk · Severomorsk · Monchegorsk · Hammerfest · North Cape · Tromsø · Barents Trough · Fugloya Bank · Nar'yan-Mar · Pechora · Usinsk · Inta · Vorkuta · Severnyy · Labytnangi · Salekhard · Ob' · Ural'skiye Gory · Urengoy · Novyy Urengoy · Tazovskiy · Taz · Pur · Kureyka

Noril'sk · Dudinka · Talnakh · Gydanskiy · Vorontsovo · Gyda · Dikson · Poluostrov Yamal · Yugorskiy Poluostrov · Ostrov Vaygach · Proliv Karskiye Vorota · Krasino · Ostrov Kolguyev · Cheshskaya Guba · Poluostrov Kanin · Mys Kanin Nos · Beloye More · Kol'skiy Poluostrov · Ostrovnoy · Murmansk Rise · Barents Sea

Karskoye More · Novaya Zemlya · East Novaya Zemlya Trough · Obskaya Guba · Baydaratskaya Guba · Yeniseyskiy Zaliv · Ostrov Belyy · Syatatya Anna Trough · Nansen Basin · Barents Plain · Svalbard · Spitsbergen · Hopen · Bjørnøya · Longyearbyen · Norwegian Sea · Voring Plateau · Norwegian Basin · Shetland Islands · Orkney Islands · Edinburgh · Glasgow · Faroe-Shetland Trough · FAROE ISLANDS (to Denmark) · Bill Baileys Bank · Faroe-Iceland Ridge

Gora Pobeda 621m · Prasina · Gakkel Ridge · Amundsen Basin · Pole Plain · North Pole · Lomonosov Ridge · Makarov Basin · Alpha Cordillera · North Magnetic Pole (2005) · ARCTIC OCEAN · Lincoln Sea · Alert · Cape Columbia · Ellesmere Island · North Geomagnetic Pole (2005) · Knud Rasmussen Land · Wandel Sea · Kap Morris Jesup · Nord · Kong Frederik VIII Land · Daneborg · Mohns Ridge · Kong Oscar Fjord · Greenland Sea · Greenland Plain · Limit of permanent ice cap · Jan Mayen Fracture Zone · Jan Mayen Ridge · Kolbeinsey Ridge · Iceland Plateau · Reykjanes Ridge

Ittoqqortoormiit · Petermann Bjerg 2940m · Gunnbjørn Fjeld 3700m · Kangikajik · Denmark Strait · Iceland · Akureyri · REYKJAVÍK · Iceland Basin · Hatton Ridge · ATLANTIC OCEAN · Reykjanes Basin

QAASUITSUP · Qaanaaq · Savissivik · Innaangaaneq · Qaarsut · Kullorsuaq · Upernavik · Uummannaq · Ilulissat · Qasigiannguit · Qeqertarsuaq · Aasiaat · Kong Frederik IX Land · Sisimiut · Maniitsoq · NUUK · Paamiut · QEQQATA · SERMERSOOQ · KUJALLEQ · Ammassalik · Kong Christian IX Land · Kong Frederik VI Kyst · Mont Forel 3360m · Nanap Isua (Kap Farvel) · Narsarsuaq · Qaqortoq · Nanortalik · Tvittut · Eirik Ridge · Limit of winter pack ice · Limit of summer pack ice

Queen Elizabeth Islands · Prince Gustaf Adolf Sea · Ellef Ringnes Island · Axel Heiberg Island · Prince Patrick Island · Melville Island · Bathurst Island · Parry Islands · Devon Island · Baffin Basin · Baffin Bay · Grise Fiord (Ausuittuq) · Qimusseriarsuaq · Baffin Island · Lancaster Sound · Resolute (Qausuittuq) · Viscount Melville Sound · Prince of Wales Island · Somerset Island · Borden Peninsula · Pond Inlet (Mittimatalik) · Brodeur Peninsula · Gulf of Boothia · Boothia Peninsula · M'Clintock Channel · Victoria Island · Cambridge Bay (Ikaluktutiak) · King William Island · Kugaaruk (Pelly Bay) · Queen Maud Gulf · Melville Peninsula · Igloolik · Prince Charles Island · Nettilling Lake · Foxe Basin · Amadjuak Lake · Hall Peninsula · Iqaluit (Frobisher Bay) · Frobisher Bay · Kimmirut (Lake Harbour) · Cape Dyer · Cumberland Peninsula · Cumberland Sound · Davis Strait · Labrador Sea · Labrador Basin · Cape Chidley · Hamilton Bank

Repulse Bay · Back · Bathurst Inlet · Reliance · Dubawnt Lake · Baker Lake · Chesterfield Inlet · Coral Harbour (Salliq) · Southampton Island · Coats Island · Mansel Island · Péninsule d'Ungava · Hudson Bay · Hudson Strait · Foxe Channel · Foxe Peninsula · Ivujivik · Quaqtaq · Ungava Bay · Kangiqsualujjuaq · Koksoak · Nain · Cartwright · Lake Melville · North West River · St. Anthony · Cape Bauld · Labrador · Grand Banks of Newfoundland · St. John's

Geographical comparisons

Largest countries

Russian Federation	6,592,735 sq miles	(17,075,200 sq km)
Canada	3,855,171 sq miles	(9,984,670 sq km)
USA	3,794,100 sq miles	(9,826,675 sq km)
China	3,705,386 sq miles	(9,596,960 sq km)
Brazil	3,286,470 sq miles	(8,511,965 sq km)
Australia	2,967,893 sq miles	(7,686,850 sq km)
India	1,269,339 sq miles	(3,287,590 sq km)
Argentina	1,068,296 sq miles	(2,766,890 sq km)
Kazakhstan	1,049,150 sq miles	(2,717,300 sq km)
Algeria	919,590 sq miles	(2,381,740 sq km)

Smallest countries

Vatican City	0.17 sq miles	(0.44 sq km)
Monaco	0.75 sq miles	(1.95 sq km)
Nauru	8.2 sq miles	(21.2 sq km)
Tuvalu	10 sq miles	(26 sq km)
San Marino	24 sq miles	(61 sq km)
Liechtenstein	62 sq miles	(160 sq km)
Marshall Islands	70 sq miles	(181 sq km)
St. Kitts & Nevis	101 sq miles	(261 sq km)
Maldives	116 sq miles	(300 sq km)
Malta	124 sq miles	(320 sq km)

Largest islands

	To the nearest 1000 – or 100,000 for the largest	
Greenland	849,400 sq miles	(2,200,000 sq km)
New Guinea	312,000 sq miles	(808,000 sq km)
Borneo	292,222 sq miles	(757,050 sq km)
Madagascar	229,300 sq miles	(594,000 sq km)
Sumatra	202,300 sq miles	(524,000 sq km)
Baffin Island	183,800 sq miles	(476,000 sq km)
Honshu	88,800 sq miles	(230,000 sq km)
Britain	88,700 sq miles	(229,800 sq km)
Victoria Island	81,900 sq miles	(212,000 sq km)
Ellesmere Island	75,700 sq miles	(196,000 sq km)

Richest countries

	GNI per capita, in US$
Monaco	186,950
Liechtenstein	136,770
Norway	102,610
Switzerland	90,760
Qatar	86,790
Luxembourg	69,900
Australia	65,390
Sweden	61,760
Denmark	61,680
Singapore	54,040

Poorest countries

	GNI per capita, in US$
Burundi	260
Malawi	270
Somalia	288
Central African Republic	320
Niger	400
Liberia	410
Dem. Rep. Congo	430
Madagascar	440
Guinea	460
Ethiopia	470
Eritrea	490
Gambia	500

Most populous countries

China	1,393,800,000
India	1,267,400,000
USA	322,600,000
Indonesia	252,800,000
Brazil	202,120,000
Pakistan	185,100,000
Nigeria	178,500,000
Bangladesh	159,000,000
Russian Federation	142,500,000
Japan	127,000,000

Least populous countries

Vatican City	842
Nauru	9488
Tuvalu	10,782
Palau	21,186
San Marino	32,742
Monaco	36,950
Liechtenstein	37,313
St Kitts & Nevis	51,538
Marshall Islands	70,983
Dominica	73,449
Andorra	85,458
Antigua & Barbuda	91,295

Most densely populated countries

Monaco	49,267 people per sq mile	(18,949 per sq km)
Singapore	23,305 people per sq mile	(9016 per sq km)
Vatican City	4953 people per sq mile	(1914 per sq km)
Bahrain	4762 people per sq mile	(1841 per sq km)
Maldives	3448 people per sq mile	(1333 per sq km)
Malta	3226 people per sq mile	(1250 per sq km)
Bangladesh	3066 people per sq mile	(1184 per sq km)
Taiwan	1879 people per sq mile	(725 per sq km)
Barbados	1807 people per sq mile	(698 per sq km)
Mauritius	1671 people per sq mile	(645 per sq km)

Most sparsely populated countries

Mongolia	5 people per sq mile	(2 per sq km)
Namibia	7 people per sq mile	(3 per sq km)
Australia	8 people per sq mile	(3 per sq km)
Suriname	8 people per sq mile	(3 per sq km)
Iceland	8 people per sq mile	(3 per sq km)
Botswana	9 people per sq mile	(4 per sq km)
Libya	9 people per sq mile	(4 per sq km)
Mauriania	10 people per sq mile	(4 per sq km)
Canada	10 people per sq mile	(4 per sq km)
Guyana	11 people per sq mile	(4 per sq km)

Most widely spoken languages

1. Chinese (Mandarin)	6. Arabic
2. English	7. Bengali
3. Hindi	8. Portuguese
4. Spanish	9. Malay-Indonesian
5. Russian	10. French

Largest conurbations

	Urban area population
Tokyo	37,800,000
Jakarta	30,500,000
Manila	24,100,000
Delhi	24,000,000
Karachi	23,500,000
Seoul	23,500,000
Shanghai	23,400,000
Beijing	21,000,000
New York City	20,600,000
Guangzhou	20,600,000
São Paulo	20,300,000
Mexico City	20,000,000
Mumbai	17,700,000
Osaka	17,400,000
Lagos	17,000,000
Moscow	16,100,000
Dhaka	15,700,000
Lahore	15,600,000
Los Angeles	15,000,000
Bangkok	15,000,000
Kolkatta	14,700,000
Buenos Aires	14,100,000
Tehran	13,500,000
Istanbul	13,300,000
Shenzhen	12,000,000

Countries with the most land borders

14: China	(Afghanistan, Bhutan, India, Kazakhstan, Kyrgyzstan, Laos, Mongolia, Myanmar (Burma), Nepal, North Korea, Pakistan, Russian Federation, Tajikistan, Vietnam)	
14: Russian Federation	(Azerbaijan, Belarus, China, Estonia, Finland, Georgia, Kazakhstan, Latvia, Lithuania, Mongolia, North Korea, Norway, Poland, Ukraine)	
10: Brazil	(Argentina, Bolivia, Colombia, French Guiana, Guyana, Paraguay, Peru, Suriname, Uruguay, Venezuela)	
9: Congo, Dem. Rep.	(Angola, Burundi, Central African Republic, Congo, Rwanda, South Sudan, Tanzania, Uganda, Zambia)	
9: Germany	(Austria, Belgium, Czech Republic, Denmark, France, Luxembourg, Netherlands, Poland, Switzerland)	
8: Austria	(Czech Republic, Germany, Hungary, Italy, Liechtenstein, Slovakia, Slovenia, Switzerland)	
8: France	(Andorra, Belgium, Germany, Italy, Luxembourg, Monaco, Spain, Switzerland)	
8: Tanzania	(Burundi, Dem. Rep. Congo, Kenya, Malawi, Mozambique, Rwanda, Uganda, Zambia)	
8: Turkey	(Armenia, Azerbaijan, Bulgaria, Georgia, Greece, Iran, Iraq, Syria)	
8: Zambia	(Angola, Botswana, Dem. Rep.Congo, Malawi, Mozambique, Namibia, Tanzania, Zimbabwe)	

Longest rivers

Nile (NE Africa)	4160 miles	(6695 km)
Amazon (South America)	4049 miles	(6516 km)
Yangtze (China)	3915 miles	(6299 km)
Mississippi/Missouri (USA)	3710 miles	(5969 km)
Ob'-Irtysh (Russian Federation)	3461 miles	(5570 km)
Yellow River (China)	3395 miles	(5464 km)
Congo (Central Africa)	2900 miles	(4667 km)
Mekong (Southeast Asia)	2749 miles	(4425 km)
Lena (Russian Federation)	2734 miles	(4400 km)
Mackenzie (Canada)	2640 miles	(4250 km)
Yenisey (Russian Federation)	2541 miles	(4090km)

Highest mountains

		Height above sea level
Everest	29,029 ft	(8848 m)
K2	28,253 ft	(8611 m)
Kangchenjunga I	28,210 ft	(8598 m)
Makalu I	27,767 ft	(8463 m)
Cho Oyu	26,907 ft	(8201 m)
Dhaulagiri I	26,796 ft	(8167 m)
Manaslu I	26,783 ft	(8163 m)
Nanga Parbat I	26,661 ft	(8126 m)
Annapurna I	26,547 ft	(8091 m)
Gasherbrum I	26,471 ft	(8068 m)

Largest bodies of inland water

		With area and depth
Caspian Sea	143,243 sq miles (371,000 sq km)	3215 ft (980 m)
Lake Superior	31,151 sq miles (83,270 sq km)	1289 ft (393 m)
Lake Victoria	26,828 sq miles (69,484 sq km)	328 ft (100 m)
Lake Huron	23,436 sq miles (60,700 sq km)	751 ft (229 m)
Lake Michigan	22,402 sq miles (58,020 sq km)	922 ft (281 m)
Lake Tanganyika	12,703 sq miles (32,900 sq km)	4700 ft (1435 m)
Great Bear Lake	12,274 sq miles (31,790 sq km)	1047 ft (319 m)
Lake Baikal	11,776 sq miles (30,500 sq km)	5712 ft (1741 m)
Great Slave Lake	10,981 sq miles (28,440 sq km)	459 ft (140 m)
Lake Erie	9,915 sq miles (25,680 sq km)	197 ft (60 m)

Deepest ocean features

Challenger Deep, Mariana Trench (Pacific)	35,827 ft	(10,920 m)
Vityaz III Depth, Tonga Trench (Pacific)	35,704 ft	(10,882 m)
Vityaz Depth, Kuril-Kamchatka Trench (Pacific)	34,588 ft	(10,542 m)
Cape Johnson Deep, Philippine Trench (Pacific)	34,441 ft	(10,497 m)
Kermadec Trench (Pacific)	32,964 ft	(10,047 m)
Ramapo Deep, Japan Trench (Pacific)	32,758 ft	(9984 m)
Milwaukee Deep, Puerto Rico Trench (Atlantic)	30,185 ft	(9200 m)
Argo Deep, Torres Trench (Pacific)	30,070 ft	(9165 m)
Meteor Depth, South Sandwich Trench (Atlantic)	30,000 ft	(9144 m)
Planet Deep, New Britain Trench (Pacific)	29,988 ft	(9140 m)

Greatest waterfalls

		Mean flow of water
Boyoma (Dem. Rep. Congo)	600,400 cu. ft/sec	(17,000 cu.m/sec)
Khône (Laos/Cambodia)	410,000 cu. ft/sec	(11,600 cu.m/sec)
Niagara (USA/Canada)	195,000 cu. ft/sec	(5500 cu.m/sec)
Grande, Salto (Uruguay)	160,000 cu. ft/sec	(4500 cu.m/sec)
Paulo Afonso (Brazil)	100,000 cu. ft/sec	(2800 cu.m/sec)
Urubupungá, Salto do (Brazil)	97,000 cu. ft/sec	(2750 cu.m/sec)
Iguaçu (Argentina/Brazil)	62,000 cu. ft/sec	(1700 cu.m/sec)
Maribondo, Cachoeira do (Brazil)	53,000 cu. ft/sec	(1500 cu.m/sec)
Victoria (Zimbabwe)	39,000 cu. ft/sec	(1100 cu.m/sec)
Murchison Falls (Uganda)	42,000 cu. ft/sec	(1200 cu.m/sec)
Churchill (Canada)	35,000 cu. ft/sec	(1000 cu.m/sec)
Kaveri Falls (India)	33,000 cu. ft/sec	(900 cu.m/sec)

Highest waterfalls

		* Indicates that the total height is a single leap
Angel (Venezuela)	3212 ft	(979 m)
Tugela (South Africa)	3110 ft	(948 m)
Utigard (Norway)	2625 ft	(800 m)
Mongefossen (Norway)	2539 ft	(774 m)
Mtarazi (Zimbabwe)	2500 ft	(762 m)
Yosemite (USA)	2425 ft	(739 m)
Ostre Mardola Foss (Norway)	2156 ft	(657 m)
Tyssestrengane (Norway)	2119 ft	(646 m)
*Cuquenan (Venezuela)	2001 ft	(610 m)
Sutherland (New Zealand)	1903 ft	(580 m)
*Kjellfossen (Norway)	1841 ft	(561 m)

Largest deserts

NB – Most of Antarctica is a polar desert, with only 50mm of precipitation annually

Sahara	3,450,000 sq miles	(9,065,000 sq km)
Gobi	500,000 sq miles	(1,295,000 sq km)
Ar Rub al Khali	289,600 sq miles	(750,000 sq km)
Great Victorian	249,800 sq miles	(647,000 sq km)
Sonoran	120,000 sq miles	(311,000 sq km)
Kalahari	120,000 sq miles	(310,800 sq km)
Kara Kum	115,800 sq miles	(300,000 sq km)
Takla Makan	100,400 sq miles	(260,000 sq km)
Namib	52,100 sq miles	(135,000 sq km)
Thar	33,670 sq miles	(130,000 sq km)

Hottest inhabited places

Djibouti (Djibouti)	86° F	(30 °C)
Tombouctou (Mali)	84.7° F	(29.3 °C)
Tirunelveli (India)		
Tuticorin (India)		
Nellore (India)	84.5° F	(29.2 °C)
Santa Marta (Colombia)		
Aden (Yemen)	84° F	(28.9 °C)
Madurai (India)		
Niamey (Niger)		
Hodeida (Yemen)	83.8° F	(28.8 °C)
Ouagadougou (Burkina Faso)		
Thanjavur (India)		
Tiruchchirappalli (India)		

Driest inhabited places

Aswân (Egypt)	0.02 in	(0.5 mm)
Luxor (Egypt)	0.03 in	(0.7 mm)
Arica (Chile)	0.04 in	(1.1 mm)
Ica (Peru)	0.1 in	(2.3 mm)
Antofagasta (Chile)	0.2 in	(4.9 mm)
Al Minya (Egypt)	0.2 in	(5.1 mm)
Asyut (Egypt)	0.2 in	(5.2 mm)
Callao (Peru)	0.5 in	(12.0 mm)
Trujillo (Peru)	0.55 in	(14.0 mm)
Al Fayyum (Egypt)	0.8 in	(19.0 mm)

Wettest inhabited places

Mawsynram (India)	467 in	(11,862 mm)
Mount Waialeale (Hawaii, USA)	460 in	(11,684 mm)
Cherrapunji (India)	450 in	(11,430 mm)
Cape Debundsha (Cameroon)	405 in	(10,290 mm)
Quibdo (Colombia)	354 in	(8892 mm)
Buenaventura (Colombia)	265 in	(6743 mm)
Monrovia (Liberia)	202 in	(5131 mm)
Pago Pago (American Samoa)	196 in	(4990 mm)
Mawlamyine (Myanmar [Burma])	191 in	(4852 mm)
Lae (Papua New Guinea)	183 in	(4645 mm)

Countries of the World

There are currently 196 independent countries in the world – more than at any previous time – and almost 60 dependencies. Antarctica is the only land area on Earth that is not officially part of, and does not belong to, any single country.

In 1950, the world comprised 82 countries. In the decades following, many more states came into being as they achieved independence from their former colonial rulers. Most recent additions were caused by the breakup of the former Soviet Union in 1991, and the former Yugoslavia in 1992, which swelled the ranks of independent states. In July 2011, South Sudan became the latest country to be formed after declaring independence from Sudan.

Country factfile key

Formation Date of independence / date current borders were established
Population Total population / population density – based on total *land* area
Languages An asterisk (*) denotes the official language(s)
Calorie consumption Average number of calories consumed daily per person

AFGHANISTAN
Central Asia

Official name Islamic Republic of Afghanistan
Formation 1919 / 1919
Capital Kabul
Population 31.3 million / 124 people per sq mile (48 people per sq km)
Total area 250,000 sq. miles (647,500 sq. km)
Languages Pashtu*, Tajik, Dari*, Farsi, Uzbek, Turkmen
Religions Sunni Muslim 80%, Shi'a Muslim 19%, Other 1%
Ethnic mix Pashtun 38%, Tajik 25%, Hazara 19%, Uzbek and Turkmen 15%, Other 3%
Government Nonparty system
Currency Afghani = 100 puls
Literacy rate rate 32%
Calorie consumption 2090 kilocalories

ALBANIA
Southeast Europe

Official name Republic of Albania
Formation 1912 / 1921
Capital Tirana
Population 3.2 million / 302 people per sq mile (117 people per sq km)
Total area 11,100 sq. miles (28,748 sq. km)
Languages Albanian*, Greek
Religions Sunni Muslim 70%, Albanian Orthodox 20%, Roman Catholic 10%
Ethnic mix Albanian 98%, Greek 1%, Other 1%
Government Parliamentary system
Currency Lek = 100 qindarka (qintars)
Literacy rate 97%
Calorie consumption 3023 kilocalories

ALGERIA
North Africa

Official name People's Democratic Republic of Algeria
Formation 1962 / 1962
Capital Algiers
Population 39.9 million / 43 people per sq mile (17 people per sq km)
Total area 919,590 sq. miles (2,381,740 sq. km)
Languages Arabic*, Tamazight (Kabyle, Shawia, Tamashek), French
Religions Sunni Muslim 99%, Christian and Jewish 1%
Ethnic mix Arab 75%, Berber 24%, European and Jewish 1%
Government Presidential system
Currency Algerian dinar = 100 centimes
Literacy rate 73%
Calorie consumption 3296 kilocalories

ANDORRA
Southwest Europe

Official name Principality of Andorra
Formation 1278 / 1278
Capital Andorra la Vella
Population 85,485 / 475 people per sq mile (184 people per sq km)
Total area 181 sq. miles (468 sq. km)
Languages Spanish, Catalan*, French, Portuguese
Religions Roman Catholic 94%, Other 6%
Ethnic mix Spanish 46%, Andorran 28%, Other 18%, French 8%
Government Parliamentary system
Currency Euro = 100 cents
Literacy rate 99%
Calorie consumption Not available

ANGOLA
Southern Africa

Official name Republic of Angola
Formation 1975 / 1975
Capital Luanda
Population 22.1 million / 46 people per sq mile (18 people per sq km)
Total area 481,351 sq. miles (1,246,700 sq. km)
Languages Portuguese*, Umbundu, Kimbundu, Kikongo
Religions Roman Catholic 68%, Protestant 20%, Indigenous beliefs 12%
Ethnic mix Ovimbundu 37%, Kimbundu 25%, Other 25%, Bakongo 13%
Government Presidential system
Currency Readjusted kwanza = 100 lwei
Literacy rate 71%
Calorie consumption 2473 kilocalories

ANTIGUA & BARBUDA
West Indies

Official name Antigua and Barbuda
Formation 1981 / 1981
Capital St. John's
Population 91,295 / 537 people per sq mile (207 people per sq km)
Total area 170 sq. miles (442 sq. km)
Languages English*, English patois
Religions Anglican 45%, Other Protestant 42%, Roman Catholic 10%, Other 2%, Rastafarian 1%
Ethnic mix Black African 95%, Other 5%
Government Parliamentary system
Currency East Caribbean dollar = 100 cents
Literacy rate 99%
Calorie consumption 2396 kilocalories

ARGENTINA
South America

Official name Argentine Republic
Formation 1816 / 1816
Capital Buenos Aires
Population 41.8 million / 40 people per sq mile (15 people per sq km)
Total area 1,068,296 sq. miles (2,766,890 sq. km)
Languages Spanish*, Italian, Amerindian languages
Religions Roman Catholic 70%, Other 18%, Protestant 9%, Muslim 2%, Jewish 1%
Ethnic mix Indo-European 97%, Mestizo 2%, Amerindian 1%
Government Presidential system
Currency Argentine peso = 100 centavos
Literacy rate 98%
Calorie consumption 3155 kilocalories

ARMENIA
Southwest Asia

Official name Republic of Armenia
Formation 1991 / 1991
Capital Yerevan
Population 3 million / 261 people per sq mile (101 people per sq km)
Total area 11,506 sq. miles (29,800 sq. km)
Languages Armenian*, Azeri, Russian
Religions Armenian Apostolic Church (Orthodox) 88%, Armenian Catholic Church 6%, Other 6%
Ethnic mix Armenian 98%, Other 1%, Yezidi 1%
Government Parliamentary system
Currency Dram = 100 luma
Literacy rate 99%
Calorie consumption 2809 kilocalories

AUSTRALIA
Australasia & Oceania

Official name Commonwealth of Australia
Formation 1901 / 1901
Capital Canberra
Population 23.6 million / 8 people per sq mile (3 people per sq km)
Total area 2,967,893 sq. miles (7,686,850 sq. km)
Languages English*, Italian, Cantonese, Greek, Arabic, Vietnamese, Aboriginal languages
Religions Roman Catholic 26%, Nonreligious 19%, Anglican 19%, Other 17%, Other Christian 13%, United Church 6%
Ethnic mix European origin 50%, Australian 25.5%, other 19%, Asian 5%, Aboriginal 0.5%
Government Parliamentary system
Currency Australian dollar = 100 cents
Literacy rate 99%
Calorie consumption 3265 kilocalories

AUSTRIA
Central Europe

Official name Republic of Austria
Formation 1918 / 1919
Capital Vienna
Population 8.5 million / 266 people per sq mile (103 people per sq km)
Total area 32,378 sq. miles (83,858 sq. km)
Languages German*, Croatian, Slovenian, Hungarian (Magyar)
Religions Roman Catholic 78%, Nonreligious 9%, Other (including Jewish and Muslim) 8%, Protestant 5%
Ethnic mix Austrian 93%, Croat, Slovene, and Hungarian 6%, Other 1%
Government Parliamentary system
Currency Euro = 100 cents
Literacy rate 99%
Calorie consumption 3784 kilocalories

AZERBAIJAN
Southwest Asia

Official name Republic of Azerbaijan
Formation 1991 / 1991
Capital Baku
Population 9.5 million / 284 people per sq mile (110 people per sq km)
Total area 33,436 sq. miles (86,600 sq. km)
Languages Azeri*, Russian
Religions Shi'a Muslim 68%, Sunni Muslim 26%, Russian Orthodox 3%, Armenian Apostolic Church (Orthodox) 2%, Other 1%
Ethnic mix Azeri 91%, Other 3%, Lazs 2%, Armenian 2%, Russian 2%
Government Presidential system
Currency New manat = 100 gopik
Literacy rate 99%
Calorie consumption 2952 kilocalories

THE BAHAMAS
West Indies

Official name Commonwealth of The Bahamas
Formation 1973 / 1973
Capital Nassau
Population 400,000 / 103 people per sq mile (40 people per sq km)
Total area 5382 sq. miles (13,940 sq. km)
Languages English*, English Creole, French Creole
Religions Baptist 32%, Anglican 20%, Roman Catholic 19%, Other 17%, Methodist 6%, Church of God 6%
Ethnic mix Black African 85%, European 12%, Asian and Hispanic 3%
Government Parliamentary system
Currency Bahamian dollar = 100 cents
Literacy rate 96%
Calorie consumption 2575 kilocalories

BAHRAIN
Southwest Asia

Official name Kingdom of Bahrain
Formation 1971 / 1971
Capital Manama
Population 1.3 million / 4762 people per sq mile (1841 people per sq km)
Total area 239 sq. miles (620 sq. km)
Languages Arabic*
Religions Muslim (mainly Shi'a) 99%, Other 1%
Ethnic mix Bahraini 63%, Asian 19%, Other Arab 10%, Iranian 8%
Government Mixed monarchical–parliamentary system
Currency Bahraini dinar = 1000 fils
Literacy rate 95%
Calorie consumption Not available

BANGLADESH
South Asia

Official name People's Republic of Bangladesh
Formation 1971 / 1971
Capital Dhaka
Population 159 million / 3066 people per sq mile (1184 people per sq km)
Total area 55,598 sq. miles (144,000 sq. km)
Languages Bengali*, Urdu, Chakma, Marma (Magh), Garo, Khasi, Santhali, Tripuri, Mro
Religions Muslim (mainly Sunni) 88%, Hindu 11%, Other 1%
Ethnic mix Bengali 98%, Other 2%
Government Parliamentary system
Currency Taka = 100 poisha
Literacy rate 59%
Calorie consumption 2450 kilocalories

BARBADOS
West Indies

Official name Barbados
Formation 1966 / 1966
Capital Bridgetown
Population 300,000 / 1807 people per sq mile (698 people per sq km)
Total area 166 sq. miles (430 sq. km)
Languages Bajan (Barbadian English), English*
Religions Anglican 40%, Other 24%, Nonreligious 17%, Pentecostal 8%, Methodist 7%, Roman Catholic 4%
Ethnic mix Black African 92%, White 3%, Other 3%, Mixed race 2%
Government Parliamentary system
Currency Barbados dollar = 100 cents
Literacy rate 99%
Calorie consumption 3047 kilocalories

BELARUS
Eastern Europe

Official name Republic of Belarus
Formation 1991 / 1991
Capital Minsk
Population 9.3 million / 116 people per sq mile (45 people per sq km)
Total area 80,154 sq. miles (207,600 sq. km)
Languages Belarussian*, Russian*
Religions Orthodox Christian 80%, Roman Catholic 14%, Other 4%, Protestant 2%
Ethnic mix Belarussian 81%, Russian 11%, Polish 4%, Ukrainian 2%, Other 2%
Government Presidential system
Currency Belarussian rouble = 100 kopeks
Literacy rate 99%
Calorie consumption 3253 kilocalories

BELGIUM
Northwest Europe

Official name Kingdom of Belgium
Formation 1830 / 1919
Capital Brussels
Population 11.1 million / 876 people per sq mile (338 people per sq km)
Total area 11,780 sq. miles (30,510 sq. km)
Languages Dutch*, French*, German*
Religions Roman Catholic 88%, Other 10%, Muslim 2%
Ethnic mix Fleming 58%, Walloon 33%, Other 6%, Italian 2%, Moroccan 1%
Government Parliamentary system
Currency Euro = 100 cents
Literacy rate 99%
Calorie consumption 3793 kilocalories

BELIZE
Central America

Official name Belize
Formation 1981 / 1981
Capital Belmopan
Population 300,000 / 34 people per sq mile (13 people per sq km)
Total area 8867 sq. miles (22,966 sq. km)
Languages English Creole, Spanish, English*, Mayan, Garifuna (Carib)
Religions Roman Catholic 62%, Other 13%, Anglican 12%, Methodist 6%, Mennonite 4%, Seventh-day Adventist 3%
Ethnic mix Mestizo 49%, Creole 25%, Maya 11%, Garifuna 6%, Other 6%, Asian Indian 3%
Government Parliamentary system
Currency Belizean dollar = 100 cents
Literacy rate 75%
Calorie consumption 2751 kilocalories

BENIN
West Africa

Official name Republic of Benin
Formation 1960 / 1960
Capital Porto-Novo
Population 10.6 million / 248 people per sq mile (96 people per sq km)
Total area 43,483 sq. miles (112,620 sq. km)
Languages Fon, Bariba, Yoruba, Adja, Houeda, Somba, French*
Religions Indigenous beliefs and Voodoo 50%, Christian 30%, Muslim 20%
Ethnic mix Fon 41%, Other 21%, Adja 16%, Yoruba 12%, Bariba 10%
Government Presidential system
Currency CFA franc = 100 centimes
Literacy rate 29%
Calorie consumption 2594 kilocalories

BHUTAN
South Asia

Official name Kingdom of Bhutan
Formation 1656 / 1865
Capital Thimphu
Population 800,000 / 44 people per sq mile (17 people per sq km)
Total area 18,147 sq. miles (47,000 sq. km)
Languages Dzongkha*, Nepali, Assamese
Religions Mahayana Buddhist 75%, Hindu 25%
Ethnic mix Drukpa 50%, Nepalese 35%, Other 15%
Government Mixed monarchical–parliamentary system
Currency Ngultrum = 100 chetrum
Literacy rate 53%
Calorie consumption Not available

BOLIVIA
South America

Official name Plurinational State of Bolivia
Formation 1825 / 1938
Capital La Paz (administrative); Sucre (judicial)
Population 10.8 million / 26 people per sq mile (10 people per sq km)
Total area 424,162 sq. miles (1,098,580 sq. km)
Languages Aymara*, Quechua*, Spanish*
Religions Roman Catholic 93%, Other 7%
Ethnic mix Quechua 37%, Aymara 32%, Mixed race 13%, European 10%, Other 8%
Government Presidential system
Currency Boliviano = 100 centavos
Literacy rate 94%
Calorie consumption 2254 kilocalories

BOSNIA & HERZEGOVINA
Southeast Europe

Official name Bosnia and Herzegovina
Formation 1992 / 1992
Capital Sarajevo
Population 3.8 million / 192 people per sq mile (74 people per sq km)
Total area 19,741 sq. miles (51,129 sq. km)
Languages Bosnian*, Serbian*, Croatian*
Religions Muslim (mainly Sunni) 40%, Orthodox Christian 31%, Roman Catholic 15%, Other 10%, Protestant 4%
Ethnic mix Bosniak 48%, Serb 34%, Croat 16%, Other 2%
Government Parliamentary system
Currency Marka = 100 pfeninga
Literacy rate 98%
Calorie consumption 3130 kilocalories

BOTSWANA
Southern Africa

Official name Republic of Botswana
Formation 1966 / 1966
Capital Gaborone
Population 2 million / 9 people per sq mile (4 people per sq km)
Total area 231,803 sq. miles (600,370 sq. km)
Languages Setswana, English*, Shona, San, Khoikhoi, isiNdebele
Religions Christian (mainly Protestant) 70%, Nonreligious 20%, Traditional beliefs 6%, Other (including Muslim) 4%
Ethnic mix Tswana 79%, Kalanga 11%, Other 10%
Government Presidential system
Currency Pula = 100 thebe
Literacy rate 87%
Calorie consumption 2285 kilocalories

BRAZIL
South America

Official name Federative Republic of Brazil
Formation 1822 / 1828
Capital Brasília
Population 202 million / 62 people per sq mile (24 people per sq km)
Total area 3,286,470 sq. miles (8,511,965 sq. km)
Languages Portuguese*, German, Italian, Spanish, Polish, Japanese, Amerindian languages
Religions Roman Catholic 74%, Protestant 15%, Atheist 7%, Other 3%, Afro-American Spiritist 1%
Ethnic mix White 54%, Mixed race 38%, Black 6%, Other 2%
Government Presidential system
Currency Real = 100 centavos
Literacy rate 91%
Calorie consumption 3263 kilocalories

BRUNEI
Southeast Asia

Official name Brunei Darussalam
Formation 1984 / 1984
Capital Bandar Seri Begawan
Population 400,000 / 197 people per sq mile (76 people per sq km)
Total area 2228 sq. miles (5770 sq. km)
Languages Malay*, English, Chinese
Religions Muslim (mainly Sunni) 66%, Buddhist 14%, Other 10%, Christian 10%
Ethnic mix Malay 67%, Chinese 16%, Other 11%, Indigenous 6%
Government Monarchy
Currency Brunei dollar = 100 cents
Literacy rate 95%
Calorie consumption 2949 kilocalories

BULGARIA
Southeast Europe

Official name Republic of Bulgaria
Formation 1908 / 1947
Capital Sofia
Population 7.2 million / 169 people per sq mile (65 people per sq km)
Total area 42,822 sq. miles (110,910 sq. km)
Languages Bulgarian*, Turkish, Romani
Religions Bulgarian Orthodox 83%, Muslim 12%, Other 4%, Roman Catholic 1%
Ethnic mix Bulgarian 84%, Turkish 9%, Roma 5%, Other 2%
Government Parliamentary system
Currency Lev = 100 stotinki
Literacy rate 98%
Calorie consumption 2877 kilocalories

BURKINA FASO
West Africa

Official name Burkina Faso
Formation 1960 / 1960
Capital Ouagadougou
Population 17.4 million / 165 people per sq mile (64 people per sq km)
Total area 105,869 sq. miles (274,200 sq. km)
Languages Mossi, Fulani, French*, Tuare g, Dyula, Songhai
Religions Muslim 55%, Christian 25%, Traditional beliefs 20%
Ethnic mix Mossi 48%, Other 21%, Peul 10%, Lobi 7%, Bobo 7%, Mandé 7%
Government Transitional regime
Currency CFA franc = 100 centimes
Literacy rate 29%
Calorie consumption 2720 kilocalories

BURUNDI
Central Africa

Official name Republic of Burundi
Formation 1962 / 1962
Capital Bujumbura
Population 10.5 million / 1060 people per sq mile (409 people per sq km)
Total area 10,745 sq. miles (27,830 sq. km)
Languages Kirundi*, French*, Kiswahili
Religions Roman Catholic 62%, Traditional beliefs 23%, Muslim 10%, Protestant 5%
Ethnic mix Hutu 85%, Tutsi 14%, Twa 1%
Government Presidential system
Currency Burundian franc = 100 centimes
Literacy rate 87%
Calorie consumption 1604 kilocalories

CAMBODIA
Southeast Asia

Official name Kingdom of Cambodia
Formation 1953 / 1953
Capital Phnom Penh
Population 15.4 million / 226 people per sq mile (87 people per sq km)
Total area 69,900 sq. miles (181,040 sq. km)
Languages Khmer*, French, Chinese, Vietnamese, Cham
Religions Buddhist 93%, Muslim 6%, Christian 1%
Ethnic mix Khmer 90%, Vietnamese 5%, Other 4%, Chinese 1%
Government Parliamentary system
Currency Riel = 100 sen
Literacy rate 74%
Calorie consumption 2411 kilocalories

CAMEROON
Central Africa

Official name Republic of Cameroon
Formation 1960 / 1961
Capital Yaoundé
Population 22.8 million / 127 people per sq mile (49 people per sq km)
Total area 183,567 sq. miles (475,400 sq. km)
Languages Bamileke, Fang, Fulani, French*, English*
Religions Roman Catholic 35%, Traditional beliefs 25%, Muslim 22%, Protestant 18%
Ethnic mix Cameroon highlanders 31%, Other 21%, Equatorial Bantu 19%, Kirdi 11%, Fulani 10%, Northwestern Bantu 8%
Government Presidential system
Currency CFA franc = 100 centimes
Literacy rate 71%
Calorie consumption 2586 kilocalories

CANADA
North America

Official name Canada
Formation 1867 / 1949
Capital Ottawa
Population 35.5 million / 10 people per sq mile (4 people per sq km)
Total area 3,855,171 sq. miles (9,984,670 sq. km)
Languages English*, French*, Chinese, Italian, German, Ukrainian, Portuguese, Inuktitut, Cree
Religions Roman Catholic 44%, Protestant 29%, Other and nonreligious 27%
Ethnic mix European origin 66%, other 27%, Asian 5%, Amerindian 2%
Government Parliamentary system
Currency Canadian dollar = 100 cents
Literacy rate 99%
Calorie consumption 3419 kilocalories

CAPE VERDE
Atlantic Ocean

Official name Republic of Cape Verde
Formation 1975 / 1975
Capital Praia
Population 500,000 / 321 people per sq mile (124 people per sq km)
Total area 1557 sq. miles (4033 sq. km)
Languages Portuguese Creole, Portuguese*
Religions Roman Catholic 97%, Other 2%, Protestant (Church of the Nazarene) 1%
Ethnic mix Mestiço 71%, African 28%, European 1%
Government Mixed presidential–parliamentary system
Currency Escudo = 100 centavos
Literacy rate 85%
Calorie consumption 2716 kilocalories

CENTRAL AFRICAN REPUBLIC
Central Africa

Official name Central African Republic
Formation 1960 / 1960
Capital Bangui
Population 4.7 million / 20 people per sq mile (8 people per sq km)
Total area 240,534 sq. miles (622,984 sq. km)
Languages Sango, Banda, Gbaya, French*
Religions Traditional beliefs 35%, Roman Catholic 25%, Protestant 25%, Muslim 15%
Ethnic mix Baya 33%, Banda 27%, Other 17%, Mandjia 13%, Sara 10%
Government Transitional regime
Currency CFA franc = 100 centimes
Literacy rate 37%
Calorie consumption 2154 kilocalories

CHAD
Central Africa

Official name Republic of Chad
Formation 1960 / 1960
Capital N'Djaména
Population 13.2 million / 27 people per sq mile (10 people per sq km)
Total area 495,752 sq. miles (1,284,000 sq. km)
Languages French*, Sara, Arabic*, Maba
Religions Muslim 51%, Christian 35%, Animist 7%, Traditional beliefs 7%
Ethnic mix Other 30%, Sara 28%, Mayo-Kebbi 12%, Arab 12%, Ouaddai 9%, Kanem-Bornou 9%
Government Presidential system
Currency CFA franc = 100 centimes
Literacy rate 37%
Calorie consumption 2110 kilocalories

CHILE
South America

Official name Republic of Chile
Formation 1818 / 1883
Capital Santiago
Population 17.8 million / 62 people per sq mile (24 people per sq km)
Total area 292,258 sq. miles (756,950 sq. km)
Languages Spanish*, Amerindian languages
Religions Roman Catholic 89%, Other and nonreligious 11%
Ethnic mix Mestizo and European 90%, Other Amerindian 9%, Mapuche 1%
Government Presidential system
Currency Chilean peso = 100 centavos
Literacy rate 99%
Calorie consumption 2989 kilocalories

CHINA
East Asia

Official name People's Republic of China
Formation 960 / 1999
Capital Beijing
Population 1.39 billion / 387 people per sq mile (149 people per sq km)
Total area 3,705,386 sq. miles (9,596,960 sq. km)
Languages Mandarin*, Wu, Cantonese, Hsiang, Min, Hakka, Kan
Religions Nonreligious 59%, Traditional beliefs 20%, Other 13%, Buddhist 6%, Muslim 2%
Ethnic mix Han 92%, Other 4%, Hui 1%, Miao 1%, Manchu 1%, Zhuang 1%
Government One-party state
Currency Renminbi (known as yuan) = 10 jiao = 100 fen
Literacy rate 95%
Calorie consumption 3108 kilocalories

COLOMBIA
South America

Official name Republic of Colombia
Formation 1819 / 1903
Capital Bogotá
Population 48.9 million / 122 people per sq mile (47 people per sq km)
Total area 439,733 sq. miles (1,138,910 sq. km)
Languages Spanish*, Wayuu, Páez, and other Amerindian languages
Religions Roman Catholic 95%, Other 5%
Ethnic mix Mestizo 58%, White 20%, European–African 14%, African 4%, African–Amerindian 3%, Amerindian 1%
Government Presidential system
Currency Colombian peso = 100 centavos
Literacy rate 94%
Calorie consumption 2804 kilocalories

COMOROS
Indian Ocean

Official name Union of the Comoros
Formation 1975 / 1975
Capital Moroni
Population 800,000 / 929 people per sq mile (359 people per sq km)
Total area 838 sq. miles (2170 sq. km)
Languages Arabic*, Comoran*, French*
Religions Muslim (mainly Sunni) 98%, Other 1%, Roman Catholic 1%
Ethnic mix Comoran 97%, Other 3%
Government Presidential system
Currency Comoros franc = 100 centimes
Literacy rate 76%
Calorie consumption 2139 kilocalories

CONGO
Central Africa

Official name Republic of the Congo
Formation 1960 / 1960
Capital Brazzaville
Population 4.6 million / 35 people per sq mile (13 people per sq km)
Total area 132,046 sq. miles (342,000 sq. km)
Languages Kongo, Teke, Lingala, French*
Religions Traditional beliefs 50%, Roman Catholic 35%, Protestant 13%, Muslim 2%
Ethnic mix Bakongo 51%, Teke 17%, Other 16%, Mbochi 11%, Mbédé 5%
Government Presidential system
Currency CFA franc = 100 centimes
Literacy rate 79%
Calorie consumption 2195 kilocalories

CONGO, DEM. REP.
Central Africa

Official name Democratic Republic of the Congo
Formation 1960 / 1960
Capital Kinshasa
Population 69.4 million / 79 people per sq mile (31 people per sq km)
Total area 905,563 sq. miles (2,345,410 sq. km)
Languages Kiswahili, Tshiluba, Kikongo, Lingala, French*
Religions Roman Catholic 50%, Protestant 20%, Traditional beliefs and other 10%, Muslim 10%, Kimbanguist 10%
Ethnic mix Other 55%, Mongo, Luba, Kongo, and Mangbetu-Azande 45%
Government Presidential system
Currency Congolese franc = 100 centimes
Literacy rate 61%
Calorie consumption 1585 kilocalories

COSTA RICA
Central America

Official name Republic of Costa Rica
Formation 1838 / 1838
Capital San José
Population 4.9 million / 249 people per sq mile (96 people per sq km)
Total area 19,730 sq. miles (51,100 sq. km)
Languages Spanish*, English Creole, Bribri, Cabecar
Religions Roman Catholic 71%, Evangelical 14%, Nonreligious 11%, Other 4%
Ethnic mix Mestizo and European 94%, Black 3%, Other 1%, Chinese 1%, Amerindian 1%
Government Presidential system
Currency Costa Rican colón = 100 céntimos
Literacy rate 97%
Calorie consumption 2898 kilocalories

CROATIA
Southeast Europe

Official name Republic of Croatia
Formation 1991 / 1991
Capital Zagreb
Population 4.3 million / 197 people per sq mile (76 people per sq km)
Total area 21,831 sq. miles (56,542 sq. km)
Languages Croatian*
Religions Roman Catholic 88%, Other 7%, Orthodox Christian 4%, Muslim 1%
Ethnic mix Croat 90%, Other 5%, Serb 5%
Government Parliamentary system
Currency Kuna = 100 lipa
Literacy rate 99%
Calorie consumption 3052 kilocalories

CUBA
West Indies

Official name Republic of Cuba
Formation 1902 / 1902
Capital Havana
Population 11.3 million / 264 people per sq mile (102 people per sq km)
Total area 42,803 sq. miles (110,860 sq. km)
Languages Spanish*
Religions Nonreligious 49%, Roman Catholic 40%, Atheist 6%, Other 4%, Protestant 1%
Ethnic mix Mulatto (mixed race) 51%, White 37%, Black 11%, Chinese 1%
Government One-party state
Currency Cuban peso = 100 centavos
Literacy rate 99%
Calorie consumption 3277 kilocalories

CYPRUS
Southeast Europe

Official name Republic of Cyprus
Formation 1960 / 1960
Capital Nicosia
Population 1.2 million / 336 people per sq mile (130 people per sq km)
Total area 3571 sq. miles (9250 sq. km)
Languages Greek*, Turkish
Religions Orthodox Christian 78%, Muslim 18%, Other 4%
Ethnic mix Greek 81%, Turkish 11%, Other 8%
Government Presidential system
Currency Euro = 100 cents; (TRNC: new Turkish lira = 100 kurus)
Literacy rate 99%
Calorie consumption 2661 kilocalories

CZECH REPUBLIC
Central Europe

Official name Czech Republic
Formation 1993 / 1993
Capital Prague
Population 10.7 million / 351 people per sq mile (136 people per sq km)
Total area 30,450 sq. miles (78,866 sq. km)
Languages Czech*, Slovak, Hungarian (Magyar)
Religions Roman Catholic 39%, Atheist 38%, Other 18%, Protestant 3%, Hussite 2%
Ethnic mix Czech 90%, Moravian 4%, Other 4%, Slovak 2%
Government Parliamentary system
Currency Czech koruna = 100 haleru
Literacy rate 99%
Calorie consumption 3292 kilocalories

DENMARK
Northern Europe

Official name Kingdom of Denmark
Formation 950 / 1944
Capital Copenhagen
Population 5.6 million / 342 people per sq mile (132 people per sq km)
Total area 16,639 sq. miles (43,094 sq. km)
Languages Danish*
Religions Evangelical Lutheran 95%, Roman Catholic 3%, Muslim 2%
Ethnic mix Danish 96%, Other (including Scandinavian and Turkish) 3%, Faeroese and Inuit 1%
Government Parliamentary system
Currency Danish krone = 100 øre
Literacy rate 99%
Calorie consumption 3363 kilocalories

DJIBOUTI
East Africa

Official name Republic of Djibouti
Formation 1977 / 1977
Capital Djibouti
Population 900,000 / 101 people per sq mile (39 people per sq km)
Total area 8494 sq. miles (22,000 sq. km)
Languages Somali, Afar, French*, Arabic*
Religions Muslim (mainly Sunni) 94%, Christian 6%
Ethnic mix Issa 60%, Afar 35%, Other 5%
Government Presidential system
Currency Djibouti franc = 100 centimes
Literacy rate 70%
Calorie consumption 2526 kilocalories

DOMINICA
West Indies

Official name Commonwealth of Dominica
Formation 1978 / 1978
Capital Roseau
Population 73,449 / 253 people per sq mile (98 people per sq km)
Total area 291 sq. miles (754 sq. km)
Languages French Creole, English*
Religions Roman Catholic 77%, Protestant 15%, Other 8%
Ethnic mix Black 87%, Mixed race 9%, Carib 3%, Other 1%
Government Parliamentary system
Currency East Caribbean dollar = 100 cents
Literacy rate 88%
Calorie consumption 3047 kilocalories

DOMINICAN REPUBLIC
West Indies

Official name Dominican Republic
Formation 1865 / 1865
Capital Santo Domingo
Population 10.5 million / 562 people per sq mile (217 people per sq km)
Total area 18,679 sq. miles (48,380 sq. km)
Languages Spanish*, French Creole
Religions Roman Catholic 95%, Other and nonreligious 5%
Ethnic mix Mixed race 73%, European 16%, Black African 11%
Government Presidential system
Currency Dominican Republic peso = 100 centavos
Literacy rate 91%
Calorie consumption 2614 kilocalories

EAST TIMOR
Southeast Asia

Official name Democratic Republic of Timor-Leste
Formation 2002 / 2002
Capital Dili
Population 1.2 million / 213 people per sq mile (82 people per sq km)
Total area 5756 sq. miles (14,874 sq. km)
Languages Tetum (Portuguese/Austronesian)*, Bahasa Indonesia, Portuguese*
Religions Roman Catholic 95%, Other (including Muslim and Protestant) 5%
Ethnic mix Papuan groups approx 85%, Indonesian approx 13%, Chinese 2%
Government Parliamentary system
Currency US dollar = 100 cents
Literacy rate 58%
Calorie consumption 2083 kilocalories

ECUADOR
South America

Official name Republic of Ecuador
Formation 1830 / 1942
Capital Quito
Population 16 million / 150 people per sq mile (58 people per sq km)
Total area 109,483 sq. miles (283,560 sq. km)
Languages Spanish*, Quechua, other Amerindian languages
Religions Roman Catholic 95%, Protestant, Jewish, and other 5%
Ethnic mix Mestizo 77%, White 11%, Amerindian 7%, Black African 5%
Government Presidential system
Currency US dollar = 100 cents
Literacy rate 93%
Calorie consumption 2477 kilocalories

EGYPT
North Africa

Official name Arab Republic of Egypt
Formation 1936 / 1982
Capital Cairo
Population 83.4 million / 217 people per sq mile (84 people per sq km)
Total area 386,660 sq. miles (1,001,450 sq. km)
Languages Arabic*, French, English, Berber
Religions Muslim (mainly Sunni) 90%, Coptic Christian and other 9%, Other Christian 1%
Ethnic mix Egyptian 99%, Nubian, Armenian, Greek, and Berber 1%
Government Transitional regime
Currency Egyptian pound = 100 piastres
Literacy rate 74%
Calorie consumption 3557 kilocalories

EL SALVADOR
Central America

Official name Republic of El Salvador
Formation 1841 / 1841
Capital San Salvador
Population 6.4 million / 800 people per sq mile (309 people per sq km)
Total area 8124 sq. miles (21,040 sq. km)
Languages Spanish*
Religions Roman Catholic 80%, Evangelical 18%, Other 2%
Ethnic mix Mestizo 90%, White 9%, Amerindian 1%
Government Presidential system
Currency Salvadorean colón = 100 centavos; and US dollar = 100 cents
Literacy rate 86%
Calorie consumption 2513 kilocalories

EQUATORIAL GUINEA
Central Africa

Official name Republic of Equatorial Guinea
Formation 1968 / 1968
Capital Malabo
Population 800,000 / 74 people per sq mile (29 people per sq km)
Total area 10,830 sq. miles (28,051 sq. km)
Languages Spanish*, Fang, Bubi, French*
Religions Roman Catholic 90%, Other 10%
Ethnic mix Fang 85%, Other 11%, Bubi 4%
Government Presidential system
Currency CFA franc = 100 centimes
Literacy rate 94%
Calorie consumption Not available

ERITREA
East Africa

Official name State of Eritrea
Formation 1993 / 2002
Capital Asmara
Population 6.5 million / 143 people per sq mile (55 people per sq km)
Total area 46,842 sq. miles (121,320 sq. km)
Languages Tigrinya*, English*, Tigre, Afar, Arabic*, Saho, Bilen, Kunama, Nara, Hadareb
Religions Christian 50%, Muslim 48%, Other 2%
Ethnic mix Tigray 50%, Tigre 31%, Other 9%, Afar 5%, Saho 5%
Government Mixed presidential–parliamentary system
Currency Nakfa = 100 cents
Literacy rate 70%
Calorie consumption 1640 kilocalories

ESTONIA
Northeast Europe

Official name Republic of Estonia
Formation 1991 / 1991
Capital Tallinn
Population 1.3 million / 75 people per sq mile (29 people per sq km)
Total area 17,462 sq. miles (45,226 sq. km)
Languages Estonian*, Russian
Religions Evangelical Lutheran 56%, Orthodox Christian 25%, Other 19%
Ethnic mix Estonian 69%, Russian 25%, Other 4%, Ukrainian 2%
Government Parliamentary system
Currency Euro = 100 cents
Literacy rate 99%
Calorie consumption 3214 kilocalories

ETHIOPIA
East Africa

Official name Federal Democratic Republic of Ethiopia
Formation 1896 / 2002
Capital Addis Ababa
Population 96.5 million / 225 people per sq mile (87 people per sq km)
Total area 435,184 sq. miles (1,127,127 sq. km)
Languages Amharic*, Tigrinya, Galla, Sidamo, Somali, English, Arabic
Religions Orthodox Christian 40%, Muslim 40%, Traditional beliefs 15%, Other 5%
Ethnic mix Oromo 40%, Amhara 25%, Other 13%, Sidama 9%, Tigray 7%, Somali 6%
Government Parliamentary system
Currency Birr = 100 cents
Literacy rate 39%
Calorie consumption 2131 kilocalories

FIJI
Australasia & Oceania

Official name Republic of Fiji
Formation 1970 / 1970
Capital Suva
Population 900,000 / 128 people per sq mile (49 people per sq km)
Total area 7054 sq. miles (18,270 sq. km)
Languages Fijian, English*, Hindi, Urdu, Tamil, Telugu
Religions Hindu 38%, Methodist 37%, Roman Catholic 9%, Muslim 8%, Other 8%
Ethnic mix Melanesian 51%, Indian 44%, Other 5%
Government Parliamentary system
Currency Fiji dollar = 100 cents
Literacy rate 94%
Calorie consumption 2930 kilocalories

FINLAND
Northern Europe

Official name Republic of Finland
Formation 1917 / 1947
Capital Helsinki
Population 5.4 million / 46 people per sq mile (18 people per sq km)
Total area 130,127 sq. miles (337,030 sq. km)
Languages Finnish*, Swedish*, Sámi
Religions Evangelical Lutheran 83%, Other 15%, Orthodox Christian 1%, Roman Catholic 1%
Ethnic mix Finnish 93%, Other (including Sámi) 7%
Government Parliamentary system
Currency Euro = 100 cents
Literacy rate 99%
Calorie consumption 3285 kilocalories

FRANCE
Western Europe

Official name French Republic
Formation 987 / 1919
Capital Paris
Population 64.6 million / 304 people per sq mile (117 people per sq km)
Total area 211,208 sq. miles (547,030 sq. km)
Languages French*, Provençal, German, Breton, Catalan, Basque
Religions Roman Catholic 88%, Muslim 8%, Protestant 2%, Buddhist 1%, Jewish 1%
Ethnic mix French 90%, North African (mainly Algerian) 6%, German (Alsace) 2%, Breton 1%, Other (including Corsicans) 1%
Government Mixed presidential–parliamentary system
Currency Euro = 100 cents
Literacy rate 99%
Calorie consumption 3524 kilocalories

GABON
Central Africa

Official name Gabonese Republic
Formation 1960 / 1960
Capital Libreville
Population 1.7 million / 17 people per sq mile (7 people per sq km)
Total area 103,346 sq. miles (267,667 sq. km)
Languages Fang, French*, Punu, Sira, Nzebi, Mpongwe
Religions Christian (mainly Roman Catholic) 55%, Traditional beliefs 40%, Other 4%, Muslim 1%
Ethnic mix Fang 26%, Shira-punu 24%, Other 16%, Foreign residents 15%, Nzabi-duma 11%, Mbédé-Teke 8%
Government Presidential system
Currency CFA franc = 100 centimes
Literacy rate 82%
Calorie consumption 2781 kilocalories

GAMBIA
West Africa

Official name Republic of the Gambia
Formation 1965 / 1965
Capital Banjul
Population 1.9 million / 492 people per sq mile (190 people per sq km)
Total area 4363 sq. miles (11,300 sq. km)
Languages Mandinka, Fulani, Wolof, Jola, Soninke, English*
Religions Sunni Muslim 90%, Christian 8%, Traditional beliefs 2%
Ethnic mix Mandinka 42%, Fulani 18%, Wolof 16%, Jola 10%, Serahuli 9%, Other 5%
Government Presidential system
Currency Dalasi = 100 butut
Literacy rate 52%
Calorie consumption 2849 kilocalories

GEORGIA
Southwest Asia

Official name Georgia
Formation 1991 / 1991
Capital Tbilisi
Population 4.3 million / 160 people per sq mile (62 people per sq km)
Total area 26,911 sq. miles (69,700 sq. km)
Languages Georgian*, Russian, Azeri, Armenian, Mingrelian, Ossetian, Abkhazian* (in Abkhazia)
Religions Georgian Orthodox 74%, Muslim 10%, Russian Orthodox 10%, Armenian Apostolic Church (Orthodox) 4%, Other 2%
Ethnic mix Georgian 84%, Azeri 6%, Armenian 6%, Russian 2%, Ossetian 1%, Other 1%
Government Presidential system
Currency Lari = 100 tetri
Literacy rate 99%
Calorie consumption 2731 kilocalories

GERMANY
Northern Europe

Official name Federal Republic of Germany
Formation 1871 / 1990
Capital Berlin
Population 82.7 million / 613 people per sq mile (237 people per sq km)
Total area 137,846 sq. miles (357,021 sq. km)
Languages German*, Turkish
Religions Protestant 34%, Roman Catholic 33%, Other 30%, Muslim 3%
Ethnic mix German 92%, Other European 3%, Other 3%, Turkish 2%
Government Parliamentary system
Currency Euro = 100 cents
Literacy rate 99%
Calorie consumption 3539 kilocalories

GHANA
West Africa

Official name Republic of Ghana
Formation 1957 / 1957
Capital Accra
Population 26.4 million / 297 people per sq mile (115 people per sq km)
Total area 92,100 sq. miles (238,540 sq. km)
Languages Twi, Fanti, Ewe, Ga, Adangbe, Gurma, Dagomba (Dagbani), English*
Religions Christian 69%, Muslim 16%, Traditional beliefs 9%, Other 6%
Ethnic mix Akan 49%, Mole-Dagbani 17%, Ewe 13%, Other 9%, Ga and Ga-Adangbe 8%, Guan 4%
Government Presidential system
Currency Cedi = 100 pesewas
Literacy rate 72%
Calorie consumption 3003 kilocalories

GREECE
Southeast Europe

Official name Hellenic Republic
Formation 1829 / 1947
Capital Athens
Population 11.1 million / 220 people per sq mile (85 people per sq km)
Total area 50,942 sq. miles (131,940 sq. km)
Languages Greek*, Turkish, Macedonian, Albanian
Religions Orthodox Christian 98%, Muslim 1%, Other 1%
Ethnic mix Greek 98%, Other 2%
Government Parliamentary system
Currency Euro = 100 cents
Literacy rate 97%
Calorie consumption 3433 kilocalories

GRENADA
West Indies

Official name Grenada
Formation 1974 / 1974
Capital St. George's
Population 110,152 / 841 people per sq mile (324 people per sq km)
Total area 131 sq. miles (340 sq. km)
Languages English*, English Creole
Religions Roman Catholic 68%, Anglican 17%, Other 15%
Ethnic mix Black African 82%, Mulatto (mixed race) 13%, East Indian 3%, Other 2%
Government Parliamentary system
Currency East Caribbean dollar = 100 cents
Literacy rate 96%
Calorie consumption 2453 kilocalories

GUATEMALA
Central America

Official name Republic of Guatemala
Formation 1838 / 1838
Capital Guatemala City
Population 15.9 million / 380 people per sq mile (147 people per sq km)
Total area 42,042 sq. miles (108,890 sq. km)
Languages Quiché, Mam, Cakchiquel, Kekchí, Spanish*
Religions Roman Catholic 65%, Protestant 33%, Other and nonreligious 2%
Ethnic mix Amerindian 60%, Mestizo 30%, Other 10%
Government Presidential system
Currency Quetzal = 100 centavos
Literacy rate 78%
Calorie consumption 2419 kilocalories

GUINEA
West Africa

Official name Republic of Guinea
Formation 1958 / 1958
Capital Conakry
Population 12 million / 126 people per sq mile (49 people per sq km)
Total area 94,925 sq. miles (245,857 sq. km)
Languages Pulaar, Malinké, Soussou, French*
Religions Muslim 85%, Christian 8%, Traditional beliefs 7%
Ethnic mix Peul 40%, Malinké 30%, Soussou 20%, Other 10%
Government Presidential system
Currency Guinea franc = 100 centimes
Literacy rate 25%
Calorie consumption 2553 kilocalories

GUINEA-BISSAU
West Africa

Official name Republic of Guinea-Bissau
Formation 1974 / 1974
Capital Bissau
Population 1.7 million / 157 people per sq mile (60 people per sq km)
Total area 13,946 sq. miles (36,120 sq. km)
Languages Portuguese Creole, Balante, Fulani, Malinké, Portuguese*
Religions Traditional beliefs 50%, Muslim 40%, Christian 10%
Ethnic mix Balante 30%, Fulani 20%, Other 16%, Mandyako 14%, Mandinka 13%, Papel 7%
Government Presidential system
Currency CFA franc = 100 centimes
Literacy rate 57%
Calorie consumption 2304 kilocalories

GUYANA
South America

Official name Cooperative Republic of Guyana
Formation 1966 / 1966
Capital Georgetown
Population 800,000 / 11 people per sq mile (4 people per sq km)
Total area 83,000 sq. miles (214,970 sq. km)
Languages English Creole, Hindi, Tamil, Amerindian languages, English*
Religions Christian 57%, Hindu 28%, Muslim 10%, Other 5%
Ethnic mix East Indian 43%, Black African 30%, Mixed race 17%, Amerindian 9%, Other 1%
Government Presidential system
Currency Guyanese dollar = 100 cents
Literacy rate 85%
Calorie consumption 2648 kilocalories

HAITI
West Indies

Official name Republic of Haiti
Formation 1804 / 1844
Capital Port-au-Prince
Population 10.5 million / 987 people per sq mile (381 people per sq km)
Total area 10,714 sq. miles (27,750 sq. km)
Languages French Creole*, French*
Religions Roman Catholic 55%, Protestant 28%, Other (including Voodoo) 16%, Nonreligious 1%
Ethnic mix Black African 95%, Mulatto (mixed race) and European 5%
Government Presidential system
Currency Gourde = 100 centimes
Literacy rate 49%
Calorie consumption 2091 kilocalories

HONDURAS
Central America

Official name Republic of Honduras
Formation 1838 / 1838
Capital Tegucigalpa
Population 8.3 million / 192 people per sq mile (74 people per sq km)
Total area 43,278 sq. miles (112,090 sq. km)
Languages Spanish*, Garífuna (Carib), English Creole
Religions Roman Catholic 97%, Protestant 3%
Ethnic mix Mestizo 90%, Black African 5%, Amerindian 4%, White 1%
Government Presidential system
Currency Lempira = 100 centavos
Literacy rate 85%
Calorie consumption 2651 kilocalories

HUNGARY
Central Europe

Official name Hungary
Formation 1918 / 1947
Capital Budapest
Population 9.9 million / 278 people per sq mile (107 people per sq km)
Total area 35,919 sq. miles (93,030 sq. km)
Languages Hungarian (Magyar)*
Religions Roman Catholic 52%, Calvinist 16%, Other 15%, Nonreligious 14%, Lutheran 3%
Ethnic mix Magyar 90%, Roma 4%, German 3%, Serb 2%, Other 1%
Government Parliamentary system
Currency Forint = 100 fillér
Literacy rate 99%
Calorie consumption 2968 kilocalories

ICELAND
Northwest Europe

Official name Republic of Iceland
Formation 1944 / 1944
Capital Reykjavík
Population 300,000 / 8 people per sq mile (3 people per sq km)
Total area 39,768 sq. miles (103,000 sq. km)
Languages Icelandic*
Religions Evangelical Lutheran 84%, Other (mostly Christian) 10%, Roman Catholic 3%, Nonreligious 3%
Ethnic mix Icelandic 94%, Other 5%, Danish 1%
Government Parliamentary system
Currency Icelandic króna = 100 aurar
Literacy rate 99%
Calorie consumption 3339 kilocalories

INDIA
South Asia

Official name Republic of India
Formation 1947 / 1947
Capital New Delhi
Population 1.27 billion / 1104 people per sq mile (426 people per sq km)
Total area 1,269,339 sq. miles (3,287,590 sq. km)
Languages Hindi*, English*, Urdu, Bengali, Marathi, Telugu, Tamil, Bihari, Gujarati, Kanarese
Religions Hindu 81%, Muslim 13%, Christian 2%, Sikh 2%, Buddhist 1%, Other 1%
Ethnic mix Indo-Aryan 72%, Dravidian 25%, Mongoloid and other 3%
Government Parliamentary system
Currency Indian rupee = 100 paise
Literacy rate 63%
Calorie consumption 2459 kilocalories

INDONESIA
Southeast Asia

Official name Republic of Indonesia
Formation 1949 / 1999
Capital Jakarta
Population 253 million / 364 people per sq mile (141 people per sq km)
Total area 741,096 sq. miles (1,919,440 sq. km)
Languages Javanese, Sundanese, Madurese, Bahasa Indonesia*, Dutch
Religions Sunni Muslim 86%, Protestant 6%, Roman Catholic 3%, Hindu 2%, Other 2%, Buddhist 1%
Ethnic mix Javanese 41%, Other 29%, Sundanese 15%, Coastal Malays 12%, Madurese 3%
Government Presidential system
Currency Rupiah = 100 sen
Literacy rate 93%
Calorie consumption 2777 kilocalories

IRAN
Southwest Asia

Official name Islamic Republic of Iran
Formation 1502 / 1990
Capital Tehran
Population 78.5 million / 124 people per sq mile (48 people per sq km)
Total area 636,293 sq. miles (1,648,000 sq. km)
Languages Farsi*, Azeri, Luri, Gilaki, Mazanderani, Kurdish, Turkmen, Arabic, Baluchi
Religions Shi'a Muslim 89%, Sunni Muslim 9%, Other 2%
Ethnic mix Persian 51%, Azari 24%, Other 10%, Lur and Bakhtiari 8%, Kurdish 7%
Government Islamic theocracy
Currency Iranian rial = 100 dinars
Literacy rate 84%
Calorie consumption 3058 kilocalories

IRAQ
Southwest Asia

Official name Republic of Iraq
Formation 1932 / 1990
Capital Baghdad
Population 34.8 million / 206 people per sq mile (80 people per sq km)
Total area 168,753 sq. miles (437,072 sq. km)
Languages Arabic*, Kurdish*, Turkic languages, Armenian, Assyrian
Religions Shi'a Muslim 60%, Sunni Muslim 35%, Other (including Christian) 5%
Ethnic mix Arab 80%, Kurdish 15%, Turkmen 3%, Other 2%
Government Parliamentary system
Currency New Iraqi dinar = 1000 fils
Literacy rate 79%
Calorie consumption 2489 kilocalories

IRELAND
Northwest Europe

Official name Ireland
Formation 1922 / 1922
Capital Dublin
Population 4.7 million / 177 people per sq mile (68 people per sq km)
Total area 27,135 sq. miles (70,280 sq. km)
Languages English*, Irish*
Religions Roman Catholic 87%, Other and nonreligious 10%, Anglican 3%
Ethnic mix Irish 99%, Other 1%
Government Parliamentary system
Currency Euro = 100 cents
Literacy rate 99%
Calorie consumption 3591 kilocalories

ISRAEL
Southwest Asia

Official name State of Israel
Formation 1948 / 1994
Capital Jerusalem (not internationally recognized)
Population 7.8 million / 994 people per sq mile (384 people per sq km)
Total area 8019 sq. miles (20,770 sq. km)
Languages Hebrew*, Arabic*, Yiddish, German, Russian, Polish, Romanian, Persian
Religions Jewish 76%, Muslim (mainly Sunni) 16%, Other 4%, Druze 2%, Christian 2%
Ethnic mix Jewish 76%, Arab 20%, Other 4%
Government Parliamentary system
Currency Shekel = 100 agorot
Literacy rate 98%
Calorie consumption 3619 kilocalories

ITALY
Southern Europe

Official name Italian Republic
Formation 1861 / 1947
Capital Rome
Population 61.1 million / 538 people per sq mile (208 people per sq km)
Total area 116,305 sq. miles (301,230 sq. km)
Languages Italian*, German, French, Rhaeto-Romanic, Sardinian
Religions Roman Catholic 85%, Other and nonreligious 13%, Muslim 2%
Ethnic mix Italian 94%, Other 4%, Sardinian 2%
Government Parliamentary system
Currency Euro = 100 cents
Literacy rate 99%
Calorie consumption 3539 kilocalories

IVORY COAST
West Africa

Official name Republic of Côte d'Ivoire
Formation 1960 / 1960
Capital Yamoussoukro
Population 20.8 million / 169 people per sq mile (65 people per sq km)
Total area 124,502 sq. miles (322,460 sq. km)
Languages Akan, French*, Krou, Voltaique
Religions Muslim 38%, Traditional beliefs 25%, Roman Catholic 25%, Other 6%, Protestant 6%
Ethnic mix Akan 42%, Voltaique 18%, Mandé du Nord 17%, Krou 11%, Mandé du Sud 10%, Other 2%
Government Presidential system
Currency CFA franc = 100 centimes
Literacy rate 41%
Calorie consumption 2799 kilocalories

JAMAICA
West Indies

Official name Jamaica
Formation 1962 / 1962
Capital Kingston
Population 2.8 million / 670 people per sq mile (259 people per sq km)
Total area 4243 sq. miles (10,990 sq. km)
Languages English Creole, English*
Religions Other and nonreligious 45%, Other Protestant 20%, Church of God 18%, Baptist 10%, Anglican 7%
Ethnic mix Black 91%, Mulatto (mixed race) 7%, European and Chinese 1%, East Indian 1%
Government Parliamentary system
Currency Jamaican dollar = 100 cents
Literacy rate 88%
Calorie consumption 2746 kilocalories

JAPAN
East Asia

Official name Japan
Formation 1590 / 1972
Capital Tokyo
Population 127 million / 874 people per sq mile (337 people per sq km)
Total area 145,882 sq. miles (377,835 sq. km)
Languages Japanese*, Korean, Chinese
Religions Shinto and Buddhist 76%, Buddhist 16%, Other (including Christian) 8%
Ethnic mix Japanese 99%, Other (mainly Korean) 1%
Government Parliamentary system
Currency Yen = 100 sen
Literacy rate 99%
Calorie consumption 2719 kilocalories

JORDAN
Southwest Asia

Official name Hashemite Kingdom of Jordan
Formation 1946 / 1967
Capital Amman
Population 7.5 million / 218 people per sq mile (84 people per sq km)
Total area 35,637 sq. miles (92,300 sq. km)
Languages Arabic*
Religions Sunni Muslim 92%, Christian 6%, Other 2%
Ethnic mix Arab 98%, Circassian 1%, Armenian 1%
Government Monarchy
Currency Jordanian dinar = 1000 fils
Literacy rate 98%
Calorie consumption 3149 kilocalories

KAZAKHSTAN
Central Asia

Official name Republic of Kazakhstan
Formation 1991 / 1991
Capital Astana
Population 16.6 million / 16 people per sq mile (6 people per sq km)
Total area 1,049,150 sq. miles (2,717,300 sq. km)
Languages Kazakh*, Russian, Ukrainian, German, Uzbek, Tatar, Uighur
Religions Muslim (mainly Sunni) 47%, Orthodox Christian 44%, Other 7%, Protestant 2%
Ethnic mix Kazakh 57%, Russian 27%, Other 8%, Uzbek 3%, Ukrainian 3%, German 2%
Government Presidential system
Currency Tenge = 100 tiyn
Literacy rate 99%
Calorie consumption 3107 kilocalories

KENYA
East Africa

Official name Republic of Kenya
Formation 1963 / 1963
Capital Nairobi
Population 45.5 million / 208 people per sq mile (80 people per sq km)
Total area 224,961 sq. miles (582,650 sq. km)
Languages Kiswahili*, English*, Kikuyu, Luo, Kalenjin, Kamba
Religions Christian 80%, Muslim 10%, Traditional beliefs 9%, Other 1%
Ethnic mix Other 28%, Kikuyu 22%, Luo 14%, Luhya 14%, Kalenjin 11%, Kamba 11%
Government Presidential system
Currency Kenya shilling = 100 cents
Literacy rate 72%
Calorie consumption 2206 kilocalories

KIRIBATI
Australasia & Oceania

Official name Republic of Kiribati
Formation 1979 / 1979
Capital Tarawa Atoll
Population 104,488 / 381 people per sq mile (147 people per sq km)
Total area 277 sq. miles (717 sq. km)
Languages English*, Kiribati
Religions Roman Catholic 55%, Kiribati Protestant Church 36%, Other 9%
Ethnic mix Micronesian 99%, Other 1%
Government Presidential system
Currency Australian dollar = 100 cents
Literacy rate 99%
Calorie consumption 3022 kilocalories

KOSOVO (not yet recognised)
Southeast Europe

Official name Republic of Kosovo
Formation 2008 / 2008
Capital Pristina
Population 1.9 million / 451 people per sq mile (174 people per sq km)
Total area 4212 sq. miles (10,908 sq. km)
Languages Albanian*, Serbian*, Bosniak, Gorani, Roma, Turkish
Religions Muslim 92%, Roman Catholic 4%, Orthodox Christian 4%
Ethnic mix Albanian 92%, Serb 4%, Bosniak and Gorani 2%, Turkish 1%, Roma 1%
Government Parliamentary system
Currency Euro = 100 cents
Literacy rate 92%
Calorie consumption Not available

KUWAIT
Southwest Asia

Official name State of Kuwait
Formation 1961 / 1961
Capital Kuwait City
Population 3.5 million / 509 people per sq mile (196 people per sq km)
Total area 6880 sq. miles (17,820 sq. km)
Languages Arabic*, English
Religions Sunni Muslim 45%, Shi'a Muslim 40%, Christian, Hindu, and other 15%
Ethnic mix Kuwaiti 45%, Other Arab 35%, South Asian 9%, Other 7%, Iranian 4%
Government Monarchy
Currency Kuwaiti dinar = 1000 fils
Literacy rate 96%
Calorie consumption 3471 kilocalories

KYRGYZSTAN
Central Asia

Official name Kyrgyz Republic
Formation 1991 / 1991
Capital Bishkek
Population 5.6 million / 73 people per sq mile (28 people per sq km)
Total area 76,641 sq. miles (198,500 sq. km)
Languages Kyrgyz*, Russian*, Uzbek, Tatar, Ukrainian
Religions Muslim (mainly Sunni) 70%, Orthodox Christian 30%
Ethnic mix Kyrgyz 69%, Uzbek 14%, Russian 9%, Other 6%, Dungan 1%, Uighur 1%
Government Presidential system
Currency Som = 100 tyiyn
Literacy rate 99%
Calorie consumption 2828 kilocalories

LAOS
Southeast Asia

Official name Lao People's Democratic Republic
Formation 1953 / 1953
Capital Vientiane
Population 6.9 million / 77 people per sq mile (30 people per sq km)
Total area 91,428 sq. miles (236,800 sq. km)
Languages Lao*, Mon-Khmer, Yao, Vietnamese, Chinese, French
Religions Buddhist 65%, Other (including animist) 34%, Christian 1%
Ethnic mix Lao Loum 66%, Lao Theung 30%, Lao Soung 2%, Other 2%
Government One-party state
Currency Kip = 100 at
Literacy rate 73%
Calorie consumption 2356 kilocalories

LATVIA
Northeast Europe

Official name Republic of Latvia
Formation 1991 / 1991
Capital Riga
Population 2 million / 80 people per sq mile (31 people per sq km)
Total area 24,938 sq. miles (64,589 sq. km)
Languages Latvian*, Russian
Religions Other 43%, Lutheran 24%, Roman Catholic 18%, O rthodox Christian 15%
Ethnic mix Latvian 62%, Russian 27%, Other 4%, Belarussian 3%, Ukrainian 2%, Polish 2%
Government Parliamentary system
Currency Euro = 100 cents
Literacy rate 99%
Calorie consumption 3293 kilocalories

LEBANON
Southwest Asia

Official name Lebanese Republic
Formation 1941 / 1941
Capital Beirut
Population 5 million / 1266 people per sq mile (489 people per sq km)
Total area 4015 sq. miles (10,400 sq. km)
Languages Arabic*, French, Armenian, Assyrian
Religions Muslim 60%, Christian 39%, Other 1%
Ethnic mix Arab 95%, Armenian 4%, Other 1%
Government Parliamentary system
Currency Lebanese pound = 100 piastres
Literacy rate 90%
Calorie consumption 3181 kilocalories

LESOTHO
Southern Africa

Official name Kingdom of Lesotho
Formation 1966 / 1966
Capital Maseru
Population 2.1 million / 179 people per sq mile (69 people per sq km)
Total area 11,720 sq. miles (30,355 sq. km)
Languages English*, Sesotho*, isiZulu
Religions Christian 90%, Traditional beliefs 10%
Ethnic mix Sotho 99%, European and Asian 1%
Government Parliamentary system
Currency Loti = 100 lisente; and South African rand = 100 cents
Literacy rate 76%
Calorie consumption 2595 kilocalories

LIBERIA
West Africa

Official name Republic of Liberia
Formation 1847 / 1847
Capital Monrovia
Population 4.4 million / 118 people per sq mile (46 people per sq km)
Total area 43,000 sq. miles (111,370 sq. km)
Languages Kpelle, Vai, Bassa, Kru, Grebo, Kissi, Gola, Loma, English*
Religions Christian 40%, Traditional beliefs 40%, Muslim 20%
Ethnic mix Indigenous tribes (12 groups) 49%, Kpellé 20%, Bassa 16%, Gio 8%, Krou 7%
Government Presidential system
Currency Liberian dollar = 100 cents
Literacy rate 43%
Calorie consumption 2251 kilocalories

LIBYA
North Africa

Official name State of Libya
Formation 1951 / 1951
Capital Tripoli
Population 6.3 million / 9 people per sq mile (4 people per sq km)
Total area 679,358 sq. miles (1,759,540 sq. km)
Languages Arabic*, Tuareg
Religions Muslim (mainly Sunni) 97%, Other 3%
Ethnic mix Arab and Berber 97%, Other 3%
Government Transitional regime
Currency Libyan dinar = 1000 dirhams
Literacy rate 90%
Calorie consumption 3211 kilocalories

LIECHTENSTEIN
Central Europe

Official name Principality of Liechtenstein
Formation 1719 / 1719
Capital Vaduz
Population 37,313 / 602 people per sq mile (233 people per sq km)
Total area 62 sq. miles (160 sq. km)
Languages German*, Alemannish dialect, Italian
Religions Roman Catholic 79%, Other 13%, Protestant 8%
Ethnic mix Liechtensteiner 66%, Other 12%, Swiss 10%, Austrian 6%, German 3%, Italian 3%
Government Parliamentary system
Currency Swiss franc = 100 rappen/centimes
Literacy rate 99%
Calorie consumption Not available

LITHUANIA
Northeast Europe

Official name Republic of Lithuania
Formation 1991 / 1991
Capital Vilnius
Population 3 million / 119 people per sq mile (46 people per sq km)
Total area 25,174 sq. miles (65,200 sq. km)
Languages Lithuanian*, Russian
Religions Roman Catholic 77%, Other 17%, Russian Orthodox 4%, Protestant 1%, Old believers 1%
Ethnic mix Lithuanian 85%, Polish 7%, Russian 6%, Belarussian 1%, Other 1%
Government Parliamentary system
Currency Euro = 100 cents
Literacy rate 99%
Calorie consumption 3463 kilocalories

LUXEMBOURG
Northwest Europe

Official name Grand Duchy of Luxembourg
Formation 1867 / 1867
Capital Luxembourg-Ville
Population 500,000 / 501 people per sq mile (193 people per sq km)
Total area 998 sq. miles (2586 sq. km)
Languages Luxembourgish*, German*, French*
Religions Roman Catholic 97%, Protestant, Orthodox Christian, and Jewish 3%
Ethnic mix Luxembourger 62%, Foreign residents 38%
Government Parliamentary system
Currency Euro = 100 cents
Literacy rate 99%
Calorie consumption 3568 kilocalories

MACEDONIA
Southeast Europe

Official name Republic of Macedonia
Formation 1991 / 1991
Capital Skopje
Population 2.1 million / 212 people per sq mile (82 people per sq km)
Total area 9781 sq. miles (25,333 sq. km)
Languages Macedonian*, Albanian*, Turkish, Romani, Serbian
Religions Orthodox Christian 65%, Muslim 29%, Roman Catholic 4%, Other 2%
Ethnic mix Macedonian 64%, Albanian 25%, Turkish 4%, Roma 3%, Serb 2%, Other 2%
Government Mixed presidential–parliamentary system
Currency Macedonian denar = 100 deni
Literacy rate 98%
Calorie consumption 2923 kilocalories

MADAGASCAR
Indian Ocean

Official name Republic of Madagascar
Formation 1960 / 1960
Capital Antananarivo
Population 23.6 million / 105 people per sq mile (41 people per sq km)
Total area 226,656 sq. miles (587,040 sq. km)
Languages Malagasy*, French*, English*
Religions Traditional beliefs 52%, Christian (mainly Roman Catholic) 41%, Muslim 7%
Ethnic mix Other Malay 46%, Merina 26%, Betsimisaraka 15%, Betsileo 12%, Other 1%
Government Mixed presidential–parliamentary system
Currency Ariary = 5 iraimbilanja
Literacy rate 64%
Calorie consumption 2052 kilocalories

MALAWI
Southern Africa

Official name Republic of Malawi
Formation 1964 / 1964
Capital Lilongwe
Population 16.8 million / 463 people per sq mile (179 people per sq km)
Total area 45,745 sq. miles (118,480 sq. km)
Languages Chewa, Lomwe, Yao, Ngoni, English*
Religions Protestant 55%, Roman Catholic 20%, Muslim 20%, Traditional beliefs 5%
Ethnic mix Bantu 99%, Other 1%
Government Presidential system
Currency Malawi kwacha = 100 tambala
Literacy rate 61%
Calorie consumption 2334 kilocalories

MALAYSIA
Southeast Asia

Official name Malaysia
Formation 1963 / 1965
Capital Kuala Lumpur; Putrajaya (administrative)
Population 30.2 million / 238 people per sq mile (92 people per sq km)
Total area 127,316 sq. miles (329,750 sq. km)
Languages Bahasa Malaysia*, Malay, Chinese, Tamil, English
Religions Muslim (mainly Sunni) 61%, Buddhist 19%, Christian 9%, Hindu 6%, Other 5%
Ethnic mix Malay 53%, Chinese 26%, Indigenous tribes 12%, Indian 8%, Other 1%
Government Parliamentary system
Currency Ringgit = 100 sen
Literacy rate 93%
Calorie consumption 2855 kilocalories

MALDIVES
Indian Ocean

Official name Republic of Maldives
Formation 1965 / 1965
Capital Male'
Population 400,000 / 3448 people per sq mile (1333 people per sq km)
Total area 116 sq. miles (300 sq. km)
Languages Dhivehi (Maldivian), Sinhala, Tamil, Arabic
Religions Sunni Muslim 100%
Ethnic mix Arab–Sinhalese–Malay 100%
Government Presidential system
Currency Rufiyaa = 100 laari
Literacy rate 98%
Calorie consumption 2722 kilocalories

MALI
West Africa

Official name Republic of Mali
Formation 1960 / 1960
Capital Bamako
Population 15.8 million / 34 people per sq mile (13 people per sq km)
Total area 478,764 sq. miles (1,240,000 sq. km)
Languages Bambara, Fulani, Senufo, Soninke, French*
Religions Muslim (mainly Sunni) 90%, Traditional beliefs 6%, Christian 4%
Ethnic mix Bambara 52%, Other 14%, Fulani 11%, Saracolé 7%, Soninka 7%, Tuareg 5%, Mianka 4%
Government Presidential system
Currency CFA franc = 100 centimes
Literacy rate 34%
Calorie consumption 2833 kilocalories

MALTA
Southern Europe

Official name Republic of Malta
Formation 1964 / 1964
Capital Valletta
Population 400,000 / 3226 people per sq mile (1250 people per sq km)
Total area 122 sq. miles (316 sq. km)
Languages Maltese*, English*
Religions Roman Catholic 98%, Other and nonreligious 2%
Ethnic mix Maltese 96%, Other 4%
Government Parliamentary system
Currency Euro = 100 cents
Literacy rate 92%
Calorie consumption 3389 kilocalories

MARSHALL ISLANDS
Australasia & Oceania

Official name Republic of the Marshall Islands
Formation 1986 / 1986
Capital Majuro
Population 70,983 / 1014 people per sq mile (392 people per sq km)
Total area 70 sq. miles (181 sq. km)
Languages Marshallese*, English*, Japanese, German
Religions Protestant 90%, Roman Catholic 8%, Other 2%
Ethnic mix Micronesian 90%, Other 10%
Government Presidential system
Currency US dollar = 100 cents
Literacy rate 91%
Calorie consumption Not available

MAURITANIA
West Africa

Official name Islamic Republic of Mauritania
Formation 1960 / 1960
Capital Nouakchott
Population 4 million / 10 people per sq mile (4 people per sq km)
Total area 397,953 sq. miles (1,030,700 sq. km)
Languages Arabic*, Hassaniyah Arabic, Wolof, French
Religions Sunni Muslim 100%
Ethnic mix Maure 81%, Wolof 7%, Tukolor 5%, Other 4%, Soninka 3%
Government Presidential system
Currency Ouguiya = 5 khoums
Literacy rate 46%
Calorie consumption 2791 kilocalories

MAURITIUS
Indian Ocean

Official name Republic of Mauritius
Formation 1968 / 1968
Capital Port Louis
Population 1.2 million / 1671 people per sq mile (645 people per sq km)
Total area 718 sq. miles (1860 sq. km)
Languages French Creole, Hindi, Urdu, Tamil, Chinese, English*, French
Religions Hindu 48%, Roman Catholic 24%, Muslim 17%, Protestant 9%, Other 2%
Ethnic mix Indo-Mauritian 68%, Creole 27%, Sino-Mauritian 3%, Franco-Mauritian 2%
Government Parliamentary system
Currency Mauritian rupee = 100 cents
Literacy rate 89%
Calorie consumption 3055 kilocalories

MEXICO
North America

Official name United Mexican States
Formation 1836 / 1848
Capital Mexico City
Population 124 million / 168 people per sq mile (65 people per sq km)
Total area 761,602 sq. miles (1,972,550 sq. km)
Languages Spanish*, Nahuatl, Mayan, Zapotec, Mixtec, Otomi, Totonac, Tzotzil, Tzeltal
Religions Roman Catholic 77%, Other 14%, Protestant 6%, Nonreligious 3%
Ethnic mix Mestizo 60%, Amerindian 30%, European 9%, Other 1%
Government Presidential system
Currency Mexican peso = 100 centavos
Literacy rate 94%
Calorie consumption 3072 kilocalories

MICRONESIA
Australasia & Oceania

Official name Federated States of Micronesia
Formation 1986 / 1986
Capital Palikir (Pohnpei Island)
Population 105,681 / 390 people per sq mile (151 people per sq km)
Total area 271 sq. miles (702 sq. km)
Languages Trukese, Pohnpeian, Kosraean, Yapese, English*
Religions Roman Catholic 50%, Protestant 47%, Other 3%
Ethnic mix Chuukese 49%, Pohnpeian 24%, Other 14%, Kosraean 6%, Yapese 5%, Asian 2%
Government Nonparty system
Currency US dollar = 100 cents
Literacy rate 81%
Calorie consumption Not available

MOLDOVA
Southeast Europe

Official name Republic of Moldova
Formation 1991 / 1991
Capital Chisinau
Population 3.5 million / 269 people per sq mile (104 people per sq km)
Total area 13,067 sq. miles (33,843 sq. km)
Languages Moldovan*, Ukrainian, Russian
Religions Orthodox Christian 93%, Other 6%, Baptist 1%
Ethnic mix Moldovan 84%, Ukrainian 7%, Gagauz 5%, Russian 2%, Bulgarian 1%, Other 1%
Government Parliamentary system
Currency Moldovan leu = 100 bani
Literacy rate 99%
Calorie consumption 2837 kilocalories

MONACO
Southern Europe

Official name Principality of Monaco
Formation 1861 / 1861
Capital Monaco-Ville
Population 36,950 / 49,267 people per sq mile (18,949 people per sq km)
Total area 0.75 sq. miles (1.95 sq. km)
Languages French*, Italian, Monégasque, English
Religions Roman Catholic 89%, Protestant 6%, Other 5%
Ethnic mix French 47%, Other 21%, Italian 16%, Monégasque 16%
Government Mixed monarchical–parliamentary system
Currency Euro = 100 cents
Literacy rate 99%
Calorie consumption Not available

MONGOLIA
East Asia

Official name Mongolia
Formation 1924 / 1924
Capital Ulan Bator
Population 2.9 million / 5 people per sq mile (2 people per sq km)
Total area 604,247 sq. miles (1,565,000 sq. km)
Languages Khalkha Mongolian, Kazakh, Chinese, Russian
Religions Tibetan Buddhist 50%, Nonreligious 40%, Shamanist and Christian 6%, Muslim 4%
Ethnic mix Khalkh 95%, Kazakh 4%, Other 1%
Government Mixed presidential–parliamentary system
Currency Tugrik (tögrög) = 100 möngö
Literacy rate 98%
Calorie consumption 2463 kilocalories

MONTENEGRO
Southeast Europe

Official name Montenegro
Formation 2006 / 2006
Capital Podgorica
Population 600,000 / 113 people per sq mile (43 people per sq km)
Total area 5332 sq. miles (13,812 sq. km)
Languages Montenegrin*, Serbian, Albanian, Bosniak, Croatian
Religions Orthodox Christian 74%, Muslim 18%, Roman Catholic 4%, Other 4%
Ethnic mix Montenegrin 43%, Serb 32%, Other 12%, Bosniak 8%, Albanian 5%
Government Parliamentary system
Currency Euro = 100 cents
Literacy rate 98%
Calorie consumption 3568 kilocalories

MOROCCO
North Africa

Official name Kingdom of Morocco
Formation 1956 / 1969
Capital Rabat
Population 35.5 million / 194 people per sq mile (75 people per sq km)
Total area 172,316 sq. miles (446,300 sq. km)
Languages Arabic*, Tamazight (Berber), French, Spanish
Religions Muslim (mainly Sunni) 99%, Other (mostly Christian) 1%
Ethnic mix Arab 70%, Berber 29%, European 1%
Government Mixed monarchical–parliamentary system
Currency Moroccan dirham = 100 centimes
Literacy rate 67%
Calorie consumption 3334 kilocalories

MOZAMBIQUE
Southern Africa

Official name Republic of Mozambique
Formation 1975 / 1975
Capital Maputo
Population 26.5 million / 88 people per sq mile (34 people per sq km)
Total area 309,494 sq. miles (801,590 sq. km)
Languages Makua, Xitsonga, Sena, Lomwe, Portuguese*
Religions Traditional beliefs 56%, Christian 30%, Muslim 14%
Ethnic mix Makua Lomwe 47%, Tsonga 23%, Malawi 12%, Shona 11%, Yao 4%, Other 3%
Government Presidential system
Currency New metical = 100 centavos
Literacy rate 51%
Calorie consumption 2283 kilocalories

MYANMAR (BURMA)
Southeast Asia

Official name Republic of the Union of Myanmar
Formation 1948 / 1948
Capital Nay Pyi Taw
Population 53.7 million / 212 people per sq mile (82 people per sq km)
Total area 261,969 sq. miles (678,500 sq. km)
Languages Myanmar (Burmese)*, Shan, Karen, Rakhine, Chin, Yangbye, Kachin, Mon
Religions Buddhist 89%, Christian 4%, Muslim 4%, Other 2%, Animist 1%
Ethnic mix Burman (Bamah) 68%, Other 12%, Shan 9%, Karen 7%, Rakhine 4%
Government Presidential system
Currency Kyat = 100 pyas
Literacy rate 93%
Calorie consumption 2571 kilocalories

NAMIBIA
Southern Africa

Official name Republic of Namibia
Formation 1990 / 1994
Capital Windhoek
Population 2.3 million / 7 people per sq mile (3 people per sq km)
Total area 318,694 sq. miles (825,418 sq. km)
Languages Ovambo, Kavango, English*, Bergdama, German, Afrikaans
Religions Christian 90%, Traditional beliefs 10%
Ethnic mix Ovambo 50%, Other tribes 22%, Kavango 9%, Damara 7%, Herero 7%, Other 5%
Government Presidential system
Currency Namibian dollar = 100 cents; and South African rand = 100 cents
Literacy rate 76%
Calorie consumption 2086 kilocalories

NAURU
Australasia & Oceania

Official name Republic of Nauru
Formation 1968 / 1968
Capital None
Population 9488 / 1171 people per sq mile (452 people per sq km)
Total area 8.1 sq. miles (21 sq. km)
Languages Nauruan*, Kiribati, Chinese, Tuvaluan, English
Religions Nauruan Congregational Church 60%, Roman Catholic 35%, Other 5%
Ethnic mix Nauruan 93%, Chinese 5%, European 1%, Other Pacific islanders 1%
Government Nonparty system
Currency Australian dollar = 100 cents
Literacy rate 95%
Calorie consumption Not available

NEPAL
South Asia

Official name Federal Democratic Republic of Nepal
Formation 1769 / 1769
Capital Kathmandu
Population 28.1 million / 532 people per sq mile (205 people per sq km)
Total area 54,363 sq. miles (140,800 sq. km)
Languages Nepali*, Maithili, Bhojpuri
Religions Hindu 81%, Buddhist 11%, Muslim 4%, Other (including Christian) 4%
Ethnic mix Other 52%, Chhetri 16%, Hill Brahman 13%, Tharu 7%, Magar 7%, Tamang 5%
Government Transitional regime
Currency Nepalese rupee = 100 paisa
Literacy rate 57%
Calorie consumption 2673 kilocalories

NETHERLANDS
Northwest Europe

Official name Kingdom of the Netherlands
Formation 1648 / 1839
Capital Amsterdam; The Hague (administrative)
Population 16.8 million / 1283 people per sq mile (495 people per sq km)
Total area 16,033 sq. miles (41,526 sq. km)
Languages Dutch*, Frisian
Religions Roman Catholic 36%, Other 34%, Protestant 27%, Muslim 3%
Ethnic mix Dutch 82%, Other 12%, Surinamese 2%, Turkish 2%, Moroccan 2%
Government Parliamentary system
Currency Euro = 100 cents
Literacy rate 99%
Calorie consumption 3147 kilocalories

NEW ZEALAND
Australasia & Oceania

Official name New Zealand
Formation 1947 / 1947
Capital Wellington
Population 4.6 million / 44 people per sq mile (17 people per sq km)
Total area 103,737 sq. miles (268,680 sq. km)
Languages English*, Maori*
Religions Anglican 24%, Other 22%, Presbyterian 18%, Nonreligious 16%, Roman Catholic 15%, Methodist 5%
Ethnic mix European 75%, Maori 15%, Other 7%, Samoan 3%
Government Parliamentary system
Currency New Zealand dollar = 100 cents
Literacy rate 99%
Calorie consumption 3170 kilocalories

NICARAGUA
Central America

Official name Republic of Nicaragua
Formation 1838 / 1838
Capital Managua
Population 6.2 million / 135 people per sq mile (52 people per sq km)
Total area 49,998 sq. miles (129,494 sq. km)
Languages Spanish*, English Creole, Miskito
Religions Roman Catholic 80%, Protestant Evangelical 17%, Other 3%
Ethnic mix Mestizo 69%, White 17%, Black 9%, Amerindian 5%
Government Presidential system
Currency Córdoba oro = 100 centavos
Literacy rate 78%
Calorie consumption 2564 kilocalories

NIGER
West Africa

Official name Republic of Niger
Formation 1960 / 1960
Capital Niamey
Population 18.5 million / 38 people per sq mile (15 people per sq km)
Total area 489,188 sq. miles (1,267,000 sq. km)
Languages Hausa, Djerma, Fulani, Tuareg, Teda, French*
Religions Muslim 99%, Other (including Christian) 1%
Ethnic mix Hausa 53%, Djerma and Songhai 21%, Tuareg 11%, Fulani 7%, Kanuri 6%, Other 2%
Government Presidential system
Currency CFA franc = 100 centimes
Literacy rate 16%
Calorie consumption 2546 kilocalories

NIGERIA
West Africa

Official name Federal Republic of Nigeria
Formation 1960 / 1961
Capital Abuja
Population 179 million / 508 people per sq mile (196 people per sq km)
Total area 356,667 sq. miles (923,768 sq. km)
Languages Hausa, English*, Yoruba, Ibo
Religions Muslim 50%, Christian 40%, Traditional beliefs 10%
Ethnic mix Other 29%, Hausa 21%, Yoruba 21%, Ibo 18%, Fulani 11%
Government Presidential system
Currency Naira = 100 kobo
Literacy rate 51%
Calorie consumption 2700 kilocalories

NORTH KOREA
East Asia

Official name Democratic People's Republic of Korea
Formation 1948 / 1953
Capital Pyongyang
Population 25 million / 538 people per sq mile (208 people per sq km)
Total area 46,540 sq. miles (120,540 sq. km)
Languages Korean*
Religions Atheist 100%
Ethnic mix Korean 100%
Government One-party state
Currency North Korean won = 100 chon
Literacy rate 99%
Calorie consumption 2094 kilocalories

NORWAY
Northern Europe

Official name Kingdom of Norway
Formation 1905 / 1905
Capital Oslo
Population 5.1 million / 43 people per sq mile (17 people per sq km)
Total area 125,181 sq. miles (324,220 sq. km)
Languages Norwegian* (Bokmål "book language" and Nynorsk "new Norsk"), Sámi
Religions Evangelical Lutheran 88%, Other and nonreligious 8%, Muslim 2%, Pentecostal 1%, Roman Catholic 1%
Ethnic mix Norwegian 93%, Other 6%, Sámi 1%
Government Parliamentary system
Currency Norwegian krone = 100 øre
Literacy rate 99%
Calorie consumption 3484 kilocalories

OMAN
Southwest Asia

Official name Sultanate of Oman
Formation 1951 / 1951
Capital Muscat
Population 3.9 million / 48 people per sq mile (18 people per sq km)
Total area 82,031 sq. miles (212,460 sq. km)
Languages Arabic*, Baluchi, Farsi, Hindi, Punjabi
Religions Ibadi Muslim 75%, Other Muslim and Hindu 25%
Ethnic mix Arab 88%, Baluchi 4%, Persian 3%, Indian and Pakistani 3%, African 2%
Government Monarchy
Currency Omani rial = 1000 baisa
Literacy rate 87%
Calorie consumption 3143 kilocalories

PAKISTAN
South Asia

Official name Islamic Republic of Pakistan
Formation 1947 / 1971
Capital Islamabad
Population 185 million / 622 people per sq mile (240 people per sq km)
Total area 310,401 sq. miles (803,940 sq. km)
Languages Punjabi, Sindhi, Pashtu, Urdu*, Baluchi, Brahui
Religions Sunni Muslim 77%, Shi'a Muslim 20%, Hindu 2%, Christian 1%
Ethnic mix Punjabi 56%, Pathan (Pashtun) 15%, Sindhi 14%, Mohajir 7%, Baluchi 4%, Other 4%
Government Parliamentary system
Currency Pakistani rupee = 100 paisa
Literacy rate 55%
Calorie consumption 2440 kilocalories

PALAU
Australasia & Oceania

Official name Republic of Palau
Formation 1994 / 1994
Capital Melekeok
Population 21,186 / 108 people per sq mile (42 people per sq km)
Total area 177 sq. miles (458 sq. km)
Languages Palauan*, English*, Japanese, Angaur, Tobi, Sonsorolese
Religions Christian 66%, Modekngei 34%
Ethnic mix Palauan 74%, Filipino 16%, Other 6%, Chinese and other Asian 4%
Government Nonparty system
Currency US dollar = 100 cents
Literacy rate 99%
Calorie consumption Not available

PANAMA
Central America

Official name Republic of Panama
Formation 1903 / 1903
Capital Panama City
Population 3.9 million / 133 people per sq mile (51 people per sq km)
Total area 30,193 sq. miles (78,200 sq. km)
Languages English Creole, Spanish*, Amerindian languages, Chibchan languages
Religions Roman Catholic 84%, Protestant 15%, Other 1%
Ethnic mix Mestizo 70%, Black 14%, White 10%, Amerindian 6%
Government Presidential system
Currency Balboa = 100 centésimos; and US dollar = 100 cents
Literacy rate 94%
Calorie consumption 2733 kilocalories

PAPUA NEW GUINEA
Australasia & Oceania

Official name Independent State of Papua New Guinea
Formation 1975 / 1975
Capital Port Moresby
Population 7.5 million / 43 people per sq mile (17 people per sq km)
Total area 178,703 sq. miles (462,840 sq. km)
Languages Pidgin English, Papuan, English*, Motu, 800 (est.) native languages
Religions Protestant 60%, Roman Catholic 37%, Other 3%
Ethnic mix Melanesian and mixed race 100%
Government Parliamentary system
Currency Kina = 100 toea
Literacy rate 63%
Calorie consumption 2193 kilocalories

PARAGUAY
South America

Official name Republic of Paraguay
Formation 1811 / 1938
Capital Asunción
Population 6.9 million / 45 people per sq mile (17 people per sq km)
Total area 157,046 sq. miles (406,750 sq. km)
Languages Guarani*, Spanish*, German
Religions Roman Catholic 90%, Protestant (including Mennonite) 10%
Ethnic mix Mestizo 91%, Other 7%, Amerindian 2%
Government Presidential system
Currency Guaraní = 100 céntimos
Literacy rate 94%
Calorie consumption 2589 kilocalories

PERU
South America

Official name Republic of Peru
Formation 1824 / 1941
Capital Lima
Population 30.8 million / 62 people per sq mile (24 people per sq km)
Total area 496,223 sq. miles (1,285,200 sq. km)
Languages Spanish*, Quechua*, Aymara
Religions Roman Catholic 81%, Other 19%
Ethnic mix Amerindian 45%, Mestizo 37%, White 15%, Other 3%
Government Presidential system
Currency New sol = 100 céntimos
Literacy rate 94%
Calorie consumption 2700 kilocalories

PHILIPPINES
Southeast Asia

Official name Republic of the Philippines
Formation 1946 / 1946
Capital Manila
Population 100 million / 870 people per sq mile (336 people per sq km)
Total area 115,830 sq. miles (300,000 sq. km)
Languages Filipino*, English*, Tagalog, Cebuano, Ilocano, Hiligaynon, many other local languages
Religions Roman Catholic 81%, Protestant 9%, Muslim 5%, Other (including Buddhist) 5%
Ethnic mix Other 34%, Tagalog 28%, Cebuano 13%, Ilocano 9%, Hiligaynon 8%, Bisaya 8%
Government Presidential system
Currency Philippine peso = 100 centavos
Literacy rate 95%
Calorie consumption 2570 kilocalories

POLAND
Northern Europe

Official name Republic of Poland
Formation 1918 / 1945
Capital Warsaw
Population 38.2 million / 325 people per sq mile (125 people per sq km)
Total area 120,728 sq. miles (312,685 sq. km)
Languages Polish*
Religions Roman Catholic 93%, Other and nonreligious 5%, Orthodox Christian 2%
Ethnic mix Polish 98%, Other 2%
Government Parliamentary system
Currency Zloty = 100 groszy
Literacy rate 99%
Calorie consumption 3485 kilocalories

PORTUGAL
Southwest Europe

Official name Portuguese Republic
Formation 1139 / 1640
Capital Lisbon
Population 10.6 million / 299 people per sq mile (115 people per sq km)
Total area 35,672 sq. miles (92,391 sq. km)
Languages Portuguese*
Religions Roman Catholic 92%, Protestant 4%, Nonreligious 3%, Other 1%
Ethnic mix Portuguese 98%, African and other 2%
Government Parliamentary system
Currency Euro = 100 cents
Literacy rate 94%
Calorie consumption 3456 kilocalories

QATAR
Southwest Asia

Official name State of Qatar
Formation 1971 / 1971
Capital Doha
Population 2.3 million / 542 people per sq mile (209 people per sq km)
Total area 4416 sq. miles (11,437 sq. km)
Languages Arabic*
Religions Muslim (mainly Sunni) 95%, Other 5%
Ethnic mix Qatari 20%, Indian 20%, Other Arab 20%, Nepalese 13%, Filipino 10%, Other 10%, Pakistani 7%
Government Monarchy
Currency Qatar riyal = 100 dirhams
Literacy rate 97%
Calorie consumption Not available

ROMANIA
Southeast Europe

Official name Romania
Formation 1878 / 1947
Capital Bucharest
Population 21.6 million / 243 people per sq mile (94 people per sq km)
Total area 91,699 sq. miles (237,500 sq. km)
Languages Romanian*, Hungarian (Magyar), Romani, German
Religions Romanian Orthodox 87%, Protestant 5%, Roman Catholic 5%, Greek Orthodox 1%, Greek Catholic (Uniate) 1%, Other 1%
Ethnic mix Romanian 89%, Magyar 7%, Roma 3%, Other 1%
Government Presidential system
Currency New Romanian leu = 100 bani
Literacy rate 99%
Calorie consumption 3363 kilocalories

RUSSIAN FEDERATION
Europe / Asia

Official name Russian Federation
Formation 1480 / 1991
Capital Moscow
Population 143 million / 22 people per sq mile (8 people per sq km)
Total area 6,592,735 sq. miles (17,075,200 sq. km)
Languages Russian*, Tatar, Ukrainian, Chavash, various other national languages
Religions Orthodox Christian 75%, Muslim 14%, Other 11%
Ethnic mix Russian 80%, Other 12%, Tatar 4%, Ukrainian 2%, Bashkir 1%, Chavash 1%
Government Mixed Presidential–Parliamentary system
Currency Russian rouble = 100 kopeks
Literacy rate 99%
Calorie consumption 3358 kilocalories

RWANDA
Central Africa

Official name Republic of Rwanda
Formation 1962 / 1962
Capital Kigali
Population 12.1 million / 1256 people per sq mile (485 people per sq km)
Total area 10,169 sq. miles (26,338 sq. km)
Languages Kinyarwanda*, French*, Kiswahili, English*
Religions Christian 94%, Muslim 5%, Traditional beliefs 1%
Ethnic mix Hutu 85%, Tutsi 14%, Other (including Twa) 1%
Government Presidential system
Currency Rwanda franc = 100 centimes
Literacy rate 66%
Calorie consumption 2148 kilocalories

ST KITTS & NEVIS
West Indies

Official name Federation of Saint Christopher and Nevis
Formation 1983 / 1983
Capital Basseterre
Population 51,538 / 371 people per sq mile (143 people per sq km)
Total area 101 sq. miles (261 sq. km)
Languages English*, English Creole
Religions Anglican 33%, Methodist 29%, Other 22%, Moravian 9%, Roman Catholic 7%
Ethnic mix Black 95%, Mixed race 3%, White 1%, Other and Amerindian 1%
Government Parliamentary system
Currency East Caribbean dollar = 100 cents
Literacy rate 98%
Calorie consumption 2507 kilocalories

ST LUCIA
West Indies

Official name Saint Lucia
Formation 1979 / 1979
Capital Castries
Population 200,000 / 847 people per sq mile (328 people per sq km)
Total area 239 sq. miles (620 sq. km)
Languages English*, French Creole
Religions Roman Catholic 90%, Other 10%
Ethnic mix Black 83%, Mulatto (mixed race) 13%, Asian 3%, Other 1%
Government Parliamentary system
Currency East Caribbean dollar = 100 cents
Literacy rate 95%
Calorie consumption 2629 kilocalories

ST VINCENT & THE GRENADINES
West Indies

Official name Saint Vincent and the Grenadines
Formation 1979 / 1979
Capital Kingstown
Population 102,918 / 786 people per sq mile (303 people per sq km)
Total area 150 sq. miles (389 sq. km)
Languages English*, English Creole
Religions Anglican 47%, Methodist 28%, Roman Catholic 13%, Other 12%
Ethnic mix Black 66%, Mulatto (mixed race) 19%, Other 12%, Carib 2%, Asian 1%
Government Parliamentary system
Currency East Caribbean dollar = 100 cents
Literacy rate 88%
Calorie consumption 2960 kilocalories

SAMOA
Australasia & Oceania

Official name Independent State of Samoa
Formation 1962 / 1962
Capital Apia
Population 200,000 / 183 people per sq mile (71 people per sq km)
Total area 1104 sq. miles (2860 sq. km)
Languages Samoan*, English*
Religions Christian 99%, Other 1%
Ethnic mix Polynesian 91%, Euronesian 7%, Other 2%
Government Parliamentary system
Currency Tala = 100 sene
Literacy rate 99%
Calorie consumption 2872 kilocalories

SAN MARINO
Southern Europe

Official name Republic of San Marino
Formation 1631 / 1631
Capital San Marino
Population 32,742 / 1364 people per sq mile (537 people per sq km)
Total area 23.6 sq. miles (61 sq. km)
Languages Italian*
Religions Roman Catholic 93%, Other and nonreligious 7%
Ethnic mix Sammarinese 88%, Italian 10%, Other 2%
Government Parliamentary system
Currency Euro = 100 cents
Literacy rate 99%
Calorie consumption Not available

SAO TOME & PRINCIPE
West Africa

Official name Democratic Republic of Sao Tome and Principe
Formation 1975 / 1975
Capital São Tomé
Population 200,000 / 539 people per sq mile (208 people per sq km)
Total area 386 sq. miles (1001 sq. km)
Languages Portuguese Creole, Portuguese*
Religions Roman Catholic 84%, Other 16%
Ethnic mix Black 90%, Portuguese and Creole 10%
Government Presidential system
Currency Dobra = 100 céntimos
Literacy rate 70%
Calorie consumption 2676 kilocalories

SAUDI ARABIA
Southwest Asia

Official name Kingdom of Saudi Arabia
Formation 1932 / 1932
Capital Riyadh
Population 29.4 million / 36 people per sq mile (14 people per sq km)
Total area 756,981 sq. miles (1,960,582 sq. km)
Languages Arabic*
Religions Sunni Muslim 85%, Shi'a Muslim 15%
Ethnic mix Arab 72%, Foreign residents (mostly south and southeast Asian) 20%, Afro-Asian 8%
Government Monarchy
Currency Saudi riyal = 100 halalat
Literacy rate 94%
Calorie consumption 3122 kilocalories

SENEGAL
West Africa

Official name Republic of Senegal
Formation 1960 / 1960
Capital Dakar
Population 14.5 million / 195 people per sq mile (75 people per sq km)
Total area 75,749 sq. miles (196,190 sq. km)
Languages Wolof, Pulaar, Serer, Diola, Mandinka, Malinké, Soninké, French*
Religions Sunni Muslim 95%, Christian (mainly Roman Catholic) 4%, Traditional beliefs 1%
Ethnic mix Wolof 43%, Serer 15%, Peul 14%, Other 14%, Toucouleur 9%, Diola 5%
Government Presidential system
Currency CFA franc = 100 centimes
Literacy rate 52%
Calorie consumption 2426 kilocalories

SERBIA
Southeast Europe

Official name Republic of Serbia
Formation 2006 / 2008
Capital Belgrade
Population 9.5 million / 318 people per sq mile (123 people per sq km)
Total area 29,905 sq. miles (77,453 sq. km)
Languages Serbian*, Hungarian (Magyar)
Religions Orthodox Christian 85%, Roman Catholic 6%, Other 6%, Muslim 3%
Ethnic mix Serb 83%, Other 10%, Magyar 4%, Bosniak 2%, Roma 1%
Government Parliamentary system
Currency Serbian dinar = 100 para
Literacy rate 98%
Calorie consumption 2724 kilocalories

SEYCHELLES
Indian Ocean

Official name Republic of Seychelles
Formation 1976 / 1976
Capital Victoria
Population 91,650 / 881 people per sq mile (339 people per sq km)
Total area 176 sq. miles (455 sq. km)
Languages French Creole*, English*, French*
Religions Roman Catholic 82%, Anglican 6%, Other (including Muslim) 6%, Other Christian 3%,
Hindu 2%, Seventh-day Adventist 1%
Ethnic mix Creole 89%, Indian 5%, Other 4%, Chinese 2%
Government Presidential system
Currency Seychelles rupee = 100 cents
Literacy rate 92%
Calorie consumption 2426 kilocalories

SIERRA LEONE
West Africa

Official name Republic of Sierra Leone
Formation 1961 / 1961
Capital Freetown
Population 6.2 million / 224 people per sq mile (87 people per sq km)
Total area 27,698 sq. miles (71,740 sq. km)
Languages Mende, Temne, Krio, English*
Religions Muslim 60%, Christian 30%, Traditional beliefs 10%
Ethnic mix Mende 35%, Temne 32%, Other 21%, Limba 8%, Kuranko 4%
Government Presidential system
Currency Leone = 100 cents
Literacy rate 44%
Calorie consumption 2333 kilocalories

SINGAPORE
Southeast Asia

Official name Republic of Singapore
Formation 1965 / 1965
Capital Singapore
Population 5.5 million / 23,305 people per sq mile (9016 people per sq km)
Total area 250 sq. miles (648 sq. km)
Languages Mandarin*, Malay*, Tamil*, English*
Religions Buddhist 55%, Taoist 22%, Muslim 16%, Hindu, Christian, and Sikh 7%
Ethnic mix Chinese 74%, Malay 14%, Indian 9%, Other 3%
Government Parliamentary system
Currency Singapore dollar = 100 cents
Literacy rate 96%
Calorie consumption Not available

SLOVAKIA
Central Europe

Official name Slovak Republic
Formation 1993 / 1993
Capital Bratislava
Population 5.5 million / 290 people per sq mile (112 people per sq km)
Total area 18,859 sq. miles (48,845 sq. km)
Languages Slovak*, Hungarian (Magyar), Czech
Religions Roman Catholic 69%, Nonreligious 13%, Other 13%, Greek Catholic (Uniate) 4%, Orthodox Christian 1%
Ethnic mix Slovak 86%, Magyar 10%, Roma 2%, Czech 1%, Other 1%
Government Parliamentary system
Currency Euro = 100 cents
Literacy rate 99%
Calorie consumption 2902 kilocalories

SLOVENIA
Central Europe

Official name Republic of Slovenia
Formation 1991 / 1991
Capital Ljubljana
Population 2.1 million / 269 people per sq mile (104 people per sq km)
Total area 7820 sq. miles (20,253 sq. km)
Languages Slovenian*
Religions Roman Catholic 58%, Other 28%, Atheist 10%, Orthodox Christian 2%, Muslim 2%
Ethnic mix Slovene 83%, Other 12%, Serb 2%, Croat 2%, Bosniak 1%
Government Parliamentary system
Currency Euro = 100 cents
Literacy rate 99%
Calorie consumption 3173 kilocalories

SOLOMON ISLANDS
Australasia & Oceania

Official name Solomon Islands
Formation 1978 / 1978
Capital Honiara
Population 600,000 / 56 people per sq mile (21 people per sq km)
Total area 10,985 sq. miles (28,450 sq. km)
Languages English*, Pidgin English, Melanesian Pidgin, 120 (est.) native languages
Religions Church of Melanesia (Anglican) 34%, Roman Catholic 19%, South Seas Evangelical Church 17%, Methodist 11%, Seventh-day Adventist 10%, Other 9%
Ethnic mix Melanesian 93%, Polynesian 4%, Micronesian 2%, Other 1%
Government Parliamentary system
Currency Solomon Islands dollar = 100 cents
Literacy rate 77%
Calorie consumption 2473 kilocalories

SOMALIA
East Africa

Official name Federal Republic of Somalia
Formation 1960 / 1960
Capital Mogadishu
Population 10.8 million / 45 people per sq mile (17 people per sq km)
Total area 246,199 sq. miles (637,657 sq. km)
Languages Somali*, Arabic*, English, Italian
Religions Sunni Muslim 99%, Christian 1%
Ethnic mix Somali 85%, Other 15%
Government Non-party system
Currency Somali shilin = 100 senti
Literacy rate 24%
Calorie consumption 1696 kilocalories

SOUTH AFRICA
Southern Africa

Official name Republic of South Africa
Formation 1934 / 1994
Capital Pretoria; Cape Town; Bloemfontein
Population 53.1 million / 113 people per sq mile (43 people per sq km)
Total area 471,008 sq. miles (1,219,912 sq. km)
Languages English, isiZulu, isiXhosa, Afrikaans, Sepedi, Setswana, Sesotho, Xitsonga, siSwati, Tshivenda, isiNdebele
Religions Christian 68%, Traditional beliefs and animist 29%, Muslim 2%, Hindu 1%
Ethnic mix Black 80%, Mixed race 9%, White 9%, Asian 2%
Government Presidential system
Currency Rand = 100 cents
Literacy rate 94%
Calorie consumption 3007 kilocalories

SOUTH KOREA
East Asia

Official name Republic of Korea
Formation 1948 / 1953
Capital Seoul; Sejong City (administrative)
Population 49.5 million / 1299 people per sq mile (501 people per sq km)
Total area 38,023 sq. miles (98,480 sq. km)
Languages Korean*
Religions Mahayana Buddhist 47%, Protestant 38%, Roman Catholic 11%, Confucianist 3%, Other 1%
Ethnic mix Korean 100%
Government Presidential system
Currency South Korean won = 100 chon
Literacy rate 99%
Calorie consumption 3329 kilocalories

SOUTH SUDAN
East Africa

Official name Republic of South Sudan
Formation 2011 / 2011
Capital Juba
Population 11.7 million / 47 people per sq mile (18 people per sq km)
Total area 248,777 sq. miles (644,329 sq. km)
Languages Arabic, Dinka, Nuer, Zande, Bari, Shilluk, Lotuko, English*
Religions Over half of the population follow Christian or traditional beliefs.
Ethnic mix Dinka 40%, Nuer 15%, Bari 10%, Shilluk/Anwak 10%, Azande 10%, Arab 10%, Other 5%
Government Transitional regime
Currency South Sudan pound = 100 piastres
Literacy rate 37%
Calorie consumption Not available

SPAIN
Southwest Europe

Official name Kingdom of Spain
Formation 1492 / 1713
Capital Madrid
Population 47.1 million / 244 people per sq mile (94 people per sq km)
Total area 194,896 sq. miles (504,782 sq. km)
Languages Spanish*, Catalan*, Galician*, Basque*
Religions Roman Catholic 96%, Other 4%
Ethnic mix Castilian Spanish 72%, Catalan 17%, Galician 6%, Basque 2%, Other 2%, Roma 1%
Government Parliamentary system
Currency Euro = 100 cents
Literacy rate 98%
Calorie consumption 3183 kilocalories

SRI LANKA
South Asia

Official name Democratic Socialist Republic of Sri Lanka
Formation 1948 / 1948
Capital Colombo; Sri Jayewardenapura Kotte
Population 21.4 million / 856 people per sq mile (331 people per sq km)
Total area 25,332 sq. miles (65,610 sq. km)
Languages Sinhala*, Tamil*, Sinhala-Tamil, English
Religions Buddhist 69%, Hindu 15%, Muslim 8%, Christian 8%
Ethnic mix Sinhalese 74%, Tamil 18%, Moor 7%, Other 1%
Government Mixed presidential–parliamentary system
Currency Sri Lanka rupee = 100 cents
Literacy rate 91%
Calorie consumption 2539 kilocalories

SUDAN
East Africa

Official name Republic of the Sudan
Formation 1956 / 2011
Capital Khartoum
Population 38.8 million / 54 people per sq mile (21 people per sq km)
Total area 718,722 sq. miles (1,861,481 sq. km)
Languages Arabic, Nubian, Beja, Fur
Religions Nearly the whole population is Muslim (mainly Sunni)
Ethnic mix Arab 60%, Other 18%, Nubian 10%, Beja 8%, Fur 3%, Zaghawa 1%
Government Presidential system
Currency New Sudanese pound = 100 piastres
Literacy rate 73%
Calorie consumption 2346 kilocalories

SURINAME
South America

Official name Republic of Suriname
Formation 1975 / 1975
Capital Paramaribo
Population 500,000 / 8 people per sq mile (3 people per sq km)
Total area 63,039 sq. miles (163,270 sq. km)
Languages Sranan (creole), Dutch*, Javanese, Sarnami Hindi, Saramaccan, Chinese, Carib
Religions Hindu 27%, Protestant 25%, Roman Catholic 23%, Muslim 20%, Traditional beliefs 5%
Ethnic mix East Indian 27%, Creole 18%, Black 15%, Javanese 15%, Mixed race 13%, Other 6%, Amerindian 4%, Chinese 2%
Government Mixed presidential–parliamentary system
Currency Surinamese dollar = 100 cents
Literacy rate 95%
Calorie consumption 2727 kilocalories

SWAZILAND
Southern Africa

Official name Kingdom of Swaziland
Formation 1968 / 1968
Capital Mbabane
Population 1.3 million / 196 people per sq mile (76 people per sq km)
Total area 6704 sq. miles (17,363 sq. km)
Languages English*, siSwati*, isiZulu, Xitsonga
Religions Traditional beliefs 40%, Other 30%, Roman Catholic 20%, Muslim 10%
Ethnic mix Swazi 97%, Other 3%
Government Monarchy
Currency Lilangeni = 100 cents
Literacy rate 83%
Calorie consumption 2275 kilocalories

SWEDEN
Northern Europe

Official name Kingdom of Sweden
Formation 1523 / 1921
Capital Stockholm
Population 9.6 million / 60 people per sq mile (23 people per sq km)
Total area 173,731 sq. miles (449,964 sq. km)
Languages Swedish*, Finnish, Sámi
Religions Evangelical Lutheran 75%, Other 13%, Muslim 5%, Other Protestant 5%, Roman Catholic 2%
Ethnic mix Swedish 86%, Foreign-born or first-generation immigrant 12%, Finnish and Sámi 2%
Government Parliamentary system
Currency Swedish krona = 100 öre
Literacy rate 99%
Calorie consumption 3160 kilocalories

SWITZERLAND
Central Europe

Official name Swiss Confederation
Formation 1291 / 1857
Capital Bern
Population 8.2 million / 534 people per sq mile (206 people per sq km)
Total area 15,942 sq. miles (41,290 sq. km)
Languages German*, Swiss-German, French*, Italian*, Romansch
Religions Roman Catholic 42%, Protestant 35%, Other and nonreligious 19%, Muslim 4%
Ethnic mix German 64%, French 20%, Other 9.5%, Italian 6%, Romansch 0.5%
Government Parliamentary system
Currency Swiss franc = 100 rappen/centimes
Literacy rate 99%
Calorie consumption 3487 kilocalories

SYRIA
Southwest Asia

Official name Syrian Arab Republic
Formation 1941 / 1967
Capital Damascus
Population 22 million / 310 people per sq mile (120 people per sq km)
Total area 71,498 sq. miles (184,180 sq. km)
Languages Arabic*, French, Kurdish, Armenian, Circassian, Turkic languages, Assyrian, Aramaic
Religions Sunni Muslim 74%, Alawi 12%, Christian 10%, Druze 3%, Other 1%
Ethnic mix Arab 90%, Kurdish 9%, Armenian, Turkmen, and Circassian 1%
Government Presidential system
Currency Syrian pound = 100 piastres
Literacy rate 85%
Calorie consumption 3106 kilocalories

TAIWAN
East Asia

Official name Republic of China (ROC)
Formation 1949 / 1949
Capital Taibei (Taipei)
Population 23.4 million / 1879 people per sq mile (725 people per sq km)
Total area 13,892 sq. miles (35,980 sq. km)
Languages Amoy Chinese, Mandarin Chinese*, Hakka Chinese
Religions Buddhist, Confucianist, and Taoist 93%, Christian 5%, Other 2%
Ethnic mix Han Chinese (pre-20th-century migration) 84%, Han Chinese (20th-century migration) 14%, Aboriginal 2%
Government Presidential system
Currency Taiwan dollar = 100 cents
Literacy rate 98%
Calorie consumption 2997 kilocalories

TAJIKISTAN
Central Asia

Official name Republic of Tajikistan
Formation 1991 / 1991
Capital Dushanbe
Population 8.4 million / 152 people per sq mile (59 people per sq km)
Total area 55,251 sq. miles (143,100 sq. km)
Languages Tajik*, Uzbek, Russian
Religions Sunni Muslim 95%, Shi'a Muslim 3%, Other 2%
Ethnic mix Tajik 80%, Uzbek 15%, Other 3%, Russian 1%, Kyrgyz 1%
Government Presidential system
Currency Somoni = 100 diram
Literacy rate 99%
Calorie consumption 2101 kilocalories

TANZANIA
East Africa

Official name United Republic of Tanzania
Formation 1964 / 1964
Capital Dodoma
Population 50.8 million / 148 people per sq mile (57 people per sq km)
Total area 364,898 sq. miles (945,087 sq. km)
Languages Kiswahili*, Sukuma, Chagga, Nyamwezi, Hehe, Makonde, Yao, Sandawe, English*
Religions Christian 63%, Muslim 35%, Other 2%
Ethnic mix Native African (over 120 tribes) 99%, European, Asian, and Arab 1%
Government Presidential system
Currency Tanzanian shilling = 100 cents
Literacy rate 68%
Calorie consumption 2208 kilocalories

THAILAND
Southeast Asia

Official name Kingdom of Thailand
Formation 1238 / 1907
Capital Bangkok
Population 67.2 million / 341 people per sq mile (132 people per sq km)
Total area 198,455 sq. miles (514,000 sq. km)
Languages Thai*, Chinese, Malay, Khmer, Mon, Karen, Miao
Religions Buddhist 95%, Muslim 4%, Other (including Christian) 1%
Ethnic mix Thai 83%, Chinese 12%, Malay 3%, Khmer and Other 2%
Government Transitional regime
Currency Baht = 100 satang
Literacy rate 96%
Calorie consumption 2784 kilocalories

TOGO
West Africa

Official name Togolese Republic
Formation 1960 / 1960
Capital Lomé
Population 7 million / 333 people per sq mile (129 people per sq km)
Total area 21,924 sq. miles (56,785 sq. km)
Languages Ewe, Kabye, Gurma, French*
Religions Christian 47%, Traditional beliefs 33%, Muslim 14%, Other 6%
Ethnic mix Ewe 46%, Other African 41%, Kabye 12%, European 1%
Government Presidential system
Currency CFA franc = 100 centimes
Literacy rate 60%
Calorie consumption 2366 kilocalories

TONGA
Australasia & Oceania

Official name Kingdom of Tonga
Formation 1970 / 1970
Capital Nuku'alofa
Population 106,440 / 383 people per sq mile (148 people per sq km)
Total area 289 sq. miles (748 sq. km)
Languages English*, Tongan*
Religions Free Wesleyan 41%, Other 17%, Roman Catholic 16%, Church of Jesus Christ of Latter-day Saints 14%, Free Church of Tonga 12%
Ethnic mix Tongan 98%, Other 2%
Government Monarchy
Currency Pa'anga (Tongan dollar) = 100 seniti
Literacy rate 99%
Calorie consumption Not available

TRINIDAD & TOBAGO
West Indies

Official name Republic of Trinidad and Tobago
Formation 1962 / 1962
Capital Port-of-Spain
Population 1.3 million / 656 people per sq mile (253 people per sq km)
Total area 1980 sq. miles (5128 sq. km)
Languages English Creole, English*, Hindi, French, Spanish
Religions Roman Catholic 26%, Hindu 23%, Other and nonreligious 23%, Anglican 8%, Baptist 7%, Pentecostal 7%, Muslim 6%
Ethnic mix East Indian 40%, Black 38%, Mixed race 20%, White and Chinese 1%, other 1%
Government Parliamentary system
Currency Trinidad and Tobago dollar = 100 cents
Literacy rate 99%
Calorie consumption 2889 kilocalories

TUNISIA
North Africa

Official name Tunisian Republic
Formation 1956 / 1956
Capital Tunis
Population 11.1 million / 185 people per sq mile (71 people per sq km)
Total area 63,169 sq. miles (163,610 sq. km)
Languages Arabic*, French
Religions Muslim (mainly Sunni) 98%, Christian 1%, Jewish 1%
Ethnic mix Arab and Berber 98%, Jewish 1%, European 1%
Government Mixed presidential–parliamentary system
Currency Tunisian dinar = 1000 millimes
Literacy rate 80%
Calorie consumption 3362 kilocalories

TURKEY
Asia / Europe

Official name Republic of Turkey
Formation 1923 / 1939
Capital Ankara
Population 75.8 million / 255 people per sq mile (98 people per sq km)
Total area 301,382 sq. miles (780,580 sq. km)
Languages Turkish*, Kurdish, Arabic, Circassian, Armenian, Greek, Georgian, Ladino
Religions Muslim (mainly Sunni) 99%, Other 1%
Ethnic mix Turkish 70%, Kurdish 20%, Other 8%, Arab 2%
Government Parliamentary system
Currency Turkish lira = 100 kurus
Literacy rate 95%
Calorie consumption 3680 kilocalories

TURKMENISTAN
Central Asia

Official name Turkmenistan
Formation 1991 / 1991
Capital Ashgabat
Population 5.3 million / 28 people per sq mile (11 people per sq km)
Total area 188,455 sq. miles (488,100 sq. km)
Languages Turkmen*, Uzbek, Russian, Kazakh, Tatar
Religions Sunni Muslim 89%, Orthodox Christian 9%, Other 2%
Ethnic mix Turkmen 85%, Other 6%, Uzbek 5%, Russian 4%
Government Presidential system
Currency New manat = 100 tenge
Literacy rate 99%
Calorie consumption 2883 kilocalories

TUVALU
Australasia & Oceania

Official name Tuvalu
Formation 1978 / 1978
Capital Funafuti Atoll
Population 10,782 / 1078 people per sq mile (415 people per sq km)
Total area 10 sq. miles (26 sq. km)
Languages Tuvaluan, Kiribati, English*
Religions Church of Tuvalu 97%, Baha'i 1%, Seventh-day Adventist 1%, Other 1%
Ethnic mix Polynesian 96%, Micronesian 4%
Government Nonparty system
Currency Australian dollar = 100 cents; and Tuvaluan dollar = 100 cents
Literacy rate 95%
Calorie consumption Not available

UGANDA
East Africa

Official name Republic of Uganda
Formation 1962 / 1962
Capital Kampala
Population 38.8 million / 504 people per sq mile (194 people per sq km)
Total area 91,135 sq. miles (236,040 sq. km)
Languages Luganda, Nkole, Chiga, Lango, Acholi, Teso, Lugbara, English*
Religions Christian 85%, Muslim (mainly Sunni) 12%, Other 3%
Ethnic mix Other 50%, Baganda 17%, Banyakole 10%, Basoga 9%, Iteso 7%, Bakiga 7%
Government Presidential system
Currency Uganda shilling = 100 cents
Literacy rate 74%
Calorie consumption 2279 kilocalories

UKRAINE
Eastern Europe

Official name Ukraine
Formation 1991 / 1991
Capital Kiev
Population 44.9 million / 193 people per sq mile (74 people per sq km)
Total area 223,089 sq. miles (603,700 sq. km)
Languages Ukrainian*, Russian, Tatar
Religions Christian (mainly Orthodox) 95%, Other 5%
Ethnic mix Ukrainian 78%, Russian 17%, Other 5%
Government Presidential system
Currency Hryvna = 100 kopiykas
Literacy rate 99%
Calorie consumption 3142 kilocalories

UNITED ARAB EMIRATES
Southwest Asia

Official name United Arab Emirates
Formation 1971 / 1972
Capital Abu Dhabi
Population 9.4 million / 291 people per sq mile (112 people per sq km)
Total area 32,000 sq. miles (82,880 sq. km)
Languages Arabic*, Farsi, Indian and Pakistani languages, English
Religions Muslim (mainly Sunni) 96%, Christian, Hindu, and other 4%
Ethnic mix Asian 60%, Emirian 25%, Other Arab 12%, European 3%
Government Monarchy
Currency UAE dirham = 100 fils
Literacy rate 90%
Calorie consumption 3215 kilocalories

UNITED KINGDOM
Northwest Europe

Official name United Kingdom of Great Britain and Northern Ireland
Formation 1707 / 1922
Capital London
Population 63.5 million / 681 people per sq mile (263 people per sq km)
Total area 94,525 sq. miles (244,820 sq. km)
Languages English*, Welsh*, Scottish Gaelic, Irish
Religions Anglican 45%, Other and nonreligious 36%, Roman Catholic 9%, Presbyterian 4%, Muslim 3%, Methodist 2%, Hindu 1%
Ethnic mix English 80%, Scottish 9%, West Indian, Asian, and other 5%, Northern Irish 3%, Welsh 3%
Government Parliamentary system
Currency Pound sterling = 100 pence
Literacy rate 99%
Calorie consumption 3414 kilocalories

UNITED STATES
North America

Official name United States of America
Formation 1776 / 1959
Capital Washington D.C.
Population 323 million / 91 people per sq mile (35 people per sq km)
Total area 3,794,100 sq. miles (9,826,675 sq. km)
Languages English*, Spanish, Chinese, French, German, Tagalog, Vietnamese, Italian, Korean, Russian, Polish
Religions Protestant 52%, Roman Catholic 25%, Other and nonreligious 20%, Jewish 2%, Muslim 1%
Ethnic mix White 60%, Hispanic 17%, Black American/African 14%, Asian 6%, American Indians & Alaksa Natives 2%, Pacific Islanders 1%
Government Presidential system
Currency US dollar = 100 cents
Literacy rate 99%
Calorie consumption 3639 kilocalories

URUGUAY
South America

Official name Oriental Republic of Uruguay
Formation 1828 / 1828
Capital Montevideo
Population 3.4 million / 50 people per sq mile (19 people per sq km)
Total area 68,039 sq. miles (176,220 sq. km)
Languages Spanish*
Religions Roman Catholic 66%, Other and nonreligious 30%, Jewish 2%, Protestant 2%
Ethnic mix White 90%, Mestizo 6%, Black 4%
Government Presidential system
Currency Uruguayan peso = 100 centésimos
Literacy rate 98%
Calorie consumption 2939 kilocalories

UZBEKISTAN
Central Asia

Official name Republic of Uzbekistan
Formation 1991 / 1991
Capital Tashkent
Population 29.3 million / 170 people per sq mile (65 people per sq km)
Total area 172,741 sq. miles (447,400 sq. km)
Languages Uzbek*, Russian, Tajik, Kazakh
Religions Sunni Muslim 88%, Orthodox Christian 9%, Other 3%
Ethnic mix Uzbek 80%, Russian 6%, Other 6%, Tajik 5%, Kazakh 3%
Government Presidential system
Currency Som = 100 tiyin
Literacy rate 99%
Calorie consumption 2675 kilocalories

VANUATU
Australasia & Oceania

Official name Republic of Vanuatu
Formation 1980 / 1980
Capital Port Vila
Population 300,000 / 64 people per sq mile (25 people per sq km)
Total area 4710 sq. miles (12,200 sq. km)
Languages Bislama (Melanesian pidgin)*, English*, French*, other indigenous languages
Religions Presbyterian 37%, Other 19%, Anglican 15%, Roman Catholic 15%, Traditional beliefs 8%, Seventh-day Adventist 6%
Ethnic mix ni-Vanuatu 94%, European 4%, Other 2%
Government Parliamentary system
Currency Vatu = 100 centimes
Literacy rate 83%
Calorie consumption 2820 kilocalories

VATICAN CITY
Southern Europe

Official name State of the Vatican City
Formation 1929 / 1929
Capital Vatican City
Population 842 / 4953 people per sq mile (1914 people per sq km)
Total area 0.17 sq. miles (0.44 sq. km)
Languages Italian*, Latin*
Religions Roman Catholic 100%
Ethnic mix The current pope is Argentinian, though most popes for the last 500 years have been Italian. Cardinals are from many nationalities, but Italians form the largest group. Most of the resident lay persons are Italian.
Government Papal state
Currency Euro = 100 cents
Literacy rate 99%
Calorie consumption Not available

VENEZUELA
South America

Official name Bolivarian Republic of Venezuela
Formation 1830 / 1830
Capital Caracas
Population 30.9 million / 91 people per sq mile (35 people per sq km)
Total area 352,143 sq. miles (912,050 sq. km)
Languages Spanish*, Amerindian languages
Religions Roman Catholic 96%, Protestant 2%, Other 2%
Ethnic mix Mestizo 69%, White 20%, Black 9%, Amerindian 2%
Government Presidential system
Currency Bolívar fuerte = 100 céntimos
Literacy rate 96%
Calorie consumption 2880 kilocalories

VIETNAM
Southeast Asia

Official name Socialist Republic of Vietnam
Formation 1976 / 1976
Capital Hanoi
Population 92.5 million / 736 people per sq mile (284 people per sq km)
Total area 127,243 sq. miles (329,560 sq. km)
Languages Vietnamese*, Chinese, Thai, Khmer, Muong, Nung, Miao, Yao, Jarai
Religions Other 74%, Buddhist 14%, Roman Catholic 7%, Cao Dai 3%, Protestant 2%
Ethnic mix Vietnamese 86%, Other 8%, Muong 2%, Tay 2%, Thai 2%
Government One-party state
Currency Dông = 10 hao = 100 xu
Literacy rate 94%
Calorie consumption 2745 kilocalories

YEMEN
Southwest Asia

Official name Republic of Yemen
Formation 1990 / 1990
Capital Sana
Population 25 million / 115 people per sq mile (44 people per sq km)
Total area 203,849 sq. miles (527,970 sq. km)
Languages Arabic*
Religions Sunni Muslim 55%, Shi'a Muslim 42%, Christian, Hindu, and Jewish 3%
Ethnic mix Arab 99%, Afro-Arab, Indian, Somali, and European 1%
Government Transitional regime
Currency Yemeni rial = 100 fils
Literacy rate 66%
Calorie consumption 2223 kilocalories

ZAMBIA
Southern Africa

Official name Republic of Zambia
Formation 1964 / 1964
Capital Lusaka
Population 15 million / 52 people per sq mile (20 people per sq km)
Total area 290,584 sq. miles (752,614 sq. km)
Languages Bemba, Tonga, Nyanja, Lozi, Lala-Bisa, Nsenga, English*
Religions Christian 63%, Traditional beliefs 36%, Muslim and Hindu 1%
Ethnic mix Bemba 34%, Other African 26%, Tonga 16%, Nyanja 14%, Lozi 9%, European 1%
Government Presidential system
Currency New Zambian kwacha = 100 ngwee
Literacy rate 61%
Calorie consumption 1930 kilocalories

ZIMBABWE
Southern Africa

Official name Republic of Zimbabwe
Formation 1980 / 1980
Capital Harare
Population 14.6 million / 98 people per sq mile (38 people per sq km)
Total area 150,803 sq. miles (390,580 sq. km)
Languages Shona, isiNdebele, English*
Religions Syncretic (Christian/traditional beliefs) 50%, Christian 25%, Traditional beliefs 24%, Other (including Muslim) 1%
Ethnic mix Shona 71%, Ndebele 16%, Other African 11%, White 1%, Asian 1%
Government Presidential system
Currency US $, South African rand, Euro, UK £, Botswana pula, Australian $, Chinese yuan, Indian rupee, and Japanese yen are legal tender
Literacy rate 84%
Calorie consumption 2110 kilocalories

Geographical names

The following glossary lists all geographical terms occurring on the maps and in main-entry names in the Index-Gazetteer. These terms may precede, follow or be run together with the proper element of the name; where they precede it the term is reversed for indexing purposes - thus Poluostrov Yamal is indexed as Yamal, Poluostrov.

Key

Geographical term
Language, Term

A

Å *Danish, Norwegian*, River
Āb *Persian*, River
Adrar *Berber*, Mountains
Agía, Ágios *Greek*, Saint
Air *Indonesian*, River
Ákra *Greek*, Cape, point
Alpen *German*, Alps
Alt- *German*, Old
Altiplanicie *Spanish*, Plateau
Älve, -älven *Swedish*, River
-ån *Swedish*, River
Anse *French*, Bay
'Aqabat *Arabic*, Pass
Archipiélago *Spanish*, Archipelago
Arcipelago *Italian*, Archipelago
Arquipélago *Portuguese*, Archipelago
Arrecife(s) *Spanish*, Reef(s)
Aru *Tamil*, River
Augstiene *Latvian*, Upland
Aukštuma *Lithuanian*, Upland
Aust- *Norwegian*, Eastern
Avtonomnyy Okrug *Russian*, Autonomous district
Āw *Kurdish*, River
'Ayn *Arabic*, Spring, well
'Ayoûn *Arabic*, Wells

B

Baelt *Danish*, Strait
Bahía *Spanish*, Bay
Baḥr *Arabic*, River
Baía *Portuguese*, Bay
Baie *French*, Bay
Bañado *Spanish*, Marshy land
Bandao *Chinese*, Peninsula
Banjaran *Malay*, Mountain range
Barajı *Turkish*, Dam
Barragem *Portuguese*, Reservoir
Bassin *French*, Basin
Batang *Malay*, Stream
Beinn, Ben *Gaelic*, Mountain
-berg *Afrikaans, Norwegian*, Mountain
Besar *Indonesian, Malay*, Big
Birkat, Birket *Arabic*, Lake, well, cistern
Boğazı *Turkish*, Lake
Boka *Serbo-Croatian*, Bay
Bol'sh-aya, -iye, -oy, -oye *Russian*, Big
Botigh(i) *Uzbek*, Depression basin
-bre(en) *Norwegian*, Glacier
Bredning *Danish*, Bay
Bucht *German*, Bay
Bugt(en) *Danish*, Bay
Buḥayrat *Arabic*, Lake, reservoir
Buḥeiret *Arabic*, Lake
Bukit *Malay*, Mountain
-bukta *Norwegian*, Bay
bukten *Swedish*, Bay
Bulag *Mongolian*, Spring
Bulak *Uighur*, Spring
Burnu *Turkish*, Cape, point
Buuraha *Somali*, Mountains

C

Cabo *Portuguese*, Cape
Caka *Tibetan*, Salt lake
Canal *Spanish*, Channel
Cap *French*, Cape
Capo *Italian*, Cape, headland
Cascada *Portuguese*, Waterfall
Cayo(s) *Spanish*, Islet(s), rock(s)
Cerro *Spanish*, Hill
Chaîne *French*, Mountain range
Chapada *Portuguese*, Hills, upland
Chau *Cantonese*, Island
Chāy *Turkish*, River
Chhâk *Cambodian*, Bay
Chhu *Tibetan*, River
-chŏsuji *Korean*, Reservoir
Chott *Arabic*, Depression, salt lake
Chŭli *Uzbek*, Grassland, steppe
Ch'ün-tao *Chinese*, Island group
Chuŏr Phnum *Cambodian*, Mountains
Ciudad *Spanish*, City, town
Co *Tibetan*, Lake
Colline(s) *French*, Hill(s)
Cordillera *Spanish*, Mountain range
Costa *Spanish*, Coast
Côte *French*, Coast
Coxilha *Portuguese*, Mountains
Cuchilla *Spanish*, Mountains

D

Daban *Mongolian, Uighur*, Pass
Dağı *Azerbaijani, Turkish*, Mountain
Dağları *Azerbaijani, Turkish*, Mountains
-dake *Japanese*, Peak
-dal(en) *Norwegian*, Valley
Danau *Indonesian*, Lake
Dao *Chinese*, Island
Đao *Vietnamese*, Island
Daryā *Persian*, River
Daryācheh *Persian*, Lake
Dasht *Persian*, Desert, plain
Dawḥat *Arabic*, Bay
Denizi *Turkish*, Sea
Dere *Turkish*, Stream
Desierto *Spanish*, Desert
Dili *Azerbaijani*, Spit
-do *Korean*, Island
Dooxo *Somali*, Valley
Düzü *Azerbaijani*, Steppe
-dwīp *Bengali*, Island

E

-eilanden *Dutch*, Islands
Embalse *Spanish*, Reservoir
Ensenada *Spanish*, Bay
Erg *Arabic*, Dunes
Estany *Catalan*, Lake
Estero *Spanish*, Inlet
Estrecho *Spanish*, Strait
Étang *French*, Lagoon, lake
-ey *Icelandic*, Island
Ezero *Bulgarian, Macedonian*, Lake
Ezers *Latvian*, Lake

F

Feng *Chinese*, Peak
Fjord *Danish*, Fjord
-fjord(en) *Danish, Norwegian, Swedish*, fjord
-fjördhur *Icelandic*, Fjord
Fleuve *French*, River
Fliegu *Maltese*, Channel
-fljór *Icelandic*, River
-flói *Icelandic*, Bay
Forêt *French*, Forest

G

-gan *Japanese*, Rock
-gang *Korean*, River
Ganga *Hindi, Nepali, Sinhala*, River
Gaoyuan *Chinese*, Plateau
Garagumy *Turkmen*, Sands
-gawa *Japanese*, River
Gebel *Arabic*, Mountain
-gebirge *German*, Mountain range
Ghadīr *Arabic*, Well
Ghubbat *Arabic*, Bay
Gjiri *Albanian*, Bay
Gol *Mongolian*, River
Golfe *French*, Gulf
Golfo *Italian, Spanish*, Gulf
Göl(ü) *Turkish*, Lake
Golyam, -a *Bulgarian*, Big
Gora *Russian, Serbo-Croatian*, Mountain
Góra *Polish*, mountain
Gory *Russian*, Mountain
Gryada *Russian*, ridge
Guba *Russian*, Bay
-gundo *Korean*, island group
Gunung *Malay*, Mountain

H

Ḥadd *Arabic*, Spit
-haehyŏp *Korean*, Strait
Haff *German*, Lagoon
Hai *Chinese*, Bay, lake, sea
Haixia *Chinese*, Strait
Hamada *Arabic*, Plateau
Ḥammādat *Arabic*, Plateau
Hāmūn *Persian*, Lake
-hantō *Japanese*, Peninsula
Har, Haré *Hebrew*, Mountain
Ḥarrat *Arabic*, Lava-field
Hav(et) *Danish, Swedish*, Sea
Hawr *Arabic*, Lake
Hāyk' *Amharic*, Lake
He *Chinese*, River
-hegység *Hungarian*, Mountain range
Heide *German*, Heath, moorland
Helodrano *Malagasy*, Bay
Higashi- *Japanese*, East(ern)
Ḥiṣā' *Arabic*, Well
Hka *Burmese*, River
-ho *Korean*, Lake
Ḫolot *Hebrew*, Dunes
Hora *Belarussian, Czech*, Mountain
Hrada *Belarussian*, Mountain, ridge
Hsi *Chinese*, River
Hu *Chinese*, Lake
Huk *Danish*, Point

I

Île(s) *French*, Island(s)
Ilha(s) *Portuguese*, Island(s)
Ilhéu(s) *Portuguese*, Islet(s)
Imeni *Russian*, In the name of
Inish- *Gaelic*, Island
Insel(n) *German*, Island(s)
Irmağı, Irmak *Turkish*, River
Isla(s) *Spanish*, Island(s)
Isola (Isole) *Italian*, Island(s)

J

Jabal *Arabic*, Mountain
Jāl *Arabic*, Ridge
-järv *Estonian*, Lake
-järvi *Finnish*, Lake
Jazā'ir *Arabic*, Islands
Jazīrat *Arabic*, Island
Jazīreh *Persian*, Island
Jebel *Arabic*, Mountain
Jezero *Serbo-Croatian*, Lake
Jezioro *Polish*, Lake
Jiang *Chinese*, River
-jima *Japanese*, Island
Jižní *Czech*, Southern
-jõgi *Estonian*, River
-joki *Finnish*, River
-jökull *Icelandic*, Glacier
Jūn *Arabic*, Bay
Juzur *Arabic*, Islands

K

Kaikyō *Japanese*, Strait
-kaise *Lappish*, Mountain
Kali *Nepali*, River
Kalnas *Lithuanian*, Mountain
Kalns *Latvian*, Mountain
Kang *Chinese*, Harbour
Kangri *Tibetan*, Mountain(s)
Kaôh *Cambodian*, Island
Kapp *Norwegian*, Cape
Káto *Greek*, Lower
Kavīr *Persian*, Desert
K'edi *Georgian*, Mountain range
Kediet *Arabic*, Mountain
Kepi *Albanian*, Cape, point
Kepulauan *Indonesian, Malay*, Island group
Khalīg, Khalīj *Arabic*, Gulf
Khawr *Arabic*, Inlet
Khola *Nepali*, River
Khrebet *Russian*, Mountain range
Ko *Thai*, Island
-ko *Japanese*, Inlet, lake
Kólpos *Greek*, Bay
-kopf *German*, Peak
Körfäzi *Azerbaijani*, Bay
Körfezi *Turkish*, Bay
Kõrgustik *Estonian*, Upland
Kosa *Russian, Ukrainian*, Spit
Koshi *Nepali*, River
Kou *Chinese*, River-mouth
Kowtal *Persian*, Pass
Kray *Russian*, Region, territory
Kryazh *Russian*, Ridge
Kuduk *Uighur*, Well
Kūh(hā) *Persian*, Mountain(s)
-kul' *Russian*, Lake
Kŭl(i) *Tajik, Uzbek*, Lake
-kundo *Korean*, Island group
-kysten *Norwegian*, Coast
Kyun *Burmese*, Island

L

Laaq *Somali*, Watercourse
Lac *French*, Lake
Lacul *Romanian*, Lake
Lagh *Somali*, Stream
Lago *Italian, Portuguese, Spanish*, Lake
Lagoa *Portuguese*, Lagoon
Laguna *Italian, Spanish*, Lagoon, lake
Laht *Estonian*, Bay
Laut *Indonesian*, Bay
Lembalemba *Malagasy*, Plateau
Lerr *Armenian*, Mountain
Lerrnashght'a *Armenian*, Mountain range
Les *Czech*, Forest
Lich *Armenian*, Lake
Liehtao *Chinese*, Island group
Liqeni *Albanian*, Lake
Límni *Greek*, Lake
Ling *Chinese*, Mountain range
Llano *Spanish*, Plain, prairie
Lumi *Albanian*, River
Lyman *Ukrainian*, Estuary

M

Madīnat *Arabic*, City, town
Mae Nam *Thai*, River
-mägi *Estonian*, Hill
Maja *Albanian*, Mountain
Mal *Albanian*, Mountains
Mal-aya, -oye, -yy, *Russian*, Small
-man *Korean*, Bay
Mar *Spanish*, Sea
Marios *Lithuanian*, Lake
Massif *French*, Mountains
Meer *German*, Lake
-meer *Dutch*, Lake
Melkosopochnik *Russian*, Plain
-meri *Estonian*, Sea
Mifraẓ *Hebrew*, Bay
Minami- *Japanese*, South(ern)
-misaki *Japanese*, Cape, point
Monkhafad *Arabic*, Depression
Montagne(s) *French*, Mountain(s)
Montañas *Spanish*, Mountains
Mont(s) *French*, Mountain(s)
Monte *Italian, Portuguese*, Mountain
More *Russian*, Sea
Mörön *Mongolian*, River
Mys *Russian*, Cape, point

N

-nada *Japanese*, Open stretch of water
Nagor'ye *Russian*, Upland
Naḥal *Hebrew*, River
Nahr *Arabic*, River
Nam *Laotian*, River
Namakzār *Persian*, Salt desert
Né-a, -on, -os *Greek*, New
Nedre- *Norwegian*, Lower
-neem *Estonian*, Cape, point
Nehri *Turkish*, River
-nes *Norwegian*, Cape, point
Nevado *Spanish*, Snow-capped mountain
Nieder- *German*, Lower
Nishi- *Japanese*, West(ern)
-nísi *Greek*, Island
Nisoi *Greek*, Islands
Nizhn-eye, -iy, -iye, -yaya *Russian*, Lower
Nizmennost' *Russian*, Lowland, plain
Nord *Danish, French, German*, North
Norte *Portuguese, Spanish*, North
Nos *Bulgarian*, Point, spit
Nosy *Malagasy*, Island
Nov-a, -i, *Bulgarian, Serbo-Croatian*, New
Nov-aya, -o, -oye, -yy, -yye *Russian*, New
Now-a, -e, -y *Polish*, New
Nur *Mongolian*, Lake
Nuruu *Mongolian*, Mountains
Nuur *Mongolian*, Lake
Nyzovyna *Ukrainian*, Lowland, plain

O

-ø *Danish*, Island
Ober- *German*, Upper
Oblast' *Russian*, Province
Órmos *Greek*, Bay
Orol(i) *Uzbek*, Island
Øster- *Norwegian*, Eastern
Ostrov(a) *Russian*, Island(s)
Otok *Serbo-Croatian*, Island
Oued *Arabic*, Watercourse
-oy *Faeroese*, Island
-øy(a) *Norwegian*, Island
Oya *Sinhala*, River
Ozero *Russian, Ukrainian*, Lake

P

Passo *Italian*, Pass
Pegunungan *Indonesian, Malay*, Mountain range
Pélagos *Greek*, Sea
Pendi *Chinese*, Basin
Penisola *Italian*, Peninsula
Pertuis *French*, Strait
Peski *Russian*, Sands
Phanom *Thai*, Mountain
Phou *Laotian*, Mountain
Pi *Chinese*, Point
Pic *Catalan, French*, Peak
Pico *Portuguese, Spanish*, Peak
-piggen *Danish*, Peak
Pik *Russian*, Peak
Pivostriv *Ukrainian*, Peninsula
Planalto *Portuguese*, Plateau
Planina, Planini *Bulgarian, Macedonian, Serbo-Croatian*, Mountain range
Plato *Russian*, Plateau
Ploskogor'ye *Russian*, Upland
Poluostrov *Russian*, Peninsula
Ponta *Portuguese*, Point
Porthmós *Greek*, Strait
Pótamos *Greek*, River
Presa *Spanish*, Dam
Prokhod *Bulgarian*, Pass
Proliv *Russian*, Strait
Pulau *Indonesian, Malay*, Island
Pulu *Malay*, Island
Punta *Spanish*, Point
Pushcha *Belorussian*, Forest
Puszcza *Polish*, Forest

Q

Qā' *Arabic*, Depression
Qalamat *Arabic*, Well
Qatorkŭh(i) *Tajik*, Mountain
Qiuling *Chinese*, Hills
Qolleh *Persian*, Mountain
Qu *Tibetan*, Stream
Quan *Chinese*, Well
Qulla(i) *Tajik*, Peak
Qundao *Chinese*, Island group

R

Raas *Somali*, Cape
-rags *Latvian*, Cape
Ramlat *Arabic*, Sands
Ra's *Arabic*, Cape, headland, point
Ravnina *Bulgarian, Russian*, Plain
Récif *French*, Reef
Recife *Portuguese*, Reef
Reka *Bulgarian*, River
Represa (Rep.) *Portuguese, Spanish*, Reservoir
Reshteh *Persian*, Mountain range
Respublika *Russian*, Republic, first-order administrative division
Respublika(si) *Uzbek*, Republic, first-order administrative division
-retsugan *Japanese*, Chain of rocks
-rettō *Japanese*, Island chain
Riacho *Spanish*, Stream
Riban' *Malagasy*, Mountains
Rio *Portuguese*, River
Río *Spanish*, River
Riu *Catalan*, River
Rivier *Dutch*, River
Rivière *French*, River
Rowd *Pashtu*, River
Rt *Serbo-Croatian*, Point
Rūd *Persian*, River
Rūdkhāneh *Persian*, River
Rudohorie *Slovak*, Mountains
Ruisseau *French*, Stream

S

-saar *Estonian*, Island
-saari *Finnish*, Island
Sabkhat *Arabic*, Salt marsh
Sāgar(a) *Hindi*, Lake, reservoir
Ṣaḥrā' *Arabic*, Desert
Saint, Sainte *French*, Saint
Salar *Spanish*, Salt-pan
Salto *Portuguese, Spanish*, Waterfall
Samudra *Sinhala*, Reservoir
-san *Japanese, Korean*, Mountain
-sanchi *Japanese*, Mountains
-sandur *Icelandic*, Beach
Sankt *German, Swedish*, Saint
-sanmaek *Korean*, Mountain range
-sanmyaku *Japanese*, Mountain range
San, Santa, Santo *Italian, Portuguese, Spanish*, Saint
São *Portuguese*, Saint
Sarīr *Arabic*, Desert
Sebkha, Sebkhet *Arabic*, Depression, salt marsh
Sedlo *Czech*, Pass
See *German*, Lake
Selat *Indonesian*, Strait
Selatan *Indonesian*, Southern
-selkä *Finnish*, Lake, ridge
Selseleh *Persian*, Mountain range
Serra *Portuguese*, Mountain
Serranía *Spanish*, Mountain
-seto *Japanese*, Channel, strait
Sever-naya, -noye, -nyy, -o *Russian*, Northern
Sha'ib *Arabic*, Watercourse
Shākh *Kurdish*, Mountain
Shamo *Chinese*, Desert
Shan *Chinese*, Mountain(s)
Shankou *Chinese*, Pass
Shanmo *Chinese*, Mountain range
Shaṭṭ *Arabic*, Distributary
Shet' *Amharic*, River
Shi *Chinese*, Municipality
-shima *Japanese*, Island
Shiqqat *Arabic*, Depression
-shotō *Japanese*, Group of islands
Shuiku *Chinese*, Reservoir
Shūrkhog(i) *Uzbek*, Salt marsh
Sierra *Spanish*, Mountains
Sint *Dutch*, Saint
-sjø(en) *Norwegian*, Lake
-sjön *Swedish*, Lake
Solonchak *Russian*, Salt lake
Solonchakovyye Vpadiny *Russian*, Salt basin, wetlands
Sơn *Vietnamese*, Mountain
Sông *Vietnamese*, River
Sør- *Norwegian*, Southern
-spitze *German*, Peak
Star-á, -é *Czech*, Old
Star-aya, -oye, -yy, -yye *Russian*, Old
Stenó *Greek*, Strait
Step' *Russian*, Steppe
Štít *Slovak*, Peak
Stœng *Cambodian*, River
Stolovaya Strana *Russian*, Plateau
Stredné *Slovak*, Middle
Střední *Czech*, Middle
Stretto *Italian*, Strait
Su Anbarı *Azerbaijani*, Reservoir
-suidō *Japanese*, Channel, strait
Sund *Swedish*, Sound, strait
Sungai *Indonesian, Malay*, River
Suu *Turkish*, River

T

Tal *Mongolian*, Plain
Tandavan' *Malagasy*, Mountain range
Tangorombohitr' *Malagasy*, Mountain massif
Tanjung *Indonesian, Malay*, Cape, point
Tao *Chinese*, Island
Ṭaraq *Arabic*, Hills
Tassili *Berber*, Mountain, plateau
Tau *Russian*, Mountain(s)
Taungdan *Burmese*, Mountain range
Techníti Límni *Greek*, Reservoir
Tekojärvi *Finnish*, Reservoir
Teluk *Indonesian, Malay*, Bay
Tengah *Indonesian*, Middle
Terara *Amharic*, Mountain
Timur *Indonesian*, Eastern
-tind(an) *Norwegian*, Peak
Tizma(si) *Uzbek*, Mountain range, ridge
-tō *Japanese*, island
Tog *Somali*, Valley
-tōge *Japanese*, pass
Togh(i) *Uzbek*, mountain
Tônlé *Cambodian*, Lake
Top *Dutch*, Peak
-tunturi *Finnish*, Mountain
Ṭurāq *Arabic*, hills
Tur'at *Arabic*, Channel

U

Udde(n) *Swedish*, Cape, point
'Uqlat *Arabic*, Well
Utara *Indonesian*, Northern
Uul *Mongolian*, Mountains

V

Väin *Estonian*, Strait
Vallée *French*, Valley
Varful *Romanian*, Peak
-vatn *Icelandic*, Lake
-vatnet *Norwegian*, Lake
Velayat *Turkmen*, Province
-vesi *Finnish*, Lake
Vestre- *Norwegian*, Western
-vidda *Norwegian*, Plateau
-vík *Icelandic*, Bay
-viken *Swedish*, Bay, inlet
Vinh *Vietnamese*, Bay
Víztárloló *Hungarian*, Reservoir
Vodaskhovishcha *Belarussian*, Reservoir
Vodokhranilishche (Vdkhr.) *Russian*, Reservoir
Vodoskhovyshche (Vdskh.) *Ukrainian*, Reservoir
Volcán *Spanish*, Volcano
Vostochn-o, yy *Russian*, Eastern
Vozvyshennost' *Russian*, Upland, plateau
Vozyera *Belarussian*, Lake
Vpadina *Russian*, Depression
Vrchovina *Czech*, Mountains
Vrh *Croat, Slovene*, Peak
Vychodné *Slovak*, Eastern
Vysochyna *Ukrainian*, Upland
Vysočina *Czech*, Upland

W

Waadi *Somali*, Watercourse
Wādī *Arabic* Watercourse
Wāḥat, Wâhat *Arabic*, Oasis
Wald *German*, Forest
Wan *Chinese*, Bay
Way *Indonesian*, River
Webi *Somali*, River
Wenz *Amharic*, River
Wiloyat(i) *Uzbek*, Province
Wyżyna *Polish*, Upland
Wzgórza *Polish*, Upland
Wzvyshsha *Belarussian*, Upland

X

Xé *Laotian*, River
Xi *Chinese*, Stream

Y

-yama *Japanese*, Mountain
Yanchi *Chinese*, Salt lake
Yanhu *Chinese*, Salt lake
Yarımadası *Azerbaijani, Turkish*, Peninsula
Yaylası *Turkish*, Plateau
Yazovir *Bulgarian*, Reservoir
Yoma *Burmese*, Mountains
Ytre- *Norwegian*, Outer
Yü *Chinese*, Island
Yunhe *Chinese*, Canal
Yuzhn-o, -yy *Russian*, Southern

Z

-zaki *Japanese*, Cape, point
Zaliv *Bulgarian, Russian*, Bay
-zan *Japanese*, Mountain
Zangbo *Tibetan*, River
Zapadn-aya, -o, -yy *Russian*, Western
Západné *Slovak*, Western
Západní *Czech*, Western
Zatoka *Polish, Ukrainian*, Bay
-zee *Dutch*, Sea
Zemlya *Russian*, Earth, land
Zizhiqu *Chinese*, Autonomous region

100 Mile House - Acerrae

INDEX

306

Index

Glossary of Abbreviations

This glossary provides a comprehensive guide to the abbreviations used in this Atlas, and in the Index.

A

abbrev. abbreviated
AD Anno Domini
Afr. Afrikaans
Alb. Albanian
Amh. Amharic
anc. ancient
approx. approximately
Ar. Arabic
Arm. Armenian
ASEAN Association of South East Asian Nations
ASSR Autonomous Soviet Socialist Republic
Aust. Australian
Az. Azerbaijani
Azerb. Azerbaijan

B

Basq. Basque
BC before Christ
Bel. Belarussian
Ben. Bengali
Ber. Berber
B-H Bosnia-Herzegovina
bn billion (one thousand million)
BP British Petroleum
Bret. Breton
Brit. British
Bul. Bulgarian
Bur. Burmese

C

C central
C. Cape
°C degrees Centigrade
CACM Central America Common Market
Cam. Cambodian
Cant. Cantonese
CAR Central African Republic
Cast. Castilian
Cat. Catalan
CEEAC Central America Common Market
Chin. Chinese
CIS Commonwealth of Independent States
cm centimetre(s)
Cro. Croat
Cz. Czech
Czech Rep. Czech Republic

D

Dan. Danish
Div. Divehi
Dom. Rep. Dominican Republic
Dut. Dutch

E

E east
EC *see* EU
EEC *see* EU
ECOWAS Economic Community of West African States
ECU European Currency Unit
EMS European Monetary System
Eng. English
est estimated
Est. Estonian
EU European Union (previously European Community [EC], European Economic Community [EEC])

F

°F degrees Fahrenheit
Faer. Faeroese
Fij. Fijian
Fin. Finnish
Fr. French
Fris. Frisian
ft foot/feet
FYROM Former Yugoslav Republic of Macedonia

G

g gram(s)
Gael. Gaelic
Gal. Galician
GDP Gross Domestic Product (the total value of goods and services produced by a country excluding income from foreign countries)
Geor. Georgian
Ger. German
Gk Greek
GNP Gross National Product (the total value of goods and services produced by a country)

H

Heb. Hebrew
HEP hydro-electric power
Hind. Hindi
hist. historical
Hung. Hungarian

I

I. Island
Icel. Icelandic
in inch(es)
In. Inuit (Eskimo)
Ind. Indonesian
Intl International
Ir. Irish
Is Islands
It. Italian

J

Jap. Japanese

K

Kaz. Kazakh
kg kilogram(s)
Kir. Kirghiz
km kilometre(s)
km² square kilometre (singular)
Kor. Korean
Kurd. Kurdish

L

L. Lake
LAIA Latin American Integration Association
Lao. Laotian
Lapp. Lappish
Lat. Latin
Latv. Latvian
Liech. Liechtenstein
Lith. Lithuanian
Lux. Luxembourg

M

m million/metre(s)
Mac. Macedonian
Maced. Macedonia
Mal. Malay
Malg. Malagasy
Malt. Maltese
mi. mile(s)
Mong. Mongolian
Mt. Mountain
Mts Mountains

N

N north
NAFTA North American Free Trade Agreement
Nep. Nepali
Neth. Netherlands
Nic. Nicaraguan
Nor. Norwegian
NZ New Zealand

P

Pash. Pashtu
PNG Papua New Guinea
Pol. Polish
Poly. Polynesian
Port. Portuguese
prev. previously

R

Rep. Republic
Res. Reservoir
Rmsch Romansch
Rom. Romanian
Rus. Russian
Russ. Fed. Russian Federation

S

S south
SADC Southern Africa Development Community
SCr. Serbo-Croatian
Sinh. Sinhala
Slvk Slovak
Slvn. Slovene
Som. Somali
Sp. Spanish
St., St Saint
Strs Straits
Swa. Swahili
Swe. Swedish
Switz. Switzerland

T

Taj. Tajik
Th. Thai
Thai. Thailand
Tib. Tibetan
Turk. Turkish
Turkm. Turkmenistan

U

UAE United Arab Emirates
Uigh. Uighur
UK United Kingdom
Ukr. Ukrainian
UN United Nations
Urd. Urdu
US/USA United States of America
USSR Union of Soviet Socialist Republics
Uzb. Uzbek

V

var. variant
Vdkhr. Vodokhranilishche (Russian for reservoir)
Vdskh. Vodokhovyshche (Ukrainian for reservoir)
Vtn. Vietnamese

W

W west
Wel. Welsh

Y

Yugo. Yugoslavia

THIS INDEX LISTS all the placenames and features shown on the regional and continental maps in this Atlas. Placenames are referenced to the largest scale map on which they appear. The policy followed throughout the Atlas is to use the local spelling or local name at regional level; commonly-used English language names may occasionally be added (in parentheses) where this is an aid to identification e.g. Firenze (Florence). English names, where they exist, have been used for all international features e.g. oceans and country names; they are also used on the continental maps and in the introductory World section; these are then fully cross-referenced to the local names found on the regional maps. The index also contains commonly-found alternative names and variant spellings, which are also fully cross-referenced.

All main entry names are those of settlements unless otherwise indicated by the use of italicized definitions or representative symbols, which are keyed at the foot of each page.

1

56 *E6* **100 Mile House** *var.* Hundred Mile House. British Columbia, SW Canada *51°39′N 121°19′W*
115 *H9* **25 de Agosto** Florida, Uruguay *34°25′S 56°24′W*
118 *G9* **25 de Mayo** La Pampa, Argentina *37°48′S 67°41′W*
25 de Mayo *see* Veinticinco de Mayo
26 Bakinskikh Komissarov *see* Häsänabad
26 Baku Komissarlary Adyndaky *see* Uzboý

A

Aa *see* Gauja
155 *D13* **Aabenraa** *var.* Åbenrå, *Ger.* Apenrade. Syddanmark, SW Denmark *55°03′N 09°26′E*
155 *D10* **Aabybro** *var.* Åbybro. Nordjylland, N Denmark *57°09′N 09°32′E*
181 *A10* **Aachen** *Dut.* Aken, *Fr.* Aix-la-Chapelle; *anc.* Aquae Grani, Aquisgranum. Nordrhein-Westfalen, W Germany *50°47′N 06°06′E*
218 *F4* **Aadchît** Lebanon *33°20′N 35°25′E*
218 *F4* **Aaïtaït** Lebanon *33°33′N 35°40′E*
Aaiún *see* Laâyoune
178 *J3* **Aakirkeby** *var.* Åkirkeby. Bornholm, E Denmark *55°04′N 14°56′E*
218 *G1* **Aakkâr el Aatîga** Lebanon
155 *D11* **Aalborg** *var.* Ålborg, Ålborg-Nørresundby; *anc.* Alburgum. Nordjylland, N Denmark *57°03′N 09°56′E*
Aalborg Bugt *see* Ålborg Bugt
179 *E12* **Aalen** Baden-Württemberg, S Germany *48°50′N 10°06′E*
155 *D11* **Aalestrup** *var.* Ålestrup. Midtjylland, NW Denmark *56°42′N 09°31′E*
218 *F3* **Aaley** Lebanon *33°48′N 35°35′E*
162 *E6* **Aalsmeer** Noord-Holland, C Netherlands *52°17′N 04°43′E*
163 *D10* **Aalst** Oost-Vlaanderen, C Belgium *50°57′N 04°03′E*
162 *F7* **Aalst** *Fr.* Alost. Noord-Brabant, S Netherlands *51°23′N 05°29′E*
162 *I7* **Aalten** Gelderland, E Netherlands *51°56′N 06°35′E*
163 *C9* **Aalter** Oost-Vlaanderen, NW Belgium *51°05′N 03°28′E*
Aanaar *see* Inari
Aanaarjävri *see* Inarijärvi
141 *H5* **Aandster** North West, South Africa *27°25′S 25°10′E*
153 *H8* **Äänekoski** Keski-Suomi, W Finland *62°34′N 25°45′E*
218 *G3* **Aanjar** *var.* ‘Anjar. C Lebanon *33°45′N 35°56′E*
218 *G3* **Aanjar** Lebanon *33°43′N 35°55′E*
138 *E7* **Aansluit** Northern Cape, N South Africa *26°41′S 22°24′E*
218 *G2* **Aaqoûra** Lebanon
218 *G2* **Aaqoûra** Lebanon *34°07′N 35°54′E*
Aar *see* Aare
176 *D4* **Aarau** Aargau, N Switzerland *47°22′N 08°00′E*
176 *C4* **Aarberg** Bern, W Switzerland *47°19′N 07°54′E*
163 *C9* **Aardenburg** Zeeland, SW Netherlands *51°16′N 03°27′E*
174 *B3* **Aare** *var.* Aar. ≈ W Switzerland
174 *B2* **Aargau** *Fr.* Argovie. ◈ *canton* N Switzerland
Aarhus *see* Århus
Aarlen *see* Arlon
155 *D11* **Aars** *var.* Års. Nordjylland, N Denmark *56°49′N 09°32′E*
163 *E10* **Aarschot** Vlaams Brabant, C Belgium *50°59′N 04°50′E*
Aassi, Nahr el *see* Orontes
Aat *see* Ath
218 *F3* **Aayoûn es Sîmâne** Lebanon *34°00′N 35°49′E*
239 *K7* **Aba** *prev.* Ngawa. Sichuan, C China *32°51′N 101°46′E*
134 *J6* **Aba** Orientale, NE Dem. Rep. Congo *03°52′N 30°14′E*
133 *K8* **Aba** Abia, S Nigeria *05°06′N 07°22′E*
220 *D3* **Abā al Qazāz, Bi’r** *well* NW Saudi Arabia
Abā as Su‘ūd *see* Najrān
107 *H4* **Abacaxis, Rio** ≈ NW Brazil
Abaco Island *see* Great Abaco/Little Abaco
Abaco Island *see* Great Abaco, N Bahamas
222 *D6* **Ābādān** Khūzestān, SW Iran *30°24′N 48°18′E*
228 *D7* **Abadan** *prev.* Bezmein, Büzmeýin, *Rus.* Byuzmeyin. Ahal Welaýaty, C Turkmenistan *38°08′N 57°53′E*
222 *F6* **Ābādeh** Fārs, C Iran *31°06′N 52°40′E*
130 *G4* **Abadla** W Algeria *31°04′N 02°39′W*
110 *E1* **Abaeté** Minas Gerais, SE Brazil *19°10′S 45°24′W*
113 *H5* **Abaí** Caazapá, S Paraguay *25°58′S 55°54′W*
Abai *see* Blue Nile
284 *B2* **Abaiang** *var.* Apia; *prev.* Charlotte Island. *atoll* Tungaru, W Kiribati
Abaj *see* Abay
133 *K7* **Abaji** Federal Capital District, C Nigeria *08°35′N 06°54′E*
79 *J5* **Abajo Peak** ▲ Utah, W USA *37°51′N 109°28′W*
133 *K8* **Abakaliki** Ebonyi, SE Nigeria *06°18′N 08°07′E*
192 *G8* **Abakan** Respublika Khakasiya, S Russian Federation *53°43′N 91°25′E*
227 *N1* **Abakan** ≈ S Russian Federation
133 *I4* **Abala** Tillabéri, SW Niger *14°55′N 03°27′E*
133 *J4* **Abalak** Tahoua, C Niger *15°28′N 06°18′E*
191 *H8* **Abalyanka** *Rus.* Obolyanka. ≈ N Belarus
192 *G7* **Aban** Krasnoyarskiy Kray, S Russian Federation *56°41′N 96°04′E*
222 *G6* **Āb Anbār-e Kān Sorkh** Yazd, C Iran *31°22′N 53°38′E*
105 *F8* **Abancay** Apurímac, SE Peru *13°37′S 72°52′W*
173 *K6* **Abanilla** Murcia, Spain *38°12′N 1°03′W*
284 *D1* **Abaokoro** *atoll* Tungaru, W Kiribati
Abariringa *see* Kanton
222 *G6* **Abarkūh** Yazd, C Iran *31°07′N 53°17′E*
252 *G3* **Abashiri** *var.* Abasiri. Hokkaidō, NE Japan *44°N 144°15′E*
252 *G3* **Abashiri-gawa** ≈ Hokkaidō, NE Japan
252 *G3* **Abashiri-ko** ◎ Hokkaidō, NE Japan
Abasiri *see* Abashiri
87 *K8* **Abasolo** Chiapas, Mexico *16°48′N 92°10′W*
85 *K5* **Abasolo** Guanajuato, Mexico
85 *M7* **Abasolo** Tamaulipas, C Mexico *24°02′N 98°18′W*
87 *H8* **Abasolo del Valle** Veracruz-Llave, Mexico *17°46′N 95°30′W*
280 *C4* **Abau** Central, S Papua New Guinea *10°04′S 148°34′E*
227 *I4* **Abay** *var.* Abaj. Karaganda, C Kazakhstan *49°38′N 72°50′E*
136 *D5* **Ābaya Hāyk’** *Eng.* Lake Margherita, *It.* Abbaia. ◎ SW Ethiopia
Ābay Wenz *see* Blue Nile
192 *G8* **Abaza** Respublika Khakasiya, S Russian Federation *52°40′N 89°58′E*
Abbaia *see* Ābaya Hāyk’
222 *G8* **Āb Bārīk** Fārs, S Iran
175 *B10* **Abbasanta** Sardegna, Italy, C Mediterranean Sea *40°08′N 08°49′E*
Abbatis Villa *see* Abbeville
72 *D4* **Abbaye, Point** *headland* Michigan, N USA *46°58′N 88°08′W*
Abbazia *see* Opatija
Abbé, Lake *see* Abhe, Lake
165 *H2* **Abbeville** *anc.* Abbatis Villa. Somme, N France *50°06′N 01°50′E*
69 *I3* **Abbeville** Alabama, S USA *31°35′N 85°16′W*
69 *K2* **Abbeville** Georgia, SE USA *31°58′N 83°18′W*
68 *C4* **Abbeville** Louisiana, S USA *29°58′N 92°08′W*
67 *H8* **Abbeville** South Carolina, SE USA *34°10′N 82°23′W*
157 *B10* **Abbeyfeale** *Ir.* Mainistir na Féile. SW Ireland *52°24′N 09°21′W*
160 *A2* **Abbeyleix** Laois, Ireland *52°54′N 7°21′W*
176 *E7* **Abbiategrasso** Lombardia, NW Italy *45°24′N 08°55′E*
154 *J1* **Abborrträsk** Norrbotten, N Sweden *65°24′N 19°33′E*
292 *F5* **Abbot Ice Shelf** *ice shelf* Antarctica
160 *G8* **Abbotsbury** United Kingdom *50°40′N 2°36′W*
140 *D9* **Abbotsdale** Western Cape, South Africa *33°29′S 18°40′E*
76 *C2* **Abbotsford** British Columbia, SW Canada *49°02′N 122°18′W*
72 *C5* **Abbotsford** Wisconsin, N USA *44°57′N 90°19′W*
231 *I4* **Abbottābād** Khyber Pakhtunkhwa, NW Pakistan *34°12′N 73°15′E*
191 *H9* **Abchuha** *Rus.* Obchuga. Minskaya Voblasts’, NW Belarus *54°30′N 29°22′E*
162 *F6* **Abcoude** Utrecht, C Netherlands *52°17′N 04°59′E*
216 *F2* **‘Abd al ‘Azīz, Jabal** ▲▲ NE Syria
221 *I9* **‘Abd al Kūrī** *island* SE Yemen
197 *K3* **Abdulino** Orenburgskaya Oblast’, W Russian Federation *53°37′N 53°39′E*
134 *F2* **Abéché** *var.* Abécher, Abeshr. Ouaddaï, SE Chad *13°49′N 20°49′E*
Abécher *see* Abéché
223 *I5* **Āb-e Garm va Sard** Khorāsān-e Janūbī, E Iran
133 *I2* **Abeïbara** Kidal, NE Mali *19°07′N 01°52′E*
171 *H3* **Abejar** Castilla y León, N Spain *41°48′N 02°47′W*
102 *B5* **Abejorral** Antioquia, W Colombia *05°48′N 75°28′W*
Abela *see* Ávila
141 *K3* **Abel Erasmuspas** *pass* Limpopo, South Africa
Abellinum *see* Avellino
152 *C5* **Abeløya** *island* Kong Karls Land, E Svalbard
136 *D4* **Ābeltī** Oromīya, C Ethiopia *08°09′N 37°31′E*
284 *B3* **Abemama** *var.* Apamama; *prev.* Roger Simpson Island. *atoll* Tungaru, W Kiribati
261 *L7* **Abemaree** *var.* Abermarre. Papua, E Indonesia *07°03′S 140°10′E*
181 *J13* **Abenberg** Bayern, Germany *49°15′N 10°58′E*
132 *G8* **Abengourou** E Ivory Coast *06°42′N 03°27′W*
172 *G5* **Abenójar** Castilla-La Manacha, Spain *38°53′N 4°21′W*
Åbenrå *see* Aabenraa
179 *G12* **Abens** ≈ SE Germany
133 *I7* **Abeokuta** Ogun, SW Nigeria *07°07′N 03°21′E*
160 *E4* **Aberaeron** SW Wales, United Kingdom *52°15′N 04°15′W*
Aberbrothock *see* Arbroath
Abercorn *see* Mbala
74 *E3* **Abercrombie** North Dakota, N USA *46°25′N 96°42′W*
160 *F6* **Aberdare** United Kingdom *51°43′N 3°27′W*
160 *D3* **Aberdaron** United Kingdom *52°49′N 4°42′W*
277 *L4* **Aberdeen** New South Wales, SE Australia *32°09′S 150°55′E*
57 *M6* **Aberdeen** Saskatchewan, S Canada *52°15′N 106°19′W*
138 *F9* **Aberdeen** Eastern Cape, S South Africa *32°30′S 24°00′E*
156 *G5* **Aberdeen** *anc.* Devana. NE Scotland, United Kingdom *57°10′N 02°04′W*
64 *E9* **Aberdeen** Maryland, NE USA *39°28′N 76°09′W*
66 *C8* **Aberdeen** Mississippi, S USA *33°49′N 88°32′W*
67 *J7* **Aberdeen** North Carolina, SE USA *35°07′N 79°25′W*
74 *D4* **Aberdeen** South Dakota, N USA *45°27′N 98°29′W*
76 *B4* **Aberdeen** Washington, NW USA *46°57′N 123°48′W*
140 *G8* **Aberdeen Kendrew** Eastern Cape, South Africa *32°33′S 24°33′E*
55 *I3* **Aberdeen Lake** ◎ Nunavut, NE Canada
140 *G9* **Aberdeen Road** Eastern Cape, South Africa *32°45′S 24°19′E*
160 *E3* **Aberdyfi** United Kingdom *52°33′N 4°02′W*
159 *H1* **Aberfeldy** C Scotland, United Kingdom *56°38′N 03°49′W*
160 *E2* **Aberffraw** United Kingdom *53°11′N 4°27′W*
158 *G2* **Aberfoyle** United Kingdom *56°11′N 4°23′W*
160 *G5* **Abergavenny** *anc.* Gobannium. SE Wales, United Kingdom *51°50′N 03°W*
160 *E5* **Abergorlech** United Kingdom *51°59′N 4°04′W*
Abergwaun *see* Fishguard
159 *I3* **Aberlady** United Kingdom *56°00′N 2°51′W*
Abermarre *see* Abemaree
70 *D3* **Abernathy** Texas, SW USA *33°49′N 101°50′W*
159 *I2* **Abernethy** C Scotland, United Kingdom
160 *E4* **Aberporth** United Kingdom *52°08′N 4°33′W*
Abersee *see* Wolfgangsee
160 *E3* **Abersoch** United Kingdom *52°49′N 4°30′W*
Abertawe *see* Swansea
Aberteifi *see* Cardigan
160 *G6* **Abertillery** United Kingdom *51°44′N 3°08′W*
76 *D8* **Abert, Lake** ◎ Oregon, NW USA
160 *E4* **Aberystwyth** W Wales, United Kingdom *52°25′N 04°05′W*
Abeshr *see* Abéché
Ábeskovvu *see* Abisko
176 *G9* **Abetone** Toscana, C Italy *44°09′N 10°42′E*
195 *L4* **Abez’** Respublika Komi, NW Russian Federation *66°32′N 61°41′E*
222 *E3* **Āb Garm** Qazvin, N Iran
220 *F7* **Abhā** ‘Asīr, SW Saudi Arabia *18°16′N 42°32′E*
222 *E3* **Abhar** Zanjān, NW Iran *36°05′N 49°18′E*
Abhé Bad/Ābhē Bid Hāyk’ *see* Abhe, Lake
136 *F3* **Abhe, Lake** *var.* Lake Abbé, *Amh.* Ābhē Bid Hāyk’, *Som.* Abhé Bad. ◎ Djibouti/Ethiopia
133 *J7* **Abia** ◈ *state* SE Nigeria
217 *K6* **‘Abīd ‘Alī** Wāsiţ, E Iraq *32°20′N 45°58′E*
191 *I10* **Abidavichy** *Rus.* Obidovichi. Mahilyowskaya Voblasts’, E Belarus *53°20′N 30°25′E*
187 *J3* **Abide** Çanakkale, NW Turkey *40°04′N 26°13′E*
132 *F8* **Abidjan** S Ivory Coast *05°19′N 04°01′W*
140 *E5* **Abierkwasputs** *salt lake* Northern Cape, South Africa
Āb-i-Istāda *see* Īstādeh-ye Moqor, Āb-e-
218 *F6* **Abila** Jordan
75 *E9* **Abilene** Kansas, C USA *38°55′N 97°14′W*
70 *F4* **Abilene** Texas, SW USA *32°27′N 99°44′W*
Abindonia *see* Abingdon
161 *I6* **Abingdon** *anc.* Abindonia. S England, United Kingdom *51°41′N 01°17′W*
73 *B10* **Abingdon** Illinois, N USA *40°48′N 90°24′W*
67 *H5* **Abingdon** Virginia, NE USA *36°42′N 81°59′W*
Abingdon *see* Pinta, Isla
65 *M3* **Abington** Massachusetts, USA *42°06′N 70°57′W*
64 *F7* **Abington** Pennsylvania, NE USA *40°06′N 75°05′W*
196 *F8* **Abinsk** Krasnodarskiy Kray, SW Russian Federation *44°51′N 38°12′E*
79 *L6* **Abiquiu Reservoir** ▣ New Mexico, SW USA
Āb-i-safed *see* Sefīd, Darya-ye
152 *F5* **Abisko** *Lapp.* Ábeskovvu. Norrbotten, N Sweden *68°21′N 18°50′E*
51 *J8* **Abitibi** ≈ Ontario, S Canada
58 *F7* **Abitibi, Lac** ◎ Ontario/Québec, S Canada
136 *E2* **Ābīy Ādī** Tigray, N Ethiopia *13°40′N 38°57′E*
190 *E5* **Abja-Paluoja** Viljandimaa, S Estonia *58°08′N 25°20′E*
Abkhazia *see* Apkhazeti
173 *I8* **Abla** Andalucía, Spain *37°08′N 2°47′W*
276 *E1* **Abminga** South Australia *26°07′S 134°49′E*
129 *I4* **Abnūb** *var.* Abnûb. C Egypt *27°18′N 31°09′E*
Abnûb *see* Abnūb
Åbo *see* Turku
232 *D4* **Abohar** Punjab, N India *30°11′N 74°14′E*
132 *G8* **Aboisso** SE Ivory Coast *05°26′N 03°13′W*
133 *N1* **Abo, Massif d’** ▲▲ NW Chad
133 *I7* **Abomey** S Benin *07°14′N 02°00′E*
167 *M10* **Abondance** Rhône-Alpes, France *46°17′N 6°44′E*
134 *C6* **Abong Mbang** Est, SE Cameroon *03°58′N 13°10′E*
183 *G12* **Abony** Pest, C Hungary *47°10′N 20°00′E*
134 *E3* **Abou-Déïa** Salamat, SE Chad *11°30′N 19°18′E*
Aboudouhour *see* Abū aḑ Ḑuhūr
Abou Kémal *see* Abū Kamāl
Abou Simbel *see* Abū Sunbul
215 *K5* **Abovyan** C Armenia *40°16′N 44°33′E*
263 *K4* **Abra** ≈ Luzon, N Philippines
129 *N9* **Abrād, Wādī** *seasonal river* W Yemen
Abraham Bay *see* The Carlton
118 *C6* **Abranquil** Maule, Chile *35°45′S 71°33′W*
172 *C4* **Abrantes** *var.* Abrántes. Santarém, C Portugal *39°28′N 08°12′W*
112 *D3* **Abra Pampa** Jujuy, N Argentina *22°47′S 65°41′W*
Abrashlare *see* Brezovo
111 *J3* **Abre Campo** Minas Gerais, Brazil *20°18′S 42°29′W*
102 *C4* **Abrego** Norte de Santander, N Colombia *08°08′N 73°14′W*
Abrene *see* Pytalovo
84 *C6* **Abreojos, Punta** *headland* NW Mexico *26°43′N 113°36′W*
169 *M3* **Abriès** Provence-Alpes-Côte d’Azur, France *44°47′N 6°56′E*
115 *J4* **Abrojal** Rivera, Uruguay *31°45′S 55°01′W*
98 *G5* **Abrolhos Bank** *undersea feature* W Atlantic Ocean *18°30′S 38°45′W*
191 *E11* **Abrova** *Rus.* Obrovo. Brestskaya Voblasts’, SW Belarus *52°30′N 25°34′E*
188 *C7* **Abrud** *Ger.* Gross-Schlatten, *Hung.* Abrudbánya. Alba, SW Romania *46°16′N 23°05′E*
Abrudbánya *see* Abrud
190 *C5* **Abruka** *island* SW Estonia
175 *F8* **Abruzzese, Appennino** ▲▲ C Italy
175 *G8* **Abruzzo** ◈ *region* C Italy
220 *F8* **‘Abs** *var.* Sūq ‘Abs. W Yemen *16°42′N 42°55′E*
77 *J6* **Absaroka Range** ▲▲ Montana/Wyoming, NW USA
64 *G9* **Absecon** New Jersey, USA *39°26′N 74°30′W*
215 *N4* **Abşeron Yarımadası** *Rus.* Apsheronskiy Poluostrov. *peninsula* E Azerbaijan
222 *F4* **Āb Shīrīn** Eşfahān, C Iran *34°17′N 51°17′E*
217 *L7* **Abtān** Maysān, SE Iraq *31°37′N 47°06′E*
177 *I3* **Abtenau** Salzburg, NW Austria *47°33′N 13°21′E*
232 *C7* **Ābu** Rājasthān, N India *24°41′N 72°50′E*
250 *D5* **Abu** Yamaguchi, Honshū, SW Japan *34°30′N 131°26′E*
216 *D3* **Abū aḑ Ḑuhūr** *Fr.* Aboudouhour. Idlib, NW Syria *35°30′N 37°00′E*
221 *J4* **Abū al Abyaḑ** *island* C United Arab Emirates
218 *H3* **Abū al ‘Aţā, Jabal** ▲▲ Syria
216 *E6* **Abū al Ḩuşayn, Khabrat** ◎ N Jordan
217 *I5* **Abū al Jīr** Al Anbār, C Iraq *33°16′N 42°55′E*
217 *M8* **Abū al Khaşīb** *var.* Abul Khasib. Al Başrah, SE Iraq *30°26′N 48°00′E*
217 *K8* **Abū at Tubrah, Thaqb** *well* S Iraq
219 *B11* **Abu Aweigila** Egypt *30°50′N 34°07′E*
Abu Balâs *see* Abu Ballās
129 *H5* **Abu Ballās** *var.* Abu Balâs. ▲ SW Egypt *24°28′N 27°36′E*
Abu Dhabi *see* Abū Z̧abī
217 *I5* **Abū Farūkh** Al Anbār, C Iraq *33°06′N 43°18′E*
134 *H3* **Abu Gabra** Eastern Darfur, W Sudan *11°02′N 26°50′E*
217 *H6* **Abū Ghār, Sha‘īb** *dry watercourse* S Iraq
129 *J8* **Abu Hamed** River Nile, N Sudan *19°32′N 33°20′E*
216 *G4* **Abū Ḩardān** *var.* Hajîne. Dayr az Zawr, E Syria *34°45′N 40°49′E*
217 *J5* **Abū Ḩasāwīyah** Diyālá, E Iraq *33°52′N 44°47′E*
216 *E6* **Abū Ḩifnah, Wādī** *dry watercourse* N Jordan
133 *K7* **Abuja** ● (Nigeria) Federal Capital District, C Nigeria *09°04′N 07°28′E*
217 *I6* **Abū Jahaf, Wādī** *dry watercourse* C Iraq
104 *E5* **Abujao, Río** ≈ E Peru
217 *J8* **Abū Jasrah** Al Muthanná, S Iraq *30°43′N 44°50′E*
219 *F12* **Abū Jurdhān** Jordan
216 *G4* **Abū Kamāl** *Fr.* Abou Kémal. Dayr az Zawr, E Syria *34°29′N 40°56′E*
260 *D5* **Abuki, Pegunungan** ▲▲ Sulawesi, C Indonesia
253 *D11* **Abukuma-gawa** ≈ Honshū, C Japan
253 *D11* **Abukuma-sanchi** ▲▲ Honshū, C Japan
Abula *see* Ávila
Abul Khasib *see* Abū al Khaşīb
134 *F6* **Abumombazi** *var.* Abumonbazi. Equateur, N Dem. Rep. Congo *03°43′N 22°06′E*
Abumonbazi *see* Abumombazi
106 *E6* **Abunã** Rondônia, W Brazil *09°41′S 65°20′W*
106 *D7* **Abunã, Rio** *var.* Río Abuná. ≈ Bolivia/Brazil
219 *G8* **Abū Nuşayr** *var.* Abu Nuseir. ‘Ammān, W Jordan *32°03′N 35°58′E*
219 *G8* **Abū Nuşayr** Jordan *32°05′N 35°52′E*
Abu Nuseir *see* Abū Nuşayr
217 *J8* **Abū Qabr** Al Muthanná, S Iraq *31°03′N 44°34′E*
218 *I2* **Abu Raḩbah** Syria *34°26′N 37°12′E*
216 *E4* **Abū Raḩbah, Jabal** ▲▲ C Syria
217 *I4* **Abū Rajāsh** Şalāḩ ad Dīn, N Iraq *34°47′N 43°36′E*
217 *K8* **Abū Raqrāq, Ghadīr** *well* S Iraq
232 *C7* **Ābu Road** Rājasthān, N India *24°29′N 72°47′E*
136 *E1* **Abu Shagara, Ras** *headland* NE Sudan *18°04′N 38°31′E*
129 *I6* **Abu Simbel** *var.* Abu Sūnbul. *headland* Egypt
Abu Simbel *see* Abū Sunbul
217 *K8* **Abū Sudayrah** Al Muthanná, S Iraq *30°55′N 44°58′E*
217 *J7* **Abū Şukhayr** Al Qādisīyah, S Iraq *31°54′N 44°27′E*
Abu Sūnbul *see* Abu Simbel
Abū Sunbul *see* Abū Sunbul
252 *D5* **Abuta** Hokkaidō, NE Japan *42°34′N 140°44′E*
219 *G12* **Abū Ţarafah, Wādī** *dry watercourse* Jordan
279 *C9* **Abut Head** *headland* South Island, New Zealand *43°06′S 170°16′E*
136 *A1* **Abu ‘Urug** Northern Kordofan, C Sudan *15°52′N 30°25′E*
136 *E4* **Ābuyē Mēda** ▲ C Ethiopia *10°28′N 39°44′E*
263 *M7* **Abuyog** Leyte, C Philippines *10°45′N 124°58′E*
134 *J2* **Abu Zabad** Western Kordofan, C Sudan *12°21′N 29°16′E*
221 *J4* **Abū Z̧abī** *var.* Abū Z̧abī, *Eng.* Abu Dhabi. ● (United Arab Emirates) Abū Z̧aby, C United Arab Emirates *24°30′N 54°20′E*
Abū Z̧abī *see* Abū Z̧abī
129 *J3* **Abu Zenīma** E Egypt *29°01′N 33°08′E*
155 *H9* **Åby** Östergötland, S Sweden *58°40′N 16°10′E*
Abyaḑ, Al Baḩr al *see* White Nile
Åbybro *see* Aabybro
134 *I3* **Abyei** Southern Kordofan, S Sudan *09°35′N 28°28′E*
134 *I3* **Abyei Area** ◈ *disputed region* Western Kordofan, S Sudan
Abyla *see* Ávila
Abymes *see* les Abymes
Abyssinia *see* Ethiopia
Açâba *see* Assaba
102 *C6* **Acacias** Meta, C Colombia *03°59′N 73°46′W*
55 *L3* **Acadia, Cape** *headland* Nunavut, C Canada
111 *I3* **Acaiaca** Minas Gerais, Brazil *20°21′S 43°09′W*
108 *C4* **Açailândia** Maranhão, E Brazil *04°51′S 47°26′W*
Acaill *see* Achill Island
88 *C5* **Acajutla** Sonsonate, W El Salvador *13°34′N 89°50′W*
172 *E7* **Acalá del Río** Andalucía, Spain *37°31′N 5°59′W*
134 *A7* **Acalayong** SW Equatorial Guinea *01°05′N 09°34′E*
86 *E6* **Acámbaro** Guanajuato, C Mexico *20°01′N 100°42′W*
86 *E6* **Acambay** México, Mexico *19°57′N 99°51′W*
102 *A4* **Acandí** Chocó, NW Colombia *08°32′N 77°20′W*
170 *C3* **A Cañiza** *var.* La Cañiza. Galicia, NW Spain *42°13′N 08°16′W*
86 *B3* **Acaponeta** Nayarit, C Mexico *22°30′N 105°21′W*
85 *I8* **Acaponeta, Río de** ≈ C Mexico
86 *E8* **Acapulco** *var.* Acapulco de Juárez. Guerrero, S Mexico *16°51′N 99°53′W*
Acapulco de Juárez *see* Acapulco
103 *J7* **Acarai Mountains** *Sp.* Serra Acaraí. ▲▲ Brazil/Guyana
Acaraí, Serra *see* Acarai Mountains
108 *G3* **Acaraú** Ceará, NE Brazil *04°35′S 37°37′W*
102 *E3* **Acarigua** Portuguesa, N Venezuela *09°35′N 69°12′W*
170 *D1* **A Carreira** Galicia, NW Spain *43°21′N 08°12′W*
86 *F5* **Acatepec** Puebla, Mexico
86 *F7* **Acatepec** Puebla, Mexico *19°02′N 98°17′W*
85 *M7* **Acatlán** Jalisco, W Mexico *23°58′N 99°08′W*
87 *H7* **Acatlán** Oaxaca, Mexico *18°32′N 96°37′W*
86 *F7* **Acatlán** *var.* Acatlán de Osorio. Puebla, S Mexico *18°12′N 98°02′W*
86 *C5* **Acatlán de Juárez** Jalisco, Mexico *20°25′N 103°35′W*
Acatlán de Osorio *see* Acatlán
87 *I7* **Acayucan** *var.* Acayucán. Veracruz-Llave, E Mexico *17°59′N 94°58′W*
Accho *see* Akko
67 *M4* **Accomac** Virginia, NE USA *37°43′N 75°41′N*
168 *D6* **Accous** Aquitaine, France *42°59′N 0°36′W*
133 *H8* **Accra** ● (Ghana)SE Ghana *05°33′N 00°15′W*
159 *J9* **Accrington** NW England, United Kingdom *53°46′N 02°21′W*
114 *B7* **Acebal** Santa Fe, C Argentina *33°14′S 60°50′W*
115 *K4* **Aceguá** Rio Grande du Sol, S Brazil *31°52′S 54°12′W*
115 *K4* **Acegua** Cerro Largo, Uruguay *31°48′S 54°18′W*
258 *C3* **Aceh** *off.* Daerah Istimewa Aceh, *var.* Acheen, Achin, Atchin, Atjeh. ◈ *autonomous district* NW Indonesia
175 *H9* **Acerenza** Basilicata, S Italy *40°46′N 15°51′E*
184 *A8* **Acerra** *anc.* Acerrae. Campania, S Italy *40°56′N 14°22′E*
Acerrae *see* Acerra

◆ Country
● Country Capital
◇ Dependent Territory
○ Dependent Territory Capital
◈ Administrative Regions
✈ International Airport
▲ Mountain
▲▲ Mountain Range
Volcano
≈ River
◎ Lake
▣ Reservoir

◆ Country   ◇ Dependent Territory   ◆ Administrative Regions   ▲ Mountain   ✕ Volcano   ☐ Lake
● Country Capital   ○ Dependent Territory Capital   ✕ International Airport   ▲ Mountain Range   ♦ River   ☐ Reservoir

Al 'Ubaydī - Angola
INDEX
310

216 G4 Al 'Ubaydī Al Anbār, W Iraq 34°22′N 41°15′E
221 I5 Al 'Ubaylah var. al-'Ubaila. Ash Sharqīyah, E Saudi Arabia 22°02′N 50°57′E
221 I5 Al 'Ubaylah spring/well E Saudi Arabia 22°02′N 50°56′E
Al Ubayyiḍ see El Obeid
221 I4 Al 'Udayd var. Al Odaid. Abū Ẓaby, W United Arab Emirates 24°34′N 51°27′E
190 F6 Alūksne Ger. Marienburg. NE Latvia 57°26′N 27°02′E
220 D3 Al 'Ulā Al Madīnah, NW Saudi Arabia 26°39′N 37°55′E
288 D6 Alula-Fartak Trench var. Illaue Fartak Trench. undersea feature W Indian Ocean 14°04′N 51°47′E
219 I9 Al 'Umarī 'Ammān, E Jordan 37°30′N 31°30′E
219 I9 Al Umarī Jordan 31°32′N 37°04′E
73 I10 Alum Creek Lake ⊠ Ohio, N USA
116 C6 Aluminé Neuquén, C Argentina 39°15′S 71°00′W
118 D10 Alumine, Lago ◎ Neuquén, Argentina
154 I7 Alunda Uppsala, C Sweden 60°04′N 18°04′E
189 J10 Alupka Avtonomna Respublika Krym, S Ukraine 44°24′N 34°01′E
128 D3 Al 'Uqaylah N Libya 30°13′N 19°12′E
Al Uqṣur see Luxor
Al Urdunn see Jordan
258 D4 Alur Panal bay Sumatera, W Indonesia
Alur Setar see Alor Setar
221 J6 Al 'Urūq al Mu'tariḍah salt lake SE Saudi Arabia
217 H4 Ālūs Al Anbār, C Iraq 34°05′N 42°27′E
189 K9 Alushta Avtonomna Respublika Krym, S Ukraine 44°41′N 34°24′E
128 B5 Al 'Uwaynāt var. Al Awaynāt. SW Libya 25°47′N 10°34′E
217 J4 Al 'Uẓaym var. Adhaim. Diyālá, E Iraq 34°12′N 44°31′E
75 D11 Alva Oklahoma, C USA 36°48′N 98°40′W
170 D5 Alva ≋ N Portugal
155 F10 Älvängen Västra Götaland, S Sweden 57°56′N 12°09′E
62 C4 Alvanley Ontario, S Canada 44°33′N 81°05′W
87 H7 Alvarado Veracruz-Llave, E Mexico 18°47′N 95°45′W
71 H5 Alvarado Texas, SW USA 32°24′N 97°12′W
87 H7 Alvarado, Laguna inlet E Mexico
106 E3 Alvarães Amazonas, NW Brazil 03°13′S 64°53′W
114 B7 Álvarez Santa Fe, Argentina 33°08′S 60°48′W
85 H4 Álvaro Obregón Chihuahua, Mexico 28°45′N 106°52′W
87 J7 Álvaro Obregón Tabasco, Mexico 18°35′N 92°39′W
84 F5 Álvaro Obregón, Presa ⊠ W Mexico
154 E5 Alvdal Hedmark, S Norway 62°07′N 10°39′E
154 G6 Älvdalen Dalarna, C Sweden 61°13′N 14°04′E
113 H7 Alvear Corrientes, NE Argentina 29°05′S 56°35′W
161 H4 Alvechurch United Kingdom 52°20′N 1°57′W
172 B5 Alverca do Ribatejo Lisboa, C Portugal 38°56′N 09°01′W
180 G7 Alverdissen Nordrhein-Westfalen, Germany 52°02′E 9°08′N
155 G11 Alvesta Kronoberg, S Sweden 56°52′N 14°34′E
161 I4 Alveston United Kingdom 52°12′N 1°39′W
172 B4 Alviela, Rio ≋ Santarém, Portugal
71 I7 Alvin Texas, SW USA 29°25′N 95°14′W
111 I3 Alvinópolis Minas Gerais, Brazil 20°06′S 43°03′W
154 I6 Älvkarleby Uppsala, C Sweden 60°34′N 17°30′E
70 G4 Alvord Texas, SW USA 33°22′N 97°39′W
154 G5 Älvros Jämtland, C Sweden 62°04′N 14°30′E
152 G6 Älvsbyn Norrbotten, N Sweden 65°41′N 21°00′E
217 M10 Al Wafrā' SE Kuwait 28°38′N 47°57′E
220 D3 Al Wajh Tabūk, NW Saudi Arabia 26°16′N 36°30′E
221 I4 Al Wakrah var. Wakra. C Qatar 25°09′N 51°36′E
216 F5 al Walaj, Sha'īb dry watercourse W Iraq
232 E6 Alwar Rājasthān, N India 27°32′N 76°35′E
220 G3 Al Warī'ah Ash Sharqīyah, N Saudi Arabia 27°54′N 47°23′E
235 D9 Alwaye var. Aluva. Kerala, SW India 10°06′N 76°23′E see also Aluva
160 F2 Alwen, Llyn ◎ United Kingdom
Alxa Zuoqi see Bayan Hot
Alx Youqi see Ehen Hudag
Al Yaman see Yemen
218 F6 Al Yarmūk Irbid, N Jordan 32°41′N 35°55′E
Alyat/Alyaty-Pristan' see Älät
187 I3 Alykí var. Aliki. Thásos, N Greece 40°36′N 24°45′E
159 I1 Alyth United Kingdom 56°37′N 3°13′W
191 D9 Alytus Pol. Olita. Alytus, S Lithuania 54°24′N 24°02′E
191 D9 Alytus ◐ province S Lithuania
179 H13 Alz ≋ SE Germany
86 C6 Alzada Colima, Mexico 19°15′N 103°31′W
77 M6 Alzada Montana, NW USA 45°00′N 104°24′W
193 H7 Alzamay Irkutskaya Oblast', S Russian Federation 55°33′N 98°36′E
181 G11 Alzenau Bayern, Germany 50°05′E 9°04′N
163 H14 Alzette ≋ S Luxembourg
181 E12 Alzey Rheinland-Pfalz, Germany 49°45′E 8°07′N
171 I6 Alzira var. Alcira ; anc. Saetabicula, Suero. Valenciana, E Spain 39°10′N 00°27′W
169 I5 Alzon Languedoc-Roussillon, France 43°58′N 3°26′E
168 G6 Alzonne Languedoc-Roussillon, France 43°16′N 2°11′E
Al Zubair see Az Zubayr
274 G5 Amadeus, Lake seasonal lake Northern Territory, C Australia
136 A5 Amadi Western Equatoria, SW South Sudan 05°32′N 30°20′E
55 M1 Amadjuak Lake ◎ Baffin Island, Nunavut, N Canada
80 D3 Amador City California, USA 38°25′N 120°49′W
155 F12 Amager island E Denmark
250 C6 Amagi var. Asakura. Fukuoka, Kyūshū, SW Japan 33°28′N 130°37′E
251 L4 Amagi-san ▲ Honshū, S Japan 34°51′N 138°57′E
260 G5 Amahai var. Masohi. Palau Seram, E Indonesia 03°19′S 128°56′E
82 G9 Amak Island island Alaska, USA
250 C7 Amakusa prev. Hondo. Kumamoto, Shimo-jima, SW Japan 32°28′N 130°12′E
250 C7 Amakusa-nada gulf SW Japan
155 F8 Åmål Västra Götaland, S Sweden 59°04′N 12°41′E
102 C4 Amalfi Antioquia, N Colombia 06°54′N 75°04′W
175 G9 Amalfi Campania, S Italy 40°37′N 14°35′E
141 H5 Amalia North-West, South Africa 27°15′S 25°03′E
186 F6 Amaliáda var. Amaliás. Dytikí Elláda, S Greece 37°48′N 21°21′E
Amaliás see Amaliáda
234 D4 Amalner Mahārāshtra, C India 21°03′N 75°04′E
261 K6 Amamapare Papua, E Indonesia 04°51′S 136°44′E
113 H3 Amambaí, Serra de var. Cordillera de Amambay, Serra de Amambay. ▲ Brazil/Paraguay see also Amambay, Cordillera de
113 H3 Amambay off. Departamento del Amambay. ◐ department E Paraguay
113 H3 Amambay, Cordillera de var. Serra de Amambaí, Serra de Amambay. ▲ Brazil/Paraguay see also Amambaí, Serra de
Amambay, Departamento del see Amambay
Amambay, Serra de see Amambaí, Serra de/ Amambay, Cordillera de
251 J9 Amami-guntō island group SW Japan
251 K9 Amami-Ō-shima island S Japan
280 A2 Amanab West Sepik, NW Papua New Guinea 03°38′S 141°16′E
167 L7 Amance Franche-Comté, France 47°48′N 6°04′E
167 L9 Amancey Franche-Comté, France 47°02′N 6°05′E
177 J10 Amandola Marche, C Italy 42°58′N 13°22′E
175 H11 Amantea Calabria, SW Italy 39°06′N 16°05′E
285 K7 Amanu island Îles Tuamotu, C French Polynesia
141 K7 Amanzimtoti Kwazulu Natal, South Africa 30°03′S 30°53′E
Amanzimtoti see eManzimtoti
108 A1 Amapá Amapá, NE Brazil 02°00′N 50°50′W
108 A1 Amapá off. Estado de Amapá; prev. Território de Amapá. ◐ state NE Brazil
Amapá, Estado de see Amapá
88 E5 Amapala Valle, S Honduras 13°16′N 87°39′W
Amapá, Território de see Amapá
136 E3 Āmara var. Amhara. ◐ N Ethiopia
Amara see Al 'Amārah
172 C1 Amarante Porto, N Portugal 41°16′N 08°05′W
256 C4 Amarapura Mandalay, C Myanmar (Burma) 21°54′N 96°01′E
Amardalay see Delgertsogt
172 D6 Amareleja Beja, S Portugal 38°12′N 07°13′W
81 H8 Amargosa Desert plain Nevada, USA
81 H8 Amargosa Range ▲▲ California, W USA
80 H7 Amargosa Valley Nevada, USA 36°39′N 116°24′W
70 E2 Amarillo Texas, SW USA 35°13′N 101°50′W
Amarinthos see Amárynthos
175 F8 Amaro, Monte ▲ C Italy 42°03′N 14°06′E
187 H6 Amárynthos var. Amarinthos. Évvoia, C Greece 38°24′N 23°53′E
219 E9 'Amasa Israel
Amasia see Amasya
214 F5 Amasya anc. Amasia. Amasya, N Turkey 40°37′N 35°50′E
214 F5 Amasya ◐ province N Turkey
87 K9 Amatenango Chiapas, Mexico 15°26′N 92°07′W
88 D3 Amatique, Bahía de bay Gulf of Honduras, W Caribbean Sea
86 C5 Amatitàn Jalisco, Mexico 20°50′N 103°43′W
88 C4 Amatitlán, Lago de ◎ S Guatemala
86 B5 Amatlàn de Cañas Nayarit, Mexico 20°48′N 104°24′W
174 F7 Amatrice Lazio, C Italy 42°38′N 13°19′E
284 F1 Amatuku atoll C Tuvalu
163 F11 Amay Liège, E Belgium 50°33′N 05°19′E
98 C3 Amazon Sp. Amazonas. ≋ Brazil/Peru
106 D4 Amazonas off. Estado do Amazonas. ◐ state N Brazil
102 D9 Amazonas off. Comisaria del Amazonas. ◐ province SE Colombia
104 C4 Amazonas off. Departamento de Amazonas. ◐ department N Peru
102 G6 Amazonas off. Territorio Amazonas. ◐ federal territory S Venezuela
Amazonas see Amazon
Amazonas, Comisaria del see Amazonas
Amazonas, Departamento de see Amazonas
Amazonas, Estado do see Amazonas
Amazonas, Territorio see Amazonas
98 D3 Amazon Basin basin N South America
95 L3 Amazon Fan undersea feature W Atlantic Ocean 05°00′N 47°30′W
108 C1 Amazon, Mouths of the delta NE Brazil
281 M2 Ambae var. Aoba, Omba. island C Vanuatu
232 E4 Ambāla Haryāna, NW India 30°19′N 76°49′E
235 G11 Ambalangoda Southern Province, SW Sri Lanka 06°14′N 80°03′E
235 G11 Ambalantota Southern Province, S Sri Lanka 06°07′N 81°01′E
139 L8 Ambalavao Fianarantsoa, C Madagascar 21°50′S 46°56′E
102 C5 Ambalema Tolima, C Colombia 04°49′N 74°48′W
134 B6 Ambam Sud, S Cameroon 02°23′N 11°17′E
139 M5 Ambanja Antsirañana, N Madagascar 13°40′S 48°27′E
193 L3 Ambarchik Respublika Sakha (Yakutiya), NE Russian Federation 69°33′N 162°08′E
116 E1 Ambargasta, Salinas de salt lake C Argentina
194 E4 Ambarnyy Respublika Kareliya, NW Russian Federation 65°53′N 33°44′E
104 C2 Ambato Tungurahua, C Ecuador 01°18′S 78°39′W
139 L8 Ambato Finandrahana Fianarantsoa, SE Madagascar
139 M7 Ambatolampy Antananarivo, C Madagascar 19°21′S 47°27′E
139 L6 Ambatomainty Mahajanga, W Madagascar 17°40′S 45°39′E
139 M6 Ambatondrazaka Toamasina, C Madagascar 17°49′S 48°28′E
168 F1 Ambazac Limousin, France 45°57′N 1°24′E
260 G6 Ambelau, Pulau island E Indonesia
179 G11 Amberg var. Amberg in der Oberpfalz. Bayern, SE Germany 49°26′N 11°52′E
Amberg in der Oberpfalz see Amberg
88 E1 Ambergris Cay island NE Belize
165 J6 Ambérieu-en-Bugey Ain, E France 45°57′N 05°21′E
279 E9 Amberley Canterbury, South Island, New Zealand 43°09′S 172°43′E
165 I6 Ambert Puy-de-Dôme, C France 45°33′N 03°45′E
Ambianum see Amiens
132 C4 Ambidédi Kayes, SW Mali 14°37′N 11°49′W
234 H3 Ambikāpur Chhattīsgarh, C India 23°09′N 83°12′E
139 M4 Ambilobe Antsirañana, N Madagascar 13°10′S 49°03′E
159 K5 Amble United Kingdom 55°20′N 1°35′W
83 H4 Ambler Alaska, USA 67°05′N 157°51′W
64 F7 Ambler Pennsylvania, USA 40°09′N 75°13′W
159 I7 Ambleside United Kingdom 54°25′N 2°58′W
Amblève see Amel
Ambo see Hāgere Hiywet
139 L10 Amboasary Toliara, S Madagascar 25°01′S 46°23′E
139 N6 Ambodifotatra var. Ambodifototra. Toamasina, E Madagascar 16°59′S 49°51′E
Amboenten see Ambunten
139 M7 Ambohidratrimo Antananarivo, C Madagascar 18°48′S 47°26′E
139 M8 Ambohimahasoa Fianarantsoa, SE Madagascar 21°07′S 47°13′E
139 N5 Ambohitralanana Antsirañana, NE Madagascar 15°13′S 50°28′E
261 K5 Amboi, Kepulauan island group E Indonesia
Amboina see Ambon
164 G4 Amboise Indre-et-Loire, C France 47°25′N 01°00′E
260 G5 Ambon prev. Amboina, Amboyna. Pulau Ambon, E Indonesia 03°41′S 128°10′E
260 G5 Ambon, Pulau island E Indonesia
137 D9 Amboseli, Lake ◎ Kenya/Tanzania
139 M8 Ambositra Fianarantsoa, SE Madagascar 20°31′S 47°15′E
139 L10 Ambovombe Toliara, S Madagascar 25°10′S 46°06′E
81 H11 Amboy California, W USA 34°33′N 115°44′W
73 C9 Amboy Illinois, N USA 41°42′N 89°19′W
Amboyna see Ambon
Ambracia see Árta
Ambre, Cap d' see Bobaomby, Tanjona
62 C8 Ambridge Pennsylvania, NE USA 40°33′N 80°11′W
Ambrim see Ambrym
135 C10 Ambriz Bengo, NW Angola 07°55′S 13°11′E
Ambrizete see N'Zeto
114 A1 Ambrosetti Santa Fe, C Argentina 30°01′S 61°33′W
281 M3 Ambrym var. Ambrim. island C Vanuatu
259 J8 Ambunten prev. Amboenten. Pulau Madura, E Indonesia 06°55′S 113°45′E
261 N6 Ambunti East Sepik, NW Papua New Guinea 04°12′S 142°49′E
235 F8 Āmbūr Tamil Nādu, SE India 12°48′N 78°44′E
218 F2 Amchit Lebanon
218 F2 Amchit Lebanon 34°09′N 35°39′E
82 B10 Amchitka Island island Aleutian Islands, Alaska, USA
82 C10 Amchitka Pass strait Aleutian Islands, Alaska, USA
221 H8 'Amd C Yemen 15°10′N 47°58′E
134 F2 Am Dam Sila, E Chad 12°46′N 20°29′E
261 I7 Amdassa Pulau Yamdena, E Indonesia 07°40′S 131°24′E
195 K1 Amderma Nenetskiy Avtonomnyy Okrug, NW Russian Federation 69°45′N 61°36′E
239 H8 Amdo Xizang Zizhiqu, W China 32°15′N 91°43′E
86 E6 Amealco Hidalgo, Mexico 20°14′N 99°33′W
86 E6 Amealco Querétaro de Arteaga, Mexico 20°11′N 100°09′W
86 B5 Ameca Jalisco, SW Mexico 20°34′N 104°03′W
86 F6 Amecameca var. Amecameca de Juárez. México, C Mexico 19°08′N 98°48′W
Amecameca de Juárez see Amecameca
217 I1 Amêdî Ar. Al 'Amādiyah. Dahūk, N Iraq 37°09′N 43°27′E
112 E10 Ameghino Buenos Aires, E Argentina 34°51′S 62°28′W
163 H11 Amel Fr. Amblève. Liège, E Belgium 50°20′N 06°13′E
162 G3 Ameland Fris. It Amelân. island Waddeneilanden, N Netherlands
Amelân, It see Ameland
174 E7 Amelia Umbria, C Italy 42°33′N 12°26′E
67 K4 Amelia Court House Virginia, NE USA 37°20′N 77°59′W
69 L4 Amelia Island island Florida, SE USA
169 H8 Amélie-les-Bains-Palalda Languedoc-Roussillon, France 42°28′N 2°40′E
65 I3 Amenia New York, NE USA 41°51′N 73°31′W
America see United States of America
291 H13 America-Antarctica Ridge undersea feature S Atlantic Ocean
America in Miniature see Maryland
110 C8 Americana São Paulo, S Brazil 22°44′S 47°19′W
80 B2 American Canyon California, USA 38°11′N 122°16′W
77 H8 American Falls Idaho, NW USA 42°47′N 112°51′W
77 H8 American Falls Reservoir ⊠ Idaho, NW USA
79 H2 American Fork Utah, W USA 40°24′N 111°47′W
286 G6 American Samoa ◇ US unincorporated territory W Polynesia
110 B6 Américo Brasiliense São Paulo, Brazil 21°43′S 48°07′W
69 J2 Americus Georgia, SE USA 32°04′N 84°13′W
162 G7 Amerongen Utrecht, C Netherlands 52°00′N 05°30′E
162 G6 Amersfoort Utrecht, C Netherlands 52°09′N 05°23′E
161 J6 Amersham SE England, United Kingdom 51°40′N 00°37′W
72 A5 Amery Wisconsin, N USA 45°18′N 92°20′W
293 L4 Amery Ice Shelf ice shelf Antarctica
74 G7 Ames Iowa, C USA 42°01′N 93°37′W
65 M1 Amesbury Massachusetts, NE USA 42°51′N 70°55′W
Amestratus see Mistretta
186 G5 Amfíkleia var. Amfíklia. Stereá Elláda, C Greece 38°38′N 22°35′E
Amfíklia see Amfíkleia
186 F5 Amfilochía var. Amfilokhía. Dytikí Elláda, C Greece 38°52′N 21°09′E
Amfilokhía see Amfilochía
187 H2 Amfípoli anc. Amphipolis. site of ancient city Kentrikí Makedonía, NE Greece
186 G5 Ámfissa Stereá Elláda, C Greece 38°32′N 22°22′E
193 K6 Amga Respublika Sakha (Yakutiya), NE Russian Federation 60°55′N 131°45′E
193 K6 Amga ≋ NE Russian Federation
247 I4 Amgalang var. Xin Barag Zuoqi. Nei Mongol Zizhiqu, N China 48°12′N 118°15′E
193 M3 Amguema ≋ NE Russian Federation
193 L7 Amgun' ≋ SE Russian Federation
Amhara see Āmara
63 N2 Amherst Nova Scotia, SE Canada 45°50′N 64°14′W
65 J2 Amherst Massachusetts, NE USA 42°22′N 72°31′W
64 A1 Amherst New York, NE USA 42°57′N 78°47′W
70 D3 Amherst Texas, SW USA 33°59′N 102°24′W
67 K4 Amherst Virginia, NE USA 37°35′N 79°04′W
Amherst see Kyaikkami
62 A6 Amherstburg Ontario, S Canada 42°05′N 83°06′W
67 H4 Amherstdale West Virginia, NE USA 37°46′N 81°46′W
62 F4 Amherst Island island Ontario, SE Canada
Amida see Diyarbakır
74 A3 Amidon North Dakota, N USA 46°29′N 103°19′W
172 C3 Amieira Castelo Branco, Portugal 39°59′N 7°55′W
172 B3 Amieira Coimbra, Portugal 40°05′N 8°46′W
165 H2 Amiens anc. Ambianum, Samarobriva. Somme, N France 49°54′N 02°18′E
216 G5 'Āmij, Wādī var. Wadi 'Amīq. dry watercourse W Iraq
214 G8 Amik Ovasī ◎ S Turkey
132 B9 Amilcar Cabral ✕ Sal, NE Cape Verde
Amilḥayt, Wādī see Umm al Ḥayt, Wādī
167 H7 Amilly Centre, France 47°59′N 2°46′E
Amíndaion/Amindeo see Amýntaio
235 C8 Amīndīvi Islands island group Lakshadweep, India, N Indian Ocean
217 K4 Amīn Ḥabīb Diyālá, E Iraq 34°17′N 45°10′E
138 D6 Aminuis Omaheke, E Namibia 23°43′S 19°21′E
218 F2 Amioun N Lebanon
218 F2 Amioun Lebanon 34°18′N 35°48′E
'Amīq, Wadi see 'Āmij, Wādī
222 D5 Amīrābād Īlām, NW Iran 33°20′N 46°16′E
Amirante Bank see Amirante Ridge
289 C8 Amirante Basin undersea feature W Indian Ocean 07°00′S 54°00′E
289 C8 Amirante Islands var. Amirantes Group. island group C Seychelles
289 D8 Amirante Ridge var. Amirante Bank. undersea feature W Indian Ocean 06°00′S 53°10′E
Amirantes Group see Amirante Islands
289 D8 Amirante Trench undersea feature W Indian Ocean 08°00′S 52°30′E
261 M6 Amisibil Papua, E Indonesia 03°59′S 140°35′E
55 H8 Amisk Lake ◎ Saskatchewan, C Canada
Amistad, Presa de la see Amistad Reservoir
70 E7 Amistad Reservoir var. Presa de la Amistad. ⊠ Mexico/USA
Amisus see Samsun
68 E4 Amite var. Amite City. Louisiana, S USA 30°40′N 90°30′W
Amite City see Amite
75 H13 Amity Arkansas, C USA 34°15′N 93°27′W
232 F9 Amla prev. Amulla. Madhya Pradesh, C India 21°53′N 78°10′E
82 D10 Amlia Island island Aleutian Islands, Alaska, USA
160 E1 Amlwch NW Wales, United Kingdom 53°25′N 04°23′W
Ammaia see Portalegre
219 G8 'Amman var. Amman; anc. Philadelphia, Bibl. Rabbah Ammon, Rabbath Ammon. ● (Jordan) 'Ammān, NW Jordan 31°57′N 35°56′E
219 G8 'Ammān off. Muḥāfazat 'Ammān; prev. Al 'Aṣimah. ◐ governorate NW Jordan
160 E5 Ammanford United Kingdom 51°48′N 3°59′W
'Ammān, Muḥāfazat see 'Ammān
152 I7 Ämmänsaari Kainuu, E Finland 64°51′N 28°58′E
152 E6 Ammarnäs Västerbotten, N Sweden 65°58′N 16°10′E
295 L7 Ammassalik var. Angmagssalik. Sermersooq, S Greenland 65°51′N 37°30′W
180 C6 Ammeloe Nordrhein-Westfalen, Germany 52°04′E 6°47′N
179 F13 Ammer ≋ SE Germany
181 I9 Ammern Thüringen, Germany 51°14′E 10°27′N
179 F13 Ammersee ◎ SE Germany
163 F8 Ammerzoden Gelderland, C Netherlands 51°46′N 05°07′E
Ammóchostos see Gazimağusa
Ammóchostos, Kólpos see Gazimağusa Körfezi
Amnok-kang see Yalu
Amoea see Portalegre
Amoentai see Amuntai
Amoerang see Amurang
222 F3 Āmol var. Amul. Māzandarān, N Iran 36°31′N 52°24′E
181 F10 Amoneburg Hessen, Germany 50°48′E 8°54′N
172 B3 Amor Leiria, Portugal 39°48′N 8°52′W
187 J8 Amorgós Amorgós, Kykládes, Greece, Aegean Sea 36°49′N 25°54′E
187 J8 Amorgós island Kykládes, Greece, Aegean Sea
66 C8 Amory Mississippi, S USA 33°58′N 88°29′W
58 F7 Amos Québec, SE Canada 48°34′N 78°08′W
154 D7 Åmot Buskerud, S Norway 59°52′N 09°55′E
155 C8 Åmot Telemark, S Norway 59°34′N 07°59′E
154 F7 Åmotfors Värmland, C Sweden 59°46′N 12°24′E
132 E3 Amourj Hodh ech Chargui, SE Mauritania 16°04′N 07°12′W
Amoy see Xiamen
139 L10 Ampanihy Toliara, SW Madagascar 24°40′S 44°45′E
235 G10 Ampara var. Amparai. Eastern Province, E Sri Lanka 07°17′N 81°41′E
139 M6 Amparafaravola Toamasina, E Madagascar 17°33′S 48°13′E
Amparai see Ampara
110 D8 Amparo São Paulo, S Brazil 22°40′S 46°49′W
111 I4 Amparo da Serra Minas Gerais, Brazil 20°31′S 42°49′W
139 M7 Ampasimanolotra Toamasina, E Madagascar 18°49′S 49°04′E
105 G9 Ampato, Nevado ▲ S Peru 15°52′S 71°51′W
179 G12 Amper ≋ SE Germany
Amphipolis see Amfípoli
262 G4 Amphitrite Group Chin. Xuande Qundao, Viet. Nhom An Vinh. island group N Paracel Islands
261 H8 Amplawas var. Emplawas. Pulau Babar, E Indonesia 08°01′S 129°42′E
169 J1 Amplepuis Rhône-Alpes, France 45°59′N 4°20′E
171 J4 Amposta Cataluña, NE Spain 40°43′N 00°34′E
59 I7 Amqui Québec, SE Canada 48°28′N 67°27′W
220 F8 'Amrān W Yemen 15°39′N 43°59′E
Amraoti see Amrāvati
234 E4 Amrāvati prev. Amraoti. Mahārāshtra, C India 20°56′N 77°45′E
219 E14 'Amrāwah Jordan
219 E14 Amrāwah Jordan 32°41′N 35°56′E
232 B9 Amreli Gujarāt, W India 21°36′N 71°20′E
176 E3 Amriswil Thurgau, NE Switzerland 47°33′N 09°18′E
216 C4 'Amrīt ruins Tarṭūs, W Syria
232 D3 Amritsar Punjab, N India 31°38′N 74°55′E
232 F5 Amroha Uttar Pradesh, N India 28°54′N 78°29′E
178 C3 Amrum island NW Germany
154 J2 Åmsele Västerbotten, N Sweden 64°31′N 19°24′E
162 F6 Amstelveen Noord-Holland, C Netherlands 52°18′N 04°50′E
162 F6 Amsterdam ● (Netherlands) Noord-Holland, C Netherlands 52°22′N 04°54′E
141 K4 Amsterdam Mpumalanga, South Africa 26°37′S 30°40′E
65 H1 Amsterdam New York, NE USA 42°56′N 74°11′W
289 F11 Amsterdam Fracture Zone tectonic feature S Indian Ocean
289 F11 Amsterdam Island island NE French Southern and Antarctic Territories
177 K2 Amstetten Niederösterreich, N Austria 48°08′N 14°52′E
134 F3 Am Timan Salamat, SE Chad 11°02′N 20°17′E
228 G7 Amu-Buxoro Kanali var. Aral-Bukhorskiy Kanal. canal C Uzbekistan
216 G1 'Āmūdah var. Amude. Al Ḥasakah, N Syria 37°06′N 40°56′E
230 G2 Amu Darya Rus. Amudar'ya, Taj. Dar"yoi Amu, Turkm. Amyderya, Uzb. Amudaryo; anc. Oxus. ≋ C Asia
Amu-Dar'ya see Amyderýa
Amudar'ya/Amudaryo/Amu, Dar"yoi see Amu Darya
Amude see 'Āmūdah
220 E1 'Amūd, Jabal al ▲ NW Saudi Arabia 30°59′N 39°17′E
82 E10 Amukta Island island Aleutian Islands, Alaska, USA
82 D10 Amukta Pass strait Aleutian Islands, Alaska, USA
Amul see Āmol
Amulla see Amla
159 H2 Amulree United Kingdom 56°30′N 3°47′W
53 H3 Amund Ringnes Island Island Nunavut, N Canada
295 I5 Amundsen Basin var. Fram Basin. undersea basin Arctic Ocean
293 L2 Amundsen Bay bay Antarctica
293 H6 Amundsen Coast physical region Antarctica
52 D6 Amundsen Gulf gulf Northwest Territories, N Canada
287 H10 Amundsen Plain undersea feature S Pacific Ocean
293 I5 Amundsen-Scott US research station Antarctica 89°59′S 10°00′E
292 F7 Amundsen Sea sea S Pacific Ocean
154 G6 Amungen ◎ C Sweden
259 K6 Amuntai prev. Amoentai. Borneo, C Indonesia 02°24′S 115°14′E
205 M4 Amur Chin. Heilong Jiang. ≋ China/Russian Federation
260 E3 Amurang prev. Amoerang. Sulawesi, C Indonesia 01°12′N 124°37′E
260 E3 Amurang, Teluk bay Sulawesi, C Indonesia
170 G2 Amurrio País Vasco, N Spain 43°03′N 03°00′W
193 L8 Amursk Khabarovskiy Kray, SE Russian Federation 50°13′N 136°34′E
193 J7 Amurskaya Oblast' ◐ province SE Russian Federation
129 J8 'Amur, Wadi ≋ NE Sudan
186 F5 Amvrakikós Kólpos gulf W Greece
Amvrosiyevka see Amvrosiyivka
189 M6 Amvrosiyivka Rus. Amvrosiyevka. Donets'ka Oblast', SE Ukraine 47°46′N 38°30′E
228 G7 Amyderýa Rus. Amu-Dar'ya. Lebap Welaýaty, NE Turkmenistan 37°58′N 65°14′E
Amyderya see Amu Darya
186 F3 Amýntaio var. Amindeo; prev. Amíndaion. Dytikí Makedonía, N Greece 40°42′N 21°42′E
72 G2 Amyot Ontario, S Canada 48°28′N 84°58′W
285 J7 Anaa atoll Îles Tuamotu, C French Polynesia
Anabanoea see Anabanua
260 C6 Anabanua prev. Anabanoea. Sulawesi, C Indonesia 03°58′S 120°07′E
283 J9 Anabar NE Nauru 0°30′S 166°56′E
193 I4 Anabar ≋ NE Russian Federation
An Abhainn Mhór see Blackwater
103 H3 Anaco Anzoátegui, NE Venezuela 09°30′N 64°28′W
77 H5 Anaconda Montana, NW USA 46°09′N 112°56′W
76 C3 Anacortes Washington, NW USA 48°30′N 122°36′W
75 D12 Anadarko Oklahoma, C USA 35°04′N 98°16′W
172 B2 Anadia Aveiro, N Portugal 40°26′N 08°27′W
Anadolu Dağları see Doğu Karadeniz Dağları
193 N3 Anadyr' Chukotskiy Avtonomnyy Okrug, NE Russian Federation 64°41′N 177°22′E
193 M3 Anadyr' ≋ NE Russian Federation
Anadyr, Gulf of see Anadyrskiy Zaliv
193 N3 Anadyrskiy Zaliv Eng. Gulf of Anadyr. gulf NE Russian Federation
187 J8 Anáfi anc. Anaphe. island Kykládes, Greece, Aegean Sea
175 F8 Anagni Lazio, C Italy 41°43′N 13°12′E
'Ānah see 'Annah
78 D8 Anaheim California, W USA 33°50′N 117°54′W
56 C5 Anahim Lake British Columbia, SW Canada 52°26′N 125°13′W
85 L5 Anáhuac Nuevo León, NE Mexico 27°13′N 100°09′W
71 I7 Anahuac Texas, SW USA 29°44′N 94°41′W
235 E9 Anai Mudi ▲ S India 10°16′N 77°08′E
Anaiza see 'Unayzah
234 H5 Anakāpalle Andhra Pradesh, E India 17°42′N 83°06′E
118 B1 Anakena, Playa de beach Easter Island, Chile, E Pacific Ocean
83 I3 Anaktuvuk Pass Alaska, USA 68°08′N 151°44′W
83 I3 Anaktuvuk River ≋ Alaska, USA
139 M5 Analalava Mahajanga, NW Madagascar 14°38′S 47°46′E
90 D4 Ana Maria, Golfo de gulf N Caribbean Sea
Anambas Islands see Anambas, Kepulauan
258 F3 Anambas, Kepulauan var. Anambas Islands. island group W Indonesia
133 K8 Anambra ◐ state SE Nigeria
74 C2 Anamoose North Dakota, N USA 47°50′N 100°14′W
74 I7 Anamosa Iowa, C USA 42°06′N 91°17′W
214 E8 Anamur İçel, S Turkey 36°06′N 32°49′E
214 E8 Anamur Burnu headland S Turkey 36°03′N 32°49′E
250 G6 Anan Tokushima, Shikoku, SW Japan 33°54′N 134°40′E
234 I4 Ānandapur var. Anandpur. Odisha, E India 21°14′N 86°10′E
Anandpur see Ānandapur
234 E7 Anantapur Andhra Pradesh, S India 14°41′N 77°36′E
232 E2 Anantnāg var. Islamabad. Jammu and Kashmir, NW India 33°44′N 75°11′E
Ananyev see Anan'yiv
188 G6 Anan'yiv Rus. Ananyev. Odes'ka Oblast', SW Ukraine 47°43′N 29°51′E
196 F8 Anapa Krasnodarskiy Kray, SW Russian Federation 44°55′N 37°20′E
Anaphe see Anáfi
109 B10 Anápolis Goiás, C Brazil 16°19′S 48°58′W
223 H6 Anār Kermān, C Iran 30°49′N 55°18′E
Anár see Inari
222 G5 Anārak Eşfahān, C Iran 33°21′N 53°43′E
Anar Dara see Anār Darreh
230 D4 Anār Darreh var. Anar Dara. Farāh, W Afghanistan 32°45′N 61°38′E
Anárjohka see Inarijoki
69 M4 Anastasia Island island Florida, SE USA
282 E3 Anatahan island C Northern Mariana Islands
204 G4 Anatolia plateau C Turkey
148 E8 Anatolian Plate tectonic feature Asia/Europe
187 I2 Anatolikí Makedonía kai Thráki Eng. Macedonia East and Thrace. ◐ region NE Greece
Anatom see Aneityum
116 F1 Añatuya Santiago del Estero, N Argentina 28°28′S 62°52′W
An Baile Meánach see Ballymena
An Bhearú see Barrow
An Bhóinn see Boyne
An Blascaod Mór see Great Blasket Island
Anbu see Chao'an
An Cabhán see Cavan
An Caisleán Nua see Newcastle
An Caisleán Riabhach see Castlerea, Ireland
An Caisleán Riabhach see Castlereagh
104 C6 Ancash off. Departamento de Ancash. ◐ department W Peru
Ancash, Departamento de see Ancash
An Cathair see Caher
164 F4 Ancenis Loire-Atlantique, NW France 47°23′N 01°10′W
167 K6 Ancerville Lorraine, France 48°38′N 5°02′E
An Chanáil Ríoga see Royal Canal
An Cheacha see Caha Mountains
111 M4 Anchieta Espírito Santo, Brazil 20°48′S 40°39′W
83 I6 Anchorage Alaska, USA 61°13′N 149°52′W
83 I6 Anchorage ✕ Alaska, USA 61°08′N 150°00′W
83 I7 Anchor Point Alaska, USA 59°46′N 151°49′W
An Chorr Chríochach see Cookstown
132 E10 Anchorstock Point headland W Tristan da Cunha 37°07′S 12°21′W
172 G4 Anchuras Castilla-La Manacha, Spain 39°29′N 4°50′W
An Clochán see Clifden
An Clochán Liath see Dunglow
69 K6 Anclote Keys island group Florida, SE USA
An Cóbh see Cobh
105 H9 Ancohuma, Nevado de ▲ W Bolivia 15°51′S 68°33′W
An Comar see Comber
104 D7 Ancón Lima, W Peru 11°45′S 77°08′W
177 J9 Ancona Marche, C Italy 43°38′N 13°30′E
159 J4 Ancroft United Kingdom 55°42′N 2°00′W
Ancuabe see Ancuabi
137 E13 Ancuabi var. Ancuabe. Cabo Delgado, NE Mozambique 13°00′S 39°50′E
117 B8 Ancud prev. San Carlos de Ancud. Los Lagos, S Chile 41°53′S 73°50′W
117 B8 Ancud, Golfo de gulf S Chile
Ancyra see Ankara
193 K9 Anda Heilongjiang, NE China 46°25′N 125°20′E
118 D8 Andacollo Neuquén, Argentina 37°11′S 70°41′W
105 F8 Andahuaylas Apurímac, S Peru 13°39′S 73°24′W
An Daingean see Dingle
233 J8 Andāl West Bengal, NE India 23°35′N 87°14′E
154 C4 Åndalsnes Møre og Romsdal, S Norway 62°33′N 07°42′E
170 F7 Andalucía Eng. Andalusia. ◐ autonomous community S Spain
130 G1 Andalucía cultural region S Spain
69 H3 Andalusia Alabama, S USA 31°18′N 86°29′W
Andalusia see Andalucía
257 A10 Andaman and Nicobar Islands var. Andamans and Nicobars. ◐ union territory India, NE Indian Ocean
288 H6 Andaman Basin undersea feature NE Indian Ocean 10°00′N 94°00′E
257 B9 Andaman Islands island group India, NE Indian Ocean
Andamans and Nicobars see Andaman and Nicobar Islands
288 H6 Andaman Sea sea NE Indian Ocean
112 C1 Andamarca Oruro, C Bolivia 18°46′S 67°31′W
261 I5 Andamata Papua, E Indonesia 02°40′S 132°30′E
276 F3 Andamooka South Australia 30°26′S 137°12′E
221 K5 'Andām, Wādī seasonal river NE Oman
139 N5 Andapa Antsirañana, NE Madagascar 14°39′S 49°40′E
231 H3 Andarāb var. Banow. Baghlān, NE Afghanistan 35°36′N 69°18′E
Andarbag see Andarbogh
229 J7 Andarbogh Rus. Andarbag, Anderbak. S Tajikistan 37°51′N 71°45′E
176 E5 Andeer Graubünden, S Switzerland 46°36′N 09°24′E
167 K6 Andelot Champagne-Ardenne, France 48°15′N 5°18′E
152 E4 Andenes Nordland, C Norway 69°18′N 16°10′E
163 F11 Andenne Namur, SE Belgium 50°29′N 05°06′E
133 I4 Andéramboukane Gao, E Mali 15°24′N 03°03′E
Anderbak see Andarbogh
163 D10 Anderlecht Brussels, C Belgium 50°50′N 04°18′E
163 D11 Anderlues Hainaut, S Belgium 50°24′N 04°16′E
176 E5 Andermatt Uri, C Switzerland 46°39′N 08°36′E
181 D11 Andernach anc. Antunnacum. Rheinland-Pfalz, SW Germany 50°26′N 07°24′E
168 C3 Andernos-les-Bains Aquitaine, France 44°44′N 1°06′W
282 B1 Andersen Air Force Base air base NE Guam 13°34′N 144°55′E
83 I5 Anderson Alaska, USA 64°20′N 149°11′W
78 B2 Anderson California, W USA 40°26′N 122°21′W
73 F10 Anderson Indiana, N USA 40°06′N 85°40′W
75 G11 Anderson Missouri, C USA 36°39′N 94°26′W
67 H7 Anderson South Carolina, SE USA 34°30′N 82°39′W
71 H6 Anderson Texas, SW USA 30°29′N 96°00′W
52 C6 Anderson ≋ Northwest Territories, NW Canada
155 F10 Anderstorp Jönköping, S Sweden 57°17′N 13°38′E
102 B5 Andes Antioquia, W Colombia 05°40′N 75°56′W
95 I5 Andes ▲▲ W South America
74 D6 Andes, Lake ◎ South Dakota, N USA
152 E4 Andfjorden fjord E Norwegian Sea
234 F6 Andhra Pradesh ◐ state E India
162 F5 Andijk Noord-Holland, NW Netherlands 52°38′N 05°00′E
229 K5 Andijon Rus. Andizhan. Andijon Viloyati, E Uzbekistan 40°46′N 72°19′E
229 K5 Andijon Viloyati Rus. Andizhanskaya Oblast'. ◐ province E Uzbekistan
Andikíthira see Antikýthira
139 M6 Andilamena Toamasina, C Madagascar 17°00′S 48°35′E
222 E5 Andīmeshk var. Andimishk; prev. Salehābād. Khūzestān, SW Iran 32°30′N 48°26′E
Andimishk see Andīmeshk
243 H4 Anding Hunan, E China 28°35′N 113°40′E
114 B6 Andino Santa Fe, Argentina 32°40′S 60°53′W
Andipáros see Antíparos
Andipaxi see Antípaxoi
Andípsara see Antípsara
214 G7 Andırın Kahramanmaraş, S Turkey 37°33′N 36°18′E
238 E6 Andirlangar Xinjiang Uygur Zizhiqu, NW China 37°38′N 83°40′E
Andírrion see Antírrio
Ándissa see Ántissa
Andizhan see Andijon
Andizhanskaya Oblast' see Andijon Viloyati
230 F2 Andkhvōy prev. Andkhvoy. Fāryāb, N Afghanistan 36°56′N 65°08′E
171 H2 Andoain País Vasco, N Spain 43°13′N 02°02′W
261 J4 Andoi Papua, E Indonesia 0°53′S 133°59′E
248 C6 Andong Jap. Antō. E South Korea 36°34′N 128°44′E
250 B2 Andong-ho prev. Andong Lake. ◎
Andong Lake see Andong-ho
245 L7 Andongwei Jiangsu, E China 34°54′N 119°06′E
177 I2 Andorf Oberösterreich, N Austria 48°22′N 13°33′E
171 I4 Andorra Aragón, NE Spain 40°59′N 00°27′W
171 J2 Andorra off. Principality of Andorra, Cat. Valls d'Andorra, Fr. Vallée d'Andorre. ◆ monarchy SW Europe
Andorra see Andorra la Vella
171 K2 Andorra la Vella var. Andorra, Fr. Andorre la Vielle, Sp. Andorra la Vieja. ● (Andorra) C Andorra 42°30′N 01°30′E
Andorra la Vieja see Andorra la Vella
Andorra, Principality of see Andorra
Andorra, Valls d'/Andorra, Vallée d' see Andorra
Andorre la Vielle see Andorra la Vella
161 I7 Andover S England, United Kingdom 51°13′N 01°28′W
75 E10 Andover Kansas, C USA 37°42′N 97°08′W
65 L2 Andover Massachusetts, USA 42°40′N 71°08′W
64 C3 Andover New York, USA 42°09′N 77°48′W
152 E4 Andøya island C Norway
113 J2 Andradina São Paulo, S Brazil 20°54′S 51°19′W
171 L6 Andratx Mallorca, Spain, W Mediterranean Sea 39°35′N 02°25′E
82 G5 Andreafsky River ≋ Alaska, USA
82 D10 Andreanof Islands island group Aleutian Islands, Alaska, USA
194 D10 Andreapol' Tverskaya Oblast', W Russian Federation 56°38′N 32°17′E
Andreas, Cape see Zafer Burnu
Andreevka see Kabanbay
110 G6 Andrelândia Minas Gerais, Brazil 21°44′S 44°18′W
114 G7 Andresito Flores, Uruguay 33°08′S 57°09′W
66 G7 Andrews North Carolina, SE USA 35°19′N 84°01′W
67 J9 Andrews South Carolina, SE USA 33°27′N 79°33′W
70 D5 Andrews Texas, SW USA 32°19′N 102°34′W
288 C7 Andrew Seamount N Indian Ocean
288 C7 Andrew Tablemount var. Gora Andryu. undersea feature W Indian Ocean 06°45′N 50°30′E
Andreyevka see Kabanbay
169 J2 Andrézieux Rhône-Alpes, France 45°32′N 4°16′E
175 H9 Andria Puglia, SE Italy 41°13′N 16°17′E
140 E4 Andriesvale Northern Cape, South Africa 26°56′S 20°39′E
184 F6 Andrijevica E Montenegro 42°45′N 19°45′E
186 F7 Andrítsaina Pelopónnisos, S Greece 37°29′N 21°52′E
Andropov see Rybinsk
187 I6 Ándros Ándros, Kykládes, Greece, Aegean Sea 37°49′N 24°54′E
187 J6 Ándros island Kykládes, Greece, Aegean Sea
63 J4 Androscoggin River ≋ Maine/New Hampshire, NE USA
61 M10 Andros Island island NW The Bahamas
197 J3 Androsovka Samarskaya Oblast', W Russian Federation 52°41′N 49°34′E
61 M10 Andros Town Andros Island, NW The Bahamas 24°40′N 77°47′W
235 C9 Āndrott Island island Lakshadweep, India, N Indian Ocean
188 G3 Andrushivka Zhytomyrs'ka Oblast', N Ukraine 50°01′N 29°02′E
183 G8 Andrychów Małopolskie, S Poland 49°51′N 19°18′E
Andryu, Gora see Andrew Tablemount
195 M8 Andryushkino Sverdlovskaya Oblast, Russian Federation
152 F4 Andselv Troms, N Norway 69°05′N 18°30′E
134 I6 Andudu Orientale, NE Dem. Rep. Congo 02°26′N 28°39′E
170 F7 Andújar anc. Illiturgis. Andalucía, SW Spain 38°02′N 04°03′W
135 D12 Andulo Bié, W Angola 11°29′S 16°43′E
165 I8 Anduze Gard, S France 44°03′N 03°59′E
An Earagail see Errigal Mountain
155 G10 Aneby Jönköping, S Sweden 57°50′N 14°45′E
Anécho see Aného
133 H3 Anéfis Kidal, NE Mali 18°05′N 00°38′E
91 K5 Anegada island NE British Virgin Islands
116 G7 Anegada, Bahía bay E Argentina
91 L5 Anegada Passage passage Anguilla/British Virgin Islands
133 I8 Aného var. Anécho; prev. Petit-Popo. S Togo 06°14′N 01°36′E
281 N6 Aneityum var. Anatom; prev. Kéamu. island S Vanuatu
118 F10 Añelo Neuquén, Argentina 38°21′S 68°47′W
188 G7 Anenii Noi Rus. Novyye Aneny. C Moldova 46°52′N 29°10′E
280 C3 Anepmete New Britain, E Papua New Guinea 05°47′S 148°37′E
168 E7 Aneto ▲ NE Spain 42°36′N 00°37′E
228 D7 Änew Rus. Annau. Ahal Welaýaty, C Turkmenistan 37°51′N 58°22′E
Änewetak see Enewetak Atoll
133 M2 Aney Agadez, NE Niger 19°22′N 13°00′E
245 L9 Anfeng Jiangsu, W China 33°05′N 120°06′E
An Fheoir see Nore
243 I5 Anfu Jiangxi, China 27°14′N 114°22′E
86 D6 Angahuàn Michoacán, Mexico 19°33′N 102°11′W
250 B3 An'gang 35°58′N 129°14′E
250 B3 An-gang prev. An'gang. Kyŏngsang-bukto, South Korea 36°00′N 129°14′E
An'gang see An-gang
193 H8 Angara ≋ C Russian Federation
193 H8 Angarsk Irkutskaya Oblast', S Russian Federation 52°31′N 103°55′E
110 A9 Angatuba São Paulo, Brazil 23°29′S 48°25′W
Angaur see Ngeaur
154 H4 Ånge Västernorrland, C Sweden 62°31′N 15°40′E
Angel see Úhlava
84 D3 Ángel de la Guarda, Isla island NW Mexico
263 K5 Angeles off. Angeles City. Luzon, N Philippines 15°16′N 120°37′E
Angeles City see Angeles
Angel Falls see Ángel, Salto
155 F11 Ängelholm Skåne, S Sweden 56°14′N 12°52′E
114 A4 Angélica Santa Fe, C Argentina 31°33′S 61°33′W
71 I5 Angelina River ≋ Texas, SW USA
103 I5 Ángel, Salto Eng. Angel Falls. waterfall E Venezuela
154 H7 Ängelsberg Västmanland, C Sweden 59°57′N 16°01′E
80 D3 Angels Camp California, W USA 38°03′N 120°31′W
177 L3 Anger Steiermark, SE Austria 47°16′N 15°41′E
Angerapp see Ozersk
Angerburg see Węgorzewo
154 H2 Ångermanälven ≋ N Sweden
178 I6 Angermünde Brandenburg, NE Germany 53°02′N 13°59′E
164 F4 Angers anc. Juliomagus. Maine-et-Loire, NW France 47°30′N 00°33′W
166 G6 Angervilliers Essonne, N France 48°35′N 02°03′E
153 G8 Ängesön island N Sweden
53 M8 Angijak Island Island Nunavut, NE Canada
55 I4 Angikuni Lake ◎ Nunavut, C Canada
Angistro see Ágkistro
187 H2 Angítis ≋ NE Greece
257 H9 Ângk Tasaôm prev. Angtassom. Takêv, S Cambodia
278 B13 Anglem, Mount ▲ Stewart Island, Southland, SW New Zealand 46°44′S 167°56′E
169 H8 Anglès Cataluña, Spain 41°57′N 2°39′E
157 F9 Anglesey cultural region NW Wales, United Kingdom
160 E2 Anglesey island NW Wales, United Kingdom
166 F9 Angles-sur-l'Anglin Poitou-Charentes, France 46°42′N 0°52′E
164 F8 Anglet Pyrénées-Atlantiques, SW France 43°29′N 01°30′W
71 I7 Angleton Texas, SW USA 29°10′N 95°27′W
Anglia see England
62 D1 Angliers Québec, SE Canada 47°33′N 79°17′W
Anglo-Egyptian Sudan see Sudan
Angmagssalik see Ammassalik
256 F6 Ang Nam Ngum ◎ C Laos
134 H6 Ango Orientale, N Dem. Rep. Congo 04°01′N 25°52′E
139 L2 Angoche Nampula, E Mozambique 16°10′S 39°58′E
118 B9 Angol Araucanía, C Chile 37°47′S 72°45′W
73 G9 Angola Indiana, N USA 41°37′N 85°00′W
135 E12 Angola off. Republic of Angola; prev. People's Republic of Angola, Portuguese West Africa. ◆ republic SW Africa

◆ Country
● Country Capital
◇ Dependent Territory
○ Dependent Territory Capital
◐ Administrative Regions
✕ International Airport
▲ Mountain
▲▲ Mountain Range
Volcano
≋ River
◎ Lake
⊠ Reservoir

◆ Country
● Country Capital
◇ Dependent Territory
○ Dependent Territory Capital
◆ Administrative Regions
✕ International Airport
▲ Mountain
▲ Mountain Range
ℜ Volcano
↗ River
◎ Lake
☷ Reservoir

INDEX

◆ Country ◇ Dependent Territory ⚑ Administrative Regions ▲ Mountain ☒ Volcano ◎ Lake
● Country Capital ○ Dependent Territory Capital ✕ International Airport ▲ Mountain Range ॐ River ☒ Reservoir

Avenio - Baia Farta
INDEX
314

Avenio see Avignon
154 C4 **Averoya** island S Norway
175 G9 **Aversa** Campania, S Italy 40°58′N 14°13′E
80 E3 **Avery** California, USA 38°12′N 120°22′W
76 F4 **Avery** Idaho, NW USA 47°14′N 115°48′W
71 I4 **Avery** Texas, SW USA 33°33′N 94°46′W
Aves, Islas de see Las Aves, Islas
Avesnes see Avesnes-sur-Helpe
167 H3 **Avesnes-le-Comte** Nord-Pas-de-Calais, France 50°17′N 2°32′E
165 I2 **Avesnes-sur-Helpe** var. Avesnes. Nord, N France 50°08′N 03°57′E
154 H7 **Avesta** Dalarna, C Sweden 60°09′N 16°10′E
165 H8 **Aveyron** ⬖ department S France
165 H8 **Aveyron** ≈ S France
175 F8 **Avezzano** Abruzzo, C Italy 42°02′N 13°26′E
186 F4 **Avgó** ▲ C Greece 39°31′N 21°24′E
Avgustov see Augustów
Avgustovskiy Kanal see Augustowski, Kanał
158 F2 **Avich, Loch** ◎ W Scotland, United Kingdom
156 F5 **Aviemore** N Scotland, United Kingdom 57°06′N 04°01′W
278 D11 **Aviemore, Lake** ◎ South Island, New Zealand
165 J8 **Avignon** anc. Avenio. Vaucluse, SE France 43°57′N 04°49′E
170 F4 **Ávila** var. Avila; anc. Abela, Abula, Abyla, Avela. Castilla y León, C Spain 40°39′N 04°42′W
170 F4 **Ávila** ⬖ province Castilla y León, C Spain
80 B7 **Avila Beach** California, USA 35°11′N 120°44′W
172 F2 **Ávila, Sierra de** ▲▲ Castilla y León, Spain
85 M8 **Avila y Urbina** Tamaulipas, Mexico 23°41′N 99°36′W
170 E1 **Avilés** Asturias, NW Spain 43°33′N 05°55′W
190 F4 **Avinurme** Ger. Awwinorm. Ida-Virumaa, NE Estonia 58°58′N 26°53′E
167 H2 **Avion** Nord-Pas-de-Calais, France 50°24′N 2°50′E
172 C5 **Avis** Portalegre, C Portugal 39°03′N 07°53′W
218 F5 **Avivim** Israel 33°05′N 35°28′E
Avlum see Aulum
277 I7 **Avoca** Victoria, SE Australia 37°09′S 143°34′E
74 F7 **Avoca** Iowa, C USA 41°27′N 95°20′W
64 C2 **Avoca** New York, USA 42°25′N 77°25′W
175 G13 **Avola** Sicilia, Italy, C Mediterranean Sea 36°54′N 15°08′E
64 C1 **Avon** New York, NE USA 42°53′N 77°41′W
74 E6 **Avon** South Dakota, N USA 43°00′N 98°03′W
161 I8 **Avon** ≈ United Kingdom
157 G12 **Avon** ≈ S England, United Kingdom
79 H9 **Avondale** Arizona, SW USA 33°25′N 112°20′W
160 G6 **Avonmouth** United Kingdom 51°31′N 2°42′W
69 M7 **Avon Park** Florida, SE USA 27°36′N 81°30′W
167 H9 **Avord** Centre, France 47°02′N 2°39′E
164 F3 **Avranches** Manche, N France 48°42′N 01°21′W
Avveel see Ivalojoki, Finland
281 J3 **Avuavu** var. Kolotambu. Guadalcanal, C Solomon Islands 09°52′S 160°25′E
165 H2 **Avure** ≈ N France
Avveel see Ivalo, Finland
Avvil see Ivalo
132 G8 **Awaaso** var. Awaso. SW Ghana 06°10′N 02°18′W
221 K4 **Awābī** var. Al 'Awābī. NE Oman 23°20′N 57°35′E
251 H5 **Awaji-shima** island SW Japan
218 G4 **Awaj, Nahr al** ≈ Syria
278 G5 **Awakino** Waikato, North Island, New Zealand 38°40′S 174°37′E
222 E9 **'Awālī** C Bahrain 26°07′N 50°33′E
163 G11 **Awans** Liège, E Belgium 50°39′N 05°30′E
278 F2 **Awanui** Northland, North Island, New Zealand 35°01′S 173°16′E
230 E8 **Awārān** Baluchistān, SW Pakistan 26°31′N 65°10′E
136 F6 **Awara Plain** plain NE Kenya
136 G4 **Āwarē** Sumalē, E Ethiopia 08°12′N 44°09′E
216 F4 **'Awāriḑ, Wādī** dry watercourse E Syria
278 B10 **Awarua Point** headland South Island, New Zealand 44°15′S 168°03′E
136 E5 **Āwasa** Southern Nationalities, S Ethiopia 06°54′N 38°26′E
136 F4 **Āwash** Āfar, NE Ethiopia 08°59′N 40°16′E
136 G5 **Āwash** var. Hawash. ≈ C Ethiopia
253 B10 **Awa-shima** island C Japan
Awaso see Awaaso
238 D4 **Awat** Xinjiang Uygur Zizhiqu, NW China 40°36′N 80°22′E
279 F8 **Awatere** ≈ South Island, New Zealand
128 C4 **Awbārī** SW Libya 26°35′N 12°46′E
128 B3 **Awbāri, Idhān** var. Edeyen d'Oubari. desert Algeria/Libya
136 G3 **Awdal** off. Gobolka Awdal. ⬖ N Somalia
134 I4 **Aweil** Northern Bahr el Ghazal, NW South Sudan 08°42′N 27°20′E
158 F2 **Awe, Loch** ◎ W Scotland, United Kingdom
133 K8 **Awka** Anambra, SW Nigeria 06°12′N 07°04′E
83 H3 **Awuna River** ≈ Alaska, USA
Awwinorm see Avinurme
Ax see Dax
Axarfjördhur see Öxarfjörður
168 G7 **Axat** Aude, S France 42°47′N 02°14′E
160 G8 **Axe** ≈ United Kingdom
163 D9 **Axel** Zeeland, SW Netherlands 51°16′N 03°55′E
295 I7 **Axel Heiberg Island** var. Axel Heiburg. island Nunavut, N Canada
Axel Heiburg see Axel Heiberg Island
132 G8 **Axim** S Ghana 04°53′N 02°14′W
186 G2 **Axiós** var. Vardar. ≈ Greece/FYR Macedonia see also Vardar
Axiós see Vardar
168 G7 **Ax-les-Thermes** Ariège, S France 42°43′N 01°49′E
160 G8 **Axminster** SW England, United Kingdom 50°47′N 03°00′W
251 H4 **Ayabe** Kyōto, Honshū, SW Japan 35°19′N 135°16′E
130 F3 **Ayachi, Jbel** ▲ C Morocco 32°30′N 05°00′W
116 H5 **Ayacucho** Buenos Aires, E Argentina 37°09′S 58°30′W
105 E8 **Ayacucho** Ayacucho, S Peru 13°10′S 74°15′W
105 E8 **Ayacucho** off. Departamento de Ayacucho. ⬖ department SW Peru
Ayacucho, Departamento de see Ayacucho
227 L5 **Ayagoz** var. Ayaguz, Kaz. Ayaköz; prev. Sergiopol. Vostochnyy Kazakhstan, E Kazakhstan 47°54′N 80°25′E
Ayaguz see Ayagoz
Ayakagytma see Oyoqog'itma
Ayakkuduk see Oyoqquduq
238 G6 **Ayakkum Hu** ◎ NW China
Ayaköz see Ayagoz
170 D8 **Ayamonte** Andalucía, S Spain 37°13′N 07°24′W
193 L7 **Ayan** Khabarovskiy Kray, E Russian Federation 56°27′N 138°09′E
214 F4 **Ayancık** Sinop, N Turkey 41°56′N 34°35′E
103 J5 **Ayanganna Mountain** ▲ C Guyana 05°21′N 59°54′W
133 K7 **Ayangba** Kogi, C Nigeria 07°36′N 07°10′E
193 M4 **Ayanka** Krasnoyarskiy Kray, E Russian Federation 63°42′N 167°31′E
102 B4 **Ayapel** Córdoba, NW Colombia 08°16′N 75°10′W
214 E5 **Ayaş** Ankara, N Turkey 40°02′N 32°21′E
105 G9 **Ayaviri** Puno, S Peru 14°53′S 70°35′W
Āybak see Aibak
229 H5 **Aydarko'l Ko'li** Rus. Ozero Aydarkul'. ◎ C Uzbekistan
Aydarkul', Ozero see Aydarko'l Ko'li
67 L6 **Ayden** North Carolina, SE USA 35°28′N 77°25′W
214 B7 **Aydın** var. Aïdin; anc. Tralles Aydin. Aydın, SW Turkey 37°51′N 27°51′E
214 B7 **Aydın** var. Aïdin. ⬖ province SW Turkey
214 E8 **Aydıncık** İçel, S Turkey 36°08′N 33°17′E
214 B6 **Aydın Dağları** ▲▲ W Turkey
238 G4 **Aydingkol Hu** ◎ NW China
187 M2 **Aydinli** İstanbul, NW Turkey 40°55′N 29°10′E
197 M3 **Aydyrlinskiy** Orenburgskaya Oblast', W Russian Federation 52°03′N 59°54′E
64 L2 **Ayer** Massachusetts, NE USA 42°33′N 71°35′W
171 I3 **Ayerbe** Aragón, NE Spain 42°16′N 00°41′W
Ayers Rock see Uluṟu
Ayeyarwady see Ayeyawady
Ayeyarwady see Irrawaddy
256 C6 **Ayeyawady** prev. Ayeyarwady var. Irrawaddy. ⬖ region SW Myanmar (Burma)
Ayiá see Agiá
Ayia Napa see Agía Nápa
Ayia Phyla see Agía Fýlaxis
Ayiásos/Ayiássos see Agiassós
Áyios Evstrátios see Ágios Efstrátios
Áyios Kírikos see Ágios Kírykos
Áyios Nikólaos see Ágios Nikólaos
Ayios Seryios see Yeniboğaziçi
136 D3 **Āykel** Āmara, N Ethiopia 12°33′N 37°01′E
193 I5 **Aykhal** Respublika Sakha (Yakutiya), NE Russian Federation 66°07′N 110°25′E
62 E3 **Aylen Lake** ◎ Ontario, SE Canada
161 J5 **Aylesbury** SE England, United Kingdom 51°50′N 00°50′W
170 G4 **Ayllón** Castilla y León, N Spain 41°25′N 03°23′W
62 C6 **Aylmer** Ontario, S Canada 42°46′N 80°57′W
62 F3 **Aylmer** Québec, SE Canada 45°23′N 75°51′W
63 J2 **Aylmer, Lac** ◎ Québec, SE Canada
54 G3 **Aylmer Lake** ◎ Northwest Territories, NW Canada
161 M3 **Aylsham** E England, United Kingdom
173 J6 **Ayna** Castilla-La Manacha, Spain 38°33′N 2°05′W
218 F3 **Aynab** Lebanon 33°46′N 35°33′E
227 K6 **Aynabulak** Kaz. Aynabulaq. Almaty, SE Kazakhstan 44°37′N 77°59′E
Aynabulaq see Aynabulak
216 E1 **'Ayn al 'Arab** Kurd. Kobani. Ḩalab, N Syria 36°55′N 38°21′E
Aynayn see 'Aynīn
217 K8 **'Ayn Ḩamūd** Dhī Qār, S Iraq 30°51′N 45°37′E
229 I6 **Ayní** prev. Varzimanor Ayni. W Tajikistan 39°24′N 68°30′E
220 E6 **'Aynīn** var. Aynayn. spring/well SW Saudi Arabia 20°52′N 41°41′E
67 K8 **Aynor** South Carolina, SE USA 33°59′N 79°11′W
217 H5 **'Ayn Zāzūh** Al Anbār, C Iraq 33°29′N 42°34′E
232 G6 **Ayodhya** Uttar Pradesh, N India 26°47′N 82°12′E
86 D5 **Ayo el Chico** Jalisco, Mexico 20°32′N 102°21′W
193 L3 **Ayon, Ostrov** island NE Russian Federation
171 I6 **Ayora** Valenciana, E Spain 39°04′N 01°04′W
133 H4 **Ayorou** Tillabéri, W Niger 14°45′N 00°54′E
134 B6 **Ayos** Centre, S Cameroon 03°53′N 12°31′E
86 B5 **Ayotitlán** Jalisco, Mexico 20°13′N 103°56′W
87 H8 **Ayotzintepec** Oaxaca, Mexico 17°40′N 96°08′W
130 E6 **'Ayoûn 'Abd el Mâlek** well N Mauritania
132 D3 **'Ayoûn el 'Atroûs** var. Aïoun el Atrous, Aïoun el Atroûss. Hodh el Gharbi, SE Mauritania 16°38′N 09°36′W
158 G4 **Ayr** W Scotland, United Kingdom 55°28′N 04°38′W
158 G4 **Ayr** ≈ United Kingdom
156 E7 **Ayr** ≈ W Scotland, United Kingdom
158 G7 **Ayre, Point of** headland N Isle of Man
160 F1 **Ayr, Point of** headland United Kingdom 53°21′N 3°19′W
156 E7 **Ayrshire** cultural region SW Scotland, United Kingdom
Aysen see Aisén
159 J7 **Aysgarth** United Kingdom 54°17′N 1°59′W
136 G3 **Āysha** Sumalē, E Ethiopia 10°36′N 42°31′E
226 F6 **Ayteke Bi** Kaz. Zhangaqazaly; prev. Novokazalinsk. Kzylorda, SW Kazakhstan 45°53′N 62°10′E
228 G4 **Aytim** Navoiy Viloyati, N Uzbekistan 42°15′N 63°25′E
275 J3 **Ayton** Queensland, NE Australia 15°54′S 145°19′E
159 L7 **Ayton** United Kingdom 54°14′N 0°29′W
185 L6 **Aytos** Burgas, E Bulgaria 42°43′N 27°14′E
166 D10 **Aytré** Poitou-Charentes, France 46°08′N 1°06′W
261 H3 **Ayu, Kepulauan** island group E Indonesia
257 I8 **A Yun Pa** prev. Cheo Reo. Gia Lai, S Vietnam 13°19′N 108°27′E
259 L5 **Ayu, Tanjung** headland Borneo, N Indonesia 0°25′N 117°34′E
86 F8 **Ayutlá** var. Ayutla de los Libres. Guerrero, S Mexico 16°51′N 99°16′W
86 B5 **Ayutla** Jalisco, C Mexico 20°07′N 104°18′W
Ayutla de los Libres see Ayutlá
257 E8 **Ayutthaya** var. Phra Nakhon Si Ayutthaya. Phra Nakhon Si Ayutthaya, C Thailand 14°20′N 100°35′E
187 K4 **Ayvacik** Çanakkale, Turkey 39°36′N 26°24′E
214 A5 **Ayvalık** Balıkesir, W Turkey 39°18′N 26°42′E
163 G11 **Aywaille** Liège, E Belgium 50°28′N 05°40′E
136 I1 **'Aywat aş Şay'ar, Wādī** seasonal river N Yemen
Azaffal see Azeffâl
171 J5 **Azahar, Costa del** coastal region E Spain
171 I4 **Azaila** Aragón, NE Spain 41°17′N 00°20′W
172 B4 **Azambuja** Lisboa, C Portugal 39°04′N 08°52′W
233 H7 **Āzamgarh** Uttar Pradesh, N India 26°03′N 83°10′E
132 G3 **Azaouâd** desert C Mali
133 I3 **Azaouagh, Vallée de l'** var. Azaouak. ≈ W Niger
Azaouak see Azaouagh, Vallée de l'
113 H6 **Azara** Misiones, NE Argentina 28°03′S 55°42′W
Azaran see Hashtrūd
Azărbaycan/Azărbaycan Respublikası see Azerbaijan
Āžarbāyjān-e Bākhtarī see Āžarbāyjān-e Gharbī
222 D3 **Āžarbāyjān-e Gharbī** off. Ostān-e Āžarbāyjān-e Gharbī, Eng. West Azerbaijan; prev. Āžarbāyjān-e Bākhtarī. NW Iran
Āžarbāyjān-e Gharbī, Ostān-e see Āžarbāyjān-e Gharbī
Āžarbāyjān-e Khāvarī see Āžarbāyjān-e Sharqī
222 D2 **Āžarbāyjān-e Sharqī** off. Ostān-e Azarbāyjān-e Sharqī, Eng. East Azerbaijan; prev. Āzarbāyjān-e Khāvarī. ⬖ province NW Iran
Āžarbāyjān-e Sharqī, Ostān-e see Āžarbāyjān-e Sharqī
133 L5 **Azare** Bauchi, N Nigeria 11°41′N 10°09′E
191 H11 **Azarychy** Rus. Ozarichi. Homyel'skaya Voblasts', SE Belarus 52°31′N 29°19′E
227 H5 **Azat, Gory** hill C Kazakhstan
166 F9 **Azay-le-Ferron** Centre, France 46°51′N 1°04′E
164 G5 **Azay-le-Rideau** Indre-et-Loire, C France 47°16′N 00°25′E
216 D2 **A'zāz** Ḩalab, NW Syria 36°35′N 37°03′E
171 M8 **Azazga** Algeria
114 D9 **Azcuénaga** Buenos Aires, Argentina 34°23′S 59°21′W
132 C1 **Azeffâl** var. Azaffal. desert Mauritania/Western Sahara
171 M8 **Azeffoun** Algeria
215 M5 **Azerbaijan** off. Azerbaijani Republic, Az. Azărbaycan, Azărbaycan Respublikası; prev. Azerbaijan SSR. ◆ republic SE Asia
Azerbaijani Republic see Azerbaijan
Azerbaijan SSR see Azerbaijan
115 J1 **Azevedo Sodré** prev. Azevedro Sodre. Rio Grande do Sul, Brazil 30°04′S 54°36′W
Azevedro Sodre see Azevedo Sodré
130 F3 **Azilal** C Morocco 31°58′N 06°53′W
Azimabad see Patna
63 J3 **Aziscohos Lake** ▣ Maine, NE USA
Azizbekov see Vayk'
Azizie see Telish
Aziziya see Al 'Azīzīyah
197 K2 **Aznakayevo** Respublika Tatarstan, W Russian Federation 54°55′N 53°15′E
172 E7 **Aznalcóllar** Andalucía, Spain 37°31′N 6°16′W
104 C3 **Azogues** Cañar, S Ecuador 02°44′S 78°48′W
86 G8 **Azompa** Oaxaca, Mexico 17°06′N 96°47′W
172 B9 **Azores** var. Açores, Ilhas dos Açores, Port. Arquipélago dos Açores. island group Portugal, NE Atlantic Ocean
290 G4 **Azores-Biscay Rise** undersea feature E Atlantic Ocean 19°00′W 42°40′N
Azotos/Azotus see Ashdod
134 F3 **Azoum, Bahr** seasonal river SE Chad
196 G7 **Azov** Rostovskaya Oblast', SW Russian Federation 47°07′N 39°26′E
189 L8 **Azov, Sea of** Rus. Azovskoye More, Ukr. Azovs'ke More. sea NE Black Sea
Azovs'ke More/Azovskoye More see Azov, Sea of
195 M4 **Azovy** Yamalo-Nenetskiy Avtonomnyy Okrug, Russian Federation
86 F8 **Azoyú** Guerrero, Mexico 16°43′N 98°44′W
168 B6 **Azpeitia** Pais Vasco, Spain 43°11′N 2°16′W
216 D7 **Azraq, Wāḩat al** oasis N Jordan
Āzro see Āzrow
130 F3 **Azrou** C Morocco 33°30′N 05°12′W
231 H4 **Āzrow** var. Āzro. Lōgar, E Afghanistan 34°11′N 69°39′E
79 K5 **Aztec** New Mexico, SW USA 36°49′N 107°59′W
79 I8 **Aztec Peak** ▲ Arizona, SW USA 33°48′N 110°54′W
91 H6 **Azua** var. Azua de Compostela. S Dominican Republic 18°29′N 70°44′W
Azua de Compostela see Azua
170 E7 **Azuaga** Extremadura, W Spain 38°16′N 05°40′W
168 C10 **Azuara** Aragón, Spain 41°15′N 0°52′W
104 C3 **Azuay** ⬖ province W Ecuador
250 B6 **Azuchi-Ō-shima** island SW Japan
170 G6 **Azuer** ≈ C Spain
89 K10 **Azuero, Península de** peninsula S Panama
105 H14 **Azufre, Volcán** var. Volcán Lastarria. ⛰ N Chile 25°16′S 68°35′W
112 B5 **Azufre, Volcn** ⛰ Chile 25°12′S 68°30′W
188 D8 **Azuga** Prahova, SE Romania 45°27′N 25°34′E
116 G5 **Azul** Buenos Aires, E Argentina 36°46′S 59°50′W
116 C1 **Azul, Cerro** ▲ NW Argentina 28°28′S 68°43′W
104 D6 **Azul, Cordillera** ▲▲ C Peru
253 C11 **Azuma-san** ▲ Honshū, C Japan 37°44′N 140°05′E
165 L9 **Azur, Côte d'** coastal region SE France
285 N2 **Azur Lagoon** ◎ Kiritimati, E Kiribati
'Azza see Gaza
Az Zāb al Kabīr see Great Zab
218 G3 **Az Zabdānī** var. Zabadani. Rīf Dimashq, W Syria 33°45′N 36°07′E
221 J5 **Aẓ Ẓāhirah** desert NW Oman
221 H3 **Aẓ Ẓahrān** Eng. Dhahran. Ash Sharqīyah, NE Saudi Arabia 26°18′N 50°02′E
221 H3 **Aẓ Ẓahrān al Khubar** var. Dhahran Al Khobar. ✈ Ash Sharqīyah, NE Saudi Arabia 26°28′N 49°42′E
218 J5 **Az Zalal** Syria 32°55′N 37°20′E
129 I3 **Az Zaqāzīq** var. Zagazig. N Egypt 30°36′N 31°32′E
219 G8 **Az Zarqā'** Jordan 32°05′N 36°06′E
219 G8 **Az Zarqā'** var. Zarqa. Az Zarqā', NW Jordan 32°04′N 36°06′E
216 C7 **Az Zarqā'** off. Muḩāfazat az Zarqā', var. Zarqa. ⬖ governorate N Jordan
128 C2 **Az Zāwiyah** var. Zawia. NW Libya 32°45′N 12°44′E
220 F8 **Az Zaydīyah** W Yemen 15°20′N 43°03′E
131 H6 **Azzel Matti, Sebkha** var. Sebkra Azz el Matti. salt flat C Algeria
220 G3 **Az Zilfī** Ar Riyāḑ, N Saudi Arabia 26°17′N 44°48′E
217 M8 **Az Zubayr** var. Al Zubair. Al Başrah, SE Iraq 30°24′N 47°45′E
Az Zuqur see Jabal Zuqar, Jazīrat

B

280 E8 **Ba** prev. Mba. Viti Levu, W Fiji 17°35′S 177°40′E
Ba see Da Răng, Sông
260 E9 **Baa** Pulau Rote, C Indonesia 10°44′S 123°06′E
280 A8 **Baaba, Île** island Îles Belep, W New Caledonia
218 F3 **Baabda** Lebanon 33°50′N 35°32′E
218 F3 **Baabdât** Lebanon 33°53′N 35°40′E
218 G3 **Baalbek** var. Ba'labakk; anc. Heliopolis. E Lebanon 34°00′N 36°15′E
218 G3 **BaaLebanonek** Lebanon 34°00′N 36°12′E
174 C3 **Baar** Zug, N Switzerland 47°12′N 08°32′E
136 F7 **Baardheere** var. Bardere, It. Bardera. Gedo, SW Somalia 02°13′N 42°19′E
Baargaal see Bargaal
163 F8 **Baarle-Hertog** Antwerpen, N Belgium 51°26′N 04°56′E
163 F8 **Baarle-Nassau** Noord-Brabant, S Netherlands 51°27′N 04°56′E
162 F6 **Baarn** Utrecht, C Netherlands 52°13′N 05°16′E
239 J2 **Baatsagaan** var. Bayansayr. Bayanhongor, C Mongolia 45°36′N 99°27′E
184 G8 **Baba** var. Buševa, Gk. Varnoús. ▲▲ FYR Macedonia/Greece
132 C3 **Bababé** Brakna, W Mauritania 16°22′N 13°57′W
214 D4 **Baba Burnu** headland NW Turkey 41°18′N 31°24′E
188 G9 **Babadag** Tulcea, SE Romania 44°53′N 28°47′E
215 M4 **Babadağ Dağı** ▲ NE Azerbaijan 41°02′N 48°04′E
228 E8 **Babadaýhan** Rus. Babadaykhan; prev. Kirovsk. Ahal Welaýaty, C Turkmenistan 37°39′N 60°17′E
Babadaykhan see Babadaýhan
228 D7 **Babadurmaz** Ahal Welaýaty, C Turkmenistan 37°39′N 59°03′E
214 B4 **Babaeski** Kırklareli, NW Turkey 41°26′N 27°06′E
217 J3 **Bāba Gurgur** Kirkūk, N Iraq 35°34′N 44°18′E
104 B2 **Babahoyo** prev. Bodegas. Los Ríos, C Ecuador 01°53′S 79°31′W
230 F3 **Bābā, Kūh-e** ▲▲ C Afghanistan
260 C5 **Babana** Sulawesi, C Indonesia 02°03′S 119°13′E
Babao see Qilian
261 H8 **Babar, Kepulauan** island group E Indonesia
261 H8 **Babar, Pulau** island Kepulauan Babar, E Indonesia
Bābāsar Pass see Babusar Pass
Babashy, Gory see Babaşy
228 B5 **Babaşy** Rus. Gory Babashy. ▲▲ W Turkmenistan
259 I8 **Babat** Jawa, S Indonesia 07°08′S 112°08′E
258 F6 **Babat** Sumatera, W Indonesia 02°45′S 104°01′E
Babatag, Khrebet see Bobotogh, Qatorkŭhi
137 D9 **Babati** Manyara, NE Tanzania 04°12′S 35°45′E
194 E8 **Babayevo** Vologodskaya Oblast', NW Russian Federation 59°23′N 35°52′E
197 I9 **Babayurt** Respublika Dagestan, SW Russian Federation 43°38′N 46°49′E
77 H3 **Babb** Montana, NW USA 48°51′N 113°26′W
160 F9 **Babbacombe Bay** bay United Kingdom
74 H2 **Babbitt** Minnesota, N USA 47°42′N 91°56′W
282 C9 **Babeldaob** var. Babeldaop, Babelthuap. island N Palau
Babeldaop see Babeldaob
212 G10 **Bab el Mandeb** strait Gulf of Aden/Red Sea
Babelthuap see Babeldaob
181 F12 **Babenhausen** Hessen, Germany 49°58′E 8°57′N
183 G9 **Babia Góra** var. Babia Hora. ▲ Poland/Slovakia 49°33′N 19°32′E
Babia Hora see Babia Góra
Babian Jiang see Black River
Babichi see Babichy
191 H11 **Babichy** Rus. Babichi. Homyel'skaya Voblasts', SE Belarus 52°17′N 30°00′E
184 E3 **Babina Greda** Vukovar-Srijem, E Croatia 45°09′N 18°33′E
56 C3 **Babine Lake** ◎ British Columbia, SW Canada
261 I5 **Babo** Papua, E Indonesia 02°29′S 133°30′E
222 G3 **Bābol** var. Babul, Balfrush, Barfrush; prev. Barfurush. Māzandarān, N Iran 36°34′N 52°39′E
222 F3 **Bābolsar** var. Babulsar; prev. Meshed-i-Sar. Māzandarān, N Iran 36°41′N 52°39′E
140 D8 **Baboon Point** cape Western Cape, South Africa
79 I10 **Baboquivari Peak** ▲ Arizona, SW USA 31°46′N 111°36′W
134 C5 **Baboua** Nana-Mambéré, W Central African Republic 05°46′N 14°47′E
191 H10 **Babruysk** Rus. Bobruysk. Mahilyowskaya Voblasts', E Belarus 53°07′N 29°13′E
Babu see Hezhou
Babul see Bābol
Babulsar see Bābolsar
185 H8 **Babuna** ▲▲ C FYR Macedonia
185 H8 **Babuna** ≈ C FYR Macedonia
231 J3 **Babusar Pass** prev. Bābāsar Pass. pass India/Pakistan
230 D4 **Bābūs, Dasht-e** Pash. Bebas, Dasht-i. ▲▲ W Afghanistan
193 H8 **Babushkin** Respublika Buryatiya, S Russian Federation 51°35′N 105°49′E
263 L3 **Babuyan Channel** channel N Philippines
263 K3 **Babuyan Islands** island N Philippines
217 J6 **Babylon** site of ancient city C Iraq
184 F3 **Bač** Ger. Batsch. Vojvodina, NW Serbia 45°24′N 19°17′E
108 E4 **Bacabal** Maranhão, E Brazil 04°15′S 44°45′W
84 F3 **Bacadéhuachi** Sonora, Mexico 29°48′N 109°08′W
87 M7 **Bacalar** Quintana Roo, SE Mexico 18°38′N 88°17′W
87 N7 **Bacalar Chico, Boca** strait SE Mexico
87 M7 **Bacalar, Laguna** ◎ SE Mexico
260 F4 **Bacan, Kepulauan** island group E Indonesia
84 F4 **Bacanora** Sonora, NW Mexico 28°59′N 109°24′W
260 G4 **Bacan, Pulau** prev. Batjan. island Maluku, E Indonesia
188 E7 **Bacău** Hung. Bákó. Bacău, NE Romania 46°36′N 26°56′E
188 E7 **Bacău** ⬖ county E Romania
Băc Bô, Vinh see Tonkin, Gulf of
256 H4 **Băc Can** var. Bach Thong. Băc Thai, N Vietnam 22°07′N 105°50′E
165 K3 **Baccarat** Meurthe-et-Moselle, NE France 48°27′N 06°46′E
277 I7 **Bacchus Marsh** Victoria, SE Australia 37°41′S 144°30′E
84 F3 **Bacerac** Sonora, NW Mexico 30°27′N 108°55′W
188 F7 **Băceşti** Vaslui, E Romania 46°50′N 27°14′E
256 H4 **Băc Giang** Ha Băc, N Vietnam 21°17′N 106°12′E
102 D3 **Bachaquero** Zulia, NW Venezuela 09°57′N 71°09′W
53 I3 **Bache Peninsula** Peninsula Nunavut, N Canada
Bacher see Pohorje
191 H8 **Bacheykava** Rus. Bocheykovo. Vitsyebskaya Voblasts', N Belarus 55°01′N 29°09′E
84 G4 **Bachíniva** Chihuahua, N Mexico 28°41′N 107°13′W
Bach Quy, Đao see Passu Keah
Bach Thong see Băc Can
238 D5 **Bachu** Xinjiang Uygur Zizhiqu, NW China 39°46′N 78°30′E
55 H3 **Back** ≈ Nunavut, N Canada
184 F3 **Bačka Palanka** prev. Palanka. Serbia, NW Serbia 44°22′N 20°57′E
184 F2 **Bačka Topola** Hung. Topolya; prev. Hung. Bácstopolya. Vojvodina, N Serbia 45°48′N 19°39′E
54 D3 **Backbone Range** ▲▲ Northwest Territories, NW Canada
155 F9 **Bäckefors** Västra Götaland, S Sweden 58°49′N 12°07′E
Bäckermühle Schulzenmühle see Żywiec
155 G8 **Bäckhammar** Värmland, C Sweden 59°09′N 14°13′E
184 F3 **Bački Petrovac** Hung. Petrőcz; prev. Petrovac, Petrovácz. Vojvodina, NW Serbia 45°22′N 19°34′E
181 G14 **Backnang** Baden-Württemberg, SW Germany 48°57′N 09°26′E
53 H9 **Back River** ≈ Nunavut, NE Canada
257 H10 **Bac Liêu** var. Vinh Loi. Minh Hai, S Vietnam 09°17′N 105°44′E
256 H4 **Băc Ninh** Ha Băc, N Vietnam 21°10′N 106°04′E
84 F3 **Bacoachi** Sonora, NW Mexico 30°36′N 110°00′W
263 L7 **Bacolod** off. Bacolod City. Negros, C Philippines 10°43′N 122°58′E
Bacolod City see Bacolod
263 K6 **Baco, Mount** ▲ Mindoro, N Philippines 12°50′N 121°08′E
183 G13 **Bácsalmás** Bács-Kiskun, S Hungary 46°07′N 19°20′E
Bácsjózseffalva see Žednik
183 G12 **Bács-Kiskun** off. Bács-Kiskun Megye. ⬖ county S Hungary
Bács-Kiskun Megye see Bács-Kiskun
Bácsszenttamás see Srbobran
Bácstopolya see Bačka Topola
Bactra see Balkh
159 J9 **Bacup** United Kingdom 53°42′N 2°11′W
180 I6 **Bad** Niedersachsen, Germany 52°03′E 10°22′N
Bada see Xilin
235 D8 **Badagara** var. Vadakara. Kerala, SW India 11°36′N 75°34′E see also Vadakara
179 G13 **Bad Aibling** Bayern, SE Germany 47°52′N 12°00′E
244 A2 **Badain Jaran Shamo** desert N China
170 D6 **Badajoz** anc. Pax Augusta. Extremadura, W Spain 38°53′N 06°58′W
170 E6 **Badajoz** ⬖ province Extremadura, W Spain
231 H2 **Badakhshān** ⬖ province NE Afghanistan
171 K3 **Badalona** anc. Baetulo. Cataluña, E Spain 41°27′N 02°15′E
169 N5 **Badalucco** Liguria, NW Italy 43°54′N 07°50′E
234 I3 **Bādāmpāhārh** var. Bādāmapāhārh. Odisha, E India 22°04′N 86°06′E
245 M1 **Badaohao** Liaoning, China 41°29′N 121°35′E
258 G4 **Badas, Kepulauan** island group W Indonesia
177 J3 **Bad Aussee** Salzburg, E Austria 47°35′N 13°44′E
72 H7 **Bad Axe** Michigan, N USA 43°48′N 83°00′W
180 C6 **Bäd Bentheim** Niedersachsen, Germany 52°19′E 7°10′N
180 E5 **Badbergen** Niedersachsen, Germany 52°38′E 7°59′N
181 E9 **Bad Berleburg** Nordrhein-Westfalen, W Germany 51°03′N 08°24′E
181 C11 **Bad Bertrich** Rheinland-Pfalz, Germany 50°04′E 7°02′N
181 J10 **Bad Blankenburg** Thüringen, C Germany 50°43′N 11°19′E
Bad Borseck see Borsec
180 H2 **Bad Bramstedt** Schleswig-Holstein, Germany 53°55′E 9°53′N
181 E11 **Bad Camberg** Hessen, W Germany 50°18′N 08°15′E
59 K8 **Baddeck** Nova Scotica, SE Canada
178 G4 **Bad Doberan** Mecklenburg-Vorpommern, N Germany 54°06′N 11°55′E
180 F7 **Bad Driburg** Nordrhein-Westfalen, Germany 51°44′E 9°01′N
179 H8 **Bad Düben** Sachsen, E Germany 51°35′N 12°34′E
181 E13 **Bad Dürkheim** Rheinland-Pfalz, Germany 49°28′E 8°10′N
243 N7 **Bade** prev. Pate. N Taiwan 24°57′N 121°17′E
181 D11 **Bad Ems** Rheinland-Pfalz, Germany 50°20′E 7°43′N
177 L2 **Baden** var. Baden bei Wien; anc. Aquae Panoniae, Thermae Pannonicae. Niederösterreich, NE Austria 48°01′N 16°14′E
176 D3 **Baden** Aargau, N Switzerland 47°28′N 08°19′E
179 C12 **Baden-Baden** anc. Aurelia Aquensis. Baden-Württemberg, SW Germany 48°46′N 08°14′E
Baden bei Wien see Baden
181 F10 **Bad Endbach** Hessen, Germany 50°45′E 8°30′N
179 D12 **Baden-Württemberg** Fr. Bade-Wurtemberg. ⬖ state SW Germany
184 A3 **Baderna** Istra, NW Croatia 45°12′N 13°45′E
180 J7 **Badersleben** Sachsen-Anhalt, Germany 51°59′E 10°54′N
180 E6 **Bad Essen** Niedersachsen, Germany 52°19′E 8°20′N
Bade-Wurtemberg see Baden-Württemberg
181 G14 **Bad Fredrichshall** Baden-Württemberg, S Germany 49°13′N 09°15′E
178 I6 **Bad Freienwalde** Brandenburg, NE Germany 52°47′N 14°04′E
180 H7 **Bad Gandersheim** Niedersachsen, Germany 51°52′E 10°02′N
177 I4 **Badgastein** var. Gastein. Salzburg, NW Austria 47°07′N 13°09′E
80 E6 **Badger** California, USA 36°38′N 119°01′W
Badger State see Wisconsin
230 E3 **Bādghīs** ⬖ province NW Afghanistan
181 C10 **Bad Godesberg** Nordrhein-Westfalen, Germany 50°41′E 7°09′N
177 J2 **Bad Hall** Oberösterreich, N Austria 48°03′N 14°13′E
136 I3 **Badhan** Maakhir, N Somalia 10°42′N 48°20′E
180 I7 **Bad Harzburg** Niedersachsen, C Germany 51°52′N 10°34′E
181 H9 **Bad Hersfeld** Hessen, C Germany 50°52′N 09°42′E
162 E6 **Badhoevedorp** Noord-Holland, C Netherlands 52°21′N 04°46′E
177 I4 **Bad Hofgastein** Salzburg, NW Austria 47°11′N 13°07′E
Bad Homburg see Bad Homburg vor der Höhe
181 F11 **Bad Homburg vor der Höhe** var. Bad Homburg. Hessen, W Germany 50°14′N 08°37′E
181 C10 **Bad Honnef** Nordrhein-Westfalen, W Germany 50°39′N 07°13′E
181 C10 **Bad Honnef am Rhein** Nordrhein-Westfalen, Germany 50°38′E 7°14′N
180 E6 **Bad Iburg** Niedersachsen, Germany 52°09′E 8°03′N
230 G10 **Badīn** Sind, SE Pakistan 24°38′N 68°53′E
67 J6 **Badin Lake** ▣ North Carolina, SE USA
84 G7 **Badiraguato** Sinaloa, C Mexico 25°21′N 107°31′W
177 J3 **Bad Ischl** Oberösterreich, N Austria 47°43′N 13°36′E
Badjawa see Bajawa
Badje-Sohppar see Övre Soppero
181 H11 **Bad Kissingen** Bayern, SE Germany 50°12′N 10°04′E
181 F12 **Bad König** Hessen, Germany 49°45′E 9°01′N
Bad Königswart see Lázně Kynžvart
181 D12 **Bad Kreuznach** Rheinland-Pfalz, SW Germany 49°51′N 07°52′E
179 C13 **Bad Krozingen** Baden-Württemberg, SW Germany 47°55′N 07°43′E
181 E9 **Bad Laasphe** Nordrhein-Westfalen, W Germany 50°57′N 08°24′E
180 E6 **Bad Laer** Niedersachsen, Germany 52°06′E 8°05′N
74 B5 **Badlands** physical region North Dakota/South Dakota, N USA
181 I9 **Bad Langensalza** Thüringen, C Germany 51°05′N 10°40′E
180 I7 **Bad Lauterberg** Niedersachsen, Germany 51°39′E 10°28′N
177 J2 **Bad Leonfelden** Oberösterreich, N Austria 48°31′N 14°17′E
181 E10 **Bad Marienberg** Rheinland-Pfalz, Germany 50°39′E 7°57′N
180 F7 **Bad Meinberg** Nordrhein-Westfalen, Germany 51°54′E 8°59′N
181 H13 **Bad Mergentheim** Baden-Württemberg, SW Germany 49°30′N 09°46′E
180 G6 **Bad Münder** Niedersachsen, Germany 52°12′E 9°28′N
181 D12 **Bad Münster-Ebernburg** Rheinland-Pfalz, Germany 49°49′E 7°51′N
181 C10 **Bad Münstereifel** Nordrhein-Westfalen, Germany 50°33′E 6°46′N
181 F11 **Bad Nauheim** Hessen, W Germany 50°22′N 08°45′E
181 C10 **Bad Neuenahr-Ahrweiler** Rheinland-Pfalz, W Germany 50°33′N 07°07′E
Bad Neustadt see Bad Neustadt an der Saale
181 I11 **Bad Neustadt an der Saale** var. Bad Neustadt. Berlin, C Germany 50°21′N 10°13′E
Badnur see Betūl
261 L7 **Bado** Papua, E Indonesia 07°06′S 139°33′E
180 F6 **Bad Oeynhausen** Nordrhein-Westfalen, NW Germany 52°12′N 08°48′E
180 I2 **Bad Oldesloe** Schleswig-Holstein, N Germany 53°49′N 10°22′E
242 F2 **Badong** Hubei, China 31°01′N 110°14′E
167 M6 **Badonviller** Lorraine, France 48°30′N 6°54′E
133 H7 **Badou** C Togo 07°37′N 00°37′E
141 K4 **Badplaas** Mpumalanga, South Africa 25°57′S 30°34′E
Badplaas see eManzana
Bad Polzin see Połczyn-Zdrój
180 G7 **Bad Pyrmont** Niedersachsen, C Germany 51°58′N 09°16′E
177 L4 **Bad Radkersburg** Steiermark, SE Austria 46°40′N 16°02′E
217 K5 **Badrah** Wāsiţ, E Iraq 33°06′N 45°58′E
Badrah see Tarialan
181 G14 **Bad Rappenau** Baden-Württemberg, Germany 49°14′E 9°06′N
180 G5 **Bad Rehburg** Niedersachsen, Germany 52°26′E 9°13′N
179 H13 **Bad Reichenhall** Bayern, SE Germany 47°43′N 12°52′E
220 D4 **Badr Ḩunayn** Al Madīnah, W Saudi Arabia 23°46′N 38°45′E
232 F4 **Badrīnāth** ▲ N India 30°44′N 79°29′E
74 C5 **Bad River** ≈ South Dakota, N USA
72 B4 **Bad River** ≈ Wisconsin, N USA
181 D11 **Bad Salzig** Rheinland-Pfalz, Germany 50°12′E 7°38′N
181 G10 **Bad Salzschlirf** Hessen, Germany 50°37′E 9°30′N
180 F6 **Bad Salzuflen** Nordrhein-Westfalen, NW Germany 52°06′N 08°45′E
181 I10 **Bad Salzungen** Thüringen, C Germany 50°48′N 10°15′E
177 K4 **Bad Sankt Leonhard im Lavanttal** Kärnten, S Austria 46°55′N 14°51′E
181 E8 **Bad Sassendorf** Nordrhein-Westfalen, Germany 51°35′E 8°10′N
180 I2 **Bad Schwartau** Schleswig-Holstein, N Germany 53°55′N 10°42′E
180 I2 **Bad Segeberg** Schleswig-Holstein, Germany 53°56′E 10°19′N
181 F11 **Bad Soden** Hessen, Germany 50°08′E 8°30′N
181 G11 **Bad Soden-Salmünster** Hessen, Germany 50°17′E 9°22′N
181 J9 **Bad Tennstedt** Thüringen, Germany 51°09′E 10°50′N
179 G13 **Bad Tölz** Bayern, SE Germany 47°44′N 11°34′E
275 I1 **Badu Island** island Queensland, NE Australia
235 G11 **Badulla** Uva Province, C Sri Lanka 06°59′N 81°03′E
181 F11 **Bad Vilbel** Hessen, Germany 50°11′E 8°45′N
177 L2 **Bad Vöslau** Niederösterreich, NE Austria 47°58′N 16°13′E
179 E13 **Bad Waldsee** Baden-Württemberg, S Germany 47°54′N 09°44′E
81 H8 **Badwater Basin** depression California, W USA
52 G2 **Bad Weather Cape** headland Nunavut, N Canada
181 G9 **Bad Wildungen** Hessen, Germany 51°07′E 9°07′N
181 I13 **Bad Windsheim** Bayern, C Germany 49°30′N 10°25′E
179 F13 **Bad Wörishofen** Bayern, S Germany 48°00′N 10°36′E
180 E4 **Bad Zwischenahn** Niedersachsen, NW Germany 53°10′N 08°01′E
170 F8 **Baena** Andalucía, S Spain 36°37′N 04°20′W
248 B4 **Baengnyong-do** prev. Paengnyŏng. island NW South Korea
110 F6 **Baependi** Minas Gerais, Brazil 21°57′S 44°53′W
181 B9 **Baesweiler** Nordrhein-Westfalen, Germany 50°54′E 6°11′N
Baeterrae/Baeterrae Septimanorum see Béziers
Baetic Cordillera/Baetic Mountains see Béticos, Sistemas
Baetulo see Badalona
104 C2 **Baeza** Napo, NE Ecuador 0°30′S 77°52′W
170 G7 **Baeza** Andalucía, S Spain 38°00′N 03°28′W
134 A5 **Bafang** Ouest, W Cameroon 05°10′N 10°11′E
132 B5 **Bafatá** C Guinea-Bissau 12°09′N 14°38′W
231 I3 **Baffa** Khyber Pakhtunkhwa, NW Pakistan 34°30′N 73°18′E
295 J8 **Baffin Basin** undersea feature N Labrador Sea
295 J8 **Baffin Bay** bay Canada/Greenland
70 G9 **Baffin Bay** inlet Texas, SW USA
295 J9 **Baffin Island** island Nunavut, NE Canada
134 B5 **Bafia** Centre, C Cameroon 04°49′N 11°14′E
133 H6 **Bafilo** NE Togo 09°23′N 01°20′E
132 D5 **Bafing** ≈ W Africa
132 D4 **Bafoulabé** Kayes, W Mali 13°43′N 10°49′W
134 A5 **Bafoussam** Ouest, W Cameroon 05°31′N 10°25′E
223 H6 **Bāfq** Yazd, C Iran 31°35′N 55°21′E
214 F4 **Bafra** Samsun, N Turkey 41°34′N 35°56′E
214 G4 **Bafra Burnu** headland N Turkey 41°42′N 36°02′E
223 H7 **Bāft** Kermān, S Iran 29°12′N 56°36′E
134 I7 **Bafwabalinga** Orientale, NE Dem. Rep. Congo 0°52′N 26°55′E
134 H7 **Bafwaboli** Orientale, NE Dem. Rep. Congo 0°36′N 26°08′E
134 I7 **Bafwasende** Orientale, NE Dem. Rep. Congo 01°00′N 27°09′E
168 G8 **Bagà** Cataluña, Spain 42°15′N 1°52′E
88 G8 **Bagaces** Guanacaste, NW Costa Rica 10°31′N 85°18′W
233 H6 **Bagaha** Bihār, N India 27°08′N 84°04′E
234 D6 **Bāgalkot** Karnātaka, W India 16°11′N 75°42′E
137 E10 **Bagamoyo** Pwani, E Tanzania 06°26′S 38°55′E
Bagan Datok see Bagan Datuk
258 D3 **Bagan Datuk** var. Bagan Datok. Perak, Peninsular Malaysia 03°58′N 100°47′E
263 N9 **Baganga** Mindanao, S Philippines 07°31′N 126°34′E
258 D4 **Bagansiapiapi** var. Pasirpangarayan. Sumatera, W Indonesia 02°06′N 100°52′E
Bagaria see Bagheria
133 J4 **Bagaroua** Tahoua, W Niger 14°34′N 04°24′E
135 E9 **Bagata** Bandundu, W Dem. Rep. Congo 03°47′S 17°57′E
Bagdad see Baghdad
193 I8 **Bagdarin** Respublika Buryatiya, S Russian Federation 54°27′N 113°34′E
115 K3 **Bagé** Rio Grande do Sul, S Brazil 31°22′S 54°06′W
167 K10 **Bâgé-le-Châtel** Rhône-Alpes, France 46°18′N 4°56′E
Bagenalstown see Muine Bheag
233 K8 **Bagerhat** var. Bagherhat. Khulna, S Bangladesh 22°40′N 89°48′E
165 I9 **Bages et de Sigean, Étang de** ◎ S France
77 K9 **Baggs** Wyoming, C USA 41°02′N 107°39′W
232 D8 **Bāgh** Madhya Pradesh, C India 22°22′N 74°49′E
217 J5 **Baghdad** var. Bagdad, Eng. Baghdad. ● Baghdād, C Iraq 33°20′N 44°26′E
217 J5 **Baghdād** ✈ (Baghdād) Baghdād, C Iraq 33°20′N 44°26′E
Baghdad see Baghdad
Bagherhat see Bagerhat
175 F12 **Bagheria** var. Bagaria. Sicilia, Italy, C Mediterranean Sea 38°05′N 13°31′E
223 H7 **Bāghīn** Kermān, C Iran 30°50′N 57°00′E
230 G2 **Baghlān** Baghlān, NE Afghanistan 36°11′N 68°44′E
231 H3 **Baghlān** var. Bāghlān. ⬖ province NE Afghanistan
Bāghlān see Baghlān
171 M8 **Baghlia** Algeria
230 E4 **Bāghrān** Helmand, S Afghanistan 32°55′N 64°57′E
74 F3 **Bagley** Minnesota, N USA 47°31′N 95°24′W
177 H8 **Bagnacavallo** Emilia-Romagna, C Italy 44°00′N 12°59′E
168 G3 **Bagnac-sur-Célé** Midi-Pyrénées, France 44°40′N 2°10′E
168 E6 **Bagnères-de-Bigorre** Hautes-Pyrénées, S France 43°04′N 00°09′E
171 J2 **Bagnères-de-Luchon** Hautes-Pyrénées, S France 42°46′N 00°34′E
176 G9 **Bagni di Lucca** Toscana, C Italy 44°01′N 10°38′E
177 H9 **Bagno di Romagna** Emilia-Romagna, C Italy 43°51′N 11°57′E
166 E6 **Bagnoles-de-l'Orne** Basse-Normandie, France 48°34′N 0°25′W
169 I4 **Bagnols-les-Bains** Languedoc-Roussillon, France 44°30′N 3°40′E
168 E7 **Bagnols-sur-Cèze** Gard, S France 44°10′N 04°37′E
244 E3 **Bag Nur** ◎ N China
256 C6 **Bago** var. Pegu. Bago, SW Myanmar (Burma) 17°18′N 96°31′E
263 L7 **Bago** off. Bago City. Negros, C Philippines 10°30′N 122°49′E
256 C6 **Bago** var. Pegu. ⬖ region S Myanmar (Burma)
Bago City see Bago
132 E5 **Bagoé** ≈ Ivory Coast/Mali
Bagong see Sansui
Bagrāmē see Bagrāmī
231 H3 **Bagrāmī** var. Bagrāmē. Kâbōl, E Afghanistan 34°29′N 69°16′E
191 A9 **Bagrationovsk** Ger. Preussisch Eylau. Kaliningradskaya Oblast', W Russian Federation 54°24′N 20°39′E
Bagrax see Bohu
Bagrax Hu see Bosten Hu
104 C4 **Bagua** Amazonas, NE Peru 05°37′S 78°36′W
263 K4 **Baguio** off. Baguio City. Luzon, N Philippines 16°25′N 120°36′E
Baguio City see Baguio
133 K3 **Bagzane, Monts** ▲ N Niger 17°48′N 08°43′E
Bāḩah, Minţaqat al see Al Bāḩah
290 C5 **Bahama Basin** C Atlantic Ocean
Bahama Islands see Bahamas, The
Bahamas see Bahamas, The
Bahamas, Commonwealth of The see Bahamas, The
90 F2 **Bahamas, The** off. Commonwealth of The Bahamas. ◆ commonwealth republic N West Indies
42 G8 **Bahamas, The** var. Bahama Islands, Bahamas. ◆ island group N West Indies
233 J8 **Baharampur** prev. Berhampore. West Bengal, NE India 24°06′N 88°19′E
228 C7 **Baharly** var. Bäherden, Rus. Bakharden; prev. Bakherden. Ahal Welaýaty, C Turkmenistan 38°30′N 57°18′E
259 K3 **Bahau, Sungai** ≈ Borneo, N Indonesia
231 I6 **Bahāwalnagar** Punjab, E Pakistan 30°00′N 73°03′E
231 I7 **Bahāwalpur** Punjab, E Pakistan 29°25′N 71°40′E
214 G7 **Bahçe** Osmaniye, S Turkey 37°14′N 36°34′E
242 C1 **Ba He** ≈ C China
219 C13 **Bahei, Wâdi el** Egypt
Bäherden see Baharly
109 F8 **Bahia** off. Estado da Bahia. ⬖ state E Brazil
84 C5 **Bahía Asunción** Baja California Sur, NW Mexico 27°08′N 114°18′W
116 F6 **Bahía Blanca** Buenos Aires, E Argentina 38°43′S 62°19′W
86 C7 **Bahía Bufadero** Michoacán, SW Mexico
117 E9 **Bahía Bustamante** Chubut, SE Argentina 45°06′S 66°30′W
84 C4 **Bahía de los Ángeles** Baja California Norte, NW Mexico
84 C5 **Bahía de Tortugas** Baja California Sur, NW Mexico 27°42′N 114°54′W
Bahia, Estado da see Bahia
88 G3 **Bahía, Islas de la** Eng. Bay Islands. island group N Honduras
84 E4 **Bahía Kino** Sonora, NW Mexico 28°48′N 111°55′W
84 D7 **Bahía Magdalena** var. Puerto Magdalena. Baja California Sur, NW Mexico 24°34′N 112°07′W
102 A5 **Bahía Solano** var. Ciudad Mutis, Solano. Chocó, W Colombia 06°13′N 77°27′W
136 D3 **Bahir Dar** var. Bahr Dar, Bahrdar Giyorgis. Āmara, N Ethiopia 11°34′N 37°23′E
221 K5 **Bahlā'** var. Bahlah, Bahlat. NW Oman 22°58′N 57°16′E
Bāhla see Bālān
Bahlah/Bahlat see Bahlā'
232 G6 **Bahraich** Uttar Pradesh, N India 27°35′N 81°36′E
221 H4 **Bahrain** off. State of Bahrain, Dawlat al Bahrayn, Ar. Al Baḩrayn, prev. Bahrein; anc. Tylos, Tyros. ◆ monarchy SW Asia
221 I3 **Bahrain** ✈ C Bahrain 26°15′N 50°39′E
222 E9 **Bahrain, Gulf of** gulf Persian Gulf, NW Arabian Sea
Bahrain, State of see Bahrain
216 D5 **Baḩrat Mallāḩah** ◎ W Syria
Bahrayn, Dawlat al see Bahrain
Bahr Dar/Bahrdar Giyorgis see Bahir Dar
Bahrein see Bahrain
Bahr el, Azraq see Blue Nile
134 D2 **Bahr el Gazel** off. Région du Bahr el Gazel. ⬖ region W Chad
Bahr el Gazel, Région du see Bahr el Gazel
Bahr el Gebel see Central Equatoria
Bahr el Jebel see Central Equatoria
136 A4 **Bahr ez Zaref** ≈ Jonglei, E South Sudan
121 J5 **Bahr Kameur** ≈ N Central African Republic
Bahr Tabariya, Sea of see Tiberias, Lake
223 K9 **Bāhū Kalāt** Sīstān va Balūchestān, SE Iran 25°42′N 61°28′E
191 H8 **Bahushewsk** Rus. Bogushëvsk. Vitsyebskaya Voblasts', NE Belarus 54°51′N 30°13′E
Bai see Tagow Bāy
188 B9 **Baia de Aramă** Mehedinţi, SW Romania 45°00′N 22°43′E
188 B7 **Baia de Criş** Ger. Altenburg, Hung. Körösbánya. Hunedoara, SW Romania 46°10′N 22°41′E
135 B14 **Baia dos Tigres** Namibe, SW Angola 16°36′S 11°44′E
135 C12 **Baia Farta** Benguela, W Angola 12°38′S 13°12′E

◆ Country
● Country Capital
◇ Dependent Territory
○ Dependent Territory Capital
⬖ Administrative Regions
✈ International Airport
▲ Mountain
▲▲ Mountain Range
⛰ Volcano
≈ River
◎ Lake
▣ Reservoir

◆ Country   **◇** Dependent Territory   **◇** Administrative Regions   **▲** Mountain   **✕** Volcano   **⊚** Lake
● Country Capital   **○** Dependent Territory Capital   **✕** International Airport   **▲** Mountain Range   **✍** River   **▨** Reservoir

◆ Country   ◇ Dependent Territory   ✪ Administrative Regions   ▲ Mountain   ▲ Volcano   ⊚ Lake
● Country Capital   ○ Dependent Territory Capital   ✈ International Airport   ▲ Mountain Range   ⊿ River   ⊠ Reservoir

Bellbrook - Beverly

INDEX

318

277 M4 **Bellbrook** New South Wales, SE Australia 30°48′S 152°32′E
168 F9 **Bellcaire d'Urgell** Cataluña, Spain 41°45′N 0°55′E
75 I10 **Belle** Missouri, C USA 38°17′N 91°43′W
67 I3 **Belle** West Virginia, NE USA 38°13′N 81°32′W
158 B7 **Belleek** W Northern Ireland, United Kingdom 54°28′N 08°06′W
73 H10 **Bellefontaine** Ohio, N USA 40°22′N 83°45′W
64 B5 **Bellefonte** Pennsylvania, NE USA 40°54′N 77°43′W
74 A5 **Belle Fourche** South Dakota, N USA 44°40′N 103°50′W
77 L7 **Belle Fourche Reservoir** ⊠ South Dakota, N USA
74 B5 **Belle Fourche River** ≈ South Dakota/Wyoming, N USA
169 J5 **Bellegarde** Languedoc-Roussillon, France 43°45′N 4°31′E
167 H7 **Bellegarde-du-Loiret** Centre, France 47°59′N 2°26′E
168 G1 **Bellegarde-en-Marche** Limousin, France 45°59′N 2°18′E
165 K6 **Bellegarde-sur-Valserine** Ain, E France 46°06′N 05°49′E
69 M8 **Belle Glade** Florida, SE USA 26°40′N 80°40′W
164 E4 **Belle Île** *island* NW France
59 L5 **Belle Isle** *island* Belle Isle, Newfoundland and Labrador, E Canada
166 A6 **Belle-Isle-en-Terre** Bretagne, France 48°33′N 3°24′W
59 L5 **Belle Isle, Strait of** *strait* Newfoundland and Labrador, E Canada
166 F6 **Bellème** Basse-Normandie, France 48°22′N 0°34′E
169 M2 **Bellentre** Rhône-Alpes, France 45°34′N 6°43′E
Bellenz *see* Bellinzona
73 A8 **Belle Plaine** Iowa, C USA 41°54′N 92°16′W
74 G5 **Belle Plaine** Minnesota, N USA 44°39′N 93°47′W
62 D1 **Belleterre** Québec, SE Canada 47°24′N 78°40′W
62 E4 **Belleville** Ontario, SE Canada 44°10′N 77°22′W
165 J6 **Belleville** Rhône, E France 46°09′N 04°42′E
73 B12 **Belleville** Illinois, N USA 38°31′N 89°58′W
75 E9 **Belleville** Kansas, C USA 39°51′N 97°38′W
65 H6 **Belleville** New Jersey, USA 40°48′N 74°09′W
167 K10 **Belleville-sur-Saône** Rhône-Alpes, France 46°06′N 4°45′E
141 H9 **Bellevue** Eastern Cape, South Africa 33°22′S 25°57′E
74 I7 **Bellevue** Iowa, C USA 42°15′N 90°25′W
75 F8 **Bellevue** Nebraska, C USA 41°08′N 95°53′W
73 H9 **Bellevue** Ohio, N USA 41°16′N 82°50′W
70 G3 **Bellevue** Texas, SW USA 33°38′N 98°00′W
76 C3 **Bellevue** Washington, NW USA 47°36′N 122°12′W
103 M6 **Bellevue de l'Inini, Montagnes** ▲▲ S French Guiana
165 J6 **Belley** Ain, E France 45°46′N 05°41′E
Bellin *see* Kangirsuk
277 M3 **Bellingen** New South Wales, SE Australia 30°27′S 152°53′E
159 J5 **Bellingham** N England, United Kingdom 55°09′N 02°16′W
76 C2 **Bellingham** Washington, NW USA 48°46′N 122°29′W
Belling Hausen Mulde *see* Southeast Pacific Basin
292 D2 **Bellingshausen** *Russian research station* South Shetland Islands, Antarctica 61°57′S 58°23′W
Bellingshausen *see* Motu One
Bellingshausen Abyssal Plain *see* Bellingshausen Plain
287 K10 **Bellingshausen Plain** *var.* Bellingshausen Abyssal Plain. *undersea feature* SE Pacific Ocean 64°00′S 90°00′W
292 E5 **Bellingshausen Sea** *sea* Antarctica
162 J4 **Bellingwolde** Groningen, NE Netherlands 53°07′N 07°10′E
176 E5 **Bellinzona** *Ger.* Bellenz. Ticino, S Switzerland 46°12′N 09°02′E
71 H5 **Bellmead** Texas, SW USA 31°36′N 97°02′W
102 B5 **Bello** Antioquia, W Colombia 06°19′N 75°34′W
158 E4 **Bellochantuy** United Kingdom 55°31′N 5°42′W
112 F10 **Bellocq** Buenos Aires, E Argentina 35°55′S 61°32′W
Bello Horizonte *see* Belo Horizonte
281 J4 **Bellona** *var.* Mungiki. *island* S Solomon Islands
53 H6 **Bellot Strait** *Sea waterway* Nunavut, N Canada
Bellovacum *see* Beauvais
64 J1 **Bellow Falls** Vermont, NE USA 43°08′N 72°26′W
276 E4 **Bell, Point** *headland* South Australia 32°13′S 133°08′E
168 F9 **Bellpuig** Cataluña, Spain 41°37′N 1°01′E
66 B6 **Bells** Tennessee, S USA 35°42′N 89°05′W
71 H3 **Bells** Texas, SW USA 33°36′N 96°24′W
159 H3 **Bellshill** United Kingdom 55°49′N 4°01′W
152 A5 **Bellsund** *inlet* SW Svalbard
177 H5 **Belluno** Veneto, NE Italy 46°08′N 12°13′E
116 F3 **Bell Ville** Córdoba, C Argentina 32°35′S 62°41′W
138 D10 **Bellville** Western Cape, SW South Africa 33°50′S 18°43′E
71 H7 **Bellville** Texas, SW USA 29°57′N 96°15′W
64 B6 **Bellwood** Pennsylvania, NE USA 40°36′N 78°19′W
180 E6 **Belm** Niedersachsen, Germany 52°18′E 8°08′N
65 H7 **Belmar** New Jersey, USA 40°11′N 74°01′W
170 F7 **Belméz** Andalucía, S Spain 38°16′N 05°12′W
74 G6 **Belmond** Iowa, C USA 42°51′N 93°36′W
80 B3 **Belmont** California, USA 37°31′N 122°17′W
80 I4 **Belmont** Nevada, USA 38°36′N 116°52′W
64 B3 **Belmont** New York, NE USA 42°14′N 78°02′W
67 I7 **Belmont** North Carolina, SE USA 35°13′N 81°01′W
109 H9 **Belmonte** Bahia, E Brazil 15°53′S 38°54′W
172 D2 **Belmonte** Castelo Branco, C Portugal 40°21′N 07°20′W
171 H6 **Belmonte** Castilla-La Mancha, C Spain 39°34′N 02°43′W
169 H5 **Belmont-sur-Rance** Midi-Pyrénées, France 43°49′N 2°46′E
88 D2 **Belmopan** ● (Belize) Cayo, C Belize 17°13′N 88°48′W
157 B8 **Belmullet** *Ir.* Béal an Mhuirhead. Mayo, W Ireland 54°14′N 09°59′W
163 C11 **Belœil** Hainaut, SW Belgium 50°33′N 03°45′E
193 K8 **Belogorsk** Amurskaya Oblast', SE Russian Federation 50°53′N 128°24′E
Belogorsk *see* Bilohirs'k
185 H5 **Belogradchik** Vidin, NW Bulgaria 43°37′N 22°42′E
139 L10 **Beloha** Toliara, S Madagascar 25°09′S 45°04′E
111 H3 **Belo Horizonte** *prev.* Bello Horizonte. *state capital* Minas Gerais, SE Brazil 19°54′S 43°54′W
75 D9 **Beloit** Kansas, C USA 39°27′N 98°06′W
73 C8 **Beloit** Wisconsin, N USA 42°31′N 89°01′W
Belokorovichi *see* Novi Bilokorovychi
192 F8 **Belokurikha** Altayskiy Kray, S Russian Federation 51°57′N 84°56′E
194 E5 **Belomorsk** Respublika Kareliya, NW Russian Federation 64°30′N 34°43′E
194 E5 **Belomorsko-Baltiyskiy Kanal** *Eng.* White Sea-Baltic Canal, White Sea Canal. *canal* NW Russian Federation
233 L8 **Belonia** Tripura, NE India 23°15′N 91°25′E
111 J2 **Belo Oriente** Minas Gerais, Brazil 19°14′S 42°28′W
Beloozersk *see* Byelaazyorsk
Belopol'ye *see* Bilopillya
170 G2 **Belorado** Castilla y León, N Spain 42°25′N 03°11′W
196 G8 **Belorechensk** Krasnodarskiy Kray, SW Russian Federation 44°46′N 39°53′E
197 M2 **Beloretsk** Respublika Bashkortostan, W Russian Federation 53°56′N 58°26′E
Belorussia/Belorussian SSR *see* Belarus
Belorusskaya Gryada *see* Byelaruskaya Hrada
Belorusskaya SSR *see* Belarus
Beloshchel'ye *see* Nar'yan-Mar
185 M5 **Beloslav** Varna, E Bulgaria 43°13′N 27°42′E
Belostok *see* Białystok
Belo-sur-Tsiribihina *see* Belo Tsiribihina
52 C7 **Belot, Lac** ◎ Northwest Territories, NW Canada
139 L7 **Belo Tsiribihina** *var.* Belo-sur-Tsiribihina. Toliara, W Madagascar 19°40′S 44°30′E
111 H4 **Belo Vale** Minas Gerais, Brazil 20°25′S 44°01′W
Belovár *see* Bjelovar
Belovezhskaya, Pushcha *see* Białowieska, Puszcza/Byelavyezhskaya, Pushcha
185 J7 **Belovo** Pazardzhik, C Bulgaria 42°10′N 24°01′E
227 M1 **Belovo** Kemerovskaya Oblast', S Russian Federation 54°25′N 86°13′E
Belovodsk *see* Bilovods'k
195 M3 **Beloyarsk** Yamalo-Nenetskiy Avtonomnyy Okrug, Russian Federation
195 M5 **Beloyarskiy** Khanty-Mansiyskiy Avtonomnyy Okrug-Yugra, N Russian Federation 63°40′N 66°31′E
194 F5 **Beloye More** *Eng.* White Sea. *sea* NW Russian Federation
194 F8 **Beloye, Ozero** ◎ NW Russian Federation
185 K7 **Belozem** Plovdiv, C Bulgaria 42°11′N 25°00′E
194 F8 **Belozërsk** Vologodskaya Oblast', NW Russian Federation 59°59′N 37°49′E
167 N9 **Belp** Bern, W Switzerland 46°54′N 07°31′E
174 B3 **Belp** ✈ (Bern) Bern, C Switzerland 46°55′N 07°29′E
175 G12 **Belpasso** Sicilia, Italy, C Mediterranean Sea 37°35′N 14°59′E
168 G6 **Belpech** Languedoc-Roussillon, France 43°12′N 1°45′E
161 I2 **Belper** United Kingdom 53°02′N 1°28′W
73 I11 **Belpre** Ohio, N USA 39°14′N 81°34′W
159 K5 **Belsay** United Kingdom 55°06′N 1°50′W
80 I7 **Belted Range** ▲▲ Nevada, USA
162 H5 **Belterwijde** ◎ N Netherlands
159 L9 **Belton** United Kingdom 53°33′N 0°49′W
75 G9 **Belton** Missouri, C USA 38°48′N 94°31′W
67 H7 **Belton** South Carolina, SE USA 34°31′N 82°29′W
70 G6 **Belton** Texas, SW USA 31°04′N 97°30′W
70 G6 **Belton Lake** ⊠ Texas, SW USA
Bel'tsy *see* Bălţi
158 C8 **Belturbet** *Ir.* Béal Tairbirt. Cavan, N Ireland 54°06′N 07°26′W
Beluchistan *see* Balochistan
227 N3 **Belukha, Gora** ▲ Kazakhstan/Russian Federation 49°50′N 86°44′E
195 J1 **Belush'ya Guba** Novaya Zemlya, Russian Federation 71°34′N 52°19′E
175 H10 **Belvedere Marittimo** Calabria, SW Italy 39°37′N 15°52′E
73 D8 **Belvidere** Illinois, N USA 42°15′N 88°50′W
64 G6 **Belvidere** New Jersey, NE USA 40°50′N 75°05′W
161 J2 **Belvoir** United Kingdom 52°53′N 0°47′W
Bely *see* Belyy
197 L4 **Belyayevka** Orenburgskaya Oblast', W Russian Federation 51°25′N 56°26′E
Belynichi *see* Byalynichy
194 D10 **Belyy** *var.* Bely, Beyj. Tverskaya Oblast', W Russian Federation 55°51′N 32°57′E
196 E3 **Belyye Berega** Bryanskaya Oblast', W Russian Federation 53°11′N 34°42′E
192 F3 **Belyy, Ostrov** *island* N Russian Federation
192 F7 **Belyy Yar** Tomskaya Oblast', C Russian Federation 58°26′N 84°57′E
166 A7 **Belz** Bretagne, France 47°41′N 3°10′W
178 H7 **Belzig** Brandenburg, NE Germany 52°09′N 12°37′E
66 A9 **Belzoni** Mississippi, S USA 33°10′N 90°29′W
139 L7 **Bemaraha** *var.* Plateau du Bemaraha. ▲▲ W Madagascar
Bemaraha, Plateau du *see* Bemaraha
135 C10 **Bembe** Uíge, NW Angola 07°03′S 14°25′E
133 I6 **Bembèrèkè** *var.* Bimbéréké. N Benin 10°10′N 02°41′E
170 E7 **Bembézar** ≈ SW Spain
170 E2 **Bembibre** Castilla y León, N Spain 42°37′N 06°25′W
161 J8 **Bembridge** S England, United Kingdom
74 F3 **Bemidji** Minnesota, N USA 47°27′N 94°53′W
162 H7 **Bemmel** Gelderland, SE Netherlands 51°53′N 05°54′E
172 D3 **Bemposta** Castelo Branco, Portugal 40°04′N 7°13′W
172 C4 **Bemposta** Santarém, Portugal 39°21′N 8°08′W
159 M8 **Bempton** United Kingdom 54°08′N 0°10′W
261 H5 **Bemu** Pulau Seram, E Indonesia 03°21′S 129°58′E
Benāb *see* Bonāb
171 J3 **Benabarre** *var.* Benavarn. Aragón, NE Spain 42°06′N 00°28′E
Benaco *see* Garda, Lago di
135 G9 **Bena-Dibele** Kasai-Oriental, C Dem. Rep. Congo 04°01′S 22°50′E
171 I5 **Benagéber, Embalse de** E Spain
173 J8 **Benahadux** Andalucía, Spain 36°56′N 2°27′W
173 N9 **Ben Ali** Algeria 36°13′N 00°41′E
277 J6 **Benalla** Victoria, SE Australia 36°33′S 146°00′E
173 H7 **Benalúa de las Villas** Andalucía, Spain 37°26′N 3°41′W
173 I7 **Benamaurel** Andalucía, Spain 37°36′N 2°42′W
170 F8 **Benamejí** Andalucía, S Spain 37°16′N 04°33′W
Benares *see* Vārānasi
168 E7 **Benasque** Aragón, Spain 42°36′N 0°31′E
Benavarn *see* Benabarre
172 B5 **Benavente** Santarém, C Portugal 38°59′N 08°49′W
170 E3 **Benavente** Castilla y León, N Spain 42°N 05°40′W
70 G9 **Benavides** Texas, SW USA 27°36′N 98°24′W
156 C5 **Benbecula** *island* NW Scotland, United Kingdom
159 H2 **Ben Chonzie** ▲ C Scotland, United Kingdom
Bencovazzo *see* Benkovac
76 C7 **Bend** Oregon, NW USA 44°04′N 121°19′W
277 H4 **Benda Range** ▲▲ South Australia
277 L4 **Bendemeer** New South Wales, SE Australia 30°54′S 151°12′E
Bender *see* Tighina
Bender Beila/Bender Beyla *see* Bandarbeyla
Bender Cassim/Bender Qaasim *see* Boosaaso
158 F2 **Benderloch** United Kingdom 56°29′N 5°24′W
Bendery *see* Tighina
277 I6 **Bendigo** Victoria, SE Australia 36°46′S 144°19′E
181 D11 **Bendorf** Rheinland-Pfalz, Germany 50°26′E 7°34′N
190 C7 **Bēne** SW Latvia 56°30′N 23°04′E
162 G7 **Beneden-Leeuwen** Gelderland, C Netherlands 51°52′N 05°32′E
179 G13 **Benediktenwand** ▲ S Germany 47°39′N 11°28′E
Benemérita de San Cristóbal *see* San Cristóbal
132 F5 **Bénéna** Ségou, S Mali 13°04′N 04°20′W
139 L9 **Benenitra** Toliara, S Madagascar 23°25′S 45°06′E
Beneschau *see* Benešov
Beneški Zaliv *see* Venice, Gulf of
183 C9 **Benešov** *Ger.* Beneschau. Středočeský Kraj, W Czech Republic 49°48′N 14°41′E
166 G10 **Bénévent-l'Abbaye** Limousin, France 46°07′N 1°38′E
175 G9 **Benevento** *anc.* Beneventum, Malventum. Campania, S Italy 41°07′N 14°45′E
Beneventum *see* Benevento
167 N6 **Benfeld** Alsace, France 48°22′N 7°36′E
288 G6 **Bengal, Bay of** *bay* N Indian Ocean
Bengalooru *see* Bangalore
Bengaluru *see* Bangalore
134 H7 **Bengamisa** Orientale, N Dem. Rep. Congo 0°58′N 25°11′E
Bengasi *see* Banghāzī
259 I8 **Bengawan, Sungai** ≈ Jawa, S Indonesia
Bengazi *see* Banghāzī
241 I3 **Bengbu** *var.* Peng-pu. Anhui, E China 32°57′N 117°17′E
76 E4 **Benge** Washington, NW USA 46°55′N 118°01′W
Benghazi *see* Banghāzī
258 E4 **Bengkalis** Pulau Bengkalis, W Indonesia 01°27′N 102°10′E
258 E4 **Bengkalis, Pulau** *island* W Indonesia
259 H4 **Bengkayang** Borneo, C Indonesia 0°45′N 109°28′E
258 E7 **Bengkulu** *prev.* Bengkoeloe, Benkoelen, Benkulen. Sumatera, W Indonesia 03°46′S 102°16′E
258 E7 **Bengkulu** *off.* Propinsi Bengkulu; *prev.* Bengkoeloe, Benkoelen, Benkulen. ◈ *province* W Indonesia
Bengkulu, Propinsi *see* Bengkulu
135 C10 **Bengo** ◈ *province* W Angola
155 F8 **Bengtsfors** Västra Götaland, S Sweden 59°03′N 12°14′E
135 C12 **Benguela** *var.* Benguella. Benguela, W Angola 12°35′S 13°30′E
135 C12 **Benguela** ◈ *province* W Angola
Benguella *see* Benguela
219 D8 **Ben Gurion** ✈ Tel Aviv, C Israel 32°04′N 34°45′E
Bengweulu, Lake *see* Bangweulu, Lake
Benha *see* Banhā
286 B4 **Benham Seamount** *undersea feature* W Philippine Sea 15°48′N 124°15′E
156 E4 **Ben Hope** ▲ N Scotland, United Kingdom 58°25′N 04°36′W
134 J7 **Beni** Nord-Kivu, NE Dem. Rep. Congo 0°31′N 29°30′E
130 G4 **Beni Abbès** W Algeria 30°07′N 02°09′W
171 N9 **Béni Aziz** Algeria
171 J5 **Benicarló** Valenciana, E Spain 40°25′N 00°25′E
171 J5 **Benicasim** *Cat.* Benicàssim. Valenciana, E Spain 40°03′N 00°03′E
Benicàssim *see* Benicasim
80 B2 **Benicia** California, USA 38°03′N 122°10′W
171 J7 **Benidorm** Valenciana, SE Spain 38°33′N 00°09′W
171 K9 **Béni Haoua** Algeria
Beni Mazâr *see* Banī Mazār
130 F3 **Beni-Mellal** C Morocco 32°20′N 06°21′W
133 I6 **Benin** *off.* Republic of Benin; *prev.* Dahomey. ◆ *republic* W Africa
133 J8 **Benin, Bight of** *gulf* W Africa
133 J8 **Benin City** Edo, SW Nigeria 06°23′N 05°40′E
Benin, Republic of *see* Benin
106 D7 **Beni, Río** ≈ N Bolivia
130 G2 **Beni Saf** *var.* Beni-Saf. NW Algeria 35°19′N 01°23′W
Beni-Saf *see* Beni Saf
Benishangul *see* Bīnshangul Gumuz
171 L9 **Beni Slimane** Algeria
173 M5 **Benissa** Valenciana, E Spain 38°43′N 00°03′E
Beni Suef *see* Banī Suwayf
55 I9 **Benito** Manitoba, S Canada 51°57′N 101°24′W
Benito *see* Uolo, Río
116 G6 **Benito Juárez** Buenos Aires, E Argentina 37°43′S 59°48′W
87 J8 **Benito Juárez** Chiapas, Mexico 16°53′N 93°11′W
85 J10 **Benito Juárez** Zacatecas, Mexico 21°29′N 103°33′W
86 F6 **Benito Juárez Internacional** ✈ (México) México, S Mexico 19°24′N 99°02′W
70 F3 **Benjamin** Texas, SW USA 33°35′N 99°49′W
106 B4 **Benjamin Constant** Amazonas, N Brazil 04°22′S 70°02′W
84 E3 **Benjamín Hill** Sonora, NW Mexico 30°13′N 111°08′W
117 B9 **Benjamín, Isla** *island* Archipiélago de los Chonos, S Chile
252 C5 **Benkei-misaki** *headland* Hokkaidō, NE Japan 42°49′N 140°10′E
75 B8 **Benkelman** Nebraska, C USA 40°04′N 101°30′W
156 F4 **Ben Klibreck** ▲ N Scotland, United Kingdom 58°15′N 04°23′W
Benkoelen/Bengkoeloe *see* Bengkulu
184 B4 **Benkovac** *It.* Bencovazzo. Zadar, SW Croatia 44°02′N 15°36′E
Benkulen *see* Bengkulu
156 F6 **Ben Lawers** ▲ C Scotland, United Kingdom 56°33′N 04°13′W
80 B4 **Ben Lomond** California, USA 37°05′N 122°05′W
158 G2 **Ben Lomond** ▲ United Kingdom 56°11′N 4°38′W
158 G2 **Ben Lui** ▲ C Scotland, United Kingdom
156 F5 **Ben Macdui** *var.* Beinn MacDuibh. ▲ C Scotland, United Kingdom 57°02′N 03°42′W
175 A13 **Ben Mehidi** Tunisia
156 D6 **Ben More** ▲ W Scotland, United Kingdom 56°26′N 06°00′W
156 F6 **Ben More** ▲ C Scotland, United Kingdom 56°22′N 04°31′W
156 E4 **Ben More Assynt** ▲ N Scotland, United Kingdom 58°09′N 04°51′W
278 D11 **Benmore, Lake** ◎ South Island, New Zealand
180 I7 **Benneckenstein** Sachsen-Anhalt, Germany 51°40′E 10°43′N
162 G7 **Bennekom** Gelderland, SE Netherlands 52°00′N 05°40′E
193 K2 **Bennetta, Ostrov** *island* Novosibirskiye Ostrova, NE Russian Federation
67 J7 **Bennettsville** South Carolina, SE USA 34°36′N 79°40′W
156 E5 **Ben Nevis** ▲ N Scotland, United Kingdom 56°80′N 05°00′W
278 G5 **Benneydale** Waikato, North Island, New Zealand 38°31′S 175°22′E
Bennichab *see* Bennichchâb
132 B2 **Bennichchâb** *var.* Bennichab. Inchiri, W Mauritania 19°26′N 15°21′W
65 I1 **Bennington** Vermont, NE USA 42°51′N 73°09′W
278 C10 **Ben Ohau Range** ▲▲ South Island, New Zealand
138 G7 **Benoni** Gauteng, NE South Africa 26°04′S 28°18′E
139 M4 **Be, Nosy** *var.* Nossi-Bé. *island* NW Madagascar
Bénoué *see* Benue
88 D2 **Benque Viejo del Carmen** Cayo, W Belize 17°04′N 89°08′W
181 C9 **Bensberg** Nordrhein-Westfalen, Germany 50°58′E 7°10′N
171 I10 **Bensekrane** Algeria
181 F12 **Bensheim** Hessen, W Germany 49°41′N 08°38′E
79 J10 **Benson** Arizona, SW USA 31°55′N 110°16′W
74 F4 **Benson** Minnesota, N USA 45°19′N 95°36′W
67 K6 **Benson** North Carolina, SE USA 35°22′N 78°33′W
171 M10 **Ben S'Rour** Algeria
260 D7 **Benteng** Pulau Selayar, C Indonesia 06°07′S 120°28′E
135 B13 **Bentiaba** Namibe, SW Angola 14°18′S 12°27′E
275 I3 **Bentinck Island** *island* Wellesley Islands, Queensland, N Australia
136 A4 **Bentiu** Unity, N South Sudan 09°14′N 29°49′E
218 F5 **Bent Jbaïl** *var.* Bint Jubayl. S Lebanon 33°07′N 35°26′E
218 F5 **Bent Jbaïl** Lebanon 33°07′N 35°25′E
57 I6 **Bentley** Alberta, SW Canada 52°27′N 114°02′W
159 L9 **Bentley** United Kingdom 53°32′N 1°09′W
113 J7 **Bento Gonçalves** Rio Grande do Sul, S Brazil 29°12′S 51°34′W
75 H13 **Benton** Arkansas, C USA 34°34′N 92°35′W
80 G5 **Benton** California, USA 37°49′N 118°29′W
73 C12 **Benton** Illinois, N USA 38°00′N 88°55′W
66 C5 **Benton** Kentucky, S USA 36°51′N 88°21′W
68 B1 **Benton** Louisiana, S USA 32°41′N 93°44′W
75 J11 **Benton** Missouri, C USA 37°05′N 89°34′W
66 F7 **Benton** Tennessee, S USA 35°10′N 84°39′W
73 E8 **Benton Harbor** Michigan, N USA 42°07′N 86°27′W
80 G5 **Benton Hot Springs** California, USA 37°48′N 118°32′W
75 G11 **Bentonville** Arkansas, C USA 36°23′N 94°13′W
159 I5 **Bentpath** United Kingdom 55°12′N 3°05′W
133 K7 **Benue** ◈ *state* SE Nigeria
134 C4 **Benue** *Fr.* Bénoué. ≈ Cameroon/Nigeria
258 E4 **Benut** Johor, Peninsular Malaysia 01°38′N 103°15′E
158 G2 **Ben Vorlich** ▲ C Scotland, United Kingdom 56°20′N 4°14′W
247 K7 **Benxi** *prev.* Pen-ch'i, Penhsihu, Penki. Liaoning, NE China 41°20′N 123°45′E
Benyakoni *see* Byenyakoni
184 F3 **Beočin** Vojvodina, N Serbia 45°13′N 19°43′E
Beodericsworth *see* Bury St Edmunds
184 G3 **Beograd** *Eng.* Belgrade, *Ger.* Belgrad; *anc.* Singidunum. ● (Serbia) Serbia, N Serbia 44°48′N 20°27′E
184 F3 **Beograd** *Eng.* Belgrade. ✈ Serbia, N Serbia 44°45′N 20°21′E
132 F7 **Béoumi** C Ivory Coast 07°40′N 05°34′W
78 F2 **Beowawe** Nevada, W USA 40°33′N 116°31′W
261 J4 **Bepondi, Pulau** *island* E Indonesia
250 D6 **Beppu** Ōita, Kyūshū, SW Japan 33°18′N 131°30′E
250 D6 **Beppu-wan** *bay* SW Japan
280 E9 **Beqa** *prev.* Mbengga. *island* W Fiji
Beqa Barrier Reef *see* Kavukavu Reef
103 I1 **Bequia** *island* C Saint Vincent and the Grenadines
217 K3 **Beranan, Shax-î** *var.* Shākh-i Barānān. ▲▲ E Iraq
184 F6 **Berane** *prev.* Ivangrad. E Montenegro 42°51′N 19°51′E
184 F9 **Berat** *var.* Berati, *SCr.* Beligrad. Berat, C Albania 40°43′N 19°58′E
184 F9 **Berat** ◈ *district* C Albania
Berătău *see* Berettyó
Berati *see* Berat
Beraun *see* Berounka, Czech Republic
Beraun *see* Beroun, Czech Republic
259 K4 **Berau, Sungai** ≈ Borneo, N Indonesia
261 I5 **Berau, Teluk** *var.* MacCluer Gulf. *bay* Papua, E Indonesia
136 C1 **Berber** River Nile, NE Sudan 18°01′N 34°00′E
136 H4 **Berbera** Woqooyi Galbeed, NW Somalia 10°24′N 45°02′E
168 A6 **Berberana** Castilla y León, Spain 42°55′N 3°03′W
134 D6 **Berbérati** Mambéré-Kadéï, SW Central African Republic 04°14′N 15°50′E
Berberia, Cabo de *see* Barbaria, Cap de
141 K5 **Berbice** Mpumalanga, South Africa 27°12′S 31°09′E
103 K5 **Berbice River** ≈ NE Guyana
Berchid *see* Berrechid
165 H1 **Berck-Plage** Pas-de-Calais, N France 50°24′N 01°35′E
166 G2 **Berck-sur-Mer** Nord-Pas-de-Calais, France 50°24′N 1°36′E
70 G8 **Berclair** Texas, SW USA 28°33′N 97°32′W
189 L6 **Berda** ≈ SE Ukraine
Berdichev *see* Berdychiv
114 C9 **Berdier** Buenos Aires, Argentina 34°24′S 60°16′W
193 J6 **Berdigestyakh** Respublika Sakha (Yakutiya), NE Russian Federation 62°02′N 127°03′E
192 F7 **Berdsk** Novosibirskaya Oblast', C Russian Federation 54°42′N 82°56′E
114 E4 **Berduc** Entre Ríos, Argentina 31°55′S 58°19′W
189 L7 **Berdyans'k** *Rus.* Berdyansk; *prev.* Osipenko. Zaporiz'ka Oblast', SE Ukraine 46°46′N 36°49′E
189 L7 **Berdyans'ka Kosa** *spit* SE Ukraine
189 L7 **Berdyans'ka Zatoka** *gulf* S Ukraine
188 G4 **Berdychiv** *Rus.* Berdichev. Zhytomyrs'ka Oblast', N Ukraine 49°54′N 28°39′E
66 F4 **Berea** Kentucky, S USA 37°34′N 84°18′W
Beregovo/Beregszász *see* Berehove
188 C5 **Berehove** *Cz.* Berehovo, *Hung.* Beregszász, *Rus.* Beregovo. Zakarpats'ka Oblast', W Ukraine 48°13′N 22°39′E
Berehovo *see* Berehove
280 B4 **Bereina** Central, S Papua New Guinea 08°29′S 146°30′E
228 C6 **Bereket** *prev. Rus.* Gazandzhyk, Kazandzhik, *Turkm.* Gazanjyk. Balkan Welaýaty, W Turkmenistan 39°17′N 55°27′E
91 L2 **Berekua** S Dominica 15°14′N 61°19′W
132 G7 **Berekum** W Ghana 07°27′N 02°35′W
80 D5 **Berenda** California, USA 37°02′N 120°09′W
Berenice *see* Baranīs
55 J8 **Berens** ≈ Manitoba/Ontario, C Canada
55 J8 **Berens River** Manitoba, C Canada 52°22′N 97°00′W
74 E6 **Beresford** South Dakota, N USA 43°02′N 96°45′W
188 D3 **Berestechko** Volyns'ka Oblast', NW Ukraine 50°21′N 25°06′E
188 F8 **Bereşti** Galaţi, E Romania 46°04′N 27°54′E
189 K4 **Berestova** ≈ E Ukraine
Beretău *see* Berettyó
183 H12 **Berettyó** *Rom.* Barcău; *prev.* Berătău, Beretău. ≈ Hungary/Romania
183 H11 **Berettyóújfalu** Hajdú-Bihar, E Hungary 47°15′N 21°33′E
Berëza/Bereza Kartuska *see* Byaroza
189 H3 **Berezan'** Kyyivs'ka Oblast', N Ukraine 50°18′N 31°30′E
189 H7 **Berezanka** Mykolayivs'ka Oblast', S Ukraine 46°51′N 31°24′E
188 D4 **Berezhany** *Pol.* Brzeżany. Ternopil's'ka Oblast', W Ukraine 49°29′N 25°00′E
Berezina *see* Byarezina
Berezino *see* Byerazino
189 H7 **Berezivka** *Rus.* Berezovka. Odes'ka Oblast', SW Ukraine 47°12′N 30°56′E
189 I2 **Berezna** Chernihivs'ka Oblast', NE Ukraine 51°35′N 31°50′E
188 E2 **Berezne** Rivnens'ka Oblast', NW Ukraine 51°00′N 26°46′E
189 I6 **Bereznehuvate** Mykolayivs'ka Oblast', S Ukraine 47°18′N 32°52′E
194 G6 **Bereznik** Arkhangel'skaya Oblast', NW Russian Federation 62°50′N 42°40′E
195 K8 **Berezniki** Permskiy Kray, NW Russian Federation 59°26′N 56°49′E
Berëzovka *see* Byarozawka, Belarus
Berezovka *see* Berezivka, Ukraine
195 M5 **Berezovo** Khanty-Mansiyskiy Avtonomnyy Okrug-Yugra, N Russian Federation 63°48′N 64°38′E
197 H5 **Berezovskaya** Volgogradskaya Oblast', SW Russian Federation 50°17′N 43°58′E
195 M9 **Berezovskiy** Sverdlovskaya Oblasť, Russian Federation
193 L8 **Berezovyy** Khabarovskiy Kray, E Russian Federation 51°42′N 135°39′E
138 D10 **Berg** ≈ W South Africa
Berg *see* Berg bei Rohrbach
171 K3 **Berga** Cataluña, NE Spain 42°06′N 01°41′E
155 H10 **Berga** Kalmar, S Sweden 57°13′N 16°03′E
214 B6 **Bergama** İzmir, W Turkey 39°08′N 27°10′E
176 F6 **Bergamo** *anc.* Bergomum. Lombardia, N Italy 45°42′N 09°40′E
171 H2 **Bergara** País Vasco, N Spain 43°05′N 02°25′W
177 J2 **Berg bei Rohrbach** *var.* Berg. Oberösterreich, N Austria 48°34′N 14°02′E
180 D5 **Berge** Niedersachsen, Germany 52°37′E 7°43′N
178 H4 **Bergen** Mecklenburg-Vorpommern, NE Germany 54°25′N 13°25′E
180 H5 **Bergen** Niedersachsen, NW Germany 52°49′N 09°57′E
162 E5 **Bergen** Noord-Holland, NW Netherlands 52°40′N 04°42′E
154 A7 **Bergen** Hordaland, S Norway 60°24′N 05°19′E
Bergen *see* Mons
103 L5 **Berg en Dal** Brokopondo, C Suriname 05°15′N 55°20′W
163 D8 **Bergen op Zoom** Noord-Brabant, S Netherlands 51°30′N 04°17′E
164 G7 **Bergerac** Dordogne, SW France 44°51′N 00°30′E
163 F9 **Bergeyk** Noord-Brabant, S Netherlands 51°19′N 05°21′E
181 B9 **Bergheim** Nordrhein-Westfalen, W Germany 50°57′N 06°39′E
103 M6 **Bergi** Sipaliwini, E Suriname 04°36′N 54°24′W
181 C9 **Bergisch Gladbach** Nordrhein-Westfalen, W Germany 50°59′N 07°09′E
181 D8 **Bergkamen** Nordrhein-Westfalen, W Germany 51°32′N 07°41′E
155 H11 **Bergkvara** Kalmar, S Sweden 56°22′N 16°04′E
140 C1 **Bergland** Hardap, Namibia 22°59′S 17°09′E
181 D9 **Bergneustadt** Nordrhein-Westfalen, Germany 51°02′E 7°39′N
Bergomum *see* Bergamo
162 G7 **Bergse Maas** ≈ S Netherlands
154 J7 **Bergshamra** Stockholm, C Sweden 59°37′N 18°40′E
154 H5 **Bergsjö** Gävleborg, C Sweden 62°00′N 17°10′E
152 G7 **Bergsviken** Norrbotten, N Sweden 65°16′N 21°24′E
181 H12 **Bergtheim** Bayern, Germany 49°54′E 10°04′N
Bergum *see* Burgum
162 H3 **Bergumer Meer** ◎ N Netherlands
154 H6 **Bergviken** ◎ C Sweden
258 F5 **Berhala, Selat** *strait* Sumatera, W Indonesia
Berhampore *see* Baharampur
171 N10 **Berhoum** Algeria
193 N5 **Beringa, Ostrov** *island* E Russian Federation
163 F9 **Beringen** Limburg, NE Belgium 51°04′N 05°14′E
83 K6 **Bering Glacier** *glacier* Alaska, USA
Beringov Proliv *see* Bering Strait
193 N3 **Beringovskiy** Chukotskiy Avtonomnyy Okrug, NE Russian Federation 63°04′N 179°09′E
286 F1 **Bering Sea** *sea* N Pacific Ocean
50 B2 **Bering Strait** *Rus.* Beringov Proliv. *strait* Bering Sea/Chukchi Sea
Berislav *see* Beryslav
114 F10 **Berisso** Buenos Aires, Argentina 34°52′S 57°53′W
170 G9 **Berja** Andalucía, S Spain 36°51′N 02°56′W
181 H9 **Berka** Thüringen, Germany 50°57′E 10°05′N
154 D4 **Berkåk** Sør-Trøndelag, S Norway 62°50′N 10°01′E
180 C7 **Berkel** ≈ Germany/Netherlands
80 B2 **Berkeley** California, W USA 37°52′N 122°16′W
117 J11 **Berkeley Sound** *sound* NE Falkland Islands
64 B8 **Berkeley Springs** *var.* Bath. West Virginia, NE USA 39°38′N 78°14′W
180 I2 **Berkenthin** Schleswig-Holstein, Germany 53°44′E 10°39′N
161 J5 **Berkhampstead** United Kingdom 51°45′N 0°33′W
292 G4 **Berkner Island** *island* Antarctica
185 I5 **Berkovitsa** Montana, NW Bulgaria 43°15′N 23°05′E
157 H11 **Berkshire** *former county* S England, United Kingdom
163 E9 **Berlaar** Antwerpen, N Belgium 51°08′N 04°39′E
172 E6 **Berlanga** Extremadura, Spain 38°17′N 5°49′W
Berlanga *see* Berlanga de Duero
170 G4 **Berlanga de Duero** *var.* Berlanga. Castilla y León, N Spain 41°28′N 02°51′W
42 E10 **Berlanga Rise** *undersea feature* E Pacific Ocean 08°30′N 93°30′W
163 D10 **Berlare** Oost-Vlaanderen, NW Belgium 51°02′N 04°01′E
170 B5 **Berlenga, Ilha da** *island* C Portugal
152 H3 **Berlevåg** *Lapp.* Bearalváhki. Finnmark, N Norway 70°51′N 29°04′E
178 H6 **Berlin** ● (Germany) Berlin, NE Germany 52°31′N 13°26′E
67 N3 **Berlin** Maryland, NE USA 38°19′N 75°13′W
63 J4 **Berlin** New Hampshire, NE USA 44°27′N 71°13′W
64 A8 **Berlin** Pennsylvania, NE USA 39°54′N 78°57′W
72 C6 **Berlin** Wisconsin, N USA 43°57′N 88°59′W
178 I6 **Berlin** ◈ *state* NE Germany
Berlinchen *see* Barlinek
53 I7 **Berlinguet Inlet** *coastal sea feature* Nunavut, NE Canada
62 B8 **Berlin Lake** ⊠ Ohio, N USA
277 L6 **Bermagui** New South Wales, SE Australia 36°26′S 150°01′E
85 J6 **Bermejillo** Durango, C Mexico 25°55′N 103°39′W
118 G1 **Bermejo** San Juan, Argentina 31°37′S 67°39′W
112 E4 **Bermejo, Río** ≈ N Argentina
116 D2 **Bermejo, Río** ≈ W Argentina
112 E4 **Bermejo viejo, Río** ≈ N Argentina
171 H1 **Bermeo** País Vasco, N Spain 43°25′N 02°44′W
170 E4 **Bermillo de Sayago** Castilla y León, N Spain 41°22′N 06°08′W
174 D4 **Bermina, Pizzo** *Rmsch.* Piz Bernina. ▲ Italy/Switzerland 46°22′N 09°52′E *see also* Bermina, Piz
140 F5 **Bermoli** Northern Cape, South Africa 28°24′S 22°53′E
67 M9 **Bermuda** *var.* Bermuda Islands, Bermudas; *prev.* Somers Islands. ◇ *UK crown colony* NW Atlantic Ocean
43 H7 **Bermuda** *var.* Great Bermuda, Long Island, Main Island. *island* Bermuda
Bermuda Islands *see* Bermuda
Bermuda-New England Seamount Arc *see* New England Seamounts
43 H7 **Bermuda Rise** *undersea feature* C Sargasso Sea 32°30′N 65°00′W
Bermudas *see* Bermuda
176 D4 **Bern** *Fr.* Berne. ● (Switzerland) Bern, W Switzerland 46°57′N 07°26′E
176 D4 **Bern** *Fr.* Berne. ◈ *canton* W Switzerland
114 G2 **Bernabe Rivera** Artigas, Uruguay 30°18′S 57°00′W
85 J9 **Bernalejo** Zacatecas, Mexico 22°42′N 104°14′W
79 L7 **Bernalillo** New Mexico, SW USA 35°18′N 106°33′W
52 D5 **Bernard Island** *Island* Northwest Territories, N Canada
62 D3 **Bernard Lake** ◎ Ontario, S Canada
114 A5 **Bernardo de Irigoyen** Santa Fe, NE Argentina 32°09′S 61°06′W
64 G6 **Bernardsville** New Jersey, NE USA 40°43′N 74°34′W
116 F6 **Bernasconi** La Pampa, C Argentina 37°55′S 63°44′W
178 H6 **Bernau** Brandenburg, NE Germany 52°41′N 13°36′E
164 G3 **Bernay** Eure, N France 49°05′N 00°36′E
178 G7 **Bernburg** Sachsen-Anhalt, C Germany 51°47′N 11°45′E
73 G10 **Berne** Indiana, N USA 40°39′N 84°57′W
Berne *see* Bern
176 D5 **Berner Alpen** *var.* Berner Oberland, *Eng.* Bernese Oberland. ▲▲ SW Switzerland
158 C1 **Berneray** *island* United Kingdom
Berner Oberland/Bernese Oberland *see* Berner Alpen
166 F10 **Berneuil** Limousin, France 46°04′N 1°06′E
167 H3 **Berneville** Nord-Pas-de-Calais, France 50°16′N 2°40′E
177 M1 **Bernhardsthal** Niederösterreich, N Austria 48°40′N 16°51′E
68 C1 **Bernice** Louisiana, S USA 32°49′N 92°39′W
75 J11 **Bernie** Missouri, C USA 36°40′N 89°58′W
53 I7 **Bernier Bay** *coastal sea feature* Nunavut, NE Canada
274 C5 **Bernier Island** *island* Western Australia
Bernina Pass *see* Bernina, Passo del
174 D3 **Bernina, Passo del** *Eng.* Bernina Pass. *pass* SE Switzerland
174 D4 **Bernina, Piz** *It.* Pizzo Bernina. ▲ Italy/Switzerland 46°22′N 09°55′E *see also* Bermina, Pizzo
Bernina, Piz *see* Bermina, Pizzo
163 C11 **Bérnissart** Hainaut, SW Belgium 50°29′N 03°37′E
181 C12 **Bernkastel-Kues** Rheinland-Pfalz, W Germany 49°55′N 07°04′E
Beroea *see* Ḥalab
181 J14 **Berolzheim** Bayern, Germany 49°01′E 10°51′N
139 L8 **Beroroha** Toliara, SW Madagascar 21°40′S 45°10′E
Béroubouay *see* Gbéroubouè
183 B8 **Beroun** *Ger.* Beraun. Středočeský Kraj, W Czech Republic 49°58′N 14°05′E
183 B8 **Berounka** *Ger.* Beraun. ≈ W Czech Republic
185 I7 **Berovo** E FYR Macedonia 41°45′N 22°50′E
218 G1 **Berqâyel** Lebanon 34°29′N 36°02′E
130 E3 **Berrechid** *var.* Berchid. W Morocco 33°16′N 07°32′W
165 J8 **Berre, Étang de** ◎ SE France
165 J8 **Berre-l'Étang** Bouches-du-Rhône, SE France 43°28′N 05°10′E
277 H5 **Berri** South Australia 34°16′S 140°35′E
73 F9 **Berrien Springs** Michigan, N USA 41°57′N 86°20′W
277 J6 **Berrigan** New South Wales, SE Australia 35°41′S 145°50′E
171 L9 **Berrouaghia** Algeria
160 G7 **Berrow** United Kingdom 51°16′N 3°01′W
165 H5 **Berry** *cultural region* C France
80 C1 **Berryessa, Lake** ◎ California, W USA
61 M9 **Berry Islands** *island group* N The Bahamas
75 H11 **Berryville** Arkansas, C USA 36°22′N 93°35′W
64 B9 **Berryville** Virginia, NE USA 39°08′N 77°59′W
140 C4 **Berseba** Karas, S Namibia 26°00′S 17°46′E
188 G5 **Bershad'** Vinnyts'ka Oblast', C Ukraine 48°20′N 29°30′E
74 B2 **Berthold** North Dakota, N USA 48°16′N 101°48′W
79 L2 **Berthoud** Colorado, C USA 40°18′N 105°04′W
79 L3 **Berthoud Pass** *pass* Colorado, C USA
110 E10 **Bertioga** São Paulo, Brazil 23°51′S 46°09′W
134 C5 **Bertoua** Est, E Cameroon 04°34′N 13°42′E
70 G6 **Bertram** Texas, SW USA 30°44′N 98°03′W
117 B12 **Bertrand, Cerro** ▲ S Argentina 50°00′S 73°27′W
163 F13 **Bertrix** Luxembourg, SE Belgium 49°52′N 05°15′E
284 B3 **Beru** *var.* Peru. *atoll* Tungaru, W Kiribati
Beruni *see* Beruniy
228 E5 **Beruniy** *var.* Biruni, *Rus.* Beruni. Qoraqalpog'iston Respublikasi, W Uzbekistan 41°48′N 60°39′E
106 G3 **Beruri** Amazonas, NW Brazil 03°44′S 61°13′W
64 E5 **Berwick** Pennsylvania, NE USA 41°03′N 76°13′W
156 G7 **Berwick** *cultural region* SE Scotland, United Kingdom
159 J3 **Berwick-upon-Tweed** N England, United Kingdom 55°46′N 02°W
189 J7 **Beryslav** *Rus.* Berislav. Khersons'ka Oblast', S Ukraine 46°51′N 33°26′E
Berytus *see* Beyrouth
172 F4 **Berzocana** Extremadura, Spain 39°26′N 5°27′W
139 K6 **Besalampy** Mahajanga, W Madagascar 16°43′S 44°29′E
169 H8 **Besalú** Cataluña, Spain 42°12′N 2°42′E
165 K5 **Besançon** *anc.* Besontium, Vesontio. Doubs, E France 47°14′N 06°01′E
165 I6 **Besbre** ≈ C France
Bescanuova *see* Baška
181 B13 **Besch** Saarland, Germany 49°30′E 6°22′N
Besdan *see* Bezdan
Besed' *see* Byesyedz'
229 J6 **Beshariq** *Rus.* Besharyk; *prev.* Kirovo. Farg'ona Viloyati, E Uzbekistan 40°26′N 70°33′E
Besharyk *see* Beshariq
228 G5 **Beshbuloq** *Rus.* Beshulak. Navoiy Viloyati, N Uzbekistan 41°55′N 64°13′E
Beshenkovichi *see* Byeshankovichy
228 G7 **Beshkent** Qashqadaryo Viloyati, S Uzbekistan 38°47′N 65°42′E
Beshulak *see* Beshbuloq
261 H4 **Besir** Papua, E Indonesia 0°25′S 130°38′E
184 F3 **Beška** Vojvodina, N Serbia 45°09′N 20°04′E
218 F3 **Beskinta** Lebanon 33°56′N 35°47′E
Beskra *see* Biskra
197 I9 **Beslan** Respublika Severnaya Osetiya, SW Russian Federation 43°12′N 44°33′E
185 H6 **Besna Kobila** ▲ SE Serbia 42°30′N 22°16′E
214 G7 **Besni** Adıyaman, S Turkey 37°42′N 37°53′E
Besontium *see* Besançon
218 C2 **Beşparmak Dağları** *Eng.* Kyrenia Mountains. ▲▲ N Cyprus
Bessarabka *see* Basarabeasca
158 D8 **Bessbrook** United Kingdom 54°12′N 6°25′W
181 G9 **Besse** Hessen, Germany 51°13′E 9°23′N
169 J4 **Bességes** Languedoc-Roussillon, France 44°17′N 4°06′E
152 B5 **Bessels, Kapp** *headland* C Svalbard 78°36′N 21°43′E
66 D9 **Bessemer** Alabama, S USA 33°24′N 86°57′W
72 C4 **Bessemer** Michigan, N USA 46°28′N 90°03′W
67 I7 **Bessemer City** North Carolina, SE USA 35°16′N 81°16′W
181 B13 **Besseringen** Saarland, Germany 49°29′E 6°36′N
169 L6 **Besse-sur-Issole** Provence-Alpes-Côte d'Azur, France 43°21′N 6°10′E
165 H6 **Bessines-sur-Gartempe** Haute-Vienne, C France 46°06′N 01°22′E
163 G8 **Best** Noord-Brabant, S Netherlands 51°31′N 05°24′E
70 E5 **Best** Texas, SW USA 31°13′N 101°34′W
217 K2 **Beste** *Ar.* Basṭah. As Sulaymānīyah, E Iraq 36°20′N 45°14′E
195 H7 **Bestuzhevo** Arkhangel'skaya Oblast', NW Russian Federation 61°36′N 43°54′E
193 J6 **Bestyakh** Respublika Sakha (Yakutiya), NE Russian Federation 61°25′N 129°05′E
Beszterce *see* Bistriţa
Besztercebánya *see* Banská Bystrica
139 L7 **Betafo** Antananarivo, C Madagascar 19°50′S 46°50′E
170 D1 **Betanzos** Galicia, NW Spain 43°17′N 08°17′W
170 C1 **Betanzos, Ría de** *estuary* NW Spain
134 C5 **Bétaré Oya** Est, E Cameroon 05°34′N 14°09′E
114 C5 **Betbeder** Entre Ríos, Argentina 32°23′S 59°55′W
219 D8 **Bet Dagan** *var.* Beit Dagan. Central, C Israel 32°00′N 34°49′E
168 B6 **Betelu** Navarra, Spain 43°01′N 1°59′W
171 I6 **Bétera** Valenciana, E Spain 39°35′N 00°28′W
133 I6 **Bétérou** C Benin 09°13′N 02°18′E
173 J2 **Beteta** Castilla-La Manacha, Spain 40°34′N 2°04′W
219 D9 **Bet Guvrin** *var.* Beit Guvrin. Southern, C Israel 31°36′N 34°54′E
141 J4 **Bethal** Mpumalanga, NE South Africa 26°27′S 29°28′E
73 B11 **Bethalto** Illinois, N USA 38°54′N 90°02′W
140 C4 **Bethanie** *var.* Bethanien, Bethany. Karas, S Namibia 26°32′S 17°11′E
Bethanien *see* Bethanie
75 G8 **Bethany** Missouri, C USA 40°15′N 94°03′W
75 E12 **Bethany** Oklahoma, C USA 35°31′N 97°37′W
Bethany *see* Bethanie
218 F6 **Bet HaShitta** *var.* Beit HaShitta. Northern, N Israel 32°33′N 35°26′E
82 G6 **Bethel** Alaska, USA 60°47′N 161°45′W
65 I4 **Bethel** Connecticut, USA 41°22′N 73°25′W
63 J4 **Bethel** Maine, NE USA 44°24′N 70°47′W
67 L6 **Bethel** North Carolina, SE USA 35°46′N 77°21′W
62 C8 **Bethel Park** Pennsylvania, NE USA 40°21′N 80°03′W
160 E2 **Bethesda** United Kingdom 53°10′N 4°05′W
64 C10 **Bethesda** Maryland, NE USA 39°00′N 77°05′W
141 J5 **Bethlehem** Free State, C South Africa 28°12′S 28°16′E
64 A5 **Bethlehem** Pennsylvania, USA 40°52′N 78°44′W
64 F6 **Bethlehem** Pennsylvania, NE USA 40°36′N 75°22′W
219 219 **Bethlehem** *var.* Beit Lekhem, *Ar.* Bayt Laḩm, *Heb.* Bet Leḥem. C West Bank 31°43′N 35°12′E
Bethlen *see* Beclean
141 H7 **Bethulie** Free State, C South Africa 30°30′S 25°59′E
165 H1 **Béthune** Pas-de-Calais, N France 50°32′N 02°38′E
165 H2 **Béthune** ≈ N France
170 F8 **Béticos, Sistemas** *var.* Sistema Penibético, *Eng.* Baetic Cordillera, Baetic Mountains. ▲▲ S Spain
102 E3 **Betijoque** Trujillo, NW Venezuela 09°25′N 70°45′W
110 G3 **Betim** Minas Gerais, SE Brazil 19°56′S 44°10′W
284 C2 **Betio** Tarawa, W Kiribati 01°21′N 172°56′E
139 L9 **Betioky** Toliara, S Madagascar 23°42′S 44°22′E
219 D9 **Bet Kama** Israel 31°27′N 34°46′E
Bet Leḥem *see* Bethlehem
218 E6 **Bet Leḥem HaGelilit** *var.* Beit Leḥem HaGelilit. Northern, N Israel 32°44′N 35°10′E
Betlen *see* Beclean
219 D9 **Bet Nir** *var.* Beit Nir. Israel 31°39′N 34°52′E
257 F12 **Betong** Yala, SW Thailand 05°45′N 101°05′E
134 E6 **Bétou** Likouala, N Congo 03°08′N 18°31′E
227 H6 **Betpakdala** *Kaz.* Betpaqdala *prev.* Betpak-Dala. *plateau* S Kazakhstan
Betpak-Dala *see* Betpakdala
Betpaqdala *see* Betpakdala
219 D9 **Bet Qama** *var.* Beit Kama. Israel
139 L9 **Betroka** Toliara, S Madagascar 23°15′S 46°07′E
Betschau *see* Bečva
167 N5 **Betschdorf** Alsace, France 48°54′N 7°54′E
218 F6 **Bet Shean** Israel 32°30′N 35°30′E
Bet She'an *see* Beit She'an
219 E8 **Bet Shemesh** *var.* Beit Shemesh. Jerusalem, C Israel 31°45′N 35°00′E
59 I7 **Betsiamites** Québec, SE Canada 48°56′N 68°40′W
59 H7 **Betsiamites** ≈ Québec, SE Canada
139 L7 **Betsiboka** ≈ N Madagascar
Betsukai *see* Bekkai
163 H14 **Bettembourg** Luxembourg, S Luxembourg 49°31′N 06°06′E
163 H13 **Bettendorf** Diekirch, NE Luxembourg 49°53′N 06°13′E
73 B9 **Bettendorf** Iowa, C USA 41°31′N 90°31′W
Bette, Pic *see* Bette, Picco
128 E6 **Bette, Picco** *var.* Bīkkū Bīttī, Pic Bette. ▲ S Libya 22°02′N 19°07′E
233 H6 **Bettiah** Bihār, N India 26°49′N 84°30′E
141 J4 **Bettiesdam** Mpumalanga, South Africa 26°27′S 29°28′E
173 L10 **Bettioua** Algeria 35°48′N 00°16′W
83 I4 **Bettles** Alaska, USA 66°54′N 151°40′W
155 H8 **Bettna** Södermanland, C Sweden 58°52′N 16°40′E
232 E9 **Betūl** *prev.* Badnur. Madhya Pradesh, C India 21°55′N 77°54′E
232 F7 **Betwa** ≈ C India
160 F2 **Betws-y-Coed** N Wales, United Kingdom 53°05′N 03°45′W
218 D7 **Bet Yiẕhaq-Shaar Ḥefer** *var.* Beit Yitskhak Sha'ar Khefer. Central, C Israel 32°19′N 34°53′E
167 H5 **Betz** Picardie, France 49°09′N 2°57′E
181 D10 **Betzdorf** Rheinland-Pfalz, W Germany 50°47′N 07°50′E
135 D10 **Béu** Uíge, NW Angola 06°15′S 15°32′E
181 F10 **Beuern** Hessen, Germany 50°38′E 8°49′N
72 F6 **Beulah** Michigan, N USA 43°35′N 83°52′W
74 B3 **Beulah** North Dakota, N USA 47°16′N 101°48′W
162 H5 **Beulakerwijde** ◎ N Netherlands
161 L7 **Beult** ≈ United Kingdom
162 G7 **Beuningen** Gelderland, SE Netherlands 51°52′N 05°47′E
Beuthen *see* Bytom
165 H4 **Beuvron** ≈ C France
163 D9 **Beveren** Oost-Vlaanderen, N Belgium 51°13′N 04°15′E
67 K6 **B. Everett Jordan Reservoir** *var.* Jordan Lake. ⊠ North Carolina, SE USA
159 M8 **Beverley** E England, United Kingdom 53°51′N 00°26′W
Beverley *see* Beverly
163 F9 **Beverlo** Limburg, NE Belgium 51°06′N 05°14′E
65 M2 **Beverly** Massachusetts, NE USA 42°33′N 70°49′W

◆ Country
● Country Capital
◇ Dependent Territory
○ Dependent Territory Capital
◈ Administrative Regions
✈ International Airport
▲ Mountain
▲▲ Mountain Range
⛰ Volcano
≈ River
◎ Lake
⊠ Reservoir

◆ Country
● Country Capital
◇ Dependent Territory
○ Dependent Territory Capital
♦ Administrative Regions
✕ International Airport
▲ Mountain
▲ Mountain Range
▲ Volcano
∅ River
⊙ Lake
⊠ Reservoir

◆ Country
● Country Capital
◇ Dependent Territory
○ Dependent Territory Capital
✕ Administrative Regions
✈ International Airport
▲ Mountain
▲ Mountain Range
🌋 Volcano
🌊 River
💧 Lake
💧 Reservoir

Chaknakdysonga - Chernavchitsy
INDEX
327

228 E9 **Chaknakdysonga** Ahal Welaýaty, S Turkmenistan 35°39′N 61°24′E
233 I8 **Chakradharpur** Jhāarkhand, N India 22°42′N 85°38′E
231 I4 **Chakwāl** Punjab, NE Pakistan 32°56′N 72°53′E
105 E9 **Chala** Arequipa, SW Peru 15°52′S 74°13′W
164 G7 **Chalais** Charente, W France 45°16′N 00°02′E
176 D5 **Chalais** Valais, SW Switzerland 46°18′N 07°37′E
187 I7 **Chalándri** *var.* Halandri; *prev.* Khalándrion. *prehistoric site* Sýros, Kykládes, Greece, Aegean Sea
283 H2 **Chalan Kanoa** Saipan, S Northern Mariana Islands 15°08′S 145°43′E
282 B2 **Chalan Pago** C Guam
Chalap Dalam/Chalap Dalan *see* Chehel Abdālān, Kūh-e
88 D4 **Chalatenango** Chalatenango, N El Salvador 14°04′N 88°53′W
88 D4 **Chalatenango** ◈ *department* NW El Salvador
139 K2 **Chalaua** Nampula, NE Mozambique 16°04′S 39°08′E
136 D7 **Chalbi Desert** *desert* N Kenya
86 G8 **Chalcatongo** Oaxaca, Mexico 17°02′N 97°35′W
85 J8 **Chalchihuites** Zacatecas, Mexico 23°29′N 103°53′W
87 I8 **Chalchijapan** Veracruz-Llave, Mexico 17°27′N 94°51′W
88 D4 **Chalchuapa** Santa Ana, W El Salvador 13°59′N 89°41′W
Chalcidice *see* Chalkidikí
Chalcis *see* Chalkída
Chālderān *see* Sīāh Chashmeh
161 I8 **Chale** United Kingdom 50°35′N 1°19′W
165 I4 **Châlette-sur-Loing** Loiret, C France 48°01′N 02°45′E
59 J7 **Chaleur Bay** *Fr.* Baie des Chaleurs. *bay* New Brunswick/Québec, E Canada
Chaleurs, Baie des *see* Chaleur Bay
105 F8 **Chalhuanca** Apurímac, S Peru 14°17′S 73°15′W
167 K7 **Chalindrey** Champagne-Ardenne, France 47°48′N 5°26′E
243 H6 **Chaling** Hunan, China 26°29′N 113°19′E
118 D1 **Chalinguita** Coquimbo, Chile 31°45′S 70°58′W
234 D4 **Chālisgaon** Mahārāshtra, C India 20°29′N 75°10′E
187 L8 **Chálki** *island* Dodekánisa, Greece, Aegean Sea
186 G4 **Chalkiádes** Thessalía, C Greece 39°24′N 22°25′E
187 H6 **Chalkída** *var.* Halkida, *prev.* Khalkís; *anc.* Chalcis. Evvoia, E Greece 38°27′N 23°38′E
187 H3 **Chalkidikí** *var.* Khalkidhikí; *anc.* Chalcidice. *peninsula* NE Greece
278 A12 **Chalky Inlet** *inlet* South Island, New Zealand
83 J4 **Chalkyitsik** Alaska, USA 66°39′N 143°43′W
164 F5 **Challans** Vendée, NW France 46°51′N 01°52′W
112 C1 **Challapata** Oruro, SW Bolivia 18°50′S 66°45′W
80 E1 **Challenge** California, USA 39°29′N 121°13′W
286 D5 **Challenger Deep** *undersea feature* W Pacific Ocean 11°20′N 142°12′E
Challenger Deep *see* Mariana Trench
287 J8 **Challenger Fracture Zone** *tectonic feature* SE Pacific Ocean
286 E8 **Challenger Plateau** *undersea feature* E Tasman Sea
76 G6 **Challis** Idaho, NW USA 44°31′N 114°14′W
158 G6 **Challoch** United Kingdom 54°58′N 4°32′W
68 E4 **Chalmette** Louisiana, S USA 29°56′N 89°57′W
194 E7 **Chalna** Respublika Kareliya, NW Russian Federation 61°53′N 33°59′E
166 D8 **Chalonnes-sur-Loire** Pays de la Loire, France 47°21′N 0°46′W
165 J3 **Châlons-en-Champagne** *prev.* Châlons-sur-Marne, *hist.* Arcae Remorum; *anc.* Carolopois. Marne, NE France 48°58′N 04°22′E
Châlons-sur-Marne *see* Châlons-en-Champagne
165 J5 **Chalon-sur-Saône** *anc.* Cabillonum. Saône-et-Loire, C France 46°47′N 04°51′E
Chaltel, Cerro *see* Fitzroy, Monte
164 G6 **Châlus** Haute-Vienne, C France 45°38′N 01°00′E
222 F3 **Chālūs** Māzandarān, N Iran 36°40′N 51°25′E
179 H11 **Cham** Bayern, SE Germany 49°13′N 12°40′E
79 L5 **Chama** New Mexico, SW USA 36°54′N 106°34′W
Cha Mai *see* Thung Song
140 B5 **Chamais Bay** Karas, Namibia
140 C5 **Chamaites** Karas, S Namibia 27°15′S 17°52′E
169 H1 **Chamaliéres** Auvergne, France 45°47′N 3°04′E
230 F6 **Chaman** Baluchistān, SW Pakistan 30°55′N 66°27′E
79 L6 **Chama, Rio** ≋ New Mexico, SW USA
232 E3 **Chamba** Himāchal Pradesh, N India 32°33′N 76°10′E
137 D12 **Chamba** Ruvuma, S Tanzania 11°33′S 37°01′E
232 E6 **Chambal** ≋ C India
57 N7 **Chamberlain** Saskatchewan, S Canada 50°49′N 105°29′W
115 H6 **Chamberlain** Uruguay 32°37′S 56°29′W
74 D5 **Chamberlain** South Dakota, N USA 43°48′N 99°19′W
63 K2 **Chamberlain Lake** ◎ Maine, NE USA
83 J2 **Chamberlin, Mount** ▲ Alaska, USA 69°16′N 144°54′W
79 J7 **Chambers** Arizona, SW USA 35°11′N 109°25′W
64 C8 **Chambersburg** Pennsylvania, NE USA 39°54′N 77°39′W
72 E5 **Chambers Island** *island* Wisconsin, N USA
165 K6 **Chambéry** *anc.* Camberia. Savoie, E France 45°34′N 05°56′E
137 B12 **Chambeshi** Muchinga, NE Zambia 10°55′S 31°07′E
137 A12 **Chambeshi** ≋ NE Zambia
131 K2 **Chambi, Jebel** *var.* Jabal ash Sha'nabī. ▲ W Tunisia 35°16′N 08°39′E
81 H11 **Chambless** California, USA 34°34′N 115°33′W
167 L5 **Chambley-Bussiéres** Lorraine, France 49°03′N 5°54′E
167 H10 **Chambon-sur-Voueize** Limousin, France 46°11′N 2°25′E
166 G8 **Chambord** Centre, France 47°37′N 1°31′E
166 F8 **Chambray** Centre, France 47°20′N 0°43′E
261 N6 **Chambri Lake** ◎ W Papua New Guinea
217 K7 **Chamcham** Al Muthanná, S Iraq 31°17′N 45°05′E
Chamchamāļ *see* Chemchemal
86 B6 **Chamela** Jalisco, SW Mexico 19°31′N 105°02′W
88 E3 **Chamelecón, Río** ≋ NW Honduras
112 D7 **Chamical** La Rioja, C Argentina 30°21′S 66°19′W
187 K9 **Chamilí** *island* Kykládes, Greece, Aegean Sea
115 H9 **Chamizo** Lavalleja, Uruguay 34°15′S 55°56′W
257 G9 **Châmnar** Kaôh Kông, SW Cambodia 11°45′N 103°32′E
232 F4 **Chamoli** Uttarakhand, N India 30°22′N 79°19′E
169 M1 **Chamonix-Mont-Blanc** Haute-Savoie, E France 45°55′N 06°52′E
234 H3 **Chāmpa** Chhattīsgarh, C India 22°02′N 82°42′E
168 E2 **Champagnac-de-Bélair** Aquitaine, France 45°24′N 0°42′E
54 B3 **Champagne** Yukon, W Canada 60°48′N 136°22′W
165 J3 **Champagne** *cultural region* N France
Champagne *see* Campania
165 J3 **Champagne-Ardenne** ◈ *region* N France
168 E1 **Champagne-Mouton** Poitou-Charentes, France 45°59′N 0°25′E
165 K5 **Champagnole** Jura, E France 46°44′N 05°55′E
167 J8 **Champagny** Bourgogne, France 47°28′N 4°46′E
73 D10 **Champaign** Illinois, N USA 40°07′N 88°15′W
256 H7 **Champasak** Champasak, S Laos 14°50′N 105°51′E
167 M6 **Champ de Feu** ▲ NE France 48°24′N 07°15′E
166 D10 **Champdeniers** Poitou-Charentes, France 46°29′N 0°24′W
59 I4 **Champdoré, Lac** ◎ Québec, NE Canada
169 H2 **Champeix** Auvergne, France 45°35′N 3°08′E
88 B4 **Champerico** Retalhuleu, SW Guatemala 14°18′N 91°54′W
176 C5 **Champéry** Valais, SW Switzerland 46°12′N 06°52′E
167 I6 **Champigny** Bourgogne, France 48°19′N 3°08′E
63 H3 **Champlain** New York, NE USA 44°58′N 73°25′W
63 H5 **Champlain Canal** *canal* New York, NE USA
63 H3 **Champlain, Lac** ◎ Québec, Canada/USA *see also* Champlain, Lake
63 H4 **Champlain, Lake** ◎ Canada/USA *see also* Champlain, Lac
165 J4 **Champlitte** Haute-Saône, E France 47°36′N 05°31′E
167 K7 **Champlitte-la-Ville** Franche-Comté, France 47°37′N 5°32′E
169 N1 **Champoluc** Valle D'Aosta, NW Italy 45°49′N 07°43′E
87 L6 **Champotón** Campeche, SE Mexico 19°18′N 90°43′W
235 E8 **Chāmrājnagar** *var.* Chamrajnagar. Karnātaka, S India 12°00′N 77°18′E
Chamrajnagar *see* Chāmrājnagar
169 L2 **Chamrousse** Rhône-Alpes, France 45°08′N 5°54′E
172 B4 **Chamusca** Santarém, C Portugal 39°21′N 08°29′W
191 I12 **Chamyarysy** *Rus.* Chemerisy. Homyel'skaya Voblasts', SE Belarus 51°42′N 30°27′E
197 I3 **Chamzinka** Respublika Mordoviya, W Russian Federation 54°22′N 45°22′E
169 I4 **Chanac** Languedoc-Roussillon, France 44°28′N 3°20′E
Chanáil Mhór, An *see* Grand Canal
Chanak *see* Çanakkale
112 B5 **Chañaral** Atacama, N Chile 26°19′S 70°34′W
170 D7 **Chança, Rio** *var.* Chanza. ≋ Portugal/Spain
114 A10 **Chancay** Buenos Aires, Argentina 35°03′S 61°15′W
104 D7 **Chancay** Lima, W Peru 11°36′S 77°14′W
Chan-chiang/Chanchiang *see* Zhanjiang
116 B5 **Chanco** Maule, C Chile 35°43′S 72°35′W
83 I3 **Chandalar** Alaska, USA 67°30′N 148°29′W
83 I4 **Chandalar River** ≋ Alaska, USA
232 G5 **Chandan Chauki** Uttar Pradesh, N India 28°32′N 80°43′E
233 J8 **Chandannagar** *prev.* Chandernagore. West Bengal, E India 22°52′N 88°21′E
232 F5 **Chandausi** Uttar Pradesh, N India 28°27′N 78°43′E
68 F5 **Chandeleur Islands** *island group* Louisiana, S USA
68 F4 **Chandeleur Sound** *sound* N Gulf of Mexico
Chandernagore *see* Chandannagar
232 E4 **Chandīgarh** *state capital* Punjab, N India 30°41′N 76°51′E
233 I8 **Chāndil** Jhārkhand, NE India 22°58′N 86°04′E
276 E1 **Chandler** South Australia 26°59′S 133°22′E
59 J7 **Chandler** Québec, SE Canada 48°21′N 64°41′W
75 E12 **Chandler** Oklahoma, C USA 35°43′N 96°54′W
71 I5 **Chandler** Texas, SW USA 32°18′N 95°28′W
83 I3 **Chandler River** ≋ Alaska, USA
104 G7 **Chandles, Río** ≋ E Peru
239 I2 **Chandmanĭ** *var.* Talshand. Govĭ-Altayĭ, C Mongolia 45°21′N 98°00′E
239 H2 **Chandmanĭ** *var.* Urdgol. Hovd, W Mongolia 47°39′N 92°46′E
62 E4 **Chandos Lake** ◎ Ontario, SE Canada
233 K8 **Chandpur** Chittagong, C Bangladesh 23°13′N 90°43′E
234 F4 **Chandrapur** Mahārāshtra, C India 19°58′N 79°21′E
138 G2 **Changa** Southern, S Zambia 16°24′S 28°27′E
250 B4 **Chang-an** 35°18′N 129°14′E
Chang'an *see* Rong'an, Guangxi Zhuangzu Zizhiqu, S China
Changan *see* Xi'an, Shaanxi, C China
235 E9 **Changanācheri** *var.* Changannassery. Kerala, SW India 09°26′N 76°31′E *see also* Changannassery
139 I5 **Changane** ≋ S Mozambique
139 I2 **Changara** Tete, NW Mozambique 16°54′S 33°15′E
248 D3 **Changbai** *var.* Changbai Chaoxianzu Zizhixian. Jilin, NE China 41°25′N 128°08′E
Changbai Chaoxianzu Zizhixian *see* Changbai
248 D2 **Changbai Shan** ▲▲ NE China
237 J4 **Changchun** *var.* Ch'angch'un, Ch'ang-ch'un; *prev.* Hsinking. *province capital* Jilin, NE China 43°53′N 125°18′E
Ch'angch'un/Ch'ang-ch'un *see* Changchun
242 G4 **Changde** Hunan, S China 29°04′N 111°42′E
243 J1 **Changfeng** Anhui, China 32°17′N 117°05′E
245 H8 **Changge** Henan, China 34°07′N 113°28′E
245 N3 **Changhai** Liaoning, China 39°10′N 122°20′E
Changhua *see* Zhanghua
258 F4 **Changi** ✈ (Singapore) E Singapore 01°22′N 103°58′E
238 F3 **Changji** Xinjiang Uygur Zizhiqu, NW China 44°02′N 87°12′E
240 F10 **Changjiang** *var.* Changjiang Lizu Zizhixian, Shiliu. Hainan, S China 19°16′N 109°09′E
243 H3 **Chang Jiang** *var.* Yangtze Kiang, *Eng.* Yangtze. ≋ C China
242 E2 **Chang Jiang** *Eng.* Yangtze. ≋ SW China
243 N1 **Changjiang Kou** *delta* E China
Changjiang Lizu Zizhixian *see* Changjiang
242 G2 **Changkang** Hubei, C China 30°36′N 112°53′E
Changkiakow *see* Zhangjiakou
257 F9 **Chang, Ko** *island* S Thailand
243 L6 **Changle** Fujan, China 25°35′N 119°19′E
243 H4 **Changle** Hunan, E China 28°51′N 113°17′E
245 K5 **Changle** Shandong, E China 36°42′N 118°50′E
243 M1 **Changlezhen** Jiangsu, E China 31°56′N 121°14′E
245 K3 **Changli** Hebei, E China 39°44′N 119°13′E
247 K5 **Changling** Jilin, NE China 44°15′N 124°03′E
242 G6 **Changning** Hunan, China 26°14′N 112°14′E
242 B4 **Changning** Sichuan, C China 28°35′N 104°57′E
Changning *see* Xunwu
250 B3 **Changnyeong** *prev.* Ch'angnyŏng. 35°32′N 128°30′E
Ch'angnyŏng *see* Changnyeong
245 J3 **Changping** Beijing Shi, N China 40°08′N 116°08′E
242 F6 **Chang-pu-tzu** Hunan, SE China 26°38′N 110°24′E
245 J6 **Changqing** Shandong, China 36°20′N 116°25′E
243 H5 **Changsha** *var.* Ch'angsha, Ch'ang-sha. *province capital* Hunan, S China 28°10′N 113°E
Ch'angsha/Ch'ang-sha *see* Changsha
243 L4 **Changshan** Zhejiang, SE China 28°54′N 118°30′E
245 N3 **Changshan Qundao** *island group* NE China
242 C3 **Changshou** Chongqing Shi, C China 29°31′N 107°01′E
243 H4 **Changshou** Hunan, E China 28°43′N 113°56′E
243 M1 **Changshu** *var.* Ch'ang-shu. Jiangsu, E China 31°39′N 120°45′E
Ch'ang-shu *see* Changshu
242 C6 **Changshun** Guizhou, S China 25°59′N 106°25′E
Changsu *see* Jangsu
243 K8 **Changtai** Fujan, China 24°22′N 117°28′E
243 J7 **Changting** Fujan, China 25°31′N 116°13′E
248 B1 **Changtu** Liaoning, NE China 42°50′N 123°59′E
89 I8 **Changuinola** Bocas del Toro, NW Panama 09°28′N 82°31′W
239 H5 **Changweiliang** Qinghai, W China 38°24′N 92°08′E
250 B4 **Ch'Ang Won** 35°14′N 128°39′E
244 D7 **Changwu** *var.* Zhaoren. Shaanxi, C China 35°12′N 107°46′E
243 L2 **Changxing** Zhejiang, China 31°01′N 119°33′E
245 M3 **Changxing Dao** *island* N China
167 J10 **Changy** Rhône-Alpes, France 46°09′N 3°54′E
242 F3 **Changyang** *var.* Longzhouping. Hubei, C China 30°45′N 111°13′E
245 L5 **Changyi** Shandong, E China 36°51′N 119°23′E
248 B4 **Changyŏn** SW North Korea 38°19′N 125°15′E
245 H6 **Changzhi** Shanxi, China 36°01′N 113°01′E
245 H6 **Changzhi** Shanxi, C China 36°10′N 113°02′E
243 L1 **Changzhou** Jiangsu, E China 31°45′N 119°58′E
187 H9 **Chaniá** *var.* Hania, Khaniá, *Eng.* Canea; *anc.* Cydonia. Kríti, Greece, E Mediterranean Sea 35°31′N 24°00′E
112 D4 **Chañi, Nevado de** ▲ NW Argentina 24°09′S 65°44′W
187 H9 **Chanión, Kólpos** *gulf* Kríti, Greece, E Mediterranean Sea
Chankiri *see* Çankırı
73 D9 **Channahon** Illinois, N USA 41°25′N 88°13′W
235 E8 **Channapatna** Karnātaka, E India 12°43′N 77°14′E
166 B4 **Channel Islands** *Fr.* Iles Normandes. *island group* S English Channel
81 B11 **Channel Islands** *island group* California, W USA
59 K7 **Channel-Port aux Basques** Newfoundland and Labrador, SE Canada 47°35′N 59°02′W
Channel, The *see* English Channel
161 N7 **Channel Tunnel** *tunnel* France/United Kingdom
70 D2 **Channing** Texas, SW USA 35°41′N 102°21′W
Chantabun/Chantaburi *see* Chanthaburi
170 D2 **Chantada** Galicia, NW Spain 42°36′N 07°46′W
167 I10 **Chantelle** Auvergne, France 46°14′N 3°09′E
86 F9 **Chantengo, Laguna de** *bay* Guerrero, S Mexico Central America
257 F8 **Chanthaburi** *var.* Chantabun, Chantaburi. Chantaburi, S Thailand 12°35′N 102°08′E
165 H3 **Chantilly** Oise, N France 49°12′N 02°28′E
166 D9 **Chantonnay** Pays de la Loire, France 46°41′N 1°03′W
53 H8 **Chantrey Inlet** *inlet* Nunavut, N Canada
217 K8 **Chanūn as Saʻūdī** Dhī Qār, S Iraq 31°04′N 46°00′E
75 F10 **Chanute** Kansas, C USA 37°40′N 95°27′W
227 J1 **Chany, Ozero** ◎ C Russian Federation
Chanza *see* Chança, Rio
243 J9 **Chao'an** *var.* Anbu. Guangdong, SE China 23°27′N 116°41′E
Ch'ao-an/Chaochow *see* Chaozhou
243 K2 **Chaohu** Anhui, E China 31°36′N 117°55′E
243 K1 **Chaohu** Anhui, China 31°22′N 117°31′E
243 K2 **Chao Hu** ◎ E China
245 M6 **Chaolian Dao** *island* Shandong, China
262 B5 **Chao Phraya, Mae Nam** ≋ C Thailand
247 K4 **Chaor Hol** *prev.* Qulin Gol. ≋ NE China
Chaouèn *see* Chefchaouen
167 J7 **Chaource** Champagne-Ardenne, France 48°04′N 4°08′E
243 J9 **Chaoyang** Guangdong, S China 23°17′N 116°33′E
243 J9 **Chaoyang** Guangdong, China 23°10′N 116°22′E
245 L1 **Chaoyang** Liaoning, NE China 41°34′N 120°29′E
Chaoyang *see* Jiayin, Heilongjiang, China
Chaoyang *see* Jiayan
Chaoyang *see* Huinan, Jilin China
243 J9 **Chaozhou** *var.* Chaoan, Chao'an, Ch'ao-an; *prev.* Chaochow. Guangdong, SE China 23°42′N 116°36′E
108 E4 **Chapadinha** Maranhão, E Brazil 03°45′S 43°23′W
58 G7 **Chapais** Québec, SE Canada 49°47′N 74°54′W
84 C3 **Chapala** Baja California Norte, NW Mexico 29°24′N 114°21′W
86 C5 **Chapala** Jalisco, SW Mexico 20°20′N 103°10′W
86 C5 **Chapala, Lago de** ◎ C Mexico
85 I10 **Chapalilla** Nayarit, Mexico
223 I2 **Chapan, Gora** ▲ C Turkmenistan 37°48′N 58°03′E
106 E10 **Chapare, Río** ≋ C Bolivia
102 B6 **Chaparral** Tolima, C Colombia 03°45′N 75°30′W
226 C3 **Chapayev** Zapadnyy Kazakhstan, NW Kazakhstan 50°12′N 51°09′E
193 I6 **Chapayevo** Respublika Sakha (Yakutiya), NE Russian Federation 60°03′N 117°19′E
197 J3 **Chapayevsk** Samarskaya Oblast', W Russian Federation 52°57′N 49°42′E
113 I6 **Chapecó** Santa Catarina, S Brazil 27°14′S 52°41′W
113 J5 **Chapecó, Rio** ≋ S Brazil
159 K10 **Chapel en le Frith** United Kingdom 53°19′N 1°54′W
66 D6 **Chapel Hill** Tennessee, S USA 35°38′N 86°40′W
90 F8 **Chapelton** C Jamaica 18°05′N 77°16′W
114 F4 **Chapicuy** Paysandú, Uruguay 31°40′S 57°55′W
58 E7 **Chapleau** Ontario, SE Canada 47°50′N 83°24′W
57 M7 **Chaplin** Saskatchewan, S Canada 50°27′N 106°37′W
196 G3 **Chaplygin** Lipetskaya Oblast', W Russian Federation 53°13′N 39°58′E
189 J7 **Chaplynka** Khersons'ka Oblast', S Ukraine 46°20′N 33°34′E
53 I8 **Chapman, Cape** *headland* Nunavut, NE Canada 69°15′N 89°09′W
70 G9 **Chapman Ranch** Texas, SW USA 27°32′N 97°25′W
Chapman's *see* Okwa
67 H4 **Chapmanville** West Virginia, NE USA 37°58′N 82°01′W
74 B7 **Chappell** Nebraska, C USA 41°05′N 102°28′W
Chapra *see* Chhapra
104 D3 **Chapuli, Río** ≋ N Peru
114 A8 **Chapuy** Santa Fe, Argentina 33°48′S 61°45′W
218 F5 **Chaqra** Lebanon 33°11′N 35°28′E
132 C1 **Châr** *well* N Mauritania
193 J7 **Chara** Zabaykal'skiy Kray, S Russian Federation 56°57′N 118°05′E
193 J7 **Chara** ≋ C Russian Federation
102 C5 **Charala** Santander, C Colombia 06°17′N 73°09′W
85 L8 **Charcas** San Luis Potosí, C Mexico 23°09′N 101°10′W
70 G8 **Charco** Texas, SW USA 28°42′N 97°35′E
85 I4 **Charco de Peña** Chihuahua, Mexico 28°31′N 104°59′W
85 J6 **Charcos de Risa** Coahuila, Mexico 26°12′N 103°06′W
292 E4 **Charcot Island** *island* Antarctica
290 H4 **Charcot Seamounts** *undersea feature* E Atlantic Ocean 11°30′W 45°00′N
160 G8 **Chard** United Kingdom 50°52′N 2°58′W
Chardara *see* Shardara
Chardarinskoye Vodokhranilishche *see* Shardarinskoye Vodokhranilishche
73 J9 **Chardon** Ohio, N USA 41°34′N 81°12′W
90 G6 **Chardonnières** SW Haiti 18°16′N 74°10′W
Chardzhev *see* Türkmenabat
Chardzhevskaya Oblast *see* Lebap Welaýaty
Chardzhou/Chardzhui *see* Türkmenabat
164 G6 **Charente** ◈ *department* W France
164 F6 **Charente** ≋ W France
164 F6 **Charente-Maritime** ◈ *department* W France
167 H9 **Charenton-du-Cher** Centre, France 46°44′N 2°39′E
215 K5 **Ch'arents'avan** C Armenia 40°23′N 44°41′E
133 N6 **Chari** *var.* Shari. ≋ Central African Republic/Chad
133 L6 **Chari-Baguirmi** *off.* Région du Chari-Baguirmi. ◈ *region* SW Chad
Chari-Baguirmi, Région du *see* Chari-Baguirmi
231 H3 **Chārīkār** Parwān, NE Afghanistan 35°01′N 69°11′E
161 L7 **Charing** United Kingdom 51°12′N 0°48′E
75 H8 **Chariton** Iowa, C USA 41°00′N 93°18′W
75 H9 **Chariton River** ≋ Missouri, C USA
103 K4 **Charity** NW Guyana 07°22′N 58°34′W
72 H7 **Charity Island** *island* Michigan, N USA
Chärjew *see* Türkmenabat
Chärjew Oblasty *see* Lebap Welaýaty
Charkhlik/Charkhliq *see* Ruoqiang
141 J4 **Charl Cilliers** Mpumalanga, South Africa 26°40′S 29°11′E
163 E11 **Charleroi** Hainaut, S Belgium 50°25′N 04°27′E
55 I7 **Charles** Manitoba, C Canada 55°27′N 100°58′W
63 I1 **Charlesbourg** Québec, SE Canada 46°50′N 71°15′W
67 M4 **Charles, Cape** *headland* Virginia, NE USA 37°09′N 75°57′W
74 H6 **Charles City** Iowa, C USA 43°04′N 92°40′W
67 L4 **Charles City** Virginia, NE USA 37°21′N 77°05′W
167 H5 **Charles de Gaulle** ✈ (Paris) Seine-et-Marne, N France 49°04′N 02°36′E
55 N2 **Charles Island** *island* Nunavut, NE Canada
Charles Island *see* Santa María, Isla
74 H5 **Charles-Lindbergh** ✈ (Minneapolis/Saint Paul) Minnesota, N USA 44°47′N 93°12′W
73 B8 **Charles Mound** *hill* Illinois, N USA
278 A11 **Charles Sound** *sound* South Island, New Zealand
279 D8 **Charleston** West Coast, South Island, New Zealand 41°54′S 171°25′E
71 J2 **Charleston** Arkansas, C USA 35°19′N 94°02′W
73 D11 **Charleston** Illinois, N USA 39°30′N 88°10′W
66 B8 **Charleston** Mississippi, S USA 34°00′N 90°03′W
66 B5 **Charleston** Missouri, C USA 36°54′N 89°22′W
67 J9 **Charleston** South Carolina, SE USA 32°48′N 79°57′W
67 I3 **Charleston** *state capital* West Virginia, NE USA 38°21′N 81°38′W
62 F4 **Charleston Lake** ◎ Ontario, SE Canada
81 I8 **Charleston Peak** ▲ Nevada, W USA 36°16′N 115°40′W
91 L6 **Charlestown** Nevis, Saint Kitts and Nevis 17°08′N 62°37′W
73 F12 **Charlestown** Indiana, N USA 38°27′N 85°40′W
63 I5 **Charlestown** New Hampshire, NE USA 43°14′N 72°23′W
64 B9 **Charles Town** West Virginia, NE USA 39°18′N 77°54′W
277 J1 **Charleville** Queensland, E Australia 26°25′S 146°18′E
165 J2 **Charleville-Mézières** Ardennes, N France 49°45′N 04°43′E
72 F5 **Charlevoix** Michigan, N USA 45°19′N 85°15′W
72 G5 **Charlevoix, Lake** ◎ Michigan, N USA
83 J4 **Charley River** ≋ Alaska, USA
290 F3 **Charlie-Gibbs Fracture Zone** *tectonic feature* N Atlantic Ocean
165 I6 **Charlieu** Loire, E France 46°11′N 04°10′E
73 G8 **Charlotte** Michigan, N USA 42°33′N 84°50′W
67 I7 **Charlotte** North Carolina, SE USA 35°14′N 80°51′W
66 D6 **Charlotte** Tennessee, S USA 36°11′N 87°18′W
70 F8 **Charlotte** Texas, SW USA 28°51′N 98°42′W
67 I7 **Charlotte** ✈ North Carolina, SE USA 35°12′N 80°54′W
91 K6 **Charlotte Amalie** *prev.* Saint Thomas. ○ (Virgin Islands (US)) Saint Thomas, N Virgin Islands (US) 18°22′N 64°56′W
67 K5 **Charlotte Court House** Virginia, NE USA 37°04′N 78°37′W
69 L8 **Charlotte Harbor** *inlet* Florida, SE USA
Charlotte Island *see* Abaiang
154 F7 **Charlottenberg** Värmland, C Sweden 59°53′N 12°17′E
Charlottenhof *see* Aegviidu
67 K4 **Charlottesville** Virginia, NE USA 38°02′N 78°29′W
59 J8 **Charlottetown** *province capital* Prince Edward Island, Prince Edward Island, SE Canada 46°14′N 63°09′W
Charlotte Town *see* Roseau, Dominica
Charlotte Town *see* Gouyave, Grenada
103 J2 **Charlotteville** Tobago, Trinidad and Tobago 11°16′N 60°33′W
277 I6 **Charlton** Victoria, SE Australia 36°18′S 143°19′E
58 F6 **Charlton Island** *island* Northwest Territories, C Canada
167 I5 **Charly-sur-Marne** Picardie, France 48°58′N 3°17′E
165 K3 **Charmes** Vosges, NE France 48°19′N 06°19′E
191 D11 **Charnawchytsy** *Rus.* Chernavchitsy. Brestskaya Voblasts', SW Belarus 52°13′N 23°44′E
167 K10 **Charnay-lès-Mâcon** Bourgogne, France 46°18′N 4°47′E
63 J2 **Charny** Québec, SE Canada 46°43′N 71°15′W
167 I7 **Charny** Bourgogne, France 47°53′N 3°06′E
167 K4 **Charny-sur-Meuse** Lorraine, France 49°12′N 5°22′E
167 J10 **Charolles** Bourgogne, France 46°26′N 4°17′E
166 G9 **Chârost** Centre, France 46°59′N 2°08′E
110 B7 **Charqueada** São Paulo, Brazil 22°30′S 47°46′W
115 N1 **Charqueadas** Rio Grande do Sul, Brazil 29°58′S 51°38′W
166 E10 **Charroux** Poitou-Charentes, France 46°09′N 0°24′E
231 I4 **Chārsadda** Khyber Pakhtunkhwa, NW Pakistan 34°12′N 71°46′E
Charshanga/Charshangngy/Charshangy *see* Köýtendag
Charsk *see* Shar
275 J4 **Charters Towers** Queensland, NE Australia 20°02′S 146°20′E
63 J3 **Chartierville** Québec, SE Canada 45°19′N 71°13′W
165 H3 **Chartres** *anc.* Autricum, Civitas Carnutum. Eure-et-Loir, C France 48°27′N 01°27′E
Charyn *see* Sharyn
Charyn *see* Sharyn
116 H4 **Chascomús** Buenos Aires, E Argentina 35°34′S 58°01′W
56 F7 **Chase** British Columbia, SW Canada 50°49′N 119°41′W
67 K5 **Chase City** Virginia, NE USA 36°48′N 78°27′W
63 K2 **Chase, Mount** ▲ Maine, NE USA 46°06′N 68°30′W
191 H8 **Chashniki** Vitsyebskaya Voblasts', N Belarus 54°52′N 29°10′E
186 F4 **Chásia** ▲▲ C Greece
74 G5 **Chaska** Minnesota, N USA 44°47′N 93°36′W
278 C13 **Chaslands Mistake** *headland* South Island, New Zealand 46°37′S 169°21′E
195 I7 **Chasovo** Respublika Komi, NW Russian Federation 61°58′N 50°34′E
Chasovo *see* Vazhgort
168 E1 **Chasseneuil** Poitou-Charentes, France 45°49′N 0°27′E
194 D9 **Chastova** Novgorodskaya Oblast', NW Russian Federation 58°37′N 32°05′E
223 H2 **Chāt** Golestān, N Iran 37°52′N 55°27′E
Chatak *see* Chhatak
Chatang *see* Zhanang
83 J4 **Chatanika** Alaska, USA 65°06′N 147°28′W
83 J4 **Chatanika River** ≋ Alaska, USA
229 K4 **Chat-Bazar** Talasskaya Oblast', NW Kyrgyzstan 42°29′N 72°37′E
91 M3 **Chateaubelair** Saint Vincent, W Saint Vincent and the Grenadines 13°15′N 61°05′W
166 C6 **Châteaubourg** Bretagne, France 48°07′N 1°24′W
114 A8 **Chateaubriand** Santa Fe, Argentina 33°38′S 61°52′W
164 F4 **Châteaubriant** Loire-Atlantique, NW France 47°43′N 01°22′W
165 I5 **Château-Chinon** Nièvre, C France 47°04′N 03°50′E
176 C5 **Château d'Oex** Vaud, W Switzerland 46°28′N 07°09′E
164 G4 **Château-du-Loir** Sarthe, NW France 47°40′N 00°25′E
165 H4 **Châteaudun** Eure-et-Loir, C France 48°04′N 01°20′E
166 C7 **Châteaugiron** Bretagne, France 48°03′N 1°30′W
164 F4 **Château-Gontier** Mayenne, NW France 47°49′N 00°42′W
63 H3 **Châteauguay** Québec, SE Canada 45°22′N 73°44′W
166 E8 **Château-la-Valliére** Centre, France 47°33′N 0°19′E
168 E2 **Château-l'Évêque** Aquitaine, France 45°15′N 0°40′E
164 D4 **Châteaulin** Finistère, NW France 48°12′N 04°07′W
165 H5 **Châteaumeillant** Cher, C France 46°33′N 02°10′E
169 I3 **Châteauneuf-de-Randon** Languedoc-Roussillon, France 44°39′N 3°40′E
166 A6 **Châteauneuf-du-Faou** Bretagne, France 48°11′N 3°49′W
166 F6 **Châteauneuf-en-Thymerais** Centre, France 48°35′N 1°15′E
168 D2 **Châteauneuf-sur-Charente** Charente, W France 45°34′N 00°33′W
167 H7 **Châteauneuf-sur-Loire** Centre, France 47°52′N 2°14′E
166 D7 **Châteauneuf-sur-Sarthe** Pays de la Loire, France 47°41′N 0°29′W
167 I8 **Châteauneuf-Val-de-Bargis** Bourgogne, France 47°17′N 3°14′E
166 F10 **Châteauponsac** Limousin, France 46°08′N 1°17′E
167 J4 **Château-Porcien** Champagne-Ardenne, France 49°32′N 4°15′E
169 J5 **Châteaurenard** Provence-Alpes-Côte d'Azur, France 43°49′N 4°50′E
166 F8 **Château-Renault** Indre-et-Loire, C France 47°34′N 00°52′E
165 H5 **Châteauroux** *prev.* Indreville. Indre, C France 46°50′N 01°43′E
165 K3 **Château-Salins** Moselle, NE France 48°50′N 06°29′E
165 I3 **Château-Thierry** Aisne, N France 49°03′N 03°24′E
167 K7 **Châteauvillain** Champagne-Ardenne, France 48°02′N 4°55′E
169 I1 **Châteldon** Auvergne, France 45°58′N 3°31′E
163 E11 **Châtelet** Hainaut, S Belgium 50°24′N 04°32′E
Châtelherault *see* Châtellerault
164 G5 **Châtellerault** *var.* Châtelherault. Vienne, W France 46°49′N 00°33′E
74 H5 **Chatfield** Minnesota, N USA 43°51′N 92°11′W
63 M1 **Chatham** New Brunswick, SE Canada 47°02′N 65°30′W
62 B6 **Chatham** Ontario, S Canada 42°24′N 82°11′W
161 L6 **Chatham** SE England, United Kingdom 51°23′N 00°31′E
73 C11 **Chatham** Illinois, N USA 39°40′N 89°42′W
67 J5 **Chatham** Virginia, NE USA 36°49′N 79°26′W
117 B12 **Chatham, Isla** *island* S Chile
Chatham Island *see* San Cristóbal, Isla
Chatham Island Rise *see* Chatham Rise
286 F9 **Chatham Islands** *island group* New Zealand, SW Pacific Ocean
267 I8 **Chatham Rise** *var.* Chatham Island Rise. *undersea feature* S Pacific Ocean
83 M7 **Chatham Strait** *strait* Alaska, USA
Chathóir, Rinn *see* Cahore Point
167 H7 **Châtillon-Coligny** Centre, France 47°50′N 2°51′E
167 I9 **Châtillon-en-Bazois** Bourgogne, France 47°03′N 3°40′E
169 K3 **Châtillon-en-Diois** Rhône-Alpes, France 44°41′N 5°28′E
167 K10 **Châtillon-sur-Chalaronne** Rhône-Alpes, France 46°07′N 4°58′E
165 H5 **Châtillon-sur-Indre** Indre, C France 46°58′N 01°10′E
167 I5 **Châtillon-sur-Marne** Champagne-Ardenne, France 49°06′N 3°45′E
165 J4 **Châtillon-sur-Seine** Côte d'Or, C France 47°51′N 04°30′E
229 J5 **Chatkal** *Uzb.* Chotqol. ≋ Kyrgyzstan/Uzbekistan
229 J5 **Chatkal Range** *Rus.* Chatkal'skiy Khrebet. ▲▲ Kyrgyzstan/Uzbekistan
Chatkal'skiy Khrebet *see* Chatkal Range
68 F3 **Chatom** Alabama, S USA 31°28′N 88°15′W
Chatrapur *see* Chhatrapur
223 H6 **Chatrūd** Kermān, C Iran 30°39′N 56°57′E
66 F7 **Chatsworth** Georgia, SE USA 34°46′N 84°46′W
Chāttagām *see* Chittagong
69 I4 **Chattahoochee** Florida, SE USA 30°40′N 84°51′W
69 I3 **Chattahoochee River** ≋ SE USA
66 F7 **Chattanooga** Tennessee, S USA 35°05′N 85°16′W
161 K4 **Chatteris** United Kingdom 52°27′N 0°03′E
229 L5 **Chatyr-Këlʼ, Ozero** ◎ C Kyrgyzstan
229 M5 **Chatyr-Tash** Narynskaya Oblast', C Kyrgyzstan 40°54′N 76°22′E
63 J3 **Chaudière** ≋ Québec, SE Canada
257 H9 **Châu Đốc** *var.* Chauphu, Chau Phu. An Giang, S Vietnam 10°53′N 105°07′E
167 J10 **Chauffailles** Bourgogne, France 46°12′N 4°20′E
232 C6 **Chauhtan** *prev.* Chohtan. Rājasthān, NW India 25°27′N 71°08′E
256 C5 **Chauk** Magway, W Myanmar (Burma) 20°52′N 94°50′E
167 H3 **Chaulnes** Picardie, France 49°49′N 2°48′E
167 K7 **Chaumant** Champagne-Ardenne, France 48°07′N 5°08′E
167 K9 **Chaumergy** Franche-Comté, France 46°51′N 5°29′E
165 J4 **Chaumont** *prev.* Chaumont-en-Bassigny. Haute-Marne, N France 48°07′N 05°08′E
Chaumont-en-Bassigny *see* Chaumont
166 G4 **Chaumont-en-Vexin** Picardie, France 49°16′N 1°53′E
167 K5 **Chaumont-sur-Aire** Lorraine, France 48°56′N 5°15′E
166 F8 **Chaumont-sur-Loire** Centre, France 47°29′N 1°11′E
193 L3 **Chaunskaya Guba** *bay* NE Russian Federation
165 I2 **Chauny** Aisne, N France 49°37′N 03°13′E
256 I7 **Châu Ô** *var.* Binh Sơn. Quang Ngai, C Vietnam 15°18′N 108°45′E
Chau Phu *see* Châu Đốc
164 F3 **Chausey, Îles** *island group* N France
167 K9 **Chaussin** Franche-Comté, France 46°58′N 5°25′E
Chausy *see* Chavusy
62 D6 **Chautauqua Lake** ◎ New York, NE USA
ChâuThanh *see* Ba Ria
166 F9 **Chauvigny** Vienne, W France 46°35′N 00°37′E
194 F4 **Chavan'ga** Murmanskaya Oblast', NW Russian Federation 66°07′N 37°44′E
62 E1 **Chavannes, Lac** ◎ Québec, SE Canada
Chavantes, Represa de *see* Xavantes, Represa de
112 G7 **Chavarría** Corrientes, NE Argentina 28°57′S 58°35′W
Chavash Respubliki *see* Chuvashskaya Respublika
170 D3 **Chaves** *anc.* Aquae Flaviae. Vila Real, N Portugal 41°44′N 07°28′W
Chávez, Isla *see* Santa Cruz, Isla
167 I4 **Chavignon** Picardie, France 49°29′N 3°31′E
138 E1 **Chavuma** North Western, NW Zambia 13°04′S 22°43′E
191 I9 **Chavusy** *Rus.* Chausy. Mahilyowskaya Voblasts', E Belarus 53°48′N 30°58′E
Chayan *see* Shayan
243 J8 **Chayang** Guangdong, SE China 24°31′N 116°40′E
229 L5 **Chayek** Narynskaya Oblast', C Kyrgyzstan 41°54′N 74°28′E
217 J4 **Chāy Khānah** Diyālá, E Iraq 34°19′N 44°33′E
195 K10 **Chaykovskiy** Permskiy Kray, NW Russian Federation 56°45′N 54°09′E
86 G9 **Chayuco** Oaxaca, Mexico 16°24′N 97°49′W
169 J2 **Chazelles-sur-Lyon** Rhône-Alpes, France 45°38′N 4°23′E
257 I8 **Chbar** Môndól Kiri, E Cambodia 12°46′N 107°10′E
161 H2 **Cheadle** United Kingdom 52°59′N 1°59′W
66 E9 **Cheaha Mountain** ▲ Alabama, S USA 33°29′N 85°48′W
Cheatharlach *see* Carlow
67 J2 **Cheat River** ≋ NE USA
183 A8 **Cheb** *Ger.* Eger. Karlovarský Kraj, W Czech Republic 50°05′N 12°23′E
218 F4 **Chebaâ** Lebanon 33°21′N 35°44′E
197 I2 **Cheboksary** Chuvashskaya Respublika, W Russian Federation 56°06′N 47°15′E
72 G5 **Cheboygan** Michigan, N USA 45°40′N 84°28′W
Chechaouèn *see* Chefchaouen
Chechenia *see* Chechenskaya Respublika
197 I9 **Chechenskaya Respublika** *Eng.* Chechenia, Chechnia, *Rus.* Chechnya. ◈ *autonomous republic* SW Russian Federation
121 H2 **Chech, Erg** *desert* Algeria/Mali
Chechevichi *see* Chachevichy
Che-chiang *see* Zhejiang
Chechnia/Chechnya *see* Chechenskaya Respublika
Chech'ŏn *see* Jecheon
182 G7 **Chęciny** Świętokrzyskie, S Poland 50°51′N 20°31′E
75 F12 **Checotah** Oklahoma, C USA 35°28′N 95°31′W
59 K8 **Chedabucto Bay** *inlet* Nova Scotia, E Canada
160 G7 **Cheddar** United Kingdom 51°17′N 2°47′W
256 B5 **Cheduba Island** *island* W Myanmar (Burma)
64 A1 **Cheektowaga** New York, USA 42°54′N 78°45′W
277 J1 **Cheepie** Queensland, Australia
79 L3 **Cheesman Lake** ▣ Colorado, C USA
293 J9 **Cheetham, Cape** *headland* Antarctica 70°26′S 162°40′E
166 E10 **Chef-Boutonne** Poitou-Charentes, France 46°06′N 0°03′W
130 F2 **Chefchaouen** *var.* Chaouèn, Chechaouèn, *Sp.* Xauen. N Morocco 35°10′N 05°16′W
Chefoo *see* Yantai
82 G7 **Chefornak** Alaska, USA 60°09′N 164°09′W
193 K8 **Chegdomyn** Khabarovskiy Kray, SE Russian Federation 51°09′N 132°58′E
130 F6 **Chegga** Tiris Zemmour, NE Mauritania 25°27′N 05°49′W
Cheghcheran *see* Chaghcharān
76 C4 **Chehalis** Washington, NW USA 46°39′N 122°57′W
76 B4 **Chehalis River** ≋ Washington, NW USA
230 E4 **Chehel Abdālān, Kūh-e** *var.* Chalap Dalam, *Pash.* Chalap Dalan. ▲▲ C Afghanistan
186 F3 **Cheimadítis, Límni** *var.* Límni Cheimadítis. ◎ N Greece
Cheimadítis, Límni *see* Cheimadítis, Límni
169 M5 **Cheiron, Mont** ▲ SE France 43°49′N 07°00′E
241 L2 **Cheju** ✈ S South Korea 33°31′N 126°29′E
Cheju *see* Jeju
Cheju-do *see* Jeju-do
Cheju-haehyeop *see* Jeju-haehyeop
Cheju Strait *see* Jeju-haehyeop
Chekiang *see* Zhejiang
Chekichler/Chekishlyar *see* Çekiçler
218 F2 **Chekka** Lebanon
218 F2 **Chekka** Lebanon 34°20′N 35°44′E
87 L7 **Chekubul** Campeche, Mexico
282 C9 **Chelab** Babeldaob, N Palau
229 H6 **Chelak** *Rus.* Chelek. Samarqand Viloyati, C Uzbekistan 39°55′N 66°45′E
76 D3 **Chelan, Lake** ▣ Washington, NW USA
Chelek *see* Chelak
Cheleken *see* Hazar
172 D6 **Cheles** Extremadura, Spain 38°31′N 7°17′W
Chélif/Chéliff *see* Chelif, Oued
131 H2 **Chelif, Oued** *var.* Chélif, Chéliff, Chellif, Shellif. ≋ N Algeria
Chelkar *see* Shalkar
Chelkar Ozero *see* Shalkar, Ozero
171 M10 **Chellal** Algeria
118 D1 **Chellepín** Coquimbo, Chile 31°53′S 70°46′W
Chellif *see* Chelif, Oued
182 J7 **Chełm** *Rus.* Kholm. Lubelskie, SE Poland 51°08′N 23°29′E
182 F4 **Chełmno** *Ger.* Culm, Kulm. Kujawski-pomorskie, C Poland 53°21′N 18°27′E
186 G6 **Chelmós** *var.* Ároania. ▲▲ S Greece
62 C2 **Chelmsford** Ontario, S Canada 46°33′N 81°16′W
161 L5 **Chelmsford** E England, United Kingdom 51°44′N 00°28′E
65 L2 **Chelmsford** Massachusetts, USA 42°36′N 71°22′W
182 F5 **Chełmża** *Ger.* Culmsee, Kulmsee. Kujawski-pomorskie, C Poland 53°11′N 18°34′E
65 L2 **Chelsea** Massachusetts, USA 42°24′N 71°02′W
75 F11 **Chelsea** Oklahoma, C USA 36°32′N 95°25′W
63 I5 **Chelsea** Vermont, NE USA 43°58′N 72°29′W
161 H5 **Cheltenham** C England, United Kingdom 51°54′N 02°04′W
171 I5 **Chelva** Valenciana, E Spain 39°45′N 01°00′W
197 N1 **Chelyabinsk** Chelyabinskaya Oblast', C Russian Federation 55°12′N 61°25′E
197 M2 **Chelyabinskaya Oblast'** ◈ *province* C Russian Federation
193 H3 **Chelyuskin, Mys** *headland* N Russian Federation 77°42′N 104°13′E
192 F8 **Chemal** Altayskiy Kray, S Russian Federation 51°22′N 85°08′E
87 N5 **Chemax** Yucatán, SE Mexico 20°41′N 87°54′W
139 J3 **Chemba** Sofala, C Mozambique 17°11′S 34°53′E
137 A12 **Chembe** Luapula, NE Zambia 11°58′S 28°45′E
217 J3 **Chemchemal** *Ar.* Juwārtā, *var.* Chamchamāļ. At Ta'mím, N Iraq 35°32′N 44°50′E
Chemenibit *see* Çemenibit
Chemerisy *see* Chamyarysy
188 E5 **Chemerivtsi** Khmel'nyts'ka Oblast', W Ukraine 49°00′N 26°21′E
164 F5 **Chemillé** Maine-et-Loire, NW France 47°15′N 00°42′W
167 K9 **Chemin** Franche-Comté, France 46°59′N 5°19′E
137 G11 **Chemin Grenier** S Mauritius 20°29′S 57°28′E
179 H9 **Chemnitz** *prev.* Karl-Marx-Stadt. Sachsen, E Germany 50°50′N 12°55′E
Chemulpo *see* Incheon
76 C7 **Chemult** Oregon, NW USA 43°14′N 121°48′W
64 D3 **Chemung River** ≋ New York/Pennsylvania, NE USA
231 I5 **Chenāb** ≋ India/Pakistan
83 J4 **Chena Hot Springs** Alaska, USA 65°06′N 146°02′W
64 F2 **Chenango River** ≋ New York, NE USA
87 L6 **Chencoh** Campeche, Mexico 19°26′N 89°49′W
87 L6 **Chencoyí** Campeche, Mexico 19°48′N 90°14′W
258 D2 **Chenderoh, Tasik** ◎ Peninsular Malaysia
166 G10 **Chénérailles** Limousin, France 46°07′N 2°10′E
63 I2 **Chêne, Rivière du** ≋ Québec, SE Canada
76 F4 **Cheney** Washington, NW USA 47°29′N 117°34′W
75 D10 **Cheney Reservoir** ▣ Kansas, C USA
245 I6 **Cheng'an** Hebei, China 36°16′N 114°25′E
242 C8 **Chengbihe Shuiku** ▣ Guangxi, China
242 F6 **Chengbu** Hunan, SE China 26°22′N 110°19′E
Chengchiatun *see* Liaoyuan
Ch'eng-chou/Chengchow *see* Zhengzhou
245 K2 **Chengde** Hebei, China 40°28′N 118°05′E
245 K2 **Chengde** *var.* Jehol. Hebei, E China 41°N 117°57′E
243 J1 **Chengdong Hu** ◎ Anhui, China
242 A2 **Chengdu** *var.* Chengtu, Ch'eng-tu. *province capital* Sichuan, C China 30°41′N 104°03′E
243 N9 **Chenggong** S Taiwan 23°06′N 121°22′E
244 D9 **Chenggu** Shaanxi, China 33°05′N 107°12′E
243 J9 **Chenghai** Guangdong, S China 23°30′N 116°42′E
Chenghsien *see* Zhengzhou
Chengjiang *see* Taihe
242 E1 **Chengkou** Chongqing Shi, C China 31°35′N 108°24′E
240 F9 **Chengmai** *var.* Jinjiang. Hainan, S China 19°45′N 109°56′E
245 N5 **Chengshan** Shandong, E China 37°22′N 122°33′E
245 N5 **Chengshan Jiao** *cape* NE China
Chengshou *see* Yangshan
Chengtu/Ch'eng-tu *see* Chengdu
Chengwen *see* Chindu
244 C8 **Chengxian** *var.* Cheng Xiang. Gansu, C China 33°42′N 105°45′E
242 E9 **Chengxiang** Guang Zhuangzu Zizhiqu, S China 23°42′N 109°17′E
Cheng Xiang *see* Chengxian
Chengyang *see* Juxian
Chengzhong *see* Ningming
245 N3 **Chengzitan** Liaoning, NE China 39°30′N 122°30′E
Chenkiang *see* Zhenjiang
234 G7 **Chennai** *prev.* Madras. *state capital* Tamil Nādu, S India 13°05′N 80°18′E
234 G7 **Chennai** ✈ Tamil Nādu, S India 13°07′N 80°13′E
166 F8 **Chenonceaux** Centre, France 47°20′N 1°04′E
165 J5 **Chenôve** Côte d'Or, C France 47°16′N 05°00′E
Chenstokhov *see* Częstochowa
242 F5 **Chenxi** *var.* Chenyang. Hunan, S China 28°02′N 110°15′E
Chen Xian/Chenxian/Chen Xiang *see* Chenzhou
245 N1 **Chenxiangtun** Liaoning, China 41°20′N 123°18′E
Chenyang *see* Chenxi
243 H7 **Chenzhou** *var.* Chenxian, Chen Xian, Chen Xiang. Hunan, S China 25°51′N 113°01′E
248 C6 **Cheonan** *Jap.* Tenan; *prev.* Ch'ŏnan. W South Korea 36°51′N 127°11′E
250 B3 **Cheongdo** *prev.* Ch'ŏngdo. Kyŏngsang-bukto, South Korea 35°39′N 128°45′E
248 B3 **Cheongju** *prev.* Chŏngju. W North Korea 39°44′N 125°13′E
Cheo Reo *see* A Yun Pa
64 L3 **Chepachet** Rhode Island, NE USA 41°54′N 71°40′W
185 J7 **Chepelare** Smolyan, S Bulgaria 41°44′N 24°41′E
185 J7 **Chepelarska Reka** ≋ S Bulgaria
104 C5 **Chepén** La Libertad, C Peru 07°15′S 79°23′W
112 C8 **Chepes** La Rioja, C Argentina 31°19′S 66°40′W
243 H10 **Chep Lap Kok** ✈ S China 22°23′N 114°11′E
89 L9 **Chepo** Panamá, C Panama 09°09′N 79°03′W
Chepping Wycombe *see* High Wycombe
195 J9 **Cheptsa** ≋ NW Russian Federation
72 C4 **Chequamegon Point** *headland* Wisconsin, N USA 46°42′N 90°45′W
165 I5 **Cher** ◈ *department* C France
165 H5 **Cher** ≋ C France
173 K4 **Chera** Valenciana, Spain 39°36′N 0°58′W
Cherangani Hills *see* Cherangany Hills
136 C7 **Cherangany Hills** *var.* Cherangani Hills. ▲▲ W Kenya
169 N3 **Cherasco** Piemonte, NW Italy 44°39′N 07°51′E
67 J7 **Cheraw** South Carolina, SE USA 34°42′N 79°52′W
164 F2 **Cherbourg** *anc.* Carusbur. Manche, N France 49°40′N 01°36′W
171 K9 **Cherchell** Algeria
197 J3 **Cherdakly** Ul'yanovskaya Oblast', W Russian Federation 54°21′N 48°54′E
195 K8 **Cherdyn'** Permskiy Kray, NW Russian Federation 60°21′N 56°39′E
194 C9 **Cherekha** ≋ W Russian Federation
193 H8 **Cheremkhovo** Irkutskaya Oblast', S Russian Federation 53°16′N 102°44′E
195 L7 **Cheremukhovo** Sverdlovskaya Oblasť, Russian Federation
192 G8 **Cheremushki** Respublika Khakastya, S Russian Federation 52°48′N 91°20′E
Cheren *see* Keren
194 F9 **Cherepovets** Vologodskaya Oblast', NW Russian Federation 59°09′N 37°50′E
195 H7 **Cherevkovo** Arkhangel'skaya Oblast', NW Russian Federation 61°45′N 45°16′E
131 H2 **Chergui, Chott ech** *salt lake* NW Algeria
Cherikov *see* Cherykaw
189 H4 **Cherkas'ka Oblast'** *var.* Cherkasy, *Rus.* Cherkasskaya Oblast'. ◈ *province* C Ukraine
Cherkasskaya Oblast' *see* Cherkas'ka Oblast'
Cherkassy *see* Cherkasy
189 I4 **Cherkasy** *Rus.* Cherkassy. Cherkas'ka Oblast', C Ukraine 49°26′N 32°05′E
Cherkasy *see* Cherkas'ka Oblast'
197 H8 **Cherkessk** Karachayevo-Cherkesskaya Respublika, SW Russian Federation 44°12′N 42°06′E
192 E7 **Cherlak** Omskaya Oblast', C Russian Federation 54°06′N 74°59′E
192 E7 **Cherlakskoye** Omskaya Oblast', C Russian Federation 53°42′N 74°23′E
195 K9 **Chermoz** Permskiy Kray, NW Russian Federation 58°49′N 56°07′E
Chernavchitsy *see* Charnawchytsy

◆ Country
● Country Capital
◇ Dependent Territory
○ Dependent Territory Capital
◈ Administrative Regions
✈ International Airport
▲ Mountain
▲▲ Mountain Range
Volcano
≋ River
◎ Lake
▣ Reservoir

◆	Country	◇	Dependent Territory	◆	Administrative Regions	▲	Mountain	◊	Volcano	◎	Lake
●	Country Capital	◇	Dependent Territory Capital	✕	International Airport	▲	Mountain Range	↗	River	⊞	Reservoir

◆ Country ◇ Dependent Territory ◇ Administrative Regions ▲ Mountain ✲ Volcano ◉ Lake
● Country Capital ○ Dependent Territory Capital ✈ International Airport ▲ Mountain Range ➣ River ◫ Reservoir

◆ Country   ◇ Dependent Territory   ◈ Administrative Regions   ▲ Mountain   ✦ Volcano   ⊡ Lake
● Country Capital   ○ Dependent Territory Capital   ✈ International Airport   ▲ Mountain Range   ♒ River   ▣ Reservoir

◆ Country   ◇ Dependent Territory   ◈ Administrative Regions   ▲ Mountain   ⊠ Volcano   ⊙ Lake
● Country Capital   ○ Dependent Territory Capital   ✕ International Airport   ▲ Mountain Range   ← River   ⊡ Reservoir

79 I3 **Fairview** Utah, W USA 39°37′N 111°26′W
80 H3 **Fairview Peak** ▲ Nevada, W USA 39°13′N 118°09′W
282 D5 **Fais** *atoll* Caroline Islands, W Micronesia
231 I5 **Faisalābād** *prev.* Lyallpur. Punjab, NE Pakistan 31°26′N 73°06′E
Faisaliya *see* Fayşalīyah
74 B4 **Faith** South Dakota, N USA 45°01′N 102°02′W
232 G6 **Faizābād** Uttar Pradesh, N India 26°46′N 82°08′E
Faizabad/Faizābād *see* Feiẕābād
91 J8 **Fajardo** E Puerto Rico 18°20′N 65°39′W
217 I6 **Fajj, Wādī al** *dry watercourse* S Iraq
220 D2 **Fajr, Bi'r** *well* NW Saudi Arabia
285 K6 **Fakahina** *atoll* Îles Tuamotu, C French Polynesia
284 E5 **Fakaofo Atoll** *island* SE Tokelau
285 J6 **Fakarava** *atoll* Îles Tuamotu, C French Polynesia
195 J9 **Fakel** Udmurtskaya Respublika, NW Russian Federation 57°35′N 53°00′E
161 L2 **Fakenham** E England, United Kingdom 52°50′N 00°51′E
261 I5 **Fakfak** Papua Barat, E Indonesia 02°55′S 132°17′E
261 I5 **Fakfak, Pegunungan** ▲▲ Papua, E Indonesia
219 I13 **Fakha, Wādī al** Saudi Arabia
233 K6 **Fakīragrām** Assam, NE India 26°22′N 90°15′E
185 L7 **Fakiyska Reka** ~ SE Bulgaria
155 E13 **Fakse** Sjælland, SE Denmark 55°16′N 12°08′E
155 F13 **Fakse Bugt** *bay* SE Denmark
155 F13 **Fakse Ladeplads** Sjælland, SE Denmark 55°14′N 12°11′E
248 B1 **Faku** Liaoning, NE China 42°30′N 123°27′E
160 D9 **Fal** ~ United Kingdom
159 I3 **Fala** United Kingdom 55°50′N 2°53′W
132 C6 **Falaba** N Sierra Leone 09°54′N 11°22′W
164 G3 **Falaise** Calvados, N France 48°52′N 00°12′W
166 F3 **Falaises d'Aval** *cave* Haute-Normandie, France
187 H2 **Falakró** ▲▲ NE Greece
282 E9 **Falalu** *island* Chuuk, C Micronesia
256 B4 **Falam** Chin State, W Myanmar (Burma) 22°58′N 93°45′E
222 F5 **Falāvarjān** Eşfahān, C Iran 32°33′N 51°28′E
188 F7 **Fălciu** Vaslui, E Romania 46°19′N 28°10′E
172 C6 **Falcoeiras** Évora, Portugal 38°29′N 7°35′W
102 E2 **Falcón** *off.* Estado Falcón. ⬖ *state* NW Venezuela
177 J9 **Falconara Marittima** Marche, C Italy 43°37′N 13°23′E
171 I10 **Falcon, Cap** *headland* Algeria
Falcone, Capo del *see* Falcone, Punta del
175 A9 **Falcone, Punta del** *var.* Capo del Falcone. *headland* Sardegna, Italy, C Mediterranean Sea 40°57′N 8°12′E
Falcón, Estado *see* Falcón
55 J10 **Falcon Lake** Manitoba, S Canada 49°44′N 95°18′W
Falcon Lake *see* Falcón, Presa/Falcon Reservoir
85 L5 **Falcón, Presa** *var.* Falcon Lake, Falcon Reservoir. ▣ Mexico/USA *see also* Falcon Reservoir
Falcón, Presa *see* Falcon Reservoir
85 L5 **Falcon Reservoir** *var.* Falcon Lake, Presa Falcón. ▣ Mexico/USA *see also* Falcón, Presa
Falcon Reservoir *see* Falcón, Presa
285 I1 **Faleālupo** Savai'i, NW Samoa 13°30′S 172°46′W
284 E2 **Falefatu** *island* Funafuti Atoll, C Tuvalu
285 I1 **Fālelima** Savai'i, NW Samoa 13°30′S 172°41′W
155 H9 **Falerum** Östergötland, S Sweden 58°07′N 16°15′E
Faleshty *see* Fălești
188 F6 **Fălești** *Rus.* Faleshty. NW Moldova 47°33′N 27°43′E
70 G9 **Falfurrias** Texas, SW USA 27°17′N 98°10′W
56 G3 **Falher** Alberta, W Canada 55°45′N 117°18′W
Falkenau an der Eger *see* Sokolov
155 F11 **Falkenberg** Halland, S Sweden 56°55′N 12°30′E
Falkenberg *see* Niemodlin
Falkenburg in Pommern *see* Złocieniec
178 H6 **Falkensee** Brandenburg, NE Germany 52°34′N 13°04′E
159 H3 **Falkirk** C Scotland, United Kingdom 56°N 03°48′W
159 I2 **Falkland** E Scotland, United Kingdom 56°15′N 3°12′E
291 D12 **Falkland Escarpment** *undersea feature* SW Atlantic Ocean 50°00′S 45°00′W
117 I10 **Falkland Islands** *var.* Falklands, Islas Malvinas. ◇ *UK dependent territory* SW Atlantic Ocean
95 M9 **Falkland Islands** *island group* SW Atlantic Ocean
291 D12 **Falkland Plateau** *var.* Argentine Rise. *undersea feature* SW Atlantic Ocean 51°00′S 50°00′W
Falklands *see* Falkland Islands
117 G13 **Falkland Sound** *var.* Estrecho de San Carlos. *strait* C Falkland Islands
Falknov nad Ohří *see* Sokolov
187 H8 **Falkonéra** *island* S Greece
155 F9 **Falköping** Västra Götaland, S Sweden 58°10′N 13°31′E
217 K6 **Fallāḥ** Wāsiţ, E Iraq 32°58′N 45°09′E
81 E12 **Fallbrook** California, W USA 33°22′N 117°15′W
80 D2 **Fallbrook** California, USA 38°25′N 121°22′W
282 E9 **Falleallej Pass** *passage* Chuuk Islands, C Micronesia
154 J1 **Fällfors** Västerbotten, N Sweden 65°07′N 20°46′E
292 E4 **Fallières Coast** *physical region* Antarctica
180 H4 **Fallingbostel** Niedersachsen, NW Germany 52°52′N 09°42′E
77 L4 **Fallon** Montana, NW USA 46°49′N 105°07′W
80 G2 **Fallon** Nevada, W USA 39°29′N 118°47′W
65 L4 **Fall River** Massachusetts, NE USA 41°42′N 71°09′W
75 F10 **Fall River Lake** ▣ Kansas, C USA
78 C2 **Fall River Mills** California, W USA 41°00′N 121°28′W
64 C10 **Falls Church** Virginia, NE USA 38°53′N 77°11′W
75 F8 **Falls City** Nebraska, C USA 40°03′N 95°36′W
70 G8 **Falls City** Texas, SW USA 28°58′N 98°01′W
67 K6 **Falls Lake** ◎ North Carolina, SE USA
Falluja *see* Al Fallūjah
133 I5 **Falmey** Dosso, SW Niger 12°29′N 02°58′E
91 M6 **Falmouth** Antigua, Antigua and Barbuda 17°02′N 61°47′W
90 E7 **Falmouth** W Jamaica 18°28′N 77°39′W
160 D10 **Falmouth** SW England, United Kingdom 50°08′N 05°04′W
66 F3 **Falmouth** Kentucky, S USA 38°40′N 84°20′W
65 M4 **Falmouth** Massachusetts, NE USA 41°31′N 70°36′W
67 L3 **Falmouth** Virginia, NE USA 38°19′N 77°28′W
160 D10 **Falmouth Bay** *bay* United Kingdom
282 E9 **Falos** *island* Chuuk, C Micronesia
138 D10 **False Bay** *Afr.* Valsbaai. *bay* SW South Africa
140 D10 **False Bay** *bay* Western Cape, South Africa
234 G6 **False Divi Point** *headland* E India 15°46′N 80°43′E
82 G9 **False Pass** Unimak Island, Alaska, USA 54°52′N 163°15′W
234 J4 **False Point** *headland* E India 20°23′N 86°52′E
171 J4 **Falset** Cataluña, NE Spain 41°08′N 00°49′E
155 F13 **Falster** *island* SE Denmark
155 F12 **Falsterbo** Skåne, S Sweden 55°22′N 12°49′E
159 J5 **Falstone** United Kingdom 55°10′N 2°26′W
188 E6 **Fălticeni** *Hung.* Falticsén. Suceava, NE Romania 47°27′N 26°20′E
Falticsén *see* Fălticeni
154 H6 **Falun** *var.* Fahlun. Kopparberg, C Sweden 60°36′N 15°36′E
110 E5 **Fama** Minas Gerais, Brazil 21°24′S 45°50′W
Famagusta *see* Gazimağusa
Famagusta Bay *see* Gazimağusa Körfezi
112 C7 **Famatina** La Rioja, NW Argentina 28°58′S 67°46′W
163 F12 **Famenne** *physical region* SE Belgium
133 L7 **Fan** ~ E Nigeria
132 E5 **Fana** Koulikoro, SW Mali 12°45′N 06°55′W
187 J6 **Fána** *ancient harbor* Chíos, SE Greece
158 C5 **Fanad Head** *headland* N Ireland
282 F10 **Fanan** *island* Chuuk, C Micronesia
282 E9 **Fanapanges** *island* Chuuk, C Micronesia
187 K6 **Fanári, Akrotírio** *headland* Ikaría, Dodekánisa, Greece, Aegean Sea 37°40′N 26°21′E
243 K2 **Fanchang** Anhui, E China 31°05′N 118°12′E
91 M3 **Fancy** Saint Vincent, Saint Vincent and the Grenadines 13°22′N 61°10′W
139 M8 **Fandriana** Fianarantsoa, SE Madagascar 20°14′S 47°21′E
256 E5 **Fang** Chiang Mai, NW Thailand 19°56′N 99°14′E
136 B4 **Fangak** Jonglei, E South Sudan 09°05′N 30°52′E
Fangatau *see* Fagatau
Fangataufa *see* Fagataufa
284 D9 **Fanga Uta** *bay* S Tonga
245 H9 **Fangcheng** Henan, C China 33°18′N 113°03′E
Fangcheng *see* Fangchenggang
242 D10 **Fangchenggang** *var.* Fangcheng Gezu Zizhixian; *prev.* Fangcheng. Guangxi Zhuangzu Zizhiqu, S China 21°49′N 108°21′E
Fangcheng Gezu Zizhixian *see* Fangchenggang
243 M10 **Fangshan** S Taiwan 22°19′N 120°41′E
243 I4 **Fangxi** Jiangxi, SE China 28°21′N 114°38′E
242 F1 **Fangxian** Hubei, China 32°02′N 110°26′E
243 M8 **Fangyuan** C Taiwan 23°55′N 120°16′E
242 G7 **Fangyuanxu** Hunan, SE China 25°39′N 112°40′E
247 M4 **Fangzheng** Heilongjiang, NE China 45°50′N 128°50′E
Fani *see* Fanit, Lumi i
191 F10 **Fanipal'** *Rus.* Fanipol'. Minskaya Voblasts', C Belarus 53°45′N 27°20′E
Fanipol' *see* Fanipal'
184 F7 **Fanit, Lumi i** *var.* Fani. ~ N Albania
168 G6 **Fanjeaux** Languedoc-Roussillon, France 43°12′N 2°03′E
242 D5 **Fanjing Shan** ▲ Guizhou, China 27°34′N 108°30′E
71 H8 **Fannin** Texas, SW USA 28°41′N 97°13′W
Fanning Island *see* Tabuaeran
154 D3 **Fannrem** Sør-Trøndelag, S Norway 63°16′N 09°48′E
177 I9 **Fano** *anc.* Colonia Julia Fanestris, Fanum Fortunae. Marche, C Italy 43°50′N 13°E
155 C12 **Fanø** *island* W Denmark
245 H3 **Fanshi** Shanxi, China 39°07′N 113°10′E
245 L9 **Fanshui** Jiangsu, E China 33°06′N 119°23′E
256 G4 **Fan Si Pan** ▲ N Vietnam 22°18′N 103°36′E
Fanum Fortunae *see* Fano
Fao *see* Al Fāw
221 J4 **Faq'** *var.* Al Faqa. Dubayy, E United Arab Emirates 24°42′N 55°37′E
219 F9 **Faqū** Jordan 31°22′N 35°41′E
Farab *see* Farap
279 E8 **Faraday, Mount** ▲ South Island, New Zealand 42°01′S 171°37′E
134 J6 **Faradje** Orientale, NE Dem. Rep. Congo 03°45′N 29°43′E
Faradofay *see* Tôlañaro
139 M9 **Farafangana** Fianarantsoa, SE Madagascar 22°50′S 47°50′E
230 D5 **Farāh** *var.* Farah, Fararud. Farāh, W Afghanistan 32°22′N 62°07′E
230 D4 **Farāh** ⬖ *province* W Afghanistan
230 D5 **Farāh Rūd** ~ W Afghanistan
218 F3 **Fâraïya** Lebanon
218 F3 **Fâraïya** Lebanon 34°01′N 35°49′E
282 E3 **Farallon de Medinilla** *island* C Northern Mariana Islands
282 E1 **Farallon de Pajaros** *var.* Uracas. *island* N Northern Mariana Islands
132 D6 **Faranah** Haute-Guinée, S Guinea 10°02′N 10°44′W
228 F7 **Farap** *Rus.* Farab. Lebap Welaýaty, NE Turkmenistan 39°15′N 63°32′E
Fararud *see* Farāh
212 F8 **Farasān, Jazā'ir** *island group* SW Saudi Arabia
139 L7 **Faratsiho** Antananarivo, C Madagascar 19°24′S 46°57′E
282 E6 **Faraulep Atoll** *atoll* Caroline Islands, C Micronesia
163 E11 **Farciennes** Hainaut, S Belgium 50°26′N 04°33′E
170 G8 **Fardes** ~ S Spain
158 B10 **Fardrum** Westmeath, Ireland 53°24′N 7°53′W
285 I7 **Fare** Huahine, W French Polynesia 16°42′S 151°01′W
161 J8 **Fareham** S England, United Kingdom 50°51′N 01°10′W
118 E3 **Farellones** Región Metropolitana, Chile 33°18′S 70°15′W
83 I6 **Farewell** Alaska, USA 62°35′N 153°59′W
278 E7 **Farewell, Cape** *headland* South Island, New Zealand 40°30′S 172°39′E
Farewell, Cape *see* Nunap Isua
278 F7 **Farewell Spit** *spit* South Island, New Zealand
155 E9 **Färgelanda** Västra Götaland, S Sweden 58°34′N 11°59′E
Farghona, Wodii/Farghona Valley *see* Fergana Valley
Farghona Wodiysi *see* Fergana Valley
69 K4 **Fargo** Georgia, SE USA 30°42′N 82°33′W
74 E3 **Fargo** North Dakota, N USA 46°53′N 96°47′W
229 J6 **Farg'ona** *Rus.* Fergana; *prev.* Novyy Margilan. Farg'ona Viloyati, E Uzbekistan 40°28′N 71°44′E
229 J6 **Farg'ona Viloyati** *Rus.* Ferganskaya Oblast'. ⬖ *province* E Uzbekistan
111 K4 **Faria Lemos** Minas Gerais, Brazil 20°48′S 42°01′W
74 H5 **Faribault** Minnesota, N USA 44°18′N 93°16′W
225 F8 **Farīdābād** Haryāna, N India 28°26′N 77°190′E
232 D4 **Farīdkot** Punjab, NW India 30°42′N 74°47′E
233 K8 **Faridpur** Dhaka, C Bangladesh 23°29′N 89°50′E
128 E3 **Fārigh, Wādī al** ~ N Libya
139 M6 **Farihy Alaotra** ◎ C Madagascar
154 H5 **Färila** Gävleborg, C Sweden 61°48′N 15°55′E
170 B5 **Farilhões** *island* C Portugal
132 B5 **Farim** NW Guinea-Bissau 12°30′N 15°09′W
Farish *see* Forish
221 I6 **Fāris, Qalamat** *well* SE Saudi Arabia
155 H11 **Färjestaden** Kalmar, S Sweden 56°38′N 16°30′E
231 H2 **Farkhār** Takhār, NE Afghanistan 36°39′N 69°43′E
229 I8 **Farkhor** *Rus.* Parkhar. SW Tajikistan 37°32′N 69°22′E
188 B8 **Fârliug** *prev.* Fîrliug, *Hung.* Furluk. Caraș-Severin, SW Romania 45°21′N 21°55′E
187 K7 **Farmakonísi** *island* Dodekánisa, Greece, Aegean Sea
73 D10 **Farmer City** Illinois, N USA 40°14′N 88°38′W
73 E11 **Farmersburg** Indiana, N USA 39°14′N 87°21′W
80 E6 **Farmersville** California, USA 36°18′N 119°12′W
71 H4 **Farmersville** Texas, SW USA 33°09′N 96°21′W
68 C1 **Farmerville** Louisiana, S USA 32°46′N 92°24′W
80 D3 **Farmington** California, USA 37°56′N 121°0′W
73 A10 **Farmington** Iowa, C USA 40°37′N 91°43′W
63 J4 **Farmington** Maine, NE USA 44°40′N 70°09′W
74 H5 **Farmington** Minnesota, N USA 44°39′N 93°09′W
75 J10 **Farmington** Missouri, C USA 37°46′N 90°26′W
63 J5 **Farmington** New Hampshire, NE USA 43°23′N 71°04′W
79 K6 **Farmington** New Mexico, SW USA 36°44′N 108°13′W
79 H2 **Farmington** Utah, W USA 40°58′N 111°53′W
67 L6 **Farmville** North Carolina, SE USA 35°37′N 77°36′W
67 K4 **Farmville** Virginia, NE USA 37°17′N 78°25′W
161 J7 **Farnborough** S England, United Kingdom 51°17′N 00°46′W
159 K4 **Farne Island** *island* United Kingdom
161 J7 **Farnham** S England, United Kingdom 51°13′N 00°49′W
159 J9 **Farnworth** United Kingdom 53°33′N 2°24′W
54 B3 **Faro** Yukon, W Canada 62°15′N 133°30′W
172 C8 **Faro** Faro, S Portugal 37°01′N 07°56′W
155 J9 **Fårö** Gotland, SE Sweden 57°55′N 19°10′E
172 C8 **Faro** ⬖ *district* S Portugal
134 C4 **Faro** ~ Cameroon/Nigeria
172 C8 **Faro** ✈ Faro, S Portugal 37°02′N 08°01′W
290 H2 **Faroe-Iceland Ridge** *var.* Faeroe-Iceland Ridge. *undersea ridge* NW Norwegian Sea
148 B5 **Faroe Islands** *var.* Faeroe Islands. *island group* N Atlantic Ocean
290 H2 **Faroe Islands** *var.* Faeroe Islands, *Dan.* Færøerne, *Faer.* Føroyar. *Self-governing territory of Denmark* N Atlantic Ocean
290 H2 **Faroe-Shetland Trough** *var.* Faeroe-Shetland Trough. *trough* NE Atlantic Ocean
Faro, Punta del *see* Peloro, Capo
155 J9 **Fårösund** Gotland, SE Sweden 57°51′N 19°02′E
289 C9 **Farquhar Group** *island group* S Seychelles
62 C7 **Farrell** Pennsylvania, NE USA 41°12′N 80°28′W
232 F6 **Farrukhābād** Uttar Pradesh, N India 27°24′N 79°34′E
222 G7 **Fārs** *off.* Ostān-e Fārs; *anc.* Persis. ⬖ *province* S Iran
186 G4 **Fársala** Thessalía, C Greece 39°17′N 22°23′E
223 H3 **Fārsīān** Golestán, N Iran
Fars, Khalīj-e *see* Persian Gulf
155 D11 **Farsø** Nordjylland, N Denmark 56°47′N 09°21′E
Fārs, Ostān-e *see* Fārs
155 B9 **Farsund** Vest-Agder, S Norway 58°05′N 06°49′E
221 J8 **Fartak, Ra's** *headland* E Yemen 15°34′N 52°19′E
113 I5 **Fartura, Serra da** ▲▲ S Brazil
Farvel, Kap *see* Nunap Isua
70 D3 **Farwell** Texas, SW USA 34°23′N 103°03′W
292 F5 **Farwell Island** *island* Antarctica
232 G5 **Far Western** ⬖ *zone* W Nepal
230 F2 **Fāryāb** ⬖ *province* N Afghanistan
222 G7 **Fasā** Fārs, S Iran 28°55′N 53°39′E
221 J7 **Fasad, Ramlat** *desert* SW Oman
175 I9 **Fasano** Puglia, SE Italy 40°50′N 17°21′E
152 D2 **Fáskrúðsfjörður** Austurland, E Iceland 64°55′N 14°01′W
180 H4 **Fassberg** Niedersachsen, Germany 52°54′E 10°10′N
188 G3 **Fastiv** *Rus.* Fastov. Kyyivs'ka Oblast', NW Ukraine 50°08′N 29°59′E
157 B11 **Fastnet Rock** *Ir.* Carraig Aonair. *island* SW Ireland
Fastov *see* Fastiv
284 F2 **Fatato** *island* Funafuti Atoll, C Tuvalu
232 F6 **Fatehgarh** Uttar Pradesh, N India 27°22′N 79°38′E
231 I4 **Fatehjang** Punjab, E Pakistan 33°33′N 72°43′E
232 D5 **Fatehpur** Rājasthān, N India 27°59′N 74°58′E
232 G7 **Fatehpur** Uttar Pradesh, N India 25°56′N 80°55′E
196 F4 **Fatezh** Kurskaya Oblast', W Russian Federation 52°01′N 35°51′E
132 B4 **Fatick** W Senegal 14°19′N 16°27′W
172 B4 **Fátima** Santarém, W Portugal 39°37′N 08°39′W
214 G5 **Fatsa** Ordu, N Turkey 41°02′N 37°31′E
Fatshan *see* Foshan
284 A9 **Fatua, Pointe** *var.* Pointe Nord. *headland* Île Futuna, S Wallis and Futuna
285 K5 **Fatu Hiva** *island* Îles Marquises, NE French Polynesia
Fatunda *see* Fatundu
135 D9 **Fatundu** *var.* Fatunda. Bandundu, W Dem. Rep. Congo 04°08′S 17°13′E
167 M7 **Faucegney** Franche-Comté, France 47°51′N 6°34′E
167 L7 **Faucogney-et-la-Mer** Haute-Saône, E France 47°50′N 06°34′E
74 D4 **Faulkton** South Dakota, N USA 45°02′N 99°07′W
167 L5 **Faulquemont** Lorraine, France 49°03′N 6°36′E
166 G2 **Fauquembergues** Nord-Pas-de-Calais, France 50°36′N 2°05′E
188 F9 **Făurei** *prev.* Filimon Sîrbu. Brăila, SE Romania 45°05′N 27°15′E
141 H6 **Fauresmith** Free State, South Africa 29°45′S 25°19′E
152 E5 **Fauske** Nordland, C Norway 67°15′N 15°27′E
57 H3 **Faust** Alberta, W Canada 55°19′N 115°33′W
166 F4 **Fauville-en-Caux** Haute-Normandie, France 49°39′N 0°34′E
163 G13 **Fauvillers** Luxembourg, SE Belgium 49°52′N 05°40′E
175 F13 **Favara** Sicilia, Italy, C Mediterranean Sea 37°19′N 13°40′E
Faventia *see* Faenza
169 L1 **Faverges** Rhône-Alpes, France 45°45′N 6°18′E
167 L7 **Faverney** Franche-Comté, France 47°46′N 6°06′E
161 M6 **Faversham** United Kingdom 51°19′N 0°54′E
175 E12 **Favignana, Isola** *island* Isole Egadi, S Italy
58 D4 **Fawn** ~ Ontario, SE Canada
Faxa Bay *see* Faxaflói
152 B2 **Faxaflói** *Eng.* Faxa Bay. *bay* W Iceland
128 E8 **Faya** *prev.* Faya-Largeau, Largeau. Borkou, N Chad 17°58′N 19°06′E
Faya-Largeau *see* Faya
280 C8 **Fayaoué** Province des Îles Loyauté, C New Caledonia 20°41′S 166°31′E
216 F4 **Fayḍāt** *hill range* E Syria
169 M5 **Fayence** Provence-Alpes-Côte d'Azur, France 43°37′N 6°41′E
66 C8 **Fayette** Alabama, S USA 33°40′N 87°49′W
72 A7 **Fayette** Iowa, C USA 42°50′N 91°48′W
68 D2 **Fayette** Mississippi, S USA 31°42′N 91°03′W
75 H9 **Fayette** Missouri, C USA 39°09′N 92°40′W
75 G12 **Fayetteville** Arkansas, C USA 36°04′N 94°10′W
67 K7 **Fayetteville** North Carolina, SE USA 35°03′N 78°53′W
66 E7 **Fayetteville** Tennessee, S USA 35°08′N 86°33′W
71 H7 **Fayetteville** Texas, SW USA 29°52′N 96°40′W
67 I4 **Fayetteville** West Virginia, NE USA 38°03′N 81°09′W
217 N9 **Faylakah** *var.* Failaka Island. *island* E Kuwait
167 K7 **Fayl-Billot** Champagne-Ardenne, France 47°47′N 5°36′E
168 E10 **Fayón** Aragón, Spain 41°15′N 0°21′E
217 J7 **Fayşalīyah** *var.* Faisaliya. Al Qādisīyah, S Iraq 31°48′N 44°36′E
282 G6 **Fayu** *var.* East Fayu. *island* Hall Islands, C Micronesia
161 I3 **Fazeley** United Kingdom 52°37′N 1°42′W
232 D4 **Fāzilka** Punjab, NW India 30°26′N 74°04′E
Fdérick *see* Fdérik
130 C7 **Fdérik** *var.* Fdérick, *Fr.* Fort Gouraud. Tiris Zemmour, NW Mauritania 22°40′N 12°41′W
Feabhail, Loch *see* Foyle, Lough
157 B10 **Feale** ~ SW Ireland
67 L8 **Fear, Cape** *headland* Bald Head Island, North Carolina, SE USA 33°50′N 77°57′W
80 D1 **Feather River** ~ California, W USA
278 G7 **Featherston** Wellington, North Island, New Zealand 41°07′S 175°28′E
159 K9 **Featherstone** United Kingdom 53°42′N 1°22′W
114 C6 **Febre** Entre Ríos, Argentina 32°28′S 59°56′W
164 G2 **Fécamp** Seine-Maritime, N France 49°45′N 00°22′E
161 H4 **Feckenham** United Kingdom 52°15′N 1°59′W
Fédala *see* Mohammedia
116 H2 **Federación** Entre Ríos, Argentina 31°12′S 58°00′W
114 F3 **Federación** Entre Ríos, E Argentina 31°00′S 57°55′W
114 F5 **Federación** Paysandú, Uruguay 32°01′S 57°20′W
114 D3 **Federal** Entre Ríos, E Argentina 30°55′S 58°45′W
133 K7 **Federal Capital District** ⬖ *capital territory* C Nigeria
Federal Capital Territory *see* Australian Capital Territory
Federal District *see* Distrito Federal
64 E10 **Federalsburg** Maryland, NE USA 38°41′N 75°46′W
131 K2 **Fedjaj, Chott el** *var.* Chott el Fejaj, Shaṭṭ al Fijāj. *salt lake* C Tunisia
154 A6 **Fedje** *island* S Norway
226 F2 **Fedorovka** Kostanay, N Kazakhstan 51°12′N 52°00′E
197 L3 **Fedorovka** Respublika Bashkortostan, W Russian Federation 53°09′N 55°07′E
Fëdory *see* Fyadory
189 K8 **Fedotova Kosa** *spit* SE Ukraine
158 C6 **Feeny** United Kingdom 54°53′N 7°01′W
159 J7 **Feetham** United Kingdom 54°22′N 2°01′W
282 F9 **Fefan** *atoll* Chuuk Islands, C Micronesia
183 I11 **Fehérgyarmat** Szabolcs-Szatmár-Bereg, E Hungary 47°59′N 22°29′E
Fehér-Körös *see* Crișul Alb
Fehértemplom *see* Bela Crkva
Fehérvölgy *see* Albac
178 F4 **Fehmarn** *island* N Germany
155 E13 **Fehmarn Belt** *Dan.* Femern Bælt, *Ger.* Fehmarnbelt. *strait* Denmark /Germany *see also* Femern Bælt
Fehmarnbelt *see* Fehmarn Belt/Femer Bælt
177 L4 **Fehring** Steiermark, SE Austria 46°56′N 16°00′E
245 J6 **Feicheng** Shandong, China 36°08′N 116°28′E
243 K1 **Feidong** Anhui, China 31°32′N 117°16′E
219 B13 **Feihi, Wâdi el** Egypt
245 K8 **Fei Huang He** ~ Jiangsu, China
106 B6 **Feijó** Acre, W Brazil 08°07′S 70°27′W
278 H7 **Feilding** Manawatu-Wanganui, North Island, New Zealand 40°15′S 175°34′E
Feira *see* Feira de Santana
109 H8 **Feira de Santana** *var.* Feira. Bahia, E Brazil 12°17′S 38°53′W
177 K3 **Feistritz** ~ SE Austria
Feistritz *see* Ilirska Bistrica
243 J1 **Feixi** *var.* Shangpai; *prev.* Shangpaihe. Anhui, E China 31°40′N 117°08′E
245 K7 **Feixian** Shandong, China 35°09′N 117°35′E
Feixin Dao *see* Flat Island
231 H2 **Feiẕābād** *var.* Faizabad, Faizābād, Feyzābād, Fyzabad; *prev.* Feyẕābād. Badakhshān, NE Afghanistan 37°06′N 70°34′E
Fejaj, Chott el *see* Fedjaj, Chott el
183 F11 **Fejér** *off.* Fejér Megye. ⬖ *county* W Hungary
Fejér Megye *see* Fejér
155 E13 **Fejø** *island* SE Denmark
214 F7 **Feke** Adana, S Turkey 37°49′N 35°55′E
218 H2 **Fékehé** Lebanon
Feketehalom *see* Codlea
Fekete-Körös *see* Crișul Negru
171 L6 **Felanitx** Mallorca, Spain, W Mediterranean Sea 39°28′N 03°08′E
177 J2 **Feldaist** ~ N Austria
177 L4 **Feldbach** Steiermark, SE Austria 46°58′N 15°53′E
179 C13 **Feldberg** ▲ SW Germany 47°52′N 08°01′E
180 H1 **Felde** Schleswig-Holstein, Germany 54°18′E 9°56′N
188 D8 **Feldioara** *Ger.* Marienburg, *Hung.* Földvár. Brașov, C Romania 45°49′N 25°36′E
176 F4 **Feldkirch** *anc.* Clunia. Vorarlberg, W Austria 47°15′N 09°38′E
177 J4 **Feldkirchen in Kärnten** *Slvn.* Trg. Kärnten, S Austria 46°42′N 14°01′E
Félegyháza *see* Kiskunfélegyháza
285 J1 **Feleolo** ✈ (Āpia) Upolu, C Samoa 13°49′S 171°59′W
172 C1 **Felgueiras** Porto, Portugal 41°22′N 8°12′W
170 C3 **Felgueiras** Porto, N Portugal 41°22′N 08°12′W
114 A3 **Felicia** Santa Fe, Argentina 31°15′S 61°13′W
Felicitas Julia *see* Lisboa
137 J8 **Félicité** *island* Inner Islands, NE Seychelles
235 C12 **Felidhu Atoll** *atoll* C Maldives
87 N6 **Felipe Carrillo Puerto** Quintana Roo, SE Mexico 19°34′N 88°02′W
52 G7 **Felix, Cape** *headland* Nunavut, N Canada
84 E3 **Félix Gómez** Sonora, Mexico 29°50′N 111°29′W
110 F1 **Felixlandia** Minas Gerais, Brazil 18°47′S 44°55′W
161 M5 **Felixstowe** E England, United Kingdom 51°58′N 01°20′E
141 L6 **Felixton** KwaZulu-Natal, South Africa 28°49′S 31°53′E
181 B12 **Fell** Rheinland-Pfalz, Germany 49°46′E 6°47′N
165 H6 **Felletin** Creuse, C France 45°53′N 02°12′E
Fellin *see* Viljandi
81 C8 **Fellows** California, USA 35°11′N 119°32′W
181 G9 **Felsberg** Hessen, Germany 51°08′E 9°25′N
Felsőbánya *see* Baia Sprie
Felsőmuzslya *see* Mužlja
Felsővisó *see* Vișeu de Sus
159 K5 **Felton** United Kingdom 55°17′N 1°42′W
80 B4 **Felton** California, W USA 37°03′N 122°04′W
177 H6 **Feltre** Veneto, NE Italy 46°01′N 11°55′E
161 L3 **Feltwell** United Kingdom 52°29′N 0°31′E
155 E13 **Femer Bælt** *Dan.* Fehmarn Belt, *Ger.* Fehmarnbelt. *strait* Denmark/Germany *see also* Fehmarn Belt
155 E13 **Femø** *island* SE Denmark
154 E5 **Femunden** ◎ S Norway
158 B8 **Fenagh** Leitrim, Ireland 54°01′N 7°50′W
170 D1 **Fene** Galicia, NW Spain 43°28′N 08°10′W
62 D4 **Fenelon Falls** Ontario, SE Canada 44°34′N 78°43′W
282 F10 **Feneppi** *atoll* Chuuk Islands, C Micronesia
215 H4 **Fener Burnu** *headland* N Turkey 41°07′N 39°26′E
Fénérive *see* Fenoarivo Atsinanana
167 M5 **Fénétrange** Lorraine, France 48°51′N 7°01′E
187 J3 **Fengári** ▲ Samothráki, E Greece 40°27′N 25°37′E
243 H8 **Fengcheng** Guangdong, SE China 24°04′N 114°12′E
243 J5 **Fengcheng** Jiangxi, China 28°07′N 115°29′E
245 N2 **Fengcheng** *var.* Feng-cheng, Fenghwangcheng. Liaoning, NE China 40°28′N 124°01′E
245 M5 **Fengcheng** Shandong, NE China 36°42′N 121°14′E
Fengcheng *see* Lianjiang
Feng-cheng *see* Fengcheng
Fengcheng *see* Xinfeng
245 H6 **Fengfeng** Hebei, E China 36°29′N 114°12′E
242 D5 **Fenggang** *var.* Longquan. Guizhou, S China 27°57′N 107°42′E
243 M3 **Fenghua** Zhejiang, SE China 29°40′N 121°25′E
242 E5 **Fenghuang** Hunan, China 27°56′N 109°35′E
243 M3 **Fenghui** Zhejiang, E China 29°57′N 120°59′E
Fenghwangcheng *see* Fengcheng
Fengjiaba *see* Wangcang
242 E2 **Fengjie** *var.* Yong'an. Sichuan, C China 31°03′N 109°31′E
242 F9 **Fengkai** *var.* Jiangkou. Guangdong, S China 23°26′N 111°28′E
243 M4 **Fenglin** Zhejiang, E China 28°19′N 120°46′E
243 N8 **Fenglin** *Jap.* Hōrin. C Taiwan 23°52′N 121°30′E
245 K3 **Fengnan** *prev.* Xugezhuang. Hebei, N China 39°34′N 118°05′E
245 J2 **Fengning** *prev.* Dagezhen. Hebei, E China 41°12′N 116°37′E
240 B7 **Fengqing** *var.* Fengshan. Yunnan, SW China 24°38′N 99°54′E
245 I7 **Fengqiu** Henan, C China 35°02′N 114°24′E
245 K3 **Fengrun** Hebei, E China 39°50′N 118°10′E
242 C8 **Fengshan** Guangxi, China 24°19′N 107°02′E
Fengshan *see* Luoyuan, Fujian, China
Fengshan *see* Fengqing, Yunnan, China
242 G6 **Fengshiyan** Hunan, SE China 26°48′N 111°59′E
243 J8 **Fengshun** Guangdong, S China 23°51′N 116°11′E
245 J9 **Fengtai** Anhui, China 32°25′N 116°26′E
245 J3 **Fengtai** Beijing Shi, N China 39°51′N 116°16′E
Fengtien *see* Liaoning, China
Fengtien *see* Shenyang, China
245 J7 **Fengxian** Jiangsu, China 34°25′N 116°20′E
244 D7 **Fengxian** Shaanxi, China 34°19′N 107°14′E
244 C8 **Fengxian** *var.* Feng Xian; *prev.* Shuangshipu. Shaanxi, C China 33°50′N 106°33′E
243 M2 **Fengxian** Shanghai Shi, China 30°33′N 121°16′E
Feng Xian *see* Fengxian
Fengxiang *see* Luobei
243 I4 **Fengxin** Jiangxi, China 28°25′N 115°13′E
243 H4 **Fengyang** *var.* Fuchang. Anhui, E China 32°52′N 117°32′E
Fengyizhen *see* Maoxian
243 M8 **Fengyuan** C Taiwan 24°13′N 120°41′E
245 H2 **Fengzhen** Nei Mongol Zizhiqu, N China 40°25′N 113°09′E
243 L7 **Fengzhou** Fujian, SE China 24°57′N 118°32′E
244 F7 **Fen He** ~ C China
244 G4 **Fenhe Shuiku** ▣ Shanxi, China
233 L8 **Feni** Chittagong, E Bangladesh 23°00′N 91°24′E
280 E2 **Feni Islands** *island group* NE Papua New Guinea
82 D10 **Fenimore Pass** *strait* Aleutian Islands, Alaska, USA
144 A5 **Feni Ridge** *undersea feature* N Atlantic Ocean 53°45′N 18°00′W
Fennern *see* Vändra
72 B7 **Fennimore** Wisconsin, N USA 42°58′N 90°39′W
139 M6 **Fenoarivo Atsinanana** *prev./Fr.* Fénérive. Toamasina, E Madagascar 20°52′S 46°52′E
244 F10 **Fenqing He** ~ C China
155 E13 **Fensmark** Sjælland, SE Denmark 55°17′N 11°48′E
157 I10 **Fens, The** *wetland* E England, United Kingdom
73 H8 **Fenton** Michigan, N USA 42°48′N 83°42′W
284 C10 **Fenuafo'ou, Île** *island* E Wallis and Futuna
159 J4 **Fenwick** United Kingdom 55°39′N 1°53′W
244 G5 **Fenyang** Shanxi, C China 37°14′N 111°40′E
243 I5 **Fenyi** Jiangxi, C China 27°29′N 114°23′E
160 D9 **Feock** United Kingdom 50°12′N 5°03′W
189 K9 **Feodosiya** *var.* Kefe, *It.* Kaffa; *anc.* Theodosia. Avtonomna Respublika Krym, S Ukraine 45°03′N 35°24′E
158 E3 **Feolin Ferry** United Kingdom 55°50′N 6°05′W
154 E4 **Feragen** ◎ S Norway
158 B10 **Ferbane** Offaly, Ireland 53°16′N 7°50′W
131 J1 **Fer, Cap de** *headland* NE Algeria 37°05′N 07°10′E
73 E12 **Ferdinand** Indiana, N USA 38°13′N 86°51′W
Ferdinand *see* Montana, Bulgaria
Ferdinand *see* Mihail Kogălniceanu, Romania
Ferdinandsberg *see* Oțelu Roșu
223 I4 **Ferdows** *var.* Firdaus; *prev.* Tūn. Khorāsān-e Jonūbī, E Iran 34°00′N 58°09′E
165 I3 **Fère-Champenoise** Marne, N France 48°45′N 03°59′E
167 I5 **Fére-en-Tardenois** Picardie, France 49°12′N 3°31′E
Ferencz-József Csúcs *see* Gerlachovský štít
175 F8 **Ferentino** Lazio, C Italy 41°40′N 13°16′E
187 J2 **Féres** Anatolikí Makedonía kai Thráki, NE Greece 40°54′N 26°12′E
Fergana *see* Farg'ona
229 J6 **Fergana Valley** *var.* Farghona Valley, *Rus.* Ferganskaya Dolina, *Taj.* Wodii Farghona, *Uzb.* Farghona Wodiysi. *basin* Tajikistan/Uzbekistan
Ferganskaya Dolina *see* Fergana Valley
Ferganskaya Oblast' *see* Farg'ona Viloyati
229 K5 **Ferganskiy Khrebet** ▲▲ C Kyrgyzstan
62 C5 **Fergus** Ontario, S Canada 43°42′N 80°22′W
74 F4 **Fergus Falls** Minnesota, N USA 46°15′N 96°02′W
280 D4 **Fergusson Island** *var.* Kaluwawa. *island* SE Papua New Guinea
183 F11 **Ferihegy** ✈ (Budapest) Budapest, C Hungary 47°25′N 19°13′E
184 G7 **Ferizaj** *Serb.* Uroševac. C Kosovo 42°23′N 21°09′E
132 F6 **Ferkessédougou** N Ivory Coast 09°36′N 05°12′W
177 J5 **Ferlach** *Slvn.* Borovlje. Kärnten, S Austria 46°31′N 14°18′E
177 J10 **Fermo** *anc.* Firmum Picenum. Marche, C Italy 43°09′N 13°44′E
170 E4 **Fermoselle** Castilla y León, N Spain 41°19′N 06°24′W
157 C10 **Fermoy** *Ir.* Mainistir Fhear Maí. SW Ireland 52°08′N 08°16′W
69 L3 **Fernandina Beach** Amelia Island, Florida, SE USA 30°40′N 81°27′W
105 A10 **Fernandina, Isla** *var.* Narborough Island. *island* Galapagos Islands, Ecuador, E Pacific Ocean
95 N3 **Fernando de Noronha** *island* E Brazil
Fernando Po/Fernando Póo *see* Bioco, Isla de
113 K2 **Fernandópolis** São Paulo, S Brazil 20°18′S 50°13′W
110 A5 **Fernando Prestes** São Paulo, Brazil 21°16′S 48°43′W
170 F8 **Fernán Núñez** Andalucía, S Spain 37°40′N 04°44′W
139 L1 **Fernão Veloso, Baia de** *bay* NE Mozambique
78 A2 **Ferndale** California, W USA 40°34′N 124°16′W
76 C2 **Ferndale** Washington, NW USA 48°51′N 122°35′W
161 I8 **Fern Down** United Kingdom 50°48′N 1°52′W
57 I9 **Fernie** British Columbia, SW Canada 49°30′N 115°00′W
80 G2 **Fernley** Nevada, W USA 39°35′N 119°15′W
160 B3 **Ferns** Wexford, Ireland 52°35′N 6°30′W
Ferozepore *see* Fīrozpur
175 H10 **Ferrandina** Basilicata, S Italy 40°30′N 16°25′E
177 H7 **Ferrara** *anc.* Forum Alieni. Emilia-Romagna, N Italy 44°50′N 11°36′E
131 H2 **Ferrat, Cap** *headland* NW Algeria 35°52′N 00°24′W
175 C10 **Ferrato, Capo** *headland* Sardegna, Italy, C Mediterranean Sea 39°18′N 09°37′E
291 F9 **Ferraz Ridge** S Atlantic Ocean
114 B8 **Ferré** Buenos Aires, Argentina 34°08′S 61°08′W
115 L1 **Ferreira** Rio Grande do Sul, Brazil 29°59′S 52°59′W
114 G3 **Ferreira** Salto, Uruguay 31°10′S 57°00′W
172 C6 **Ferreira do Alentejo** Beja, S Portugal 38°04′N 08°06′W
104 B5 **Ferreñafe** Lambayeque, W Peru 06°42′S 79°45′W
176 C6 **Ferret** Valais, SW Switzerland 45°57′N 07°04′E
168 C3 **Ferret, Cap** *headland* W France 44°37′N 01°15′W
167 M8 **Ferrette** Alsace, France 47°30′N 7°19′E
68 D2 **Ferriday** Louisiana, S USA 31°37′N 91°33′W
167 H7 **Ferriéres** Centre, France 48°05′N 2°47′E
169 H2 **Ferrieres-St-Mary** Auvergne, France 45°11′N 3°04′E
Ferro *see* Hierro
175 C9 **Ferro, Capo** *headland* Sardegna, Italy, C Mediterranean Sea 41°09′N 09°31′E
170 D1 **Ferrol** *var.* El Ferrol; *prev.* El Ferrol del Caudillo. Galicia, NW Spain 43°29′N 08°14′W
104 C6 **Ferrol, Península de** *peninsula* W Peru
79 I4 **Ferron** Utah, W USA 39°05′N 111°07′W
111 H1 **Ferros** Minas Gerais, Brazil 19°14′S 43°02′W
67 J5 **Ferrum** Virginia, NE USA 36°54′N 80°01′W
160 B3 **Ferrybank** Wicklow, Ireland 52°48′N 6°09′W
159 K6 **Ferryhill** United Kingdom 54°41′N 1°33′W
68 G4 **Ferry Pass** Florida, SE USA 30°30′N 87°12′W
Ferryville *see* Menzel Bourguiba
74 F3 **Fertile** Minnesota, N USA 47°32′N 96°16′W
Fertő *see* Neusiedler See
111 J4 **Fervedouro** Minas Gerais, Brazil 20°44′S 42°17′W
Ferwerd *see* Ferwert
162 G3 **Ferwert** *Dutch.* Ferwerd. Fryslân, N Netherlands 53°21′N 05°47′E
130 F2 **Fès** *Eng.* Fez. N Morocco 34°06′N 04°57′W
135 E10 **Feshi** Bandundu, SW Dem. Rep. Congo 06°08′S 18°12′E
74 D2 **Fessenden** North Dakota, N USA 47°36′N 99°37′W
Festenberg *see* Twardogóra
66 B3 **Festus** Missouri, C USA 38°13′N 90°24′W
188 F10 **Fetești** Ialomița, SE Romania 44°22′N 27°51′E
160 A4 **Fethard** Wexford, SE Ireland
214 C8 **Fethiye** Muğla, SW Turkey 36°37′N 29°08′E
156 H1 **Fetlar** *island* NE Scotland, United Kingdom
154 E7 **Fetsund** Akershus, S Norway 59°55′N 11°03′E
159 I1 **Fettercairn** United Kingdom 56°51′N 2°34′W
181 J13 **Feucht** Bayern, Germany 49°23′E 11°13′N
181 I14 **Feuchtwangen** Bayern, Germany 49°10′E 10°20′N
167 J6 **Feuges** Champagne-Ardenne, France 48°24′N 4°07′E
51 J6 **Feuilles, Lac aux** ◎ Québec, E Canada
59 H3 **Feuilles, Rivière aux** ~ Québec, E Canada
163 H13 **Feulen** Diekirch, C Luxembourg 49°52′N 06°03′E
165 J6 **Feurs** Loire, E France 45°44′N 04°14′E
155 C9 **Fevik** Aust-Agder, S Norway 58°22′N 08°40′E
193 K8 **Fevral'sk** Amurskaya Oblast', SE Russian Federation 52°25′N 131°06′E
159 K8 **Fewston** United Kingdom 53°59′N 1°42′W
Feyzābād *see* Feiẕābād
Feyẕābād *see* Feiẕābād
Fez *see* Fès
128 E5 **Fezzan** ⬖ *cultural region* C Libya
160 E2 **Ffestiniog** NW Wales, United Kingdom 52°55′N 03°54′W
Fhóid Duibh, Cuan an *see* Blacksod Bay
112 C6 **Fiambalá** Catamarca, NW Argentina 27°45′S 67°37′W
139 L8 **Fianarantsoa** Fianarantsoa, C Madagascar 21°27′S 47°05′E
139 M8 **Fianarantsoa** ⬖ *province* SE Madagascar
134 C3 **Fianga** Mayo-Kébbi Est, SW Chad 09°57′N 15°09′E
Ficce *see* Fichē
136 E4 **Fichē** *It.* Ficce. Oromīya, C Ethiopia 09°48′N 38°43′E
179 H9 **Fichtelberg** ▲ Czech Republic/Germany 50°26′N 12°57′E
179 G10 **Fichtelgebirge** ▲▲ SE Germany
179 G10 **Fichtelnaab** ~ SE Germany
181 H14 **Fichtenberg** Baden-Württemberg, Germany 48°59′E 9°43′N
141 I6 **Ficksburg** Free State, South Africa 28°53′S 27°53′E
219 E11 **Fidān, Wādī al** *dry watercourse* Jordan
80 D2 **Fiddletown** California, USA 38°30′N 120°45′W
176 F7 **Fidenza** Emilia-Romagna, N Italy 44°52′N 10°04′E
184 F9 **Fier** *var.* Fieri. Fier, SW Albania 40°44′N 19°34′E
184 F9 **Fier** ⬖ *district* W Albania
Fieri *see* Fier
Fierza *see* Fierzë
184 F7 **Fierzë** *var.* Fierza. Shkodër, N Albania 42°15′N 20°02′E
184 F7 **Fierzës, Liqeni i** ◎ N Albania
176 D5 **Fiesch** Valais, SW Switzerland 46°25′N 08°09′E
176 G9 **Fiesole** Toscana, C Italy 43°50′N 11°18′E
219 F10 **Fīfah** Jordan 30°56′N 35°26′E
219 F10 **Fīfah** Aṭ Ṭafīlah, W Jordan 30°55′N 35°25′E
159 I2 **Fife** *var.* Kingdom of Fife. *cultural region* E Scotland, United Kingdom
Fife, Kingdom of *see* Fife
159 J2 **Fife Ness** *headland* E Scotland, United Kingdom 56°16′N 02°35′W
Fifteen Twenty Fracture Zone *see* Barracuda Fracture Zone
171 I10 **Figalo, Cap** *headland* Algeria
165 H7 **Figeac** Lot, S France 44°37′N 02°01′E
155 H10 **Figeholm** Kalmar, SE Sweden 57°12′N 16°34′E
114 B7 **Fighiera** Santa Fe, Argentina 33°12′S 60°28′W
Figig *see* Figuig
138 G4 **Figtree** Matabeleland South, SW Zimbabwe 20°24′S 28°21′E
172 B3 **Figueira da Foz** Coimbra, W Portugal 40°09′N 08°51′W
171 L3 **Figueres** Cataluña, E Spain 42°16′N 02°57′E
131 H3 **Figuig** *var.* Figig. E Morocco 32°09′N 01°13′W
Fijājj, Shaṭṭ al *see* Fedjaj, Chott el
281 M7 **Fiji** *off.* Republic of Fiji, *prev.* Sovereign Democratic Republic of Fiji, *prev.* Republic of the Fiji Islands, *Fij.* Viti. ◆ *republic* SW Pacific Ocean
280 E10 **Fiji** *island group* SW Pacific Ocean
Fiji Islands, Republic of the *see* Fiji
267 I6 **Fiji Plate** *tectonic feature*
Fiji, Republic of *see* Fiji
Fiji, Sovereign Democratic Republic of *see* Fiji
171 H8 **Filabres, Sierra de los** ▲▲ SE Spain
138 G4 **Filabusi** Matabeleland South, S Zimbabwe 20°34′S 29°20′E
88 G8 **Filadelfia** Guanacaste, W Costa Rica 10°28′N 85°33′W
183 G10 **Fil'akovo** *Hung.* Fülek. Banskobystrický Kraj, C Slovakia 48°15′N 19°53′E
293 H3 **Filchner Ice Shelf** *ice shelf* Antarctica
62 E2 **Fildegrand** ~ Québec, SE Canada
76 G8 **Filer** Idaho, NW USA 42°34′N 114°36′W
Filevo *see* Varbitsa
159 M7 **Filey** United Kingdom 54°13′N 0°17′W
159 M8 **Filey Bay** *bay* United Kingdom
188 C9 **Filiași** Dolj, SW Romania 44°32′N 23°31′E
186 E4 **Filiátes** Ípeiros, W Greece 39°38′N 20°16′E
186 F7 **Filiatrá** Pelopónnisos, S Greece 37°09′N 21°35′E
175 G11 **Filicudi, Isola** *island* Isole Eolie, S Italy
221 K6 **Filim** E Oman 20°37′N 58°11′E
Filimon Sîrbu *see* Făurei
133 I4 **Filingué** Tillabéri, W Niger 14°21′N 03°22′E
Filiouri *see* Líssos
187 I2 **Fílippoi** *anc.* Philippi. *site of ancient city* Anatolikí Makedonía kai Thráki, NE Greece
154 G7 **Filipstad** Värmland, C Sweden 59°44′N 14°10′E
87 I8 **Filisola** Veracruz-Llave, Mexico 17°50′N 94°19′W
176 F5 **Filisur** Graubünden, S Switzerland 46°40′N 09°43′E
154 C6 **Fillefjell** ▲▲ S Norway
78 D8 **Fillmore** California, W USA 34°23′N 118°56′W
79 H4 **Fillmore** Utah, W USA 38°57′N 112°19′W
86 E8 **Filo de los Caballos** Guerrero, S Mexico 17°39′N 99°50′W
87 M6 **Filomeno Mata** Quintana Roo, Mexico 19°52′N 88°24′W
85 N10 **Filomeno Mata** Veracruz-Llave, SE Mexico 20°12′N 97°41′W
62 E2 **Fils, Lac du** ◎ Québec, SE Canada
180 D3 **Filsum** Niedersachsen, Germany 53°14′E 7°38′N
Filyos Çayı *see* Yenice Çayı
Fimbul Ice Shelf *see* Fimbulisen
293 I2 **Fimbutheimen** *physical region* Antarctica
176 G7 **Finale Emilia** Emilia-Romagna, C Italy 44°50′N 11°17′E
176 D8 **Finale Ligure** Liguria, NW Italy 44°11′N 08°22′E
170 G8 **Fiñana** Andalucía, S Spain 37°09′N 02°47′W
67 J4 **Fincastle** Virginia, NE USA 37°30′N 79°54′W
161 L3 **Fincham** United Kingdom 52°37′N 0°30′E
161 L5 **Finchingfield** SE England, United Kingdom 51°58′N 00°18′E
161 K6 **Finchley** United Kingdom 51°36′N 0°10′W
163 H13 **Findel** ✈ (Luxembourg) Luxembourg, C Luxembourg 49°39′N 06°16′E
156 F5 **Findhorn** ~ N Scotland, United Kingdom
73 H9 **Findlay** Ohio, N USA 41°02′N 83°40′W
161 K8 **Findon** United Kingdom 50°51′N 0°24′W
158 E2 **Fingal's Cave** *cave* W Scotland, United Kingdom
64 D2 **Finger Lakes** ◎ New York, NE USA
139 H2 **Fingoè** Tete, NW Mozambique 15°10′S 31°51′E
214 C8 **Finike** Antalya, SW Turkey 36°18′N 30°08′E
164 E3 **Finistère** ⬖ *department* NW France
280 B3 **Finisterre Range** ▲▲ N Papua New Guinea
276 E1 **Finke** Northern Territory, N Australia 25°37′S 134°35′E
177 J5 **Finkenstein** Kärnten, S Austria 46°34′N 13°53′E
283 H10 **Finkol, Mount** *var.* Mount Crozer. ▲ Kosrae, E Micronesia 05°18′N 163°00′E
153 I8 **Finland** *off.* Republic of Finland, *Fin.* Suomen Tasavalta, Suomi. ◆ *republic* N Europe
153 H10 **Finland, Gulf of** *Est.* Soome Laht, *Fin.* Suomenlahti, *Ger.* Finnischer Meerbusen, *Rus.* Finskiy Zaliv, *Swe.* Finska Viken. *gulf* E Baltic Sea
Finland, Republic of *see* Finland
56 D1 **Finlay** ~ British Columbia, W Canada
277 J6 **Finley** New South Wales, SE Australia 35°41′S 145°33′E
80 B1 **Finley** California, USA 39°0′N 122°53′W
74 E3 **Finley** North Dakota, N USA 47°30′N 97°50′W
72 C2 **Finmark** Ontario, S Canada 48°35′N 89°45′W
158 C6 **Finn** ~ Ireland
181 E9 **Finnentrop** Nordrhein-Westfalen, Germany 51°10′E 7°58′N
55 L2 **Finnie Bay** *bay* NW Atlantic Ocean
Finnischer Meerbusen *see* Finland, Gulf of
152 G4 **Finnmark** ⬖ *county* N Norway
152 G4 **Finnmarksvidda** *physical region* N Norway
152 F4 **Finnsnes** Troms, N Norway 69°16′N 18°00′E
280 C3 **Finschhafen** Morobe, C Papua New Guinea 06°35′S 147°51′E
154 C6 **Finse** Hordaland, S Norway 60°35′N 07°33′E
Finska Viken/Finskiy Zaliv *see* Finland, Gulf of
155 H9 **Finspång** Östergötland, S Sweden 58°42′N 15°45′E
174 B3 **Finsteraarhorn** ▲ Switzerland 46°33′N 08°07′E
179 I8 **Finsterwalde** Brandenburg, E Germany 51°38′N 13°43′E
180 H4 **Fintel** Niedersachsen, Germany 53°10′E 9°41′N
158 C7 **Fintona** United Kingdom 54°30′N 7°19′W
158 B6 **Fintown** Donegal, Ireland 54°52′N 8°07′W
158 G3 **Fintry** United Kingdom 56°03′N 4°12′W
158 E1 **Fionnphort** W Scotland, United Kingdom 56°20′N 06°17′W
278 A12 **Fiordland** *physical region* South Island, New Zealand
176 F7 **Fiorenzuola d'Arda** Emilia-Romagna, C Italy 44°57′N 09°53′E
Firat Nehri *see* Euphrates
Firdaus *see* Ferdows
80 C5 **Firebaugh** California, USA 36°52′N 120°27′W
55 H4 **Firedrake Lake** ◎ Northwest Territories, NW Canada
65 J6 **Fire Island** *island* New York, NE USA
176 G9 **Firenze** *Eng.* Florence; *anc.* Florentia. Toscana, C Italy 43°47′N 11°15′E
176 G9 **Firenzuola** Toscana, C Italy 44°07′N 11°22′E
72 H2 **Fire River** Ontario, S Canada 48°46′N 83°34′W
Fîrliug *see* Fârliug
114 A7 **Firmat** Santa Fe, C Argentina 33°29′S 61°29′W
165 J6 **Firminy** Loire, E France 45°22′N 04°18′E
Firmum Picenum *see* Fermo
232 F6 **Fīrozābād** Uttar Pradesh, N India 27°09′N 78°24′E
232 D4 **Fīrozpur** *var.* Ferozepore. Punjab, NW India 30°55′N 74°38′E
First State *see* Delaware
222 F7 **Fīrūzābād** Fārs, S Iran 28°51′N 52°35′E
287 H3 **Fisberling Seamount** N Pacific Ocean
Fischamend *see* Fischamend Markt
177 L2 **Fischamend Markt** *var.* Fischamend. Niederösterreich, NE Austria 48°08′N 16°37′E
181 D12 **Fischbach** Rheinland-Pfalz, Germany 49°44′E 7°24′N
181 D14 **Fischbach** Rheinland-Pfalz, Germany 49°05′E 7°43′N
177 L3 **Fischbacher Alpen** ▲▲ E Austria
180 G6 **Fischbeck** Niedersachsen, Germany 52°08′E 9°18′N
Fischhausen *see* Primorsk
140 C2 **Fish** *var.* Vis. ~ S Namibia
138 D9 **Fish** *Afr.* Vis. ~ SW South Africa
140 E7 **Fish** *Afr.* Vis. ~ SW South Africa
80 E5 **Fish Camp** California, USA 37°29′N 119°38′W
55 J9 **Fisher Branch** Manitoba, S Canada 51°09′N 97°34′W
55 J9 **Fisher River** Manitoba, S Canada 51°25′N 97°23′W
65 K5 **Fishers Island** *island* New York, NE USA
79 M5 **Fishers Peak** ▲ Colorado, C USA 37°06′N 104°27′W
55 L3 **Fisher Strait** *strait* Nunavut, N Canada
160 D5 **Fishguard** *Wel.* Abergwaun. SW Wales, United Kingdom 51°59′N 04°49′W

◆ Country
● Country Capital
◇ Dependent Territory
○ Dependent Territory Capital
⬖ Administrative Regions
✈ International Airport
▲ Mountain
▲▲ Mountain Range
Volcano
~ River
◎ Lake
▣ Reservoir

G

◆ Country   ◇ Dependent Territory   ● Administrative Regions   ▲ Mountain   ❖ Volcano   ⊚ Lake
● Country Capital   ○ Dependent Territory Capital   ✈ International Airport   ▲ Mountain Range   ♣ River   ▨ Reservoir

197 I9 **Gudermes** Chechenskaya Respublika, SW Russian Federation 43°23´N 46°06´E
180 I3 **Gudow** Schleswig-Holstein, Germany 53°33´E 10°47´N
234 F7 **Gūdūr** Andhra Pradesh, E India 14°10´N 79°51´E
228 B7 **Gudurolum** Balkan Welaýaty, W Turkmenistan 37°28´N 54°30´E
154 B6 **Gudvangen** Sogn Og Fjordane, S Norway 60°54´N 06°49´E
167 M7 **Guebwiller** Haut-Rhin, NE France 47°55´N 07°13´E
Guéckédou see Guékédou
132 D7 **Guékédou** var. Gueckédou. Guinée-Forestière, S Guinea 08°33´N 10°08´W
87 H8 **Guelatao** Oaxaca, SE Mexico
Guelders see Gelderland
134 C4 **Guélengdeng** Mayo-Kébbi Est, W Chad 10°55´N 15°31´E
131 J1 **Guelma** var. Gâlma. NE Algeria 36°29´N 07°25´E
130 D3 **Guelmime** var. Goulimine. SW Morocco 28°59´N 10°10´W
62 C5 **Guelph** Ontario, S Canada 43°34´N 80°16´W
166 C7 **Guémené-Penfao** Loire-Atlantique, NW France 47°37´N 01°49´W
166 A6 **Guémené-sur-Scorff** Bretagne, France 48°04´N 3°12´W
171 N9 **Guenzet** Algeria
134 F4 **Guér** Morbihan, NW France 47°54´N 02°07´W
134 E4 **Guéra** off. Région du Guéra. ◆ region S Chad
164 E4 **Guérande** Loire-Atlantique, NW France 47°20´N 02°25´W
Guéra, Région du see Guéra
165 H6 **Guéréda** Wadi Fira, E Chad 14°30´N 01°52´E
165 H6 **Guéret** Creuse, C France 46°10´N 01°52´E
167 J9 **Guérigny** Bourgogne, France 47°05´N 3°12´E
80 B1 **Guerneville** California, USA 38°30´N 123°00´W
Guernica/Guernica y Lumo see Gernika-Lumo
80 E4 **Guerney** California, USA 36°13´N 119°38´W
77 M8 **Guernsey** Wyoming, C USA 42°16´N 104°44´W
198 D1 **Guernsey** ◇ British Crown Dependency Channel Islands, NW Europe
166 B4 **Guernsey** island Channel Islands, NW Europe
132 C4 **Guéroù** Assaba, S Mauritania 16°48´N 11°40´W
75 F4 **Guerrero** Texas, SW USA 28°54´N 98°53´W
85 L4 **Guerrero** Coahuila, Mexico 28°18´N 100°23´W
85 J6 **Guerrero** San Luis Potosí, C Mexico 21°43´N 99°48´W
85 M9 **Guerrero** Tamaulipas, NE Mexico 23°33´N 99°48´W
86 D7 **Guerrero** ◆ state S Mexico
84 C5 **Guerrero Negro** Baja California Sur, NW Mexico 27°56´N 114°04´W
114 B8 **Guerrico** Buenos Aires, Argentina 33°40´S 60°24´W
165 N3 **Gueugnon** Saône-et-Loire, C France 46°36´N 04°03´E
132 E5 **Guéyo** S Ivory Coast 05°25´N 06°04´W
243 H5 **Gugang** Jiangxi, SE China 28°17´N 113°45´E
175 G8 **Guglionesi** Molise, C Italy 41°54´N 14°54´E
282 E3 **Gugan** island C Northern Mariana Islands
Guhrau see Góra
Gui see Guangxi Zhuangzu Zizhiqu
172 B3 **Guia** Leiria, Portugal 39°57´N 8°47´W
Guiana see French Guiana
95 **Guiana Basin** undersea feature W Atlantic Ocean 11°00´N 52°00´W
98 D2 **Guiana Highlands** var. Macizo de las Guayanas. ▲ N South America
Guiba see Juba
164 E4 **Guichen** Ille-et-Vilaine, NW France 47°57´N 01°47´W
243 K2 **Guichi** prev. Guichi. Anhui, SE China 30°39´N 117°29´E
Guichi see Guichi
114 B8 **Guichón** Paysandú, W Uruguay 32°30´S 57°13´W
133 K5 **Guidan-Roumji** Maradi, S Niger 13°40´N 06°41´E
Guidder see Guider
134 C3 **Guider** var. Guidder. Nord, N Cameroon 09°55´N 13°59´E
132 C4 **Guidimaka** ◆ region S Mauritania
131 J3 **Guidimouni** Zinder, S Niger 13°40´N 09°31´E
242 C6 **Guiding** Guizhou, China 26°08´N 107°07´E
111 I5 **Guidoval** Minas Gerais, Brazil 21°09´S 42°48´W
132 B3 **Guier, Lac de** var. Lac de Guiers. ⊗ N Senegal
Guiers, Lac de see Guier, Lac de
242 E9 **Guigang** var. Guixian, Gui Xian. Guangxi Zhuangzu Zizhiqu, S China 23°06´N 109°36´E
132 E8 **Guiglo** W Ivory Coast 06°33´N 07°29´W
167 H6 **Guignes** Île-de-France, France 48°38´N 2°48´E
102 F3 **Güigüe** Carabobo, N Venezuela 10°05´N 67°48´W
88 C4 **Güija, Lago de** ⊗ El Salvador/Guatemala
29 F4 **Gui Jiang** var. Gui Shui. ↗ S China
243 M3 **Guiji Shan** Zhejiang, China
170 E4 **Guijuelo** Castilla y León, N Spain 40°34´N 05°40´W
Guilan see Gīlān
161 J7 **Guildford** SE England, United Kingdom 51°14´N 00°35´W
63 K3 **Guildhall** Vermont, NE USA 44°34´N 71°36´W
159 H2 **Guildtown** United Kingdom 56°28´N 3°24´W
63 J5 **Guilford** Connecticut, USA
165 J5 **Guilford** Maine, NE USA 45°10´N 69°22´W
Guillaume-Delisle, Lac ⊗ Québec, NE Canada
169 M4 **Guillaumes** Provence-Alpes-Côte d'Azur, France 44°05´N 6°51´E
84 B3 **Guillermo Prieto** Durango, Mexico 30°26´N 112°40´W
72 D7 **Guillermo Prieto** Durango, Mexico 24°26´N 105°0´W
165 K7 **Guillestre** Hautes-Alpes, SE France 44°41´N 06°38´E
160 G3 **Guilsfield** United Kingdom 52°43´N 3°08´W
173 G2 **Guimarães** var. Guimaráes. Braga, N Portugal 41°26´N 08°19´W
Guimarães see Guimarães
102 G8 **Guimarães Rosas, Pico** ▲ NW Brazil
111 L1 **Guimarinia** Minas Gerais, Brazil 18°51´S 46°47´W
245 K7 **Guimeng Ding** ▲ Shandong, China 35°36´N 117°30´E
66 C8 **Guin** Alabama, S USA 33°58´N 87°54´W
80 C1 **Guinda** California, USA 38°50´N 122°12´W
132 D6 **Guinea** off. Republic of Guinea. var. Guinée; prev. French Guinea, People's Revolutionary Republic of Guinea. ◆ republic W Africa
291 I0 **Guinea Basin** undersea feature E Atlantic Ocean 0°00´N 05°00´W
132 A5 **Guinea-Bissau** off. Republic of Guinea-Bissau, Fr. Guinée-Bissau, Port. Guiné-Bissau; prev. Portuguese Guinea. ◆ republic W Africa
Guinea-Bissau, Republic of see Guinea-Bissau
120 F4 **Guinea Fracture Zone** tectonic feature E Atlantic Ocean
133 H9 **Guinea, Gulf of** Fr. Golfe de Guinée. gulf E Atlantic Ocean
Guinea, People's Revolutionary Republic of see Guinea
Guinea, Republic of see Guinea-Bissau
Guinée see Guinea
Guinée-Bissau see Guinea-Bissau
Guinée, Golfe de see Guinea, Gulf of
61 L10 **Güines** La Habana, W Cuba 22°50´N 82°02´W
166 G1 **Guînes** Nord-Pas-de-Calais, France 50°52´N 1°52´E
133 I9 **Guiniamp** Côtes d'Armor, NW France 48°34´N 03°09´W
242 E9 **Guiping** Guangxi, China 23°14´N 110°02´E
Guipúzcoa see Gipuzkoa
90 C4 **Güira de Melena** La Habana, W Cuba 22°47´N 82°33´W
130 D4 **Guir, Hamada du** desert Algeria/Morocco
103 I2 **Güiria** Sucre, NE Venezuela 10°37´N 62°21´W
111 J5 **Guiricema** Minas Gerais, Brazil 21°00´S 42°43´W
157 L7 **Guisborough** United Kingdom 54°32´N 1°09´W
113 F4 **Guise** Picardie, France 49°54´N 3°38´E
159 K8 **Guiseley** United Kingdom 53°52´N 1°43´W
242 F8 **Gui Shui** ↗ Guang Zhuangzu Zizhiqu, China
Gui Shui see Gui Jiang
170 D1 **Guitiriz** Galicia, NW Spain 43°10´N 07°52´W
247 I3 **Guixi** Jiangxi, C China 28°18´N 117°12´E
263 M7 **Guiuan** Samar, C Philippines 11°02´N 125°45´E
243 K4 **Guixi** Jiangxi, SE China 28°18´N 117°12´E
Gui Xian/Guixian see Guigang
242 C6 **Guiyang** var. Kuei-Yang, Kuei-yang, Kueyang, Kweiyang; prev. Kweichu. province capital Guizhou, S China 26°35´N 106°45´E
242 D6 **Guiyang** Hunan, China 24°26´N 112°26´E
242 H9 **Guiyang** var. Guizhou Sheng, Kuei-chou, Kuei-chou, Gui. province S China
242 C6 **Guizhou** var. Guizhou Sheng, Kuei-chou, Kweichow, Qian. ◆ province S China
Guizhou Sheng see Guizhou
168 G3 **Gujan-Mestras** Gironde, SW France 44°39´N 01°04´W
234 E3 **Gujarāt** var. Gujerat. ◆ state W India
231 I4 **Gūjar Khān** Punjab, E Pakistan 33°19´N 73°23´E
234 D5 **Gujarāt** see Gujarāt
242 C5 **Gujiang** Jiangxi, SE China 27°11´N 114°47´E
231 J4 **Gujrānwāla** Punjab, NE Pakistan 32°11´N 74°09´E
231 J5 **Gujrāt** Punjab, E Pakistan 32°34´N 74°04´E
228 B5 **Gulandag** Rus. Gory Kulandag. ▲ Balkan Welaýaty, W Turkmenistan

229 K6 **Gul'cha** Kir. Gülchö. Oshskaya Oblast', SW Kyrgyzstan 40°16´N 73°27´E
Gülchö see Gul'cha
289 H10 **Gulden Draak Seamount** undersea feature E Indian Ocean 33°S 101°00´E
214 H4 **Gülek Boğazı** var. Cilician Gates. pass S Turkey
280 B4 **Gulf** ◇ province S Papua New Guinea
68 G4 **Gulf Breeze** Florida, SE USA 30°21´N 87°09´W
Gulf of Liaotung see Liaodong Wan
69 K4 **Gulfport** Florida, SE USA 27°45´N 82°42´W
68 F4 **Gulfport** Mississippi, S USA 30°22´N 89°06´W
277 K4 **Gulf Shores** Alabama, S USA 30°15´N 87°40´W
277 K4 **Gulgong** New South Wales, SE Australia 32°22´S 149°31´E
242 B4 **Gulin** Sichuan, C China 28°06´N 105°47´E
261 I6 **Gulir** Pulau Kundi, E Indonesia 04°27´S 131°41´E
229 I6 **Gulistan** Rus. Gulistan. Sirdaryo Viloyati, E Uzbekistan 40°29´N 68°46´E
Gulja see Yining
83 J6 **Gulkana** Alaska, USA 62°17´N 145°25´W
159 I3 **Gullane** United Kingdom 56°02´N 2°49´W
57 L8 **Gull Lake** Saskatchewan, S Canada 50°05´N 108°30´W
55 C5 **Gull Lake** ⊗ Michigan, N USA
74 G3 **Gull Lake** ⊗ Minnesota, N USA
243 J6 **Gullonggang** Jiangxi, SE China 26°26´N 115°42´E
155 G8 **Gullspång** Västra Götaland, S Sweden 58°58´N 14°04´E
214 D7 **Güllük Körfezi** prev. Akbük Limanı. bay SW Turkey
232 D2 **Gulmarg** Jammu and Kashmir, NW India 34°04´N 74°25´E
Gulpaigan see Golpāyegān
163 H10 **Gulpen** Limburg, SE Netherlands 50°48´N 05°53´E
Gul'shad see Gul'shat
227 I6 **Gul'shat** Kir. Gul'shad. Karaganda, E Kazakhstan 46°37´N 74°22´E
136 B7 **Gulu** N Uganda 02°46´N 32°21´E
Gülübovo see Galabovo
185 I9 **Gulyantsi** Pleven, N Bulgaria 43°37´N 24°40´E
Gulyaypole see Hulyaypole
Gumal see Pishan
Gümai see Darlag
134 F6 **Gumba** Équateur, W Dem. Rep. Congo 02°58´N 21°23´E
Gumbinnen see Gusev
137 D12 **Gumbiro** Ruvuma, S Tanzania 10°19´S 35°40´E
228 B6 **Gumdag** prev. Kum-Dag. Balkan Welaýaty, W Turkmenistan 39°13´N 54°35´E
133 L5 **Gumel** Jigawa, N Nigeria 12°35´N 09°23´E
250 A1 **Gumi** prev. Kumi. 36°07´N 128°20´E
170 G3 **Gumiel de Hizán** Castilla y León, N Spain 41°46´N 03°42´W
233 I8 **Gumla** Jhārkhand, N India 23°03´N 84°36´E
Gumma see Gunma
181 D9 **Gummersbach** Nordrhein-Westfalen, W Germany 51°01´N 07°34´E
133 J5 **Gummi** Zamfara, NW Nigeria 12°07´N 05°07´E
181 I10 **Gumpelstadt** Thüringen, Germany 50°50´E 10°18´N
Gumpolds see Humpolec
Gumti see Gomati
141 I6 **Gumtree** Free State, South Africa 28°27´S 27°52´E
Gümülcine/Gümüljina see Komotini
215 H5 **Gümüşhane** var. Gümüşane, Gumushkhane. 41°13´N 39°31´E
215 H5 **Gümüşhane** var. Gümüşane, Gumushane, Gumushkhane. ◆ province NE Turkey
Gümüşhane see Gümüşhane
Gumushkhane see Gümüşhane
261 J9 **Gumzai** Pulau Kola, E Indonesia 05°27´S 134°38´E
232 E7 **Guna** Madhya Pradesh, C India 24°39´N 77°18´E
Gunabad see Gonābād
Gunan see Qijiang
Gunbad-i-Qawus see Gonbad-e Kāvūs
277 J5 **Gunbar** New South Wales, SE Australia 34°03´N 145°33´E
277 K6 **Gundagai** New South Wales, SE Australia 35°06´S 148°03´E
181 G13 **Gundelsheim** Baden-Württemberg, Germany 49°17´E 9°10´N
134 F5 **Gundji** Équateur, C Dem. Rep. Congo 00°20´N 21°31´E
235 E8 **Gundlupet** Karnātaka, W India 11°48´N 76°42´E
181 D8 **Gündoğmuş** Antalya, S Turkey 36°50´N 32°07´E
221 M5 **Güney Doğu Toroslar** ▲ SE Turkey
135 E9 **Gungu** Bandundu, SW Dem. Rep. Congo 05°43´S 19°21´E
197 J9 **Gunib** Respublika Dagestan, SW Russian Federation
181 E14 **Gunia** Vukovar-Srijem, E Croatia 45°16´N 18°51´E
73 F8 **Gun Lake** ⊗ Michigan, N USA
253 B12 **Gunma** var. Gumma. ◆ prefecture Honshū, S Japan
295 L6 **Gunnbjørn Fjeld** var. Gunnbjörns Bjerge. ▲ C Greenland 69°03´N 29°36´W
Gunnbjörns Bjerge see Gunnbjørn Fjeld
181 B8 **Gunne** Nordrhein-Westfalen, W Germany 51°30´E 8°02´N
114 C5 **Gunnedah** New South Wales, SE Australia 30°59´S 150°15´E
137 G12 **Gunner's Quoin** var. Coin de Mire. island N Mauritius
79 H4 **Gunnison** Colorado, C USA 38°33´N 106°55´W
73 H4 **Gunnison** Utah, W USA 39°09´N 111°49´W
79 H4 **Gunnison River** ↗ Colorado, C USA
151 K7 **Gunpowder River** ↗ Maryland, NE USA
Guns see Kőszeg
248 B4 **Gunsan** var. Gunsan, Jap. Gunzan; prev. Kunsan. S South Korea 35°59´N 126°42´E
Gunsan see Gunsan
177 I3 **Gunskirchen** Oberösterreich, N Austria 48°07´N 13°54´E
181 E12 **Gunsterblum** Rheinland-Pfalz, Germany 49°48´S 8°21´N
66 E6 **Guntersville** Alabama, S USA 34°21´N 86°17´W
66 D8 **Guntersville Lake** ⊗ Alabama, S USA
234 G6 **Guntur** var. Guntur. Andhra Pradesh, SE India 16°20´N 80°27´E
258 E4 **Gunungsitoli** Pulau Nias, W Indonesia 01°11´N 97°35´E
234 H5 **Gunupur** Odisha, E India 19°04´N 83°52´E
169 C10 **Gunwalloe** United Kingdom 50°02´N 5°16´W
250 A1 **Gunwi** prev. Kunwi. Kyŏngsang-bukto, South Korea 36°11´N 128°34´E
Gurganj see Kōneürgench
Gurgan/Gurgān see Gorgān
232 E3 **Gurgaon** Haryāna, E India 28°27´N 77°01´E
103 H4 **Gurguéia, Rio** ↗ E Venezuela
177 J4 **Guri, Embalse de** ⊞ E Venezuela
103 K1 **Guria** Rus. Gurdzhaani. C Georgia 41°42´N 45°47´E
177 J4 **Gurk** Kärnten, S Austria 46°52´N 14°13´E
190 **Gurk** var. Urt. Ömnögovĭ, S Mongolia 43°16´N 100°00´E
Gurk-aani see Gurdzhaani
177 I4 **Gurkfeld** see Krško
185 K9 **Gurkovo** prev. Kolupchii. Stara Zagora, C Bulgaria 42°42´N 25°46´E
177 J4 **Gurktaler Alpen** ▲ S Austria
235 E5 **Gürün** Kars. Gurlen. Xorazm Viloyati, W Uzbekistan 41°54´N 60°18´E
214 G3 **Guru** Manica, C Mozambique 14°27´N 33°18´E
107 H4 **Gurupá** Pará, NE Brazil 01°25´S 51°39´W
107 K6 **Gurupi** Tocantins, C Brazil 11°44´N 49°04´W
219 C11 **Gurūn, Wādī el** ↗ NE Brazil
232 E7 **Guru Sikhar** ▲ NW India 24°45´N 72°51´E
261 L8 **Gurvanbulag** var. Höhrin Am. Bayanhongor, C Mongolia 47°02´N 98°41´E
187 K8 **Gurvanbulag** var. Urt. Ömnögovĭ, S Mongolia 43°16´N 100°00´E
239 K1 **Gurvantes** var. Urt. Ömnögovĭ, S Mongolia 43°16´N 100°00´E
Gur'yev/Gur'yevskaya Oblast' see Atyrau
133 L5 **Gusau** Zamfara, NW Nigeria 12°18´N 06°27´E
191 D9 **Gusev** Ger. Gumbinnen. Kaliningradskaya Oblast', W Russian Federation 54°36´N 22°11´E
245 N2 **Gushan** Liaoning, China 39°32´N 123°23´E

228 F9 **Gushgy** Rus. Kushka. ↗ Mary Welaýaty, S Turkmenistan
Gush Halav see Jish
243 I6 **Gushi** Henan, China 32°06´N 115°25´E
133 H4 **Gushiagu** var. Gushiago. NE Ghana 09°54´N 00°12´W
251 I10 **Gushikawa** Okinawa, Okinawa, SW Japan 26°21´N 127°50´E
193 J6 **Gusinje** E Montenegro 42°34´N 19°51´E
191 I8 **Gusino** Smolenskaya Oblast', Russian Federation
193 I8 **Gusinoozersk** Respublika Buryatiya, S Russian Federation 51°18´N 106°28´E
196 G2 **Gus'-Khrustal'nyy** Vladimirskaya Oblast', W Russian Federation 55°39´N 40°42´E
175 B10 **Guspini** Sardegna, Italy, C Mediterranean Sea 39°33´N08°39´E
177 K3 **Güssing** Burgenland, SE Austria 47°03´N 16°19´E
177 K3 **Gusswerk** Steiermark, E Austria 47°43´N 15°15´E
152 B5 **Gustav Adolf Land** physical region NE Svalbard
293 L3 **Gustav Bull Mountains** ▲ Antarctica
88 B4 **Gustavo Sotelo** Sonora, Mexico 31°33´N 113°34´W
83 L7 **Gustavus** Alaska, USA 58°24´N 135°44´W
78 B4 **Gustav V Land** physical region NE Svalbard
80 C4 **Gustine** California, W USA 37°14´N 121°00´W
178 I3 **Güstrow** Mecklenburg-Vorpommern, NE Germany 53°48´N 12°12´E
155 H9 **Gusum** Östergötland, S Sweden 58°15´N 16°30´E
Guta/Gúta see Kolárovo
Gutenstein see Ravne na Koroškem
180 I7 **Gütersloh** Nordrhein-Westfalen, W Germany
244 E4 **Gutian** Fujan, China 26°20´N 118°26´E
243 L6 **Gutian Shuiku** ⊞ Fujan, China
86 G5 **Gutiérrez Zamora** Veracruz-Llave, E Mexico 20°29´N 97°07´W
Guting see Yutai
Gutta see Kolárovo
176 F4 **Guttenberg** Iowa, C USA 42°47´N 91°06´W
291 D4 **Guttentag** see Dobrodzień
Guttstadt see Dobre Miasto
239 I2 **Guulin** Govĭ-Altay, C Mongolia 46°33´N 97°21´E
233 I6 **Guwāhāti** prev. Gauhati. Assam, NE India 26°09´N 91°42´E
217 I2 **Guwēr** var. Al Kuwayr, Al Quwayr, Quwair. Arbil, N Iraq 36°03´N 43°30´E
228 A5 **Guwlumaýak** Rus. Kuuli-Mayak. Balkan Welaýaty, NW Turkmenistan 41°04´N 52°43´E
103 **Guyana** off. Co-operative Republic of Guyana; prev. British Guiana. ◆ republic N South America
Guyana, Co-operative Republic of see Guyana
67 H4 **Guyandotte River** ↗ West Virginia, NE USA
Guyane see French Guiana
244 G3 **Guyang** Inner Mongolia, China 41°01´N 110°02´E
161 K3 **Güyhırn** United Kingdom 52°37´N 0°06´E
Guyi see Sanjiang
75 C9 **Guymon** Oklahoma, C USA 36°42´N 101°30´W
228 F7 **Guýmök** Lebap Welaýaty, NE Turkmenistan
Guiyong see Jiangle
66 G6 **Guyot, Mount** ▲ North Carolina/Tennessee, SE USA 35°42´N 83°15´W
277 L5 **Guyra** New South Wales, SE Australia 30°13´S 151°42´E
245 I8 **Guyuan** Hebei, China 41°24´N 115°25´E
244 C6 **Guyuan** Ningxia, N China 35°57´N 106°13´E
218 B2 **Güzelyurt** Gk. Kólpos Mórfou, Morphou. W Cyprus 35°12´N 33°E
218 B2 **Güzelyurt Körfezi** var. Morfou Bay, Morphou Bay, Gk. Kólpos Mórfou. bay W Cyprus
245 K9 **Guzhen** Anhui, China 38°11´N 117°11´E
Guzhou see Rongjiang
84 G2 **Guzmán** Chihuahua, N Mexico 31°13´N 107°27´W
84 G2 **Guzmán, Laguna de** ⊗ Chihuahua, Mexico
229 H7 **G'uzor** Rus. Guzar. Qashqadaryo Viloyati, S Uzbekistan 38°41´N 66°12´E
230 F9 **Gwādar** var. Gwadur. Baluchistān, SW Pakistan 25°09´N 62°21´E
230 F9 **Gwādar East Bay** bay SW Pakistan
230 F9 **Gwādar West Bay** bay SW Pakistan
Gwadur see Gwādar
138 G3 **Gwai** Matabeleland North, W Zimbabwe 19°17´S 27°32´E
160 F2 **Gwalchmai** United Kingdom 53°15´N 4°25´W
234 F4 **Gwalior** Madhya Pradesh, C India 26°16´N 78°12´E
138 G3 **Gwanda** Matabeleland South, SW Zimbabwe 20°56´S 29°0´E
134 F5 **Gwane** Orientale, N Dem. Rep. Congo 04°40´N 25°51´E
248 C4 **Gwangju** off. Kwangju-gwangyoksi, var. Guangju, Kwangchu, Jap. Kōshū; prev. Kwangju. SW South Korea 35°09´N 126°53´E
250 A1 **Gwangju** prev. Kwangju. Kyŏnggi-do, South Korea 35°17´N 127°16´E
250 A4 **Gwangyang** prev. Kwangyang. 34°58´N 127°35´E
184 B3 **Gwayi** ↗ W Zimbabwe
184 F2 **Gwda** var. Glda. Ger. Küddow. ↗ NW Poland
157 B8 **Gweebarra Bay** Ir. Béal an Bheara. inlet W Ireland
158 B5 **Gweedore** Ir. Gaoth Dobhair. Donegal, NW Ireland 55°03´N 08°14´W
Gwelo see Gweru
157 G11 **Gwent** cultural region S Wales, United Kingdom
139 H4 **Gweru** prev. Gwelo. Midlands, C Zimbabwe 19°27´S 29°49´E
133 L6 **Gwoza** Borno, NE Nigeria 11°07´N 13°40´E
160 F1 **Gwy** see Wye
277 J3 **Gwydir River** ↗ New South Wales, SE Australia
157 F10 **Gwynedd** var. Gwyneth. cultural region NW Wales, United Kingdom
Gwyneth see Gwynedd
167 F6 **Gy** Franche-Comté, France 47°24´N 5°49´E
239 J2 **Gyaca** var. Ngarrab. Xizang Zizhiqu, W China 29°06´N 92°37´E
Gya'gya see Saga
Gyaisi/Jiacuojhangg'ê see Zhidoi
Gyali see Yiali
187 K8 **Gyali** var. Yialí. island Dodekánisa, Greece, Aegean Sea
Gyamotang see Dêngqên
Gyandzha see Gäncä
Gyandzhe see Gäncä
Gyangkar see Dinggyê
245 J9 **Gyangkar** Anhui, E China 34°07´N 117°12´E
238 E9 **Gyangzê** Xizang Zizhiqu, W China 28°50´N 89°38´E
238 G9 **Gyaring Co** ⊗ W China
239 I7 **Gyaring Hu** ⊗ C China
187 J3 **Gyáros** var. Yioúra. island Kykládes, Greece, Aegean Sea
192 F4 **Gyda** Yamalo-Nenetskiy Avtonomnyy Okrug, N Russian Federation 70°52´N 78°29´E
192 F4 **Gydanskiy Poluostrov** Eng. Gyda Peninsula. peninsula N Russian Federation
Gyda Peninsula see Gydanskiy Poluostrov
171 I10 **Gydel** Algeria
140 D9 **Gydo Pass** pass Western Cape, South Africa
Gyêgu see Yushu
250 A1 **Gyeonggi-do** prev. Kyŏnggi-do, South Korea
248 C4 **Gyeonggi-man** prev. Kyŏnggi-man. bay NW South Korea 35°49´N 129°09´E
250 B2 **Gyeongju** Jap. Keishū; prev. Kyŏngju. SE South Korea 35°49´N 129°09´E
250 B2 **Gyeongsangbuk-do** prev. Kyŏngsangbuk-do. ◆ province SE South Korea
250 A3 **Gyeongsangnam-Do** prev. Kyŏngsangnam-do. ◆ province SE South Korea
Gyéres see Câmpia Turzii
Gyergyószentmiklós see Gheorgheni
Gyergyótölgyes see Tulgheş
Gyeva see Déva
Gyigang see Zayü
Gyixong see Gonggar
178 G9 **Gyldenløveshøy** hill range C Denmark
250 A4 **Gyöko-ri** 34°30´N 127°05´E
106 D4 **Gylppo** SW Myanmar (Burma) 18°14´N 95°39´E
183 H12 **Gyömaendrőd** Békés, SE Hungary 46°56´N 20°50´E
Gyömbér see Dumbier
183 G13 **Gyömöre** Győr-Moson-Sopron, NW Hungary 47°34´N 17°49´E
183 E11 **Győr** Ger. Raab, Lat. Arrabona. Győr-Moson-Sopron, NW Hungary 47°41´N 17°40´E
183 F11 **Győr-Moson-Sopron** off. Győr-Moson-Sopron Megye. ◆ county NW Hungary
Győr-Moson-Sopron Megye see Győr-Moson-Sopron
77 F4 **Gypsum Point** headland Northwest Territories, NW Canada
55 I3 **Gypsumville** Manitoba, S Canada 51°47´N 98°38´W
59 H2 **Gyrfalcon Islands** island group Northwest Territories, N Canada
193 J8 **Gytheio** var. Githio; prev. Yíthion. Pelopónnisos, S Greece 36°46´N 22°34´E
155 E9 **Gysinge** Gävleborg, C Sweden 60°16´N 16°55´E
186 I7 **Gythio** see Gytheio

228 G7 **Gyuichbirleshik** Lebap Welaýaty, E Turkmenistan 38°10´N 64°33´E
183 H12 **Gyula** Rom. Jula. Békés, SE Hungary 46°39´N 21°17´E
Gyulafehérvár see Alba Iulia
Gyulovo see Roza
215 K5 **Gyumri** var. Giumri, Rus. Kumayri; prev. Aleksandropol', Leninakan. W Armenia 40°48´N 43°51´E
228 F8 **Gyunuzyndag, Gora** ▲ Ahal Welaýaty, W Turkmenistan 38°51´N 56°25´E
228 F5 **Gyzylbaydak** Rus. Krasnoye Znamya. Mary Welaýaty, S Turkmenistan 36°51´N 62°24´E
Gyzyletrek see Etrek
228 C5 **Gyzylgaýa** Rus. Balkan Welaýaty, NW Turkmenistan 40°37´N 55°15´E
228 A6 **Gyzylsuw** Rus. Kizyl-Sua. Balkan Welaýaty, W Turkmenistan 39°49´N 53°00´E
Gyzyrlabat see Serdar
Gzhatsk see Gagarin

H

233 K6 **Ha** Bhutan 27°17´N 89°22´E
Haabai see Ha'apai Group
163 D10 **Haacht** Vlaams Brabant, C Belgium 50°58´N 04°38´E
177 I2 **Haag** Niederösterreich, NE Austria 48°07´N 14°32´E
292 F5 **Haag Nunataks** ▲ Antarctica
141 J2 **Haäkdoring** Limpopo, South Africa 24°26´S 28°49´E
152 A3 **Haakon VII Land** physical region NW Svalbard
162 F3 **Haaksbergen** Overijssel, E Netherlands 52°09´N 06°45´E
162 D5 **Haamstede** Zeeland, SW Netherlands 51°43´N 03°45´E
284 D7 **Ha'ano** island Ha'apai Group, C Tonga
152 I7 **Ha'apai Group** var. Haabai. island group C Tonga
153 H9 **Haapajärvi** Pohjois-Pohjanmaa, C Finland 63°45´N 25°20´E
152 I6 **Haapamäki** Pirkanmaa, C Finland 62°11´N 24°32´E
190 D4 **Haapsalu** Ger. Hapsal. Läänemaa, W Estonia 58°58´N 23°32´E
155 D13 **Haarby** var. Hårby. Syddjylland, C Denmark 55°13´N 10°07´E
181 H8 **Haaren** Nordrhein-Westfalen, Germany 51°34´E 8°44´N
162 E6 **Haarlem** prev. Harlem. Noord-Holland, W Netherlands 52°23´N 04°38´E
278 C10 **Haast** West Coast, South Island, New Zealand 43°53´S 169°02´E
278 C10 **Haast** South Island, New Zealand
284 E10 **Ha'atua** Tutuila, Tonga 21°23´S 174°57´W
230 F9 **Hab** ↗ SW Pakistan
221 I4 **Haba** var. Al Haba. Dubayy, NE United Arab Emirates 25°01´N 55°37´E
238 F3 **Habahe** var. Kaba. Xinjiang Uygur Zizhiqu, NW China 48°04´N 86°27´E
221 I7 **Ḩabarūt** var. Habrut. SW Oman 17°19´N 52°45´E
136 F7 **Habaswein** Isiolo, NE Kenya 01°01´N 39°27´E
163 G13 **Habay-la-Neuve** Luxembourg, SE Belgium 49°43´N 05°38´E
217 K7 **Ḩabbānīyah, Buḩayrat** ⊗ C Iraq
131 I3 **Habbouch** Lebanon 33°24´N 35°34´E
233 J7 **Habeshawerd** see Bystrzyca Kłodzka
233 J7 **Ḩabīganj** Sylhet, NE Bangladesh 24°23´N 91°25´E
155 G10 **Habo** Västra Götaland, S Sweden 57°55´N 14°05´E
252 D3 **Haboma Islands** island group Kuril'skiye Ostrova, SE Russian Federation
218 C6 **HaBonim** Israel 32°38´N 34°56´E
251 K3 **Habororo** Hokkaidō, NE Japan 44°19´N 141°42´E
161 K1 **Habrough** United Kingdom 53°36´N 0°15´W
Habrut see Ḩabarūt
221 J4 **Ḩabshān** Abū Ẕaby, C United Arab Emirates 23°51´N 53°34´E
102 B8 **Hacha** Putumayo, S Colombia 0°02´S 75°30´W
221 I7 **Ḩachīnan** Gifu, Honshū, SW Japan 35°46´N 136°57´E
253 C8 **Hachiman** Akita, Honshū, C Japan 40°22´N 139°59´E
253 D9 **Hachinohe** Aomori, Honshū, C Japan 40°30´N 141°29´E
251 M8 **Hachijō** Tōkyō, Hachijō-jima, SE Japan 35°40´N 139°20´E
253 B12 **Hachiōji** Tōkyō, Honshū, S Japan 35°40´N 139°20´E
86 F2 **Hacienda de la Mesa** Tamaulipas, Mexico 24°14´N 99°15´W
81 H9 **Hacienda Heights** California, USA 34°00´N 117°58´W
215 M5 **Hacıqabal** prev. Qazımämmäd. SE Azerbaijan 40°03´N 48°56´E
154 A4 **Hackås** Jämtland, C Sweden 62°55´N 14°31´E
65 J5 **Hackensack** New Jersey, NE USA 40°51´N 74°03´W
64 A6 **Hackettstown** New Jersey, NE USA 40°51´N 74°49´W
128 C6 **Ḩadabat al Jilf al Kabīr** var. Gilf Kebir Plateau. plateau SW Egypt
Hadama see Nazrēt
253 D10 **Hadano** Kanagawa, Honshū, S Japan 35°22´N 139°14´E
221 J5 **Ḩadbaram** S Oman 17°27´N 55°13´E
218 C6 **Hadchit** Lebanon
217 H7 **Ḩadīthah** Al Anbār, SW Iraq 34°07´N 42°23´E
218 C6 **Ḩaddādiyeh** well S Iraq
159 H2 **Haddington** SE Scotland, United Kingdom 55°59´N 02°46´W
161 K6 **Haddiscoe** United Kingdom 52°31´N 1°35´E
221 L5 **Ḩadd, Ra's al** headland NE Oman 22°28´N 59°58´E
133 L5 **Hadejia** Jigawa, N Nigeria 12°22´N 10°02´E
133 L5 **Hadejia** ↗ N Nigeria
218 C6 **Hadera** Israel 32°26´N 34°55´E
218 C6 **Hadera** var. Khadera; prev. Ḩadera. Haifa, N Israel 32°26´N 34°55´E
155 D13 **Haderslev** Ger. Hadersleben. Syddanmark, SW Denmark 55°15´N 09°30´E
Hadersleben see Haderslev
235 M5 **Hadhdhunmathi Atoll** atoll S Maldives
221 I7 **Ḩadīboh** Suquţrā, SE Yemen 12°38´N 54°05´E
238 F6 **Hadilik** Xinjiang Uygur Zizhiqu, W China
217 H7 **Ḩadīthah** Al Anbār, SW Iraq 34°07´N 42°23´E
220 E5 **Ḩadiyah** Al Madīnah, W Saudi Arabia 25°36´N 38°31´E
184 E5 **Hadjer-Lamis** off. Région du Hadjer-Lamis. ◆ region SW Chad
161 L9 **Hadjout** Algeria
161 L8 **Hadleigh** United Kingdom 51°32´N 0°36´E
161 L8 **Hadley** United Kingdom 52°42´N 0°58´E
59 I2 **Hadley Bay** bay Victoria Island, Nunavut, N Canada
180 H7 **Hadmersleben** Sachsen-Anhalt, Germany 51°59´E 11°18´N
160 G3 **Hadnall** United Kingdom 52°46´N 2°42´W
256 H6 **Ha Đông** var. Hadong. Ha Tây, N Vietnam 20°58´N 105°46´E
Hadong see Ha Đông
Hadramaut see Ḩaḑramawt
136 J7 **Ḩaḑramawt** var. Hadramaut. ▲ S Yemen
Hadria see Adria
Hadrianopolis see Edirne
159 **Hadrian's Wall** ancient wall N England, United Kingdom
Hadria Picena see Apricena
155 D11 **Hadsten** Midtjylland, C Denmark 56°19´N 10°03´E
155 D11 **Hadsund** Nordjylland, N Denmark 56°43´N 10°07´E
189 J7 **Hadyach** Rus. Gadyach. Poltavs'ka Oblast', NE Ukraine 50°21´N 34°00´E
185 M7 **Hadzhiyska Reka** var. Khadzhiyska Reka. ↗ E Bulgaria
184 F5 **Ḩaḑīći** Federacija Bosne I Hercegovine, SE Bosnia and Herzegovina 43°48´N 18°12´E
248 B5 **Haeju** S North Korea 38°04´N 125°40´E
Haeju-man see Gyeonggi-man
220 E3 **Ḩafar al Bāţin** Ash Sharqīyah, N Saudi Arabia 28°25´N 45°59´E
180 H3 **Haffkrug** Schleswig-Holstein, Germany 54°03´E 10°44´N
215 K7 **Ḩakkârı** var. Çölemerik, Hakâri. Hakkâri, SE Turkey 37°36´N 43°45´E
219 F13 **Ḩafīr, Wādī** dry watercourse Jordan
253 D11 **Ḩafizābād** Punjab, NE Pakistan 32°03´N 73°42´E
152 B2 **Hafnarfjörður** Höfuðborgarsvæðið, W Iceland 64°03´N 21°57´W
Hafnia see København
Hafnia see København
Hafren see Severn
Hafun see Xaafuun
Hafun, Ras see Xaafuun, Raas
136 J7 **Hag 'Abdullah** Sinnar, E Sudan 13°59´N 33°35´E
181 D8 **Hagakke** Niedersachsen, Germany 51°57´E 8°44´N
183 E11 **Hagari** Győr-Moson-Sopron, NW Hungary 47°44´N 19°49´E
178 I5 **Hagall** Eng. Galilee. ▲ N Israel
55 J4 **Hagar** Ontario, S Canada 46°27´N 80°02´W
282 A2 **Hagåtña** var. Agaña. ◯ (Guam) NW Guam 13°27´N 144°45´E

56 C5 **Hagensborg** British Columbia, SW Canada 52°24´N 126°24´W
136 C6 **Hāgere Hiywet** var. Agere Hiywet, Ambo. Oromīya, C Ethiopia 09°00´N 37°55´E
78 G4 **Hagerman** Idaho, NW USA 42°48´N 114°53´W
79 I6 **Hagerman** New Mexico, SW USA 33°07´N 104°19´W
64 B6 **Hagerstown** Maryland, NE USA 39°39´N 77°44´W
62 C6 **Hagersville** Ontario, S Canada 42°58´N 80°03´W
154 B5 **Hagetmau** Landes, SW France 43°40´N 00°36´W
154 G2 **Hagfors** Värmland, C Sweden 60°03´N 13°45´E
153 G8 **Häggenås** Jämtland, C Sweden 63°24´N 14°56´E
250 D5 **Hagi** Yamaguchi, Honshū, SW Japan 34°25´N 131°22´E
256 H4 **Ha Giang** Ha Giang, N Vietnam 22°50´N 104°58´E
Hagios Evstrátios see Ágios Efstrátios
HaGolan see Golan Heights
165 K3 **Hagondange** Moselle, NE France 49°16´N 06°06´E
157 B10 **Hag's Head** Ir. Ceann Caillí. headland W Ireland 52°56´N 09°29´W
165 J1 **Hague, Cap de la** headland N France 49°43´N 01°56´W
165 L3 **Haguenau** Bas-Rhin, NE France 48°49´N 07°47´E
281 N9 **Hajima-retto** island group C Japan
139 I7 **Hahaya** ✈ (Moroni) Grande Comore, NW Comoros
181 D9 **Hahn** Nordrhein-Westfalen, Germany 51°58´E 7°27´N
180 I7 **Hahndorf** Niedersachsen, Germany 51°58´E 10°25´N
140 D5 **Haib** Karas, S Namibia 28°12´S 18°19´E
248 A4 **Haibo** 42°15´N 124°12´E
Haibowan see Wuhai
245 N2 **Haicheng** Liaoning, NE China 40°53´N 122°45´E
Haicheng see Haifeng
Haicheng see Haiyuan
Haida see Nový Bor
Haïdara see Aïdar
256 D4 **Hai Dương** Hai Hưng, N Vietnam 20°56´N 106°21´E
218 D6 **Haifa** ◆ district NW Israel
Haifa see Hefa
Haifa, Bay of see Mifrats Hefa
243 I9 **Haifeng** var. Haicheng. Guangdong, S China 22°58´N 115°19´E
245 J4 **Hai He** ↗ E China
Haikang see Leizhou
240 D4 **Haikou** var. Hai-k'ou, Hoihow, Fr. Hoï-Hao. province capital Hainan, S China 20°05´N 110°17´E
Hai-k'ou see Haikou
220 F4 **Ḩā'il** NW Saudi Arabia 27°31´N 41°45´E
239 H5 **Hailar** var. Hai-la-erh. ↗ province N Saudi Arabia 27°33´N 42°50´E
247 I3 **Hailar He** ↗ N China
76 D4 **Hailey** Idaho, NW USA 43°31´N 114°18´W
62 D1 **Haileybury** Ontario, S Canada 47°27´N 79°39´W
248 E1 **Hailin** Heilongjiang, NE China 44°37´N 129°24´E
Ḩā'il, Mintaqah see Ḩā'il
Hailong see Meihekou
161 L3 **Hailsham** United Kingdom 50°52´N 0°16´E
247 L3 **Hailun** Heilongjiang, NE China 48°16´N 126°29´E
152 F7 **Hailuoto** Swe. Karlö. island W Finland
Haima see Haymā'
243 J9 **Haimen** Guangdong, SE China 23°12´N 116°37´E
243 M1 **Haimen** Jiangsu, China 31°52´N 121°05´E
240 F9 **Hainan** var. Hainan Sheng, Qiong. ◆ province S China
243 H5 **Hainan Dao** island S China
243 H5 **Hainan Sheng** see Hainan
Hainan Strait see Qiongzhou Haixia
Hainasch see Ainaži
163 C11 **Hainaut** ◆ province SW Belgium
177 M2 **Hainburg an der Donau** var. Hainburg. Niederösterreich, NE Austria 48°09´N 16°57´E
83 L7 **Haines** Alaska, USA 59°13´N 135°27´W
58 B2 **Haines** Oregon, NW USA 44°53´N 117°56´W
69 M7 **Haines City** Florida, SE USA 28°06´N 81°36´W
177 L2 **Haines Junction** Yukon, W Canada
179 H9 **Hainfeld** Niederösterreich, NE Austria 48°03´N 15°47´E
179 H9 **Hainichen** Sachsen, E Germany 50°58´N 13°08´E
243 M2 **Haining** Zhejiang, China 30°19´N 120°25´E
Hai Ninh see Mong Cai
256 H5 **Hai Phong** var. Haifong, Haiphong. N Vietnam 20°50´N 106°41´E
Haiphong see Hai Phong
243 J9 **Haitan Dao** island SE China
90 G5 **Haiti** off. Republic of Haiti. ◆ republic C West Indies
Haiti, Republic of see Haiti
137 F9 **Haiwee Reservoir** ⊞ California, W USA
128 J3 **Haiya** Red Sea, NE Sudan 18°17´N 36°21´E
239 J7 **Haiyan** var. Sanjiaocheng. Qinghai, W China
243 M2 **Haiyang** Zhejiang, E China 30°31´N 120°56´E
244 C6 **Haiyuan** var. Haicheng. Ningxia, N China 36°35´N 105°35´E
183 H11 **Hajdú-Bihar** off. Hajdú-Bihar Megye. ◆ county E Hungary
Hajdúböszörmény see Hajdú-Bihar
183 H11 **Hajdúböszörmény** E Hungary 47°39´N 21°32´E
183 H11 **Hajdúhadház** Hajdú-Bihar, E Hungary 47°39´N 21°40´E
183 H11 **Hajdúnánás** Hajdú-Bihar, E Hungary 47°50´N 21°26´E
183 H11 **Hajdúszoboszló** Hajdú-Bihar, E Hungary 47°27´N 21°24´E
175 B14 **Haïb El Ayoun** Tunisia
217 J3 **Ḩājjī Ebrāhīm, Kūh-e** ▲ Iran/Iraq 36°53´N 44°56´E
253 B10 **Hajiki-zaki** headland Sado, C Japan 38°19´N 138°28´E
233 I8 **Hājipur** Bihār, N India 25°41´N 85°13´E
220 F8 **Ḩajjah** W Yemen 15°43´N 43°32´E
221 J4 **Ḩajjīābād** Al Muthanná, S Iraq 31°24´N 45°20´E
217 L8 **Ḩajjīābād** Hormozgān, C Iran
217 K8 **Ḩajji, Thaqb al** well S Iraq
184 F5 **Ḩajla** E Montenegro
183 J4 **Hajnówka** Ger. Hermhausen. Podlaskie, NE Poland 52°46´N 23°32´E
Haka see Hakha
Hakapehi see Punaauia
Hakâri see Hakkâri
219 D11 **HaLmakhtesh** see HaMakhtesh
HaQatan see
180 J6 **Hakenstedt** Sachsen-Anhalt, Germany 52°11´E 11°16´N
256 B4 **Hakha** var. Haka. Chin State, W Myanmar (Burma)
215 K7 **Hakkâri** var. Çölemerik, Hakâri. Hakkâri, SE Turkey 37°36´N 43°45´E
215 K7 **Hakkâri** var. Hakâri. ◆ province SE Turkey
Hakkâri see Hakkâri
152 G5 **Hakkas** Norrbotten, N Sweden 66°51´N 21°36´E
251 J7 **Hakken-zan** ▲ Honshū, SW Japan 34°11´N 135°57´E
253 D8 **Hakkōda-san** ▲ Honshū, C Japan 40°40´N 140°49´E
252 D3 **Hako-dake** ▲ Hokkaidō, NE Japan 41°46´N 140°43´E
253 C9 **Hakodate** Hokkaidō, NE Japan 41°46´N 140°43´E
140 E4 **Hakskeenpan** salt lake Northern Cape, South Africa
251 J2 **Haku-san** ▲ Honshū, SW Japan 36°07´N 136°45´E
Hakupu see Mātū
Hal see Halle
196 J5 **Hala** Sind, SE Pakistan 25°47´N 68°28´E
216 C3 **Ḩalab** Eng. Aleppo, Fr. Alep; anc. Beroea. Ḩalab, NW Syria 36°14´N 37°10´E
216 C2 **Ḩalab** off. Muḩāfaẕat Ḩalab, var. Halab, Halab. ◆ governorate NW Syria
218 G1 **Ḩalabā** ✈ Ḩalab, NW Syria
217 L5 **Ḩalabjah** var. Halabja. NE Iraq
Halab, Muḩāfaẕat see Ḩalab
180 J6 **Halaç** Rus. Khalach. Lebap Welaýaty, E Turkmenistan 38°02´N 64°49´E
128 J4 **Ḩalā'ib Triangle** ◆ disputed region S Egypt / N Sudan
128 C10 **Halba** Lebanon
131 L9 **Hal'b bi Uvea** N Wallis and Futuna 13°17´S 176°07´W
256 H5 **Ha Lâm Quảng Nam-Đà Nẵng, C Vietnam
57 H4 **Halbrite** Saskatchewan, S Canada
221 J7 **Ḩalānīyat, Juzur al** var. Jazā'ir Bin Ghalfân, Eng. Kuria Muria Islands. island group S Oman
221 J7 **Ḩalānīyat, Khalīj al** Eng. Kuria Muria Bay. bay S Oman
151 **Halas** see Kiskunhalas
181 G11 **Halberg** Rheinland-Pfalz, Germany
180 H7 **Halberstadt** Sachsen-Anhalt, C Germany 51°54´N 11°03´E
216 C4 **Ḩalbūn** Syria 33°40´N 36°15´E

◆ Country	◇ Dependent Territory	◆ Administrative Regions	▲ Mountain	⛰ Volcano	◎ Lake
● Country Capital	○ Dependent Territory Capital	✕ International Airport	▲ Mountain Range	◢ River	◉ Reservoir

I

◆ Country
● Country Capital
◇ Dependent Territory
○ Dependent Territory Capital
✕ International Airport
▲ Mountain
▲ Mountain Range
🌋 Volcano
✍ River
◆ Lake
◇ Reservoir

◆ Country ● Country Capital ◇ Dependent Territory ○ Dependent Territory Capital ◈ Administrative Regions ✕ International Airport ▲ Mountain ▲ Mountain Range ◉ Volcano ◣ River ◎ Lake ◙ Reservoir

Iténez, Río - Jekyll Island
INDEX
354

Iténez, Río *see* Guaporé, Rio
102 D6 **Iteviate, Río** ≈ C Colombia
178 E7 **Ith** *hill range* C Germany
72 G7 **Ithaca** Michigan, N USA 43°17′N 84°36′W
64 D2 **Ithaca** New York, NE USA 42°26′N 76°30′W
186 E6 **Itháki** *island* Iónia Nísiá, Greece, C Mediterranean Sea
Itháki *see* Vathy
It Hearrenfean *see* Heerenveen
Itihara *see* Ichihara
134 G6 **Itimbiri** ≈ N Dem. Rep. Congo
Itinomiya *see* Ichinomiya
Itinoseki *see* Ichinoseki
110 B7 **Itirapina** São Paulo, Brazil 22°15′S 47°49′W
110 C4 **Itirapuá** São Paulo, Brazil 20°38′S 47°14′W
83 I2 **Itkilik River** ≈ Alaska, USA
251 L4 **Itō** Shizuoka, Honshū, S Japan 34°58′N 139°04′E
110 C6 **Itobi** Sao Paulo, Brazil 21°42′S 46°54′W
253 A11 **Itoigawa** Niigata, Honshū, C Japan 37°02′N 137°53′E
251 I10 **Itoman** Okinawa, SW Japan 26°05′N 127°40′E
164 G3 **Iton** ≈ N France
106 E8 **Itonamas Río** ≈ NE Bolivia
Itoupé, Mont *see* Sommet Tabulaire
163 D13 **Ítreaupont** Aisne, N France 49°55′N 03°56′E
Itseqqortoormiit *see* Ittoqqortoormiit
66 A8 **Itta Bena** Mississippi, S USA 33°30′N 90°19′W
175 B9 **Ittiri** Sardegna, Italy, C Mediterranean Sea 40°36′N 08°34′E
167 N5 **Ittlenheim** Alsace, France 48°40′N 7°33′E
295 L6 **Ittoqqortoormiit** *var.* Itseqqortoormiit, *Dan.* Scoresbysund, *Eng.* Scoresby Sound. Sermersooq, C Greenland 70°33′N 21°52′W
110 C9 **Itu** São Paulo, S Brazil 23°17′S 47°16′W
263 H7 **Itu Aba Island** *Chin.* Taiping Dao. *island* W Spratly Islands
102 B4 **Ituango** Antioquia, NW Colombia 07°07′N 75°46′W
111 L2 **Itueta** Minas Gerais, Brazil 19°23′S 41°11′W
106 B4 **Ituí, Rio** ≈ NW Brazil
135 I9 **Itula** Sud-Kivu, E Dem. Rep. Congo 03°30′S 27°50′E
109 B11 **Itumbiara** Goiás, C Brazil 18°25′S 49°15′W
103 K5 **Ituni** E Guyana 05°24′N 58°18′W
110 C8 **Itupeva** São Paulo, Brazil 23°09′S 47°04′W
87 M6 **Iturbide** Campeche, SE Mexico 19°41′N 89°29′W
85 L7 **Iturbide** Nuevo León, N Mexico 24°44′N 99°54′W
Ituri *see* Aruwimi
249 M2 **Iturup, Ostrov** *island* Kuril'skiye Ostrova, SE Russian Federation
110 G5 **Itutinga** Minas Gerais, Brazil 21°18′S 44°40′W
110 B4 **Ituverava** São Paulo, S Brazil 20°22′S 47°48′W
106 D6 **Ituxi, Rio** ≈ W Brazil
113 H6 **Ituzaingó** Corrientes, NE Argentina 27°34′S 56°44′W
181 J11 **Itz** ≈ C Germany
180 G2 **Itzehoe** Schleswig-Holstein, N Germany 53°56′N 09°31′E
66 C7 **Iuka** Mississippi, S USA 34°48′N 88°11′W
111 K4 **Iúna** Esperíto Santo, Brazil 20°24′S 41°36′W
113 J4 **Ivaiporã** Paraná, S Brazil 24°16′S 51°46′W
113 I3 **Ivaí, Rio** ≈ S Brazil
152 H4 **Ivalo** *Lapp.* Avveel, Avvil. Lappi, N Finland 68°34′N 27°29′E
152 H4 **Ivalojoki** *Lapp.* Avveel. ≈ N Finland
191 E11 **Ivanava** *Pol.* Janów, Janów Poleski, *Rus.* Ivanovo. Brestskaya Voblasts', SW Belarus 52°09′N 25°32′E
134 B7 **Ivando** *var.* Djidji. ≈ Congo/Gabon
190 G3 **Ivangorod** *Est.* Jaanilinn. Leningradskaya Oblast', NW Russian Federation 59°22′N 28°14′E
Ivangorod *see* Dęblin
Ivangrad *see* Berane
277 I4 **Ivanhoe** New South Wales, SE Australia 32°55′S 144°21′E
80 E6 **Ivanhoe** California, USA 36°23′N 119°13′W
74 F5 **Ivanhoe** Minnesota, N USA 44°27′N 96°15′W
184 C2 **Ivanić-Grad** Sisak-Moslavina, N Croatia 45°43′N 16°23′E
189 K7 **Ivanivka** Khersons'ka Oblast', S Ukraine 46°43′N 34°28′E
189 H7 **Ivanivka** Odes'ka Oblast', SW Ukraine 46°57′N 30°26′E
184 F5 **Ivanjica** Serbia, C Serbia 43°36′N 20°14′E
184 D3 **Ivanjska** *var.* Potkozarje. Republika Srpska, NW Bosnia and Herzegovina 44°54′N 17°04′E
183 E10 **Ivanka** ✈ (Bratislava) Bratislavský Kraj, W Slovakia 48°10′N 17°13′E
188 G2 **Ivankiv** *Rus.* Ivankov. Kyyivs'ka Oblast', N Ukraine 50°55′N 29°53′E
Ivankov *see* Ivankiv
82 G9 **Ivanof Bay** Alaska, USA 55°55′N 159°28′W
188 D5 **Ivano-Frankivs'k** *Ger.* Stanislau, *Pol.* Stanisławów, *Rus.* Ivano-Frankovsk; *prev.* Stanislav. Ivano-Frankivs'ka Oblast', W Ukraine 48°55′N 24°45′E
Ivano-Frankivs'k *see* Ivano-Frankivs'ka Oblast'
188 D5 **Ivano-Frankivs'ka Oblast'** *var.* Ivano-Frankivs'k, *Rus.* Ivano-Frankovskaya Oblast'; *prev.* Stanislavskaya Oblast'. ◈ *province* W Ukraine
Ivano-Frankovsk *see* Ivano-Frankivs'k
Ivano-Frankovskaya Oblast' *see* Ivano-Frankivs'ka Oblast'
194 G10 **Ivanovo** Ivanovskaya Oblast', W Russian Federation 57°02′N 40°58′E
190 H7 **Ivanovo** Pskovskaya Oblast', Russian Federation
Ivanovo *see* Ivanava
194 G10 **Ivanovskaya Oblast'** ◈ *province* W Russian Federation
81 I10 **Ivanpah** California, USA 35°20′N 115°19′W
81 I10 **Ivanpah Lake** ◎ California, W USA
184 C1 **Ivanščica** ▲▲ NE Croatia
185 L5 **Ivanski** Shumen, NE Bulgaria 43°09′N 27°02′E
197 J4 **Ivanteyevka** Saratovskaya Oblast', W Russian Federation 52°13′N 49°06′E
Ivantsevichi/Ivatsevichi *see* Ivatsevichy
188 D3 **Ivanychi** Volyns'ka Oblast', NW Ukraine 50°37′N 24°22′E
191 E11 **Ivatsevichy** *Pol.* Iwacewicze, *Rus.* Ivantsevichi, Ivatsevichi. Brestskaya Voblasts', SW Belarus 52°43′N 25°21′E
185 L7 **Ivaylovgrad** Haskovo, S Bulgaria 41°32′N 26°06′E
185 L7 **Ivaylovgrad, Yazovir** ⊠ S Bulgaria
195 L7 **Ivdel'** Sverdlovskaya Oblast', C Russian Federation 60°42′N 60°07′E
Ivenets *see* Ivyanyets
188 F8 **Iveşti** Galaţi, E Romania 45°27′N 28°00′E
Ivgovuotna *see* Lyngen
171 J9 **Ivi, Cap** *headland* Algeria
Ivigtut *see* Ivittuut
134 C7 **Ivindo** ≈ Gabon
161 J5 **Ivinghoe** United Kingdom 51°50′N 0°37′W
113 I3 **Ivinheima** Mato Grosso do Sul, SW Brazil 22°16′S 53°52′W
295 M8 **Ivittuut** *var.* Ivigtut. Sermersooq, S Greenland 61°12′N 48°10′W
Iviza *see* Eivissa\Ibiza
139 L9 **Ivohibe** Fianarantsoa, SE Madagascar 22°28′S 46°53′E
Ivoire, Côte d' *see* Ivory Coast
132 F7 **Ivory Coast** *off.* Republic of the Ivory Coast, *Fr.* Côte d'Ivoire, République de la Côte d'Ivoire. ◆ *republic* W Africa
124 B5 **Ivory Coast** *Fr.* Côte d'Ivoire. *coastal region* S Ivory Coast
Ivory Coast, Republic of the *see* Ivory Coast
155 G12 **Ivösjön** ◎ S Sweden
176 D6 **Ivrea** *anc.* Eporedia. Piemonte, NW Italy 45°28′N 07°52′E
166 G5 **Ivry-la-Bataille** Haute-Normandie, France 48°53′N 1°28′E
58 G1 **Ivujivik** Québec, NE Canada 62°26′N 77°49′W
191 F10 **Ivyanyets** *Rus.* Ivenets. Minskaya Voblasts', C Belarus 53°53′N 26°45′E
Iv'ye *see* Iwye
Iwacewicze *see* Ivatsevichy
253 E8 **Iwaizumi** Iwate, Honshū, NE Japan 39°48′N 141°46′E
253 D12 **Iwaki** Fukushima, Honshū, N Japan 37°01′N 140°52′E
252 D7 **Iwaki-san** ▲ Honshū, C Japan 40°39′N 140°20′E
250 E5 **Iwakuni** Yamaguchi, Honshū, SW Japan 34°08′N 132°06′E
252 E4 **Iwamizawa** Hokkaidō, NE Japan 43°12′N 141°47′E
252 D4 **Iwanai** Hokkaidō, NE Japan 42°51′N 140°21′E
253 D10 **Iwanuma** Miyagi, Honshū, C Japan 38°06′N 140°51′E
251 K5 **Iwata** Shizuoka, Honshū, S Japan 34°42′N 137°51′E
253 D8 **Iwate** Iwate, Honshū, N Japan 40°03′N 141°12′E
253 E8 **Iwate** *off.* Iwate-ken. ◈ *prefecture* Honshū, C Japan
Iwate-ken *see* Iwate
253 D8 **Iwate-san** ▲ Honshū, C Japan 39°52′N 140°59′E
133 J7 **Iwo** Oyo, SW Nigeria 07°21′N 03°58′E
Iwojima *see* Iō-jima
191 E9 **Iwye** *Pol.* Iwje, *Rus.* Iv'ye. Hrodzyenskaya Voblasts', W Belarus 53°56′N 25°46′E
86 F7 **Ixcamilpa** Puebla, Mexico 18°02′N 98°42′W
87 K9 **Ixcán, Río** ≈ Guatemala/Mexico
86 F8 **Ixcuinatoyac** Guerrero, Mexico 17°15′N 98°28′W
163 E10 **Ixelles** *Dut.* Elsene. Brussels, C Belgium 50°49′N 04°21′E
87 I9 **Ixhuatán** Oaxaca, Mexico 16°21′N 94°29′W
86 G5 **Ixhuatlán** Veracruz-Llave, Mexico 20°42′N 98°01′W
106 C8 **Ixiamas** La Paz, NW Bolivia 13°45′S 68°10′W
86 F5 **Ixmiquilpan** *var.* Ixmiquilpán. Hidalgo, C Mexico 20°30′N 99°15′W
Ixmiquilpán *see* Ixmiquilpan
141 K7 **Ixopo** KwaZulu-Natal, South Africa 30°09′S 30°05′E
Ixopo *see* eXobho
Ixtaccíhuatal, Volcán *see* Iztaccíhuati, Volcán
86 D8 **Ixtapa** Guerrero, S Mexico 17°38′N 101°29′W
87 J8 **Ixtapangajoya** Chiapas, Mexico 17°29′N 92°59′W
86 G9 **Ixtayutla** Oaxaca, Mexico 16°34′N 97°39′W
87 I9 **Ixtepec** Oaxaca, SE Mexico 16°32′N 95°03′W
85 I10 **Ixtlán** *var.* Ixtlán del Río. Nayarit, C Mexico 21°02′N 104°21′W
87 H8 **Ixtlán de Juárez** Oaxaca, Mexico 17°20′N 96°29′W
Ixtlán del Río *see* Ixtlán
195 N9 **Iyevlevo** Tyumenskaya Oblast', C Russian Federation 57°36′N 67°20′E
250 F6 **Iyo** Ehime, Shikoku, SW Japan 33°43′N 132°42′E
250 F5 **Iyomishima** *var.* Iyomisima. Ehime, Shikoku, SW Japan 33°58′N 133°30′E
Iyomisima *see* Iyomishima
250 E6 **Iyo-nada** *sea* S Japan
88 D3 **Izabal** *off.* Departamento de Izabal. ◈ *department* E Guatemala
Izabal, Departamento de *see* Izabal
88 D3 **Izabal, Lago de** *prev.* Golfo Dulce. ◎ E Guatemala
222 F6 **Īzad Khvāst** Fārs, C Iran 31°31′N 52°09′E
87 M5 **Izamal** Yucatán, SE Mexico 20°58′N 89°00′W
218 I7 **Izam, Jibal Abu al** ▲▲ Jordan
197 J9 **Izberbash** Respublika Dagestan, SW Russian Federation 42°32′N 47°51′E
190 G5 **Izborsk** Pskovskaya Oblast', Russian Federation
163 B10 **Izegem** *prev.* Iseghem. West-Vlaanderen, W Belgium 50°55′N 03°13′E
222 E6 **Īzeh** Khūzestān, SW Iran 31°48′N 49°49′E
251 I9 **Izena-jima** *island* Nansei-shotō, SW Japan
185 M7 **Izgrev** Burgas, E Bulgaria 42°09′N 27°49′E
195 J10 **Izhevsk** *prev.* Ustinov. Udmurtskaya Respublika, NW Russian Federation 56°48′N 53°12′E
195 J5 **Izhma** Respublika Komi, NW Russian Federation 64°56′N 53°52′E
195 J5 **Izhma** ≈ NW Russian Federation
221 K5 **Izkī** NE Oman 22°45′N 57°36′E
Izmail *see* Izmayil
188 G9 **Izmayil** *Rus.* Izmail. Odes'ka Oblast', SW Ukraine 45°19′N 28°49′E
214 B6 **İzmir** *prev.* Smyrna. İzmir, W Turkey 38°25′N 27°10′E
214 B6 **İzmir** *prev.* Smyrna. ◈ *province* W Turkey
214 C4 **İzmit** *var.* Ismid; *anc.* Astacus. Kocaeli, NW Turkey 40°47′N 29°55′E
170 F8 **Iznájar** Andalucía, S Spain 37°17′N 04°16′W
170 F8 **Iznajar, Embalse de** ⊠ S Spain
170 G8 **Iznalloz** Andalucía, S Spain 37°23′N 03°31′W
214 C5 **İznik** Bursa, NW Turkey 40°27′N 29°43′E
214 C5 **İznik Gölü** ◎ NW Turkey
197 H8 **Izobil'nyy** Stavropol'skiy Kray, SW Russian Federation 45°22′N 41°40′E
177 J6 **Izola** *It.* Isola d'Istria. SW Slovenia 45°31′N 13°40′E
218 G6 **Izra'** *var.* Ezra, Ezraa. Dar'ā, S Syria 32°52′N 36°15′E
218 G6 **Izra** Syria 32°51′N 36°15′E
86 F6 **Iztaccíhuati, Volcán** *var.* Volcán Ixtaccíhuatal. ⛰ S Mexico 19°07′N 98°37′W
86 F6 **Iztaccíhuatl, Volcàn** ⛰ Mexico 19°11′N 98°39′W
88 C4 **Iztapa** Escuintla, SE Guatemala 13°58′N 90°42′W
Izúcar de Matamoros *see* Matamoros
251 L4 **Izu-hantō** *peninsula* Honshū, S Japan
Izuhara *see* Tsushima
250 C8 **Izumi** Kagoshima, Kyūshū, SW Japan 32°02′N 130°20′E
251 H5 **Izumiōtsu** Ōsaka, Honshū, SW Japan 34°29′N 135°25′E
251 H5 **Izumisano** Ōsaka, Honshū, SW Japan 34°23′N 135°18′E
250 F4 **Izumo** Shimane, Honshū, SW Japan 35°22′N 132°46′E
Izu Shichito *see* Izu-shotō
251 M8 **Izu-shotō** *var.* Izu Shichito. *island group* S Japan
286 D3 **Izu Trench** *undersea feature* NW Pacific Ocean
192 G3 **Izvestiy TsIK, Ostrova** *island* N Russian Federation
185 I6 **Izvor** Pernik, W Bulgaria 42°27′N 22°53′E
188 E3 **Izyaslav** Khmel'nyts'ka Oblast', W Ukraine 50°08′N 26°53′E
189 L4 **Izyum** Kharkivs'ka Oblast', E Ukraine 49°12′N 37°19′E

J

153 I9 **Jaala** Kymenlaakso, S Finland 61°04′N 26°30′E
Jaanilinn *see* Ivangorod
218 E7 **Jaba** West Bank 32°19′N 35°13′E
218 H6 **Jabal ad Durūz** *dry watercourse* Syria
218 I6 **Jabal al Arab** ▲ Syria 32°40′N 36°47′E
218 I6 **Jabal al Arab** ▲▲ Syria
218 F4 **Jabal ash Shaykh** ▲ Lebanon 33°26′N 35°51′E
212 E5 **Jabal ash Shifā** *desert* NW Saudi Arabia
221 I4 **Jabal aẓ Ẓannah** *var.* Jebel Dhanna. Abū Ẓaby, W United Arab Emirates 24°10′N 52°36′E
219 E14 **Jabal Bāqir** *dry watercourse* Jordan
219 C9 **Jabalīya** *var.* Jabāliyah. NE Gaza Strip 31°32′N 34°29′E
Jabāliyah *see* Jabalīya
218 J1 **Jab al Jarraḥ** Ḥimṣ, C Syria 34°50′N 37°18′E
218 F3 **Jabal Lubnan** ◈ *governorate* Lebanon
173 H5 **Jabalón** ≈ C Spain
232 F8 **Jabalpur** *prev.* Jubbulpore. Madhya Pradesh, C India 23°10′N 79°59′E
129 J3 **Jabal Ramm** ▲ Jordan
136 F2 **Jabal Zuqar, Jazīrat** *var.* Az Zuqur. *island* SW Yemen
Jabar *see* Jawa Barat
Jabat *see* Jabwot
216 D2 **Jabbūl, Sabkhat al** *sabkha* NW Syria
274 G2 **Jabiru** Northern Territory, N Australia 12°44′S 132°48′E
216 C3 **Jablah** *var.* Jeble, *Fr.* Djéblé. Al Lādhiqīyah, W Syria 35°00′N 36°00′E
184 B3 **Jablanac** Lika-Senj, W Croatia 44°43′N 14°54′E
184 D5 **Jablanica** Federacija Bosne I Hercegovine, SW Bosnia and Herzegovina 43°39′N 17°43′E
184 G8 **Jablanica** *Alb.* Mali i Jabllanicës, *var.* Malet e Jabllanicës. ▲ Albania/FYR Macedonia *see also* Jabllanicës, Mali i
Jabllanicës, Malet e *see* Jablanica/Jabllanicës, Mali i
184 F8 **Jabllanicës, Mali i**, *Mac.* Jablanica. ▲ Albania/FYR Macedonia *see also* Jablanica
182 C7 **Jablonec nad Nisou** *Ger.* Gablonz an der Neisse. Liberecký Kraj, N Czech Republic 50°44′N 15°10′E
Jabłonków/Jablunkau *see* Jablunkov
182 F4 **Jabłonowo Pomorskie** Kujawski-pomorskie, C Poland 53°24′N 19°08′E
183 F9 **Jablunkov** *Ger.* Jablunkau, *Pol.* Jabłonków. Moravskoslezský Kraj, E Czech Republic 49°35′N 18°46′E
108 J6 **Jaboatão** Pernambuco, E Brazil 08°05′S 35°W
110 A4 **Jaborandi** São Paulo, Brazil 20°40′S 48°25′W
110 A5 **Jaboticabal** São Paulo, S Brazil 21°15′S 48°17′W
172 D7 **Jabugo** Andalucía, Spain 37°55′N 6°44′W
283 L6 **Jabwot** *var.* Jabat, Jebat, Jōwat. *island* Ralik Chain, S Marshall Islands
171 I2 **Jaca** Aragón, NE Spain 42°34′N 00°33′W
88 B3 **Jacaltenango** Huehuetenango, W Guatemala 15°39′N 91°46′W
115 I1 **Jacaquá** Rio Grande do Sul, Brazil 29°41′S 55°12′W
107 H4 **Jacaré-a-Canga** Pará, NE Brazil 05°59′S 57°32′W
110 E9 **Jacareí** São Paulo, S Brazil 23°18′S 45°55′W
107 J10 **Jaciara** Mato Grosso, W Brazil 15°59′S 54°57′W
106 E6 **Jaciparaná** Rondônia, W Brazil 09°20′S 64°28′W
63 J3 **Jackman** Maine, NE USA 45°35′N 70°14′W
78 G1 **Jackpot** Nevada, W USA 41°57′N 114°41′W
66 F5 **Jacksboro** Tennessee, S USA 36°19′N 84°11′W
70 G4 **Jacksboro** Texas, SW USA 33°13′N 98°11′W
68 G3 **Jackson** Alabama, S USA 31°30′N 87°53′W
80 D3 **Jackson** California, W USA 38°19′N 120°46′W
66 G9 **Jackson** Georgia, SE USA 33°17′N 83°58′W
66 G4 **Jackson** Kentucky, S USA 37°32′N 83°24′W
68 D3 **Jackson** Louisiana, S USA 30°50′N 91°13′W
73 G8 **Jackson** Michigan, N USA 42°15′N 84°24′W
74 F6 **Jackson** Minnesota, N USA 43°38′N 95°00′W
68 E2 **Jackson** *state capital* Mississippi, S USA 32°19′N 90°12′W
75 J10 **Jackson** Missouri, C USA 37°23′N 89°40′W
67 L5 **Jackson** North Carolina, SE USA 36°24′N 77°25′W
73 I11 **Jackson** Ohio, NE USA 39°03′N 82°40′W
66 C6 **Jackson** Tennessee, S USA 35°37′N 88°50′W
77 I7 **Jackson** Wyoming, C USA 43°28′N 110°45′W
278 B10 **Jackson Bay** *bay* South Island, New Zealand
280 C4 **Jackson Field** ✈ (Port Moresby) Central/National Capital District, S Papua New Guinea 09°28′S 147°12′E
278 B10 **Jackson Head** *headland* South Island, New Zealand 43°57′S 168°38′E
69 J4 **Jackson, Lake** ◎ Florida, SE USA
77 J7 **Jackson Lake** ◎ Wyoming, C USA
292 F4 **Jackson, Mount** ▲ Antarctica 71°43′S 63°45′W
79 M2 **Jackson Reservoir** ◎ Colorado, C USA
66 E8 **Jacksonville** Alabama, S USA 33°48′N 85°45′W
75 I13 **Jacksonville** Arkansas, C USA 34°52′N 92°08′W
69 L4 **Jacksonville** Florida, SE USA 30°20′N 81°39′W
73 B11 **Jacksonville** Illinois, N USA 39°43′N 90°13′W
67 L7 **Jacksonville** North Carolina, SE USA 34°45′N 77°26′W
71 I5 **Jacksonville** Texas, SW USA 31°57′N 95°16′W
65 J1 **Jacksonville** Vermont, USA 42°48′N 72°49′W
69 L4 **Jacksonville Beach** Florida, SE USA 30°17′N 81°23′W
90 G6 **Jacmel** *var.* Jaquemel. S Haiti 18°13′N 72°33′W
Jacob *see* Nkayi
230 G7 **Jacobābād** Sind, SE Pakistan 28°16′N 68°30′E
141 H6 **Jacobsdal** Free State, South Africa 29°08′S 24°46′E
103 K7 **Jacobs Ladder Falls** *waterfall* S Guyana
86 D6 **Jacona de Plancarte** Michoacán, Mexico 19°57′N 102°16′W
91 L1 **Jaco, Pointe** *headland* N Dominica 15°38′N 61°25′W
111 H2 **Jacoticatubas** Minas Gerais, Brazil 19°30′S 43°49′W
63 I1 **Jacques-Cartier** ≈ Québec, SE Canada
59 J6 **Jacques-Cartier, Détroit de** *var.* Jacques-Cartier Passage. *strait* Gulf of St. Lawrence/St. Lawrence River, Canada
59 J7 **Jacques-Cartier, Mont** ▲ Québec, SE Canada 48°58′N 66°00′W
Jacques-Cartier Passage *see* Jacques-Cartier, Détroit de
110 G8 **Jacuecanga** Rio de Janeiro, Brazil 23°01′S 44°13′W
110 D5 **Jacuí** Minas Gerais, Brazil 21°00′S 46°48′W
115 M1 **Jacuí, Rio** ≈ S Brazil
81 F13 **Jacumba** California, USA 32°37′N 116°11′W
111 M2 **Jacupemba** Esperíto Santo, Brazil 19°36′S 40°12′W
113 K4 **Jacupiranga** São Paulo, S Brazil 24°42′S 48°00′W
110 D7 **Jacutinga** Minas Gerais, Brazil 22°17′S 46°37′W
180 E3 **Jade** ≈ NW Germany
178 C5 **Jadebusen** *bay* NW Germany
Jadotville *see* Likasi
Jadransko More/Jadransko Morje *see* Adriatic Sea
170 G4 **Jadraque** Castilla-La Mancha, C Spain 40°55′N 02°55′W
178 G2 **Jægerspris** Hovedstaden, E Denmark 55°52′N 11°59′E
104 C4 **Jaén** Cajamarca, N Peru 05°45′S 78°51′W
170 G8 **Jaén** Andalucía, SW Spain 37°46′N 03°48′W
170 G7 **Jaén** ◈ *province* Andalucía, S Spain
155 A9 **Jæren** *physical region* S Norway
235 F9 **Jaffna** Northern Province, N Sri Lanka 09°42′N 80°03′E
235 G9 **Jaffna Lagoon** *lagoon* N Sri Lanka
65 K1 **Jaffrey** New Hampshire, NE USA 42°46′N 72°00′W
216 C8 **Jafr, Qāʻ al** *var.* El Jafr. *salt pan* S Jordan
232 E4 **Jagādhri** Haryāna, N India 30°11′N 77°18′E
190 E3 **Jägala** *var.* Jägala Jõgi, *Ger.* Jaggowal. ≈ NW Estonia
Jägala Jõgi *see* Jägala
Jagannath *see* Puri
234 G5 **Jagdalpur** Chhattīsgarh, C India 19°07′N 82°04′E
237 J2 **Jagdaqi** Nei Mongol Zizhiqu, N China 50°26′N 124°03′E
Jägerndorf *see* Krnov
141 H6 **Jagersfontein** Free State, South Africa 29°46′S 25°25′E
Jaggowal *see* Jägala
216 G1 **Jaghjaghah, Nahr** ≈ N Syria
140 E6 **Jaght Drift** Northern Cape, South Africa 94°24′S 21°04′E
263 M8 **Jagna** Bohol, C Philippines 09°37′N 124°16′E
184 G4 **Jagodina** *prev.* Svetozarevo. Serbia, C Serbia 43°59′N 21°15′E
184 F4 **Jagodnja** ▲▲ W Serbia
181 H14 **Jagst** ≈ SW Germany
181 G13 **Jagsthausen** Baden-Württemberg, Germany 49°19′E 9°28′N
181 H14 **Jagstheim** Baden-Württemberg, Germany 49°06′E 10°04′N
181 H14 **Jagstzell** Baden-Württemberg, Germany 49°02′E 10°06′N
234 F5 **Jagtiāl** Telangana, C India 18°49′N 78°53′E
111 I2 **Jaguaraçu** Minas Gerais, Brazil 19°40′S 42°45′W
115 L6 **Jaguarão** Rio Grande do Sul, S Brazil 32°30′S 53°25′W
115 K5 **Jaguarão, Rio** *var.* Río Yaguarón. ≈ Brazil/Uruguay
113 K4 **Jaguariaíva** Paraná, S Brazil 24°15′S 49°44′W
289 B10 **Jaguar Seamount** *undersea feature* SW Indian Ocean 21°50′S 39°27′E
90 C3 **Jagüey Grande** Matanzas, W Cuba 22°31′N 81°07′W
233 I7 **Jahānābād** Bihār, N India 25°13′N 84°59′E
219 G13 **Jahdānīyah, Wādī al** *dry watercourse* Jordan
219 G13 **Jahdhānīyah, Wādī al** Jordan
Jahra *see* Al Jahrā'
222 G8 **Jahrom** *var.* Jahrum. Fārs, S Iran 28°35′N 53°32′E
Jahrum *see* Jahrom
118 D2 **Jahuel** Valparaíso, Chile 32°43′S 70°39′W
Jailolo *see* Halmahera, Pulau
261 H3 **Jailolo, Selat** *strait* E Indonesia
87 L5 **Jaina** Campeche, Mexico 20°14′N 90°28′W
Jainat *see* Chai Nat
244 A6 **Jainca** Qinghai, China 35°36′N 102°02′E
Jainti *see* Jayanti
232 E6 **Jaipur** *prev.* Jeypore. *state capital* Rājasthān, N India 26°54′N 75°47′E
233 K7 **Jaipurhat** *var.* Joypurhat. Rajshahi, NW Bangladesh 25°04′N 89°06′E
232 C6 **Jaisalmer** Rājasthān, NW India 26°55′N 70°56′E
234 I4 **Jājapur** *var.* Jajpur, Panikoilli. Odisha, E India 18°54′N 82°36′E
223 H3 **Jājarm** Khorāsān-e Shemālī, NE Iran 36°58′N 56°26′E
184 D4 **Jajce** Federacija Bosne I Hercegovine, W Bosnia and Herzegovina 44°20′N 17°16′E
Jajı *see* 'Alī Khēl
Jajpur *see* Jājapur
138 C4 **Jakalsberg** Otjozondjupa, N Namibia 19°23′S 17°28′E
258 G8 **Jakarta** *prev.* Djakarta, *Dut.* Batavia. ● (Indonesia) Jawa, C Indonesia 06°08′S 106°45′E
258 G8 **Jakarta Raya** *off.* Daerah Khusus Ibukota Jakarta, *var.* Djakarta Raya. ◈ *metropolitan district* Jakarta Raya, S Indonesia Asia
54 B4 **Jakes Corner** Yukon, W Canada 60°18′N 134°00′W
232 E4 **Jākhal** Haryāna, NW India 29°46′N 75°51′E
Jakobeny *see* Iacobeni
153 G8 **Jakobstad** *Fin.* Pietarsaari. Österbotten, W Finland 63°41′N 22°40′E
Jakobstadt *see* Jēkabpils
184 G7 **Jakupica** ▲▲ C FYR Macedonia
79 N9 **Jal** New Mexico, SW USA 32°07′N 103°10′W
220 G4 **Jalājil** *var.* Ğalāğıl. Ar Riyāḍ, C Saudi Arabia 25°43′N 45°22′E
231 H3 **Jalālābād** *var.* Jalalabad, Jelalabad. Nangarhār, E Afghanistan 34°26′N 70°28′E
Jalal-Abad *see* Dzhalal-Abad, Dzhalal-Abadskaya Oblast', Kyrgyzstan
Jalal-Abad Oblasty *see* Dzhalal-Abadskaya Oblast'
231 J4 **Jalālpur** Punjab, E Pakistan 32°39′N 74°11′E
231 I6 **Jalālpur Pīrwāla** Punjab, E Pakistan 29°30′N 71°20′E
232 E4 **Jalandhar** *prev.* Jullundur. Punjab, N India 31°20′N 75°37′E
88 G4 **Jalán, Río** ≈ S Honduras
88 C4 **Jalapa** Jalapa, C Guatemala 14°39′N 89°59′W
87 J8 **Jalapa** Tabasco, Mexico 17°44′N 92°49′W
88 G5 **Jalapa** Nueva Segovia, NW Nicaragua 13°56′N 86°11′W
88 C4 **Jalapa** *off.* Departamento de Jalapa. ◈ *department* SE Guatemala
Jalapa, Departamento de *see* Jalapa
223 K8 **Jālaq** Sīstān va Balūchestān, SE Iran
153 G8 **Jalasjärvi** Etelä-Pohjanmaa, W Finland 62°30′N 22°50′E
230 F5 **Jaldak** Zābul, SE Afghanistan 32°00′N 66°45′E
113 J2 **Jales** São Paulo, S Brazil 20°15′S 50°34′W
234 J3 **Jaleshwar** *var.* Jaleswar. Odisha, NE India 21°51′N 87°15′E
Jaleswar *see* Jaleshwar
234 D4 **Jalgaon** Mahārāshtra, C India 21°01′N 75°34′E
217 L8 **Jalībah** Dhī Qār, S Iraq 30°37′N 46°31′E
217 L8 **Jalīb Shahāb** Al Muthannā, S Iraq 30°26′N 46°09′E
167 I10 **Jaligny** Auvergne, France 46°23′N 3°35′E
133 L7 **Jalingo** Taraba, E Nigeria 08°54′N 11°22′E
86 B5 **Jalisco** ◈ *state* SW Mexico
234 D4 **Jālna** Mahārāshtra, W India 19°50′N 75°53′N
Jalomitsa *see* Ialomiţa
171 H3 **Jalón** ≈ N Spain
232 C7 **Jālor** Rājasthān, N India 25°21′N 72°43′E
86 C5 **Jalostotilan** Jalisco, Mexico 21°12′N 102°28′W
85 K10 **Jalostotitlán** Jalisco, W Mexico 21°11′N 102°29′W
184 F4 **Jalovik** Serbia, W Serbia 44°37′N 19°48′E
87 J7 **Jalpa** Tabasco, Mexico 18°08′N 93°05′W
85 J10 **Jalpa** Zacatecas, C Mexico 21°40′N 103°W
233 J6 **Jalpāiguri** West Bengal, NE India 26°43′N 88°24′E
86 F5 **Jalpán** *var.* Jalpan. Querétaro de Arteaga, C Mexico 21°13′N 99°28′W
Jalpan *see* Jalpán
218 E7 **Jalqamūs** West Bank 32°25′N 35°21′E
121 I1 **Jalta** *island* N Tunisia
87 J9 **Jaltenango** Chiapas, Mexico 15°55′N 92°43′W
86 G8 **Jaltepec** Oaxaca, Mexico 17°18′N 97°12′W
128 F3 **Jālū** *var.* Jūlā. NE Libya 29°02′N 21°33′E
283 L6 **Jaluit Atoll** *var.* Jālwōj. *atoll* Ralik Chain, S Marshall Islands
Jālwōj *see* Jaluit Atoll
137 G8 **Jamaame** *It.* Giamame; *prev.* Margherita. Jubbada Hoose, S Somalia 0°00′N 42°43′E
133 L5 **Jamaare** ≈ NE Nigeria
90 D4 **Jamaica** ◆ *commonwealth republic* W West Indies
95 I2 **Jamaica** *island* W West Indies
90 G7 **Jamaica Channel** *channel* Haiti/Jamaica
233 K7 **Jamalpur** Dhaka, N Bangladesh 24°54′N 89°57′E
233 I7 **Jamālpur** Bihār, NE India 25°19′N 86°30′E
258 F4 **Jamaluang** *var.* Jemaluang. Johor, Peninsular Malaysia 02°15′N 103°50′E
107 I4 **Jamanxim, Rio** ≈ C Brazil
87 H7 **Jamapa** Veracruz-Llave, Mexico 19°02′N 96°14′W
110 E9 **Jambeiro** São Paulo, Brazil 23°16′S 45°41′W
104 B3 **Jambeli, Canal de** *channel* S Ecuador
163 F11 **Jambes** Namur, SE Belgium 50°26′N 04°51′E
258 E6 **Jambi** *var.* Telanaipura; *prev.* Djambi. Sumatera, W Indonesia 01°34′S 103°37′E
258 E5 **Jambi** *off.* Propinsi Jambi, *var.* Djambi. ◈ *province* W Indonesia
141 K4 **Jambila** Mpumalanga, South Africa 25°45′S 30°49′E
Jambi, Propinsi *see* Jambi
Jamdena *see* Yamdena, Pulau
180 J4 **Jameln** Niedersachsen, Germany 53°03′E 11°05′N
58 F4 **James Bay** *bay* Ontario/Québec, E Canada
80 B5 **Jamesburg** California, USA 36°22′N 121°35′W
64 G7 **Jamesburg** New Jersey, USA 40°21′N 74°26′W
117 A9 **James, Isla** *island* Archipiélago de los Chonos, S Chile
274 G5 **James Ranges** ▲▲ Northern Territory, C Australia
52 E8 **James River** ≈ Nunavut, N Canada
74 E6 **James River** ≈ North Dakota/South Dakota, N USA
67 L4 **James River** ≈ Virginia, NE USA
292 E3 **James Ross Island** *island* Antarctica
276 G5 **Jamestown** South Australia 33°13′S 138°36′E
132 G9 **Jamestown** ○ (Saint Helena) NW Saint Helena 15°56′S 05°44′W
80 D3 **Jamestown** California, W USA 37°57′N 120°25′W
66 F5 **Jamestown** Kentucky, S USA 36°58′N 85°03′W
62 D7 **Jamestown** New York, NE USA 42°05′N 79°15′W
74 D3 **Jamestown** North Dakota, N USA 46°54′N 98°42′W
66 F5 **Jamestown** Tennessee, S USA 36°24′N 84°58′W
Jamestown *see* Holetown
62 G2 **Jamet** ≈ Québec, SE Canada
86 G9 **Jamiltepec** *var.* Santiago Jamiltepec. Oaxaca, SE Mexico 16°18′N 97°51′W
155 D10 **Jammerbugten** *bay* Skagerrak, E North Sea
141 I6 **Jammerdrif** Free State, South Africa 29°42′S 26°58′E
232 D3 **Jammu** *prev.* Jummoo. *state capital* Jammu and Kashmir, NW India 32°43′N 74°54′E
232 E2 **Jammu and Kashmir** *var.* Jammu-Kashmir, Kashmir. ◈ *state* NW India
231 J3 **Jammu and Kashmīr** *disputed region* India/Pakistan
Jammu-Kashmir *see* Jammu and Kashmir
232 B8 **Jāmnagar** *prev.* Navanagar. Gujarāt, W India 22°28′N 70°06′E
231 H6 **Jāmpur** Punjab, E Pakistan 29°38′N 70°40′E
153 H9 **Jämsä** Keski-Suomi, C Finland 61°51′N 25°10′E
153 H9 **Jämsänkoski** Keski-Suomi, C Finland 61°54′N 25°10′E
233 I8 **Jamshedpur** Jhārkhand, NE India 22°47′N 86°12′E
154 G3 **Jämtland** ◈ *county* C Sweden
233 I7 **Jamūi** Bihār, NE India 24°57′N 86°14′E
Jamuna *see* Brahmaputra
233 K7 **Jamuna Nadi** ≈ N Bangladesh
Jamundá *see* Nhamundá, Rio
102 B6 **Jamundí** Valle del Cauca, SW Colombia 03°16′N 76°31′W
219 H8 **Janāb, Wādī al** *dry watercourse* Jordan
233 I6 **Janakpur** Central, C Nepal 26°45′N 85°55′E
109 E9 **Janaúba** Minas Gerais, SE Brazil 15°47′S 43°16′W
107 L1 **Janaucu, Ilha** *island* NE Brazil
222 G4 **Jandaq** Eşfahān, C Iran 34°04′N 54°26′E
170 B10 **Jandia, Punta de** *headland* Fuerteventura, Islas Canarias, Spain, NE Atlantic Ocean 28°03′N 14°32′W
106 B4 **Jandiatuba, Rio** ≈ NW Brazil
173 H6 **Jándula** ≈ S Spain
74 G5 **Janesville** Minnesota, N USA 44°07′N 93°43′W
73 C8 **Janesville** Wisconsin, N USA 42°42′N 89°02′W
139 J6 **Jangamo** Inhambane, SE Mozambique 24°04′S 35°25′E
234 F5 **Jangaon** Telangana, C India 18°47′N 79°25′E
233 J7 **Jangīpur** West Bengal, NE India 24°31′N 88°03′E
250 A3 **Jangsu** *prev.* Changsu. 35°41′N 127°54′E
Janina *see* Ioánnina
Janischken *see* Joniškis
184 F4 **Janja** NE Bosnia and Herzegovina 44°40′N 19°15′E
Jankovac *see* Jánoshalma
295 L5 **Jan Mayen** ◇ *constituent part of Norway* N Atlantic Ocean
144 B3 **Jan Mayen** *island* N Atlantic Ocean
295 L5 **Jan Mayen Fracture Zone** *tectonic feature* Greenland Sea/Norwegian Sea
295 L5 **Jan Mayen Ridge** *undersea feature* Greenland Sea/Norwegian Sea 69°00′N 08°00′W
84 G2 **Janos** Chihuahua, N Mexico 30°50′N 108°10′W
183 G13 **Jánoshalma** *SCr.* Jankovac. Bács-Kiskun, S Hungary 46°19′N 19°16′E
Janów *see* Ivanava, Belarus
182 E5 **Janowiec Wielkopolski** *Ger.* Janowitz. Kujawski-pomorskie, C Poland 52°47′N 17°30′E
Janowitz *see* Janowiec Wielkopolski
Janow/Janów *see* Jonava, Lithuania
182 I7 **Janów Lubelski** Lubelski, E Poland 50°42′N 22°24′E
Janów Poleski *see* Ivanava
140 G9 **Jansenville** Eastern Cape, S South Africa 32°56′S 24°40′E
261 J6 **Jantan** Papua, E Indonesia 03°53′S 134°20′E
109 E9 **Januária** Minas Gerais, SE Brazil 15°28′S 44°23′W
Janūbīyah, Al Bādiyah al *see* Ash Shāmīyah
166 G6 **Janville** Centre, France 48°12′N 1°53′E
164 F4 **Janzé** Ille-et-Vilaine, NW France 47°55′N 01°28′W
232 D8 **Jaora** Madhya Pradesh, C India 23°40′N 75°10′E
253 D14 **Japan** *var.* Nippon, *Jap.* Nihon. ◆ *monarchy* E Asia
205 N5 **Japan** *island group* E Asia
286 D2 **Japan Basin** *undersea feature* N Sea of Japan 40°00′N 135°00′E
205 N5 **Japan, Sea of** *var.* East Sea, *Rus.* Yapanskoye More. *sea* NW Pacific Ocean *see also* East Sea
286 D3 **Japan Trench** *undersea feature* NW Pacific Ocean 37°00′N 143°00′E
110 E3 **Japaraíba** Minas Gerais, Brazil 20°08′S 45°30′W
Japen *see* Yapen, Pulau
111 H7 **Japeri** Rio de Janeiro, Brazil 22°39′S 43°40′W
106 A5 **Japiim** *var.* Máncio Lima. Acre, W Brazil 08°00′S 73°39′W
106 D2 **Japurá** Amazonas, N Brazil 01°43′S 66°14′W
106 C2 **Japurá, Rio** *var.* Río Caquetá, Yapurá. ≈ Brazil/Colombia *see also* Caquetá, Río
Japurá, Rio *see* Caquetá, Río
89 M10 **Jaqué** Darién, SE Panama 07°31′N 78°09′W
Jaquemel *see* Jacmel
Jarablos *see* Jarābulus
216 E1 **Jarābulus** *var.* Jarablos, Jerablus, *Fr.* Djérablous. Ḥalab, N Syria 36°51′N 38°02′E
113 K5 **Jaraguá do Sul** Santa Catarina, S Brazil 26°29′S 49°07′W
170 E5 **Jaraicejo** Extremadura, W Spain 39°40′N 05°49′W
170 E5 **Jaráiz de la Vera** Extremadura, W Spain 40°04′N 05°45′W
170 G4 **Jarama** ≈ C Spain
117 D10 **Jaramillo** Santa Cruz, SE Argentina 47°10′S 67°07′W
Jarandilla de la Vega *see* Jarandilla de la Vera
170 E5 **Jarandilla de la Vera** *var.* Jarandilla de la Vega. Extremadura, W Spain 40°08′N 05°39′W
231 J5 **Jarānwāla** Punjab, E Pakistan 31°20′N 73°26′E
218 G7 **Jarash** *var.* Jerash; *anc.* Gerasa. Jarash, NW Jordan 32°17′N 35°54′E
219 G7 **Jarash** *off.* Muḥāfa at Jarash. ◈ *governorate* N Jordan
Jarash, Muḥāfa at *see* Jarash
Jarbah, Jazīrat *see* Jerba, Île de
154 H6 **Järbo** Gävleborg, C Sweden 60°43′N 16°40′E
Jardan *see* Yordon
90 D4 **Jardines de la Reina, Archipiélago de los** *island group* C Cuba
110 B5 **Jardinópolis** São Paulo, Brazil 21°02′S 47°46′W
239 J2 **Jargalant** Bayanhongor, C Mongolia 47°14′N 99°43′E
239 K1 **Jargalant** Bulgan, N Mongolia 49°09′N 104°19′E
239 I2 **Jargalant** *var.* Buyanbat. Govĭ-Altay, W Mongolia 47°00′N 95°57′E
239 I1 **Jargalant** *var.* Orgil. Hövsgöl, C Mongolia 48°31′N 99°19′E
Jargalant *see* Battsengel
Jargalant *see* Bulgan, Bayan-Ölgiy, Mongolia
Jargalant *see* Biger, Govĭ-Altay, Mongolia
Jarīd, Shaṭṭ al *see* Jerid, Chott el
110 D8 **Jarinu** Sao Paulo, Brazil 23°06′S 46°42′W
108 A2 **Jari, Rio** *var.* Jary. ≈ N Brazil
222 A9 **Jarīr, Wādī al** *dry watercourse* C Saudi Arabia
Jarja *see* Yur'ya
154 G7 **Järna** *var.* Dala-Järna. Dalarna, C Sweden 60°31′N 14°22′E
155 I8 **Järna** Stockholm, C Sweden 59°05′N 17°35′E
164 G6 **Jarnac** Charente, W France 45°41′N 00°10′W
167 L5 **Jarny** Lorraine, France 49°09′N 5°53′E
182 E6 **Jarocin** Wielkopolskie, C Poland 51°59′N 17°30′E
183 D8 **Jaroměř** *Ger.* Jermer. Královéhradecký Kraj, N Czech Republic 50°22′N 15°55′E
Jaroslau *see* Jarosław
183 I8 **Jarosław** *Ger.* Jaroslau, *Rus.* Yaroslav. Podkarpackie, SE Poland 50°01′N 22°41′E
154 F3 **Järpen** Jämtland, C Sweden 63°21′N 13°30′E
229 H8 **Jarqo'rg'on** *Rus.* Dzharkurgan. Surkhondaryo Viloyati, S Uzbekistan 37°31′N 67°20′E
216 G1 **Jarrāh, Wadi** *dry watercourse* NE Syria
Jars, Plain of *see* Xiangkhoang, Plateau de
244 C3 **Jartai Yanchi** ◎ N China
106 F7 **Jaru** Rondônia, W Brazil 10°24′S 62°45′W
Jarud Qi *see* Lubei
190 E4 **Järva-Jaani** *Ger.* Sankt-Johannis. Järvamaa, N Estonia 59°03′N 25°54′E
190 E4 **Järvakandi** *Ger.* Jerwakant. Raplamaa, NW Estonia 58°45′N 24°49′E
190 E4 **Järvamaa** *var.* Järva Maakond. ◈ *province* N Estonia
Järva Maakond *see* Järvamaa
153 H10 **Järvenpää** Uusimaa, S Finland 60°29′N 25°06′E
62 C6 **Jarvis** Ontario, S Canada 42°53′N 80°06′W
284 G3 **Jarvis Island** ◇ *US unincorporated territory* C Pacific Ocean
154 H5 **Järvsö** Gävleborg, C Sweden 61°43′N 16°25′E
Jary *see* Jari, Rio
184 G3 **Jaša Tomić** Vojvodina, NE Serbia 45°27′N 20°51′E
184 B4 **Jasenice** Zadar, SW Croatia 44°15′N 15°33′E
219 I9 **Jashshat al 'Adlah, Wādī al** *dry watercourse* C Jordan
133 H7 **Jasikan** E Ghana 07°24′N 00°28′E
Jāsk *see* Bandar-e Jāsk
228 D3 **Jasliq** *Rus.* Zhaslyk. Qoraqalpog'iston Respublikasi, NW Uzbekistan 43°57′N 57°30′E
183 H9 **Jasło** Podkarpackie, SE Poland 49°45′N 21°28′E
55 H9 **Jasmin** Saskatchewan, S Canada 51°11′N 103°34′W
117 G10 **Jason Islands** *island group* NW Falkland Islands
292 E3 **Jason Peninsula** *peninsula* Antarctica
73 E11 **Jasonville** Indiana, N USA 39°09′N 87°12′W
56 G5 **Jasper** Alberta, SW Canada 52°55′N 118°05′W
62 F4 **Jasper** Ontario, SE Canada 44°50′N 75°57′W
66 D8 **Jasper** Alabama, S USA 33°49′N 87°16′W
75 H12 **Jasper** Arkansas, C USA 36°00′N 93°11′W
69 K4 **Jasper** Florida, SE USA 30°31′N 82°57′W
73 E12 **Jasper** Indiana, N USA 38°22′N 86°57′W
74 F5 **Jasper** Minnesota, N USA 43°51′N 96°24′W
75 G11 **Jasper** Missouri, C USA 37°20′N 94°18′W
64 C3 **Jasper** New York, USA 42°07′N 77°30′W
66 E7 **Jasper** Tennessee, S USA 35°04′N 85°36′W
71 J6 **Jasper** Texas, SW USA 30°55′N 94°00′W
56 G6 **Jasper National Park** *national park* Alberta/British Columbia, SW Canada
Jassy *see* Iaşi
185 H6 **Jastrebac** ▲▲ SE Serbia
184 C2 **Jastrebarsko** Zagreb, N Croatia 45°40′N 15°40′E
Jastrow *see* Jastrowie
182 E4 **Jastrowie** *Ger.* Jastrow. Wielkopolskie, C Poland 53°25′N 16°48′E
183 F8 **Jastrzębie-Zdrój** Śląskie, S Poland 49°58′N 18°34′E
183 G11 **Jászapáti** Jász-Nagykun-Szolnok, E Hungary 47°30′N 20°10′E
183 G11 **Jászberény** Jász-Nagykun-Szolnok, E Hungary 47°30′N 19°56′E
183 G11 **Jász-Nagykun-Szolnok** *off.* Jász-Nagykun-Szolnok Megye. ◈ *county* E Hungary
Jász-Nagykun-Szolnok Megye *see* Jász-Nagykun-Szolnok
109 A10 **Jataí** Goiás, C Brazil 17°58′S 51°45′W
107 H2 **Jatapu, Serra do** ▲▲ N Brazil
87 K9 **Jatate, Río** ≈ SE Mexico
Jateng *see* Jawa Tengah
230 G10 **Jāti** Sind, SE Pakistan 24°20′N 68°18′E
90 D4 **Jatibonico** Sancti Spíritus, C Cuba 21°56′N 79°11′W
258 G8 **Jatiluhur, Danau** ◎ Jawa, S Indonesia
Jatim *see* Jawa Timur
Jativa *see* Xàtiva
231 H7 **Jatoi** *prev.* Jattoi. Punjab, E Pakistan 29°29′N 70°58′E
Jattoi *see* Jatoi
110 A7 **Jaú** São Paulo, S Brazil 22°11′S 48°35′W
103 J8 **Jauaperi, Rio** ≈ N Brazil
163 F10 **Jauche** Walloon Brabant, C Belgium 50°42′N 04°55′E
Jauer *see* Jawor
Jauf *see* Al Jawf
231 I5 **Jauharābād** Punjab, E Pakistan 32°16′N 72°17′E
104 E7 **Jauja** Junín, C Peru 11°48′S 75°30′W
85 M8 **Jaumave** Tamaulipas, C Mexico 23°28′N 99°22′W
166 E9 **Jaunay-Clan** Poitou-Charentes, France 46°41′N 0°22′E
190 E6 **Jaunjelgava** *Ger.* Friedrichstadt. S Latvia 56°38′N 25°03′E
Jaunlatgale *see* Pytalovo
190 E6 **Jaunpiebalga** NE Latvia 57°10′N 26°02′E
190 C6 **Jaunpils** C Latvia 56°45′N 23°03′E
233 H7 **Jaunpur** Uttar Pradesh, N India 25°44′N 82°41′E
114 D9 **Jàuregui** Buenos Aires, Argentina 34°36′S 59°10′W
169 M5 **Jaussiers** Provence-Alpes-Côte d'Azur, France 43°56′N 6°42′E
74 C4 **Java** South Dakota, N USA 45°29′N 99°54′W
Java *see* Jawa
171 I5 **Javalambre** ▲ E Spain 40°02′N 01°06′W
289 I8 **Java Ridge** *undersea feature* E Indian Ocean
104 F4 **Javari, Rio** *var.* Yavarí. ≈ Brazil/Peru
Javarthushuu *see* Bayan-Uul
259 H7 **Java Sea** *Ind.* Laut Jawa. *sea* W Indonesia
289 I8 **Java Trench** *var.* Sunda Trench. *undersea feature* E Indian Ocean
223 H6 **Javazm** *var.* Jowzam. Kermān, C Iran 30°31′N 55°01′E
171 J6 **Jávea** *Cat.* Xàbia. Valenciana, E Spain 38°48′N 00°10′E
Javhlant *see* Bayan-Ovoo
114 G2 **Javier de Viana** Artigas, Uruguay 30°24′S 56°47′W
117 B10 **Javier, Isla** *island* S Chile
184 E4 **Javor** ▲▲ Bosnia and Herzegovina/Serbia
183 F10 **Javorie** *Hung.* Jávoros. ▲ S Slovakia 48°26′N 19°16′E
Jávoros *see* Javorie
152 G7 **Jävre** Norrbotten, N Sweden 65°07′N 21°31′E
259 H8 **Jawa** *Eng.* Java; *prev.* Djawa. *island* C Indonesia
258 G8 **Jawa Barat** *off.* Propinsi Jawa Barat, *var.* Jabar, *Eng.* West Java. ◈ *province* S Indonesia
Jawa Barat, Propinsi *see* Jawa Barat
Jawa, Laut *see* Java Sea
217 I2 **Jawān** Nīnawýa, NW Iraq 35°57′N 43°03′E
259 H8 **Jawa Tengah** *off.* Propinsi Jawa Tengah, *var.* Jateng, *Eng.* Central Java. ◈ *province* S Indonesia
Jawa Tengah, Propinsi *see* Jawa Tengah
259 I9 **Jawa Timur** *off.* Propinsi Jawa Timur, *var.* Jatim, *Eng.* East Java. ◈ *province* S Indonesia
Jawa Timur, Propinsi *see* Jawa Timur
136 H7 **Jawhar** *var.* Jowhar, *It.* Giohar. Shabeellaha Dhexe, S Somalia 02°37′N 45°30′E
182 D7 **Jawor** *Ger.* Jauer. Dolnośląskie, SW Poland 51°01′N 16°11′E
Jaworów *see* Yavoriv
183 F8 **Jaworzno** Śląskie, S Poland 50°13′N 19°11′E
Jaxartes *see* Syr Darya
75 G11 **Jay** Oklahoma, C USA 36°25′N 94°49′W
Jayabum *see* Chaiyaphum
86 G8 **Jayacatlán** Oaxaca, Mexico 17°26′N 96°50′W
Jayanath *see* Chai Nat
233 K6 **Jayanti** *prev.* Jainti. West Bengal, NE India 26°45′N 89°40′E
261 K6 **Jaya, Puncak** *prev.* Puntjak Carstensz, Puntjak Sukarno. ▲ Papua, E Indonesia 04°00′S 137°10′E
261 M5 **Jayapura** *var.* Djajapura, *Dut.* Hollandia; *prev.* Kotabaru, Sukarnapura. Papua, E Indonesia 02°37′S 140°39′E
261 L6 **Jayawijaya, Pegunungan** ▲▲ Papua, E Indonesia
Jay Dairen *see* Dalian
Jayhawker State *see* Kansas
229 J6 **Jayilgan** *Rus.* Dzhailgan, Dzhayilgan. C Tajikistan 39°17′N 71°32′E
234 H5 **Jaypur** *var.* Jeypore, Jeypur. Odisha, E India 18°54′N 82°36′E
70 E4 **Jayton** Texas, SW USA 33°16′N 100°35′W
223 I8 **Jaz Mūriān, Hāmūn-e** ◎ SE Iran
216 F3 **Jazrah** Ar Raqqah, C Syria 35°56′N 39°02′E
218 F4 **Jbâa Ech Chuôf** Lebanon 33°40′N 35°36′E
218 F2 **Jbaïl** *var.* Jebeil, Jubayl, Jubeil; *anc.* Biblical Gebal, Bybles. W Lebanon 34°00′N 35°45′E
218 F2 **Jbaïl** Lebanon 34°07′N 35°39′E
70 E4 **J. B. Thomas, Lake** ⊠ Texas, SW USA
Jdaïdé *see* Judaydah
171 J9 **Jdiouia** Algeria
81 I9 **Jean** Nevada, W USA 35°45′N 115°20′W
68 D4 **Jeanerette** Louisiana, S USA 29°54′N 91°39′W
52 C10 **Jean Marie River** Northwest Territories, Canada
90 G5 **Jean-Rabel** NW Haiti 19°48′N 73°05′W
223 I8 **Jebāl Bārez, Kūh-e** ▲▲ SE Iran
Jebat *see* Jabwot
133 J7 **Jebba** Kwara, W Nigeria 09°04′N 04°50′E
Jebeil *see* Jbaïl
188 A8 **Jebel** *Hung.* Széphely; *prev. Hung.* Zsebely. Timiş, W Romania 45°33′N 21°14′E
228 B6 **Jebel** *Rus.* Dzhebel. Balkan Welaýaty, W Turkmenistan 39°42′N 54°10′E
Jebel, Bahr el *see* White Nile
Jebel Dhanna *see* Jabal aẓ Ẓannah
Jeble *see* Jablah
111 H4 **Jeceaba** Minas Gerais, Brazil 20°32′S 43°59′W
248 C5 **Jecheon** *Jap.* Teisen; *prev.* Chech'ŏn. N South Korea 37°06′N 128°15′E
159 J4 **Jedburgh** SE Scotland, United Kingdom 55°29′N 02°34′W
Jedda *see* Jiddah
183 G8 **Jędrzejów** *Ger.* Endersdorf. Świętokrzyskie, C Poland 50°39′N 20°18′E
180 J5 **Jeetze** *var.* Jeetzel. ≈ C Germany
Jeetzel *see* Jeetze
67 I5 **Jefferson** North Carolina, SE USA 36°24′N 81°33′W
71 J4 **Jefferson** Texas, SW USA 32°45′N 94°21′W
72 D7 **Jefferson** Wisconsin, N USA 43°01′N 88°48′W
75 H10 **Jefferson City** *state capital* Missouri, C USA 38°33′N 92°13′W
77 I5 **Jefferson City** Montana, NW USA 46°24′N 112°01′W
66 G6 **Jefferson City** Tennessee, S USA 36°07′N 83°29′W
80 I4 **Jefferson, Mount** ▲ Nevada, W USA 38°49′N 116°54′W
76 C6 **Jefferson, Mount** ▲ Oregon, NW USA 44°40′N 121°48′W
66 E3 **Jeffersontown** Kentucky, S USA 38°11′N 85°33′W
65 H5 **Jefferson Valley** New York, USA 41°20′N 73°47′W
73 F12 **Jeffersonville** Indiana, N USA 38°16′N 85°45′W
77 K8 **Jeffrey City** Wyoming, C USA 42°29′N 107°49′W
141 H10 **Jeffrey's Bay** Eastern Cape, South Africa 34°02′S 24°54′E
133 J5 **Jega** Kebbi, NW Nigeria 12°15′N 04°21′E
Jehol *see* Chengde
243 J9 **Jeixi** *var.* Hepo. Guangdong, SE China 23°34′N 115°22′E
248 B8 **Jeju** *Jap.* Saishū; *prev.* Cheju. S South Korea 33°31′N 126°34′E
237 J6 **Jeju-do** *Jap.* Saishū; *prev.* Cheju-do, Quelpart. *island* S South Korea
237 J6 **Jeju-haehyeop** *Eng.* Cheju Strait; *prev.* Cheju-haehyeop. *strait* S South Korea
113 H4 **Jejui-Guazú, Río** ≈ E Paraguay
190 E7 **Jēkabpils** *Ger.* Jakobstadt. S Latvia 56°30′N 25°56′E
69 L3 **Jekyll Island** *island* Georgia, SE USA

◆ Country
● Country Capital
◇ Dependent Territory
○ Dependent Territory Capital
◈ Administrative Regions
✈ International Airport
▲ Mountain
▲▲ Mountain Range
⛰ Volcano
≈ River
◎ Lake
⊠ Reservoir

Jelai, Sungai - Juma He

258 I6 **Jelai, Sungai** 〰 Borneo, N Indonesia
Jelalabad *see* Jalālābād
182 E7 **Jelcz-Laskowice** Dolnośląskie, SW Poland 51°01′N 17°24′E
182 D7 **Jelenia Góra** *Ger.* Hirschberg, Hirschberg im Riesengebirge, Hirschberg in Riesengebirge, Hirschberg in Schlesien. Dolnośląskie, SW Poland 50°54′N 15°48′E
233 J6 **Jelep La** *pass* N India
190 D6 **Jelgava** *Ger.* Mitau. C Latvia 56°38′N 23°47′E
184 F5 **Jelica** ▲▲ C Serbia
66 G5 **Jellico** Tennessee, S USA 36°33′N 84°06′W
155 D12 **Jelling** Syddanmark, C Denmark 55°45′N 09°24′E
258 F3 **Jemaja, Pulau** *island* W Indonesia
Jemaluang *see* Jamaluang
163 D11 **Jemappes** Hainaut, S Belgium 50°27′N 03°53′E
259 J9 **Jember** *prev.* Djember. Jawa, C Indonesia 08°07′S 113°45′E
163 E11 **Jemeppe-sur-Sambre** Namur, S Belgium 50°27′N 04°41′E
79 L6 **Jemez Pueblo** New Mexico, SW USA 35°36′N 106°43′W
180 D3 **Jemgum** Niedersachsen, Germany 53°16′E 7°23′N
238 F2 **Jeminay** *var.* Tuotiereke. Xinjiang Uygur Zizhiqu, NW China 47°28′N 85°49′E
175 C14 **Jemmel** Tunisia
283 L5 **Jemo Island** *atoll* Ratak Chain, C Marshall Islands
259 K5 **Jempang, Danau** ◎ Borneo, N Indonesia
179 G9 **Jena** Thüringen, C Germany 50°56′N 11°35′E
68 C2 **Jena** Louisiana, S USA 31°40′N 92°07′W
176 F4 **Jenaz** Graubünden, SE Switzerland 46°56′N 09°43′E
177 H3 **Jenbach** Tirol, W Austria 47°24′N 11°47′E
175 B13 **Jendouba** NW Tunisia Africa 36°29′N 08°46′E
260 C6 **Jeneponto** *prev.* Djeneponto. Sulawesi, C Indonesia 05°41′S 119°42′E
218 E7 **Jenin** N West Bank 32°28′N 35°17′E
67 H5 **Jenkins** Kentucky, S USA 37°10′N 82°37′W
75 F12 **Jenks** Oklahoma, C USA 36°01′N 95°58′W
Jenné *see* Djenné
80 A1 **Jenner** California, USA 38°27′N 123°07′W
177 L4 **Jennersdorf** Burgenland, SE Austria 46°57′N 16°08′E
68 C4 **Jennings** Louisiana, S USA 30°13′N 92°39′W
52 G8 **Jenny Lind Island** *island* Nunavut, N Canada
69 N7 **Jensen Beach** Florida, SE USA 27°15′N 80°13′W
53 J7 **Jens Munk Island** *island* Nunavut, NE Canada
250 A2 **Jeollabuk-do** *prev.* Chŏllabuk-Do. ◈
250 B1 **Jeongseon** *prev.* Chŏngsŏn. 37°22′N 128°39′E
241 L1 **Jeonju** *Jap.* Zenshū; *prev.* Chŏnju. SW South Korea 35°51′N 127°08′E
111 J4 **Jequeri** Minas Gerais, Brazil 20°27′S 42°40′W
109 G8 **Jequié** Bahia, E Brazil 13°52′S 40°06′W
111 H4 **Jequitibà** Minas Gerais, Brazil 20°38′S 43°28′W
111 H2 **Jequitibá** Minas Gerais, Brazil 19°16′S 44°01′W
109 G9 **Jequitinhonha, Rio** 〰 E Brazil
Jerablus *see* Jarābulus
130 G2 **Jerada** NE Morocco 34°16′N 02°07′W
Jerash *see* Jarash
131 L2 **Jerba, Île de** *var.* Djerba, Jazīrat Jarbah. *island* E Tunisia
86 E6 **Jerécuaro** Guanajuato, Mexico
90 G6 **Jérémie** SW Haiti 18°39′N 74°11′W
111 L4 **Jerénimo Monteiro** Espírito Santo, Brazil 20°47′S 41°24′W
Jeréz *see* Jerez de García Salinas, Mexico
Jerez *see* Jeréz de la Frontera, Spain
85 J9 **Jerez de García Salinas** *var.* Jeréz. Zacatecas, C Mexico 22°40′N 103°00′W
172 E9 **Jerez de la Frontera** *var.* Jerez; *prev.* Xeres. Andalucía, SW Spain 36°41′N 06°08′W
170 D7 **Jerez de los Caballeros** Extremadura, W Spain 38°20′N 06°45′W
Jergucati *see* Jorgucat
173 L3 **Jérica** Valenciana, Spain 39°55′N 0°34′W
219 F8 **Jericho** *Ar.* Arīḩā, *Heb.* Yeriḥo. E West Bank 31°51′N 35°27′E
219 F8 **Jericho Plain** West Bank
131 K2 **Jerid, Chott el** *var.* Shaṭṭ al Jarīd. *salt lake* SW Tunisia
277 J6 **Jerilderie** New South Wales, SE Australia 35°24′S 145°43′E
110 B3 **Jeriquara** São Paulo, Brazil 20°19′S 47°36′W
Jerischmarkt *see* Câmpia Turzii
152 G5 **Jerisjärvi** ◎ NW Finland
175 A14 **Jerissa** Tunisia
Jermak *see* Aksu
Jermentau *see* Yereymentau
Jermer *see* Jaroměř
79 H7 **Jerome** Arizona, SW USA 34°45′N 112°06′W
76 G8 **Jerome** Idaho, NW USA 42°43′N 114°31′W
198 D1 **Jersey** ◇ *British Crown Dependency* Channel Islands, NW Europe
166 B4 **Jersey** *island* Channel Islands, NW Europe
65 H6 **Jersey City** New Jersey, NE USA 40°42′N 74°01′W
64 C5 **Jersey Shore** Pennsylvania, NE USA 41°12′N 77°13′W
73 B11 **Jerseyville** Illinois, N USA 39°07′N 90°19′W
172 F3 **Jerte** Extremadura, Spain 40°13′N 5°45′W
170 E5 **Jerte** 〰 W Spain
219 E8 **Jerusalem** *Ar.* Al Quds, Al Quds ash Sharīf, *Heb.* Yerushalayim; *anc.* Hierosolyma. ● (Israel) Jerusalem, NE Israel 31°47′N 35°13′E
219 D8 **Jerusalem** ◈ *district* E Israel
277 L6 **Jervis Bay** New South Wales, SE Australia 35°09′S 150°42′E
277 L6 **Jervis Bay Territory** ◈ *territory* SE Australia
Jerwakant *see* Järvakandi
180 J6 **Jerxheim** Niedersachsen, Germany 52°05′E 10°54′N
177 J5 **Jesenice** *Ger.* Assling. NW Slovenia 46°26′N 14°01′E
183 E8 **Jeseník** *Ger.* Freiwaldau. Olomoucký Kraj, E Czech Republic 50°14′N 17°12′E
Jesi *see* Iesi
177 I6 **Jesolo** *var.* Iesolo. Veneto, NE Italy 45°32′N 12°37′E
Jesselton *see* Kota Kinabalu
141 L5 **Jesser Point** *point* KwaZulu-Natal, South Africa
154 E7 **Jessheim** Akershus, S Norway 60°07′N 11°10′E
233 K8 **Jessore** Khulna, W Bangladesh 23°10′N 89°12′E
180 H3 **Jesteburg** Niedersachsen, Germany 53°18′E 9°58′N
110 F6 **Jesuânia** Minas Gerais, Brazil 22°00′S 45°18′W
69 L2 **Jesup** Georgia, SE USA 31°36′N 81°54′W
64 D9 **Jessup** Maryland, USA 39°09′N 76°47′W
87 I8 **Jesús Carranza** Veracruz-Llave, SE Mexico 17°30′N 95°01′W
116 E2 **Jesús María** Córdoba, C Argentina 30°59′S 64°05′W
85 K9 **Jesús María** Baja California Sur, NW Mexico 22°15′N 101°48′W
86 D5 **Jesús María** Jalisco, Mexico 20°37′N 102°07′W
85 N7 **Jesús María, Boca** *inlet* Tamaulipas, NE Mexico
75 C10 **Jetmore** Kansas, C USA 38°05′N 99°55′W
163 D12 **Jeumont** Nord, N France 50°18′N 04°06′E
250 A1 **Jeungpyeong** *prev.* Chŭngp'yŏng. Ch'ungch'ŏng-bukto, South Korea 36°47′N 127°35′E
180 E3 **Jever** Niedersachsen, Germany 53°35′E 7°54′N
154 D7 **Jevnaker** Oppland, S Norway 60°15′N 10°25′E
Jewe *see* Jõhvi
71 H5 **Jewett** Texas, SW USA 31°21′N 96°08′W
65 K4 **Jewett City** Connecticut, NE USA 41°36′N 71°58′W
Jewish Autonomous Oblast *see* Yevreyskaya Avtonomnaya Oblast'
Jeypore *see* Jaipur, Rājasthān, India
Jeypore/Jeypur *see* Jaypur, Orissa, India
184 F6 **Jezercës, Maja e** ▲ N Albania 42°27′N 19°49′E
183 A9 **Jezerní Hora** ▲ SW Czech Republic 49°10′N 13°11′E
218 F4 **Jezzine** Lebanon 33°32′N 35°35′E
232 D8 **Jhābua** Madhya Pradesh, C India 22°44′N 74°37′E
232 E7 **Jhālāwār** Rājasthān, N India 24°37′N 76°12′E
Jhang/Jhang Sadar *see* Jhang Sadr
231 I5 **Jhang Sadr** *var.* Jhang, Jhang Sadar. Punjab, NE Pakistan 31°16′N 72°19′E
232 F7 **Jhānsi** Uttar Pradesh, N India 25°27′N 78°34′E
234 I3 **Jhārkhand** ◈ *state* NE India
234 H3 **Jhārsuguda** Odisha, E India 21°56′N 84°04′E
231 J4 **Jhelum** Punjab, NE Pakistan 32°55′N 73°42′E
205 I5 **Jhelum** 〰 E Pakistan
Jhenaidaha *see* Jhenida
233 K8 **Jhenida** *var.* Jhenaidaha. Dhaka, W Bangladesh 23°34′N 89°39′E
230 G9 **Jhimpīr** Sind, SE Pakistan 25°00′N 68°01′E
Jhind *see* Jīnd
231 H9 **Jhudo** Sind, SE Pakistan 24°58′N 69°18′E
Jhumra *see* Chak Jhumra
232 D5 **Jhunjhunūn** Rājasthān, N India 28°05′N 75°30′E
Ji *see* Hebei, China
Ji *see* Jilin, China
Jiading *see* Xinfeng
233 J7 **Jiāganj** West Bengal, NE India 24°18′N 88°07′E
242 G7 **Jiahe** Hunan, S China 25°33′N 112°15′E
Jiaji *see* Qionghai
242 A3 **Jiajiang** Sichuan, C China 29°45′N 103°35′E
244 C8 **Jialing Jiang** 〰 C China
245 I8 **Jialu He** 〰 Henan, China
237 K3 **Jiamusi** *var.* Chia-mu-ssu, Kiamusze. Heilongjiang, NE China 46°46′N 130°19′E
243 I6 **Ji'an** Jiangxi, S China 27°08′N 115°00′E
243 I6 **Ji'an** Jiangxi, China 27°02′N 114°32′E
248 C3 **Ji'an** Jilin, NE China 41°04′N 126°07′E
245 L2 **Jianchang** Liaoning, NE China 40°48′N 119°51′E
Jianchang *see* Nancheng
Jiancheng *see* Jianyang
240 B6 **Jianchuan** *var.* Jinhuan. Yunnan, SW China 26°28′N 99°49′E
243 L3 **Jiande** Zhejiang, China 29°17′N 119°10′E
242 B4 **Jiang'an** Sichuan, C China 28°47′N 105°07′E
243 H9 **Jiangcun** Guangdong, SE China 23°17′N 113°14′E
243 L1 **Jiangdu** Jiangsu, China 32°16′N 119°19′E
245 K3 **Jianggezhuang** Tianjin Shi, N China 39°27′N 119°07′E
242 F7 **Jianghua** Hunan, China 25°07′N 111°20′E
242 C3 **Jiangjin** Chongqing Shi, C China 29°11′N 106°10′E
238 G3 **Jiangjunmiao** Xinjiang Uygur Zizhiqu, W China 44°42′N 90°06′E
245 J9 **Jiangkou** Anhui, E China 32°48′N 116°07′E
242 E9 **Jiangkou** Guang Zhuangzu Zizhiqu, S China 23°35′N 110°13′E
242 E5 **Jiangkou** *var.* Shuangjiang. Guizhou, S China 27°46′N 108°53′E
242 G3 **Jiangkou** Hubei, C China 30°26′N 111°50′E
242 F5 **Jiangkou** Hunan, S China 27°52′N 110°24′E
Jiangkou *see* Fengkai
243 K6 **Jiangle** *var.* Guyong. Fujian, SE China 26°44′N 117°26′E
243 H10 **Jiangmen** Guangdong, S China 22°35′N 113°02′E
Jiangna *see* Yanshan
243 L4 **Jiangshan** Zhejiang, SE China 28°41′N 118°33′E
245 K9 **Jiangsu** *var.* Chiang-su, Jiangsu Sheng, Kiangsu, Su. ◈ *province* E China
Jiangsu *see* Nanjing
Jiangsu Sheng *see* Jiangsu
243 H8 **Jiangwei** Guangdong, SE China 24°27′N 114°08′E
243 I5 **Jiangxi** *var.* Chiang-hsi, Gan, Jiangxi Sheng, Kiangsi. ◈ *province* S China
Jiangxi Sheng *see* Jiangxi
243 L1 **Jiangyan** Jiangsu, China 32°19′N 120°05′E
243 M1 **Jiangyin** Jiangsu, China 31°32′N 120°10′E
242 F7 **Jiangyong** Hunan, China 25°10′N 111°12′E
242 B1 **Jiangyou** *prev.* Zhongba. Sichuan, C China 31°52′N 104°52′E
242 D6 **Jianhe** Guizhou, China 26°26′N 108°17′E
245 L8 **Jianhu** Jiangsu, E China 33°29′N 119°48′E
242 G3 **Jianli** *var.* Rongcheng. Hubei, C China 29°51′N 112°50′E
243 J6 **Jianning** Jiangxi, SE China 26°48′N 116°50′E
243 K5 **Jian'ou** Fujian, SE China 27°04′N 118°20′E
245 L1 **Jianping** *var.* Yebaishou. Liaoning, NE China 41°13′N 119°37′E
Jianshe *see* Baiyü
242 E2 **Jianshi** *var.* Yezhou. Hubei, C China 30°37′N 109°42′E
Jiantang *see* Xamgyi'nyilha
205 L7 **Jian Xi** 〰 SE China
243 K5 **Jianyang** Fujian, SE China 27°20′N 118°01′E
242 B2 **Jianyang** *var.* Jiancheng. Sichuan, C China 30°22′N 104°31′E
245 L2 **Jiaoding Shan** ▲ Liaoning, China 41°07′N 120°01′E
245 J4 **Jiaohe** Hebei, E China 38°01′N 116°17′E
248 D1 **Jiaohe** Jilin, NE China 43°41′N 127°20′E
Jiaojiang *see* Taizhou
243 K8 **Jiaomei** Fujian, SE China 24°32′N 117°54′E
245 L6 **Jiaonan** Shandong, China 35°32′N 119°35′E
Jiaoxian *see* Jiaozhou
245 L6 **Jiaozhou** *prev.* Jiaoxian. Shandong, E China 36°17′N 120°00′E
245 L6 **Jiaozhou Wan** *bay* Shandong, China
245 H7 **Jiaozuo** Henan, C China 35°14′N 113°13′E
Jiashan *see* Mingguang
238 C5 **Jiashi** *var.* Baren, Payzawat. Xinjiang Uygur Zizhiqu, NW China 39°27′N 76°45′E
232 G7 **Jiāwān** Madhya Pradesh, C India 24°20′N 82°17′E
245 K8 **Jiawang** Jiangsu, E China 34°26′N 117°27′E
245 H8 **Jia Xian** Henan, E China 33°58′N 113°12′E
245 J7 **Jiaxiang** Shandong, China 35°14′N 116°12′E
243 M2 **Jiaxing** Zhejiang, SE China 30°44′N 120°46′E
247 M3 **Jiayan** *var.* Chaoyang. Heilongjiang, NE China 48°55′N 130°25′E
243 M9 **Jiayi** *var.* Chia-i, Chiai, Chiayi, Kiayi, *Jap.* Kagi. C Taiwan 23°29′N 120°27′E
247 M3 **Jiayin** *var.* Chaoyang. Heilongjiang, NE China 48°51′N 130°24′E
243 H3 **Jiayu** Hubei, China 29°34′N 113°32′E
239 I5 **Jiayuguan** Gansu, N China 39°47′N 98°14′E
243 J9 **Jiazi** Guizhou, SE China 22°53′N 116°04′E
Jibhalanta *see* Uliastay
216 F3 **Jibli** Ar Raqqah, C Syria 35°49′N 39°23′E
188 C6 **Jibou** *Hung.* Zsibó. Sălaj, NW Romania 47°15′N 23°17′E
221 L5 **Jibsh, Ra's al** *headland* E Oman 21°20′N 59°23′E
Jibuti *see* Djibouti
87 H7 **Jícaro** Veracruz-Llave, Mexico 18°33′N 96°15′W
242 B6 **Jichang** Guizhou, S China 26°30′N 105°22′E
183 C8 **Jičín** *Ger.* Jitschin. Královéhradecký Kraj, N Czech Republic 50°27′N 15°20′E
220 D5 **Jiddah** *Eng.* Jedda. (Saudi Arabia) Makkah, W Saudi Arabia 21°34′N 39°13′E
221 J6 **Jiddat al Ḩarāsīs** *desert* C Oman
Jiehu *see* Yinan
243 J9 **Jieshi** Guangdong, SE China 22°49′N 115°48′E
243 J9 **Jieshi Wan** *bay* Guangdong, China
245 I9 **Jieshou** Anhui, China 33°09′N 115°13′E
243 H6 **Jieshou** Hunan, SE China 26°36′N 113°24′E
Jiesjavrre *see* Iešjavri
243 I9 **Jiexi** Guangdong, China 23°16′N 115°30′E
244 G5 **Jiexiu** Shanxi, C China 37°00′N 111°50′E
243 J9 **Jieyang** Guangdong, S China 23°32′N 116°20′E
191 D9 **Jieznas** Kaunas, S Lithuania 54°37′N 24°10′E
Jīfa', Bi'r *see* Jif'īyah, Bi'r
220 G8 **Jif'īyah, Bi'r** *var.* Bi'r Jīfa'. *well* C Yemen
133 L5 **Jigawa** ◈ *state* N Nigeria
228 F5 **Jigerbent** *Rus.* Dzhigirbent. Lebap Welaýaty, NE Turkmenistan 40°44′N 61°56′E
90 F5 **Jiguaní** Granma, E Cuba 20°24′N 76°26′W
239 K7 **Jigzhi** *var.* Chugqênsumdo. Qinghai, C China 33°23′N 101°25′E
Jih-k'a-tse *see* Xigazê
183 C9 **Jihlava** *Ger.* Iglau, *Pol.* Igława. Vysocina, S Czech Republic 49°22′N 15°36′E
183 D9 **Jihlava** *var.* Igel, *Ger.* Iglawa. 〰 Vysočina, C Czech Republic
Jihlavský Kraj *see* Vysočina
183 B9 **Jihočeský Kraj** *prev.* Budějovický Kraj. ◈ *region* Czech Republic
183 D9 **Jihomoravský Kraj** *prev.* Brněnský Kraj. ◈ *region* SE Czech Republic
131 J1 **Jijel** *var.* Djidjel; *prev.* Djidjelli. NE Algeria 36°50′N 05°43′E
188 F6 **Jijia** 〰 N Romania
136 G4 **Jijiga** *It.* Giggiga. Sumalē, E Ethiopia 09°21′N 42°53′E
171 I7 **Jijona** *var.* Xixona. Valenciana, E Spain 38°34′N 00°29′W
137 G8 **Jilib** *It.* Gelib. Jubbada Dhexe, S Somalia 0°18′N 42°48′E
237 J4 **Jilin** *var.* Chi-lin, Girin, Kirin; *prev.* Yungki, Yunki. Jilin, NE China 43°46′N 126°32′E
248 C1 **Jilin** *var.* Chi-lin, Girin, Ji, Jilin Sheng, Kirin. ◈ *province* NE China
248 C2 **Jilin Hada Ling** ▲▲ NE China
Jilin Sheng *see* Jilin
247 J2 **Jiliu He** 〰 NE China
171 H4 **Jiloca** 〰 N Spain
168 B10 **Jiloca** 〰 Spain
243 N7 **Jilong** *var.* Keelung, *Jap.* Kirun, Kirun', *prev.* Chilung, *prev. Sp.* Santissima Trinidad. N Taiwan 25°10′N 121°43′E
86 E6 **Jilotepec de Abasolo** México, Mexico 19°57′N 99°32′W
136 D5 **Jīma** *var.* Jimma, *It.* Gimma. Oromīya, C Ethiopia 07°42′N 36°51′E
91 H6 **Jimaní** W Dominican Republic 18°29′N 71°49′W
188 A8 **Jimbolia** *Ger.* Hatzfeld, *Hung.* Zsombolya. Timiș, W Romania 45°47′N 20°43′E
170 E9 **Jimena de la Frontera** Andalucía, S Spain 36°27′N 05°28′W
85 I5 **Jiménez** Chihuahua, N Mexico 27°09′N 104°54′W
85 L4 **Jiménez** Coahuila, NE Mexico 29°05′N 100°40′W
85 M7 **Jiménez** *var.* Santander Jiménez. Tamaulipas, C Mexico 24°11′N 98°29′W
85 J8 **Jiménez del Teul** Zacatecas, C Mexico 23°13′N 103°46′W
133 M6 **Jimeta** Adamawa, E Nigeria 09°16′N 12°25′E
Jimma *see* Jīma
245 L6 **Jimo** Shandong, China 36°13′N 120°17′E
238 G3 **Jimsar** Xinjiang Uygur Zizhiqu, NW China 44°05′N 88°48′E
64 E6 **Jim Thorpe** Pennsylvania, NE USA 40°51′N 75°43′W
Jin *see* Shanxi
Jin *see* Tianjin Shi
245 J6 **Jinan** *var.* Chinan, Chi-nan, Tsinan. *province capital* Shandong, E China 36°43′N 116°58′E
Jin'an *see* Songpan
Jinbi *see* Dayao
244 A4 **Jinchang** Gansu, N China 38°31′N 102°07′E
245 H7 **Jincheng** Shanxi, C China 35°30′N 112°52′E
Jincheng *see* Wuding
Jinchengjiang *see* Hechi
232 E5 **Jīnd** *prev.* Jhind. Haryāna, NW India 29°29′N 76°22′E
277 K6 **Jindabyne** New South Wales, SE Australia 36°28′S 148°36′E
248 B7 **Jin-do** *Jap.* Chin-tō; *prev.* Chin-do. *island* SW South Korea
183 C9 **Jindřichův Hradec** *Ger.* Neuhaus. Jihočeský Kraj, S Czech Republic 49°09′N 15°01′E
Jing *see* Beijing Shi
Jing *see* Jinghe, China
243 I4 **Jing'an** Jiangxi, SE China 28°52′N 115°22′E
244 E5 **Jingbian** Shaanxi, China 37°22′N 108°29′E
244 D7 **Jingchuan** Gansu, C China 35°20′N 107°45′E
243 K3 **Jingde** *var.* Jingyang. Anhui, E China 30°18′N 118°33′E
243 K4 **Jingdezhen** Jiangxi, S China 29°18′N 117°18′E
243 H6 **Jinggangshan** Jiangxi, S China 26°36′N 114°11′E
243 H6 **Jinggang Shan** ▲ Jiangxi, China 26°25′N 114°04′E
245 J4 **Jinghai** Tianjin Shi, E China 38°53′N 116°45′E
238 E3 **Jinghe** *var.* Jing. Xinjiang Uygur Zizhiqu, NW China 44°35′N 82°55′E
244 D7 **Jing He** 〰 C China
243 L5 **Jinghing** *var.* Hexi. Zhejiang, SE China 27°58′N 119°39′E
240 B8 **Jinghong** *var.* Yunjinghong. Yunnan, SW China 22°03′N 100°56′E
243 M1 **Jingjiang** Jiangsu, China 32°01′N 120°10′E
242 G2 **Jingmen** Hubei, C China 30°58′N 112°09′E
244 C6 **Jingning** Gansu, China 35°19′N 105°26′E
248 D1 **Jingpo Hu** ◎ NE China
243 H2 **Jingshan** Hubei, China 31°01′N 113°04′E
242 F1 **Jing Shan** ▲▲ C China
244 B5 **Jingtai** *var.* Yitiaoshan. Gansu, C China 37°12′N 104°06′E
242 B9 **Jingxi** *var.* Xinjing. Guangxi Zhuangzu Zizhiqu, S China 23°10′N 106°22′E
245 J5 **Jingxian** Hebei, E China 37°41′N 116°16′E
Jing Xian *see* Jingzhou, Hunan, China
242 A3 **Jingyan** Sichuan, China 29°25′N 104°02′E
244 E7 **Jingyang** Shaanxi, China 34°19′N 108°30′E
Jingyang *see* Jingde
248 C2 **Jingyu** Jilin, NE China 42°23′N 126°48′E
244 B5 **Jingyuan** *var.* Wulan. Gansu, C China 36°35′N 104°40′E
244 C6 **Jingyuan** Ningxia, China 35°19′N 106°11′E
245 L6 **Jingzhi** Shandong, E China 36°19′N 119°23′E
242 G3 **Jingzhou** *prev.* Shashi, Sha-shih, Shasi. Hubei, C China 30°21′N 112°09′E
242 E6 **Jingzhou** *var.* Jing Xian, Jingzhou Miaouzu Dongzu Zizhixian, Quyang. Hunan, S China 26°35′N 109°40′E
Jingzhou Miaouzu Dongzu Zizhixian *see* Jingzhou
248 C7 **Jinhae** *Jap.* Chinkai; *prev.* Chinhae. S South Korea 35°06′N 128°48′E
Jinhe *see* Jinping
Jinhong Dao *see* Sin Cowe Island
245 L9 **Jinhu** *var.* Licheng. Jiangsu, E China 33°01′N 119°01′E
243 L4 **Jinhua** Zhejiang, SE China 29°15′N 119°36′E
Jinhuan *see* Jianchuan
245 J7 **Jining** Shandong, E China 35°25′N 116°35′E
Jining *see* Ulan Qab
137 C8 **Jinja** S Uganda 0°27′N 33°14′E
243 L7 **Jinjiang** *var.* Qingyang. Fujian, SE China 24°53′N 118°36′E
243 J4 **Jinjiang** Jiangxi, SE China 28°24′N 116°53′E
243 K7 **Jin Jiang** 〰 Fujan, China
242 E5 **Jin Jiang** 〰 Guizhou, China
243 I4 **Jin Jiang** 〰 S China
Jinjiang *see* Chengmai
248 C7 **Jinju** *prev.* Chinju, *Jap.* Shinshū. S South Korea 35°12′N 128°06′E
261 J7 **Jin, Kepulauan** *island group* E Indonesia
243 H3 **Jinkou** Hubei, C China 30°20′N 114°07′E
245 K5 **Jinlingzhen** Shandong, E China 36°48′N 118°12′E
243 L8 **Jinmen Dao** *var.* Chinmen Tao, Quemoy. *island* W Taiwan
88 G5 **Jinotega** Jinotega, NW Nicaragua 13°03′N 85°59′W
88 G5 **Jinotega** ◈ *department* N Nicaragua
88 G6 **Jinotepe** Carazo, SW Nicaragua 11°50′N 86°10′W
242 E6 **Jinping** *var.* Sanjiang. Guizhou, S China 26°42′N 109°13′E
240 C8 **Jinping** *var.* Jinhe. Yunnan, SW China 22°47′N 103°12′E
Jinsen *see* Incheon
242 C5 **Jinsha** Guizhou, S China 27°24′N 106°16′E
245 K1 **Jinshan** Inner Mongolia, China 41°34′N 118°25′E
243 M2 **Jinshan** Shanghai Shi, E China 30°43′N 121°19′E
242 G3 **Jinshi** Hunan, S China 29°42′N 111°46′E
Jinshi *see* Xinning
218 F4 **Jinsnâya** Lebanon 33°32′N 35°25′E
239 J2 **Jinst** *var.* Bodī. Bayanhongor, C Mongolia 45°25′N 100°33′E
239 J5 **Jinta** Gansu, N China 40°01′N 98°57′E
243 L1 **Jintan** Jiangsu, China 31°28′N 119°20′E
242 B2 **Jintang** Sichuan, China 30°30′N 104°14′E
263 L7 **Jintotolo Channel** *channel* C Philippines
245 K3 **Jinxi** Henan, NE China 39°35′N 118°38′E
243 J5 **Jinxi** Jiangxi, China 27°32′N 116°28′E
243 K6 **Jin Xi** 〰 SE China
Jinxi *see* Huludao
243 J4 **Jinxian** *var.* Minhe. Fujian, SE China 28°22′N 116°14′E
Jinxian *see* Jinzhou
245 J7 **Jinxiang** Shandong, E China 35°08′N 116°19′E
242 E8 **Jinxiu** Guangxi, China 24°05′N 110°07′E
250 A4 **Jinyang-ho** *prev.* Chinyang Lake. ◎
250 B3 **Jinyeong** *prev.* Chinyŏng. 35°18′N 128°41′E
243 M4 **Jinyun** Zhejiang, China 28°23′N 120°02′E
243 I1 **Jinzhai** *var.* Meishan. Anhui, E China 31°42′N 115°47′E
245 H5 **Jinzhong** *var.* Yuci. Shanxi, C China 37°34′N 112°45′E
245 I4 **Jinzhou** Hebei, China 38°01′N 115°01′E
245 M3 **Jinzhou** *prev.* Jinxian. Liaoning, NE China 39°04′N 121°45′E
245 M2 **Jinzhou** *var.* Chin-chou, Chinchow; *prev.* Chinhsien. Liaoning, NE China 41°07′N 121°06′E
245 M2 **Jinzhou Wan** *bay* Liaoning, China
Jinzhu *see* Daocheng
216 C8 **Jinz, Qā' al** ◎ C Jordan
95 K5 **Jiparaná, Rio** 〰 W Brazil
104 B2 **Jipijapa** Manabí, W Ecuador 01°23′S 80°35′W
88 D5 **Jiquilisco** Usulután, S El Salvador 13°19′N 88°35′W
86 C6 **Jiquilpan de Juárez** Michoacán, Mexico 19°59′N 102°43′W
86 E6 **Jiquipilco** México, Mexico 19°33′N 99°37′W
Jirgalanta *see* Hovd
229 J7 **Jirgatol** *Rus.* Dzhirgatal'. C Tajikistan 39°13′N 71°09′E
250 A3 **Jiri-san** *prev.* Chiri-san. ▲ 35°16′N 127°39′E
129 I5 **Jirjā** *var.* Girga, Girgeh, Jirjā. C Egypt 26°17′N 31°58′E
Jirjā *see* Jirjā
183 B8 **Jirkov** *Ger.* Görkau. Ústecký Kraj, NW Czech Republic 50°30′N 13°27′E
223 I8 **Jīroft** *var.* Sabzawaran, Sabzvārān. Kermān, SE Iran 28°40′N 57°40′E
136 I5 **Jirriiban** Mudug, E Somalia 07°15′N 48°55′E
218 F5 **Jīsh** *var.* Gush Ḫalav. Northern, N Israel 33°01′N 35°26′E
244 F7 **Jishan** Shanxi, China 35°22′N 110°35′E
242 E4 **Jishou** Hunan, S China 28°20′N 109°43′E
243 I5 **Jishui** Jiangxi, China 27°08′N 115°05′E
Jisr ash Shadadi *see* Ash Shadādah
188 D9 **Jitaru** Olt, S Romania 44°27′N 24°32′E
Jitschin *see* Jičín
188 C10 **Jiu** *Ger.* Schil, Schyl, *Hung.* Zsil, Zsily. 〰 S Romania
245 K3 **Jiucheng** Hebei, NE China 40°10′N 118°20′E
242 A1 **Jiuding Shan** ▲ Sichuan, China 31°22′N 103°32′E
243 K8 **Jiufeng** Fujian, SE China 24°15′N 117°01′E
243 L5 **Jiufeng Shan** ▲▲ SE China
243 I4 **Jiugong Shan** ▲ Hubei, China 29°14′N 114°23′E
243 J3 **Jiujiang** Jiangxi, S China 29°45′N 115°59′E
243 J3 **Jiujiang** Jiangxi, China 29°22′N 115°32′E
242 G2 **Jiukou** Hebei, C China 30°51′N 112°37′E
243 J8 **Jiulian Shan** ▲▲ Guangdong, China
243 I4 **Jiuling Shan** ▲▲ S China
245 K5 **Jiulinzi** Shandong, E China 36°51′N 118°21′E
240 C5 **Jiulong** *var.* Garba, *Tib.* Gyaisi. Sichuan, C China 29°00′N 101°30′E
243 K7 **Jiulong Jiang** 〰 SE China
243 J6 **Jiulong Xi** 〰 SE China
245 J5 **Jiuqihe** Shandong, E China 36°52′N 117°05′E
239 J5 **Jiuquan** *var.* Suzhou. Gansu, N China 39°47′N 98°30′E
240 F10 **Jiusuo** Hainan, S China 18°25′N 109°55′E
248 C1 **Jiutai** Jilin, NE China 44°01′N 125°51′E
242 D7 **Jiuwan Dashan** ▲▲ S China
245 J3 **Jiuwuqing** Tianjin Shi, N China 39°31′N 116°52′E
244 B8 **Jiuzhaigou** *var.* Nongle; *prev.* Nanping. Sichuan, C China 33°25′N 104°05′E
245 K5 **Jiuzhangqiu** Shandong, E China 36°48′N 117°28′E
280 B3 **Jiwaka** ◈ *province* C Papua New Guinea
230 D9 **Jīwani** Baluchistān, SW Pakistan 25°05′N 61°46′E
237 K3 **Jixi** Heilongjiang, NE China 45°17′N 131°01′E
247 M4 **Jixian** *var.* Fuli. Heilongjiang, NE China 46°38′N 131°04′E
244 F6 **Jixian** *var.* Ji Xian. Shanxi, C China 36°15′N 110°41′E
245 J3 **Jixian** Tianjin Shi, China 40°01′N 117°14′E
Jixian *see* Weihui
Ji Xian *see* Jixian
245 J5 **Jiyang** Shandong, China 36°35′N 117°07′E
244 G7 **Jiyuan** Henan, China 35°02′N 112°21′E
Jiza *see* Al Jīzah
220 F7 **Jīzān** *var.* Qīzān. Jīzān, SW Saudi Arabia 17°50′N 42°50′E
220 F7 **Jīzān** *var.* Minṭaqat Jīzān. ◈ *province* SW Saudi Arabia
Jīzān, Minṭaqat *see* Jīzān
245 I5 **Jizhou** Hebei, China 37°20′N 115°20′E
129 K5 **Jizl, Wādī al** *dry watercourse* W Saudi Arabia
250 F3 **Jizō-zaki** *headland* Honshū, SW Japan 35°34′N 133°16′E
136 J1 **Jiz', Wādī al** *dry watercourse* E Yemen
229 I6 **Jizzax** *Rus.* Dzhizak. Jizzax Viloyati, C Uzbekistan 40°08′N 67°47′E
229 H6 **Jizzax Viloyati** *Rus.* Dzhizakskaya Oblast'. ◈ *province* C Uzbekistan
113 J6 **Joaçaba** Santa Catarina, S Brazil 27°08′S 51°30′W
87 H7 **Joachín** Veracruz-Llave, Mexico 18°38′N 96°14′W
Joal *see* Joal-Fadiout
132 A4 **Joal-Fadiout** *prev.* Joal. W Senegal 14°09′N 16°50′W
111 J1 **Joanésia** Minas Gerais, Brazil 19°12′S 42°40′W
110 D8 **Joanópolis** São Paulo, Brazil 22°56′S 46°17′W
132 B9 **João Barrosa** Boa Vista, E Cape Verde 16°01′N 22°44′W
João Belo *see* Xai-Xai
João de Almeida *see* Chibia
111 I3 **João Monlevade** Minas Gerais, Brazil 19°50′S 44°08′W
111 M2 **João Neiva** Espírito Santo, Brazil 19°45′S 40°24′W
108 J5 **João Pessoa** *prev.* Paraíba. *state capital* Paraíba, E Brazil 07°06′S 34°53′W
71 J5 **Joaquin** Texas, SW USA 31°58′N 94°03′W
115 H9 **Joaquín Suárez** Canelones, Uruguay 34°44′S 56°02′W
112 D5 **Joaquín V. González** Salta, N Argentina 25°06′S 64°07′W
Joazeiro *see* Juazeiro
277 J2 **Jobs Gate** Queensland, Australia
Job'urg *see* Johannesburg
183 A11 **Jochberger Ache** 〰 W Austria
Jo-ch'iang *see* Ruoqiang
152 G6 **Jock** Norrbotten, N Sweden 66°40′N 22°45′E
118 F2 **Jocolí** Mendoza, Argentina 32°35′S 68°41′W
88 G3 **Jocón** Yoro, N Honduras 15°17′N 86°55′W
86 C5 **Jocotepec** Jalisco, Mexico
170 G7 **Jódar** Andalucía, S Spain 37°51′N 03°18′W
232 C6 **Jodhpur** Rājasthān, NW India 26°17′N 73°02′E
163 F10 **Jodoigne** Walloon Brabant, C Belgium 50°43′N 04°52′E
153 J8 **Joensuu** Pohjois-Karjala, SE Finland 62°36′N 29°45′E
79 N3 **Joes** Colorado, C USA 39°36′N 102°40′W
285 N2 **Joe's Hill** *hill* Kiritimati, NE Kiribati
253 A11 **Jōetsu** *var.* Zyôetu. Niigata, Honshū, C Japan 37°09′N 138°13′E
139 I4 **Jofane** Inhambane, S Mozambique 21°16′S 34°21′E
234 D7 **Jog** *waterfall* W India
233 J6 **Jogbani** Bihār, NE India 26°23′N 87°16′E
190 F4 **Jõgeva** *Ger.* Laisholm. Jõgevamaa, E Estonia 58°48′N 26°28′E
190 F4 **Jõgevamaa** *off.* Jõgeva Maakond. ◈ *province* E Estonia
Jõgeva Maakond *see* Jõgevamaa
234 C7 **Jog Falls** *Waterfall* Karnātaka, W India
223 H3 **Joghatāy** Khorāsān-e Razavī, NE Iran 36°34′N 57°00′E
233 K6 **Jogīghopa** Assam, NE India 26°14′N 90°35′E
232 E3 **Jogindarnagar** Himāchal Pradesh, N India 31°51′N 76°47′E
Jogjakarta *see* Yogyakarta
251 J2 **Jōhana** Toyama, Honshū, SW Japan 36°30′N 136°53′E
138 G6 **Johannesburg** *var.* Egoli, Erautini, Gauteng, *abbrev.* Job'urg. Gauteng, NE South Africa 26°10′S 28°02′E
78 E7 **Johannesburg** California, W USA 35°20′N 117°37′W
Johannisburg *see* Pisz
230 G8 **Johi** Sind, SE Pakistan 26°46′N 67°28′E
103 K7 **Johi Village** S Guyana 01°48′N 58°33′W
91 M6 **John A. Osborne** ✈ (Plymouth) E Montserrat 16°45′N 62°09′W
76 E6 **John Day** Oregon, NW USA 44°25′N 118°57′W
76 D6 **John Day River** 〰 Oregon, NW USA
52 G6 **John Dyer, Cape** *headland* Nunavut, N Canada
65 H6 **John F Kennedy** ✈ (New York) Long Island, New York, NE USA 40°39′N 73°45′W
67 K5 **John H. Kerr Reservoir** *var.* Buggs Island Lake, Kerr Lake. ⊠ North Carolina/Virginia, SE USA
79 N4 **John Martin Reservoir** ⊠ Colorado, C USA
156 F3 **John o'Groats** N Scotland, United Kingdom 58°38′N 03°03′W
75 F10 **John Redmond Reservoir** ⊠ Kansas, C USA
83 I3 **John River** 〰 Alaska, USA
159 J1 **Johnshaven** United Kingdom 56°47′N 2°20′W
75 B10 **Johnson** Kansas, C USA 37°33′N 101°46′W
63 I4 **Johnson** Vermont, NE USA 44°39′N 72°40′W
64 A4 **Johnsonburg** Pennsylvania, NE USA 41°28′N 78°37′W
64 E3 **Johnson City** New York, NE USA 42°06′N 75°54′W
67 H5 **Johnson City** Tennessee, S USA 36°18′N 82°21′W
70 G6 **Johnson City** Texas, SW USA 30°17′N 98°27′W
80 E7 **Johnsondale** California, W USA 35°58′N 118°32′W
84 D4 **Johnson, Pico de** ▲ Massachusetts, NE USA 29°12′N 112°03′W
54 B4 **Johnsons Crossing** Yukon, W Canada 60°30′N 133°15′W
67 J8 **Johnsonville** South Carolina, SE USA 33°50′N 79°26′W
160 D5 **Johnston** United Kingdom 51°45′N 5°00′W
67 H8 **Johnston** South Carolina, SE USA 33°49′N 81°48′W
286 G4 **Johnston Atoll** ◇ *US unincorporated territory* C Pacific Ocean
267 K2 **Johnston Atoll** *atoll* C Pacific Ocean
73 C13 **Johnston City** Illinois, N USA 37°49′N 88°55′W
158 G3 **Johnstone** United Kingdom 55°50′N 4°30′W
274 E7 **Johnston, Lake** *salt lake* Western Australia
65 H1 **Johnstown** New York, USA 43°0′N 74°22′W
73 I10 **Johnstown** Ohio, N USA 40°08′N 82°39′W
64 A7 **Johnstown** Pennsylvania, NE USA 40°20′N 78°56′W
258 E4 **Johor** *var.* Johore. ◈ *state* Peninsular Malaysia
Johor Baharu *see* Johor Bahru
258 E4 **Johor Bahru** *var.* Johor Baharu, Johore Bahru. Johor, Peninsular Malaysia 01°29′N 103°44′E
Johore *see* Johor
Johore Bahru *see* Johor Bahru
190 F3 **Jõhvi** *Ger.* Jewe. Ida-Virumaa, NE Estonia 59°21′N 27°25′E
165 I4 **Joigny** Yonne, C France 47°58′N 03°24′E
Joinvile *see* Joinville
113 K5 **Joinville** *var.* Joinvile. Santa Catarina, S Brazil 26°20′S 48°55′W
165 J3 **Joinville** Haute-Marne, N France 48°26′N 05°07′E
292 E2 **Joinville Island** *island* Antarctica
86 F7 **Jojutla** *var.* Jojutla de Juárez. Morelos, S Mexico 18°38′N 99°10′W
Jojutla de Juárez *see* Jojutla
152 F6 **Jokkmokk** *Lapp.* Dálvvadis. Norrbotten, N Sweden 66°35′N 19°57′E
152 D2 **Jökuldalur** 〰 E Iceland
152 D2 **Jökulsá á Fjöllum** 〰 NE Iceland
Jokyakarta *see* Yogyakarta
73 D9 **Joliet** Illinois, N USA 41°33′N 88°05′W
63 H2 **Joliette** Québec, SE Canada 46°02′N 73°27′W
263 K9 **Jolo** Jolo Island, SW Philippines 06°02′N 121°00′E
80 B6 **Jolon** California, USA 35°58′N 121°11′W
154 B5 **Jølstervatnet** ◎ S Norway
259 I8 **Jombang** *prev.* Djombang. Jawa, S Indonesia 07°33′S 112°14′E
239 J8 **Jomda** Xizang Zizhiqu, W China 31°26′N 98°09′E
104 A2 **Jome, Punta de** *headland* W Ecuador 0°57′S 80°49′W
191 D8 **Jonava** *Ger.* Janow, *Pol.* Janów. Kaunas, C Lithuania 55°05′N 24°19′E
228 G6 **Jondor** *Rus.* Zhondor. Buxoro Viloyati, C Uzbekistan 39°46′N 64°11′E
244 A7 **Jonê** *var.* Liulin. Gansu, C China 34°36′N 103°39′E
75 J12 **Jonesboro** Arkansas, C USA 35°50′N 90°42′W
66 F9 **Jonesboro** Georgia, SE USA 33°31′N 84°21′W
73 C13 **Jonesboro** Illinois, N USA 37°25′N 89°19′W
68 C2 **Jonesboro** Louisiana, S USA 32°14′N 92°43′W
67 H6 **Jonesboro** Tennessee, S USA 36°17′N 82°28′W
63 L4 **Jonesport** Maine, NE USA 44°33′N 67°35′W
42 E3 **Jones Sound** *channel* Nunavut, N Canada
68 C2 **Jonesville** Louisiana, S USA 31°37′N 91°49′W
73 G9 **Jonesville** Michigan, N USA 41°58′N 84°39′W
67 I7 **Jonesville** South Carolina, SE USA 34°49′N 81°38′W
228 F5 **Jongeldi** *Rus.* Dzhankel'dy. Buxoro Viloyati, C Uzbekistan 40°50′N 63°16′E
136 B5 **Jonglei** Jonglei, E South Sudan 06°54′N 31°19′E
136 B5 **Jonglei** *var.* Gongoleh State. ◈ *state* E South Sudan
136 A5 **Jonglei Canal** *canal* E South Sudan
190 D7 **Joniškėlis** Panevėžys, N Lithuania 56°02′N 24°10′E
190 D7 **Joniškis** *Ger.* Janischken. Šiaulai, N Lithuania 56°15′N 23°36′E
155 G10 **Jönköping** Jönköping, S Sweden 57°45′N 14°10′E
155 G10 **Jönköping** ◈ *county* S Sweden
59 H7 **Jonquière** Québec, SE Canada 48°25′N 71°16′W
87 K7 **Jonuta** Tabasco, SE Mexico 18°04′N 92°03′W
164 F6 **Jonzac** Charente-Maritime, W France 45°26′N 00°25′W
75 G11 **Joplin** Missouri, C USA 37°04′N 94°31′W
64 E9 **Joppatowne** Maryland, USA 39°27′N 76°21′W
77 K4 **Jordan** Montana, NW USA 47°18′N 106°54′W
216 C7 **Jordan** *off.* Hashemite Kingdom of Jordan, *Ar.* Al Mamlaka al Urduniya al Hashemīyah, Al Urdunn; *prev.* Transjordan. ◆ *monarchy* SW Asia
218 F7 **Jordan** *Ar.* Urdunn, *Heb.* HaYarden. 〰 SW Asia
Jordan Lake *see* B. Everett Jordan Reservoir
183 G9 **Jordanów** Małopolskie, S Poland 49°39′N 19°51′E
218 F4 **Jordanub Jannine** Lebanon 33°37′N 35°47′E
218 F3 **Jordanûnié** Lebanon 33°58′N 35°37′E
76 F8 **Jordan Valley** Oregon, NW USA 42°59′N 117°03′W
216 B6 **Jordan Valley** *valley* N Israel
104 D7 **Jorge Chávez Internacional** *var.* Lima. ✈ (Lima) Lima, W Peru 12°07′S 77°01′W
184 F10 **Jorgucat** *var.* Jergucati, Jorgucati. Gjirokastër, S Albania 39°57′N 20°14′E
Jorgucati *see* Jorgucat
233 M6 **Jorhāt** Assam, NE India 26°45′N 94°09′E
180 H3 **Jork** Niedersachsen, Germany 53°32′E 9°41′N
154 J1 **Jörn** Västerbotten, N Sweden 65°03′N 20°04′E
79 L9 **Jornada Del Muerto** *valley* New Mexico, SW USA
153 I8 **Joroinen** Etelä-Savo, E Finland 62°11′N 27°50′E
155 B8 **Jørpeland** Rogaland, S Norway 59°01′N 06°04′E
133 K6 **Jos** Plateau, C Nigeria 09°59′N 08°57′E
263 M10 **Jose Abad Santos** *var.* Trinidad. Mindanao, S Philippines 05°51′N 125°35′E
115 I7 **José Batlle y Ordóñez** *var.* Batlle y Ordóñez. Florida, C Uruguay 33°28′S 55°08′W
116 I4 **José Batlle y Ordóñez** Lavalleja, Uruguay 33°30′S 54°54′W
117 C9 **José de San Martín** Chubut, S Argentina 44°04′S 70°29′W
116 H4 **José Enrique Rodó** Soriano , Uruguay 33°48′S 57°30′W
114 F8 **José Enrique Rodó** *var.* Rodó, José E.Rodo; *prev.* Drabble, Drable. Soriano, SW Uruguay 33°43′S 57°33′W
José E.Rodo *see* José Enrique Rodó
114 F10 **José Ferrari** Buenos Aires, Argentina 35°17′S 57°52′W
Josefsdorf *see* Žabalj
61 L10 **José Martí** ✈ (La Habana) Cuidad de La Habana, N Cuba 23°03′N 82°22′W
115 J7 **José Pedro Varela** *var.* José P.Varela. Lavalleja, S Uruguay 33°30′S 54°28′W
274 F2 **Joseph Bonaparte Gulf** *gulf* N Australia
79 I7 **Joseph City** Arizona, SW USA 34°56′N 110°18′W
59 I5 **Joseph, Lake** ◎ Newfoundland and Labrador, E Canada
62 C3 **Joseph, Lake** ◎ Ontario, S Canada
280 B2 **Josephstaal** Madang, N Papua New Guinea 04°42′S 144°55′E
José P.Varela *see* José Pedro Varela
107 L4 **José Rodrigues** Pará, N Brazil 05°45′S 51°20′W
232 F4 **Joshīmath** Uttarakhand, N India 30°33′N 79°35′E
70 G4 **Joshua** Texas, SW USA 32°27′N 97°23′W
81 G11 **Joshua Tree** California, W USA 34°07′N 116°18′W
133 K6 **Jos Plateau** *plateau* C Nigeria
164 E4 **Josselin** Morbihan, NW France 47°57′N 02°35′W
Jos Sudarso *see* Yos Sudarso, Pulau
154 C5 **Jostedalsbreen** *glacier* S Norway
154 C6 **Jotunheimen** ▲▲ S Norway
218 E5 **Jouaiya** Lebanon 33°14′N 35°19′E
218 F4 **Joub Jannine** Lebanon 33°37′N 35°47′E
218 F3 **Joûnié** *var.* Junīyah. W Lebanon 33°54′N 33°36′E
70 G8 **Jourdanton** Texas, SW USA 28°55′N 98°34′W
162 G4 **Joure** *Fris.* De Jouwer. Fryslân, N Netherlands 52°58′N 05°48′E
153 H9 **Joutsa** Keski-Suomi, C Finland 61°46′N 26°09′E
153 I9 **Joutseno** Etelä-Karjala, SE Finland 61°06′N 28°30′E
152 I5 **Joutsijärvi** Lappi, NE Finland 66°40′N 28°00′E
167 L9 **Joux, Lac de** ◎ W Switzerland
Jovakān *see* Jowkān
90 C3 **Jovellanos** Matanzas, W Cuba 22°49′N 81°11′W
233 L7 **Jowai** Meghālaya, NE India 25°25′N 92°21′E
Jōwat *see* Jabwot
Jowhar *see* Jawhar
222 F7 **Jowkān** *var.* Jovakān. Fārs, S Iran
Jowzam *see* Javazm
230 F3 **Jowzjān** ◈ *province* N Afghanistan
169 J4 **Joyeuse** Rhône-Alpes, France 44°29′N 4°14′E
Joypurhat *see* Jaipurhat
141 L5 **Jozini** KwaZulu-Natal, South Africa 27°26′S 32°04′E
Józseffalva *see* Žabalj
J.Storm Thurmond Reservoir *see* Clark Hill Lake
91 I8 **Juana Díaz** C Puerto Rico 18°03′N 66°30′W
85 J7 **Juan Aldama** Zacatecas, C Mexico 24°20′N 103°23′W
114 A9 **Juan B. Alberdi** Buenos Aires, Argentina 34°24′S 61°42′W
85 I8 **Juan B. Ceballos** Durango, Mexico
114 B7 **Juan Bernabé Molina** Santa Fe, Argentina 33°29′S 60°30′W
42 C5 **Juan de Fuca Plate** *tectonic feature*
76 B3 **Juan de Fuca, Strait of** *strait* Canada/USA
87 I7 **Juan Díaz Covarrubias** Veracruz-Llave, Mexico 18°10′N 95°11′W
Juan Fernandez Islands *see* Juan Fernández, Islas
287 K8 **Juan Fernández, Islas** *Eng.* Juan Fernandez Islands. *island group* W Chile
103 H2 **Juangriego** Nueva Esparta, NE Venezuela 11°06′N 63°59′W
114 D10 **Juan José Almeyra** Buenos Aires, Argentina 34°55′S 59°34′W
104 D5 **Juanjuí** *var.* Juanjuy. San Martín, N Peru 07°10′S 76°44′W
Juanjuy *see* Juanjuí
153 I8 **Juankoski** Pohjois-Savo, C Finland 63°01′N 28°24′E
Juan Lacaze *see* Juan L. Lacaze
114 F9 **Juan L. Lacaze** *var.* Juan Lacaze, Puerto Sauce; *prev.* Sauce. Colonia, SW Uruguay 34°26′S 57°25′W
114 F2 **Juan Pujol** Corrientes, Argentina 30°25′S 57°52′W
87 H7 **Juan Rodríguez Clara** Veracruz-Llave, Mexico 17°59′N 95°25′W
112 E4 **Juan Solá** Salta, N Argentina 23°30′S 62°42′W
114 G9 **Juan Soler** Uruguay 34°19′S 56°48′W
117 A11 **Juan Stuven, Isla** *island* S Chile
107 I7 **Juará** Mato Grosso, W Brazil 11°10′S 57°28′W
84 G3 **Juárez** Chihuahua, Mexico 30°19′N 108°05′W
85 L5 **Juárez** *var.* Villa Juárez. Coahuila, NE Mexico 27°39′N 100°43′W
86 B6 **Juárez** Colima, Mexico 19°10′N 103°55′W
84 B1 **Juárez, Sierra de** ▲▲ NW Mexico
110 G3 **Juatuba** Minas Gerais, Brazil 19°57′S 44°20′W
108 G6 **Juazeiro** *prev.* Joazeiro. Bahia, E Brazil 09°25′S 40°30′W
108 G5 **Juazeiro do Norte** Ceará, E Brazil 07°10′S 39°18′W
136 B6 **Juba** *var.* Jūbā. ● Central Equatoria, S South Sudan 04°50′N 31°35′E
136 F7 **Juba** *Amh.* Genalē Wenz, *It.* Guiba, *Som.* Ganaane, Webi Jubba. 〰 Ethiopia/Somalia
292 D2 **Jubany** *Argentinian research station* Antarctica 61°57′S 58°23′W
Jubayl *see* Jbaïl
136 G7 **Jubbada Dhexe** *off.* Gobolka Jubbada Dhexe. ◈ *region* SW Somalia
Jubbada Dhexe, Gobolka *see* Jubbada Dhexe
137 F8 **Jubbada Hoose** ◈ *region* SW Somalia
Jubba, Webi *see* Juba
Jubbulpore *see* Jabalpur
Jubeil *see* Jbaïl
276 B3 **Jubilee Lake** ◎ Western Australia, S Australia
114 E4 **Jubileo** Entre Ríos, Argentina 31°44′S 58°37′W
130 C5 **Juby, Cap** *headland* SW Morocco 27°58′N 12°56′W
171 H5 **Júcar** *var.* Jucar. 〰 C Spain
86 G9 **Juchatengo** Oaxaca, Mexico 16°21′N 97°06′W
181 B9 **Jüchen** Nordrhein-Westfalen, Germany 51°06′E 6°30′N
85 J10 **Juchipila** Zacatecas, C Mexico 21°25′N 103°06′W
87 I9 **Juchitán** *var.* Juchitán de Zaragosa. Oaxaca, SE Mexico 16°27′N 95°W
Juchitán de Zaragosa *see* Juchitán
86 B5 **Juchitlán** Jalisco, Mexico 20°05′N 104°07′W
181 I10 **Jüchsen** Thüringen, Germany 50°29′E 10°29′N
216 B7 **Judaea** *cultural region* Israel/West Bank
219 E9 **Judaean Hills** *Heb.* Haré Yehuda. *hill range* E Israel
218 G4 **Judaydah** Syria
218 G4 **Judaydah** *Fr.* Jdaïdé. Rīf Dimashq, W Syria 33°17′N 36°15′E
217 H7 **Judayyidat Hāmir** Al Anbār, S Iraq 31°50′N 41°50′E
177 K4 **Judenburg** Steiermark, C Austria 47°09′N 14°43′E
172 G5 **Judío** ▲ Castilla-La Manacha, Spain 38°35′N 4°31′W
77 J4 **Judith River** 〰 Montana, NW USA
71 K2 **Judsonia** Arkansas, C USA 35°16′N 91°38′W
129 N9 **Jufrah, Wādī al** *dry watercourse* NW Yemen
Jugar *see* Sêrxü
245 H7 **Jugongxian** Henan, E China 34°49′N 113°05′E
Jugoslavija *see* Serbia
245 L2 **Juhua Dao** *island* Liaoning, China
88 G6 **Juigalpa** Chontales, S Nicaragua 12°04′N 85°21′W
168 F2 **Juillac** Limousin, France 45°19′N 1°19′E
178 B5 **Juist** *island* NW Germany
Juisui *see* Ruisui
111 I6 **Juiz de Fora** Minas Gerais, SE Brazil 21°47′S 43°23′W
112 C4 **Jujuy** *off.* Provincia de Jujuy. ◈ *province* N Argentina
Jujuy *see* San Salvador de Jujuy
Jujuy, Provincia de *see* Jujuy
152 F5 **Jukkasjärvi** *Lapp.* Čohkkiras. Norrbotten, N Sweden 67°52′N 20°39′E
Jula *see* Gyula, Hungary
Jūlā *see* Jālū, Libya
79 N2 **Julesburg** Colorado, C USA 40°59′N 102°15′W
Julia Beterrae *see* Béziers
105 G9 **Juliaca** Puno, SE Peru 15°32′S 70°10′W
275 I4 **Julia Creek** Queensland, C Australia 20°40′S 141°49′E
81 F13 **Julian** California, W USA 33°04′N 116°36′W
162 E4 **Julianadorp** Noord-Holland, NW Netherlands 52°53′N 04°43′E
177 J5 **Julian Alps** *Ger.* Julische Alpen, *It.* Alpi Giulie, *Slvn.* Julijske Alpe. ▲▲ Italy/Slovenia
103 L6 **Juliana Top** ▲ C Suriname 03°39′N 56°36′W
Julianehåb *see* Qaqortoq
181 B9 **Jülich** Nordrhein-Westfalen, Germany 50°56′E 6°22′N
Julijske Alpe *see* Julian Alps
85 I4 **Julimes** Chihuahua, N Mexico 28°29′N 105°21′W
114 F10 **Julio Arditi** Buenos Aires, Argentina 35°08′S 57°39′W
Julio Briga *see* Bragança
Juliobriga *see* Logroño
113 I7 **Júlio de Castilhos** Rio Grande do Sul, S Brazil 29°14′S 53°42′W
Juliomagus *see* Angers
115 K7 **Julio María Sanz** Uruguay 33°09′S 54°09′W
Julische Alpen *see* Julian Alps
245 M1 **Juliuhe** Liaoning, NE China 42°03′N 122°55′E
Jullundur *see* Jalandhar
229 H6 **Juma** *Rus.* Dzhuma. Samarqand Viloyati, C Uzbekistan 39°43′N 66°37′E
245 J4 **Juma He** 〰 E China

◆ Country
● Country Capital
◇ Dependent Territory
○ Dependent Territory Capital
◈ Administrative Regions
✈ International Airport
▲ Mountain
▲▲ Mountain Range
🌋 Volcano
〰 River
◎ Lake
⊠ Reservoir

INDEX
355

◆ Country ◇ Dependent Territory ◉ Administrative Regions ▲ Mountain ⌖ Volcano ☉ Lake
● Country Capital ○ Dependent Territory Capital ✈ International Airport ▲ Mountain Range ≈ River ▨ Reservoir

◆ Country ● Country Capital ◇ Dependent Territory ○ Dependent Territory Capital ◆ Administrative Regions ✕ International Airport ▲ Mountain ▲ Mountain Range ▼ Volcano ♣ River ⊚ Lake ⊡ Reservoir

◆ Country	◇ Dependent Territory	♦ Administrative Regions	▲ Mountain	☰ Volcano	◎ Lake
● Country Capital	○ Dependent Territory Capital	✈ International Airport	▲ Mountain Range	♣ River	▣ Reservoir

◆ Country
● Country Capital
◇ Dependent Territory
○ Dependent Territory Capital
◆ Administrative Regions
✕ International Airport
▲ Mountain
▲ Mountain Range
☆ Volcano
ॐ River
⊚ Lake
⊡ Reservoir

Lampasas River - Laurens
INDEX
365

70 G6 **Lampasas River** ∽ Texas, SW USA
85 L5 **Lampazos** *var.* Lampazos de Naranjo. Nuevo León, NE Mexico 27°00′N 100°28′W
Lampazos de Naranjo *see* Lampazos
186 F6 **Lámpeia** Dytikí Elláda, S Greece 37°51′N 21°48′E
181 E13 **Lampertheim** Hessen, W Germany 49°36′N 08°28′E
160 E5 **Lampeter** SW Wales, United Kingdom 52°08′N 04°04′W
256 E6 **Lamphun** *var.* Lampun, Muang Lamphun. Lamphun, NW Thailand 18°36′N 99°02′E
55 J6 **Lamprey** Manitoba, C Canada 58°18′N 94°06′W
Lampun *see* Lamphun
258 F7 **Lampung** *off.* Propinsi Lampung. ◈ *province* SW Indonesia
Lampung, Propinsi *see* Lampung
258 F8 **Lampung, Teluk** *bay* Sumatera, SW Indonesia
196 F3 **Lamskoye** Lipetskaya Oblast', W Russian Federation 52°57′N 38°04′E
180 G2 **Lamstedt** Niedersachsen, Germany 53°38′E 9°06′N
137 F9 **Lamu** Lamu, SE Kenya 02°17′S 40°54′E
137 E9 **Lamu** ◈ *county* SE Kenya
89 H8 **La Muerte, Cerro** ▲ C Costa Rica 09°33′N 83°47′W
165 K7 **la Mure** Isère, E France 44°54′N 05°48′E
167 J10 **Lamure-sur-Azergues** Rhône-Alpes, France 46°04′N 4°30′E
79 L7 **Lamy** New Mexico, SW USA 35°27′N 105°52′W
191 F11 **Lan'** ∽ C Belarus
82 C2 **Lāna'i** *var.* Lanai. *island* Hawai'i, USA, C Pacific Ocean
82 C2 **Lāna'i City** *var.* Lanai City. Lanai, Hawaii, USA, C Pacific Ocean 20°49′N 156°55′W
Lanai City *see* Lāna'i City
168 D9 **Lanaja** Aragón, Spain 41°46′N 0°20′W
163 G10 **Lanaken** Limburg, NE Belgium 50°53′N 05°39′E
118 B9 **Lanalhue, Lago** ◎ Bío-Bío, Chile
263 M9 **Lanao, Lake** *var.* Lake Sultan Alonto. ◎ Mindanao, S Philippines
169 M5 **La Napoule** Provence-Alpes-Côte d'Azur, France 43°31′N 6°56′E
159 H4 **Lanark** S Scotland, United Kingdom 55°38′N 04°25′W
156 F7 **Lanark** *cultural region* C Scotland, United Kingdom
170 F5 **La Nava de Ricomalillo** Castilla-La Mancha, C Spain 39°40′N 04°59′W
Lanba *see* Shuicheng
257 D9 **Lanbi Kyun** *prev.* Sullivan Island. *island* Mergui Archipelago, S Myanmar (Burma)
Lancang Jiang *see* Mekong
159 I9 **Lancashire** *cultural region* N England, United Kingdom
62 G3 **Lancaster** Ontario, SE Canada 45°10′N 74°31′W
159 I8 **Lancaster** NW England, United Kingdom 54°03′N 02°48′W
81 E9 **Lancaster** California, W USA 34°42′N 118°08′W
66 F4 **Lancaster** Kentucky, S USA 37°35′N 84°34′W
75 H8 **Lancaster** Missouri, C USA 40°32′N 92°31′W
63 I4 **Lancaster** New Hampshire, NE USA 44°29′N 71°34′W
64 A1 **Lancaster** New York, NE USA 42°54′N 78°40′W
73 I11 **Lancaster** Ohio, N USA 39°42′N 82°36′W
64 E7 **Lancaster** Pennsylvania, NE USA 40°03′N 76°18′W
67 I7 **Lancaster** South Carolina, SE USA 34°43′N 80°47′W
71 H4 **Lancaster** Texas, SW USA 32°35′N 96°45′W
67 M4 **Lancaster** Virginia, NE USA 37°48′N 76°30′W
72 B7 **Lancaster** Wisconsin, N USA 42°52′N 90°43′W
53 I5 **Lancaster Sound** *sound* Nunavut, N Canada
159 K6 **Lanchester** United Kingdom 54°49′N 1°44′W
Lan-chou/Lan-chow/Lanchow *see* Lanzhou
175 G8 **Lanciano** Abruzzo, C Italy 42°13′N 14°23′E
245 L6 **Lancun** Shandong, E China 36°25′N 120°10′E
183 I8 **Łańcut** Podkarpackie, SE Poland 50°04′N 22°14′E
259 H5 **Landak, Sungai** ∽ Borneo, N Indonesia
Landao *see* Lantau Island
Landau *see* Landau an der Isar
Landau *see* Landau in der Pfalz
179 H12 **Landau an der Isar** *var.* Landau. Bayern, SE Germany 48°40′N 12°41′E
181 E14 **Landau in der Pfalz** *var.* Landau. Rheinland-Pfalz, SW Germany 49°12′N 08°07′E
Land Burgenland *see* Burgenland
179 F14 **Landeck** Tirol, W Austria 47°09′N 10°35′E
163 F10 **Landen** Vlaams Brabant, C Belgium 50°45′N 05°05′E
77 K8 **Lander** Wyoming, C USA 42°49′N 108°43′W
164 D3 **Landerneau** Finistère, NW France 48°27′N 04°16′W
81 G11 **Landers** California, USA 34°16′N 116°24′W
155 F11 **Landeryd** Halland, S Sweden 57°04′N 13°15′E
164 F8 **Landes** ◈ *department* SW France
180 G5 **Landesbergen** Niedersachsen, Germany 52°34′E 9°08′N
Landeshut/Landeshut in Schlesien *see* Kamienna Góra
171 I5 **Landete** Castilla-La Mancha, C Spain 39°54′N 01°22′W
166 A7 **Landévant** Bretagne, France 47°46′N 3°07′W
163 H10 **Landgraaf** Limburg, SE Netherlands 50°55′N 06°04′E
164 D3 **Landivisiau** Finistère, NW France 48°31′N 04°03′W
166 D6 **Landivy** Pays de la Loire, France 48°28′N 1°02′W
Land Kärten *see* Kärnten
Land of Enchantment *see* New Mexico
The Land of Opportunity *see* Arkansas
Land of Steady Habits *see* Connecticut
Land of the Midnight Sun *see* Alaska
176 F4 **Landquart** Graubünden, SE Switzerland 46°58′N 09°35′E
176 F4 **Landquart** ∽ Austria/Switzerland
67 H7 **Landrum** South Carolina, SE USA 35°10′N 82°11′W
Landsberg *see* Gorzów Wielkopolski, Lubuskie, Poland
Landsberg *see* Górowo Iławeckie, Warmińsko-Mazurskie, NE Poland
179 F13 **Landsberg am Lech** Bayern, S Germany 48°03′N 10°52′E
Landsberg an der Warthe *see* Gorzów Wielkopolski
52 E4 **Lands End** *headland* Northwest Territories, N Canada
160 C10 **Land's End** *headland* SW England, United Kingdom 50°02′N 05°41′W
179 G12 **Landshut** Bayern, SE Germany 48°32′N 12°09′E
Landskron *see* Lanškroun
155 F12 **Landskrona** Skåne, S Sweden 55°52′N 12°52′E
162 F6 **Landsmeer** Noord-Holland, C Netherlands 52°26′N 04°55′E
155 F10 **Landvetter** ✈ (Göteborg) Västra Götaland, S Sweden 57°39′N 12°22′E
Landwarów *see* Lentvaris
158 B9 **Lanesborough** Longford, Ireland 53°40′N 7°59′W
166 A7 **Lanester** Bretagne, France 47°45′N 3°21′W
69 I1 **Lanett** Alabama, S USA 32°52′N 85°11′W
166 F5 **La Neuve-Lyre** Haute-Normandie, France 48°54′N 0°45′E
176 C4 **La Neuveville** *var.* Neuveville, *Ger.* Neuenstadt. Neuchâtel, W Switzerland 47°05′N 07°03′E
155 D11 **Langå** *var.* Langaa. Midtjylland, C Denmark 56°23′N 09°55′E
Langaa *see* Langå
238 D8 **La'nga Co** ◎ W China
Langada *see* Lagkáda
Langades/Langadhás *see* Lagkadás
Langádhia/ *see* Lagkádia
229 K8 **Langar** *Rus.* Lyangar. SE Tajikistan 37°04′N 72°39′E
229 H6 **Langar** *Rus.* Lyangar. Navoiy Viloyati, C Uzbekistan 40°27′N 65°54′E
222 E3 **Langarūd** Gīlān, NW Iran 37°10′N 50°09′E
55 I10 **Langbank** Saskatchewan, S Canada 50°01′N 102°16′W
Langchi *see* Yangshan
74 D2 **Langdon** North Dakota, N USA 48°45′N 98°22′W
169 I3 **Langeac** Haute-Loire, C France 45°06′N 03°31′E
164 G5 **Langeais** Indre-et-Loire, C France 47°22′N 00°27′E
140 D9 **Langebaan** Western Cape, South Africa 33°06′S 18°02′E
140 F6 **Langeberg** ▲▲ Northern Cape, South Africa
140 E10 **Langeberg** ▲▲ Western Cape, South Africa
129 K8 **Langeb, Wadi** ∽ NE Sudan
Långed *see* Dals Långed
155 E13 **Langeland** *island* S Denmark
180 I7 **Langelsheim** Niedersachsen, Germany 51°56′E 10°20′N
163 B10 **Langemark** West-Vlaanderen, W Belgium 50°55′N 02°55′E
181 F12 **Langen** Hessen, W Germany 49°58′N 08°40′E
180 F2 **Langen** Niedersachsen, Germany 53°37′E 8°36′N
179 E12 **Langenau** Baden-Württemberg, S Germany 48°30′N 10°08′E
181 C8 **Langenberg** Nordrhein-Westfalen, Germany 51°21′E 7°08′N
55 I9 **Langenburg** Saskatchewan, S Canada 50°50′N 101°43′W
181 H14 **Langenburg** Baden-Württemberg, Germany 49°15′E 9°51′N
180 G5 **Langendamm** Niedersachsen, Germany 52°37′E 9°15′N
176 G4 **Längenfeld** Tirol, W Austria 47°04′N 10°59′E
181 C9 **Langenfeld** Nordrhein-Westfalen, W Germany 51°06′N 06°57′E
180 H5 **Langenhagen** Niedersachsen, N Germany 52°26′N 09°45′E
180 H5 **Langenhagen** ✈ (Hannover) Niedersachsen, NW Germany 52°28′N 09°40′E
180 F6 **Langenholzhausen** Nordrhein-Westfalen, Germany 52°09′E 8°58′N
177 K2 **Langenlois** Niederösterreich, NE Austria 48°29′N 15°42′E
176 D4 **Langenthal** Bern, NW Switzerland 47°13′N 07°48′E
180 D2 **Langeoog** Niedersachsen, Germany 53°45′E 7°29′N
180 D2 **Langeoog** *island* NW Germany
192 F6 **Langepas** Khanty-Mansiyskiy Avtonomnyy Okrug-Yugra, C Russian Federation 61°12′N 75°24′E
155 D12 **Langeskov** Syddtjylland, C Denmark 55°22′N 10°36′E
155 D8 **Langesund** Telemark, S Norway 59°00′N 09°43′E
155 D8 **Langesundsfjorden** *fjord* S Norway
154 B4 **Langevågen** Møre og Romsdal, S Norway 62°26′N 06°15′E
245 J3 **Langfang** Hebei, E China 39°30′N 116°39′E
154 C4 **Langfjorden** *fjord* S Norway
74 E4 **Langford** South Dakota, N USA 45°36′N 97°48′W
180 E5 **Langförden** Niedersachsen, Germany 52°47′E 8°16′N
258 D4 **Langgapayung** Sumatera, W Indonesia 01°42′N 99°57′E
176 F8 **Langhirano** Emilia-Romagna, C Italy 44°37′N 10°16′E
159 I5 **Langholm** S Scotland, United Kingdom 55°14′N 03°11′W
152 B2 **Langjökull** *glacier* C Iceland
258 D2 **Langkawi, Pulau** *island* Peninsular Malaysia
260 E6 **Langkesi, Kepulauan** *island group* C Indonesia
257 D10 **Langkha Tuk, Khao** ▲ SW Thailand 09°19′N 98°39′E
140 E5 **Langklip** Northern Cape, South Africa 31°01′S 17°41′E
140 F5 **Langkloof Pass** Northern Cape, South Africa
56 D9 **Langley** British Columbia, SW Canada 49°07′N 122°39′W
161 K5 **Langley** United Kingdom 51°59′N 0°06′E
256 H5 **Lang Mô** Thanh Hoa, N Vietnam 19°36′N 105°30′E
Langnau *see* Langnau im Emmental
167 N9 **Langnau im Emmental** *var.* Langnau. Bern, W Switzerland 46°57′N 07°47′E
169 I3 **Langogne** Lozère, S France 44°40′N 03°52′E
164 G7 **Langon** Gironde, SW France 44°33′N 00°14′W
La Ngounié *see* Ngounié
152 E4 **Langøya** *island* C Norway
238 D8 **Langqên Zangbo** ∽ China/India
243 L6 **Langqi** Fujian, SE China 26°05′N 119°34′E
Langreo *see* Llangréu
165 J4 **Langres** Haute-Marne, N France 47°53′N 05°20′E
165 J4 **Langres, Plateau de** *plateau* C France
258 C3 **Langsa** Sumatera, W Indonesia 04°30′N 97°53′E
154 H3 **Långsele** Västernorrland, C Sweden 63°11′N 17°05′E
244 D1 **Lang Shan** ▲▲ N China
154 H7 **Långshyttan** Dalarna, C Sweden 60°26′N 16°02′E
256 H4 **Lang Sơn** *var.* Langson. Lang Sơn, N Vietnam 21°50′N 106°45′E
Langson *see* Lang Sơn
257 E10 **Lang Suan** Chumphon, SW Thailand 09°55′N 99°07′E
159 J7 **Langthwaite** United Kingdom 54°25′N 1°59′W
154 J1 **Långträsk** Norrbotten, N Sweden 65°22′N 20°19′E
70 D7 **Langtry** Texas, SW USA 29°46′N 101°25′W
165 I8 **Languedoc** *cultural region* S France
165 H9 **Languedoc-Roussillon** ◈ *region* S France
66 A6 **L'Anguille River** ∽ Arkansas, C USA
154 I3 **Långviksmon** Västernorrland, N Sweden 63°39′N 18°45′E
180 E3 **Langwarden** Niedersachsen, Germany 53°36′E 8°19′N
180 G4 **Langwedel** Niedersachsen, Germany 52°58′E 9°13′N
179 F12 **Langweid** Bayern, S Germany 48°29′N 10°50′E
243 L2 **Langxi** Anhui, E China 31°08′N 119°10′E
242 C1 **Langzhong** Sichuan, C China 31°46′N 105°55′E
Lan Hsü *see* Lan Yu
57 N6 **Lanigan** Saskatchewan, S Canada 51°50′N 105°01′W
188 E4 **Lanivtsi** Ternopil's'ka Oblast', W Ukraine 49°52′N 26°05′E
215 M6 **Länkäran** *Rus.* Lenkoran'. S Azerbaijan 38°46′N 48°51′E
164 G9 **Lannemezan** Hautes-Pyrénées, S France 43°08′N 00°22′E
164 E3 **Lannion** Côtes d'Armor, NW France 48°44′N 03°27′W
62 G2 **L'Annonciation** Québec, SE Canada 46°22′N 74°51′W
171 K3 **L'Anoia** ∽ NE Spain
86 A3 **La Noria** Sinaloa, Mexico 23°30′N 106°18′W
84 G3 **La Norteña** Chihuahua, Mexico 29°39′N 108°25′W
168 F2 **Lanouaille** Aquitaine, France 45°23′N 1°08′E
64 F7 **Lansdale** Pennsylvania, NE USA 40°14′N 75°13′W
62 F4 **Lansdowne** Ontario, SE Canada 44°25′N 76°00′W
232 F4 **Lansdowne** Uttarakhand, N India 29°50′N 78°42′E
58 D5 **Lansdowne** ◎ Ontario, C Canada
55 L8 **Lansdowne House** Ontario, Canada 52°11′N 87°52′W
263 H8 **Lansdowne Reef** *Chin.* Qiong Jiao, *Viet.* Đa Len Đao. *island* S Spratly Islands
72 D4 **L'Anse** Michigan, N USA 46°45′N 88°27′W
64 E6 **Lansford** Pennsylvania, NE USA 40°49′N 75°52′W
242 G7 **Lanshan** Hunan, S China 25°18′N 112°06′E
74 I6 **Lansing** Iowa, C USA 43°22′N 91°11′W
73 G8 **Lansing** *state capital* Michigan, N USA 42°44′N 84°33′W
152 G6 **Lansjärv** Norrbotten, N Sweden 66°39′N 22°10′E
183 D8 **Lanškroun** *Ger.* Landskron. Pardubický Kraj, E Czech Republic 49°55′N 16°38′E
169 M2 **Lanslebourg-Mont-Cenis** Rhône-Alpes, France 45°17′N 6°52′E
168 G5 **Lanta** Midi-Pyrénées, France 43°34′N 1°40′E
257 D11 **Lanta, Ko** *island* S Thailand
243 I9 **Lantang** Guangdong, SE China 23°25′N 114°57′E
243 H10 **Lantau Island** *Cant.* Tai Yue Shan, *Chin.* Landao. *island* Hong Kong, S China
166 G8 **Lanthenay** Centre, France 47°23′N 1°44′E
244 E8 **Lantian** Shaanxi, China 34°05′N 109°11′E
Lantian *see* Lianyuan
Lan-ts'ang Chiang *see* Mekong
Lantung, Gulf of *see* Liaodong Wan
260 D3 **Lanu** Sulawesi, N Indonesia 01°00′N 121°33′E
114 E9 **Lanús** Buenos Aires, Argentina 34°43′S 58°24′W
175 C10 **Lanusei** Sardegna, Italy, C Mediterranean Sea 39°55′N 09°31′E
85 N6 **La Nutria** Tamaulipas, Mexico
164 E4 **Lanvaux, Landes de** *physical region* NW France
166 B5 **Lanvollon** Bretagne, France 48°38′N 2°59′W
247 L4 **Lanxi** Heilongjiang, NE China 46°18′N 126°!5′E
243 L3 **Lanxi** Zhejiang, SE China 29°12′N 119°27′E
La Nyanga *see* Nyanga
244 G4 **Lanyi He** ∽ Shanxi, China
243 N10 **Lan Yu** *var.* Huoshao Tao, Hungt'ou, Lan Hsü, Lanyü, *Eng.* Orchid Island; *prev.* Kotosho, Koto Sho, Lan Yü. *island* SE Taiwan
Lanyü *see* Lan Yu
170 B9 **Lanzarote** *island* Islas Canarias, Spain, NE Atlantic Ocean
244 B6 **Lanzhou** *var.* Lan-chou, Lanchow, Lan-chow; *prev.* Kaolan. *province capital* Gansu, C China 36°01′N 103°52′E
176 C7 **Lanzo Torinese** Piemonte, NE Italy 45°18′N 07°26′E
263 K3 **Laoag** Luzon, N Philippines 18°11′N 120°34′E
263 M6 **Laoang** Samar, C Philippines 12°29′N 125°01′E
256 G4 **Lao Cai** Lao Cai, N Vietnam 22°29′N 104°00′E
85 J8 **La Ochoa** Durango, Mexico 24°01′N 103°55′W
Laodicea/Laodicea ad Mare *see* Al Lādhiqīyah
Laoet *see* Laut, Pulau
248 A1 **Laoha He** ∽ NE China
242 G1 **Laohekou** *var.* Guanghua. Hubei, C China 32°20′N 111°42′E
Laoi, An *see* Lee
244 G8 **Laojun Shan** ▲ Henan, China 33°27′N 111°23′E
248 C3 **Lao Ling** ▲▲ N China
170 B10 **La Oliva** *var.* Oliva. Fuerteventura, Islas Canarias, Spain, NE Atlantic Ocean 28°36′N 13°53′W
Lao, Loch *see* Belfast Lough
Laolong *see* Longchuan
Lao Mangnai *see* Mangnai
165 I2 **Laon** *var.* la Laon; *anc.* Laudunum. Aisne, N France 49°34′N 03°37′E
Lao People's Democratic Republic *see* Laos
102 G2 **La Orchila, Isla** *island* N Venezuela
114 B9 **La Oriental** Buenos Aires, Argentina 34°34′S 60°49′W
170 B10 **La Orotava** Tenerife, Islas Canarias, Spain, NE Atlantic Ocean 28°23′N 16°32′W
104 D7 **La Oroya** Junín, C Peru 11°36′S 75°54′W
256 G5 **Laos** *off.* Lao People's Democratic Republic. ◆ *republic* SE Asia
245 M6 **Laoshan Wan** *bay* E China
244 F7 **Laotongguan** Shaanxi, C China 34°36′N 110°17′E
248 E2 **Laoye Ling** ▲▲ NE China
113 K5 **Lapa** Paraná, S Brazil 25°46′S 49°44′W
173 J7 **La Paca** Murcia, Spain 37°52′N 1°50′W
167 I10 **La Pacaudiére** Rhône-Alpes, France 46°11′N 3°52′E
165 I6 **Lapalisse** Allier, C France 46°13′N 03°39′E
118 C5 **La Palma** Libertador General Bernardo O'Higgins, Chile 34°43′S 71°42′W
102 C5 **La Palma** Cundinamarca, C Colombia 05°23′N 74°24′W
88 D4 **La Palma** Chalatenango, N El Salvador 14°19′N 89°10′W
84 E4 **La Palma** Guerrero, S Mexico 28°17′N 110°47′W
89 M9 **La Palma** Darién, SE Panama 08°24′N 78°09′W
170 A9 **La Palma** *island* Islas Canarias, Spain, NE Atlantic Ocean
170 D8 **La Palma del Condado** Andalucía, S Spain 37°23′N 06°33′W
115 K9 **La Paloma** Durazno, C Uruguay 32°54′S 55°36′W
115 I6 **La Paloma** Rocha, E Uruguay 34°37′S 54°08′W
116 E5 **La Pampa** *off.* Provincia de La Pampa. ◈ *province* C Argentina
La Pampa, Provincia de *see* La Pampa
103 H4 **La Paragua** Bolívar, E Venezuela 06°53′N 63°16′W
84 E7 **La Partida, Isla** *island* Baja California Sur, NW Mexico
191 I10 **Lapatsichy** *Rus.* Lopatichi. Mahilyowskaya Voblasts', E Belarus 53°34′N 30°53′E
114 C3 **La Paz** Entre Ríos, E Argentina 30°45′S 59°36′W
118 G3 **La Paz** Mendoza, C Argentina 33°30′S 67°36′W
106 D10 **La Paz** *var.* La Paz de Ayacucho. ● (Bolivia-seat of government) La Paz, W Bolivia 16°30′S 68°13′W
88 E4 **La Paz** La Paz, SW Honduras 14°20′N 87°40′W
84 E8 **La Paz** Baja California Sur, NW Mexico 24°07′N 110°18′W
115 H9 **La Paz** Canelones, S Uruguay 34°46′S 56°13′W
105 H9 **La Paz** ◈ *department* W Bolivia
88 D5 **La Paz** ◈ *department* S El Salvador
88 E4 **La Paz** ◈ *department* SW Honduras
La Paz *see* El Alto, Bolivia
La Paz *see* Robles, Colombia
La Paz *see* La Paz Centro
84 F7 **La Paz, Bahía de** *bay* NW Mexico
88 G6 **La Paz Centro** *var.* La Paz. León, W Nicaragua 12°20′N 86°41′W
La Paz de Ayacucho *see* La Paz
102 E9 **La Pedrera** Amazonas, SE Colombia 01°19′S 69°31′W
115 K5 **La Pedrera** Cerro Largo, Uruguay 32°21′S 54°08′W
72 H7 **Lapeer** Michigan, N USA 43°03′N 83°19′W
114 A3 **La Pelada** Santa Fe, Argentina 30°52′S 60°59′W
114 B2 **La Penca** Santa Fe, Argentina 30°20′S 60°34′W
85 I4 **La Perla** Chihuahua, N Mexico 28°18′N 104°34′W
252 E1 **La Pérouse Strait** *Jap.* Sōya-kaikyō, *Rus.* Proliv Laperuza. *strait* Japan/Russian Federation
118 G9 **La Perra, Salitral de** *salt lake* C Argentina
Laperuza, Proliv *see* La Pérouse Strait
85 N7 **La Pesca** Tamaulipas, C Mexico 23°49′N 97°45′W
167 M5 **La Petite-Pierre** Alsace, France 48°52′N 7°19′E
160 F8 **Lapford** United Kingdom 50°52′N 3°48′W
114 B4 **La Picada** Entre Ríos, Argentina 31°44′S 60°18′W
86 D5 **La Piedad Cavadas** Michoacán, C Mexico 20°20′N 102°01′W
Lapines *see* Lafnitz
153 I8 **Lapinlahti** Pohjois-Savo, C Finland 63°21′N 27°25′E
Lápithos *see* Lapta
68 E4 **Laplace** Louisiana, S USA 30°04′N 90°28′W
114 B9 **Laplacette** Buenos Aires, Argentina 34°43′S 61°10′W
91 L2 **La Plaine** SE Dominica 15°20′N 61°15′W
137 I10 **La Plaine-des-Palmistes** C Réunion
152 G5 **Lapland** *Fin.* Lappi, *Swe.* Lappland. *cultural region* N Europe
Lapland *see* Lappi
74 C4 **La Plant** South Dakota, N USA 45°06′N 100°40′W
114 F10 **La Plata** Buenos Aires, E Argentina 34°56′S 57°55′W
102 B7 **La Plata** Huila, SW Colombia 02°33′N 75°55′W
67 L3 **La Plata** Maryland, NE USA 38°32′N 76°59′W
La Plata *see* Sucre
91 I8 **la Plata, Río de** ∽ C Puerto Rico
168 E4 **Laplume** Aquitaine, France 44°07′N 0°31′E
171 K3 **La Pobla de Lillet** Cataluña, NE Spain 42°15′N 01°57′E
171 J3 **La Pobla de Segur** Cataluña, NE Spain 42°15′N 00°58′E
63 J1 **La Pocatière** Québec, SE Canada 47°21′N 70°04′W
170 E2 **La Pola** *prev.* Pola de Lena. Asturias, N Spain 43°10′N 05°49′W
170 E2 **La Pola de Gordón** Castilla y León, N Spain 42°50′N 05°38′W
170 E1 **La Pola Siero** *prev.* Pola de Siero. Asturias, N Spain 43°24′N 05°39′W
80 E1 **La Porte** California, USA 39°41′N 120°59′W
73 E9 **La Porte** Indiana, N USA 41°36′N 86°43′W
64 D4 **Laporte** Pennsylvania, NE USA 41°25′N 76°29′W
73 A8 **La Porte City** Iowa, C USA 42°19′N 92°11′W
112 D6 **La Posta** Catamarca, C Argentina 27°59′S 65°32′W
84 F4 **La Poza** Sonora, Mexico 29°24′N 110°12′W
84 D6 **La Poza Grande** Baja California Sur, NW Mexico 25°50′N 112°00′W
153 H8 **Lappajärvi** Etelä-Pohjanmaa, W Finland 63°13′N 23°40′E
153 H8 **Lappajärvi** ◎ W Finland
153 I9 **Lappeenranta** *Swe.* Villmanstrand. Etelä-Karjala, SE Finland 61°04′N 28°15′E
153 G9 **Lappfjärd** *Fin.* Lapväärtti. Österbotten, W Finland 62°14′N 21°30′E
152 H5 **Lappi** *Swe.* Lappland, *Eng.* Lapland. ◈ *region* N Finland
Lappi/Lappland *see* Lapland
Lappo *see* Lapua
116 G5 **Laprida** Buenos Aires, E Argentina 37°34′S 60°45′W
70 F8 **La Pryor** Texas, SW USA 28°56′N 99°51′W
214 B5 **Lâpseki** Çanakkale, NW Turkey 40°22′N 26°42′E
218 B2 **Lapta** *Gk.* Lápithos. NW Cyprus 35°20′N 33°11′E
190 G5 **Laptevo** Pskovskaya Oblast', Russian Federation
Laptev Sea *see* Laptevykh, More
193 I3 **Laptevykh, More** *Eng.* Laptev Sea. *sea* Arctic Ocean
153 G8 **Lapua** *Swe.* Lappo. Etelä-Pohjanmaa, W Finland 62°57′N 23°E
168 A7 **La Puebla de Arganzón** Castilla y León, Spain 42°46′N 2°50′W
170 G2 **La Puebla de Arganzón** País Vasco, N Spain 42°45′N 02°49′W
170 E8 **La Puebla de Cazalla** Andalucía, S Spain 37°14′N 05°18′W
172 F7 **La Puebla de los Infantes** Andalucía, Spain 37°47′N 5°23′W
170 F5 **La Puebla de Montalbán** Castilla-La Mancha, C Spain 39°52′N 04°22′W
173 K3 **La Puebla de Valverde** Aragón, Spain 40°13′N 0°55′W
115 J4 **Lapuente** Rivera, Uruguay 31°25′S 54°55′W
118 D7 **La Puente, Río** ∽ Maule, Chile
84 F3 **La Puerta** Baja California Norte, NW Mexico 30°18′N 109°51′W
102 E3 **La Puerta** Trujillo, NW Venezuela 09°08′N 70°46′W
263 M7 **Lapu-Lapu** C Philippines 10°18′N 123°58′E
Lapurdum *see* Bayonne
84 D6 **La Purísima** Baja California Sur, NW Mexico 26°10′N 112°05′W
Lapväärtti *see* Lappfjärd
182 I5 **Łapy** Podlaskie, NE Poland 53°N 22°54′E
129 H7 **Laqiya Arba'in** Northern, NW Sudan 20°01′N 28°01′E
115 I9 **La Querencia** Canelones, Uruguay 34°38′S 55°37′W
112 D3 **La Quiaca** Jujuy, N Argentina 22°12′S 65°36′W
175 F8 **L'Aquila** *var.* Aquila, Aquila degli Abruzzi. Abruzzo, C Italy 42°21′N 13°24′E
81 D10 **La Quinta** California, USA 34°20′N 118°38′W
81 F12 **La Quinta** California, USA 33°40′N 116°19′W
222 G8 **Lār** Fārs, S Iran 27°42′N 54°19′E
102 E3 **Lara** *off.* Estado Lara. ◈ *state* NW Venezuela
173 I9 **La Ràbita** Andalucía, Spain 36°45′N 3°10′W
170 C1 **Laracha** Galicia, NW Spain 43°14′N 08°34′W
130 F2 **Larache** *var.* al Araïch, El Araïch, *prev.* El Araïche; *anc.* Lixus. NW Morocco 35°12′N 06°10′W
158 D9 **Laracor** Meath, Ireland 53°32′N 6°47′W
Lara, Estado *see* Lara
160 B2 **Laragh** Wicklow, Ireland 53°0′N 6°18′W
165 K8 **Laragne-Montéglin** Hautes-Alpes, SE France 44°21′N 05°46′E
170 F8 **La Rambla** Andalucía, S Spain 37°37′N 04°44′W
77 M9 **Laramie** Wyoming, C USA 41°18′N 105°35′W
77 L8 **Laramie Mountains** ▲▲ Wyoming, C USA
77 M8 **Laramie River** ∽ Wyoming, C USA
111 L3 **Laranja da Terra** Espírito Santo, Brazil 19°54′S 41°04′W
111 J5 **Laranjal** Minas Gerais, Brazil 21°22′S 42°28′W
110 B8 **Laranjal Paulista** São Paulo, Brazil 23°04′S 47°51′W
113 J5 **Laranjeiras do Sul** Paraná, S Brazil 25°23′S 52°23′W
Larantoeka *see* Larantuka
260 E8 **Larantuka** *prev.* Larantoeka. Flores, C Indonesia 08°20′S 123°00′E
118 B8 **Laraquete** Bío-Bío, Chile 37°10′S 73°11′W
261 I7 **Larat** Pulau Larat, E Indonesia 07°07′S 131°46′E
261 I7 **Larat, Pulau** *island* Kepulauan Tanimbar, E Indonesia
171 L9 **Larba** Algeria
169 J1 **L'Arbresle** Rhône-Alpes, France 45°50′N 4°37′E
155 J10 **Lärbro** Gotland, SE Sweden 57°46′N 18°49′E
169 M4 **Larche, Col de** *pass* France/Italy
58 F7 **Larder Lake** Ontario, S Canada 48°06′N 79°44′W
170 G1 **Laredo** Cantabria, N Spain 43°23′N 03°22′W
70 F9 **Laredo** Texas, SW USA 27°30′N 99°30′W
84 G7 **La Reforma** Sinaloa, W Mexico 25°05′N 108°03′W
162 H6 **Laren** Gelderland, E Netherlands 52°12′N 06°22′E
162 F6 **Laren** Noord-Holland, C Netherlands 52°15′N 05°13′E
164 G7 **la Réole** Gironde, SW France 44°34′N 00°00′W
La Réunion *see* Réunion
Largeau *see* Faya
169 J4 **Largentiére** Rhône-Alpes, France 44°32′N 4°18′E
176 C7 **l'Argentière-la-Bessée** Hautes-Alpes, SE France 44°49′N 06°34′E
230 F3 **Lar Gerd** *var.* Largird. Balkh, N Afghanistan 35°36′N 66°48′E
Largird *see* Lar Gerd
69 K7 **Largo** Florida, SE USA 27°55′N 82°47′W
79 K6 **Largo, Canon** *valley* New Mexico, SW USA
90 C4 **Largo, Cayo** *island* W Cuba
69 N9 **Largo, Key** *island* Florida Keys, Florida, SE USA
159 I2 **Largo Ward** United Kingdom 56°15′N 2°51′W
158 F3 **Largs** W Scotland, United Kingdom 55°48′N 04°50′W
114 A6 **Larguía** Santa Fe, Argentina 32°34′S 61°13′W
168 B6 **la Rhune** *var.* Larrún. ▲ France/Spain 43°19′N 01°36′W *see also* Larrún
la Rhune *see* Larrún
175 C13 **L'Ariana** Tunisia 36°54′N 10°11′E
114 C10 **La Rica** Buenos Aires, Argentina 34°59′S 59°52′W
la Riege *see* Ariège
74 E2 **Larimore** North Dakota, N USA 47°54′N 97°37′W
175 G8 **Larino** Molise, C Italy 41°46′N 14°50′E
Lario *see* Como, Lago di
116 D1 **La Rioja** La Rioja, NW Argentina 29°26′S 66°50′W
116 D2 **La Rioja** *off.* Provincia de La Rioja. ◈ *province* NW Argentina
171 H3 **La Rioja** ◈ *autonomous community* N Spain
La Rioja, Provincia de *see* La Rioja
186 G4 **Lárisa** *var.* Larissa. Thessalía, C Greece 39°38′N 22°27′E
Larissa *see* Lárisa
161 L4 **Lark** ∽ United Kingdom
230 G8 **Lārkāna** *var.* Larkhana. Sind, SE Pakistan 27°32′N 68°18′E
159 H4 **Larkhall** United Kingdom 55°44′N 3°58′W
Larkhana *see* Lārkāna
166 A7 **Larmor-Plage** Bretagne, France 47°42′N 3°23′W
Larnaca *see* Lárnaka
218 C2 **Lárnaka** *var.* Larnaca, Larnax. SE Cyprus 34°55′N 33°39′E
218 C2 **Lárnaka** ✈ SE Cyprus 34°52′N 33°38′E
Larnax *see* Lárnaka
158 E6 **Larne** *Ir.* Latharna. E Northern Ireland, United Kingdom 54°51′N 05°49′W
75 D10 **Larned** Kansas, C USA 38°12′N 99°05′W
170 E2 **La Robla** Castilla y León, N Spain 42°48′N 05°37′W
170 D6 **La Roca de la Sierra** Extremadura, W Spain 39°06′N 06°41′W
168 E2 **La Rochebeaucourt-et-Argentine** Aquitaine, France 45°29′N 0°22′E
166 B8 **La Roche-Bernard** Bretagne, France 47°31′N 2°18′W
168 G2 **La Roche-Canillac** Limousin, France 45°12′N 1°58′E
169 M3 **La Roche-de-Rame** Provence-Alpes-Côte d'Azur, France 44°45′N 6°35′E
163 G12 **La Roche-en-Ardenne** Luxembourg, SE Belgium 50°11′N 05°35′E
164 G6 **la Rochefoucauld** Charente, W France 45°43′N 00°23′E
164 F6 **la Rochelle** *anc.* Rupella. Charente-Maritime, W France 46°09′N 01°07′W
166 F9 **La Roche-Posay** Poitou-Charentes, France 46°47′N 0°49′E
164 F5 **la Roche-sur-Yon** *prev.* Bourbon Vendée, Napoléon-Vendée. Vendée, NW France 46°40′N 01°26′W
171 H6 **La Roda** Castilla-La Mancha, C Spain 39°13′N 02°10′W
170 F8 **La Roda de Andalucía** Andalucía, S Spain 37°12′N 04°45′W
91 I6 **La Romana** E Dominican Republic 18°25′N 69°00′W
57 M3 **La Ronge** Saskatchewan, C Canada 55°07′N 105°18′W
57 M3 **La Ronge, Lac** ◎ Saskatchewan, C Canada
168 G3 **Laroquebrou** Auvergne, France 44°58′N 2°12′E
169 L6 **La Roquebrussanne** Provence-Alpes-Côte d'Azur, France 43°20′N 5°59′E
68 E5 **Larose** Louisiana, S USA 29°34′N 90°22′W
85 K4 **La Rosita** Coahuila, Mexico 28°20′N 101°40′W
89 H5 **La Rosita** Región Autónoma Atlántico Norte, NE Nicaragua 13°55′N 84°23′W
115 J5 **Larrayos** Uruguay 32°10′S 55°01′W
114 A5 **Larrechea** Santa Fe, Argentina 31°56′S 61°03′W
274 G3 **Larrimah** Northern Territory, N Australia 15°30′S 133°12′E
114 D6 **Larroque** Entre Ríos, E Argentina 33°05′S 59°06′W
168 B6 **Larrún** *Fr.* la Rhune. ▲ France/Spain 43°18′N 01°35′W *see also* la Rhune
Larrún *see* la Rhune
293 L4 **Lars Christensen Coast** *physical region* Antarctica
83 I8 **Larsen Bay** Kodiak Island, Alaska, USA 57°32′N 153°58′W
292 E3 **Larsen Ice Shelf** *ice shelf* Antarctica
52 G7 **Larsen Sound** *sound* Nunavut, N Canada
La Rúa *see* A Rúa de Valdeorras
168 D6 **Laruns** Pyrénées-Atlantiques, SW France 43°00′N 00°25′W
155 D8 **Larvik** Vestfold, S Norway 59°04′N 10°02′E
192 F6 **Lar'yak** Khanty-Mansiyskiy Avtonomnyy Okrug-Yugra, C Russian Federation 61°09′N 80°01′E
La-sa *see* Lhasa
173 I7 **La Sagra** ▲ Andalucía, Spain 37°57′N 2°34′W
260 G5 **Lasahata** Pulau Seram, E Indonesia 02°52′S 128°27′E
Lasahau *see* Lasihao
79 J4 **La Sal** Utah, W USA 38°19′N 109°14′W
118 E8 **La Salada** Neuquén, Argentina 37°19′S 70°15′W
84 B2 **La Salina** Baja California Norte, NW Mexico
62 A6 **La Salle** Ontario, S Canada 42°13′N 83°05′W
73 C9 **La Salle** Illinois, N USA 41°19′N 89°06′W
173 I8 **Las Alpujarras** ▲▲ Andalucía, Spain
168 G4 **La Salvetat-Payralés** Midi-Pyrénées, France 44°13′N 2°12′E
91 I6 **Las Americas** ✈ (Santo Domingo) S Dominican Republic 18°24′N 69°38′W
84 E3 **La Sangre** Sonora, Mexico
79 N4 **Las Animas** Colorado, C USA 38°04′N 103°13′W
115 J5 **Las Arenas** Uruguay 32°01′S 55°04′W
176 C5 **La Sarine** *var.* Sarine. ∽ SW Switzerland
176 C5 **La Sarraz** Vaud, W Switzerland 46°40′N 06°32′E
58 F7 **La Sarre** Québec, SE Canada 48°49′N 79°12′W
102 F2 **Las Aves, Islas** *var.* Islas de Aves. *island group* N Venezuela
114 A5 **Las Bandurrias** Santa Fe, Argentina 32°11′S 61°29′W
84 F6 **Las Bocas** Sonora, Mexico 26°35′N 109°20′W
102 G4 **Las Bonitas** Bolívar, C Venezuela 07°50′N 65°40′W
170 E8 **Las Cabezas de San Juan** Andalucía, S Spain 36°59′N 05°56′W
114 B3 **Las Cañas** Santa Fe, Argentina 31°01′S 60°18′W
115 K8 **Lascano** Rocha, E Uruguay 33°40′S 54°12′W
105 I13 **Lascar, Volcán** 🌋 N Chile 23°22′S 67°33′W
112 C4 **Lascar, Volcn** 🌋 Chile 23°18′S 67°36′W
118 G1 **Las Casuarinas** San Juan, Argentina 31°49′S 68°20′W
118 G3 **Las Catitas** Mendoza, Argentina 33°18′S 68°02′W
85 I8 **Las Cebollas** Durango, Mexico 23°23′N 104°50′W
118 C4 **Las Chacras** Libertador General Bernardo O'Higgins, Chile 34°15′S 71°33′W
87 I7 **Las Choapas** *var.* Choapas. Veracruz-Llave, SE Mexico 17°51′N 94°00′W
84 F8 **Las Cruces** Baja California Sur, NW Mexico 24°12′N 110°05′W
81 B8 **Las Cruces** California, USA 34°30′N 120°14′W
79 L9 **Las Cruces** New Mexico, SW USA 32°19′N 106°49′W
114 D6 **Las Cuevas** Entre Ríos, Argentina 32°22′S 60°22′W
Lasdehnen *see* Krasnoznamensk
La Selle *see* Selle, Pic de la
116 B2 **La Serena** Coquimbo, C Chile 29°54′S 71°18′W
172 F5 **La Serena** *physical region* W Spain
171 K2 **La Seu d'Urgell** *var.* La Seu d'Urgell; *prev.* Seo de Urgel. Cataluña, NE Spain 42°22′N 01°27′E
La Seu d'Urgell *see* La Seu d'Urgell
165 K9 **la Seyne-sur-Mer** Var, SE France 43°07′N 05°53′E
116 H5 **Las Flores** Buenos Aires, E Argentina 36°03′S 59°08′W
112 B7 **Las Flores** San Juan, W Argentina 30°14′S 69°10′W
114 C4 **Las Garaas** Entre Rios, Argentina 31°24′S 59°42′W
84 F6 **Las Grullas** Sinaloa, Mexico 25°53′N 109°18′W
114 D5 **Las Guachas** Entre Ríos, Argentina 32°28′S 59°10′W
161 J7 **Lasham** United Kingdom 51°10′N 1°02′W
57 K5 **Lashburn** Saskatchewan, S Canada 53°09′N 109°37′W
118 F2 **Las Heras** Mendoza, W Argentina 32°48′S 68°50′W
256 D4 **Lashio** Shan State, E Myanmar (Burma) 22°58′N 97°48′E
230 E5 **Lashkar Gāh** *var.* Lash-Kar-Gar'. Helmand, S Afghanistan 31°35′N 64°21′E
Lash-Kar-Gar' *see* Lashkar Gāh
172 E2 **Las Hurdes** Extremadura, W Spain 40°27′N 06°03′W
260 D6 **Lasihao** *var.* Lasahau. Pulau Muna, C Indonesia 05°01′S 122°23′E
175 H11 **La Sila** ▲▲ SW Italy
117 C13 **La Silueta, Cerro** ▲ S Chile 52°22′S 72°09′W
88 G5 **La Sirena** Región Autónoma Atlántico Sur, E Nicaragua 12°59′N 84°35′W
182 F7 **Łask** Lódzkie, C Poland 51°36′N 19°06′E
177 K5 **Laško** *Ger.* Tüffer. C Slovenia 46°08′N 15°13′E
118 E10 **Las Lajas** Neuquén, W Argentina 38°31′S 70°22′W
118 D10 **Las Lajas, Cerro** ▲ W Argentina 38°49′S 70°42′W
112 F4 **Las Lomitas** Formosa, N Argentina 24°45′S 60°35′W
118 G5 **Las Malvinas** Mendoza, Argentina 34°50′S 68°15′W
86 B2 **Las Margaritas** Chiapas, SE Mexico 16°15′N 91°58′W
114 D10 **Las Marianas** Buenos Aires, Argentina 35°04′S 59°31′W
Las Marismas *see* Guadalquivir, Marismas del
118 E4 **Las Melosas Refugio** Región Metropolitana, Chile 33°51′S 70°14′W
114 D7 **Las Mercedes** Entre Ríos, Argentina 33°10′S 58°54′W
102 G3 **Las Mercedes** Guárico, N Venezuela 09°08′N 66°27′W
88 D4 **Las Minas, Cerro** ▲ W Honduras 14°33′N 88°41′W
114 D2 **Las Mulitas** Entre Ríos, Argentina 30°22′S 58°59′W
172 F7 **Las Navas de la Concepción** Andalucía, Spain 37°56′N 5°28′W
118 D5 **Las Nieves** Libertador General Bernardo O'Higgins, Chile 34°33′S 70°42′W
85 I6 **Las Nieves** Durango, Mexico 26°24′N 105°22′W
170 G6 **La Solana** Castilla-La Mancha, C Spain 38°56′N 03°14′W
84 D7 **La Soledad, Boca** *coastal sea feature* Baja California Sur, NW Mexico
91 M3 **La Soufrière** ▲ Saint Vincent, Saint Vincent and the Grenadines 13°20′N 61°11′W
165 H6 **la Souterraine** Creuse, C France 46°15′N 01°28′E
112 G6 **Las Palmas** Chaco, N Argentina 27°08′S 58°45′W
89 J10 **Las Palmas** Veraguas, W Panama 08°09′N 81°28′W
170 B10 **Las Palmas** *var.* Las Palmas de Gran Canaria. Gran Canaria, Islas Canarias, Spain, NE Atlantic Ocean 28°08′N 15°27′W
170 C10 **Las Palmas** ◈ *province* Islas Canarias, Spain, NE Atlantic Ocean
170 B10 **Las Palmas** ✈ Gran Canaria, Islas Canarias, Spain, NE Atlantic Ocean
84 F8 **Las Palmas, Bahía de** *bay* Baja California Sur, NW Mexico
Las Palmas de Gran Canaria *see* Las Palmas
114 A2 **Las Palmeras** Santa Fe, Argentina 30°30′S 61°54′W
84 D5 **Las Palomas** Baja California Norte, W Mexico 31°44′N 107°37′W
84 G2 **Las Palomas** Chihuahua, Mexico 31°44′N 107°37′W
85 I5 **Las Parras** Baja California Sur, NW Mexico
118 C4 **Las Pataguas** Libertador General Bernardo O'Higgins, Chile 34°25′S 71°38′W
171 H6 **Las Pedroñeras** Castilla-La Mancha, C Spain 39°27′N 02°41′W
176 F9 **La Spezia** Liguria, NW Italy 44°08′N 09°50′E
115 H9 **Las Piedras** Canelones, S Uruguay 34°42′S 56°14′W
117 D8 **Las Plumas** Chubut, S Argentina 43°46′S 67°15′W
118 B7 **Las Raíces** Bío-Bío, Chile 36°18′S 72°48′W
114 A6 **Las Rosas** Santa Fe, C Argentina 32°27′S 61°30′W
87 K9 **Las Rosas** Chiapas, Mexico 16°24′N 92°23′W
173 H2 **Las Rozas de Madrid** Madrid, Spain 40°29′N 3°52′W
Lassa *see* Lhasa
180 J2 **Lassahn** Mecklenburg-Vorpommern, Germany 53°36′E 10°57′N
166 D6 **Lassay** Pays de la Loire, France 48°26′N 0°30′W
78 C2 **Lassen Peak** ▲ California, W USA 40°27′N 121°28′W
167 H4 **Lassigny** Picardie, France 49°35′N 2°51′E
292 F4 **Lassiter Coast** *physical region* Antarctica
177 K4 **Lassnitz** ∽ SE Austria
63 H3 **L'Assomption** Québec, SE Canada 45°48′N 73°27′W
63 H2 **L'Assomption** ∽ Québec, SE Canada
89 K10 **Las Tablas** Los Santos, S Panama 07°45′N 80°17′W
Lastarria, Volcán *see* Azufre, Volcán
79 M3 **Last Chance** Colorado, C USA 39°41′N 103°34′W
Last Frontier, The *see* Alaska
159 L7 **Lastingham** United Kingdom 54°17′N 0°53′W
57 N7 **Last Mountain Lake** ◎ Saskatchewan, S Canada
116 C2 **Las Tórtolas, Cerro** ▲ W Argentina 29°57′S 69°49′W
112 F6 **Las Toscas** Santa Fe, C Argentina 28°22′S 59°20′W
115 J5 **Las Toscas** Uruguay 32°09′S 55°03′W
135 B8 **Lastoursville** Ogooué-Lolo, E Gabon 0°50′S 12°43′E
184 C6 **Lastovo** *It.* Lagosta. *island* SW Croatia
184 C6 **Lastovski Kanal** *channel* SW Croatia
84 D5 **Las Tres Vírgenes, Volcán** 🌋 NW Mexico 27°27′N 112°34′W
84 E3 **Las Trincheras** Sonora, NW Mexico 30°21′N 111°27′W
103 H4 **Las Trincheras** Bolívar, E Venezuela 06°57′N 64°49′W
114 B4 **Las Tunas** Santa Fe, Argentina 31°34′S 60°59′W
90 E4 **Las Tunas** *var.* Victoria de las Tunas. Las Tunas, E Cuba 20°58′N 76°59′W
La Suisse *see* Switzerland
166 E7 **La Suze-sur-Sarthe** Pays de la Loire, France 47°54′N 0°01′E
118 C1 **Las Vacas** Coquimbo, Chile 31°50′S 71°22′W
84 G3 **Las Varas** Chihuahua, N Mexico 29°35′N 108°01′W
85 I10 **Las Varas** Nayarit, C Mexico 21°12′N 105°10′W
116 F3 **Las Varillas** Córdoba, E Argentina 31°54′S 62°45′W
81 J9 **Las Vegas** Nevada, W USA 36°09′N 115°10′W
79 M6 **Las Vegas** New Mexico, SW USA 35°35′N 105°15′W
281 I5 **Lata** Nendö, Solomon Islands 10°45′S 165°43′E
59 K6 **La Tabatière** Québec, E Canada 50°51′N 58°59′W
104 C2 **Latacunga** Cotopaxi, C Ecuador 0°58′S 78°36′W
292 E4 **Latady Island** *island* Antarctica
102 C8 **La Tagua** Putumayo, S Colombia 0°05′S 74°39′W
Latakia *see* Al Lādhiqīyah
152 G4 **Lätäseno** ∽ NW Finland
62 D1 **Latchford** Ontario, S Canada 47°20′N 79°45′W
62 E3 **Latchford Bridge** Ontario, SE Canada 45°16′N 77°29′W
284 D7 **Late** *island* Vava'u Group, N Tonga
233 H8 **Lātehār** Jhārkhand, N India 23°48′N 84°28′E
114 F6 **La Tentación** Paysandú, Uruguay 32°31′S 57°45′W
164 F7 **la Teste** Gironde, SW France 44°38′N 01°04′W
71 I5 **Latexo** Texas, SW USA 31°24′N 95°28′W
65 H1 **Latham** New York, NE USA 42°45′N 73°45′W
Latharna *see* Larne
180 D4 **Lathen** Niedersachsen, Germany 52°52′E 7°19′N
176 C4 **La Thielle** *var.* Thièle. ∽ W Switzerland
80 C3 **Lathrop** California, USA 37°49′N 121°17′W
75 G9 **Lathrop** Missouri, C USA 39°33′N 94°19′W
169 M1 **La Thuile** Valle D'Aosta, NW Italy 45°42′N 06°56′E
175 F9 **Latina** *prev.* Littoria. Lazio, C Italy 41°28′N 12°53′E
87 H7 **La Tinaja** Veracruz-Llave, S Mexico
177 I6 **Latisana** Friuli-Venezia Giulia, NE Italy 45°47′N 13°01′E
Latium *see* Lazio
187 J10 **Latò** *site of ancient city* Kríti, Greece, E Mediterranean Sea
80 D6 **Laton** California, USA 36°26′N 119°41′W
280 B9 **La Tontouta** ✈ (Noumea) Province Sud, S New Caledonia 22°06′S 166°12′E
102 G2 **La Tortuga, Isla** *var.* Isla Tortuga. *island* N Venezuela
169 H2 **La Tour-d'Auvergne** Auvergne, France 45°32′N 2°42′E
169 H7 **Latour-de-France** Languedoc-Roussillon, France 42°46′N 2°39′E
167 M10 **La Tour-de-Peilz** *var.* La Tour de Peilz. Vaud, SW Switzerland 46°28′N 06°52′E
La Tour de Peilz *see* La Tour-de-Peilz
165 J6 **la Tour-du-Pin** Isère, E France 45°34′N 05°25′E
168 C1 **la Tremblade** Charente-Maritime, W France 45°45′N 01°07′W
164 G5 **la Trimouille** Vienne, W France 46°27′N 01°02′E
114 B8 **La Trinidad** Buenos Aires, Argentina 34°12′S 61°12′W
88 G5 **La Trinidad** Estelí, NW Nicaragua 12°57′N 86°15′W
263 K4 **La Trinidad** Luzon, N Philippines 16°30′N 120°39′E
87 K9 **La Trinitaria** Chiapas, SE Mexico 16°02′N 92°00′W
91 N1 **la Trinité** E Martinique 14°44′N 60°58′W
166 B6 **La Trinité-Porhoët** Bretagne, France 48°06′N 2°33′W
62 D8 **Latrobe** Pennsylvania, NE USA 40°18′N 79°19′W
277 J7 **La Trobe River** ∽ Victoria, SE Australia
Lattakia/Lattaquié *see* Al Lādhiqīyah
260 G5 **Latu** Pulau Seram, E Indonesia 03°24′S 128°37′E
63 H1 **La Tuque** Québec, SE Canada 47°26′N 72°47′W
234 E5 **Lātūr** Mahārāshtra, C India 18°24′N 76°34′E
190 E4 **Latvia** *off.* Republic of Latvia, *Ger.* Lettland, *Latv.* Latvija, Latvijas Republika; *prev.* Latvian SSR, *Rus.* Latviyskaya SSR. ◆ *republic* NE Europe
Latvian SSR/Latvija/Latvijas Republika/Latviyskaya SSR *see* Latvia
Latvia, Republic of *see* Latvia
280 D3 **Lau** New Britain, E Papua New Guinea 05°46′S 151°21′E
181 F10 **Laubach** Hessen, Germany 50°33′E 8°59′N
267 J6 **Lau Basin** *undersea feature* S Pacific Ocean
169 I4 **Laubert** Languedoc-Roussillon, France 44°33′N 3°38′E
179 I8 **Lauchhammer** Brandenburg, E Germany 51°30′N 13°48′E
181 H13 **Lauda** Baden-Württemberg, Germany 49°34′E 9°42′N
159 I4 **Lauder** United Kingdom 55°42′N 2°45′W
170 G2 **Laudio** *var.* Llodio. País Vasco, N Spain 43°08′N 02°59′W
Laudunum *see* Laon
Laudus *see* St-Lô
180 H7 **Lauenberg** Niedersachsen, Germany 51°46′E 9°45′N
180 G4 **Lauenbrück** Niedersachsen, Germany 53°12′E 9°34′N
180 I3 **Lauenburg** Schleswig-Holstein, Germany 53°22′E 10°34′N
Lauenburg/Lauenburg in Pommern *see* Lębork
180 G6 **Lauenstein** Niedersachsen, Germany 52°05′E 9°33′N
181 J13 **Lauf** Bayern, Germany 49°28′E 11°10′N
179 F11 **Lauf an der Pegnitz** Bayern, SE Germany 49°31′N 11°16′E
176 D3 **Laufen** Basel, NW Switzerland 47°26′N 07°31′E
177 I3 **Lauffen** Salzburg, NW Austria 47°54′N 12°57′E
152 B2 **Laugarbakki** Norðurland Vestra, N Iceland 65°18′N 20°51′W
152 B3 **Laugarvatn** Suðurland, SW Iceland 64°09′N 20°43′W
72 E4 **Laughing Fish Point** *headland* Michigan, N USA 46°31′N 87°01′W
81 J11 **Laughlin** Nevada, USA 35°10′N 114°34′W
280 G9 **Lau Group** *island group* E Fiji
Lauis *see* Lugano
153 H8 **Laukaa** Keski-Suomi, C Finland 62°27′N 25°58′E
190 C7 **Laukuva** Tauragė, W Lithuania 55°37′N 22°12′E
Laun *see* Louny
277 J9 **Launceston** Tasmania, SE Australia 41°25′S 147°07′E
160 E8 **Launceston** *anc.* Dunheved. SW England, United Kingdom 50°38′N 04°21′W
102 B7 **La Unión** Nariño, SW Colombia 01°35′N 77°09′W
88 E5 **La Unión** La Unión, SE El Salvador 13°20′N 87°50′W
88 G4 **La Unión** Olancho, C Honduras 15°02′N 86°40′W
85 J7 **La Unión** Durango, Mexico 25°17′N 103°30′W
85 K7 **La Unión** Durango, Mexico 22°28′N 101°53′W
86 D7 **La Unión** Guerrero, S Mexico 17°58′N 101°49′W
87 M7 **La Unión** Quintana Roo, E Mexico 18°00′N 101°48′W
171 I8 **La Unión** Murcia, SE Spain 37°37′N 00°54′W
102 F4 **La Unión** Barinas, C Venezuela 08°15′N 67°46′W
88 E5 **La Unión** ◈ *department* E El Salvador
82 D3 **Laupāhoehoe** *var.* Laupahoehoe. Hawaii, USA, C Pacific Ocean 20°00′N 155°15′W
Laupahoehoe *see* Laupāhoehoe
179 E12 **Laupheim** Baden-Württemberg, S Germany 48°13′N 09°54′E
275 J3 **Laura** Queensland, NE Australia 15°37′S 144°34′E
283 M10 **Laura** *atoll* Majuro Atoll, SE Marshall Islands
Laurana *see* Lovran
102 F4 **La Urbana** Bolívar, C Venezuela 07°05′N 66°58′W
67 M3 **Laurel** Delaware, NE USA 38°33′N 75°34′W
69 L7 **Laurel** Florida, SE USA 27°07′N 82°27′W
64 D9 **Laurel** Maryland, NE USA 39°06′N 76°51′W
68 F2 **Laurel** Mississippi, S USA 31°41′N 89°10′W
77 J5 **Laurel** Montana, NW USA 45°40′N 108°46′W
74 E7 **Laurel** Nebraska, C USA 42°25′N 97°04′W
64 E7 **Laureldale** Pennsylvania, NE USA 40°24′N 75°52′W
85 K5 **Laureles** Coahuila, Mexico 27°20′N 102°02′W
114 F4 **Laureles** Salto, Uruguay 31°31′S 57°32′W
62 D9 **Laurel Hill** *ridge* Pennsylvania, NE USA
159 J1 **Laurencekirk** United Kingdom 56°49′N 2°27′W
158 B10 **Laurencetown** Galway, Ireland 53°14′N 8°11′W
74 G6 **Laurens** Iowa, C USA 42°51′N 94°51′W
67 H7 **Laurens** South Carolina, SE USA 34°29′N 82°01′W

◆ Country
● Country Capital
◇ Dependent Territory
○ Dependent Territory Capital
◈ Administrative Regions
✈ International Airport
▲ Mountain
▲▲ Mountain Range
🌋 Volcano
∽ River
◎ Lake
⊡ Reservoir

Country ◇ Dependent Territory ◆ Administrative Regions ▲ Mountain ◐ Volcano ◎ Lake
● Country Capital ○ Dependent Territory Capital ✈ International Airport ▲ Mountain Range ♔ River ◫ Reservoir

◆ Country   ◇ Dependent Territory   ◆ Administrative Regions   ▲ Mountain   ☆ Volcano   ● Lake
● Country Capital   ○ Dependent Territory Capital   ✕ International Airport   ▲ Mountain Range   ✍ River   ◉ Reservoir

M

Madre, Sierra - Maluu

79 K2 **Madre, Sierra** ▲▲ Colorado/Wyoming, C USA
170 G5 **Madrid** ● (Spain) Madrid, C Spain 40°25′N 03°43′W
74 G7 **Madrid** Iowa, C USA 41°52′N 93°49′W
170 G4 **Madrid** ◐ *autonomous community* C Spain
170 G6 **Madridejos** Castilla-La Mancha, C Spain 39°29′N 03°32′W
170 F4 **Madrigal de las Altas Torres** Castilla y León, N Spain 41°05′N 05°00′W
170 E6 **Madrigalejo** Extremadura, W Spain 39°08′N 05°36′W
78 A2 **Mad River** California, W USA
88 G5 **Madriz** ◐ *department* NW Nicaragua
173 H1 **Madrona** Castilla y León, Spain 40°54′N 4°10′W
170 E6 **Madroñera** Extremadura, W Spain 39°25′N 05°46′W
276 B4 **Madura** Western Australia 31°52′S 127°01′E
Madura *see* Madurai
235 E9 **Madurai** *prev.* Madura, Mathurai. Tamil Nādu, S India 09°55′N 78°07′E
259 J8 **Madura, Pulau** *prev.* Madoera. *island* C Indonesia
259 J8 **Madura, Selat** *strait* C Indonesia
197 J10 **Madzhalis** Respublika Dagestan, SW Russian Federation 42°12′N 47°46′E
185 K7 **Madzharovo** Haskovo, S Bulgaria 41°36′N 25°52′E
139 I1 **Madzimoyo** Eastern, E Zambia 13°42′S 32°34′E
253 B12 **Maebashi** *var.* Maebasi, Mayebashi. Gunma, Honshū, S Japan 36°24′N 139°02′E
Maebasi *see* Maebashi
256 E5 **Mae Chan** Chiang Rai, NW Thailand 20°13′N 99°52′E
256 D5 **Mae Hong Son** *var.* Maehongson, Muai To. Mae Hong Son, NW Thailand 19°16′N 97°56′E
Maehongson *see* Mae Hong Son
166 A6 **Maël-Carhaix** Bretagne, France 48°17′N 3°25′W
168 E10 **Maella** Aragón, Spain 41°08′N 0°09′E
Mae Nam Khong *see* Mekong
256 E5 **Mae Nam Nan** NW Thailand
256 E7 **Mae Nam Tha Chin** W Thailand
256 E5 **Mae Nam Yom** W Thailand
160 E2 **Maentwrog** United Kingdom 52°56′N 3°59′W
250 B1 **Maepo** *prev.* Maep'o. 37°02′N 128°17′E
Maep'o *see* Maepo
256 D6 **Mae Sariang** Mae Hong Son, NW Thailand 18°08′N 97°57′E
79 J2 **Maeser** Utah, W USA 40°28′N 109°35′W
Maeseyck *see* Maaseik
256 D6 **Mae Sot** *var.* Ban Mae Sot. Tak, W Thailand 16°44′N 98°32′E
160 F6 **Maesteg** United Kingdom 51°36′N 3°39′W
90 E5 **Maestra, Sierra** ▲▲ E Cuba
Maestricht *see* Maastricht
256 E5 **Mae Suai** *var.* Ban Mae Suai. Chiang Rai, NW Thailand 19°43′N 99°30′E
256 E5 **Mae Tho, Doi** ▲ NW Thailand 18°56′N 99°20′E
139 L6 **Maevatanana** Mahajanga, C Madagascar 16°57′S 46°50′E
281 M2 **Maéwo** *prev.* Aurora. *island* C Vanuatu
260 G4 **Mafa** Pulau Halmahera, E Indonesia 0°01′N 127°50′E
141 I6 **Mafeteng** W Lesotho 29°48′S 27°15′E
163 F11 **Maffe** Namur, SE Belgium 50°21′N 05°19′E
261 L5 **Maffin** Papua, E Indonesia 01°57′S 138°48′E
277 J7 **Maffra** Victoria, SE Australia 37°59′S 147°03′E
137 F11 **Mafia** *island* E Tanzania
137 E11 **Mafia Channel** *sea waterway* E Tanzania
141 H4 **Mafikeng** *off.* Mahikeng. North-West, N South Africa 25°53′S 25°39′E
113 K5 **Mafra** Santa Catarina, S Brazil 26°08′S 49°47′W
172 A5 **Mafra** Lisboa, C Portugal 38°57′N 09°19′W
221 J4 **Mafraq** Abū Ẓaby, C United Arab Emirates 24°21′N 54°33′E
Mafraq/Muḥāfaẓat al Mafraq *see* Al Mafraq
141 J7 **Mafube** Eastern Cape, South Africa 30°10′S 28°43′E
193 L5 **Magadan** Magadanskaya Oblast', E Russian Federation 59°38′N 150°50′E
193 L5 **Magadanskaya Oblast'** ◐ *province* E Russian Federation
176 E5 **Magadino** Ticino, S Switzerland 46°09′N 08°50′E
117 C13 **Magallanes** *var.* Región de Magallanes y de la Antártica Chilena. ◐ *region* S Chile
Magallanes *see* Punta Arenas
Magallanes, Estrecho de *see* Magellan, Strait of
Magallanes y de la Antártica Chilena, Región de *see* Magallanes
62 E2 **Maganasipi, Lac** ◎ Québec, SE Canada
102 C3 **Magangué** Bolívar, N Colombia 09°14′N 74°46′W
285 L8 **Magareva** *var.* Mangareva. *island* Îles Tuamotu, SE French Polynesia
133 K5 **Magaria** Zinder, S Niger 13°00′N 08°55′E
280 C4 **Magarida** Central, SW Papua New Guinea 10°10′S 149°21′E
263 L4 **Magat** Luzon, N Philippines
75 H12 **Magazine Mountain** ▲ Arkansas, C USA 35°10′N 93°38′W
132 C7 **Magburaka** C Sierra Leone 08°44′N 11°57′W
193 J8 **Magdagachi** Amurskaya Oblast', SE Russian Federation 53°25′N 125°41′E
114 F10 **Magdalena** Buenos Aires, E Argentina 35°05′S 57°30′W
106 E8 **Magdalena** El Beni, N Bolivia 13°22′S 64°07′W
85 H3 **Magdalena** Jalisco, W Mexico 30°33′N 106°31′W
84 E3 **Magdalena** Sonora, NW Mexico 30°38′N 110°59′W
79 L8 **Magdalena** New Mexico, SW USA 34°07′N 107°14′W
102 C3 **Magdalena** *off.* Departamento del Magdalena. ◐ *province* N Colombia
84 D7 **Magdalena, Bahía** *bay* W Mexico
Magdalena, Departamento del *see* Magdalena
117 B9 **Magdalena, Isla** *island* Archipiélago de los Chonos, S Chile
84 D7 **Magdalena, Isla** *island* NW Mexico
95 I3 **Magdalena, Río** C Colombia
84 E3 **Magdalena, Río** NW Mexico
Magdalen Islands *see* Madeleine, Îles de la
229 H8 **Magdanly** *Rus.* Govurdak; *prev.* gowurdak, Guardak. Lebap Welaýaty, E Turkmenistan 37°50′N 66°06′E
178 G7 **Magdeburg** Sachsen-Anhalt, C Germany 52°08′N 11°39′E
68 E2 **Magee** Mississippi, S USA 31°52′N 89°43′W
158 E6 **Magee, Island** *island* United Kingdom
259 H8 **Magelang** Jawa, C Indonesia 07°28′S 110°11′E
286 F5 **Magellan Rise** *undersea feature* C Pacific Ocean
286 D4 **Magellan Seamounts** *undersea feature* W Pacific Ocean
117 C13 **Magellan, Strait of** *Sp.* Estrecho de Magallanes. *strait* Argentina/Chile
219 C9 **Magen** Israel 31°17′N 34°25′E
176 E7 **Magenta** Lombardia, NW Italy 45°28′N 08°52′E
Magerøy *see* Magerøya
152 G3 **Magerøya** *var.* Magerøy, *Lapp.* Máhkarávju. *island* N Norway
250 C10 **Mage-shima** *island* Nansei-shotō, SW Japan
176 E5 **Maggia** Ticino, S Switzerland 46°15′N 08°42′E
176 E5 **Maggia** SW Switzerland
Maggiore, Lago *see* Maggiore, Lake
174 C4 **Maggiore, Lake** *It.* Lago Maggiore. ◎ Italy/Switzerland
90 E8 **Maggotty** W Jamaica 18°09′N 77°46′W
132 C4 **Maghama** Gorgol, S Mauritania 15°31′N 12°50′W
219 H10 **Maghār, Wādī** *dry watercourse* Jordan
218 G7 **Maghayyir as Sarḥān** Jordan 32°28′N 36°12′W
158 D6 **Maghera** *Ir.* Machaire Rátha. C Northern Ireland, United Kingdom 54°51′N 06°40′W
158 D6 **Magherafelt** *Ir.* Machaire Fíolta. C Northern Ireland, United Kingdom 54°45′N 06°36′W
158 D7 **Magheralin** United Kingdom 54°28′N 6°16′W
283 H2 **Magicienne Bay** *bay* Saipan, S Northern Mariana Islands
170 G8 **Magina** ▲ S Spain 37°43′N 03°24′W
137 C12 **Magingo** Ruvuma, S Tanzania 09°57′S 35°23′E
184 D4 **Maglaj** ◐ Federacija Bosne I Hercegovine, N Bosnia and Herzegovina
175 J10 **Maglie** Puglia, SE Italy 40°07′N 18°18′E
79 H2 **Magna** Utah, W USA 40°42′N 112°06′W
166 F10 **Magnac-Laval** Limousin, France 46°13′N 1°10′E
Magnesia *see* Manisa
62 C3 **Magnetawan** Ontario, S Canada
197 M3 **Magnitogorsk** Chelyabinskaya Oblast', C Russian Federation 53°28′N 59°06′E
75 H14 **Magnolia** Arkansas, C USA 33°17′N 93°16′W
68 E3 **Magnolia** Mississippi, S USA 31°08′N 90°27′W
71 I6 **Magnolia** Texas, SW USA 30°12′N 95°46′W
Magnolia State *see* Mississippi
154 F7 **Magnor** Hedmark, S Norway 59°57′N 12°14′E
166 G5 **Magny-en-Vexin** Val-d'Oise, France 49°09′N 1°47′E
280 G8 **Mago** *prev.* Mango. *island* Lau Group, E Fiji
139 H2 **Màgoé** Tete, NW Mozambique 15°50′S 31°42′E
63 I3 **Magog** Québec, SE Canada 45°16′N 72°09′W
141 K2 **Magoro** Limpopo, South Africa 23°18′S 30°19′E
138 G2 **Magoye** Southern, S Zambia 16°00′S 27°34′E
86 G4 **Magozal** Veracruz-Llave, C Mexico 21°33′N 97°57′W
59 J6 **Magpie** Québec, E Canada
171 N10 **Magra** Algeria
57 J9 **Magrath** Alberta, SW Canada 49°27′N 112°52′W
171 I6 **Magre** Valenciana, E Spain
173 N9 **Magroua, Cap** *cape* Algeria
80 G6 **Magruder Mountain** ▲ Nevada, USA 37°25′N 117°33′W
132 C3 **Magṭa' Laḥjar** *var.* Magta Lahjar, Magta' Lahjar, Magtá Lahjar. Brakna, SW Mauritania 17°27′N 13°07′W
228 C7 **Magtymguly** *prev.* Garrygala, *Rus.* Kara-Kala. Balkan Welaýaty, W Turkmenistan 38°27′N 56°15′E
242 A9 **Maguan** Yunnan, China 23°00′N 104°14′E
139 I6 **Magude** Maputo, S Mozambique 25°02′S 32°40′E
141 L5 **Magudu** KwaZulu-Natal, South Africa 27°32′S 31°39′E
158 C7 **Maguires Bridge** United Kingdom 54°18′N 7°28′W
133 M5 **Magumeri** Borno, NE Nigeria 12°07′N 12°48′E
282 F6 **Magur Islands** *island group* Caroline Islands, C Micronesia
55 J4 **Maguse Lake** ◎ Nunavut, E Canada
256 C5 **Magway** *var.* Magwe. Magway, W Myanmar (Burma) 20°08′N 94°55′E
256 C5 **Magway** *var.* Magwe. ◐ *region* C Myanmar (Burma)
Magwe *see* Magway
Magyar-Becse *see* Bečej
Magyarkanizsa *see* Kanjiža
Magyarország *see* Hungary
Magyarzsombor *see* Zimbor
222 D3 **Mahābād** *var.* Mehabad; *prev.* Sāūjbulāgh. Āẕarbāyjān-e Gharbī, NW Iran 36°44′N 45°44′E
288 E7 **Mahabiss Fracture Zone** *tectonic feature* N Indian Ocean
139 L8 **Mahabo** Toliara, W Madagascar 20°22′S 44°39′E
Maha Chai *see* Samut Sakhon
234 C5 **Mahād** Mahārāshtra, W India 18°04′N 73°21′E
136 H7 **Mahadday Weyne** Shabeellaha Dhexe, C Somalia 02°55′N 45°30′E
134 J6 **Mahagi** Orientale, NE Dem. Rep. Congo 02°16′N 30°59′E
Mahāil *see* Muhāyil
139 M6 **Mahajamba** *seasonal river* NW Madagascar
232 D5 **Mahājan** Rājasthān, NW India 28°47′N 73°50′E
139 L5 **Mahajanga** *var.* Majunga. Mahajanga, NW Madagascar 15°40′S 46°20′E
139 M5 **Mahajanga** ◐ *province* W Madagascar
139 L5 **Mahajanga** ✈ Mahajanga, NW Madagascar 15°40′S 46°20′E
259 K4 **Mahakam, Sungai** *var.* Koetai, Kutai. Borneo, C Indonesia
141 I2 **Mahalapye** *var.* Mahalatswe. Central, SE Botswana 23°02′S 26°53′E
Mahalatswe *see* Mahalapye
260 D5 **Mahalona** Sulawesi, C Indonesia 02°37′S 121°26′E
Mahameru *see* Semeru, Gunung
223 I7 **Māhān** Kermān, E Iran 30°00′N 57°00′E
234 I4 **Mahanādi** E India
234 G3 **Mahānadi** Chhattásgarh, India
139 M7 **Mahanoro** Toamasina, E Madagascar 19°53′S 48°48′E
64 E6 **Mahanoy City** Pennsylvania, USA 40°49′N 76°09′W
233 H6 **Mahārājganj** Bihār, N India 26°07′N 84°31′E
234 D4 **Mahārāshtra** ◐ *state* W India
139 L6 **Mahavavy** *seasonal river* N Madagascar
235 G10 **Mahaweli Ganga** C Sri Lanka
Mahbés *see* El Mahbas
234 F5 **Mahbūbābād** Telangana, E India 17°35′N 80°00′E
234 E6 **Mahbūbnagar** Telangana, C India 16°46′N 78°01′E
220 E4 **Mahd adh Dhahab** Al Madīnah, W Saudi Arabia 23°33′N 40°56′E
103 J5 **Mahdia** C Guyana 05°16′N 59°08′W
131 L2 **Mahdia** *var.* Al Mahdīyah, Mehdia. NE Tunisia 35°14′N 11°06′E
235 D8 **Mahe** *Fr.* Mahé; *prev.* Mayyali. Puducherry, SW India 11°41′N 75°31′E
137 I9 **Mahé** ✈ Mahé, NE Seychelles 04°37′S 55°27′E
137 I9 **Mahé** *island* Inner Islands, NE Seychelles
Mahé *see* Mahe
137 G11 **Mahebourg** SE Mauritius 20°24′S 57°42′E
232 E5 **Mahendragarh** *prev.* Mohendergarh. Haryāna, N India 28°17′N 76°14′E
231 M6 **Mahendranagar** Far Western, W Nepal 28°58′N 80°13′E
137 D11 **Mahenge** Morogoro, SE Tanzania 08°41′S 36°41′E
278 D11 **Maheno** Otago, South Island, New Zealand 45°10′S 170°51′E
232 C8 **Mahesāna** Gujarāt, W India 23°37′N 72°28′E
232 D8 **Maheshwar** Madhya Pradesh, C India 22°11′N 75°40′E
233 L9 **Maheskhali Island** *var.* Maiskhal Island. *island* SE Bangladesh
232 D8 **Mahi** N India
278 J6 **Mahia Peninsula** *peninsula* North Island, New Zealand
Mahikeng *see* Mafikeng
191 H9 **Mahilyow** *Rus.* Mogilëv. Mahilyowskaya Voblasts', E Belarus 53°55′N 30°23′E
191 I9 **Mahilyowskaya Voblasts'** *Rus.* Mogilëvskaya Oblast'. ◐ *province* E Belarus
285 M3 **Mahina** Tahiti, W French Polynesia 17°29′S 149°27′W
278 C12 **Mahinerangi, Lake** ◎ South Island, New Zealand
Máhkarávju *see* Magerøya
139 H7 **Mahlabatini** KwaZulu/Natal, E South Africa 28°15′S 31°28′E
256 C4 **Mahlaing** Mandalay, C Myanmar (Burma) 21°03′N 95°44′E
177 L4 **Mahldorf** Steiermark, SE Austria 46°54′N 15°55′E
Mahmūd-e 'Erāqī *see* Maḥmūd-e Rāqī
231 H3 **Maḥmūd-e Rāqī** *var.* Mahmūd-e 'Erāqī. Kāpīsā, NE Afghanistan 35°01′N 69°20′E
Mahmudiya *see* Al Maḥmūdīyah
187 M2 **Mahmutbey** İstanbul, NW Turkey 41°03′N 28°50′E
74 F3 **Mahnomen** Minnesota, N USA 47°19′N 95°58′W
232 F7 **Mahoba** Uttar Pradesh, N India 25°18′N 79°53′E
Mahón *see* Maó
137 E10 **Mahonda** Zanzibar North, E Tanzania 6°00′S 39°14′E
62 D8 **Mahoning Creek Lake** ⊠ Pennsylvania, NE USA
54 D2 **Mahony Lake** ◎ Northwest Territories, NW Canada
171 H6 **Mahora** Castilla-La Mancha, C Spain 39°13′N 01°44′W
Mähren *see* Moravia
Mährisch-Budwitz *see* Moravské Budějovice
Mährisch-Kromau *see* Moravský Krumlov
Mährisch-Neustadt *see* Uničov
Mährisch-Schönberg *see* Šumperk
Mährisch-Trübau *see* Moravská Třebová
Mährisch-Weisskirchen *see* Hranice
Māh-Shahr *see* Bandar-e Māhshahr
Mahuan Dao *see* Nanshan Island
135 I8 **Mahulu** Maniema, E Dem. Rep. Congo 01°04′S 27°10′E
232 C9 **Mahuva** Gujarāt, W India 21°06′N 71°46′E
141 J2 **Mahwelereng** Limpopo, South Africa 24°09′S 28°59′E
187 L1 **Mahya Dağı** ▲ NW Turkey 41°47′N 27°34′E
171 J4 **Maials** *var.* Mayals. Cataluña, NE Spain 41°22′N 00°30′E
284 B2 **Maiana** *prev.* Hall Island. *atoll* Tungaru, W Kiribati
285 I7 **Maiao** *var.* Tapuaemanu, Tubuai-Manu. *island* Îles du Vent, W French Polynesia
102 D2 **Maicao** La Guajira, N Colombia 11°23′N 72°16′W
Mai Ceu/Mai Chio *see* Maych'ew
165 K5 **Maîche** Doubs, E France 47°15′N 06°43′E
230 G3 **Maīdān Shahr** *var.* Maydān Shahr; *prev.* Meydān Shahr. Wardak, E Afghanistan 34°27′N 68°48′E
161 H7 **Maiden Bradley** United Kingdom 51°09′N 2°17′W
161 J6 **Maidenhead** S England, United Kingdom 51°32′N 00°44′W
158 F5 **Maidens** United Kingdom 55°20′N 4°49′W
57 L5 **Maidstone** Saskatchewan, S Canada 53°06′N 109°21′W
161 L7 **Maidstone** SE England, United Kingdom 51°17′N 00°31′E
133 M5 **Maiduguri** Borno, NE Nigeria 11°51′N 13°10′E
176 F4 **Maienfeld** Sankt Gallen, NE Switzerland 47°01′N 09°30′E
188 D8 **Măieruș** *Hung.* Szászmagyarós. Brașov, C Romania 45°55′N 25°30′E
Maigh Chromtha *see* Macroom
Maigh Eo *see* Mayo
167 H4 **Maignelay** Picardie, France 49°33′N 2°31′E
102 G5 **Maigualida, Sierra** ▲▲ S Venezuela
232 G7 **Maihar** Madhya Pradesh, C India 24°18′N 80°46′E
232 G9 **Maikala Range** ▲▲ C India
121 K6 **Maiko** W Dem. Rep. Congo
Mailand *see* Milano
232 G5 **Mailāni** Uttar Pradesh, N India 28°17′N 80°20′E
167 K10 **Maillat** Rhône-Alpes, France 46°08′N 5°33′E
167 J6 **Mailly-le-Camp** Champagne-Ardenne, France 48°40′N 4°13′E
231 I6 **Māilsi** Punjab, E Pakistan 29°46′N 72°15′E
227 I8 **Maimak** Talasskaya Oblast', NW Kyrgyzstan 42°40′N 71°12′E
Maimāna *see* Maīmanah
230 E2 **Maīmanah** *var.* Maimāna, Maymana; *prev.* Meymaneh. Fāryāb, NW Afghanistan 35°57′N 64°48′E
261 J5 **Maimawa** Papua Barat, E Indonesia 03°21′S 133°36′E
Maimuna *see* Al Maymūnah
219 G8 **Mā'īn** Mādabā, NW Jordan 31°40′N 35°44′E
181 E12 **Main** C Germany
186 G8 **Maïna** *ancient monument* Pelopónnisos, S Greece
186 F7 **Maínalo** ▲▲ S Greece
179 G12 **Mainburg** Bayern, SE Germany 48°40′N 11°48′E
Main Camp *see* Banana
62 B3 **Main Channel** *lake channel* Ontario, S Canada
135 E8 **Mai-Ndombe, Lac** *prev.* Lac Léopold II. ◎ W Dem. Rep. Congo
181 J13 **Main-Donau-Kanal** *canal* SE Germany
63 K2 **Maine** *off.* State of Maine, *also known as* Lumber State, Pine Tree State. ◐ *state* NE USA
164 G4 **Maine** *cultural region* NW France
164 G4 **Maine-et-Loire** ◐ *department* NW France
63 K5 **Maine, Gulf of** *gulf* NE USA
133 M5 **Maïné-Soroa** Diffa, SE Niger 13°14′N 12°00′E
256 D2 **Maingkwan** *var.* Mungkawn. Kachin State, N Myanmar (Burma) 26°20′N 96°37′E
181 G14 **Mainhardt** Baden-Württemberg, Germany 49°05′E 9°33′N
Main Island *see* Bermuda
Mainistir Fhear Maí *see* Fermoy
Mainistirna Búille *see* Boyle
Mainistir na Corann *see* Midleton
Mainistir na Féile *see* Abbeyfeale
156 F3 **Mainland** *island* N Scotland, United Kingdom
156 G1 **Mainland** *island* NE Scotland, United Kingdom
239 H9 **Mainling** *var.* Tungdor. Xizang Zizhiqu, W China 29°12′N 94°06′E
232 F6 **Mainpuri** Uttar Pradesh, N India 27°14′N 79°01′E
165 H3 **Maintenon** Eure-et-Loir, C France 48°35′N 01°34′E
139 K7 **Maintirano** Mahajanga, W Madagascar 18°01′S 44°03′E
152 I7 **Mainua** Kainuu, C Finland 64°05′N 27°28′E
181 E12 **Mainz** *Fr.* Mayence. Rheinland-Pfalz, SW Germany 50°00′N 08°16′E
132 B10 **Maio** *var.* Vila do Maio. Maio, S Cape Verde 15°07′N 23°12′W
132 B10 **Maio** *var.* Mayo. *island* Ilhas de Sotavento, SE Cape Verde
118 C3 **Maipo, Río** Chile
116 B4 **Maipo, Río** C Chile
118 E4 **Maipo, Volcán** W Argentina 34°09′S 69°51′W
112 B10 **Maipo, Volcn** Chile 34°12′S 69°48′W
116 H5 **Maipú** Buenos Aires, E Argentina 36°52′S 57°52′W
118 F3 **Maipú** Mendoza, E Argentina 33°00′S 68°46′W
116 B4 **Maipú** Santiago, C Chile 33°30′S 70°52′W
176 C8 **Maira** NW Italy
176 F5 **Maira** *It.* Mera. Italy/Switzerland
233 L6 **Mairābari** Assam, NE India 26°28′N 92°22′E
172 F8 **Mairena del Alcor** Andalucía, Spain 37°22′N 5°45′W
90 G5 **Maisí** Guantánamo, E Cuba 20°13′N 74°08′W
191 E8 **Maišiagala** Vilnius, SE Lithuania 54°52′N 25°03′E
Maiskhal Island *see* Maheskhali Island
257 E9 **Mai Sombun** Chumphon, SW Thailand 10°49′N 99°13′E
Mai Son *see* Hat Lot
167 H6 **Maisse** Île-de-France, France 48°23′N 02°22′E
Maisur *see* Mysore, India
Maisur *see* Karnātaka, India
118 C6 **Maitenes** Maule, Chile 35°37′S 71°23′W
277 L5 **Maitland** New South Wales, SE Australia 32°33′S 151°33′E
276 G5 **Maitland** South Australia 34°21′S 137°42′E
62 C4 **Maitland** Ontario, S Canada
293 I1 **Maitri** *Indian research station* Antarctica 70°03′S 08°59′E
243 H3 **Maiwang** Hubei, C China 30°31′N 113°36′E
239 H9 **Maizhokunggar** Xizang Zizhiqu, W China 29°50′N 91°40′E
89 I6 **Maíz, Islas del** *var.* Corn Islands. *island group* SE Nicaragua
251 H3 **Maizuru** Kyōto, Honshū, SW Japan 35°30′N 135°20′E
102 C3 **Majagual** Sucre, N Colombia 08°36′N 74°39′W
87 N6 **Majahual** Quintana Roo, E Mexico 18°43′N 87°43′W
Majardah, Wādī *see* Medjerda, Oued/Mejerda
218 E7 **Majdal Banī Fāḍel** West Bank 32°04′N 35°21′E
218 E5 **Majd el-Kurūm** Israel 32°55′N 35°16′E
111 I7 **Majé** Rio de Janeiro, Brazil 22°39′S 43°01′W
Mājeej *see* Mejit Island
260 C5 **Majene** *prev.* Madjene. Sulawesi, C Indonesia 03°33′S 118°59′E
89 L9 **Majé, Serranía de** ▲▲ E Panama
184 E4 **Majevica** ▲▲ NE Bosnia and Herzegovina
136 D5 **Majī** Southern Nationalities, S Ethiopia 06°11′N 35°32′E
242 D6 **Majiang** Guizhou, China 26°17′N 107°22′E
244 D5 **Majiatan** Ningxia, China 37°28′N 106°29′E
Majiazhou *see* Mashi
221 K4 **Majīs** NW Oman 24°25′N 56°34′E
Majorca *see* Mallorca
Mājro *see* Majuro Atoll
Majunga *see* Mahajanga
283 N10 **Majuro** ✈ Majuro Atoll, SE Marshall Islands 07°05′N 171°08′E
283 N10 **Majuro Atoll** *var.* Mājro. *atoll* Ratak Chain, SE Marshall Islands
283 M10 **Majuro Lagoon** *lagoon* Majuro Atoll, SE Marshall Islands
132 B4 **Maka** C Senegal 13°40′N 14°12′W
135 B9 **Makabana** Niari, SW Congo 03°28′S 12°36′E
82 B2 **Makahū'ena Point** *var.* Makahuena Point. *headland* Kaua'i, Hawai'i, USA 21°52′N 159°28′W
82 B2 **Makakilo City** O'ahu, Hawaii, USA, C Pacific Ocean 21°21′N 158°05′W
138 E4 **Makalamabedi** Central, C Botswana 20°19′S 23°51′E
Makale *see* Mek'elē
233 J6 **Makalu** *Chin.* Makaru Shan. ▲ China/Nepal 27°53′N 87°09′E
137 C11 **Makampi** Mbeya, S Tanzania 08°00′S 33°17′E
Makanchi *see* Makanshy
227 L5 **Makanshy** *prev.* Makanchi. Vostochnyy Kazakhstan, E Kazakhstan 46°47′N 82°00′E
89 H5 **Makantaka** Región Autónoma Atlántico Norte, NE Nicaragua 13°13′N 84°04′W
285 H10 **Makapu Point** *headland* W Niue 18°59′S 169°56′E
278 B12 **Makarewa** Southland, South Island, New Zealand 46°17′S 168°16′E
188 G3 **Makariv** Kyyivs'ka Oblast', N Ukraine 50°28′N 29°49′E
278 C10 **Makarora** South Island, New Zealand
193 M8 **Makarov** Ostrov Sakhalin, Sakhalinskaya Oblast', SE Russian Federation 48°24′N 142°37′E
295 H5 **Makarov Basin** *undersea feature* Arctic Ocean
286 D3 **Makarov Seamount** *undersea feature* W Pacific Ocean 29°30′N 153°30′E
184 C5 **Makarska** *It.* Macarsca. Split-Dalmacija, SE Croatia 43°18′N 17°00′E
Makaru Shan *see* Makalu
195 H9 **Makar'yev** Kostromskaya Oblast', NW Russian Federation 57°52′N 43°46′E
137 B11 **Makasa** Northern, NE Zambia 09°42′S 31°54′E
Makasar *see* Makassar
Makasar, Selat *see* Makassar Straits
260 C6 **Makassar** *var.* Macassar, Makasar; *prev.* Ujungpandang. Sulawesi, C Indonesia 05°09′S 119°28′E
286 B6 **Makassar Straits** *Ind.* Makasar Selat. *strait* C Indonesia
226 C5 **Makat** *Kaz.* Maqat. Atyrau, SW Kazakhstan 47°40′N 53°28′E
285 J6 **Makatea** *island* Îles Tuamotu, C French Polynesia
217 K4 **Makātū** Diyālá, E Iraq 33°55′N 45°25′E
139 L8 **Makay** *var.* Massif du Makay. ▲▲ SW Madagascar
Makay, Massif du *see* Makay
187 J2 **Makaza** *pass* Bulgaria/Greece
261 I4 **Makbon** Papua, E Indonesia 0°43′S 131°30′E
Makedonija *see* Macedonia, FYR
285 H9 **Makefu** W Niue 18°59′S 169°55′W
285 K6 **Makemo** *atoll* Îles Tuamotu, C French Polynesia
132 C6 **Makeni** C Sierra Leone 08°57′N 12°02′W
Makenzen *see* Orlyak
Makeyevka *see* Makiyivka
197 J9 **Makhachkala** *prev.* Petrovsk-Port. Respublika Dagestan, SW Russian Federation 42°58′N 47°30′E
Makhado *see* Louis Trichardt
226 C5 **Makhambet** Atyrau, W Kazakhstan 47°35′N 51°35′E
Makharadze *see* Ozurgeti
217 L9 **Makhfar al Buṣayyah** Al Muthanná, S Iraq 30°09′N 46°09′E
Makhmūr *see* Mexmûr
195 M8 **Makhnevo** Sverdlovskaya Oblast', Russian Federation
190 H5 **Makhnovka** Pskovskaya Oblast', Russian Federation
219 I10 **Makhrūq, Wadi al** *dry watercourse* E Jordan
219 D11 **Makhtésh Ramon** ▲▲ Israel
217 I3 **Makḥūl, Jabal** ▲▲ C Iraq
136 I1 **Makhyah, Wādī** *dry watercourse* N Yemen
261 J5 **Maki** Papua Barat, E Indonesia 03°00′S 134°10′E
260 F3 **Makian, Pulau** *island* Maluku, E Indonesia
278 D11 **Makikihi** Canterbury, South Island, New Zealand 44°36′S 171°09′E
284 B2 **Makin** *prev.* Pitt Island. *atoll* Tungaru, W Kiribati
137 E9 **Makindu** Makueni, S Kenya 02°15′S 37°49′E
227 I2 **Makinsk** Akmola, N Kazakhstan 52°40′N 70°28′E
Makira *see* Makira-Ulawa
Makira *see* San Cristobal
281 K4 **Makira-Ulawa** *prev.* Makira. ◐ *province* SE Solomon Islands
189 M5 **Makiyivka** *Rus.* Makeyevka; *prev.* Dmitriyevsk. Donets'ka Oblast', E Ukraine 47°57′N 37°47′E
220 E5 **Makkah** *Eng.* Mecca. Makkah, W Saudi Arabia 21°28′N 39°50′E
220 E5 **Makkah** *var.* Minṭaqat Makkah. ◐ *province* W Saudi Arabia
Makkah, Minṭaqat *see* Makkah
59 K4 **Makkovik** Newfoundland and Labrador, NE Canada 55°06′N 59°07′W
162 G4 **Makkum** Fryslân, N Netherlands 53°03′N 05°25′E
183 H13 **Makó** *Rom.* Macău. Csongrád, SE Hungary 46°14′N 20°28′E
Mako *see* Makung
62 C1 **Makobe Lake** ◎ Ontario, S Canada
280 E8 **Makogai** *island* C Fiji
134 B7 **Makokou** Ogooué-Ivindo, NE Gabon 0°38′N 12°47′E
137 C11 **Makongolosi** Mbeya, S Tanzania 08°24′S 33°09′E
58 C4 **Makoop Lake** ◎ Ontario, SE Canada
140 F3 **Makopong** Kgalagadi, Botswana 25°20′S 22°59′E
137 A8 **Makota** SW Uganda 0°37′S 30°12′E
134 D7 **Makoua** Cuvette, C Congo 0°01′S 15°40′E
171 M8 **Makouda** Algeria
182 H5 **Maków Mazowiecki** Mazowieckie, C Poland 52°51′N 21°06′E
183 G9 **Maków Podhalański** Małopolskie, S Poland 49°43′N 19°40′E
223 J9 **Makran** *cultural region* Iran/Pakistan
232 D6 **Makrāna** Rājasthān, N India 27°02′N 74°44′E
223 J10 **Makran Coast** *coastal region* SE Iran
191 D12 **Makrany** *Rus.* Mokrany. Brestskaya Voblasts', SW Belarus 51°50′N 24°15′E
Makrinoros *see* Makrynóros
187 H6 **Makrónisosi** *island* Kykládes, Greece, Aegean Sea
186 F5 **Makrynóros** *var.* Makrinoros. ▲▲ C Greece
187 H6 **Makryplági** ▲ C Greece 38°00′N 23°06′E
Maksamaa *see* Maxmo
Maksatha *see* Maksatikha
Maksaticha *see* Maksatikha
194 E9 **Maksatikha** *var.* Maksatha, Maksaticha. Tverskaya Oblast', W Russian Federation 57°49′N 35°46′E
218 G3 **Maksé** Lebanon 33°49′N 35°50′E
232 E8 **Maksi** Madhya Pradesh, C India 23°20′N 76°36′E
252 B1 **Maksumovka** Primorskiy Kray, SE Russian Federation 46°05′N 137°54′E
175 B14 **Maktar** Tunisia
222 C1 **Mākū** Āẕarbāyjān-e Gharbī, NW Iran 39°20′N 44°50′E
137 D8 **Makueni** ◐ *county* S Kenya
256 C2 **Mākum** Assam, NE India 27°28′N 95°28′E
Makun *see* Makung
243 M9 **Makung** *prev.* Mako, Makun. W Taiwan 23°35′N 119°35′E
250 C9 **Makurazaki** Kagoshima, Kyūshū, SW Japan 31°16′N 130°18′E
133 K7 **Makurdi** Benue, C Nigeria 07°42′N 08°36′E
227 H1 **Makushino** Kurganskaya Oblast', C Russian Federation 55°11′N 67°16′E
82 F9 **Makushin Volcano** ▲ Unalaska Island, Alaska, USA 53°53′N 166°55′W
141 H5 **Makwassie** North West, South Africa 27°19′S 25°59′E
139 H3 **Makwiro** Mashonaland West, N Zimbabwe 17°58′S 30°25′E
105 D8 **Mala** Lima, W Peru 12°40′S 76°36′W
154 I1 **Malå** Västerbotten, N Sweden 65°12′N 18°45′E
Mala *see* Mallow, Ireland
Mala *see* Malaita, Solomon Islands
263 M9 **Malabang** Mindanao, S Philippines
235 D8 **Malabār Coast** *coast* SW India
134 A6 **Malabo** *prev.* Santa Isabel. ● (Equatorial Guinea) Isla de Bioco, NW Equatorial Guinea 03°43′N 08°52′E
134 A6 **Malabo** ✈ Isla de Bioco, N Equatorial Guinea 03°44′N 08°51′E
114 G8 **Mal Abrigo** Uruguay 34°09′S 56°57′W
Malaca *see* Málaga
Malacca *see* Melaka
258 C2 **Malacca, Strait of** *Ind.* Selat Malaka. *strait* Indonesia/Malaysia
Malacka *see* Malacky
183 E10 **Malacky** *Hung.* Malacka. Bratislavský Kraj, W Slovakia 48°26′N 17°01′E
77 I8 **Malad City** Idaho, NW USA 42°10′N 112°16′W
189 I3 **Mala Divytsya** Chernihivs'ka Oblast', N Ukraine 50°40′N 32°13′E
191 F9 **Maladzyechna** *Pol.* Molodeczno, *Rus.* Molodechno. Minskaya Voblasts', C Belarus 54°19′N 26°51′E
284 C10 **Mala'etoli** Île Uvea, E Wallis and Futuna
102 D4 **Málaga** Santander, C Colombia 06°44′N 72°45′W
170 F9 **Málaga** *anc.* Malaca. Andalucía, S Spain 36°43′N 04°25′W
80 D6 **Malaga** California, USA 36°41′N 119°44′W
79 N9 **Malaga** New Mexico, SW USA 32°10′N 104°04′W
170 F8 **Málaga** ◐ *province* Andalucía, S Spain
170 F9 **Málaga** ✈ Andalucía, S Spain 36°38′N 04°36′W
Malagasy Republic *see* Madagascar
170 G6 **Malagón** Castilla-La Mancha, C Spain 39°10′N 03°51′W
158 E10 **Malahide** *Ir.* Mullach Íde. Dublin, E Ireland 53°27′N 06°09′W
281 K2 **Malaita** *off.* Malaita Province. ◐ *province* N Solomon Islands
281 J2 **Malaita** *var.* Mala. *island* N Solomon Islands
Malaita Province *see* Malaita
136 B4 **Malakal** Upper Nile, NE South Sudan 09°31′N 31°40′E
184 B3 **Mala Kapela** ▲▲ NW Croatia
71 H5 **Malakoff** Texas, SW USA 32°10′N 96°00′W
Malakula *see* Malekula
231 J5 **Malakwāl** *var.* Mālikwāla. Punjab, E Pakistan 32°32′N 73°18′E
280 C3 **Malalamai** Madang, W Papua New Guinea 05°49′S 146°44′E
275 K1 **Malalaua** Gulf, S Papua New Guinea 08°05′S 146°09′E
260 D5 **Malamala** Sulawesi, C Indonesia 03°21′S 120°58′E
259 J9 **Malang** Jawa, C Indonesia 07°59′S 112°45′E
137 D13 **Malanga** Niassa, N Mozambique 13°27′S 36°05′E
Malange *see* Malanje
152 F4 **Malangen** *sound* N Norway
135 D11 **Malanje** *var.* Malange. Malanje, NW Angola 09°34′S 16°25′E
135 D11 **Malanje** *var.* Malange. ◐ *province* N Angola
230 E9 **Malān, Rās** *cape* SW Pakistan
133 I5 **Malanville** NE Benin 11°50′N 03°23′E
Malapane *see* Ozimek
235 D8 **Malappuram** Kerala, SW India 11°00′N 76°02′E
89 K10 **Mala, Punta** *headland* S Panama 07°28′N 79°58′W
230 E9 **Malar** Baluchistān, SW Pakistan 26°19′N 64°55′E
155 H8 **Mälaren** ◎ C Sweden
112 B10 **Malargüe** Mendoza, W Argentina 35°32′S 69°35′W
58 F7 **Malartic** Québec, SE Canada 48°09′N 78°09′W
191 D12 **Malaryta** *Pol.* Maloryta, *Rus.* Malorita. Brestskaya Voblasts', SW Belarus 51°47′N 24°05′E
117 D9 **Malaspina** Chubut, SE Argentina 44°56′S 66°52′W
83 K6 **Malaspina Glacier** *glacier* Yukon, W Canada
83 K6 **Malaspina Glacier** *glacier* Alaska, USA
215 H7 **Malatya** *anc.* Melitene. Malatya, SE Turkey 38°22′N 38°18′E
215 H6 **Malatya** ◐ *province* C Turkey
169 K4 **Malaucéne** Provence-Alpes-Côte d'Azur, France 44°10′N 5°08′E
Malavate *see* Pédima
189 I5 **Mala Vyska** *Rus.* Malaya Viska. Kirovohrads'ka Oblast', S Ukraine 48°37′N 31°36′E
137 C13 **Malawi** *off.* Republic of Malaŵi; *prev.* Nyasaland, Nyasaland Protectorate. ◆ *republic* S Africa
Malawi, Lake *see* Nyasa, Lake
Malaŵi, Republic of *see* Malawi
153 G8 **Malax** *Fin.* Maalahti. Österbotten, W Finland 62°55′N 21°30′E
194 D9 **Malaya Vishera** Novgorodskaya Oblast', W Russian Federation 58°52′N 32°12′E
Malaya Viska *see* Mala Vyska
263 M8 **Malaybalay** Mindanao, S Philippines 08°10′N 125°08′E
222 E4 **Malāyer** *prev.* Daulatabad. Hamadān, W Iran 34°20′N 48°47′E
258 D2 **Malay Peninsula** *peninsula* Malaysia/Thailand
258 F2 **Malaysia** *off.* Malaysia, *var.* Federation of Malaysia; *prev.* the separate territories of Federation of Malaya, Sarawak and Sabah (North Borneo) and Singapore. ◆ *monarchy* SE Asia
Malaysia, Federation of *see* Malaysia
215 J6 **Malazgirt** Muş, E Turkey 39°09′N 42°30′E
133 J4 **Malbaza** Tahoua, S Niger 13°57′N 05°32′E
182 F4 **Malbork** *Ger.* Marienburg, Marienburg in Westpreussen. Pomorskie, N Poland 54°01′N 19°03′E
181 C12 **Malborn** Rheinland-Pfalz, Germany 49°43′E 6°59′N
160 F9 **Malborough** United Kingdom 50°14′N 3°49′W
178 H5 **Malchin** Mecklenburg-Vorpommern, N Germany 53°43′N 12°46′E
178 H5 **Malchiner See** ◎ NE Germany
163 C9 **Maldegem** Oost-Vlaanderen, NW Belgium 51°12′N 03°27′E
162 G7 **Malden** Gelderland, SE Netherlands 51°47′N 05°51′E
65 L2 **Malden** Massachusetts, NE USA 42°25′N 71°04′W
75 J11 **Malden** Missouri, C USA 36°33′N 89°58′W
285 H4 **Malden Island** *prev.* Independence Island. *atoll* E Kiribati
225 E13 **Maldive Islands** *island group* Maldive Islands, NW Maldives
288 F7 **Maldives** *off.* Maldivian Divehi, Republic of Maldives. ◆ *republic* N Indian Ocean
Maldives, Republic of *see* Maldives
Maldivian Divehi *see* Maldives
161 L5 **Maldon** E England, United Kingdom 51°44′N 00°40′E
115 J10 **Maldonado** Maldonado, S Uruguay 34°57′S 54°59′W
115 J9 **Maldonado** ◐ *department* S Uruguay
86 F9 **Maldonado, Punta** *headland* S Mexico 16°18′N 98°31′W
176 G5 **Malè** Trentino-Alto Adige, N Italy 46°21′N 10°51′E
235 C12 **Male'** *Div.* Maale. ● (Maldives) Male' Atoll, C Maldives 04°10′N 73°29′E
132 D5 **Maléa** *var.* Maléya. NE Guinea 11°46′N 09°43′W
Maléas, Ákra *see* Agriliá, Akrotírio
187 H8 **Maléas, Akrotírio** *headland* S Greece 36°25′N 23°11′E
235 C12 **Male' Atoll** *var.* Kaafu Atoll. *atoll* C Maldives
Malebo, Pool *see* Stanley Pool
234 D4 **Mālegaon** Mahārāshtra, W India 20°33′N 74°32′E
136 B5 **Malek** Jonglei, E South Sudan 06°04′N 31°36′E
281 M3 **Malekula** *var.* Malakula; *prev.* Mallicolo. *island* W Vanuatu
283 I10 **Malem** Kosrae, E Micronesia 05°16′N 163°01′E
139 K2 **Malema** Nampula, N Mozambique 14°57′S 37°28′E
135 H10 **Malemba-Nkulu** Katanga, SE Dem. Rep. Congo 08°01′S 26°48′E
217 J1 **Małe Mela** *Ar.* Marī Mīlā, *var.* Mārī Mīlah. Arbīl, E Iraq 36°58′N 44°42′E
194 F6 **Malen'ga** Respublika Kareliya, NW Russian Federation 63°50′N 36°21′E
180 I1 **Malente** Schleswig-Holstein, Germany 54°11′E 10°34′N
155 H11 **Målerås** Kalmar, S Sweden 56°55′N 15°34′E
165 H4 **Malesherbes** Loiret, C France 48°18′N 02°25′E
187 H5 **Malesína** Stereá Elláda, E Greece 38°37′N 23°15′E
166 B7 **Malestroit** Bretagne, France 47°49′N 2°23′W
Maléya *see* Maléa
197 I9 **Malgobek** Respublika Ingushetiya, SW Russian Federation 43°34′N 44°34′E
171 L3 **Malgrat de Mar** Cataluña, NE Spain 41°39′N 02°45′E
134 H1 **Malha** Northern Darfur, W Sudan 15°07′N 26°00′E
217 I4 **Malḥah** *var.* Malḥāt. Ṣalāḥ ad Dīn, C Iraq 34°44′N 42°41′E
159 J8 **Malham** United Kingdom 54°03′N 2°09′W
Malḥāt *see* Malḥah
76 E7 **Malheur Lake** ◎ Oregon, NW USA
76 E7 **Malheur River** Oregon, NW USA
132 C5 **Mali** NW Guinea 12°08′N 12°29′W
132 G3 **Mali** *off.* Republic of Mali, *Fr.* République du Mali; *prev.* French Sudan, Sudanese Republic. ◆ *republic* W Africa
260 F8 **Maliana** W East Timor 08°57′S 125°25′E
244 D6 **Malian He** Gansu, China
81 C10 **Malibu** California, USA 34°0′N 118°49′W
166 E7 **Malicorne-sur-Sarthe** Pays de la Loire, France 47°49′N 0°05′W
218 H6 **Māliḥah** Syria 32°44′N 36°12′E
256 D2 **Mali Hka** N Myanmar (Burma)
Mali Idjoš *see* Mali Iđoš
184 F2 **Mali Iđoš** *var.* Mali Idjoš, *Hung.* Kishegyes; *prev.* Krivaja. Vojvodina, N Serbia 45°43′N 19°40′E
184 G7 **Mali i Sharrit** *Serb.* Šar Planina. ▲▲ FYR Macedonia/Serbia
Mali i Zi *see* Crna Gora
184 F2 **Mali Kanal** *canal* N Serbia
260 E4 **Maliku** Sulawesi, N Indonesia 0°36′S 123°13′E
Malik, Wadi al *see* Milk, Wadi el
Mālikwāla *see* Malakwāl
257 D8 **Mali Kyun** *var.* Tavoy Island. *island* Mergui Archipelago, S Myanmar (Burma)
155 H10 **Målilla** Kalmar, S Sweden 57°24′N 15°49′E
184 A4 **Mali Lošinj** *It.* Lussinpiccolo. Primorje-Gorski Kotar, W Croatia 44°31′N 14°28′E
158 C5 **Malin** Donegal, Ireland 55°18′N 7°16′W
Malin *see* Malyn
86 E7 **Malinalco** México, Mexico 18°57′N 99°30′W
263 L8 **Malindang, Mount** ▲ Mindanao, S Philippines 08°12′N 123°37′E
137 F9 **Malindi** Kilifi, SE Kenya 03°14′S 40°05′E
Malines *see* Mechelen
156 C7 **Malin Head** *Ir.* Cionn Mhálanna. *headland* NW Ireland 55°37′N 07°37′W
158 A6 **Malin More** Donegal, Ireland 54°42′N 8°46′W
260 D3 **Malino, Gunung** ▲ Sulawesi, N Indonesia 0°44′N 120°45′E
184 G9 **Maliq** *var.* Maliqi. Korçë, SE Albania 40°45′N 20°45′E
Maliqi *see* Maliq
Mali, Republic of *see* Mali
Mali, République du *see* Mali
263 M9 **Malita** Mindanao, S Philippines 06°13′N 125°39′E
Maljovica *see* Malyovitsa
234 E4 **Malkāpur** Mahārāshtra, C India 20°52′N 76°18′E
214 B4 **Malkara** Tekirdağ, NW Turkey 40°54′N 26°54′E
191 F11 **Mal'kavichy** *Rus.* Mal'kovichi. Brestskaya Voblasts', SW Belarus 52°31′N 26°36′E
Malkiye *see* Al Mālikīyah
218 F5 **Malkiyya** Israel 33°06′N 35°31′E
187 K1 **Malko Sharkovo, Yazovir** ⊠ SE Bulgaria
185 M7 **Malko Tarnovo** *var.* Malko Tŭrnovo. Burgas, E Bulgaria 42°00′N 27°33′E
Malko Tŭrnovo *see* Malko Tarnovo
Mal'kovichi *see* Mal'kavichy
277 K7 **Mallacoota** Victoria, SE Australia 37°34′S 149°45′E
218 I6 **Mallāḥ** Syria 32°30′N 36°51′E
156 E5 **Mallaig** N Scotland, United Kingdom 57°04′N 05°48′W
276 G5 **Mallala** South Australia 34°29′S 138°30′E
129 I4 **Mallawī** *var.* Mallawi. C Egypt 27°44′N 30°50′E
Mallawi *see* Mallawī
118 B9 **Malleco, Río** Araucanía, Chile
171 I3 **Mallén** Aragón, NE Spain 41°53′N 01°25′W
176 G5 **Malles Venosta** *Ger.* Mals im Vinschgau. Trentino-Alto Adige, N Italy 46°40′N 10°37′E
Mallicolo *see* Malekula
180 J3 **Malliss** Mecklenburg-Vorpommern, Germany 53°13′E 11°19′N
177 I4 **Mallnitz** Salzburg, S Austria 46°58′N 13°09′E
171 L5 **Mallorca** *Eng.* Majorca; *anc.* Baleares Major. *island* Islas Baleares, Spain, W Mediterranean Sea
157 B10 **Mallow** *Ir.* Mala. SW Ireland 52°08′N 08°39′W
160 F3 **Mallwyd** United Kingdom 52°41′N 3°41′W
154 E2 **Malm** Nord-Trøndelag, C Norway 64°04′N 11°12′E
155 G10 **Malmbäck** Jönköping, S Sweden 57°34′N 14°30′E
152 F5 **Malmberget** *Lapp.* Malmivaara. Norrbotten, N Sweden 67°09′N 20°39′E
163 H11 **Malmédy** Liège, E Belgium 50°26′N 06°02′E
138 D10 **Malmesbury** Western Cape, SW South Africa 33°28′S 18°43′E
Malmivaara *see* Malmberget
155 H8 **Malmköping** Södermanland, C Sweden 59°08′N 16°49′E
155 F12 **Malmö** Skåne, S Sweden 55°36′N 13°E
155 F12 **Malmo** ✈ Skåne, S Sweden 55°33′N 13°23′E
102 F2 **Malmok** *headland* N Bonaire 12°16′N 68°21′W
155 H9 **Malmslätt** Östergötland, S Sweden 58°25′N 15°30′E
195 J10 **Malmyzh** Kirovskaya Oblast', NW Russian Federation 56°30′N 50°37′E
104 B4 **Mal Nomre, Punta** *headland* NW Peru 06°03′S 81°08′W
281 L3 **Malo** *island* W Vanuatu
196 F4 **Maloarkhangel'sk** Orlovskaya Oblast', W Russian Federation 52°25′N 36°37′E
Maļoeļap *see* Maloelap Atoll
283 M6 **Maloelap Atoll** *var.* Maļoeļap. *atoll* E Marshall Islands
Maloenda *see* Malunda
176 F5 **Maloja** Graubünden, S Switzerland 46°25′N 09°42′E
137 B12 **Malole** Northern, NE Zambia 10°05′S 31°37′E
167 H1 **Malo-les-Bains** Nord-Pas-de-Calais, France 51°03′N 2°24′E
280 D9 **Malolo** *island* Mamanuca Group, W Fiji
280 D9 **Malolo Barrier Reef** *var.* Ro Ro Reef. *reef* W Fiji
263 K5 **Malolos** Luzon, N Philippines 14°51′N 120°49′E
62 G4 **Malone** New York, NE USA 44°51′N 74°18′W
135 G11 **Malonga** Katanga, S Dem. Rep. Congo 10°26′S 23°10′E
183 G8 **Małopolska** *plateau* S Poland
183 G8 **Małopolskie** ◐ *province* SE Poland
Malorita/Maloryta *see* Malaryta
194 F6 **Maloshuyka** Arkhangel'skaya Oblast', NW Russian Federation 63°43′N 37°20′E
Mal'ovitsa *see* Malyovitsa
227 K7 **Malovodnoye** Almaty, SE Kazakhstan 43°31′N 77°42′E
154 A5 **Måløy** Sogn Og Fjordane, S Norway 61°57′N 05°06′E
196 F2 **Maloyaroslavets** Kaluzhskaya Oblast', W Russian Federation 55°03′N 36°31′E
195 I3 **Malozemel'skaya Tundra** *physical region* NW Russian Federation
170 E6 **Malpartida de Cáceres** Extremadura, W Spain 39°26′N 06°30′W
170 E5 **Malpartida de Plasencia** Extremadura, W Spain 39°59′N 06°03′W
160 G2 **Malpas** United Kingdom 53°01′N 2°46′W
85 K9 **Malpaso** Zacatecas, Mexico 22°37′N 102°46′W
174 C4 **Malpensa** ✈ (Milano) Lombardia, N Italy 45°41′N 08°40′E
130 D7 **Malqteïr** *desert* N Mauritania
181 E14 **Malsch** Baden-Württemberg, Germany 48°53′E 8°20′N
181 G9 **Malsfeld** Hessen, Germany 51°05′E 9°32′N
Mals im Vinschgau *see* Malles Venosta
190 F7 **Malta** SE Latvia 56°19′N 27°11′E
172 D2 **Malta** Guarda, Portugal 40°44′N 7°06′W
77 K3 **Malta** Montana, NW USA 48°21′N 107°52′W
175 J13 **Malta** *off.* Republic of Malta. ◆ *republic* C Mediterranean Sea
177 I4 **Malta** *var.* Maltabach. S Austria
175 J13 **Malta** *island* Malta, C Mediterranean Sea
Maltabach *see* Malta
Malta, Canale di *see* Malta Channel
175 G13 **Malta Channel** *It.* Canale di Malta. *strait* Italy/Malta
140 C3 **Maltahöhe** Hardap, SW Namibia 24°50′S 17°00′E
Malta, Republic of *see* Malta
159 L10 **Maltby** United Kingdom 53°25′N 1°12′W
159 M10 **Maltby le Marsh** United Kingdom 53°18′N 0°12′E
187 M2 **Maltepe** İstanbul, NW Turkey 40°56′N 29°09′E
159 L8 **Malton** N England, United Kingdom 54°07′N 00°50′W
260 G5 **Maluku** *off.* Propinsi Maluku, *Dut.* Molukken, *Eng.* Moluccas. ◐ *province* E Indonesia
260 G5 **Maluku** *Dut.* Molukken, *Eng.* Moluccas; *prev.* Spice Islands. *island group* E Indonesia
Maluku, Laut *see* Molucca Sea
Maluku, Propinsi *see* Maluku
260 F4 **Maluku Utara** *off.* Propinsi Maluku Utara. ◐ *province* E Indonesia
Maluku Utara, Propinsi *see* Maluku Utara
218 H3 **Malūlā** Syria 33°50′N 36°33′E
133 K5 **Malumfashi** Katsina, N Nigeria 11°51′N 07°39′E
260 C5 **Malunda** *prev.* Maloenda. Sulawesi, C Indonesia 02°58′S 118°52′E
154 F6 **Malung** Dalarna, C Sweden 60°40′N 13°45′E
154 F6 **Malungsfors** Dalarna, C Sweden 60°43′N 13°34′E
141 J6 **Maluti Mountains** ▲▲ Lesotho
281 J2 **Maluu** *var.* Malu'u. Malaita, N Solomon Islands 08°22′S 160°39′E

◆ Country
● Country Capital
◇ Dependent Territory
○ Dependent Territory Capital
◐ Administrative Regions
✈ International Airport
▲ Mountain
▲▲ Mountain Range
Volcano
River
◎ Lake
⊠ Reservoir

INDEX

◆ Country	◇ Dependent Territory	◈ Administrative Regions	▲ Mountain	✦ Volcano	⊗ Lake
● Country Capital	○ Dependent Territory Capital	✈ International Airport	▲ Mountain Range	♣ River	⊠ Reservoir

◆ Country   ◇ Dependent Territory   ◇ Administrative Regions   ▲ Mountain   ☒ Volcano   ◎ Lake
● Country Capital   ○ Dependent Territory Capital   ✕ International Airport   ▲ Mountain Range   ◈ River   ▨ Reservoir

◆ Country
◇ Dependent Territory
● Country Capital
◇ Dependent Territory Capital
◇ Administrative Regions
✈ International Airport
▲ Mountain
▲▲ Mountain Range
✦ Volcano
✍ River
⊗ Lake
⊡ Reservoir

◆ Country   ◇ Dependent Territory   ◈ Administrative Regions   ▲ Mountain   ▲ Volcano   ◉ Lake
◆ Country Capital   ◉ Dependent Territory Capital   ✈ International Airport   ▲ Mountain Range   ♦ River   ◙ Reservoir

◆ Country ◇ Dependent Territory ◈ Administrative Regions ▲ Mountain ◉ Volcano ◎ Lake
● Country Capital ◉ Dependent Territory Capital ✗ International Airport ▲ Mountain Range ✎ River ▣ Reservoir

Nikunau - North Plainfield
INDEX
383

284 *B3* **Nikunau** *var.* Nukunau; *prev.* Byron Island. *atoll* Tungaru, W Kiribati
118 *C5* **Nilahue** Libertador General Bernardo O'Higgins, Chile *34°42′S 71°45′W*
235 *D8* **Nilambūr** Kerala, SW India *11°17′N 76°15′E*
81 *G13* **Niland** California, W USA *33°14′N 115°31′W*
136 *B6* **Nile** *former province* NW Uganda
121 *K2* **Nile** *Ar.* Nahr an Nīl. ≈ N Africa
129 *I2* **Nile Delta** *delta* N Egypt
121 *K2* **Nile Fan** *undersea feature* E Mediterranean Sea *33°00′N 31°00′E*
73 *F9* **Niles** Michigan, N USA *41°49′N 86°15′W*
73 *J9* **Niles** Ohio, N USA *41°10′N 80°46′W*
235 *D8* **Nīleswaram** Kerala, SW India *12°18′N 75°07′E*
62 *E2* **Nilgaut, Lac** ◎ Québec, SE Canada
223 *M4* **Nīlī** Dāykundī, C Afghanistan *33°43′N 66°07′E*
238 *E3* **Nilka** Xinjiang Uygur Zizhiqu, NW China *43°46′N 82°33′E*
Nīl, Nahr an *see* Nile
111 *I8* **Nilópolis** Rio de Janeiro, Brazil *22°48′S 43°25′W*
153 *I8* **Nilsiä** Pohjois-Savo, C Finland *63°13′N 28°00′E*
232 *D7* **Nimach** Madhya Pradesh, C India *24°27′N 74°56′E*
232 *D7* **Nimbāhera** Rājasthān, N India *24°38′N 74°45′E*
132 *E7* **Nimba, Monts** *var.* Nimba Mountains. ▲▲ W Africa
Nimba Mountains *see* Nimba, Monts
Nimburg *see* Nymburk
165 *J8* **Nîmes** *anc.* Nemausus, Nismes. Gard, S France *43°49′N 04°20′E*
232 *E5* **Nīm ka Thāna** Rājasthān, N India *27°42′N 75°50′E*
277 *K6* **Nimmitabel** New South Wales, SE Australia *36°34′S 149°18′E*
Nimptsch *see* Niemcza
293 *I6* **Nimrod Glacier** *glacier* Antarctica
230 *D5* **Nīmrōz** *var.* Nimroze; *prev.* Chakhānsūr, Nīmrūz. ◈ *province* SW Afghanistan
Nimroze *see* Nīmrōz
Nīmrūz *see* Nīmrōz
136 *B6* **Nimule** Eastern Equatoria, S South Sudan *03°35′N 32°03′E*
Nimwegen *see* Nijmegen
53 *J7* **Nina Bang Lake** ◎ Nunavut, NE Canada
235 *C9* **Nine Degree Channel** *channel* India/Maldives
80 *J4* **Ninemile Peak** ▲ Nevada, USA *39°09′N 116°15′W*
62 *F5* **Ninemile Point** *headland* New York, NE USA *43°31′N 76°22′W*
289 *G8* **Ninetyeast Ridge** *undersea feature* E Indian Ocean *04°00′S 90°00′E*
277 *J7* **Ninety Mile Beach** *beach* Victoria, SE Australia
278 *E2* **Ninety Mile Beach** *beach* North Island, New Zealand
67 *H8* **Ninety Six** South Carolina, SE USA *34°10′N 82°01′W*
161 *L8* **Ninfield** United Kingdom *50°52′N 0°25′E*
248 *E1* **Ning'an** Heilongjiang, NE China *44°20′N 129°28′E*
243 *N3* **Ningbo** *var.* Ning-po, Yin-hsien; *prev.* Ninghsien. Zhejiang, SE China *29°54′N 121°33′E*
245 *K1* **Ningcheng** Inner Mongolia, China *41°22′N 119°11′E*
243 *L6* **Ningde** Fujian, SE China *26°48′N 119°33′E*
243 *J6* **Ningdu** *var.* Meijiang. Jiangxi, S China *26°28′N 115°53′E*
Ning'er *see* Pu'er
280 *A3* **Ningerum** Western, SW Papua New Guinea *05°40′S 141°10′E*
Ninggang *see* Longshi
243 *L2* **Ningguo** Anhui, E China *30°33′N 118°58′E*
243 *M3* **Ninghai** Zhejiang, SE China *29°18′N 121°26′E*
245 *K3* **Ninghe** *var.* Lutai. Tianjin Shi, N China *39°19′N 117°49′E*
Ning-hsia *see* Ningxia
Ninghsien *see* Ningbo
243 *J6* **Ninghua** Fujan, China *26°10′N 116°23′E*
245 *I5* **Ningjin** Hebei, China *37°22′N 114°33′E*
245 *J5* **Ningjin** Shandong, E China *37°39′N 116°48′E*
242 *C10* **Ningming** *var.* Chengzhong. Guangxi Zhuangzu Zizhiqu, S China *22°07′N 106°43′E*
240 *C6* **Ningnan** *var.* Pisha. Sichuan, C China *26°59′N 102°49′E*
Ning-po *see* Ningbo
244 *B9* **Ningqiang** Shaanxi, China *32°49′N 106°15′E*
Ningsia/Ningsia Hui/Ningsia Hui Autonomous Region *see* Ningxia
243 *M4* **Ningxi** Zhejiang, SE China *28°35′N 120°58′E*
244 *C5* **Ningxia** *off.* Ningxia Huizu Zizhiqu, *var.* Ning-hsia, Ningsia, *Eng.* Ningsia Hui, Ningsia Hui Autonomous Region. ◈ *autonomous region* N China
Ningxia Huizu Zizhiqu *see* Ningxia
244 *D7* **Ningxian** *var.* Xinning. Gansu, N China *35°30′N 108°05′E*
242 *G5* **Ningxiang** Hunan, S China *28°15′N 112°33′E*
245 *J6* **Ningyang** Shandong, China *35°28′N 116°28′E*
242 *G7* **Ningyuan** Hunan, S China *25°36′N 111°54′E*
256 *H5* **Ninh Binh** Ninh Binh, N Vietnam *20°14′N 106°00′E*
257 *J8* **Ninh Hoa** Khanh Hoa, S Vietnam *12°28′N 109°07′E*
280 *B1* **Ninigo Group** *island group* N Papua New Guinea
83 *I7* **Ninilchik** Alaska, USA *60°03′N 151°40′W*
75 *D10* **Ninnescah River** ≈ Kansas, C USA
293 *K9* **Ninnis Glacier** *glacier* Antarctica
253 *E8* **Ninohe** Iwate, Honshū, C Japan *40°16′N 141°18′E*
163 *D10* **Ninove** Oost-Vlaanderen, C Belgium *50°50′N 04°02′E*
263 *K5* **Ninoy Aquino** ✈ (Manila) Luzon, N Philippines *14°26′N 121°00′E*
Nio *see* Íos
74 *E6* **Niobrara** Nebraska, C USA *42°43′N 97°59′W*
74 *D6* **Niobrara River** ≈ Nebraska/Wyoming, C USA
135 *E8* **Nioki** Bandundu, W Dem. Rep. Congo *02°44′S 17°42′E*
132 *F4* **Niono** Ségou, C Mali *14°18′N 05°59′W*
132 *D4* **Nioro** *var.* Nioro du Sahel. Kayes, W Mali *15°13′N 09°39′W*
132 *B4* **Nioro du Rip** SW Senegal *13°44′N 15°48′W*
Nioro du Sahel *see* Nioro
164 *G6* **Niort** Deux-Sèvres, W France *46°21′N 00°25′W*
139 *I10* **Nioumachoua** Mohéli, S Comoros *12°21′S 43°43′E*
280 *A3* **Nipa** Southern Highlands, W Papua New Guinea *06°11′S 143°27′E*
57 *N4* **Nipawin** Saskatchewan, S Canada *52°23′N 104°01′W*
58 *C7* **Nipigon** Ontario, S Canada *49°02′N 88°15′W*
58 *C6* **Nipigon, Lake** ◎ Ontario, S Canada
57 *K3* **Nipin** ≈ Saskatchewan, C Canada
62 *D2* **Nipissing, Lake** ◎ Ontario, S Canada
81 *B8* **Nipomo** California, W USA *35°02′N 120°28′W*
Nippon *see* Japan
81 *I10* **Nipton** California, USA *35°28′N 115°16′W*
216 *E4* **Niqniqīyah, Jabal an** ▲▲ C Syria
118 *C7* **Niquen** Bío-Bío, Chile *36°14′S 72°04′W*
112 *C8* **Niquivil** San Juan, W Argentina *30°25′S 68°42′W*
261 *L5* **Nirabotong** Papua, E Indonesia *02°35′S 140°08′E*
253 *A13* **Nirasaki** Yamanashi, Honshū, S Japan *35°45′N 138°27′E*
Niriz *see* Neyrīz
247 *K3* **Nirji** *var.* Morin Dawa Daurzu Zizhiqi. Nei Mongol Zizhiqu, N China *48°21′N 124°32′E*
234 *F5* **Nirmal** Telangana, C India *19°04′N 78°21′E*
233 *I6* **Nirmāli** Bihār, NE India *26°18′N 86°35′E*
185 *H5* **Niš** *Eng.* Nish, *Ger.* Nisch; *anc.* Naissus. Serbia, SE Serbia *43°21′N 21°53′E*
172 *C4* **Nisa** Portalegre, C Portugal *39°31′N 07°39′W*
Nisa *see* Neisse
220 *G2* **Niṣab** Al Ḥudūd ash Shamālīyah, N Saudi Arabia *29°11′N 44°43′E*
220 *G9* **Niṣāb** *var.* Anṣāb. SW Yemen *14°24′N 46°47′E*
185 *H5* **Nišava** *Bul.* Nishava. ≈ Bulgaria/Serbia *see also* Nishava
Nišava *see* Nishava
175 *G13* **Niscemi** Sicilia, Italy, C Mediterranean Sea *37°09′N 14°23′E*
Nisch/Nish *see* Niš
252 *D5* **Niseko** Hokkaidō, NE Japan *42°50′N 140°43′E*
Nishapur *see* Neyshābūr
185 *I6* **Nishava** *var.* Nišava. ≈ Bulgaria/Serbia *see also* Nišava
Nishava *see* Nišava
190 *G7* **Nishcha** ≈ N Belarus
252 *H4* **Nishibetsu-gawa** ≈ Hokkaidō, NE Japan
250 *E5* **Nishi-gawa** ≈ Honshū, SW Japan
250 *E5* **Nishi-Nōmi-jima** *var.* Nōmi-jima. *island* SW Japan
250 *D10* **Nishinoomote** Kagoshima, Tanega-shima, SW Japan *30°42′N 130°59′E*
251 *M9* **Nishino-shima** *Eng.* Rosario. *island* Ogasawara-shotō, SE Japan
251 *J4* **Nishio** *var.* Nisio. Aichi, Honshū, SW Japan *34°52′N 137°02′E*
250 *C6* **Nishi-Sonogi-hantō**
250 *B7* **Nishi-Sonogi-hantō** *peninsula* Kyūshū, SW Japan
251 *H4* **Nishiwaki** *var.* Nisiwaki. Hyōgo, Honshū, SW Japan *34°59′N 134°58′E*
221 *I8* **Nishtūn** SE Yemen *15°47′N 52°08′E*
Nisibin *see* Nusaybin
Nisio *see* Nishio
Nisiros *see* Nísyros
Nisiwaki *see* Nishiwaki
Niska *see* Niesky
185 *H5* **Niška Banja** Serbia, SE Serbia *43°18′N 22°01′E*
58 *D3* **Niskibi** ≈ Ontario, C Canada
183 *I8* **Nisko** Podkrapackie, SE Poland *50°31′N 22°09′E*
54 *A2* **Nisling** ≈ Yukon, W Canada
163 *E12* **Nismes** Namur, S Belgium *50°04′N 04°31′E*
Nismes *see* Nîmes
188 *F7* **Nisporeni** *Rus.* Nisporeny. W Moldova *47°04′N 28°10′E*
Nisporeny *see* Nisporeni
155 *G10* **Nissan** ≈ S Sweden
Nissan Islands *see* Green Islands
155 *C8* **Nisser** ◎ S Norway
155 *C11* **Nissum Bredning** *inlet* NW Denmark
74 *G3* **Nisswa** Minnesota, N USA *46°31′N 94°17′W*
Nistru *see* Dniester
187 *L8* **Nísyros** *var.* Nisiros. *island* Dodekánisa, Greece, Aegean Sea
190 *E6* **Nītaure** C Latvia *57°05′N 25°12′E*
111 *I8* **Niterói** *prev.* Nictheroy. Rio de Janeiro, SE Brazil *22°54′S 43°06′W*
156 *F7* **Nith** ≈ S Scotland, United Kingdom
Nitinan *see* Nichinan
183 *E10* **Nitra** *Ger.* Neutra, *Hung.* Nyitra. Nitriansky Kraj, SW Slovakia *48°20′N 18°05′E*
183 *F10* **Nitra** *Ger.* Neutra, *Hung.* Nyitra. ≈ W Slovakia
183 *E10* **Nitriansky Kraj** ◈ *region* SW Slovakia
67 *H3* **Nitro** West Virginia, NE USA *38°24′N 81°50′W*
195 *N9* **Nitsa** ≈ C Russian Federation
Nitsana *see* Nizzana
Nitsanei Oz *see* Nizzané 'Oz
Nitsanim *see* Nizzanim
181 *B12* **Nittel** Rheinland-Pfalz, Germany *49°39′E 6°27′N*
Niuatobutabu *see* Niuatoputapu
284 *D6* **Niuatoputapu** *var.* Niuatobutabu; *prev.* Keppel Island. *island* N Tonga
284 *D9* **Niu'Aunofa** *headland* Tongatapu, S Tonga *21°03′S 175°19′W*
245 *J3* **Niubaotun** Beijing Shi, N China *39°45′N 116°41′E*
Niuchwang *see* Yingkou
285 *H4* **Niue** ◇ *self-governing territory in free association with New Zealand* S Pacific Ocean
284 *C5* **Niulakita** *var.* Nurakita. *atoll* S Tuvalu
245 *M6* **Niuqibu** Shandong, E China *36°33′N 120°33′E*
284 *B4* **Niutao** *atoll* NW Tuvalu
245 *M2* **Niuzhuang** Liaoning, NE China *40°57′N 122°32′E*
152 *H7* **Nivala** Pohjois-Pohjanmaa, C Finland *63°56′N 25°00′E*
164 *F9* **Nive** ≈ SW France
163 *D10* **Nivelles** Walloon Brabant, C Belgium *50°36′N 04°04′E*
165 *I5* **Nivernais** *cultural region* C France
75 *H11* **Nixa** Missouri, C USA *37°02′N 93°17′W*
80 *G1* **Nixon** Nevada, W USA *39°48′N 119°24′W*
70 *G7* **Nixon** Texas, SW USA *29°16′N 97°45′W*
Niya *see* Minfeng
Niyazov *see* Nyýazow
234 *F5* **Nizāmābād** Telangana, C India *18°40′N 78°05′E*
234 *F5* **Nizām Sāgar** ◎ C India
197 *H2* **Nizhegorodskaya Oblast'** ◈ *province* W Russian Federation
193 *I7* **Nizhneangarsk** Respublika Buryatiya, S Russian Federation *55°47′N 109°39′E*
Nizhnegorskiy *see* Nyzhn'ohirs'kyy
197 *K2* **Nizhnekamsk** Respublika Tatarstan, W Russian Federation *55°36′N 51°45′E*
197 *L1* **Nizhnekamskoye Vodokhranilishche** ⊠ W Russian Federation
193 *L3* **Nizhnekolymsk** Respublika Sakha (Yakutiya), NE Russian Federation *68°32′N 161°00′E*
193 *H8* **Nizhneudinsk** Irkutskaya Oblast', S Russian Federation *54°48′N 98°51′E*
192 *F6* **Nizhnevartovsk** Khanty-Mansiyskiy Avtonomnyy Okrug-Yugra, C Russian Federation *60°57′N 76°40′E*
193 *J4* **Nizhneyansk** Respublika Sakha (Yakutiya), NE Russian Federation *71°25′N 135°59′E*
197 *I6* **Nizhniy Baskunchak** Astrakhanskaya Oblast', SW Russian Federation *48°15′N 46°49′E*
195 *L10* **Nizhniye Sergi** Sverdlovskaya Oblast', Russian Federation
197 *H3* **Nizhniy Lomov** Penzenskaya Oblast', W Russian Federation *53°32′N 43°39′E*
197 *H1* **Nizhniy Novgorod** *prev.* Gor'kiy. Nizhegorodskaya Oblast', W Russian Federation *56°17′N 44°E*
195 *J6* **Nizhniy Odes** Respublika Komi, NW Russian Federation *63°42′N 54°59′E*
Nizhniy Pyandzh *see* Panji Poyon
195 *L9* **Nizhniy Tagil** Sverdlovskaya Oblast', C Russian Federation *57°57′N 59°51′E*
195 *K6* **Nizhnyaya-Omra** Respublika Komi, NW Russian Federation *62°46′N 55°54′E*
195 *I4* **Nizhnyaya Pësha** Nenetskiy Avtonomnyy Okrug, NW Russian Federation *66°54′N 47°37′E*
195 *M9* **Nizhnyaya Salda** Sverdlovskaya Oblast', Russian Federation
195 *N9* **Nizhnyaya Tavda** Tyumenskaya Oblast', C Russian Federation *57°41′N 65°54′E*
192 *G5* **Nizhnyaya Tunguska** *Eng.* Lower Tunguska. ≈ N Russian Federation
189 *I2* **Nizhyn** *Rus.* Nezhin. Chernihivs'ka Oblast', NE Ukraine *51°03′N 31°54′E*
214 *G8* **Nizip** Gaziantep, S Turkey *37°02′N 37°47′E*
221 *K5* **Nizwa** *var.* Nazwāh. NE Oman *23°00′N 57°50′E*
Nizza *see* Nice
176 *D8* **Nizza Monferrato** Piemonte, NE Italy *44°47′N 08°22′E*
219 *C11* **Nizzana** *var.* Nitsana. Southern, S Israel *30°53′N 34°25′E*
218 *E7* **Nizzané 'Oz** *var.* Nitsanei Oz. Central, C Israel *32°18′N 35°00′E*
219 *D8* **Nizzanim** *var.* Nitsanim. Southern, S Israel *31°43′N 34°38′E*
Njávdám *see* Näätämöjoki
Njazidja *see* Ngazidja
Njellim *see* Nellim
137 *C11* **Njombe** Njombe, S Tanzania *09°20′S 34°47′E*
137 *C11* **Njombe** *off.* Mkoa wa Njombe. ◈ *region* S Tanzania
137 *C11* **Njombe** ≈ C Tanzania
Njombe, Mkoa wa *see* Njombe
Njuk, Ozero *see* Nyuk, Ozero
Njuksenica *see* Nyuksenitsa
152 *F4* **Njunis** ▲ N Norway *68°47′N 19°24′E*
Njurunda *see* Njurundabommen
154 *I4* **Njurundabommen** *prev.* Njurunda. Västernorrland, C Sweden *62°15′N 17°24′E*
154 *I5* **Njutånger** Gävleborg, C Sweden *61°37′N 17°04′E*
134 *B5* **Nkambe** Nord-Ouest, NW Cameroon *06°35′N 10°44′E*
141 *K6* **Nkandla** KwaZulu-Natal, South Africa *28°37′S 31°05′E*
Nkata Bay *see* Nkhata Bay
135 *C9* **Nkayi** *prev.* Jacob. Bouenza, S Congo *04°11′S 13°17′E*
138 *G3* **Nkayi** Matabeleland North, W Zimbabwe *19°00′S 28°54′E*
137 *C12* **Nkhata Bay** *var.* Nkata Bay. Northern, N Malawi *11°37′S 34°20′E*
137 *A10* **Nkonde** Kigoma, N Tanzania *06°16′S 30°17′E*
134 *A5* **Nkongsamba** *var.* N'Kongsamba. Littoral, W Cameroon *04°59′N 09°53′E*
N'Kongsamba *see* Nkongsamba
135 *E14* **Nkurenkuru** Okavango, N Namibia *17°38′S 18°39′E*
141 *L6* **Nkwalini** KwaZulu-Natal, South Africa *28°44′S 31°31′E*
133 *H7* **Nkwanta** E Ghana *08°18′N 00°27′E*
256 *D2* **Nmai Hka** *var.* Me Hka. ≈ N Myanmar (Burma)
166 *G4* **Noailles** Picardie, France *49°20′N 2°12′E*
Noardwâlde *see* Noordwolde
169 *M2* **Noasca** Valle D'Aosta, NW Italy *45°27′N 07°18′E*
82 *G3* **Noatak** Alaska, USA *67°34′N 162°58′W*
83 *H3* **Noatak River** ≈ Alaska, USA
158 *D8* **Nobber** Meath, Ireland *53°49′N 6°45′W*
Nobeji *see* Noheji
250 *D7* **Nobeoka** Miyazaki, Kyūshū, SW Japan *32°34′N 131°37′E*
75 *E12* **Noble** Oklahoma, C USA *35°08′N 97°23′W*
73 *F11* **Noblesville** Indiana, N USA *40°03′N 86°00′W*
252 *D5* **Noboribetsu** *var.* Noboribetu. Hokkaidō, NE Japan *42°27′N 141°08′E*
Noboribetu *see* Noboribetsu
107 *I9* **Nobres** Mato Grosso, W Brazil *14°44′S 56°15′W*
175 *H11* **Nocera Terinese** Calabria, S Italy *39°03′N 16°10′E*
85 *K10* **Nochistlán** Zacatecas, Mexico *21°22′N 102°51′W*
86 *G8* **Nochixtlán** *var.* Asunción Nochixtlán. Oaxaca, SE Mexico *17°29′N 97°17′W*
70 *G3* **Nocona** Texas, SW USA *33°47′N 97°43′W*
86 *D6* **Nocupétaro** Michoacán, Mexico
117 *E11* **Nodales, Bahía de los** *bay* S Argentina
75 *F8* **Nodaway River** ≈ Iowa/Missouri, C USA
167 *L8* **Nods** Franche-Comté, France *47°06′N 6°20′E*
75 *G11* **Noel** Missouri, C USA *36°33′N 94°29′W*
140 *E5* **Noenieput** Northern Cape, South Africa *27°31′S 20°08′E*
84 *G2* **Nogales** Chihuahua, NW Mexico *18°50′N 97°12′W*
84 *E2* **Nogales** Sonora, NW Mexico *31°17′N 110°53′W*
79 *I10* **Nogales** Arizona, SW USA *31°20′N 110°55′W*
Nogal Valley *see* Dooxo Nugaaleed
164 *G8* **Nogaro** Gers, S France *43°46′N 00°01′W*
182 *F4* **Nogat** ≈ N Poland
250 *D6* **Nōgata** Fukuoka, Kyūshū, SW Japan *33°46′N 130°42′E*
197 *I8* **Nogayskaya Step'** *steppe* SW Russian Federation
167 *K7* **Nogent-en-Bassigny** Champagne-Ardenne, France *48°02′N 5°21′E*
166 *G6* **Nogent-le-Roi** Centre, France *48°39′N 1°32′E*
164 *G4* **Nogent-le-Rotrou** Eure-et-Loir, C France *48°19′N 00°50′E*
167 *H5* **Nogent-sur-Marne** Île-de-France, France *48°50′N 02°29′E*
165 *H3* **Nogent-sur-Oise** Oise, N France *49°16′N 02°28′E*
165 *I3* **Nogent-sur-Seine** Aube, N France *48°30′N 03°31′E*
192 *G6* **Noginsk** Krasnoyarskiy Kray, N Russian Federation *64°28′N 91°09′E*
196 *G2* **Noginsk** Moskovskaya Oblast', W Russian Federation *55°51′N 38°23′E*
193 *L7* **Nogliki** Ostrov Sakhalin, Sakhalinskaya Oblast', SE Russian Federation *51°44′N 143°14′E*
251 *I3* **Nōgōhaku-san** ▲ Honshū, SW Japan *35°46′N 136°30′E*
238 *G1* **Nogoonnuur** Bayan-Ölgiy, NW Mongolia *49°31′N 89°48′E*
114 *C5* **Nogoyá** Entre Ríos, E Argentina *32°25′S 59°50′W*
183 *G11* **Nógrád** *off.* Nógrád Megye. ◈ *county* N Hungary
Nógrád Megye *see* Nógrád
168 *F8* **Noguera Pallaresa** ≈ NE Spain
171 *J3* **Noguera Ribagorçana** ≈ NE Spain
166 *G9* **Nohant-Vic** Centre, France *46°38′N 1°59′E*
252 *D7* **Noheji** *var.* Nobeji. Aomori, Honshū, C Japan *40°51′N 141°07′E*
181 *C13* **Nohfelden** Saarland, SW Germany *49°35′N 07°08′E*
170 *C2* **Noia** Galicia, NW Spain *42°48′N 08°52′W*
165 *H8* **Noire, Montagne** ▲▲ S France
62 *E2* **Noire, Rivière** ≈ Québec, SE Canada
63 *I3* **Noire, Rivière** ≈ Québec, SE Canada
Noire, Rivi`ere *see* Black River
164 *E4* **Noires, Montagnes** ▲▲ NW France
169 *I1* **Noirétable** Rhône-Alpes, France *45°49′N 3°46′E*
164 *E5* **Noirmoutier-en-l'Île** Vendée, NW France *47°00′N 02°15′W*
164 *E5* **Noirmoutier, Île de** *island* NW France
253 *C14* **Nojima-zaki** *headland* Honshū, S Japan *34°54′N 139°54′E*
281 *I5* **Noka** Nendö, E Solomon Islands *10°42′S 165°57′E*
138 *E4* **Nokaneng** North West, NW Botswana *19°40′S 22°12′E*
153 *H9* **Nokia** Pirkanmaa, W Finland *61°29′N 23°30′E*
252 *H4* **Nokke-suidō** *prev.* Notsuke-suidō. *strait* Japan/Russian Federation
252 *H4* **Nokke-zaki** *prev.* Notsuke-zaki. *headland* Hokkaidō, NE Japan *43°33′N 145°18′E*
230 *D7* **Nok Kundi** Baluchistān, SW Pakistan *28°49′N 62°39′E*
73 *C11* **Nokomis** Illinois, N USA *39°18′N 89°17′W*
72 *C5* **Nokomis, Lake** ⊠ Wisconsin, N USA
133 *N4* **Nokou** Kanem, W Chad *14°36′N 14°45′E*
281 *L2* **Nokuku** Espiritu Santo, W Vanuatu *14°56′S 166°34′E*
155 *F10* **Nol** Västra Götaland, S Sweden *57°55′N 12°03′E*
134 *D6* **Nola** Sangha-Mbaéré, SW Central African Republic *03°29′N 16°05′E*
70 *E5* **Nolan** Texas, SW USA *32°15′N 100°15′W*
195 *I9* **Nolinsk** Kirovskaya Oblast', NW Russian Federation *57°35′N 49°54′E*
Nolsø *see* Nólsoy
154 *B2* **Nólsoy** *Dan.* Nolsø. *island* E Faroe Islands
280 *A3* **Nomad** Western, SW Papau New Guinea *06°11′S 142°13′E*
55 *N7* **Nomansland Point** *headland* Ontario, C Canada
250 *C9* **Noma-zaki** Kyūshū, SW Japan
85 *J8* **Nombre de Dios** Durango, C Mexico *23°51′N 104°14′W*
88 *G3* **Nombre de Dios, Cordillera** ▲▲ N Honduras
82 *F5* **Nome** Alaska, USA *64°30′N 165°24′W*
74 *E3* **Nome** North Dakota, N USA *46°39′N 97°49′W*
82 *G5* **Nome, Cape** *headland* Alaska, USA *64°25′N 165°00′W*
167 *L5* **Nomény** Lorraine, France *48°54′N 6°14′E*
239 *K3* **Nomgon** *var.* Sangiyn Dalay. Ömnögovĭ, S Mongolia *42°50′N 105°04′E*
Nōmi-jima *see* Nishi-Nōmi-jima
62 *G2* **Nominingue, Lac** ◎ Québec, SE Canada
Nomoi Islands *see* Mortlock Islands
250 *B7* **Nomo-zaki** *headland* Kyūshū, SW Japan *32°34′N 129°45′E*
239 *I1* **Nömrög** *var.* Hödrögö. Dzavhan, N Mongolia *48°51′N 96°48′E*
140 *C3* **Nomtsas** Hardap, Namibia *24°26′S 16°50′E*
284 *D7* **Nomuka Group** *island group* W Tonga
282 *G6* **Nomwin Atoll** *atoll* Hall Islands, C Micronesia
54 *G4* **Nonacho Lake** ◎ Northwest Territories, NW Canada
166 *F5* **Nonancourt** Haute-Normandie, France *48°46′N 1°12′E*
Nondaburi *see* Nonthaburi
83 *H7* **Nondalton** Alaska, USA *59°58′N 154°51′W*
141 *K5* **Nondweni** KwaZulu-Natal, South Africa *28°11′S 30°48′E*
247 *L5* **Nong'an** Jilin, NE China *44°25′N 125°10′E*
256 *F7* **Nong Bua Khok** Nakhon Ratchasima, C Thailand *15°23′N 101°51′E*
256 *F6* **Nong Bua Lamphu** Udon Thani, E Thailand *17°11′N 102°27′E*
256 *G5* **Nông Hèt** Xiangkhoang, N Laos *19°27′N 104°02′E*
Nongkaya *see* Nong Khai
262 *C3* **Nong Khai** *var.* Mi Chai, Nongkaya. Nong Khai, E Thailand *17°52′N 102°44′E*
Nongle *see* Jiuzhaigou
257 *E10* **Nong Met** Surat Thani, SW Thailand *09°27′N 99°09′E*
139 *H7* **Nongoma** KwaZulu/Natal, E South Africa *27°54′S 31°40′E*
141 *L5* **Nongoma** KwaZulu-Natal, South Africa *27°54′S 31°39′E*
256 *F7* **Nong Phai** Phetchabun, C Thailand *15°58′N 101°02′E*
233 *L7* **Nongstoin** Meghālaya, NE India *25°24′N 91°19′E*
140 *A1* **Nonidas** Erongo, N Namibia *22°36′S 14°40′E*
Nonni *see* Nen Jiang
181 *C13* **Nonnweiler** Saarland, Germany *49°37′E 6°57′N*
85 *H5* **Nonoava** Chihuahua, N Mexico *27°24′N 106°18′W*
284 *B3* **Nonouti** *prev.* Sydenham Island. *atoll* Tungaru, W Kiribati
257 *E8* **Nonthaburi** *var.* Nondaburi, Nontha Buri. Nonthaburi, C Thailand *13°48′N 100°11′E*
Nontha Buri *see* Nonthaburi
164 *G6* **Nontron** Dordogne, SW France *45°34′N 00°41′E*
229 *K6* **Nookat** *var.* Iski-Nauket; *prev.* Eski-Nookat. Oshskaya Oblast', SW Kyrgyzstan *40°18′N 72°29′E*
274 *G2* **Noonamah** Northern Territory, N Australia *12°46′S 131°08′E*
74 *B1* **Noonan** North Dakota, N USA *48°51′N 102°57′W*
Noonu *see* South Miladhunmadulu Atoll
163 *C8* **Noord-Beveland** *var.* North Beveland. *island* SW Netherlands
163 *F8* **Noord-Brabant** *Eng.* North Brabant. ◈ *province* S Netherlands
162 *E4* **Noorder Haaks** *spit* NW Netherlands
162 *E5* **Noord-Holland** *Eng.* North Holland. ◈ *province* NW Netherlands
Noordhollandsch Kanaal *see* Noordhollands Kanaal
162 *E4* **Noordhollands Kanaal** *var.* Noordhollandsch Kanaal. *canal* NW Netherlands
Noord-Kaap *see* Northern Cape
140 *D8* **Noordkuil** Western Cape, South Africa *32°31′S 18°22′E*
162 *G5* **Noordoostpolder** *island* N Netherlands
102 *E2* **Noordpunt** *headland* N Curaçao *12°21′N 69°08′W*
162 *E5* **Noord-Scharwoude** Noord-Holland, NW Netherlands *52°42′N 04°48′E*
Noordwes *see* North-West
162 *E6* **Noordwijk aan Zee** Zuid-Holland, W Netherlands *52°15′N 04°25′E*
162 *E6* **Noordwijkerhout** Zuid-Holland, W Netherlands *52°16′N 04°30′E*
162 *H4* **Noordwolde** *Fris.* Noardwâlde. Fryslân, N Netherlands *52°54′N 06°10′E*
Noordzee *see* North Sea
162 *E6* **Noordzee-Kanaal** *canal* NW Netherlands
190 *C1* **Noormarkku** *Swe.* Norrmark. Satakunta, SW Finland *61°35′N 21°54′E*
82 *G4* **Noorvik** Alaska, USA *66°50′N 161°01′W*
277 *M1* **Noosa Heads** Queensland, Australia
56 *B8* **Nootka Sound** *inlet* British Columbia, W Canada
81 *H9* **Nopah Range** *ridge* California, USA
135 *C10* **Nóqui** Zaire Province, NW Angola *05°54′S 13°30′E*
155 *G8* **Nora** Örebro, C Sweden *59°31′N 15°02′E*
229 *I7* **Norak** *Rus.* Nurek. W Tajikistan *38°23′N 69°14′E*
74 *H6* **Nora Springs** Iowa, C USA *43°08′N 93°00′W*
154 *H7* **Norberg** Västmanland, C Sweden *60°04′N 15°56′E*
114 *C10* **Norberto de la Riestra** Buenos Aires, Argentina *35°16′S 59°46′W*
62 *F3* **Norcan Lake** ◎ Ontario, SE Canada
81 *E11* **Norco** California, USA *33°56′N 117°33′W*
295 *J5* **Nord** N Greenland *81°38′N 12°51′W*
134 *C4* **Nord** *Eng.* North. ◈ *province* N Cameroon
165 *H1* **Nord** ◈ *department* N France
152 *B4* **Nordaustlandet** *island* NE Svalbard
155 *D13* **Nordborg** *Ger.* Nordburg. Syddanmark, SW Denmark *55°04′N 09°41′E*
Nordburg *see* Nordborg
155 *C12* **Nordby** Syddtjylland, W Denmark *55°27′N 08°25′E*
180 *C2* **Norddeich** Niedersachsen, Germany *53°37′E 7°09′N*
57 *H6* **Nordegg** Alberta, SW Canada *52°27′N 116°06′W*
180 *C2* **Norden** Niedersachsen, NW Germany *53°36′N 07°12′E*
80 *E2* **Norden** California, USA *39°19′N 120°21′W*
180 *F3* **Nordenham** Niedersachsen, NW Germany *53°30′N 08°29′E*
193 *H3* **Nordenshel'da, Arkhipelag** *island group* N Russian Federation
152 *B5* **Nordenskiold Land** *physical region* W Svalbard
178 *B5* **Norderney** *island* NW Germany
180 *H2* **Norderstedt** Schleswig-Holstein, N Germany *53°42′N 09°59′E*
154 *B5* **Nordfjord** *fjord* S Norway
154 *B5* **Nordfjord** *physical region* S Norway
154 *B5* **Nordfjordeid** Sogn og Fjordane, S Norway *61°54′N 06°E*
152 *E5* **Nordfold** Nordland, C Norway *67°48′N 15°16′E*
Nordfriesische Inseln *see* North Frisian Islands
178 *D3* **Nordfriesland** *cultural region* N Germany
180 *G1* **Nordhastedt** Schleswig-Holstein, Germany *54°10′E 9°11′N*
181 *J8* **Nordhausen** Thüringen, C Germany *51°31′N 10°48′E*
181 *E12* **Nordheim** Hessen, Germany *49°41′E 8°23′N*
70 *G8* **Nordheim** Texas, SW USA *28°55′N 97°36′W*
181 *I10* **Nordheim vor der Rhön** Bayern, Germany *50°29′E 10°11′N*
154 *A6* **Nordhordland** *physical region* S Norway
180 *C5* **Nordhorn** Niedersachsen, NW Germany *52°26′N 07°04′E*
137 *I8* **Nord, Île du** *island* Inner Islands, NE Seychelles
155 *D11* **Nordjylland** ◈ *county* N Denmark
152 *H2* **Nordkapp** *Eng.* North Cape. *headland* N Norway *25°47′E 71°10′N*
152 *B4* **Nordkapp** *headland* N Svalbard *80°31′N 19°58′E*
135 *I8* **Nord-Kivu** *off.* Région du Nord Kivu. ◈ *region* E Dem. Rep. Congo
Nord Kivu, Région du *see* Nord-Kivu
154 *F1* **Nordland** ◈ *county* C Norway
179 *F12* **Nördlingen** Bayern, S Germany *48°49′N 10°28′E*
154 *J3* **Nordmaling** Västerbotten, N Sweden *63°35′N 19°30′E*
154 *G7* **Nordmark** Värmland, C Sweden *59°52′N 14°04′E*
Nord, Mer du *see* North Sea
180 *G1* **Nord-Ostee-Kanal** *canal* N Germany
42 *F2* **Nordøstrundingen** *cape* NE Greenland
134 *A5* **Nord-Ouest** *Eng.* North-West. ◈ *province* NW Cameroon
Nord-Ouest, Territoires du *see* Northwest Territories
165 *H1* **Nord-Pas-de-Calais** ◈ *region* N France
181 *D13* **Nordpfälzer Bergland** ▲▲ W Germany
Nord, Pointe *see* Fatua, Pointe
280 *B9* **Nord, Province** ◈ *province* C New Caledonia
152 *F4* **Nordreisa** Troms, N Norway *69°45′N 21°03′E*
180 *C7* **Nordrhein-Westfalen** *Eng.* North Rhine-Westphalia, *Fr.* Rhénanie du Nord-Westphalie. ◈ *state* W Germany
Nordsee/Nordsjøen/Nordsøen *see* North Sea
180 *H6* **Nordstemmen** Niedersachsen, Germany *52°10′E 9°47′N*
178 *D4* **Nordstrand** *island* N Germany
154 *F2* **Nord-Trøndelag** ◈ *county* C Norway
152 *B1* **Norðurfjörður** Vestfirðir, NW Iceland *66°01′N 21°33′W*
152 *C1* **Norðurland Eystra** ◈ *region* N Iceland
152 *C2* **Norðurland Vestra** ◈ *region* N Iceland
157 *C10* **Nore** *Ir.* An Fheoir. ≈ S Ireland
160 *A4* **Nore** ≈ Kilkenny, Ireland
65 *I3* **Norfolk** Connecticut, USA *42°00′N 73°12′W*
74 *E7* **Norfolk** Nebraska, C USA *42°01′N 97°25′W*
67 *M5* **Norfolk** Virginia, NE USA *36°51′N 76°17′W*
161 *L3* **Norfolk** *cultural region* E England, United Kingdom
286 *E8* **Norfolk Island** ◈ *Australian self-governing territory* SW Pacific Ocean
267 *H6* **Norfolk Ridge** *undersea feature* W Pacific Ocean
75 *I11* **Norfork Lake** ⊠ Arkansas/Missouri, C USA
162 *I4* **Norg** Drenthe, NE Netherlands *53°04′N 06°28′E*
Norge *see* Norway
159 *J4* **Norham** United Kingdom *55°43′N 2°09′W*
154 *B7* **Norheimsund** Hordaland, S Norway *60°22′N 06°09′E*
195 *N3* **Nori** Yamalo-Nenetskiy Avtonomnyy Okrug, Russian Federation
70 *G10* **Norias** Texas, SW USA *26°47′N 97°45′W*
251 *J3* **Norikura-dake** ▲ Honshū, S Japan *36°06′N 137°33′E*
192 *G5* **Noril'sk** Krasnoyarskiy Kray, N Russian Federation *69°21′N 88°02′E*
62 *D4* **Norland** Ontario, SE Canada *44°46′N 78°48′W*
67 *K5* **Norlina** North Carolina, SE USA *36°26′N 78°11′W*
73 *C10* **Normal** Illinois, N USA *40°30′N 88°59′W*
75 *E12* **Norman** Oklahoma, C USA *35°13′N 97°27′W*
Norman *see* Tulita
280 *D4* **Normanby Island** *island* SE Papua New Guinea
Normandes, Îles *see* Channel Islands
103 *J6* **Normandia** Roraima, N Brazil *03°57′N 59°39′W*
164 *G3* **Normandie** *Eng.* Normandy. *cultural region* N France
164 *F3* **Normandie, Collines de** *hill range* NW France
141 *K5* **Normandien** KwaZulu-Natal, South Africa *27°59′S 29°47′E*
Normandy *see* Normandie
71 *H6* **Normangee** Texas, SW USA *31°01′N 96°06′W*
67 *I6* **Norman, Lake** ⊠ North Carolina, SE USA
90 *F8* **Norman Manley** ✈ (Kingston) E Jamaica *17°55′N 76°46′W*
275 *I3* **Norman River** ≈ Queensland, NE Australia
275 *I3* **Normanton** Queensland, NE Australia *17°49′S 141°08′E*
159 *K9* **Normanton** United Kingdom *53°42′N 1°25′W*
54 *D2* **Norman Wells** Northwest Territories, NW Canada *65°18′N 126°42′W*
58 *F7* **Normétal** Québec, S Canada *48°59′N 79°23′W*
85 *H5* **Norogachic** Chihuahua, Mexico *27°15′N 107°07′W*
247 *H4* **Norovlin** *var.* Uldz. Hentiy, NE Mongolia *48°47′N 112°01′E*
55 *I9* **Norquay** Saskatchewan, S Canada *51°51′N 102°04′W*
Norra Karelen *see* Pohjois-Karjala
154 *G2* **Norråker** Jämtland, C Sweden *64°25′N 15°40′E*
154 *H6* **Norrala** Gävleborg, C Sweden *61°22′N 17°04′E*
Norra Ny *see* Stöllet
Norra Österbotten *see* Pohjois-Pohjanmaa
Norra Savolax *see* Pohjois-Savo
152 *E6* **Norra Storjfjället** ▲ N Sweden *65°57′N 15°15′E*
152 *F5* **Norrbotten** ◈ *county* N Sweden
154 *H5* **Norrdellen** ◎ C Sweden
155 *D12* **Nørre Aaby** *var.* Nørre Åby. Syddtjylland, C Denmark *55°28′N 09°53′E*
Nørre Åby *see* Nørre Aaby
155 *E13* **Nørre Alslev** Sjælland, SE Denmark *54°54′N 11°53′E*
155 *C12* **Nørre Nebel** Syddtjylland, W Denmark *55°45′N 08°16′E*
155 *D10* **Nørresundby** Nordjylland, N Denmark *57°05′N 09°55′E*
66 *G5* **Norris Lake** ⊠ Tennessee, S USA
64 *F7* **Norristown** Pennsylvania, NE USA *40°07′N 75°20′W*
155 *H9* **Norrköping** Östergötland, S Sweden *58°35′N 16°10′E*
Norrmark *see* Noormarkku
154 *I6* **Norrsundet** Gävleborg, C Sweden *60°55′N 17°09′E*
154 *J7* **Norrtälje** Stockholm, C Sweden *59°46′N 18°42′E*
274 *E7* **Norseman** Western Australia *32°16′S 121°46′E*
154 *J1* **Norsjö** Västerbotten, N Sweden *64°55′N 19°30′E*
155 *D8* **Norsjø** ◎ S Norway
193 *K8* **Norsk** Amurskaya Oblast', SE Russian Federation *52°20′N 129°57′E*
Norske Havet *see* Norwegian Sea
281 *M3* **Norsup** Malekula, C Vanuatu *16°05′S 167°24′E*
118 *A1* **Norte, Cabo** *headland* Easter Island, Chile, E Pacific Ocean *27°03′S 109°24′W*
102 *C4* **Norte de Santander** *off.* Departamento de Norte de Santander. ◈ *province* N Colombia
Norte de Santander, Departamento de *see* Norte de Santander
70 *C8* **Norte, Meseta del** *plain* N Mexico
116 *I5* **Norte, Punta** *headland* E Argentina *36°17′S 56°46′W*
67 *I8* **North** South Carolina, SE USA *33°37′N 81°06′W*
North *see* Nord
65 *I2* **North Adams** Massachusetts, NE USA *42°40′N 73°06′W*
184 *F6* **North Albanian Alps** *Alb.* Bjeshkët e Namuna, *SCr.* Prokletije. ▲▲ SE Europe
159 *K7* **Northallerton** N England, United Kingdom *54°20′N 01°26′W*
274 *D7* **Northam** Western Australia *31°40′S 116°40′E*
141 *I3* **Northam** Limpopo, South Africa *24°57′S 27°16′E*
138 *G6* **Northam** Northern, N South Africa *24°56′S 27°18′E*
42 **North America** *continent*
43 *H7* **North American Basin** *undersea feature* W Sargasso Sea *30°00′N 60°00′W*
42 *B3* **North American Plate** *tectonic feature*
65 *J2* **North Amherst** Massachusetts, NE USA *42°24′N 72°31′W*
161 *J4* **Northampton** C England, United Kingdom *52°14′N 00°54′W*
65 *J2* **Northampton** Massachusetts, USA *42°20′N 72°38′W*
64 *F6* **Northampton** Pennsylvania, USA *40°41′N 75°30′W*
161 *J4* **Northamptonshire** *cultural region* C England, United Kingdom
257 *A8* **North Andaman** *island* Andaman Islands, India, NE Indian Ocean
65 *L1* **North Andover** Massachusetts, USA *42°42′N 71°08′W*
117 *I11* **North Arm** East Falkland, Falkland Islands *52°06′S 59°21′W*
65 *L3* **North Attleboro** Massachusetts, USA *41°59′N 71°20′W*
67 *H9* **North Augusta** South Carolina, SE USA *33°30′N 81°58′W*
289 *J9* **North Australian Basin** *Fr.* Bassin Nord de l' Australie. *undersea feature* E Indian Ocean
158 *F1* **North Ballachulish** United Kingdom *56°42′N 5°11′W*
73 *H9* **North Baltimore** Ohio, N USA *41°10′N 83°40′W*
57 *L5* **North Battleford** Saskatchewan, S Canada *52°47′N 108°19′W*
62 *D2* **North Bay** Ontario, S Canada *46°20′N 79°28′W*
58 *F3* **North Belcher Islands** *island group* Belcher Islands, Nunavut, C Canada
74 *E7* **North Bend** Nebraska, C USA *41°27′N 96°46′W*
76 *B7* **North Bend** Oregon, NW USA *43°24′N 124°13′W*
65 *I1* **North Bennington** Vermont, USA *42°56′N 73°15′W*
159 *I3* **North Berwick** SE Scotland, United Kingdom *56°04′N 02°44′W*
North Beveland *see* Noord-Beveland
North Borneo *see* Sabah
65 *L2* **Northborough** Massachusetts, USA *42°19′N 71°38′W*
277 *J3* **North Bourke** New South Wales, SE Australia *30°03′S 145°56′E*
North Brabant *see* Noord-Brabant
276 *E2* **North Branch Neales** *seasonal river* South Australia
65 *J5* **North Branford** Connecticut, USA *41°20′N 72°46′W*
91 *H4* **North Caicos** *island* NW Turks and Caicos Islands
75 *D12* **North Canadian River** ≈ Oklahoma, C USA
73 *J10* **North Canton** Ohio, N USA *40°52′N 81°24′W*
59 *K7* **North, Cape** *headland* Cape Breton Island, Nova Scotia, SE Canada *47°06′N 60°24′W*
278 *F1* **North Cape** *headland* North Island, New Zealand *34°23′S 173°02′E*
280 *D2* **North Cape** *headland* New Ireland, NE Papua New Guinea *02°33′S 150°48′E*
North Cape *see* Nordkapp
64 *G10* **North Cape May** New Jersey, NE USA *38°59′N 74°55′W*
58 *C5* **North Caribou Lake** ◎ Ontario, C Canada
67 *K6* **North Carolina** *off.* State of North Carolina, *also known as* Old North State, Tar Heel State, Turpentine State. ◈ *state* SE USA
159 *L9* **North Cave** United Kingdom *53°46′N 0°39′W*
North Celebes *see* Sulawesi Utara
235 *G10* **North Central** ◈ *province* N Sri Lanka
58 *E8* **North Channel** *lake channel* Canada/USA
156 *E7* **North Channel** *strait* Northern Ireland/Scotland, United Kingdom
67 *J9* **North Charleston** South Carolina, SE USA *32°53′N 79°59′W*
159 *K4* **North Charlton** United Kingdom *55°29′N 1°44′W*
73 *D8* **North Chicago** Illinois, N USA *42°19′N 87°50′W*
293 *L6* **Northcliffe Glacier** *glacier* Antarctica
73 *G11* **North College Hill** Ohio, N USA *39°13′N 84°33′W*
70 *E5* **North Concho River** ≈ Texas, SW USA
63 *J4* **North Conway** New Hampshire, NE USA *44°03′N 71°06′W*
161 *L2* **North Creake** United Kingdom *52°54′N 0°46′E*
75 *I14* **North Crossett** Arkansas, C USA *33°10′N 91°56′W*
74 *D3* **North Dakota** *off.* State of North Dakota, *also known as* Flickertail State, Peace Garden State, Sioux State. ◈ *state* N USA
North Devon Island *see* Devon Island
161 *K7* **North Downs** *hill range* SE England, United Kingdom
62 *C6* **North East** Pennsylvania, NE USA *42°13′N 79°49′W*
138 *G4* **North East** ◈ *district* NE Botswana
132 *C9* **North East Bay** *bay* Ascension Island, C Atlantic Ocean
82 *F5* **Northeast Cape** *headland* Saint Lawrence Island, Alaska, USA *63°16′N 168°50′W*
263 *H7* **Northeast Cay** *Chin.* Beizi Dao, *alt.* Parola. *island* NW Spratly Islands
North East Frontier Agency/North East Frontier Agency of Assam *see* Arunāchal Pradesh
282 *F9* **Northeast Island** *island* Chuuk, C Micronesia
90 *G8* **North East Point** *headland* E Jamaica *18°09′N 76°19′W*
285 *N1* **Northeast Point** *headland* Kiritimati, E Kiribati *10°23′S 105°45′E*
90 *G4* **Northeast Point** *headland* Great Inagua, S The Bahamas *21°18′N 73°01′W*
90 *G3* **Northeast Point** *headland* Acklins Island, SE The Bahamas *22°43′N 73°50′W*
61 *M9* **Northeast Providence Channel** *channel* N The Bahamas
81 *E9* **North Edwards** California, USA *35°01′N 117°50′W*
180 *H7* **Northeim** Niedersachsen, C Germany *51°42′N 10°E*
73 *A9* **North English** Iowa, C USA *41°30′N 92°04′W*
218 *E5* **Northern** ◈ *district* N Israel
137 *C12* **Northern** ◈ *region* N Malawi
280 *C4* **Northern** *var.* Oro. ◈ *province* S Papua New Guinea
235 *G9* **Northern** ◈ *province* N Sri Lanka
136 *A1* **Northern** ◈ *state* N Sudan
137 *B12* **Northern** ◈ *province* NE Zambia
Northern *see* Limpopo
134 *I4* **Northern Bahr el Ghazal** ◈ *state* NW South Sudan
Northern Border Region *see* Al Ḥudūd ash Shamālīyah
64 *A6* **Northern Cambria** Pennsylvania, NE USA *40°39′N 78°46′W*
138 *E8* **Northern Cape** *off.* Northern Cape Province, *Afr.* Noord-Kaap. ◈ *province* W South Africa
Northern Cape Province *see* Northern Cape
284 *G5* **Northern Cook Islands** *island group* N Cook Islands
134 *H1* **Northern Darfur** ◈ *state* NW Sudan
Northern Dvina *see* Severnaya Dvina
55 *I6* **Northern Indian Lake** ◎ Manitoba, C Canada
158 *C7* **Northern Ireland** *var.* The Six Counties. ◈ *political division* Northern Ireland, United Kingdom
158 *C7* **Northern Ireland** *var.* The Six Counties. *cultural region* Northern Ireland, United Kingdom
136 *A2* **Northern Kordofan** ◈ *state* C Sudan
280 *G8* **Northern Lau Group** *island group* Lau Group, NE Fiji
282 *E2* **Northern Mariana Islands** ◇ *US commonwealth territory* W Pacific Ocean
Northern Rhodesia *see* Zambia
Northern Sporades *see* Vóreies Sporádes
275 *H4* **Northern Territory** ◈ *territory* N Australia
Northern Transvaal *see* Limpopo
Northern Ural Hills *see* Severnyye Uvaly
159 *I1* **North Esk** ≈ United Kingdom
144 *E6* **North European Plain** *plain* N Europe
75 *I8* **North Fabius River** ≈ Missouri, C USA
117 *I10* **North Falkland Sound** *sound* N Falkland Islands
64 *M4* **North Falmouth** Massachusetts, NE USA *41°38′N 70°37′W*
74 *H5* **Northfield** Minnesota, N USA *44°27′N 93°10′W*
63 *I5* **Northfield** New Hampshire, NE USA *43°26′N 71°34′W*
267 *I5* **North Fiji Basin** *undersea feature* N Coral Sea
161 *N6* **North Foreland** *headland* SE England, United Kingdom *51°22′N 01°26′E*
80 *E5* **North Fork** California, USA *37°14′N 119°31′W*
80 *E1* **North Fork American River** ≈ California, W USA
83 *I3* **North Fork Chandalar River** ≈ Alaska, USA
74 *B4* **North Fork Grand River** ≈ North Dakota/South Dakota, N USA
66 *G4* **North Fork Kentucky River** ≈ Kentucky, S USA
83 *I3* **North Fork Koyukuk River** ≈ Alaska, USA
75 *C12* **North Fork Red River** ≈ Oklahoma/Texas, SW USA
75 *C9* **North Fork Solomon River** ≈ Kansas, C USA
69 *L8* **North Fort Myers** Florida, SE USA *26°40′N 81°52′W*
72 *F5* **North Fox Island** *island* Michigan, N USA
178 *C3* **North Frisian Islands** *var.* Nordfriesische Inseln. *island group* N Germany
295 *I8* **North Geomagnetic Pole** *pole* Arctic Ocean
64 *J2* **North Hatfield** Massachusetts, NE USA *42°24′N 72°37′W*
65 *J4* **North Haven** Connecticut, NE USA *41°25′N 72°51′W*
278 *F3* **North Head** *headland* North Island, New Zealand *36°23′S 174°01′E*
55 *I4* **North Henik Lake** ◎ Nunavut, E Canada
63 *H4* **North Hero** Vermont, NE USA *44°49′N 73°14′W*
80 *D2* **North Highlands** California, W USA *38°40′N 121°25′W*
160 *E9* **North Hill** United Kingdom *50°33′N 4°26′W*
North Holland *see* Noord-Holland
136 *D6* **North Horr** Marsabit, N Kenya *03°17′N 37°08′E*
235 *C14* **North Huvadhu Atoll** *var.* Gaafu Alifu Atoll. *atoll* S Maldives
278 *H3* **North Island** *island* N New Zealand
67 *K9* **North Island** *island* South Carolina, SE USA
73 *E9* **North Judson** Indiana, N USA *41°12′N 86°44′W*
North Kalimantan *see* Kalimantan Utara
North Karelia *see* Pohjois-Karjala
North Kazakhstan *see* Severnyy Kazakhstan
73 *J9* **North Kingsville** Ohio, N USA *41°54′N 80°41′W*
55 *J6* **North Knife River** ≈ Manitoba, C Canada
248 *C4* **North Korea** *off.* Democratic People's Republic of Korea, *Kor.* Chosŏn-minjujuŭi-inmin-kanghwaguk. ◆ *republic* E Asia
233 *M6* **North Lakhimpur** Assam, NE India *27°10′N 94°00′E*
278 *F2* **Northland** *off.* Northland Region. ◈ *region* North Island, New Zealand
286 *F8* **Northland Plateau** *undersea feature* S Pacific Ocean
Northland Region *see* Northland
81 *J9* **North Las Vegas** Nevada, W USA *36°12′N 115°07′W*
161 *I5* **Northleach** United Kingdom *51°50′N 1°49′W*
73 *E9* **North Liberty** Indiana, N USA *41°36′N 86°22′W*
74 *I7* **North Liberty** Iowa, C USA *41°45′N 91°36′W*
75 *I13* **North Little Rock** Arkansas, C USA *34°46′N 92°15′W*
74 *C6* **North Loup River** ≈ Nebraska, C USA
235 *B11* **North Maalhosmadulu Atoll** *var.* North Malosmadulu Atoll, Raa Atoll. *atoll* N Maldives
73 *J9* **North Madison** Ohio, N USA *41°48′N 81°03′W*
North Malosmadulu Atoll *see* North Maalhosmadulu Atoll
73 *F10* **North Manchester** Indiana, N USA *41°00′N 85°45′W*
72 *F5* **North Manitou Island** *island* Michigan, N USA
74 *G5* **North Mankato** Minnesota, N USA *44°11′N 94°03′W*
69 *N9* **North Miami** Florida, SE USA *25°54′N 80°11′W*
235 *C11* **North Miladhunmadulu Atoll** *var.* Shaviyani Atoll. *atoll* N Maldives
North Minch *see* Minch, The
52 *B10* **North Nahanni River** ≈ Northwest Territories, W Canada
69 *L8* **North Naples** Florida, SE USA *26°13′N 81°47′W*
267 *H5* **North New Hebrides Trench** *undersea feature* N Coral Sea
69 *M8* **North New River Canal** ≈ Florida, SE USA
235 *B12* **North Nilandhe Atoll** *atoll* C Maldives
79 *H2* **North Ogden** Utah, W USA *41°18′N 111°57′W*
North Ossetia *see* Severnaya Osetiya-Alaniya, Respublika
North Ostrobothnia *see* Pohjois-Pohjanmaa
80 *F6* **North Palisade** ▲ California, W USA *37°06′N 118°31′W*
282 *F8* **North Pass** *passage* Chuuk Islands, C Micronesia
160 *G7* **North Petherton** United Kingdom *51°05′N 3°01′W*
64 *G6* **North Plainfield** New Jersey, USA *40°38′N 74°26′W*

◆ Country
● Country Capital
◇ Dependent Territory
○ Dependent Territory Capital
◈ Administrative Regions
✈ International Airport
▲ Mountain
▲▲ Mountain Range
Volcano
≈ River
◎ Lake
⊠ Reservoir

Nyasaland/Nyasaland Protectorate - Olderfjord

INDEX

385

Nyasaland/Nyasaland Protectorate *see* Malawi
Nyassa, Lago *see* Nyasa, Lake
191 F10 **Nyasvizh** *Pol.* Nieśwież, *Rus.* Nesvizh. Minskaya Voblasts', C Belarus 53°13′N 26°40′E
256 D6 **Nyaunglebin** Bago, SW Myanmar (Burma) 17°59′N 94°44′E
256 C4 **Nyaung-u** Magway, C Myanmar (Burma) 21°03′N 95°44′E
195 L6 **Nyays** ≈ Khanty-Mansiyskiy Avtonomnyy Okrug-Yugra, Russian Federation
195 L10 **Nyazepetrovsk** Chelyabinskaya Oblast', Russian Federation
155 E13 **Nyborg** Syddtjylland, C Denmark 55°19′N 10°48′E
155 H11 **Nybro** Kalmar, S Sweden 56°45′N 15°54′E
195 N3 **Nyda** Yamalo-Nenetskiy Avtonomnyy Okrug, Russian Federation
191 F10 **Nyeharelaye** *Rus.* Negoreloye. Minskaya Voblasts', C Belarus 53°36′N 27°04′E
293 L2 **Nye Mountains** ▲▲ Antarctica
137 D8 **Nyeri** Nyeri, C Kenya 0°25′S 36°56′E
137 D8 **Nyeri** ⬖ *county* C Kenya
190 H7 **Nyeshcharda, Vozyera** *Rus.* Ozero Neshcherdo. ◎ N Belarus
152 B5 **Ny-Friesland** *physical region* N Svalbard
154 G7 **Nyhammar** Dalarna, C Sweden 60°19′N 14°55′E
240 C3 **Nyikog Qu** ≈ C China
238 F9 **Nyima** Xizang Zizhiqu, W China 31°56′N 87°16′E
139 H1 **Nyimba** Eastern, E Zambia 14°33′S 30°49′E
239 I9 **Nyingchi** *var.* Pula. Xizang Zizhiqu, W China 29°34′N 94°33′E
Nyinma *see* Maqu
183 I11 **Nyírbátor** Szabolcs-Szatmár-Bereg, E Hungary 47°50′N 22°09′E
183 I11 **Nyíregyháza** Szabolcs-Szatmár-Bereg, NE Hungary 47°57′N 21°43′E
Nyitra *see* Nitra
Nyitrabánya *see* Handlová
153 G8 **Nykarleby** *Fin.* Uusikaarlepyy. Österbotten, W Finland 63°22′N 22°30′E
155 C11 **Nykøbing** Midtjylland, NW Denmark 56°48′N 08°52′E
155 E13 **Nykøbing** Sjælland, SE Denmark 54°47′N 11°53′E
155 E12 **Nykøbing** Sjælland, C Denmark 55°56′N 11°41′E
155 I9 **Nyköping** Södermanland, S Sweden 58°45′N 17°03′E
155 G8 **Nykroppa** Värmland, C Sweden 59°37′N 14°18′E
Nyland *see* Uusimaa
138 G6 **Nylstroom** Northern, NE South Africa 24°39′S 28°24′E
Nylstroom *see* Modimolle
277 J4 **Nymagee** New South Wales, SE Australia 32°06′S 146°19′E
277 M3 **Nymboida** New South Wales, SE Australia 29°57′S 152°45′E
277 M3 **Nymboida River** ≈ New South Wales, SE Australia
183 C8 **Nymburk** *var.* Neuenburg an der Elbe, *Ger.* Nimburg. Středočesky Kraj, C Czech Republic 50°12′N 15°00′E
155 I8 **Nynäshamn** Stockholm, C Sweden 58°54′N 17°55′E
277 K4 **Nyngan** New South Wales, SE Australia 31°36′S 147°07′E
Nyoman *see* Neman
176 B5 **Nyon** *Ger.* Neuss; *anc.* Noviodunum. Vaud, SW Switzerland 46°23′N 06°15′E
134 A6 **Nyong** ≈ SW Cameroon
165 J7 **Nyons** Drôme, E France 44°22′N 05°06′E
134 A5 **Nyos, Lac** *Eng.* Lake Nyos. ◎ NW Cameroon
Nyos, Lake *see* Nyos, Lac
195 K7 **Nyrob** *var.* Nyrov. Permskiy Kray, NW Russian Federation 60°41′N 56°42′E
Nyrov *see* Nyrob
183 E8 **Nysa** *Ger.* Neisse. Opolskie, S Poland 50°28′N 17°20′E
76 F7 **Nyssa** Oregon, NW USA 43°52′N 116°59′W
Nysa Łużycka *see* Neisse
Nyslott *see* Savonlinna
Nystad *see* Uusikaupunki
155 E13 **Nysted** Sjælland, SE Denmark 54°40′N 11°41′E
195 K9 **Nytva** Permskiy Kray, NW Russian Federation 57°56′N 55°22′E
253 C8 **Nyūdō-zaki** *headland* Honshū, C Japan 39°59′N 139°40′E
195 H6 **Nyukhcha** Arkhangel'skaya Oblast', NW Russian Federation 63°24′N 46°34′E
194 E5 **Nyuk, Ozero** *var.* Ozero Njuk. ◎ NW Russian Federation
195 H8 **Nyuksenitsa** *var.* Njuksenica. Vologodskaya Oblast', NW Russian Federation 60°25′N 44°12′E
135 I10 **Nyunzu** Katanga, SE Dem. Rep. Congo 05°55′S 28°00′E
193 I6 **Nyurba** Respublika Sakha (Yakutiya), NE Russian Federation 63°17′N 118°15′E
193 I6 **Nyuya** Respublika Sakha (Yakutiya), NE Russian Federation 60°33′N 116°10′E
193 I6 **Nyuya** ≈ NE Russian Federation
228 F7 **Nyýazow** *Rus.* Niyazov. Lebap Welaýaty, NE Turkmenistan 39°13′N 63°16′E
189 J7 **Nyzhni Sirohozy** Khersons'ka Oblast', S Ukraine 46°49′N 34°21′E
189 K8 **Nyzhn'ohirs'kyy** *Rus.* Nizhnegorskiy. Avtonomna Respublika Krym, S Ukraine 45°26′N 34°42′E
NZ *see* New Zealand
137 C9 **Nzega** Tabora, C Tanzania 04°13′S 33°11′E
132 D7 **Nzérékoré** SE Guinea 07°45′N 08°49′W
135 B10 **N'Zeto** *prev.* Ambrizete. Zaire Province, NW Angola 07°14′S 12°52′E
141 K1 **Nzhelele** ≈ Limpopo, South Africa
135 H11 **Nzilo, Lac** *prev.* Lac Delcommune. ◎ SE Dem. Rep. Congo
139 I9 **Nzwani** *Fr.* Anjouan, *var.* Ndzouani. *island* SE Comoros

O

74 D5 **Oacoma** South Dakota, N USA 43°49′N 99°25′W
161 J3 **Oadby** United Kingdom 52°36′N 1°05′W
74 C5 **Oahe Dam** *dam* South Dakota, N USA
74 C5 **Oahe, Lake** ⊞ North Dakota/South Dakota, N USA
82 B2 **Oa'hu** *var.* Oahu. *island* Hawai'ian Islands, Hawai'i, USA
252 G4 **O-Akan-dake** ▲ Hokkaidō, NE Japan 43°26′N 144°09′E
277 H5 **Oakbank** South Australia 33°07′S 140°36′E
65 M4 **Oak Bluffs** Martha's Vineyard, New York, NE USA 41°25′N 70°32′W
79 H3 **Oak City** Utah, W USA 39°22′N 112°19′W
79 K2 **Oak Creek** Colorado, C USA 40°16′N 106°57′W
80 D3 **Oakdale** California, W USA 37°46′N 120°51′W
68 C3 **Oakdale** Louisiana, S USA 30°49′N 92°39′W
161 H3 **Oakengates** United Kingdom 52°42′N 2°26′W
74 E4 **Oakes** North Dakota, N USA 46°08′N 98°05′W
160 F7 **Oakford** United Kingdom 50°59′N 3°33′W
80 D3 **Oak Grove** California, USA 38°07′N 120°53′W
80 E6 **Oak Grove** California, USA 36°27′N 118°48′W
81 F12 **Oak Grove** California, USA 33°23′N 116°47′W
68 D1 **Oak Grove** Louisiana, S USA 32°51′N 91°25′W
161 J3 **Oakham** C England, United Kingdom 52°41′N 00°45′W
76 C3 **Oak Harbor** Washington, NW USA 48°17′N 122°38′W
67 I4 **Oak Hill** West Virginia, NE USA 37°59′N 81°09′W
161 K4 **Oakington** United Kingdom 52°16′N 0°04′E
80 B3 **Oakland** California, W USA 37°48′N 122°16′W
74 F7 **Oakland** Iowa, C USA 41°18′N 95°22′W
63 K4 **Oakland** Maine, NE USA 44°32′N 69°43′W
67 J2 **Oakland** Maryland, NE USA 39°24′N 79°25′W
74 F7 **Oakland** Nebraska, C USA 41°50′N 96°28′W
73 D9 **Oak Lawn** Illinois, N USA 41°43′N 87°45′W
80 C3 **Oakley** California, USA 38°00′N 121°43′W
77 H8 **Oakley** Idaho, NW USA 42°13′N 113°54′W
75 C9 **Oakley** Kansas, C USA 39°08′N 100°53′W
73 D9 **Oak Park** Illinois, N USA 41°53′N 87°46′W
55 J9 **Oak Point** Manitoba, S Canada 50°23′N 97°00′W
76 C7 **Oakridge** Oregon, NW USA 43°45′N 122°27′W
66 F6 **Oak Ridge** Tennessee, S USA 36°02′N 84°12′W
278 F6 **Oakura** Taranaki, North Island, New Zealand 39°07′S 173°58′E
68 E3 **Oak Vale** Mississippi, S USA 31°26′N 89°57′W
81 C9 **Oak View** California, USA 34°24′N 119°18′W
62 D5 **Oakville** Ontario, S Canada 43°27′N 79°41′W
71 H5 **Oakwood** Texas, SW USA 31°34′N 95°51′W
278 D11 **Oamaru** Otago, South Island, New Zealand 45°10′S 170°51′E
158 E4 **Oa, Mull of** *headland* W Scotland, United Kingdom 55°35′N 06°20′W
260 D3 **Oan** Sulawesi, N Indonesia 01°16′N 121°25′E
279 F9 **Oaro** Canterbury, South Island, New Zealand 42°29′S 173°30′E
81 F12 **Oasis** California, USA 33°28′N 116°06′W
78 G2 **Oasis** Nevada, W USA 41°01′N 114°29′W
293 J8 **Oates Land** *physical region* Antarctica
277 J10 **Oatlands** Tasmania, SE Australia 42°21′S 147°23′E
81 J11 **Oatman** Arizona, SW USA 35°03′N 114°19′W
86 G8 **Oaxaca** *var.* Oaxaca de Juárez; *prev.* Antequera. Oaxaca, SE Mexico 17°04′N 96°41′W
86 G8 **Oaxaca** ⬖ *state* SE Mexico
Oaxaca de Juárez *see* Oaxaca
236 C1 **Ob'** ≈ C Russian Federation
195 M6 **Ob'** ≈ Khanty-Mansiyskiy Avtonomnyy Okrug-Yugra, Russian Federation
195 M4 **Ob'** ≈ Yamalo-Nenetskiy Avtonomnyy Okrug, Russian Federation
72 G1 **Oba** Ontario, S Canada 49°03′N 84°06′W
227 L3 **Oba** *prev.* Uba. ≈ E Kazakhstan
62 C1 **Obabika Lake** ◎ Ontario, S Canada
Obagan *see* Ubagan
191 H8 **Obal'** *Rus.* Obol'. Vitsyebskaya Voblasts', N Belarus 55°22′N 29°17′E
134 B6 **Obala** Centre, SW Cameroon 04°09′N 11°32′E
72 G2 **Oba Lake** ◎ Ontario, S Canada
251 I4 **Obama** Fukui, Honshū, SW Japan 35°32′N 135°45′E
158 F2 **Oban** W Scotland, United Kingdom 56°25′N 05°29′W
Oban *see* Halfmoon Bay
253 D10 **Obanazawa** Yamagata, Honshū, C Japan 38°37′N 140°22′E
Obando *see* Puerto Inírida
170 D2 **O Barco** *var.* El Barco, El Barco de Valdeorras, O Barco de Valdeorras. Galicia, NW Spain 42°24′N 07°00′W
O Barco de Valdeorras *see* O Barco
Obbia *see* Hobyo
154 J3 **Obbola** Västerbotten, N Sweden 63°41′N 20°16′E
Obbrovazzo *see* Obrovac
Obchuga *see* Abchuha
Obdorsk *see* Salekhard
Óbecse *see* Bečej
190 E7 **Obeliai** Panevėžys, NE Lithuania 55°57′N 25°47′E
113 H6 **Oberá** Misiones, NE Argentina 27°29′S 55°08′W
181 G9 **Oberaula** Hessen, Germany 50°51′E 9°28′N
181 H11 **Oberbach** Bayern, Germany 50°21′E 9°53′N
176 D4 **Oberburg** Bern, W Switzerland 47°00′N 07°37′E
181 I13 **Oberdachstetten** Bayern, Germany 49°25′E 10°26′N
177 I4 **Oberdrauburg** Salzburg, S Austria 46°45′N 12°59′E
181 H11 **Oberelsbach** Bayern, Germany 50°26′E 10°08′N
Oberglogau *see* Głogówek
177 K2 **Ober Grafendorf** Niederösterreich, NE Austria 48°09′N 15°33′E
181 C8 **Oberhausen** Nordrhein-Westfalen, W Germany 51°27′N 06°50′E
181 J10 **Oberhof** Thüringen, Germany 50°43′E 10°44′N
Oberhollabrunn *see* Tulln
181 D11 **Oberlahnstein** Rheinland-Pfalz, Germany 50°18′E 7°37′N
Oberlaibach *see* Vrhnika
179 I8 **Oberlausitz** *var.* Hornja Łužica. *physical region* E Germany
75 C9 **Oberlin** Kansas, C USA 39°49′N 100°33′W
68 C4 **Oberlin** Louisiana, S USA 30°37′N 92°45′W
73 I9 **Oberlin** Ohio, N USA 41°17′N 82°13′W
181 I10 **Obermassfeld-Grimmenthal** Thüringen, Germany 50°32′E 10°26′N
181 I8 **Obermehler** Thüringen, Germany 51°16′E 10°36′N
165 K3 **Obernai** Bas-Rhin, NE France 48°28′N 07°30′E
177 I2 **Obernberg am Inn** Oberösterreich, N Austria 48°19′N 13°20′E
181 E10 **Oberndorf** Hessen, Germany 50°45′E 8°24′N
180 G2 **Oberndorf** Niedersachsen, Germany 53°45′E 9°09′N
Oberndorf *see* Oberndorf am Neckar
179 D12 **Oberndorf am Neckar** *var.* Oberndorf. Baden-Württemberg, SW Germany 48°18′N 08°32′E
177 I3 **Oberndorf bei Salzburg** Salzburg, W Austria 47°57′N 12°57′E
Oberneustadtl *see* Kysucké Nové Mesto
180 G6 **Obernkirchen** Niedersachsen, Germany 52°16′E 9°08′N
277 L5 **Oberon** New South Wales, SE Australia 33°42′S 149°50′E
179 I13 **Oberösterreich** *off.* Land Oberösterreich, *Eng.* Upper Austria. ⬖ *state* NW Austria
Oberösterreich, Land *see* Oberösterreich
Oberpahlen *see* Põltsamaa
179 G10 **Oberpfälzer Wald** ▲▲ SE Germany
177 L3 **Oberpullendorf** Burgenland, E Austria 47°32′N 16°30′E
Oberradkersburg *see* Gornja Radgona
181 F12 **Ober-Ramstadt** Hessen, Germany 49°50′E 8°45′N
181 J8 **Oberröblingen** Sachsen-Anhalt, Germany 51°26′E 11°19′N
181 G14 **Oberrot** Baden-Württemberg, Germany 49°01′E 9°40′N
181 H11 **Obersinn** Bayern, Germany 50°13′E 9°37′N
181 G14 **Oberstenfeld** Baden-Württemberg, Germany 49°02′E 9°19′N
181 I12 **Obertheres** Bayern, Germany 50°01′E 10°26′N
181 E10 **Obertiefenbach** Hessen, Germany 50°28′E 8°07′N
181 F11 **Oberursel** Hessen, W Germany 50°12′N 08°34′E
177 I4 **Obervellach** Salzburg, S Austria 46°56′N 13°10′E
177 L3 **Oberwart** Burgenland, SE Austria 47°18′N 16°12′E
181 J10 **Oberweissbach** Thüringen, Germany 50°35′E 11°09′N
181 D11 **Oberwesel** Rheinland-Pfalz, Germany 50°06′E 7°44′N
Oberwischau *see* Vişeu de Sus
177 J4 **Oberwölz** *var.* Oberwölz-Stadt. Steiermark, SE Austria 47°12′N 14°20′E
Oberwölz-Stadt *see* Oberwölz
181 H11 **Oberzell** Hessen, Germany 50°20′E 9°43′N
73 H11 **Obetz** Ohio, N USA 39°52′N 82°57′W
Ob', Gulf of *see* Obskaya Guba
107 J2 **Óbidos** Pará, NE Brazil 01°52′S 55°30′W
172 A4 **Óbidos** Leiria, C Portugal 39°21′N 09°09′W
Obidovichi *see* Abidavichy
229 I7 **Obigarm** W Tajikistan 38°42′N 69°34′E
252 F5 **Obihiro** Hokkaidō, NE Japan 42°56′N 143°10′E
Obi-Khingou *see* Khingov
229 I7 **Obikiik** SW Tajikistan 38°07′N 68°36′E
Obilić *see* Obiliq
184 G6 **Obiliq** *Serb.* Obilić. N Kosovo 42°50′N 20°57′E
197 I7 **Obil'noye** Respublika Kalmykiya, SW Russian Federation 47°31′N 44°24′E
66 B5 **Obion** Tennessee, S USA 36°15′N 89°11′W
66 B5 **Obion River** ≈ Tennessee, S USA
260 G4 **Obi, Pulau** *island* Maluku, E Indonesia
252 E3 **Obira** Hokkaidō, NE Japan 44°01′N 141°39′E
260 G4 **Obi, Selat** *strait* Maluku, E Indonesia
168 F2 **Objat** Limousin, France 45°16′N 1°25′E
197 H6 **Oblivskaya** Rostovskaya Oblast', SW Russian Federation 48°34′N 42°31′E
193 K8 **Obluch'ye** Yevreyskaya Avtonomnaya Oblast', SE Russian Federation 48°59′N 131°18′E
196 F2 **Obninsk** Kaluzhskaya Oblast', W Russian Federation 55°06′N 36°40′E
185 K5 **Obnova** Pleven, N Bulgaria 43°26′N 25°04′E
134 H5 **Obo** Haut-Mbomou, E Central African Republic 05°20′N 26°29′E
140 E5 **Obobogorab** Northern Cape, South Africa 72°31′S 20°05′E
136 G3 **Obock** E Djibouti 11°57′N 43°09′E
Obol' *see* Obal'
Obolyanka *see* Abalyanka
261 I6 **Obome** Papua Barat, E Indonesia 03°42′S 133°21′E
182 E5 **Oborniki** Wielkolpolskie, W Poland 52°38′N 16°48′E
135 D8 **Obouya** Cuvette, C Congo 0°56′S 15°41′E
196 F4 **Oboyan'** Kurskaya Oblast', W Russian Federation 51°12′N 36°15′E
194 G6 **Obozerskiy** Arkhangel'skaya Oblast', NW Russian Federation 63°26′N 40°20′E
184 F4 **Obrenovac** Serbia, N Serbia 44°39′N 20°12′E
181 G13 **Obrigheim** Baden-Württemberg, Germany 49°21′E 9°05′N
184 B4 **Obrovac** *It.* Obbrovazzo. Zadar, SW Croatia 44°12′N 15°40′E
Obrovo *see* Abrova
78 C2 **Observation Peak** ▲ California, W USA 40°48′N 120°07′W
195 N2 **Obskaya Guba** *Eng.* Gulf of Ob. *gulf* N Russian Federation
289 C12 **Ob' Tablemount** *undersea feature* S Indian Ocean 50°16′S 51°59′E
289 H11 **Ob' Trench** *undersea feature* E Indian Ocean
132 G8 **Obuasi** S Ghana 06°15′N 01°36′W
189 H3 **Obukhiv** *Rus.* Obukhov. Kyyivs'ka Oblast', N Ukraine 50°05′N 30°37′E
Obukhov *see* Obukhiv
195 K9 **Obva** ≈ NW Russian Federation
189 L7 **Obytichna Kosa** *spit* SE Ukraine
189 K7 **Obytichna Zatoka** *gulf* SE Ukraine
185 M6 **Obzor** Burgas, E Bulgaria 42°52′N 27°53′E
170 G2 **Oca** ≈ N Spain
69 L5 **Ocala** Florida, SE USA 29°11′N 82°08′W
84 G4 **Ocampo** Chihuahua, Mexico 28°11′N 108°23′W
85 K5 **Ocampo** Coahuila, NE Mexico 27°18′N 102°24′W
86 D4 **Ocampo** Guanajuato, Mexico 21°39′N 101°30′W
85 M8 **Ocampo** Tamaulipas, Mexico 22°50′N 99°20′W
102 C4 **Ocaña** Norte de Santander, N Colombia 08°16′N 73°21′W
170 G5 **Ocaña** Castilla-La Mancha, C Spain 39°57′N 03°30′W
170 D2 **O Carballiño** *Cast.* Carballiño. Galicia, NW Spain 42°26′N 08°05′W
79 M6 **Ocate** New Mexico, SW USA 36°09′N 105°03′W
Ocavango *see* Okavango
104 D7 **Occidental, Cordillera** ▲▲ W South America
112 B2 **Occidental, Cordillera** ▲▲ Bolivia/Chile
67 I4 **Oceana** West Virginia, NE USA 37°41′N 81°37′W
67 N3 **Ocean City** Maryland, NE USA 38°20′N 75°05′W
64 G9 **Ocean City** New Jersey, NE USA 39°15′N 74°33′W
56 B5 **Ocean Falls** British Columbia, SW Canada 52°24′N 127°42′W
Ocean Island *see* Banaba
Ocean Island *see* Kure Atoll
80 B7 **Oceano** California, USA 35°06′N 120°37′W
290 F4 **Oceanographer Fracture Zone** *tectonic feature* NW Atlantic Ocean
81 E12 **Oceanside** California, W USA 33°12′N 117°23′W
68 F4 **Ocean Springs** Mississippi, S USA 30°24′N 88°49′W
Ocean State *see* Rhode Island
70 E5 **O C Fisher Lake** ⊞ Texas, SW USA
189 I7 **Ochakiv** *Rus.* Ochakov. Mykolayivs'ka Oblast', S Ukraine 46°36′N 31°33′E
Ochakov *see* Ochakiv
Ochamchira *see* Ochamchire
215 I3 **Ochamchire** *Rus.* Ochamchira; *prev.* Och'amch'ire. W Georgia 42°45′N 41°30′E
Och'amch'ire *see* Ochamchire
Ochansk *see* Okhansk
195 K9 **Ochër** Permskiy Kray, NW Russian Federation 57°54′N 54°40′E
187 I6 **Óchi** ▲ Évvoia, C Greece 38°03′N 24°27′E
252 I4 **Ochiishi-misaki** *headland* Hokkaidō, NE Japan 43°10′N 145°29′E
159 H2 **Ochil Hills** ▲▲ United Kingdom
158 G4 **Ochiltree** United Kingdom 55°27′N 4°22′W
69 I4 **Ochlockonee River** ≈ Florida/Georgia, SE USA
250 A2 **Och'on** 36°17′N 127°33′E
90 F7 **Ocho Rios** C Jamaica 18°24′N 77°06′W
Ochrida *see* Ohrid
Ochrida, Lake *see* Ohrid, Lake
181 H13 **Ochsenfurt** Bayern, C Germany 49°39′N 10°03′E
181 D11 **Ochtendung** Rheinland-Pfalz, Germany 50°21′E 7°23′N
180 C6 **Ochtrup** Nordrhein-Westfalen, Germany 52°13′E 7°11′N
69 K3 **Ocilla** Georgia, SE USA 31°35′N 83°15′W
154 H6 **Ockelbo** Gävleborg, C Sweden 60°51′N 16°46′E
Ocker *see* Oker
155 E10 **Öckerö** Västra Götaland, S Sweden 57°43′N 11°39′E
69 K2 **Ocmulgee River** ≈ Georgia, SE USA
188 C7 **Ocna Mureş** *Hung.* Marosújvár; *prev.* Ocna Mureşului, *prev. Hung.* Marosújvárakna. Alba, C Romania 46°25′N 23°53′E
Ocna Mureşului *see* Ocna Mureş
188 C8 **Ocna Sibiului** *Ger.* Salzburg, *Hung.* Vizakna. Sibiu, C Romania 45°52′N 23°59′E
188 C9 **Ocnele Mari** *prev.* Vioara. Vâlcea, S Romania 45°03′N 24°18′E
188 F5 **Ocniţa** *Rus.* Oknitsa. N Moldova 48°25′N 27°30′E
66 G9 **Oconee, Lake** ⊞ Georgia, SE USA
69 K2 **Oconee River** ≈ Georgia, SE USA
72 D7 **Oconomowoc** Wisconsin, N USA 43°06′N 88°29′W
72 D6 **Oconto** Wisconsin, N USA 44°55′N 87°52′W
72 D6 **Oconto Falls** Wisconsin, N USA 44°52′N 88°06′W
72 D6 **Oconto River** ≈ Wisconsin, N USA
170 D2 **O Corgo** Galicia, NW Spain 42°56′N 07°25′W
87 K8 **Ocosingo** Chiapas, SE Mexico 17°04′N 92°15′W
88 G5 **Ocotal** Nueva Segovia, NW Nicaragua 13°38′N 86°28′W
86 G8 **Ocotepec** Oaxaca, Mexico 17°09′N 97°46′W
88 D4 **Ocotepeque** ⬖ *department* W Honduras
Ocotepeque *see* Nueva Ocotepeque
81 F13 **Ocotillo** California, USA 32°44′N 116°00′W
81 F13 **Ocotillo Wells** California, USA 33°09′N 116°08′W
86 C5 **Ocotlán** Jalisco, SW Mexico 20°21′N 102°42′W
87 H8 **Ocotlán** *var.* Ocotlán de Morelos. Oaxaca, SE Mexico 16°49′N 96°49′W
Ocotlán de Morelos *see* Ocotlán
87 J8 **Ocozocuautla** Chiapas, SE Mexico 16°46′N 93°22′W
67 M7 **Ocracoke Island** *island* North Carolina, SE USA
164 F2 **Octeville** Manche, N France 49°37′N 01°39′W
October Revolution Island *see* Oktyabr'skoy Revolyutsii, Ostrov
89 K10 **Ocú** Herrera, S Panama 07°55′N 80°43′W
137 E13 **Ocua** Cabo Delgado, NE Mozambique 13°37′S 39°44′E
Ocumare *see* Ocumare del Tuy
102 G3 **Ocumare del Tuy** *var.* Ocumare. Miranda, N Venezuela 10°07′N 66°47′W
133 H8 **Oda** SE Ghana 05°55′N 00°56′W
250 E4 **Ōda** *var.* Oda. Shimane, Honshū, SW Japan 35°10′N 132°29′E
152 D2 **Ódaðahraun** *lava flow* C Iceland
261 L7 **Odammun** ≈ Papua, E Indonesia
253 D8 **Ōdate** Akita, Honshū, C Japan 40°18′N 140°34′E
253 B14 **Odawara** Kanagawa, Honshū, S Japan 35°15′N 139°08′E
154 B7 **Odda** Hordaland, S Norway 60°03′N 06°34′E
155 D12 **Odder** Midtjylland, C Denmark 55°59′N 10°10′E
Oddur *see* Xuddur
74 F7 **Odebolt** Iowa, C USA 42°18′N 95°15′W
172 B7 **Odeceixe** Faro, Portugal 37°26′N 8°46′W
168 G7 **Odeillo-Via** Languedoc-Roussillon, France 42°30′N 2°02′E
172 C8 **Odeleite** Faro, S Portugal 37°02′N 07°29′W
172 C8 **Odeleite, Ribeira de** ≈ Portugal
70 F3 **Odell** Texas, SW USA 34°19′N 99°24′W
172 B8 **Odelouca, Ribeira de** ≈ Portugal
70 G8 **Odem** Texas, SW USA 27°57′N 97°34′W
172 B7 **Odemira** Beja, S Portugal 37°35′N 08°38′W
214 B6 **Ödemiş** İzmir, SW Turkey 38°11′N 27°58′E
Ödenburg *see* Sopron
138 G7 **Odendaalsrus** Free State, C South Africa 27°52′S 26°42′E
181 F14 **Odenheim** Baden-Württemberg, Germany 49°11′E 8°45′N
Odenpäh *see* Otepää
155 D12 **Odense** Syddtjylland, C Denmark 55°24′N 10°23′E
64 D9 **Odenton** Maryland, USA 39°05′N 76°42′W
181 F12 **Odenwald** ▲▲ W Germany
144 D6 **Oder** *Cz./Pol.* Odra. ≈ C Europe
178 I6 **Oderbruch** *wetland* Germany/Poland
Oderhaff *see* Szczeciński, Zalew
178 I6 **Oder-Havel-Kanal** *canal* NE Germany
Oderhellen *see* Odorheiu Secuiesc
181 D12 **Odernheim** Rheinland-Pfalz, Germany 49°46′E 7°42′N
178 I7 **Oder-Spree-Kanal** *canal* NE Germany
Odertal *see* Zdzieszowice
177 I6 **Oderzo** Veneto, NE Italy 45°48′N 12°33′E
189 H7 **Odesa** *Rus.* Odessa. Odes'ka Oblast', SW Ukraine 46°29′N 30°44′E
70 D5 **Odessa** Texas, SW USA 31°51′N 102°22′W
76 E4 **Odessa** Washington, NW USA 47°19′N 118°41′W
Odesa *see* Odes'ka Oblast'
155 G9 **Ödeshög** Östergötland, S Sweden 58°13′N 14°40′E
189 H7 **Odes'ka Oblast'** *var.* Odesa, *Rus.* Odesskaya Oblast'. ⬖ *province* SW Ukraine
Odessa *see* Odesa
Odesskaya Oblast' *see* Odes'ka Oblast'
192 E7 **Odesskoye** Omskaya Oblast', C Russian Federation 54°15′N 72°45′E
Odessus *see* Varna
164 D4 **Odet** ≈ NW France
170 D8 **Odiel** ≈ SW Spain
132 E6 **Odienné** NW Ivory Coast 09°32′N 07°35′W
263 L6 **Odiongan** Tablas Island, C Philippines 12°23′N 122°01′E
234 I4 **Odisha** *prev.* Orissa. ⬖ *state* NE India
172 C6 **Odivelas** Beja, Portugal 38°10′N 8°09′W
172 A5 **Odivelas** Lisboa, Portugal 38°47′N 9°11′W
188 E8 **Odobeşti** Vrancea, E Romania 45°46′N 27°06′E
182 E7 **Odolanów** *Ger.* Adelnau. Wielkopolskie, C Poland 51°35′N 17°42′E
257 H9 **Ôdôngk** Kâmpóng Spoe, S Cambodia 11°48′N 104°45′E
70 D4 **O'donnell** Texas, SW USA 32°57′N 101°49′W
162 I4 **Odoorn** Drenthe, NE Netherlands 52°52′N 06°49′E
Odorhei *see* Odorheiu Secuiesc
188 D7 **Odorheiu Secuiesc** *Ger.* Oderhellen, *Hung.* Vámosudvarhely; *prev.* Odorhei, *Ger.* Hofmarkt. Harghita, C Romania 46°18′N 25°19′E
Odra *see* Oder
136 H4 **Odweyne** ⬖ *region* NW Somalia
184 F2 **Odžaci** *Ger.* Hodschag, *Hung.* Hodság. Vojvodina, NW Serbia 45°31′N 19°15′E
181 G8 **Oedelsheim** Hessen, Germany 51°36′E 9°36′N
250 B3 **Oedong** 35°41′N 129°19′E
108 F5 **Oeiras** Piauí, E Brazil 07°00′S 42°07′W
172 A5 **Oeiras** Lisboa, C Portugal 38°41′N 09°18′W
180 E7 **Oelde** Nordrhein-Westfalen, W Germany 51°49′N 08°09′E
74 A6 **Oelrichs** South Dakota, N USA 43°08′N 103°13′W
179 G9 **Oelsnitz** Sachsen, E Germany 50°22′N 12°12′E
Oels/Oels in Schlesien *see* Oleśnica
74 H6 **Oelwein** Iowa, C USA 42°40′N 91°54′W
Oeniadae *see* Oiniádes
285 M8 **Oeno Island** *atoll* Pitcairn Group of Islands, C Pacific Ocean
180 F7 **Oerlinghausen** Nordrhein-Westfalen, Germany 51°58′E 8°40′N
Oesel *see* Saaremaa
176 G4 **Oetz** *var.* Ötz. Tirol, W Austria 47°15′N 10°56′E
215 I5 **Of** Trabzon, NE Turkey 40°57′N 40°17′E
Ofakim *see* Ofaqim
73 B12 **O'Fallon** Illinois, N USA 38°35′N 89°54′W
66 A2 **O'Fallon** Missouri, C USA 38°54′N 90°31′W
175 H9 **Ofanto** ≈ S Italy
219 D9 **Ofaqim** *var.* Ofakim. Southern, S Israel 31°17′N 34°37′E
141 K2 **Ofcolaco** Limpopo, South Africa 24°06′S 30°23′E
158 B10 **Offaly** *Ir.* Ua Uíbh Fhailí; *prev.* King's County. *cultural region* C Ireland
181 F11 **Offenbach** *var.* Offenbach am Main. Hessen, W Germany 50°06′N 08°46′E
Offenbach am Main *see* Offenbach
181 D13 **Offenbach-Hundheim** Rheinland-Pfalz, Germany 49°37′E 7°33′N
179 C12 **Offenburg** Baden-Württemberg, SW Germany 48°28′N 07°57′E
276 D2 **Officer Creek** *seasonal river* South Australia
166 F3 **Offranville** Haute-Normandie, France 49°52′N 1°03′E
Oficina María Elena *see* María Elena
Oficina Pedro de Valdivia *see* Pedro de Valdivia
187 J8 **Ofidoússa** *island* Kykládes, Greece, Aegean Sea
Ofiral *see* Sharm ash Shaykh
152 E4 **Ofotfjorden** *fjord* N Norway
285 L2 **Ofu** *island* Manua Islands, E American Samoa
253 E9 **Ōfunato** Iwate, Honshū, C Japan 39°04′N 141°43′E
253 C8 **Oga** Akita, Honshū, C Japan 39°56′N 139°47′E
Ogaadeen *see* Ogaden
253 D9 **Ogachi** Akita, Honshū, C Japan 39°03′N 140°26′E
253 D9 **Ogachi-tōge** *pass* Honshū, C Japan
136 H5 **Ogaden** *Som.* Ogaadeen. *plateau* Ethiopia/Somalia
253 C8 **Oga-hantō** *peninsula* Honshū, C Japan
251 I4 **Ōgaki** Gifu, Honshū, SW Japan 35°22′N 136°35′E
74 B7 **Ogallala** Nebraska, C USA 41°09′N 101°44′W
258 F7 **Ogan, Air** ≈ Sumatera, W Indonesia
251 N9 **Ogasawara-shotō** *Eng.* Bonin Islands. *island group* SE Japan
62 E1 **Ogascanane, Lac** ◎ Québec, SE Canada
252 E7 **Ogawara-ko** ◎ Honshū, C Japan
133 J7 **Ogbomosho** *var.* Ogmoboso. Oyo, W Nigeria 08°10′N 04°16′E
161 I6 **Ogbourne Saint George** United Kingdom 51°27′N 1°43′W
74 G7 **Ogden** Iowa, C USA 42°03′N 94°01′W
79 H2 **Ogden** Utah, W USA 41°09′N 111°58′W
62 G4 **Ogdensburg** New York, NE USA 44°42′N 75°25′W
280 G10 **Ogea Driki** *island* Lau Group, E Fiji
280 G10 **Ogea Levu** *island* Lau Group, E Fiji
69 L2 **Ogeechee River** ≈ Georgia, SE USA
Oger *see* Ogre
253 A11 **Ogi** Niigata, Sado, C Japan 37°49′N 138°16′E
54 B1 **Ogilvie** Yukon, NW Canada 63°34′N 139°44′W
54 B1 **Ogilvie** ≈ Yukon, NW Canada
54 B1 **Ogilvie Mountains** ▲▲ Yukon, NW Canada
Oginskiy Kanal *see* Ahinski Kanal
239 J1 **Ögiynuur** *var.* Dzegstey. Arhangay, C Mongolia 47°38′N 102°31′E
228 D3 **Og'iyon Sho'rxogi** *wetland* NW Uzbekistan
228 B6 **Oglanly** Balkan Welaýaty, W Turkmenistan 39°56′N 54°25′E
69 J2 **Oglethorpe** Georgia, SE USA 32°17′N 84°03′W
66 F7 **Oglethorpe, Mount** ▲ Georgia, SE USA 34°29′N 84°19′W
176 F6 **Oglio** *anc.* Ollius. ≈ N Italy
Ogmoboso *see* Ogbomosho
160 F6 **Ogmore-by-Sea** United Kingdom 51°28′N 3°38′W
165 K4 **Ognon** ≈ E France
260 D3 **Ogoamas, Pegunungan** ▲▲ Sulawesi, N Indonesia
193 K7 **Ogodzha** Amurskaya Oblast', S Russian Federation 52°51′N 132°49′E
133 K8 **Ogoja** Cross River, S Nigeria 06°37′N 08°48′E
58 D6 **Ogoki** ≈ Ontario, S Canada
58 D6 **Ogoki Lake** ◎ Ontario, C Canada
Ögöömör *see* Hanhongor
134 B7 **Ogooué** ≈ Congo/Gabon
134 C7 **Ogooué-Ivindo** *off.* Province de l'Ogooué-Ivindo, *var.* L'Ogooué-Ivindo. ⬖ *province* N Gabon
Ogooué-Ivindo, Province de l' *see* Ogooué-Ivindo
135 B8 **Ogooué-Lolo** *off.* Province de l'Ogooué-Lolo, *var.* L'Ogooué-Lolo. ⬖ *province* C Gabon
Ogooué-Lolo, Province de l' *see* Ogooué-Lolo
135 A8 **Ogooué-Maritime** *off.* Province de l'Ogooué-Maritime, *var.* L'Ogooué-Maritime. ⬖ *province* W Gabon
Ogooué-Maritime, Province de l' *see* Ogooué-Maritime
190 I3 **Ogoreli** Leningradskaya Oblast', Russian Federation
250 C6 **Ogōri** Fukuoka, Kyūshū, SW Japan 33°24′N 130°34′E
250 D5 **Ogōri** Yamaguchi, Honshū, SW Japan 34°05′N 131°20′E
185 I5 **Ogosta** ≈ NW Bulgaria
185 I8 **Ogražden** *Bul.* Ograzhden. ▲ Bulgaria/FYR Macedonia *see also* Ograzhden
185 I8 **Ograzhden** *Mac.* Orgražden. ▲ Bulgaria/FYR Macedonia *see also* Orgražden
Ograzhden *see* Ogražden
190 D6 **Ogre** *Ger.* Oger. C Latvia 56°49′N 24°36′E
190 E6 **Ogre** ≈ C Latvia
184 B3 **Ogulin** Karlovac, NW Croatia 45°15′N 15°13′E
133 I7 **Ogun** ⬖ *state* SW Nigeria
Ogurdzhaly, Ostrov *see* OgurjalyAdasy
228 A6 **OgurjalyAdasy** *Rus.* Ogurdzhaly, Ostrov. *island* W Turkmenistan
133 K8 **Ogwashi-Uku** Delta, S Nigeria 06°08′N 06°38′E
278 B12 **Ohai** Southland, South Island, New Zealand 45°56′S 167°59′E
229 I5 **Ohangaron** *Rus.* Akhangaran. Toshkent Viloyati, E Uzbekistan 40°56′N 69°37′E
229 I5 **Ohangaron** *Rus.* Akhangaran. ≈ E Uzbekistan
138 C3 **Ohangwena** ⬖ *district* N Namibia
253 C14 **Ōhara** Chiba, Honshū, S Japan 35°14′N 140°19′E
73 D8 **O'Hare** ✈ (Chicago) Illinois, N USA 41°59′N 87°56′W
252 D6 **Ōhata** Aomori, Honshū, C Japan 41°23′N 141°09′E
278 G7 **Ohau** Manawatu-Wanganui, North Island, New Zealand 40°40′S 175°15′E
278 C10 **Ohau, Lake** ◎ South Island, New Zealand
Ohcejohka *see* Utsjoki
163 F11 **Ohey** Namur, SE Belgium 50°26′N 05°07′E
114 B9 **O'Higgins** Buenos Aires, Argentina 34°35′S 60°42′W
118 B1 **O'Higgins, Cabo** *headland* Easter Island, Chile, E Pacific Ocean 27°05′S 109°15′W
O'Higgins, Lago *see* San Martín, Lago
73 H10 **Ohio** *off.* State of Ohio, *also known as* Buckeye State. ⬖ *state* N USA
42 F6 **Ohio River** ≈ N USA
Ohlau *see* Oława
181 G10 **Ohm** ≈ C Germany
284 E10 **Ohonua** 'Eua, E Tonga 21°20′S 174°57′W
69 K2 **Ohoopee River** ≈ Georgia, SE USA
179 G10 **Ohre** *Ger.* Eger. ≈ Czech Republic/Germany
Ohri *see* Ohrid
184 G8 **Ohrid** *Turk.* Ochrida, Ohri. SW FYR Macedonia 41°07′N 20°48′E
184 G8 **Ohrid, Lake** *var.* Lake Ochrida, *Alb.* Liqeni i Ohrit, *Mac.* Ohridsko Ezero. ◎ Albania/FYR Macedonia
Ohridsko Ezero/Ohrit, Liqeni i *see* Ohrid, Lake
141 K3 **Ohrigstad** Mpumalanga, South Africa 24°45′S 30°34′E
141 K3 **Ohrigstaddam** ⊞ Mpumalanga, South Africa
181 G14 **Öhringen** Baden-Württemberg, Germany 49°12′E 9°30′N
278 G5 **Ohura** Manawatu-Wanganui, North Island, New Zealand 38°51′S 174°58′E
103 N6 **Oiapoque** Amapá, E Brazil 03°54′N 51°46′W
103 N6 **Oiapoque, Rio** *var.* Fleuve l'Oyapok, Oyapock. ≈ Brazil/French Guiana *see also* Oyapok, Fleuve l'
Oiapoque, Rio *see* Oyapok, Fleuve l'
102 C5 **Oiba** Santander, C Colombia 06°16′N 73°18′W
63 J1 **Oies, Île aux** *island* Québec, SE Canada
152 H6 **Oijärvi** Pohjois-Pohjanmaa, C Finland 65°38′N 26°05′E
152 H6 **Oikarainen** Lappi, N Finland 66°30′N 25°46′E
282 C10 **Oikuul** Babeldaob, N Palau
62 C7 **Oil City** Pennsylvania, NE USA 41°25′N 79°42′W
64 A6 **Oil City** Pennsylvania, USA 40°24′N 78°38′W
62 C7 **Oil Creek** ≈ Pennsylvania, NE USA
81 D8 **Oildale** California, W USA 35°25′N 119°01′W
Oileán Ciarral *see* Castleisland
Oil Islands *see* Chagos Archipelago
186 F6 **Oiniádes** *anc.* Oeniadae. *site of ancient city* Dytikí Elláda, W Greece
187 K5 **Oinoússes** *island* E Greece
Oírr, Inis *see* Inisheer
163 F8 **Oirschot** Noord-Brabant, S Netherlands 51°30′N 05°18′E
165 H2 **Oise** ⬖ *department* N France
165 H3 **Oise** ≈ N France
166 G3 **Oisemont** Picardie, France 49°57′N 1°46′E
163 F8 **Oisterwijk** Noord-Brabant, S Netherlands 51°35′N 05°12′E
91 K3 **Oistins** S Barbados 13°04′N 59°33′W
250 D6 **Ōita** Ōita, Kyūshū, SW Japan 33°15′N 131°35′E
250 D6 **Ōita** *off.* Ōita-ken. ⬖ *prefecture* Kyūshū, SW Japan
Ōita-ken *see* Ōita
186 G5 **Oíti** ▲ C Greece 38°48′N 22°12′E
252 E5 **Oiwake** Hokkaidō, NE Japan 42°54′N 141°49′E
81 C9 **Ojai** California, W USA 34°25′N 119°15′W
154 G6 **Öje** Dalarna, C Sweden 60°49′N 13°54′E
152 G6 **Öjebyn** Norrbotten, N Sweden 65°20′N 21°26′E
250 B6 **Ojika-jima** *island* SW Japan
85 I3 **Ojinaga** Chihuahua, N Mexico 29°31′N 104°26′W
253 B11 **Ojiya** *var.* Oziya. Niigata, Honshū, C Japan 37°18′N 138°49′E
85 H3 **Ojo Caliente** Chihuahua, Mexico 30°43′N 106°55′W
85 K9 **Ojo Caliente** *var.* Ojocaliente. Zacatecas, C Mexico 22°35′N 102°18′W
Ojocaliente *see* Ojo Caliente
86 G8 **Ojo de Agua** Oaxaca, Mexico 17°07′N 97°14′W
84 E2 **Ojo De Agua** Sonora, Mexico 30°42′N 110°41′W
84 F5 **Ojo de Agua** Sonora, Mexico 28°06′N 109°14′W
85 I3 **Ojo del Carrizo** Chihuahua, Mexico 29°57′N 105°16′W
84 C5 **Ojo de Liebre, Laguna** *var.* Laguna Scammon, Scammon Lagoon. *lagoon* NW Mexico
112 B6 **Ojos del Salado, Cerro** ▲ W Argentina 27°04′S 68°34′W
84 B1 **Ojos Negros** Baja California Norte, NW Mexico 31°52′N 116°16′W
171 H4 **Ojos Negros** Aragón, NE Spain 40°43′N 01°30′W
86 D4 **Ojuelos de Jalisco** Aguascalientes, C Mexico 21°52′N 101°40′W
196 G2 **Oka** ≈ W Russian Federation
140 C1 **Okahandja** Otjozondjupa, C Namibia 21°58′S 16°55′E
278 G5 **Okahukura** Manawatu-Wanganui, North Island, New Zealand 38°48′S 175°13′E
278 F2 **Okaihau** Northland, North Island, New Zealand 35°19′S 173°45′E
138 C4 **Okakarara** Otjozondjupa, N Namibia 20°33′S 17°20′E
59 J3 **Okak Islands** *island group* Newfoundland and Labrador, E Canada
76 E2 **Okanagan** ≈ British Columbia, SW Canada
56 F8 **Okanagan Lake** ◎ British Columbia, SW Canada
Ókanizsa *see* Kanjiža
138 C3 **Okankolo** Oshikoto, N Namibia 17°57′S 16°28′E
54 D10 **Okanogan River** ≈ Washington, NW USA
138 C4 **Okaputa** Otjozondjupa, N Namibia 20°09′S 16°56′E
231 J6 **Okāra** Punjab, E Pakistan 30°49′N 73°31′E
75 D12 **Okarche** Oklahoma, C USA 35°43′N 97°58′W
Okarem *see* Ekerem
283 H10 **Okat Harbor** *harbor* Kosrae, E Micronesia
68 F2 **Okatibbee Creek** ≈ Mississippi, S USA
138 B4 **Okaukuejo** Kunene, N Namibia 19°10′S 15°23′E
Okavanggo *see* Cubango/Okavango
138 D3 **Okavango** ⬖ *district* NW Namibia
138 E3 **Okavango** *var.* Cubango, Kavango, Kavengo, Kubango, Okavanggo, *Port.* Ocavango. ≈ S Africa *see also* Cubango
Okavango *see* Cubango
138 E3 **Okavango Delta** *wetland* N Botswana
251 K3 **Okaya** Nagano, Honshū, S Japan 36°03′N 138°00′E
250 G5 **Okayama** Okayama, Honshū, SW Japan 34°40′N 133°54′E
250 G4 **Okayama** *off.* Okayama-ken. ⬖ *prefecture* Honshū, SW Japan
Okayama-ken *see* Okayama
251 J4 **Okazaki** Aichi, Honshū, C Japan 34°58′N 137°10′E
250 A2 **Okcheon** *prev.* Okch'on. Ch'ungch'ŏng-bukto, South Korea 36°07′N 127°53′E
Okch'on *see* Okcheon
182 H6 **Okęcie** ✈ (Warszawa) Mazowieckie, C Poland 52°08′N 20°57′E
69 M7 **Okeechobee** Florida, SE USA 27°14′N 80°49′W
69 M8 **Okeechobee, Lake** ◎ Florida, SE USA
75 D12 **Okeene** Oklahoma, C USA 36°07′N 98°19′W
69 L3 **Okefenokee Swamp** *wetland* Georgia, SE USA
160 E8 **Okehampton** SW England, United Kingdom 50°44′N 04°W
75 F12 **Okemah** Oklahoma, C USA 35°25′N 96°20′W
133 J7 **Okene** Kogi, S Nigeria 07°31′N 06°15′E
180 I5 **Oker** *var.* Ocker. ≈ NW Germany
180 I7 **Oker-Stausee** ◎ C Germany
193 L7 **Okha** Ostrov Sakhalin, Sakhalinskaya Oblast', SE Russian Federation 53°33′N 142°55′E
195 K9 **Okhansk** *var.* Ochansk. Permskiy Kray, NW Russian Federation 57°44′N 55°20′E
193 K6 **Okhota** ≈ E Russian Federation
193 L6 **Okhotsk** Khabarovskiy Kray, E Russian Federation 59°21′N 143°15′E
286 D1 **Okhotsk, Sea of** *sea* NW Pacific Ocean
189 K3 **Okhtyrka** *Rus.* Akhtyrka. Sums'ka Oblast', NE Ukraine 50°19′N 34°54′E
190 J5 **Okhvat** Tverskaya Oblast', Russian Federation
286 C4 **Oki-Daitō Ridge** *undersea feature* W Pacific Ocean 23°50′N 133°00′E
140 C6 **Okiep** Northern Cape, W South Africa 29°39′S 17°53′E
Oki-guntō *see* Oki-shotō
250 F3 **Oki-kaikyō** *strait* SW Japan
250 F10 **Okinawa** Okinawa, SW Japan 26°20′N 127°47′E
251 I9 **Okinawa** *off.* Okinawa-ken. ⬖ *prefecture* Okinawa, SW Japan
251 I10 **Okinawa** *island* SW Japan
Okinawa-ken *see* Okinawa
255 I1 **Okinawa-shotō** *island group* Nansei-shotō, SW Japan Asia
251 J9 **Okinoerabu-jima** *island* Nansei-shotō, SW Japan
250 E7 **Okino-shima** *island* SW Japan
250 F3 **Oki-shotō** *var.* Oki-guntō. *island group* SW Japan
133 J8 **Okitipupa** Ondo, SW Nigeria 06°33′N 04°43′E
256 C6 **Okkan** Bago, SW Myanmar (Burma) 17°30′N 95°52′E
190 J3 **Okladnevo** Novgorodskaya Oblast', Russian Federation
75 D12 **Oklahoma** *off.* State of Oklahoma, *also known as* The Sooner State. ⬖ *state* C USA
75 E12 **Oklahoma City** *state capital* Oklahoma, C USA 35°28′N 97°31′W
70 F3 **Oklaunion** Texas, SW USA 34°07′N 99°07′W
69 L5 **Oklawaha River** ≈ Florida, SE USA
75 F12 **Okmulgee** Oklahoma, C USA 35°38′N 95°59′W
Oknitsa *see* Ocniţa
66 C8 **Okolona** Mississippi, S USA 34°00′N 88°45′W
252 F3 **Okoppe** Hokkaidō, NE Japan 44°27′N 143°06′E
57 I7 **Okotoks** Alberta, SW Canada 50°46′N 113°57′W
129 K7 **Oko, Wadi** ≈ NE Sudan
135 C8 **Okoyo** Cuvette, W Congo 01°28′S 15°04′E
133 I7 **Okpara** ≈ Benin/Nigeria
152 G3 **Øksfjord** Finnmark, N Norway 70°13′N 22°22′E
261 M6 **Oksibil** Papua, E Indonesia 04°52′S 140°32′E
195 J3 **Oksino** Nenetskiy Avtonomnyy Okrug, NW Russian Federation 67°33′N 52°15′E
152 E6 **Oksskolten** ▲ C Norway 66°00′N 14°18′E
Oksu *see* Oqsu
226 F2 **Oktaybr'skiy** Kostanay, N Kazakhstan
280 A3 **Ok Tedi** Western, W Papua New Guinea
Oktemberyan *see* Armavir
256 C5 **Oktwin** Bago, C Myanmar (Burma) 18°47′N 96°21′E
197 J3 **Oktyabr'sk** Samarskaya Oblast', W Russian Federation 53°13′N 48°36′E
Oktyab'rsk *see* Kandyagash
194 G7 **Oktyabr'skiy** Arkhangel'skaya Oblast', NW Russian Federation 61°03′N 43°16′E
193 M6 **Oktyabr'skiy** Kamchatskiy Kray, E Russian Federation 52°35′N 156°18′E
197 K2 **Oktyabr'skiy** Respublika Bashkortostan, W Russian Federation 54°28′N 53°29′E
197 H6 **Oktyabr'skiy** Volgogradskaya Oblast', SW Russian Federation 48°00′N 43°35′E
Oktyabr'skiy *see* Aktsyabrski
195 M6 **Oktyabr'skoye** Khanty-Mansiyskiy Avtonomnyy Okrug-Yugra, Russian Federation
197 L3 **Oktyabr'skoye** Orenburgskaya Oblast', W Russian Federation 52°22′N 55°39′E
193 H2 **Oktyabr'skoy Revolyutsii, Ostrov** *Eng.* October Revolution Island. *island* Severnaya Zemlya, N Russian Federation
Ōkuchi *see* Isa
194 E9 **Okulovka** *var.* Okulovka. Novgorodskaya Oblast', W Russian Federation 58°24′N 33°16′E
Okulovka *see* Okulovka
252 C5 **Okushiri-tō** *var.* Okusiri Tô. *island* NE Japan
Okusiri Tô *see* Okushiri-tō
133 I6 **Okuta** Kwara, W Nigeria 09°18′N 03°09′E
Ōkuti *see* Isa
140 F1 **Okwa** *var.* Chapman's. ≈ Botswana/Namibia
193 L5 **Ola** Magadanskaya Oblast', E Russian Federation 59°36′N 151°18′E
75 H12 **Ola** Arkansas, C USA 35°01′N 93°13′W
Ola *see* Ala
80 F7 **Olacha Peak** ▲ California, W USA 36°15′N 118°07′W
152 C1 **Ólafsfjörður** Norðurland Eystra, N Iceland 66°04′N 18°36′W
152 A2 **Ólafsvík** Vesturland, W Iceland 64°52′N 23°45′W
Oláhbrettye *see* Bretea-Română
Oláhszentgyőrgy *see* Sângeorz-Băi
Oláh-Toplicza *see* Topliţa
190 D6 **Olaine** C Latvia 56°47′N 23°56′E
80 F7 **Olancha** California, W USA 36°16′N 118°00′W
88 G3 **Olanchito** Yoro, C Honduras 15°30′N 86°34′W
88 G4 **Olancho** ⬖ *department* E Honduras
155 I11 **Öland** *island* S Sweden
155 I10 **Ölands norra udde** *headland* S Sweden 57°21′N 17°06′E
155 I11 **Ölands södra udde** *headland* S Sweden 56°12′N 16°26′E
169 H5 **Olargues** Languedoc-Roussillon, France 43°33′N 2°55′E
111 H6 **Olaria** Minas Gerais, Brazil 21°52′S 43°56′W
277 H4 **Olary** South Australia 32°18′S 140°16′E
114 B10 **Olascoaga** Buenos Aires, Argentina 35°12′S 60°36′W
75 G9 **Olathe** Kansas, C USA 38°52′N 94°50′W
116 G5 **Olavarría** Buenos Aires, E Argentina 36°57′S 60°20′W
152 B5 **Olav V Land** *physical region* C Svalbard
182 E7 **Oława** *Ger.* Ohlau. Dolnośląskie, SW Poland 50°57′N 17°18′E
175 C9 **Olbia** *prev.* Terranova Pausania. Sardegna, Italy, C Mediterranean Sea 40°55′N 09°30′E
90 E3 **Old Bahama Channel** *channel* The Bahamas/Cuba
Old Bay State/Old Colony State *see* Massachusetts
161 H4 **Oldbury** United Kingdom 52°30′N 2°01′W
158 C9 **Oldcastle** Meath, Ireland 53°46′N 7°10′W
65 I2 **Old Chatham** New York, USA 42°26′N 73°34′W
52 A6 **Old Crow** Yukon, NW Canada 67°34′N 139°55′W
52 A6 **Old Crow Flats** *Wetland* Yukon, NW Canada
Old Dominion *see* Virginia
Oldeberkeap *see* Oldeberkoop
162 H4 **Oldeberkoop** *Fris.* Oldeberkeap. Fryslân, N Netherlands 52°55′N 06°07′E
162 H6 **Oldebroek** Gelderland, E Netherlands 52°27′N 05°54′E
162 H4 **Oldemarkt** Overijssel, N Netherlands 52°49′N 05°58′E
154 B5 **Olden** Sogn Og Fjordane, C Norway 61°52′N 06°44′E
180 E3 **Oldenbrok** Niedersachsen, Germany 53°17′E 8°24′N
180 E4 **Oldenburg** Niedersachsen, NW Germany 53°09′N 08°13′E
180 J1 **Oldenburg** *var.* Oldenburg in Holstein. Schleswig-Holstein, N Germany 54°17′N 10°55′E
Oldenburg in Holstein *see* Oldenburg
162 I6 **Oldenzaal** Overijssel, E Netherlands 52°19′N 06°53′E
152 F4 **Olderdalen** Troms, N Norway 69°36′N 20°34′E
Olderfjord *see* Leaibevuotna

◆ Country
● Country Capital
◇ Dependent Territory
○ Dependent Territory Capital
⬖ Administrative Regions
✈ International Airport
▲ Mountain
▲▲ Mountain Range
Volcano
≈ River
◎ Lake
⊞ Reservoir

◆ Country ◇ Dependent Territory ▲ Administrative Regions ▲ Mountain ℞ Volcano ⊚ Lake
● Country Capital ○ Dependent Territory Capital ✕ International Airport ▲ Mountain Range ← River ⊠ Reservoir

◆ Country   ◇ Dependent Territory   ◆ Administrative Regions   ▲ Mountain   ✕ Volcano   ◎ Lake
● Country Capital   ○ Dependent Territory Capital   ✕ International Airport   ▲ Mountain Range   ✦ River   ▨ Reservoir

◆ Country	◇ Administrative Regions	▲ Mountain	⦵ Volcano	◎ Lake
● Country Capital	○ Dependent Territory Capital	✈ International Airport	▲ Mountain Range	◈ River	◻ Reservoir

Portland Point - Provincetown
INDEX
393

90 F8 **Portland Point** *headland* C Jamaica 17°42′N 77°10′W
165 I9 **Port-la-Nouvelle** Aude, S France 43°01′N 03°04′E
Portlaoighise *see* Port Laoise
160 A2 **Port Laoise** *var.* Portlaoise, *Ir.* Portlaoighise; *prev.* Maryborough. C Ireland 53°02′N 07°17′W
Portlaoise *see* Port Laoise
71 H8 **Port Lavaca** Texas, SW USA 28°36′N 96°39′W
276 F5 **Port Lincoln** South Australia 34°43′S 135°49′E
83 I8 **Port Lions** Kodiak Island, Alaska, USA 57°55′N 152°48′W
158 F6 **Port Logan** United Kingdom 54°42′N 4°57′W
132 C7 **Port Loko** W Sierra Leone 08°50′N 12°50′W
166 A7 **Port-Louis** Bretagne, France 47°43′N 3°21′W
91 J1 **Port-Louis** Grande Terre, N Guadeloupe 16°25′N 61°32′W
137 G10 **Port Louis** ● (Mauritius) NW Mauritius 20°10′S 57°30′E
Port Louis *see* Scarborough
Port-Lyautey *see* Kénitra
277 H7 **Port MacDonnell** South Australia 38°04′S 140°40′E
277 M4 **Port Macquarie** New South Wales, SE Australia 31°26′S 152°55′E
Portmadoc *see* Porthmadog
Port Mahon *see* Maó
63 M4 **Port Maitland** Nova Scotia, SE Canada 43°58′N 66°08′W
90 F7 **Port Maria** C Jamaica 18°22′N 76°54′W
56 B7 **Port McNeill** Vancouver Island, British Columbia, SW Canada 50°34′N 127°06′W
59 J6 **Port-Menier** Île d'Anticosti, Québec, E Canada 49°49′N 64°19′W
82 G8 **Port Moller** Alaska, USA 56°00′N 160°31′W
90 G8 **Port Morant** E Jamaica 17°53′N 76°20′W
90 F8 **Portmore** C Jamaica 17°58′N 76°52′W
280 C4 **Port Moresby** ● (Papua New Guinea) Central/National Capital District, SW Papua New Guinea 09°28′S 147°12′E
158 F1 **Portnacroish** United Kingdom 56°34′N 5°22′W
158 D4 **Portnahaven** United Kingdom 55°41′N 6°30′W
Port Natal *see* Durban
166 B7 **Port-Navalo** Bretagne, France 47°33′N 2°54′W
71 J7 **Port Neches** Texas, SW USA 29°59′N 93°57′W
276 F5 **Port Neill** South Australia 34°06′S 136°19′E
140 C6 **Port Nolloth** Northern Cape, W South Africa 29°17′S 16°51′E
64 G9 **Port Norris** New Jersey, NE USA 39°13′N 75°00′W
Port-Nouveau-Québec *see* Kangiqsualujjuaq
172 B1 **Porto** *Eng.* Oporto; *anc.* Portus Cale. Porto, NW Portugal 41°09′N 08°37′W
172 B1 **Porto** *var.* Pôrto. ⬖ *district* N Portugal
170 C4 **Porto** ✈ Porto, W Portugal 41°09′N 08°37′W
Pôrto *see* Porto
113 J7 **Porto Alegre** *var.* Pôrto Alegre. *state capital* Rio Grande do Sul, S Brazil 30°03′S 51°10′W
Porto Alexandre *see* Tombua
135 C11 **Porto Amboim** Kwanza Sul, NW Angola 10°47′S 13°43′E
Porto Amélia *see* Pemba
Porto Bello *see* Portobelo
89 K8 **Portobelo** *var.* Porto Bello, Puerto Bello. Colón, N Panama 09°33′N 79°37′W
113 I4 **Pôrto Camargo** Paraná, S Brazil 23°23′S 53°47′W
71 H8 **Port O'Connor** Texas, SW USA 28°26′N 96°26′W
172 B7 **Porto Covo** Setúbal, Portugal 37°51′N 8°48′W
Pôrto de Mós *see* Porto de Moz
107 K2 **Porto de Moz** *var.* Pôrto de Mós. Pará, NE Brazil 01°45′S 52°15′W
170 A8 **Porto do Moniz** Madeira, Portugal, NE Atlantic Ocean
172 B6 **Porto do Rei** Setúbal, Portugal 38°17′N 8°24′W
107 I7 **Porto dos Gaúchos** Mato Grosso, W Brazil 11°32′S 57°16′W
Porto Edda *see* Sarandë
175 F13 **Porto Empedocle** Sicilia, Italy, C Mediterranean Sea 37°18′N 13°32′E
112 G2 **Porto Esperança** Mato Grosso do Sul, SW Brazil 19°36′S 57°24′W
174 D7 **Portoferraio** Toscana, C Italy 42°49′N 10°18′E
110 C6 **Porto Ferreira** São Paulo, Brazil 21°51′S 47°28′W
158 G2 **Port of Menteith** United Kingdom 56°11′N 4°16′W
156 D3 **Port of Ness** NW Scotland, United Kingdom 58°29′N 06°15′W
91 L7 **Port-of-Spain** ● (Trinidad and Tobago) Trinidad, Trinidad and Tobago 10°39′N 61°30′W
Port of Spain *see* Piarco
174 B7 **Porto, Golfe de** *gulf* Corse, France, C Mediterranean Sea
Porto Grande *see* Mindelo
177 I6 **Portogruaro** Veneto, NE Italy 45°46′N 12°50′E
78 C3 **Portola** California, W USA 39°48′N 120°28′W
281 L2 **Port-Olry** Espiritu Santo, C Vanuatu 15°03′S 167°04′E
153 G8 **Pörtom** *Fin.* Pirttikylä. Österbotten, W Finland 62°42′N 21°40′E
Port Omna *see* Portumna
112 G3 **Porto Murtinho** Mato Grosso do Sul, SW Brazil 21°42′S 57°52′W
108 C7 **Porto Nacional** Tocantins, C Brazil 10°41′S 48°19′W
133 I8 **Porto-Novo** ● (Benin) S Benin 06°29′N 02°37′E
69 M5 **Port Orange** Florida, SE USA 29°06′N 80°59′W
76 C3 **Port Orchard** Washington, NW USA 47°32′N 122°38′W
Porto Re *see* Kraljevica
76 A7 **Port Orford** Oregon, NW USA 42°45′N 124°30′W
Porto Rico *see* Puerto Rico
177 J10 **Porto San Giorgio** Marche, C Italy 43°10′N 13°47′E
165 N9 **Porto San Stefano** Toscana, C Italy 42°26′N 11°09′E
170 B8 **Porto Santo** *var.* Vila Baleira. Porto Santo, Madeira, Portugal, NE Atlantic Ocean 33°04′N 16°20′W
170 B8 **Porto Santo** ✈ Porto Santo, Madeira, Portugal, NE Atlantic Ocean 33°04′N 16°20′W
170 B8 **Porto Santo** *var.* Ilha do Porto Santo. *island* Madeira, Portugal, NE Atlantic Ocean
Porto Santo, Ilha do *see* Porto Santo
113 I3 **Porto São José** Paraná, S Brazil 22°43′S 53°10′W
109 H10 **Porto Seguro** Bahia, E Brazil 16°25′S 39°07′W
175 B9 **Porto Torres** Sardegna, Italy, C Mediterranean Sea 40°50′N 08°23′E
113 J5 **Porto União** Santa Catarina, S Brazil 26°15′S 51°04′W
175 C8 **Porto-Vecchio** Corse, France, C Mediterranean Sea 41°35′N 09°17′E
106 E6 **Porto Velho** *var.* Velho. *state capital* Rondônia, W Brazil 08°45′S 63°54′W
104 B2 **Portoviejo** *var.* Puertoviejo. Manabí, W Ecuador 01°03′S 80°31′W
158 F6 **Portpatrick** United Kingdom 54°51′N 5°07′W
278 B13 **Port Pegasus** *bay* Stewart Island, New Zealand
62 D4 **Port Perry** Ontario, SE Canada 44°08′N 78°57′W
277 I7 **Port Phillip Bay** *harbor* Victoria, SE Australia
276 G5 **Port Pirie** South Australia 33°11′S 138°01′E
160 C9 **Portreath** United Kingdom 50°16′N 5°18′W
156 D5 **Portree** N Scotland, United Kingdom 57°26′N 06°12′W
56 C9 **Port Renfrew** Vancouver Island, British Columbia, SW Canada 48°33′N 124°24′W
Port Rex *see* East London
Port Rois *see* Portrush
90 F8 **Port Royal** E Jamaica 17°55′N 76°52′W
67 I10 **Port Royal** South Carolina, SE USA 32°22′N 80°41′W
67 I10 **Port Royal Sound** *inlet* South Carolina, SE USA
158 D5 **Portrush** *Ir.* Port Rois. N Northern Ireland, United Kingdom 55°12′N 06°40′W
Port Said *see* Būr Saʿīd
168 E4 **Port-Sainte-Marie** Aquitaine, France 44°15′N 0°24′E
69 I5 **Port Saint Joe** Florida, SE USA 29°49′N 85°18′W
69 M6 **Port Saint John** Florida, SE USA 28°28′N 80°46′W
169 J6 **Port-St-Louis-du-Rhône** Bouches-du-Rhône, SE France 43°22′N 04°48′E
158 G8 **Port St Mary** SW Isle of Man
158 C5 **Portsalon** Donegal, Ireland 55°12′N 7°37′W
90 G6 **Port Salut** SW Haiti 18°04′N 73°55′W
117 I11 **Port Salvador** *inlet* East Falkland, Falkland Islands
117 I11 **Port San Carlos** East Falkland, Falkland Islands 51°30′S 58°59′W
59 L6 **Port Saunders** Newfoundland, Newfoundland and Labrador, SE Canada 50°40′N 57°17′W
139 H8 **Port Shepstone** KwaZulu/Natal, E South Africa 30°44′S 30°28′E
91 L1 **Portsmouth** *var.* Grand-Anse. NW Dominica 15°34′N 61°27′W
161 J8 **Portsmouth** S England, United Kingdom 50°48′N 01°05′W
65 M1 **Portsmouth** New Hampshire, NE USA 43°04′N 70°47′W
66 G3 **Portsmouth** Ohio, N USA 38°43′N 83°00′W
67 M5 **Portsmouth** Virginia, NE USA 36°50′N 76°18′W
62 B6 **Port Stanley** Ontario, S Canada 42°39′N 81°12′W
Port Stanley *see* Stanley
117 H11 **Port Stephens** *inlet* West Falkland, Falkland Islands
117 H11 **Port Stephens Settlement** West Falkland, Falkland Islands
158 D5 **Portstewart** *Ir.* Port Stíobhaird. N Northern Ireland, United Kingdom 55°11′N 06°43′W
Port Stíobhaird *see* Portstewart
139 H9 **Port St. Johns** Eastern Cape, SE South Africa 31°37′S 29°32′E
141 J8 **Port St. Johns** Eastern Cape, South Africa 31°36′S 29°32′E
129 K7 **Port Sudan** Red Sea, NE Sudan 19°37′N 37°14′E
68 E5 **Port Sulphur** Louisiana, S USA 29°28′N 89°41′W
167 L7 **Port-sur-Saéne** Franche-Comté, France 47°41′N 6°03′E
Port Swettenham *see* Klang/Pelabuhan Klang
160 F6 **Port Talbot** S Wales, United Kingdom 51°36′N 03°47′W
152 H5 **Porttipahdan Tekojärvi** ◎ N Finland
76 C3 **Port Townsend** Washington, NW USA 48°07′N 122°45′W
170 D5 **Portugal** *off.* Portuguese Republic. ◆ *republic* SW Europe
114 A2 **Portugalete** Santa Fe, Argentina 30°16′S 61°22′W
170 G1 **Portugalete** País Vasco, N Spain 43°19′N 03°01′W
102 F3 **Portuguesa** *off.* Estado Portuguesa. ⬖ *state* N Venezuela
Portuguesa, Estado *see* Portuguesa
Portuguese East Africa *see* Mozambique
Portuguese Guinea *see* Guinea-Bissau
Portuguese Republic *see* Portugal
Portuguese Timor *see* East Timor
Portuguese West Africa *see* Angola
157 C9 **Portumna** *Ir.* Port Omna. Galway, W Ireland 53°06′N 08°13′W
172 B2 **Portunhos** Coimbra, Portugal 40°18′N 8°33′W
Portus Cale *see* Porto
Portus Magnus *see* Almería
Portus Magonis *see* Maó
165 I9 **Port-Vendres** *var.* Port Vendres. Pyrénées-Orientales, S France 42°31′N 03°06′E
276 G5 **Port Victoria** South Australia 34°34′S 137°31′E
281 M4 **Port-Vila** *var.* Vila. ● (Vanuatu) Éfaté, C Vanuatu 17°45′S 168°21′E
Port Vila *see* Bauer Field
64 B3 **Portville** Pennsylvania, NE USA 42°02′N 78°20′W
65 I6 **Port Washington** New York, USA 40°50′N 73°42′W
72 D7 **Port Washington** Wisconsin, N USA 43°23′N 87°54′W
158 G6 **Port William** SW Scotland, United Kingdom 54°45′N 04°41′W
106 C7 **Porvenir** Pando, NW Bolivia 11°15′S 68°43′W
117 C13 **Porvenir** Magallanes, S Chile 53°18′S 70°22′W
114 F5 **Porvenir** Paysandú, W Uruguay 32°23′S 57°59′W
153 H10 **Porvoo** *Swe.* Borgå. Uusimaa, S Finland 60°25′N 25°40′E
Porzecze *see* Parechcha
170 F6 **Porzuna** Castilla-La Mancha, C Spain 39°10′N 04°10′W
190 I3 **Posad** Novgorodskaya Oblast', Russian Federation
113 H6 **Posadas** Misiones, NE Argentina 27°27′S 55°52′W
170 F7 **Posadas** Andalucía, S Spain 37°48′N 05°06′W
Poschega *see* Požega
176 F5 **Poschiavino** ≈ Italy/Switzerland
176 F5 **Poschiavo** *Ger.* Puschlav. Graubünden, S Switzerland 46°19′N 10°02′E
184 B4 **Posedarje** Zadar, SW Croatia 44°12′N 15°27′E
Posen *see* Poznań
194 F9 **Poshekhon'ye** Yaroslavskaya Oblast', W Russian Federation 58°31′N 39°07′E
152 I6 **Posio** Lappi, NE Finland 66°06′N 28°16′E
Poskam *see* Zepu
Posnania *see* Poznań
260 D4 **Poso** Sulawesi, C Indonesia 01°23′S 120°45′E
260 C4 **Poso, Danau** ◎ Sulawesi, C Indonesia
215 J4 **Posof** Ardahan, NE Turkey 41°30′N 42°33′E
195 N4 **Pospoluy** Yamalo-Nenetskiy Avtonomnyy Okrug, Russian Federation
140 B5 **Possession Island** Karas, Namibia
70 G4 **Possum Kingdom Lake** ⊠ Texas, SW USA
70 E4 **Post** Texas, SW USA 33°14′N 101°24′W
Postavy/Postawy *see* Pastavy
Poste-de-la-Baleine *see* Kuujjuarapik
163 H9 **Posterholt** Limburg, SE Netherlands 51°07′N 06°02′E
140 F5 **Postmasburg** Northern Cape, N South Africa 28°20′S 23°05′E
Pôsto Diuarum *see* Campo de Diauarum
107 J8 **Pôsto Jacaré** Mato Grosso, W Brazil 12°S 53°27′W
177 J6 **Postojna** *Ger.* Adelsberg, *It.* Postumia. SW Slovenia 45°48′N 14°12′E
81 I12 **Poston** Nevada, USA 33°59′N 114°24′W
Postumia *see* Postojna
74 I6 **Postville** Iowa, C USA 43°04′N 91°34′W
Pöstyén *see* Piešt'any
184 D5 **Posušje** Federacija Bosne I Hercegovine, SW Bosnia and Herzegovina 43°28′N 17°20′E
260 D8 **Pota** Flores, C Indonesia 08°21′S 120°50′E
187 H9 **Potamós** Antikýthira, S Greece 35°53′N 23°17′E
243 M4 **Potan** Zhejiang, SE China 28°44′N 120°30′E
190 J2 **Potanino** Leningradskaya Oblast', Russian Federation
103 J5 **Potaru River** ≈ C Guyana
219 F10 **Potash City** Al Karak, W Jordan
141 I4 **Potchefstroom** North-West, N South Africa 26°42′S 27°06′E
75 G12 **Poteau** Oklahoma, C USA 35°03′N 94°36′W
70 G8 **Poteet** Texas, SW USA 29°02′N 98°34′W
187 H3 **Poteídaia** *site of ancient city* Kentrikí Makedonía, N Greece
Potentia *see* Potenza
175 H9 **Potenza** *anc.* Potentia. Basilicata, S Italy 40°40′N 15°50′E
278 B12 **Poteriteri, Lake** ◎ South Island, New Zealand
170 F2 **Potes** Cantabria, N Spain 43°10′N 04°41′W
140 G7 **Potfontein** Northern Cape, South Africa 30°12′S 24°07′E
Potgietersrus *see* Mokopane
70 G7 **Poth** Texas, SW USA 29°04′N 98°04′W
76 D4 **Potholes Reservoir** ⊠ Washington, NW USA
215 I4 **Poti** *prev.* P'ot'i. W Georgia 42°10′N 41°42′E
P'ot'i *see* Poti
133 L5 **Potiskum** Yobe, NE Nigeria 11°38′N 11°07′E
Potkozarje *see* Ivanjska
76 F4 **Potlatch** Idaho, NW USA 46°55′N 116°51′W
76 G4 **Pot Mountain** ▲ Idaho, NW USA 46°44′N 115°24′W
184 D5 **Potoci** Federacija Bosne I Hercegovine, S Bosnia and Herzegovina 43°24′N 17°52′E
61 L5 **Potomac River** ≈ NE USA
Pòtoprens *see* Port-au-Prince
112 C2 **Potosí** Potosí, S Bolivia 19°35′S 65°51′W
88 E5 **Potosí** Chinandega, NW Nicaragua 12°58′N 87°30′W
75 I10 **Potosi** Missouri, C USA 37°57′N 90°49′W
112 C2 **Potosí** ⬖ *department* SW Bolivia
86 D8 **Potosí, Bahía del** *bay* Guerrero, Mexico
118 F2 **Potrerillos** Mendoza, Argentina 32°57′S 69°11′W
112 B5 **Potrerillos** Atacama, N Chile 26°30′S 69°25′W
88 E3 **Potrerillos** Cortés, NW Honduras 15°10′N 87°58′W
84 B2 **Potrero** Veracruz-Llave, E Mexico 31°08′N 115°57′W
116 C1 **Potro, Cerro del** ▲ N Chile 28°22′S 69°34′W
178 H7 **Potsdam** Brandenburg, NE Germany 52°24′N 13°04′E
62 G4 **Potsdam** New York, NE USA 44°40′N 74°58′W
177 L2 **Pottendorf** Niederösterreich, E Austria 47°55′N 16°23′E
177 L2 **Pottenstein** Niederösterreich, E Austria 47°58′N 16°07′E
59 I1 **Potter Island** *Island* Nunavut, NE Canada
161 K6 **Potters Bar** United Kingdom 51°41′N 0°10′W
280 A8 **Pott, Île** *island* Îles Belep, W New Caledonia
161 K4 **Potton** United Kingdom 52°07′N 0°12′W
64 F7 **Pottstown** Pennsylvania, NE USA 40°15′N 75°39′W
64 E6 **Pottsville** Pennsylvania, NE USA 40°40′N 76°10′W
235 G11 **Pottuvil** Eastern Province, SE Sri Lanka 06°53′N 81°49′E
231 I4 **Potwar Plateau** *plateau* NE Pakistan
164 F4 **Pouancé** Maine-et-Loire, W France 47°46′N 01°11′W
280 A8 **Pouébo** Province Nord, C New Caledonia 20°40′S 164°02′E
280 B9 **Pouembout** Province Nord, W New Caledonia 21°09′S 164°52′E
65 H4 **Poughkeepsie** New York, USA 41°42′N 73°55′W
167 I9 **Pougues-les-Eaux** Bourgogne, France 47°04′N 3°06′E
167 J8 **Pouilly-en-Auxois** Bourgogne, France 47°16′N 4°33′E
167 J10 **Pouilly-sous-Charlieu** Rhône-Alpes, France 46°09′N 4°07′E
160 B2 **Poulaphouca Reservoir** ⊠ Ireland
63 H5 **Poultney** Vermont, NE USA 43°31′N 73°12′W
161 H6 **Poulton** United Kingdom 51°42′N 1°51′W
159 I8 **Poulton le Fylde** United Kingdom 53°50′N 2°59′W
280 A8 **Poum** Province Nord, W New Caledonia 20°15′S 164°03′E
Poŭn *see* Boeun
160 E8 **Poundstock** United Kingdom 50°46′N 4°33′W
140 G7 **Poupan** Northern Cape, South Africa 26°38′S 22°29′E
172 D2 **Pousade** Guarda, Portugal 40°33′N 7°08′W
110 E7 **Pouso Alegre** Minas Gerais, NE Brazil 22°13′S 45°56′W
285 J2 **Poutasi** Upolu, SE Samoa 14°00′S 171°43′W
257 G8 **Poŭthĭsăt** *prev.* Pursat. Poŭthĭsăt, W Cambodia 12°32′N 103°55′E
257 G9 **Poŭthĭsăt, Stœ̆ng** *prev.* Pursat. ≈ W Cambodia
168 E6 **Pouyastruc** Midi-Pyrénées, France 43°16′N 0°10′E
166 D9 **Pouzauges** Vendée, NW France 46°47′N 00°54′W
Po, Valle del *see* Po Valley
176 G7 **Po Valley** *It.* Valle del Po. *valley* N Italy
183 F9 **Považská Bystrica** *Ger.* Waagbistritz, *Hung.* Vágbeszterce. Trenčiansky Kraj, W Slovakia 49°07′N 18°26′E
194 E6 **Povenets** Respublika Kareliya, NW Russian Federation 62°50′N 34°47′E
278 J5 **Poverty Bay** *inlet* North Island, New Zealand
184 F4 **Povlen** ▲▲ W Serbia
170 C3 **Póvoa de Varzim** Porto, NW Portugal 41°22′N 08°46′W
197 H4 **Povorino** Voronezhskaya Oblast', W Russian Federation 51°10′N 42°16′E
Povungnituk *see* Puvirnituq
Rivière de Povungnituk *see* Puvirnituq, Riviere de
62 D2 **Powassan** Ontario, S Canada 46°05′N 79°21′W
81 E12 **Poway** California, W USA 32°57′N 117°02′W
159 J4 **Powburn** N England, United Kingdom 55°27′N 01°57′W
77 L7 **Powder River** Wyoming, C USA 43°01′N 106°57′W
77 L7 **Powder River** ≈ Montana/Wyoming, NW USA
76 E6 **Powder River** ≈ Oregon, NW USA
77 K7 **Powder River Pass** *pass* Wyoming, C USA
77 J6 **Powell** Wyoming, C USA 44°45′N 108°45′W
79 I5 **Powell, Lake** ◎ Utah, W USA
79 L3 **Powell, Mount** ▲ Colorado, C USA 39°25′N 106°20′W
80 G4 **Powell Mountain** ▲ Nevada, USA 38°20′N 118°42′W
56 C8 **Powell River** British Columbia, SW Canada 49°54′N 124°34′W
72 E5 **Powers** Michigan, N USA 45°40′N 87°29′W
74 B2 **Powers Lake** North Dakota, N USA 48°33′N 102°37′W
67 K4 **Powhatan** Virginia, NE USA 37°33′N 77°56′W
73 J11 **Powhatan Point** Ohio, N USA 39°49′N 80°49′W
280 B9 **Poya** Province Nord, C New Caledonia 21°19′S 165°07′E
243 J4 **Poyang** Jiangxi, China 29°00′N 116°25′E
243 J4 **Poyang Hu** ◎ S China
72 D6 **Poygan, Lake** ◎ Wisconsin, N USA
177 L1 **Poysdorf** Niederösterreich, NE Austria 48°40′N 16°38′E
184 G4 **Požarevac** *Ger.* Passarowitz. Serbia, NE Serbia 44°37′N 21°11′E
86 G5 **Poza Rica** *var.* Poza Rica de Hidalgo. Veracruz-Llave, E Mexico 20°34′N 97°26′W
Poza Rica de Hidalgo *see* Poza Rica
184 D3 **Požega** *prev.* Slavonska Požega, *Ger.* Poschega, *Hung.* Pozsega. Požega-Slavonija, NE Croatia 45°19′N 17°42′E
184 D3 **Požega-Slavonija** *off.* Požeško-Slavonska Županija. ⬖ *province* NE Croatia
Požeško-Slavonska Županija *see* Požega-Slavonija
190 H5 **Pozherevitsy** Pskovskaya Oblast', Russian Federation
190 I5 **Pozhnya** Tverskaya Oblast', Russian Federation
195 K8 **Pozhva** Komi-Permyatskiy Okrug, NW Russian Federation 59°07′N 56°04′E
182 E6 **Poznań** *Ger.* Posen, Posnania. Wielkolpolskie, C Poland 52°24′N 16°56′E
170 G8 **Pozo Alcón** Andalucía, S Spain 37°43′N 02°55′W
112 B2 **Pozo Almonte** Tarapacá, N Chile 20°16′S 69°50′W
170 F7 **Pozoblanco** Andalucía, S Spain 38°23′N 04°48′W
171 H6 **Pozo Cañada** Castilla-La Mancha, C Spain 38°49′N 01°45′W
112 G4 **Pozo Colorado** Presidente Hayes, C Paraguay 23°26′S 58°51′W
85 I9 **Pozo de Higueras** Nayarit, Mexico 22°17′N 105°29′W
117 E10 **Pozos, Punta** *headland* S Argentina 47°55′S 65°46′W
Pozsega *see* Požega
Pozsony *see* Bratislava
91 L10 **Pozuelos** Anzoátegui, NE Venezuela 10°11′N 64°39′W
175 G13 **Pozzallo** Sicilia, Italy, C Mediterranean Sea 36°44′N 14°51′E
175 F9 **Pozzuoli** *anc.* Puteoli. Campania, S Italy 40°49′N 14°07′E
132 G8 **Pra** ≈ S Ghana
Prabumulih *see* Perabumulih
183 B9 **Prachatice** *Ger.* Prachatitz. Jihočeský Kraj, S Czech Republic 49°01′N 14°02′E
Prachatitz *see* Prachatice
257 F8 **Prachin Buri** *var.* Prachinburi. Prachin Buri, C Thailand 14°05′N 101°23′E
Prachinburi *see* Prachin Buri
181 D10 **Pracht** Rheinland-Pfalz, Germany 50°45′E 7°39′N
Prachuab Girikhand *see* Prachuap Khiri Khan
257 E9 **Prachuap Khiri Khan** *var.* Prachuab Girikhand. Prachuap Khiri Khan, SW Thailand 11°50′N 99°49′E
183 E8 **Praděd** *Ger.* Altvater. ▲ NE Czech Republic 50°06′N 17°14′E
169 I3 **Pradelles** Auvergne, France 44°46′N 3°53′E
102 B6 **Pradera** Valle del Cauca, SW Colombia 03°23′N 76°11′W
168 G7 **Prades** Pyrénées-Orientales, S France 42°36′N 02°22′E
169 M4 **Pradleves** Piemonte, NW Italy 44°25′N 07°16′E
109 G10 **Prado** Bahia, SE Brazil 17°13′S 39°15′W
102 C6 **Prado** Tolima, C Colombia 03°45′N 74°55′W
Prado del Ganso *see* Goose Green
110 B5 **Pradópolis** São Paulo, Brazil 21°22′S 48°04′W
110 G5 **Prados** Minas Gerais, Brazil 21°03′S 44°05′W
Prae *see* Phrae
155 E13 **Præstø** Sjælland, SE Denmark 55°08′N 12°03′E
169 M3 **Pragelato** Piemonte, NW Italy 45°00′N 06°56′E
Prag/Praga/Prague *see* Praha
75 E12 **Prague** Oklahoma, C USA 35°29′N 96°40′W
183 C8 **Praha** *Eng.* Prague, *Ger.* Prag, *Pol.* Praga. ● (Czech Republic) Středočeský Kraj, NW Czech Republic 50°06′N 14°26′E
166 D10 **Prahecq** Poitou-Charentes, France 46°15′N 0°21′W
196 B7 **Prahova** ⬖ *county* SE Romania
188 D9 **Prahova** ≈ S Romania
132 B10 **Praia** ● (Cape Verde) Santiago, S Cape Verde 14°55′N 23°31′W
172 A5 **Praia das Maçãs** Lisboa, Portugal 38°49′N 9°28′W
172 B3 **Praia da Vieira** Leiria, Portugal 39°52′N 8°58′W
172 B7 **Praia de Odeceixe** Faro, Portugal 37°26′N 8°48′W
139 I6 **Praia do Bilene** Gaza, S Mozambique 25°18′S 33°10′E
139 I6 **Praia do Xai-Xai** Gaza, S Mozambique 25°04′S 33°43′E
110 D10 **Praia Grande** São Paulo, Brazil 24°00′S 46°24′W
188 D7 **Praid** *Hung.* Parajd. Harghita, C Romania 46°33′N 25°06′E
75 C9 **Prairie Dog Creek** ≈ Kansas/Nebraska, C USA
72 B7 **Prairie du Chien** Wisconsin, N USA 43°02′N 91°08′W
75 G12 **Prairie Grove** Arkansas, C USA 35°58′N 94°19′W
73 F9 **Prairie River** ≈ Michigan, N USA
Prairie State *see* Illinois
71 H7 **Prairie View** Texas, SW USA 30°05′N 95°59′W
256 G7 **Prakhon Chai** Buri Ram, E Thailand 14°36′N 103°04′E
169 M4 **Pra-Loup** Provence-Alpes-Côte d'Azur, France 44°22′N 6°37′E
177 I2 **Pram** ≈ N Austria
257 G9 **Prâmaôy** *prev.* Phumĭ Prâmaôy. Poŭthĭsăt, W Cambodia 12°13′N 103°05′E
177 J2 **Prambachkirchen** Oberösterreich, N Austria 48°18′N 13°50′E
190 E3 **Prangli** *island* N Estonia
234 F4 **Prānhita** ≈ C India
137 I8 **Praslin** *island* Inner Islands, NE Seychelles
187 L9 **Prasonísi, Akrotírio** *cape* Ródos, Dodekánisa, Greece, Aegean Sea
182 F7 **Praszka** Opolskie, S Poland 51°05′N 18°29′E
110 A8 **Pratânia** São Paulo, Brazil 22°49′S 48°39′W
110 C4 **Pratápolis** Minas Gerais, Brazil 20°45′S 46°52′W
Pratas Island *see* Tungsha Tao
191 H11 **Pratasy** *Rus.* Protasy. Homyel'skaya Voblasts', SE Belarus 52°47′N 29°05′E
256 F7 **Prathai** Nakhon Ratchasima, E Thailand 15°31′N 102°42′E
80 E5 **Prather** California, USA 37°02′N 119°31′W
Prathet Thai *see* Thailand
Prathum Thani *see* Pathum Thani
110 D2 **Pratinha** Minas Gerais, Brazil 19°46′S 46°24′W
117 B11 **Prat, Isla** *island* S Chile
176 G9 **Prato** Toscana, C Italy 43°53′N 11°05′E
168 G8 **Prats de Lluçanès** *prev.* Prats de Llusanés. Cataluña, Spain 42°01′N 2°02′E
Prats de Llusanés *see* Prats de Lluçanès
169 H8 **Prats-de-Mollo-la-Preste** Pyrénées-Orientales, S France 42°25′N 02°28′E
75 D10 **Pratt** Kansas, C USA 37°40′N 98°45′W
176 D3 **Pratteln** Basel Landschaft, NW Switzerland 47°32′N 07°42′E
287 H1 **Pratt Seamount** *undersea feature* N Pacific Ocean 56°09′N 142°30′W
69 H2 **Prattville** Alabama, S USA 32°27′N 86°27′W
Praust *see* Pruszcz Gdański
167 K7 **Prauthoy** Champagne-Ardenne, France 47°40′N 5°17′E
185 L4 **Pravda** *prev.* Dogrular. Silistra, NE Bulgaria 43°53′N 26°58′E
191 B9 **Pravdinsk** *Ger.* Friedland. Kaliningradskaya Oblast', W Russian Federation 54°26′N 21°01′E
170 E1 **Pravia** Asturias, N Spain 43°30′N 06°06′W
160 F9 **Prawle Point** *headland* United Kingdom 50°13′N 3°42′W
168 F4 **Prayssac** Midi-Pyrénées, France 44°30′N 1°12′E
191 G8 **Prazaroki** *Rus.* Prozoroki. Vitsyebskaya Voblasts', N Belarus 55°18′N 28°13′E
Prázsmár *see* Prejmer
257 H8 **Preăh Vihéar** Preăh Vihéar, N Cambodia 13°57′N 104°48′E
191 J8 **Prechistoye** Smolenskaya Oblast', Russian Federation
188 D8 **Predeal** *Hung.* Predeál. Braşov, C Romania 45°30′N 25°31′E
177 J4 **Predlitz** Steiermark, SE Austria 47°04′N 13°54′E
55 H9 **Preeceville** Saskatchewan, S Canada 51°58′N 102°40′W
Preenkuln *see* Priekule
164 G3 **Pré-en-Pail** Mayenne, NW France 48°27′N 00°15′W
160 G3 **Prees** United Kingdom 52°53′N 2°40′W
159 I8 **Preesall** United Kingdom 53°55′N 2°58′W
180 I1 **Preetz** Schleswig-Holstein, Germany 54°14′E 10°17′N
177 J2 **Pregarten** Oberösterreich, N Austria 48°21′N 14°31′E
102 D4 **Pregonero** Táchira, NW Venezuela 08°02′N 71°35′W
190 F7 **Preiļi** *Ger.* Preli. SE Latvia 56°17′N 26°52′E
188 E8 **Prejmer** *Ger.* Tartlau, *Hung.* Prázsmár. Braşov, S Romania 45°42′N 25°49′E
184 E6 **Prekornica** ▲▲ C Montenegro
Preli *see* Preiļi
167 I8 **Prémery** Bourgogne, France 47°10′N 3°20′E
Prëmet *see* Përmet
178 G6 **Premnitz** Brandenburg, NE Germany 52°33′N 12°22′E
70 G9 **Premont** Texas, SW USA 27°21′N 98°07′W
184 D5 **Prenj** ▲▲ S Bosnia and Herzegovina
Prenjas/Prenjasi *see* Përrenjas
68 E3 **Prentiss** Mississippi, S USA 31°36′N 89°52′W
Preny *see* Prienai
178 I5 **Prenzlau** Brandenburg, NE Germany 53°19′N 13°52′E
193 I7 **Preobrazhenka** Irkutskaya Oblast', C Russian Federation 60°01′N 108°00′E
256 B7 **Preparis Island** *island* SW Myanmar (Burma)
Prerau *see* Přerov
183 E9 **Přerov** *Ger.* Prerau. Olomoucký Kraj, E Czech Republic 49°27′N 17°27′E
85 L8 **Presa** San Luis Potosí, Mexico 22°51′N 100°09′W
85 J4 **Presa** San Luis Potosí, Mexico 28°28′N 103°15′W
84 F5 **Presa Adolfo Ruiz Cortines** Sonora, Mexico 27°16′N 109°02′W
85 K6 **Presa Rodríguez** Baja California Sur, NW Mexico
Preschau *see* Prešov
62 G4 **Prescott** Ontario, SE Canada 44°43′N 75°33′W
79 H8 **Prescott** Arizona, SW USA 34°33′N 112°26′W
75 H13 **Prescott** Arkansas, C USA 33°49′N 93°25′W
76 E5 **Prescott** Washington, NW USA 46°17′N 118°21′W
72 A5 **Prescott** Wisconsin, N USA 44°46′N 92°45′W
278 A12 **Preservation Inlet** *inlet* South Island, New Zealand
185 H7 **Preševo** Serbia, SE Serbia 42°20′N 21°38′E
74 C5 **Presho** South Dakota, N USA 43°54′N 100°03′W
108 E4 **Presidente Dutra** Maranhão, E Brazil 05°17′S 44°30′W
113 J3 **Presidente Epitácio** São Paulo, S Brazil 21°45′S 52°07′W
112 G4 **Presidente Hayes** *off.* Departamento de Presidente Hayes. ⬖ *department* C Paraguay
Presidente Hayes, Departamento de *see* Presidente Hayes
113 J3 **Presidente Prudente** São Paulo, S Brazil 22°09′S 51°24′W
114 C10 **Presidente Quintana** Buenos Aires, Argentina 35°15′S 60°11′W
Presidente Stroessner *see* Ciudad del Este
Presidente Vargas *see* Itabira
113 J3 **Presidente Venceslau** São Paulo, S Brazil 21°52′S 51°51′W
287 H7 **President Thiers Seamount** *undersea feature* C Pacific Ocean 24°39′S 145°50′W
70 B7 **Presidio** Texas, SW USA 29°33′N 104°22′W
85 H8 **Presidio, Río** ≈ Sinaloa, Mexico
Preslav *see* Veliki Preslav
183 H9 **Prešov** *var.* Preschau, *Ger.* Eperies, *Hung.* Eperjes. Prešovský Kraj, E Slovakia 49°N 21°14′E
183 H9 **Prešovský Kraj** ⬖ *region* E Slovakia
184 G8 **Prespa, Lake** *Alb.* Liqen i Prespës, *Gk.* Límni Megáli Préspa, Límni Prespa, *Mac.* Prespansko Ezero, *Serb.* Prespansko Jezero. ◎ SE Europe
Prespa, Limni/Prespansko Ezero/Prespansko Jezero/Prespës, Liqen i *see* Prespa, Lake
63 L1 **Presque Isle** Maine, NE USA 46°40′N 68°01′W
62 C6 **Presque Isle** *headland* Pennsylvania, NE USA 42°09′N 80°06′W
Pressburg *see* Bratislava
160 F1 **Prestatyn** United Kingdom 53°20′N 3°25′W
159 J10 **Prestbury** United Kingdom 53°17′N 2°09′W
132 G8 **Prestea** SW Ghana 05°22′N 02°07′W
160 G4 **Presteigne** United Kingdom 52°17′N 3°00′W
183 B9 **Přeštice** *Ger.* Pschestitz. Plzeňský Kraj, W Czech Republic 49°36′N 13°19′E
172 C2 **Préstimo** Aveiro, Portugal 40°37′N 8°20′W
159 I9 **Preston** NW England, United Kingdom 53°46′N 02°42′W
69 J2 **Preston** Georgia, SE USA 32°03′N 84°32′W
77 I8 **Preston** Idaho, NW USA 42°06′N 111°52′W
74 I7 **Preston** Iowa, C USA 42°03′N 90°24′W
74 H6 **Preston** Minnesota, N USA 43°41′N 92°06′W
161 J7 **Preston Candover** United Kingdom 51°10′N 1°08′W
67 H4 **Prestonsburg** Kentucky, S USA 37°40′N 82°46′W
158 G4 **Prestwick** W Scotland, United Kingdom 55°31′N 04°39′W
141 J4 **Pretoria** *var.* Epitoli. ● (South Africa) Gauteng, NE South Africa 25°41′S 28°12′E
Pretoria-Witwatersrand-Vereeniging *see* Gauteng
Pretusha *see* Pretushë
184 G9 **Pretushë** *var.* Pretusha. Korçë, SE Albania 40°50′N 20°45′E
166 F9 **Preuilly-sur-Claise** Centre, France 46°51′N 0°55′E
Preussisch Eylau *see* Bagrationovsk
Preußisch Holland *see* Pasłęk
180 F6 **Preussisch Oldendorf** Nordrhein-Westfalen, Germany 52°18′E 8°30′N
Preussisch-Stargard *see* Starogard Gdański
186 E5 **Préveza** Ípeiros, W Greece 38°59′N 20°44′E
79 N2 **Prewitt Reservoir** ⊠ Colorado, C USA
257 H9 **Prey Vêng** Prey Vêng, S Cambodia 11°30′N 105°20′E
Priaral'skiye Karakumy, Peski *see* Priaral'skiy Karakum
226 G5 **Priaral'skiy Karakum** *prev.* Priaral'skiye Karakumy, Peski. *desert* SW Kazakhstan
193 J8 **Priargunsk** Zabaykal'skiy Kray, S Russian Federation 50°25′N 119°12′E
82 E8 **Pribilof Islands** *island group* Alaska, USA
184 F5 **Priboj** Serbia, W Serbia 43°34′N 19°33′E
183 B9 **Příbram** *Ger.* Pibrans. Středočeský Kraj, W Czech Republic 49°41′N 14°02′E
79 I3 **Price** Utah, W USA 39°35′N 110°49′W
79 I4 **Price River** ≈ Utah, W USA
68 G4 **Prichard** Alabama, S USA 30°44′N 88°04′W
181 I12 **Prichsenstadt** Bayern, Germany 49°49′E 10°21′N
70 G5 **Priddy** Texas, SW USA 31°39′N 98°30′W
171 H5 **Priego** Castilla-La Mancha, C Spain 40°26′N 02°19′W
170 F8 **Priego de Córdoba** Andalucía, S Spain 37°27′N 04°12′W
190 B7 **Priekule** *Ger.* Preenkuln. SW Latvia 56°26′N 21°36′E
190 B7 **Priekulė** *Ger.* Prökuls. Klaipėda, W Lithuania 55°36′N 21°16′E
191 D9 **Prienai** *Pol.* Preny. Kaunas, S Lithuania 54°37′N 23°56′E
140 F6 **Prieska** Northern Cape, C South Africa 29°40′S 22°45′E
140 F6 **Prieskapoort** *pass* Northern Cape, South Africa
76 F3 **Priest Lake** ⊠ Idaho, NW USA
76 F3 **Priest River** Idaho, NW USA 48°10′N 117°02′W
170 F2 **Prieta, Peña** ▲ N Spain 43°01′N 04°42′W
85 I8 **Prieto, Cerro** ▲ C Mexico 24°10′N 105°21′W
183 F10 **Prievidza** *var.* Priewitz, *Ger.* Priwitz, *Hung.* Privigye. Trenčiansky Kraj, W Slovakia 48°47′N 18°35′E
Priewitz *see* Prievidza
184 C3 **Prijedor** ⬖ Republika Srpska, NW Bosnia and Herzegovina
184 F5 **Prijepolje** Serbia, W Serbia 43°24′N 19°39′E
Prikaspiyskaya Nizmennost' *see* Caspian Depression
184 G8 **Prilep** *Turk.* Perlepe. S FYR Macedonia 41°21′N 21°34′E
176 C5 **Prilly** Vaud, SW Switzerland 46°32′N 06°38′E
Priluki *see* Pryluky
116 F2 **Primero, Río** ≈ C Argentina
74 F6 **Primghar** Iowa, C USA 43°05′N 95°37′W
81 I10 **Primm** Nevada, USA 35°37′N 115°23′W
184 B2 **Primorje-Gorski Kotar** *off.* Primorsko-Goranska Županija. ⬖ *province* NW Croatia
191 A8 **Primorsk** *Ger.* Fischhausen. Kaliningradskaya Oblast', W Russian Federation 54°45′N 20°00′E
194 D8 **Primorsk** *Fin.* Koivisto. Leningradskaya Oblast', NW Russian Federation 60°20′N 28°39′E
248 G1 **Primorskiy Kray** *prev. Eng.* Maritime Territory. ⬖ *territory* SE Russian Federation
185 M6 **Primorsko** *prev.* Keupriya. Burgas, E Bulgaria 42°15′N 27°45′E
196 F7 **Primorsko-Akhtarsk** Krasnodarskiy Kray, SW Russian Federation 46°03′N 38°44′E
Primorsko-Goranska Županija *see* Primorje-Gorski Kotar
Primorsk/Primorskoye *see* Prymors'k
184 C5 **Primošten** Šibenik-Knin, S Croatia 43°34′N 15°57′E
57 K3 **Primrose Lake** ◎ Saskatchewan, C Canada
181 C13 **Primstal** Saarland, Germany 49°32′E 6°59′N
57 M5 **Prince Albert** Saskatchewan, S Canada 53°09′N 105°43′W
138 E10 **Prince Albert** Western Cape, SW South Africa 33°13′S 22°03′E
52 E6 **Prince Albert Peninsula** *peninsula* Victoria Island, Northwest Territories, NW Canada
52 E7 **Prince Albert Sound** *inlet* Northwest Territories, N Canada
52 D4 **Prince Alfred, Cape** *headland* Northwest Territories, NW Canada
140 G9 **Prince Alfred's Pass** *pass* Western Cape, South Africa
53 K8 **Prince Charles Island** *island* Nunavut, NE Canada
293 K4 **Prince Charles Mountains** ▲▲ Antarctica
Prince-Édouard, Île-du *see* Prince Edward Island
289 B12 **Prince Edward Fracture Zone** *tectonic feature* SW Indian Ocean
59 K8 **Prince Edward Island** *Fr.* Île-du Prince-Édouard. ⬖ *province* SE Canada
59 K8 **Prince Edward Island** *Fr.* Île-du Prince-Édouard. *island* SE Canada
289 C12 **Prince Edward Islands** *island group* S South Africa
67 L3 **Prince Frederick** Maryland, NE USA 38°32′N 76°35′W
56 E4 **Prince George** British Columbia, SW Canada 53°55′N 122°49′W
67 L4 **Prince George** Virginia, NE USA 37°13′N 77°13′W
295 H7 **Prince Gustaf Adolf Sea** *sea* Nunavut, N Canada
294 E7 **Prince of Wales, Cape** *headland* Alaska, USA 65°39′N 168°12′W
275 J1 **Prince of Wales Island** *island* Queensland, E Australia
52 G6 **Prince of Wales Island** *island* Queen Elizabeth Islands, Nunavut, NW Canada
83 M8 **Prince of Wales Island** *island* Alexander Archipelago, Alaska, USA
Prince of Wales Island *see* Pinang, Pulau
52 E6 **Prince of Wales Strait** *strait* Northwest Territories, N Canada
295 H7 **Prince Patrick Island** *island* Parry Islands, Northwest Territories, NW Canada
53 H6 **Prince Regent Inlet** *channel* Nunavut, N Canada
56 A3 **Prince Rupert** British Columbia, SW Canada 54°18′N 130°17′W
Prince's Island *see* Príncipe
67 M3 **Princess Anne** Maryland, NE USA 38°12′N 75°42′W
Princess Astrid Coast *see* Prinsesse Astrid Kyst
275 J2 **Princess Charlotte Bay** *bay* Queensland, NE Australia
293 K4 **Princess Elizabeth Land** *physical region* Antarctica
53 I3 **Princess Marie Bay** *coastal sea feature* Nunavut, N Canada
56 B5 **Princess Royal Island** *island* British Columbia, SW Canada
91 L8 **Princes Town** Trinidad, Trinidad and Tobago 10°16′N 61°23′W
56 F8 **Princeton** British Columbia, SW Canada 49°25′N 120°35′W
80 A3 **Princeton** California, USA 37°30′N 122°29′W
73 C9 **Princeton** Illinois, N USA 41°22′N 89°27′W
73 D12 **Princeton** Indiana, N USA 38°21′N 87°33′W
74 J7 **Princeton** Iowa, C USA 41°40′N 90°21′W
66 D5 **Princeton** Kentucky, S USA 37°06′N 87°52′W
74 G4 **Princeton** Minnesota, N USA 45°34′N 93°34′W
75 G8 **Princeton** Missouri, C USA 40°22′N 93°37′W
64 G7 **Princeton** New Jersey, NE USA 40°21′N 74°39′W
67 I4 **Princeton** West Virginia, NE USA 37°23′N 81°06′W
160 E9 **Princetown** United Kingdom 50°33′N 4°00′W
83 J6 **Prince William Sound** *inlet* Alaska, USA
121 I6 **Príncipe** *var.* Príncipe Island, *Eng.* Prince's Island. *island* N Sao Tome and Principe
Príncipe Island *see* Príncipe
76 D6 **Prineville** Oregon, NW USA 44°19′N 120°50′W
74 A5 **Pringle** South Dakota, N USA 43°34′N 103°34′W
70 E1 **Pringle** Texas, SW USA 35°55′N 101°28′W
163 E8 **Prinsenbeek** Noord-Brabant, S Netherlands 51°36′N 04°42′E
162 G4 **Prinses Margriet Kanaal** *canal* N Netherlands
293 I2 **Prinsesse Astrid Kyst** *Eng.* Princess Astrid Coast. *physical region* Antarctica
293 J2 **Prinsesse Ragnhild Kyst** *physical region* Antarctica
293 K2 **Prins Harald Kyst** *physical region* Antarctica
152 A5 **Prins Karls Forland** *island* W Svalbard
89 H5 **Prinzapolka** Región Autónoma Atlántico Norte, NE Nicaragua 13°19′N 83°35′W
88 G5 **Prinzapolka, Río** ≈ NE Nicaragua
195 M6 **Priob'ye** Khanty-Mansiyskiy Avtonomnyy Okrug-Yugra, N Russian Federation 62°25′N 65°36′E
170 C1 **Prior, Cabo** *headland* NW Spain 43°33′N 08°21′W
74 G5 **Prior Lake** Minnesota, N USA 44°42′N 93°25′W
141 H7 **Priors** Free State, South Africa 30°25′S 25°36′E
227 J6 **Priozersk** Karaganda, SE Kazakhstan 46°02′N 73°42′E
194 D7 **Priozersk** *Fin.* Käkisalmi. Leningradskaya Oblast', NW Russian Federation 61°02′N 30°07′E
191 F11 **Pripet** *Bel.* Prypyats', *Ukr.* Pryp"yat'. ≈ Belarus/Ukraine
188 E1 **Pripet Marshes** *wetland* Belarus/Ukraine
184 G6 **Prishtinë** , *Serb.* Priština. C Kosovo 42°40′N 21°10′E
196 F4 **Pristen'** Kurskaya Oblast', W Russian Federation 51°15′N 36°47′E
Priština *see* Prishtinë
178 G5 **Pritzwalk** Brandenburg, NE Germany 53°10′N 12°11′E
165 J7 **Privas** Ardèche, E France 44°45′N 04°35′E
175 F9 **Priverno** Lazio, C Italy 41°28′N 13°11′E
Privigye *see* Prievidza
184 B4 **Privlaka** Zadar, SW Croatia 44°15′N 15°07′E
194 G10 **Privolzhsk** Ivanovskaya Oblast', NW Russian Federation 57°24′N 41°16′E
197 I4 **Privolzhskaya Vozvyshennost'** *var.* Volga Uplands. ▲▲ W Russian Federation
197 I4 **Privolzhskoye** Saratovskaya Oblast', W Russian Federation 51°08′N 45°57′E
Priwitz *see* Prievidza
197 H7 **Priyutnoye** Respublika Kalmykiya, SW Russian Federation 46°08′N 43°33′E
184 G7 **Prizren** S Kosovo 42°14′N 20°46′E
175 F12 **Prizzi** Sicilia, Italy, C Mediterranean Sea 37°44′N 13°26′E
185 H7 **Probištip** NE FYR Macedonia 42°00′N 22°06′E
259 J9 **Probolinggo** Jawa, C Indonesia 07°45′S 113°12′E
Probstberg *see* Wyszków
182 D7 **Prochowice** *Ger.* Parchwitz. Dolnośląskie, SW Poland 51°15′N 16°22′E
74 H3 **Proctor** Minnesota, N USA 46°46′N 92°13′W
70 G5 **Proctor** Texas, SW USA 31°57′N 98°25′W
70 G5 **Proctor Lake** ⊠ Texas, SW USA
234 F7 **Proddatūr** Andhra Pradesh, E India 14°45′N 78°34′E
172 C3 **Proença-a-Nova** *var.* Proença a Nova. Castelo Branco, C Portugal 39°45′N 07°56′W
Proença a Nova *see* Proença-a-Nova
163 F11 **Profondeville** Namur, SE Belgium 50°22′N 04°52′E
114 A3 **Progresso** Santa Fe, Argentina 31°08′S 60°59′W
85 K5 **Progreso** Coahuila, Mexico 27°25′N 100°59′W
87 L5 **Progreso** Yucatán, SE Mexico 21°14′N 89°41′W
193 K8 **Progress** Amurskaya Oblast', SE Russian Federation 49°40′N 129°30′E
197 H9 **Prokhladnyy** Kabardino-Balkarskaya Respublika, SW Russian Federation 43°48′N 44°02′E
Prokletije *see* North Albanian Alps
227 M1 **Prokop'yevsk** Kemerovskaya Oblast', S Russian Federation 53°56′N 86°48′E
Prökuls *see* Priekulė
184 G5 **Prokuplje** Serbia, SE Serbia 43°15′N 21°35′E
194 D9 **Proletariy** Novgorodskaya Oblast', W Russian Federation 58°24′N 31°40′E
197 H7 **Proletarsk** Rostovskaya Oblast', SW Russian Federation 46°42′N 41°48′E
197 H7 **Proletarskoye Vodokhranilishche** *salt lake* SW Russian Federation
Prome *see* Pyay
113 K3 **Promissão** São Paulo, S Brazil 21°33′S 49°51′W
109 B12 **Promissão, Represa de** ⊠ S Brazil
195 L3 **Promyshlennyy** Respublika Komi, NW Russian Federation 67°36′N 64°E
180 I2 **Pronstorf** Schleswig-Holstein, Germany 53°57′E 10°28′N
114 E5 **Pronunciamiento** Entre Ríos, Argentina 32°21′S 58°26′W
191 I10 **Pronya** ≈ E Belarus
54 D5 **Prophet River** British Columbia, W Canada 58°07′N 122°39′W
73 C9 **Prophetstown** Illinois, N USA 41°40′N 89°56′W
Propinsi Kepulauan Riau *see* Kepulauan Riau
Propinsi Papua Barat *see* Papua Barat
Propinsi Sulawesi Barat *see* Sulawesi Barat
108 I7 **Propriá** Sergipe, E Brazil 10°15′S 36°51′W
175 B8 **Propriano** Corse, France, C Mediterranean Sea 41°41′N 08°54′E
Prościejów *see* Prostějov
Proskurov *see* Khmel 'nyts'kyy
187 H2 **Prosotsáni** Anatolikí Makedonía kai Thráki, NE Greece 41°11′N 23°59′E
263 N8 **Prosperidad** Mindanao, S Philippines 08°36′N 125°54′E
76 D5 **Prosser** Washington, NW USA 46°12′N 119°46′W
Prossnitz *see* Prostějov
183 E9 **Prostějov** *Ger.* Prossnitz, *Pol.* Prościejów. Olomoucký Kraj, E Czech Republic 49°29′N 17°08′E
189 L5 **Prosyana** Dnipropetrovs'ka Oblast', E Ukraine 48°07′N 36°22′E
183 G8 **Proszowice** Malopolskie, S Poland 50°12′N 20°15′E
Protasy *see* Pratasy
289 A11 **Protea Seamount** *undersea feature* SW Indian Ocean 36°50′S 18°05′E
291 D13 **Protector Basin** S Atlantic Ocean
140 E10 **Protem** Western Cape, South Africa 34°16′S 20°05′E
186 F7 **Próti** *island* S Greece
185 M5 **Provadia** , Provadiya. Varna, E Bulgaria 43°10′N 27°29′E
Provadiya *see* Provadia
165 K8 **Provence** *cultural region* SE France
169 K6 **Provence** *prev.* Marseille-Marigrane. ✈ (Marseille) Bouches-du-Rhône, SE France 43°25′N 05°15′E
165 K8 **Provence-Alpes-Côte d'Azur** ⬖ *region* SE France
66 D4 **Providence** Kentucky, S USA 37°23′N 87°47′W
65 L3 **Providence** *state capital* Rhode Island, NE USA 41°50′N 71°26′W
79 H1 **Providence** Utah, W USA 41°42′N 111°49′W
Providence *see* Fort Providence
Providence *see* Providence Atoll
121 M6 **Providence Atoll** *var.* Providence. *atoll* S Seychelles
62 B2 **Providence Bay** Manitoulin Island, Ontario, S Canada 45°39′N 82°16′W
69 I2 **Providence Canyon** *valley* Alabama/Georgia, S USA
68 D1 **Providence, Lake** ◎ Louisiana, S USA
81 H10 **Providence Mountains** ▲▲ California, W USA
114 A3 **Providencia** Santa Fe, Argentina 30°59′S 61°01′W
89 I5 **Providencia, Isla de** *island* NW Colombia, Caribbean Sea
91 H4 **Providenciales** *island* W Turks and Caicos Islands
50 B2 **Provideniya** Chukotskiy Avtonomnyy Okrug, NE Russian Federation 64°22′N 173°14′W
65 N3 **Provincetown** Massachusetts, NE USA 42°01′N 70°10′W

◆ Country
● Country Capital
◇ Dependent Territory
○ Dependent Territory Capital
⬖ Administrative Regions
✈ International Airport
▲ Mountain
▲▲ Mountain Range
Volcano
≈ River
◎ Lake
⊠ Reservoir

◆ Country   ◇ Dependent Territory   ◆ Administrative Regions   ▲ Mountain   ☒ Volcano   ☺ Lake
● Country Capital   ○ Dependent Territory Capital   ✕ International Airport   ▲ Mountain Range   ✍ River   ☐ Reservoir

Reichenbach - Rivas

179 G9 **Reichenbach** *var.* Reichenbach im Vogtland. Sachsen, E Germany *50°36′N 12°18′E*
Reichenbach *see* Dzierżoniów
Reichenbach im Vogtland *see* Reichenbach
181 D13 **Reichenbach-Steegen** Rheinland-Pfalz, Germany *49°30′E 7°33′N*
181 H12 **Reichenberg** Bayern, Germany *49°44′E 9°55′N*
Reichenberg *see* Liberec
181 G12 **Reicholzheim** Baden-Württemberg, Germany *49°44′E 9°32′N*
276 C4 **Reid** Western Australia *30°49′S 128°24′E*
69 L2 **Reidsville** Georgia, SE USA *32°05′N 82°07′W*
67 J5 **Reidsville** North Carolina, SE USA *36°21′N 79°39′W*
Reifnitz *see* Ribnica
161 K7 **Reigate** SE England, United Kingdom *51°14′N 00°13′W*
159 M8 **Reighton** United Kingdom *54°09′N 0°16′W*
Reikjavik *see* Reykjavík
164 F6 **Ré, Île de** *island* W France
79 J9 **Reiley Peak** ▲ Arizona, SW USA *32°24′N 110°09′W*
219 C9 **Re'im** Israel *31°23′N 34°27′E*
165 I3 **Reims** *Eng.* Rheims; *anc.* Durocortorum, Remi. Marne, N France *49°16′N 04°01′E*
117 B13 **Reina Adelaida, Archipiélago** *island group* S Chile
102 E1 **Reina Beatrix** ✈ (Oranjestad) C Aruba *12°30′N 69°57′W*
176 D4 **Reinach** Aargau, W Switzerland *47°16′N 08°12′E*
176 D3 **Reinach** Basel Landschaft, NW Switzerland *47°30′N 07°36′E*
170 A10 **Reina Sofía** ✈ (Tenerife) Tenerife, Islas Canarias, Spain, NE Atlantic Ocean
74 H7 **Reinbeck** Iowa, C USA *42°19′N 92°36′W*
180 I3 **Reinbek** Schleswig-Holstein, N Germany *53°31′N 10°15′E*
57 N2 **Reindeer** ≈ Saskatchewan, C Canada
57 N1 **Reindeer Lake** ◎ Manitoba/Saskatchewan, C Canada
Reine-Charlotte, Îles de la *see* Queen Charlotte Islands
Reine-Élisabeth, Îles de la *see* Queen Elizabeth Islands
154 C6 **Reineskarvet** ▲ S Norway *60°38′N 07°48′E*
278 E1 **Reinga, Cape** *headland* North Island, New Zealand *34°24′S 172°40′E*
181 F12 **Reinheim** Hessen, Germany *49°50′E 8°50′N*
170 G2 **Reinosa** Cantabria, N Spain *43°01′N 04°09′W*
181 C12 **Reinsfeld** Rheinland-Pfalz, Germany *49°41′E 6°53′N*
181 F10 **Reiskirchen** Hessen, Germany *50°30′E 8°31′N*
179 I14 **Reisseck** ▲ S Austria *46°57′N 13°21′E*
181 E8 **Reiste** Nordrhein-Westfalen, Germany *51°16′E 8°15′N*
64 D9 **Reisterstown** Maryland, NE USA *39°27′N 76°46′W*
Reisui *see* Yeosu
162 H3 **Reitdiep** ≈ NE Netherlands
285 K7 **Reitoru** *atoll* Îles Tuamotu, C French Polynesia
141 J5 **Reitz** Free State, South Africa *27°48′S 28°26′E*
155 H8 **Rejmyre** Östergötland, S Sweden *58°49′N 15°55′E*
Reka *see* Rijeka
Reka Ili *see* Ile/Ili He
Rekarne *see* Tumbo
Rekhovot *see* Rehovot
118 B8 **Relbun, Río** ≈ Bío-Bío, Chile
54 G4 **Reliance** Northwest Territories, C Canada *62°45′N 109°08′W*
77 J9 **Reliance** Wyoming, C USA *41°42′N 109°13′W*
131 H2 **Relizane** *var.* Ghelîzâne, Ghilizane. NW Algeria *35°45′N 00°33′E*
180 H2 **Rellingen** Schleswig-Holstein, Germany *53°39′E 9°49′N*
166 F6 **Rémalard** Basse-Normandie, France *48°26′N 0°46′E*
276 G4 **Remarkable, Mount** ▲ South Australia *32°46′S 138°08′E*
171 I10 **Remchi** Algeria
181 J10 **Remda** Thüringen, Germany *50°46′E 11°13′N*
102 C4 **Remedios** Antioquia, N Colombia *07°02′N 74°42′W*
89 J9 **Remedios** Veraguas, W Panama *08°13′N 81°48′W*
88 C5 **Remedios, Punta** *headland* SW El Salvador *13°31′N 89°48′W*
180 D3 **Remels** Niedersachsen, Germany *53°18′E 7°45′N*
Remi *see* Reims
163 I14 **Remich** Grevenmacher, SE Luxembourg *49°33′N 06°23′E*
163 F10 **Remicourt** Liège, E Belgium *50°40′N 05°19′E*
103 N5 **Rémire** NE French Guiana *04°52′N 52°16′W*
167 M7 **Remiremont** Lorraine, France *48°01′N 6°35′E*
197 H7 **Remontnoye** Rostovskaya Oblast', SW Russian Federation *46°35′N 43°38′E*
261 I6 **Remoon** Pulau Kur, E Indonesia *05°18′S 131°59′E*
163 G11 **Remouchamps** Liège, E Belgium *50°29′N 05°43′E*
165 J8 **Remoulins** Gard, S France *43°56′N 04°34′E*
137 G11 **Rempart, Mont du** *hill* W Mauritius
181 C9 **Remscheid** Nordrhein-Westfalen, W Germany *51°10′N 07°11′E*
74 F6 **Remsen** Iowa, C USA *42°48′N 95°58′W*
169 K4 **Rémuzat** Rhône-Alpes, France *44°24′N 5°21′E*
154 E6 **Rena** Hedmark, S Norway *61°08′N 11°21′E*
154 E5 **Renåa** ≈ S Norway
118 B9 **Renaico, Río** ≈ Chile
Renaix *see* Ronse
166 D7 **Renazé** Pays de la Loire, France *47°48′N 1°03′W*
190 E5 **Rencēni** N Latvia *57°43′N 25°25′E*
190 C6 **Renda** W Latvia *57°04′N 22°18′E*
175 H11 **Rende** Calabria, SW Italy *39°19′N 16°10′E*
163 G12 **Rendeux** Luxembourg, SE Belgium *50°15′N 05°28′E*
Rendina *see* Rentína
73 C12 **Rend Lake** ⊠ Illinois, N USA
281 H2 **Rendova** *island* New Georgia Islands, NW Solomon Islands
180 H1 **Rendsburg** Schleswig-Holstein, N Germany *54°18′N 09°40′E*
176 C5 **Renens** Vaud, SW Switzerland *46°32′N 06°36′E*
62 F3 **Renfrew** Ontario, SE Canada *45°28′N 76°44′W*
158 G3 **Renfrew** United Kingdom *55°52′N 4°22′W*
156 E7 **Renfrew** *cultural region* SW Scotland, United Kingdom
258 E5 **Rengat** Sumatera, W Indonesia *0°26′S 102°38′E*
233 M6 **Rengma Hills** ▲▲ NE India
116 B4 **Rengo** Libertador, C Chile *34°24′S 70°50′W*
181 D10 **Rengsdorf** Rheinland-Pfalz, Germany *50°30′E 7°30′N*
242 C5 **Renhuai** Guizhou, China *27°29′N 106°14′E*
188 F8 **Reni** Odes'ka Oblast', SW Ukraine *45°30′N 28°18′E*
243 J8 **Renju** Guangdong, SE China *24°49′N 115°53′E*
136 B3 **Renk** Upper Nile, NE South Sudan *11°48′N 32°49′E*
190 D2 **Renko** Kanta-Häme, S Finland *60°52′N 24°16′E*
162 G7 **Renkum** Gelderland, SE Netherlands *51°58′N 05°43′E*
277 H5 **Renmark** South Australia *34°12′S 140°43′E*
281 J4 **Rennell** *var.* Mu Nggava. *island* S Solomon Islands
281 I2 **Rennell and Bellona** *prev.* Central. ◈ *province* S Solomon Islands
275 H3 **Renner Springs Roadhouse** Northern Territory, N Australia *18°12′S 133°48′E*
164 F4 **Rennes** *Bret.* Roazon; *anc.* Condate. Ille-et-Vilaine, NW France *48°08′N 01°40′W*
293 J8 **Rennick Glacier** *glacier* Antarctica
55 J9 **Rennie** Manitoba, S Canada *49°51′N 95°28′W*
80 F2 **Reno** Nevada, W USA *39°32′N 119°49′W*
177 H8 **Reno** ≈ N Italy
80 F2 **Reno-Cannon** ✈ Nevada, W USA *39°26′N 119°42′W*
141 I5 **Renoster** ≈ Free State, South Africa
138 E9 **Renoster** ≈ SW South Africa
140 F8 **Renosterkop** Western Cape, South Africa *32°12′S 22°51′E*
140 E9 **Renosterleegte** ≈ Western Cape, South Africa
63 M1 **Renous** New Brunswick, SE Canada *46°49′N 65°48′W*
64 C4 **Renovo** Pennsylvania, NE USA *41°19′N 77°42′W*
245 J4 **Renqiu** Hebei, E China *38°49′N 116°02′E*
242 A3 **Renshou** Sichuan, C China *30°02′N 104°09′E*
73 E10 **Rensselaer** Indiana, N USA *40°57′N 87°09′W*
65 H2 **Rensselaer** New York, NE USA *42°38′N 73°44′W*
Rentería *see* Errenteria
186 G5 **Rentína** *var.* Rendina. Thessalía, C Greece *39°04′N 21°58′E*
74 F5 **Renville** Minnesota, N USA *44°48′N 95°13′W*
159 I6 **Renwick** United Kingdom *54°46′N 2°37′W*
132 G5 **Réo** W Burkina Faso *12°20′N 02°28′W*
63 H3 **Repentigny** Québec, SE Canada *45°42′N 73°28′W*
228 F7 **Repetek** Lebap Welaýaty, E Turkmenistan *38°40′N 63°12′E*
153 G8 **Replot** *Fin.* Raippaluoto. *island* W Finland
Repola *see* Reboly
Reppen *see* Rzepin
172 C5 **Represa** Évora, Portugal *38°42′N 8°06′W*
110 A9 **Represa Armando** *mountains* Sao Paulo, Brazil
110 E4 **Represa de Furnas** *mountains* Minas Gerais, Brazil
110 D4 **Represa dos Peixotos** *mountains* Minas Gerais, Brazil
Reps *see* Rupea
75 H11 **Republic** Missouri, C USA *37°07′N 93°28′W*
76 E3 **Republic** Washington, NW USA *48°39′N 118°44′W*
75 E9 **Republican River** ≈ Kansas/Nebraska, C USA
53 I9 **Repulse Bay** Northwest Terretories, N Canada *66°35′N 86°20′W*
104 F4 **Requena** Loreto, NE Peru *05°05′S 73°52′W*
171 I6 **Requena** Valenciana, E Spain *39°29′N 01°08′W*
165 H8 **Réquista** Aveyron, S France *44°00′N 02°31′E*
214 G5 **Reşadiye** Tokat, N Turkey *40°24′N 37°19′E*
Reschenpass *see* Resia, Passo di
Reschitza *see* Reşiţa
290 C5 **Researcher Seamount** *undersea feature* C Atlantic Ocean
184 G8 **Resen** *Turk.* Resne. SW FYR Macedonia *41°07′N 21°00′E*
113 J4 **Reserva** Paraná, S Brazil *24°40′S 50°52′W*
55 H9 **Reserve** Saskatchewan, S Canada *52°24′N 102°37′W*
79 K8 **Reserve** New Mexico, SW USA *33°42′N 108°45′W*
118 D2 **Resguardo de los Patos** Valparaíso, Chile *32°29′S 70°35′W*
Reshetilovka *see* Reshetylivka
190 G5 **Reshety** Pskovskaya Oblast', Russian Federation
189 J4 **Reshetylivka** *Rus.* Reshetilovka. Poltavs'ka Oblast', NE Ukraine *49°34′N 34°05′E*
Resht *see* Rasht
217 I2 **Reshwan** *Ar.* Rashwān, *var.* Ŗashwan. Arbīl, N Iraq *36°28′N 43°54′E*
174 D3 **Resia, Passo di** *Ger.* Reschenpass. *pass* Austria/Italy
Resicabánya *see* Reşiţa
112 F6 **Resistencia** Chaco, NE Argentina *27°27′S 58°56′W*
188 B8 **Reşiţa** *Ger.* Reschitza, *Hung.* Resicabánya. Caraş-Severin, W Romania *45°14′N 21°58′E*
Resne *see* Resen
118 G5 **Resolana** Mendoza, Argentina *34°32′S 68°06′W*
295 I8 **Resolute** *Inuit* Qausuittuq. Cornwallis Island, Nunavut, N Canada *74°41′N 94°54′W*
Resolution *see* Fort Resolution
59 I1 **Resolution Island** *island* Nunavut, NE Canada
278 A12 **Resolution Island** *island* SW New Zealand
111 L2 **Resplendor** Minas Gerais, Brazil *19°20′S 41°15′W*
111 H5 **Ressaguinha** Minas Gerais, Brazil *21°04′S 43°46′W*
167 H4 **Ressons-sur-Matz** Picardie, France *49°33′N 2°45′E*
Resta *see* Rasta
115 L1 **Restinga Sèca** Rio Grande do Sul, Brazil *29°49′S 53°23′W*
55 I10 **Reston** Manitoba, S Canada *49°33′N 101°03′W*
62 D2 **Restoule Lake** ◎ Ontario, S Canada
102 C6 **Restrepo** Meta, C Colombia *04°20′N 73°29′W*
140 F8 **Restvale** Western Cape, South Africa *32°05′S 23°00′E*
88 B4 **Retalhuleu** Retalhuleu, SW Guatemala *14°31′N 91°40′W*
88 B4 **Retalhuleu** *off.* Departamento de Retalhuleu. ◈ *department* SW Guatemala
Retalhuleu, Departamento de *see* Retalhuleu
118 F1 **Retamito** San Juan, Argentina *32°06′S 68°36′W*
159 L10 **Retford** C England, United Kingdom *53°18′N 00°52′W*
165 I2 **Rethel** Ardennes, N France *49°31′N 04°22′E*
Rethimno/Réthimnon *see* Réthymno
187 I10 **Réthymno** *prev.* Rethimno, Réthimnon. Kríti, Greece, E Mediterranean Sea *35°21′N 24°29′E*
Retiche, Alpi *see* Rhaetian Alps
163 F9 **Retie** Antwerpen, N Belgium *51°18′N 05°05′E*
169 I2 **Retournac** Auvergne, France *45°12′N 4°02′E*
183 F11 **Rétság** Nógrád, N Hungary *47°57′N 19°08′E*
172 G4 **Retuerta de Bullaque** Castilla-La Manacha, Spain *39°27′N 4°24′W*
177 L1 **Retz** Niederösterreich, NE Austria *48°46′N 15°58′E*
166 G9 **Reuilly** Centre, France *47°05′N 2°03′E*
137 I10 **Réunion** *off.* La Réunion. ◇ *French overseas department* W Indian Ocean
204 G10 **Réunion** *island* W Indian Ocean
171 J4 **Reus** Cataluña, E Spain *41°10′N 01°06′E*
163 F9 **Reusel** Noord-Brabant, S Netherlands *51°21′N 05°10′E*
Reuţel *see* Ciuhuru
179 D12 **Reutlingen** Baden-Württemberg, S Germany *48°30′N 09°13′E*
176 G3 **Reutte** Tirol, W Austria *47°30′N 10°44′E*
163 H9 **Reuver** Limburg, SE Netherlands *51°17′N 06°05′E*
74 B4 **Reva** South Dakota, N USA *45°30′N 103°03′W*
Revakha *see* Rewaha
Reval/Revel *see* Tallinn
194 F3 **Revda** Murmanskaya Oblast', NW Russian Federation *67°57′N 34°29′E*
195 L10 **Revda** Sverdlovskaya Oblast', C Russian Federation *56°48′N 59°42′E*
80 J6 **Reveille Range** ▲▲ Nevada, USA
171 K1 **Revel** Haute-Garonne, S France *43°27′N 01°59′E*
56 G7 **Revelstoke** British Columbia, SW Canada *51°02′N 118°12′W*
89 H8 **Reventazón, Río** ≈ E Costa Rica
176 G7 **Revere** Lombardia, N Italy *45°03′N 11°07′E*
159 M10 **Revesby** United Kingdom *53°08′N 0°03′W*
167 K5 **Revigny-sur-Ornain** Lorraine, France *48°50′N 4°59′E*
83 N8 **Revillagigedo Island** *island* Alexander Archipelago, Alaska, USA
165 J2 **Revin** Ardennes, N France *49°57′N 04°39′E*
219 D10 **Revivim** Israel *31°02′N 34°43′E*
152 B5 **Revnosa** *headland* C Svalbard *78°03′N 18°52′E*
Revolyutsii, Pik *see* Revolyutsiya, Qullai
229 K7 **Revolyutsiya, Qullai** *Rus.* Pik Revolyutsii. ▲ SE Tajikistan *38°40′N 72°26′E*
183 G10 **Revúca** *Ger.* Grossrauschenbach, *Hung.* Nagyrőce. Banskobystrický Kraj, C Slovakia *48°40′N 20°10′E*
232 G7 **Rewa** Madhya Pradesh, C India *24°32′N 81°18′E*
219 D9 **Rewaha** *var.* Revakha. Southern, S Israel *31°39′N 34°44′E*
217 J2 **Rewanduz** *Ar.* Rawāndūz, *var.* Ŗawāndiz. Arbīl, N Iraq *36°38′N 44°32′E*
232 E5 **Rewāri** Haryāna, N India *28°14′N 76°38′E*
77 I7 **Rexburg** Idaho, NW USA *43°49′N 111°47′W*
134 C4 **Rey Bouba** Nord, NE Cameroon *08°40′N 14°11′E*
152 D2 **Reyðarfjörður** Austurland, E Iceland *65°02′N 14°12′W*
106 D9 **Reyes** El Beni, NW Bolivia *14°17′S 67°18′W*
86 G6 **Reyes de Vallarta** Puebla, Mexico *20°06′N 97°32′W*
80 A2 **Reyes, Point** *headland* California, W USA *37°59′N 123°01′W*
102 A7 **Reyes, Punta** *headland* SW Colombia *02°44′N 78°08′W*
214 G8 **Reyhanlı** Hatay, S Turkey *36°16′N 36°35′E*
89 L9 **Rey, Isla del** *island* Archipiélago de las Perlas, SE Panama
152 B2 **Reykhólar** Vestfirðir, W Iceland *65°28′N 22°12′W*
152 D2 **Reykjahlíð** Norðurland Eystra, NE Iceland *65°37′N 16°54′W*
295 M7 **Reykjanes Basin** *var.* Irminger Basin. *undersea feature* N Atlantic Ocean *62°30′N 33°30′W*
295 M7 **Reykjanes Ridge** *undersea feature* N Atlantic Ocean *62°00′N 27°00′W*
152 B3 **Reykjavík** *var.* Reikjavik. ● (Iceland) Höfudhborgarsvaedhi, W Iceland *64°08′N 21°54′W*
85 J5 **Rey, Laguna del** ◎ Coahuila, Mexico
64 A5 **Reynoldsville** Pennsylvania, NE USA *41°04′N 78°51′W*
85 M6 **Reynosa** Tamaulipas, C Mexico *26°03′N 98°19′W*
Reza'iyeh *see* Orūmīyeh
Reẕā'īyeh, Daryācheh-ye *see* Orūmīyeh, Daryācheh-ye
164 F5 **Rezé** Loire-Atlantique, NW France *47°10′N 01°36′W*
190 F7 **Rēzekne** *Ger.* Rositten; *prev. Rus.* Rezhitsa. SE Latvia *56°31′N 27°22′E*
217 J2 **Rezge** *Ar.* Razkah, *var.* Ŗazga. As Sulaymānīyah, E Iraq *36°25′N 45°06′E*
195 M9 **Rezh** Sverdlovskaya Oblast, Russian Federation
195 M9 **Rezh** ≈ Sverdlovskaya Oblast, Russian Federation
Rezhitsa *see* Rēzekne
188 G6 **Rezina** NE Moldova *47°44′N 28°58′E*
185 M7 **Rezovo** *Turk.* Rezve. Burgas, E Bulgaria *42°00′N 28°00′E*
185 M7 **Rezovska Reka** *Turk.* Rezve Deresi. ≈ Bulgaria/Turkey *see also* Rezve Deresi
Rezovska Reka *see* Rezve Deresi
Rezve *see* Rezovo
187 L1 **Rezve Deresi** *Bul.* Rezovska Reka. ≈ Bulgaria/Turkey *see also* Rezovska Reka
Rezve Deresi *see* Rezovska Reka
169 N4 **Rezzo** Liguria, NW Italy *44°01′N 07°52′E*
Rhadames *see* Ghadāmis
180 C7 **Rhade** Nordrhein-Westfalen, Germany *51°45′E 6°56′N*
Rhaedestus *see* Tekirdağ
176 F5 **Rhaetian Alps** *Fr.* Alpes Rhétiques, *Ger.* Rätische Alpen, *It.* Alpi Retiche. ▲▲ C Europe
176 F4 **Rhätikon** ▲▲ C Europe
181 D12 **Rhaunen** Rheinland-Pfalz, Germany *49°51′E 7°21′N*
160 F4 **Rhayader** United Kingdom *52°19′N 3°30′W*
180 E7 **Rheda-Wiedenbrück** Nordrhein-Westfalen, W Germany *51°51′N 08°17′E*
180 D4 **Rhede** Niedersachsen, Germany *53°04′E 7°16′N*
180 B7 **Rhede** Nordrhein-Westfalen, Germany *51°50′E 6°42′N*
162 H7 **Rheden** Gelderland, E Netherlands *52°01′N 06°03′E*
Rhegion/Rhegium *see* Reggio di Calabria
Rheims *see* Reims
Rhein *see* Rhine
181 C10 **Rheinbach** Nordrhein-Westfalen, W Germany *50°37′N 06°57′E*
181 B8 **Rheinberg** Nordrhein-Westfalen, Germany *51°33′E 6°36′N*
180 D6 **Rheine** *var.* Rheine in Westfalen. Nordrhein-Westfalen, NW Germany *52°17′N 07°27′E*
Rheine in Westfalen *see* Rheine
Rheinfeld *see* Rheinfelden
176 D3 **Rheinfelden** Baden-Württemberg, S Germany *47°34′N 07°46′E*
176 D3 **Rheinfelden** *var.* Rheinfeld. Aargau, N Switzerland *47°33′N 07°47′E*
181 D10 **Rheinisches Schiefergebirge** *var.* Rhine State Uplands, *Eng.* Rhenish Slate Mountains. ▲▲ W Germany
181 C11 **Rheinland-Pfalz** *Eng.* Rhineland-Palatinate, *Fr.* Rhénanie-Palatinat. ◈ *state* W Germany
181 F12 **Rhein Main** ✈ (Frankfurt am Main) Hessen, W Germany *50°03′N 08°33′E*
181 E14 **Rheinzabern** Rheinland-Pfalz, Germany *49°07′E 8°17′N*
Rhénanie du Nord-Westphalie *see* Nordrhein-Westfalen
Rhénanie-Palatinat *see* Rheinland-Pfalz
162 G7 **Rhenen** Utrecht, C Netherlands *52°01′N 06°02′E*
Rhenish Slate Mountains *see* Rheinisches Schiefergebirge
181 D11 **Rhens** Rheinland-Pfalz, Germany *50°17′E 7°37′N*
Rhétiques, Alpes *see* Rhaetian Alps
181 B9 **Rheydt** Nordrhein-Westfalen, Germany *51°10′E 6°27′N*
178 H6 **Rhin** ≈ NE Germany
Rhin *see* Rhine
144 D6 **Rhine** *Dut.* Rijn, *Fr.* Rhin, *Ger.* Rhein. ≈ W Europe
64 H3 **Rhinebeck** New York, NE USA *41°55′N 73°54′W*
72 C5 **Rhinelander** Wisconsin, N USA *45°39′N 89°23′W*
Rhineland-Palatinate *see* Rheinland-Pfalz
Rhine State Uplands *see* Rheinisches Schiefergebirge
178 H6 **Rhinkanal** *canal* NE Germany
136 B7 **Rhino Camp** NW Uganda *02°58′N 31°24′E*
130 D4 **Rhir, Cap** *headland* W Morocco *30°40′N 09°54′W*
160 D3 **Rhiw** United Kingdom *52°49′N 4°38′W*
176 E6 **Rho** Lombardia, N Italy *45°32′N 09°02′E*
158 C10 **Rhode** Offaly, Ireland *53°21′N 7°12′W*
65 L4 **Rhode Island** *off.* State of Rhode Island and Providence Plantations, *also known as* Little Rhody, Ocean State. ◈ *state* NE USA
65 L4 **Rhode Island** *island* Rhode Island, NE USA
65 L4 **Rhode Island Sound** *sound* Maine/Rhode Island, NE USA
Rhodes *see* Ródos
Rhode-Saint-Genèse *see* Sint-Genesius-Rode
144 F9 **Rhodes Basin** *undersea feature* E Mediterranean Sea *35°55′N 28°30′E*
Rhodesia *see* Zimbabwe
187 I2 **Rhodope Mountains** *var.* Rodhópi Óri, *Bul.* Rhodope Planina, Rodopi, *Gk.* Orosirá Rodhópis, *Turk.* Dospad Dagh. ▲▲ Bulgaria/Greece
Rhodope Planina *see* Rhodope Mountains
181 H11 **Rhön** ▲▲ C Germany
165 J6 **Rhône** ◈ *department* E France
148 B7 **Rhône** ≈ France/Switzerland
165 J6 **Rhône-Alpes** ◈ *region* E France
162 E7 **Rhoon** Zuid-Holland, SW Netherlands *51°52′N 04°25′E*
160 F6 **Rhoose** United Kingdom *51°23′N 3°21′W*
160 E5 **Rhos** United Kingdom *52°00′N 4°22′W*
160 G2 **Rhosllanerchrugog** United Kingdom *53°00′N 3°04′W*
160 E6 **Rhossili** United Kingdom *51°34′N 4°17′W*
159 H10 **Rhuddlan** United Kingdom *53°17′N 3°28′W*
156 D5 **Rhum** *var.* Rum. *island* W Scotland, United Kingdom
181 I8 **Rhumspringe** Niedersachsen, Germany *51°35′E 10°18′N*
168 B6 **Rhune** ▲ Spain *43°18′N 1°38′W*
Rhuthun *see* Ruthin
160 F1 **Rhyl** NE Wales, United Kingdom *53°19′N 03°28′W*
166 C8 **Riaillé** Pays de la Loire, France *47°31′N 1°17′W*
109 B9 **Rialma** Goiás, S Brazil *15°22′S 49°35′W*
81 E11 **Rialto** California, USA *34°06′N 117°22′W*
170 F2 **Riaño** Castilla y León, N Spain *42°59′N 05°00′W*
169 L5 **Rians** Provence-Alpes-Côte d'Azur, France *43°37′N 5°45′E*
173 I3 **Riansáres** ≈ C Spain
232 D3 **Riāsi** Jammu and Kashmir, NW India *33°03′N 74°51′E*
258 E5 **Riau** *off.* Propinsi Riau. ◈ *province* W Indonesia
Riau Archipelago *see* Riau, Kepulauan
258 F4 **Riau, Kepulauan** *var.* Riau Archipelago, *Dut.* Riouw-Archipel. *island group* W Indonesia
Riau, Propinsi *see* Riau
170 G4 **Riaza** Castilla y León, N Spain *41°17′N 03°29′W*
170 G3 **Riaza** ≈ N Spain
136 F7 **Riba** *spring/well* NE Kenya *01°56′N 40°38′E*
170 D2 **Ribadavia** Galicia, NW Spain *42°17′N 08°08′W*
170 D1 **Ribadeo** Galicia, NW Spain *43°32′N 07°04′W*
170 F1 **Ribadesella** *var.* Ribeseya. Asturias, N Spain *43°27′N 05°04′W*
172 A4 **Ribamar** Lisboa, Portugal *39°12′N 9°20′W*
168 G8 **Ribas de Freser** Cataluña, Spain *42°18′N 2°10′E*
170 C6 **Ribatejo** *former province* C Portugal
139 K2 **Ribáuè** Nampula, N Mozambique *14°56′S 38°19′E*
159 I9 **Ribble** ≈ United Kingdom
157 G9 **Ribble** ≈ NW England, United Kingdom
159 J9 **Ribchester** United Kingdom *53°48′N 2°32′W*
155 C12 **Ribe** Syddtjylland, W Denmark *55°20′N 08°47′E*
167 M6 **Ribeauvillé** Alsace, France *48°12′N 7°19′E*
167 H4 **Ribécourt** Picardie, France *49°31′N 2°55′E*
Ribeira *see* Sanata Uxía de Ribeira
170 A8 **Ribeira Brava** Madeira, Portugal, NE Atlantic Ocean *32°39′N 17°04′W*
172 C10 **Ribeira Grande** São Miguel, Azores, Portugal, NE Atlantic Ocean *37°34′N 25°32′W*
110 B7 **Ribeirão Bonito** São Paulo, Brazil *22°04′S 48°10′W*
110 A10 **Ribeirão Branco** São Paulo, Brazil *24°13′S 48°47′W*
110 D9 **Ribeirão Pires** São Paulo, Brazil *23°43′S 46°25′W*
110 B5 **Ribeirão Preto** São Paulo, S Brazil *21°09′S 47°48′W*
110 F5 **Ribeirão Vermelho** Minas Gerais, Brazil *21°11′S 45°03′W*
113 K4 **Ribeira, Rio** ≈ S Brazil
167 I3 **Ribemont** Picardie, France *49°48′N 3°28′E*
175 F12 **Ribera** Sicilia, Italy, C Mediterranean Sea *37°31′N 13°16′E*
168 E2 **Ribérac** Aquitaine, France *45°15′N 0°19′E*
172 E5 **Ribera del Fresno** Extremadura, Spain *38°33′N 6°14′W*
106 D7 **Riberalta** El Beni, N Bolivia *11°01′S 66°04′W*
110 B4 **Riberão Corrente** São Paulo, Brazil *20°27′S 47°35′W*
171 K3 **Ribes de Freser** Cataluña, NE Spain *42°18′N 02°11′E*
Ribeseya *see* Ribadesella
169 L4 **Ribiers** Provence-Alpes-Côte d'Azur, France *44°14′N 5°52′E*
72 C5 **Rib Mountain** ▲ Wisconsin, N USA *44°55′N 89°41′W*
177 K6 **Ribnica** *Ger.* Reifnitz. S Slovenia *45°46′N 14°40′E*
188 G6 **Rîbniţa** *var.* Râbniţa, *Rus.* Rybnitsa. NE Moldova *47°46′N 29°01′E*
178 G4 **Ribnitz-Damgarten** Mecklenburg-Vorpommern, NE Germany *54°14′N 12°25′E*
183 C8 **Říčany** *Ger.* Ritschan. Středočesky Kraj, W Czech Republic *49°59′N 14°40′E*
85 H3 **Ricardo Flores Magón** Chihuahua, Mexico *29°56′N 106°57′W*
114 B6 **Ricardone** Santa Fe, Argentina *32°46′S 60°47′W*
81 I12 **Rice** California, USA *34°05′N 114°51′W*
74 G4 **Rice** Minnesota, N USA *45°42′N 94°10′W*
72 A5 **Rice Lake** Wisconsin, N USA *45°33′N 91°43′W*
62 E4 **Rice Lake** ◎ Ontario, SE Canada
72 I3 **Rice Lake** ◎ Ontario, S Canada
67 H8 **Richard B. Russell Lake** ⊠ Georgia, SE USA
52 G6 **Richard Collinson, Cape** *headland* Nunavut, NW Canada *72°46′N 120°47′W*
52 F6 **Richard Collinson Inlet** *coastal sea feature* Northwest Territories, NW Canada
141 L6 **Richards Bay** KwaZulu-Natal, South Africa *28°48′S 32°06′E*
71 H4 **Richardson** Texas, SW USA *32°55′N 96°44′W*
57 J1 **Richardson** ≈ Alberta, C Canada
52 A6 **Richardson Mountains** ▲▲ Yukon, NW Canada
278 B11 **Richardson Mountains** ▲▲ South Island, New Zealand
88 D2 **Richardson Peak** ▲ SE Belize *16°34′N 88°46′W*
132 B3 **Richard Toll** N Senegal *16°28′N 15°44′W*
74 B3 **Richardton** North Dakota, N USA *46°52′N 102°19′W*
64 G7 **Richboro** Pennsylvania, USA *40°13′N 75°01′W*
62 C4 **Rich, Cape** *headland* Ontario, S Canada *44°42′N 80°37′W*
164 G5 **Richelieu** Indre-et-Loire, C France *47°01′N 00°18′E*
181 F14 **Richen** Baden-Württemberg, Germany *49°10′E 8°56′N*
76 G8 **Richfield** Idaho, NW USA *43°03′N 114°11′W*
79 H4 **Richfield** Utah, W USA *38°45′N 112°05′W*
64 G1 **Richfield Springs** New York, NE USA *42°52′N 74°57′W*
63 I3 **Richford** Vermont, NE USA *44°59′N 72°37′W*
80 D7 **Richgrove** California, USA *35°48′N 119°06′W*
158 D7 **Rich Hill** United Kingdom *54°23′N 6°33′W*
75 G10 **Rich Hill** Missouri, C USA *38°06′N 94°22′W*
63 N1 **Richibucto** New Brunswick, SE Canada *46°42′N 64°54′W*
176 E4 **Richisau** Glarus, NE Switzerland *47°00′N 08°54′E*
69 J2 **Richland** Georgia, SE USA *32°05′N 84°40′W*
75 H10 **Richland** Missouri, C USA *37°51′N 92°24′W*
71 H5 **Richland** Texas, SW USA *31°55′N 96°25′W*
76 E5 **Richland** Washington, NW USA *46°17′N 119°16′W*
72 B7 **Richland Center** Wisconsin, N USA *43°22′N 90°24′W*
67 L7 **Richlands** North Carolina, SE USA *34°52′N 77°33′W*
67 H5 **Richlands** Virginia, NE USA *37°05′N 81°47′W*
70 F6 **Richland Springs** Texas, SW USA *31°16′N 98°56′W*
277 L5 **Richmond** New South Wales, SE Australia *33°36′S 150°44′E*
56 D9 **Richmond** British Columbia, SW Canada *49°07′N 123°09′W*
62 F3 **Richmond** Ontario, SE Canada *45°12′N 75°49′W*
63 I3 **Richmond** Québec, SE Canada *45°39′N 72°07′W*
279 F8 **Richmond** Tasman, South Island, New Zealand *41°25′S 173°04′E*
140 G8 **Richmond** Northern Cape, South Africa *31°25′S 23°56′E*
159 K7 **Richmond** United Kingdom *54°24′N 1°44′W*
80 B2 **Richmond** California, W USA *37°57′N 122°22′W*
73 G11 **Richmond** Indiana, N USA *39°50′N 84°51′W*
66 F4 **Richmond** Kentucky, S USA *37°45′N 84°19′W*
75 G9 **Richmond** Missouri, C USA *39°15′N 93°59′W*
71 I7 **Richmond** Texas, SW USA *29°36′N 95°48′W*
79 H1 **Richmond** Utah, W USA *41°55′N 111°51′W*
67 L4 **Richmond** *state capital* Virginia, NE USA *37°33′N 77°28′W*
62 D5 **Richmond Hill** Ontario, S Canada *43°51′N 79°24′W*
279 F8 **Richmond Range** ▲▲ South Island, New Zealand
161 K6 **Richmond upon Thames** United Kingdom *51°26′N 0°18′W*
75 G13 **Rich Mountain** ▲ Arkansas, C USA *34°37′N 94°17′W*
140 C5 **Richtersveld National Park** Northern Cape, South Africa
73 H10 **Richwood** Ohio, N USA *40°25′N 83°18′W*
67 I3 **Richwood** West Virginia, NE USA *38°13′N 80°31′W*
180 H1 **Rickling** Schleswig-Holstein, Germany *54°01′E 10°10′N*
161 K6 **Rickmansworth** United Kingdom *51°38′N 0°28′W*
168 C9 **Ricla** Aragón, Spain *41°31′N 1°24′W*
170 E3 **Ricobayo, Embalse de** ⊠ NW Spain
Ricomagus *see* Riom
Ridā' *see* Radā'
197 K2 **Ridder** Respublika Tatarstan, W Russian Federation *54°34′N 52°27′E*
162 E7 **Ridderkerk** Zuid-Holland, SW Netherlands *51°52′N 04°35′E*
76 F8 **Riddle** Idaho, NW USA *42°07′N 116°09′W*
76 B7 **Riddle** Oregon, NW USA *42°57′N 123°21′W*
62 F3 **Rideau** ≈ Ontario, SE Canada
64 J5 **Ridge** New York, NE USA *40°53′N 72°53′W*
81 F8 **Ridgecrest** California, W USA *35°37′N 117°40′W*
65 I5 **Ridgefield** Connecticut, NE USA *41°16′N 73°30′W*
68 E2 **Ridgeland** Mississippi, S USA *32°25′N 90°07′W*
67 I10 **Ridgeland** South Carolina, SE USA *32°30′N 80°59′W*
66 B5 **Ridgely** Tennessee, S USA *36°15′N 89°29′W*
62 B6 **Ridgetown** Ontario, S Canada *42°27′N 81°52′W*
67 I8 **Ridgeway** South Carolina, SE USA *34°17′N 80°56′W*
Ridgeway *see* Ridgway
161 L5 **Ridgewell** United Kingdom *52°03′N 0°31′E*
161 J5 **Ridgmont** United Kingdom *52°00′N 0°35′W*
64 A4 **Ridgway** *var.* Ridgeway. Pennsylvania, NE USA *41°24′N 78°40′W*
159 J6 **Ridingmill** United Kingdom *54°56′N 1°58′W*
55 I9 **Riding Mountain** ▲▲ Manitoba, S Canada
159 J5 **Ridsdale** United Kingdom *55°09′N 2°08′W*
141 I5 **Riebeeckstad** Free State, South Africa *27°55′S 26°49′E*
140 D9 **Riebeek-Wes** Western Cape, South Africa *33°21′S 18°52′E*
Ried *see* Ried im Innkreis
180 F4 **Riede** Niedersachsen, Germany *52°58′E 8°57′N*
177 I2 **Ried im Innkreis** *var.* Ried. Oberösterreich, NW Austria *48°13′N 13°29′E*
181 C13 **Riegelsberg** Saarland, Germany *49°18′E 6°56′N*
177 L4 **Riegersburg** Steiermark, SE Austria *47°03′N 15°52′E*
176 D3 **Riehen** Basel-Stadt, NW Switzerland *47°35′N 07°39′E*
152 G4 **Riehppegáisá** *var.* Rieppe. ▲ N Norway *69°38′N 21°31′E*
180 E6 **Riemsloh** Niedersachsen, Germany *52°11′E 8°25′N*
163 G10 **Riemst** Limburg, NE Belgium *50°49′N 05°36′E*
181 H11 **Rieneck** Bayern, Germany *50°06′E 9°39′N*
Rieppe *see* Riehppegáisá
180 J1 **Riepsdorf** Schleswig-Holstein, Germany *54°14′E 10°58′N*
179 H8 **Riesa** Sachsen, E Germany *51°18′N 13°18′E*
117 C13 **Riesco, Isla** *island* S Chile
175 F13 **Riesi** Sicilia, Italy, C Mediterranean Sea *37°17′N 14°05′E*
180 E5 **Rieste** Niedersachsen, Germany *52°29′E 8°01′N*
140 E8 **Riet** ≈ SW South Africa
190 B7 **Rietavas** Telšiai, W Lithuania *55°43′N 21°56′E*
180 E7 **Rietberg** Nordrhein-Westfalen, Germany *51°47′E 8°26′N*
140 G9 **Rietbron** Eastern Cape, South Africa *32°54′S 23°09′E*
140 E1 **Rietfontein** Omaheke, E Namibia *21°58′S 20°58′E*
141 J3 **Rietfontein** Gauteng, South Africa *25°42′S 28°13′E*
140 E4 **Rietfontein** Northern Cape, South Africa *26°44′S 20°02′E*
175 F8 **Rieti** *anc.* Reate. Lazio, C Italy *42°24′N 12°51′E*
140 D7 **Rietpoort** Western Cape, South Africa *30°57′S 18°03′E*
140 E7 **Riet se Vloer** *salt lake* Northern Cape, South Africa
168 F6 **Rieumes** Midi-Pyrénées, France *43°25′N 1°07′E*
168 G4 **Rieupeyroux** Midi-Pyrénées, France *44°18′N 2°14′E*
169 L5 **Riez** Provence-Alpes-Côte d'Azur, France *43°49′N 6°06′E*
144 B9 **Rif** *var.* Riff, Er Rif, Er Riff. ▲▲ N Morocco
110 C3 **Rifaina** São Paulo, Brazil *20°05′S 47°26′W*
218 I3 **Rīf Dimashq** *off.* Muḩāfaẓat Dimashq, *var.* Damascus, *Ar.* Ash Sham, Ash Shām, Damasco, Esh Sham, *Fr.* Damas. ◈ *governorate* S Syria
Riff *see* Rif
79 K3 **Rifle** Colorado, C USA *39°30′N 107°46′W*
72 G6 **Rifle River** ≈ Michigan, N USA
Rift Valley *see* Great Rift Valley
190 D6 **Rīga** *Eng.* Riga. ● C Latvia *56°57′N 24°08′E*
Rigaer Bucht *see* Riga, Gulf of
190 D5 **Riga, Gulf of** *Est.* Liivi Laht, *Ger.* Rigaer Bucht, *Latv.* Rīgas Jūras Līcis, *Rus.* Rizhskiy Zaliv; *prev. Est.* Riia Laht. *gulf* Estonia/Latvia
Rīgas Jūras Līcis *see* Riga, Gulf of
62 G3 **Rigaud** ≈ Ontario/Québec, SE Canada
77 I7 **Rigby** Idaho, NW USA *43°40′N 111°54′W*
Rīgestān *see* Rēgestān
76 F6 **Riggins** Idaho, NW USA *45°24′N 116°18′W*
168 G4 **Rignac** Midi-Pyrénées, France *44°25′N 2°18′E*
59 K4 **Rigolet** Newfoundland and Labrador, NE Canada *51°10′N 58°25′W*
133 M4 **Rig-Rig** Kanem, W Chad *14°16′N 14°21′E*
190 D4 **Riguldi** Läänemaa, W Estonia *59°07′N 23°34′E*
218 F5 **Rīḩāniya** Israel *33°03′N 35°29′E*
Riia Laht *see* Riga, Gulf of
153 H9 **Riihimäki** Kanta-Häme, S Finland *60°45′N 24°45′E*
293 K2 **Riiser-Larsen Peninsula** *peninsula* Antarctica
291 J14 **Riiser-Larsen Sea** *sea* Antarctica
Riisser Larsen Ice Shelf *see* Riiser-Larsenisen
84 C1 **Riíto** Sonora, NW Mexico *32°06′N 114°57′W*
184 A2 **Rijeka** *Ger.* Sankt Veit am Flaum, *It.* Fiume, *Slvn.* Reka; *anc.* Tarsatica. Primorje-Gorski Kotar, NW Croatia *45°20′N 14°26′E*
163 F8 **Rijen** Noord-Brabant, S Netherlands *51°35′N 04°55′E*
163 E9 **Rijkevorsel** Antwerpen, N Belgium *51°23′N 04°43′E*
Rijn *see* Rhine
162 E6 **Rijnsburg** Zuid-Holland, W Netherlands *52°12′N 04°27′E*
Rijssel *see* Lille
162 I6 **Rijssen** Overijssel, E Netherlands *52°19′N 06°30′E*
162 E7 **Rijswijk** *Eng.* Ryswick. Zuid-Holland, W Netherlands *52°03′N 04°20′E*
152 F4 **Riksgränsen** Norrbotten, N Sweden *68°24′N 18°15′E*
252 G4 **Rikubetsu** Hokkaidō, NE Japan *43°30′N 143°43′E*
253 E9 **Rikuzen-Takata** Iwate, Honshū, C Japan *39°03′N 141°38′E*
75 E9 **Riley** Kansas, C USA *39°18′N 96°49′W*
163 F10 **Rillaar** Vlaams Brabant, C Belgium *50°58′N 04°58′E*
159 L8 **Rillington** United Kingdom *54°09′N 0°42′W*
Rí, Loch *see* Ree, Lough
185 I7 **Rilska Reka** ≈ W Bulgaria
133 J5 **Rima** ≈ N Nigeria
222 A9 **Rimah, Wādī ar** *var.* Wādī ar Rummah. *dry watercourse* C Saudi Arabia
169 N1 **Rimasco** Piemonte, NW Italy *45°51′N 08°03′E*
Rimaszombat *see* Rimavská Sobota
285 I8 **Rimatara** *island* Îles Australes, SW French Polynesia
183 G10 **Rimavská Sobota** *Ger.* Gross-Steffelsdorf, *Hung.* Rimaszombat. Bankobystrický Kraj, C Slovakia *48°24′N 20°01′E*
57 I6 **Rimbey** Alberta, SW Canada *52°39′N 114°10′W*
154 I7 **Rimbo** Stockholm, C Sweden *59°44′N 18°21′E*
169 N1 **Rimella** Piemonte, NW Italy *45°54′N 08°11′E*
155 H9 **Rimforsa** Östergötland, S Sweden *58°06′N 15°40′E*
177 I9 **Rimini** *anc.* Ariminum. Emilia-Romagna, N Italy *44°03′N 12°33′E*
Rîmnicu-Sărat *see* Râmnicu Sărat
Rîmnicu Vîlcea *see* Râmnicu Vâlcea
231 K3 **Rimo Muztāgh** ▲▲ India/Pakistan
59 I7 **Rimouski** Québec, SE Canada *48°26′N 68°32′W*
238 G9 **Rinbung** Xizang Zizhiqu, W China *29°15′N 89°40′E*
110 B6 **Rincão** São Paulo, Brazil *21°35′S 48°05′W*
115 K6 **Rincón** Uruguay *32°47′S 53°50′W*
112 C4 **Rincón, Cerro** ▲ N Chile *24°01′S 67°19′W*
85 I10 **Rincón de Guayabitos** Nayarit, Mexico
170 F9 **Rincón de la Victoria** Andalucía, S Spain *36°43′N 04°18′W*
115 H6 **Rincón del Bonete** Durazno, Uruguay *32°48′S 56°25′W*
Rincón del Bonete, Lago Artificial de *see* Río Negro, Embalse del
114 F6 **Rincón del Cololó** Uruguay *33°05′S 57°26′W*
114 B5 **Rincón del Doll** Entre Ríos, Argentina *32°26′S 60°23′W*
114 G9 **Rincón del Pino** Uruguay *34°29′S 56°51′W*
85 K9 **Rincón de Romos** Aguascalientes, Mexico *22°14′N 102°18′W*
171 H3 **Rincón de Soto** La Rioja, N Spain *42°15′N 01°50′W*
87 I9 **Rincón Juárez** Oaxaca, Mexico *16°17′N 94°22′W*
154 D4 **Rindal** Møre og Romsdal, S Norway *63°02′N 09°09′E*
181 H14 **Rindelbach** Baden-Württemberg, Germany *48°59′E 10°08′N*
187 I7 **Ríneia** *island* Kykládes, Greece, Aegean Sea
232 E6 **Rīngas** *prev.* Reengus, Rīngus. Rājasthān, N India *27°18′N 75°27′E*
155 D13 **Ringe** Syddtjylland, C Denmark *55°14′N 10°30′E*
154 D5 **Ringebu** Oppland, S Norway *61°31′N 10°09′E*
Ringen *see* Rõngu
159 H6 **Ringford** United Kingdom *54°53′N 4°02′W*
281 H2 **Ringgi** Kolombangara, NW Solomon Islands *08°03′S 157°08′E*
66 F7 **Ringgold** Georgia, SE USA *34°55′N 85°06′W*
68 B2 **Ringgold** Louisiana, S USA *32°19′N 93°16′W*
70 G3 **Ringgold** Texas, SW USA *33°47′N 97°56′W*
155 C12 **Ringkøbing** Midtjylland, W Denmark *56°04′N 08°22′E*
155 C12 **Ringkøbing Fjord** *fjord* W Denmark
77 I5 **Ringling** Montana, NW USA *46°15′N 110°48′W*
75 E13 **Ringling** Oklahoma, C USA *34°12′N 97°35′W*
161 K8 **Ringmer** United Kingdom *50°53′N 0°04′E*
154 E6 **Ringsaker** Hedmark, S Norway *60°54′N 10°45′E*
158 D5 **Ringsend** United Kingdom *55°02′N 6°45′W*
161 L2 **Ringstead** United Kingdom *52°56′N 0°33′E*
155 E12 **Ringsted** Sjælland, E Denmark *55°28′N 11°48′E*
Rīngus *see* Rīngas
152 F3 **Ringvassøya** *Lapp.* Ránes. *island* N Norway
161 I8 **Ringwood** United Kingdom *50°51′N 1°47′W*
65 H5 **Ringwood** New Jersey, NE USA *41°06′N 74°15′W*
180 D7 **Rinkerode** Nordrhein-Westfalen, Germany *51°51′E 7°41′N*
Rinn Duáin *see* Hook Head
180 G6 **Rinteln** Niedersachsen, NW Germany *52°10′N 09°04′E*
186 F6 **Río** Dytikí Elláda, S Greece *38°18′N 21°48′E*
Rio *see* Rio de Janeiro
111 H3 **Rio Acima** Minas Gerais, Brazil *20°05′S 43°47′W*
104 C2 **Riobamba** Chimborazo, C Ecuador *01°44′S 78°40′W*
111 M2 **Rio Bananal** Espírito Santo, Brazil *19°16′S 40°20′W*
111 J8 **Rio Bonito** Rio de Janeiro, SE Brazil *22°42′S 42°38′W*
106 D6 **Rio Branco** *state capital* Acre, W Brazil *09°59′S 67°49′W*
115 L6 **Río Branco** Cerro Largo, NE Uruguay *32°32′S 53°28′W*
Rio Branco, Território de *see* Roraima
85 M6 **Río Bravo** Tamaulipas, C Mexico *25°57′N 98°03′W*
116 B7 **Río Bueno** Los Ríos, C Chile *40°20′S 72°55′W*
106 2B **Rio Caqueta** ≈ Guainia, Colombia
103 H2 **Río Caribe** Sucre, NE Venezuela *10°43′N 63°06′W*
111 J3 **Rio Casca** Minas Gerais, Brazil *20°13′S 42°39′W*
102 G3 **Río Chico** Miranda, N Venezuela *10°18′N 66°00′W*
117 C9 **Río Cisnes** Aisén, S Chile *44°29′S 71°15′W*
110 B7 **Rio Claro** São Paulo, S Brazil *22°19′S 47°35′W*
91 L8 **Rio Claro** Trinidad, Trinidad and Tobago *10°18′N 61°11′W*
102 E3 **Río Claro** Lara, N Venezuela *09°54′N 69°23′W*
116 E6 **Río Colorado** Río Negro, E Argentina *39°01′S 64°05′W*
85 H5 **Río Conchos** ≈ Chihuahua, Mexico
116 E3 **Río Cuarto** Córdoba, C Argentina *33°06′S 64°20′W*
111 H7 **Rio das Flores** Rio de Janeiro, Brazil *22°10′S 43°35′W*
111 K7 **Rio das Ostras** Rio de Janeiro, Brazil *22°32′S 41°57′W*
110 B8 **Rio das Pedras** São Paulo, Brazil *22°50′S 47°36′W*
111 I8 **Rio de Janeiro** *var.* Rio. *state capital* Rio de Janeiro, SE Brazil *22°53′S 43°17′W*
111 K6 **Rio de Janeiro** *off.* Estado do Rio de Janeiro. ◈ *state* SE Brazil
Rio de Janeiro, Estado do *see* Rio de Janeiro
89 J10 **Río de Jesús** Veraguas, S Panama *07°58′N 81°01′W*
78 A2 **Rio Dell** California, W USA *40°30′N 124°07′W*
111 I3 **Rio Doce** Minas Gerais, Brazil *20°15′S 42°54′W*
113 K6 **Rio do Sul** Santa Catarina, S Brazil *27°15′S 49°37′W*
111 H4 **Rio Espera** Minas Gerais, Brazil *20°51′S 43°29′W*
117 D12 **Río Gallegos** *var.* Gallegos, Puerto Gallegos. Santa Cruz, S Argentina *51°40′S 69°21′W*
117 D13 **Río Grande** Tierra del Fuego, S Argentina *53°45′S 67°46′W*
115 N5 **Rio Grande** *var.* São Pedro do Rio Grande do Sul. Rio Grande do Sul, S Brazil *32°03′S 52°08′W*
85 J8 **Río Grande** Zacatecas, C Mexico *23°50′N 103°20′W*
88 G5 **Río Grande** León, NW Nicaragua *12°59′N 86°34′W*
91 J7 **Río Grande** E Puerto Rico *18°23′N 65°51′W*
70 F9 **Rio Grande** ≈ Texas, SW USA
70 F10 **Rio Grande City** Texas, SW USA *26°24′N 98°50′W*
108 I4 **Rio Grande do Norte** *off.* Estado do Rio Grande do Norte. ◈ *state* E Brazil
Rio Grande do Norte, Estado do *see* Rio Grande do Norte
115 K2 **Rio Grande do Sul** *off.* Estado do Rio Grande do Sul. ◈ *state* S Brazil
Rio Grande do Sul, Estado do *see* Rio Grande do Sul
291 G10 **Rio Grande Fracture Zone** *tectonic feature* C Atlantic Ocean
291 F10 **Rio Grande Gap** *undersea feature* S Atlantic Ocean
Rio Grande Plateau *see* Rio Grande Rise
291 F10 **Rio Grande Rise** *var.* Rio Grande Plateau. *undersea feature* SW Atlantic Ocean *31°00′S 35°00′W*
102 D2 **Ríohacha** La Guajira, N Colombia *11°23′N 72°47′W*
89 K9 **Río Hato** Coclé, C Panama *08°21′N 80°10′W*
70 G10 **Rio Hondo** Texas, SW USA *26°14′N 97°34′W*
104 D4 **Rioja** San Martín, N Peru *06°02′S 77°10′W*
87 M4 **Río Lagartos** Yucatán, SE Mexico *21°35′N 88°08′W*
165 I6 **Riom** *anc.* Ricomagus. Puy-de-Dôme, C France *45°54′N 03°06′E*
172 B4 **Rio Maior** Santarém, C Portugal *39°20′N 08°55′W*
110 G3 **Rio Manso** Minas Gerais, Brazil *20°16′S 44°19′W*
165 H7 **Riom-ès-Montagnes** Cantal, C France *45°15′N 02°39′E*
113 K5 **Rio Negro** Paraná, S Brazil *26°06′S 49°46′W*
116 D7 **Río Negro** *off.* Provincia de Río Negro. ◈ *province* C Argentina
114 F6 **Río Negro** ◈ *department* W Uruguay
95 L7 **Río Negro, Embalse del** *var.* Lago Artificial de Rincón del Bonete. ⊠ C Uruguay
Río Negro, Provincia de *see* Río Negro
175 H9 **Rionero in Vulture** Basilicata, S Italy *40°55′N 15°40′E*
215 J3 **Rioni** ≈ W Georgia
111 I5 **Rio Novo** Minas Gerais, Brazil *21°29′S 43°08′W*
171 H7 **Riópar** Castilla-La Mancha, C Spain *38°31′N 02°27′W*
115 M1 **Rio Pardo** Rio Grande do Sul, S Brazil *29°41′S 52°25′W*
111 I3 **Rio Piracicaba** Minas Gerais, Brazil *19°55′S 43°11′W*
111 H6 **Rio Prèto** Minas Gerais, Brazil *22°05′S 43°50′W*
79 L7 **Rio Rancho Estates** New Mexico, SW USA *35°14′N 106°40′W*
88 G7 **Río San Juan** ◈ *department* S Nicaragua
102 B5 **Ríosucio** Caldas, W Colombia *05°26′N 75°44′W*
102 B4 **Ríosucio** Chocó, NW Colombia *07°25′N 77°05′W*
114 C8 **Río Tala** Buenos Aires, Argentina *33°45′S 59°38′W*
116 E3 **Río Tercero** Córdoba, C Argentina *32°15′S 64°08′W*
88 G3 **Río Tinto, Sierra** ▲▲ NE Honduras
102 E3 **Río Tocuyo** Lara, N Venezuela *10°18′N 70°00′W*
Riouw-Archipel *see* Riau, Kepulauan
109 A10 **Rio Verde** Goiás, C Brazil *17°50′S 50°55′W*
85 L9 **Río Verde** *var.* Rioverde. San Luis Potosí, C Mexico *21°58′N 100°W*
Rioverde *see* Río Verde
80 C2 **Rio Vista** California, W USA *38°09′N 121°42′W*
167 L8 **Rioz** Franche-Comté, France *47°25′N 6°04′E*
184 G4 **Ripanj** Serbia, N Serbia *44°37′N 20°30′E*
177 J10 **Ripatransone** Marche, C Italy *43°00′N 13°45′E*
161 I2 **Ripley** United Kingdom *53°02′N 1°24′W*
81 H13 **Ripley** California, USA *33°32′N 114°39′W*
66 C7 **Ripley** Mississippi, S USA *34°43′N 88°57′W*
73 H12 **Ripley** Ohio, N USA *38°45′N 83°51′W*
66 B6 **Ripley** Tennessee, S USA *35°43′N 89°30′W*
67 H3 **Ripley** West Virginia, NE USA *38°49′N 81°44′W*
171 K3 **Ripoll** Cataluña, NE Spain *42°12′N 02°12′E*
159 K8 **Ripon** N England, United Kingdom *54°07′N 01°31′W*
80 C3 **Ripon** California, USA *37°45′N 121°07′W*
72 D7 **Ripon** Wisconsin, N USA *43°52′N 88°48′W*
175 G12 **Riposto** Sicilia, Italy, C Mediterranean Sea *37°44′N 15°13′E*
80 D5 **Ripperdan** California, USA *36°51′N 120°03′W*
163 G8 **Rips** Noord-Brabant, SE Netherlands *51°31′N 05°49′E*
167 M6 **Riquewihr** Alsace, France *48°10′N 7°18′E*
102 B5 **Risaralda** *off.* Departamento de Risaralda. ◈ *province* C Colombia
Risaralda, Departamento de *see* Risaralda
160 G6 **Risca** United Kingdom *51°37′N 3°06′W*
188 F6 **Rîşcani** *var.* Râşcani, *Rus.* Ryshkany. NW Moldova *47°55′N 27°31′E*
180 G7 **Rischenau** Nordrhein-Westfalen, Germany *51°53′E 9°17′N*
168 D5 **Riscle** Midi-Pyrénées, France *43°40′N 0°05′W*
232 F4 **Rishikesh** Uttarakhand, N India *30°06′N 78°16′E*
252 E2 **Rishiri-suidō** *strait* E Sea of Japan
252 D2 **Rishiri-tō** *var.* Risiri Tô. *island* NE Japan
252 E2 **Rishiri-yama** ⛰ Rishiri-tō, NE Japan *45°11′N 141°11′E*
Rishon Le-Tsiyon *see* Rishon LeZiyyon
219 D8 **Rishon LeZiyyon** *var.* Rishon Le-Tsiyon. Central, C Israel *31°57′N 34°48′E*
70 F5 **Rising Star** Texas, SW USA *32°06′N 98°57′W*
73 G12 **Rising Sun** Indiana, N USA *38°58′N 84°53′W*
Risiri Tô *see* Rishiri-tō
164 G3 **Risle** ≈ N France
Rîşno *see* Râşnov
75 I13 **Rison** Arkansas, C USA *33°58′N 92°11′W*
155 D9 **Risør** Aust-Agder, S Norway *58°44′N 09°15′E*
152 E4 **Risøyhamn** Nordland, C Norway *69°00′N 15°37′E*
219 D8 **RIsraelhon LeZiyyon** Israel *31°57′N 34°48′E*
179 E13 **Riss** ≈ S Germany
114 F7 **Risso** Soriano, Uruguay *33°30′S 57°36′W*
190 D4 **Risti** *Ger.* Kreuz. Läänemaa, W Estonia *59°01′N 24°01′E*
153 I9 **Ristiina** Etelä-Savo, E Finland *61°32′N 27°15′E*
152 I7 **Ristijärvi** Kainuu, C Finland *64°30′N 28°15′E*
282 B1 **Ritidian Point** *headland* N Guam *13°39′N 144°51′E*
118 C2 **Ritoque, Punta** *point* Valparaíso, Chile
Ritschan *see* Říčany
180 F4 **Ritterhude** Niedersachsen, Germany *53°11′E 8°45′N*
80 F5 **Ritter, Mount** ▲ California, W USA *37°40′N 119°10′W*
73 I10 **Rittman** Ohio, N USA *40°58′N 81°46′W*
76 E4 **Ritzville** Washington, NW USA *47°07′N 118°22′W*
Riva *see* Riva del Garda
112 E10 **Rivadavia** Buenos Aires, E Argentina *35°29′S 62°59′W*
176 G6 **Riva del Garda** *var.* Riva. Trentino-Alto Adige, N Italy *45°54′N 10°50′E*
85 H4 **Riva Palacio** Chihuahua, Mexico *28°33′N 106°30′W*
176 D7 **Rivarolo Canavese** Piemonte, W Italy *45°21′N 07°42′E*
114 C9 **Rivas** Buenos Aires, Argentina *34°37′S 59°46′W*
88 G7 **Rivas** Rivas, SW Nicaragua *11°26′N 85°50′W*
88 G7 **Rivas** ◈ *department* SW Nicaragua

◆ Country
● Country Capital
◇ Dependent Territory
○ Dependent Territory Capital
◈ Administrative Regions
✈ International Airport
▲ Mountain
▲▲ Mountain Range
⛰ Volcano
≈ River
◎ Lake
⊠ Reservoir

INDEX

◆ Country   ◇ Dependent Territory   ✶ Administrative Regions   ▲ Mountain   ⊛ Volcano   ○ Lake
● Country Capital   ○ Dependent Territory Capital   ✈ International Airport   ▲ Mountain Range   ❧ River   ⬚ Reservoir

S

◆ Country	◇ Dependent Territory	◆ Administrative Regions	▲ Mountain	◇ Volcano	◎ Lake
● Country Capital	○ Dependent Territory Capital	✈ International Airport	▲ Mountain Range	♒ River	⬚ Reservoir

◆ Country
● Country Capital
◇ Dependent Territory
○ Dependent Territory Capital
◈ Administrative Regions
✈ International Airport
▲ Mountain
▲ Mountain Range
◈ Volcano
↩ River
⊙ Lake
☐ Reservoir

◆ Country
◇ Dependent Territory
● Country Capital
◉ Dependent Territory Capital
◆ Administrative Regions
✕ International Airport
▲ Mountain
▲ Mountain Range
▲ Volcano
◆ River
☐ Lake
▣ Reservoir

◆ Country
● Country Capital
◇ Dependent Territory
○ Dependent Territory Capital
◇ Administrative Regions
✕ International Airport
▲ Mountain
▲ Mountain Range
🌋 Volcano
◇ River
◎ Lake
◎ Reservoir

◆ Country	◇ Dependent Territory	◇ Administrative Regions	▲ Mountain	☒ Volcano	⊚ Lake
● Country Capital	○ Dependent Territory Capital	✈ International Airport	▲ Mountain Range	♠ River	⊠ Reservoir

Ushtobe - Varrel
INDEX
421

227 K6 **Ushtobe** *Kaz.* Üshtöbe. Almaty, SE Kazakhstan 45°15′N 77°59′E
Üshtöbe *see* Ushtobe
117 D14 **Ushuaia** Tierra del Fuego, S Argentina 54°48′S 68°19′W
83 I5 **Usibelli** Alaska, USA 63°54′N 148°41′W
Usibuka *see* Ushibuka
181 F11 **Usingen** Hessen, Germany 50°20′E 8°32′N
280 B3 **Usino** Madang, N Papua New Guinea 05°40′S 145°31′E
195 K4 **Usinsk** Respublika Komi, NW Russian Federation 66°01′N 57°37′E
160 G6 **Usk** United Kingdom 51°42′N 2°54′W
160 G6 **Usk** ≈ United Kingdom
157 G11 **Usk** *Wel.* Wysg. ≈ SE Wales, United Kingdom
Uskočke Planine/Uskokengebirge *see* Gorjanci
Uskoplje *see* Gornji Vakuf
Üsküb/Üsküp *see* Skopje
196 G4 **Usman'** Lipetskaya Oblast', W Russian Federation 52°04′N 39°41′E
190 B6 **Usmas Ezers** ◎ NW Latvia
190 I7 **Usmyn'** Pskovskaya Oblast', Russian Federation
195 K8 **Usol'ye** Permskiy Kray, NW Russian Federation 59°27′N 56°33′E
193 H8 **Usol'ye-Sibirskoye** Irkutskaya Oblast', C Russian Federation 52°48′N 103°40′E
118 F2 **Uspallata** Mendoza, Argentina 32°35′S 69°20′W
118 F2 **Uspallata, Sierra de** ▲▲ Mendoza, Argentina
87 I8 **Uspanapa, Río** ≈ SE Mexico
Uspenskiy *see* Upenskoye
86 C6 **Uspero** Michoacán, Mexico
165 H6 **Ussel** Corrèze, C France 45°33′N 02°18′E
247 N3 **Ussuri** *var.* Usuri, Wusuri, *Chin.* Wusuli Jiang. ≈ China/Russian Federation
193 L9 **Ussuriysk** *prev.* Nikol'sk, Nikol'sk-Ussuriyskiy, Voroshilov. Primorskiy Kray, SE Russian Federation 43°48′N 131°59′E
214 F4 **Usta Burnu** *headland* N Turkey 41°58′N 34°30′E
230 G7 **Usta Muhammad** Baluchistān, SW Pakistan 28°07′N 68°00′E
168 C6 **Ustaritz** Aquitaine, France 43°24′N 1°27′W
195 M10 **Ust'-Bagaryak** Chelyabinskaya Oblast', Russian Federation
193 I8 **Ust'-Barguzin** Respublika Buryatiya, S Russian Federation 53°28′N 109°00′E
193 M6 **Ust'-Bol'sheretsk** Kamchatskiy Kray, E Russian Federation 52°48′N 156°12′E
197 H5 **Ust'-Buzulukskaya** Volgogradskaya Oblast', SW Russian Federation 50°12′N 42°06′E
190 H7 **Ust'-Dolyssy** Pskovskaya Oblast', Russian Federation
183 B8 **Ustecký Kraj** ◈ *region* NW Czech Republic
175 F11 **Ustica, Isola d'** *island* S Italy
193 H7 **Ust'-Ilimsk** Irkutskaya Oblast', C Russian Federation 57°57′N 102°30′E
183 B8 **Ústí nad Labem** *Ger.* Aussig. Ústecký Kraj, NW Czech Republic 50°41′N 14°04′E
183 D8 **Ústí nad Orlicí** *Ger.* Wildenschwert. Pardubický Kraj, C Czech Republic 49°58′N 16°24′E
Ustinov *see* Izhevsk
184 E5 **Ustiprača** ◈ Republika Srpska, SE Bosnia and Herzegovina
192 E6 **Ust'-Ishim** Omskaya Oblast', C Russian Federation 57°42′N 70°58′E
182 E3 **Ustka** *Ger.* Stolpmünde. Pomorskie, N Poland 54°35′N 16°50′E
193 N5 **Ust'-Kamchatsk** Kamchatskiy Kray, E Russian Federation 56°14′N 162°28′E
227 L3 **Ust'-Kamenogorsk** *Kaz.* Öskemen. Vostochnyy Kazakhstan, E Kazakhstan 49°58′N 82°36′E
193 M6 **Ust'-Khayryuzovo** Krasnoyarskiy Kray, E Russian Federation 57°07′N 156°37′E
227 M3 **Ust'-Koksa** Respublika Altay, S Russian Federation 50°15′N 85°45′E
195 J7 **Ust'-Kulom** Respublika Komi, NW Russian Federation 61°42′N 53°42′E
193 H7 **Ust'-Kut** Irkutskaya Oblast', C Russian Federation 56°49′N 105°32′E
193 J4 **Ust'-Kuyga** Respublika Sakha (Yakutiya), NE Russian Federation 69°59′N 135°27′E
196 G8 **Ust'-Labinsk** Krasnodarskiy Kray, SW Russian Federation 44°40′N 40°46′E
193 K6 **Ust'-Maya** Respublika Sakha (Yakutiya), NE Russian Federation 60°27′N 134°28′E
193 K5 **Ust'-Nera** Respublika Sakha (Yakutiya), NE Russian Federation 64°28′N 143°01′E
193 J7 **Ust'-Nyukzha** Amurskaya Oblast', S Russian Federation 56°30′N 121°32′E
193 I4 **Ust'-Olenëk** Respublika Sakha (Yakutiya), NE Russian Federation 73°03′N 119°34′E
193 L5 **Ust'-Omchug** Magadanskaya Oblast', E Russian Federation 61°07′N 149°17′E
193 H8 **Ust'-Ordynskiy** Irkutskaya Oblast', S Russian Federation 52°50′N 104°42′E
195 H8 **Ust'-Ordynskiy Buryatskiy Okrug** ◈ *autonomous district* S Russian Federation
194 G6 **Ust'-Pinega** Arkhangel'skaya Oblast', NW Russian Federation 64°09′N 41°55′E
192 G4 **Ust'-Port** Krasnoyarskiy Kray, N Russian Federation 69°42′N 84°25′E
185 L7 **Ustrem** *prev.* Vakav. Yambol, E Bulgaria 42°01′N 26°28′E
183 I9 **Ustrzyki Dolne** Podkarpackie, SE Poland 49°26′N 22°36′E
195 K5 **Ust'-Shchuger** Respublika Komi, NW Russian Federation 64°15′N 57°45′W
Ust'-Sysol'sk *see* Syktyvkar
195 J5 **Ust'-Tsil'ma** Respublika Komi, NW Russian Federation 65°25′N 52°09′E
Ust Urt *see* Ustyurt Plateau
195 M4 **Ust' -Voykar** Yamalo-Nenetskiy Avtonomnyy Okrug, Russian Federation
195 H7 **Ust'ya** ≈ NW Russian Federation
194 F4 **Ust'ye Varzugi** Murmanskaya Oblast', NW Russian Federation 66°16′N 36°47′E
193 M6 **Ust'yevoye** *prev.* Kirovskiy. Kamchatskiy Kray, E Russian Federation 54°06′N 155°48′E
189 I6 **Ustynivka** Kirovohrads'ka Oblast', C Ukraine 47°58′N 32°32′E
226 D7 **Ustyurt Plateau** *var.* Ust Urt, *Uzb.* Ustyurt Platosi. *plateau* Kazakhstan/Uzbekistan
Ustyurt Platosi *see* Ustyurt Plateau
194 F9 **Ustyuzhna** Vologodskaya Oblast', NW Russian Federation 58°50′N 36°25′E
238 F3 **Usu** Xinjiang Uygur Zizhiqu, NW China 44°27′N 84°37′E
260 D5 **Usu** Sulawesi, C Indonesia 02°34′S 120°58′E
250 E7 **Usuki** Ōita, Kyūshū, SW Japan 33°07′N 131°48′E
88 E5 **Usulután** Usulután, SE El Salvador 13°20′N 88°26′W
88 D5 **Usulután** ◈ *department* SE El Salvador
87 J7 **Usumacinta, Río** ≈ Guatemala/Mexico
Usumbura *see* Bujumbura
Usuri *see* Ussuri
U.S./USA *see* United States of America
141 J1 **Usutu** Limpopo, South Africa 22°34′S 28°35′E
190 I7 **Usvyaty** Pskovskaya Oblast', Russian Federation
261 K6 **Uta** Papua, E Indonesia 04°28′S 136°03′E
79 I4 **Utah** *off.* State of Utah, *also known as* Beehive State, Mormon State. ◈ *state* W USA
79 H3 **Utah Lake** ◎ Utah, W USA
Utaidhani *see* Uthai Thani
152 H7 **Utajärvi** Pohjois-Pohjanmaa, C Finland 64°45′N 26°25′E
Utamboni *see* Mitemele, Río
Utaradit *see* Uttaradit
258 D2 **Utara, Selat** *strait* Peninsular Malaysia
252 E4 **Utashinai** *var.* Utasinai. Hokkaidō, NE Japan 43°32′N 142°03′E
Utasinai *see* Utashinai
261 K6 **Uta, Sungai** ≈ Papua, E Indonesia
284 D7 **'Uta Vava'u** *island* Vava'u Group, N Tonga
79 N6 **Ute Creek** ≈ New Mexico, SW USA
191 E8 **Utena** Utena, E Lithuania 55°30′N 25°34′E
191 E8 **Utena** ◈ *province* Lithuania
79 N6 **Ute Reservoir** ⊠ New Mexico, SW USA
256 E7 **Uthai Thani** *var.* Muang Uthai Thani, Udayadhani, Utaidhani. Uthai Thani, W Thailand 15°22′N 100°03′E
230 F9 **Uthal** Baluchistān, SW Pakistan 25°53′N 66°37′E
uThongathi *see* oThongathi
64 F1 **Utica** New York, NE USA 43°06′N 75°15′W
171 I6 **Utiel** Valenciana, E Spain 39°33′N 01°13′W
57 H3 **Utikuma Lake** ◎ Alberta, W Canada
88 G3 **Utila, Isla de** *island* Islas de la Bahía, N Honduras
Utina *see* Udine
109 F8 **Utinga** Bahia, E Brazil 12°05′S 41°07′W
Utirik *see* Utrik Atoll
189 K7 **Utlyuts'kyy Lyman** *bay* S Ukraine
250 C7 **Uto** Kumamoto, Kyūshū, SW Japan 32°42′N 130°40′E
155 I8 **Utö** Stockholm, C Sweden 58°55′N 18°19′E
70 F7 **Utopia** Texas, SW USA 29°30′N 99°31′W
190 H3 **Utorgosh** Novgorodskaya Oblast', Russian Federation
162 F7 **Utrecht** *Lat.* Trajectum ad Rhenum. Utrecht, C Netherlands 52°06′N 05°07′E
139 H7 **Utrecht** KwaZulu/Natal, E South Africa 27°40′S 30°20′E
162 F6 **Utrecht** ◈ *province* C Netherlands
170 E8 **Utrera** Andalucía, S Spain 37°10′N 05°47′W
283 L5 **Utrik Atoll** *var.* Utirik, Utrōk, Utrōnk. *atoll* Ratak Chain, N Marshall Islands
173 K2 **Utrillas** Aragón, Spain 40°49′N 0°51′W
Utrōk/Utrōnk *see* Utrik Atoll
155 A8 **Utsira** *island* SW Norway
152 H3 **Utsjoki** *var.* Ohcejohka. Lappi, N Finland 69°54′N 27°01′E
253 C12 **Utsunomiya** *var.* Utunomiya. Tochigi, Honshū, S Japan 36°36′N 139°53′E
197 I7 **Utta** Respublika Kalmykiya, SW Russian Federation 46°22′N 46°03′E
256 E6 **Uttaradit** *var.* Utaradit. Uttaradit, N Thailand 17°38′N 100°05′E
232 F5 **Uttarakhand** ◈ *state* N India
232 F4 **Uttarkāshi** Uttarakhand, N India 30°45′N 78°19′E
232 F5 **Uttar Pradesh** *prev.* United Provinces, United Provinces of Agra and Oudh. ◈ *state* N India
161 H2 **Uttoxeter** United Kingdom 52°54′N 1°51′W
91 I8 **Utuado** C Puerto Rico 18°17′N 66°41′W
238 F2 **Utubulak** Xinjiang Uygur Zizhiqu, W China 46°50′N 86°15′E
82 G2 **Utukok River** ≈ Alaska, USA
Utunomiya *see* Utsunomiya
281 I5 **Utupua** *island* Santa Cruz Islands, E Solomon Islands
Utva *see* Shyngyrlau
283 H10 **Utwe** Kosrae, E Micronesia
283 H10 **Utwe Harbor** *harbor* Kosrae, E Micronesia
Uubulan *see* Hayrhan
239 M1 **Uulbayan** *var.* Dzüünbulag. Sühbaatar, E Mongolia 46°30′N 112°22′E
194 B9 **Uulu** Estonia
190 D5 **Uulu** Pärnumaa, SW Estonia 58°15′N 24°32′E
295 K8 **Uummannaq** *var.* Umanak, Umanaq. ◈ Qaasuitsup, C Greenland
Uummannarsuaq *see* Nunap Isua
246 E2 **Üüreg Nuur** ◎ NW Mongolia
Uusikaarlepyy *see* Nykarleby
153 G10 **Uusikaupunki** *Swe.* Nystad. Varsinais-Suomi, SW Finland 60°48′N 21°25′E
153 H10 **Uusimaa** *Swe.* Nyland. ◈ *region* S Finland
195 J10 **Uva** Udmurtskaya Respublika, NW Russian Federation 56°41′N 52°15′E
235 G11 **Uva** ◈ *province* SE Sri Lanka
184 F5 **Uvac** ≈ W Serbia
70 F7 **Uvalde** Texas, SW USA 29°14′N 99°49′W
191 I11 **Uvaravichy** *Rus.* Uvarovichi. Homyel'skaya Voblasts', SE Belarus 52°36′N 30°44′E
102 E6 **Uvá, Río** ≈ E Colombia
Uvarovichi *see* Uvaravichy
197 H4 **Uvarovo** Tambovskaya Oblast', W Russian Federation 51°58′N 42°13′E
192 E6 **Uvat** Tyumenskaya Oblast', C Russian Federation 59°11′N 68°37′E
284 C10 **Uvea, Île** *island* N Wallis and Futuna
137 A10 **Uvinza** Kigoma, W Tanzania 05°08′S 30°23′E
135 I9 **Uvira** Sud-Kivu, E Dem. Rep. Congo 03°24′S 29°05′E
141 K7 **Uvongo** KwaZulu-Natal, South Africa 30°50′S 30°24′E
239 H1 **Uvs** ◈ *province* NW Mongolia
192 G8 **Uvs Nuur** *var.* Ozero Ubsu-Nur. ◎ Mongolia/Russian Federation
250 E6 **Uwa** *var.* Seiyo. Ehime, Shikoku, SW Japan 33°22′N 132°29′E
250 E6 **Uwajima** *var.* Uwazima. Ehime, Shikoku, SW Japan 33°13′N 132°32′E
128 G7 **'Uwaynāt, Jabal al** *var.* Jebel Uweinat. ▲ Libya/Sudan 21°51′N 25°01′E
Uwazima *see* Uwajima
64 F7 **Uwchlan** Pennsylvania, NE USA 40°03′N 75°40′W
Uweinat, Jebel *see* 'Uwaynāt, Jabal al
261 M7 **Uwimmerah, Sungai** ≈ Papua, E Indonesia
62 D4 **Uxbridge** Ontario, S Canada 44°07′N 79°07′W
161 J6 **Uxbridge** United Kingdom 51°33′N 0°29′W
Uxellodunum *see* Issoudun
Uxin Qi *see* Dabqig, N China
87 L5 **Uxmal, Ruinas** *ruins* Yucatán, SE Mexico
205 J3 **Uy** ≈ Kazakhstan/Russian Federation
226 F7 **Uyaly** Kzylorda, S Kazakhstan 44°22′N 61°16′E
193 K4 **Uyandina** ≈ NE Russian Federation
239 J2 **Uyanga** *var.* Ongi. Övörhangay, C Mongolia 46°30′N 102°18′E
Üydzen *see* Manlay
Uyeda *see* Ueda
192 G2 **Uyedineniya, Ostrov** *island* N Russian Federation
133 K8 **Uyo** Akwa Ibom, S Nigeria 05°00′N 07°57′E
239 H2 **Üyönch** Hovd, W Mongolia 46°04′N 92°05′E
Uyuk *see* Oyyk
221 J7 **'Uyūn** SW Oman 17°19′N 53°50′E
112 C2 **Uyuni** Potosí, W Bolivia 20°27′S 66°48′W
112 C2 **Uyuni, Salar de** *wetland* SW Bolivia
228 F5 **Uzbekistan** *off.* Republic of Uzbekistan. ◆ *republic* C Asia
Uzbekistan, Republic of *see* Uzbekistan
229 K7 **Uzbel Shankou** *Rus.* Pereval Kyzyl-Dzhiik. *pass* China/Tajikistan
228 B6 **Uzboý** *prev. Rus.* Imeni 26 Bakinskikh Komissarov, *Turkm.* 26 Baku Komissarlary Adyndaky. Balkan Welaýaty, W Turkmenistan 39°24′N 54°04′E
191 F10 **Uzda** Minskaya Voblasts', C Belarus 53°29′N 27°10′E
168 F2 **Uzerche** Corrèze, C France 45°24′N 01°35′E
165 J8 **Uzès** Gard, S France 44°00′N 04°25′E
229 K5 **Uzgen** *Kir.* Özgön. Oshskaya Oblast', SW Kyrgyzstan 40°42′N 73°17′E
188 G2 **Uzh** ≈ N Ukraine
Uzhgorod *see* Uzhhorod
188 B5 **Uzhhorod** *Rus.* Uzhgorod; *prev.* Ungvár. Zakarpats'ka Oblast', W Ukraine 48°36′N 22°19′E
192 G7 **Uzhur** Krasnoyarskiy Kray, S Russian Federation 55°18′N 89°36′E
Uzi *see* Uji
184 F5 **Užice** *prev.* Titovo Užice. Serbia, W Serbia 43°52′N 19°51′E
Uzin *see* Uzyn
196 G3 **Uzlovaya** Tul'skaya Oblast', W Russian Federation 54°01′N 38°15′E
179 D14 **Uznach** Sankt Gallen, NE Switzerland 47°12′N 09°00′E
Uzunagach *see* Uzynagash
214 B4 **Uzunköprü** Edirne, NW Turkey 41°18′N 26°40′E
190 C7 **Užventis** Šiaulai, C Lithuania 55°49′N 22°38′E
189 H4 **Uzyn** *Rus.* Uzin. Kyyivs'ka Oblast', N Ukraine 49°52′N 30°27′E
227 K7 **Uzynagash** *prev.* Uzunagach. Almaty, SE Kazakhstan 43°08′N 76°20′E
226 G2 **Uzynkol'** *prev.* Lenīn, Leninskoye. Kustanay, N Kazakhstan 54°05′N 65°23′E

V

Vääksy *see* Asikkala
138 F8 **Vaal** ≈ C South Africa
152 H7 **Vaala** Kainuu, C Finland 64°34′N 26°49′E
141 I4 **Vaal Dam** ⊠ South Africa
141 H5 **Vaalhartsdam** ◎ Northern Cape, South Africa
153 I9 **Vaalimaa** Etelä-Karjala, SE Finland 60°34′N 27°49′E
163 H10 **Vaals** Limburg, SE Netherlands 50°46′N 06°01′E
141 J2 **Vaalwater** Limpopo, South Africa 24°10′S 28°06′E
153 G8 **Vaasa** *Swe.* Vasa; *prev.* Nikolainkaupunki. Österbotten, W Finland 63°07′N 21°39′E
162 H6 **Vaassen** Gelderland, E Netherlands 52°18′N 05°59′E
190 D7 **Vabalninkas** Panevėžys, NE Lithuania 55°59′N 24°45′E
Vabkent *see* Vobkent
168 G5 **Vabre** Midi-Pyrénées, France 43°41′N 2°26′E
183 F11 **Vác** *Ger.* Waitzen. Pest, N Hungary 47°46′N 19°08′E
113 J6 **Vacaria** Rio Grande do Sul, S Brazil 28°31′S 50°52′W
80 C2 **Vacaville** California, W USA 38°21′N 121°59′W
Vaccacahy *see* Vaccacaí
115 K2 **Vaccacaí** *prev.* Vaccacahy. Rio Grande do Sul, Brazil 30°26′S 54°22′W
165 J8 **Vaccarès, Étang de** ◎ SE France
185 J7 **Vacha** *var.* Vŭcha. ≈ SW Bulgaria
90 G6 **Vache, Île à** *island* SW Haiti
137 G11 **Vacoas** W Mauritius 20°18′S 57°29′E
76 C4 **Vader** Washington, NW USA 46°23′N 122°58′W
154 B6 **Vadheim** Sogn Og Fjordane, S Norway 61°12′N 05°48′E
218 C2 **Vadili** *Gk.* Vatilí. C Cyprus 35°09′N 33°39′E
232 C8 **Vadodara** *prev.* Baroda. Gujarāt, W India 22°19′N 73°14′E
152 I3 **Vadsø** *Fin.* Vesisaari. Finnmark, N Norway 70°07′N 29°47′E
155 G9 **Vadstena** Östergötland, S Sweden 58°26′N 14°55′E
176 F4 **Vaduz** ● (Liechtenstein) W Liechtenstein 47°08′N 09°32′E
Vág *see* Váh
194 G7 **Vaga** ≈ NW Russian Federation
154 D5 **Vågåmo** Oppland, S Norway 61°52′N 09°06′E
184 B4 **Vaganski Vrh** ▲ W Croatia 44°24′N 15°32′E
154 A1 **Vágar** *Dan.* Vågø. *island* W Faroe Islands
Vágbeszterce *see* Považská Bystrica
155 G10 **Vaggeryd** Jönköping, S Sweden 57°30′N 14°10′E
215 K5 **Vagharshapat** *var.* Ejmiadzin, Ejmiatsin, Etchmiadzin, *Rus.* Echmiadzin. W Armenia 40°10′N 44°17′E
155 I8 **Vagnhärad** Södermanland, C Sweden 58°57′N 17°32′E
Vågø *see* Vágar
172 B2 **Vagos** Aveiro, N Portugal 40°33′N 08°42′W
Vágsellye *see* Sal'a
152 E4 **Vågsfjorden** *fjord* N Norway
154 A5 **Vågsøy** *island* S Norway
Vágújhely *see* Nové Mesto nad Váhom
183 E11 **Váh** *Ger.* Waag, *Hung.* Vág. ≈ W Slovakia
153 G8 **Vähäkyrö** Österbotten, W Finland 63°04′N 22°05′E
285 K7 **Vahitahi** *atoll* Îles Tuamotu, E French Polynesia
Váhtjer *see* Gällivare
284 F1 **Vaiaku** Funafuti Atoll, SE Tuvalu 08°31′N 179°11′E
172 C4 **Vaiamonte** Portalegre, Portugal 39°06′N 7°31′W
Vaidei *see* Vulcan
66 B9 **Vaiden** Mississippi, S USA 33°19′N 89°42′W
235 F9 **Vaigai** ≈ SE India
166 D7 **Vaiges** Pays de la Loire, France 48°03′N 0°28′W
181 F14 **Vaihingen** Baden-Württemberg, Germany 48°56′E 8°58′N
118 A2 **Vaihu** Easter Island, Chile, E Pacific Ocean 27°10′S 109°22′W
190 E5 **Väike Emajõgi** ≈ S Estonia
190 E4 **Väike-Maarja** *Ger.* Klein-Marien. Lääne-Virumaa, NE Estonia 59°07′N 26°16′E
Väike-Salatsi *see* Mazsalaca
79 L3 **Vail** Colorado, C USA 39°36′N 106°20′W
167 H9 **Vailly** Centre, France 46°52′N 2°44′E
167 I4 **Vailly-sur-Aisne** Picardie, France 49°25′N 3°31′E
284 D9 **Vaina** Tongatapu, S Tonga 21°12′S 175°10′W
190 C4 **Väinameri** *prev.* Muhu Väin, *Ger.* Moon-Sund. *sea* E Baltic Sea
153 I9 **Vainikkala** Etelä-Karjala, SE Finland 60°54′N 28°18′E
190 B7 **Vaiņode** SW Latvia 56°25′N 21°52′E
235 E9 **Vaippār** ≈ SE India
285 K7 **Vairaatea** *atoll* Îles Tuamotu, C French Polynesia
285 N4 **Vairao** Tahiti, W French Polynesia 17°48′S 149°17′W
165 J8 **Vaison-la-Romaine** Vaucluse, SE France 44°15′N 05°04′E
284 C9 **Vaitupu** Île Uvea, E Wallis and Futuna 13°14′S 176°09′W
284 C4 **Vaitupu** *atoll* C Tuvalu
Vajdahunyad *see* Hunedoara
Vajdej *see* Vulcan
134 G3 **Vakaga** ◈ *prefecture* NE Central African Republic
185 I6 **Vakarel** Sofia, W Bulgaria 42°35′N 23°40′E
Vakav *see* Ustrem
196 G10 **Vakfikebir** Trabzon, NE Turkey 41°03′N 39°19′E
192 F6 **Vakh** ≈ C Russian Federation
Vakhon, Qatorkŭhi *see* Nicholas Range
229 I8 **Vakhsh** SW Tajikistan 37°46′N 68°48′E
229 I8 **Vakhsh** ≈ SW Tajikistan
195 H9 **Vakhtan** Nizhegorodskaya Oblast', W Russian Federation 58°00′N 46°43′E
154 B6 **Vaksdal** Hordaland, S Norway 60°29′N 05°45′E
Valachia *see* Wallachia
172 B4 **Valada** Santarém, Portugal 39°05′N 8°46′W
174 B3 **Valais** *Ger.* Wallis. ◈ *canton* SW Switzerland
184 G9 **Valamarës, Mali i** ▲ SE Albania 40°48′N 20°31′E
195 J9 **Valamaz** Udmurtskaya Respublika, NW Russian Federation 57°36′N 52°07′E
185 H8 **Valandovo** SE FYR Macedonia 41°20′N 22°33′E
183 E9 **Valašské Meziříčí** *Ger.* Wallachisch-Meseritsch, *Pol.* Wałeckie Międzyrzecze. Zlínský Kraj, E Czech Republic 49°29′N 17°57′E
187 I5 **Valáxa** *island* Vóreies Sporádes, Greece, Aegean Sea
169 M4 **Valberg** Provence-Alpes-Côte d'Azur, France 44°06′N 6°56′E
155 F8 **Vålberg** Värmland, C Sweden 59°24′N 13°12′E
172 C3 **Valbom** Castelo Branco, Portugal 39°58′N 7°40′W
172 D2 **Valbom** Guarda, Portugal 40°47′N 7°08′W
169 L3 **Valbonnais** Rhône-Alpes, France 44°54′N 5°54′E
168 C6 **Valcarlos** Navarra, Spain 43°06′N 1°18′W
188 C8 **Vâlcea** *prev.* Vîlcea. ◈ *county* SW Romania
Vălčedrăm *see* Valchedram
185 I5 **Valchedram** *var.* Vălčedrăm, Vŭlchedrŭm. Montana, NW Bulgaria 43°42′N 23°25′E
116 E7 **Valcheta** Río Negro, E Argentina 40°42′S 66°08′W
185 M5 **Valchi Dol** *var.* Vŭlchidol; *prev.* Kurt-Dere. Varna, E Bulgaria 43°25′N 27°33′E
63 I3 **Valcourt** Québec, SE Canada 45°28′N 72°18′W
Valdai Hills *see* Valdayskaya Vozvyshennost'
170 F2 **Valdavia** ≈ N Spain
194 E9 **Valday** Novgorodskaya Oblast', W Russian Federation 57°57′N 33°20′E
194 D9 **Valdayskaya Vozvyshennost'** *var.* Valdai Hills. *hill range* W Russian Federation
170 E5 **Valdecañas, Embalse de** ⊠ W Spain
172 F3 **Valdecañas, Embalse de** ⊠ Spain
172 E4 **Valdefuentes** Extremadura, Spain 39°16′N 6°07′W
190 C6 **Valdemārpils** *Ger.* Sassmacken. NW Latvia 57°23′N 22°36′E
155 H9 **Valdemarsvik** Östergötland, S Sweden 58°13′N 16°35′E
170 G5 **Valdemoro** Madrid, C Spain 40°12′N 03°40′W
170 G6 **Valdepeñas** Castilla-La Mancha, C Spain 38°46′N 03°24′W
173 H7 **Valdepeñas de Jaén** Andalucía, Spain 37°35′N 3°49′W
170 E3 **Valderaduey** ≈ NE Spain
170 F3 **Valderas** Castilla y León, N Spain 42°05′N 05°27′W
171 J4 **Valderrobres** *var.* Vall-de-roures. Aragón, NE Spain 40°53′N 00°08′E
117 F8 **Valdés, Península** *peninsula* SE Argentina
104 C1 **Valdéz** *var.* Limones. Esmeraldas, NW Ecuador 01°13′N 79°00′W
83 J6 **Valdez** Alaska, USA 61°08′N 146°21′W
Valdia *see* Weldiya
169 M2 **Val d'Isère** Savoie, E France 45°23′N 07°03′E
116 B7 **Valdivia** Los Ríos, C Chile 39°50′S 73°13′W
118 D4 **Valdivia de Paine** Región Metropolitana, Chile 33°49′S 70°52′W
165 H3 **Val-d'Oise** ◈ *department* N France
58 F7 **Val-d'Or** Québec, SE Canada 48°06′N 77°42′W
69 K3 **Valdosta** Georgia, SE USA 30°49′N 83°16′W
154 D6 **Valdres** *physical region* S Norway
76 F7 **Vale** Oregon, NW USA 43°59′N 117°15′W
188 B6 **Valea lui Mihai** *Hung.* Érmihályfalva. Bihor, NW Romania 47°31′N 22°08′E
172 C4 **Vale de Açor** Portalegre, Portugal 39°15′N 7°55′W
172 B6 **Vale de Água** Setúbal, Portugal 37°54′N 8°35′W
172 B1 **Vale de Cambra** Aveiro, Portugal 40°51′N 8°24′W
172 B3 **Vale de Paredes** Leiria, Portugal 39°42′N 9°03′W
172 B7 **Vale de Santiago** Beja, Portugal 37°45′N 8°25′W
172 C6 **Vale de Vargo** Beja, Portugal 37°59′N 7°25′W
172 C4 **Vale do Peso** Portalegre, Portugal 39°21′N 7°39′W
172 D3 **Vale Feitoso** Castelo Branco, Portugal 40°04′N 6°59′W
56 F6 **Valemount** British Columbia, SW Canada 52°46′N 119°17′W
109 H8 **Valença** Bahia, E Brazil 13°22′S 39°06′W
170 C3 **Valença do Minho** Viana do Castelo, N Portugal 42°02′N 08°38′W
108 F5 **Valença do Piauí** Piauí, E Brazil 06°26′S 41°46′W
165 H5 **Valençay** Indre, C France 47°10′N 01°31′E
165 J7 **Valence** *anc.* Valentia, Valentia Julia, Ventia. Drôme, E France 44°56′N 04°54′E
171 I6 **Valencia** Valenciana, E Spain 39°29′N 00°24′W
81 D9 **Valencia** California, USA 34°27′N 118°37′W
102 F3 **Valencia** Carabobo, N Venezuela 10°12′N 68°02′W
171 I6 **Valencia** *Cat.* València. ◈ *province* Valenciana, E Spain
171 I6 **Valencia** ✈ Valencia, E Spain
170 D6 **Valencia de Alcántara** Extremadura, W Spain 39°25′N 07°14′W
170 F3 **Valencia de Don Juan** Castilla y León, N Spain 42°17′N 05°31′W
171 J6 **Valencia, Golfo de** *var.* Gulf of Valencia. *gulf* E Spain
Valencia, Gulf of *see* Valencia, Golfo de
157 A11 **Valencia Island** *Ir.* Dairbhre. *island* SW Ireland
171 I6 **Valenciana** *var.* Valencia, *Cat.* València; *anc.* Valentia. ◈ *autonomous community* NE Spain
Valencia/València *see* Valenciana
165 I1 **Valenciennes** Nord, N France 50°21′N 03°32′E
188 E9 **Vălenii de Munte** Prahova, SE Romania 45°11′N 26°02′E
169 L5 **Valensole** Provence-Alpes-Côte d'Azur, France 43°50′N 5°59′E
Valentia *see* Valence, France
Valentia *see* Valenciana
Valentia Julia *see* Valence
165 K4 **Valentigney** Doubs, E France 47°27′N 06°49′E
74 C6 **Valentine** Nebraska, C USA 42°53′N 100°31′W
70 B6 **Valentine** Texas, SW USA 30°35′N 104°30′W
115 I7 **Valentines** Uruguay 33°16′S 55°07′W
Valentine State *see* Oregon
176 E7 **Valenza** Piemonte, NW Italy 45°01′N 08°37′E
154 E6 **Våler** Hedmark, S Norway 60°39′N 11°52′E
102 E3 **Valera** Trujillo, NW Venezuela 09°21′N 70°38′W
286 G8 **Valerie Guyot** S Pacific Ocean 33°00′S 164°00′W
172 C7 **Vales Mortos** Beja, Portugal 37°48′N 7°30′W
Valetta *see* Valletta
190 E5 **Valga** *Ger.* Walk, *Latv.* Valka. Valgamaa, S Estonia 57°48′N 26°04′E
190 E5 **Valgamaa** *var.* Valga Maakond. ◈ *province* S Estonia
Valga Maakond *see* Valgamaa
169 J4 **Valgorge** Rhône-Alpes, France 44°35′N 4°07′E
169 M1 **Valgrisanche** Valle D'Aosta, NW Italy 45°37′N 07°03′E
172 D2 **Valhelhas** Guarda, Portugal 40°24′N 7°24′W
89 J9 **Valiente, Península** *peninsula* NW Panama
175 B8 **Valinco, Golfe de** *gulf* Corse, France, C Mediterranean Sea
184 F4 **Valjevo** Serbia, W Serbia 44°17′N 19°54′E
Valjok *see* Válljohka
173 M1 **Valjunquera** Aragón, Spain 40°57′N 0°01′E
190 E5 **Valka** *Ger.* Walk. N Latvia 57°48′N 26°01′E
Valka *see* Valga
153 H9 **Valkeakoski** Pirkanmaa, W Finland 61°17′N 24°05′E
153 I9 **Valkeala** Kymenlaakso, S Finland 60°55′N 26°49′E
163 G10 **Valkenburg** Limburg, SE Netherlands 50°52′N 05°50′E
163 G9 **Valkenswaard** Noord-Brabant, S Netherlands 51°21′N 05°29′E
191 D9 **Valkininkai** Alytus, S Lithuania 54°22′N 24°51′E
189 K4 **Valky** Kharkivs'ka Oblast', E Ukraine 49°51′N 35°40′E
87 M5 **Valladolid** Yucatán, SE Mexico 20°39′N 88°13′W
170 F3 **Valladolid** Castilla y León, NW Spain 41°39′N 04°45′W
170 F3 **Valladolid** ◈ *province* Castilla y León, N Spain
165 K8 **Vallauris** Alpes-Maritimes, SE France 43°34′N 07°03′E
Vall-de-roures *see* Valderrobres
Vall D'Uxó *see* La Vall d'Uixó
155 C8 **Valle** Aust-Agder, S Norway 59°13′N 07°33′E
170 F2 **Valle** Cantabria, N Spain 43°14′N 04°16′W
88 E5 **Valle** ◈ *department* S Honduras
170 G5 **Vallecas** Madrid, C Spain 40°23′N 03°37′W
85 L5 **Vallecillo** Nuevo León, Mexico 26°40′N 100°00′W
79 K5 **Vallecito Reservoir** ⊠ Colorado, C USA
174 B4 **Valle d'Aosta** *Fr.* Vallée d'Aoste. ◈ *region* NW Italy
172 G8 **Valle de Abdalajís** Andalucía, Spain 36°56′N 4°41′W
86 E6 **Valle de Bravo** México, S Mexico 19°19′N 100°08′W
84 B1 **Valle de Guadalupe** Jalisco, W Mexico 32°04′N 116°37′W
102 G3 **Valle de Guanape** Anzoátegui, N Venezuela 09°54′N 65°41′W
86 C6 **Valle de Juárez** Jalisco, Mexico 19°56′N 102°56′W
102 G3 **Valle de La Pascua** Guárico, N Venezuela 09°15′N 66°00′W
102 B6 **Valle del Cauca** *off.* Departamento del Valle del Cauca. ◈ *province* W Colombia
Valle del Cauca, Departamento del *see* Valle del Cauca
85 H5 **Valle de Olivos** Chihuahua, Mexico 27°12′N 106°17′W
86 D5 **Valle de Santiago** Guanajuato, C Mexico 20°25′N 101°15′W
85 H5 **Valle de Zaragoza** Chihuahua, N Mexico 27°25′N 105°50′W
102 C2 **Valledupar** Cesar, N Colombia 10°31′N 73°16′W
Vallée d'Aoste *see* Valle d'Aosta
132 B3 **Vallée de Ferlo** ≈ NW Senegal
112 D1 **Vallegrande** Santa Cruz, C Bolivia 18°30′S 64°06′W
118 C2 **Valle Hermoso** Valparaíso, Chile 32°26′S 71°14′W
87 M6 **Valle Hermoso** Quintana Roo, Mexico 19°11′N 88°31′W
86 G1 **Valle Hermoso** Tamaulipas, C Mexico 25°39′N 97°49′W
80 B2 **Vallejo** California, W USA 38°08′N 122°16′W
114 B5 **Valle María** Entre Ríos, Argentina 31°59′S 60°35′W
169 N1 **Valle Mosso** Piemonte, NW Italy 45°38′N 08°08′E
87 H8 **Valle Nacional** Oaxaca, Mexico 17°47′N 96°19′W
116 B1 **Vallenar** Atacama, N Chile 28°35′S 70°44′W
155 I8 **Vallentuna** Stockholm, C Sweden 59°32′N 18°05′E
169 I4 **Valleraugue** Languedoc-Roussillon, France 44°05′N 3°38′E
166 C8 **Vallet** Pays de la Loire, France 47°10′N 1°16′W
175 J13 **Valletta** *prev.* Valetta. ● (Malta) E Malta 35°54′N 14°31′E
81 F11 **Valle Vista** California, USA 33°45′N 116°54′W
81 C8 **Valley Acres** California, USA 35°12′N 119°24′W
81 E12 **Valley Center** California, USA 33°13′N 117°02′W
75 E10 **Valley Center** Kansas, C USA 37°49′N 97°22′W
74 E3 **Valley City** North Dakota, N USA 46°57′N 97°58′W
76 D8 **Valley Falls** Oregon, NW USA 42°28′N 120°16′W
Valleyfield *see* Salaberry-de-Valleyfield
67 J3 **Valley Head** West Virginia, NE USA 38°33′N 80°01′W
80 D3 **Valley Home** California, USA 37°50′N 120°55′W
70 G5 **Valley Mills** Texas, SW USA 31°36′N 97°27′W
129 I5 **Valley of the Kings** *headland* Egypt
129 I5 **Valley of the Kings** *ancient monument* E Egypt
74 F6 **Valley Springs** South Dakota, N USA 43°34′N 96°28′W
66 E3 **Valley Station** Kentucky, S USA 38°06′N 85°52′W
65 I6 **Valley Stream** New York, USA 40°40′N 73°43′W
57 H3 **Valleyview** Alberta, W Canada 55°02′N 117°17′W
71 H4 **Valley View** Texas, SW USA 33°27′N 97°08′W
116 G5 **Vallimanca, Arroyo** ≈ E Argentina
152 H4 **Válljohka** *var.* Valjok. Finnmark, N Norway 69°40′N 25°52′E
175 G10 **Vallo della Lucania** Campania, S Italy 40°13′N 15°15′E
169 L2 **Valloire** Rhône-Alpes, France 45°10′N 6°26′E
169 J4 **Vallon-Pont-d'Arc** Rhône-Alpes, France 44°24′N 4°24′E
176 B4 **Vallorbe** Vaud, W Switzerland 46°43′N 06°21′E
169 L3 **Vallouise** Provence-Alpes-Côte d'Azur, France 44°51′N 6°29′E
171 K4 **Valls** Cataluña, NE Spain 41°18′N 01°15′E
154 H5 **Vallsta** Gävleborg, C Sweden 61°30′N 16°25′E
154 I6 **Vallvik** Gävleborg, C Sweden 61°10′N 17°15′E
57 M9 **Val Marie** Saskatchewan, S Canada 49°15′N 107°44′W
190 E5 **Valmiera** *Est.* Volmari, *Ger.* Wolmar. N Latvia 57°34′N 25°26′E
80 J1 **Valmy** Nevada, USA 40°48′N 117°08′W
170 G2 **Valnera** ▲ N Spain 43°08′N 03°39′W
164 F2 **Valognes** Manche, N France 49°31′N 01°28′W
Valona *see* Vlorë
Valona Bay *see* Vlorës, Gjiri i
170 C4 **Valongo** *var.* Valongo de Gaia. Porto, N Portugal 41°11′N 08°30′W
Valongo de Gaia *see* Valongo
170 F3 **Valoria la Buena** Castilla y León, N Spain 41°48′N 04°33′W
191 F9 **Valozhyn** *Pol.* Wołożyn, *Rus.* Volozhin. Minskaya Voblasts', C Belarus 54°05′N 26°32′E
170 D3 **Valpaços** Vila Real, N Portugal 41°36′N 07°17′W
116 B3 **Valparaíso** Valparaíso, C Chile 33°05′S 71°18′W
85 J9 **Valparaíso** Zacatecas, C Mexico 22°49′N 103°28′W
69 H4 **Valparaiso** Florida, SE USA 30°30′N 86°28′W
73 E9 **Valparaiso** Indiana, N USA 41°28′N 87°04′W
116 B3 **Valparaíso** *off.* Región de Valparaíso. ◈ *region* C Chile
Valparaíso, Región de *see* Valparaíso
169 M1 **Valpelline** Valle D'Aosta, NW Italy 45°49′N 07°19′E
Valpo *see* Valpovo
184 E2 **Valpovo** *Hung.* Valpo. Osijek-Baranja, E Croatia 45°40′N 18°25′E
169 I6 **Valras-Plage** Languedoc-Roussillon, France 43°15′N 3°17′E
165 J7 **Valréas** Vaucluse, SE France 44°22′N 05°00′E
141 I5 **Vals** ≈ Free State, South Africa
Vals *see* Vals-Platz
232 C9 **Valsād** *prev.* Bulsar. Gujarāt, W India 20°40′N 72°55′E
Valsbaai *see* False Bay
261 H5 **Valse Pisang, Kepulauan** *island group* E Indonesia
176 E5 **Vals-Platz** *var.* Vals. Graubünden, S Switzerland 46°39′N 09°09′E
141 I5 **Valsrivier** Free State, South Africa 28°09′S 28°06′E
261 K8 **Vals, Tanjung** *headland* Papua, SE Indonesia 08°26′S 137°35′E
152 I7 **Valtimo** Pohjois-Karjala, E Finland 63°39′N 28°49′E
186 F5 **Váltou** ▲▲ C Greece
169 N1 **Valtournanche** Valle D'Aosta, NW Italy 45°52′N 07°37′E
197 H7 **Valuyevka** Rostovskaya Oblast', SW Russian Federation 46°48′N 43°49′E
196 F5 **Valuyki** Belgorodskaya Oblast', W Russian Federation 50°11′N 38°07′E
79 H2 **Val Verda** Utah, W USA 40°51′N 111°53′W
170 A10 **Valverde** Hierro, Islas Canarias, Spain, NE Atlantic Ocean 27°48′N 17°55′W
81 E11 **Val Verde** California, USA 33°51′N 117°15′W
170 D8 **Valverde del Camino** Andalucía, S Spain 37°35′N 06°45′W
172 D3 **Valverde del Fresno** Extremadura, Spain 40°13′N 6°52′W
81 E10 **Valyermo** California, USA 34°27′N 117°51′W
155 D12 **Vamdrup** Syddanmark, C Denmark 55°26′N 09°18′E
154 G6 **Våmhus** Dalarna, C Sweden 61°07′N 14°30′E
153 G9 **Vammala** Pirkanmaa, SW Finland 61°20′N 22°55′E
Vámosudvarhely *see* Odorheiu Secuiesc
215 J6 **Van** Van, E Turkey 38°30′N 43°23′E
71 I4 **Van** Texas, SW USA 32°31′N 95°38′W
215 K6 **Van** ◈ *province* E Turkey
215 K5 **Vanadzor** *prev.* Kirovakan. N Armenia 40°49′N 44°29′E
71 H4 **Van Alstyne** Texas, SW USA 33°25′N 96°34′W
77 K5 **Vananda** Montana, NW USA 46°22′N 106°58′W
185 J1 **Vânători** *Hung.* Héjjasfalva; *prev.* Vînători. Mureș, C Romania 46°14′N 24°56′E
285 K7 **Vanavana** *atoll* Îles Tuamotu, SE French Polynesia
Vana-Vändra *see* Vändra
193 H6 **Vanavara** Krasnoyarskiy Kray, C Russian Federation 60°19′N 102°19′E
75 G12 **Van Buren** Arkansas, C USA 35°28′N 94°25′W
63 L1 **Van Buren** Maine, NE USA 47°07′N 67°57′W
75 I11 **Van Buren** Missouri, C USA 37°00′N 91°00′W
63 L3 **Vanceboro** Maine, NE USA 45°31′N 67°25′W
67 L6 **Vanceboro** North Carolina, SE USA 35°16′N 77°06′W
66 G3 **Vanceburg** Kentucky, S USA 38°36′N 84°40′W
91 L6 **Vance W. Amory** ✈ Nevis, Saint Kitts and Nevis 17°08′N 62°36′W
Vanch *see* Vanj
56 D8 **Vancouver** British Columbia, SW Canada 49°13′N 123°06′W
76 C5 **Vancouver** Washington, NW USA 45°38′N 122°39′W
56 D9 **Vancouver** ✈ British Columbia, SW Canada 49°03′N 123°00′W
54 B9 **Vancouver Island** *island* British Columbia, SW Canada
Vanda *see* Vantaa
261 K5 **Van Daalen** ≈ Papua, E Indonesia
73 C11 **Vandalia** Illinois, N USA 38°57′N 89°05′W
75 I9 **Vandalia** Missouri, C USA 39°18′N 91°29′W
73 G11 **Vandalia** Ohio, N USA 39°53′N 84°12′W
141 I4 **Vanderbijlpark** Gauteng, South Africa 26°42′S 27°49′E
71 H8 **Vanderbilt** Texas, SW USA 28°45′N 96°37′W
73 G8 **Vandercook Lake** Michigan, N USA 42°11′N 84°23′W
56 D4 **Vanderhoof** British Columbia, SW Canada 53°54′N 124°00′W
141 H7 **Vanderkloof** Free State, South Africa 30°00′S 24°44′E
141 H7 **Vanderkloof Damm** ◎ Free State, South Africa
63 H5 **Vanderwhacker Mountain** ▲ New York, NE USA 43°54′N 74°06′W
274 G2 **Van Diemen Gulf** *gulf* Northern Territory, N Australia
Van Diemen's Land *see* Tasmania
190 E4 **Vändra** *Ger.* Fennern; *prev.* Vana-Vändra. Pärnumaa, SW Estonia 58°39′N 25°00′E
Vandsburg *see* Więcbork
76 B10 **Van Duzen River** ≈ California, W USA
141 J4 **Vandyksdrif** Mpumalanga, South Africa 26°05′S 29°19′E
191 D8 **Vandžiogala** Kaunas, C Lithuania 55°07′N 23°55′E
85 L8 **Vanegas** San Luis Potosí, C Mexico 23°53′N 100°55′W
Vaner, Lake *see* Vänern
155 F8 **Vänern** *Eng.* Lake Vaner; *prev.* Lake Vener. ◎ S Sweden
155 F9 **Vänersborg** Västra Götaland, S Sweden 58°16′N 12°22′E
154 C6 **Vang** Oppland, S Norway 61°07′N 08°34′E
139 M9 **Vangaindrano** Fianarantsoa, SE Madagascar 23°21′S 47°35′E
215 J6 **Van Gölü** *Eng.* Lake Van; *anc.* Thospitis. *salt lake* E Turkey
281 I2 **Vangunu** *island* New Georgia Islands, NW Solomon Islands
70 B5 **Van Horn** Texas, SW USA 31°03′N 104°51′W
281 I5 **Vanikolo** *var.* Vanikoro. *island* Santa Cruz Islands, E Solomon Islands
Vanikoro *see* Vanikolo
280 A2 **Vanimo** West Sepik, NW Papua New Guinea 02°40′S 141°17′E
193 L8 **Vanino** Khabarovskiy Kray, SE Russian Federation 49°10′N 140°18′E
234 D7 **Vānīvilāsa Sāgara** ◎ SW India
229 J7 **Vanj** *Rus.* Vanch. S Tajikistan 38°22′N 71°27′E
188 B9 **Vânju Mare** *prev.* Vînju Mare. Mehedinți, SW Romania 44°25′N 22°52′E
62 G3 **Vankleek Hill** Ontario, SE Canada 45°32′N 74°39′W
Van, Lake *see* Van Gölü
154 J2 **Vännäs** Västerbotten, N Sweden 63°54′N 19°43′E
154 J2 **Vännäsby** Västerbotten, N Sweden 63°55′N 19°53′E
164 E4 **Vannes** *anc.* Dariorigum. Morbihan, NW France 47°40′N 02°45′W
152 F3 **Vannøya** *island* N Norway
165 K6 **Vanoise, Massif de la** ▲▲ E France
141 J5 **Van Reenen** Free State, South Africa 28°22′S 29°23′E
141 J6 **Van Reenens Pass** *pass* South Africa
261 K5 **Van Rees, Pegunungan** ▲▲ Papua, E Indonesia
140 D8 **Vanrhynsdorp** Western Cape, SW South Africa 31°36′S 18°45′E
140 D8 **Vanrhynspas** *pass* Northern Cape, South Africa
67 H4 **Vansant** Virginia, NE USA 37°13′N 82°03′W
154 G7 **Vansbro** Dalarna, C Sweden 60°32′N 14°15′E
155 B9 **Vanse** Vest-Agder, S Norway 58°04′N 06°40′E
53 J9 **Vansittart Island** *island* Nunavut, NE Canada
141 I7 **Van Stadensrus** Free State, South Africa 29°59′S 27°01′E
153 H10 **Vantaa** *Swe.* Vanda. Uusimaa, S Finland 60°18′N 25°01′E
153 H10 **Vantaa** ✈ (Helsinki) Uusimaa, S Finland 60°18′N 25°01′E
76 D4 **Vantage** Washington, NW USA 46°55′N 119°55′W
280 G8 **Vanua Balavu** *prev.* Vanua Mbalavu. *island* Lau Group, E Fiji
281 M1 **Vanua Lava** *island* Banks Islands, N Vanuatu
280 F8 **Vanua Levu** *island* N Fiji
280 E8 **Vanua Levu Barrier Reef** *reef* C Fiji
Vanua Mbalavu *see* Vanua Balavu
281 J6 **Vanuatu** *off.* Republic of Vanuatu; *prev.* New Hebrides. ◆ *republic* SW Pacific Ocean
267 I5 **Vanuatu** *island group* SW Pacific Ocean
Vanuatu, Republic of *see* Vanuatu
280 G9 **Vanua Vatu** *island* Lau Group, E Fiji
73 G10 **Van Wert** Ohio, N USA 40°52′N 84°34′W
140 F7 **Van Wyksvlei** Northern Cape, South Africa 30°21′S 21°49′E
140 F7 **Vanwyksvleidam** ⊠ Northern Cape, South Africa
169 N1 **Vanzone Con San Carlo** Piemonte, NW Italy 45°58′N 08°05′E
140 F4 **Van Zylsrust** Northern Cape, South Africa 26°52′S 22°03′E
280 C10 **Vao** Province Sud, S New Caledonia 22°40′S 167°29′E
168 G5 **Vaour** Midi-Pyrénées, France 44°04′N 1°48′E
Vapincum *see* Gap
188 G5 **Vapnyarka** Vinnyts'ka Oblast', C Ukraine 48°31′S 28°44′E
172 C7 **Vaqueiros** Faro, Portugal 37°23′N 7°44′W
85 M9 **Vaqueras** San Luis Potosí, Mexico 21°42′N 99°45′W
165 K8 **Var** ◈ *department* SE France
165 K8 **Var** ≈ SE France
155 F9 **Vara** Västra Götaland, S Sweden 58°16′N 12°57′E
Varaďinska Županija *see* Varaždin
190 F6 **Varakļāni** C Latvia 56°36′N 26°40′E
176 D6 **Varallo** Piemonte, NE Italy 45°51′N 08°16′E
222 F4 **Varāmīn** *var.* Veramin. Tehrān, N Iran 35°19′N 51°40′E
233 H7 **Vārānasi** *prev.* Banaras, Benares, *hist.* Kasi. Uttar Pradesh, N India 25°20′N 83°E
195 K2 **Varandey** Nenetskiy Avtonomnyy Okrug, NW Russian Federation 68°48′N 57°54′E
152 H3 **Varangerbotn** *Lapp.* Vuonnabahta. Finnmark, N Norway 70°09′N 28°28′E
152 I3 **Varangerfjorden** *Lapp.* Várjjatvuotna. *fjord* N Norway
152 H3 **Varangerhalvøya** *Lapp.* Várnjárga. *peninsula* N Norway
Varannó *see* Vranov nad Topl'ou
175 G8 **Varano, Lago di** ◎ SE Italy
191 F8 **Varapayeva** *Rus.* Voropayevo. Vitsyebskaya Voblasts', NW Belarus 55°09′N 27°13′E
Varasd *see* Varaždin
184 C1 **Varaždin** *Ger.* Warasdin, *Hung.* Varasd. Varaždin, N Croatia 46°18′N 16°21′E
184 C1 **Varaždin** *off.* Varaďinska Županija. ◈ *province* N Croatia
176 E8 **Varazze** Liguria, NW Italy 44°21′N 08°35′E
155 F10 **Varberg** Halland, S Sweden 57°06′N 12°15′E
185 K7 **Varbitsa** *var,* Vŭrbitsa; *prev.* Filevo. Haskovo, S Bulgaria 42°02′N 25°25′E
185 K8 **Varbitsa** ≈ S Bulgaria
185 H7 **Vardar** *Gk.* Axiós. ≈ FYR Macedonia/Greece *see also* Axiós
Vardar *see* Axiós
155 C12 **Varde** Syddtjylland, W Denmark 55°38′N 08°31′E
215 L5 **Vardenis** E Armenia 40°11′N 45°43′E
152 I3 **Vardø** *Fin.* Vuoreija. Finnmark, N Norway 70°22′N 31°06′E
186 G5 **Vardoúsia** ▲▲ C Greece
Vareia *see* Logroño
180 E3 **Varel** Niedersachsen, NW Germany 53°24′N 08°07′E
191 D9 **Varėna** *Pol.* Orany. Alytus, S Lithuania 54°13′N 24°35′E
63 H3 **Varennes** Québec, SE Canada 45°42′N 73°25′W
167 K4 **Varennes-en-Argonne** Lorraine, France 49°14′N 5°02′E
165 I6 **Varennes-sur-Allier** Allier, C France 46°17′N 03°24′E
184 E4 **Vareš** Federacija Bosne I Hercegovine, E Bosnia and Herzegovina 44°12′N 18°19′E
176 E6 **Varese** Lombardia, N Italy 45°49′N 08°50′E
188 D8 **Vârful Moldoveanu** *var.* Moldoveanul; *prev.* Vîrful Moldoveanu. ▲ C Romania 45°35′N 24°48′E
Varganzi *see* Warganza
155 F9 **Vårgårda** Västra Götaland, S Sweden 58°00′N 12°49′E
102 G2 **Vargas** *off.* Estado Vargas. ◈ *state* N Venezuela
195 N10 **Vargashi** Kurganskaya Oblast', C Russian Federation 55°22′N 65°39′E
110 D8 **Vargem** São Paulo, Brazil 22°53′S 46°25′W
111 J2 **Vargem Alegre** Minas Gerais, Brazil 19°35′S 42°18′W
111 L4 **Vargem Alta** Espírito Santo, Brazil 20°40′S 41°00′W
110 D3 **Vargem Bonita** Minas Gerais, Brazil 20°20′S 46°22′W
110 C6 **Vargem Grande do Sul** São Paulo, Brazil 21°50′S 46°53′W
110 F6 **Varginha** Minas Gerais, Brazil 21°33′S 45°26′W
155 F9 **Vargön** Västra Götaland, S Sweden 58°21′N 12°22′E
155 A9 **Varhaug** Rogaland, S Norway 58°37′N 05°39′E
168 G6 **Varilhes** Midi-Pyrénées, France 43°03′N 1°38′E
Várjjatvuotna *see* Varangerfjorden
153 I8 **Varkaus** Pohjois-Savo, C Finland 62°20′N 27°50′E
152 C2 **Varmahlíð** Norðurland Vestra, N Iceland 65°32′N 19°33′W
155 F8 **Värmland** ◈ *county* C Sweden
185 M5 **Varna** *prev.* Stalin; *anc.* Odessus. Varna, E Bulgaria 43°14′N 27°56′E
185 M5 **Varna** ◈ *province* E Bulgaria
185 M5 **Varna** ✈ Varna, E Bulgaria 43°16′N 27°52′E
155 G10 **Värnamo** Jönköping, S Sweden 57°11′N 14°03′E
185 M5 **Varnenski Zaliv** *prev.* Stalinski Zaliv. *bay* E Bulgaria
185 M5 **Varnensko Ezero** *estuary* E Bulgaria
190 C7 **Varniai** Telšiai, W Lithuania 55°45′N 22°22′E
Várnjárga *see* Varangerhalvøya
Varnoús *see* Baba
182 C7 **Varnsdorf** *Ger.* Warnsdorf. Ústecký Kraj, NW Czech Republic 50°57′N 14°35′E
183 F12 **Várpalota** Veszprém, W Hungary 47°12′N 18°08′E
180 F5 **Varrel** Niedersachsen, Germany 52°37′E 8°44′N

◆ Country
● Country Capital
◇ Dependent Territory
○ Dependent Territory Capital
◈ Administrative Regions
✈ International Airport
▲ Mountain
▲▲ Mountain Range
⛰ Volcano
≈ River
◎ Lake
⊠ Reservoir

◆ Country	◇ Dependent Territory	◆ Administrative Regions
● Country Capital	○ Dependent Territory Capital	✕ International Airport

▲ Mountain	▲ Volcano	☒ Lake
▲ Mountain Range	❧ River	☒ Reservoir

◆ Country
● Country Capital
◇ Dependent Territory
○ Dependent Territory Capital
◈ Administrative Regions
✈ International Airport
▲ Mountain
▲ Mountain Range
♺ Volcano
♦ River
◆ Lake
■ Reservoir

Picture credits

♦ Country   ◇ Dependent Territory   ◈ Administrative Regions   ▲ Mountain   🜨 Volcano   ◎ Lake
● Country Capital   ○ Dependent Territory Capital   ✕ International Airport   ▲ Mountain Range   ⚓ River   ▨ Reservoir

Key to Map Pages

	Small scale above 1:10,000,000
	Medium scale 1: 10,000,000 to 1:4,000,000
	Large scale below 1:4,000,000

United States of America

Europe

76
74
72
62
78
64
80
66
70
68

50
52
THE ARCTIC
294-295

82
54
152
58
156
56
164
60
198
170
130
NORTH
AMERICA
42-93
132
84
86
ATLANTIC OCEAN
290-291
132
90
82
88
PACIFIC OCEAN
286-287
102
284
105
104
106
108
SOUTH
AMERICA
94-119
112
110
116
114
118

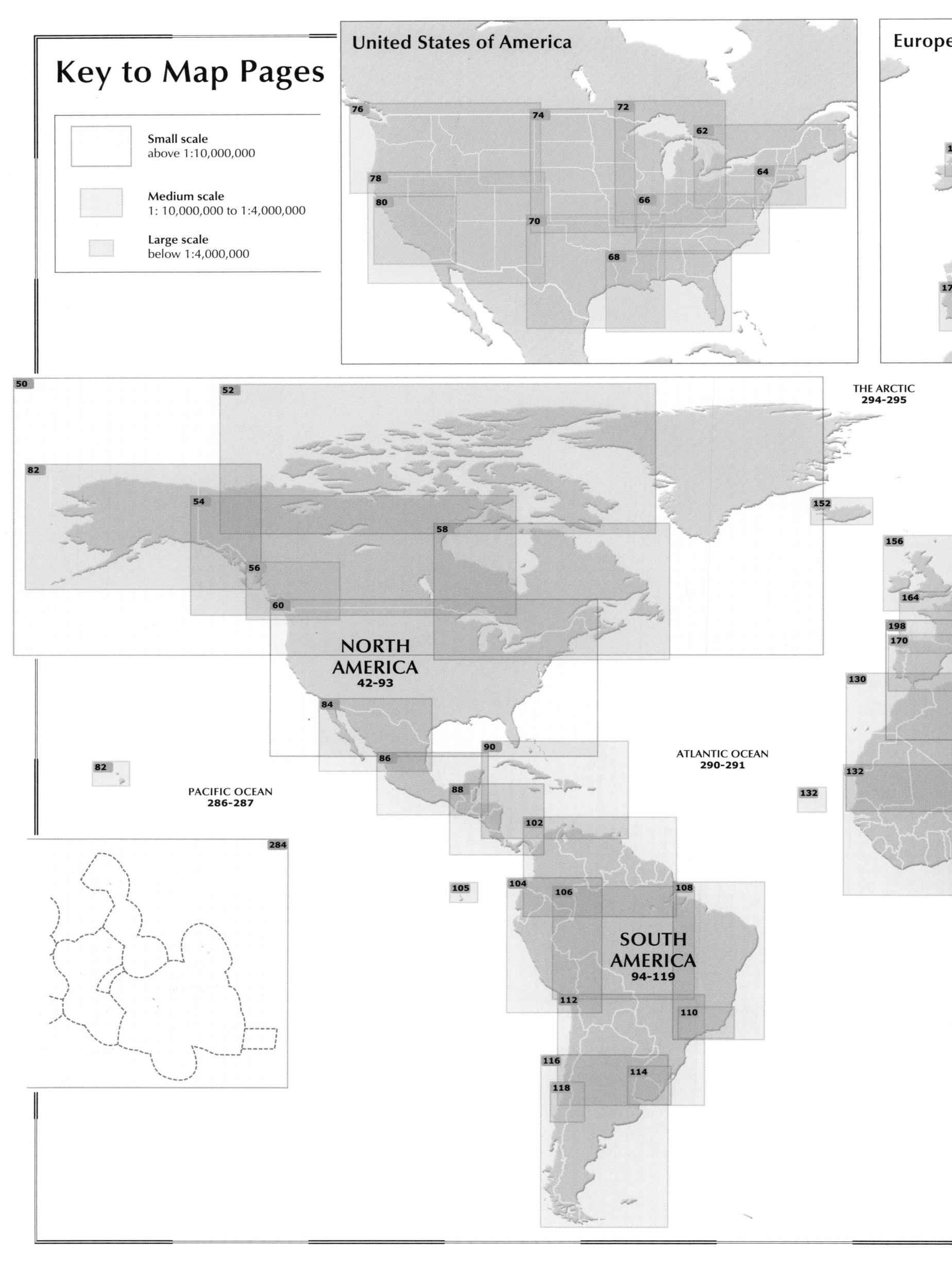